DATE DUE

DEMCO 38-296

HANDBOOK OF FLUID DYNAMICS AND FLUID MACHINERY

HANDBOOK OF FLUID DYNAMICS AND FLUID MACHINERY

VOLUME I: FUNDAMENTALS OF FLUID DYNAMICS

Edited by

Joseph A. Schetz and Allen E. Fuhs

A WILEY–INTERSCIENCE PUBLICATION

JOHN WILEY & SONS, INC.

New York • Chichester • Brisbane • Toronto • Singapore

This publication is designed to provide accurate and
authoritative information in regard to the subject
matter covered. It is sold with the understanding that
the publisher is not engaged in rendering professional
services. If legal accounting, medical, psychological, or any other
expert assistance is required, the services of a competent
professional person should be sought.

Library of Congress Cataloging in Publication Data:
Handbook of fluid dynamics and fluid machinery/editors, Joseph A.
  Schetz and Allen E. Fuhs.
      p.   cm.
    Includes index.
    ISBN: 0-471-12598-9 (Volume I)
    ISBN: 0-471-87352-7 (set)
    1. Fluid mechanics. 2. Hydraulic machinery. I. Schetz, Joseph A.
  II. Fuhs, Allen E.
  TA357.H286   1996
  620.1′06—dc20   95-5671
Printed in the United States of America

10 9 8 7 6 5 4 3 2 1

EDITORIAL REVIEW BOARD

LIST OF CONTRIBUTORS

DANIEL F. ANCONA, U.S. Department of Energy, Washington, D.C.

JOHN D. ANDERSON, JR., University of Maryland, College Park, MD

W. KYLE ANDERSON, NASA Langley Research Center, Hampton, VA

TAKAKAGE ARAI, Muroran Institute of Technology, Muroran, Japan

GEORGE D. ASHTON, U.S. Army Cold Regions Research and Engineering Laboratory, Hanover, NH

FRITZ H. BARK, The Royal Institute of Technology, Stockholm, Sweden

ALAN BEERBOWER (Deceased), University of California at San Diego, La Jolla, CA

MAREK BEHR, University of Minnesota, Minneapolis, MN

R. BYRON BIRD, University of Wisconsin–Madison, Madison, WI

RODNEY D. W. BOWERSOX, Air Force Institute of Technology, Dayton, OH

BOBBIE CARR, Naval Postgraduate School, Monterey, CA

CAMPBELL D. CARTER, Systems Research Laboratories, Dayton, OH

MICHAEL V. CASEY, Sulzer Innotec/Sulzer Hydro, Winterthur/Zurich, Switzerland

JACK E. CERMAK, Colorado State University, Fort Collins, CO

CHAU-LYAN CHANG, High Technology Corporation, Hampton, VA

HERBERT S. CHENG, Northwestern University, Evanston, IL

PASQUALE CINNELLA, Mississippi State University, Starkville, MS

FAYETTE S. COLLIER, JR., NASA Langley Research Center, Hampton, VA

J. ROBERT COOKE, Cornell University, Ithaca, NY

MALCOLM J. CROCKER, Auburn University, Auburn, AL

CLAYTON T. CROWE, Washington State University, Pullman, WA

NICHOLAS A. CUMPSTY, University of Cambridge, Cambridge, UK

WILLIAM G. DAY, JR., David Taylor Model Basin, Carderock, MD

REINER DECHER, University of Washington, Seattle, WA

JAMES D. DE LAURIER, University of Toronto, Toronto, Canada

Note: All Contributors to Volumes I, II, and III are mentioned in this Section of the handbook.

SERGE T. DEMETRIADES,   STD Research Incorporated, Arcadia, CA

AYODEJI O. DEMUREN,   Old Dominion University, Norfolk, VA

THOMAS E. DILLER,   Virginia Polytechnic Institute and State University, Blacksburg, VA

PANAYIOTIS DIPLAS,   Virginia Polytechnic Institute and State University, Blacksburg, VA

LOUIS V. DIVONE,   U.S. Department of Energy, Washington, D.C.

FRANKLIN T. DODGE,   Southwest Research Institute, San Antonio, TX

EARL H. DOWELL,   Duke University, Durham, NC

DONALD J. DUSA (Retired),   General Electric Aircraft Engines, Evandale, OH

HARRY A. DWYER,   University of California, Davis, CA

PETER R. EISEMAN,   Program Development Corporation, White Plains, NY

GEORGE EMANUEL,   University of Oklahoma, Norman, OK

JOHN F. FOSS,   Michigan State University, East Lansing, MI

ALLEN E. FUHS (Emeritus),   U.S. Naval Postgraduate School, Monterey, CA

SUSAN E. FUHS,   Rand Corporation, Santa Monica, CA

WALTER S. GEARHART (Retired),   Pennsylvania State University, State College, PA

WILLIAM K. GEORGE,   State University of New York at Buffalo, Buffalo, NY

ALFRED GESSOW,   University of Maryland, College Park, MD

K. N. GHIA,   University of Cincinnati, Cincinnati, OH

U. GHIA,   University of Cincinnati, Cincinnati, OH

CARL H. GIBSON,   University of California—San Diego, La Jolla, CA

RICHARD J. GOLDSTEIN,   University of Minnesota, Minneapolis, MI

SANFORD GORDON,   Sanford Gordon & Associates, Cleveland, OH

ROBERT A. GREENKORN,   Purdue University, West Lafayette, IN

EDWARD M. GREITZER,   Massachusetts Institute of Technology, Cambridge, MA

BERNARD GROSSMAN,   Virginia Polytechnic Institute and State University, Blacksburg, VA

FREDRICK G. HAMMITT (Deceased),   University of Michigan, Ann Arbor, MI

EVERETT J. HARDGRAVE, JR. (Retired),   Applied Physics Laboratory, The Johns Hopkins University, Laurel, MD

A. GEORGE HAVENER,   U.S. Air Force Academy, Colorado Springs, CO

STEPHEN HEISTER,   Purdue University, West Lafayette, IN

ROBERT E. HENDERSON,   Pennsylvania State University, State College, PA

JACKSON R. HERRING, National Center for Atmospheric Research, Boulder, CO

JOHN L. HESS (Retired), Douglas Aircraft Company, Long Beach, CA

RAYMOND M. HICKS, NASA Ames Research Center, Moffett Field, CA

GUSTAVE J. HOKENSON (Deceased), Air Force Institute of Technology, Dayton, OH

TERRY L. HOLST, NASA Ames Research Center, Moffett Field, CA

DAVID P. HOULT, Massachusetts Institute of Technology, Cambridge, MA

THOMAS J. R. HUGHES, Stanford University, Stanford, CA

EVANGELOS HYTOPOULOS, Automated Analysis Corp., Ann Arbor, MI

TAKAO INAMURA, Tohoku University, Sendai, Japan

DAVID JAPIKSE, Concepts ETI, Incorporated, Wilder, VT

SAM P. JONES, TCOM, L.P., Columbia, MD

HELMUT KECK, Sulzer Innotec/Sulzer Hydro, Winterthur/Zurich, Switzerland

JAMES L. KEIRSEY (Retired), Applied Physics Laboratory, The Johns Hopkins University, Laurel, MD

LAWRENCE A. KENNEDY, University of Illinois at Chicago, Chicago, IL

JOHN KIM, University of California Los Angeles, Los Angeles, CA

ANJANEYULU KROTHAPALLI, Florida State University, Tallahassee, FL

PAUL KUTLER, NASA Ames Research Center, Moffett Field, CA

K. KUWAHARA, Institute of Space and Astronautical Science, Sagamihara, Japan

E. EUGENE LARRABEE (Emeritus), Massachusetts Institute of Technology, Cambridge, MA

J. GORDON LEISHMAN, University of Maryland, College Park, MD

PETER E. LILEY (Retired), Purdue University, West Lafayette, IN

RAINALD LÖHNER, The George Mason University, Fairfax, VA

LUIZ M. LOURENCO, Florida State University, Tallahassee, FL

MUJEEB R. MALIK, High Technology Corporation, Hampton, VA

JAMES F. MARCHMAN, III, Virginia Polytechnic Institute and State University, Blacksburg, VA

CRAIG SAMUEL MARTIN, Georgia Institute of Technology, Atlanta, GA

HUGH R. MARTIN, University of Waterloo, Waterloo, Ontario, Canada

C. D. MAXWELL, STD Research Incorporated, Arcadia, CA

UNMEEL B. MEHTA, NASA Ames Research Center, Moffett Field, CA

JOHN E. MINARDI, University of Dayton, Dayton, OH

ALBERT M. MOMENTHY,   Boeing Commercial Airplane Company, Seattle, WA

THOMAS B. MORROW,   Southwest Research Institute, San Antonio, TX

HANY MOUSTAPHA,   Pratt and Whitney Canada, Montreal, Canada

S. NAKAMURA,   Ohio State University, Columbus, OH

RICHARD NEERKEN (Retired),   The Ralph M. Parsons Company, Pasadena, CA

WAYNE L. NEU,   Virginia Polytechnic Institute and State University, Blacksburg, VA

NEIL OLIEN,   National Institute of Standards and Technology, Boulder, CO

BHARATAN R. PATEL,   Fluent Incorporated, Lebanon, NH

VICTOR L. PETERSON (Retired),   NASA Ames Research Center, Moffett Field, CA

J. LEITH POTTER (Retired),   Vanderbilt University, Nashville, TN

THOMAS H. PULLIAM,   NASA Ames Research Center, Moffett Field, CA

SAAD RAGAB,   Virginia Polytechnic Institute and State University, Blacksburg, VA

RICHARD H. RAND,   Cornell University, Ithaca, NY

EVERETT V. RICHARDSON (Emeritus),   Colorado State University, Fort Collins, CO

DONALD O. ROCKWELL,   Lehigh University, Bethlehem, PA

COLIN RODGERS,   Los Angeles, CA

JOHN P. ROLLINS,   Clarkson University, Potsdam, NY

PHILIP G. SAFFMAN,   California Institute of Technology, Pasadena, CA

MANUEL D. SALAS,   NASA Langley Research Center, Hampton, VA

P. SAMPATH,   Pratt and Whitney Canada, Montreal, Canada

TURGUT SARPKAYA,   Naval Postgraduate School, Monterey, CA

JOSEPH A. SCHETZ,   Virginia Polytechnic Institute and State University, Blacksburg, VA

LEON H. SCHINDEL,   Naval Surface Warfare Center, Silver Spring, MD

JOHN E. SCHMIDT (Retired),   Boeing Commercial Airplane Company, Seattle, WA

WILLIAM B. SHIPPEN (Retired),   Applied Physics Laboratory, The Johns Hopkins University, Laurel, MD

TERRY W. SIMON,   University of Minnesota, Minneapolis, MI

HELMUT SOCKEL,   Technical University of Vienna, Vienna, Austria

GEOFFREY R. SPEDDING,   University of Southern California, Los Angeles, CA

PHILIP C. STEIN, JR.,   Stein Seal Company, Kulpsville, PA

WILLIAM G. STELTZ (Retired),   Westinghouse Electric Company, Orlando, FL

KENNETH G. STEVENS,   NASA Ames Research Center, Moffett Field, CA

PAUL N. SWARZTRAUBER, National Center for Atmospheric Research, Boulder, CO

ROLAND A. SWEET, University of Colorado at Denver, Denver, CO

JULIAN SZEKELY, Massachusetts Institute of Technology, Cambridge, MA

JIMMY TAN-ATICHAT, California State University, Chico, Chico, CA

RICHARD S. TANKIN, Northwestern University, Evanston, IL

TAYFUN E. TEZDUYAR, University of Minnesota, Minneapolis, MN

JAMES L. THOMAS, NASA Langley Research Center, Hampton, VA

CHANG LIN TIEN, University of California, Berkeley, CA

EUGENE D. TRAGANZA, Naval Postgraduate School, Monterey, CA

STEVENS P. TUCKER, Naval Postgraduate School, Monterey, CA

ERNEST W. UPTON, Bloomfield Hills, MI

MICHAEL W. VOLK, Oakland, CA

JAMES WALLACE, University of Maryland, College Park, MD

CANDACE WARK, Illinois Institute of Technology, Chicago, IL

FRANK M. WHITE, University of Rhode Island, Kingston, RI

JOHN M. WIEST, Purdue University, West Lafayette, IN

JAMES C. WILLIAMS, III, Auburn University, Auburn, AL

SCOTT WOODWARD, State University of New York at Buffalo, Buffalo, NY

TERRY WRIGHT, University of Alabama at Birmingham, Birmingham, AL

GEORGE T. YATES, Consultant, Boardman, OH

H. C. YEE, NASA Ames Research Center, Moffett Field, CA

HIDEO YOSHIHARA (Retired), Boeing Company, Seattle, WA

TSUKASA YOSHINAKA, Concepts ETI, Incorporated, Wilder, VT

VIRGINIA E. YOUNG, Virginia Polytechnic Institute and State University, Blacksburg, VA

JAMES L. YOUNGHANS, General Electric Aircraft Engines, Evandale, OH

HENRY C. YUEN, TRW Space and Technology Group, Redondo Beach, CA

EDWARD E. ZUKOSKI, California Institute of Technology, Pasadena, CA

CONTENTS

PREFACE

More than a quarter century has elapsed since the publication of a major handbook on the subject of fluid dynamics and fluid machinery. We, and many others, learned much of our fluid dynamics from the volume edited by Victor Streeter and the relevant volumes of the *Handbuch der Physik*. While fluid dynamics is viewed by some as a mature science, during the past three decades many significant advances have occurred. Building on a glorious history and the recent rapid pace of progress, the future for fluid dynamics is equally exciting. The advances in fluid dynamics and associated equipment are the result of new and powerful forms of instrumentation, e.g., laser velocimeters, and the impressive achievements and rapidly growing capability of computational fluid dynamics. A new handbook is needed to document and record the advanced state of knowledge; hopefully, this handbook fills that need. In addition to the new generation of fluid dynamic information and understanding, the classical topics are presented here also. Finally, a number of exciting topics such as the fluid dynamics of green plants which have not traditionally been considered in such a handbook have been included.

The intended reader and user for this handbook is an engineer with a bachelor's degree in a technical major with basic coursework in fluid dynamics. The authors have written chapters and sections which summarize important engineering facts and data. The material in the handbook is not intended to be a status report concerning ongoing research, but rather the information was selected primarily to provide useful answers for the practicing engineer. The answer may be in the form of a chart, an equation, a constant, a performance parameter, or guidance to the appropriate engineering literature. With the widespread availability of the personal computer, the need to produce pages of tabulated data, e.g., flow properties across a normal shock wave, has declined sharply. In many cases in this handbook, the equation which can generate a table is given in lieu of the many pages of tables.

Although this handbook is part of the Wiley Series of Handbooks for Mechanical Engineers, the information contained within these volumes is of value not only to mechanical engineers but also to aerospace, civil, chemical, environmental, materials, and nuclear engineers. In addition, meteorologists and oceanographers will find the handbook to be a useful tool in their personal library. Physicists with a bent toward fluid dynamics (and there are several thousand such scientists in the world), likewise, will be interested in this handbook.

Looking at the list of contributors, you will notice a large, international group of talented engineers and scientists. As editors, we express our thanks to the many cooperative individuals who have contributed freely of their talents and who wrote sections of the handbook. Such people are always busy, and taking on this additional task was often a burden. We, and the technical community, appreciate their dedication and service. Sadly, several of the authors passed away before seeing this handbook in print.

The Editorial Review Board for the handbook consists of very distinguished and knowledgeable people. Their helpful suggestions and insights are deeply appreciated.

JOSEPH A. SCHETZ
ALLEN E. FUHS

1 Equations for Fluid Statics and Dynamics

FRANK M. WHITE
University of Rhode Island
Kingston, RI

JOHN L. HESS (Retired)
Douglas Aircraft Company
Long Beach, CA

CARL H. GIBSON
University of California
La Jolla, CA

CONTENTS

Handbook of Fluid Dynamics and Fluid Machinery, Edited by Joseph A. Schetz and Allen E. Fuhs
ISBN 0-471-12598-9 Copyright © 1996 John Wiley & Sons, Inc.

1.1   FLUIDS IN STATIC EQUILIBRIUM
Frank M. White

When a fluid is in static equilibrium, its viscous stresses vanish, and its velocity and acceleration are identically zero. The equation of motion reduces to a balance between the static pressure gradient and the body force distribution in the fluid

$$\nabla p = \rho \overline{\mathbf{X}} \qquad (1.1)$$

In general, the body force may be due to gravity, electromagnetic, or other force fields acting on the fluid. By far the most common case is the effect of gravity alone, where $\overline{\mathbf{X}} = \overline{\mathbf{g}}$ and is a known function of space.

From Eq. (1.1), one notes that the only two fluid properties involved in static equilibrium problems are the static pressure p and the mass density ρ. It is assumed that both are known thermodynamic state functions of temperature and concentration. For a single-phase fluid, it is assumed that $\rho = \rho(p, T)$ only.

It is customary to define the coordinate z as *upward* or opposite to the acceleration of gravity, as in Fig. 1.1. Thus, $\overline{\mathbf{g}} = -g\overline{\mathbf{k}}$, so that pressure varies only with z, and Eq. (1.1) becomes

$$\frac{dp}{dz} = -\rho g \, dz,$$

or

$$p_2 - p_1 = -\int_1^2 \rho g \, dz \qquad (1.2)$$

This is the *hydrostatic pressure formula* and holds between any points 1 and 2 in a fluid of uniformly varying density. Since transport properties such as viscosity and thermal conductivity are eliminated from the formula, it follows that Eq. (1.2) is valid for *any* type of fluid in static equilibrium, including strongly non-Newtonian substances.

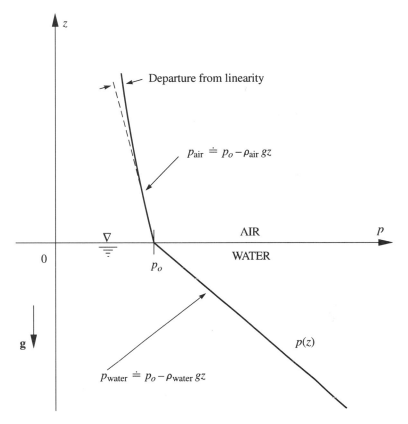

FIGURE 1.1 Hydrostatic pressure distribution in gases and liquids.

For problems involving the earth's atmosphere or oceans, the overall variation in g is less than $\pm 1\%$. Thus, g is commonly considered to be constant at approximately 9.81 m/s². Further, for liquids or gases over short changes in altitude, density is approximately constant, and Eq. (1.2) may be integrated to yield

$$p_2 - p_1 \doteq -\rho_o g(z_2 - z_1) \tag{1.3}$$

where ρ_o is the average fluid density. Equation (1.3) predicts a linear distribution of pressure, increasing downward, as shown in Fig. 1.1. Over large altitude changes, $p(z)$ may depart from linearity; a rule of thumb is that the departure will be less than 5% error if the change in altitude $|z_2 - z_1| < 0.1a^2/g$, where a is the speed of sound of the fluid. Typical hydrostatic pressure distributions are shown in Fig. 1.1.

1.1.1 The Standard Atmosphere

Equation (1.3) is adequate for liquids, which are nearly incompressible; for example, it holds with 1% accuracy over the entire ocean. However, in the atmosphere, density varies considerably with altitude. Assuming an ideal gas,

$$p = \rho R T, \quad R = \text{specific gas constant}$$

the equation of state may be substituted into Eq. (1.2) and integrated with the result

$$p_2/p_1 = \exp\left[-\frac{g}{R}\int_{z_1}^{z_2}\frac{dz}{T(z)}\right] \qquad (1.4)$$

where, for air, $R = 287$ m^2/s$^2 \cdot$K. The *standard atmosphere* assumes the following temperature variation with altitude

a) troposphere, $0 < z < 11.0$ km: $T = 288.16$ K $-$ 0.00650 z(m)

b) stratosphere, $11.0 < z < 20.1$ km: $T = 216.66$ K

Substitution into Eq. (1.4) yields the following formulas for pressure distribution in the standard atmosphere

a) troposphere: $p = 101350[1 - z/44332]^{5.26}$ $\qquad$ (1.5)

b) stratosphere: $p = 22612 \exp[(11000 - z)/6339]$

with p in pascals and z in meters. Tabulated values of pressure, density, and temperature in the standard atmosphere are given in Sec. 2.3 of this handbook.

1.1.2 Forces on Submerged Surfaces

Assuming constant density and gravity, Eq. (1.3) may be integrated to find the hydrostatic pressure force on any surface submerged in a fluid. The most important case is the flat surface shown in Fig. 1.2. Viewed from the side, the surface makes an angle θ with the horizontal.

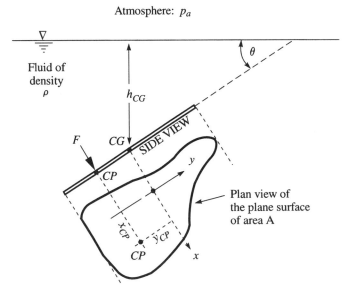

FIGURE 1.2 Location and magnitude of the hydrostatic force on one side of a plane submerged surface.

By integrating the elemental force $dF = pdA$ over the surface, one finds that the total force F on one side of the flat surface is given by

$$F = p_{CG} A = (p_a + \rho g h_{CG}) A \qquad (1.6)$$

where A is the area of the flat surface, and h_{CG} is the depth down to the centroid of the surface, as shown in Fig. 1.2.

However, this force F is not at the centroidal point CG. Rather, it acts at the *center of pressure CP*, which is below the centroid and has centroidal coordinates, in the plane of the flat surface, given by

$$x_{CP} = -\rho g \sin \theta \, I_{xy}/F$$
$$y_{CP} = -\rho g \sin \theta \, I_{xx}/F \qquad (1.7)$$

Here I_{xx} and I_{xy} denote the moment of inertia and product of inertia, respectively, of the area of the plane surface. The coordinate system is shown in Fig. 1.2.

1.1.3 Buoyancy—Archimedes Principles

Consider a body completely submerged in one or more fluids. Using Eq. (1.3) for the net integrated pressure on the body yields the two principles of Archimedes:

I. A body immersed in a fluid(s) experiences a vertical buoyant force B equal to the weight(s) of the fluid(s) it displaces.

II. A floating body displaces its own weight in the fluid(s) in which it floats.

More than one fluid may be involved; for example, a floating body may be partially in air and partially in water.

These two principles of buoyancy are illustrated in Fig. 1.3. For the floating body in Fig. 1.3a, the body weight $W = B = (\rho_1 g v_1 + \rho_2 g v_2)$, where v_1 and v_2 are the body volumes displaced in each fluid. If fluid 1 is a gas and fluid 2 a liquid, the term $(\rho_1 g v_1)$ is often negligible.

For the completely submerged body in Fig. 1.3b, the upward buoyant force $B = \rho_2 g v$, where v is the total displaced volume. Generally B does not equal W, and an additional support force is required. If $B = W$, the body is said to be *neutrally buoyant*.

The true position of a floating body is determined by a stability analysis. See Sec. 3.6 of Fox and McDonald (1992) or Sec. 2.8 of White (1994). A body floating in an unstable position will soon overturn.

1.1.4 Surface Tension

An important boundary condition occurring at the interface between two fluids involves the *coefficient of surface tension*, σ, defined as the surface energy per unit interfacial area. A cut of length dL made in a surface will expose a tension force σdL normal to the cut and parallel to the surface. If the surface is curved, these surface forces can cause a pressure jump across the surface.

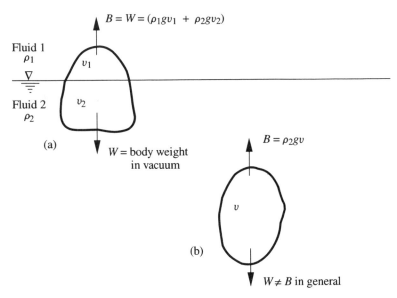

FIGURE 1.3 The two buoyancy principles of Archimedes: (a) a floating body and (b) a completely submerged body.

Figure 1.4 shows a doubly curved element of surface with the surface tension forces on the cut sides. These forces have a net component normal to the center of the surface. A force balance, as shown in Fig. 1.4, reveals that the pressure difference is given by

$$\Delta p = \sigma\left(\frac{1}{R_1} + \frac{1}{R_2}\right) \tag{1.8}$$

where R_1 and R_2 are the principal radii of curvature of the surface. From this relation, one deduces three special cases for particular shapes

a) solid liquid cylinder: $\Delta p = \sigma/R$

b) solid liquid droplet: $\Delta p = 2\sigma/R$ (1.9)

c) thin-walled bubble: $\Delta p = 4\sigma/R$

where R is the body radius.

When a fluid interface strikes a solid surface, its behavior is characterized by the *contact angle* θ between the heavier fluid and the solid as shown in Fig. 1.5. When the contact angle is less than 90°, as in Fig. 1.5a, the fluid is said to *wet* the surface; an example is the contact between air/water and clean glass, $\theta \doteq 0°$. If θ is greater than 90°, the fluid is *nonwetting*, Fig. 1.5b; an example is air/mercury and clean glass, $\theta \doteq 130°$.

When a tube is inserted in a pool of fluid, as in Fig. 1.5c, the wetting action of surface tension causes a capillary rise of fluid in the tube to a height h given by

$$h = \frac{2\sigma \cos \theta}{\rho g R}, \tag{1.10}$$

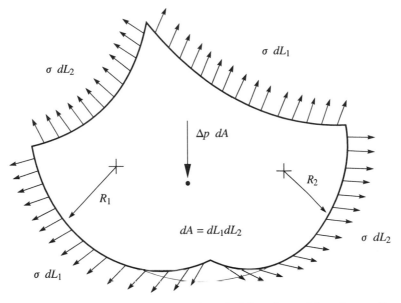

FIGURE 1.4 An element of doubly curved interfacial surface subjected to surface tension forces and a resulting pressure difference.

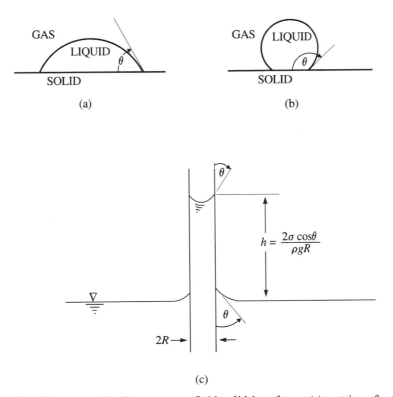

FIGURE 1.5 Contact angle phenomena at fluid-solid interfaces: (a) wetting, $\theta < 90°$, (b) nonwetting, $\theta > 90°$, and (c) capillary rise in a tube for a wetting liquid.

where R is the tube radius. This rise can be very large for small diameter tubes. If the fluid, like mercury on glass, is nonwetting, $\cos\theta$ will be negative, resulting in a capillary depression in the tube. Equation (1.10) is valid for wetting or nonwetting fluids.

Further data on surface tension and contact angle of common fluid/fluid and fluid/solid interfaces are given in Chap. 2.

1.2 FLUID DYNAMICS
Frank M. White

The term *fluid dynamics* encompasses the study of the laws of conservation of mass, momentum, and energy as they apply to flow of fluids. In general, the fluid may be viscous, heat-conducting, non-Newtonian, compressible, and multiphase or multi-component. However, if all these effects occur simultaneously, the flow is so complex and the number of parameters so large that it is essentially impossible to study, even by experimental means. Therefore, to gain any insight into fluid dynamics, it is necessary to limit the range of parameters in any given study. The general field of fluid dynamics has been divided into sub-problems where only a few parameters vary: inviscid potential flow (Sec. 1.3), non-Newtonian flow (Chap. 3), compressible flow (Chaps. 8, 9, and 10), viscous heat-conducting flow (Chap. 4), unsteady flow (Chap. 12), and multiphase flow (Chap. 15). Even thus limited, flow analysis of the basic fluid dynamic equations is usually confined to relatively simple geometries and flow patterns. More complex flows can be studied by either computational schemes (Chaps. 18 to 22) or by experiment (Chaps. 15 and 16). In this section the basic equations governing inviscid fluid dynamics are discussed.

1.2.1 Relevant Fluid Properties

For inviscid flow of a single-phase fluid, the relevant fluid properties are the density ρ, static pressure p, temperature T, and the velocity vector $\overline{V}$ or its Cartesian components (u, v, w). These four unknowns are computed from the three conservation relations of mass, momentum, and energy plus the thermodynamic state relation of the particular fluid.

1.2.2 Eulerian and Lagrangian Coordinates

When analyzing the motion of a fluid, one may adopt either of two different viewpoints. One is the *Eulerian* system, in which all variables are described in terms of a single set of fixed coordinates plus time. Of special interest is the Eulerian velocity vector

$$\begin{aligned} \overline{V} &= \overline{V}(x, y, z, t) \\ &= \bar{\mathbf{i}}u(x, y, z, t) + \bar{\mathbf{j}}v(x, y, z, t) + \overline{\mathbf{k}}w(x, y, z, t) \end{aligned} \tag{1.11}$$

An analogy is as if the entire flow field were watched through a glass window on which is scribed a set of fixed coordinate lines (x, y, z). This system is very appropriate to fluid mechanics which involves continuous variations in space of a deformable material.

The second common viewpoint is the *Lagrangian* system, in which the position $\overline{\mathbf{X}}$ of each particle is studied as a function of time and the particle's initial position $\overline{\mathbf{X}}_0$ at $t = 0$

$$\overline{\mathbf{X}} = \overline{\mathbf{X}}(\overline{\mathbf{X}}_0, t)$$
$$= \overline{\mathbf{i}}x(x_0, y_0, z_0, t) + \overline{\mathbf{j}}y(x_0, y_0, z_0, t) + \overline{\mathbf{k}}z(x_0, y_0, z_0, t) \qquad (1.12)$$

The Lagrangian velocity and acceleration are given by

$$\overline{\mathbf{V}}(\overline{\mathbf{X}}_0, t) = \frac{\partial}{\partial t}[\overline{\mathbf{X}}(\overline{\mathbf{X}}_0, t)]$$
$$\overline{\mathbf{a}}(\overline{\mathbf{X}}_0, t) = \frac{\partial^2}{\partial t^2}[\overline{\mathbf{X}}(\overline{\mathbf{X}}_0, t)] \qquad (1.13)$$

and also follow each specific particle. The Lagrangian system is appropriate to problems involving particle physics and solid mechanics. Only a few fluid dynamics problems are suitable for Lagrangian coordinates; an example is the analysis of particle motions beneath surface waves in the ocean. In this handbook, Eulerian system is used almost exclusively. A very good description of the two systems and the transformations which relate them is given in the classic text by Lamb (1945).

The Eulerian acceleration is found by differentiation of Eq. (1.11)

$$\overline{\mathbf{a}} = \frac{d\overline{\mathbf{V}}}{dt}$$
$$= \frac{\partial\overline{\mathbf{V}}}{\partial t} + u\frac{\partial\overline{\mathbf{V}}}{\partial x} + v\frac{\partial\overline{\mathbf{V}}}{\partial y} + w\frac{\partial\overline{\mathbf{V}}}{\partial z} \qquad (1.14)$$

This expression is nonlinear because of the products of velocity components and spatial velocity derivatives. Expanded in scalar form, Eq. (1.14) contains twelve terms and is the principal reason why exact mathematical analysis of fluid dynamics problems is very difficult.

1.2.3 One-Dimensional Steady Inviscid Flow

The general analysis of three-dimensional inviscid flow is rather complex and is treated in Sec. 1.3. As a first approximation, one can obtain surprisingly good results from a one-dimensional approximation.

Figure 1.6(a) shows property profiles for flow of a real fluid through a duct. The velocity is generally large in the middle of the duct but falls to zero at the walls due to the no-slip condition for viscous fluids. The temperature may be hot or cold in the middle of the duct but always approaches T_w at the walls due to the no-temperature-jump condition. The density also varies across the duct and attains a value ρ_w at the walls corresponding to T_w and the local wall pressure. In a real fluid, then, all property profiles vary both downstream and across the duct.

In the one-dimensional approximation, sketched in Fig. 1.6(b), these profiles are averaged across the duct to yield mean properties $V(x)$, $T(x)$, $\rho(x)$, and $p(x)$ which vary only in the flow direction. The theory is greatly simplified.

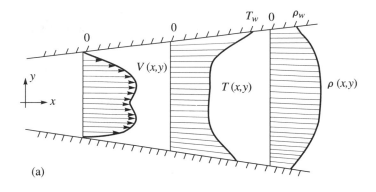

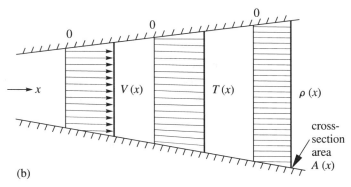

FIGURE 1.6 Compressible flow through a duct: (a) real-fluid velocity, temperature, and density profiles and (b) the one-dimensional approximation.

Continuity. For one-dimensional unsteady flow, the mass conservation or *continuity* equation is

$$A \frac{\partial \rho}{\partial t} + \frac{\partial}{\partial x} (\rho A V) = 0 \tag{1.15}$$

where $A(x)$ is the duct cross-section area which is assumed to be known. For steady flow, Eq. (1.15) reduces to the case of constant mass flow $\dot{m}$ through the duct

$$\rho(x) A(x) V(x) = \dot{m} = \text{constant} \tag{1.16}$$

When density is highly variable, as in a compressible gas flow, this equation has many interesting properties, as discussed in Sec. 8.3.

Momentum. Assuming inviscid flow, the x-momentum relation for unsteady flow through a duct is the *Euler equation,*

$$\frac{\partial V}{\partial t} + \frac{1}{\rho} \frac{\partial p}{\partial x} + V \frac{\partial V}{\partial t} + g \frac{\partial z}{\partial x} = 0 \tag{1.17}$$

where $z(x)$ is the vertical elevation of any point along the axis of the duct. Note that Eq. (1.17) does not explicitly contain the duct area $A(x)$. For steady flow, Eq. (1.17) integrates to the celebrated Bernoulli equation, which is valid along a streamline (here the duct axis) for any inviscid flow

$$\int \frac{dp}{\rho} + \tfrac{1}{2} V^2 + gz = \text{constant} \tag{1.18}$$

The integral cannot be evaluated until a pressure-density relation is assumed. The most common assumption is an *incompressible* fluid, $\rho = \text{constant}$, which is valid for any flow, liquid or gas, moving at Mach numbers less than about 0.3. For incompressible flow, the Bernoulli relation becomes

$$p + \tfrac{1}{2}\rho V^2 + \rho g z = \text{constant} \tag{1.19}$$

The constant may be evaluated from known properties at any convenient point along the flow.

If the fluid is compressible, it is customary to assume isentropic flow for which the enthalpy and pressure changes are related by

$$dh = dp/\rho$$

Substitution into Eq. (1.18) and integration yields the compressible form of the Bernoulli relation

$$h + \tfrac{1}{2} V^2 + gz = \text{constant} \tag{1.20}$$

This can be transformed into some useful algebraic forms by making the ideal-gas assumption, as in Sec. 8.3 of this handbook. It is interesting that Eq. (1.20) is identical to the *steady flow energy equation* for any flow (liquid or gas) with zero heat flux and zero shaft work in the duct. A conclusion is that Eq. (1.20) is not valid for conditions involving heat flux or shaft work.

Applications of the Bernoulli Equation. The Bernoulli equation has many practical applications. Four of these are shown in Fig. 1.7: a) stagnation pressure, b) the Pitot tube, c) the venturi meter, and d) the orifice meter. Each application is discussed in order assuming an incompressible flow with negligible variations in the gravity term gz (e.g., a horizontal flow).

Since the Bernoulli "constant" holds at any point along a streamline of the flow, it must hold at the *stagnation point* which occurs at the front of any body immersed in a freestream, as shown in Fig. 1.7(a). This point divides the flow into an upper part which moves over the body and a lower part which moves beneath the body. The velocity at a stagnation point is zero and, from Eq. (1.19), the pressure at this point rises to a large value called the *stagnation pressure*, p_0

$$p_0 = p_\infty + \tfrac{1}{2}\rho V_\infty^2 \tag{1.21}$$

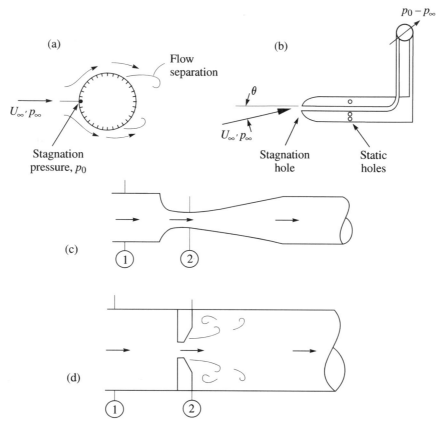

FIGURE 1.7 Four applications of Bernoulli's equation: (a) stagnation pressure, (b) Pitot tube, (c) venturi tube, and (d) orifice plate.

where p_∞ and V_∞ are the flow properties in the approaching freestream. The stagnation pressure, p_0, can be interpreted as the sum of the *static pressure* p_∞ and the *dynamic pressure* $\frac{1}{2}\rho V_\infty^2$ in the freestream. The dynamic pressure can be quite large if the fluid is dense and moving at a high velocity. For example, in water flow, the dynamic pressure equals one atmosphere when the flow velocity is about 14 m/s. The determination of stagnation pressure in a high-Mach-number (compressible) gas flow is discussed in Sec. 8.3.

As a second application, Fig. 1.7(b) shows a velocity-measuring instrument called a *Pitot tube*, named after its inventor, Henri de Pitot (1695–1771). The *stagnation* hole at the front of the tube measures p_0 while the *static* holes on the side sense p_∞. A differential pressure gage placed across the output of these two values registers the pressure difference $(p_0 - p_\infty)$, when Eq. (1.21) predicts

$$V_\infty \doteq [2(p_0 - p_\infty)/\rho]^{1/2} \tag{1.22}$$

The friction losses of this device are very small, but a substantial error may occur if the flow is yawed at an angle θ with respect to the tube axis as shown in Fig. 1.7(b). Further details about Pitot tubes are given in Sec. 15.3.

Another application of Bernoulli's equation is to flow measurement in a duct. Since Eq. (1.19) holds at every point along an (inviscid) streamline, it implies that an increase in velocity will cause a decrease in pressure. Thus, if a duct is fitted with a local obstruction, it will cause a local pressure drop which can be correlated with the flow rate through the duct. Two examples of such *Bernoulli obstruction* flow meters are the Venturi tube in Fig. 1.7(c) and the orifice plate in Fig. 1.7(d).

Let point 1 be upstream of the obstruction and point 2 be in the constricted high-velocity flow through the obstruction. Then, for either the venturi tube or the orifice plate, the continuity and Bernoulli equations may be used as an estimate of the flow changes between 1 and 2

$$\text{a) Continuity:} \quad \dot{m} = \rho_1 \frac{\pi}{4} D_1^2 V_1 = \rho_2 \frac{\pi}{4} D_2^2 V_2$$

$$\text{b) Bernoulli:} \quad p_0 = p_1 + \frac{1}{2} \rho V_1^2 = p_2 + \frac{1}{2} \rho V_2^2$$

Assuming $\rho_1 = \rho_2$ (incompressible flow), V_1 may be eliminated between these two to obtain an estimate of V_2

$$V_2^2 = \frac{2(p_1 - p_2)}{\rho(1 - D_2^4/D_1^4)}$$

The corresponding volume flow would be $Q = V_2 A_2$. This is an idealized estimate. In practice, friction losses in the obstruction cause the formula to be modified by a fractional (i.e., $C_d < 1.0$) *discharge coefficient, C_d*. Then the recommended correlation for volume flow through an obstruction meter is

$$Q = C_d A_2 \left[\frac{2(p_1 - p_2)}{\rho(1 - \beta^4)} \right]^{1/2} \tag{1.23}$$

where $\beta = D_2/D_1$. The experimental values of discharge coefficient are primarily affected by Reynolds number, diameter ratio, and the geometry of the device

$$C_d = f(\rho V_1 D_1/\mu, \, D_2/D_1, \, \text{geometry}) \tag{1.24}$$

Geometric effects include the shape of the obstruction, the location of the pressure taps, and the surface roughness. There are also compressibility effects if the throat Mach number M_2 is larger than about 0.2.

The venturi tube in Fig. 1.7(b) is designed to have very small friction losses; hence, its measured C_d is of the order of 0.98. The orifice plate in Fig. 1.7(d) causes large flow distortion and has $C_d \doteq 0.6$. One advantage of the orifice plate is that it is inexpensive to construct and install. For further details about obstruction flow meters, see Chap. 15 of this handbook or Chap. H of the text by Ward–Smith (1980).

The Energy Equation. Application of the First Law of Thermodynamics to steady one-dimensional flow between section 1 (upstream) and section 2 (downstream) leads to the energy equation

$$\left(h_1 + \frac{1}{2}V_1^2 + gz_1\right) + \frac{dQ}{dm} = \left(h_2 + \frac{1}{2}V_2^2 + gz_2\right) + \frac{dW_s}{dm} \qquad (1.25)$$

where dQ/dm is the heat added per unit mass, and dW_s/dm is the shaft work done by the flow per unit mass between sections 1 and 2. For adiabatic flow with no shaft work, Eq. (1.25) reduces to the form of Bernoulli's equation given as Eq. (1.20), as mentioned before.

Equation (1.25) is often called the *steady flow energy equation*. It assumes that viscous work on the fluid is negligible, so it should be applied only to a duct flow with fixed solid walls or to a streamtube with negligible surface viscous stresses. It also requires that the flow properties at sections 1 and 2 be approximately one-dimensional.

State Relations. In problems involving shaft work, heat transfer, and compressibility, all three of the basic thermodynamic properties of pressure, temperature, and density are variables. In this case, they should be related by the thermodynamic *state relation* for the particular fluid substance. Assume a single-phase substance. If the fluid is complex—steam, for example—use of charts and tables of the form

$$\rho = \rho(p, T) \qquad h = h(p, T) \qquad (1.26)$$

may be required. However, there are at least three useful approximate formulas commonly used for analysis of fluid flow. One is the incompressible fluid assumption

$$\rho \doteq \text{constant}; \qquad dh \doteq c_p dT \qquad (1.27)$$

Compressible gas flows are often analyzed with the ideal-gas assumption

$$p = \rho RT; \qquad dh = c_p dT \qquad (1.28)$$

where R is the specific gas constant. For air, $R \doteq 287 \text{ m}^2/(\text{s}^2 \cdot \text{K})$.

Finally, for liquid flow with large pressure changes, compressibility can be modeled by the empirical relation [Cole (1948)]

$$\frac{p}{p_a} \doteq (B + 1)\left(\frac{\rho}{\rho_a}\right)^n - B \qquad (1.29)$$

where B and n are dimensionless constants to fit the particular liquid and p_a and ρ_a are any convenient reference properties, say, at atmospheric pressure. For water at 20°C with $p_a = 1$ atm and $\rho_a = 999 \text{ kg/m}^3$, experiments at pressures up to 1000 atm yield $B \doteq 3000$ and $n \doteq 7$ [Cole (1948)].

The one-dimensional flow approximation is useful for making simple design-type estimates of flow behavior in many problems, such as compressible flow (Chap. 8), internal flows (Chap. 5), and fluid meters (Chap. 15). One should note, however, that it is a very poor approximation in other problems, for example, boundary layer flow (Chap. 4).

1.2.4 Three-Dimensional Control Volume Relations

Many fluid flow problems are inherently two or three dimensional, and the one-dimensional approximations of Sec. 1.2.3 do not apply. In three-dimensional flows, one has the choice of analyzing either a) a fixed region of the flow, called a *control volume*, or b) a fixed mass of fluid, called a *system*. The system approach deals with a differential-sized particle and will be treated in Sec. 1.2.5. In this section, control volumes are considered.

System analyses may be directly applied to fluid flows, because the basic laws of conservation of mass, momentum, and energy are commonly written for systems. To apply these laws to a region or control volume, through which many fluid systems pass, requires a transformation between system derivatives and control volume rates of change.

Consider the arbitrary fixed control volume *CV* in Fig. 1.8, with an arbitrary control surface *CS*. The flow passes through the control volume—from left to right for convenience—and may have both spatial and time variations.

Let *B* be any extensive property, i.e., proportional to mass, of the fluid system which occupies the control volume at time t. At time $t + dt$, this system has begun to move out of the control volume, as shown in Fig. 1.8. Let $\beta(x, y, z, t) = dB/dm$ be the local value of *B* per unit mass at any point in the control volume. Then Fig. 1.8 indicates that, in control volume terms, the rate of change of *B* consists of the changes within the control volume, plus the flux of *B* out of the control surface, minus the flux of *B* into the control surface. The two flux terms can be combined

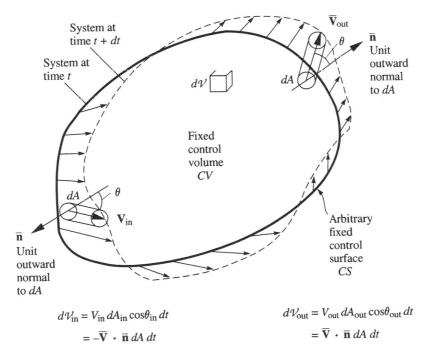

$$dV_{in} = V_{in}\, dA_{in}\, \cos\theta_{in}\, dt$$

$$= -\overline{\mathbf{V}} \cdot \overline{\mathbf{n}}\, dA\, dt$$

$$dV_{out} = V_{out}\, dA_{out}\, \cos\theta_{out}\, dt$$

$$= \overline{\mathbf{V}} \cdot \overline{\mathbf{n}}\, dA\, dt$$

FIGURE 1.8 Definition sketch for an arbitrary fixed control volume subjected to an arbitrary flow pattern.

by defining an outward normal unit vector $\bar{\mathbf{n}}$ at every point on the control surface, as shown in Fig. 1.8. Then the normal velocity component around the surface, $V_n = \overline{\mathbf{V}} \cdot \bar{\mathbf{n}}$, will be positive for outflow and negative for inflow.

The result is that the system rate of change of B can be written as the sum of two integrals involving the intensive property β and the fluid velocity vector

$$\left.\frac{dB}{dt}\right|_{\text{system}} = \frac{\partial}{\partial t} \iiint_{CV} \beta\rho \, d\mathcal{V} + \iint_{CS} \beta\rho(\overline{\mathbf{V}} \cdot \bar{\mathbf{n}}) \, dA \tag{1.30}$$

where $d\mathcal{V}$ and dA are the volume and area elements associated with the control volume and control surface, respectively, as shown in Fig. 1.8.

Equation (1.30) is a general transformation for a fixed control volume called the *Reynolds transport equation*. It may be applied successively to mass, momentum, angular momentum, and energy in the next sections.

In a more general case, the control volume is not fixed but instead moves and deforms with a variable velocity $\overline{\mathbf{V}}_s(x, y, z, t)$. In this case the Reynolds transport equation may be generalized to

$$\left.\frac{dB}{dt}\right|_{\text{system}} = \frac{d}{dt} \iiint_{CV} \beta\rho \, d\mathcal{V} + \iint_{CS} \beta\rho(\overline{\mathbf{V}}_r \cdot \bar{\mathbf{n}}) \, dA \tag{1.31}$$

where $\overline{\mathbf{V}}_r = \overline{\mathbf{V}} - \overline{\mathbf{V}}_s$ is the relative velocity vector between the fluid and the moving control surface. Some examples of deformable control volume analyses are given by White (1994) and in the text by Potter and Foss (1975).

Continuity. For mass conservation, the appropriate property B is the system mass m, with β equal to unity. The system law requires that $dm/dt = 0$, hence, Eq. (1.30) becomes the integral continuity relation for a fixed control volume

$$\frac{\partial}{\partial t} \iiint_{CV} \rho \, d\mathcal{V} + \iint_{CS} \rho(\overline{\mathbf{V}} \cdot \bar{\mathbf{n}}) \, dA = 0 \tag{1.32}$$

For steady flow with a number of one-dimensional inlets and exits, this relation reduces to

$$\left(\sum_j \rho_j A_j V_j\right)_{\text{out}} - \left(\sum_j \rho_j A_j V_j\right)_{\text{in}} = 0 \tag{1.33}$$

Finally, if there is only one inlet and one exit, Eq. (1.33) reduces to the simple one-dimensional relation, Eq. (1.16), discussed earlier.

Linear Momentum. For conservation of linear momentum (Newton's second law), B is a vector, the linear momentum $m\overline{\mathbf{V}}$, and $\beta = \overline{\mathbf{V}}$. Newton's law requires that the system change $d(m\overline{\mathbf{V}})/dt = \Sigma\overline{\mathbf{F}}$, so that Eq. (1.30) becomes the *integral momentum relation* for a fixed control volume

$$\frac{\partial}{\partial t} \iiint_{CV} \overline{\mathbf{V}}\rho \, d\mathcal{V} + \iint_{CS} \overline{\mathbf{V}}\rho(\overline{\mathbf{V}} \cdot \overline{\mathbf{n}}) \, dA = \Sigma \overline{\mathbf{F}} \tag{1.34}$$

where $\Sigma \overline{\mathbf{F}}$ denotes the vector sum of all forces on any type acting on the control volume or its surface. These forces are generally two types: a) *body forces* $\overline{\mathbf{B}}$ which act on each mass element dm within the control volume, and b) *surface tractions* $\overline{\mathbf{T}}$ which act on each element dA of the surface. Then one could evaluate the total force as

$$\Sigma \overline{\mathbf{F}} = \iiint_{CV} \overline{\mathbf{B}}\rho \, d\mathcal{V} + \iint_{CS} \overline{\mathbf{T}} \, dA \tag{1.35}$$

The most common body force is the gravity vector, $\overline{\mathbf{B}} = -g\overline{\mathbf{k}}$.

For the special case of steady flow with a finite number of one-dimensional inlets and outlets, Eq. (1.34) reduces to

$$\left(\sum_j \overline{\mathbf{V}}_j \dot{m}_j \right)_{\text{out}} - \left(\sum_j \overline{\mathbf{V}}_j \dot{m}_j \right)_{\text{in}} = \Sigma \overline{\mathbf{F}} \tag{1.36}$$

where $\dot{m}_j = \rho_j A_j V_j$ is the mass flow at each section j. Finally, for a single inlet and single outlet, the one-dimensional vector momentum relation is

$$\dot{m}(\overline{\mathbf{V}}_{\text{out}} - \overline{\mathbf{V}}_{\text{in}}) = \Sigma \overline{\mathbf{F}} \tag{1.37}$$

This relation can be used, for example, to estimate the fluid-induced force on a turning vane or turbomachinery blade.

In some cases, notably the theory of rotating fluids (Sec. 13.2) or turbomachinery (Chap. 27), it is convenient to write the integral momentum equation in a noninertial coordinate system. Further details about noninertial control volumes can be found in the texts by White (1994), Potter and Foss (1975), and Shames (1982).

Angular Momentum. The Reynolds transport relation, Eq. (1.30), can be applied to conservation of angular momentum by letting the property B be the angular momentum vector $\overline{\mathbf{r}}_O \times (m\overline{\mathbf{V}})$, where $\overline{\mathbf{r}}_O$ is the position vector to the mass m from any point O about which applied moments are computed. Then $\beta = \overline{\mathbf{r}}_O \times \overline{\mathbf{V}}$, and the angular momentum integral relation becomes

$$\frac{\partial}{\partial t} \iiint_{CV} (\overline{\mathbf{r}}_O \times \overline{\mathbf{V}})\rho \, d\mathcal{V} + \iint_{CS} (\overline{\mathbf{r}}_O \times \overline{\mathbf{V}})\rho(\overline{\mathbf{V}} \cdot \overline{\mathbf{n}}) \, dA = \Sigma \overline{\mathbf{M}}_O \tag{1.38}$$

where $\Sigma \overline{\mathbf{M}}_O$ denotes the sum of all applied moments of the control volume body and surface forces about point O. The net moment may be written in terms of the force fields $\overline{\mathbf{B}}$ and $\overline{\mathbf{T}}$ discussed in Eq. (1.35)

$$\Sigma \overline{\mathbf{M}}_O = \iiint_{CV} (\overline{\mathbf{r}}_O \times \overline{\mathbf{B}})\rho \, d\mathcal{V} + \iint_{CS} (\overline{\mathbf{r}}_O \times \overline{\mathbf{T}}) \, dA \tag{1.39}$$

Equation (1.38), when applied to the axis of a turbomachine, relates the machine torque to its flow rate and geometry as shown in Chap. 27.

The noninertial coordinate form of Eq. (1.38) is discussed in detail on pp. 166–170 of Shames (1982).

First Law of Thermodynamics. The first law of thermodynamics for a system is

$$\frac{dE}{dt} = \frac{dQ}{dt} - \frac{dW}{dt} \tag{1.40}$$

where E is the system energy, Q is the heat added to the system, and W is the work done by the system. Dimensions for E, Q, and W are Joules. To apply the Reynolds transport relation, Eq. (1.30), interpret B as E, with $\beta = dE/dm = e$. Then the integral energy relation is

$$\frac{\partial}{\partial t} \iiint_{CV} e_t \rho \, d\mathcal{V} + \iint_{CS} e_t \rho (\overline{\mathbf{V}} \cdot \overline{\mathbf{n}}) \, dA = \frac{dQ}{dt} - \frac{dW}{dt} \tag{1.41}$$

for a fixed control volume. For a moving fluid, interpret e_t to be the sum of internal, kinetic, and potential energy

$$e_t = e + \tfrac{1}{2}V^2 + gz \tag{1.42}$$

where z is the vertical coordinate.

The work term in Eq. (1.41) includes shaft work W_s, surface traction work, and the work of body forces—*not* including the work of gravity forces, which are already given as the potential energy term in Eq. (1.42)

$$\frac{dW}{dt} = \frac{dW_s}{dt} - \iiint_{CV} (\overline{\mathbf{B}}' \cdot \overline{\mathbf{V}}) \rho \, d\mathcal{V} - \iint_{CS} (\overline{\mathbf{T}} \cdot \overline{\mathbf{V}}) \, dA \tag{1.43}$$

where $\overline{\mathbf{B}}'$ denotes the nongravitational body forces per unit mass.

For steady flow through a control volume with solid walls, the viscous work terms are negligible because of the no-slip condition. The work of surface tractions is due only to pressure forces at the boundary, which can be combined with the control surface energy flux terms. Then the energy equation for steady flow becomes

$$\iint_{CS} \left(h + \frac{1}{2}V^2 + gz \right) \rho (\overline{\mathbf{V}} \cdot \overline{\mathbf{n}}) \, dA$$
$$= \frac{dQ}{dt} - \frac{dW_s}{dt} + \iiint_{CV} (\overline{\mathbf{B}}' \cdot \overline{\mathbf{V}}) \rho \, d\mathcal{V} \tag{1.44}$$

If there are no body forces except gravity, $\overline{\mathbf{B}}' = 0$. If, further, the control volume has only a finite number of one-dimensional inlets and exits, Eq. (1.44) becomes

$$\left[\sum_j (h_j + \tfrac{1}{2}V_j^2 + gz_j)\dot{m}_j \right]_{\text{out}} - \left[\sum_j \left(h_j + \tfrac{1}{2}V_j^2 + gz_j \right)\dot{m}_j \right]_{\text{in}} = \frac{dQ}{dt} - \frac{dW_s}{dt} \tag{1.45}$$

Finally, if there is only one inlet and one exit, then $\dot{m}_{out} = \dot{m}_{in}$ from the continuity relation Eq. (1.16) which yields the one-dimensional steady flow energy equation

$$\left(h + \frac{1}{2} V^2 + gz\right)_{out} - \left(h + \frac{1}{2} V^2 + gz\right)_{in} = \frac{dQ}{dm} - \frac{dW_s}{dm} \tag{1.46}$$

This is identical to our earlier Eq. (1.25), in which subscript 1 meant inflow and subscript 2 outflow.

For additional discussion of thermodynamics, see Chap. 7.

Second Law of Thermodynamics. The second law, being an inequality, is rarely used in flow analysis except as a test of the validity of a particular flow pattern. For example, in Sec. 8.4, the second law is used to eliminate a mathematical solution for a shock wave which is impossible physically. For a system, the second law of thermodynamics takes the form

$$\frac{1}{T} \frac{dQ}{dt} \leq \frac{dS}{dt} \tag{1.47}$$

where S is the system entropy, J/K, and T its absolute temperature. For use in the Reynolds transport relation Eq. (1.30), interpret B as S and β as $dS/dm = s$. The integral form of the second law becomes

$$\iint_{CS} \frac{1}{T} \frac{dQ}{dt} \leq \frac{\partial}{\partial t} \iiint_{CV} s\rho \, d\mathcal{V} + \iint_{CS} s\rho(\overline{V} \cdot \overline{n}) \, dA \tag{1.48}$$

For steady flow with a number of one-dimensional inlets and outlets, this reduces to

$$\iint_{CS} \frac{1}{T} \frac{dQ}{dt} \leq \left(\sum_j s_j \dot{m}_j\right)_{out} - \left(\sum_j s_j \dot{m}_j\right)_{in}$$

If there is only one inlet and one outlet, the result is

$$\iint_{CS} \frac{1}{T} \frac{dQ}{dt} \leq \dot{m}(s_{out} - s_{in})$$

Finally, for one-dimensional adiabatic flow $(dQ = 0)$, this reduces to

$$s_{out} \geq s_{in} \tag{1.49}$$

It is this condition which is used as a test in the theory of shock waves in Sec. 8.4.

These five basic control volume integral relations—continuity, Eq. (1.32), linear momentum, Eq. (1.34), angular momentum, Eq. (1.38), first law, Eq. (1.41), and second law, Eq. (1.48)—are used in many different types of analyses of fluid flow and fluid machinery, as seen in subsequent sections of this handbook. The monograph by Bejan (1982) gives an interesting application of the entropy-generation concept to fluid mechanics.

1.2.5 Three-Dimensional Inviscid Differential Relations

As mentioned in the introduction to Sec. 1.2.4, one may take a *system approach* to fluid mechanics by examining the motion of a differential particle of mass $dm = \rho\, d\mathcal{V}$ moving in a flow. Application of conservation of mass, linear momentum, and energy leads to three basic partial differential equations of fluid motion which, when integrated with the proper boundary conditions, yield the flow field properties as a function of space and time.

In differential analysis, except for rarefied gases, one makes the *continuum hypothesis* that all fluid properties are continuous functions of space and time. Therefore, the concepts of the differential calculus apply, and all properties change by a differential amount across the breadth of a differential-sized element. Some examples are shown in Fig. 1.9. Between the two "x" faces on the left and right sides of the element, properties will change only if they have a nonzero partial derivative with respect to x. These differences cause the element to change size, accelerate, receive heat by conduction, etc., and the balances of these differences result in the basic differential equations.

Continuity. Referring to Fig. 1.9, the excess velocity $[(\partial u/\partial x)dx]$ of the right face will change the volume of the element, as will similar velocity differences $[(\partial v/\partial y)dy]$ and $[(\partial w/\partial z)dz]$ in the y- and z-directions, respectively. To conserve the mass of the element during these volume changes, the density of the element must also

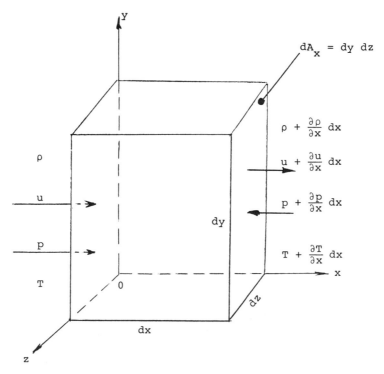

FIGURE 1.9 Schematic of a differential fluid particle, showing differential variations of fluid properties in the x direction. Variations in y and z directions are omitted.

change. This balance results in the *continuity* or conservation of mass differential equation for flow of a single-phase fluid

$$\frac{\partial \rho}{\partial t} + \frac{\partial}{\partial x}(\rho u) + \frac{\partial}{\partial y}(\rho v) + \frac{\partial}{\partial z}(\rho w) = 0$$

or

$$\frac{\partial \rho}{\partial t} + \nabla \cdot (\rho \overline{\mathbf{V}}) = 0 \tag{1.50}$$

This relation is quite general and applies to steady or unsteady flow of viscous or inviscid fluids. If the flow is *steady*, drop the term $(\partial \rho / \partial t)$.

If the flow is *incompressible*, then density drops out entirely from the continuity equation, which becomes

$$\nabla \cdot \overline{\mathbf{V}} = \frac{\partial u}{\partial x} + \frac{\partial v}{\partial y} + \frac{\partial w}{\partial z} = 0 \tag{1.51}$$

The flow may be either steady or unsteady.

Cylindrical Coordinates. Besides the Cartesian system, there are two curvilinear orthogonal coordinate systems in use: a) cylindrical, and b) spherical polar. Both are illustrated in Fig. 1.10.

The cylindrical coordinates, as shown in Fig. 1.10(a), are defined by the Cartesian transformation

$$x = r \cos \theta; \qquad y = r \sin \theta; \qquad z = z \tag{1.52}$$

where the baseline $(\theta = 0)$ is parallel to the x axis. The equation of continuity Eq. (1.50) in these coordinates becomes

$$\frac{\partial \rho}{\partial t} + \frac{1}{r}\frac{\partial}{\partial r}(r \rho v_r) + \frac{1}{r}\frac{\partial}{\partial \theta}(\rho v_\theta) + \frac{\partial}{\partial z}(\rho v_z) = 0 \tag{1.53}$$

where the radial, tangential, and axial velocity components (v_r, v_θ, v_z) are sketched in Fig. 1.10a. The incompressible form of this relation is

$$\frac{1}{r}\frac{\partial}{\partial r}(r v_r) + \frac{1}{r}\frac{\partial}{\partial \theta}(v_\theta) + \frac{\partial}{\partial z}(v_z) = 0 \tag{1.54}$$

Spherical Coordinates. Spherical polar coordinates are illustrated in Fig. 1.10(b) and are defined by the Cartesian transformation

$$x = r \sin \theta \cos \phi$$
$$y = r \sin \theta \sin \phi \tag{1.55}$$
$$z = r \cos \theta$$

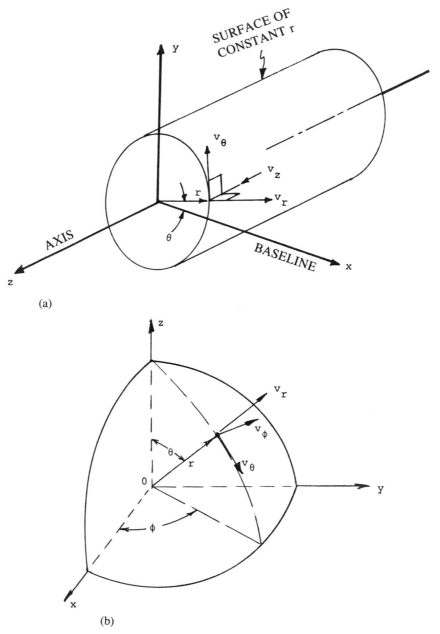

FIGURE 1.10   Illustration of orthogonal coordinate systems: (a) cylindrical and (b) spherical polar coordinates.

The azimuthal angle is ϕ, and the polar angle is θ. The general equation of continuity becomes

$$\frac{\partial \rho}{\partial t} + \frac{1}{r^2} \frac{\partial}{\partial r} (r^2 \rho v_r) + \frac{1}{r \sin \theta} \frac{\partial}{\partial \theta} (\rho v_\theta \sin \theta) + \frac{1}{r \sin \theta} \frac{\partial}{\partial \phi} (\rho v_\phi) = 0 \quad (1.56)$$

and the incompressible form is obtained by equating ρ to a constant.

Other curvilinear orthogonal coordinate systems are discussed in Appendix 2 of the text by Yih (1969).

The Stream Function. Whenever the general equation of continuity, Eq. (1.50), can be reduced to only two coordinates—such as (x, y) or (r, θ) or (x, z)—a very useful mathematical device called the *stream function* exists and is denoted by ψ. Its purpose is to satisfy conservation of mass identically and thus eliminate the continuity relation from the analysis.

Take, for example, steady incompressible flow in the x–y plane, for which Eq. (1.51) reduces to

$$\frac{\partial u}{\partial x} + \frac{\partial v}{\partial y} = 0$$

This relation is satisfied identically by a stream function $\psi(x, y)$ such that

$$u = \frac{\partial \psi}{\partial y} \qquad v = -\frac{\partial \psi}{\partial x} \tag{1.57}$$

Further, there is a geometric interpretation: lines of constant ψ are *streamlines* of the flow, that is, they are everywhere tangential to the local velocity vector. Also, the volume flow Q per unit depth between any points 1 and 2 in the flow field is given by

$$Q = \int_1^2 (\overline{\mathbf{V}} \cdot \overline{\mathbf{n}}) \, dA = \psi_2 - \psi_1 \tag{1.58}$$

Finally, by substituting for u and v from Eq. (1.57) in the momentum equation of the next section, one obtains a single partial differential equation in $\psi(x, y)$. Other useful forms of the stream function for various coordinate systems are as follows:

a) compressible flow (u, v) in the x–y plane

$$u = \frac{1}{\rho}\frac{\partial \psi}{\partial y} \qquad v = -\frac{1}{\rho}\frac{\partial \psi}{\partial x} \tag{1.59a}$$

b) incompressible plane flow (v_r, v_θ) in polar coordinates

$$v_r = \frac{1}{r}\frac{\partial \psi}{\partial \theta} \qquad v_\theta = -\frac{\partial \psi}{\partial r} \tag{1.59b}$$

c) incompressible axisymmetric flow (v_r, v_z)

$$v_r = -\frac{1}{r}\frac{\partial \psi}{\partial z} \qquad v_z = \frac{1}{r}\frac{\partial \psi}{\partial r} \tag{1.59c}$$

d) spherical polar flow (v_r, v_θ)

$$v_r = \frac{1}{r^2 \sin\theta}\frac{\partial \psi}{\partial \theta} \qquad v = -\frac{1}{r \sin\theta}\frac{\partial \psi}{\partial r} \tag{1.59d}$$

For all of these cases, lines of constant ψ are streamlines of the flow. Equation (1.58) is valid for case (b), but for case (a) $\psi_2 - \psi_1$ equals the mass flow $\dot{m}$ between 1 and 2. For cases (c) and (d), usually called *Stokes stream functions*, the volume flow between 1 and 2 equals $2\pi(\psi_2 - \psi_1)$.

A stream function exists in principle for unsteady flow in one dimension, but it is seldom used. For fully three-dimensional, three-coordinate flow, a pair of stream functions exists, as discussed in the text by Yih (1969). Again, stream function pairs are rather complex and seldom used.

The Inviscid Momentum Equation. The momentum equation results from the application of Newton's law, $\overline{\mathbf{F}} = m\overline{\mathbf{a}}$, to a differential fluid particle. If viscous stresses are absent, the only applied forces are due to pressure and a body force field. The result is the three-dimensional *Euler equation*

$$\rho \frac{d\overline{\mathbf{V}}}{dt} = -\nabla p + \rho\overline{\mathbf{X}} \tag{1.60}$$

where $\overline{\mathbf{X}}$ is the body force per unit volume; in most cases, $\overline{\mathbf{X}} = \overline{\mathbf{g}}$, the acceleration of gravity. One must be careful to interpret $(d\overline{\mathbf{V}}/dt)$ as the four-term Eulerian acceleration vector from Eq. (1.14).

If the density is constant or a function only of pressure (a *barotropic* fluid) and $\overline{\mathbf{X}} = -g\overline{\mathbf{k}}$, Euler's equation may be rewritten in the form

$$\frac{d\overline{\mathbf{V}}}{dt} = -\nabla\left[\int \frac{dp}{\rho} + gz\right] \tag{1.61}$$

Equation (1.61) may be integrated by making certain assumptions about the form of the velocity vector.

Boundary Conditions for Incompressible Flow. If the density is constant, Eqs. (1.51) and (1.60) are two simultaneous partial differential equations for $\overline{\mathbf{V}}$ and p. The proper velocity condition is that the flow must be purely tangential at any solid surface

$$V_n = 0 \tag{1.62}$$

In addition, the velocity vector must be known all around a boundary enclosing the flow field, and the pressure must be known at one point in the flow. It is not possible to satisfy the no-slip condition ($\overline{\mathbf{V}} = 0$) in an inviscid flow.

The Inviscid Energy Equation. In the absence of viscous stresses, the energy equation or first law of thermodynamics becomes a balance among internal energy, pressure work, and heat transfer. When expressed for a differential element, the result is the inviscid energy equation

$$\rho \frac{de}{dt} + p\nabla \cdot \overline{\mathbf{V}} + \nabla \cdot \overline{\mathbf{q}} = 0 \tag{1.63}$$

where e is the specific internal energy, J/kg, and $\overline{q}$ is the heat transfer rate vector, Jm^2s. The Eulerian time derivative (d/dt) is given by Eq. (1.14).

For the special case of adiabatic steady flow $(\overline{q} = 0, \partial/\partial t = 0)$, Eq. (1.63) reduces to the algebraic energy relation given earlier as Eq. (1.20).

1.2.6 Rotational and Irrotational Flow

Although there was an angular momentum integral relation Eq. (1.38), there was no angular momentum differential relation in Sec. 1.2.5. The reason is that, in the absence of concentrated ''body moments,'' conservation of angular momentum for a differential element simply requires that the stress tensor by symmetric, $\tau_{ij} = \tau_{ji}$; see pp. 207–208 of White (1994).

In spite of the fact that angular acceleration is not a factor in flow of a continuous fluid, the angular velocity $\overline{\Omega}$ is of interest and is equal to one-half the curl of the local velocity vector; see p. 224 of White (1994)

$$\overline{\Omega} = \frac{1}{2} \nabla \times \overline{V} = \frac{1}{2} \begin{vmatrix} \overline{i} & \overline{j} & \overline{k} \\ \dfrac{\partial}{\partial x} & \dfrac{\partial}{\partial y} & \dfrac{\partial}{\partial z} \\ u & v & w \end{vmatrix} \tag{1.64}$$

The distribution of angular velocity in a flow field is of great interest. If $\overline{\Omega}$ is everywhere zero, the flow is termed *irrotational* and has many interesting properties.

Velocity Potential. Mathematically, in order for a flow to be irrotational $(\nabla \times \overline{V} \equiv 0)$, the velocity must be the gradient of a scalar function called the velocity potential, Φ

$$\overline{V} = \nabla\Phi \tag{1.65}$$

The form of the gradient function in various coordinate systems may be listed as follows:

a) Cartesian

$$u = \frac{\partial\Phi}{\partial x} \qquad v = \frac{\partial\Phi}{\partial y} \qquad w = \frac{\partial\Phi}{\partial z} \tag{1.66a}$$

b) cylindrical polar (Fig. 1.10a)

$$v_r = \frac{\partial\Phi}{\partial r} \qquad v_\theta = \frac{1}{r}\frac{\partial\Phi}{\partial\theta} \qquad v_z = \frac{\partial\Phi}{\partial z} \tag{1.66b}$$

c) spherical polar (Fig. 1.10b)

$$v_r = \frac{\partial\Phi}{\partial r} \qquad v_\theta = \frac{1}{r}\frac{\partial\Phi}{\partial\theta} \qquad v_\phi = \frac{1}{r\sin\theta}\frac{\partial\Phi}{\partial\phi} \tag{1.66c}$$

Note that $\Phi(x, y, z, t)$ is a fully three-dimensional function, in contrast to the stream function ψ, which is limited to two coordinates as in Eq. (1.59).

If Eq. (1.65) is substituted into the incompressible continuity relation Eq. (1.51), a single partial differential equation results for the velocity potential

$$\nabla^2 \Phi = 0 \tag{1.67}$$

This is *Laplace's equation*, probably the most widely studied boundary value problem in mathematical physics. Many types of solution to this *potential flow* problem are discussed in Sec. 1.3 of this handbook. Compressible potential flow is discussed in Chap. 8.

If Eq. (1.65) is substituted into the inviscid, compressible momentum Eq. (1.61) there results

$$\frac{\partial \Phi}{\partial t} + \int \frac{dp}{\rho} + \frac{1}{2} V^2 + gz = \text{constant} \tag{1.68}$$

This is the unsteady, irrotational form of Bernoulli's relation. For steady, incompressible flow it reduces to Eq. (1.19).

If the curl of the velocity is not zero, the flow is termed *rotational*, and a velocity potential does not exist. Nevertheless, if inviscid, it satisfies a form of the unsteady Bernoulli relation along any given streamline direction ds

$$\int \frac{\partial V}{\partial t} ds + \int \frac{dp}{\rho} + \frac{1}{2} V^2 + gz = \text{constant} \tag{1.69}$$

but the constant may vary across the streamlines. In Eqs. (1.19) and (1.68), the same constant is valid everywhere.

1.2.7 Circulation

An important concept is inviscid flow analysis is the *circulation* of a flow. As illustrated in Fig. 1.11, the circulation Γ is defined as the line integral around a closed path of the velocity component tangential to the path, times the arc length along the path

$$\Gamma = \int_C \overline{V} \cdot d\overline{S} = \int_C V \cos \alpha \, dS \tag{1.70}$$

The angle α is defined by the geometry illustrated in Fig. 1.11. The circulation is a function of the path and the type of flow field and may vary with time. By Stokes' theorem [see p. 499 of White (1994)] it is identically related to the area integral of the normal component of angular velocity within the closed curve, also shown in Fig. 1.11

$$\int_C \overline{V} \cdot d\overline{S} = \iint_A (\nabla \times \overline{V}) \cdot \overline{n} \, dA \tag{1.71}$$

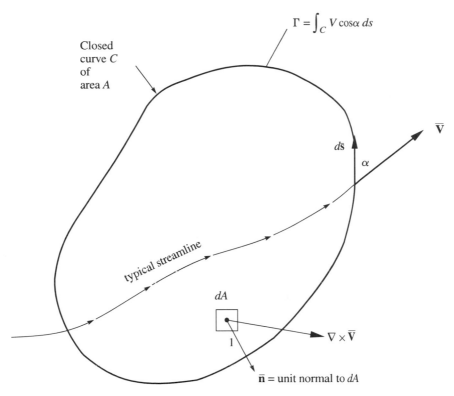

FIGURE 1.11 Definition of the circulation of a fluid.

It follows that the circulation in an irrotational flow is zero for any path which does not enclose a singularity or *vortex center*, where $\nabla \times \overline{V}$ would be infinite.

Kelvin's Theorem. The rate of change of circulation is found by substituting for $d\overline{V}/dt$ from the momentum equation into Eq. (1.70). For an inviscid, incompressible fluid, the result is striking

$$\frac{d\Gamma}{dt} = 0 \tag{1.72}$$

This theorem is due to Lord Kelvin (1869) and states that the circulation is invariant for any closed path. From Stokes' theorem Eq. (1.71), one can infer that the angular velocity distribution is also invariant in an inviscid, constant density flow.

Bjerknes' Theorem. Kelvin's theorem was extended to variable density inviscid flow by V. Bjerknes in 1900. It is assumed that the body forces are conservative (given by the gradient of a potential) and hence do not contribute to circulation. Bjerknes' theorem then results

$$\frac{d\Gamma}{dt} = \int\int_{A} \frac{1}{\rho^2} (\nabla\rho \times \nabla p) \cdot \overline{n} \, dA \tag{1.73}$$

It follows that for a barotropic fluid, where $\rho = \rho(p)$ only, the circulation change is still zero. However, circulation may be generated in nonbarotropic situations where the density gradient is not parallel to the pressure gradient. Two common cases are the stratified flow patterns in the atmosphere and in the ocean, as discussed by Yih (1980) and Sec. 13.4. Another example is *thermal blooming* of a horizontal laser beam in the absence of wind.

Forces in Two-Dimensional Inviscid Flow. The circulation concept is directly related to the force induced by a uniform inviscid incompressible stream flow past a two-dimensional body, as shown in Fig. 1.12. Regardless of the shape of the body, integration of the pressure forces around the surface of the body yields a force normal to the stream (lift) and a force parallel to the stream (drag) given by

$$\text{Lift} = -\rho U \Gamma$$

$$\text{Drag} = 0$$

$$(1.74)$$

per unit depth into the paper. The lift result was found by W. M. Kutta (1910) and independently by N. E. Joukowsky (1905). The problem of lifting-body theory is thus reduced to the determination of the circulation about the body as a function of velocity and geometry.

The zero drag result in Eq. (1.74) is due to d'Alembert (1752) and is now called the *d'Alembert paradox* since all experiments show it to be false. The reason of course is that a real fluid is not truly inviscid. Viscous stresses generated at the body surface create both frictional drag and pressure or *form drag* due to flow separation at the rear of the body.

The analysis of fluid forces in potential flow is treated in more detail in Sec. 1.3 and also in a very interesting discussion by Batchelor (1967).

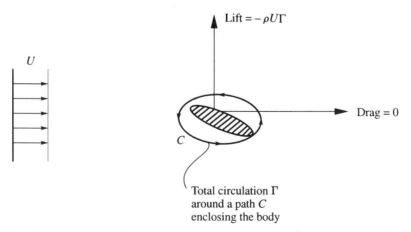

FIGURE 1.12 Lift and drag forces in inviscid incompressible flow about a two-dimensional body.

1.2.8 Vorticity and Vortex Motion

The factor of one-half in Eq. (1.64) is inconvenient, and it has been customary to define a quantity called *vorticity*, $\overline{\omega}$, as twice the angular velocity vector

$$\overline{\omega} = 2\overline{\Omega} = \nabla \times \overline{V}$$

For an irrotational flow field, the vorticity is zero. If the flow is rotational, the vorticity is a vector function of space and time compatible with the three conservation laws. In particular, for inviscid incompressible flow, take the curl of the momentum Eq. (1.60) to yield the inviscid *vorticity dynamics equation*

$$\frac{d\overline{\omega}}{dt} = (\overline{\omega} \cdot \nabla)\overline{V} \tag{1.75}$$

From this, note that if the fluid is initially irrotational everywhere, its rate of change is zero everywhere and the flow remains irrotational. This is a corollary of Kelvin's theorem Eq. (1.72).

The Line Vortex. A classic example of vorticity-related flow is the circulating motion induced by a *line vortex* as shown in Fig. 1.13. The streamlines are circular and the tangential velocity v_θ varies inversely with radius r from the vortex center, $v_\theta = K/r$. Evaluation of $\nabla \times \overline{V}$ reveals that the vorticity is infinite at the vortex center and zero elsewhere. For any closed path about the vortex center, evaluation of the circulation from Eq. (1.70) gives the result $\Gamma = 2\pi K$. The tangential velocity may thus be written in the form

$$v_\theta = \frac{\Gamma}{2\pi r} \tag{1.76}$$

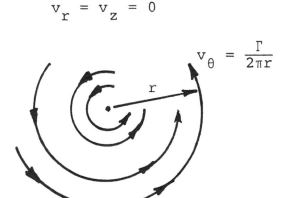

FIGURE 1.13 Illustration of the circular streamline pattern induced by a line vortex of circulation Γ.

and Γ is said to be the *strength* of the vortex. If the vortex is immersed in a uniform stream U, it will cause a lift $\rho U \Gamma$ as in Fig. 1.12.

The Law of Biot and Savart. Equation (1.76) is for the case of a vortex line of infinite length into and out of the paper, as in Fig. 1.13. Think of a vortex line as being composed of many short *vortex filaments* of length ds, as in Fig. 1.14. The velocity induced at any point P a distance R from this filament is perpendicular to the plane of $d\overline{\mathbf{s}}$ and $\overline{\mathbf{R}}$, as shown in Fig. 1.14, and is given by

$$d\overline{\mathbf{V}} = \frac{\Gamma \, d\overline{\mathbf{s}} \times \overline{\mathbf{R}}}{4\pi R^3} = \frac{\Gamma \, ds \sin \alpha}{4\pi R^2} \, \overline{\mathbf{n}} \tag{1.77}$$

where $\overline{\mathbf{n}}$ is a unit vector normal to $d\overline{\mathbf{s}}$ and $\overline{\mathbf{R}}$. This is the law of Biot and Savart (1820), who discovered experimentally that the same relation holds for the magnetic field induced by a current flowing in a wire.

Given the strength and position in space of the vortex filament, integrate Eq. (1.77) to obtain the total velocity at P induced by the whole filament. If the vortex line is straight and infinitely long, the integration yields the potential vortex result of Eq. (1.76).

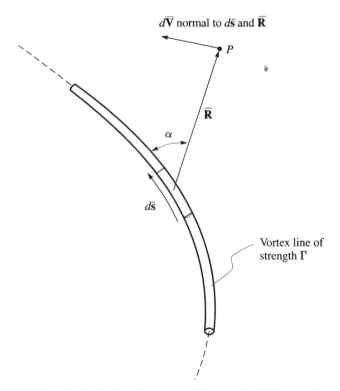

FIGURE 1.14   Fluid velocity induced at any arbitrary point P by an element of a vortex line: the Biot–Savart law.

The Three Helmholtz Vortex Laws. In a classic paper in 1858, H. von Helmholtz stated three basic laws for motion of a vortex filament in an otherwise irrotational fluid. Essentially, the laws are corollaries to Kelvin's theorem Eq. (1.72) and are as follows:

1. A vortex filament can neither begin nor end in the fluid; it must either form a closed loop or else end on the fluid boundary.
2. Vortex tubes always contain the same fluid particles.
3. The strength Γ of a vortex line is constant. Hence, from Stokes' theorem Eq. (1.71), the product of its vorticity and its cross-sectional area must be constant.

A good discussion of these and related theorems is given in the text by Robertson (1965).

Some surprisingly complex flows can be modeled numerically by the superposition of multiple vortex filaments and other vortex surface elements, as discussed in Sec. 1.3 of this handbook.

The Vortex Sheet. A surface of distributed vorticity is called a *vortex sheet*, and a schematic is sketched in Fig. 1.15. It may be visualized as a row of line vortices very closely spaced; see p. 455 of White (1994). The local sheet strength is denoted by γ and varies with s, the arc length along the sheet. A sheet element of length ds has circulation $d\Gamma = \gamma\,ds$, as shown in Fig. 1.15.

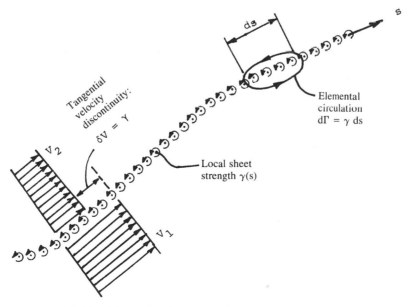

FIGURE 1.15   Schematic of a vortex sheet and its kinematic properties.

By integration of its total induced velocity, it may be shown [p. 97 of Batchelor (1967)] that there is a discontinuity in tangential velocity across a vortex sheet equal to the local sheet strength, as in Fig. 1.15

$$\delta V = V_1 - V_2 = \gamma(s) \tag{1.78}$$

The concept of local sheet vorticity can be generalized to a strength which varies with both coordinates in the sheet surface. The sheet strength can then be varied to match almost any boundary condition or body shape, as shown in Sec. 1.3. This is the basis of the panel method used in three-dimensional aerodynamics [Kraus (1978)]. The one-dimensional sheet strength model of Fig. 1.15 is used in the theory of thin airfoils; see pp. 474–477 of White (1994). An interesting analysis of the instability of a vortex sheet is given in Sec. 7.1 of Batchelor (1967).

1.2.9 Generalized Bernoulli Relations

Two different forms of the Bernoulli equation have been discussed as follows: 1) Eq. (1.68), which is valid everywhere for an irrotational flow, and 2) Eq. (1.69), which is valid along any given streamline for a rotational flow.

The Bernoulli relations express a balance among the work done by pressure forces, gravity, and inertia terms in the momentum equation. Other types of energy are excluded. Therefore, all Bernoulli relations are subject to the following limitations: 1) inviscid flow, 2) no shaft work, and 3) no heat transfer. Even within these limitations, certain other forms of the Bernoulli relation can be derived; refer to pp. 69–77 of Yih (1969). Three of these are listed here:

1. Steady two-dimensional flow with constant vorticity ω

$$\int \frac{dp}{\rho} + \frac{1}{2} V^2 + gz + \omega\psi = \text{constant everywhere}$$

2. Irrotational flow which is steady with respect to a frame of reference rotating at constant angular velocity Ω

$$\int \frac{dp}{\rho} + \frac{1}{2} V^2 + gz - \frac{1}{2} \Omega^2 r^2 = \text{constant along a streamline,}$$

   where r denotes distance along the axis of rotation.

3. Flow with constant vorticity ω, which is steady with respect to a frame rotating at constant angular velocity Ω

$$\int \frac{dp}{\rho} + \frac{1}{2} V^2 + gz - \frac{1}{2} \Omega^2 r^2 + (\omega + 2\Omega)\psi = \text{constant everywhere}$$

Unsteady forms are also derived by Yih (1969).

1.3 POTENTIAL FLOW
John L. Hess

1.3.1 Mathematical Formulation

The equations of motion for inviscid-incompressible flow consist of the zero divergence condition on the fluid velocity plus the vector Euler equation. This set represents four scalar partial differential equations for four unknowns: pressure and three components of velocity. Three of the four equations are nonlinear, and thus solution by ordinary means is quite difficult. However, if the additional condition of irrotationality of the velocity field is imposed, which Kelvin's theorem states will hold under fairly general circumstances, the only partial differential equation that need be solved is the linear Laplace equation for a single scalar unknown—the velocity potential. The only nonlinearity that remains has been relegated to an algebraic equation, the Bernoulli equation, which yields the pressure in terms of the velocity.

For definiteness the problem considered is that of steady flow past a stationary body immersed in an onset flow, which in most cases is a uniform stream. This may be denoted the *wind-tunnel problem*, and it clearly also represents the case of a body translating through fluid at rest. It is convenient to consider the fluid velocity $\bar{\mathbf{V}}$ at any point as the sum of the onset flow velocity $\bar{\mathbf{U}}$ and the disturbance velocity $\bar{\mathbf{v}}$ due to the body (see Fig. 1.16). The condition that the velocity field $\bar{\mathbf{v}}$ be irrotational means that $\bar{\mathbf{v}}$ may be taken as the gradient of a scalar velocity potential ϕ

$$\bar{\mathbf{v}} = \text{grad } \phi \tag{1.79}$$

Applying the zero divergence condition to Eq. (1.79) then given Laplace's equation for ϕ

$$\nabla^2 \phi = 0 \text{ in the field} \tag{1.80}$$

At infinity, the disturbance velocity must vanish so that the velocity approaches the onset flow, i.e.,

$$|\text{grad } \phi| \to 0 \text{ at } \infty \tag{1.81}$$

The body surface is assumed to be impervious to fluid so that the normal component of the total velocity must vanish there, or

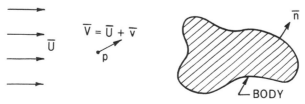

FIGURE 1.16 The basic problem of potential flow.

$$\overline{\mathbf{v}} \cdot \overline{\mathbf{n}} = \frac{\partial \phi}{\partial n} = -\overline{\mathbf{U}} \cdot \overline{\mathbf{n}} \text{ on the body} \tag{1.82}$$

Equations (1.80) to (1.82) define what may be called the basic problem of potential flow. In many important applications, particularly lifting flows to be discussed later, certain auxiliary conditions may be required in addition to the above three. Nevertheless, the basic problem is mathematically well defined and is a very useful starting point for a study of potential flow. Once it has been solved for ϕ, $\overline{\mathbf{v}}$ is obtained from Eq. (1.79) and the pressure p is calculated from the Bernoulli equation. Define

$$V = |\overline{\mathbf{U}} + \overline{\mathbf{v}}|, \ U = |\overline{\mathbf{U}}| \tag{1.83}$$

Then the Bernoulli equation is

$$p + \tfrac{1}{2}\rho V^2 = \text{const} = p_\infty + \tfrac{1}{2}\rho U^2 \tag{1.84}$$

where p_∞ is the pressure at infinity, and ρ is the constant density of the fluid. In applications where the onset flow $\overline{\mathbf{U}}$ is a uniform stream, it is customary to express p by means of the nondimensional pressure coefficient

$$C_p = (p - p_\infty)/0.5\rho U^2 = 1 - V^2/U^2 \tag{1.85}$$

1.3.2 Streamlines and Stream Functions

The external flow field is characterized by the streamline pattern of the flow. Moreover, streamlines and their associated streamfunctions are useful for constructing analytic solutions (Sec. 1.3.4). Accordingly, these concepts are introduced here.

In the flow field there is a vector velocity at each point as shown in Fig. 1.17. It is possible to begin a curve at a particular point and proceed in such a way that at any point the curve is tangent to the local velocity vector. Such a curve is called a streamline of the flow. In steady flow, it also represents the path of all fluid particles that pass through the initial point. Both conceptually and numerically the process of generating a streamline is similar to the numerical solution of a differential equation. Let dx, dy, dz represent infinitesimal movement along a streamline. A tangent vector

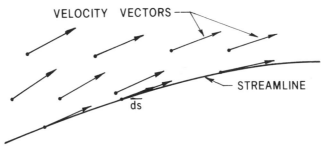

FIGURE 1.17   Streamline of the flow.

to the streamline is

$$\overline{\mathbf{T}} = \frac{dx}{dt} \overline{\mathbf{i}} + \frac{dy}{dt} \overline{\mathbf{j}} + \frac{dz}{dt} \overline{\mathbf{k}} \qquad (1.86)$$

where t is any convenient parameter. This vector will be parallel to the velocity with components u, v, w if

$$\frac{dx}{dt} = \lambda u, \qquad \frac{dy}{dt} = \lambda v, \qquad \frac{dz}{dt} = \lambda w \qquad (1.87)$$

This is the differential equation of a streamline in a form that is convenient for numerical integration, because it is not invalidated by local zero values of any velocity component. One popular numerical integration scheme takes t as time and $\lambda = 1$, while another takes t as arc length and λ as $1/V$, where V is the local velocity magnitude. It is efficient to use a variable integration increment based on local streamline curvature. The differential equation [Eq. (1.87)] may be written in the more compact form

$$dx/u = dy/v = dz/w \qquad (1.88)$$

but this requires special handling for local zero values of a velocity component.

A concept that is useful in two-dimensional or axisymmetric flow is that of the stream function (see Sec. 1.2.5). In its broadest sense it is simply a function that is constant along each streamline with a value that changes from streamline to streamline. In most cases a normalization is used to give the stream function a physical significance in terms of velocity flux. Two-dimensional and axisymmetric flows have the property that in those cases there is a single nontrivial stream function. In three-dimensional flow there are two independent stream functions, and it is difficult to make use of this concept to produce meaningful results.

Two-dimensional Flow. Two-dimensional flow has the property that w and all derivatives with respect to z are zero so that attention may be restricted to the xy-plane. Let the stream function be denoted $\psi(x, y)$. The condition that this be constant along streamlines may be written

$$\overline{\mathbf{V}} \cdot \operatorname{grad} \psi = 0 \qquad (1.89)$$

Clearly grad ψ is determined only to within a scalar multiplicative function of position. In incompressible flow the relation of the stream function to the velocity is

$$u = \frac{\partial \psi}{\partial y} \qquad v = -\frac{\partial \psi}{\partial x} \qquad (1.90)$$

Consider an arbitrary curve c as shown in Fig. 1.18. The flux of fluid of unit density crossing c from left to right between points 1 and 2 is

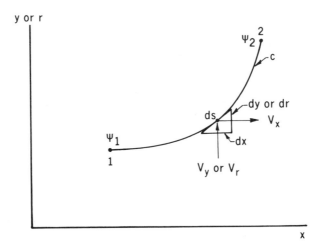

FIGURE 1.18 Velocity flux across a curve.

$$\text{Flux}(1, 2) = \int_1^2 (u \, dy - v \, dx) = \int_1^2 \left(\frac{\partial \psi}{\partial y} \, dy + \frac{\partial \psi}{\partial x} \, dx \right)$$

$$\text{Flux}(1, 2) = \int_1^2 d\psi = \psi(2) - \psi(1) \tag{1.91}$$

Thus, the difference in the values of the stream function at two points equals the flux of velocity across a curve joining the points, and the flux is the same for all curves joining the points.

Axisymmetric Flow. Flows symmetric about the x-axis are most conveniently described in cylindrical coordinates x, r, θ where

$$r = \sqrt{y^2 + z^2} \qquad \theta = \tan^{-1} \left(\frac{z}{y} \right) \tag{1.92}$$

and the velocity components are u, u_r, u_θ. Axisymmetric flow has u_θ and all θ derivatives equal to zero. The stream function $\psi(x, r)$ again satisfies Eq. (1.89) but the usual normalization gives

$$ru = \frac{\partial \psi}{\partial r} \qquad ru_r = -\frac{\partial \psi}{\partial x} \tag{1.93}$$

With this definition the flux across the circular annulus whose trace in the xy-plane is the curve of Fig. 1.18 is

$$\text{Flux}(1, 2) = \int_1^2 (u \, dr - u_r \, dx)2\pi r = 2\pi \int_1^2 \left(\frac{\partial \psi}{\partial r} \, dr + \frac{\partial \psi}{\partial x} \, dx \right)$$

$$= 2\pi[\psi(2) - \psi(1)] \tag{1.94}$$

The flux between the points equals 2π times the difference in stream function values, but this is preferred to the alternative, which is using a factor of 2π in Eq. (1.93).

1.3.3 Singular Solutions

It can be proved that every nonconstant solution of Laplace's equation is singular somewhere—either at infinity or in the finite part of the field. In constructing flows about bodies immersed in a uniform onset flow, the uniform-stream solution is the only one allowed to be singular at infinity. (Its potential is singular, not its velocity.) The other allowable solutions, which represent the effect of the body, fall to zero at infinity as required by Eq. (1.81) and thus are singular at some finite point. The "building blocks" used to construct solutions are the flows for the source, doublet and vortex singularities. The process by which these are superposed is discussed in subsequent subsections. Here a library of singular solutions is presented.

In the equations below it is assumed that in two dimensions the singularity is located at x_0, y_0 and the effects are being evaluated at x, y. The two-dimensional distance between the points is

$$R_2 = \sqrt{(x - x_0)^2 + (y - y_0)^2} \tag{1.95}$$

In three-dimensions, the singularity is at x_0, y_0, z_0 and the effects are evaluated at x, y, z. The three-dimensional distance is

$$R_3 = \sqrt{(x - x_0)^2 + (y - y_0)^2 + (z - z_0)^2} \tag{1.96}$$

Note below the special treatment that must be accorded the axisymmetric stream function.

Two-Dimensional Uniform Stream (Speed U, Angle of Attack α)

Potential:	$\phi_\infty = U(x \cos \alpha + y \sin \alpha)$	(1.97a)
Velocity components:	$u = U \cos \alpha, \quad v = U \sin \alpha$	(1.97b)
Stream function:	$\psi = U(-x \sin \alpha + y \cos \alpha)$	(1.97c)

Three-Dimensional Uniform Stream (Speed U, Direction cosines, l, m, n)

Potential:	$\phi_\infty = U(lx + my + nz)$	(1.98a)
Velocity components:	$u = Ul, \quad v = Um, \quad w = Un$	(1.98b)
Axisymmetric stream function:	$\psi = U(y^2 + z^2)/2$	(1.98c)

(valid only for $m = n = y_0 = z_0 = 0$)

Two-Dimensional source (Strength m, Volume Flux 4πm)

Potential:	$\phi = 2m \ln R_2$	(1.99a)
Velocity components:	$u = 2m(x - x_0)/R_2^2$	(1.99b)
	$v = 2m(y - y_0)/R_2^2$	(1.99c)
Stream function:	$\psi = 2m \tan^{-1} [(y - y_0)/(x - x_0)]$	(1.99d)

Three-Dimensional Source (Strength m, Volume Flux 4πm)

Potential:	$\phi = -m/R_3$	(1.100a)
Velocity components:	$u = m(x - x_0)/R_3^3$	(1.100b)
	$v = m(y - y_0)/R_3^3$	(1.100c)
	$w = m(z - z_0)R_3^3$	(1.100d)
Axisymmetric stream function:	$\psi = -m(x - x_0)/R_3$	(1.100e)

(valid only for $y_0 = z_0 = 0$)

Two-Dimensional Doublet (Strength μ)

Doublet axis:	$\bar{t} = l\bar{i} + m\bar{j}, \quad l^2 + m^2 = 1$	(1.101a)
Potential:	$\phi = -2\mu[l(x - x_0) + m(y - y_0)]/R_2^2$	(1.101b)
Velocity components:	$u = -2[\mu l + (x - x_0)\phi]/R_2^2$	(1.101c)
	$v = -2[\mu m + (y - y_0)\phi]/R_2^2$	(1.101d)
Stream function:	$\psi = 2 \, \text{sgn} \, (l)\mu(y - y_0)/R_2^2$	(1.101e)

(valid only for $m = 0$)

Three-Dimensional Doublet (Strength μ)

Doublet axis:	$\bar{t} = l\bar{i} + m\bar{j} + n\bar{k}, \, l^2 + m^2 + n^2 = 1$	(1.102a)
Potential:	$\phi = -\mu[l(x - x_0) + m(y - y_0) + n(z - z_0)]/R_3^3$	(1.102b)
Velocity components:	$u = -[\mu l + 3R_3(x - x_0)\phi]/R_3^3$	(1.102c)
	$v = -[\mu m + 3R_3(y - y_0)\phi]/R_3^3$	(1.102d)
	$w = -[\mu n + 3R_3(z - z_0)\phi]/R_3^3$	(1.102e)
Axisymmetric stream function:	$\psi = \text{sgn} \, (l)\mu(y^2 + z^2)/R_3^3$	(1.102f)

(valid only for $y_0 = z_0 = m = n = 0$)

Two-Dimensional Vortex (Circulation Γ)

Potential:	$\phi = -(\Gamma/2\pi) \tan^{-1} [(y - y_0)/(x - x_0)]$	(1.103a)
Velocity components:	$u = (\Gamma/2\pi) (y - y_0)/R_2^2$	(1.103b)
	$v = -(\Gamma/2\pi) (x - x_0)/R_2^2$	(1.103c)
Stream function:	$\psi = (\Gamma/2\pi)\ln R_2$	(1.103d)

It may be noted above that formulas are not given for a three-dimensional vortex, because there is no three-dimensional analogy of a point vortex singularity. Three-dimensional vortices must be of constant strength and either form closed loops or extend to infinity, otherwise the associated velocity field is not irrotational.

1.3.4 Construction of Solutions by Superposition. The Indirect Method

The equations of Sec. (1.3.1) define the so-called direct problem of potential flow: given a body shape, find the flow field and the surface pressure distribution. While it can be shown that a solution always exists, construction of a solution that satisfies the boundary condition on the body surface, Eq. (1.82), is usually quite difficult. As one means of avoiding the problems inherent in accommodating arbitrary geometric boundaries, recourse has been made to generating analytic solutions by a so-called indirect method. Such a method cannot produce the flow about an arbitrarily specified body, but the analytic solutions generated illustrate the properties of potential flow, thus providing a certain "feel" for the subject, and also provides a means of evaluating the accuracy of numerical methods.

The indirect method, as well as the direct numerical methods to be considered subsequently, utilizes the linearity of Laplace's equation to construct a flow solution as a linear combination of simpler solutions. Specifically, the method assumes a particular uniform stream and a particular set of singularities of the type given in Sec. 1.3.3. Both the strengths and locations of the singularities are specified. Clearly, the velocity and pressure at any point in space can be calculated immediately. The more difficult task is to determine the body shape that gives rise to the flow. In two-dimensional or axisymmetric flow, the total stream function of the flow is simply the sum of the stream function of the uniform stream and those of the assumed singularities. Curves of constant values of total stream function are streamlines, any of which may be considered a solid wall. If the singularities have been properly arranged, there is one streamline that has the appearance of a body. The determining feature is the existence on the streamline of an upstream stagnation point (point where the fluid velocity is zero) downstream of which the streamline bifurcates. The region between the branches of this so-called *dividing streamline* may be taken as the body. If the body is finite and closed, as opposed to semi-infinite, the two branches of the dividing streamline will rejoin at a downstream stagnation point. Body closure requires that the net source strength of the specified singularities vanish.

Source in a Uniform Stream. Schematically the situation is as sketched in Fig. 1.19. A uniform stream of speed U is flowing from left to right parallel to the x-axis. A source of strength m is located on the x-axis at x_0. Clearly, the source could be located anywhere along the x-axis. For definiteness the source location x_0 will be chosen so that the upstream stagnation point, which is the "nose" of the body, is at the origin. The appropriate equations of Sec. 1.3.3 are superposed to give the solution.

Two-Dimensional Flow. The combined potential, stream function, and velocity components are

$$\Phi = Ux + 2m \ln D \tag{1.104a}$$

$$\Psi = Uy + 2m \tan^{-1} [y/(x - x_0)] \tag{1.104b}$$

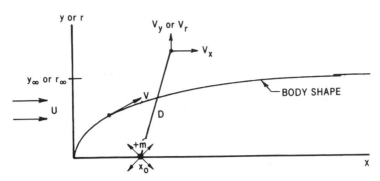

FIGURE 1.19   Point source in a uniform stream.

$$u = U + 2m(x - x_0)/D, \qquad v = 2my/D \tag{1.104c}$$

$$D^2 = (x - x_0)^2 + y^2 \tag{1.104d}$$

The inverse tangent must be evaluated in the range $-\pi$ to $+\pi$ by accounting for the individual signs of the numerator and denominator of its argument.

By symmetry or by examination of v it is clear that the stagnation point is on the x-axis. If it is to be at the origin, the following relation, obtained by setting $x = y = 0$ in $u = 0$, must exist between the parameters of the problem

$$x_0 = 2m/U \tag{1.105}$$

Since $x_0 > 0$, the inverse trigonometric function in the stream function has a value of π at the origin, and the stream function on the dividing streamline has value $2m\pi$. Thus the equation of the body profile is

$$Uy + 2m \tan^{-1}[y/(x - x_0)] = 2m\pi \tag{1.106}$$

As $x \to +\infty$, the $\tan^{-1} \to 0$, and y approaches an asymptotic value y_∞ proportional to m. Thus m and x_0 may be eliminated in terms of the more physically meaningful y_∞

$$m = Uy_\infty/2\pi, \qquad x_0 = y_\infty/\pi \tag{1.107}$$

Now the solution may be written in a nondimensional form. The body shape is given by

$$x/y_\infty = 1/\pi - (y/y_\infty)/\tan[\pi(y/y_\infty)] \tag{1.108}$$

and the velocity components are

$$u/U = 1 + (x/y_\infty - 1/\pi)/\pi(D/y_\infty)^2 \tag{1.109a}$$

$$v/U = (y/y_\infty)/\pi(D/y_\infty)^2 \tag{1.109b}$$

The body shape and the total surface velocity distribution are shown in Fig. 1.19.

Axisymmetric Flow. The combined potential, stream function, and velocity components are

$$\Phi = ux - m/D, \qquad \Psi = Ur^2/2 - m(x - x_0)/D$$
$$u = U + m(x - x_0)/D^3, \qquad u_r = mr/D^3 \qquad (1.110)$$
$$D^2 = (x - x_0)^2 + r^2, \qquad r^2 = y^2 + z^2$$

The *r*-component of velocity is zero for $r = 0$. Setting $x = r = 0$ in $u = 0$ gives the relation

$$x_0^2 = m/U \qquad (1.111)$$

The stream function assumes the value m at the origin, so the body shape is given by

$$Ur^2/2 - m(x - x_0)/D = m \qquad (1.112)$$

As $x \to +\infty$ the radius approaches an asymptotic value r_∞ given by

$$Ur_\infty^2 = 4m \qquad (1.113)$$

which gives from Eq. (1.111)

$$x_0 = r_\infty/2 \qquad (1.114)$$

If these are inserted into Eq. (1.112), the resulting equation may be solved for x/r_∞ to give the equation of the body shape in nondimensional form

$$\frac{x}{r_\infty} = \frac{1}{2} + [(r/r_\infty)^2 - 1/2]/\sqrt{1 - (r/r_\infty)^2} \qquad (1.115)$$

The velocity components are

$$u/U = 1 + [(x/r_\infty) - 1/2]/4(D/r_\infty)^3$$
$$u_r/U = (r/r_\infty)/4(D/r_{/\infty})^3 \qquad (1.116)$$

The body shape and the total surface velocity distribution are shown in Fig. 1.20, which compares the two-dimensional and axisymmetric single-source bodies having the same asymptotic thickness. Even though the two-dimensional body is less full at the nose, it disturbs the flow more. That this is generally true is illustrated even more strongly by the results for closed bodies given below.

Source and Equal Sink in a Uniform Stream. The indirect method may be used to obtain the flow about a closed body by placing a source of strength $+m$ on the *x*-axis at $x = -d$ and a sink of strength $-m$ on the *x*-axis at $x = +d$, as illustrated in Fig. 1.21. Again the uniform stream is from left to right parallel to the *x*-axis

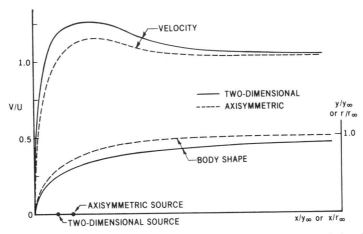

FIGURE 1.20 Body shapes and surface velocity distributions for semi-infinite single-source bodies.

with velocity U. The stagnation points on the dividing streamline lie on the x-axis at $x = \pm x_s$. It turns out that the resulting body profiles are symmetric about the y-axis or r-axis ("fore-and-aft" symmetry) and thus attain their maximum thickness at $x = 0$. Only the thickness ratio, y_m/x_s or r_m/x_s, is physically significant. By varying m and d thickness ratios from zero to unity can be obtained. Bodies of this type are known as *Rankine bodies*.

Two-dimensional flow. The total potential, stream function, and velocity components are obtained from the formulas of Sec. 1.3.3 as follows

$$\Phi = Ux + 2m \ln D_1 - 2m \ln D_2 \tag{1.117a}$$

$$\Psi = Uy + 2m \tan^{-1}[y/(x + d)] - 2m \tan^{-1}[y/(x - d)] \tag{1.117b}$$

$$u = U + 2m(x + d)/D_1^2 - 2m(x - d)/D_2^2 \tag{1.117c}$$

$$v = 2my/D_1^2 - 2my/D_2^2 \tag{1.117d}$$

$$D_1^2 = (x + d)^2 + y^2, \qquad D_2^2 = (x - d)^2 + y^2 \tag{1.117e}$$

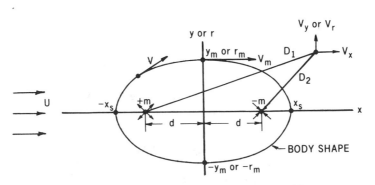

FIGURE 1.21 Point source and equal sink in a uniform stream.

By inspection, the stagnation points are on the x-axis. Clearly, the stream function is zero on the x-axis for $|x| > d$, so that $\Psi = 0$ gives the equation of the body profile. Let a prime denote normalization with respect to d, so that

$$x' = x/d, \quad y' = y/d, \quad y'_m = y_m/d, \quad x'_s = x_s/d \tag{1.118}$$

and define the parameter

$$K = 4m/Ud \tag{1.119}$$

The equation of the $\Psi = 0$ streamline is

$$y' + \left(\frac{K}{2}\right)\left\{ \tan^{-1}\left[\frac{y'}{(x' + 1)}\right] - \tan^{-1}\left[\frac{y'}{(x' - 1)}\right] \right\} = 0 \tag{1.120}$$

Setting $y = 0$ in $u = 0$ gives, after some manipulation, the nondimensional stagnation point location as

$$x'_s = \pm\sqrt{1 + K} \tag{1.121}$$

Setting $x' = 0$, $y' = y'_m$ in Eq. (1.120) gives

$$K = 2y'_m/[\pi - 2\tan^{-1}(y'_m)] \tag{1.122}$$

The flow is characterized by the velocity at the maximum half-breath on the body surface. Denote this velocity V_m. Then $V_m = u(x = 0, y = y_m)$, which from Eq. (1.117) is

$$V_m = U\{1 + K/[1 + (y'_m)^2]\} \tag{1.123}$$

Equations (1.121) to (1.123) implicitly define the velocity ratio V_m/U as a function of the thickness ratio $\tau = y'_m/x'_s$. This relation is graphed in Fig. 1.22. Equation (1.120) may be put into the form

$$(x')^2 = 1 - (y')^2 + 2y'/\tan(2y'/K) \tag{1.124}$$

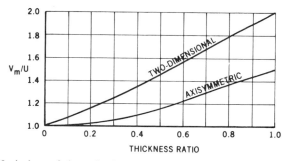

FIGURE 1.22 Variation of the velocity at maximum body thickness with thickness ratio for Rankine bodies.

from which the body shape may be calculated after y'_m is selected. The velocity components are

$$u/U = 1 + K[(y')^2 - (x')^2 + 1]/Q, \quad v/U = -2Kx'y'/Q \qquad (1.125a)$$

$$Q = [(x' - 1)^2 + (y')^2] [(x' + 1)^2 + (y')^2] \qquad (1.125b)$$

By way of example, Fig. 1.23 shows the body shape and total surface velocity distribution for the body of thickness ratio 0.415 that corresponds to a value $y'_m = 0.5$. Note that the maximum velocity does not occur at $x = 0$, so that to denote V_m as the maximum velocity is incorrect.

Axisymmetric Flow. Figure 1.21 still describes the situation. The formulas of Sec. 1.3.3 give the total potential, stream function and velocity components as

$$\Phi = Ux - m/D_1 + m/D_2 \qquad (1.126a)$$

$$\Psi = Ur^2/2 - m(x + d)/D_1 + m(x - d)/D_2 \qquad (1.126b)$$

$$u = U + m(x + d)/D_1^3 - m(x - d)/D_2^3 \qquad (1.126c)$$

$$u_r = mr/D_1^3 - mr/D_2^3, \quad r^2 = y^2 + z^2 \qquad (1.126d)$$

$$D_1^2 = (x + d)^2 + r^2, \quad D_2^2 = (x - d)^2 + r^2 \qquad (1.126e)$$

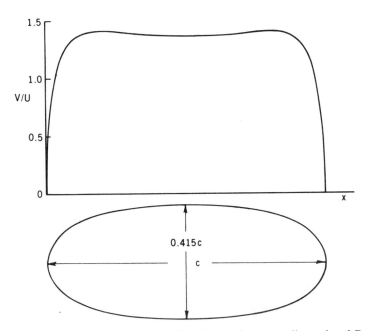

FIGURE 1.23 Shape and surface velocity distribution for a two-dimensional Rankine body of thickness ratio 0.415.

As before the stagnation points are on the x-axis, and $\Psi = 0$ gives the equation of the body. In nondimensional form, the body shape is

$$(r')^2 = q\left\{\frac{(x' + 1)}{[(x' + 1)^2 + (r')^2]^{1/2}} - \frac{(x' - 1)}{[(x' - 1)^2 + (r')^2]^{1/2}}\right\}, \quad (1.127)$$

where Eq. (1.118) still applies and

$$r' = r/d, \qquad r'_m = r_m/d, \qquad q = 2m/Ud^2 \quad (1.128)$$

This equation must be solved by numerical techniques. The maximum radius occurs at $x' = 0$ and is given by

$$(r'_m)^2 \sqrt{1 + (r'_m)^2} = 2q \quad (1.129)$$

The nondimensional stagnation point location is obtained by setting $r - 0$ in $u = 0$. In nondimensional form

$$2qx'_s/[(x'_s)^2 - 1]^2 = \pm 1 \quad (1.130)$$

The velocity V_m at the maximum body thickness is

$$\begin{aligned} V_m &= U\{1 + q/[(r'_m)^2 + 1]^{3/2}\} \\ &= U\{1 + (r'_m)^2/2[(r'_m)^2 + 1]\} \end{aligned} \quad (1.131)$$

Equations (1.129) to (1.131) define the velocity ratio V_m/U as a function of the thickness ratio $\tau = r'_m/x'_s$. This relation is shown in Fig. 1.22, where it may be compared with the analogous two-dimensional result. Note that the disturbance to the uniform stream is less for the axisymmetric than for the two-dimensional body. This is particularly true for small thickness ratios where the limiting forms are

$$(V_m/U) \ (2D) \sim 1 + (2/\pi)\tau,$$

$$\tau \to 0 \quad (1.132)$$

$$(V_m/U) \ (\text{Axi}) \sim 1 + (1/2)\tau^2$$

The axisymmetric disturbance is an order of magnitude smaller. Sufficiently thin source-sink bodies of either type have maximum values of surface velocity near their ends and a local minimum V_m at the location of maximum thickness (Fig. 1.23). However, V_m is larger than freestream velocity (Fig. 1.22).

Doublet in a Uniform Stream. Circular Cylinder and Sphere. Unexpectedly, the superposition of a uniform stream and a doublet, a more complicated singularity, gives rise to a simpler body shape than the Rankine bodies if the doublet axis is parallel to the uniform stream. The situation is shown in Fig. 1.24. With a uniform stream U parallel to the positive x-axis, the doublet, which is located at the origin,

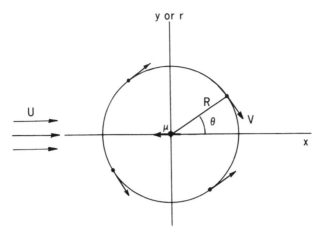

FIGURE 1.24   Point doublet in a uniform stream: circular cylinder and sphere.

must have its axis pointing along the negative x-axis if the combined flow is to have stagnation points on the x-axis for a positive value of doublet strength μ. (As can be seen from the doublet equations Sec. 1.3.3, with $m = n = 0$, it is the sign of the product μl that determines flow direction.) Thus, the doublet formulas of Sec. 1.3.3 are used with $l = -1$, $m = n = x_0 = y_0 = z_0 = 0$.

Two-Dimensional Flow. The Circular Cylinder. The stream function of the combined flow is

$$\Psi = Uy - 2\mu y/(x^2 + y^2) \tag{1.133}$$

Clearly, this is zero on the x-axis and on the circle centered at the origin of radius R where

$$R^2 = 2\mu/U \tag{1.134}$$

Thus, the body is a circular cylinder. Let Eq. (1.134) define μ in terms of R. Then the total potential, stream function, and velocity components at any point are

$$\Phi = Ux[1 + R^2/(x^2 + y^2)] \tag{1.135a}$$

$$\Psi = Uy[1 - R^2/(x^2 + y^2)] \tag{1.135b}$$

$$u = U[1 + R^2(y^2 - x^2)/(x^2 + y^2)^2] \tag{1.135c}$$

$$v = -2UR^2xy/(x^2 + y^2)^2 \tag{1.135d}$$

It is readily verified that the stagnation points are at $y = 0$, $x = \pm R$. On the body $x = R \cos \theta$, $y = R \sin \theta$, where θ is the angle measured from the positive x-axis (Fig. 1.24). Using these in Eq. (1.135), the total surface velocity V is found to be

$$V = 2U \sin \theta \tag{1.136}$$

a very simple and useful result. As can be inferred from Eq. (1.135), the flow direction is as shown in Fig. 1.24, namely a positive value of V corresponds to a clockwise velocity, i.e., in the direction of decreasing θ.

Axisymmetric Flow. The Sphere. The total stream function of the combined flow is

$$\Psi = Ur^2/2 - \mu r^2/(x^2 + r^2)^{3/2} \tag{1.137}$$

where as usual $r^2 = y^2 + z^2$. This is zero on the x-axis and on the circle $x^2 + r^2 = R^2$ where

$$R^3 = 2\mu/U \tag{1.138}$$

Thus, the body is a sphere. Using this value for μ, the total potential, stream function, and velocity components are

$$\Phi = Ux[1 + R^3/2(x^2 + r^2)^{3/2}] \tag{1.139a}$$

$$\Psi = (1/2)Ur^2[1 - R^3/(x^2 + r^2)^{3/2}] \tag{1.139b}$$

$$u = U[1 + R^3(r^2 - 2x^2)/2(x^2 + r^2)^{5/2}] \tag{1.139c}$$

$$u_r = 3UR^3rx/2(x^2 + r^2)^{5/2} \tag{1.139d}$$

The stagnation points are at $r = 0$, $x = \pm R$. On the body $x = R \cos \theta$, $r = R \sin \theta$ (Fig. 1.24), and the total surface velocity turns out to be

$$V = 1.5U \sin \theta \tag{1.140}$$

a result that can be compared with Eq. (1.136) and with Fig. 1.22.

Linearly Varying Line Source in a Uniform Stream. Ellipsoid of Revolution. The indirect method illustrated above can be greatly elaborated using an ensemble of singularities of various types. Generally, singularities on the x-axis are preferred for the axisymmetric case, although ring sources could also be used. However, there is no restriction to point singularities. Variable strength distributions on the axis yield a wide variety of bodies. In Sec. 1.3.6 a simple approximate method for adjusting such a variation to solve the direct problem, i.e., a given body shape, is outlined. Here the indirect method is used.

An interesting example is that of a linearly varying three-dimensional source distribution on the x-axis. Specifically, let the source strength occupy the portion of the x-axis $|x| \leq c$ and have the strength per unit length

$$\lambda = -K\xi, \qquad -c \leq \xi \leq c \tag{1.141}$$

Clearly, the net strength is zero, so the body is closed. The negative sign insures the source strength is positive on the upstream portion of the axis, which is necessary if stagnation points are to be achieved for a positive value of K. The potential, stream function and velocity components due to this source distribution are obtained by integrating the relevant formulas of Sec. 1.3.3. For example, the potential and stream function are

$$\phi = -\int_{-c}^{c} (-K\xi d\xi)/\sqrt{(x - \xi)^2 + r^2} \qquad (1.142a)$$

$$\psi = -\int_{-c}^{c} [-K\xi(x - \xi)]d\xi/\sqrt{(x - \xi)^2 + r^2} \qquad (1.142b)$$

Let a point (x, r) have a distance D_1 from the point $(-c, 0)$ and a distance D_2 from the point $(+c, 0)$ (see Fig. 1.25). Then the above integrals are

$$\phi = K(D_2 - D_1 + x \ln Q] \qquad (1.143a)$$

$$\psi = -K[c(D_1 + D_2)/2 + x(D_2 - D_1)/2 - (r^2/2) \ln Q] \qquad (1.143b)$$

$$Q = (D_1 + D_2 + 2c)/(D_1 + D_2 - 2c) \qquad (1.143c)$$

The stream function Ψ of the combined flow is obtained by adding that of the uniform stream U to ψ from Eq. (1.143). It turns out by sheer luck that $\Psi = 0$ on the x-axis and on the ellipse $D_1 + D_2 = 2a$, where $x = \pm a$ are the stagnation point locations (Fig. 1.25) and the semi-minor axis b is the maximum radius. The points $x = \pm c$ are the foci of the ellipse. Let $\tau = b/a$ be the thickness ratio of the ellipse, then the source gradient K is

$$K = \tau^2 V_m(\tau)/2(1 - \tau^2)^{3/2} \qquad (1.144)$$

where

$$V_m(\tau) = U(1 - \tau^2)^{3/2}/[\sqrt{1 - \tau^2} - (\tau^2/2) \ln P] \qquad (1.145a)$$

$$P = [1 + \sqrt{1 - \tau^2}]/[1 - \sqrt{1 - \tau^2}] \qquad (1.145b)$$

This can be used in Eq. (1.143) to get ϕ, ψ and, by differentiation, the velocity components at any point. Principal interest is in flow on the body surface, and the

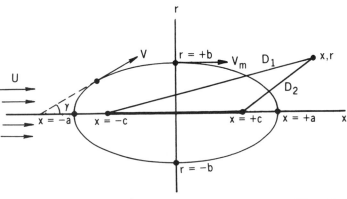

FIGURE 1.25 Linearly varying line source in a uniform stream: ellipsoid of revolution.

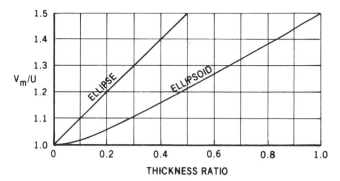

FIGURE 1.26 Maximum velocities on two-dimensional ellipses and ellipsoids of revolution with freestreams parallel to the major axis.

total surface velocity V, including the freestream contribution, can be written

$$V = V_m(\tau) \cos \gamma \qquad (1.146)$$

where γ is the local slope angle of the body profile. Thus, the velocity $V_m(\tau)$, which is attained at the point of maximum thickness, is the maximum velocity, and the velocity at any other point is its projection along the local tangent. The function $V_m(\tau)$ is shown in Fig. 1.26. For comparison, the corresponding quantity for the two-dimensional elliptic cylinder, which is derived in Sec. 1.3.5 below, is also shown. The similarity to Fig. 1.22 for the Rankine bodies is striking, particularly the behavior for small thickness.

Lifting Flow About a Circular Cylinder. From Eq. (1.103) it can be shown that a point vortex at the origin having circulation Γ gives rise to a velocity field whose magnitude is inversely proportional to distance from the origin and whose direction is perpendicular to the radius from the origin. Thus, any such flow may be added to the circular cylinder flow of Fig. 1.24, and the circle will remain a streamline. For the case of a freestream parallel to the x-axis (Fig. 1.24), the combined surface velocity is

$$V = 2U \sin \theta + \Gamma/2\pi R \qquad (1.147)$$

which is the positive in the clockwise sense. The force on the body is

$$\overline{\mathbf{F}} = D\overline{\mathbf{i}} + L\overline{\mathbf{j}} = \overline{\mathbf{i}} \oint p \, dy + \overline{\mathbf{j}} \oint p \, dx \qquad (1.148)$$

where the pressure p is obtained from Bernoulli's equation (1.84), and the integrals are performed clockwise around the body. The result is

$$L = \rho U \Gamma \qquad D = 0 \qquad (1.149)$$

It can be shown that Eq. (1.149) is true for arbitrary two-dimensional bodies [Milne–Thompson (1950) and Karamcheti (1966)]. This is often referred to as the *Kutta–*

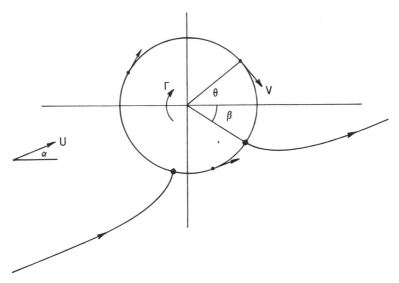

FIGURE 1.27   Lifting flow about a circular cylinder at angle of attack.

Joukowski formula for the lift on a two-dimensional body. For $0 < \Gamma < 4\pi RU$, the stagnation points are on the cylinder and are displaced below the x-axis. The flow pattern is symmetric about the y-axis ("fore and aft" symmetry). If the uniform stream is at angle of attack α (Fig. 1.27), the flow pattern is simply rotated by α, and the surface velocity is obtained from Eq. (1.147) by replacing θ by $\theta - \alpha$. So far the circulation Γ has been undetermined. In many applications its value is adjusted to place the rear stagnation point at a desired location, say $\theta = -\beta$ (Fig. 1.27). This implies

$$2U \sin(-\beta - \alpha) + \Gamma/2\pi R = 0$$

$$\Gamma = 4\pi RU \sin(\alpha + \beta) \qquad (1.150)$$

and the surface velocity is

$$V = 2U[\sin(\theta - \alpha) + \sin(\alpha + \beta)] \qquad (1.151)$$

Equation 1.149 still holds, but L is now the force component perpendicular to the uniform stream and D the component parallel to it. Discussion of the physical significance of the condition for determining the otherwise arbitrary value of Γ is given below.

1.3.5   Solution of Two-Dimensional Flows by Conformal Mapping

General Statement. The Joukowski Transformation.   A different method for solving two-dimensional potential flow problems is through the use of *conformal mapping*. The power of this technique is based on the fact that two-dimensional potential flow is closely related to the theory of functions of a complex variable. This relationship does not exist for axisymmetric (or three-dimensional) flow. The basis of the connection is that the real and imaginary parts of an analytic function satisfy the Cauchy–Riemann equations, which are the two-dimensional forms of the zero curl and divergence conditions. Thus, these functions are solutions of Laplace's equation. Moreover, Laplace's equation is invariant under a conformal transformation. The relevant complex analysis is discussed extensively in the literature [Milne–Thompson (1950) and Karamcheti (1966)], and derivations should be pursued there. The present discussion assumes those developments, as well as familiarity with complex notation, and it presents some results that appear to be of interest.

Consider complex variables $z = x + iy$ and $\zeta = \xi + i\eta$. In the ζ-plane, there is a circle of radius R with center at $\zeta = P$, where P is complex (Fig. 1.28). The complex equation of the circle is

$$\zeta = P + Re^{i\theta} \qquad (1.152)$$

and the parameters are chosen so that the circle intersects the positive real axis at ζ_T. This point defines a value $\theta = -\beta$. The circle is immersed in a uniform stream U at angle of attack α, and the circulation is chosen so that $\theta = -\beta$ is a stagnation point. Suppose the transformation $z(\zeta)$ takes the circle to a body in the z-plane (shown as an airfoil in Fig. 1.28). Let the mapping be such that points at infinity are not disturbed, i.e., $z(\zeta) \to \zeta$ as $|\zeta| \to \infty$. Then the uniform onset flow is the same in both planes, as is the circulation about the body contour. Velocities at corresponding points of the two planes are related by the reciprocal of the mapping derivative. Specifically, for points on the body the mapping $z(\zeta)$ and Eq. (1.152) give

$$x = x(\theta), \qquad y = y(\theta), \qquad dz/d\zeta = f(\theta) \qquad (1.153)$$

The velocity magnitude V on the body surface is

$$V(\text{body}) = V(\text{circle})/|f(\theta)| \qquad (1.154)$$

where $V(\text{circle})$ is given by Eq. (1.151).

Consider an airfoil-type body with a sharp corner at its "trailing edge," as illustrated in Fig. 1.28. It is known [Milne–Thompson (1950) and Karamcheti (1966)]

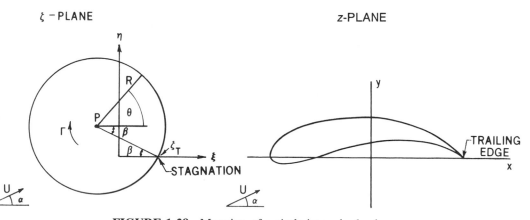

FIGURE 1.28 Mapping of a circle into a body shape.

that the mapping $z(\zeta)$ that transforms a circle to such a body must have a zero value of mapping derivative $dz/d\zeta$ at the point that maps to the corner. Thus, from Eq. (1.154) the velocity at the corner will be infinite unless the velocity at the corresponding point on the circle is zero. The well-known Kutta condition states that such an infinity must be avoided and thus, that the point $\theta = -\beta$ on the circle must map to the trailing edge of the airfoil. As the angle of attack α changes, the circulation Γ, which is the same for both circle and airfoil, changes according to Eq. (1.150) and thus the lift, Eq. (1.149), is proportional to $\sin(\alpha + \beta)$, where $\alpha = -\beta$ is the angle of attack for zero lift. With the mapping defined as above, any value of circulation other than that of Eq. (1.150) would move the stagnation point on the circle away from $\theta = -\beta$ and give an infinite velocity at the airfoil trailing edge. Thus, in physical terms the condition of finite trailing-edge velocity defines a unique value of circulation. (More generally, for multiple airfoils the Kutta condition determines a unique value for each airfoil.) Since the flow leaves the circle at the point $\theta = -\beta$, the flow leaves the airfoil at the trailing edge. The order of the zero of $dz/d\zeta$ depends on the size of the included angle of the trailing edge. It is known [Milne–Thompson (1950) and Karamcheti (1966)] that if the trailing-edge angle is nonzero, the velocity there as given by Eq. (1.154) is zero, but if the trailing-edge angle is zero (cusped trailing edge), the velocity there is finite. Obviously the above discussion does not apply to a body without a sharp corner. Nevertheless, it is customary in analytic work to designate some point as the "trailing edge" to fix the circulation. This point has zero velocity, and the flow leaves the body there.

The simplest meaningful transformation is the Joukowsky transformation,

$$z = \zeta + C^2/\zeta, \qquad dz/d\zeta = 1 - C^2/\zeta^2 \qquad (1.155)$$

where C is real and positive. It can generate an interesting variety of shapes depending on where the center of the circle is located in the ζ-plane. Since $dz/d\zeta = 0$ at $\zeta = \pm C$, P and R must be chosen so that these points are not exterior to the circle in order to avoid infinite velocities at exterior points in the z-plane.

Elliptic Cylinder. If the center of the circle is taken at the origin, then $P = 0$ and $\beta = 0$. To avoid singularities $R \geq C$. Then Eqs. (1.152) and (1.155) give the ellipse

$$x = a \cos\theta \qquad\qquad y = b \sin\theta$$
$$a = R(1 + C^2/R^2) \qquad b = R(1 - C^2/R^2) \qquad (1.156)$$
$$a = 2R/(1 + \tau) \qquad b = 2R\tau/(1 + \tau)$$

where $\tau = a/b$ is the thickness ratio. After some manipulation, the magnitude of the mapping derivative on the body can be written

$$\left\|\frac{dz}{d\zeta}\right\| = \frac{2}{(1 + \tau)} \sqrt{1 - (1 - \tau^2)\cos^2\theta} \qquad (2.157)$$

The "trailing edge" of the ellipse is on the positive x-axis. Since $\beta = 0$, Eqs. (1.154) and (1.155) then give the surface velocity on the lifting ellipse as

$$V = U(1 + \tau) \frac{\sin \alpha + \sin (\theta - \alpha)}{\sqrt{1 - (1 - \tau^2) \cos^2 \theta}} \tag{1.158}$$

which together with Eq. (1.156) gives the velocity distribution. The circulation is given by Eq. (1.150) with $\beta = 0$, and since the *chord* of the ellipse is $2a$, the lift coefficient is

$$C_L = \rho U \Gamma / (0.5 \rho U^2 2a) = 2\pi (1 + \tau) \sin \alpha \tag{1.159}$$

The above results are valid for $0 \leq \tau \leq 1$. Setting $\tau = 0$ gives the velocity distribution on a lifting flat plate lying on $-2R \leq x \leq 2R$ as

$$V(\tau = 0) = U\{\cos \alpha \pm \sin \alpha \sqrt{[1 - (x/2R)]/[1 + (x/2R)]}\} \tag{1.160}$$

where the plus sign applies to the upper surface and the minus to the lower. This is the usual thin-airfoil result. Finally, taking $\alpha = 0$ in Eq. (1.158) gives the nonlifting flow about an ellipse at zero angle of attack as

$$V(\alpha = 0) = U(1 + \tau) \cos \gamma \tag{1.161}$$

where γ is the local slope angle (Fig. 1.25). The maximum velocity occurs at the middle, $x = 0$, and has the value $V_m = U(1 + \tau)$. This is compared with V_m for an ellipsoid of revolution in Fig. 1.26.

Joukowsky Airfoils. To obtain a *Joukowsky airfoil* with a sharp trailing edge, the center and radius of the circle in the ζ-plane must be such that $\zeta_T = C$, i.e., the circle must intersect the positive real axis at the point where the mapping derivative is zero. (For the ellipse, this point was inside the circle.) The point $\zeta_T = C$, which is the stagnation point of the flow about the circle, maps to the airfoil trailing edge. The airfoil shape is completely determined by the location P of the center of the circle, since the radius R is determined from P and the $\zeta_T = C$ condition. To prevent the other zero of $dz/d\zeta$ at $\zeta = -C$ from lying outside the circle, P must have a nonpositive real part. A center on the positive imaginary axis gives an "airfoil" that is a circular arc of zero thickness with the degree of camber dependent on the height of the center above the origin. If the center lies on the negative real axis, the airfoil is symmetric, i.e., uncambered, but with finite thickness. More general locations of the center yield airfoils with both camber and thickness. It can be shown that all Joukowsky airfoils have cusped trailing edges. The theory for symmetric Joukowsky airfoils is given below.

Let the center of the circle in the ζ-plane be on the real axis at a point $\zeta = P = -\epsilon C$, where ϵ is positive. Referring to Fig. 1.28, $\beta = 0$ and $R = (1 - \epsilon)C$. Separating real and imaginary parts in Eq. (1.162) gives the airfoil shape as

$$x = C[(1 + \epsilon) \cos \theta - \epsilon] [1 + 1/D] \tag{1.162a}$$

$$y = C(1 + \epsilon) \sin \theta [1 - 1/D] \tag{1.162b}$$

$$D = \epsilon^2 + (1 + \epsilon)^2 - 2\epsilon(1 + \epsilon) \cos \theta \tag{1.162c}$$

The chord of the airfoil is

$$c = 4C(\xi + 1)^2/(2\xi + 1) \tag{1.163}$$

After some manipulation, the absolute value of the mapping derivative on the body is

$$\left\| \frac{dz}{d\zeta} \right\| = \frac{2(1 + \epsilon)}{D} \sqrt{(1 - \cos\theta)\,[1 + \epsilon^2 + (1 - \epsilon^2)\cos\theta]} \tag{1.164}$$

The above, together with Eqs. (1.151) and (1.154) with $\beta = 0$, define the surface velocity distribution. It can be shown that for small values of thickness ratio τ, the relation $\tau = 1.3\epsilon$ is a good approximation. At the trailing edge the velocity approaches

$$V_{TE} = U\cos\alpha/(1 + \epsilon) \tag{1.165}$$

for all values of ϵ. Figure 1.29 shows the airfoil shape and surface velocity distribution at $\alpha = 10°$ for the 13%-thick airfoil obtained by setting $\epsilon = 0.1$.

As outlined above, the mapping technique is an indirect method in the sense of Sec. 1.3.4: Given a mapping, find the body and the velocity distribution. Just as for the indirect method based on singularities, the indirect mapping method can be greatly elaborated. James (1972) presents an extensive study, which includes airfoils of complicated shapes. The mappings are simpler for shapes with cusped trailing edges, but airfoils with nonzero trailing edge angles can also be generated, such as the well-known *Karman–Trefftz airfoils* [Milne–Thompson (1950) and Karamcheti (1966)]. Using numerical techniques, the mapping procedure may be adapted to solve the direct problem, which of course is far more useful. Halsey (1979) has

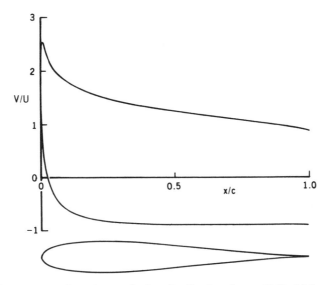

FIGURE 1.29 Shape and surface velocity distribution for a 13% thick symmetric Joukowsky airfoil at 10° angle of attack.

constructed a very powerful method, which gives highly accurate solutions for virtually any two-dimensional configuration including multiple airfoils.

1.3.6 Direct Methods of Solution Based on Numerical Discretization of Integral Equations

The superposition principle for singular solutions, which was used in Sec. 1.3.4 to generate solutions indirectly, may also be used to solve the direct problem for arbitrarily prescribed bodies. Because of the greatly increased complexity of the mathematics, analytic solutions are not possible. Instead, numerical solutions are obtained, but this is accomplished in a routine manner, and the results are quite useful. The basic philosophy of the various "singularity methods" consists of three steps. First, a continuously variable singularity distribution is assumed whose location is known but whose strength variation is unknown. Second, application of the zero normal-velocity boundary condition, Eq. 1.82, leads to an integral equation for the singularity distribution. Third, the numerical discretization replaces the integral equation by a set of linear algebraic equations for a finite number of values of the singularity. The last set of equations then determine the flow field. Stream functions or velocities due to distributed singularities are obtained by integrating the corresponding point singularity formulas of Sec. 1.3.3 over the domain of the distribution. Numerically, the two main tasks are: (1) calculating the coefficient matrix of the linear equations, and (2) solving the system.

Von Karman's Axial Source Method. A classical example of this type of method is due to Von Karman (1930) and is applicable to axisymmetric flow. Suppose there is a uniform stream U parallel to the x-axis, immersed in which is an axisymmetric body whose profile curve has the equation $r = r(x)$ with $r = 0$ at $x = d$ and $x = g$ (Fig. 1.30). A line source distribution of variable strength per unit length $\lambda(x)$ is assumed to lie along the x-axis between $x = e$ and $x = f$. (Clearly, the inequalities $d < e < f < g$ must hold.) Von Karman applied the zero normal velocity boundary condition in terms of the stream function instead of the velocity. Using the formulas of Sec. 1.3.3 the combined stream function is set equal to zero on the curve $r = r(x)$, i.e.,

$$\Psi = U[r(x)]^2/2 - \int_e^f (x - \xi)\lambda(\xi)d\xi/\sqrt{(x - \xi)^2 + [r(x)]^2} = 0$$

$$d \le x \le g \tag{1.66}$$

since all functions are known except $\lambda(\xi)$, this is an integral equation for λ.

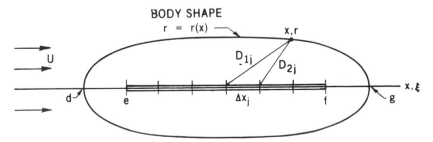

FIGURE 1.30 von Karman's axial source method.

The numerical discretization is as follows. Divide the line $e \le x \le f$ into N increments Δx_j, and assume that over each increment $\lambda(x)$ may be approximated by a constant λ_j. Then for each increment the constant λ_j may be taken out of the integral of Eq. (1.166), and the resulting integral over Δx_j of known geometric quantities can be accomplished analytically. The various integrals over Δx_j, each of which is the stream function of a constant-strength line source, comprise the desired coefficient matrix. Specifically, the linear equations that approximate Eq. (1.166) are

$$\sum_{j=1}^{N} A_{ij}\lambda_j = C_i, \qquad i = 1, 2, \ldots, N \tag{1.167}$$

where

$$A_{ij} = D_{2j}(x_i) - D_{1j}(x_i), \quad C_i = U[r(x_i)]^2/2 \tag{1.168}$$

Note (Fig. 1.30), that D_2 and D_1 are the distances from the point $[x_i, r(x_i)]$ to the two ends of the segment Δx_j. The very simple coefficient matrix is the main advantage of this method.

While it is simple, Von Karman's method has rather severe limitations both theoretically and practically. The basic restriction is stated by Von Karman (1930). "This [representation by an axial line source] is possible only in the exceptional case when the analytical continuation of the potential function, free from singularities in the space outside the body, can be extended to the axis of symmetry without encountering singular spots." The implications of this condition are difficult to assess. It is not necessarily true that the body must be slender, as shown by the solution for the ellipsoid of revolution in Sec. 1.3.4. As a practical matter the body probably should be fairly thin and very smooth. Another unanswered question is the extent of the axial source distribution, i.e., what values should be given e and f. Clearly, e cannot equal d, but how big should the gap be? The line source of Sec. 1.3.4 extends between the foci for the ellipsoid of revolution, but no general result is known. Because current computers have made computing costs for axisymmetric flow so small, the more general methods of the type discussed below have rendered this method largely obsolete. Nevertheless, interest in the method continues [Zedan and Dalton (1980)].

Surface Singularity Distributions. The Surface Source Method. In order to solve the direct problem of calculating the velocity field about a given body of arbitrarily prescribed shape, singularity distributions must be placed on the surface of the body. It is known [Lamb (1945)] that the desired flow solution for any body can be expressed in terms of surface distributions of source and doublet. The perturbation velocity at any point in the field is

$$\overline{v}(x, y, z) = \int\int_S [\overline{V}(\text{source})\sigma(x_0, y_0, z_0) + \overline{V}(\text{doublet})\mu(x_0, y_0, z_0)] \, dS \tag{1.169}$$

where: (1) σ and μ are, respectively, the source and doublet strength per unit area, (2) (x_0, y_0, z_0) is a point on the body surface S, (3) the doublet axis is along the local normal to S, and (4) the point source and doublet velocities are given by the appropriate equations in Sec. 1.3.3. Because it is based on the source and doublet solutions, the velocity of Eq.(1.169) is a potential flow, i.e., has zero curl and divergence, and approaches zero at infinity for arbitrary functions σ and μ. These last two functions σ and μ are available to satisfy the zero normal velocity boundary condition. Since there are two functions and only one condition, there is a "degree of freedom" in the problem. Either σ or μ or a relation between them may be arbitrarily prescribed. In addition, internal singularities may be used. The choice of singularity should be made to achieve favorable numerics. Different investigators have made different choices, and descriptions of the various methods have been given in several general reviews [Hess (1975a) and Shaw (1980)]. In all cases, the integral equation resulting from application of the boundary condition has the body surface as its domain.

By way of example, the surface source method is described here. It forms the basis of many of the existing flow-calculation methods [Hess (1975a) and Shaw (1980)] and provides the most direct analogy of Von Karman's method. In this approach, the source strength σ is adjusted to satisfy the normal velocity boundary condition. If the flow is nonlifting, the doublet strength μ is set equal to zero. In lifting flow, an auxiliary doublet distribution is used to satisfy the Kutta condition and provide the lift. There are various numerical discretizations of the surface source method applicable to two-dimensional, axisymmetric, and three-dimensional flows [Hess and Smith (1966) and Hess (1974)]. Moreover, there are first-order and second-order methods [Hess (1975a), Shaw (1980), Hess (1973), and Hess (1975b)], as well as various assumptions for the auxiliary doublet distribution [Hess (1975a) and Shaw (1980)]. For definiteness, a two-dimensional, first-order source method is described below using surface vorticity (equivalent to surface doublets) to obtain lift.

Numerical Discretization in Two Dimensions. In two dimensions the basic velocity influences are those due to a straight-line element of constant strength. These influences are most conveniently calculated in a coordinate system based on the element. Let a source distribution of constant strength unity lie along the x-axis from $-\Delta/2$ to $+\Delta/2$. Integrating the point-source formulas of Sec. 1.3.3 gives the velocity components at a point (x, y) as

$$
\begin{aligned}
u &= \int_{-\Delta/2}^{\Delta/2} \frac{(x - \xi)d\xi}{(x - \xi)^2 + y^2} = \frac{\ln[(x + \Delta/2)^2 + y^2]}{\ln[(x - \Delta/2)^2 + y^2]} \\
v &= \int_{-\Delta/2}^{\Delta/2} \frac{y\,d\xi}{[(x - \xi)^2 + y^2]} = 2 \tan^{-1}\left(\frac{y\Delta}{[x^2 + y^2 - (\Delta/2)^2]}\right)
\end{aligned}
\tag{1.170}
$$

where the $\tan^{-1}$ must be evaluated in the range $-\pi$ to $+\pi$ by accounting for the individual signs of the numerator and denominator. The equations in Sec. 1.3.3. may be used to show that the two-dimensional velocity field due to a doublet distribution on a closed curve is identical to that of a vorticity distribution on the same curve whose strength is proportional to the derivative of the doublet strength. (An analogous relation in terms of doublet gradient exists in three dimensions [Hess (1974)].) It is more convenient to generate two-dimensional lift by means of surface vorticity. The equations in Sec. 1.3.3 also show that the velocity components due to a point vortex of circulation $\Gamma = 2\pi m$ have the same magnitude as those of the source of strength m, but that the x and y components are interchanged with one sign changed. By inspection it is evident that if this interchange is made in the integrands of Eq. (1.170), the integrated components of source and vorticity have the same relation.

A two-dimensional body contour is defined by a set of $N + 1$ points whose coordinates are input to the computer. If the body is an airfoil, by far the most common case, both the first and last points are at the trailing edge. Points are input in clockwise order around the contour, Fig. 1.31. For numerical purposes, successive input points are joined by straight-line elements. The midpoints of the elements are designated control points. It is only at these points that the boundary condition is applied, and the fluid velocity and pressure are ultimately calculated.

Let subscripts i and j denote quantities associated with particular elements. Clearly i and j range from 1 to N. For each element certain geometric quantities are calculated, namely

$$\bar{x}_j = \tfrac{1}{2}(x_j + x_{j+1}), \qquad \bar{y}_j = \tfrac{1}{2}(y_j + y_{j+1})$$

$$C_j = x_{j+1} - x_j \qquad S_j = y_{j+1} - y_j \qquad (1.171)$$

$$\Delta_j = \sqrt{C_j^2 + S_j^2}$$

$$\cos \gamma_j = C_j/\Delta_j \qquad \sin \gamma_j = S_j/\Delta_j$$

These are the control point coordinates $\bar{x}, \bar{y}$, the element length Δ, and the sine and cosine of the element inclination angle γ, where this last quantity may be in any of the four quadrants.

To compute the effect of the elements on each other, the basic task is to compute the normal velocity A_{ij} and tangential velocity B_{ij} induced at the control point of the ith element by a unit source density on the jth element. This is accomplished by transforming the ith control point, along with its unit normal and tangent vectors, into a coordinate system based on the jth element, which has its x-axis lying in the jth element, y-axis normal to it, and origin at the ith control point. In this coordinate system, Eq. (1.170) may be used to obtain Cartesian velocity components and dot

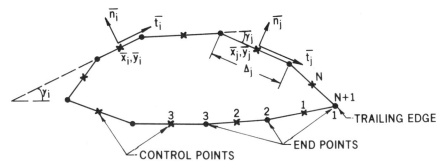

FIGURE 1.31 Numerical discretization in two dimensions.

products may be taken to get normal and tangential components. Specifically, calculate

$$x = (\bar{x}_i - \bar{x}_j) \cos \gamma_j + (\bar{y}_i - \bar{y}_j) \sin \gamma_j$$

$$y = -(\bar{x}_i - \bar{x}_j) \sin \gamma_j + (\bar{y}_i - \bar{y}_j) \cos \gamma_j \qquad (1.172)$$

These are used in Eq. (1.170) together with Δ_j to find u and v after which the desired quantities are

$$A_{ij} = -u \sin (\gamma_i - \gamma_j) + v \cos (\gamma_i - \gamma_j)$$

$$B_{ij} = u \cos (\gamma_i - \gamma_j) + v \sin (\gamma_i - \gamma_j) \qquad (1.173)$$

The self-effect of an element is

$$A_{ii} = 2\pi \qquad B_{ii} = 0 \qquad (1.174)$$

In general, the uniform stream is at angle of attack α. It simplifies the logic to generate three independent flow solutions and combine them later. These are: (1) noncirculatory flow for $\alpha = 0$, (2) noncirculatory flow for $\alpha = 90°$, and (3) pure circulatory flow due to a unit vorticity distribution ω on all elements. The normal and tangential components of these flows at the ith control point are

$$\alpha = 0°: \quad N_i^{(0)} = -\sin \gamma_i, \; T_i^{(0)} = \cos \gamma_i$$

$$\alpha = 90°: \quad N_i^{(90)} = \cos \gamma_i, \quad T_i^{(90)} = \sin \gamma_i \qquad (1.175)$$

$$\omega = 1: \quad N_i^{(\omega)} = \sum_{j=1}^{N} B_{ij}, \; T_i^{(\omega)} = \sum_{j=1}^{N} A_{ij}$$

The components for the circulatory flow have been obtained based on the above-mentioned relation between source and vorticity velocity components.

The set of linear algebraic equations for the values of source density on the elements is

$$\sum_{j=1}^{N} A_{ij}\sigma_j = -N_i, \qquad i = 1, 2, \ldots, N \qquad (1.176)$$

When this has been solved, surface velocities at the control points are obtained from

$$V_i = \sum_{j=1}^{N} B_{ij}\sigma_j + T_i \qquad (1.177)$$

Equations (1.176) and (1.177) are solved for each of the flows of Eq. (1.175). Thus, three complete surface velocity distributions are obtained: $V_i^{(0)}$, $V_i^{(90)}$, $V_i^{(\omega)}$. The combined velocity distribution for any angle of attack α is

$$V_i^{(\alpha)} = V_i^{(0)} \cos \alpha + V_i^{(90)} \sin \alpha + \omega V_i^{(\omega)}, \quad i = 1, \ldots, N \qquad (1.178)$$

The sign convention is that a positive tangential velocity is clockwise around the contour, and a negative one is counterclockwise. The value of ω is determined from the Kutta condition, whose numerical approximation requires that the velocities at the first and last control points are equal in magnitude and opposite in sign. (The velocity is negative at $i = 1$ and positive at $i = N$ to insure flow off the trailing edge.) The equation is

$$\omega = -\frac{[V_1^{(0)} + V_N^{(0)}]\cos\alpha + [V_1^{(90)} + V_N^{(90)}]\sin\alpha}{V_1^{(\omega)} + V_N^{(\omega)}} \qquad (1.179)$$

Besides its logical simplicity, the above scheme is computationally efficient because the three solutions of Eq. (1.176) can be obtained simultaneously and because solutions of the form of Eq. (1.178) can be obtained for a set of values of α with very little additional effort.

The Three-Dimensional Problem. The flow calculation that was noted as the basic problem of potential flow in Sec. 1.3.1 may be discretized for three-dimensional problems in a manner completely analogous to the above two-dimensional procedure. Input points on the surface define the body to the computer. These are associated in groups of four to form N four-sided surface elements. In a first-order method, the elements are plane quadrilaterals as shown in Fig. 1.32. The appearance of the elements in Fig. 1.32 has led to their designation as *panels* and this approach as a *panel method*. The centroid of a quadrilateral serves as the control point where the normal velocity is set equal to zero and where the pressure is calculated. Source strength is taken constant on each element, and the boundary condition yields N linear equations in the N unknown values of source strength. Computation of the velocity influence matrices of the elements on each others' control points requires integrating the three-dimensional point-source equations in Sec. 1.3.3 over a general

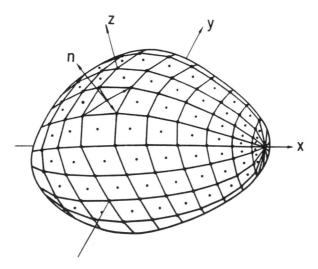

FIGURE 1.32 Approximation of a three-dimensional body by plane quadrilateral surface elements or "panels."

plane quadrilateral. This can be accomplished analytically if the quadrilateral is in the xy-plane [Hess and Smith (1966)]. Thus, just as in two dimensions, the velocity induced at the ith control point by a unit source density on the jth element is accomplished using a coordinate system based on the jth element. Its z-axis is the element normal vector, and its origin is the jth control point. Again, the induced normal velocity is the dot product of the induced velocity vector with the normal vector of the ith element. The complete set of induced normal velocities comprises the coefficient matrix of the linear equations for the source densities. Because three-dimensional computing costs are typically two orders of magnitude higher than two-dimensional ones, a number of efficiency features are used in three-dimensional programs. The most important are the use of simple approximate induced velocity formulas when the ith and jth elements are far apart and the use of an iterative matrix solution if the number of elements is very large. Details of this somewhat complicated calculation are given in Hess and Smith (1966).

While the addition of lift in two dimensions causes no significant increase in the complexity of the problem, three-dimensional lifting flow is not only considerably more complicated than nonlifting flow but requires a set of more-or-less arbitrary assumptions, which, while reasonable in many cases, are difficult to justify in a fundamental way. The reasoning used to construct a potential flow model of three-dimensional lift is described in Hess (1974). Figure 1.33 illustrates the main features of the model. Wings or other lifting portions of the configuration are characterized by having trailing edges which cause trailing vortex wakes. So-called *bound vorticity* is hypothesized to lie on or within the wing surface with strength varying in the direction generally parallel to the trailing edge (*spanwise* in aerodynamic jargon). The bound vorticity strength is adjusted to satisfy a condition of smooth flow off the entire trailing edge, while the trailing vorticity has constant strength in the stream direction and an initial strength at the trailing edge equal to the ''spanwise'' deriv-

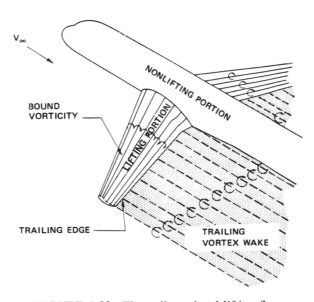

FIGURE 1.33 Three-dimensional lifting flow.

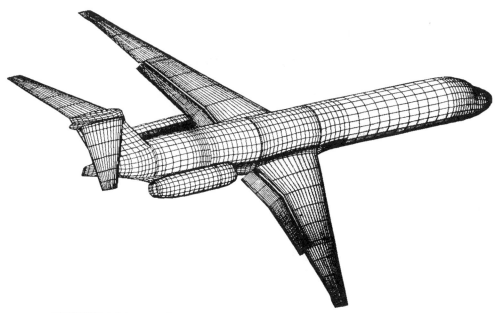

FIGURE 1.34 Paneling of a complicated three-dimensional configuration.

ative of bound vorticity. The location of the wake once it leaves the trailing edge is initially unknown, which introduces a nonlinearity. Usually, the wake location is assumed, because it has a relatively minor effect on surface pressures. Various first-order numerical implementations [Hess (1975a), Shaw (1980), and Hess (1974)] have been devised. All of them panel the configuration of Fig. 1.33, lifting and nonlifting portions and wake, with quadrilaterials like those of Fig. 1.32. First-order surface source methods for many years were the standard design tool throughout the world. The source methods proved fast and reliable. Now, more elaborate methods are available of both the source and dipole type [Hess, Friedman, and Clark (1985), Carmichael and Erickson (1981)], and very complicated configurations can be analyzed, e.g., Fig. 1.34. It is remarkable how well these potential flow results agree with experiment.

1.4 INTRODUCTION TO FLOW OF VISCOUS FLUIDS
Frank M. White

The previous sections of this chapter have been restricted to inviscid flow. In reality, all fluids exhibit frictional or *viscous* effects which can have profound consequences on the flow pattern. Whereas inviscid fluids are devoid of shear stress everywhere, a real fluid can support finite shear stresses when in motion.

1.4.1 Viscous Stress Versus Strain Rate

A real fluid at rest, as in Sec. 1.1 (hydrostatics), has no shear stress because its weight can be balanced entirely by pressure (normal) stresses, as in Eq. (1.1). But, a real fluid in motion will develop shear stresses if it is subjected to *spatial gradients* in the velocity field. Such gradients are commonly caused by the presence of a solid boundary to which a viscous fluid will *stick* and not slip.

A classic case of shear flow is developed by placing a viscous fluid between a fixed and a moving plate, as in Fig. 1.35(a). It takes a finite force F to maintain the upper plate ($y = H$) at a steady velocity V while the lower plate ($y = 0$) is fixed. The fluid develops a *velocity profile* $u(y)$ which has no-slip at the upper surface ($u = V$) and also at the lower surface ($u = 0$). Regardless of the type of fluid placed between the plates, the velocity profile will be linear in y, assuming no edge effects

$$u = V\frac{y}{H} \tag{1.180}$$

where H is the spacing between plates. The linear profile reflects the fact that the shear stress $\tau_{xy} = F/A$ is constant between the plates—a very special case.

If one varies the force F on the plate in Fig. 1.35(a), the plate velocity V will

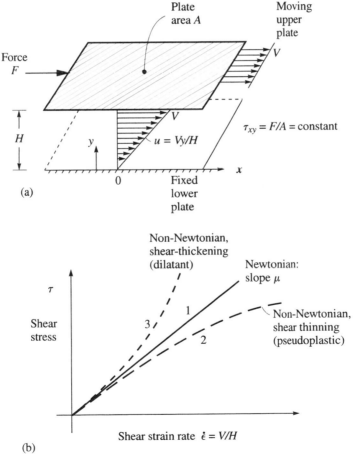

FIGURE 1.35 Fluid sheared between two plates. (a) Linear velocity profile and (b) linear or nonlinear stress/strain rate behavior.

vary, possibly in a nonlinear manner. Each viscous fluid will exhibit a unique relation between shear stress and the fluid strain rate $\dot{\epsilon} = \partial u / \partial y = V/H$ between the plates. Three typical curves are shown in Fig. 1.35(b). A straight line (curve 1) is characteristic of a linear or *Newtonian* viscous fluid

$$\tau = \mu \dot{\epsilon} \tag{1.181}$$

The slope of the curve is called the *coefficient of viscosity*, μ, with units of stress time: Pa · s or kg/(m · s) in SI units. All gases, fresh water, seawater, mercury, and most common hydrocarbon liquids (gasoline, kerosene, lubricating oils) are Newtonian fluids. This section only treats the Newtonian fluids.

Figure 1.35(b) also illustrates two non-Newtonian fluids: shear-thinning or *pseudoplastic* (curve 2), and shear-thickening or *dilatant* (curve 3). The general analysis of non-Newtonian fluids is given in Chap. 3.

The Reynolds Number. Consider dynamic motion of a viscous flow with velocity scale U and length scale L. The two most important fluid parameters influencing the motion are the density ρ and the viscosity μ. These four parameters (U, L, ρ, μ) may be combined into a single dimensionless group called the *Reynolds number*, Re, after Osborne Reynolds (1883)

$$\text{Re} = \rho UL/\mu = UL/\nu \tag{1.182}$$

where $\nu = \mu/\rho$ is a convenient ratio called the *kinematic viscosity* of the fluid. The Reynolds number is the dominant parameter affecting almost all viscous flows. Experiments used to predict a prototype viscous flow from a model test should be scaled by the Reynolds number, that is, Re(model) equal to Re(prototype), as discussed in Chap. 16. Theoretical analyses of viscous flow (Chap. 4) also reveal the Reynolds number to be the primary parameter of such flows.

Laminar Versus Turbulent Flow. The importance of the Reynolds number was beautifully illustrated in a classic experiment by Reynolds himself, using a dye streak to visualize flow through a smooth pipe, as in Fig. 1.36. If the Reynolds number was low, the dye streak remained straight and smooth [Fig. 1.36(a)], a condition called *laminar* or *streamline* flow. In an intermediate Reynolds number range [Fig. 1.36(b)], the dye streak exhibited erratic behavior, and a point measurement of, say, velocity versus time showed irregular "bursts" of activity. This intermediate range is called *transition flow*.

At high Reynolds numbers [Fig. 1.36(c)], the dye streak breaks up and mixes at an intense rate, filling the tube with color. Point velocity measurements show a continuous random fluctuation called *turbulence*, and instantaneous streamlines entwine like spaghetti. This is turbulent flow, and it has friction and heat transfer of quite a different character compared to laminar flow.

For flow in a pipe, the proper length scale for the Reynolds number is the diameter, D, and approximate flow ranges are

a)   laminar flow: $\text{Re}_D < 2000$

b)   transition flow: $2000 < \text{Re}_D < 4000$

c)   turbulent flow: $\text{Re}_D > 4000$

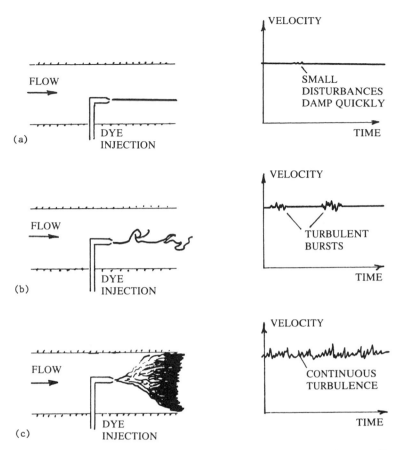

FIGURE 1.36 Dye streak visualization and velocity measurements in duct flow (after a famous experiment by Osborne Reynolds in 1883): (a) laminar flow, low Re, (b) transition flow, moderate Re, and (c) turbulent flow, large Re.

All viscous flows are characterized by these three regimes of motion, but the numerical values for the transition are dependent upon the geometry of the flow (Chap. 4).

1.4.2 The Navier–Stokes Equation

The inviscid momentum Eq. (1.60) may be readily modified to account for viscous stresses. The viscous stress tensor τ_{ij} has nine components as illustrated in Fig. 1.37. Gradients in these components give rise to a net force on a differential element, and the momentum equation becomes

$$\rho \, \frac{d\overline{\mathbf{V}}}{dt} = -\nabla p + \rho \overline{\mathbf{X}} + \nabla \cdot \tau_{ij} \qquad (1.183)$$

This modification was first given by L. M. H. Navier in 1827.

Meanwhile, the simple Newtonian shear relation Eq. (1.181) was generalized to a three-dimensional flow by George G. Stokes in 1845, with the result

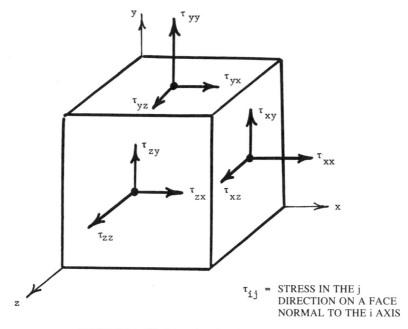

$$\tau_{ij} = \text{STRESS IN THE } j \text{ DIRECTION ON A FACE NORMAL TO THE } i \text{ AXIS}$$

FIGURE 1.37 Notation for Stresses.

$$\tau_{ij} = \mu \left(\frac{\partial u_i}{\partial x_j} + \frac{\partial u_j}{\partial x_i} \right) + \delta_{ij} \lambda \, \nabla \cdot \overline{\mathbf{V}} \qquad (1.184)$$

where λ is called the second or *bulk coefficient of viscosity*, and δ_{ij} is the Kroneker delta function. Since $\nabla \cdot \overline{\mathbf{V}}$ is very small for most flows, the term involving λ is usually neglected. Stokes hypothesized that the two viscosity coefficients were directly related,

$$\lambda \doteq -\frac{2}{3} \mu$$

but experiments show this to be accurate only for monatomic gases.

The combination of Eqs. (1.183) and (1.184) are called the *Navier–Stokes equation* for a Newtonian fluid. The equations are quite complex mathematically, as shown when the three scalar components are written in full for Cartesian coordinates

$$\rho \left(\frac{\partial u}{\partial t} + u \frac{\partial u}{\partial x} + v \frac{\partial u}{\partial y} + w \frac{\partial u}{\partial z} \right) = -\frac{\partial p}{\partial x} + \rho g_x + \frac{\partial}{\partial x} \left(2\mu \frac{\partial u}{\partial x} + \lambda \nabla \cdot \overline{\mathbf{V}} \right)$$

$$+ \frac{\partial}{\partial y} \left[\mu \left(\frac{\partial u}{\partial y} + \frac{\partial v}{\partial x} \right) \right] + \frac{\partial}{\partial z} \left[\mu \left(\frac{\partial w}{\partial x} + \frac{\partial u}{\partial z} \right) \right] \qquad (1.185a)$$

$$\rho \left(\frac{\partial v}{\partial t} + u \frac{\partial v}{\partial x} + v \frac{\partial v}{\partial y} + w \frac{\partial v}{\partial z} \right) = -\frac{\partial p}{\partial y} + \rho g_y + \frac{\partial}{\partial x} \left[\mu \left(\frac{\partial v}{\partial x} + \frac{\partial u}{\partial y} \right) \right]$$

$$+ \frac{\partial}{\partial y} \left(2\mu \frac{\partial v}{\partial y} + \lambda \nabla \cdot \overline{\mathbf{V}} \right) + \frac{\partial}{\partial z} \left[\mu \left(\frac{\partial v}{\partial z} + \frac{\partial w}{\partial y} \right) \right] \qquad (1.185b)$$

$$\rho\left(\frac{\partial w}{\partial t} + u\,\frac{\partial w}{\partial x} + v\,\frac{\partial w}{\partial y} + w\,\frac{\partial w}{\partial z}\right) = -\frac{\partial p}{\partial z} + \rho g_z + \frac{\partial}{\partial x}\left[\mu\left(\frac{\partial w}{\partial x} + \frac{\partial u}{\partial z}\right)\right]$$

$$+ \frac{\partial}{\partial y}\left[\mu\left(\frac{\partial v}{\partial z} + \frac{\partial w}{\partial y}\right)\right] + \frac{\partial}{\partial z}\left(2\mu\frac{\partial w}{\partial z} + \lambda\nabla\cdot\overline{\mathbf{V}}\right) \tag{1.185c}$$

This is a formidable set of nonlinear partial differential equations whose general properties are still not known. As shown in Chap. 4, certain special case solutions have been found, and many successful digital computer simulations have been reported.

Incompressible Flow. Most of the known solutions to the Navier–Stokes equation are for the special case of incompressible flow with constant viscosity, for which Eq. (1.185) reduces to

$$\rho\left(\frac{\partial\overline{\mathbf{V}}}{\partial t} + u\,\frac{\partial\overline{\mathbf{V}}}{\partial x} + v\,\frac{\partial\overline{\mathbf{V}}}{\partial y} + w\,\frac{\partial\overline{\mathbf{V}}}{\partial z}\right) = -\nabla p + \rho\mathbf{g} + \mu\,\nabla^2\overline{\mathbf{V}} \tag{1.186}$$

While simplified, this equation is sufficiently general to reveal many characteristics of viscous flows, such as boundary layer behavior, separation, flow instability, and turbulence. A detailed study of this equation and its consequences is given in Chap. 4 and in White (1992) and Schetz (1993).

Cylindrical Coordinates. Assuming incompressible flow with constant viscosity, Eq. (1.186) has the following three components for the cylindrical coordinates of Fig. 1.10(a)

$$\frac{\partial v_r}{\partial t} + (\overline{\mathbf{V}}\cdot\nabla)v_r - \frac{v_\theta^2}{r} = -\frac{1}{\rho}\frac{\partial p}{\partial r} + g_r + \nu\left(\nabla^2 v_r - \frac{v_r}{r^2} - \frac{2}{r^2}\frac{\partial v_\theta}{\partial\theta}\right)$$

$$\frac{\partial v_\theta}{\partial t} + (\overline{\mathbf{V}}\cdot\nabla)v_\theta + \frac{v_\theta v_r}{r} = -\frac{1}{\rho r}\frac{\partial p}{\partial\theta} + g_\theta + \nu\left(\nabla^2 v_\theta + \frac{2}{r^2}\frac{\partial v_r}{\partial\theta} - \frac{v_\theta}{r^2}\right)$$

$$\frac{\partial v_z}{\partial t} + (\overline{\mathbf{V}}\cdot\nabla)v_z = -\frac{1}{\rho}\frac{\partial p}{\partial z} + g_z + \nu\,\nabla^2 v_z \tag{1.187}$$

where

$$\overline{\mathbf{V}}\cdot\nabla = v_r\frac{\partial}{\partial r} + \frac{v_\theta}{r}\frac{\partial}{\partial\theta} + v_z\frac{\partial}{\partial z}$$

$$\nabla^2 = \frac{1}{r}\frac{\partial}{\partial r}\left(r\frac{\partial}{\partial r}\right) + \frac{1}{r^2}\frac{\partial^2}{\partial\theta^2} + \frac{\partial^2}{\partial z^2}$$

These may be further reduced to the special case of *axisymmetric flow*, where there are no circumferential variations, $(\partial/\partial\theta)\equiv 0$.

The Navier–Stokes equation in spherical polar coordinates [Fig. 1.10(b)] is given in Appendix B [White (1992)].

1.4.3 The General Energy Equation

The energy equation or first law of thermodynamics was given as Eq. (1.63) for an inviscid fluid. This may be extended to a viscous, heat-conducting fluid by adding the work done by viscous stresses and by assuming that the fluid is isotropic and follows *Fourier's Law of Conduction*

$$\overline{q} = -k\nabla T \tag{1.188}$$

where k is the thermal conductivity of the fluid. The general energy equation thus becomes

$$\rho \frac{de}{dt} + p\nabla \cdot \overline{V} = \nabla \cdot (k\nabla T) + \mu\Phi \tag{1.189}$$

where Φ is called the *viscous dissipation function* and is a positive definite relation depending on the strain rates

$$\Phi = 2\left(\frac{\partial u}{\partial x}\right)^2 + 2\left(\frac{\partial v}{\partial y}\right)^2 + 2\left(\frac{\partial w}{\partial z}\right)^2 + \left(\frac{\partial v}{\partial x} + \frac{\partial u}{\partial y}\right)^2$$

$$+ \left(\frac{\partial w}{\partial y} + \frac{\partial v}{\partial z}\right)^2 + \left(\frac{\partial u}{\partial z} + \frac{\partial w}{\partial x}\right)^2 + \frac{\lambda}{\mu}(\nabla \cdot \overline{V})^2$$

An alternate form of Eq. (1.189) uses the fluid enthalpy

$$\rho \frac{dh}{dt} = \frac{dp}{dt} + \nabla \cdot (k\nabla T) + \mu\Phi$$

The only restrictions on this relation are that the fluid must be single-phase, Newtonian, satisfy Fourier's Law, and have no internal sources of heat generation within the element.

1.4.4 Boundary Conditions

The continuity, Eq. (1.50), Navier–Stokes, Eq. (1.185), and energy, Eq. (1.189) relations constitute three simultaneous partial differential equations in the five variables ρ, $\overline{V}$, p, T, and e (or h). The system is closed mathematically if the thermodynamic state equations for the fluid are used to relate (ρ, p, T, e).

As far as is known, this system of equations is well posed and can be solved for realistic viscous flow patterns by using the following boundary conditions:

1) at any solid surface with velocity $\overline{V}_w$ and temperature T_w

$$\overline{V} = \overline{V}_w \quad \text{(no-slip condition)}$$

$$T = T_w \quad \text{(no-temperature-jump condition)}$$

2) at any ''open'' boundary across a fluid inlet or exit

known distributions of $\overline{V}$, p, and T

3) at an interface $\eta(x, y)$ between fluids 1 and 2

$$(w, \dot{m}, \tau_{ij}, q_z)_1 = (w, \dot{m}, \tau_{ij}, q_z)_2$$

$$p_2 = p_1 - \sigma \left(\frac{1}{R_1} + \frac{1}{R_2} \right)$$

The coefficient of surface tension, σ, is discussed in Sec. 1.1.4.

1.4.5 Turbulence—The Reynolds-Averaged Equations

The energy and Navier–Stokes equations are valid for either laminar or turbulent flow of a Newtonian fluid. For turbulent flow, as in Fig. 1.36(c), the fluid properties develop rapid random fluctuations superimposed on a more or less steady mean value. Osborne Reynolds, in 1895, exploited this idea by defining every property P to be the sum of a time-mean value $\overline{P}$ plus a fluctuation P', such that

$$\overline{P} = \frac{1}{T} \int_0^T P \, dt \tag{1.190}$$

where T is much longer than any significant period of the fluctuations. Assuming incompressible flow with constant μ, c_p, and k, the following mean and fluctuating variables are defined

$$u = U + u' \qquad v = V + v' \qquad w = W + w'$$
$$p = \overline{p} + p' \qquad T = \overline{T} + T'$$

The statistical implications of this splitting of variables are profound and will not be discussed here. Full details are given in the text by Hinze (1978).

Substituting these split variables in the Navier–Stokes Eq. (1.186) and taking the time-mean of the entire equation gives the *Reynolds-averaged* momentum equation

$$\rho \frac{d\overline{V}}{dt} = -\nabla \overline{p} + \rho \overline{g} + \mu \nabla^2 \overline{V} + \nabla \cdot \tau_{ij} \text{ (turb)} \tag{1.191}$$

where $\overline{V} = \overline{i}U + \overline{j}V + \overline{k}W$ is now the *time-mean velocity* and τ_{ij}(turb) is the turbulent *Reynolds-stress tensor*

$$\tau_{ij}(\text{turb}) = -\rho \overline{u_i' u_j'}$$

Thus, the onset of turbulence introduces nine new ''stress'' terms into the analysis, each of which must be modeled as functions of time and space to allow a solution to proceed. Fortunately, under the two-dimensional boundary layer approximations, Sec. 1.4.9, only one stress term dominates.

Similarly, substitution of the split variables into the energy Eq. (1.189) and time-averaging yields the mean-flow energy equation

$$\rho c_v \frac{d\overline{T}}{dt} = \nabla \cdot (k\nabla\overline{T}) - \nabla \cdot q_i(\text{turb}) + \mu\overline{\Phi} \qquad (1.192)$$

where $\overline{\Phi}$ is a combination of both laminar and turbulent viscous dissipation,

$$\overline{\Phi} = \frac{1}{2}\sum\left(\frac{\partial u_i}{\partial x_j} + \frac{\partial u_i'}{\partial x_j} + \frac{\partial u_j}{\partial x_i} + \frac{\partial u_j'}{\partial x_i}\right)^2$$

and $q_i(\text{turb})$ denotes the *turbulent heat flux* vector

$$q_i(\text{turb}) = \rho c_v \overline{u_i' T'}$$

Again, in boundary layer flow, a single flux term dominates. The Reynolds-averaged Eqs. (1.191) and (1.192) are subject to the same boundary conditions as for laminar flow, Sec. 1.4.4. The turbulent flux terms must also be modeled realistically to complete the analysis.

Further details of Reynolds-averaging and turbulence modeling are given in Secs. 4.5 and 21.2 of this handbook.

1.4.6 Laminar Pipe Flow

A classic problem, dating back to the work of Hagen (1839) and Poiseuille (1840), is the steady laminar flow of a Newtonian fluid through a very long pipe of radius R, as shown in Fig. 1.38(a). Let x be the axial coordinate.

For a location far downstream of the entrance, the flow will be *fully-developed*, that is, the axial velocity $u = u(r)$ only while the in-plane velocities $v = w = 0$. Then continuity is satisfied identically and the momentum equation reduces to

$$\frac{\mu}{r}\frac{d}{dr}\left(r\frac{du}{dt}\right) = \frac{dp}{dx} = \text{constant}$$

subject to the no-slip condition $u(R) = 0$.

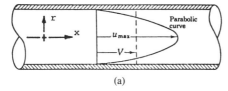
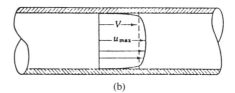

(a)   (b)

FIGURE 1.38 Comparison of fully-developed pipe flow velocity profiles for the same average velocity V. (a) laminar and (b) turbulent.

The solution is the Poiseuille paraboloid distribution

$$u = u_{max}(1 - r^2/R^2) \tag{1.193}$$

where

$$u_{max} = \frac{R^2}{4\mu}\left(-\frac{dp}{dx}\right)$$

The average velocity V can be computed from the volume flow

$$V = \frac{Q}{A}, \quad Q = \int u \, dA \tag{1.194}$$

Combining Eqs. (1.193) and (1.194) yields the laminar result

$$V = \frac{1}{2} u_{max} = \frac{R^2}{8\mu}\left(-\frac{dp}{dx}\right)$$

The solution Eq. (1.193) also yields the wall shear stress and the pressure gradient

$$\tau_w = 4\mu V/R = \frac{1}{2} R\left(-\frac{dp}{dx}\right)$$

$$-\frac{dp}{dx} = \frac{\Delta p}{L} = 8\mu Q/(\pi R^4) = 8\mu V/R^2$$

where Δp denotes the pressure drop through a pipe of length L.

It is customary to write these results in dimensionless form. The pipe Reynolds number is defined as

$$Re_D = \rho VD/\mu \tag{1.195}$$

and the Darcy friction factor is defined as

$$f = \frac{2D}{\rho V^2}\frac{\Delta p}{L} \tag{1.196}$$

Use of these variables enables the pressure drop to be written in the dimensionless form

$$f = \frac{64}{Re_D} \tag{1.197}$$

for laminar fully-developed pipe flow. The results of this section are valid up to $Re_D \doteq 2000$, after which the flow becomes unstable and undergoes transition to turbulence.

1.4.7 Turbulent Pipe Flow

The analysis of turbulent pipe flow is simplified by the fact that the mean velocity profile is correlated over the complete range of Reynolds numbers by a simple formula called the *Law-of-the-Wall*. Both axial velocity U and distance y from the wall are nondimensionalized by wall shear stress, density, and viscosity

$$u^+ = U/v^*, \qquad v^* = (\tau_w/\rho)^{1/2}$$
$$y^+ = \rho v^* y/\mu$$

where $y = R - r$ for pipe flow. R is radius of pipe. For smooth wall flow with modest or negligible pressure gradient, turbulent velocity data forms a nearly unique curve $u^+ (y^+)$, as shown in Fig. 1.39. Except very near the wall and in the center of the pipe, the data is well approximated by a logarithmic law

$$u^+ \doteq \frac{1}{\kappa} \ln (y^+) + 5.0, \qquad \kappa \doteq 0.41 \tag{1.198}$$

This correlation was developed in the 1930's by L. Prandtl and by T. von Kármán; the constant κ is now called von Kármán's constant and is insensitive to wall or

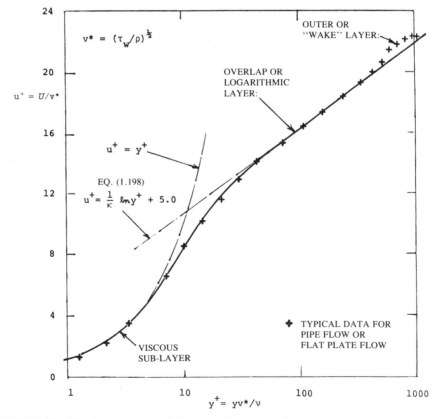

FIGURE 1.39 The law-of-the-wall for turbulent mean flow past a smooth, impermeable surface with modest pressure gradient.

external flow conditions. Note in Fig. 1.39, the *viscous sublayer* very near the wall ($y^+ \leq 5$), where $u^+ = y^+$ and turbulence effects on the profile are completely damped out.

If Eq. (1.198) is substituted for $U(r)$ in Eq. (1.194), the result is a friction factor formula for fully-developed smooth-wall turbulent pipe flow

$$1/\sqrt{f} = 2.0 \log_{10} (\mathrm{Re}_D \sqrt{f}/2.51)$$

where f is also defined by Eq. (1.196).

The turbulent velocity profile Eq. (1.198) may be generalized to include the effect of rough walls of average roughness height ϵ. Then the friction factor becomes a function of Re_D and roughness ratio ϵ/D. The universally accepted formula is the correlation of Colebrook and White [Colebrook (1939)]

$$\frac{1}{\sqrt{f}} = -2.0 \log_{10} \left(\frac{\epsilon/D}{3.7} + \frac{2.51}{\mathrm{Re}_D \sqrt{f}} \right) \qquad (1.199)$$

This formula was plotted in 1944 by Moody (1944) as the celebrated Moody friction factor chart, shown in Fig. 1.40.

Equation (1.199) is implicit in f and thus requires iteration if f is unknown. An alternate explicit formula was recently proposed by Haaland (1983)

$$\frac{1}{\sqrt{f}} \doteq -1.8 \log_{10} \left[\left(\frac{\epsilon/D}{3.7} \right)^{1.11} + \frac{6.9}{\mathrm{Re}_D} \right]$$

The maximum deviation from Eq. (1.199) is less than 1.5%.

When plotted in linear coordinates, the velocity profile $U(r)$ for turbulent flow has a steep wall gradient and a flat central core, as shown in Fig. 1.38(b). The average velocity is much greater than one-half $\bar{u}_{\max}$, and an accurate formula is

$$V/\bar{u}_{\max} = (1 + 1.3\sqrt{f})^{-1}$$

valid for either smooth or rough walls if f is computed from Eq. (1.199).

The Moody chart may be extended to uniform ducts of noncircular cross section if D in Eqs. (1.195) and (1.196) is replaced by the *hydraulic diameter*,

$$D_h = 4A/P \qquad (1.200)$$

where A is the cross-sectional area and P is the perimeter wetted by the fluid. The accuracy of this approximation is excellent for turbulent flow, but only fair for laminar flow. For better accuracy in laminar flow, one should seek the exact solution of the Navier–Stokes equation for the given cross section.

1.4.8 Open Channel Flow

A classic problem in fluids engineering is the determination of the flow rate and surface elevation in an *open channel* such as a river or flume or spillway. If the bottom slope and cross section are highly variable and there are obstructions such

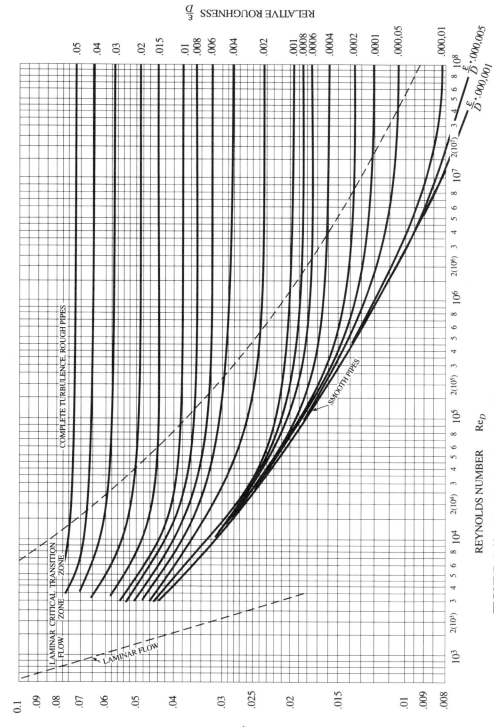

FIGURE 1.40 The Moody chart for fully developed pipe friction, from Eq. (2.199) (Reprinted with permission, Moody, L. F., "Friction Factors for Pipe Flow," *ASME Transactions,* Vol. 66, 1994.)

as dams or weirs, the analysis and flow patterns can be quite complex. The general study of open channel flow is treated in specialized texts such as Chow (1959).

A simple but instructive special case is the flow of a liquid down a long, straight channel of constant bottom slope S_0 and uniform cross-section area A, as shown in Fig. 1.41. The flow is steady, and the free surface slope also equals S_0. This condition is called *uniform flow*, wherein the gravity force is exactly balanced by the fluid shear along the bottom and sides of the channel. For a channel of length L,

$$\rho g A L S_0 = \tau_w P L,$$

or

$$\tau_w = \rho g (A/P) S_0 = \tfrac{1}{8} \rho V^2 f$$

where the friction factor f has been introduced from Eq. (1.196). The ratio of cross-section area A to wetted perimeter P (see Fig. 1.41) is called the *hydraulic radius*, R_h. Solving for average flow velocity, one obtains

$$V = (g/f)^{1/2} (R_h S_0)^{1/2} \tag{1.201}$$

This is the *Chézy formula* for uniform channel flow, named after its discoverer, Antoine Chézy (1768). Since the geometry resembles pipe flow, one may estimate

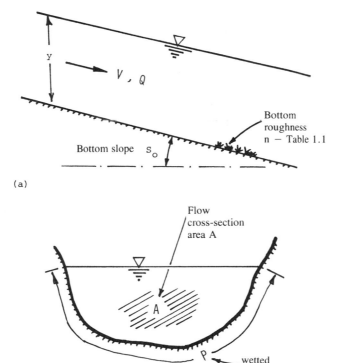

FIGURE 1.41 Uniform flow in a straight open channel. (a) side view and (b) cross section.

f from the Moody chart, Fig. 1.40, basing Reynolds number and roughness ratio on the hydraulic diameter $D_h = 4R_h = 4A/P$. A compilation of estimated roughness heights for various channel types is given in Table 1.1.

In practice, the flow parameters in natural channels are sufficiently uncertain that the Moody chart amounts to overkill, and one customarily uses the power-law correlation of Robert Manning (1889)

$$V \text{ (m/s)} \doteq \frac{1.0}{n} [R_h(\text{m})]^{2/3} S_0^{1/2} \tag{1.202}$$

The dimensional constant *n* is called *Manning's roughness parameter*, with typical values listed in Table 1.1. Equation (1.202) is dimensional and is restricted to SI

TABLE 1.1   Experimental Values of Manning's *n* Factor[†]

		Average roughness height, ϵ	
	n	ft	mm
Artificial lined channels:			
Glass	0.010 ± 0.002	0.0011	0.3
Brass	0.011 ± 0.002	0.0019	0.6
Steel, smooth	0.012 ± 0.002	0.0032	1.0
Painted	0.014 ± 0.003	0.0080	2.4
Riveted	0.015 ± 0.002	0.012	3.7
Cast iron	0.013 ± 0.003	0.0051	1.6
Cement, finished	0.012 ± 0.002	0.0032	1.0
Unfinished	0.014 ± 0.002	0.0080	2.4
Planed wood	0.012 ± 0.002	0.0032	1.0
Clay tile	0.014 ± 0.003	0.0080	2.4
Brick work	0.015 ± 0.002	0.012	3.7
Asphalt	0.016 ± 0.003	0.018	5.4
Corrugated metal	0.022 ± 0.005	0.12	37
Rubble masonry	0.025 ± 0.005	0.26	80
Excavated earth channels:			
Clean	0.022 ± 0.004	0.12	37
Gravelly	0.025 ± 0.005	0.26	80
Weedy	0.030 ± 0.005	0.8	240
Stony, cobbles	0.035 ± 0.010	1.5	500
Natural channels:			
Clean and straight	0.030 ± 0.005	0.8	240
Sluggish, deep pools	0.040 ± 0.010	3	900
Major rivers	0.035 ± 0.010	1.5	500
Floodplains:			
Pasture, farmland	0.035 ± 0.010	1.5	500
Light brush	0.05 ± 0.02	6	2000
Heavy brush	0.075 ± 0.025	15	5000
Trees	0.15 ± 0.05	?	?

[†]A more complete list is given in Chow (1959), pp. 110–113.
(Reprinted with permission, White, F., *Fluid Mechanics*, 3rd ed., McGraw-Hill, New York, 1994.)

units for V, m/s, and R_h, m. Manning's formula is equivalent to a bottom friction correlation and is widely used in modeling unsteady and variable open channel flows.

1.4.9 The Boundary Layer Equations

Most fluids engineering problems involve low viscosity fluids such as air and water, plus human-scale velocities (a few m/s) and sizes (a meter or so). Thus, the characteristic Reynolds numbers are quite large, 10^4 or more. For these conditions it is appropriate to use the *boundary layer approximations* proposed in 1904 by Ludwig Prandtl wherein viscous and heat-conduction effects are confined to thin layers near solid boundaries. The resulting boundary layer equations are probably the most important tools ever developed in fluid flow analysis.

Figure 1.42 shows a schematic of steady two-dimensional boundary layer flow near a solid surface, using curvilinear coordinates x along the surface and y normal to the wall. It is assumed that the Reynolds number is large

$$\mathrm{Re}_x = \rho U_\infty x/\mu \gg 1$$

For this condition, Prandtl showed that streamwise velocities and cross-stream gradients will dominate the flow

$$u \gg v$$

$$\frac{\partial}{\partial y} \gg \frac{\partial}{\partial x}$$

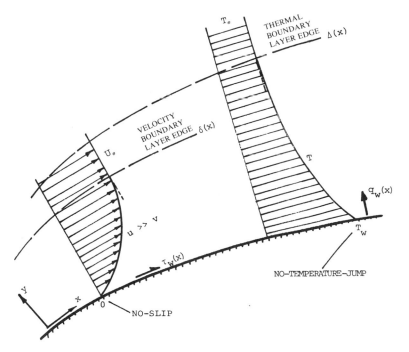

FIGURE 1.42 Schematic of velocity and temperature distributions in a two-dimensional boundary layer near a fixed hot solid surface.

Shear effects will be confined to a thin *velocity boundary layer* of width $\delta(x) \ll x$. Heat-conduction effects will be confined to a *thermal boundary layer* of width $\Delta(x) \ll x$. The immediate consequence of this is that the static pressure has a negligible variation across the stream

$$\frac{\partial p}{\partial y} \doteq 0$$

The y-momentum equation thus vanishes, as do many other terms in the Navier–Stokes and energy equations. Pressure is assumed to be known from an inviscid flow Bernoulli relation

$$\frac{dp}{dx} = -\rho U_e \frac{dU_e}{dx} \tag{1.203}$$

The problem reduces to a set of three approximate equations in the three unknowns (u, v, T) for steady incompressible flow:

Continuity

$$\frac{\partial u}{\partial x} + \frac{\partial v}{\partial y} = 0 \tag{1.204a}$$

x-momentum

$$u \frac{\partial u}{\partial x} + v \frac{\partial u}{\partial y} \doteq U_e \frac{dU_e}{dx} + g_x + \frac{1}{\rho} \frac{\partial \tau}{\partial y} \tag{1.204b}$$

energy

$$\rho c_p \left(u \frac{\partial T}{\partial x} + v \frac{\partial T}{\partial y} \right) \doteq -\frac{\partial q}{\partial y} + \tau \frac{\partial u}{\partial y} \tag{1.204c}$$

where

$$\tau = \mu \frac{\partial u}{\partial y} - \rho \overline{u'v'}$$

and

$$q = -k \frac{\partial T}{\partial y} + \rho c_p \overline{v'T'}$$

Written in this form, the equations are valid for either laminar or turbulent flow.

The boundary conditions for an impermeable surface are

$$\text{at the wall, } y = 0: \qquad u = v = 0, \quad T = T_w$$

$$\text{in the stream, } y \to\to \infty: \quad u \to\to U_e, \quad T \to\to T_e$$

Note that a condition for v in the outer stream is not specified. These equations are extended to compressible flow in Sec. 4.6 and to three-dimensional flow in Sec. 4.8. Many more details about boundary layer analysis are given in Chap. 4.

Integral Relations. The boundary layer Eqs. (1.204) and (1.205) are *parabolic* in character and many different analytical and numerical techniques have been devised for their solution (see Chap. 4 for further details). There is also an approximate technique, developed by von Kármán in 1921, of integrating the momentum and energy relations across the boundary layer, leaving an ordinary differential equation for the streamwise variation of *integral* parameters such as δ and τ_w. For two-dimensional flow, the integral relations are
    momentum integral

$$\tau_w = \frac{d}{dx}\left[\int_0^\infty \rho u(U_e - u)\, dy \right] + \rho \frac{dU_e}{dx} \int_0^\infty (U_e - u)\, dy \qquad (1.206a)$$

energy integral

$$q_w = \frac{d}{dx}\left[\int_0^\infty \rho u\left(c_p T + \frac{u^2}{2} - c_p T_e - \frac{U_e^2}{2}\right) dy \right] \qquad (1.206b)$$

Integral relations for three-dimensional boundary layer flow are given Schetz (1993).

Although Eqs. (1.206) are exact in principle, they are normally solved by introducing approximate profile shapes $u(y)$ and $T(y)$. Many particular solutions are given White (1992) and Schetz (1993).

1.4.10 The Flat Plate Boundary Layer

A classic case illustrating boundary layer analysis is a sharp flat plate at zero incidence, as shown in Fig. 1.43. Since the boundary layer is thin at large Re_x, one may assume that $U_e = U_\infty$ is constant. Also, assume constant stream and wall temperatures. The flow near the leading edge will be laminar. Transition occurs typically at about $\mathrm{Re}_x \doteq 5 \times 10^5$ and the flow then becomes turbulent. This problem has been extensively analyzed in the literature [White (1992), Schetz (1993), Cebeci and Bradshaw (1977), and Kays and Crawford (1980)].

Laminar Flow. The laminar momentum solution was given by H. Blasius in 1908; details are discussed in Sec. 4.2. The results for thickness, shear, and total drag are

$$\delta/x = 5.0/\mathrm{Re}_x^{1/2}$$

$$c_f = 2\tau_w/(\rho U_\infty^2) = 0.664/\mathrm{Re}_x^{1/2} \qquad (1.207)$$

$$C_D = 2(\text{Drag})/(\rho U_\infty^2 bL) = 1.328/\mathrm{Re}_L^{1/2}$$

where b is the width and L the length of the plate.

The heat transfer may be written in terms of either the *Nusselt number* or the *Stanton number*

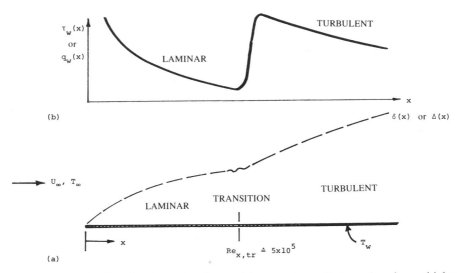

FIGURE 1.43 The flat plate boundary layer. (a) geometry, with boundary layer thickness greatly exaggerated and (b) typical shear and heat flux distributions.

$$\text{Nu}_x = \frac{q_w x}{(T_w - T_\infty)k} \tag{1.208a}$$

$$\text{St}_x = \frac{q_w}{(T_w - T_\infty)\rho U_\infty c_p} = \frac{\text{Nu}_x}{\text{Re}_x \text{Pr}} \tag{1.208b}$$

where $\text{Pr} = \mu c_p/k$ is the *Prandtl number* of the fluid. The average Nusselt number over the plate length is defined by

$$\overline{\text{Nu}_L} = \frac{1}{L} \int_0^L \text{Nu}_x \, dx$$

The laminar heat transfer solution was given by E. Pohlhausen in 1921

$$\text{Nu}_x = 0.332 \, \text{Re}_x^{1/2} \text{Pr}^{1/3}$$

$$\overline{\text{Nu}_L} = 0.664 \, \text{Re}_L^{1/2} \text{Pr}^{1/3} \tag{1.209}$$

Note that the friction increases as $U_\infty^{3/2}$ while heat transfer only increases as $U_\infty^{1/2}$. The above results are limited to the laminar flow range $10^3 \leq \text{Re}_x \leq 5 \times 10^5$.

Turbulent Flow. Unless the plate is extremely smooth and the freestream unusually quiet, transition may be assumed to occur on the plate at about $\text{Re}_x = 5 \times 10^5$. The resulting turbulent flow cannot be analyzed exactly, but there are two excellent approximate theories available: 1) numerical integration of Eq. (1.204) with an empirical correlation for the turbulent sheat and heat flux terms, and 2) use of turbulent profile correlations in the integral relations Eq. (1.206). Some details of these analyses are given in Chap. 4.

The results of these analyses for smooth-wall turbulent flow may be summarized in the following equations

$$\delta/x = \frac{0.0598}{\log_{10} \text{Re}_x - 3.17}$$

$$c_f = 0.455/\ln^2 (0.06 \text{ Re}_x) \tag{1.210}$$

$$C_D = 0.523/\ln^2 (0.06 \text{ Re}_L)$$

Here, either the laminar portion of the flow is neglected or x is taken as the point of *apparent origin* of the turbulent portion of the boundary layer. If $\text{Re}_{tr} = 5 \times 10^5$, the following two equations provide a smoothly varying transition zone between laminar and turbulent flow

$$c_f \doteq \frac{0.455}{\ln^2 (0.06 \text{ Re}_x)} - \frac{1671}{\text{Re}_x}$$

$$C_D \doteq \frac{0.523}{\ln^2 (0.06 \text{ Re}_L)} - \frac{1522}{\text{Re}_L} \tag{1.211}$$

where x denotes distance from the leading edge. Some engineering equations for turbulent heat flux are

$$\text{Nu}_x \doteq 0.029 \text{ Re}_x^{0.8} \text{Pr}^{0.6}$$

$$\overline{\text{Nu}_L} \doteq 0.036 \text{ Re}_L^{0.8} \text{Pr}^{0.6} \tag{1.212}$$

Here, note that turbulent friction increases as $U_\infty^{1.8}$ while turbulent heat flux only increases as $U_\infty^{0.8}$.

Many additional parametric effects on turbulent boundary layers—such as wall roughness, compressibility, wall curvature, blowing, and suction—are discussed later.

1.4.11 Stability of Laminar Flow

All laminar flows become unstable at a finite Reynolds number which depends on geometry and flow conditions. The theory of the initial instability is now well developed and many results have been reported in a specialized text [Betchov and Criminale (1967)]. The subsequent growth of the unstable waves and their breakdown into turbulence is not so well understood, but there have been many experiments and transition correlations. Summaries of the overall subject of stability and transition are given in various viscous flow texts: Chap. 5 [White (1992)], Chap. 6 [Schetz (1993)], Chap. 9 [Cebeci and Bradshaw (1977)].

The basic idea of stability theory is to assume that a known exact solution (U, V) of the Navier–Stokes Eq. (1.185) undergoes a small disturbance (u', v') for an assumed two-dimensional flow pattern. Substitute the slightly disturbed velocities $(U + u', V + v')$ into Eq. (1.185), cancel out the known exact solution terms, and neglect higher powers of (u', v'). It is convenient to eliminate u' in terms of v' from the continuity equation. Finally, assume a solution in the form of a streamwise traveling wave whose amplitude varies with y

$$v'(x, y, t) = v(y) \exp [i(\alpha x - \omega t)]$$

where $i = (-1)^{1/2}$. The result is a fourth-order linear ordinary differential equation in $v(y)$ which depends upon both the streamwise velocity profile $U(y)$ and its curvature $U''(y)$

$$\left(U - \frac{\omega}{\alpha}\right) (v'' - \alpha^2 v) - U'' v + i \frac{\nu}{\alpha} (v'''' - 2\alpha^2 v'' + \alpha^4 v) = 0 \quad (1.213)$$

where primes denote differentiation with respect to y. The basic flow $U(y)$ is assumed incompressible and laminar. Equation (1.213) was found by Orr in 1907 and independently by Sommerfeld in 1908. The boundary conditions are that the disturbances $v(y)$ should vanish at the wall and in the stream

$$v(0) = v'(0) = 0 \qquad v(\infty) = v'(\infty) = 0 \qquad (1.214)$$

The Orr–Sommerfeld equation is thus an eigenvalue problem for the wavenumber α and frequency ω of the disturbance. The *neutral curve* between stability and instability is the locus of points for which the imaginary part of ω vanishes. Two examples are shown in Fig. 1.44. The neutral curves are called *thumb curves* because of their shape. Traveling waves occurring within the thumb are unstable, hence, there is a minimum or *critical* Reynolds number Re_{crit} below which the flow is en-

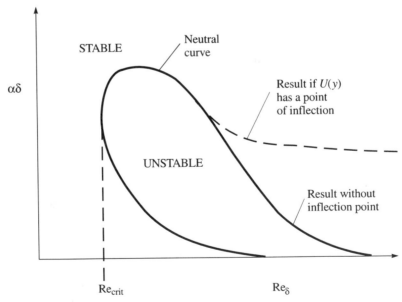

FIGURE 1.44 Typical ''thumb curve'' solution boundaries of the Orr–Sommerfeld equation for the stability of boundary layer flow.

tirely stable. Above Re_{crit}, there is a range of wavenumbers and frequencies which will grow unstably to large amplitudes and eventually break up into turbulence downstream. The point where turbulence ensues everywhere is usually called the "transition" point and often Re_{tr} is much greater than Re_{crit}.

An accurate numerical analysis of Eq. (1.213) is a difficult task because the four linearly independent solutions are extremely sensitive to small numerical errors. Details of the solution techniques are given in Sec. 4.3.

An interesting fact, shown in Fig. 1.44, is that profiles $U(y)$ which have a point of inflection ($U'' = 0$) within the boundary layer, have a much broader thumb curve which extends to infinite Reynolds number ("inviscid" instability). Such flows are very unstable and include 1) jet flow, 2) wake flow, and 3) adverse pressure gradients. Flows without a point of inflection yield a much thinner thumb curve and are inherently more stable, e.g., 1) flat plate flow, 2) pipe or channel flow, and 3) boundary layers with favorable pressure gradient.

Further discussion of stability and transition of laminar flows is given in Chap. 4.

1.5 INTRODUCTION TO TURBULENT FLOW AND MIXING
Carl H. Gibson

Turbulent flows are more complex than viscous or potential flows because they are dominated by a nonlinear force that randomly scrambles the motion on all length scales permitted by other forces that tend to damp out the turbulence, such as viscous, buoyancy, Coriolis, and electromagnetic forces. Turbulence is strongly diffusive. Heat, chemical species, and momentum transports and mixing are generally dominated by turbulence in flows where turbulence exists. Most industrial flows take advantage of this strongly diffusive property of turbulence. Flows and fluxes in the atmosphere, ocean, sun, interstellar medium, and intergalactic medium are affected, if not dominated, by turbulence, either past or present.

The dynamics of turbulence is best illustrated by the conservation of momentum equations written in the form

$$\frac{\partial \bar{v}}{\partial t} = \bar{v} \times \overline{\omega} - \nabla B + \nabla \cdot (\bar{\bar{\tau}}/\rho) \qquad (1.215)$$

where $\bar{v}$ is the velocity, t is time, $\overline{\omega} = \text{curl } \bar{v}$ is the vorticity, $\bar{\bar{\tau}}$ is the viscous stress tensor, and the density ρ is assumed constant. The Bernoulli group $B = v^2/2 + p/\rho + gx_3$, where p is the pressure, g is gravity, and x_3 is up. Coriolis and electromagnetic forces are omitted here for simplicity. Turbulence occurs when the inertial-vortex forces $\bar{v} \times \overline{\omega}$ per unit mass exceed the viscous forces $\nabla \cdot (\bar{\bar{\tau}}/\rho)$ per unit mass. The ratio of inertial vortex forces to viscous forces is the Reynolds number $Re = UL/\nu$, where U is a characteristic velocity and L is a characteristic length scale of the flow. Re may also be considered the ratio of the eddy diffusivity, UL, to the molecular diffusivity, ν.

Turbulence first appears as viscous eddies that form on thin shear layers such as boundary layers on solid surfaces in a flow, or after such vortex sheets are injected into the flow from separation lines or jets and break up into random eddies. Figure

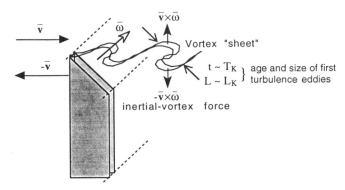

FIGURE 1.45   Formation of turbulence by the instability of a shear layer.

1.45 shows a sharp edged blade moving at right angles through a fluid to produce a *vortex sheet*, or shear layer. A segment of the blade is shown in the figure shedding the shear layer, which thickens by viscous diffusion until its Reynolds number, $v\delta/\nu$, exceeds a critical value, where δ is the thickness of the shear layer, v is half the velocity of the fluid with respect to the blade, and ν is the kinematic viscosity of the fluid. The vorticity $\overline{\omega} = \nabla \times \overline{v}$ is directed into the paper. The shear layer is absolutely unstable because vortex forces $\overline{v} \times \overline{\omega}$ arise in the direction of any perturbation of the layer from a plane. The growing perturbation produces an opposite perturbation on the shear layer with oppositely directed vortex force, and the two perturbations then induce motions so that they rotate each other forming an eddy. The size of the eddy formed is proportional to the Kolmogorov length scale, $L_K = (\nu^3/\epsilon)^{1/4}$, where ϵ is the viscous dissipation rate of the shear layer, and the time of rotation is proportional to the Kolmogorov time, $T_K = (\nu/\epsilon)^{1/2}$, with universal proportionality constants, according to the universal similarity theory of Kolmogorov (1941) that implies the existence of a universal critical Reynolds number at these scales. The eddies interact with each other to form pairs; the pairs of eddies interact with other pairs of eddies to form pairs of pairs, and so on, forming an upward cascade of interacting, nested eddy clusters of larger and larger length and time scales as more and more external fluid is entrained. Contrary to a common misconception and the well-known Richardson poem,* the natural direction of the turbulence cascade is up: that is, from small scales to large, with feedback; not down from large scales to small, one way. Turbulence extracts energy from larger scale laminar shear flows by sucking it into crevices between the growing turbulent eddies. This induced laminar cascade of energy from large scales to small is not turbulence because $\overline{\omega} = 0$ in the laminar fluid. Turbulence eddies are constrained at large scales by the solid walls of the container or duct, or by buoyancy, Coriolis, or electromagnetic forces that prevent further growth in the vertical direction for buoyancy, or in directions perpendicular to the angular rotation and magnetic field vectors for Coriolis and electromagnetic forces, respectively.

*Big whorls have little whorls, which feed on their velocity; And little whorls have lesser whorls, And so on to viscosity (in the molecular sense). L. F. Richardson (1922). Corrected version: Little whorls by vortex forces, form and pair with more of, whorls that grow by vortex forces (Slava Kolmogorov!).

The random motions excited by the damping of turbulence at small scales by viscosity and at large scales by buoyancy, Coriolis, or electromagnetic forces should not be called turbulence, but viscous laminar flow or forced waves. Neither should the inviscid, ideal flow of irrotational fluid as it is pulled into the interstices of a turbulent flow, and eventually crosses the superlayer interface separating turbulence from nonturbulence in the final stage of entrainment, by viscosity.

Definition: Turbulence is defined as an eddy-like state of fluid motion where the inertial-vortex forces are larger than any other forces that tend to damp them out.

Therefore, in order for a flow to be considered turbulence by this definition, it should possess eddy-like motions with Reynolds numbers, Froude numbers, Rossby numbers, inverse Chapman numbers of the first kind, etc., that are greater than universal critical values. Many other definitions of turbulence are in current use, some of which are so broad that they fail to differentiate between waves and turbulence, or only list various distinctive characteristics of turbulence (e.g., random, rotational, dissipative, diffusive) as one might define a disease, or syndrome. Perhaps the most common procedure is to avoid any precise definition of turbulence, and apply the term to any random velocity field, or even any random scalar field. This can lead to trouble. Just as in geology where all that glitters is not gold, in fluid mechanics all that wiggles is not turbulence. Treating the dissipation rates of turbulence and temperature in most ocean microstructure patches as representative of the active turbulence that does most of the mixing can lead to vast underestimates of the average values if such patches are *fossil turbulence* (remnants in any hydrophysical field of previous turbulence in fluid that is no longer turbulent on the scale of the remnant). This is a possible explanation of the so-called *dark mixing* phenomenon of the ocean (T. Dillon, unpublished 1993); that is, the unobserved mixing that must exist to balance mass, momentum, and energy budgets of the ocean. The unobserved *dark mixing* is analogous to the unobserved *dark matter* that must exist to bind galaxies together.

The existence of a universal critical Reynolds number for shear instability plus the tendency of turbulence to scramble away any preferred directions, form the physical bases of the two 1941 Kolmogorov universal similarity hypotheses. By the first Kolmogorov hypothesis, the statistical laws, F, for velocity differences between points separated by distances smaller than the external scale, L_0, should be locally isotropic and homogeneous in space and time, and should be universal when all lengths and times are normalized by the Kolmogorov length, L_K, and time, T_K. It is easy to show that $\delta_{crit} \approx L_K$ and $t_{crit} \approx T_K$ for turbulent boundary layers or separated shear layers such as in Fig. 1.45. This is a very powerful hypothesis. It means that all turbulent flows should be statistically identical in Kolmogorov space and time. All spectra should have the same universal form, no matter whether the turbulence is in air or water, in the ocean or the atmosphere, or in outer space. By the second hypothesis, F becomes independent of ν for scales larger than L_K. Thus, by the two K-41 hypotheses, not only is all turbulence statistically identical in K space, but it depends only on ϵ and L when $L_0 \gg L \gg L_K$.

Turbulence is turbulence according to Kolmogorov (1941), and, remarkably, also by numerical, laboratory, and field experiments. Numerous attempts have been made to test the consequences of the K-41 hypotheses, and first-order statistical parameters such as power spectra of velocity components show excellent agreement with universal Kolmogorov similarity and local isotropy. The power spectrum, Φ_u, of the streamwise velocity fluctuation u' as a function of the streamwise wavenumber k should have an inertial subrange

$$\Phi_u = \alpha \epsilon^{2/3} k^{-5/3} \text{ for } L_0^{-1} \gg k \gg L_K^{-1} \tag{1.216}$$

(by dimensional analysis) according to the second K-41 hypothesis, where $\int_0^\infty \Phi_u \, dk$ is U'^2 and α is a universal constant. The value of the constant α is 0.55 ± 0.05, and the exponential viscous cutoff of the spectrum occurs at $kL_K = 0.1$ for a wide variety of flows and Reynolds numbers.

However, the dissipation rate $\epsilon = 2\nu e_{ij}^2$ is not constant in space but an intermittent random variable, where $e_{ij} = (\partial v_i/\partial x_j + \partial v_j/\partial x_i)/2$ is the rate of strain tensor. The temperature dissipation rate $\chi = 2D(\partial T/\partial x_i)^2$ for temperature, T, with molecular diffusivity, D, mixed by turbulence is also intermittent. For higher order statistics of turbulence such as spectra of squared velocity derivatives, this intermittency of ϵ and γ must be taken into account. In 1962, Kolmogorov offered a refinement to the K-41 hypotheses consisting of a third universal similarity hypothesis that ϵ_r averaged over scales, r, should be lognormal with variance

$$\sigma_{\ln \epsilon r}^2 = \mu \ln (L_0/cr), \tag{1.217}$$

where μ is a universal constant and c is a constant (probably in the range 10–100 and depending somewhat on the external flow). This third *intermittency* hypothesis of Kolmogorov is readily extended to scalar fields like temperature [Gibson (1991)] and both versions have been subjected to experimental tests. Such tests and their interpretation are difficult, and the results are a matter of current debate. Measurements of ϵ and χ in the ocean and atmosphere generally show excellent agreement with intermittent lognormal distributions, with *intermittency factors* $\sigma_{\ln \epsilon}^2$ and $\sigma_{\ln \chi}^2$ values in the range 3–7 [Baker and Gibson (1987)] and μ values for both velocity and temperature dissipation fields in the range 0.4–0.5 [Gibson, Stegen and Mc-Connell (1970)]. Analysis of doppler velocity inferred from spectral line broadening in dense molecular clouds of the interstellar medium [Falgarone and Phillips (1990)] also give μ values of 0.4–0.5. Other investigators, for example, Sreenivasan and Kailasnath (1993), have inferred values of μ in the range 0.2 from laboratory and smaller Reynolds number field data. However, such small μ values would require astronomically large values of L_0 to explain the measured values of $\sigma_{\ln \epsilon}^2$ and $\sigma_{\ln \chi}^2$ using Kolmogorov's expression, Eq. (1.217). Assuming a horizontal cascade between the viscous and Coriolis-inertial scales, L_0 must be thousands or millions of times the earth's circumference for $\mu = 0.2$ rather than reasonable values of only tens or hundreds of kilometers for $\mu = 0.4$–0.5.

Most natural flows such as in the ocean and atmosphere are stably stratified by gravitational forces that strongly inhibit the formation of turbulence and limit the vertical extent of its eddy motions. The limiting size $L_R = (\epsilon/N^3)^{1/2}$ is the Ozmidov length scale, where inertial forces $F_I \approx \rho U^2 L^2 \approx \rho \epsilon^{2/3} L^{8/3}$ are balanced by buoyancy forces $F_B = (\delta m)gL^3 \approx g(\partial \rho/\partial z)L^4 \approx \rho N^2 L^4$, where $U \approx (\epsilon L)^{1/3}$ for the turbulence by the second K-41 hypothesis, L is the scale of the motion U, g is gravity, z is depth, and N is the stratification frequency. Turbulence appears in the ocean or atmosphere interior as intermittent, isolated patches that are rapidly converted to forced internal wave motions by the buoyancy. These patches persist as fossil or active-fossil turbulence long after the turbulence is damped, with ϵ values less than the maximum previous value ϵ_0, and with temperature and density fields scrambled on vertical scales up to $L_{R0} \approx (\epsilon_0/N^3)^{1/2}$. All turbulence ceases in the patch when ϵ decreases to values below about $\epsilon_F \approx 30\nu N^2$ [Gibson (1980, 1986)].

Turbulence can grow to much larger scales in the horizontal direction than in the vertical in a stably stratified medium. For a rotating system such as the earth, inertial forces $F_I \approx \rho U^2 Lh$ are balanced by Coriolis forces $F_C \approx \rho U \Omega L^2 h$ at the Hopfinger scale $L_H \approx (\epsilon/\Omega^3)^{1/2}$, where $U \approx (\epsilon L)^{1/3}$ for the turbulence, h is the thickness of the turbulent layer, and Ω is the vertical component of the angular velocity. When such *two-dimensional turbulence* reaches scales as large as L_H, it ceases to grow and is fossilized to form forced Coriolis-inertial waves, or eddies, that persist for long periods and preserve information about their maximum previous ϵ_0 at fossilization by preserving the length scale $L_{H0} \approx (\epsilon_0/\Omega^3)^{1/2}$. An example may be the *meddies* formed when warm, saline, sea water is injected at depth from the Mediterrenian Sea into the Atlantic at Gibraltar. Fifty kilometer meddies with $\Omega \approx 2.5 \times 10^{-5}$ radian/s imply $\epsilon_0 \approx 0.3$ cm^2 s^{-3}, which is not inconsistent with measurements through the strait. Another example may be the arctic eddies observed under the north polar ice cap. These are only about ten kilometers in diameter, which is consistent with $\Omega \approx 7 \times 10^{-5}$ radians/s and a slightly smaller value of $\epsilon_0 \approx 0.29$ cm^{2}s^{-3} at the point of origin of the eddies, presumably the Bering strait.

Scalar fields like temperature mixed by turbulence also are locally isotropic, and they obey universal similarity hypotheses similar to those for turbulent velocity fields, depending on the molecular diffusivity ratio, or Prandtl number, Pr $= \nu/D$. At large scales, the relevant parameters are only χ, ϵ, and the size, L, of the scalar fluctuations, so that the scalar spectrum of temperature fluctuations

$$\Phi_T = \beta \chi \epsilon^{-1/3} k^{-5/3} \tag{1.218}$$

by dimensional analysis, where β is a universal constant, $\int_0^\infty \Phi_T \, dk = T^2$, and T is the temperature fluctuation. For Pr $\ll 1$, the smallest scale fluctuations are mixed by uniform straining on scales smaller than L_K, and depend on the rate of strain parameter $\gamma = (\epsilon/\nu)^{1/2}$ [Batchelor (1959)]. For Pr $\ll 1$, molecular diffusion effects first occur at the Obukhov–Corrsin scale $L_C = (D^3/\epsilon)^{1/4}$.

The mixing mechanism for scales smaller than L_C is controversial for Pr $\ll 1$. According to Batchelor, Howells, and Townsend (1959) the rate of strain γ is irrelevant for scales smaller than L_C because $L_C \gg L_K$, leading to a $\Phi_T \sim k^{-17/3}$ subrange for $L_C^{-1} \ll k \ll L_K^{-1}$. According to Gibson (1968a,b) the zero gradient point stretching mechanism of small scale scalar mixing is independent of Pr, giving a $\Phi_T \sim k^{-3}$ subrange for $L_C^{-1} \ll k \ll L_B^{-1}$ for Pr $\ll 1$, and confirming the $\Phi_T \sim k^{-1}$ subrange for $L_K^{-1} \ll k \ll L_B^{-1}$ for Pr $\gg 1$ predicted by Batchelor (1959), where $L_B = (D/\gamma)^{1/2}$ is the Batchelor length scale. Points of maximum and minimum scalar value are produced by turbulence overturning uniform scalar gradients on scales L_K for Pr $\gg 1$ and L_C for Pr $\ll 1$, and these hot and cold spots are stretched and split to smaller scales by local straining and diffusive instability until a local equilibrium is reached where the extrema have radii of curvature L_B. Thus, the smallest scale cascade of scalar turbulent mixing is downward from large scales to small, contrary to the upward direction of the cascade of turbulence discussed previously.

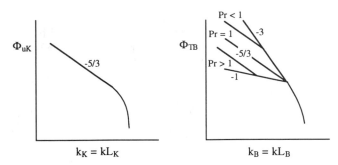

FIGURE 1.46 Spectral consequences of universal similarity hypotheses for turbulent velocity (left) and turbulent temperature (right). Log–log plots of power spectra Φ are plotted versus wavenumber k, normalized by Kolmogorov length and time scales (K subscripts) and Batchelor length, time, and temperature scales (B subscripts).

This cascade requires a brief mixing transition time, t_D, between the onset of turbulence and the onset of mixing. For $Pr \gg 1$, $t_D\gamma \approx (1/2) \ln Pr$. For $Pr \ll 1$, $t_D\gamma \approx Pr^{-1/2}$. Note that $\gamma = T_K^{-1}$. Figure 1.46 summarizes the spectral consequences of universal similarity theories for turbulent velocity fields and turbulent temperature fields for various Pr values. Power law subranges of Eqs. (1.216) and (1.218) appear as straight lines with slopes indicated for the various universal similarity theories of turbulent velocity and temperature (and other scalar fields) mixed by turbulent velocity, as discussed in Gibson (1991). All turbulent velocity spectra should collapse to the same universal spectral function shown on the left when normalized by Kolmogorov length, L_K, and time, T_K, scales according to Kolmogorov's 1941 theory. All turbulent scalar fields like temperature, with the appropriate Pr value, should collapse to the various universal scalar spectral forms shown on the right when normalized by Batchelor length, $L_B = L_K/Pr^{1/2}$, Batchelor time, $T_B = \gamma^1 = T_K$, and Batchelor temperature $T_B = (\chi T_B)^{1/2}$ according to the analogous turbulent mixing theories.

REFERENCES

Baker, M. A. and Gibson, C. H., "Sampling Turbulence in the Stratified Ocean: Statistical Consequences of Strong Intermittency," *J. Phys. Oceanogr.*, Vol. 17, pp. 1817–1837, 1987.

Batchelor, G. K., "Small-Scale Variation of Convected Quantities Like Temperature in Turbulent Fluids," *J. Fluid Mech.*, Vol. 5, pp. 113–133, 1959.

Batchelor, G. K., *An Introduction to Fluid Dynamics*, Cambridge University Press, Cambridge, 1967.

Batchelor, G. K., Howells, I. D., and Townsend, A. A., "Small-Scale Variation of Convected Quantities Like Temperature in Turbulent Fluid. Part 2. The Case of Large Conductivity," *J. Fluid Mech.*, Vol. 5, pp. 134–139, 1959.

Bejan, A., *Entropy Generation through Heat and Fluid Flow*, Wiley, New York, 1982.

Betchov, R. and Criminale, W. O., *Stability of Parallel Flows*, Academic Press, New York, 1967.

Carmichael, R. L. and Erickson, L. L., "PAN-AIR—A Higher-Order Panel Method for Predicting Subsonic or Supersonic Linear Potential Flows About Arbitrary Configurations," *AIAA Paper No. 81-1255*, June 1981.

Cebeci, T. and Bradshaw, P., *Momentum Transfer in Boundary Layers*, Hemisphere Pub. Corp., New York, 1977.

Chow, V. T., *Open Channel Hydraulics*, McGraw-Hill, New York, 1959.

Cole, R. H., *Underwater Explosions*, Princeton University Press, Princeton, NJ, 1948.

Colebrook, C. F., "Turbulent Flow in Pipes, with Particular Reference to the Transition Region between the Smooth and Rough Pipe Laws," *J. Inst. Civil Eng.*, London, Vol. 11, pp. 133–156, 1939.

Falgarone, E. and Phillips, T. G., "A Signature of the Intermittency of Interstellar Turbulence: The Wings of Molecular Line Profiles," *The Astrophysical Journal*, Vol. 359, pp. 344–354, 1990.

Fox, R. W. and McDonald, A. T., *Introduction to Fluid Mechanics*, 3rd ed., John Wiley & Sons, New York, 1992.

Gibson, C. H., "Fine Structure of Scalar Fields Mixed by Turbulence: I. Zero-Gradient Points and Minimal Gradient Surfaces," *Phys. Fluids*, Vol. 11, pp. 2305–2315, 1968a.

Gibson, C. H., "Fine Structure of Scalar Fields Mixed by Turbulence: II. Spectral Theory," *Phys. Fluids*, Vol. 11, pp. 2316–2327, 1968b.

Gibson, C. H., "Fossil Temperature, Salinity, and Vorticity Turbulence in the Ocean," *Marine Turbulence*, Nihou, J. (Ed.), Elsevier Publishing Co., Amsterdam, pp. 221–257, 1980.

Gibson, C. H., "Internal Waves, Fossil Turbulence, and Composite Ocean Microstructure Spectra," *J. Fluid Mech.*, Vol. 168, pp. 89–117, 1986.

Gibson, C. H., "Kolmogorov Similarity Hypotheses for Scalar Fields: Sampling Intermittent Turbulent Mixing in the Ocean and Galaxy," *Turbulence and Stochastic Processes: Kolmogorov's Ideas 50 Years on, Proc. Roy. Soc. Lond. A*, Vol. 433 (No. 1890), pp. 149–164, 1991.

Gibson, C. H., Stegen, G. R., and McConnell, S., "Measurements of the Universal Constant in Kolmogorov's Third Hypothesis for High Reynolds Number Turbulence," *Phys. Fluids*, Vol. 13, pp. 2448–2451, 1970.

Haaland, S. E., "Simple and Explicit Formulas for the Friction Factor in Turbulent Pipe Flow," *J. Fluids Engrg.*, Vol. 105, pp. 89–90, 1983.

Halsey, N. D., "Potential Flow Analysis of Multielement Airfoils Using Conformal Mapping," *AIAA Journal*, p. 1281, 1979.

Hess, J. L., "Higher Order Numerical Solution of the Integral Equation for the Two-Dimensional Neumann Problem," *Computer Methods in Applied Mechanics and Engineering*, pp. 1–15, 1973.

Hess, J. L., "The Problem of Three-Dimsnional Lifting Potential Flow and Its Solution by Means of Surface Singularity Distribution," *Computer Methods in Applied Mechanics and Engineering*, pp. 283–319, Nov. 1974.

Hess, J. L., "Review of Integral-Equation Techniques for Solving Potential Flow Problems with Emphasis on the Surface Source Methods," *Computer Methods in Applied Mechanics and Engineering*, pp. 145–196, 1975a.

Hess, J. L., "Improved Solution for Potential Flow About Arbitrary Axisymmetric Bodies by the Use of a Higher-Order Surface Source Method," *Computer Methods in Applied Mechanics and Engineering*, pp. 297–305, 1975b.

Hess, J. L. and Smith, A. M. O., "Calculation of Potential Flow About Arbitrary Bodies," *Progress in Aeronautical Science*, Vol. 8, Pergamon Press, Oxford, pp. 1–138, 1966.

Hess, J. L., Friedman, D. M., and Clark, R. W., "Calculation of Compressible Flow about Three-Dimensional Inlets with Auxilliary Inlets, Slats and Vanes by Means of a Panel Method," *NASA CR-174975*, June 1985.

Hinze, J. O., *Turbulence*, 2nd ed., McGraw-Hill, New York, 1978.

James, R. M., "A General Class of Exact Airfoil Solutions," *Journal of Aircraft*, pp. 574–580, 1972.

Karamcheti, K., *Principles of Ideal-Fluid Aerodynamics*, John Wiley & Sons, New York, 1966.

Kays, W. M. and Crawford, M. E., *Convective Heat and Mass Transfer*, 2nd ed., McGraw-Hill, New York, 1980.

Kolmogorov, A. N., "The Local Structure of Turbulence in Incompressible Viscous Fluid for Very Large Reynolds Numbers," *Dokl. Akad. Nauk SSSR*, Vol. 30, pp. 301–305, 1941.

Kolmogorov, A. N., "A Refinement of Previous Hypotheses Concerning the Local Structure of Turbulence in a Viscous Imcompressible Fluid at High Reynolds Numbers," *J. Fluid Mech.*, Vol. 13, pp. 82–85, 1962.

Kraus, W., "Panel Methods in Aerodynamics," Chap. 4 of *Numerical Methods in Fluid Dynamics*, Wirz, H. J. and Smolderen, J. J. (Eds.), McGraw-Hill/Hemisphere, New York, 1978.

Lamb, H., *Hydrodynamics*, Dover, New York, 1945.

Milne-Thomson, L. M., *Theoretical Hydrodynamics*, Macmillan, New York, 1950.

Moody, L. F., "Friction Factors for Pipe Flow," *ASME Transactions*, Vol. 66, pp. 671–684, 1944.

Potter, M. C. and Foss, J. F., *Fluid Mechanics*, Ronald Press, New York, 1975.

Richardson, L. F., *Weather Prediction by Numerical Process*. Cambridge University Press, Cambridge, 1922.

Robertson, J. M., Hydrodynamics in Theory and Application, Prentice-Hall, Englewood Cliffs, NJ, 1965.

Schetz, J. A., *Boundary Layer Analysis*, Prentice Hall, Englewood Cliffs, NJ, 1993.

Shames, I. H., *Mechanics of Fluids*, 2nd ed., McGraw-Hill, New York, 1982.

Shaw, R., Pilkey, W., Pilkey, B., Wilson, R., Lakis, A., Chaudouet, A., and Marino, C. (Eds.), *Innovative Numerical Analysis for Applied Engineering Sciences*, University Press of Virginia, Charlottesville, pp. 131–143, 1980.

Sreenivasan, K. R. and Kailasnath, P., "An Update on the Intermittency Exponent in Turbulence," *Phys. Fluids A*, Vol. 5, No. 2, pp. 512–514, 1993.

von Karman, T., "Calculation of Pressure Distribution on Airship Hulls," *NACA TM 574*, 1930.

Ward-Smith, A. J., *Internal Fluid Flow*, Clarendon Press, Oxford, 1980.

White, F. M., *Fluid Mechanics*, 3rd ed., McGraw-Hill, New York, 1994.

White, F. M., *Viscous Fluid Flow*, 2nd ed., McGraw-Hill, New York, 1992.

Yih, C. S., *Fluid Mechanics*, McGraw-Hill, New York, 1969.

Yih, C. S., *Stratified Flows*, Academic Press, New York, 1980.

Zedan, M. F. and Dalton, C. "Higher Order Axial Singularity Distributions for Potential Flow About Bodies of Revolution," *Computer Methods in Applied Mechanics and Engineering*, Vol. 21, pp. 295–314, 1980.

2 Properties of Fluids

PETER E. LILEY
Purdue University
West Lafayette, IN

EUGENE D. TRAGANZA and STEVENS P. TUCKER
Naval Postgraduate School
Monterey, CA

SANFORD GORDON
Sanford Gordon & Associates
Cleveland, OH

ALAN BEERBOWER (Deceased)
University of California at San Diego
La Jolla, CA

JOHN E. SCHMIDT (Retired) and ALBERT M. MOMENTHY
Boeing Commercial Airplane Company
Seattle, WA

ALLEN E. FUHS
Naval Postgraduate School
Monterey, CA

NEIL OLIEN
National Institute of Standards and Technology
Boulder, CO

JOSEPH A. SCHETZ
Virginia Polytechnic Institute and State University
Blacksburg, VA

SUSAN E. FUHS
Rand Corporation
Santa Monica, CA

Handbook of Fluid Dynamics and Fluid Machinery, Edited by Joseph A. Schetz and Allen E. Fuhs
ISBN 0-471-12598-9 Copyright © 1996 John Wiley & Sons, Inc.

CONTENTS

2.1 WATER AND STEAM
Peter E. Liley

The data in this section are based upon Juza (1966), Schmidt (1969), Kesselman and Blank (1968), and Reynard (1967).

TABLE 2.1 Specific Volume (m³/kg) of Water as a Function of Pressure and Temperature at Low Pressure

Temp. (K)	Pressure (kN/m²)								
	100	1000	2000	3000	4000	5000	6000	8000	10000
273.15	1.000.−3	1.000.−3	1.000.−3	9.987.−4	9.982.−4	9.977.−4	9.972.−4	9.962.−4	9.953.−4
280	1.000.−3	1.000.−3	1.000.−3	9.988.−4	9.983.−4	9.978.−4	9.973.−4	9.964.−4	9.955.−4
290	1.001.−3	1.001.−3	1.000.−3	9.999.−4	9.994.−4	9.990.−4	9.985.−4	9.976.−4	9.967.−4
300	1.003.−3	1.003.−3	1.003.−3	1.002.−3	1.002.−3	1.001.−3	1.001.−3	9.999.−4	9.990.−4
310	1.007.−3	1.006.−3	1.006.−3	1.005.−3	1.005.−3	1.005.−3	1.004.−3	1.003.−3	1.002.−3
320	1.011.−3	1.010.−3	1.010.−3	1.009.−3	1.009.−3	1.009.−3	1.008.−3	1.007.−3	1.006.−3
330	1.016.−3	1.015.−3	1.015.−3	1.014.−3	1.014.−3	1.014.−3	1.013.−3	1.012.−3	1.011.−3
340	1.021.−3	1.021.−3	1.020.−3	1.020.−3	1.019.−3	1.019.−3	1.018.−3	1.017.−3	1.017.−3
350	1.027.−3	1.027.−3	1.026.−3	1.026.−3	1.025.−3	1.025.−3	1.024.−3	1.023.−3	1.023.−3
360	1.034.−3	1.034.−3	1.033.−3	1.033.−3	1.032.−3	1.032.−3	1.031.−3	1.030.−3	1.029.−3
370	1.041.−3	1.041.−3	1.040.−3	1.040.−3	1.039.−3	1.039.−3	1.038.−3	1.037.−3	1.036.−3
380	1.729	1.049.−3	1.048.−3	1.048.−3	1.047.−3	1.047.−3	1.046.−3	1.045.−3	1.044.−3
390	1.778	1.057.−3	1.057.−3	1.056.−3	1.055.−3	1.055.−3	1.054.−3	1.053.−3	1.052.−3
400	1.827	1.067.−3	1.066.−3	1.065.−3	1.065.−3	1.064.−3	1.064.−3	1.063.−3	1.061.−3
420	1.921	1.087.−3	1.086.−3	1.086.−3	1.085.−3	1.085.−3	1.084.−3	1.082.−3	1.081.−3
440	2.016	1.110.−3	1.110.−3	1.108.−3	1.108.−3	1.107.−3	1.107.−3	1.105.−3	1.104.−3
460	2.110	0.198	1.137.−3	1.136.−3	1.135.−3	1.134.−3	1.131.−3	1.131.−3	1.129.−3
480	2.204	0.210	1.167.−3	1.166.−3	1.165.−3	1.164.−3	1.163.−3	1.161.−3	1.159.−3
500	2.298	0.221	0.104	1.202.−3	1.201.−3	1.200.−3	1.198.−3	1.196.−3	1.193.−3
550	2.531	0.246	0.119	0.0765	0.0549	0.0417	0.0327	1.316.−3	1.313.−3
600	2.763	0.271	0.133	0.0861	0.0630	0.0490	0.0396	0.0276	0.0201
650	2.996	0.295	0.145	0.0958	0.0703	0.0551	0.0451	0.0324	0.0247
700	3.227	0.319	0.158	0.104	0.0769	0.0608	0.0500	0.0346	0.0283
750	3.459	0.343	0.170	0.112	0.0834	0.0661	0.0545	0.0401	0.0314
800	3.690	0.367	0.182	0.120	0.0889	0.0713	0.0589	0.0436	0.0343
900	4.152	0.414	0.206	0.137	0.102	0.0813	0.0674	0.0501	0.0398
1000	4.614	0.460	0.230	0.153	0.114	0.0911	0.0757	0.0564	0.0449
1200	5.538	0.553	0.276	0.184	0.138	0.1102	0.0918	0.0687	0.0549
1400	6.461	0.646	0.323	0.215	0.161	0.1291	0.1075	0.0800	0.0645
1500	6.924	0.692	0.346	0.231	0.173	0.1384	0.1153	0.0865	0.0692
Pressure	1	10	20	30	40	50	60	80	100
									(bar)

Interpolated or extrapolated from the tables of Juza, J., "An equation of state for water and steam," Academia, Prague, Czechoslovakia, 144 pp., 1966; Schmidt, E., "Properties of water and steam in S.I. Units," Springer–Verlag, 205 pp., 1969.

TABLE 2.2   Specific Volume (m³/kg) of Water as a Function of Pressure and Temperature at Medium Pressures

Temp. (K)	Pressure (kN/m²)							
	1×10^4	1.5×10^4	2×10^4	3×10^4	4×10^4	6×10^4	8×10^4	10^5
273.15	9.953.−4	9.928.−4	9.904.−4	9.857.−4	9.811.−4	9.723.−4	9.641.−4	9.567.−4
280	9.955.−4	9.931.−4	9.908.−4	9.863.−4	9.819.−4	9.735.−4	9.655.−4	9.586.−4
290	9.967.−4	9.944.−4	9.923.−4	9.880.−4	9.838.−4	9.757.−4	9.680.−4	9.613.−4
300	9.990.−4	9.968.−4	9.947.−4	9.906.−4	9.865.−4	9.786.−4	9.712.−4	9.642.−4
310	1.002.−4	1.000.−3	9.981.−4	9.940.−4	9.899.−4	9.825.−4	9.748.−4	9.681.−4
320	1.006.−3	1.004.−3	1.002.−3	9.980.−4	9.939.−4	9.863.−4	9.789.−4	9.721.−4
330	1.011.−3	1.009.−3	1.007.−3	1.003.−3	9.986.−4	9.908.−4	9.834.−4	9.767.−4
340	1.017.−3	1.014.−3	1.012.−3	1.008.−3	1.004.−3	9.959.−4	9.884.−4	9.816.−4
350	1.023.−3	1.020.−3	0.018.−3	1.014.−3	1.009.−3	1.001.−3	9.930.−4	9.867.−4
360	1.029.−3	1.027.−3	1.024.−3	1.019.−3	1.016.−3	1.007.−3	0.999.−3	9.925.−4
370	1.036.−3	1.034.−3	1.031.−3	1.027.−3	1.022.−3	1.014.−3	1.006.−3	9.984.−4
380	1.044.−3	1.041.−3	1.039.−3	1.034.−3	1.029.−3	1.021.−3	1.012.−3	1.004.−3
390	1.052.−3	1.050.−3	1.047.−3	1.042.−3	1.037.−3	1.028.−3	1.019.−3	1.011.−3
400	1.061.−3	1.059.−3	1.056.−3	1.050.−3	1.045.−3	1.035.−3	1.027.−3	1.018.−3
420	1.081.−3	1.078.−3	1.075.−3	1.069.−3	1.063.−3	1.052.−3	1.043.−3	1.033.−3
440	1.104.−3	1.100.−3	1.096.−3	1.093.−3	1.083.−3	1.071.−3	1.060.−3	1.050.−3
460	1.129.−3	1.125.−3	1.121.−3	1.113.−3	1.106.−3	1.092.−3	1.080.−3	1.069.−3
480	1.159.−3	1.154.−3	1.149.−3	1.144.−3	1.133.−3	1.116.−3	1.102.−3	1.089.−3
500	1.193.−3	1.187.−3	1.181.−3	1.170.−3	1.160.−3	1.142.−3	1.126.−3	1.112.−3
550	1.313.−3	1.300.−3	1.289.−3	1.269.−3	1.252.−3	1.233.−3	1.199.−3	1.179.−3
600	0.0201	1.519.−3	1.483.−3	1.428.−3	1.392.−3	1.337.−3	1.296.−3	1.265.−3
650	0.0247	0.0140	7.706.−3	1.828.−3	1.658.−3	1.514.−3	1.435.−3	1.381.−3
700	0.0283	0.0172	0.0116	5.416.−3	2.630.−3	1.831.−3	1.639.−3	1.536.−3
750	0.0314	0.0198	0.0139	7.864.−3	4.821.−3	2.509.−3	1.973.−3	1.760.−3
800	0.0343	0.0220	0.0158	9.512.−3	6.391.−3	3.496.−3	2.484.−3	2.078.−3
900	0.0398	0.0259	0.0190	0.0120	8.619.−3	5.257.−3	3.704.−3	2.907.−3
1000	0.0449	0.0295	0.0219	0.0142	1.038.−2	6.605.−3	4.793.−3	3.763.−3
1200	0.0549	0.0364	0.0272	0.0179	0.0133	0.00876	0.00652	0.00519
1400	0.0645	0.0429	0.0322	0.0214	0.0156	0.0104	0.00777	0.00645
1500	0.0692	0.0461	0.0346	0.0231	0.0168	0.0114	0.00868	0.00700
	100	150	200	300	400	600	800	1000
	Pressure (bar)							

Interpolated or extrapolated from the tables of Kesselman, P. H. and Blank, Yu, I., Paper B-11, 7th Int. Conf. Properties of Steam, Tokyo, Japan, 1968; Reynard, K. W., Eng. Sci. Data Unit, London, England item 68008, 33 pp, 1968.

TABLE 2.3 Specific Volume (m³/kg) of Water as a Function of Pressure and Temperature at High Pressure

Pressure (kN/m²)

Temp. (K)	1×10^5	1.5×10^5	2×10^5	2.5×10^5	3×10^5	4×10^5	5×10^5	6×10^5	8×10^5	1×10^6
273.15	9.567.–4	9.395.–4	9.244.–4	9.111.–4	8.991.–4	8.785.–4	8.609.–4	8.455.–4		
280	9.586.–4	9.416.–4	9.267.–4	9.136.–4	9.016.–4	8.810.–4	8.633.–4	8.477.–4		
290	9.613.–4	9.449.–4	9.286.–4	9.172.–4	9.054.–4	8.846.–4	8.667.–4	8.510.–4	8.245.–4	
300	9.642.–4	9.482.–4	9.340.–4	9.210.–4	9.092.–4	8.888.–4	8.703.–4	8.545.–4	8.279.–4	
310	9.681.–4	9.523.–4	9.380.–4	9.250.–4	9.132.–4	8.924.–4	8.744.–4	8.587.–4	8.323.–4	8.118.–4
320	9.721.–4	9.563.–4	9.421.–4	9.291.–4	9.173.–4	8.965.–4	8.786.–4	8.630.–4	8.366.–4	8.154.–4
330	9.767.–4	9.608.–4	9.466.–4	9.336.–4	9.216.–4	9.006.–4	8.826.–4	8.669.–4	8.405.–4	8.190.–4
340	9.816.–4	9.656.–4	9.512.–4	9.381.–4	9.259.–4	9.047.–4	8.866.–4	8.708.–4	8.443.–4	8.226.–4
350	9.867.–4	9.704.–4	9.559.–4	9.426.–4	9.304.–4	9.089.–4	8.905.–4	8.746.–4	8.479.–4	8.262.–4
360	9.925.–4	9.758.–4	9.610.–4	9.475.–4	9.351.–4	9.132.–4	8.946.–4	8.784.–4	8.513.–4	8.293.–4
370	9.984.–4	9.813.–4	9.662.–4	9.525.–4	9.399.–4	9.176.–4	8.986.–4	8.821.–4	8.547.–4	8.325.–4
380	1.004.–3	9.872.–4	9.717.–4	9.577.–4	9.449.–4	9.222.–4	9.028.–4	8.860.–4	8.581.–4	8.356.–4
390	1.011.–3	9.934.–4	9.774.–4	9.631.–4	9.500.–4	9.268.–4	9.070.–4	8.898.–4	8.615.–4	8.386.–4
400	1.018.–3	9.997.–4	9.863.–4	9.686.–4	9.546.–4	9.315.–4	9.112.–4	8.937.–4	8.648.–4	8.416.–4
420	1.033.–3	1.011.–3	9.957.–4	9.800.–4	9.661.–4	9.414.–4	9.201.–4	9.018.–4	8.717.–4	8.476.–4
440	1.050.–3	1.028.–3	1.009.–3	9.926.–4	9.776.–4	9.517.–4	9.295.–4	9.102.–4	8.768.–4	8.538.–4
460	1.069.–3	1.045.–3	1.024.–3	1.006.–3	9.898.–4	9.623.–4	9.391.–4	9.189.–4	8.860.–4	8.600.–4
480	1.089.–3	1.063.–3	1.040.–3	1.020.–3	1.003.–3	9.735.–4	9.491.–4	9.279.–4	8.935.–4	8.664.–4
500	1.112.–3	1.082.–3	1.057.–3	1.035.–3	1.017.–3	9.853.–4	9.595.–4	9.373.–4	9.012.–4	8.730.–4
550	1.179.–3	1.137.–3	1.104.–3	1.073.–3	1.054.–3	1.017.–3	9.865.–4	9.616.–4	9.203.–4	8.899.–4
600	1.265.–3	1.205.–3	1.160.–3	1.125.–3	1.096.–3	1.051.–3	1.016.–3	9.875.–4	9.425.–4	9.078.–4
650	1.381.–3	1.288.–3	1.226.–3	1.181.–3	1.145.–3	1.090.–3	1.048.–3	1.015.–3	9.648.–4	9.267.–4
700	1.536.–3	1.446.–3	1.305.–3	1.245.–3	1.199.–3	1.132.–3	1.083.–3	1.045.–3	9.882.–4	9.463.–4
750	1.760.–3	1.522.–3	1.398.–3	1.319.–3	1.261.–3	1.178.–3	1.121.–3	1.077.–3	1.013.–3	9.677.–4
800	2.072.–3	1.682.–3	1.508.–3	1.403.–3	1.329.–3	1.229.–3	1.161.–3	1.111.–3	1.039.–3	9.878.–4
900	2.907.–3	2.090.–3	1.773.–3	1.601.–3	1.486.–3	1.342.–3	1.224.–3	1.185.–3	1.094.–3	1.032.–3
1000	3.763.–3	2.575.–3	2.088.–3	1.831.–3	1.667.–3	1.469.–3	1.349.–3	1.265.–3	1.182.–3	1.080.–3
1200	0.00519	3.537.–3	2.679.–3	2.341.–3	2.070.–3	1.751.–3	1.565.–3	1.441.–3	1.30.–3	1.18.–3
1400	0.00645	4.35.–3	3.87.–3	2.87.–3	2.50.–3	2.03.–3	1.76.–3	1.59.–3	1.42.–3	1.28.–3
1500	0.00700	4.89.–3	3.80.–3	3.18.–3	2.72.–3	2.19.–3	1.85.–3	1.66.–3	1.49.–3	1.33.–3

TABLE 2.4 Specific Heat (kJ/kgK) of Water as a Function of Pressure and Temperature

Pressure (kN/m²)

Temp. (K)	100	1000	2000	3000	4000	5000	6000	8000	10000	15000	20000	30000	40000	60000	80000	100000
273.15	4.217	4.212	4.207	4.201	4.196	4.191	4.186	4.178	4.170	4.141	4.117	4.073	4.032	3.957	3.883	3.801
280	4.200	4.197	4.191	4.188	4.182	4.178	4.175	4.167	4.158	4.140	4.118	4.080	4.044	3.978	3.912	3.845
290	4.186	4.183	4.179	4.176	4.173	4.170	4.167	4.160	4.152	4.138	4.120	4.091	4.061	4.008	3.955	3.910
300	4.179	4.177	4.174	4.171	4.168	4.165	4.162	4.156	4.150	4.137	4.123	4.101	4.077	4.036	3.993	3.966
310	4.179	4.177	4.175	4.172	4.169	4.166	4.164	4.158	4.153	4.142	4.129	4.108	4.086	4.048	4.011	3.985
320	4.180	4.178	4.176	4.173	4.171	4.168	4.166	4.160	4.156	4.146	4.135	4.115	4.095	4.060	4.029	4.005
330	4.183	4.181	4.179	4.176	4.174	4.171	4.170	4.165	4.161	4.149	4.141	4.121	4.102	4.069	4.039	4.013
340	4.187	4.186	4.183	4.181	4.178	4.176	4.174	4.169	4.165	4.154	4.146	4.127	4.108	4.075	4.045	4.018
350	4.193	4.192	4.189	4.187	4.184	4.182	4.180	4.175	4.171	4.162	4.153	4.134	4.115	4.082	4.051	4.023
360	4.203	4.202	4.199	4.197	4.194	4.192	4.190	4.185	4.181	4.171	4.161	4.142	4.123	4.090	4.059	4.030
370	4.214	4.213	4.210	4.207	4.205	4.203	4.201	4.196	4.192	4.180	4.170	4.150	4.132	4.098	4.066	4.037
380	2.016	4.225	4.222	4.219	4.217	4.215	4.213	4.208	4.204	4.192	4.182	4.161	4.142	4.107	4.075	4.044
390	2.004	4.239	4.237	4.234	4.232	4.230	4.228	4.223	4.219	4.205	4.195	4.173	4.152	4.118	4.085	4.052
400	1.996	4.256	4.253	4.250	4.247	4.245	4.243	4.238	4.235	4.221	4.209	4.186	4.163	4.128	4.095	4.060
420	1.986	4.304	4.291	4.298	4.293	4.290	4.287	4.282	4.278	4.261	4.246	4.221	4.197	4.153	4.115	4.078
440	1.982	4.351	4.357	4.353	4.349	4.347	4.340	4.332	4.326	4.308	4.293	4.264	4.238	4.184	4.139	4.101
460	1.980	2.518				4.443	4.428	4.416	4.406	4.385	4.355	4.318	4.286	4.223	4.170	4.126
480	1.980	2.411						4.503	4.488	4.412	4.483	4.386	4.343	4.266	4.211	4.153
500	1.983	2.324	2.807							4.525	4.511	4.44	4.416	4.332	4.264	4.195
550	2.000	2.178	2.412	2.742	3.191	3.820					4.94	4.81	4.70	4.53	4.41	4.32
600	2.024	2.126	2.260	2.417	2.608	2.850	3.105	3.910				5.60	5.31	4.92	4.67	4.47
650	2.054	2.116	2.204	2.285	2.421	2.530	2.677	2.996	3.409			9.17	6.92	5.62	5.28	5.02
700	2.085	2.129	2.182	2.246	2.313	2.379	2.456	2.633	2.853	3.55	4.67	10.2	13.2	6.93	5.60	5.08
750	2.118	2.151	2.189	2.231	2.272	2.321	2.390	2.498	2.610	2.99	3.49	5.00	7.36	7.79	6.09	5.31
800	2.151	2.178	2.208	2.239	2.272	2.306	2.344	2.419	2.496	2.74	3.04	3.82	4.86	6.38	6.09	5.51
900	2.219	2.236	2.259	2.281	2.303	2.325	2.346	2.392	2.440	2.57	2.71	3.03	3.39	4.19	4.77	4.88
1000	2.286	2.301	2.316	2.332	2.349	2.365	2.380	2.414	2.447	2.53	2.62	2.81	3.01	3.38	3.75	3.96
1200	2.43	2.44	2.45	2.45	2.46	2.47	2.48	2.50	2.51	2.54	2.57	2.65	2.70	2.88	3.01	3.15
1400	2.58	2.58	2.59	2.59	2.59	2.60	2.60	2.62	2.63	2.65	2.66	2.72	2.77	2.87	2.95	3.05
1500	2.65	2.66	2.67	2.67	2.67	2.67	2.67	2.68	2.68	2.70	2.70	2.75	2.79	2.86	2.93	3.01
Pressure (bar)	1	10	20	30	40	50	60	80	100	150	200	300	400	600	800	1000

Interpolated or extrapolated from the tables of Kesselman, P. H. and Blank, Yu. I., Paper B-11, 7th Int. Conf. Properties of Steam, Tokyo, Japan, 1968; Reynard, K. W., Eng. Sci. Data Unit, London, England item 68009, 42 pp., 1968.

TABLE 2.5 Thermal Conductivity (W/mK) of Water as a Function of Pressure and Temperature

Temp. (K)	Pressure (kN/m²)															
	100	1000	2000	3000	4000	5000	6000	8000	10000	15000	20000	30000	40000	60000	80000	100000
273.15	0.563	0.564	0.565	0.566	0.566	0.567	0.568	0.570	0.571	0.573	0.574	0.578	0.583	0.590	0.599	0.609
280	0.576	0.577	0.578	0.579	0.579	0.580	0.580	0.582	0.583	0.585	0.586	0.590	0.594	0.602	0.612	0.620
290	0.595	0.596	0.596	0.597	0.597	0.598	0.598	0.599	0.601	0.603	0.604	0.608	0.612	0.620	0.630	0.637
300	0.612	0.613	0.613	0.614	0.614	0.615	0.615	0.616	0.617	0.619	0.622	0.626	0.629	0.638	0.647	0.653
310	0.626	0.626	0.626	0.627	0.627	0.628	0.629	0.631	0.632	0.633	0.635	0.640	0.643	0.652	0.660	0.667
320	0.639	0.639	0.639	0.640	0.640	0.641	0.642	0.643	0.644	0.646	0.649	0.654	0.657	0.666	0.673	0.681
330	0.649	0.649	0.649	0.650	0.650	0.651	0.652	0.653	0.654	0.656	0.659	0.664	0.669	0.677	0.685	0.694
340	0.658	0.658	0.658	0.659	0.659	0.660	0.661	0.662	0.664	0.666	0.668	0.674	0.678	0.688	0.697	0.706
350	0.667	0.667	0.667	0.668	0.669	0.669	0.669	0.670	0.670	0.674	0.677	0.682	0.687	0.698	0.708	0.717
360	0.674	0.674	0.674	0.675	0.675	0.676	0.677	0.677	0.677	0.681	0.683	0.689	0.694	0.705	0.715	0.725
370	0.679	0.679	0.680	0.680	0.680	0.681	0.682	0.683	0.684	0.687	0.689	0.695	0.700	0.711	0.721	0.733
380	0.026	0.682	0.682	0.683	0.683	0.684	0.684	0.686	0.687	0.691	0.693	0.699	0.704	0.716	0.726	0.737
390	0.026	0.684	0.684	0.685	0.685	0.686	0.686	0.688	0.689	0.694	0.697	0.703	0.709	0.720	0.730	0.740
400	0.027	0.686	0.686	0.687	0.687	0.688	0.688	0.690	0.691	0.697	0.700	0.706	0.712	0.724	0.734	0.743
420	0.029	0.688	0.688	0.690	0.690	0.691	0.690	0.693	0.694	0.700	0.703	0.709	0.716	0.728	0.738	0.748
440	0.030	0.681	0.682	0.683	0.683	0.686	0.686	0.689	0.691	0.694	0.698	0.705	0.710	0.726	0.743	0.747
460	0.032	0.035	0.673	0.674	0.675	0.677	0.677	0.681	0.683	0.686	0.690	0.698	0.705	0.720	0.732	0.746
480	0.034	0.037	0.662	0.663	0.664	0.666	0.666	0.669	0.670	0.675	0.679	0.687	0.693	0.711	0.725	0.740
500	0.036	0.038	0.042	0.643	0.644	0.649	0.649	0.653	0.654	0.660	0.664	0.673	0.674	0.699	0.714	0.731
550	0.041	0.042	0.044	0.048	0.050	0.052	0.054	0.589	0.595	0.603	0.611	0.624	0.637	0.658	0.676	0.694
600	0.046	0.047	0.048	0.050	0.051	0.053	0.055	0.066	0.071	0.507	0.519	0.545	0.565	0.596	0.624	0.650
650	0.052	0.053	0.054	0.055	0.056	0.057	0.059	0.062	0.065	0.085	0.142	0.330	0.466	0.521	0.561	0.592
700	0.058	0.059	0.059	0.060	0.060	0.061	0.062	0.065	0.067	0.078	0.094	0.172	0.326	0.419	0.477	0.516
750	0.064	0.065	0.065	0.066	0.066	0.066	0.068	0.071	0.072	0.079	0.090	0.119	0.151	0.318	0.390	0.410
800	0.070	0.071	0.071	0.072	0.072	0.072	0.074	0.077	0.080	0.085	0.094	0.112	0.144	0.239	0.320	0.372
900	0.083	0.085	0.085	0.086	0.086	0.086	0.089	0.091	0.094	0.097	0.103	0.112	0.129	0.170	0.221	0.270
1000	0.097	0.098	0.098	0.098	0.098	0.100	0.101	0.105	0.107	0.110	0.113	0.123	0.134	0.159	0.193	0.228
1200	0.13	0.13	0.13	0.13	0.13	0.13	0.13	0.13	0.13	0.14	0.14	0.14	0.15			
1400	0.16	0.16·	0.16	0.16	0.16	0.16	0.16	0.16	0.16	0.16	0.16	0.16	0.17			
1500	0.18	0.18	0.18	0.18	0.18	0.18	0.18	0.18	0.18	0.18	0.18	0.18	0.18			
Pressure (bar)	1	10	20	30	40	50	60	80	100	150	200	300	400	600	800	1000

Interpolated or extrapolated from the tables of Kesselman, P. H. and Blank, Yu. I., Paper B-11, 7th Int. Conf. Properties of Steam, Tokyo, Japan, 1968; Reynard, K. W., Eng. Sci. Data Unit, London, England item 68009, 42 pp., 1968.

TABLE 2.6 Viscosity (10^{-4} Pa · s) of Water as a Function of Pressure and Temperature

Temp. (K)	\[Pressure (kN/m²)\] 100	1000	2000	3000	4000	5000	6000	8000	10000	15000	20000	30000	40000	60000	80000	100000
273.15	17.92	17.90	17.88	17.85	17.83	17.80	17.78	17.73	17.69	17.59	17.49	17.31	17.14	17.04	16.87	16.52
280	14.61	14.59	14.58	14.55	14.54	14.51	14.50	14.46	14.42	14.34	14.26	14.11	13.97	13.75	13.59	13.47
290	11.04	11.03	11.01	11.00	10.98	10.97	10.95	10.92	10.90	10.84	10.78	10.66	10.56	10.39	10.27	10.18
300	8.57	8.57	8.57	8.56	8.56	8.56	8.55	8.55	8.55	8.54	8.53	8.52	8.52	8.52	8.53	8.56
310	7.10	7.10	7.10	7.09	7.09	7.09	7.08	7.08	7.08	7.08	7.07	7.06	7.06	7.06	7.07	7.09
320	5.71	5.71	5.71	5.70	5.70	5.70	5.70	5.70	5.70	5.69	5.68	5.68	5.68	5.68	5.68	5.70
330	4.92	4.92	4.92	4.93	4.93	4.93	4.93	4.93	4.94	4.95	4.95	4.96	4.98	5.03	5.06	5.11
340	4.24	4.24	4.24	4.25	4.25	4.25	4.25	4.25	4.26	4.26	4.26	4.28	4.29	4.33	4.36	4.40
350	3.69	3.70	3.70	3.70	3.71	3.71	3.71	3.71	3.72	3.73	3.74	3.77	3.80	3.85	3.90	3.96
360	3.27	3.28	3.28	3.28	3.29	3.29	3.29	3.29	3.30	3.30	3.31	3.34	3.37	3.41	3.45	3.50
370	2.92	2.93	2.93	2.93	2.94	2.94	2.94	2.94	2.94	2.95	2.96	2.98	3.00	3.04	3.08	3.13
380	0.125	2.64	2.64	2.65	2.65	2.66	2.66	2.66	2.67	2.68	2.70	2.71	2.74	2.79	2.85	2.89
390	0.130	2.34	2.34	2.35	2.35	2.36	2.36	2.36	2.36	2.37	2.39	2.41	2.43	2.47	2.52	2.56
400	0.134	2.19	2.19	2.20	2.20	2.21	2.21	2.21	2.22	2.24	2.25	2.28	2.30	2.32	2.34	2.36
420	0.141	1.86	1.86	1.87	1.87	1.88	1.88	1.88	1.89	1.90	1.91	1.94	1.95	1.97	1.99	2.00
440	0.149	1.64	1.64	1.64	1.65	1.65	1.65	1.66	1.66	1.68	1.68	1.71	1.73	1.77	1.82	1.86
460	0.157	0.153	1.44	1.44	1.45	1.45	1.45	1.45	1.46	1.48	1.49	1.51	1.53	1.57	1.62	1.65
480	0.165	0.162	1.29	1.29	1.29	1.30	1.30	1.30	1.31	1.32	1.33	1.36	1.38	1.42	1.46	1.50
500	0.173	0.171	0.169	1.18	1.18	1.19	1.19	1.19	1.20	1.21	1.22	1.24	1.26	1.31	1.36	1.39
550	0.193	0.192	0.192	0.191	0.190	0.189	0.188	0.98	0.99	1.00	1.01	1.03	1.04	1.09	1.13	1.17
600	0.214	0.213	0.213	0.212	0.212	0.211	0.211	0.210	0.209	0.82	0.83	0.85	0.87	0.93	0.98	1.02
650	0.235	0.235	0.235	0.233	0.234	0.234	0.234	0.234	0.235	0.239	0.249	0.58	0.70	0.79	0.85	0.90
700	0.256	0.256	0.256	0.256	0.256	0.256	0.256	0.256	0.257	0.261	0.269	0.37	0.54	0.65	0.73	0.79
750	0.277	0.277	0.277	0.277	0.277	0.277	0.277	0.278	0.279	0.283	0.289	0.36	0.39	0.52	0.63	0.69
800	0.297	0.297	0.297	0.298	0.298	0.298	0.298	0.299	0.300	0.304	0.309	0.33	0.35	0.45	0.54	0.62
900	0.337	0.337	0.337	0.338	0.338	0.339	0.340	0.341	0.342	0.346	0.350	0.36	0.38	0.42	0.48	0.54
1000	0.376	0.376	0.376	0.377	0.377	0.378	0.379	0.381	0.382	0.386	0.390	0.40	0.41	0.44	0.48	0.51
1200	0.443	0.443	0.444	0.445	0.445	0.446	0.446	0.446	0.447	0.449	0.451	0.46	0.46	0.49		
1400	0.505	0.505	0.505	0.506	0.506	0.506	0.506	0.506	0.506	0.507	0.509	0.51	0.52	0.51		
1500	0.535	0.535	0.535	0.535	0.535	0.535	0.536	0.536	0.536	0.537	0.539	0.54	0.53	0.52		
Pressure (bar)	1	10	20	30	40	50	60	80	100	150	200	300	400	600	800	1000

Interpolated or extrapolated from the tables of Kesselman, P. H. and Blank, Yu. I., Paper B-11, 7th Int. Conf. Properties of Steam, Tokyo, Japan, 1968; Reynard, K. W., Eng. Sci. Data Unit, London, England item 68009, 42 pp., 1968.

TABLE 2.7 Prandtl Number of Water as a Function of Pressure and Temperature

Temp. (K)	Pressure (kN/m²)															
	100	1000	2000	3000	4000	5000	6000	8000	10000	15000	20000	30000	40000	60000	80000	100000
273.15	13.4	13.4	13.3	13.3	13.2	13.2	13.1	13.0	12.9	12.7	12.5	12.2	11.9	11.3	10.8	10.3
280	10.6	10.6	10.5	10.5	10.4	10.4	10.3	10.3	10.2	10.0	9.8	9.5	9.2	8.6	8.1	7.6
290	7.7	7.6	7.6	7.6	7.5	7.5	7.5	7.4	7.4	7.2	7.0	6.8	6.7	6.5	6.4	6.3
300	5.9	5.8	5.8	5.8	5.8	5.8	5.8	5.8	5.7	5.7	5.7	5.6	5.6	5.4	5.3	5.2
310	4.7	4.7	4.6	4.6	4.6	4.6	4.5	4.5	4.5	4.5	4.5	4.4	4.4	4.3	4.3	4.3
320	3.77	3.75	3.73	3.71	3.69	3.67	3.65	3.64	3.62	3.6	3.6	3.5	3.5	3.4	3.4	3.4
330	3.18	3.17	3.16	3.14	3.13	3.12	3.11	3.09	3.08	3.0	3.0	2.9	2.9	2.9	2.8	2.8
340	2.70	2.70	2.69	2.69	2.68	2.67	2.66	2.65	2.64	2.6	2.6	2.6	2.6	2.5	2.5	2.5
350	2.34	2.34	2.33	2.33	2.33	2.33	2.32	2.32	2.31	2.30	2.29	2.28	2.27	2.25	2.23	2.21
360	2.05	2.05	2.05	2.05	2.04	2.04	2.04	2.04	2.03	2.03	2.03	2.03	2.02	2.00	1.99	1.97
370	1.80	1.80	1.80	1.80	1.80	1.79	1.79	1.79	1.79	1.79	1.79	1.78	1.78	1.76	1.75	1.73
380	0.99	1.62	1.62	1.62	1.62	1.62	1.61	1.61	1.61	1.61	1.61	1.60	1.60	1.59	1.58	1.58
390	0.99	1.47	1.47	1.47	1.47	1.47	1.47	1.47	1.47	1.47	1.47	1.47	1.47	1.46	1.44	1.43
400	0.98	1.35	1.35	1.35	1.35	1.35	1.35	1.35	1.35	1.35	1.35	1.34	1.34	1.32	1.30	1.28
420	0.97	1.16	1.16	1.16	1.16	1.16	1.15	1.15	1.15	1.15	1.15	1.15	1.15	1.14	1.14	1.14
440	0.97	1.04	1.04	1.04	1.04	1.04	1.04	1.04	1.04	1.04	1.04	1.04	1.04	1.04	1.04	1.04
460	0.97	1.15	0.95	0.95	0.95	0.95	0.95	0.95	0.95	0.95	0.95	0.94	0.94	0.93	0.93	0.93
480	0.96	1.08	0.89	0.89	0.89	0.89	0.89	0.89	0.89	0.88	0.87	0.87	0.86	0.86	0.85	0.85
500	0.96	1.02	1.13	0.85	0.85	0.85	0.84	0.84	0.84	0.84	0.83	0.83	0.82	0.82	0.81	0.80
550	0.95	0.99	1.06								1.05	0.93	0.89	0.81	0.79	0.77
600	0.94	0.97	1.01	1.05								1.05	0.96	0.82	0.78	0.74
650	0.93	0.95	0.97	1.00	1.03	1.04	1.08	1.13	1.24	1.51	2.31	1.30	1.03	0.86	0.79	0.74
700	0.92	0.93	0.94	0.95	0.97	0.99	1.01	1.05	1.09			1.47	1.30	1.01	0.88	0.73
750	0.91	0.91	0.92	0.93	0.97	0.97	0.98	1.00	1.02	1.05	1.12	1.32	1.55	1.26	0.97	0.77
800	0.91	0.91	0.92	0.92	0.94	0.95	0.96	0.97	0.97	0.98	1.01	1.12	1.23	1.24	1.07	0.82
900	0.90	0.90	0.90	0.91	0.91	0.92	0.92	0.92	0.92	0.92	0.93	0.96	1.00	1.04	1.02	0.91
1000	0.89	0.89	0.89	0.89	0.89	0.89	0.89	0.89	0.89	0.89	0.90	0.91	0.92	0.94	1.04	.95
1200	0.84	0.84	0.85	0.85	0.85	0.85	0.85	0.85	0.85	0.85	0.85	0.86	0.87	0.93	0.93	0.91
1400	0.80	0.81	0.82	0.82	0.82	0.82	0.82	0.83	0.83	0.83	0.83	0.84	0.85			
1500	0.78	0.78	0.79	0.79	0.80	0.80	0.81	0.81	0.81	0.82	0.82	0.83	0.84			
Pressure (bar)	1	10	20	30	40	50	60	80	100	150	200	300	400	600	800	1000

Calculated and smoothed from Tables 2.4 to 2.6 of Specific Heat at Constant Pressure, Thermal Conductivity and Viscosity.

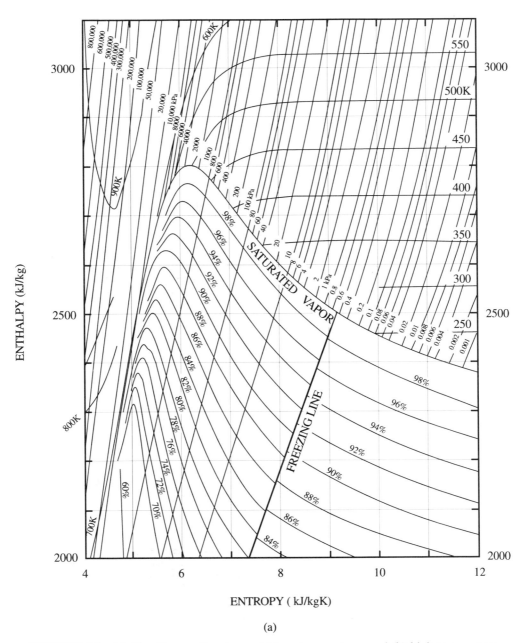

(a)

FIGURE 2.1 Mollier diagram for steam: (a) low temperatures and (b) high temperatures.

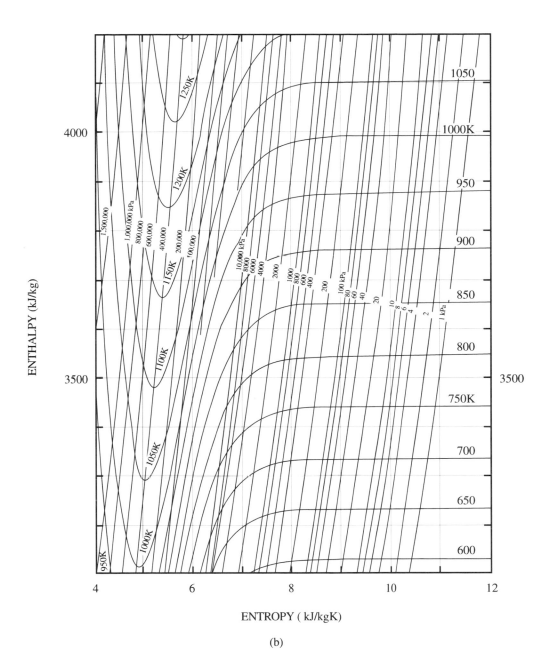

(b)

FIGURE 2.1   (*Continued*)

2.2 SEAWATER

E. D. Traganza and S. P. Tucker

2.2.1 Composition of Seawater

Seawater is a moderately concentrated (~ 0.5 molal), multicomponent, aqueous, electrolytic solution [Horne (1969)]. In the open ocean, the ratios or proportions of the major elements (Table 2.8) remain relatively constant even though the total salt content varies from ~ 0.033–0.038 kg/kg seawater. For the major elements, variation in concentration in the open ocean is in principle by the addition or removal of pure water to a saline solution of fixed elemental composition. Therefore, knowing the concentration of one of the major elements is sufficient to closely approximate the total salt content or salinity of seawater. A total salt content of 0.033 kg/kg seawater is expressed by a dimensionless quantity called *Practical Salinity*, $S = 33$, for 0.038 kg/kg, $S = 38$, etc. Both have numerically identical major elements to total salt content ratios, assuming they are open ocean seawater. Compositional constancy generally does not hold for the minor or trace elements (Table 2.9), nor for major elements where there is a source or sink and mixing is insufficient, e.g., near rivers, submarine springs, hydrothermal vents, submarine volcanos, ice edges, in anoxic basins, shallow tropical seas, etc.

Excluding dissolved gases and elements with boundary effects or impacted by

TABLE 2.8 Concentrations of the Major Constituents in Seawater

Constituent	g/kg	Molality	% of each species
Chloride	19.353	0.56241	Cl^-
Sodium	10.76	0.48284	Na^+ (97.7%), $NaSO_4^-$ (2.2%), $NaHCO_3^\circ$ (0.03%)
Magnesium	1.297	0.05440	Mg^{2+} (89%), $MgSO_4^\circ$ (10%), $MgHCO_3^+$ (0.6%), $MgCO_3^\circ$ (0.1%)
Sulfate	2.712	0.02909	SO_4^{2-} (39%), $NaSO_4^-$ (37%), $MgSO_4^\circ$ (19%), $CaSO_4^\circ$ (4%)
Calcium	0.4119	0.01059	Ca^{2+} (88%), $CaSO_4^\circ$ (11%), $CaHCO_3^+$ (0.6%), $CaCO_3^\circ$ (0.1%)
Potassium	0.399	0.01052	K^+ (98.8%), KSO_4^- (1.2%)
Bicarbonate*	0.145	0.00245	HCO_3^- (64%), $MgHCO_3^+$ (16%), $NaHCO_3^\circ$ (8%), $CaHCO_3^+$ (3%), CO_3^{2-} (0.8%), $MgCO_3^\circ$ (6%), $NaCO_3^-$ (1%), $CaCO_3^\circ$ (0.5%)
Bromide	0.0673	0.00087	Br^-
Boron	0.0046	0.00044	$B(OH)_3$ (84%), $B(OH)_4^-$ (16%)
Strontium	0.0078	0.00009	Sr^{2+}
Fluoride	0.0013		F^- (50–80%), MgF^+ (20–50%)

The bicarbonate ion, denoted by *, is actually Titration Alkalinity. Concentrations of the other constituents, in g/kg, are average values for seawater with a total salt content of 0.0350 kg/kg seawater, or (salinity) $S = 35$. Molalities are for $S = 34.8$. Percentages are for species of each element in seawater, $S = 35$, and pH = 8 at 25°C and 1 atmosphere. This table is compiled from tables by Kestor and Pytkowicz [Hood and Pytkowicz (1974)]. See also Millero (1983) and Millero and Schreiber (1982) for ion pairing, activity coefficients, and ionic strengths.

TABLE 2.9 Observed Elemental Concentrations in Seawater and Predicted Mean Oceanic Concentration

Atomic Number	Element	Species	Surface Concentration (Sample Depth)	Deep Water Concentration (Sample Depth)	Predicted Mean Water Concentration
1	Hydrogen	H_2	1.7 nmol/kg (2 m)	1.9 nmol/kg (927 m)	1.9 nmol/kg
2	Helium		178 μg/kg (2 m)	c	178 μg/kg
3	Lithium		36 pg/kg (40 m)	144 pg/kg (789 m)	0.2 ng/kg
4	Beryllium		4.4 mg/kg	c	4.4 mg/kg
5	Boron	Inorganic Boron			
6	Carbon	ΣCO_2	2039 μmol/kg (1 m)	2264 μmol/kg (714 m)	2200 μmol/kg
7	Nitrogen	N_2		560 μmol/kg (995 m)	590 μmol/kg
		NO_3	0 μmol/kg (22 m)	41 μmol/kg (891 m)	30 μmol/kg
8	Oxygen	Dissolved O_2	205 μmol/kg (4.7 mL/L) (22 m)	47 μmol/kg (1.1 mL/L) (891 m)	150 μmol/kg
9	Fluorine		1.3 mg/kg	c	1.3 mg/kg
10	Neon		7.0 nmol/kg (10 m)	8.0 nmol/kg (1086 m)	8 nmol/kg (est)
11	Sodium		10.781 g/kg	c	10.781 g/kg
12	Magnesium		1.284 g/kg	c	1.28 g/kg
13	Aluminum		1.0 μg/kg (0 m)	0.6* μg/kg (1000 m)	1 μg/kg
14	Silicon	Silicate	2.0 μmol/kg (22 m)	100.8 μmol/kg (891 m)	110 μmol/kg
15	Phosphorous	Reactive phosphate	0.08 μmol/kg (22 m)	2.84 μmol/kg (891 m)	2 μmol/kg
16	Sulfur	Sulfate	2.712 g/kg	c	2.712 g/kg
17	Chlorine	Chloride	19.353 g/kg	c	19.353 g/kg
18	Argon		10.0 μmol/kg (10 m)	15.6 μmol/kg (1086 m)	15.6 μmol/kg
19	Potassium		399 mg/kg	c	399 mg/kg
20	Calcium		417.6 mg/kg (30 m)	413.6 mg/kg (1000 m)	415 mg/kg (C.A.)
					412 mg/kg (Ca/CL)
21	Scandium		0.4–1.0 ng/kg (all depths)		<1 ng/kg
22	Titanium		0.9 μg/kg		<1 ng/kg
23	Vanadium		1.0 μg/kg (0 m)	1.2* μg/kg (1000 m)	<1 μg/kg

TABLE 2.9 *(Continued)*

Atomic Number	Element	Species	Surface Concentration (Sample Depth)	Deep Water Concentration (Sample Depth)	Predicted Mean Water Concentration
24	Chromium	Cr(tot)	268 ng/kg (0 m)	295 ng/kg (1000 m)	330 ng/kg (Si) 330 ng/kg (Si, PO$_4$) 350 ng/kg (Si, NO$_3$)
25	Manganese	Dissolved Mn	34 ng/kg (<1 m)	38 ng/kg (985 m)	10 ng/kg
26	Iron		8 ng/kg (0 m)	45 ng/kg	40 ng/kg
27	Cobalt		7 ng/kg (0.2 m)	2 ng/kg (980 m)	2 ng/kg
28	Nickel		146 ng/kg (0 m)	566 ng/kg (985 m)	480 ng/kg
29	Copper		34 ng/kg (0 m)	130 ng/kg (985 m)	120 ng/kg
30	Zinc		6–7 ng/kg (0 m)	438 ng/kg (985 m)	390 ng/kg
31	Gallium		15–30 ng/kg		10–20 ng/kg
32	Germanium		<0.5 ng/kg (0 m)	6 ng/kg (1000 m)	5 ng/kg
33	Arsenic	As(V)	1.1 µg/kg (0 m)	1.8 µg/kg (1000 m)	2 µg/kg
		Dimethylarsenate	0.1 µg/kg (1 m)		
34	Selenium	[Se(tot)]	27 ng/kg (1 m)	85* ng/kg (732 m)	170 ng/kg
		[Se IV]	<2 ng/kg (1 m)	20 ng/kg (732 m)	—
		[Se VI]	25 ng/kg (1 m)	65 ng/kg (732 m)	—
35	Bromine	Bromide	67 mg/kg	c	67 mg/kg
36	Krypton		2.0 nmol/kg (10 m)	3.7 nmol/kg (989 m)	3.7 nmol/kg
37	Rubidium		124 µg/kg (10 m)	c	124 µg/kg
38	Strontium		7.404 mg/kg (10 m)	7.720 mg/kg (700 m)	7.8 mg/kg (PO$_4$) 7.7 mg/kg (Sr/CL)
39	Yttrium			13 ng/kg ('deep-water')	13 ng/kg
40	Zirconium		~ 1 µg/kg (0 m)		<1 µg/kg
41	Niobium		<10 ng/kg (0 m)		1 ng/kg
42	Molybdenum		11 µg/kg (0 m)	c	11 µg/kg
44	Ruthenium		0.7 ng/kg (10 m)	0.5* ng/kg (880 m)	0.5 ng/kg
45	Rhodium				
46	Palladium				

#	Element			
47	Silver	0.3 ng/kg (0 m)	3* ng/kg (1100 m)	3 ng/kg
48	Cadmium	0.3 ng/kg (2 m)	117 ng/kg (985 m)	70 ng/kg
49	Indium	1.0 ng/kg (0 m)	0.1* ng/kg (1000 m)	0.2 ng/kg
50	Tin	0.2 μg/kg		0.5 ng/kg
51	Antimony		0.2 μg/kg	0.2 μg/kg
52	Tellurium			
53	Iodine	48 μg/kg (2 m)	60* μg/kg (1036 m)	59 μg/kg (PO$_4$ corr.)
				60 μg/kg (NO$_3$ corr.)
54	Xenon	0.28 nmol/kg (10 m)		0.5 nmol/kg
55	Cesium	0.3 μg/kg (0 m)	c	0.3 ng/kg
56	Barium	4.8 μg/kg (13 m)	13.3 μg/kg (997 m)	11.7 μg/kg
57	Lanthanum	1.80 ng/kg (100 m)	3.50* ng/kg (700 m)	4 ng/kg
58	Cerium	2.35 ng/kg (100 m)	3.46* ng/kg (700 m)	4 ng/kg
59	Praeseodymium		0.6* ng/kg (deep)	0.6 ng/kg
60	Neodymium	1.85 ng/kg (100 m)	3.04* ng/kg (700 m)	4 ng/kg
61	Promethium			
62	Samarium	0.40 ng/kg (100 m)	0.65* ng/kg (900 m)	0.6 ng/kg
63	Europeum	0.10 ng/kg (100 m)	0.12* ng/kg (900 m)	0.1 ng/kg
64	Gadolinium	0.54 ng/kg (100 m)	0.82* ng/kg (900 m)	0.8 ng/kg
65	Terbium		0.1* ng/kg (deep)	0.1 ng/kg
66	Dysprosium	0.78 ng/kg (100 m)	0.91 ng/kg (900 m)	1 ng/kg
67	Holmium		0.2* ng/kg (deep)	0.2 ng/kg
68	Erbium	0.8 ng/kg (100 m)	0.83* ng/kg (900 m)	0.9 ng/kg
69	Thulium		0.2* ng/kg (deep)	0.2 ng/kg
70	Ytterbium	0.61 ng/kg (100 m)	0.81* ng/kg (00 m)	0.9 ng/kg
71	Lutetium		0.2* ng/kg (deep)	0.2 ng/kg
72	Hafnium	<8 ng/kg		<8 ng/kg
73	Tantalum	<2.5 ng/kg		<2.5 ng/kg
74	Tungsten	0.1 μg/kg		<1 ng/kg
75	Rhenium	5.6 ng/kg (0 m)	4.3* ng/kg (900 m)	4 ng/kg
76	Osmium			
77	Iridium	0.1		
78	Platinum			

TABLE 2.9 (*Continued*)

Atomic Number	Element	Species	Surface Concentration (Sample Depth)	Deep Water Concentration (Sample Depth)	Predicted Mean Water Concentration
79	Gold		4–27 ng/kg (all depths)		11 ng/kg
80	Mercury		3 ng/kg (above thermocline)	4* ng/kg (1000 m)	6 ng/kg
81	Thallium		11–13 ng/kg (0 m)		12 ng/kg
82	Lead		13.6 ng/kg (0.2 m)	4.5 ng/kg (1000 m)	1 ng/kg
83	Bismuth		20 ng/kg (0 m)		10 ng/kg
84	Polonium				
86	Radon				
88	Radium				
89	Actinium				
90	Thorium		<0.7 ng/kg		<0.7 ng/kg
91	Proactinium				
92	Uranium		3.2 µg/kg	c	3.2 µg/kg

These data give a general range of surface and deep water concentrations from different oceans. Where possible, deep water data are from the Pacific; *
denotes Atlantic deep water data. In addition, it may be important to see actual profiles of concentration vs. depth, such as those referenced in this table.
[Reprinted with permission, Quinby–Hunt, M. S. and Turekian, K. K., "Distribution of Elements in Seawater," *EOS*, Vol. 64, No. 14, 1983.]

man, variations in the composition of open ocean seawater arises principally from the removal of elements by organisms in the upper ocean, and subsequent destruction of organisms and organism-produced particles as they move downward [Broecker (1974)]. Surface waters generally are depleted in these biologically active elements, and deep waters have higher concentrations. This process is well known in plant nutrient elements, nitrogen, phosphorous, and silicon. Some trace elements, e.g., Sr and Cd, appear to be biologically active in that they follow a pattern similar to the nutrient elements [Quinby–Hunt and Turekian (1983)]. Calcium is a major element which varies slightly ($\sim 1\%$) due to removal at the surface to form shells and other hard parts of organisms which sink and undergo dissolution with depth. Marine organisms are constantly removing inorganic and organic chemicals from seawater and releasing others into it. Regions where deeper waters resupply the surface with nutrients and other elements are biologically productive, given adequate sunlight, temperature, and seed populations of marine algae. Dynamically stratified regions, in which deep waters are not upwelled or mixed to the surface layers, are nutrient depleted and low in biological production.

The dissolved gases have a more complex distribution [Quinby–Hunt and Turekian (1983)]. Initially, the levels are determined by the solubility of the atmospheric gases (Henry's law) and by bubble trapping at the surface. In principle, the concentration of each gas could be determined by the temperature and salinity of water in contact with the atmosphere and the relative abundance of atmospheric gases, with corrections for bubble trapping. Supersaturation can occur at the surface by turbulent mixing and with heating of the water as it sinks and moves away from a cold surface site. However, greater changes occur for biologically active gases as a result of metabolism. Oxygen is produced by photosynthesis in surface waters and removed by respiration of plants and animals, and by microbial and chemical oxidation of organic matter. Deep and bottom waters are also relatively high in oxygen as a result of dense, cold, oxygen-bearing waters which sink in high latitudes and advect at depth to lower latitudes. An intermediate oxygen minimum layer develops in a balance between biological consumption and the net of advection and diffusion. A very low oxygen minimum exists over a large ocean region of the eastern subequatorial Pacific. In areas where circulation is poor, such as deep trenches, oxygen will disappear, and sulfate (a major element species) will diminish as it is the next available oxidant. CO_2, N_2, N_2O, H_2S, H_2, and CO are also affected by biological activity.

Quinby–Hunt and Turekian (1983) point to the impact of man as seen by the presence of bomb produced nuclides such as ^{90}Sr, ^{137}Cs, ^{3}H, (bomb) ^{14}C, and plutonium. Lead, Pb, is another element which has been shown to owe its distribution to inputs by man. The distribution of these nuclides is controlled by supply from the atmosphere and coastal sources, thus generally showing decreasing concentration with depth. Fluxes of some nuclides from the ocean boundaries can modify the distribution pattern of these nuclides. One example is the degassing of ^{3}H from the earth's interior at oceanic sea floor spreading centers. Removal processes, such as precipitation, are also important.

2.2.2 Comparison with Pure Water; Similarities and Differences

For many practical purposes, representing seawater by an aqueous 0.5 M NaCl solution might be permissible [Horne (1969)]. Table 2.10 compares a number of prop-

TABLE 2.10   Comparison of Pure Water and Seawater Properties

Property	Seawater, 35‰ S	Aqueous NaCl Solution, 0.50 M	Aqueous NaCl Solution, 0.6 M	Pure Water
Density, g/cm³, 25 C	1.02412	1.10752	1.02172	1.0029
Equivalent conductivity, 25 C, cm² ohm⁻¹ equiv⁻¹	—	93.62	91.58	—
Specific conductivity, 25 C, ohm⁻¹ cm⁻¹	0.0532	0.0468	0.0458	—
Viscosity, 25 C, millipoise	9.02	9.32	9.41	8.90
Vapor pressure, mm Hg at 20 C	17.4	17.27	17.18	17.34
Isothermal compressibility, 0 C, unit vol/atm	46.4×10^{-6}	46.6×10^{-6}	45.9×10^{-6}	50.3×10^{-6}
Temperature of maximum density, C	−3.52	—	—	+3.98
Freezing point, C	−1.91	−1.72	−2.04	0.00
Surface tension, 25 C, dyne/cm	72.74	72.79	72.95	71.97
Velocity of sound, 0 C, m/sec	1450	—	—	1407
Specific heat, 17.5 C, joule/g C	3.898	4.019	3.998	4.182

Reprinted with permission from Horne, R. A., "Comparison of Pure Water and Seawater Properties," *Marine Chemistry*, 1969. See also Millero (1974).

erties of typical seawater (salinity, $S = 35$) with pure water and an aqueous NaCl solution. However, there are notable exceptions, e.g., electrical conductivity, freezing point depression, and sound absorption. The exceptions arise from the fact that seawater is not a dilute but a moderately concentrated electrolytic solution (i.e., not an *ideal solution*), that it contains not a simple 1:1 electrolyte but rather a mix of electrolytes some of which alter the structure of the system. In addition, there are various materials of terrestrial origin, living and dead organisms, and their products.

Dissolved organic compounds may influence the state of inorganic substances in seawater and *vice versa* [Williams (1975)]. However, the organic content of seawater lies in the range 0.5–2.0 mg/l. The present knowledge of organic species and their concentrations is very sparse because of the great difficulty of measuring them. On the average, every gram of dissolved organic matter is dwarfed by 35,000 grams of salts in 900,000 grams of water. Since very few direct, nondestructive techniques with sufficient selectivity, sensitivity, and accuracy exist, the problem is to remove the relatively enormous amount of salts while concentrating the minute amount of organic matter to measurable quantities. Dissolved organic material in seawater is being studied, but despite the increasing effort, many aspects of the subject are still in a rudimentary state [Williams (1975)].

In recent years, many fundamental thermodynamic properties of seawater have been carefully measured by Millero (1982), (1983), and (1992) and others. However, because the system is not ideal, it is multicomponent, and it is rarely at equilibrium. In many cases it is necessary to develop empirical equations.

2.2.3 Equation of State for Seawater

The *Equation of State for Seawater, 1980* is the mathematical expression to calculate density from measurements of temperature, pressure, and salinity, or parameters dependent upon these [UNESCO (1981)]. This equation is based on measurements of the density of standard seawater solutions obtained by weight dilution with distilled water and by evaporation. As the absolute density of pure water is not known with enough accuracy, the density of distilled water used for these measurements was determined from the equation of the *Standard Mean Ocean Water* (SMOW) whose isotopic composition is well defined. To determine absolute densities from the relative density equation, it is necessary to know the density of SMOW [Millero and Poisson (1981)].

The One Atmosphere International Equation of State of Seawater, 1980 [UNESCO (1981)]. The density (ρ, kg/m^{-3}) of seawater at one standard atmosphere ($p = 0$) is to be computed from the Practical Salinity (S) and the temperature (T, C) with the following equation

$$\rho(S, T, 0) = \rho_w + (8.244.93 \times 10^{-1} - 4.0899 \times 10^{-3}T$$
$$+ 7.6438 \times 10^{-5}T^2 - 8.2467 \times 10^{-7}T^3 + 5.3875 \times 10^{-9}T^4)S$$
$$+ (-5.72466 \times 10^{-3} + 1.0227 \times 10^{-4}T$$
$$- 1.6546 \times 10^{-6}T^2)S^{3/2} + 4.8314 \times 10^{-4}S^2 \qquad (2.1)$$

where ρ_w, the density of the Standard Mean Ocean Water (SMOW) taken as pure water reference, is given by

$$\rho_w = 999.842594 + 6.793952 \times 10^{-2}T - 9.095290 \times 10^{-3}T^2$$
$$+ 1.001685 \times 10^{-4}T^3 - 1.120083 \times 10^{-6}T^4$$
$$+ 6.536332 \times 10^{-9}T^5 \qquad (2.2)$$

The One Atmosphere International Equation of State of Seawater, 1980 is valid for Practical Salinity from 0–42 and temperature from -2–40 C.

The High Pressure International Equation of State of Seawater, 1980 [UNESCO (1981)]. The density (ρ, kg/m^{-3}) of seawater at high pressure is to be computed from the Practical Salinity (S), the temperature (T, C) and the applied pressure (p, bars) with the following equation

$$\rho(S, T, p) = \frac{\rho(S, T, 0)}{1 - p/K(S, T, p)} \qquad (2.3)$$

where $\rho(S, T, 0)$ is the One Atmosphere International Equation of State, 1980, given above, and $K(S, T, p)$ is the secant bulk modulus given by

$$K(S, T, p) = K(S, T, 0) + Ap + Bp^2 \qquad (2.4)$$

where

$$K(S, T, 0) = K_w + (54.6746 - 0.603459T + 1.09987 \times 10^{-2}T^2$$
$$- 6.1670 \times 10^{-5}T^3)S + (7.944 \times 10^{-2} + 1.6483 \times 10^{-2}T$$
$$- 5.3009 \times 10^{-4}T^2)\, S^{3/2}, \qquad (2.5)$$
$$A = A_w + (2.2838 \times 10^{-3} - 1.0981 \times 10^{-5}T - 1.6078 \times 10^{-6}T^2)S$$
$$+ 1.91075 \times 10^{-4}S^{3/2} \qquad (2.6)$$
$$B = B_w + (-9.9348 \times 10^{-7} + 2.0816 \times 10^{-8}T$$
$$+ 9.1697 \times 10^{-10}T^2)S \qquad (2.7)$$

the pure water terms K_w, A_w, and B_w of the secant bulk modulus are given by

$$K_w = 19652.21 + 148.4206T - 2.327105T^2 + 1.360477 \times 10^{-2}T^3$$
$$- 5.155288 \times 10^{-5}T^4 \qquad (2.8)$$
$$A_w = 3.239908 + 1.43713 \times 10^{-3}T + 1.16092 \times 10^{-4}T^2$$
$$- 5.77905 \times 10^{-7}T^3 \qquad (2.9)$$
$$B_w = 8.50935 \times 10^{-5} - 6.12293 \times 10^{-6}T + 5.2787 \times 10^{-8}T^2 \qquad (2.10)$$

The High Pressure International Equation of State of Seawater, 1980, is valid for Practical Salinity from 0–42, temperature from -2–40 C and applied pressure from 0–1000 bars.

Practical Salinity. Practical Salinity (S) is defined in terms of the ratio K_{15} of the electrical conductivity of seawater (under a pressure of one standard atmosphere) to that of a KCl solution having exactly a concentration of 32.43356 g KCl/kg of solution (in vacuum), both samples being at 15 C. Practical Salinity can be computed from the equation

$$S = .0080 - 0.1692K_{15}^{1/2} + 25.3851K_{15} + 14.0941K_{15}^{3/2}$$
$$- 7.0261K_{15}^2 + 2.7081K_{15}^{5/2} \qquad (2.11)$$

This equation can be used directly to give salinities from conductivity ratios measured by salinometers thermostated at 15 C. This equation may also be used to compute an uncorrected salinity from the ratio R_t measured by a salinometer at any temperature between -2 and $+35$ C. The uncorrected salinity obtained is then corrected by addition of

$$\hat{S} = \frac{(T - 15)}{1 + 0.0162(T - 15)} (0.0005 - 0.0056R_t^{1/2} - 0.0066R_t$$
$$- 0.0375R_t^{3/2} + 0.0636R_t^2 - 0.0144R_t^{5/2}) \tag{2.12}$$

to give the Practical Salinity to 0.001. The principle of calibrating standard seawater in electrical conductivity with a potassium chloride solution is given in UNESCO (1981).

Algorithms and sample FORTRAN subprograms are given in an UNESCO paper [Fofonoff and Millard (1983)] for the conversion of conductivity to salinity, salinity to conductivity, and algorithms and subprograms for density, specific volume, and pressure to depth conversion.

2.2.4 Thermodynamic Properties of Seawater

Density. For density, see Sec. 2.2.3 and Table 2.10.

Specific Heat. The UNESCO Division of Marine Sciences published the extensive "Algorithms for Computation of Fundamental Properties of Seawater" [Fofonoff and Millard (1983)]. The following selected text on specific heat and freezing point temperature is for the most part a direct quote. For additional information, including adiabatic lapse rate and potential temperature, consult that report which includes a sample FORTRAN subprogram for each formula, together with a summary of typical values.

Specific heat of seawater c_p (J/kg C), defined to be the heat in Joules required to raise the temperature of one kilogram of seawater one degree Celsius at constant pressure, is a function of salinity, S, temperature, T, and pressure, p. For seawater of oceanic salinities, the specific heat increases with temperature and decreases with salinity and pressure.

Millero, Perron, and Desnoyers (1973) and Millero and Leung (1976) measured c_p for Standard Seawater, diluted with pure water or concentrated by evaporation, over the chlorinity range of 0–22 (salinity 0.4–40) and temperature range 5–35 C (International Practical Temperature Scale-68, ITPS-68), relative to pure water with a precision of 0.0001 J/g C.

For c_p computation, specific heats are fitted to an equation (S.D. 0.0003 J/g C). The empirical formula given by Millero *et al.* (1973) has been selected by UNESCO because it is in agreement with the earlier work of Bromley *et al.* (1967) and because of their use of Standard Seawater for the measurements.

Specific heat values differ slightly on the International Practical Temperature Scale-48 (IPTS-48) and IPTS-68 temperature scales because 1 C intervals on the two scales are different,

$$dT_{48}/dT_{68} = 1.00044 \quad \text{at 0 C} \tag{2.13}$$

The differences introduced by correcting for the temperature scale change are comparable to the accuracy of determination of c_p, and may be neglected.

The formula for c_p, converted to SI units and salinity, is

$$c_p(S, T, 0) = c_p(0, T, 0) + AS + BS^{3/2} \tag{2.14}$$

where

$$c_p(0, T, 0) = C_0 + C_1T + C_2T^2 + C_3T^3 + C_4T^4 \tag{2.15}$$

$$A = a_0 + a_1T + a_2T^2 \tag{2.16}$$

$$B = b_0 + b_1T + b_2T^2 \tag{2.17}$$

$C_0 = +4217.4$	$a_0 = -7.643575$	$b_0 = +0.1770383$
$C_1 = -3.720283$	$a_1 = +0.1072763$	$b_1 = -4.07718$ E-3
$C_2 = +0.1412855$	$a_2 = -1.38385$ E-3	$b_2 = +5.148$ E-5
$C_3 = -2.654387$ E-3		
$C_4 = 2.093236$ E-5		

Range of Validity: $S = 0$–40; $T = 0$–35 C

Check: $c_p = 3980.051$ J/(kg°) for $S = 40$, $T = 40$ C

Std. Dev. = 0.5 J/(kg C)

Specific Heat of Seawater c_p: Pressure Dependence. Direct measurements of the specific heat of seawater are not available. The pressure dependence is computed from the thermodynamic equation

$$\frac{\partial c_p}{\partial p} = -T_K \frac{\partial^2 V}{\partial T^2} \tag{2.18}$$

where V is specific volume (m³/kg), T_K absolute temperature (K), and p pressure (P_a). For pressure in bars, the equation can be integrated to yield

$$c_p(S, T, p) = c_p(S, T, 0) - 10^5 \int_0^p (T + 273.15) \frac{\partial^2 V}{\partial T^2} \, dp \tag{2.19}$$

Because the pressure dependence is not economically evaluated from the exact integrals, a polynomial expression has been fitted to a table of values generated from the exact formulas. These least squares polynomials combined with the polynomials for c_p (S, T, 0) provide specific heat estimates over the full range of salinity, temperature, and pressure.

Specific heat of seawater c_p

$$c_p(S, T, p) = c_p(S, T, 0) + \Delta_1 c_p(0, T, p) + \Delta_2 c_p(S, T, p) \tag{2.20}$$

Polynomials for $S = 0$

$$\Delta_1 c_p(0, T, p) = (a_0 + a_1T + a_2T^2 + a_3T^3 + a_4T^4) \, p$$
$$+ (b_0 + b_1T + b_2T^2 + b_3T^3 + b_4T^4) \, p^2$$
$$+ (c_0 + c_1T + c_2T^2 + c_3T^3) \, p^3 \tag{2.21}$$

$$a_0 = -4.9592 \text{ E-1} \quad b_0 = +2.4931 \text{ E-4} \quad c_0 = -5.422 \text{ E-8}$$

$$a_1 = +1.45747 \text{ E-2} \quad b_1 = -1.08645 \text{ E-5} \quad c_1 = +2.6380 \text{ E-9}$$

$$a_2 = -3.13885 \text{ E-4} \quad b_2 = +2.87533 \text{ E-7} \quad c_2 = -6.5637 \text{ E-11}$$

$$a_3 = +2.0357 \text{ E-6} \quad b_3 = -4.0027 \text{ E-9} \quad c_3 = +6.136 \text{ E-13}$$

$$a_4 = +1.7168 \text{ E-8} \quad b_4 = +2.2956 \text{ E-11}$$

CHECK VALUES

$$c_p(S, T, 0) = 3980.051 \text{ J/(kg C)}$$

$$\Delta_1 c_p(0, T, P) = -177.985 \text{ J/(kg C)}$$

$$p = 10{,}000 \text{ decibars, } T = 40 \text{ C } S = 40$$

Standard Deviation: $\Delta_1 c_p = 0.074$ J/(kg C)

Polynomials for $S > 0$

$$
\begin{aligned}
\Delta_2 c_p(S, T, P) = {} & [(d_0 + d_1 T + d_2 T^2 + d_3 T^3 + d_4 T^4) \, S \\
& + (e_0 + e_1 T + e_2 T^2) \, S^{3/2}] \, p \\
& + [(f_0 + f_1 T + f_2 T^2 + f_3 T^3) \, S \\
& + (g_0) \, S^{3/2}] \, p^2 \\
& + [(h_0 + h_1 T + h_2 T^2) \, S \\
& + J_1 T \, S^{3/2}] \, p^3
\end{aligned}
\tag{2.22}
$$

$$d_0 = +4.9247 \text{ E-3} \qquad e_1 = -1.517 \text{ E-6}$$

$$d_1 = -1.28315 \text{ E-4} \qquad e_2 = +3.122 \text{ E-8}$$

$$d_2 = +9.802 \text{ E-7} \qquad f_0 = -2.9558 \text{ E-6}$$

$$d_3 = +2.5941 \text{ E-8} \qquad f_1 = +1.17054 \text{ E-7}$$

$$d_4 = -2.9179 \text{ E-10} \qquad f_2 = +-2.3905 \text{ E-9}$$

$$e_0 = -1.2331 \text{ E-4} \qquad f_3 = +1.8448 \text{ E-11}$$

$$g_0 = +9.971 \text{ E-8}$$

$$h_0 = +5.540 \text{ E-10} \qquad h_1 = -1.7682 \text{ E-11}$$

$$h_2 = +3.513 \text{ E-13} \qquad J_1 = -1.4300 \text{ E-12}$$

CHECK VALUES

$$\Delta_2 c_p(S,\, T,\, p) = 47.433 \text{ J/(kg C)}$$

$$c_p(S,\, T,\, p) = 3849.500 \text{ J/(kg C)}$$

$$p = 10{,}000 \text{ decibars},\ T = 40 \text{ C } S = 40$$

$$\text{Standard Deviation: } \Delta_2 c_p = .062 \text{ J/(kg C)}$$

The effects of pressure on heat capacity are discussed theoretically by Leyendekkers (1980).

Freezing Point. The freezing point temperature T_f is given as a function of salinity, S, and pressure, p, by [Fofonoff and Millard (1983)]

$$T_f = a_0 S + a_1 S^{3/2} + a_2 S^2 + bp \qquad (2.23)$$

where

$$a_0 = -0.0575$$

$$a_1 = +1.710523 \quad \text{E-3}$$

$$a_2 = -2.154996 \quad \text{E-4}$$

$$b = -7.53 \quad \text{E-4}$$

Check Value: $T_f = -2.588567$ C for $S = 40$, $p = 500$ decibars.

The formula is valid in the range 4–40 Practical Salinity at atmospheric pressure. Measurements at elevated pressures showed no significant dependence of the pressure coefficient on salinity in the range $S = 27$–35. The estimated error to pressure of 500 decibars is 0.003 C.

Adiabatic Lapse Rate. For adiabatic lapse rate see Fofonoff and Millard (1983).

Potential Temperature. For potential temperature see Fofonoff and Millard (1983).

2.2.5 . Transport Properties of Seawater

Viscosity. *Dynamic viscosity* of seawater is a measure of the internal fluid friction or the forces of drag which its molecular and ionic constituents exert on one another. It is independent of the state of motion and is a characteristic property of the fluid comparable to the elasticity of a solid body. A second and fundamentally different quantity is the *eddy viscosity* which depends upon the state of motion and is not a characteristic physical property of the fluid. The numerical value of the *eddy viscosity* varies within very wide limits according to the type of motion, and, as far as ocean currents are concerned, only the order of magnitude has been ascertained [Sverdrup *et al.* (1942)].

The determination of absolute dynamic viscosity is difficult, but the measurement of relative viscosity, the ratio of the viscosity of a liquid to that of some standard liquid such as water, is simple and adequate for most purposes. The viscosity of seawater at various salinities and temperatures can be computed from values for distilled water using Millero's equation [Millero (1974)].

Viscosity of pure water μ_T at temperature T C is given by

$$\log \mu_T/\mu_{20} = \frac{1.1709(20 - T) - 0.001827(T - 20)^2}{T + 89.93} \qquad (2.24)$$

where μ_{20} is the viscosity at 20 C.

Viscosity of seawater μ is calculated from

$$\mu/\mu^\circ = 1 + ACl_v^{1/2} + BCl_v \qquad (2.25)$$

where μ° is the viscosity of pure water, Cl_v is the volume chlorinity (Cl_v = Chlorinity $\times$ density), and $A = 0.000366, 0.0014303$ and $B = 0.002756, 0.003416$ at 5 and 25 C; constants at other temperatures are obtained by linear interpolation or extrapolation.

The change in dynamic viscosity ($\hat{\mu}$ centipoises) produced by an increase in pressure (p, kg/cm^{-2}) at temperature, (T C) can be calculated from the expression

$$\hat{\mu}_p = -1.7913 \times 10^{-4}p + 9.5182 \times 10^{-8}p^2 + p(1.3550 \times 10^{-5}T$$

$$- 2.5853 \times 10^{-7}T^2) - p^2(6.0833 \times 10^{-9}T - 1.1652 \times 10^{-10}T^2)$$

$$(2.26)$$

Chlorinity, Cl, can be obtained from the relationship between salinity and chlorinity

$$S = 1.80655 \, Cl \qquad (2.27)$$

This conversion factor is based on the concept of constancy of the composition of seawater. The discrepancy between the defined and the chemically determined chlorinity (chloride, bromide, and iodide content in one kg of seawater) is negligible in *normal* oceanic water. (Prior to 1978, when salinity was redefined as Practical Salinity, S and Cl were written $S‰$ and $Cl‰$ signifying units of parts per thousand or g/kg. S and Cl are now taken to mean kg $\times 10^3$ kg and, therefore, are dimensionless, e.g., 0.035001 kg salt/kg seawater is equivalent to Practical Salinity, $S = 35.001$.)

Transport Phenomena in Pure Water and Seawater. For transport phenomena see Table 2.11.

2.2.6 Acoustics

Speed of Sound in Seawater. The empirical equation for the speed of sound, $U = U(S, T, P)$, which is given here is due to Chen and Millero (1977). It is based on observations of Wilson (1959), Chen and Millero (1976), and Millero and Kubinski (1975). Over the range $5 \le S \le 40$ g/kg for salinity, $0 \le T \le 30$ C for temper-

TABLE 2.11 Comparison of Transport Phenomena in Pure Water and Seawater at One Atmosphere

Name, Symbol, Units	Pure Water		Seawater Salinity 35/mille	
	0 C	20 C	0 C	20 C
Dynamic viscosity, μ, g cm^{-1} sec^{-1} = poise	0.01787	0.01022	0.01877	0.01075
Thermal conductivity, k, watt cm^{-1} C^{-1}	0.00566	0.00599	0.00563	0.00596
Kinematic viscosity, $\nu = \mu/\rho$, cm^2 sec^{-1}	0.01787	0.01004	0.01826	0.01049
Thermal diffusivity, $\alpha = k/c_p\rho$ cm^2 sec^{-1}	0.00134	0.00143	0.00139	0.00149
Diffusivity, D, cm^2 sec^{-1}				
NaCl	0.0000074	0.0000141	0.0000068	0.0000129
N$_2$	0.0000106	0.0000169	—	—
O$_2$		0.000021	—	—
Prandtl number, Pr $= \nu/\alpha$	13.3	7.0	13.1	7.0

Reprinted with permission from Horne, R. A., "Comparison of Pure Water and Seawater Properties," *Marine Chemistry*, 1969.

ature, and $0 \leq P \leq 1000$ bars pressure above one atmosphere the uncertainty in sound speed is generally much less than 0.5 m/s, especially within the more restricted range that is applicable to most of the world's oceans. Below $S = 5$ g/kg the maximum uncertainty is probably not more than 0.8 m/s, depending on pressure. For a discussion of these uncertainties, see Chen and Millero (1977), and for a general discussion of sound speed measurements see Mackenzie (1981). The units of U will be meters per second if the salinity is in g/kg, the temperature is in C and the pressure in bars. (See Fofonoff and Millard [1983] for a FORTRAN algorithm for the computation of sound speed based on Chen and Millero [1977].)

$$U^p(S, T, p) = AS + BS^{3/2} + CS^2 + (U^0 - U^0_{H_2O}) + U^p_{H_2O} \qquad (2.28)$$

where

$$
\begin{aligned}
A = {}& (9.4742 \times 10^{-5} - 1.2580 \times 10^{-5}T - 6.4885 \times 10^{-8}T^2 \\
& + 1.0507 \times 10^{-8}T^3 - 2.0122 \times 10^{-10}T^4)p + (-3.9064 \times 10^{-7} \\
& + 9.1041 \times 10^{-9}T - 1.6002 \times 10^{-10}T^2 + 7.988 \times 10^{-12}T^3)p^2 \\
& + (1.100 \times 10^{-10} + 6.649 \times 10^{-12}T \\
& - 3.389 \times 10^{-13}T^2)p^3 \qquad\qquad\qquad\qquad\qquad\qquad\qquad (2.29)
\end{aligned}
$$

$$B = (7.3637 \times 10^{-5} + 1.7945 \times 10^{-7}t)p \qquad (2.30)$$

$$C = -7.9836 \times 10^{-6}p \qquad (2.31)$$

$$U^0 - U^0_{H_2O} = (1.389 - 1.262 \times 10^{-2}T + 7.164 \times 10^{-5}T^2$$
$$+ 2.006 \times 10^{-6}T^3 - 3.21 \times 10^{-8}T^4)S(‰)$$
$$+ (-1.922 \times 10^{-2} - 4.42 \times 10^{-5}T)S(‰)^{3/2}$$
$$+ 1.727 \times 10^{-3}S(‰)^2 \tag{2.32}$$

$$U^p_{H_2O} = 1402.388 + 5.03711T - 5.80852 \times 10^{-2}T^2 + 3.3420 \times 10^{-4}T^3$$
$$- 1.47800 \times 10^{-6}T^4 + 3.1464 \times 10^{-9}T^5$$
$$+ (0.153563 + 6.8982 \times 10^{-4}T - 8.1788 \times 10^{-6}T^2$$
$$+ 1.3621 \times 10^{-7}T^3 - 6.1185 \times 10^{-10}T^4)p$$
$$+ (3.1260 \times 10^{-5} - 1.7107 \times 10^{-6}T + 2.5974 \times 10^{-8}T^2$$
$$- 2.5335 \times 10^{-10}T^3 + 1.0405 \times 10^{-12}T^4)p^2$$
$$+ (-9.7729 \times 10^{-9} + 3.8504 \times 10^{-10}T - 2.3643 \times 10^{-12}T^2)p^3 \tag{2.33}$$

Absorption of Sound in Seawater. The absorption of sound, α, in seawater may be considered due to contributions from boric acid and magnesium sulfate as well as from the water itself. Francois and Garrison (1982) give the following empirical equations for the absorption coefficient. Within the salinity range $25 \leq S \leq 40$ g/kg and the temperature range $-2 \leq T \leq 30$ C, that is the temperature–salinity domain found throughout the greater part of the world's oceans, and for frequencies in the range 200 Hz $\leq f \leq 1$ MHz, α is estimated to $\pm 5\%$ by the equations.

$$\begin{matrix} \text{Total} \\ \text{Absorption} \end{matrix} = \begin{matrix} \text{Boric} \\ \text{Acid} \\ \text{Contrib.} \end{matrix} + \begin{matrix} \text{MgSO}_4 \\ \text{Contrib.} \end{matrix} + \begin{matrix} \text{Pure} \\ \text{Water} \\ \text{Contrib.} \end{matrix}$$

$$\alpha = \frac{A_1 P_1 f_1 f^2}{f^2 + f_1^2} + \frac{A_2 P_2 f_2 f^2}{f^2 + f_2^2} + A_3 P_3 f^2 \ \text{dB km}^{-1} \tag{2.34}$$

for frequency f in kilohertz.

Boric Acid Contribution

$$A_1 = \frac{8.86}{c} \times 10^{(0.78\,\text{pH} - 5)}$$

$$P_1 = 1$$

$$f_1 = 2.8 \, (S/35)^{0.5} 10^{(4 - 1245/\theta)}, \ \text{kHz} \tag{2.35}$$

where c is the sound speed (m/s), given approximately by

$$c = 1412 + 3.21T + 1.19S + 0.0167D \tag{2.36}$$

T is the temperature in C, $\theta = 273 + T$, S is the salinity in g/kg, and D is the depth in meters. pH $= -\log[H^+]$, where $[H^+]$ is the hydrogen ion concentration of the water, which may be estimated according to a procedure devised by Lovett (1980).

MgSO$_4$ Contribution

$$A_2 = 21.44 \, \frac{S}{c} \, (1 + 0.025 \, T) \qquad \text{dB km}^{-1} \text{ kHz}^{-1}$$

$$P_2 = 1 - 1.37 \times 10^{-4}D + 6.2 \times 10^{-9}D^2$$

$$f_2 = \frac{8.17 \times 10^{(8 - 1990/\theta)}}{1 + 0.0018 \, (S - 35)} \qquad \text{kHz} \qquad (2.37)$$

Pure Water Contribution

for $T \le 20$ C,

$$A_3 = 4.937 \times 10^{-4} - 2.59 \times 10^{-5}T$$
$$+ 9.11 \times 10^{-7}T^2 - 1.50 \times 10^{-8}T^3 \quad \text{dB km}^{-1} \text{ kHz}^{-2}$$

$$(2.38)$$

for $T > 20$ C,

$$A_3 = 3.964 \times 10^{-4} - 1.146 \times 10^{-5}T$$
$$+ 1.45 \times 10^{-7}T^2 - 6.5 \times 10^{-10}T^3 \quad \text{dB km}^{-1} \text{ kHz}^{-2}$$

$$(2.39)$$

$$P_3 = 1 - 3.83 \times 10^{-5}D + 4.9 \times 10^{-10}D^2 \qquad (2.40)$$

2.2.7 Optical Properties of Seawater

Refractive Index. Austin and Halikas (1976) have developed a series of tables which give the optical refractive index $n = n(\lambda, S, T, p)$ for wavelengths in the range 400 $\le \lambda \le$ 643.8 nm, for salinities in the range $0 \le S \le 43$ g/kg, for temperatures in the range $0 \le T \le 30$ C and for pressures in the range $0 \le p \le 1100$ kg/cm^2 to within an absolute value of approximately 0.00003. Their work is based on observations by Mehu and Johannin–Gilles (1968) giving the dependence of n on T, S, and λ at one atmosphere, Stanley (1971) giving the dependence of n on T, p, and λ at $S = 36$ g/kg, and Tilton and Taylor (1938) giving n for pure water as a function of λ and T at one atmosphere.

The equations given below for pure water, $n = n(\lambda, T, p)$, and for seawater, $n = n(\lambda, S, T, p)$, were developed by McNeil (1980) and are based on the tables of Austin and Halikas (1976) with which they agree to within ± 0.00001 within the more restricted ranges of parameters: namely, for $n = n(\lambda, T, p)$ wavelengths in the range $470 \le \lambda \le 550$ nm, temperature in the range $0 \le T \le 30$ C, and pressure in the range $0 \le p \le 200$ kg force/cm^2; and for $n = n(\lambda, S, T, p)$ the same ranges for S and T but an expanded pressure range, $0 \le P \le 1200$ kg force/cm^2.

For pure water: $n = n(\lambda, T, p) = A + B + C + D + E$ with

$$A = 1.33792 + 1.51083 \times 10^{-4}\lambda - 5.125$$
$$\times 10^{-7}\lambda^2 + 4.1667 \times 10^{-10}\lambda^3$$
$$B = 1.605 \times 10^{-5}P - 1.556 \times 10^{-9}p^2$$
$$C = -3.8 \times 10^{-6}T - 2.55 \times 10^{-6}T^2$$
$$+ 1.33 \times 10^{-8}T^3$$
$$D = (-7.2 \times 10^{-6}T + 10^{-7}T^2)$$
$$\times (2.1 \times 10^{-2}p - 1.7 \times 10^{-5}p^2)$$
$$E = (500 - \lambda) \times 1.2 \times 10^{-9}p \qquad (2.41)$$

For seawater: $n = n(\lambda, S, T, p) = A + B + C + D + E + F$ with

$$A = 1.40392 - 1.8775 \times 10^{-4}\lambda + 1.375 \times 10^{-7}\lambda^2$$
$$B = 1.5538 \times 10^{-5}p - 1.375 \times 10^{-9}p^2 - 7.8 \times 10^{-14}p^3$$
$$C = -3.82 \times 10^{-5}T - 2 \times 10^{-6}T^2 + 1.16 \times 10^{-8}T^3$$
$$D = (-9 \times 10^{-6}T + 1.2 \times 10^{-7}T^2)$$
$$\times 10^{-2}p - 2.06 \times 10^{-6}p^2)$$
$$E = (500 - \lambda) \times 1.2 \times 10^{-9}p$$
$$F = (S - 35) (2 \times 10^{-4} - 1.1 \times 10^{-8}p - 10^{-6}T$$
$$+ 1.5 \times 10^{-8}T^2) \qquad (2.42)$$

λ = wavelength in nm, S_2 = salinity in g/kg, T = temperature in C, and p = pressure in kg force/cm^2 above one atmosphere. (p = 0 corresponds to the sea surface.)

Absorption and Scattering of Light. The loss of flux ΔF from a beam of light incident at right angles to a thin slab of homogeneous material having a thickness Δx may be described by the relation $\Delta F = -c_\lambda F/\Delta x$, where c_λ, an *inherent* property of the water [Preisendorfer (1976)], is defined as the *beam attenuation coefficient*, and the incident flux is F. Integration of the defining relation gives the exponential relation $F_x = F_0 e^{-c_\lambda x}$, in which F_x is the flux at distance x, F_0 is the flux at $x = 0$, and c_λ is the beam attenuation coefficient for the medium. The subscript λ is used here to emphasize that the beam attenuation depends strongly on wavelength. It is also to be noted, that in general, natural bodies of water, i.e., the oceans, lakes, etc., are not homogeneous or stationary: Their optical properties are functions of horizontal and vertical position in the water mass and of time. c_λ may be considered to be due to contributions due to scattering, b_λ, and absorption, a_λ: $c_\lambda = a_\lambda + b_\lambda$. a_λ is called the *absorption coefficient* and b_λ is called the *total scattering coefficient*. The *volume scattering function*, $\beta(\theta)$, represents the scattered radiant intensity, $dI(\theta)$, in direction θ per unit scattering volume, dV, divided by the incident irradiance, E:

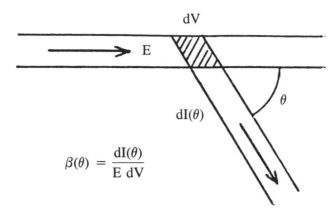

FIGURE 2.2   Definition of the volume scattering function.

$\beta(\theta) = (dI(\theta)/E\,dV)$ m^{-1}-steradian^{-1} (see Fig. 2.2). The total scattering coefficient is given by

$$b_\lambda = \iint\limits_{4\pi} \beta(\theta)\,d\Omega = 2\pi \int_0^\pi \beta(\theta)\sin\theta\,d\theta \qquad (2.43)$$

where Ω is the solid angle. In natural waters, there are generally contributions to the beam attenuation coefficient due to the presence of particles (p) and dissolved *yellow substance* (y) as well as to the water molecules themselves. Thus, dropping the subscript λ which should be taken as understood present, $c = c_w + a_y + a_p + b_p$, where c_w is in fact composed of the molecular absorption term for pure water and a molecular scattering term, and a scattering term due to dissolved matter is ignored because of its smallness with respect to the other scattering terms. The purpose of the tables and figures given below is to present representative values found in various ocean waters and to give an idea of the ranges of the parameters involved. The tables and figures do not indicate the scales of spatial and temporal variability which may be actually encountered.

Figure 2.3 [Morel (1974)] shows c_λ for pure water for $200 \le \lambda \le 2800$ nm.

Table 2.12 gives scattering and attenuation parameters, for pure water and for representative filtered seawater. Table 2.13 gives an indication of the variability of inherent properties at selected wavelengths for various ocean regions (Gordon, Smith, and Zaneveld, 1979). Figure 2.4 [Petzold (1972)] shows the volume scattering function $\beta(\theta)$ for three widely varying water types, namely, Tongue of the Ocean, Bahamas, two stations off the coast of California, and the harbor at San Diego, California, all during 1971. Figure 2.5 [Morel (1966)] illustrates the variation in shape of the volume scattering function for samples taken at various depths in the Western Mediterranean and for pure water (symmetric curve).

The *diffuse attenuation coefficient* for downwelling irradiance, K_d (e.g., attenuation of daylight in the ocean) is an *apparent* optical property [Préisendorfer (1976)], that is one which—unlike the *inherent* properties described above—depends

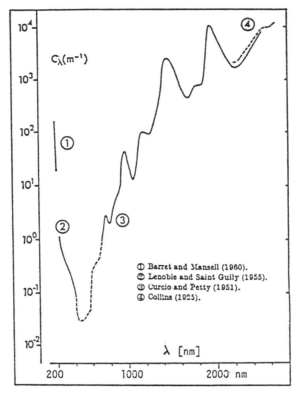

$C_\lambda (m^{-1})$

① Barret and Mansell (1960).
② Lenoble and Saint Guily (1955).
③ Curcio and Petty (1951).
④ Collins (1925).

λ [nm]

FIGURE 2.3 Attenuation curve for water between 200 and 2800 nm. (From Morel, 1974.)

on the geometric structure of the light field. In magnitude, it falls between the absorption and beam attenuation coefficients: $a < K_d < c$. For homogeneous water the downwelling irradiance for a particular wavelength at depth z may be given by $E_d(z) = E_d(0) \, e^{-K_d z}$ where $E_d(0)$ is the irradiance just beneath the surface ($z = 0$) and K_d is the irradiance attenuation coefficient. K_d forms the basis of the Jerlov (1976) optical classification scheme for ocean waters. In Table 2.14 and Fig. 2.6, the Roman numerals refer to ocean waters which are generally far from land and particle free. The progression of numbers is towards increasing turbidity, starting with I. Coastal waters are characterized using Arabic numerals, again starting with 1 and progressing toward increasing turbidity. The values given in Table 2.14 correspond very closely to values which would be observed for K_d in the various kinds of water represented and may be used for engineering purposes.

2.2.8 Electrical Properties of Seawater

Electrical Conductivity. The specific conductance of seawater at one atmosphere, $\chi_{T,S}$, is given by Poisson (1980) in units of ohm^{-1}-cm^{-1} as the following polynomial based on laboratory observations. The absolute value of the mean deviation between values calculated with this polynomial and the measured values is 3×10^{-6} ohm^{-1}-cm^{-1}.

TABLE 2.12 Scattering and Attenuation Parameters for Pure Water and Pure Seawater [from Gordon, Smith, and Zaneveld, 1979]

Wavelength	350	375	400	425	450	475	500	525	550	575	600	625	650	675	700	725	750	775	800
Pure water transmittance in %/m		95.6	95.8	96.8	98.1	98.2	96.5	96.0	93.3	91.3 / 89.7	83.3 / 75.2	79.6 / 73.7	75.0 / 70.4	69.3 / 64.5	60.7 / 52.3	29 / 17	9 / 7	9 / 7	18
Pure water attenuation coefficient in 10^{-3} m^{-1}, c		45	43	33	19	18	36	41	69	91 / 109	186 / 272	228 / 305	288 / 351	367 / 438	500 / 648	1,240 / 1,750	2,400 / 2,680	2,400 / 2,630	2,050
Pure water scattering coefficient in 10^{-3} m^{-1}, b	10.35	7.68	5.81	4.47	3.49	2.76	2.22	1.79	1.49	1.25	1.09								
Pure water volume scattering coefficient at 90° in m^{-1} sr, $b\gamma$ (90°)	.647	.480	.363	.280	.218	.173	.138	.112	.093	.078	.068								
Pure seawater scattering coefficient in 10^{-3} m^{-1}, b	13.45	9.98	7.75	5.81	4.54	3.59	2.88	2.33	1.93	1.62	1.41								
Pure seawater volume scattering coefficient at 90° in m^{-1} sr, β (90°)	.841	.624	.472	.363	.284	.225	.180	.146	.121	.101	.038								

TABLE 2.13 A Comparison of Inherent Optical Properties. Units of m^{-1} $c_{observed}$ = $c_w + a_p + b_p + a_y$ [from Gordon, Smith, and Zaneveld, 1979]

Location	Wavelength (nm)	$c-c_w$	a_p	b_p	c_p	a_y
Sargasso Sea	440	0.05		0.04		
Caribbean Sea	655	0.06	0	0.06	0.06	
	440	0.09		0.06		
Equator						
Central Pacific	440	0.09		0.06		
Romanche Deep	440	0.12		0.07		
Galapagos	655	0.11	0.04	0.07	0.11	
	440	0.24		0.08		
Off Peru (64 miles)	700	0.39			0.39	0
	400	0.73			0.64	0.09
Galapagos	700	0.16			0.16	0
	400	0.25			0.21	0.04
Northwestern	700	0.07			0.07	0
Galapagos	400	0.11			0.08	0.03
Continential slope	668	0.05				0
	365	0.10			0.08	0.02
Bermuda	655	0.10			0.10	0
	380	0.20			0.17	0.03
Kattegat	655	0.23	0.08	0.15	0.23	0
	380	0.54	0.27	0.16	0.44	0.11
South Baltic Sea	655	0.27	0.07	0.20	0.27	0
	380	1.15	0.28	0.21	0.49	0.68
Bothnian Gulf	655	0.38	0.10	0.28	0.38	0
	380	1.72	0.33	0.31	0.64	1.08
North Atlantic	665					0
	420					0.03
North Sea	665					0.01
	420					0.10
Baltic Sea	665					0.02
	420					0.33
Sargasso Sea	633	0.03	0.01	0.02	0.03	
	484	0.03	0.01			
	440	0.05	0.01	0.04		
	375			0.02		
Western Mediterranean area	655	0.08	0.04	0.04	0.08	0
	375	0.06		0.04		

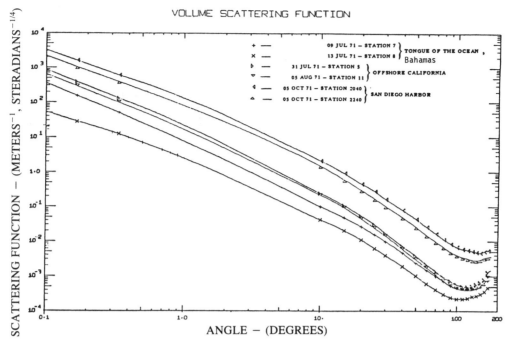

FIGURE 2.4 Volume scattering function of various water types. (From Petzold, 1972.)

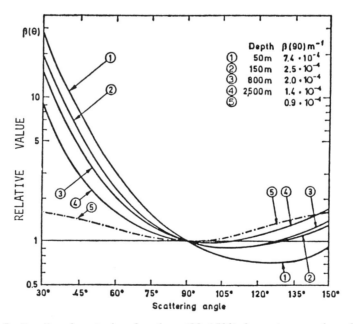

FIGURE 2.5 Family of scattering functions (30–150°) for water samples of the western Mediterranean (1–4) compared with the theoretical function for pure water (5). The curves are normalized at 90° so that they may be compared. (From Morel, 1966.)

TABLE 2.14 Downward Irradiance Attenuation Coefficient $K_d \cdot 10^2 \text{m}^{-1}$ (0–10 m) [from Jerlov, *Marine Optics*, 1976]

Water Type	Wavelength (nm)															
	310	350	375	400	425	450	475	500	525	550	575	600	625	650	675	700
I	15	6.2	3.8	2.8	2.2	1.9	1.8	2.7	4.3	6.3	8.9	23.5	30.5	36	42	56
IA	18	7.8	5.2	3.8	3.1	2.6	2.5	3.2	4.8	6.7	9.4	24	31	37	43	57
IB	22	10	6.6	5.1	4.2	3.6	3.3	4.2	5.4	7.2	9.9	24.5	31.5	37.5	43.5	58
II	37	17.5	12.2	9.6	8.1	6.8	6.2	7.0	7.6	8.9	11.5	26	33.5	40	46.5	61
III	65	32	22	18.5	16	13.5	11.6	11.5	11.6	12.0	14.8	29.5	37.5	44.5	52	66
1	180	120	80	51	36	25	17	14	13	12	15	30	37	45	51	65
3	240	170	110	78	54	39	29	22	20	19	21	33	40	46	56	71
5	350	230	160	110	78	56	43	36	31	30	33	40	48	54	65	80
7		300	210	160	120	89	71	58	49	46	46	48	54	63	78	92
9		390	300	240	190	160	123	99	78	63	58	60	65	76	92	110

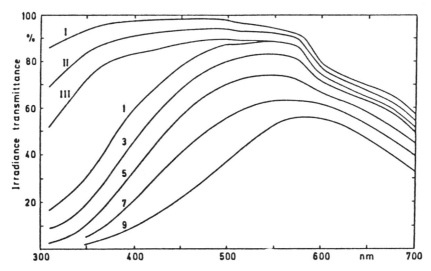

FIGURE 2.6   Transmittance per meter of downward irradiance in the surface layer for water types. Oceanic types I, II, III and coastal types 1, 3, 5, 7, 9. (Transmittance is given by $E_d(z)/E_d(0) = e^{-K_d z}$ and percent transmittance is $100 \times e^{-K_d z}$.) (From Jerlov, 1976.)

$$\chi_{T,S} = \frac{S}{35}(0.042933 R_D) + S(S - 35)(B_0 + B_1 S^{1/2} + B_2 T$$

$$+ B_3 S + B_4 S^{1/2} T + B_5 T^2 + B_6 S^{3/2} + B_7 T S$$

$$+ B_8 T^2 S^{1/2}) \tag{2.44}$$

where

$$B_0 = -8.647 \times 10^{-6} \qquad B_5 = 1.08 \times 10^{-9}$$

$$B_1 = 2.752 \times 10^{-6} \qquad B_6 = 2.61 \times 10^{-8}$$

$$B_2 = -2.70 \times 10^{-7} \qquad B_7 = -3.9 \times 10^{-9}$$

$$B_3 = -4.37 \times 10^{-7} \qquad B_8 = 1.2 \times 10^{-10}$$

$$B_4 = 5.29 \times 10^{-8} \tag{2.45}$$

$$R_D = \frac{\chi_{T,35}}{\chi_{15,35}} = 0.6765836 + 2.005294 \times 10^{-2} T$$

$$+ 1.110990 \times 10^{-4} T^2 - 7.26684 \times 10^{-7} T^3$$

$$+ 1.3587 \times 10^{-9} T^4$$

$\chi_{T,S}$ will have units of ohm^{-1}-cm^{-1} if the salinity S is in g/kg and the temperature T is in C.

For pressures above one atmosphere, the following expression due to Ribe and

Howe (1975) may be used. Its accuracy is approximately 0.007 g/kg (in terms of S assuming χ is known) for pressures from 0–7000 decibars, temperatures from 0–25 C, and for salinities between 20 and 40 g/kg. At pressures below 1000 decibars the expression may be used with temperatures up to 30 C. For salinities below 20 g/kg correction terms are given by Ribe and Howe (1975).

$$\chi_{S,T,p} = \chi_{35,15,0} \{R_D(1 + r) + (1 + Dr) [H(1 + Ar)$$
$$\cdot (35 - S)^2 + G(35 - S)]\} \qquad (2.46)$$

where r, A, G, D, and H are taken as the following [Ribe and Howe (1975)]

$$r = [(1.8993 \times 10^{-6})p - (5.71 \times 10^{-11})p^2]$$
$$\times \left[(4.498 \times 10^{-4}) \exp \frac{2651.5}{273.165 + T} + 1 \right]$$
$$A = 3.389 - (2.99 \times 10^{-2})T - (1.162 \times 10^{-3})T^2$$
$$+ (2.50 \times 10^{-5})T^3$$
$$H = -(3.67 \times 10^{-5}) - (9.683 \times 10^{-7})T$$
$$- (2.973 \times 10^{-8})T^2 + (3.59 \times 10^{-10})T^3$$
$$G = -0.017483 - (5.0058 \times 10^{-4})T$$
$$- (2.453 \times 10^{-6})T^2 + (10.05 \times 10^{-9})T^3$$
$$D = 0.731 + 0.0028T \qquad (2.47)$$

and the Poisson (1980) value for $\chi_{35,15,0} = 0.042933$ ohm^{-1}-cm^{-1} is used with his polynomial for R_D which is given above for the one-atmosphere case. S, T, and p must be in units of g/kg, C, and decibars, respectively.

Dielectric Constant of Seawater. According to the model of Klein and Swift (1977), the complex dielectric constant of seawater may be determined at any frequency within the microwave band below 5.2 GHz using the relation

$$\kappa = \kappa_\infty + \frac{\kappa_s - \kappa_\infty}{1 + (j\omega\tau)^{1-\alpha}} - j\frac{\sigma}{\omega\epsilon_0} \qquad (2.48)$$

where $\omega = 2\pi f$ is the radian frequency with f in hertz, κ_∞ is the dielectric constant at infinite frequency, κ_s is the static dielectric constant, τ is the relaxation time in seconds, σ is the conductivity in mhos/meter, α is an empirical parameter that describes the distribution of relaxation times, $j = \sqrt{-1}$, and $\epsilon_0 = 8.854 \times 10^{-12}$ farads/meter is the permittivity of free space. A choice of $\kappa_\infty = 4.9$ and $\alpha = 0$ yields values for κ with an uncertainty of about 0.3 percent if the following equations are used to evaluate κ_s, τ, and σ for temperatures and salinities in the ranges $4 \leq T \leq 30$ C and $5 \leq S \leq 36$ g/kg, respectively. T must be in degrees Celsius, and S must be in grams of salt per kilogram of seawater.

$$\kappa_s(T, S) = \kappa_s(T)a(S, T)$$

$$\kappa_s(T) = 87.134 - 1.949 \times 10^{-1}T - 1.276 \times 10^{-2}T^2 + 2.491 \times 10^{-4}T^3$$

$$a(S, T) = 1.000 + 1.613 \times 10^{-5}ST - 3.656 \times 10^{-3}S + 3.210 \times 10^{-5}S^2$$

$$- 4.232 \times 10^{-7}S^3 \tag{2.51}$$

$$\tau(T, S) = \tau(T, 0)b(S, T) \tag{2.52}$$

where $\tau(T, 0)$ is the distilled water value of the relaxation time, whose regression equation is

$$\tau(T, 0) = 1.768 \times 10^{-11} - 6.086$$
$$\times 10^{-13}T + 1.104 \times 10^{-14}T^2$$
$$- 8.111 \times 10^{-17}T^3$$

and

$$b(S, T) = 1.000 + 2.282 \times 10^{-5}ST - 7.638$$
$$\times 10^{-4}S - 7.760 \times 10^{-6}S^2$$
$$+ 1.105 \times 10^{-8}S^3.$$

$$\sigma(T, S) = \sigma(25, S) \exp(-\Delta\beta)$$

$$\Delta = 25 - T$$

$$\beta = 2.033 \times 10^{-2} + 1.266 \times 10^{-4}\Delta + 2.464 \times 10^{-6}\Delta^2$$
$$- S(1.849 \times 10^{-5} - 2.551 \times 10^{-7}\Delta + 2.551 \times 10^{-8}\Delta^2)$$

and

$$\sigma(25, S) = S(0.182521 - 1.46192 \times 10^{-3}S$$
$$+ 2.09324 \times 10^{-5}S^2 - 1.28205 \times 10^{-7}S^3)$$

2.3 AIR
Sanford Gordon

The ideal thermodynamic and transport properties of dry air presented in this section are taken from Gordon (1982a). Table 2.15 gives properties based on the assumption that the composition of air remains constant (undissociated) even at high temperatures. Table 2.16 gives properties based on the assumption that composition is in chemical equilibrium at each temperature. The assumed undissociated composition is a simplified version of the U.S. Standard Atmosphere, 1976 and consists of four components. This composition and the chemical equilibrium composition at 3000 K and 101325 Pa (1 atmosphere) are given in Table 2.17.

TABLE 2.15 Properties of Air Based on Constant Gaseous Compositions

T K	Density (P = 1.0) G/CM³	Density (P = 50.) G/CM³	H J/G	Entropy (P = .01) J/G K	(P = .10) J/G K	(P = 1.0) J/G K	(P = 10.) J/G K	(P = 50.) J/G K	CP J/G K	GAM	VS M/S	Visc. Micropoise	Cond. Micro W/CM K	Pr
200	1.7649-3	8.8247-2	-103.0	7.7810	7.1200	6.4591	5.7981	5.3361	1.0079	1.3982	283.3	133	183	0.7339
210	1.6809-3	8.4045-2	-92.9	7.8301	7.1692	6.5082	5.8473	5.3853	1.0069	1.3987	290.4	139	191	0.7321
220	1.6045-3	8.0225-2	-82.8	7.8770	7.2160	6.5550	5.8941	5.4321	1.0062	1.3992	297.3	144	199	0.7301
230	1.5347-3	7.6737-2	-72.8	7.9217	7.2607	6.5998	5.9388	5.4768	1.0055	1.3995	304.0	150	207	0.7282
240	1.4708-3	7.3539-2	-62.7	7.9644	7.3035	6.6425	5.9816	5.5196	1.0050	1.3998	310.5	155	215	0.7263
250	1.4120-3	7.0598-2	-52.7	8.0055	7.3445	6.6836	6.0226	5.5606	1.0047	1.4000	317.0	160	222	0.7245
260	1.3576-3	6.7882-2	-42.6	8.0449	7.3839	6.7230	6.0620	5.6000	1.0045	1.4001	323.3	165	230	0.7230
270	1.3074-3	6.5368-2	-32.6	8.0828	7.4218	6.7609	6.0999	5.6379	1.0044	1.4002	329.4	170	237	0.7216
280	1.2607-3	6.3034-2	-22.5	8.1193	7.4583	6.7974	6.1364	5.6744	1.0044	1.4001	335.5	175	245	0.7206
290	1.2172-3	6.0860-2	-12.5	8.1545	7.4936	6.8326	6.1717	5.7097	1.0046	1.4001	341.4	180	252	0.7197
298	1.1839-3	5.9196-2	-4.3	8.1824	7.5214	6.8605	6.1995	5.7375	1.0048	1.3999	346.1	184	257	0.7191
300	1.1766-3	5.8831-2	-2.4	8.1886	7.5277	6.8667	6.2057	5.7438	1.0048	1.3999	347.2	185	259	0.7190
310	1.1387-3	5.6934-2	7.6	8.2216	7.5606	6.8997	6.2387	5.7767	1.0052	1.3997	352.9	190	265	0.7188
320	1.1031-3	5.5154-2	17.7	8.2535	7.5925	6.9316	6.2706	5.8086	1.0057	1.3994	358.5	194	272	0.7188
330	1.0697-3	5.3483-2	27.7	8.2844	7.6535	6.9625	6.3016	5.8696	1.0063	1.3991	364.0	199	279	0.7188
340	1.0382-3	5.1910-2	37.8	8.3145	7.6535	6.9926	6.3316	5.8696	1.0071	1.3987	369.5	204	285	0.7188
350	1.0085-3	5.0427-2	47.9	8.3437	7.6827	7.0218	6.3608	5.8988	1.0079	1.3982	374.8	208	292	0.7189
360	9.8052-4	4.9026-2	57.9	8.3721	7.7112	7.0502	6.3892	5.9273	1.0088	1.3977	380.0	212	298	0.7189
370	9.5402-4	4.7701-2	68.0	8.3998	7.7388	7.0778	6.4169	5.9549	1.0098	1.3972	385.2	217	304	0.7189
380	9.2892-4	4.6446-2	78.1	8.4267	7.7657	7.1048	6.4438	5.9818	1.0109	1.3966	390.3	221	311	0.7189
390	9.0510-4	4.5255-2	88.3	8.4530	7.7920	7.1311	6.4701	6.0081	1.0120	1.3959	395.3	225	317	0.7188
400	8.8247-4	4.4124-2	98.4	8.4786	7.8177	7.1567	6.4957	6.0338	1.0133	1.3952	400.3	229	323	0.7188
410	8.6095-4	4.3047-2	108.5	8.5036	7.8427	7.1817	6.5208	6.0588	1.0146	1.3945	405.1	234	330	0.7187
420	8.4045-4	4.2022-2	118.7	8.5281	7.8672	7.2062	6.5453	6.0833	1.0161	1.3938	409.9	238	336	0.7186
430	8.2090-4	4.1045-2	128.8	8.5520	7.8911	7.2301	6.5692	6.1072	1.0175	1.3930	414.6	242	342	0.7184
440	8.0225-4	4.0112-2	139.0	8.5755	7.9145	7.2535	6.5926	6.1306	1.0191	1.3921	419.3	246	348	0.7183

TABLE 2.15 *(Continued)*

T K	Density (P = 1.0) G/CM³	(P = 50.) G/CM³	H J/G	Entropy (P = .01) J/G K	(P = .10) J/G K	(P = 1.0) J/G K	(P = 10.) J/G K	(P = 50.) J/G K	CP J/G K	GAM	VS M/S	Visc. Micro-poise	Cond. Micro W/CM K	Pr
450	7.8442-4	3.9221-2	149.2	8.5984	7.9374	7.2765	6.6155	6.1535	1.0207	1.3913	423.9	250	355	0.7182
460	7.6737-4	3.8368-2	159.4	8.6208	7.9599	7.2989	6.6380	6.1760	1.0224	1.3904	428.5	253	361	0.7181
470	7.5104-4	3.7552-2	169.7	8.6428	7.9819	7.3209	6.6600	6.1980	1.0241	1.3894	433.0	257	367	0.7180
480	7.3539-4	3.6770-2	179.9	8.6644	8.0035	7.3425	6.6815	6.2196	1.0259	1.3885	437.4	261	373	0.7179
490	7.2038-4	3.6019-2	190.2	8.6856	8.0246	7.3637	6.7027	6.2407	1.0278	1.3875	441.8	265	379	0.7178
500	7.0598-4	3.5299-2	200.5	8.7064	8.0454	7.3845	6.7235	6.2615	1.0297	1.3865	446.1	269	385	0.7178
510	6.9213-4	3.4607-2	210.8	8.7268	8.0658	7.4049	6.7439	6.2819	1.0317	1.3855	450.4	272	392	0.7178
520	6.7882-4	3.3941-2	221.1	8.7468	8.0859	7.4249	6.7640	6.3020	1.0337	1.3845	454.6	276	398	0.7178
530	6.6602-4	3.3301-2	231.5	8.7665	8.1056	7.4446	6.7837	6.3217	1.0357	1.3834	458.8	280	404	0.7178
540	6.5368-4	3.2684-2	241.8	8.7859	8.1250	7.4640	6.8031	6.3411	1.0378	1.3823	462.9	283	410	0.7179
550	6.4180-4	3.2090-2	252.2	8.8050	8.1440	7.4831	6.8221	6.3601	1.0399	1.3813	467.0	287	416	0.7179
560	6.3034-4	3.1517-2	262.6	8.8237	8.1628	7.5018	6.8409	6.3789	1.0421	1.3802	471.0	291	422	0.7179
570	6.1928-4	3.0964-2	273.1	8.8422	8.1812	7.5203	6.8593	6.3974	1.0443	1.3791	475.0	294	428	0.7179
580	6.0860-4	3.0430-2	283.5	8.8604	8.1994	7.5385	6.8775	6.4155	1.0465	1.3780	479.0	298	434	0.7179
590	5.9828-4	2.9914-2	294.0	8.8783	8.2173	7.5564	6.8954	6.4334	1.0488	1.3768	482.9	301	440	0.7179
600	5.8831-4	2.9416-2	304.5	8.8959	8.2350	7.5740	6.9131	6.4511	1.0510	1.3757	486.8	304	446	0.7179
610	5.7867-4	2.8933-2	315.0	8.9133	8.2524	7.5914	6.9305	6.4685	1.0533	1.3746	490.6	308	452	0.7178
620	5.6934-4	2.8467-2	325.6	8.9305	8.2695	7.6086	6.9476	6.4856	1.0557	1.3735	494.4	311	458	0.7177
630	5.6030-4	2.8015-2	336.1	8.9474	8.2864	7.6255	6.9645	6.5025	1.0580	1.3723	498.2	315	464	0.7176
640	5.5154-4	2.7577-2	346.7	8.9641	8.3031	7.6422	6.9812	6.5192	1.0604	1.3712	501.9	318	470	0.7175
650	5.4306-4	2.7153-2	357.3	8.9805	8.3196	7.6586	6.9977	6.5357	1.0627	1.3701	505.6	322	476	0.7173
660	5.3483-4	2.6742-2	368.0	8.9968	8.3358	7.6749	7.0139	6.5519	1.0651	1.3689	509.3	325	482	0.7172
670	5.2685-4	2.6342-2	378.6	9.0128	8.3519	7.6909	7.0299	6.5680	1.0675	1.3678	512.9	328	489	0.7171
680	5.1910-4	2.5955-2	389.3	9.0286	8.3677	7.7067	7.0458	6.5838	1.0699	1.3667	516.5	332	495	0.7169
690	5.1158-4	2.5579-2	400.0	9.0443	8.3833	7.7224	7.0614	6.5994	1.0723	1.3655	520.1	335	501	0.7168

700	0.7167	507	338	523.6	1.3644	1.0747	6.6149	7.0769	7.7378	8.3988	9.0597	410.8	2.5213−2	5.0427−4
710	0.7165	513	341	527.1	1.3633	1.0771	6.6301	7.0921	7.7531	8.4140	9.0750	421.5	2.4858−2	4.9717−4
720	0.7164	519	345	530.6	1.3622	1.0795	6.6452	7.1072	7.7682	8.4291	9.0901	432.3	2.4513−2	4.9026−4
730	0.7163	525	348	534.1	1.3611	1.0819	6.6601	7.1221	7.7831	8.4440	9.1050	443.1	2.4177−2	4.8355−4
740	0.7161	531	351	537.5	1.3600	1.0843	6.6749	7.1368	7.7978	8.4588	9.1197	453.9	2.3851−2	4.7701−4
750	0.7160	537	354	540.9	1.3589	1.0867	6.6894	7.1514	7.8124	8.4733	9.1343	464.8	2.3533−2	4.7065−4
760	0.7159	544	357	544.3	1.3579	1.0891	6.7038	7.1658	7.8268	8.4877	9.1487	475.7	2.3223−2	4.6446−4
770	0.7158	550	360	547.6	1.3568	1.0915	6.7181	7.1801	7.8410	8.5020	9.1629	486.6	2.2921−2	4.5843−4
780	0.7157	556	364	551.0	1.3558	1.0939	6.7322	7.1942	7.8551	8.5161	9.1770	497.5	2.2627−2	4.5255−4
790	0.7156	562	367	554.3	1.3547	1.0963	6.7461	7.2081	7.8691	8.5300	9.1910	508.5	2.2341−2	4.4682−4
800	0.7155	568	370	557.6	1.3537	1.0986	6.7599	7.2219	7.8829	8.5438	9.2048	519.4	2.2062−2	4.4124−4
810	0.7154	574	373	560.8	1.3527	1.1009	6.7736	7.2356	7.8966	8.5575	9.2185	530.4	2.1789−2	4.3579−4
820	0.7153	580	376	564.1	1.3517	1.1033	6.7871	7.2491	7.9101	8.5710	9.2320	541.5	2.1524−2	4.3047−4
830	0.7152	586	379	567.3	1.3507	1.1056	6.8005	7.2625	7.9235	8.5844	9.2454	552.5	2.1264−2	4.2529−4
840	0.7151	592	382	570.5	1.3497	1.1078	6.8138	7.2758	7.9367	8.5977	9.2586	563.6	2.1011−2	4.2022−4
850	0.7150	598	385	573.7	1.3488	1.1101	6.8269	7.2889	7.9498	8.6108	9.2717	574.7	2.0764−2	4.1528−4
860	0.7149	603	388	576.8	1.3478	1.1123	6.8399	7.3019	7.9628	8.6238	9.2847	585.8	2.0523−2	4.1045−4
870	0.7148	609	391	580.0	1.3469	1.1146	6.8528	7.3148	7.9757	8.6367	9.2976	596.9	2.0287−2	4.0573−4
880	0.7148	615	394	583.1	1.3460	1.1168	6.8655	7.3275	7.9885	8.6494	9.3104	608.1	2.0056−2	4.0112−4
890	0.7147	621	397	586.2	1.3451	1.1189	6.8781	7.3401	8.0011	8.6620	9.3230	619.2	1.9831−2	3.9662−4
900	0.7146	627	400	589.3	1.3442	1.1211	6.8907	7.3526	8.0136	8.6746	9.3355	630.4	1.9610−2	3.9221−4
910	0.7145	633	403	592.4	1.3433	1.1232	6.9031	7.3650	8.0260	8.6870	9.3479	641.7	1.9395−2	3.8790−4
920	0.7145	639	405	595.4	1.3424	1.1253	6.9153	7.3773	8.0383	8.6992	9.3602	652.9	1.9184−2	3.8368−4
930	0.7144	644	408	598.5	1.3416	1.1274	6.9275	7.3895	8.0505	8.7114	9.3724	664.2	1.8978−2	3.7956−4
940	0.7143	650	411	601.5	1.3408	1.1294	6.9396	7.4016	8.0625	8.7235	9.3844	675.5	1.8776−2	3.7552−4
950	0.7143	656	414	604.5	1.3400	1.1314	6.9516	7.4135	8.0745	8.7355	9.3964	686.8	1.8578−2	3.7157−4
960	0.7142	662	417	607.5	1.3392	1.1334	6.9634	7.4254	8.0864	8.7473	9.4083	698.1	1.8385−2	3.6770−4
970	0.7142	667	420	610.5	1.3384	1.1353	6.9752	7.4372	8.0981	8.7591	9.4200	709.4	1.8195−2	3.6391−4
980	0.7141	673	423	613.4	1.3376	1.1372	6.9868	7.4488	8.1098	8.7707	9.4317	720.8	1.8010−2	3.6019−4
990	0.7141	679	425	616.4	1.3369	1.1391	6.9984	7.4604	8.1213	8.7823	9.4432	732.2	1.7828−2	3.5655−4

TABLE 2.15 (*Continued*)

T K	Density (P = 1.0) G/CM³	Density (P = 50.) G/CM³	H J/G	Entropy (P = .01) J/G K	(P = .10) J/G K	(P = 1.0) J/G K	(P = 10.) J/G K	(P = 50.) J/G K	CP J/G K	GAM	VS M/S	Visc. Micropoise	Cond. Micro W/CM K	Pr
1000	3.5299-4	1.7649-2	743.6	9.4547	8.7937	8.1328	7.4718	7.0098	1.1410	1.3362	619.3	428	684	0.7140
1050	3.3618-4	1.6809-2	800.8	9.5106	8.8496	8.1887	7.5277	7.0657	1.1495	1.3328	633.8	442	711	0.7142
1100	3.2090-4	1.6045-2	858.5	9.5642	8.9033	8.2423	7.5814	7.1194	1.1576	1.3297	648.0	455	738	0.7145
1150	3.0695-4	1.5347-2	916.6	9.6159	8.9549	8.2939	7.6330	7.1710	1.1653	1.3269	661.8	468	764	0.7148
1200	2.9416-4	14708-2	975.0	9.6656	9.0046	8.3437	7.6827	7.2207	1.1726	1.3241	675.4	481	789	0.7152
1250	2.8239-4	1.4120-2	1033.8	9.7136	9.0527	8.3917	7.7308	7.2688	1.1796	1.3216	688.6	494	814	0.7156
1300	2.7153-4	1.3576-2	1093.0	9.7600	9.0991	8.4381	7.7771	7.3152	1.1863	1.3192	701.6	506	839	0.7160
1350	2.6147-4	1.3074-2	1152.5	9.8049	9.1439	8.4830	7.8220	7.3600	1.1926	1.3170	714.4	519	863	0.7165
1400	2.5213-4	1.2607-2	1212.2	9.8484	9.1874	8.5265	7.8655	7.4035	1.1987	1.3149	726.9	531	887	0.7169
1450	2.4344-4	1.2172-2	1272.3	9.8906	9.2296	8.5686	7.9077	7.4457	1.2044	1.3129	739.2	542	911	0.7172
1500	2.3533-4	1.1766-2	1332.7	9.9315	9.2705	8.6096	7.9486	7.4866	1.2099	1.3111	751.3	554	934	0.7175
1550	2.2773-4	1.1387-2	1393.3	9.9712	9.3103	8.6493	7.9884	7.5264	1.2151	1.3093	763.2	566	958	0.7175
1600	2.2062-4	1.1031-2	1454.2	10.0099	9.3489	8.6880	8.0270	7.5650	1.2200	1.3077	775.0	577	981	0.7174
1650	2.1393-4	1.0697-2	1515.3	10.0475	9.3865	8.7256	8.0646	7.6027	1.2247	1.3061	786.5	588	1004	0.7173
1700	2.0764-4	1.0382-2	1576.7	10.0841	9.4232	8.7622	8.1013	7.6393	1.2291	1.3047	797.9	599	1027	0.7172
1750	2.0171-4	1.0085-2	1638.2	10.1198	9.4589	8.7979	8.1370	7.6750	1.2334	1.3033	809.1	610	1050	0.7170
1800	1.9610-4	9.8052-3	1700.0	10.1546	9.4937	8.8327	8.1718	7.7098	1.2374	1.3020	820.2	621	1072	0.7169
1850	1.9080-4	9.5402-3	1762.0	10.1886	9.5276	8.8667	8.2057	7.7437	1.2412	1.3008	831.1	632	1094	0.7168
1900	1.8578-4	9.2892-3	1824.1	10.2217	9.5608	8.8998	8.2389	7.7769	1.2448	1.2997	841.9	643	1116	0.7167
1950	1.8102-4	9.0510-3	1886.4	10.2541	9.5932	8.9322	8.2712	7.8093	1.2483	1.2986	842.6	653	1138	0.7167
2000	1.7649-4	8.8247-3	1948.9	10.2858	9.6248	8.9638	8.3029	7.8409	1.2515	1.2976	863.1	664	1159	0.7166
2050	1.7219-4	8.6095-3	2011.6	10.3167	9.6557	8.9948	8.3338	7.8718	1.2546	1.2967	873.5	674	1180	0.7168
2100	1.6809-4	8.4045-3	2074.4	10.3470	9.6860	9.0251	8.3641	7.9021	1.2576	1.2958	883.8	684	1200	0.7171
2150	1.6418-4	8.2090-3	2137.3	10.3766	9.7156	9.0547	8.3937	7.9317	1.2604	1.2949	894.0	695	1220	0.7173
2200	1.6045-4	8.0225-3	2200.4	10.4056	9.7446	9.0837	8.4227	7.9607	1.2631	1.2941	904.0	705	1240	0.7176

2250	1.5688-4	7.8442-3	2263.6	10.4340	9.7731	9.1121	8.4511	7.9892	1.2656	1.2933	914.0	715	1260	0.7179
2300	1.5347-4	7.6737-3	2327.0	10.4619	9.8009	9.1399	8.4790	8.0170	1.2681	1.2926	923.8	725	1279	0.7183
2350	1.5021-4	7.5104-3	2390.5	10.4892	9.8282	9.1672	8.5063	8.0443	1.2704	1.2919	933.5	735	1298	0.7187
2400	1.4708-4	7.3539-3	2454.0	10.5159	9.8550	9.1940	8.5331	8.0711	1.2726	1.2913	943.2	744	1317	0.7191
2450	1.4408-4	7.2038-3	2517.7	10.5422	9.8812	9.2203	8.5593	8.0973	1.2747	1.2906	952.7	754	1336	0.7195
2500	1.4120-4	7.0598-3	2581.5	10.5680	9.9070	9.2460	8.5851	8.1231	1.2767	1.2900	962.2	764	1354	0.7200
2550	1.3843-4	6.9213-3	2645.4	10.5933	9.9323	9.2713	8.6104	8.1484	1.2787	1.2895	971.5	773	1373	0.7202
2600	1.3576-4	6.7882-3	2709.4	10.6181	9.9571	9.2962	8.6352	8.1733	1.2806	1.2889	980.8	783	1391	0.7204
2650	1.3320-4	6.6602-3	2773.4	10.6425	9.9816	9.3206	8.6596	8.1977	1.2824	1.2884	990.0	792	1410	0.7206
2700	1.3074-4	6.5368-3	2837.6	10.6665	10.0055	9.3446	8.6836	8.2216	1.2841	1.2879	999.1	802	1428	0.7208
2750	1.2836-4	6.4180-3	2901.8	10.6901	10.0291	9.3682	8.7072	8.2452	1.2858	1.2874	1008.1	811	1447	0.7210
2800	1.2607-4	6.3034-3	2966.2	10.7133	10.0523	9.3914	8.7304	8.2684	1.2874	1.2869	1017.0	820	1465	0.7211
2850	1.2386-4	6.1928-3	3030.6	10.7361	10.0751	9.4142	8.7532	8.2912	1.2890	1.2865	1025.9	830	1483	0.7212
2900	1.2172-4	6.0860-3	3095.1	10.7585	10.0975	9.4366	8.7756	8.3136	1.2905	1.2861	1034.7	839	1501	0.7213
2950	1.1966-4	5.9829-3	3159.6	10.7806	10.1196	9.4587	8.7977	8.3357	1.2920	1.2856	1043.4	848	1519	0.7213
3000	1.1766-4	5.8831-3	3224.3	10.8023	10.1413	9.4804	8.8194	8.3574	1.2935	1.2852	1052.0	857	1537	0.7213

Dry Air Only; F/A = 0; Equiv. Ratio = 0; Chem. Equiv. Ratio = 0.0015; MW = 28.9651; Gaseous Composition: N_2 = .78084; O_2 = .20948; AR = .00937; CO_2 = .00032

TABLE 2.16 Properties of Air Based on Eqiulibrium Compositions

								Reacting Compositions					Frozen Compositions				
T K	Density G/CM³	H J/G	Entropy J/G K	MW	Visc. Micro-poise	DLVDLT	DLVDLP	CP J/G K	(GAM)S	VS M/S	Cond. Micro W/CM K	Pr	CP J/G K	GAM	VS M/S	Cond. Micro W/CM K	Pr
900	3.9221-4	630.5	8.0136	28.965	400	1.0000	−1.0000	1.1215	1.3440	589.3	627	0.715	1.1211	1.3442	589.3	627	0.715
950	3.7157-4	686.8	8.0746	28.965	414	1.0000	−1.0000	1.1322	1.3397	604.4	656	0.714	1.1314	1.3400	604.5	656	0.714
1000	3.5299-4	743.7	8.1329	28.965	428	1.0000	−1.0000	1.1422	1.3357	619.2	685	0.714	1.1410	1.3362	619.3	684	0.714
1050	3.3618-4	801.0	8.1888	28.965	442	1.0000	−1.0000	1.1513	1.3321	633.6	712	0.714	1.1495	1.3328	633.8	711	0.714
1100	3.2090-4	858.8	8.2426	28.965	455	1.0000	−1.0000	1.1602	1.3287	647.7	739	0.715	1.1576	1.3297	648.0	738	0.714
1150	3.0695-4	917.0	8.2944	28.965	468	1.0000	−1.0000	1.1690	1.3255	661.5	766	0.715	1.1653	1.3269	661.8	764	0.715
1200	2.9416-4	975.7	8.3443	28.965	481	1.0000	−1.0000	1.1777	1.3223	674.9	792	0.715	1.1726	1.3241	675.4	789	0.715
1250	2.8239-4	1034.8	8.3926	28.965	494	1.0000	−1.0000	1.1864	1.3192	688.0	819	0.716	1.1796	1.3216	688.6	814	0.716
1300	2.7153-4	1094.3	8.4393	28.965	506	1.0000	−1.0000	1.1950	1.3161	700.8	845	0.716	1.1863	1.3192	701.6	839	0.716
1350	2.6147-4	1154.3	8.4845	28.965	519	1.0000	−1.0000	1.2036	1.3132	713.4	871	0.717	1.1926	1.3170	714.4	863	0.716
1400	2.5214-4	1214.7	8.5285	28.965	531	1.0000	−1.0000	1.2123	1.3102	725.6	897	0.717	1.1987	1.3149	726.9	887	0.717
1450	2.4344-4	1275.5	8.5711	28.965	542	1.0000	−1.0000	1.2211	1.3073	737.7	923	0.717	1.2044	1.3129	739.2	911	0.717
1500	2.3533-4	1336.8	8.6127	28.965	554	1.0000	−1.0000	1.2300	1.3044	749.4	950	0.718	1.2099	1.3110	751.3	934	0.718
1550	2.2773-4	1398.5	8.6532	28.965	566	1.0000	−1.0000	1.2389	1.3016	761.0	977	0.718	1.2151	1.3093	763.2	958	0.717
1600	2.2062-4	1460.7	8.6926	28.965	577	1.0001	−1.0000	1.2481	1.2987	772.3	1004	0.717	1.2200	1.3077	775.0	981	0.717
1650	2.1393-4	1523.4	8.7312	28.965	588	1.0001	−1.0000	1.2575	1.2959	783.4	1031	0.717	1.2247	1.3061	786.5	1004	0.717
1700	2.0764-4	1586.5	8.7689	28.965	599	1.0002	−1.0000	1.2671	1.2930	794.3	1059	0.717	1.2291	1.3047	797.9	1027	0.717
1750	2.0171-4	1650.1	8.8058	28.965	610	1.0003	−1.0000	1.2772	1.2901	805.0	1088	0.717	1.2334	1.3033	809.1	1050	0.717

1800	1.9610-4	1714.2	8.8419	28.964	621	1.0005	−1.0000	1.2877	1.2872	815.5	1117	0.716	1.2374	1.3021	820.2	1072	0.717
1850	1.9080-4	1778.8	8.8773	28.964	632	1.0007	−1.0000	1.2989	1.2842	825.8	1147	0.716	1.2412	1.3009	831.2	1095	0.717
1900	1.8577-4	1844.1	8.9121	28.963	643	1.0011	−1.0000	1.3110	1.2811	835.9	1179	0.715	1.2448	1.2997	842.0	1116	0.717
1950	1.8100-4	1910.0	8.9463	28.962	653	1.0016	−1.0001	1.3242	1.2778	845.8	1212	0.714	1.2482	1.2987	852.6	1138	0.717
2000	1.7647-4	1976.5	8.9800	28.961	664	1.0023	−1.0001	1.3388	1.2744	855.4	1247	0.713	1.2515	1.2977	863.2	1159	0.717
2050	1.7215-4	2043.9	9.0133	28.959	674	1.0033	−1.0001	1.3552	1.2709	864.9	1285	0.711	1.2546	1.2968	873.6	1180	0.717
2100	1.6804-4	2112.1	9.0462	28.956	685	1.0045	−1.0002	1.3738	1.2670	874.1	1326	0.710	1.2575	1.2959	884.0	1201	0.717
2150	1.6411-4	2181.3	9.0787	28.953	695	1.0062	−1.0002	1.3950	1.2630	883.0	1371	0.707	1.2603	1.2951	894.2	1221	0.717
2200	1.6035-4	2251.6	9.1111	28.948	705	1.0084	−1.0003	1.4194	1.2586	891.8	1421	0.704	1.2630	1.2944	904.4	1241	0.718
2250	1.5676-4	2323.3	9.1433	28.941	715	1.0112	−1.0004	1.4476	1.2539	900.3	1478	0.701	1.2655	1.2937	914.4	1261	0.718
2300	1.5330-4	2396.5	9.1755	28.933	725	1.0147	−1.0006	1.4802	1.2490	908.6	1542	0.697	1.2679	1.2931	924.5	1280	0.718
2350	1.4999-4	2471.4	9.2077	28.923	735	1.0190	−1.0007	1.5178	1.2437	916.6	1614	0.692	1.2702	1.2925	934.4	1300	0.719
2400	1.4680-4	2548.4	9.2401	28.910	745	1.0244	−1.0010	1.5611	1.2382	924.5	1696	0.686	1.2724	1.2920	944.4	1319	0.719
2450	1.4372-4	2627.6	9.2728	28.893	755	1.0308	−1.0012	1.6110	1.2324	932.1	1790	0.680	1.2746	1.2916	954.3	1338	0.719
2500	1.4075-4	2709.6	9.3059	28.873	765	1.0386	−1.0016	1.6679	1.2265	939.7	1897	0.673	1.2766	1.2913	964.2	1357	0.720
2550	1.3787-4	2794.6	9.3395	28.848	775	1.0478	−1.0020	1.7327	1.2205	947.1	2018	0.665	1.2785	1.2910	974.1	1376	0.720
2600	1.3508-4	2883.0	9.3739	28.819	784	1.0587	−1.0025	1.8059	1.2144	954.4	2155	0.657	1.2804	1.2909	984.0	1395	0.720
2650	1.3237-4	2975.3	9.4090	28.783	794	1.0713	−1.0031	1.8880	1.2084	961.8	2309	0.649	1.2822	1.2908	994.0	1414	0.720
2700	1.2972-4	3071.9	9.4452	28.741	804	1.0848	−1.0038	1.9796	1.2026	969.2	2482	0.641	1.2840	1.2908	1004.1	1434	0.720
2750	1.2715-4	3173.4	9.4824	28.692	814	1.1024	−1.0046	2.0808	1.1970	976.7	2673	0.633	1.2857	1.2910	1014.3	1453	0.720
2800	1.2463-4	3280.2	9.5209	28.634	823	1.1210	−1.0056	2.1917	1.1918	984.3	2883	0.626	1.2873	1.2913	1024.6	1472	0.720
2850	1.2216-4	3392.7	9.5607	28.568	833	1.1418	−1.0067	2.3122	1.1869	992.2	3112	0.619	1.2890	1.2916	1035.1	1492	0.719
2900	1.1973-4	3511.6	9.6020	28.492	842	1.1647	−1.0079	2.4417	1.1824	1000.3	3360	0.612	1.2906	1.2922	1045.7	1512	0.719
2950	1.1735-4	3637.0	9.6449	28.406	852	1.1896	−1.0092	2.5795	1.1783	1008.7	3623	0.607	1.2921	1.2929	1056.6	1532	0.719
3000	1.1500-4	3769.6	9.6895	28.309	862	1.2164	−1.0107	2.7243	1.1748	1017.4	3901	0.602	1.2937	1.2937	1067.7	1552	0.718

TABLE 2.17   Simplified Version of Composition of the U.S. Standard Atmosphere

	Mole Fractions	
Species	4—Component Air	Equilibrium Air, $T = 3000$ K, $P = 1$ atm
N_2	0.78084	0.74203
O_2	0.209476	0.16109
A_r	0.009365	0.00915
CO_2	0.000319	0.00017
CO		0.00014
N		0.00001
NO		0.04220
NO_2		0.00002
O		0.04518

Section 2.4 gives the sources of the transport property data for pure species. Section 2.7 gives the sources of the ideal thermodynamic data for the pure species, the method used to calculate equilibrium compositions, and the equations used for obtaining mixture properties.

2.4   COMMON GASES
Sanford Gordon

Tables of viscosity and thermal conductivity are given for seventeen gaseous species which occur as combustion products of hydrocarbons with air in Table 2.18. This table was taken directly from Gordon (1982a), and the original sources are detailed there. For the range of combustion conditions given, sixteen of these species appeared with a mole fraction greater than 0.000005 for at least some of the conditions [Gordon (1982b)]. These species are: Ar, CH_4, CO, CO_2, H, H_2, H_2O, N, NH_3, NO, NO_2, N_2, N_2O, O, OH, and O_2. Atomic carbon gas C was included for completeness.

The tables are given for every 100 K from 200–5000 K. Viscosity is given in units of micropoise (1 poise = 1 g/cm s), and thermal conductivity is given in units of microwatts per centimeter per unit Kelvin. These units were selected to permit the tables to be given in whole numbers.

Experimental transport data are usually available for a considerably shorter temperature range than the 200–5000 K range given in the tables. Properties were extrapolated to the higher temperatures and may be in error.

The transport property values selected are in excellent agreement with the values selected in Touloukian *et al.* (1975) and Touloukian *et al.* (1979), except for a few species. For example, for N_2, there is excellent agreement at some temperatures but as much as 11 percent difference at other temperatures. Property values for N_2 from several sources are compared in Table 2.19.

It is not surprising that there is lack of agreement at high temperatures inasmuch as

"Most of the data for the thermal conductivity coefficient for both nitrogen and oxygen seem unreliable outside the range of about 150 to 600 K" [Hanley and Fly (1973)].

2.5 LUBRICATING OILS
Alan Beerbower

2.5.1 Introduction

Types of Oils. Lubricating oils used to be classified, according to the principal by-product from the crude, as *paraffinic* and *asphaltic*. When crudes containing both occurred, the crudes were all called *mixed base*. The more asphaltic were called *naphthenic*, alternative to the modern *cycloparaffinic*. Although these terms are still in use, the terms are quite misleading as cycloparaffins are the largest component in all oils. The variation is largely in the proportions of branched chain *isoparaffins* and *aromatics*. A more meaningful classification is by viscosity index, a scale organized with Pennsylvania paraffinic at 100 and a typical asphaltic at 0 VI. Regardless of crude source, the modern classes are High VI (90–105), Medium VI (55–75) and Low VI (negative to 50). A number of synthetic oils are now in use. The most important are diesters and polyol esters, *poly-alpha-olefin* isoparaffinic and alkyl-aromatic hydrocarbons, and polyglycols.

By far the largest volume of lubricants is now the high VI class, and this section will be devoted entirely to their properties. If a user has need for the properties of other petroleum oils or if a synthetic is required, the supplier will normally provide the data.

The applications covered by *fluid machinery* can be logically argued to include just about everything with moving parts. However, to keep this discussion within space limitations it will be confined to the following classes: pumps, fans, blowers, and air compressors; steam, water, and gas turbines; and air motors. This excludes the very large subject of internal combustion engines, for which the lubricants are so heavily dosed with additives that the base oil properties are masked. Automatic transmission fluids are also in that category. All the aviation turbine oils are also excluded, as these oils are universally synthetic esters. Hydraulic fluids are given in Sec. 2.8, though in many cases lubricating oils are used for hydraulic fluids. Oils for piston steam engines are very specialized, as are those for compressing refrigerants and other gases, and so are omitted. Again, the oil supplier can be very helpful on data for these varieties.

2.5.2 Physical Properties

Viscosity is the most critical property in most lubricant application (Sec. 23.1). The definition is still that proposed by Newton: the force transmitted to one square unit of plane surface by a parallel surface at unit distance moving at unit velocity by the fluid between them. The practice has grown of calling this the *dynamic* or *absolute* viscosity, but that is not really needed as the original word is quite explicit. It is already absolute, while dynamic implies *under accelerated motion*, in which case, Newton's definition may not be appropriate. These modifiers were probably introduced to emphasize that it is not the more frequently discussed kinematic viscosity.

The *kinematic viscosity* is a unit derived from the basic definition by dividing by the density. The special virtue of this concept is that it is directly related to the efflux time from a viscometer under standard gravitational force. Most of the common laboratory instruments operate on this principle (i.e., ASTM D 445). However, there are a few devices that measure the force on a rotating member which yields

TABLE 2.18 Transport Properties of Pure Species

T K	AR Visc. Micro-poise	AR Cond. Micro W/CM K	CH₄ Visc. Micro-poise	CH₄ Cond. Micro W/CM K	CO Visc. Micro-poise	CO Cond. Micro W/CM K	CO₂ Visc. Micro-poise	CO₂ Cond. Micro W/CM K
200	159	124	77	217	129	186	101	97
300	228	178	112	346	177	254	150	167
400	288	225	141	496	218	315	197	242
500	342	267	168	667	254	373	240	322
600	391	305	192	856	287	429	280	401
700	436	340	214	1056	317	486	317	478
800	478	373	235	1260	346	543	350	551
900	518	404	255	1465	373	598	382	621
1000	556	434	274	1669	399	652	411	689
1100	592	462	292	1866	425	704	438	753
1200	628	490	309	2061	449	754	465	813
1300	661	516	326	2252	473	804	489	871
1400	694	542	342	2439	496	853	513	927
1500	726	567	368	2622	519	901	535	979
1600	757	591	374	2801	541	949	557	1030
1700	788	615	389	2975	564	996	577	1078
1800	818	638	403	3145	586	1043	597	1124
1900	847	661	418	3310	608	1090	616	1168
2000	875	683	432	3470	631	1136	635	1211
2100	903	704	446	3626	653	1182	653	1251
2200	930	725	460	3778	675	1227	670	1291
2300	956	746	473	3925	697	1272	687	1328
2400	982	766	486	4068	719	1317	704	1365
2500	1007	786	499	4207	742	1362	720	1400
2600	1032	806	512	4343	764	1408	735	1433
2700	1057	825	525	4475	786	1453	750	1466
2800	1082	844	538	4604	808	1497	765	1497
2900	1106	863	550	4730	831	1542	780	1528
3000	1129	881	562	4854	853	1587	794	1558
3100	1153	900	574	4977	874	1631	808	1588
3200	1176	918	586	5097	896	1676	822	1617
3300	1199	936	598	5216	917	1720	835	1646
3400	1221	953	610	5333	939	1764	848	1675
3500	1244	970	621	5450	961	1808	861	1703
3600	1266	988	633	5565	982	1852	873	1732
3700	1287	1005	644	5679	1004	1896	886	1760
3800	1309	1021	655	5793	1026	1941	898	1787
3900	1330	1038	667	5905	1048	1985	910	1815
4000	1351	1054	678	6017	1070	2029	922	1842
4100	1372	1071	689	6128	1091	2073	934	1869
4200	1393	1087	699	6238	1113	2117	945	1894
4300	1413	1103	710	6347	1134	2162	956	1919
4400	1433	1119	721	6455	1155	2206	967	1942
4500	1453	1134	731	6561	1176	2250	978	1965
4600	1473	1150	742	6664	1197	2293	989	1987
4700	1493	1165	752	6765	1218	2337	1000	2008
4800	1512	1180	762	6863	1239	2380	1010	2029
4900	1532	1195	773	6958	1259	2422	1021	2050
5000	1551	1210	783	7047	1280	2464	1031	2072

TABLE 2.18 (*Continued*)

H		H$_2$		H$_2$O		N		NH$_3$	
Visc. Micro- poise	Cond. Micro W/CM K	Visc. Micro- poise	Cond. Micro W/CM K	Visc. Micro- poise	Cond. Micro W/CM K	Visc. Micro- poise	Cond. Micro W/CM K	Visc. Micro- poise	Cond. Micro W/CM K
38	1197	67	1293	51	96	125	279	68	152
53	1661	89	1835	91	180	173	387	103	243
67	2094	109	2256	133	271	218	486	137	366
81	2507	127	2634	172	359	260	580	172	509
93	2906	144	2985	214	464	301	670	206	658
106	3292	160	3326	255	579	339	756	239	816
118	3670	175	3660	296	704	377	839	271	984
130	4040	190	3991	336	835	413	920	301	1149
142	4403	205	4321	376	971	448	999	331	1317
153	4761	218	4684	413	1111	483	1076	359	1493
165	5114	232	5051	450	1254	417	1151	386	1671
176	5462	245	5420	487	1402	550	1225	412	1847
187	5807	258	5790	524	1553	582	1298	438	2022
198	6149	271	6160	560	1707	614	1369	462	2195
209	6487	283	6511	595	1860	646	1439	486	2365
220	6823	295	6864	630	2015	677	1508	509	2531
231	7156	307	7221	664	2170	708	1576	531	2695
242	7488	318	7585	698	2327	738	1644	553	2855
252	7817	330	7957	731	2485	768	1710	574	3011
263	8144	341	8321	763	2642	798	1776	595	3164
273	8470	352	8701	795	2799	827	1842	616	3314
284	8794	363	9101	826	2958	856	1907	636	3459
294	9117	374	9519	857	3118	885	1972	655	3602
305	9438	385	9952	887	3278	914	2037	674	3741
315	9758	395	10415	917	3443	942	2102	693	3876
325	10077	406	10849	946	3605	970	2167	712	4009
336	10395	416	11256	974	3764	998	2232	730	4138
346	10711	426	11643	1003	3919	1026	2298	748	4265
356	11027	436	12015	1030	4070	1053	2364	766	4388
366	11347	446	12377	1058	4213	1081	2431	783	4509
377	11664	456	12735	1085	4352	1108	2498	801	4626
387	11978	466	13089	1111	4488	1135	2566	818	4742
397	12288	475	13440	1137	4621	1162	2635	834	4855
407	12592	485	13789	1163	4752	1188	2705	851	4965
416	12884	494	14136	1189	4880	1215	2775	867	5074
425	13171	504	14481	1214	5006	1241	2848	883	5181
434	13455	513	14824	1239	5130	1268	2921	899	5285
444	13735	522	15166	1264	5252	1294	2995	915	5388
452	14012	532	15506	1288	5372	1320	3071	931	5489
461	14290	541	15846	1312	5491	1346	3149	946	5589
470	14566	550	16185	1336	5608	1371	3227	962	5686
479	14842	559	16524	1359	5724	1397	3308	977	5782
488	15116	568	16863	1383	5838	1422	3390	992	5875
497	15389	576	17204	1406	5951	1448	3474	1007	5967
506	15661	585	17547	1429	6063	1473	3559	1022	6056
515	15933	594	17893	1451	6173	1498	3647	1036	6142
523	16204	603	18242	1474	6283	1524	3737	1051	6225
532	16477	611	18595	1496	6391	1549	3828	1065	6305
541	16750	620	18952	1518	6497	1574	3922	1080	6382

TABLE 2.18 (*Continued*)

T K	NO Visc. Micro-poise	NO Cond. Micro W/CM K	NO$_2$ Visc. Micro-poise	NO$_2$ Cond. Micro W/CM K	N$_2$ Visc. Micro-poise	N$_2$ Cond. Micro W/CM K	N$_2$O Visc. Micro-poise	N$_2$O Cond. Micro W/CM K
200	136	189	111	112	129	182	99	106
300	192	262	165	180	179	259	149	180
400	239	330	213	253	222	323	194	257
500	282	395	257	327	260	383	236	334
600	320	459	298	400	295	442	274	409
700	356	522	335	471	327	502	309	481
800	389	585	370	539	358	562	341	550
900	421	646	403	603	387	621	372	616
1000	452	704	434	663	414	678	402	680
1100	481	758	463	719	441	732	430	738
1200	509	812	492	773	467	784	457	795
1300	536	864	520	826	492	835	483	851
1400	563	914	546	877	516	885	508	905
1500	589	964	572	926	540	933	532	957
1600	614	1012	597	974	563	981	556	1008
1700	639	1059	622	1021	585	1028	579	1058
1800	663	1105	646	1066	607	1075	601	1106
1900	686	1150	669	1110	629	1120	623	1152
2000	710	1194	692	1153	650	1164	645	1198
2100	733	1237	715	1194	670	1207	666	1242
2200	755	1279	737	1234	691	1249	686	1284
2300	777	1320	758	1273	711	1291	706	1326
2400	799	1361	779	1312	730	1332	726	1367
2500	820	1401	800	1349	750	1372	746	1406
2600	842	1439	821	1386	769	1412	765	1445
2700	862	1478	842	1422	788	1451	784	1483
2800	883	1515	862	1457	806	1490	803	1521
2900	903	1552	882	1492	825	1528	821	1558
3000	923	1589	901	1527	843	1567	840	1594
3100	943	1625	921	1561	860	1605	858	1630
3200	963	1661	940	1595	878	1644	876	1666
3300	982	1697	959	1628	896	1682	893	1701
3400	1001	1732	978	1661	913	1719	911	1736
3500	1020	1767	996	1694	930	1757	928	1771
3600	1039	1801	1015	1727	947	1793	945	1805
3700	1058	1836	1033	1759	963	1829	962	1839
3800	1076	1870	1051	1791	980	1865	979	1873
3900	1094	1904	1069	1823	996	1900	996	1907
4000	1112	1937	1086	1855	1012	1935	1012	1941
4100	1130	1971	1104	1887	1029	1968	1029	1974
4200	1148	2004	1121	1918	1044	2000	1045	2007
4300	1166	2037	1139	1949	1060	2033	1061	2040
4400	1183	2070	1156	1980	1076	2066	1077	2072
4500	1200	2102	1173	2011	1091	2100	1093	2104
4600	1217	2135	1189	2040	1107	2134	1109	2135
4700	1234	2166	1206	2069	1122	2168	1124	2166
4800	1251	2197	1223	2098	1137	2203	1140	2196
4900	1268	2228	1239	2125	1152	2238	1155	2225
5000	1285	2258	1256	2152	1167	2272	1170	2253

TABLE 2.18 (*Continued*)

O		OH		O$_2$	
Visc. Micropoise	Cond. Micro W/CM K	Visc. Micropoise	Cond. Micro W/CM K	Visc. Micropoise	Cond. Micro W/CM K
160	333	73	182	146	185
206	421	115	288	206	263
249	500	158	392	257	336
289	576	199	493	303	410
328	649	240	594	344	481
365	720	279	694	383	549
401	789	317	795	420	614
436	855	354	896	455	678
470	921	390	999	488	740
503	985	425	1103	521	797
535	1048	458	1207	552	852
567	1110	491	1309	582	907
599	1171	522	1410	612	959
630	1231	553	1510	640	1011
660	1290	582	1608	668	1064
690	1348	611	1704	696	1116
719	1405	639	1799	723	1166
749	1462	666	1892	749	1216
777	1518	693	1984	775	1263
806	1573	719	2074	801	1307
834	1628	744	2163	826	1349
862	1683	769	2250	851	1388
889	1738	793	2336	875	1424
917	1792	817	2420	899	1457
944	1846	841	2503	922	1492
971	1900	864	2585	946	1527
997	1954	886	2666	969	1561
1024	2007	909	2746	992	1595
1050	2061	931	2824	1014	1629
1076	2115	952	2901	1036	1661
1102	2169	974	2977	1058	1692
1128	2222	995	3051	1080	1724
1154	2276	1016	3125	1101	1755
1179	2330	1036	3198	1123	1785
1205	2385	1056	3269	1144	1816
1230	2439	1076	3340	1165	1847
1255	2493	1096	3410	1185	1877
1280	2548	1115	3479	1206	1907
1305	2603	1135	3548	1226	1937
1330	2657	1154	3616	1246	1965
1354	2713	1173	3684	1266	1993
1379	2768	1192	3751	1286	2020
1403	2823	1211	3817	1306	2046
1428	2878	1229	3884	1325	2073
1452	2934	1248	3950	1345	2099
1476	2990	1266	4017	1364	2125
1500	3045	1284	4083	1383	2151
1524	3101	1301	4149	1402	2176
1548	3157	1319	4216	1421	2201

TABLE 2.19 Comparison of Values for N_z from Several Sources

Temperature T K	Viscosity, Micropoise			Thermal Conductivity, microwatt/cm K	
	These tables (From Hanley and Fly, 1973)	Maitland and Smith (1972)	Touloukian et al. (1975)	This report (From Hanley and Fly, 1973)	Touloukian et al. (1979)
300	179	179	179	259	260
500	260	260	259	383	386
1000	414	417	401	678	631
1500	540	541	512	933	842
2000	650	650	601	1164	1146
2500	—	—	—	1372	1406
3000	—	—	—	1567	1640

the dynamic viscosity. The viscosities are related by the equation

$$\nu = \mu/\rho \tag{2.53}$$

where ν is the kinematic viscosity, μ the dynamic viscosity, and ρ the density. The SI unit for ν is m^2/s, and for μ it is Pa · s, but the customary units are *centistokes* (mm^2/s) and *centipoises* (mPa · s). The customary units are abbreviated *cSt* and *cP*.

Another use of the word is in *apparent viscosity* which is related to non-Newtonian lubricants such as greases and highly compounded engine oils. It is simply the viscosity of the Newtonian oil which would give the same force reading under the same conditions. See Chap. 3 for a discussion of non-Newtonian fluids.

The International Standards Organization has a system for classifying industrial lubricating oils by the viscosities of the oils at 40 C (ISO Std. 3448). The same classes are also used in ASTM D 2422 and by several organizations in other nations. ISO lists 18 viscosity grades, from 2–1500 cSt at 40 C. Fourteen of the most commonly used are listed in Table 2.20 along with many other properties discussed below. Properties of the others may be estimated by extrapolation.

Density used to be considered a key property of lubricants because it provided a clue as to the *paraffinicity*. Today, it is most often needed for converting cSt to cP units, as bearings are like rotary viscometers. It is usually reported at 15 C. Many suppliers report the closely related *specific gravity*, the ratio of density to that of water at the same temperature. For most purposes, the two are interchangeable. In the older literature one may still find references to the *API gravity*, a term related to specific gravity [API (1970)] by

$$API = (141.5/sp\ gr) - 131.5 \tag{2.54}$$

where both are at 60 F (15.4 C).

The variation of density with viscosity is rather complicated because it depends on the paraffinicity. However, by limiting this discussion to paraffinic oils as defined above, it is possible to express the relation simply as entries in Table 2.20. Likewise, the variation in density with temperature can be shown as a single entry for each

TABLE 2.20 Properties of Lubricating Oils

ISO Grade cSt, 40 C	Density 15 C kg/L	$d\rho/dT$ g/L · K	Visc. Temp. A	Visc. Temp. B	Specif. Heat kJ/ kg · K	Visc./P Expon. 1/MPa	Visc./P Slope 1/GPa · K	Surface Tension mN/m
10	0.829	0.716	9.56	3.83	1.921	17.6	73	28.0
15	0.843	0.702	9.57	3.80	1.900	19.4	94	29.0
22	0.851	0.795	9.59	3.79	1.892	21.4	118	29.7
32	0.858	0.689	9.58	3.77	1.883	23.4	141	30.3
46	0.863	0.685	9.46	3.70	1.875	25.0	158	30.7
68	0.869	0.681	9.29	3.62	1.867	26.7	174	31.0
100	0.874	0.677	9.07	3.51	1.858	28.0	183	31.6
150	0.876	0.676	8.87	3.42	1.852	29.5	194	31.8
220	0.884	0.670	8.72	3.35	1.846	30.4	196	32.1
320	0.889	0.667	8.60	3.29	1.838	31.3	195	32.4
460	0.893	0.665	8.49	3.23	1.833	31.9	190	32.7
680	0.898	0.662	8.40	3.18	1.825	32.0	176	32.9
1000	0.903	0.660	8.33	3.15	1.821	31.0	142	33.2
1500	0.908	0.658	8.28	3.12	1.813	30.9	123	33.9

viscosity grade. The values are those of the density–temperature slope $d\rho/dT$ which is essentially independent of temperature, rather than the coefficient of thermal expansion $d\rho/\rho dT$ which is not. These values are used in

$$\rho_T = \rho - (T - 15)\, d\rho/dT \qquad (2.55)$$

Viscosity–temperature relations are quite another matter, both in terms of importance and complication. The trade unit for this is the *Viscosity Index*, which is defined in ASTM D 2272. In limiting the scope of this section to paraffinic oils, the discussion is automatically limited to those of 100 ± 5 VI—in other words, comparable to the Pennsylvania lubes used in defining the system. Thus, it is possible to express this property as a single constant for each ISO grade. The unit chosen is that from ASTM D 341, the parameter defined by B in the equation

$$A - B \log T = \log \log (\nu + 0.7) \qquad (2.56)$$

where T is in Kelvin. The value of A can be computed if values of ν, B, and T are assigned, but for convenience both A and B are listed in Table 2.20.

Pour point is the term used for the lowest temperature at which flow is detectable under test conditions rigidly specified in ASTM D 97. In paraffinic lubricants the cessation of flow is caused by precipitation of wax (normal paraffins) into a gel structure. This is controlled by removing a major part of the wax and can be further controlled by additives. Thus, pour point is a property that can be selected on a cost basis, and so it is not shown in Table 2.20. Unless a premium is paid for special dewaxing, the levels available are about -18 C (0 F) for the first three grades, -12 C (10 F) for the second three, and -6 C (20 F) for the last three.

Specific heat capacity of lubricating oils at a given temperature depends only on the density, so it is possible to include values at 15 C for all the grades in Table 2.20 as c_p.

The variation of specific heat capacity is nearly logarithmic with temperature. c_{pT} can be estimated by

$$c_{pT} = c_p [1 + 0.0013(9T - 15)] \tag{2.57}$$

where T is in Celsius.

Thermal conductivity is essentially independent of density and viscosity and may be estimated quite accurately by

$$k = 463.3 - 0.511T \tag{2.58}$$

where T is in Celsius and k in J/hr · m · K.

The *Prandtl number* is a composite parameter that is sometimes used as a convenient expression of the heat transfer capability of a fluid because it contains all the relevant fluid properties. It may be calculated for any of the oils at any desired temperature by the equation

$$\text{Pr} = c_p \mu / k \tag{2.59}$$

Compressibility modulus (or *bulk modulus*) is the reciprocal of compressibility. While that definition seems to be quite straightforward, there are a number of complications in how it is to be used. Part of this is related to the fact that as the amount of *free volume* left in the liquid decreases, the pressure required for a further decrease rises sharply. Thus, the pressure–density curve is essentially hyperbolic. The modulus can be interpreted as the total pressure required for unit volume decrease, or the incremental pressure increase for the same change. The former is known as the *secant bulk modulus* because it cuts across the hyperbola, and the latter as the *tangent bulk modulus* because it is tangent to the pressure–density curve. The equations are

$$B_S = p\rho_p / (\rho_p - \rho) \tag{2.60a}$$

and

$$B_T = \rho_p \, dp/d\rho \tag{2.60b}$$

where ρ_p is the density at the desired pressure. Similar equations using volume rather than density yield negative B_S and B_T.

This complex situation is best approached in two steps. First, the secant bulk modulus to any reference pressure is a function only of the density at atmospheric pressure and the temperature of interest. Figure 2.7 can be used to estimate this for any temperature and 137.9 MPa (20,000 psig). The resulting modulus is then entered into Fig. 2.8 to obtain B_S at any desired pressure. That is the value most often required, but if B_T is needed it can be closely approximated by B_S at half the pressure. A third sort of bulk modulus is that found when a fluid is so rapidly compressed that the heat cannot escape. That is known as the *adiabatic bulk modulus* and its

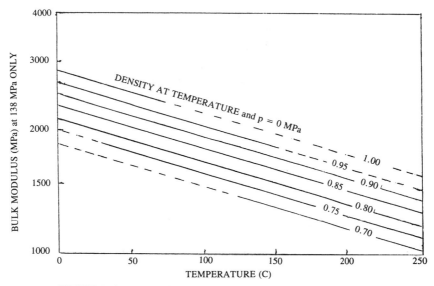

FIGURE 2.7 Isothermal secant bulk modulus at 138 MPa.

symbol is K_T as it is normally reported only in the tangent sense. It can best be calculated from the sonic velocity in the fluid, and is useful in the analysis of hydraulic systems. It is less in demand for lubes though it has potential in considering rapid asperity encounters.

The *pressure–viscosity exponent* is often desired on oils to be used in elastohydrodynamic lubrication (see Sec. 23.1). This is the parameter α_p defined by the equation

$$\mu_p = \mu \exp\,(\alpha_p p) \tag{2.61}$$

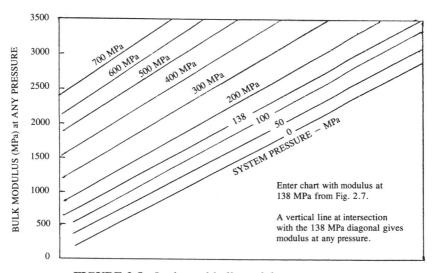

FIGURE 2.8 Isothermal bulk modulus at any pressure.

Estimated values [So and Klaus (1980)] of this exponent are listed in Table 2.20, along with values for its temperature slope.

Surface tension (more properly, *surface free energy*) of paraffinic oils does not vary strongly with viscosity. Values for the various grades are shown in Table 2.20. The variation with temperature is almost exactly linear and can be represented by this equation

$$\sigma_T = \sigma + 0.10(T - 25) \tag{2.62}$$

where T is in Celsius.

2.5.3 Chemical Properties

Toxicity does not present any problem with these oils. Numerous tests have shown the oils to be noncarcinogenic. Some sensitive individuals have developed dermatitis after prolonged exposure, but proper personal hygiene is the only precaution that is usually needed.

Although gear oils are outside the scope of this section, it must be mentioned that *gear lubricants* often contain toxic compounds of lead, chlorine, and phosphorus. If such oils are used indiscriminately as *all-purpose* lubricants there can be some danger of long-term health problems.

Oxidation stability is a property that should be of interest as it (along with contamination) controls how frequently the oil must be changed. Unfortunately, paraffinic oils do not have very good oxidation resistance without inhibitors, but paraffinic oils do respond very well to small doses of the proper additives. For low temperature uses, such as cold water pumps, the base oils may be good enough, but steam turbine lubrication requires all the best technology. Obviously, stability is a property that the customer can dictate, and so it cannot be shown in Table 2.20. As a rough guide, the minimum standard for steam turbine oils is 1000 hours in the ASTM D 943 oxidation test, in which the oil is in contact with steel, copper, and water at 95 C while being blown with oxygen until the acidity rises to the incipient danger level. (Some customers are now demanding, and receiving, oils of 2000 and even 5000 hours life which is considered equivalent to between 20 and 50 years in the turbines.) Taking as baseline the life of 1000 hours for an ISO Grade 32 oil, the life of a Grade 3 would be about 1500 hours, and that of a Grade 1500 about 400.

2.6 COMMON PETROLEUM BASED FUELS
John E. Schmidt and Albert M. Momenthy

The common petroleum based fuels (gasolines, turbine fuels, diesel fuels, and heating oils) all are mixtures of a large number of hydrocarbon species, each species with its peculiar set of properties. The compositions of these fuels vary widely among samples, depending on crude oil sources, refinery processing, and blending, with resultant significant variations in properties, subject to the limitations of the specifications for the various fuels. Typical values of physical and thermodynamic properties are tabulated along with an indication of the range over which each may be expected to vary. In general, excursions outside of these ranges can be expected only for one to two percent of the samples for each fuel type.

Tables 2.21, 2.22, and 2.23 summarize properties of aviation turbine fuels Jet A, JP-5, and JP-4, respectively. The typical values and ranges are based primarily on the inspection data reported by the U.S. Department of Energy (DOE) [Dickson (1992), (1993)]. Most properties not reported in the inspection data were obtained by correlations with distillation curves and relative density. The properties of aviation gasoline in Table 2.24 are based on previously published typical distillation curves and properties [Barnett and Hibbard (1956), CRC (1983)] and limited inspection data. The properties given in Tables 2.25, 2.26, and 2.27 for automobile gasoline and 1-D and 2-D diesel fuels, respectively, are also based on recent inspection data [Shelton (1981), (1982a), (1982b)]. The property ranges were estimated on the basis of these data along with distillation and volatility specification limits.

The properties tabulated for Jet A fuel also apply to Jet A-1 and JP-8 fuels which meet the same specification [ASTM (1990a)] and MIL (1989), except for a maximum freezing point of −47 C as compared to −40 C for Jet A. Similarly, the properties tabulated for JP-4 apply equally to Jet B fuel which has nearly the same specifications [MK (1989b)]. The major difference is that Jet B has a −50 C maximum freezing point instead of the −58 C specified for JP-4. Although the ASTM specifications for aviation gasolines [ASTM (1990b)] include three different grades, these differ only in required octane rating and lead content, and the variations in properties found within each grade can be expected to be much greater than the differences between the average values for the several grades. Therefore, only one set of average properties is presented for aviation gasolines. The same rationale applies to automobile gasolines, for which five volatility classes are specified for both leaded and unleaded gasolines covering several ranges of octane rating [ASTM (1990c)]. However, significant variations in the volatility of automobile gasolines occur from season to season and from one geographical area of the United States to another, as climate dictates. The properties reported represent overall averages for all seasons and all areas of the contiguous United States. The tabulations for 1-D and 2-D diesel fuels also represent average properties for all seasons and all areas of the country, although pour point and viscosity requirements may vary significantly due to climatic conditions. The properties given for 1-D and 2-D diesel fuels should also apply approximately to No. 1 and No. 2 heating oils which have similar specifications [ASTM (1990d), (1990e)].

The tabulated properties are based on data assembled in the past and do not reflect changes in properties resulting from reformulations of gasoline and diesel fuels which are presently being introduced or planned for the future to meet environmental requirements. Scant data are available for the properties of these reformulated products, but it should be recognized that the changes involved (e.g., inclusion of oxygenated constituents, lowered aromatics and sulfur contents, lowered vapor pressures, and altered boiling ranges) can be expected to affect the properties of these fuels. Moreover, the seasonal and geographical variations of the reformulations make it impossible to generalize as to the effects. While these reformulations will primarily affect gasoline and diesel fuel, there may be some indirect effects on jet fuels due to changes in refinery operations and processes.

TABLE 2.21 Properties of JET A Aircraft Turbine Fuel

DISTILLATION CURVE

VOLUME %	TEMPERATURE, C TYPICAL	RANGE
0	168	137–187
10	190	162–202
50	214	179–232
90	246	203–269
100	268	218–293

CHARACTERISTIC PROPERTIES

	TYPICAL	RANGE
REL. DENSITY	0.815	0.785–0.840
AVG. MOL. WT	170	154–178
SPECIES MOL. WT.		140–230
CRIT. TEMP., K	680	652–696
CRIT. PRES., kPa	2310	2200–2520
FREEZING PT., C	−45	−40, −65

COMBUSTION PROPERTIES

	TYPICAL	RANGE
MASS % HYDROGEN	13.8	13.6–14.1
MASS C/H RATIO	6.25	6.1–6.4
HEAT OF COMBUSTION		
GROSS, MJ/kg	46.12	45.76–46.42
NET, MJ/kg	43.15	42.84–43.38

SAFETY PARAMETERS

	TYPICAL	RANGE
FLASH POINT, C	55	38–71
AUTO IGNITION, C	257	250–300
FLAMMABILITY LIMITS		
LOWER VOLUME %	0.60	0.57–0.62
UPPER VOLUME %	4.53	4.4–4.7
PROLONGED EXPOSURE LIMIT: 500 PPM		

TEMP. C	DENSITY kg/m³	COEF. THERMAL EXPN. 10³/K	SECANT BULK MODULUS GPa	KINEMATIC VISCOSITY 10⁶×m²/s	DYNAMIC VISCOSITY mPa×s	SURFACE TENSION N/m	VAPOR PRESSURE kPa	SPECIFIC HEAT kJ/kgK	HEAT OF VAPORIZATION kJ/kg	DIELECT. COEF.	WATER SOLUBILITY MASS %	OSTWALD COEFFICIENT O₂	N₂	AIR
−40	852.0	0.829	1.948	12.79	10.90	0.0311		1.674		2.21		0.184	0.079	0.105
−20	837.8	0.855	1.771	5.44	4.56	0.0294		1.772	376	2.18		0.199	0.097	0.123
0	823.3	0.881	1.606	3.03	2.49	0.0278	0.05	1.867	365	2.16	0.002	0.212	0.116	0.140
20	808.7	0.907	1.452	1.98	1.60	0.0261	0.18	1.960	356	2.13	0.006	0.224	0.135	0.157
40	794.0	0.934	1.308	1.43	1.14	0.0245	0.54	2.050	346	2.11	0.015	0.235	0.154	0.173
60	779.1	0.960	1.175	1.09	0.85	0.0229	1.42	2.138	335	2.08	0.032	0.246	0.173	0.189
80	764.1	0.986	1.050	0.86	0.66	0.0214	3.32	2.224	323	2.06	0.064	0.255	0.191	0.204
100	749.0	1.012	0.935	0.71	0.53	0.0198	7.10	2.307	311	2.03	0.12	0.264	0.210	.0218
120	733.8	1.038	0.828	0.59	0.43	0.0183	14.1	2.388	300	2.01	0.20	0.272	0.228	0.232
140	718.5	1.064	0.728	0.50	0.36	0.0167	26.0	2.467	289	1.98	0.34	0.280	0.245	0.245
160	703.2	1.090	0.636	0.43	0.30	0.0152	45.6	2.543	281	1.96	0.52	0.287	0.262	0.258
180	687.9	1.116	0.550				75.9	2.617	270	1.93	0.77	0.294	0.279	0.270
200	675.2	1.142	0.470				121.0	2.688	258	1.91	1.16	0.300	0.295	0.282
RANGE	+30 / −40	+12% / −8%	±10%	+40% / −60%	+40% / −60%	±0.0015	+120% / −35%	±0.06	±10	±0.06	±50%	+22% / −14%	+22% / −14%	+22% / −14%

TABLE 2.22 Properties of JP-5 Aircraft Turbine Fuel

DISTILLATION CURVE

VOLUME %	TEMPERATURE, C TYPICAL	RANGE
0	179	171–194
10	199	189–204
50	215	206–223
90	240	233–250
100	260	232–280

CHARACTERISTIC PROPERTIES

	TYPICAL	RANGE
REL. DENSITY	0.821	0.807–0.845
AVG. MOL. WT	172	167–176
SPECIES MOL. WT.		140–230
CRIT. TEMP., K	685	630–698
CRIT. PRES., kPa	2300	2200–2430
FREEZING PT., C	−49	−46, −60

COMBUSTION PROPERTIES

	TYPICAL	RANGE
MASS % HYDROGEN	13.6	13.5–13.7
MASS C/H RATIO	6.35	6.3–6.4
HEAT OF COMBUSTION		
GROSS, MJ/kg	45.93	45.67–46.09
NET, MJ/kg	43.05	42.81–43.18

SAFETY PARAMETERS

	TYPICAL	RANGE
FLASH POINT, C	66	60–71
AUTO IGNITION, C	257	245–300
FLAMMABILITY LIMITS		
LOWER VOLUME %	0.60	0.57–0.62
UPPER VOLUME %	4.53	4.4–4.7
PROLONGED EXPOSURE LIMIT: 500 PPM		

TEMP. C	DENSITY kg/m³	COEF. THERMAL EXPN. 10³/K	SECANT BULK MODULUS GPa	KINEMATIC VISCOSITY 10⁶×m²/s	DYNAMIC VISCOSITY mPa x s	SURFACE TENSION N/m	VAPOR PRESSURE kPa	SPECIFIC HEAT kJ/kgK	HEAT OF VAPORIZATION kJ/kg	DIELECT. COEF.	WATER SOLUBILITY MASS %	OSTWALD COEFFICIENT O₂	N₂	AIR
−40	859.6	0.815	1.980	13.53	11.63	0.0314		1.654		2.22		0.177	0.076	0.101
−20	845.5	0.840	1.801	5.95	5.03	0.0298		1.751	377	2.19		0.192	0.094	0.119
0	831.2	0.865	1.634	3.30	2.75	0.0281	0.03	1.846	365	2.17	0.002	0.206	0.113	0.136
20	816.8	0.890	1.478	2.12	1.73	0.0265	0.11	1.939	355	2.15	0.006	0.219	0.132	0.153
40	802.2	0.915	1.333	1.50	1.21	0.0249	0.35	2.029	346	2.12	0.015	0.231	0.151	0.170
60	787.4	0.940	1.198	1.13	0.89	0.0233	0.95	2.117	336	2.10	0.032	0.242	0.170	0.186
80	772.6	0.965	1.073	0.89	0.69	0.0217	2.29	2.202	325	2.07	0.064	0.252	0.189	0.201
100	757.6	0.990	0.956	0.72	0.55	0.0201	5.05	2.285	316	2.05	0.12	0.261	0.208	.216
120	742.6	1.015	0.848	0.60	0.45	0.0186	10.3	2.366	307	2.02	0.20	0.270	0.227	0.231
140	727.5	1.040	0.747	0.51	0.37	0.0171	19.5	2.445	296	2.00	0.34	0.278	0.245	0.245
160	712.3	1.065	0.653	0.44	0.31	0.0156	34.8	2.521	286	1.97	0.52	0.286	0.263	0.258
180	697.1	1.090	0.567	0.38	0.27	0.0141	59.1	2.594	274	1.95	0.77	0.293	0.280	0.271
200	681.9	1.115	0.486	0.34	0.23	0.0127	96.1	2.666	263	1.92	1.20	0.300	0.297	0.283
RANGE	+25	+10%	+8%	±35%	+40%	+0.0010	+100%	+0.04%	+10	±0.05	±50%	+18%	+18%	+18%
	−35	−7%	−10%			−0.0005	−20%	−0.06%				−13%	−13%	−13%

149

TABLE 2.23 Properties of JP-4 Aircraft Turbine Fuel

DISTILLATION CURVE			CHARACTERISTIC PROPERTIES			COMBUSTION PROPERTIES			SAFETY PARAMETERS		
VOLUME %	TEMPERATURE, C TYPICAL	RANGE		TYPICAL	RANGE		TYPICAL	RANGE		TYPICAL	RANGE
0	61	49–78	REL. DENSITY	0.764	0.75–0.80	MASS % HYDROGEN	13.8	13.6–14.5	FLASH POINT, C	2	–10, +21
10	97	82–125	AVG. MOL. WT.	120	103–145	MASS C/H RATIO	6.25	5.9–6.4	AUTO IGNITION, C	300	250–350
50	141	109–182	SPECIES MOL. WT.		80–200	HEAT OF COMBUSTION			FLAMMABILITY LIMITS		
90	207	150–236	CRIT. TEMP., K	594	567–636	GROSS, MJ/kg	46.40	45.93–46.85	LOWER VOLUME %	0.80	0.74–0.90
100	240	187–260	CRIT. PRES., kPa	3220	3000–3560	NET, MJ/kg	43.50	43.05–43.78	UPPER VOLUME %	5.63	5.3–6.2
			FREEZING PT., C	–63	–58, –80				PROLONGED EXPOSURE LIMIT: 500 PPM		

TEMP. C	DENSITY kg/m^3	COEF. THERMAL EXPN. 10^3/K	SECANT BULK MODULUS GPa	KINEMATIC VISCOSITY 10^6×m^2/s	DYNAMIC VISCOSITY mPa × s	SURFACE TENSION N/m	VAPOR PRESSURE kPa	SPECIFIC HEAT kJ/kgK	HEAT OF VAPORIZATION kJ/kg	DIELECT. COEF.	WATER SOLUBILITY MASS %	OSTWALD COEFFICIENT O$_2$	N$_2$	AIR
–40	811.5	1.048	1.79	3.14	2.55	0.0281	0.48	1.777		2.14	0.0015	0.237	0.103	0.137
–20	794.3	1.092	1.61	1.90	1.51	0.0263	1.52	1.878	375	2.11	0.003	0.246	0.121	0.153
0	776.8	1.136	1.45	1.30	1.01	0.0244	4.07	1.965	365	2.08	0.006	0.253	0.139	0.167
20	759.0	1.180	1.30	0.96	0.73	0.0226	9.52	2.056	356	2.05	0.011	0.260	0.156	0.181
40	741.0	1.224	1.15	0.74	0.55	0.0208	20.0	2.144	346	2.02	0.019	0.267	0.172	0.195
60	722.8	1.268	1.02	0.59	0.43	0.0190	3.84	2.230	334	1.99	0.032	0.272	0.188	0.207
80	704.4	1.311	0.90	0.49	0.34	0.0173	68.4	2.314	323	1.96		0.277	0.204	0.219
100	685.8	1.355	0.79	0.41	0.28	0.0156	114	2.395	311	1.93		0.282	0.219	.230
120	667.2	1.399	0.69	0.35	0.24	0.0139	182	2.474	300	1.90		0.286	0.233	0.240
140	648.5	1.443	0.59			0.0123	277	2.551	290	1.87		0.290	0.247	0.250
160	629.8	1.487	0.50			0.0107	405	2.626	281	1.83		0.293	0.260	0.259
180	611.1	1.531	0.42			0.0091	574	2.698	270	1.80		0.297	0.273	0.267
200	592.4	1.574	0.35			0.0076	788	2.768	258	1.77		0.300	0.285	0.276
RANGE	+40	+3%	+15%	+65%	+70%	+0.0025	+22%	+0.03%	+10	+0.07	±50%	+7%	+7%	+7%
	–15	–10%	–5%	–35%	–35%	–0.0020		–0.08%		–0.02		–19%	–19%	–19%

TABLE 2.24 Properties of Aviation Gasoline

DISTILLATION CURVE		
VOLUME %	TEMPERATURE, C TYPICAL	RANGE
0	41	32–44
10	60	50–75
50	95	85–105
90	127	112–139
100	158	136–170

CHARACTERISTIC PROPERTIES	TYPICAL	RANGE
REL. DENSITY	0.71	0.69–0.73
AVG. MOL. WT.	98	93–103
SPECIES MOL. Wt.		72–128
CRIT. TEMP., K	539	523–556
CRIT. PRES., kPa	3480	3140–3620
FREEZING PT., C		–60, –90

COMBUSTION PROPERTIES	TYPICAL	RANGE
MASS % HYDROGEN	14.9	14.3–15.4
MASS C/H RATIO	5.71	5.5–6.0
HEAT OF COMBUSTION		
GROSS, MJ/kg	47.22	46.86–47.48
NET, MJ/kg	44.06	43.83–44.21

SAFETY PARAMETERS	TYPICAL	RANGE
FLASH POINT, C	–33	–20, –45
AUTO IGNITION, C	482	455–510
FLAMMABILITY LIMITS		
LOWER VOLUME %	1.34	1.27–1.40
UPPER VOLUME %	7.4	7.1–7.7
PROLONGED EXPOSURE LIMIT: 500 PPM		

TEMP. C	DENSITY kg/m³	COEF. THERMAL EXPN. 10^3/K	SECANT BULK MODULUS GPa	KINEMATIC VISCOSITY $10^{-6} \times m^2/s$	DYNAMIC VISCOSITY mPa×s	SURFACE TENSION N/m	VAPOR PRESSURE kPa	SPECIFIC HEAT kJ/kgK	HEAT OF VAPORIZATION kJ/kg	DIELECT. COEF.	WATER SOLUBILITY MASS %	OSTWALD COEFFICIENT O_2	N_2	AIR
–40	760.2	1.154	1.60	1.35	1.02	0.0251	1.50	1.912		2.05		0.309	0.137	0.180
–20	742.5	1.208	1.44	0.99	0.74	0.0231	4.42	2.006	386	2.02	0.0010	0.308	0.153	0.192
0	724.3	1.263	1.28	0.77	0.56	0.0212	11.2	2.098	374	1.99	0.0030	0.307	0.168	0.203
20	705.9	1.317	1.14	0.62	0.44	0.0193	29.8	2.188	362	1.96	0.0075	0.306	0.182	0.213
40	687.1	1.372	1.01	0.51	0.35	0.0174	49.8	2.276	349	1.93	0.0160	0.306	0.196	0.222
60	668.2	1.426	0.89	0.44	0.29	0.0156	92.0	2.361	335	1.90		0.306	0.208	0.230
80	649.0	1.481	0.78	0.38	0.24	0.0138	158.0	2.444	321	1.87		0.305	0.220	0.238
100	629.8	1.535	0.67	0.33	0.21	0.0120	258.0	2.525	305	1.83		0.305	0.231	0.245
120	610.4	1.590	0.58	0.29	0.18	0.0103	398.0	2.603	288	1.80		0.305	0.242	0.251
140	590.9	1.645	0.49			0.0086	591.0	2.680	271	1.77		0.304	0.252	0.257
160	571.5	1.699	0.41			0.0070	845.0	2.754	251	1.74		0.304	0.261	0.263
180	552.1	1.754	0.33			0.0055	1170.0	2.825	226	1.71		0.304	0.270	0.268
200	532.8	1.808	0.26			0.0040	1580.0	2.895	198	1.67		0.303	0.278	0.273
RANGE ±20		±5%	±7%	+30% / –20%	+35% / –20%	±0.0001	+15%	±0.050%	±10	±0.04	±50%	+11% / –10%	+11% / –10%	+11% / –10%

TABLE 2.25 Properties of Automobile Gasoline

DISTILLATION CURVE

VOLUME %	TEMPERATURE, C TYPICAL	RANGE
0	30	18–40
10	47	32–60
50	102	79–120
90	168	142–190
100	210	181–225

CHARACTERISTIC PROPERTIES

	TYPICAL	RANGE
REL. DENSITY	0.74	0.705–0.775
AVG. MOL. WT.	97	88–104
SPECIES MOL. Wt.		58–142
CRIT. TEMP., K	556	535–580
CRIT. PRES., kPa	4050	3770–4330
FREEZING PT., C		–70, –90

COMBUSTION PROPERTIES

	TYPICAL	RANGE
MASS % HYDROGEN	13.7	12.4–14.4
MASS C/H RATIO	6.30	5.9–7.1
HEAT OF COMBUSTION		
GROSS, MJ/kg	46.55	45.67–47.11
NET, MJ/kg	43.63	43.04–44.06

SAFETY PARAMETERS

	TYPICAL	RANGE
FLASH POINT, C	–43	–30, –50
AUTO IGNITION, C		257–300
FLAMMABILITY LIMITS		
LOWER VOLUME %	1.3	1.21–1.43
UPPER VOLUME %	7.6	6.7–7.7
PROLONGED EXPOSURE LIMIT:	500 PPM	

TEMP.	DENSITY	COEF. THERMAL EXPN.	SECANT BULK MODULUS	KINEMATIC VISCOSITY	DYNAMIC VISCOSITY	SURFACE TENSION	VAPOR PRESSURE	SPECIFIC HEAT	HEAT OF VAPOR- IZATION	DIELECT. COEF.	WATER SOLUBILITY	OSTWALD COEFFICIENT		
C	kg/m³	10^3/K	GPa	$10^6 \times$ m²/s	mPa × s	N/m	kPa	kJ/kgK	kJ/kg		MASS %	O_2	N_2	AIR
–40	789.2	1.092	1.70	1.55	1.23	0.0268	3.6	1.833		2.10		0.266	0.117	0.154
–20	771.8	1.140	1.53	1.08	0.84	0.0248	9.7	1.927	395	2.07		0.271	0.134	0.169
0	754.0	1.188	1.37	0.81	0.61	0.0228	23.0	2.018	384	2.04	0.003	0.276	0.151	0.182
20	736.0	1.236	1.23	0.63	0.47	0.0209	48.3	2.107	371	2.01	0.007	0.280	0.167	0.194
40	717.6	1.284	1.09	0.51	0.37	0.0191	92.3	2.194	358	1.98	0.020	0.283	0.182	0.206
60	699.1	1.332	0.96	0.42	0.30	0.0172	163	2.278	345	1.95		0.286	0.197	0.217
80	680.4	1.381	0.85	0.36	0.25	0.0155	271	2.360	332	1.92		0.289	0.211	0.227
100	661.5	1.429	0.74	0.31	0.21	0.0135	425	2.440	316	1.89		0.291	0.224	0.236
120	642.6	1.477	0.64	0.28	0.18	0.0118	637	2.517	300	1.86		0.294	0.237	0.245
140	623.6	1.525	0.55			0.0101	919	2.592	284	1.82		0.296	0.249	0.253
160	604.6	1.573	0.46			0.0085	1280	2.665	267	1.79		0.297	0.260	0.260
180	585.6	1.621	0.38			0.0068	1730	2.736	248	1.76		0.299	0.271	0.267
200	566.6	1.669	0.31			0.0052	2290	2.805	225	1.73		0.301	0.282	0.274
RANGE	±35	±10%	±15%	+25% / –15%	+30% / –20%	±0.002	+40% / –20%	±0.10%	±10	±0.06	±50%	+20% / –17%	+20% / –17%	+20% / –17%

TABLE 2.26 Properties of 1-D Diesel Fuels

DISTILLATION CURVE

VOLUME %	TEMPERATURE, C TYPICAL	RANGE
0	176	154–208
10	193	162–224
50	219	199–256
90	251	229–293
100	274	246–321

CHARACTERISTIC PROPERTIES

	TYPICAL	RANGE
REL. DENSITY	0.814	0.789–0.836
AVG. MOL. WT.	175	157–205
SPECIES MOL. WT.		150–226
CRIT. TEMP., K	685	663–720
CRIT. PRES., kPa	2250	2050–2410
POUR POINT, C	−40	−18, −54

COMBUSTION PROPERTIES

	TYPICAL	RANGE
MASS % HYDROGEN	13.5	12.8–14.3
MASS C/H RATIO	6.4	6.0–6.8
HEAT OF COMBUSTION		
GROSS, MJ/kg	46.09	45.58–46.65
NET, MJ/kg	43.23	42.85–43.62

SAFETY PARAMETERS

	TYPICAL	RANGE
FLASH POINT, C	59	40,80
AUTO IGNITION, C	255	
FLAMMABILITY LIMITS		
LOWER VOLUME %	0.58	0.53–0.61
UPPER VOLUME %	4.45	4.2–4.6
PROLONGED EXPOSURE LIMIT: 500 PPM		

TEMP C	DENSITY kg/m³	COEF. THERMAL EXPN. 10³/K	SECANT BULK MODULUS GPa	KINEMATIC VISCOSITY 10⁶ m²/s	DYNAMIC VISCOSITY mPa×s	SURFACE TENSION N/m	VAPOR PRESSURE kPa	SPECIFIC HEAT kJ/kgK	HEAT OF VAPORIZATION kJ/kg	DIELECT. COEF.	WATER SOLUBILITY MASS %	OSTWALD COEFFICIENT O₂	N₂	AIR
−20	839.4	0.852	1.78	6.71	5.63	0.0295		1.771	373	2.18	0.001	0.197	0.096	0.122
0	825.0	0.878	1.62	3.61	2.98	0.0279	0.044	1.867	363	2.16	0.003	0.210	0.115	0.139
20	810.4	0.904	1.46	2.28	1.84	0.0262	0.155	1.960	353	2.13	0.008	0.223	0.134	0.156
40	795.7	0.929	1.31	1.59	1.27	0.0247	0.460	2.050	343	2.11	0.019	0.234	0.153	0.172
60	780.9	0.955	1.18	1.19	0.93	0.0230	1.210	2.139	332	2.09	0.040	0.245	0.172	0.188
80	765.9	0.981	1.06	0.93	0.71	0.0215	2.83	2.225	321	2.06	0.080	0.254	0.191	0.203
100	750.8	1.007	0.94	0.75	0.56	0.0200	6.05	2.308	310	2.04	0.15	0.263	0.209	0.218
120	735.7	1.033	0.83	0.62	0.46	0.0184	12.00	2.389	300	2.01	0.25	0.272	0.227	0.232
140	720.4	1.059	0.73	0.53	0.38	0.0169	22.2	2.468	290	1.99	0.42	0.279	0.245	0.245
160	705.2	1.085	0.64	0.45	0.32	0.0154	38.9	2.545	280	1.96	0.66	0.287	0.262	0.258
180	689.9	1.110	0.55	0.39	0.27	0.0140	64.8	2.619	269	1.94	1.00	0.293	0.279	0.270
200	674.5	1.136	0.47	0.35	0.23	0.0126	103	2.690	258	1.91	1.50	0.300	0.296	0.282
220	659.2	1.162	0.40	0.31	0.21	0.0111	159	2.760	245	1.88		0.306	0.311	0.293
RANGE	±30	±7%	±7%	+65% / −20%	+70% / −25%	+0.0015 / −0.0010	+100% / −60%	±0.07%	±10	±0.04	±30%	+15% / −10%	+15% / −10%	+15% / −10%

153

TABLE 2.27 Properties of 2-D Diesel Fuels

DISTILLATION CURVE

VOLUME %	TEMPERATURE, C TYPICAL	RANGE
0	192	160–250
10	222	191–272
50	261	228–290
90	308	256–337
100	335	303–382

CHARACTERISTIC PROPERTIES

	TYPICAL	RANGE
REL. DENSITY	0.850	0.810–0.885
AVG. MOL. WT.	203	185–227
SPECIES MOL. Wt.		165–240
CRIT. TEMP., K	731	712–756
CRIT. PRES., kPa	2190	2030–2400
POUR POINT, C	−21	+30, −40

COMBUSTION PROPERTIES

	TYPICAL	RANGE
MASS % HYDROGEN	12.8	11.9–14.1
MASS C/H RATIO	6.8	6.1–7.4
HEAT OF COMBUSTION		
GROSS, MJ/kg	45.50	44.64–46.58
NET, MJ/kg	42.79	42.13–43.59

SAFETY PARAMETERS

	TYPICAL	RANGE
FLASH POINT, C	74	55,116
AUTO IGNITION, C	259	
FLAMMABILITY LIMITS		
LOWER VOLUME %	0.52	0.47–0.57
UPPER VOLUME %	4.10	3.8–4.4
PROLONGED EXPOSURE LIMIT:	500 PPM	

TEMP. C	DENSITY kg/m³	COEF. THERMAL EXPN. 10^3/K	SECANT BULK MODULUS GPa	KINEMATIC VISCOSITY $10^6 \times m^2/s$	DYNAMIC VISCOSITY mPa × s	SURFACE TENSION N/m	VAPOR PRESSURE kPa	SPECIFIC HEAT kJ/kgK	HEAT OF VAPORIZATION kJ/kg	DIELECT. COEF.	WATER SOLUBILITY MASS %	OSTWALD COEFFICIENT O_2	N_2	AIR
−20	875.3	0.791	1.92	18.20	15.92	0.0314		1.695	375	2.24	0.001	0.168	0.082	0.104
0	861.3	0.813	1.74	8.08	6.96	0.0298	0.02	1.791	365	2.22	0.003	0.185	0.101	0.122
20	847.3	0.835	1.58	4.46	3.78	0.0282	0.07	1.884	354	2.20	0.008	0.200	0.121	0.140
40	833.0	0.857	1.43	2.83	2.36	0.0267	0.21	1.975	344	2.17	0.019	0.214	0.141	0.158
60	818.7	0.879	1.29	1.98	1.62	0.0252	0.57	2.063	334	2.15	0.040	0.228	0.162	0.176
80	804.3	0.901	1.16	1.48	1.19	0.0237	1.36	2.149	325	2.13	0.080	0.240	0.183	0.193
100	789.7	0.923	1.04	1.16	0.92	0.0222	3.00	2.233	317	2.10	0.14	0.252	0.204	0.211
120	775.1	0.945	0.92	0.94	0.73	0.0207	6.00	2.314	309	2.08	0.24	0.264	0.225	0.227
140	760.4	0.967	0.82	0.78	0.59	0.0192	11.30	2.393	299	2.05	0.39	0.274	0.245	0.243
160	745.7	0.989	0.72	0.66	0.49	0.0178	20.1	2.470	288	2.03	0.61	0.284	0.266	0.259
180	730.9	1.012	0.63	0.57	0.42	0.0164	34.0	2.544	278	2.00	0.94	0.293	0.286	0.274
200	716.1	1.034	0.55	0.50	0.36	0.0150	55.0	2.616	267	1.98	1.40	0.302	0.306	0.289
220	701.3	1.056	0.47	0.44	0.31	0.0136	85.0	2.686	257	1.95		0.311	0.325	0.303
RANGE	±40	±8%	±10%	+60% / −45%	+65% / −50%	±0.001	+100% / −75%	±0.10	+10 / −20	±0.07	±30%	+25% / −17%	+25% / −17%	+25% / −17%

2.6.1 Derivation and Correlation of Tabulated Fuel Properties

For each of the fuel properties tabulated, the origin of the data or of the correlation from which it was derived is indicated below. The correlations are generally based on distillation characteristics and relative density (15.6 C/15.6 C). General equation forms to represent the properties tabulated as functions of temperature are given. The appropriate constants for these equations to represent typical values and envelopes of these properties for each fuel are given in Table 2.28. These equations are generally valid over the range of the tabulated data, some of which have already been extrapolated beyond the range of proven validity. The errors involved in these extrapolations are generally small relative to the variation of values among different samples of each fuel.

Distillation Characteristics. The typical distillation curves and the envelopes for all of the petroleum fuels, except aviation gasoline, have been derived from ASTM D 86 [ASTM (1990f)] distillation data reported by the DOE. The typical distillation curve for aviation gasoline is based on those published in several sources, and the envelope has been estimated on the basis of specification limits. Distillation curve slopes and average boiling points needed in some property correlations were also obtained from the distillation curves using methods in the API Technical Data Book [API (1966)].

Relative Density. The typical value and range of relative density (15.6 C/15.6 C) for each fuel was obtained from the sources cited above. The range of relative density for aviation gasoline was estimated on the basis of information from aviation gasoline suppliers.

Fuel Molecular Weight. The typical mean molecular weight and its range of variation for each petroleum fuel were determined from mean average boiling points and relative densities using an API nomograph [API (1966)]. The indicated ranges of species molecular weights within each fuel are based on published data [Panghorn and Gillis (1974)] for gasoline, kerosene, and diesel fuel, and on limits of the fuel distillation curves as compared to boiling points of pure hydrocarbons.

Critical Properties. The critical temperatures and pressures were obtained by methods suggested in the API Technical Data Handbook [API (1966)].

The critical temperature in deg. K is given by the equation

$$T_c = 358.75 + 0.9261\Delta - 0.0003959\Delta^2 \tag{2.63}$$

where $\Delta = \rho_{rel}(1.8\text{VABP} + 132)$, and VABP is the volume average boiling point in C. The critical pressures were obtained by a correlation with the VABP, the slope of the distillation curve, and ρ_{rel}, the relative density at 15.6 C.

Freezing and Pour Points. Typical values and ranges of freezing points of aviation turbine fuels and pour points of diesel fuels were obtained from the DOE inspection data for these fuels. The ranges of freezing point given for aviation and automobile gasolines are rough estimates based on a loose correlation between distillation end point and freezing point.

TABLE 2.28 Constants for Fuel Property Equations

FUEL	PROPERTY	DENSITY		KINEMATIC VISCOSITY		SURFACE TENSION		VAPOR PRESSURE		SPECIFIC HEAT			HEAT OF VAPORIZATION	
		ρ_T	$\beta_T \times 10^3$	A_1	$A_2 \times 10^{-10}$	T_c	K_f	$A_3 \times 10^{-6}$	A_4	A_5	$A_6 \times 10^3$	$A_7 \times 10^6$	A_8	A_9
JET A	TYPICAL	812	0.902	-4.13	1.500	680	11.77	4.78	-5007	0.364	6.30	-2.95	473	0.52
	MAXIMUM	840*	0.843	-4.13	1.820	696	11.43	4.10	-4748	0.413	6.30	-2.95	485	0.52
	MINIMUM	775*	0.990	-4.13	0.632	652	12.01	5.36	-5141	0.320	6.21	-2.90	454	0.46
JP-5	TYPICAL	820	0.884	-4.10	1.350	685	11.68	5.69	-5199	0.347	6.29	-2.94	466	0.49
	MAXIMUM	845*	0.833	-4.10	1.540	698	11.39	4.97	-4985	0.369	6.33	-2.96	487	0.52
	MINIMUM	788*	0.957	-4.10	1.090	680	11.86	5.58	-5234	0.301	6.22	-2.91	471	0.52
JP-4	TYPICAL	763	1.170	-4.14	0.850	594	11.83	1.05	-3404	0.497	6.15	-2.88	450	0.36
	MAXIMUM	802*	1.086	-4.14	1.190	636	11.55	1.28	-3404	0.494	6.26	-2.93	501	0.50
	MINIMUM	751*	1.199	-4.14	0.612	567	12.10	0.80	-3404	0.415	6.15	-2.87	473	0.39
AVIATION GASOLINE	TYPICAL	710	1.305	-3.50	0.0139	539	12.22	1.36	-3200	0.641	6.11	-2.86	518	0.45
	MAXIMUM	730	1.251	-3.50	0.0187	556	12.02	1.54	-3200	0.692	6.11	-2.86	534	0.48
	MINIMUM	690	1.364	-3.50	0.0108	523	12.43	1.17	-3200	0.592	6.11	-2.86	540	0.49
AUTOMOBILE GASOLINE	TYPICAL	740	1.226	-3.94	0.173	556	11.77	1.22	-2970	0.579	6.03	-2.82	514	0.44
	MAXIMUM	775	1.143	-3.94	0.227	579	11.37	1.98	-3010	0.665	6.06	-2.83	526	0.47
	MINIMUM	705	1.319	-3.94	0.140	535	12.16	1.34	-3140	0.497	5.98	-2.80	518	0.44
1D-DIESEL	TYPICAL	814	0.898	-4.14	1.780	685	11.80	4.12	-5010	0.360	6.32	-2.95	468	0.51
	MAXIMUM	836	0.851	-4.14	2.560	721	11.75	4.59	-4810	0.408	6.37	-2.98	484	0.56
	MINIMUM	789	0.955	-4.14	1.460	663	11.85	4.31	-5440	0.328	6.24	-2.91	457	0.49
2D-DIESEL	TYPICAL	850	0.830	-3.97	1.020	731	11.59	2.90	-5150	0.281	6.33	-2.96	460	0.52
	MAXIMUM	885	0.788	-3.97	1.330	756	11.58	1.38	-4520	0.350	6.47	-3.02	466	0.52
	MINIMUM	810	0.885	-3.97	0.656	712	11.62	6.34	-6010	0.216	6.25	-2.92	432	0.53

*SPECIFICATION LIMITS

Hydrogen Content and C/H Ratio. The typical values and ranges of mass percent hydrogen in the petroleum fuels were estimated from a correlation with fuel characterization factor and average boiling point [Hougen, *et al.* (1954)]. The minimum values for aviation turbine fuels are based on specification limits on API gravity and mid-boiling point [ASTM (1990a) and MIL (1989a)]. The C/H (carbon to hydrogen) mass ratios were obtained either by calculation from the hydrogen content or from a correlation with relative density and mean boiling point [API (1966)].

Heat of Combustion. Typical values and ranges of net heat of combustion for aviation turbine fuels and diesel fuels were obtained from the DOE inspection data; those for aviation and automobile gasolines are derived from a correlation with fuel relative density and characterization factor [API (1966)]. The gross heats of combustion for all of the fuels were calculated from the net heats and hydrogen contents by correcting for the heat of vaporization of the product water.

Flash Point. Typical values and ranges of flash points for JP-5, Jet A and diesel fuels were obtained from the DOE inspection data, while those for JP-4 and aviation and automobile gasolines are estimates based on a correlation with the 5% distillation point [Van Winkle (1954)].

Auto-Ignition Temperature. The auto-ignition temperatures tabulated here are based on data from several sources [Barnett and Hibbard (1956), Pangborn and Gillis (1974)]. Because only minimum values are usually reported, the typical and maximum values may not be representative.

Flammability Limits. Typical flammability limits were obtained from several sources, and the ranges were determined from correlations with molecular weight [Barnett and Hibbard (1956) and Pangborn and Gillis (1974)].

Toxicity. The maximum concentration of vapors allowable for prolonged exposure has been established at 500 parts per million in air for the various petroleum fuels [Pangborn and Gillis (1974)].

Density and Coefficient of Thermal Expansion. The densities and coefficients of thermal expansion were determined as a function of temperature using an ASTM procedure [ASTM (1980a)]. The coefficient of expansion in reciprocal degrees K is given by

$$\beta = \beta_T(1 + 1.6\beta_T(T - T_B)) \qquad (2.64)$$

where β_T is the value of β at any base temperature T_B and is a function of the density, ρ_T at T_B. The density in kg/m^3 is given by

$$\rho = \rho_T \exp\left[-\beta_T(T - T_B)(1 + 0.8\beta_T(T - T_B))\right] \qquad (2.65)$$

The appropriate values of β_T and ρ_T for a base temperature of $T_B = 288.6$ K are given in Table 2.28 for the various fuels.

Bulk Modulus. The isothermal secant bulk moduli at atmospheric pressure are tabulated for the various fuels. The values were determined from an API procedure [API (1966)] which may be represented by the following equations

$$B_S = m_B \frac{(B_R - 100{,}000)}{23170} + B_I \qquad (2.66)$$

where B_S is the isothermal bulk modulus in GPa.

$$B_R = 10^{-0.001098T + 5.235 + 0.0007133\rho}$$

$$m_B = 0.14924 + 0.0734p + 0.02098p^2$$

$$B_I = 0.1048 + 4.704p - 3.743p^2 + 2.2232p^3 \qquad (2.66a)$$

where ρ is the density in kg/m^3 at atmospheric pressure. $T(K)$ and $p(GPa)$ are the temperature and pressure of interest.

Viscosity. The tabulated kinematic viscosities of the various fuels as a function of temperature represent straight lines on ASTM D 341 viscosity charts [ASTM (1990g)]. The position of typical, minimum, and maximum curves for the aviation turbine fuels and diesel fuels were established from the DOE inspection data at specified temperatures. The position of the curves for aviation and automobile gasolines are based on correlations of the viscosity at specified temperatures with specific gravity and characterization factor [API (1966)]. These data agree well with the curves presented in other sources [CRC (1983) and Smith (1970)]. The slopes for all fuels are typical of each type [Barnett and Hibbard (1956)]. The data can be represented approximately by an equation of the form

$$\nu = (Z - \exp[-0.7487 - 3.295Z + 0.6119Z^2 - 0.3193Z^3]) \times 10^{-6} \qquad (2.67)$$

with

$$Z = \exp(A_2 T^{A_1}) - 0.7 \qquad (2.68)$$

and inversely,

$$Z = \nu \times 10^6 + \exp[-1.47 - 184 \times 10^6 \nu - 0.51 \times 10^{12} \nu^2]$$

where ν is the kinematic viscosity in m^2/s at a temperature $T(K)$ and A_1 and A_2 are constants. The appropriate values of A_1 and A_2 for typical and envelope curves for each fuel are given in Table 2.28. A better fit to the tabular data at the lower viscosities is obtained using the factors of ASTM D 341.

   The dynamic viscosities, μ (Pa $\cdot$ s), are the product of density (kg/m^3) and kinematic viscosity.

$$\mu = \rho \nu \qquad (2.69)$$

Surface Tension. The surface tensions tabulated for the various petroleum fuels were determined from a correlation [API (1966)] which may be represented by the equation

$$\sigma = \frac{0.605}{K_f} \left(1 - \frac{T}{T_c} \right)^{1.2}$$ (2.70)

where σ is the surface tension in N/m, T and T_c are the temperature of interest and critical temperature in K, and K_f is the fuel characterization factor. $K_f = 1.2164$ (MABP)$^{1/3}/\rho_{rel}$, where MABP(K) is the mean average boiling point.

For petroleum products of undefined species, MABP is empirically correlated with the volume average boiling point VABP(C). Based on an API chart [API (1966)]

$$\text{MABP} = \frac{5}{9} [\text{VABP} - (D_1 M + D_2 M^2 + D_3 M^3)] + 273.2$$ (2.71)

where $M = 1.8(T_{90}-T_{10})/80$ is the slope of the ASTM D 86 distillation curve between the 10% and 90% distillation temperatures, T_{10}(C) and T_{90}(C). The constants D_1, D_2, and D_3 are calculated as functions of VABP

$$\log_{10} D_1 = 0.724 - 0.0004565(\text{VABP})$$

$$\log_{10} D_2 = 0.00865 - 0.001702(\text{VABP})$$

$$D_3 = 0.08$$ (2.72)

The appropriate values of T_c and K_f for typical and envelope curves for the various fuels are included in Table 2.28. These values pertain to pure hydrocarbon fuels, and certain additives and contaminants in relatively small amounts can have order of magnitude effects on the surface tension.

The above equation is appropriate for experimentally determined static surface tension. For design calculations, a dynamic value, obtained by multiplying the static value by 1.125, should be used.

Vapor Pressure. The typical and envelope vapor pressures (p_v) for the JP-5 and Jet A aviation turbine fuels and diesel fuels were obtained from the ASTM D 86 distillation curves (reported in the DOE inspection data for the fuels) by determining the zero evaporation point of the equilibrium flash vaporization curves at various pressures according to an API procedure [API (1966)]. Pressures above atmospheric are extrapolations of log p_v versus $1/T$. Vapor pressure–temperature relationships for JP-4 and gasolines are based on Reid vapor pressure averages and extremes reported in the DOE inspection data and specification limits using a procedure from Barnett and Hibbard (1956). In all cases, the data may be approximated by an equation of the form

$$p_v = A_3 \exp \left(\frac{A_4}{T} \right)$$ (2.73)

Values of A_3 and A_4 to represent typical and envelope curves for p_v (kPa) and $T(K)$ for the various fuels are included in Table 2.28.

Specific Heat. The tabulated specific heats at constant pressure as a function of temperature result from an equation of the form

$$c_p = A_5 + A_6 T + A_7 T^2 \qquad (2.74)$$

where the constants A_5, A_6, and A_7 for c_p (kJ/kg · K) and $T(K)$ were evaluated by an API procedure [API (1966)] and are included in Table 2.28 for typical and envelope curves for the various fuels.

Heat of Vaporization. The tabulated heats of vaporization were derived from generalized enthalpy temperature charts [API (1966)] as a function of relative density and characterization factors. These data have been fitted to an equation of the following form [Hougen *et al.* (1954)].

$$h_v = A_8 \left(1 - \frac{T}{T_c} \right)^{A_9} \qquad (2.75)$$

For h_v (kJ/kg) and T and T_c in K, A_8 and A_9 are included in Table 2.28 for typical and envelope curves for the various fuels. At the high end of the temperature range, h_v's from the equation tend to be 5–10% higher than the tabulated values.

Thermal Conductivity. No consistent correlation of the thermal conductivity of petroleum fuels with other properties has been developed, but, in general, the following equation has been accurate within $\pm 5\%$. [API (1966)]

$$k = 0.1604 - 0.0000766T \qquad (2.76)$$

where the thermal conductivity, k is in $W/(m · K)$ and T is in K.

Dielectric Coefficient. The dielectric coefficient (ratio of dielectric constant of the fuel to that of a vacuum) correlates simply with the fuel density, within an accuracy of about $+2\%$, according to the equation [Barnett and Hibbard (1956)]

$$K = 0.001667\rho + 0.785 \qquad (2.77)$$

where ρ is the fuel density in kg/m³ at the temperature of interest. The tabulated values were obtained from this relationship.

Electrical Conductivity. Data for the electrical conductivity of petroleum fuels is sparse. The available data [CRC (1983), Smith (1970)] suggests that conductivities for a given fuel type at a given temperature may vary over orders of magnitude and that the dependence on temperature is roughly exponential. The conductivity, $\sigma(pS/M)$, may be expressed in the form

$$\sigma = A_{10} \exp (A_{11}T) \qquad (2.78)$$

where T is in K. Reasonable typical and envelope values of A_{10} and A_{11} for aviation turbine fuels in general and for gasolines are listed below:

	Turbine Fuels		Gasolines	
	A_{10}	A_{11}	A_{10}	A_{11}
Typical	0.019	0.017	0.155	0.018
Maximum	0.004	0.027	0.75	0.018
Minimum	0.008	0.017	0.08	0.018

The values for gasoline are much higher than those for turbine fuels due to additives in gasolines. At times, the conductivity of turbine fuels may also be greatly increased with additives. Conductivities of very clean diesel fuels would be comparable to those of turbine fuels, but the less stringent specification limits on contaminants can be expected to result in significantly higher values.

Water Solubility. The tabulated data for water solubility are derived from several sources [Barnett and Hibbard (1956), API (1966), and Smith (1970)]. Only limited data are available, but discrepancies suggest solubilities may be very sensitive to fuel composition, especially aromatic content and contaminants, thus the wide range of uncertainty.

The solubility is generally exponentially related to the reciprocal of the temperature. The following equation has been suggested in the absence of other data [API (1966)].

$$S_w = \frac{1800}{M} \times 10^{-[4200(H/C) + 1050][(1/T) - 0.0016]} \tag{2.79}$$

where the solubility, S_w, is in mass % of water, T is K, and M and H/C are the molecular weight and hydrogen to carbon mass ratio of the fuel. This relationship fits some of the tabulated data well and others, notably those for JP-4, poorly.

Gas Solubility. The solubility of nonreactive gases in hydrocarbon fuels can be characterized by the Ostwald coefficient, which is the ratio of the volume of gas dissolved to the volume of solvent, the gas volume being measured at the prevailing condition of gas partial pressure. Defined thus, the Ostwald coefficient is independent of pressure so long as Henry's Law remains valid. The Ostwald coefficients tabulated for oxygen, nitrogen, and air in the various fuels are based on the following correlation [ASTM (1990f)] which expresses Ostwald coefficient, L, as a function of temperature $T(K)$

$$L = A_{13} \exp\left[(0.0395(A_{12} - A_{14})^2 - 2.66)\left(1 - \frac{273}{T}\right) \right.$$
$$\left. - 0.303A_{12} - 0.024(17.6 - A_{14})^2 + 5.731 \right] \tag{2.80}$$

where the constant A_{12} is related to the fuel density at 15 C, $\rho_o(kg/m^3)$, by

$$A_{12} = 0.01203\rho_o + 7.36 \tag{2.81}$$

The constants A_{13} and A_{14} are functions of the gas species. Expressions for A_{13} and A_{14} for oxygen, nitrogen, and air are tabulated below

Solute Gas	A_{13}	A_{14}
Oxygen	1.28	7.75
Nitrogen	1.70	6.04
Air	1.44	6.67

These equations apply for all fuels, but a standard error of 18% is reported for the correlation.

2.7   THERMODYNAMIC AND TRANSPORT COMBUSTION PROPERTIES OF HYDROCARBONS WITH AIR
Sanford Gordon

Combustion compositions and properties for various conditions may be found in the literature [Gordon (1982a,b), GE (1955), Poferl and Svehla (1973), and Keenan (1980)]. For example, Gordon (1982a), (1982b) presents tables for a range of fuel hydrogen-to-carbon atom ratio, fuel-to-air mass ratios, pressures, and temperatures. The material presented in this section is a condensed version of Gordon's work. Only a few tables for stoichiometric conditions are reproduced here for purposes of illustration. Combustion compositions are presented in Tables 2.29 and 2.30. Properties are given in Tables 2.31 to 2.33. The properties presented are composition, density, molecular weight, enthalpy, entropy, specific heat at constant pressure, volume derivatives, isentropic exponent, velocity of sound, viscosity, thermal conductivity, and Prandtl number.

2.7.1   Calculation of Combustion Products

The usefulness of theoretical thermodynamic and transport properties in predicting results of physical processes depends to a large extent on how closely the assumptions used in the calculations approximate physical reality. In some cases, significantly different property values may be calculated for different assumptions. The property data presented here are calculated for three different assumptions concerning combustion compositions. These assumptions are: (1) constant composition for all temperatures and pressures, (2) chemical equilibrium with only homogeneous gas-phase species present, and (3) chemical equilibrium with heterogeneous-phase species.

Equilibrium compositions were obtained by means of a computer program [Gordon and McBride (1976)] which uses a free-energy minimization method. This program assumes that all gases are ideal and that interactions among phases can be neglected. An example of equilibrium compositions is given in Table 2.29. These compositions were obtained for the following conditions: a hydrogen-to-carbon atom ratio of 2, the 4-component air given in Sec. 2.3, stoichiometric fuel–air ratio, and a pressure of 1 atmosphere. For these conditions, out of 55 gaseous species included in the calculations, only the following 13 had a mole fraction greater than 0.000005

TABLE 2.29 Equilibrium Compositions (Mole Fractions)

T K	AR	CO	CO_2	H	HO_2	H_2	H_2O	N	NO	N_2	O	OH	O_2
900	0.00875	0.00000	0.13083	0.00000	0.00000	0.00000	0.13054	0.00000	0.00000	0.72988	0.00000	0.00000	0.00000
950	0.00875	0.00000	0.13083	0.00000	0.00000	0.00000	0.13054	0.00000	0.00000	0.72988	0.00000	0.00000	0.00000
1000	0.00875	0.00000	0.13083	0.00000	0.00000	0.00000	0.13054	0.00000	0.00000	0.72988	0.00000	0.00000	0.00000
1050	0.00875	0.00000	0.13083	0.00000	0.00000	0.00000	0.13054	0.00000	0.00000	0.72988	0.00000	0.00000	0.00000
1100	0.00875	0.00000	0.13083	0.00000	0.00000	0.00000	0.13054	0.00000	0.00000	0.72988	0.00000	0.00000	0.00000
1150	0.00875	0.00000	0.13083	0.00000	0.00000	0.00000	0.13054	0.00000	0.00000	0.72988	0.00000	0.00000	0.00000
1200	0.00875	0.00000	0.13083	0.00000	0.00000	0.00000	0.13053	0.00000	0.00000	0.72987	0.00000	0.00000	0.00000
1250	0.00875	0.00000	0.13083	0.00000	0.00000	0.00000	0.13053	0.00000	0.00000	0.72987	0.00000	0.00000	0.00000
1300	0.00875	0.00001	0.13082	0.00000	0.00000	0.00000	0.13053	0.00000	0.00000	0.72987	0.00000	0.00000	0.00001
1350	0.00875	0.00002	0.13082	0.00000	0.00000	0.00001	0.13052	0.00000	0.00000	0.72987	0.00000	0.00000	0.00001
1400	0.00875	0.00003	0.13080	0.00000	0.00000	0.00001	0.13052	0.00000	0.00001	0.72986	0.00000	0.00000	0.00002
1450	0.00875	0.00005	0.13078	0.00000	0.00000	0.00002	0.13051	0.00000	0.00001	0.72984	0.00000	0.00001	0.00003
1500	0.00875	0.00009	0.13074	0.00000	0.00000	0.00003	0.13049	0.00000	0.00002	0.72982	0.00000	0.00001	0.00005
1550	0.00875	0.00015	0.13067	0.00000	0.00000	0.00005	0.13046	0.00000	0.00003	0.72978	0.00000	0.00002	0.00008
1600	0.00875	0.00023	0.13058	0.00000	0.00000	0.00008	0.13042	0.00000	0.00005	0.72973	0.00000	0.00003	0.00012
1650	0.00875	0.00036	0.13044	0.00000	0.00000	0.00011	0.13037	0.00000	0.00007	0.72966	0.00000	0.00005	0.00019
1700	0.00875	0.00053	0.13025	0.00000	0.00000	0.00016	0.13029	0.00000	0.00011	0.72955	0.00000	0.00008	0.00027
1750	0.00875	0.00078	0.12998	0.00000	0.00000	0.00022	0.13019	0.00000	0.00016	0.72941	0.00000	0.00012	0.00039
1800	0.00875	0.00112	0.12961	0.00001	0.00000	0.00030	0.13005	0.00000	0.00022	0.72921	0.00000	0.00018	0.00055
1850	0.00874	0.00157	0.12913	0.00001	0.00000	0.00040	0.12986	0.00000	0.00030	0.72895	0.00001	0.00026	0.00076
1900	0.00874	0.00216	0.12849	0.00002	0.00000	0.00052	0.12963	0.00000	0.00041	0.72861	0.00001	0.00038	0.00104
1950	0.00874	0.00292	0.12766	0.00003	0.00000	0.00068	0.12933	0.00000	0.00055	0.72817	0.00002	0.00053	0.00139
2000	0.00873	0.00387	0.12662	0.00005	0.00000	0.00087	0.12895	0.00000	0.00073	0.72762	0.00003	0.00072	0.00182
2050	0.00872	0.00506	0.12533	0.00007	0.00000	0.00110	0.12847	0.00000	0.00094	0.72692	0.00005	0.00098	0.00236
2100	0.00872	0.00652	0.12374	0.00011	0.00000	0.00137	0.12788	0.00000	0.00121	0.72606	0.00008	0.00130	0.00301
2150	0.00870	0.00828	0.12182	0.00017	0.00000	0.00170	0.12716	0.00000	0.00153	0.72502	0.00012	0.00170	0.00378
2200	0.00869	0.01037	0.11954	0.00026	0.00000	0.00209	0.12629	0.00000	0.00191	0.72376	0.00018	0.00220	0.00470
2250	0.00868	0.01282	0.11686	0.00037	0.00000	0.00254	0.12525	0.00000	0.00236	0.72227	0.00028	0.00281	0.00576
2300	0.00866	0.01565	0.11376	0.00053	0.00000	0.00307	0.12401	0.00000	0.00288	0.72052	0.00041	0.00354	0.00696
2350	0.00864	0.01887	0.11024	0.00075	0.00000	0.00368	0.12255	0.00000	0.00347	0.71848	0.00060	0.00441	0.00832

TABLE 2.29 (*Continued*)

T K	AR	CO	CO₂	H	HO₂	H₂	H₂O	N	NO	N₂	O	OH	O₂
2400	0.00861	0.02247	0.10628	0.00104	0.00000	0.00437	0.12085	0.00000	0.00415	0.71614	0.00085	0.00543	0.00982
2450	0.00859	0.02644	0.10189	0.00143	0.00000	0.00515	0.11887	0.00000	0.00490	0.71347	0.00120	0.00661	0.01145
2500	0.00856	0.03076	0.09710	0.00194	0.00000	0.00602	0.11661	0.00000	0.00574	0.71047	0.00165	0.00795	0.01319
2550	0.00852	0.03539	0.09196	0.00260	0.00000	0.00700	0.11402	0.00000	0.00665	0.70711	0.00225	0.00946	0.01503
2600	0.00848	0.04026	0.08651	0.00343	0.00000	0.00809	0.11110	0.00000	0.00764	0.70340	0.00301	0.01115	0.01692
2650	0.00844	0.04531	0.08083	0.00449	0.00001	0.00928	0.10783	0.00000	0.00869	0.69933	0.00397	0.01299	0.01883
2700	0.00839	0.05046	0.07498	0.00581	0.00001	0.01056	0.10419	0.00000	0.00981	0.69490	0.00517	0.01499	0.02072
2750	0.00834	0.05562	0.06907	0.00743	0.00001	0.01195	0.10018	0.00000	0.01097	0.69012	0.00664	0.01712	0.02255
2800	0.00829	0.06071	0.06316	0.00942	0.00001	0.01341	0.09580	0.00000	0.01217	0.68498	0.00841	0.01935	0.02428
2850	0.00823	0.06565	0.05736	0.01182	0.00001	0.01494	0.09104	0.00000	0.01338	0.67951	0.01053	0.02165	0.02587
2900	0.00817	0.07035	0.05173	0.01468	0.00001	0.01651	0.08595	0.00001	0.01461	0.67370	0.01304	0.02398	0.02727
2950	0.00810	0.07474	0.04634	0.01806	0.00001	0.01809	0.08054	0.00001	0.01583	0.66757	0.01595	0.02629	0.02846
3000	0.00803	0.07879	0.04126	0.02200	0.00001	0.01965	0.07486	0.00001	0.01702	0.66115	0.01931	0.02852	0.02941

Fuel H/C Atom Ratio = 2.000; F/A = 0.067628; Equiv. Ratio = 1.000; Chem. Equiv. Ratio = 1.0000; P = 101.325 KPA (1.00 ATM) Dry Air

TABLE 2.30   Equilibrium Compositions (Mole Fractions)

T K	AR	CO_2	$H_2O(S)$	$H_2O(L)$	H_2O	N_2
			Pressure = 0.01 ATM			
200	0.00875	0.13083	0.13040	0.00000	0.00014	0.72988
220	0.00875	0.13083	0.12826	0.00000	0.00227	0.72988
240	0.00875	0.13083	0.10659	0.00000	0.02395	0.72988
			Pressure = 0.10 ATM			
200	0.00875	0.13083	0.13052	0.00000	0.00001	0.72988
220	0.00875	0.13083	0.13031	0.00000	0.00023	0.72988
240	0.00875	0.13083	0.12820	0.00000	0.00234	0.72988
260	0.00875	0.13083	0.11347	0.00000	0.01707	0.72988
280	0.00875	0.13083	0.00000	0.03665	0.09388	0.72988
			Pressure = 1.00 ATM			
200	0.00875	0.13083	0.13053	0.00000	0.00000	0.72988
220	0.00875	0.13083	0.13051	0.00000	0.00002	0.72988
240	0.00875	0.13083	0.13030	0.00000	0.00023	0.72988
260	0.00875	0.13083	0.12866	0.00000	0.00168	0.72988
280	0.00875	0.13083	0.00000	0.12198	0.00856	0.72988
298	0.00875	0.13083	0.00000	0.10261	0.02793	0.72988
300	0.00875	0.13083	0.00000	0.09926	0.03127	0.72988
320	0.00875	0.13083	0.00000	0.03034	0.10019	0.72988
			Pressure = 10.00 ATM			
200	0.00875	0.13083	0.13054	0.00000	0.00000	0.72988
220	0.00875	0.13083	0.13053	0.00000	0.00000	0.72988
240	0.00875	0.13083	0.13051	0.00000	0.00002	0.72988
260	0.00875	0.13083	0.13037	0.00000	0.00017	0.72988
280	0.00875	0.13083	0.00000	0.12969	0.00085	0.72988
298	0.00875	0.13083	0.00000	0.12782	0.00271	0.72988
300	0.00875	0.13083	0.00000	0.12751	0.00303	0.72988
320	0.00875	0.13083	0.00000	0.12146	0.00908	0.72988
340	0.00875	0.13083	0.00000	0.10682	0.02372	0.72988
360	0.00875	0.13083	0.00000	0.07458	0.05595	0.72988
380	0.00875	0.13083	0.00000	0.00701	0.12352	0.72988
			Pressure = 50.00 ATM			
200	0.00875	0.13083	0.13054	0.00000	0.00000	0.72988
220	0.00875	0.13083	0.13054	0.00000	0.00000	0.72988
240	0.00875	0.13083	0.13053	0.00000	0.00000	0.72988
260	0.00875	0.13083	0.13050	0.00000	0.00003	0.72988
280	0.00875	0.13083	0.00000	0.13037	0.00017	0.72988
298	0.00875	0.13083	0.00000	0.12999	0.00054	0.72988
300	0.00875	0.13083	0.00000	0.12993	0.00060	0.72988
320	0.00875	0.13083	0.00000	0.12874	0.00180	0.72988
340	0.00875	0.13083	0.00000	0.12589	0.00464	0.72988
360	0.00875	0.13083	0.00000	0.11989	0.01064	0.72988
380	0.00875	0.13083	0.00000	0.10835	0.02218	0.72988
400	0.00875	0.13083	0.00000	0.08766	0.04287	0.72988
420	0.00875	0.13083	0.00000	0.05232	0.07822	0.72988

Fuel H/C Atom Ratio = 2.000; F/A = 0.067628; Equiv. Ratio = 1.000; Chem. Equiv. Ratio = 1.0000; Dry Air.

TABLE 2.31 Properties Based on Constant Gaseous Compositions

T K	Density (P = 1.0) G/CM³	Density (P = 50.0) G/CM³	H J/G	Entropy (P = .10) J/G K	Entropy (P = .10) J/G K	Entropy (P = 1.0) J/G K	Entropy (P = 1.0) J/G K	Entropy (P = 50.0) J/G K	CP J/G K	GAM	VS M/S	Visc. Micropoise	Cond. Micro W/CM K	Pr
200	1.7613-3	8.8066-2	−2976.4	7.8377	7.1754	6.5131	5.8508	5.3879	1.0436	1.3805	281.8	114	152	0.7843
210	1.6774-3	8.3872-2	−2965.9	7.8887	7.2264	6.5641	5.9017	5.4388	1.0451	1.3797	288.7	119	160	0.7797
220	1.6012-3	8.0060-2	−2955.5	7.9373	7.2750	6.6127	5.9504	5.4875	1.0467	1.3789	295.4	125	169	0.7757
230	1.5316-3	7.6579-2	−2945.0	7.9839	7.3216	6.6593	5.9970	5.5340	1.0484	1.3781	301.9	130	177	0.7722
240	1.4678-3	7.3388-2	−2934.5	8.0286	7.3663	6.7039	6.0416	5.5787	1.0502	1.3772	308.3	136	185	0.7692
250	1.4091-3	7.0453-2	−2924.0	8.0715	7.4092	6.7469	6.0845	5.6216	1.0520	1.3763	314.6	141	193	0.7667
260	1.3549-3	6.7743-2	−2913.5	8.1128	7.4505	6.7881	6.1258	5.6629	1.0539	1.3754	320.7	146	201	0.7646
270	1.3047-3	6.5234-2	−2902.9	8.1526	7.4903	6.8280	6.1656	5.7027	1.0559	1.3744	326.7	151	209	0.7629
280	1.2581-3	6.2904-2	−2892.3	8.1910	7.5287	6.8664	6.2041	5.7411	1.0579	1.3734	332.6	156	217	0.7616
290	1.2147-3	6.0735-2	−2881.7	8.2282	7.5659	6.9036	6.2412	5.7783	1.0600	1.3724	338.4	161	225	0.7606
298	1.1815-3	5.9075-2	−2873.1	8.2576	7.5953	6.9330	6.2706	5.8077	1.0617	1.3716	343.0	165	231	0.7601
300	1.1742-3	5.8711-2	−2871.1	8.2642	7.6018	6.9395	6.2772	5.8143	1.0621	1.3714	344.0	166	232	0.7600
310	1.1363-3	5.6817-2	−2860.5	8.2990	7.6367	6.9744	6.3121	5.8491	1.0643	1.3703	349.6	171	239	0.7599
320	1.1008-3	5.5041-2	−2849.8	8.3328	7.6705	7.0082	6.3459	5.8830	1.0666	1.3693	355.0	176	247	0.7600
330	1.0675-3	5.3373-2	−2839.2	8.3657	7.7034	7.0411	6.3788	5.9158	1.0689	1.3682	360.4	181	254	0.7602
340	1.0361-3	5.1803-2	−2828.5	8.3976	7.7353	7.0730	6.4107	5.9478	1.0712	1.3671	365.6	185	261	0.7604
350	1.0065-3	5.0323-2	−2817.7	8.4287	7.7664	7.1041	6.4418	5.9788	1.0736	1.3660	370.8	190	268	0.7606
360	9.7851-4	4.8925-2	−2807.0	8.4590	7.7967	7.1344	6.4721	6.0091	1.0761	1.3648	375.9	194	275	0.7600
370	9.5206-4	4.7603-2	−2796.2	8.4885	7.8262	7.1639	6.5016	6.0386	1.0786	1.3637	381.0	199	283	0.7594
380	9.2701-4	4.6350-2	−2785.4	8.5173	7.8550	7.1927	6.5304	6.0674	1.0811	1.3625	385.9	203	290	0.7587
390	9.0324-4	4.5162-2	−2774.6	8.5454	7.8831	7.2208	6.5585	6.0956	1.0837	1.3613	390.8	208	297	0.7579
400	8.8066-4	4.4033-2	−2763.8	8.5729	7.9106	7.2483	6.5860	6.1230	1.0863	1.3601	395.6	212	305	0.7572
410	8.5918-4	4.2959-2	−2752.9	8.5998	7.9374	7.2751	6.6128	6.1499	1.0890	1.3589	400.3	217	312	0.7569
420	8.3872-4	4.1936-2	−2742.0	8.6260	7.9637	7.3014	6.6391	6.1762	1.0917	1.3577	405.0	221	319	0.7566
430	8.1922-4	4.0961-2	−2731.0	8.6518	7.9894	7.3271	6.6648	6.2019	1.0944	1.3565	409.6	225	326	0.7564
440	8.0060-4	4.0030-2	−2720.1	8.6770	8.0146	7.3523	6.6900	6.2271	1.0972	1.3553	414.2	229	333	0.7563

0.7562	340	233	418.7	1.3541	1.1000	6.2518	6.7147	7.3770	8.0393	8.7016	−2709.1	3.9140−2	7.8281−4	450
0.7561	346	238	423.1	1.3528	1.1029	6.2750	6.7389	7.4012	8.0635	8.7258	−2698.1	3.8289−2	7.6579−4	460
0.7561	353	242	427.5	1.3516	1.1057	6.2997	6.7627	7.4250	8.0873	8.7496	−2687.0	3.7475−2	7.4960−4	470
0.7562	360	246	431.8	1.3504	1.1086	6.3230	6.7860	7.4483	8.1106	8.7729	−2675.0	3.6694−2	7.3388−4	480
0.7653	367	250	436.1	1.3491	1.1115	6.3459	6.8089	7.4712	8.1335	8.7958	−2664.9	3.5945−2	7.1890−4	490
0.7563	374	254	440.3	1.3479	1.1145	6.3684	6.8313	7.4937	8.1560	8.8183	−2653.7	3.5226−2	7.0453−4	500
0.7561	380	257	444.5	1.3466	1.1175	6.3905	6.8534	7.5158	8.1781	8.8404	−2642.6	3.4536−2	6.9071−4	510
0.7559	387	261	448.6	1.3454	1.1205	6.4122	6.8752	7.5375	8.1998	8.8621	−2631.4	3.3871−2	6.7743−4	520
0.7556	394	265	452.7	1.3441	1.1235	6.4336	6.8965	7.5588	8.2212	8.8835	−2620.2	3.3232−2	6.6465−4	530
0.7553	401	269	456.7	1.3429	1.1265	6.4546	6.9176	7.5799	8.2422	8.9045	−2608.9	3.2617−2	6.5234−4	540
0.7550	408	273	460.7	1.3417	1.1296	6.4753	6.9383	7.6006	8.2629	8.9252	−2597.6	3.2024−2	6.4048−4	550
0.7465	415	277	464.7	1.3404	1.1326	6.4957	6.9586	7.6210	8.2833	8.9456	−2586.3	3.1452−2	6.2904−4	560
0.7543	422	280	468.6	1.3392	1.1357	6.5158	6.9787	7.6410	8.3033	8.9657	−2575.0	3.0900−2	6.1801−4	570
0.7539	429	284	472.5	1.3379	1.1388	6.5356	6.9985	7.6608	8.3231	8.9854	−2563.6	3.0368−2	6.0735−4	580
0.7535	536	288	476.3	1.3367	1.1419	6.5550	7.0180	7.6803	8.3426	9.0049	−2552.2	2.9853−2	5.9706−4	590
0.7531	443	292	480.1	1.3355	1.1450	6.5743	7.0372	7.6995	8.3618	9.0242	−2540.8	2.9355−2	5.8711−4	600
0.7526	450	295	483.8	1.3343	1.1482	6.5932	7.0562	7.7185	8.3808	9.0431	−2529.3	2.8874−2	5.7748−4	610
0.7522	457	299	487.6	1.3331	1.1513	6.6119	7.0749	7.7372	8.3995	9.0618	−2517.8	2.8408−2	5.6817−4	620
0.7517	465	302	491.3	1.3318	1.1544	6.6304	7.0933	7.7556	8.4179	9.0802	−2506.3	2.7957−2	5.5915−4	630
0.7512	472	306	494.9	1.3306	1.1576	6.6486	7.1115	7.7738	8.4361	9.0985	−2494.7	2.7521−2	5.5041−4	640
0.7507	479	310	498.6	1.3294	1.1607	6.6665	7.1295	7.7918	8.4541	9.1164	−2483.1	2.7097−2	5.4194−4	650
0.7502	486	313	502.2	1.3283	1.1639	6.6843	7.1472	7.8095	8.4719	9.1342	−2471.5	2.6687−2	5.3373−4	660
0.7497	493	317	505.7	1.3271	1.1670	6.7018	7.1647	7.8271	8.4894	9.1517	−2459.9	2.6288−2	5.2577−4	670
0.7492	500	320	509.3	1.3259	1.1702	6.7191	7.1821	7.8444	8.5067	9.1690	−2448.2	2.5902−2	5.1803−4	680
0.7487	507	324	512.8	1.3248	1.1734	6.7362	7.1992	7.8615	8.5238	9.1861	−2436.4	2.5526−2	5.1053−4	690
0.7483	514	327	516.2	1.3236	1.1765	6.7531	7.2161	7.8784	8.5407	9.2030	−2424.7	2.5162−2	5.0323−4	700
0.7478	521	331	519.7	1.3225	1.1796	6.7698	7.2328	7.8951	8.5574	9.2197	−2412.9	2.4807−2	4.9615−4	710
0.7473	529	334	523.1	1.3213	1.1828	6.7864	7.2493	7.9116	8.5739	9.2362	−2401.1	2.4463−2	4.8925−4	720
0.7468	536	337	526.5	1.3202	1.1859	6.8027	7.2656	7.9280	8.5903	9.2526	−2389.3	2.4128−2	4.8255−4	730
0.7464	543	341	529.9	1.3191	1.1891	6.8189	7.2818	7.9441	8.6064	9.2687	−2377.4	2.3802−2	4.7603−4	740

TABLE 2.31 (Continued)

T K	Density (P = 1.0) G/CM³	Density (P = 50.0) G/CM³	H J/G	Entropy (P = .10) J/G K	Entropy (P = .10) J/G K	Entropy (P = 1.0) J/G K	Entropy (P = 10.0) J/G K	Entropy (P = 50.0) J/G K	CP J/G K	GAM	VS M/S	Visc. Micropoise	Cond. Micro W/CM K	Pr
750	4.6968-4	2.3484-2	−2365.5	9.2847	8.6224	7.9601	7.2978	6.8348	1.1922	1.3180	533.2	344	550	0.7459
760	4.6350-4	2.3175-2	−2353.5	9.3005	8.6382	7.9759	7.3136	6.8507	1.1953	1.3169	536.5	347	557	0.7455
770	4.5748-4	2.2874-2	−2341.6	9.3162	8.6539	7.9916	7.3292	6.8663	1.1984	1.3158	539.8	351	564	0.7451
780	4.5162-4	2.2581-2	−2329.6	9.3317	8.6693	8.0070	7.3447	6.8818	1.2015	1.3148	543.1	354	571	0.7446
790	4.4590-4	2.2295-2	−2317.5	9.3470	8.6847	8.0224	7.3600	6.8971	1.2045	1.3137	546.4	357	578	0.7442
800	4.4033-4	2.2016-2	−2305.5	9.3622	8.6998	8.0375	7.3752	6.9123	1.2076	1.3127	549.6	361	585	0.7438
810	4.3489-4	2.1745-2	−2293.4	9.3772	8.7149	8.0526	7.3902	6.9273	1.2106	1.3116	552.8	364	592	0.7435
820	4.2959-4	2.1479-2	−2281.3	9.3921	8.7297	8.0674	7.4051	6.9422	1.2137	1.3106	556.0	367	599	0.7432
830	4.2441-4	2.1221-2	−2269.1	9.4068	8.7445	8.0822	7.4198	6.9569	1.2167	1.3096	559.2	370	606	0.7429
840	4.1936-4	2.0968-2	−2256.9	9.4214	8.7591	8.0967	7.4344	6.9715	1.2197	1.3086	562.3	373	613	0.7426
850	4.1443-4	2.0721-2	−2244.7	9.4358	8.7735	8.1112	7.4489	6.9859	1.2226	1.3076	565.4	377	620	0.7423
860	4.0961-4	2.0480-2	−2232.5	9.4501	8.7878	8.1255	7.4632	7.0003	1.2256	1.3067	568.5	380	627	0.7420
870	4.0490-4	2.0245-2	−2220.2	9.4643	8.8020	8.1397	7.4774	7.0144	1.2285	1.3057	571.6	383	634	0.7418
880	4.0030-4	2.0015-2	−2207.9	9.4784	8.8161	8.1538	7.4914	7.0285	1.2314	1.3048	574.7	386	641	0.7415
890	3.9580-4	1.9790-2	−2195.6	9.4923	8.8300	8.1677	7.5054	7.0424	1.2343	1.3038	577.7	389	658	0.7412
900	3.9140-4	1.9570-2	−2183.2	9.5061	8.8438	8.1815	7.5192	7.0562	1.2372	1.3029	580.8	392	655	0.7410
910	3.8710-4	1.9355-2	−2170.8	9.5198	8.8575	8.1952	7.5329	7.0699	1.2400	1.3020	583.8	395	662	0.7408
920	3.8289-4	1.9145-2	−2158.4	9.5334	8.8711	8.2087	7.5464	7.0835	1.2428	1.3011	586.8	398	669	0.7405
930	3.7878-4	1.8939-2	−2146.0	9.5468	8.8845	8.2222	7.5599	7.0969	1.2456	1.3003	589.8	401	675	0.7403
940	3.7475-4	1.8737-2	−2133.5	9.5602	8.8978	8.2355	7.5732	7.1103	1.2483	1.2994	592.7	405	682	0.7401
950	3.7080-4	1.8540-2	−2121.0	9.5734	8.9111	8.2488	7.5864	7.1235	1.2510	1.2986	595.7	408	689	0.7398
960	3.6694-4	1.8347-2	−2108.5	9.5865	8.9242	8.2619	7.5996	7.1366	1.2537	1.2977	598.6	411	696	0.7796
970	3.6316-4	1.8158-2	−2096.0	9.5995	8.9372	8.2749	7.6126	7.1496	1.2563	1.2969	601.5	414	703	0.7394
980	3.5945-4	1.7973-2	−2083.4	9.6124	8.9501	8.2878	7.6255	7.1625	1.2590	1.2961	604.5	417	709	0.7392
990	3.5582-4	1.7791-2	−2070.8	9.6252	8.9629	8.3006	7.6383	7.1753	1.2616	1.2953	607.3	419	716	0.7390

0.7388	723	422	610.2	1.2946	1.2641	7.1880	7.6509	8.3133	8.9756	9.6379	−2058.1	1.7613−2	3.5226−4	1000
0.7379	756	437	624.4	1.2910	1.2762	7.2500	7.7129	8.3752	9.0375	9.6999	−1994.6	1.6774−2	3.3549−4	1050
0.7371	788	451	638.3	1.2876	1.2877	7.3096	7.7726	8.4349	9.0972	9.7595	−1930.5	1.6012−2	3.2024−4	1100
0.7363	821	465	651.8	1.2845	1.2986	7.3671	7.8300	8.4923	9.1547	9.8170	−1865.9	1.5316−2	3.0632−4	1150
0.7357	852	479	665.1	1.2816	1.3091	7.4226	7.8855	8.5478	9.2102	9.8725	−1800.7	1.4678−2	2.9355−4	1200
0.7350	884	493	678.1	1.2789	1.3190	7.4762	7.9392	8.6015	9.2638	9.9261	−1735.0	1.4091−2	2.8181−4	1250
0.7344	915	506	690.8	1.2764	1.3284	7.5281	7.9911	8.6534	9.3157	9.9780	−1668.8	1.3549−2	2.7097−4	1300
0.7339	946	519	703.4	1.2740	1.3374	7.5784	8.0414	8.7037	9.3660	10.0283	−1602.1	1.3047−2	2.6094−4	1350
0.7333	977	532	715.6	1.2718	1.3460	7.6272	8.0902	8.7525	9.4148	10.0771	−1535.1	1.2581−2	2.5162−4	1400
0.7327	1007	545	727.7	1.2697	1.3541	7.6746	8.1376	8.7999	9.4622	10.1245	−1467.6	1.2147−2	2.4294−4	1450
0.7321	1037	558	739.6	1.2678	1.3618	7.7207	8.1836	8.8459	9.5082	10.1705	−1399.7	1.1742−2	2.3484−4	1500
0.7314	1067	570	751.3	1.2660	1.3692	7.7654	8.2284	8.8907	9.5530	10.2153	−1331.4	1.1363−2	2.2727−4	1550
0.7307	1097	583	762.8	1.2643	1.3761	7.8090	8.2720	8.9343	9.5966	10.2589	−1262.7	1.1008−2	2.2016−4	1600
0.7300	1127	595	774.1	1.2627	1.3827	7.8515	8.3144	8.9767	9.6390	10.3013	−1193.8	1.0675−2	2.1349−4	1650
0.7292	1156	607	785.3	1.2612	1.3889	7.8928	8.3558	9.0181	9.6804	10.3427	−1124.5	1.0361−2	2.0721−4	1700
0.7285	1185	619	796.3	1.2598	1.3949	7.9332	8.3961	9.0584	9.7207	10.3831	−1054.9	1.0065−2	2.0129−4	1750
0.7277	1214	631	807.2	1.2585	1.4005	7.9726	8.4355	9.0978	9.7601	10.4224	−985.0	9.7851−3	1.9570−4	1800
0.7270	1242	642	817.9	1.2572	1.4058	8.0110	8.4739	9.1363	9.7986	10.4609	−914.8	9.5206−3	1.9041−4	1850
0.7262	1270	654	828.5	1.2561	1.4108	8.0486	8.5115	9.1738	9.8361	10.4984	−844.4	9.2701−3	1.8540−4	1900
0.7254	1298	665	839.0	1.2550	1.4156	8.0853	8.5482	9.2105	9.8728	10.5351	−773.8	9.0324−3	1.8065−4	1950
0.7246	1326	677	849.4	1.2540	1.4201	8.1212	8.5841	9.2464	9.9087	10.5710	−702.9	8.8066−3	1.7613−4	2000
0.7239	1354	688	859.6	1.2530	1.4243	8.1563	8.6192	9.2815	9.9438	10.6062	−631.8	8.5918−3	1.7184−4	2050
0.7232	1381	699	869.7	1.2522	1.4284	8.1906	8.6536	9.3159	9.9782	10.6405	−560.4	8.3872−3	1.6774−4	2100
0.7225	1408	710	879.7	1.2513	1.4322	8.2243	8.6872	9.3496	10.0119	10.6742	−488.9	8.1922−3	1.6384−4	2150
0.7217	1434	721	889.6	1.2505	1.4358	8.2573	8.7202	9.3825	10.0448	10.7072	−417.2	8.0060−3	1.6012−4	2200
0.7209	1461	732	899.4	1.2498	1.4392	8.2896	8.7525	9.4148	10.0771	10.7395	−345.4	7.8281−3	1.5656−4	2250
0.7201	1487	742	909.0	1.2491	1.4425	8.3212	8.7842	9.4465	10.1088	10.7711	−273.3	7.6579−3	1.5316−4	2300
0.7193	1513	753	918.6	1.2484	1.4455	8.3523	8.8152	9.4776	10.1399	10.8022	−201.1	7.4950−3	1.4990−4	2350
0.7184	1540	764	928.1	1.2478	1.4485	8.3828	8.8457	9.5080	10.1703	10.8326	−128.8	7.3388−3	1.4678−4	2400
0.7176	1565	774	937.5	1.2472	1.4512	8.4127	8.8756	9.5379	10.2002	10.8625	−56.3	7.1890−3	1.4378−4	2450

TABLE 2.31 (*Continued*)

T	Density (P = 1.0)	Density (P = 50.0)	H	Entropy (P = .10)	Entropy (P = .10)	Entropy (P = 1.0)	Entropy (P = 10.0)	Entropy (P = 50.0)	CP	GAM	VS	Visc. Micro-	Cond. Micro-	Pr
K	G/CM³	G/CM³	J/G	J/G K	J/G K	J/G K	J/G K	J/G K	J/G K		M/S	poise	W/CM K	
2500	1.4091-4	7.0453-3	16.4	10.8919	10.2296	9.5673	8.9049	8.4420	1.4539	1.2466	946.8	784	1591	0.7167
2550	1.3814-4	6.9071-3	89.1	10.9207	10.2584	9.5961	8.9338	8.4708	1.4564	1.2461	956.0	795	1617	0.7157
2600	1.3549-4	6.7743-3	162.0	10.9490	10.2867	9.6244	8.9621	8.4991	1.4587	1.2456	965.0	805	1643	0.7146
2650	1.3293-4	6.6465-3	235.0	10.9768	10.3145	9.6522	8.9899	8.5269	1.4610	1.2451	974.2	815	1668	0.7137
2700	1.3047-4	6.5234-3	308.1	11.0041	10.3418	9.6795	9.0172	8.5543	1.4632	1.2447	983.2	825	1693	0.7127
2750	1.2810-4	6.4048-3	381.3	11.0310	10.3687	9.7064	9.0441	8.5811	1.4653	1.2443	992.1	835	1718	0.7118
2800	1.2581-4	6.2904-3	454.6	11.0574	10.3951	9.7328	9.0705	8.6075	1.4673	1.2438	1000.9	845	1743	0.7108
2850	1.2360-4	6.1801-3	528.0	11.0834	10.4211	9.7588	9.0965	8.6335	1.4692	1.2434	1009.6	854	1768	0.7100
2900	1.2147-4	6.0735-3	601.5	11.1090	10.4467	9.7844	9.1220	8.6591	1.4710	1.2431	1018.3	864	1793	0.7091
2950	1.1941-4	5.9706-3	675.1	11.1341	10.4718	9.8095	9.1472	8.6843	1.4728	1.2427	1026.9	874	1817	0.7083
3000	1.1742-4	5.8711-3	748.8	11.1589	10.4966	9.8343	9.1720	8.7090	1.4745	1.2424	1035.4	883	1841	0.7075

Fuel H/C Atom Ratio = 2.000; F/A = 0.067628; Equiv. Ratio = 1.000; Chem. Equiv. Ratio = 1.0000; MW = 28.9056; Gaseous Composition: CO_2 = .13083; H_2O = .13054; N_2 = .72988; O_2 = .00000; AR = .00875.

TABLE 2.32 Properties Based on Equilibrium Compositions

| | | | | | | Reacting Compositions | | | | | | | Frozen Compositions | | | | |
T K	Density G/CM³	H J/G	Entropy J/G K	MW	Visc. Micro-poise	DLVDLT	DLVDLP	CP J/G K	(GAM)S	VS M/S	Cond. Micro W/CM K	Pran.	CP J/G K	GAM	VS M/S	Cond. Micro W/CM K	Pr
900	3.9140-4	-2183.2	8.1815	28.906	392	1.0000	-1.0000	1.2372	1.3029	580.8	655	0.741	1.2372	1.3029	580.8	655	0.741
950	3.7080-4	-2121.0	8.2487	28.906	408	1.0000	-1.0000	1.2510	1.2986	595.7	689	0.740	1.2510	1.2986	595.7	689	0.740
1000	3.5226-4	-2058.1	8.3133	28.906	422	1.0000	-1.0000	1.2641	1.2946	610.2	723	0.739	1.2641	1.2946	610.2	723	0.739
1050	3.3549-4	-1994.6	8.3752	28.906	437	1.0000	-1.0000	1.2762	1.2910	624.4	756	0.738	1.2762	1.2910	624.4	756	0.738
1100	3.2024-4	-1930.3	8.4349	28.906	451	1.0000	-1.0000	1.2878	1.2876	638.3	789	0.737	1.2877	1.2876	638.3	788	0.737
1150	3.0632-4	-1865.9	8.4924	28.906	465	1.0000	-1.0000	1.2989	1.2845	651.8	821	0.736	1.2986	1.2845	651.8	821	0.736
1200	2.9355-4	-1800.7	8.5479	28.906	479	1.0000	-1.0000	1.3096	1.2815	665.1	853	0.736	1.3091	1.2816	665.1	852	0.736
1250	2.8181-4	-1734.9	8.6015	28.906	493	1.0001	-1.0000	1.3200	1.2787	678.0	885	0.735	1.3190	1.2789	678.1	884	0.735
1300	2.7097-4	-1668.7	8.6535	28.905	506	1.0001	-1.0000	1.3302	1.2760	690.7	917	0.734	1.3284	1.2764	690.9	915	0.734
1350	2.6093-4	-1601.9	8.7039	28.905	519	1.0002	-1.0000	1.3405	1.2733	703.2	949	0.733	1.3374	1.2740	703.4	946	0.734
1400	2.5161-4	-1534.6	8.7528	28.905	532	1.0004	-1.0000	1.3511	1.2707	715.3	981	0.733	1.3460	1.2718	715.7	977	0.733
1450	2.4293-4	-1466.8	8.8004	28.904	545	1.0006	-1.0000	1.3623	1.2681	727.3	1015	0.732	1.3541	1.2697	727.7	1007	0.733
1500	2.3483-4	-1398.4	8.8468	28.904	558	1.0010	-1.0000	1.3746	1.2653	738.9	1049	0.731	1.3618	1.2678	739.6	1037	0.732
1550	2.2724-4	-1329.3	8.8921	28.903	570	1.0015	-1.0000	1.3884	1.2623	750.2	1086	0.729	1.3691	1.2660	751.3	1067	0.731
1600	2.2013-4	-1259.5	8.9364	28.901	582	1.0023	-1.0001	1.4043	1.2590	761.3	1124	0.728	1.3761	1.2643	762.9	1097	0.731
1650	2.1344-4	-1188.8	8.9799	28.898	595	1.0033	-1.0001	1.4230	1.2554	772.0	1166	0.726	1.3826	1.2628	774.3	1127	0.730
1700	2.0714-4	-1117.1	9.0227	28.895	607	1.0048	-1.0001	1.4454	1.2514	782.4	1212	0.723	1.3889	1.2613	785.5	1156	0.729
1750	2.0119-4	-1044.2	9.0650	28.890	619	1.0068	-1.0002	1.4723	1.2468	792.4	1264	0.721	1.3947	1.2600	796.6	1185	0.728
1800	1.9555-4	-969.8	9.1069	28.884	630	1.0094	-1.0002	1.5045	1.2417	802.1	1322	0.717	1.4003	1.2588	807.6	1214	0.727
1850	1.9021-4	-893.6	9.1487	28.875	642	1.0127	-1.0003	1.5432	1.2361	811.5	1388	0.714	1.4055	1.2577	818.5	1242	0.727
1900	1.8513-4	-815.4	9.1904	28.864	654	1.0170	-1.0005	1.5894	1.2300	820.5	1465	0.709	1.4104	1.2566	829.3	1270	0.726
1950	1.8029-4	-734.6	9.2324	28.849	665	1.0223	-1.0006	1.6440	1.2233	829.2	1553	0.704	1.4151	1.2558	840.1	1298	0.725
2000	1.7567-4	-650.8	9.2728	28.830	676	1.0288	-1.0018	1.7081	1.2163	837.6	1655	0.698	1.4194	1.2550	850.8	1326	0.724
2050	1.7125-4	-563.6	9.3179	28.807	687	1.0368	-1.0011	1.7827	1.2091	845.8	1773	0.691	1.4235	1.2543	861.5	1353	0.723
2100	1.6701-4	-472.3	9.3618	28.779	698	1.0463	-1.0014	1.8685	1.2017	853.9	1910	0.683	1.4272	1.2538	872.2	1380	0.722
2150	1.6292-4	-376.5	9.4069	28.743	709	1.0575	-1.0018	1.9661	1.1943	861.8	2069	0.674	1.4307	1.2534	882.9	1407	0.721
2200	1.5899-4	-275.5	9.4534	28.701	719	1.0706	-1.0023	2.0760	1.1871	869.8	2253	0.663	1.4340	1.2532	893.7	1434	0.719
2250	1.5518-4	-168.7	9.5014	28.651	730	1.0857	-1.0029	2.1983	1.1803	877.9	2464	0.651	1.4370	1.2531	904.5	1461	0.718
2300	1.5150-4	-55.5	9.5511	28.592	740	1.1028	-1.0035	2.3329	1.1738	886.0	2705	0.638	1.4397	1.2531	915.5	1487	0.717
2350	1.4792-4	64.8	9.6028	28.523	750	1.1220	-1.0043	2.4794	1.1678	894.0	2981	0.624	1.4422	1.2533	926.6	1513	0.715

TABLE 2.32 *(Continued)*

						Reacting Compositions							Frozen Compositions			
T K	Density G/CM³	H J/G	Entropy J/G K	MW	Visc. Micro-poise	DLVDLT	DLVDLP	CP J/G K	(GAM)S	VS M/S	Cond. Micro W/CM K	Pran.	CP J/G K	GAM	VS M/S	Cond. Micro W/CM K	Pr
2400	1.4443-4	192.6	9.6567	28.444	760	1.1433	-1.0052	2.6371	1.1624	903.1	3294	0.609	1.4445	1.2537	937.8	1540	0.713
2450	1.4103-4	328.6	9.7128	28.353	770	1.1667	-1.0061	2.8048	1.1576	912.0	3647	0.592	1.4465	1.2543	949.3	1566	0.711
2500	1.3771-4	473.3	9.7712	28.251	780	1.1920	-1.0073	2.9816	1.1534	921.2	4045	0.575	1.4484	1.2550	960.9	1593	0.709
2550	1.3446-4	626.9	9.8320	28.136	789	1.2193	-1.0085	3.1662	1.1498	930.8	4490	0.556	1.4501	1.2559	972.8	1620	0.706
2600	1.3128-4	790.0	9.8954	28.008	798	1.2484	-1.0099	3.3573	1.1467	940.8	4986	0.537	1.4516	1.2571	985.0	1648	0.703
2650	1.2816-4	962.8	9.9612	27.868	807	1.2792	-1.0114	3.5538	1.1442	951.1	5535	0.518	1.4531	1.2584	997.4	1676	0.700
2700	1.2509-4	1145.5	10.0295	27.715	816	1.3115	-1.0130	3.7548	1.1421	961.8	6139	0.499	1.4544	1.2599	1010.2	1705	0.696
2750	1.2208-4	1338.3	10.1002	27.548	825	1.3455	-1.0148	3.9597	1.1405	972.9	6798	0.481	1.4556	1.2616	1023.3	1735	0.692
2800	1.1912-4	1541.5	10.1734	27.369	834	1.3810	-1.0167	4.1680	1.1393	984.5	7511	0.463	1.4568	1.2635	1036.7	1766	0.688
2850	1.1621-4	1755.2	10.2491	27.176	842	1.4179	-1.0188	4.3795	1.1385	996.4	8274	0.446	1.4580	1.2656	1050.5	1798	0.683
2900	1.1334-4	1979.5	10.3271	26.970	851	1.4562	-1.0210	4.5938	1.1380	1008.7	9080	0.430	1.4592	1.2679	1064.7	1832	0.677
2950	1.1051-4	2214.6	10.4075	26.752	859	1.4957	-1.0234	4.8101	1.1379	1021.4	9920	0.417	1.4604	1.2704	1079.2	1868	0.672
3000	1.0773-4	2460.5	10.4901	26.521	867	1.5361	-1.0259	5.0271	1.1380	1034.5	10777	0.405	1.4617	1.2730	1094.2	1906	0.665

TABLE 2.33 Low Temperature Properties Based on Equilibrium Compositions

	Heterogeneous Phase Properties					Gas Phase Properties								Cond.	
T K	Density G/CM^3	H J/G	Entropy J/G K	MW	CP J/G K	Density G/CM^3	MW	Visc. Micro-poise	DLVDLT	DLVDLP	CP J/G K	(GAM)S	VS M/S	Micro W/CM K	Pr
								Pressure = 0.01 ATM							
200	2.025–5	−3206.7	6.9031	28.906	1.0581	1.861–5	30.539	129	1.000	−1.000	0.9723	1.3889	275	170	0.734
220	1.837–5	−3182.3	7.0187	28.906	1.5478	1.690–5	30.508	139	1.000	−1.000	0.9769	1.3869	288	186	0.731
240	1.643–5	−3123.1	7.2739	28.906	5.7018	1.534–5	30.205	147	1.000	−1.000	0.9931	1.3835	302	199	0.736
								Pressure = 0.10 ATM							
200	2.026–4	−3206.9	6.3261	28.906	1.0242	1.861–4	30.540	129	1.000	−1.000	0.9722	1.3889	275	170	0.734
220	1.841–4	−3186.0	6.4258	28.906	1.0864	1.692–4	30.537	140	1.000	−1.000	0.9757	1.3870	288	187	0.730
240	1.684–4	−3161.4	6.5325	28.906	1.4937	1.549–4	30.507	150	1.000	−1.000	0.9805	1.3849	301	202	0.728
260	1.528–4	−3114.1	6.7203	28.906	3.8642	1.420–4	30.300	159	1.000	−1.000	0.9929	1.3819	314	216	0.730
280	1.306–4	−2949.1	7.3276	28.906	12.1790	1.276–4	29.320	160	1.000	−1.000	1.0392	1.3753	330	221	0.752
								Pressure = 1.00 ATM							
200	2.025–3	−3206.9	5.7501	28.906	1.0209	1.861–3	30.541	129	1.000	−1.000	0.9722	1.3889	275	170	0.734
220	1.841–3	−3186.3	5.8482	28.906	1.0405	1.692–3	30.540	140	1.000	−1.000	0.9756	1.3871	288	187	0.730
240	1.687–3	−3165.1	5.9406	28.906	1.0949	1.551–3	30.537	151	1.000	−1.000	0.9793	1.3851	301	203	0.727
260	1.555–3	−3141.4	6.0354	28.906	1.3368	1.430–3	30.517	161	1.000	−1.000	0.9839	1.3829	313	218	0.725
280	1.433–3	−3081.3	6.2565	28.906	2.1555	1.324–3	30.419	170	1.000	−1.000	0.9919	1.3804	325	233	0.726
298	1.316–3	−3029.3	6.4357	28.906	3.8234	1.232–3	30.151	177	1.000	−1.000	1.0069	1.3772	336	243	0.731
300	1.304–3	−3022.0	6.4602	28.906	4.0965	1.223–3	30.106	177	1.000	−1.000	1.0092	1.3768	338	244	0.733
320	1.135–3	−2895.1	6.8679	28.906	9.5454	1.114–3	29.246	179	1.000	−1.000	1.0511	1.3708	353	250	0.753

TABLE 2.33 (*Continued*)

T K	Heterogeneous Phase Properties					Gas Phase Properties									
	Density G/CM³	H J/G	Entropy J/G K	MW	CP J/G K	Density G/CM³	MW	Visc. Micro-poise	DLVDLT	DLVDLP	CP J/G K	(GAM)S	VS M/S	Cond. Micro W/CM K	Pr
Pressure = 10.00 ATM															
200	2.022-2	-3206.9	5.1742	28.906	1.0205	1.861-2	30.541	129	1.000	-1.000	0.9722	1.3889	275	170	0.734
220	1.839-2	-3186.4	5.2722	28.906	1.0359	1.692-2	30.541	140	1.000	-1.000	0.9756	1.3871	288	187	0.730
240	1.686-2	-3165.5	5.3631	28.906	1.0552	1.551-2	30.540	151	1.000	-1.000	0.9792	1.3851	301	203	0.727
260	1.556-2	-3144.0	5.4488	28.906	1.0936	1.431-2	30.538	161	1.000	-1.000	0.9830	1.3831	313	219	0.724
280	1.444-2	-3093.2	5.6358	28.906	1.3377	1.329-2	30.528	171	1.000	-1.000	0.9874	1.3809	325	234	0.723
298	1.353-2	-3067.7	5.7240	28.906	1.4932	1.247-2	30.502	180	1.000	-1.000	0.9922	1.3788	335	247	0.723
300	1.344-2	-3064.9	5.7333	28.906	1.5174	1.239-2	30.497	181	1.000	-1.000	0.9928	1.3785	336	248	0.723
320	1.252-2	-3031.0	5.8428	28.906	1.9335	1.158-2	30.411	190	1.000	-1.000	1.0007	1.3759	347	261	0.726
340	1.159-2	-2984.7	5.9828	28.906	2.7921	1.083-2	30.208	197	1.000	-1.000	1.0137	1.3727	358	273	0.731
360	1.057-2	-2913.9	6.1846	28.906	4.4673	1.008-2	29.783	202	1.000	-1.000	1.0366	1.3685	371	283	0.741
380	9.335-3	-2795.3	6.5045	28.906	7.7926	9.295-3	28.983	204	1.000	-1.000	1.0775	1.3628	385	291	0.757
Pressure = 50.00 ATM															
200	1.004-1	-3206.9	4.7717	28.906	1.0205	9.305-2	30.541	129	1.000	-1.000	0.9722	1.3889	275	170	0.734
220	9.133-2	-3186.4	4.8697	28.906	1.0355	8.459-2	30.541	140	1.000	-1.000	0.9756	1.3871	288	187	0.730
240	8.378-2	-3165.5	4.9605	28.906	1.0517	7.754-2	30.541	151	1.000	-1.000	0.9792	1.3851	301	203	0.727
260	7.738-2	-3144.3	5.0454	28.906	1.0721	7.157-2	30.540	161	1.000	-1.000	0.9830	1.3831	313	219	0.724
280	7.191-2	-3094.3	5.2293	28.906	1.2664	6.646-2	30.538	171	1.000	-1.000	0.9870	1.3809	324	234	0.723
298	6.753-2	-3071.0	5.3097	28.906	1.2986	6.240-2	30.533	180	1.000	-1.000	0.9910	1.3789	335	247	0.722
300	6.711-2	-3068.6	5.3178	28.906	1.3036	6.201-2	30.532	181	1.000	-1.000	0.9914	1.3787	336	249	0.722
320	6.285-2	-3041.8	5.4043	28.906	1.3877	5.811-2	30.515	191	1.000	-1.000	0.9964	1.3764	346	262	0.723
340	5.898-2	-3012.6	5.4928	28.906	1.5524	5.461-2	30.474	199	1.000	-1.000	1.0026	1.3739	357	276	0.725
360	5.535-2	-2978.9	5.5891	28.906	1.8432	5.144-2	30.389	208	1.000	-1.000	1.0107	1.3712	367	288	0.728
380	5.179-2	-2937.6	5.7005	28.906	2.3194	4.847-2	30.229	215	1.000	-1.000	1.0223	1.3681	378	301	0.732
400	4.813-2	-2884.3	5.8370	28.906	3.0644	4.563-2	29.952	221	1.000	-1.000	1.0392	1.3645	389	312	0.737
420	4.418-2	-2812.4	6.0123	28.906	4.2129	4.281-2	29.507	226	1.000	-1.000	1.0640	1.3602	401	323	0.746

for at least some of the temperatures in the range of 900–3000 K: Ar, CO, CO_2, H, HO_2, H_2, H_2O, N, NO, N_2, O, OH, and O_2. Thermodynamic and transport properties based on the compositions of Table 2.29 are given in Table 2.32. At low temperatures, condensation of water may occur, producing either ice or liquid water. Table 2.30 gives the heterogeneous equilibrium composition at various temperatures from 200–420 K for pressures of 0.01, 0.1, 1, 10, and 50 atmospheres. Properties based on these compositions are given in Table 2.33.

If there is no dissociation, then for stoichiometric conditions the following mole fractions are obtained for Ar, CO_2, H_2O, and N_2, respectively: 0.00875, 0.13083, 0.13054, and 0.72988. This composition was used to obtain the mixture properties of Table 2.31.

2.7.2 Equations for Calculating Mixture Properties

The equations used to obtain the thermodynamic and transport mixture properties are given in this section. Some mixture properties, such as density, enthalpy, molecular weight, and viscosity, depend on composition only. Entropy depends on composition and pressure. The assumptions concerning the calculation of composition have been discussed in the previous section. Other properties, such as thermodynamic derivatives and thermal conductivity, depend not only on composition but also on assumptions concerning the change of composition during some process in which temperature, pressure, or some other thermodynamic state functions may be changing. The two assumptions used to calculate change of composition with respect to temperature or pressure are: (1) instantaneous change of compositions to the new equilibrium condition, referred to here as *reacting compositions* and (2) zero change of compositions, referred to here as *frozen compositions*. Certain mixture properties (such as isentropic exponent, velocity of sound, viscosity, and conductivity) are difficult to define unambiguously for heterogeneous-phase mixtures. For this reason, some properties of heterogeneous systems (Table 2.33) are based on heterogeneous compositions, and others are based on compositions that have been normalized to the gas phase only. The parameters calculated for heterogeneous and gas-phase compositions are given in Table 2.34.

TABLE 2.34 Listing of Properties Based on Heterogeneous Composition or Gas Phase Only

Properties Based on Heterogeneous Compositions	Properties Based on Compositions Normalized to Gas Phase Only
Density	Density
Enthalpy	Molecular weight
Entropy	Viscosity
Molecular weight	Volume derivatives
Specific heat	Specific heat
	Isentropic exponent
	Velocity of sound
	Thermal conductivity
	Prandtl number

Note that density, molecular weight, and specific heat are given for both the heterogeneous mixture and the mixture normalized to the gas phase only (Table 2.33). The gas-phase values for density and molecular weight are the appropriate ones for the equation of state for gases which follows.

Equation of State for Gases. If it is assumed that all gases are ideal the equation of state for a gaseous mixture is

$$pv_g = n_g RT \tag{2.82}$$

where p is the pressure, R is the gas constant, T is the temperature, v_g is the specific volume of the gas, and n_g is the moles of the gas per unit mass.

Equation (2.82) applies only to gas-phase species even for those conditions where condensed species also exist.

Molecular Weight for Gases Only. Two alternative definitions will be given for the molecular weight of the gas phase only M_g. One definition is in terms of the mole fraction of the combustion species x_j, which can be defined as

$$x_j = \frac{n_j}{\displaystyle\sum_{j=1}^{NS} n_j} \tag{2.83}$$

where n_j is the number of moles of species j per unit mass of the mixture (and denotes the composition variables of Gordon and McBride (1976) and NS is an index indicating the total number of species present including both gas- and condensed-phase species. Molecular weight M_g is defined as

$$M_g = \frac{\displaystyle\sum_{j=1}^{NG} x_j M_j}{\displaystyle\sum_{j=1}^{NG} x_j} \tag{2.84}$$

where M_j is the molecular weight of species j, and NG is an index indicating the number of gases present. For heterogeneous mixtures, the denominator of Eq. (2.84) is less than 1. Alternatively M_g can be defined as the reciprocal of n_g appearing in Eq. (2.82).

$$M_g = \frac{1}{n_g} \tag{2.85}$$

Molecular Weight for Heterogeneous Mixtures. For heterogeneous mixtures, the molecular weight of the mixture M is defined as

$$M = \sum_{j=1}^{NS} x_j M_j \tag{2.86}$$

where

$$\sum_{j=1}^{NS} x_j = 1 \tag{2.87}$$

Density. The density of the gas phase ρ_g can be defined as the reciprocal of v_g appearing in Eq. (2.82).

$$\rho_g = \frac{1}{v_g} \tag{2.88}$$

Gordon (1982a) derived the following expression for the density of heterogeneous-phase mixtures

$$\rho = \frac{M}{\dfrac{RT}{p} \displaystyle\sum_{j=1}^{NG} x_j + \displaystyle\sum_{j=1}^{NC} \dfrac{x_j M_j}{\rho_j}} \tag{2.89}$$

where NC is an index indicating the number of condensed species present. When only gases are present ($NC = 0$), Eq. (2.89) reduces to the same form as Eq. (2.82).

Enthalpy. The expression for enthalpy is

$$h = \sum_{j=1}^{NS} n_j(H_T^o)_j = \frac{\displaystyle\sum_{j=1}^{NS} x_j(H_T^o)_j}{M} \tag{2.90}$$

where h is the enthalpy of the mixture and $(H_T^o)_j$ is the standard-state enthalpy for species j. The enthalpy base selected assigns a value of $H_{298.15} = 0$ to the reference elements Ar, C(s), H_2, O_2, and N_2.

Entropy. The expression for entropy is

$$s = \sum_{j=1}^{NS} n_j S_j = \frac{\displaystyle\sum_{j=1}^{NS} x_j S_j}{M} \tag{2.91}$$

where

$$S_j = \begin{cases} (S_T^o)_j - R \ln\left(\dfrac{x_j}{\displaystyle\sum_{j=1}^{NG} x_j}\right) - R \ln p_{atm} & j = 1, \ldots, NG \\[4mm] (S_T^o)_j & j = NG + 1, \ldots, NS \end{cases} \tag{2.92}$$

and where s is the entropy of the mixture, $(S_T^o)_j$ is the standard-state entropy of species j, and p_{atm} is the pressure in atmospheres.

Specific Heat at Constant Pressure. All thermodynamic first derivatives can be expressed in terms of any three independent first derivatives. The Bridgman tables express first derivatives in terms of $(\partial V/\partial T)_p$, $(\partial V/\partial p)_T$, and $(\partial n/\partial T)_p = c_p$.

In Sec. 2.7.1, it was pointed out that various assumptions can be used in calculating reaction compositions. In the calculation of thermodynamic derivatives, such as specific heat, additional assumptions can be used to specify the rate of change of composition with respect to temperature and pressure. As a result of various assumptions concerning compositions and rate of change of composition, several values of specific heat and other derivatives can be found in Tables 2.31 to 2.33 for the same conditions of temperature, pressure, and fuel-air ratio.

The following expression is obtained from the definition of enthalpy [Eq. (2.90)] in terms of n_j.

$$c_p = \left(\frac{\partial h}{\partial T}\right)_p = \sum_{j=1}^{NS} n_j (C_p^\circ)_j + \sum_{j=1}^{NS} (H_T^\circ)_j \left(\frac{\partial n_j}{\partial T}\right)_p \qquad (2.93)$$

where c_p is the specific heat at constant pressure and $(C_p^\circ)_j$ is the heat capacity of species j. The composition derivatives are obtained by the method of Gordon and McBride (1976).

If one assumes that a change in temperature produces a change in composition (reacting compositions), both terms on the right side of Eq. (2.93) contribute to specific heat. However, if one assumes that compositions remain fixed with changes in temperature (frozen compositions), only the first term contributes. An illustration of the large effects that changing compositions can make on specific heat is given in Fig. 2.9. The three curves were plotted from data in Tables 2.31 to 2.33. The large spike at low temperatures is due to the heat of transition from condensed water (liquid or solid) to gaseous water.

In addition to the two possible values of specific heat that may be obtained by means of Eq. (2.93), an additional value is given in Table 2.33 for heterogeneous mixtures. As will be discussed under Gas-phase Transport Properties, the transport mixture properties of viscosity and thermal conductivity are calculated for the gas phase only. Therefore, before calculating transport mixture properties, equilibrium compositions for heterogeneous mixtures are first normalized to gas-phase compositions only. To have specific heat consistent with the other transport properties, additional values are calculated for compositions normalized to gas-phase compositions only. In this case, specific heat becomes

$$(c_p)_g = \sum_{j=1}^{NG} n_j (C_p^\circ)_j + \sum_{j=1}^{NG} (H_T^\circ)_j \left(\frac{\partial n_j}{\partial T}\right)_p \qquad (2.94)$$

Note that Eq. (2.94) differs from Eq. (2.93) in that the summation index in Eq. (2.94) is for gases only, and n_j are relative to the mass of gases only.

Volume Derivatives. These derivatives are calculated for gas-phase compositions only. When heterogeneous-phase compositions are present, mole fractions are first normalized to gases only before the volume derivatives are calculated. From Eq. (2.82)

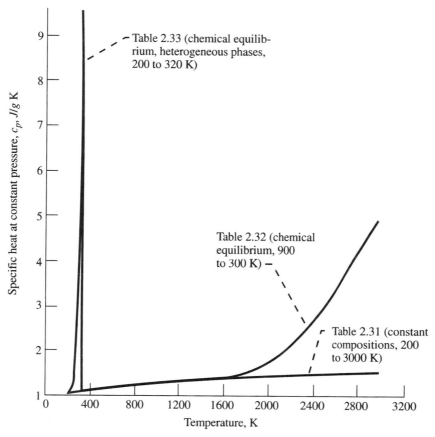

FIGURE 2.9 Specific heat, as a function of temperature, based on several assumptions concerning combustion. Stoichiometric equivalence ratio: pressure, $p_{atm} = 1$ atm, hydrogen–carbon atom ratio, H/C = 2, dry air.

$$\left(\frac{\partial \ln v_g}{\partial \ln T}\right)_p = 1 + \left(\frac{\partial \ln n_g}{\partial \ln T}\right)_p \tag{2.95}$$

$$\left(\frac{\partial \ln v_g}{\partial \ln p}\right)_T = -1 + \left(\frac{\partial \ln n_g}{\partial \ln p}\right)_T \tag{2.96}$$

The derivatives of n_g in Eq. (2.95) and (2.96) are obtained by matrix methods [Gordon and McBride (1976)].

Isentropic Exponent and Velocity of Sound. In Tables 2.31 to 2.33, only the velocity of sound of a gas is calculated even when the system is heterogeneous. Therefore, the isentropic exponent γ_s, which is related to the velocity of sound a, is calculated for the gas phase only. For heterogeneous mixtures, compositions are first normalized to gases only before γ_s is calculated. The isentropic exponent is defined to be

$$\gamma_s = \left(\frac{\partial \ln p}{\partial \ln \rho_g}\right)_s = -\left(\frac{\partial \ln p}{\partial \ln v_g}\right)_s \tag{2.97}$$

This expression for γ_s is related to the velocity of sound a as follows

$$a^2 = \left(\frac{\partial p}{\partial \rho_g}\right)_s = \frac{p}{\rho_g}\left(\frac{\partial \ln p}{\partial \ln \rho_g}\right)_s = \frac{p\gamma_s}{\rho_g} \tag{2.98}$$

By using the equation of state given in Eq. (2.82), the familiar expression for velocity of sound can be obtained.

$$a = \sqrt{\gamma_s n_g RT} = \sqrt{\frac{\gamma_s RT}{M_g}} \tag{2.99}$$

Specific Heat Ratio. The isentropic exponent γ_s is often confused with the specific heat ratio, which is defined to be

$$\gamma \equiv \frac{c_p}{c_v} \tag{2.100}$$

In this report, γ is assumed to apply to gases only.

The two *gammas* defined in Eqs. (2.97) and (2.100) are related as follows

$$\gamma_s = \frac{-\gamma}{\left(\dfrac{\partial \ln v_g}{\partial \ln p}\right)_T} \tag{2.101}$$

The two *gammas* are equal if and only if $(\partial \ln v_g/\partial \ln p)_T = -1$. Note that γ_s, defined by Eq. (2.97), is required in Eq. (2.99) and not the specific heat ratio, defined by Eq. (2.100).

Gas-Phase Transport Properties. The equations describing the calculation of gas-phase transport properties of mixtures are given in Svehla and McBride (1973). Because of the large amount of detail and special terminology used to present these equations, the equations will not be reproduced here. Only a very simple form of these equations will be given. Note that the method of Svehla and McBride (1973) is restricted to the gas phase, and allowances are not made for the effect of *condensed species* in *heterogeneous* systems. Only gaseous species are used in the transport property calculations. Viscosity can be written as

$$\mu = \frac{\displaystyle\sum_{j=1}^{NG} x_j \mu_j}{\displaystyle\sum_{j=1}^{NG} x_j} \tag{2.102}$$

where μ is the mixture viscosity and μ_j are variables whose values are obtained by the solution of a set of simultaneous algebraic equations. It should be emphasized that the μ_j in Eq. (2.102) are not the viscosities of the pure species.

Thermal conductivity consists of several terms that can be written as

$$k = k_{\text{trans}} + k_{\text{int}} + k_{\text{reaction}} = k_{\text{frozen}} + k_{\text{reaction}} \tag{2.103}$$

where k is the mixture thermal conductivity and where the subscripts refer to the translational, internal, reaction, and frozen contributions to thermal conductivity. The equations for obtaining these contributions are given in Svehla and McBride (1973). Values for k_{frozen} are given in Table 2.32 under the heading "Frozen Compositions"; values for k are given under the heading "Reacting Compositions."

Prandtl number is defined as

$$\text{Pr} = \frac{c_p \mu}{k} \tag{2.104}$$

In Table 2.32, the values of Prandtl number under the headings "Reacting Compositions" are obtained by means of Eq. (2.104) with the appropriate values for c_p and k together with the value given for μ.

These tables are from computer outputs which did not permit use of Greek letters, lower-case letters, subscripts, or superscripts. Therefore, some symbols in the tables differ from the text.

The following sources were used for physical constants [Cohen and Taylor (1973)], atomic weights [Holden (1979)], and ideal thermodynamic properties of pure species [JANNAF (1979)]. The sources for transport properties of pure species are given in Sec. 2.4.

2.8 CRYOGENIC GASES
Neil Olien

TABLE 2.35 Properties of Helium

Thermophysical Properties of Liquid Helium Along the Saturation Line

Temperature, K	Pressure, MPa	Liquid Density, kg/m³	Vapor Density, kg/m³	Enthalpy, kJ/kg	Entropy, kJ/kg · K	Heat of Vaporization, kJ/kg	Specific Heat Cp, kJ/kg · K	$\frac{1}{V}\left(\frac{\partial V}{\partial T}\right)_P$ 1/K	$-\frac{1}{V}\left(\frac{\partial V}{\partial P}\right)_T$ 1/MPa	Sonic Velocity m/s	Dielectric Constant	Viscosity kg/m · s	Thermal Conductivity mW/m · K	Prandtl Number	Thermal Diffusivity m²/h
2.177	0.00504	146.2	1.177	—	—	—	—	—	—	—	—	—	—	—	—
2.2	0.00533	146.1	1.235	−11.45	1.782	22.22	3.16	0.0238	0.149	216.1	1.05500	36.1×10^{-7}	14.8	0.770	1.156×10^{-4}
2.4	0.00834	145.3	1.805	−10.91	2.008	22.47	2.25	0.0374	0.156	217.4	1.05468	37.1×10^{-7}	16.0	0.521	1.764×10^{-4}
2.5	0.01021	144.8	2.144	−10.68	2.097	22.62	2.13	0.0430	0.161	217.9	1.05448	37.3×10^{-7}	16.5	0.482	1.926×10^{-4}
2.6	0.01235	144.2	2.521	−10.46	2.179	22.76	2.10	0.0482	0.166	217.7	1.05424	37.3×10^{-7}	16.8	0.468	1.994×10^{-4}
2.8	0.01753	142.8	3.401	−9.99	2.336	22.97	2.22	0.0582	0.181	215.5	1.05371	37.2×10^{-7}	17.4	0.475	1.973×10^{-4}
3.0	0.02402	141.1	4.463	−9.503	2.493	23.09	2.42	0.0688	0.200	211.5	1.05309	36.8×10^{-7}	17.8	0.500	1.876×10^{-4}
3.2	0.03197	139.3	5.731	−8.951	2.653	23.08	2.67	0.0808	0.224	206.5	1.05238	36.2×10^{-7}	18.2	0.529	1.768×10^{-4}
3.4	0.04155	137.2	7.235	−8.341	2.817	22.93	2.95	0.0950	0.256	201.0	1.05157	35.5×10^{-7}	18.6	0.562	1.656×10^{-4}
3.5	0.04699	136.0	8.085	−8.013	2.900	22.80	3.10	0.103	0.276	198.0	1.05113	35.1×10^{-7}	18.8	0.580	1.602×10^{-4}
3.6	0.05289	134.8	9.008	−7.667	2.985	22.63	3.28	0.113	0.298	194.9	1.05066	34.7×10^{-7}	19.0	0.599	1.544×10^{-4}
3.8	0.06614	132.1	11.09	−6.919	3.161	22.15	3.68	0.135	0.355	188.4	1.04963	33.8×10^{-7}	19.2	0.646	1.426×10^{-4}
4.0	0.08147	129.0	13.56	−6.084	3.345	21.46	4.19	0.164	0.435	181.2	1.04845	32.8×10^{-7}	19.5	0.707	1.296×10^{-4}
4.2	0.09902	125.4	16.49	−5.146	3.540	20.55	4.88	0.204	0.551	173.0	1.04710	31.8×10^{-7}	19.6	0.791	1.156×10^{-4}
4.224	0.10133	125.0	16.89	−5.023	3.564	20.41	4.98	0.210	0.571	172.0	1.04692	31.7×10^{-7}	19.6	0.804	1.134×10^{-4}
4.4	0.1190	121.3	20.05	−4.078	3.750	19.32	5.86	0.264	0.737	163.8	1.04553	30.7×10^{-7}	19.7	0.916	0.997×10^{-4}

4.5	0.1299	118.9	22.13	-3.484	3.863	18.56	6.55	0.307	0.881	158.6	1.04463	30.1×10^{-7}	19.9	0.994	0.918×10^{-4}
4.6	0.1416	116.3	24.49	-2.838	3.984	17.68	7.44	0.365	1.074	153.1	1.04364	29.5×10^{-7}	20.0	1.10	0.832×10^{-4}
4.8	0.1670	110.1	30.37	-1.339	4.255	15.40	10.5	0.574	1.821	140.7	1.04125	28.2×10^{-7}	20.5	1.45	0.634×10^{-4}
5.0	0.1954	101.1	39.30	0.6496	4.605	11.89	19.9	1.23	4.406	126.0	1.03785	26.4×10^{-7}	22.2	2.37	0.396×10^{-4}
5.1	0.2109	94.09	46.80	2.083	4.858	9.017	38.5	2.59	10.20	117.3	1.03520	25.1×10^{-7}	26.3	3.68	0.262×10^{-4}
5.2014	0.2275	69.54	69.54	6.620	5.600	—	—	—	—	0	—	—	—	—	—

Source: McCarty, R.D., *J. Phys. Chem. Ref. Data* **2**, pp. 923–1042, 1973.

Fixed Point Properties

Molecular Weight = 4.0026

Lambda Point:	Temperature = 2.177 K;	Pressure = 0.005035 MPa;	Liquid Density = 146.2 kg/m³;
Normal Boiling Point:	Temperature = 4.224 K;	Pressure = 0.10133 MPa;	Liquid Density = 125.0 kg/m³;
Critical Point:	Temperature = 5.2014 K;	Pressure = 0.22796 MPa;	Density = 69.64 kg/m³;

Standard Conditions:

STP	Temperature = 273.15 K;	Pressure = 0.10133 MPa;	Gas Density = 0.1785 kg/m³;
NTP	Temperature = 298.15 K;	Pressure = 0.10133 MPa;	Gas Density = 0.1637 kg/m³;

Vapor Density = 1.177 kg/m³
Vapor Density = 16.89 kg/m³

Other Properties

Refractive Index *n*: at NBP (4.224 K) $n = 1.02461$ for $\lambda = 435.8$ nm; $n = 1.02449$ for $\lambda = 546.2$ nm; $n = 1.02440$ for $\lambda = 693.9$ nm

Surface Tension σ: at 3 K, $\sigma = 0.2247 \times 10^{-3}$ N/m; at 4 K, $\sigma = 0.1226 \times 10^{-3}$ N/m; at 4.224 K, $\sigma = 0.0997 \times 10^{-3}$ N/m; at 5 K, $\sigma = 0.02055 \times 10^{-3}$ N/m

Flammability Limits (in Air): Lower limit = ; Upper Limit = Helium is nonflammable

Toxicity: Helium is nontoxic and therefore is rated as a simple asphyxiant

TABLE 2.36 Properties of Hydrogen (Parahydrogen)

Thermophysical Properties of Liquid Parahydrogen Along the Saturation Line

Temperature, K	Pressure, MPa	Liquid Density, kg/m^3	Vapor Density, kg/m^3	Enthalpy, kJ/kg	Entropy, kJ/kg · K	Heat of Vaporization kJ/kg	Specific Heat Cp, kJ/kg · K	$\frac{1}{V}\left(\frac{\partial V}{\partial T}\right)_p$ 1/K	$-\frac{1}{V}\left(\frac{\partial V}{\partial P}\right)_T$ 1/MPa	Sonic Velocity m/s	Dielectric Constant	Viscosity kg/m · s	Thermal Conductivity mW/m · K	Prandtl Number	Thermal Diffusivity m^2/h
13.803	0.0070	77.0235	0.1254	−308.9	4.964	449.2	6.36	0.01022	0.01107	1264	1.25160	255.05×10^{-7}	72.59	2.2359	0.00053
14.0	0.0079	76.8554	0.1388	−307.5	5.057	449.6	6.47	0.01046	0.01132	1256	1.25101	248.01×10^{-7}	74.62	2.1517	0.00054
15.0	0.0134	75.9963	0.2228	−300.7	5.525	451.8	6.91	0.01111	0.01226	1228	1.24799	217.84×10^{-7}	83.02	1.8128	0.00057
16.0	0.0216	75.1030	0.3385	−293.4	5.989	453.0	7.36	0.01181	0.01328	1200	1.24486	194.15×10^{-7}	88.86	1.6074	0.00058
17.0	0.0329	74.1710	0.4922	−285.6	6.453	453.2	7.88	0.01286	0.01469	1168	1.24159	175.07×10^{-7}	92.71	1.4884	0.00057
18.0	0.0482	73.1950	0.6902	−277.3	6.917	452.3	8.42	0.01394	0.01619	1140	1.23818	159.34×10^{-7}	95.43	1.4063	0.00056
19.0	0.0682	72.1693	0.9395	−268.4	7.382	450.1	8.93	0.01486	0.01756	1120	1.23460	146.10×10^{-7}	97.00	1.3451	0.00054
20.0	0.0935	71.0870	1.2474	−258.9	7.851	446.5	9.45	0.01585	0.01910	1100	1.23083	134.76×10^{-7}	98.40	1.2942	0.00053
20.268	0.10133	70.7864	1.3378	−256.2	7.977	445.5	9.66	0.01644	0.01992	1089	1.22978	131.98×10^{-7}	98.92	1.2893	0.00052
21.0	0.1250	69.9403	1.6189	−248.8	8.323	441.8	10.13	0.01754	0.02160	1070	1.22684	124.87×10^{-7}	100.05	1.2648	0.00051
22.0	0.1634	68.7200	2.0711	−237.9	8.802	435.1	10.82	0.01921	0.02428	1044	1.22261	116.11×10^{-7}	100.95	1.2444	0.00049
23.0	0.2096	67.4149	2.6119	−226.3	9.287	426.8	11.69	0.02162	0.02818	1010	1.21810	108.22×10^{-7}	101.20	1.2499	0.00046
24.0	0.2645	66.0112	3.2548	−213.8	9.783	416.4	12.52	0.02394	0.03241	980	1.21325	101.01×10^{-7}	100.84	1.2543	0.00044
25.0	0.3288	64.4917	4.0171	−200.4	10.291	404.0	13.44	0.02665	0.03774	948	1.20802	94.32×10^{-7}	99.93	1.2682	0.00042
26.0	0.4035	62.8337	4.9215	−185.9	10.813	389.0	14.80	0.03116	0.04668	903	1.20233	88.03×10^{-7}	98.43	1.3240	0.00038
27.0	0.4892	61.0065	5.9999	−170.1	11.356	370.9	16.17	0.03585	0.05736	860	1.19608	82.02×10^{-7}	96.38	1.3764	0.00035

28.0	0.5871	58.9665	7.2979	−152.8	11.926	18.48	0.04417	0.07623	806	1.18913	76.19 × 10⁻⁷	93.78	1.5012	0.00031
29.0	0.6978	56.6460	8.8866	−133.6	12.535	22.05	0.05763	0.10873	747	1.18125	70.42 × 10⁻⁷	90.55	1.7146	0.00026
30.0	0.8225	53.9303	10.8872	−111.7	13.200	26.59	0.07585	0.15938	689	1.17208	64.57 × 10⁻⁷	86.57	1.9834	0.00022
31.0	0.9627	50.5892	13.5411	−85.8	13.960	36.55	0.11759	0.28343	619	1.16087	58.40 × 10⁻⁷	85.85	2.4889	0.00017
32.0	1.1198	45.9927	17.4983	−52.4	14.917	65.37	0.24599	0.72015	534	1.14555	51.31 × 10⁻⁷	91.46	3.6672	0.00011
32.976	1.2928	31.4285	31.4285	38.3	17.565	—	—	—	0	1.09795	35.43 × 10⁻⁷	—	—	—

Source: McCarty, R. D., Hord, J., and Roder, H. M., *Nat. Bur. Stand. Monograph 168*, 1981.

Fixed Point Properties

Molecular Weight = 2.01594

Triple Point:	Temperature = 13.803 K;	Pressure = 7042.1 Pa;	Liquid Density = 77.03 kg/m³;	Vapor Density = 0.126 kg/m³
Normal Boiling Point:	Temperature = 20.268 K;	Pressure = 0.10133 MPa;	Liquid Density = 70.78 kg/m³;	Vapor Density = 1.338 kg/m³
Critical Point:	Temperature = 32.976 K;	Pressure = 1.2928 MPa;	Density = 31.43 kg/m³;	

Standard Conditions:

| STP | Temperature = 273.15 K; | Pressure = 0.10133 MPa; | Gas Density = 0.08963 kg/m³; |
| NTP | Temperature = 298.15 K; | Pressure = 0.10133 MPa; | Gas Density = 0.08376 kg/m³; |

Other Properties

Refractive Index n: at NBP (20.268 K) n = 1.1136 for λ = 435.8 nm; n = 1.1109 for λ = 546.2 nm; n = 1.1007 for λ = 693.9 nm

Surface Tension σ: at 15 K, σ = 2.7921 × 10⁻³ N/m; at 20.268 K, σ = 1.9299 × 10⁻³ N/m at 25 K, σ = 1.1751 × 10⁻³ N/m; at 30 K, σ = 0.4112 × 10⁻³ N/m

Heat of Fusion: at Triple Point (13.803 K) ΔH = 58.17 kJ/kg

Flammability Limits (in Air): Lower Limit = 4%; Upper Limit = 75%

Toxicity: Hydrogen is nontoxic and therefore is rated as a simple asphyxiant

186

TABLE 2.37 Properties of Methane

Thermophysical Properties of Liquid Methane Along the Saturation Line

Temperature, K	Pressure, MPa	Liquid Density, kg/m³	Vapor Density, kg/m³	Enthalpy, kJ/kg	Entropy, kJ/kg · K	Heat of Vaporization kJ/kg	Specific Heat Cp, kJ/kg · K	$\frac{1}{V}\left(\frac{\partial V}{\partial T}\right)_P$ 1/K	$-\frac{1}{V}\left(\frac{\partial V}{\partial P}\right)_T$ 1/MPa	Sonic Velocity m/s	Dielectric Constant	Viscosity kg/m · s	Thermal Conductivity mW/m · K	Prandtl Number	Thermal Diffusivity m²/h
90.680	0.01174	451.56	0.2515	−360.1	4.213	544.2	3.346	0.002925	0.00144	1544	1.6782	2036×10^{-7}	223.9	3.0425	5.34×10^{-4}
95.0	0.0199	445.83	0.407	−345.5	4.370	538.5	3.407	0.003008	0.00156	1501	1.6681	1793×10^{-7}	215.5	2.8344	5.11×10^{-4}
100.0	0.0345	439.07	0.668	−328.4	4.546	531.2	3.447	0.003124	0.00172	1453	1.6561	1564×10^{-7}	206.2	2.6146	4.90×10^{-4}
105.0	0.0566	432.12	1.046	−311.0	4.715	523.3	3.475	0.003259	0.00192	1405	1.6438	1377×10^{-7}	197.3	2.4258	4.73×10^{-4}
110.0	0.0884	424.96	1.570	−293.5	4.877	514.7	3.501	0.003412	0.00214	1355	1.6313	1223×10^{-7}	188.8	2.2674	4.57×10^{-4}
111.632	0.10133	422.57	1.778	−287.7	4.929	511.7	3.510	0.003466	0.00223	1338	1.6271	1178×10^{-7}	186.0	2.2218	4.52×10^{-4}
115.0	0.1326	417.56	2.273	−275.8	5.034	505.4	3.530	0.003585	0.00241	1304	1.6184	1092×10^{-7}	180.5	2.1363	4.41×10^{-4}
120.0	0.1919	409.92	3.194	−257.9	5.185	495.3	3.565	0.003784	0.00273	1251	1.6051	981×10^{-7}	172.5	2.0285	4.25×10^{-4}
125.0	0.2693	402.00	4.373	−239.7	5.331	484.3	3.609	0.004014	0.00313	1197	1.5914	886×10^{-7}	164.7	1.9411	4.09×10^{-4}
130.0	0.3681	393.76	5.855	−221.3	5.474	472.4	3.664	0.004284	0.00360	1141	1.5773	802×10^{-7}	157.1	1.8716	3.92×10^{-4}
135.0	0.4912	385.18	7.691	−202.5	5.613	459.4	3.733	0.004606	0.00420	1084	1.5627	729×10^{-7}	149.6	1.8185	3.75×10^{-4}
140.0	0.6422	376.20	9.942	−183.4	5.749	445.2	3.821	0.004997	0.00496	1025	1.5474	663×10^{-7}	142.3	1.7810	3.56×10^{-4}
145.0	0.8244	366.76	12.681	−163.7	5.883	429.5	3.932	0.005483	0.00595	965	1.5315	605×10^{-7}	135.1	1.7595	3.37×10^{-4}
150.0	1.0414	356.78	15.998	−143.6	6.016	412.3	4.074	0.006098	0.00727	904	1.5148	551×10^{-7}	128.0	1.7550	3.17×10^{-4}
155.0	1.2966	346.17	20.011	−122.7	6.147	393.1	4.259	0.006898	0.00909	841	1.4972	502×10^{-7}	120.9	1.7703	2.95×10^{-4}
160.0	1.5939	334.77	24.879	−101.1	6.279	371.7	4.507	0.007972	0.01169	776	1.4784	457×10^{-7}	113.9	1.8104	2.72×10^{-4}
165.0	1.9373	322.38	30.829	−78.5	6.411	347.5	4.849	0.009478	0.01564	710	1.4582	415×10^{-7}	106.8	1.8841	2.46×10^{-4}
170.0	2.3308	308.66	38.211	−54.5	6.546	319.7	5.349	0.011729	0.02204	641	1.4360	375×10^{-7}	99.9	2.0073	2.18×10^{-4}
175.0	2.7793	293.05	47.614	−28.7	6.687	287.0	6.151	0.015448	0.03371	569	1.4111	336×10^{-7}	93.3	2.2133	1.86×10^{-4}

180.0	3.2883	274.37	60.214	0.1	6.839	246.6	7.681	0.022820	0.05957	491	1.3817	296×10^{-7}	87.6	2.5965	1.50×10^{-4}
185.0	3.8646	249.41	79.163	34.8	7.017	191.6	12.027	0.044852	0.14876	403	1.3430	252×10^{-7}	83.6	3.6262	1.00×10^{-4}
190.0	4.5199	195.42	126.558	96.4	7.329	73.7	126.212	0.686638	3.69514	285	1.2622	182×10^{-7}	93.1	24.6721	0.14×10^{-4}
190.555	4.5988	160.43	160.43	132.4	7.516	0.0	—	—	—	0	1.2118	149×10^{-7}	—	—	—

Source: Goodwin, R. D., *Nat. Bur. Stand. Tech. Note 653*, 1974.

Fixed Point Properties

Molecular Weight = 16.043

Triple Point:	Temperature = 90.680 K;	Pressure = 0.01174 MPa;	Liquid Density = 451.56 kg/m^3
Normal Boiling Point:	Temperature = 111.532 K;	Pressure = 0.10133 MPa;	Liquid Density = 422.57 kg/m^3
Critical Point:	Temperature = 190.555 K;	Pressure = 4.5988 MPa;	Density = 160.43 kg/m^3

Standard Conditions:

STP	Temperature = 273.15 K;	Pressure = 0.10133 MPa;	Gas Density = 0.04472 kg/m^3
NTP	Temperature = 298.15 K;	Pressure = 0.10133 MPa;	Gas Density = 0.04095 kg/m^3

Other Properties

Refractive Index n: at 90.680; $n = 1.29581$ for $\lambda = 546.2$ nm; $n = 1.27761$ at 110 K; $n = 1.23226$ at 150 K; $n = 1.10333$ at 190.555 K

Heat of Fusion: at Triple Point (90.680 K, $\Delta H = 57.87$ kJ/kg

Flammability Limits (in Air): Lower limit = 5.3%; Upper Limit = 15%

Toxicity: Methane is nontoxic and therefore is rated as a simple asphyxiant

2.9 HYDRAULIC FLUIDS
Allen E. Fuhs

TABLE 2.38 Specifications for Hydraulic Fluids

Type of Hydraulic Fluid	Petroleum Base				Synthetic Fire-Resistant Non-neurotoxic	Synthetic	Petroleum Base Aircraft,
Application	Machine Tools				Accumulator Loaded Systems Operating in Excess of 600 psi (4 MPa)	Aircraft	Aircraft, Missile, Ordnance
Reference	MIL (1982)				MIL (1981)	MIL (1986)	MIL (1978)
	GRADE 1	GRADE 2	GRADE 3	GRADE 4			
Kinematic Viscosity, Centistokes							
@ 204 C	—	—	—	—	—	2.5	—
@ 100 C	—	—	—	—	4.8 (min)	—	4.9
@ 40 C	28.8 to 35.2	41.4 to 50.6	61.2 to 74.8	135 to 165	38.5 to 45.5	—	13.2
@ −40 C	—	—	—	—	—	—	600
@ −54 C	—	—	—	—	—	2500	2500
Viscosity Index (min)	90	90	90	90	—	—	—
Pour Point, C	−12	−12	−12	−6	−18	−60 (max)	−60 (max)
Flash Point, C (min)	188	196	196	220	—	202	82
Fire Point, C (min)	216	218	218	246	—	371	—
Acid Number, (max)	—	—	—	—	0.1	0.2	0.2
API Gravity, degrees	30 to 33	28 to 32	29 to 31	27 to 46	—	—	—
Water, percent (max)	none	none	none	none	none	0.01	100 parts/million

For definition of terms, see MIL (1978), (1981), (1982), (1986), Hatton (1962), and Holzbock (1968).

2.10 HEAT TRANSFER FLUIDS
Peter E. Liley

TABLE 2.39 Saturated Mercury

T (K)	P (kPa)	v_f (m³/kg)	Δh_v (kJ/kg)	c_{pf} (kJ/kgK)	μ_f (10^{-4} Pa · s)	k_f (W/mK)	Pr_f (−)
250	2.21.−6	7.325.−5	307.58	0.1415	18.81	7.84	0.0339
260	6.93.−6	7.339.−5	307.31	0.1409	17.91	8.00	0.0315
270	1.99.−5	7.352.−5	306.94	0.1405	17.12	8.14	0.0295
280	4.96.−5	7.365.−5	306.57	0.1401	16.44	8.27	0.0279
290	1.24.−4	7.378.−5	306.21	0.1397	15.86	8.41	0.0263
300	2.91.−4	7.392.−5	305.84	0.1393	15.19	8.54	0.0248
350	0.01022	7.459.−5	304.11	0.1377	13.06	9.18	0.0196
400	0.1394	7.526.−5	302.43	0.1366	11.71	9.80	0.0163
450	1.053	7.595.−5	300.80	0.1357	10.93	10.27	0.0144
500	5.261	7.664.−5	299.19	0.1353	10.60	10.93	0.0131
600	56.95	7.807.−5	295.92	0.1356	9.11	11.9	0.0104
700	315.3	7.957.−5	291.28	0.1372	8.50	12.8	0.0091
800	1118	8.118.−5	287.81	0.1398	8.08	13.6	0.0083
900	2990	8.292.−5	282.08	0.1439	7.78	14.2	0.0079
1000	6575	8.482.−5	274.66	0.1492	7.55	14.7	0.0077

Interpolated, extrapolated, or converted from the tables of Vukalovich, M. P., Ivanov, A. I., Fokin, L. R., and Yakovlev, A. T., "Thermophysical Properties of Mercury," *Izd. Standartov*, Moscow, U.S.S.R., 312 pp., 1971.

TABLE 2.40 Saturated Potassium

T (K)	P (kPa)	v_f (m³/kg)	Δh_v (kJ/kg)	c_{pf} (kJ/kgK)	μ_f (10^{-4} Pa · s)	k_f (W/mK)	Pr_f (−)
336	1.37.−7	1.208.−3	2333	0.82	5.44	49.9	0.0089
350	3.82.−6	1.213.−3	2230	0.82	5.04	49.5	0.0083
400	1.84.−5	1.229.−3	2196	0.81	4.13	48.0	0.0070
450	3.21.−4	1.247.−3	2180	0.79	3.53	46.6	0.0060
500	3.13.−3	1.266.−3	2165	0.79	3.01	45.2	0.0053
600	9.26.−2	1.304.−3	2130	0.77	2.38	42.3	0.0043
700	1.022	1.346.−3	2088	0.76	1.98	39.4	0.0038
800	6.116	1.389.−3	2042	0.76	1.71	36.5	0.0036
900	24.41	1.437.−3	1994	0.77	1.51	33.6	0.0034
1000	73.22	1.488.−3	1942	0.79	1.35	30.7	0.0035
1100	186.4	1.546.−3	1890	0.82	1.23	27.8	0.0036
1200	396.3	1.605.−3	1836	0.85	1.14		
1300	730.4	1.672.−3	1783	0.87	1.05		
1400	1244	1.742.−3	1731	0.90	0.98		
1500	2000	1.816.−3	1678	0.92	0.93		

Interpolated, extrapolated, or converted from tables of Vargaftik, N. B., *Tables of the Thermophysical Properties of Liquids and Gases*, Halstead Press, 758 pp. New York, 1975.

TABLE 2.41   Saturated Potassium/Sodium Eutectic

T (K)	P (kPa)	v_f (m³/kg)	Δh_v (kJ/kg)	c_{pf} (kJ/kgK)	μ_f (10^{-4} Pa · s)	k_f (W/mK)	Pr_f (−)
262	2.83.−11	1.137.−3		1.35	9.99	21.5	0.063
270	8.84.−11	1.140.−3		1.34	9.45	21.6	0.059
280	3.34.−10	1.143.−3		1.32	8.83	21.8	0.054
290	1.15.−9	1.146.−3		1.30	8.27	22.1	0.049
300	3.66.−9	1.149.−3		1.29	7.77	22.2	0.045
310	1.08.−8	1.152.−3		1.27	7.32	22.4	0.042
320	2.97.−8	1.155.−3		1.25	6.90	22.6	0.038
330	7.70.−8	1.158.−3		1.22	6.53	22.8	0.035
340	1.89.−7	1.162.−3		1.20	6.21	23.9	0.032
350	4.40.−7	1.165.−3		1.18	5.85	23.0	0.030
360	9.76.−7	1.169.−3		1.16	5.61	23.2	0.028
370	2.08.−6	1.172.−3		1.15	5.39	23.3	0.027
380	4.25.−6	1.175.−3		1.14	5.17	23.4	0.025
390	8.37.−6	1.178.−3		1.13	4.96	23.5	0.024
400	1.59.−5	1.182.−3		1.12	4.76	23.6	0.023
450	2.60.−5	1.200.−3		1.11	4.01	24.2	0.018
500	2.43.−3	1.218.−3		1.06	3.47	24.8	0.015
550	0.0151	1.237.−3		1.05	3.06	25.4	0.013
600	0.0694	1.257.−3		1.03	2.75	26.1	0.011
650	0.2521	1.277.−3		1.01	2.49	26.7	0.009
700	0.7610	1.297.−3		0.992	2.27	27.4	0.008
750	1.983	1.319.−3		0.975	2.08	28.1	0.007
800	4.583	1.339.−3		0.958	1.93	28.7	0.006
850	9.600	1.361.−3		0.942	1.79	29.3	0.006
900	18.52	1.384.−3		0.925	1.67	29.9	0.005
950	33.35	1.407.−3		0.908	1.56	30.6	0.005
1000	56.6	1.431.−3		0.891	1.45	31.2	0.004

Interpolated, extrapolated, or converted from tables of Vargaftik, N. B., *Tables of the Thermophysical Properties of Liquids and Gases*, Halstead Press, 758 pp. New York, N.Y., 1975.

TABLE 2.42   Saturated Sodium

T (K)	P (kPa)	v_f (m³/kg)	Δh_v (kJ/kg)	c_{pf} (kJ/kgK)	μ_f (10^{-4} Pa · s)	k_f (W/mK)	Pr_f (−)
380	2.63.−10	1.081.−3	4497	1.39	6.93	85.5	0.0113
400	1.39.−9	1.086.−3	4493	1.37	6.08	84.7	0.0098
450	4.59.−8	1.100.−3	4467	1.35	5.08	82.3	0.0083
500	7.52.−5	1.114.−3	4439	1.33	4.24	80.0	0.0070
600	4.93.−3	1.145.−3	4375	1.30	3.28	75.4	0.0057
700	0.0965	1.177.−3	4298	1.27	2.69	70.7	0.0048
800	0.8904	1.211.−3	4212	1.26	2.30	64.9	0.0045
900	4.980	1.247.−3	4118	1.25	2.02	61.4	0.0041

TABLE 2.42   *(Continued)*

T (K)	P (kPa)	v_f (m³/kg)	Δh_v (kJ/kg)	c_{pf} (kJ/kgK)	μ_f (10^{-4} Pa·s)	k_f (W/mK)	Pr_f (−)
1000	19.63	1.286.−3	4023	1.26	1.81	56.7	0.0040
1100	60.02	1.327.−3	3926	1.27	1.65	52.1	0.0040
1200	150.4	1.372.−3	3832	1.29	1.51	47.6	0.0041
1300	324.5	1.419.−3	3735	1.31	1.41		
1400	625.4	1.469.−3	3633	1.33	1.32		
1500	1101	1.523.−3	3522	1.35	1.24		
1600	1802	1.581.−3	3400				
1700	2778	1.643.−3	3269				
1800	4090	1.713.−3	3130				
1900	5800	1.792.−3	2982				
2000	7450	1.884.−3	2824				
2100	10600	1.993.−3	2653				
2200	13790	2.123.−3	2465				
2300	17560	2.285.−3	2256				
2400	21950	2.495.−3	2017				
2500	26980	2.781.−3	1733				
2600	32700	3.230.−3	1365				
2700	39090	4.201.−3	753				
2733	41360	5.501.−3	0				

Interpolated and/or converted from tables of Padilla, A., Argonne National Laboratory Report ANL 8095, 1974.

TABLE 2.43   **Saturated Benzene**

T (K)	P (kPa)	v_f (m³/kg)	Δh_v (kJ/kg)	c_{pf} (kJ/kgK)	μ_f (10^{-4} Pa·s)	k_f (W/mK)	Pr_f (−)
290	9	1.133.−3	439.2	1.72	6.75	0.147	7.90
300	14	1.147.−3	432.1	1.75	5.80	0.144	7.05
320	32	1.176.−3	417.7	1.80	4.52	0.138	5.90
340	66	1.207.−3	402.8	1.87	3.55	0.132	5.03
360	124	1.241.−3	388.0	1.92	2.99	0.126	4.56
380	216	1.277.−3	372.9	1.99	2.46	0.120	4.08
400	352	1.318.−3	357.3	2.07	2.05	0.114	3.72
450	971	1.444.−3	311.9	2.32	1.37	0.098	3.24
500	2165	1.640.−3	246.4	2.67	0.98	0.083	3.15
562	4900	3.290.−3	0				

Converted from the tables of Counsell, J. F., Lawrenson, I. J., and Lees, E. B., National Physical Laboratory Report Chem. 52, 13 pp., 1976.

TABLE 2.44 Saturated Ethylene Glycol

T (K)	P (kPa)	v_f (m³/kg)	Δh_v (kJ/kg)	c_{pf} (kJ/kgK)	μ_f (10^{-4} Pa · s)	k_f (W/mK)	Pr_f (−)
260	3.2.−6	8.82.−4		2.20	0.130	0.246	1160
270	9.3.−6	8.87.−4		2.24	0.0702	0.248	640
280	2.5.−5	8.92.−4		2.28	0.0407	0.250	370
290	6.7.−5	8.97.−4		2.32	0.0247	0.252	230
300	1.61.−4	9.02.−4	1093	2.36	0.0162	0.254	150
310	3.67.−4	9.07.−4	1082	2.40	0.0112	0.256	105
320	8.06.−4	9.14.−4	1070	2.45	0.0081	0.258	77
330	1.62.−3	9.20.−4	1059	2.49	0.0060	0.259	58
340	3.07.−3	9.27.−4	1047	2.53	0.0045	0.260	44
350	5.76.−3	9.33.−4	1036	2.59	0.0035	0.261	35
360	1.05.−2	9.39.−4	1024	2.64	0.0028	0.261	28
370	1.73.−2	9.45.−4	1013	2.69	0.0022	0.262	23
380	2.84.−2	9.52.−4	1001	2.74	0.0018	0.261	19
390	4.60.−2	9.58.−4	990	2.79	0.0015	0.261	16
400	7.50.−2	9.64.−4	978	2.84	0.00130	0.260	14
410	0.116	9.70.−4	966	2.90	0.00111	0.259	12
420	0.171	9.76.−4	954	2.96	0.00097	0.258	11
430	0.253	9.82.−4	943	3.01	0.00086	0.257	10
440	0.368	9.88.−4	931	3.07	0.00078	0.256	9
450	0.523	9.96.−4	919	3.13	0.00071	0.255	9

Interpolated and converted from the graphical values of *Glycols*, Union Carbide Corp. Publ. F-41515B, 74 pp., 1978.

TABLE 2.45 Saturated Propylene Glycol

T (K)	P (kPa)	v_f (m³/kg)	Δh_v (kJ/kg)	c_{pf} (kJ/kgK)	μ_f (10^{-4} Pa · s)	k_f (W/mK)	Pr_f (−)
260	2.9.−5	9.45.−4		2.28	0.990	0.200	11300
270	5.84.−5	9.51.−4		2.34	0.360	0.200	4200
280	1.12.−4	9.57.−4		2.39	0.1530	0.200	1830
290	2.06.−4	9.63.−4		2.42	0.0675	0.200	820
300	3.63.−4	9.70.−4	908	2.51	0.0330	0.200	415
310	6.17.−4	9.77.−4	896	2.57	0.0200	0.200	255
320	1.49.−3	9.85.−4	885	2.63	0.0125	0.200	165
330	2.78.−3	9.92.−4	873	2.69	0.0080	0.200	110
340	5.21.−3	0.001000	861	2.74	0.0056	0.200	77
350	9.42.−3	0.001008	849	2.80	0.0041	0.199	58
360	1.63.−2	0.001017	838	2.86	0.0032	0.199	46
370	2.56.−2	0.001026	826	2.92	0.0025	0.198	37
380	4.24.−2	0.001035	813	2.98	0.0020	0.197	30
390	6.90.−2	0.001044	800	3.04	0.00166	0.196	26
400	0.107	0.001053	787	3.09	0.00136	0.194	22
410	0.168	0.001063	774	3.15	0.00155	0.193	19
420	0.249	0.001073	761	3.21	0.00097	0.192	16
430	0.370	0.001084	748	3.28	0.00082	0.190	14
440	0.527	0.001094	735	3.34	0.00071	0.189	13
450	0.741	0.001105	722	3.40	0.00061	0.187	11

Interpolated and converted from the graphical values of *Glycols*, Union Carbide Corp. Publ. F-41515B, 74 pp., 1978.

TABLE 2.46   Saturated Toluene

T (K)	P (kPa)	v_f (m³/kg)	Δh_v (kJ/kg)	c_{p_f} (kJ/kgK)	μ_f (10^{-4} Pa · s)	k_f (W/mK)	Pr_f (−)
270	0.8	1.127.−3	429.0	1.64	8.02	0.141	9.33
280	1.4	1.138.−3	423.1	1.66	6.96	0.138	8.37
290	2.5	1.150.−3	417.2	1.68	6.10	0.136	7.54
300	4.2	1.162.−3	411.3	1.71	5.41	0.133	6.96
310	6.8	1.175.−3	405.5	1.74	4.83	0.131	6.46
320	10.7	1.188.−3	399.6	1.78	4.34	0.128	6.04
330	16.3	1.201.−3	393.3	1.81	3.93	0.126	5.66
340	24.2	1.215.−3	387.8	1.84	3.58	0.124	5.36
350	34.8	1.230.−3	381.8	1.88	3.28	0.121	5.10
360	48.9	1.245.−3	375.6	1.92	3.01	0.119	4.89
370	67.4	1.261.−3	369.2	1.96	2.78	0.117	4.69
380	90.9	1.277.−3	362.8	2.01	2.56	0.114	4.51
390	121	1.294.−3	356.2	2.05	2.37	0.112	4.34
400	157	1.312.−3	349.4	2.09	2.19	0.110	4.18
420	226	1.350.−3	335.0	2.17	1.89	0.105	3.91
440	397	1.393.−3	319.7	2.24	1.64	0.101	3.66
460	589	1.443.−3	303.2	2.31			
480	845	1.499.−3	285.6	2.38			
500	1176	1.567.−3	266.9	2.45			
592	4104	3.452.−3	0				

Converted from the tables of Counsell, J. F., Lawrenson, I. J., and Lees, E. B., National Physical Laboratory Report Chem. 52, 13 pp., 1976.

2.11   ACOUSTICS
Allen E. Fuhs

2.11.1   Introduction

The acoustical properties of water and seawater are discussed in Sec. 2.2. Consequently, this section focuses on acoustical properties of gases. Data are given for three aspects of acoustics as follows: speed of sound in gases, characteristic impedance, and the attentuation of sound.

2.11.2   Speed of Sound in Gases

The speed of sound in gases is discussed in Chaps. 6, 7, and 8 and in Sec. 23.2. The variation of sound speed with temperature and molecular weight of the gas is discussed in Chap. 4. Table 2.47 provides the sound speed for common gases.

2.11.3   Characteristic Impedance

As discussed in Sec. 23.2, the characteristic impedance for sound (longitudinal) waves is given by

$$(\text{characteristic impedance}) = \rho c \qquad (2.105)$$

TABLE 2.47   Speed of Sound in Gases at 273 K

Gas	Chemical Formula	Speed of Sound m/s	Molecular Weight kg/kilomole
Air	—	331.5	28.966
Ammonia	NH_3	415	17.03
Argon	Ar	307.8	39.95
Carbon monoxide	CO	337.1	28.01
Carbon dioxide	CO_2	258*	44.01
		268.6**	44.01
Helium	He	970	4.003
Hydrogen	H_2	1269.5	1.008
Methane	CH_4	432	16.04
Neon	Ne	435	20.18
Nitrogen	N_2	337	28.01
Oxygen	O_2	317.2	32.00
Steam (@ 100 C)	H_2O	404.8	18.015

*Sufficiently low frequency so that thermal relaxation occurs.
**Sufficiently high frequency so that thermal relaxation does not have time to occur.

The symbols frequently used for speed of sound are a or c; ρ is the mass density. Transmission and reflection of waves at an interface depends on the values of ρc in each media.

Table 2.48 provides values of the characteristic impedance for several gases.

2.11.4   Attenuation of Acoustical Waves

The equation which relates sound intensity, W/m^2, to the instantaneous value of the pressure amplitude of the wave, is repeated here for convenience

$$I = p^2/\rho c \tag{2.106}$$

TABLE 2.48   Characteristic Impedance, ρc, of Gases at 273 K and 1.0 Atmosphere Pressure

Gas	Chemical Formula	ρc N $\cdot$ s/m^3
Air	—	430
Argon	Ar	569
Carbon monoxide	CO	421
Carbon dioxide	CO_2	508
Helium	He	173
Hydrogen	H_2	114
Neon	Ne	385
Nitrogen	N_2	421
Oxygen	O_2	453

For a sinusoidal wave, the average intensity is given by

$$\langle I \rangle = p_1^2/\rho c \tag{2.107}$$

where p_1 is the amplitude of the sinusoidal wave.

Define m as the attenuation coefficient with units of per meter. Define α as the dB/m attentuation. Both m and α refer to acoustic intensity. Hence,

$$dB(x) = 10 \log_{10} I(x)/I_0 \tag{2.108}$$

where x is distance from a point on a plane wave with intensity I_0, and where

$$dB(x) = \alpha x \tag{2.108a}$$

Equations (2.108) and (2.108a) apply to a plane wave.
    Solving for the intensity ratio yields

$$I(x)/I_0 = 10^{-\alpha x/10} \tag{2.109}$$

By definition, the intensity ratio in terms of m is

$$I(x)/I_0 = \exp(-mx) \tag{2.109a}$$

Equating the intensity ratios in Eqs. (2.109) and (2.109a) gives

$$\alpha = 10m \log_{10} e = 4.34m \tag{2.110}$$

Attenuation of acoustic waves is discussed in Strutt (1945), Beranek (1949), Knudsen and Harris (1950), Humphreys (1964), and Athey (1970).

Sound can be attenuated or dispersed by wind [Humphreys (1964)], turbulence in the atmosphere [Athey (1970)], and temperature gradients [Wood (1966)]. These mechanisms are not a characteristic of the medium—in this case, air. Two properties of air, which are discussed here, attenuate sound. One cause of attenuation is molecular absorption and dispersion due to transfer of energy between translation and molecular internal degrees of freedom; hence, molecular absorption does not occur in mono-atomic gases such as helium. The other cause of attenuation discussed here is heat conduction and viscosity. Absorption by this mechanism was first studied by Kirchhoff and Stokes [Strutt (1945)] who derived the following equation

$$m_c = \frac{\omega^2}{2\rho c^3} \left[\frac{4\mu}{3} + (\gamma - 1) \frac{k}{C_p} \right] \tag{2.111}$$

where m_c is the attenuation due to thermal conductivity and viscous dissipation, $1/m$, ω is the frequency of the sound, radians/second, ρ is mass density, kg/m^3, c is speed of sound, m/s, μ is dynamic viscosity coefficient, N s/m^2, γ is ratio of heat capacities, dimensionless, k is thermal conductivity, J/smK, and C_p is specific heat capacity at constant pressure, J/kg K. The subscript c for m denotes attenuation due to conductivity and viscosity.

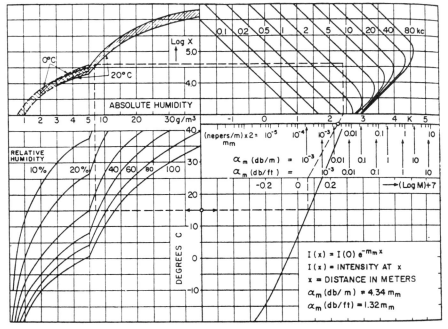

FIGURE 2.10 Nomogram for estimating the attenuation in humid air caused by molecular absorption. (Reproduced with permission, Beranek, L. L., *Acoustic Measurements*, John Wiley & Sons, Inc., New York, 1949.)

The attenuation due to molecular absorption in humid air [Beranek (1949)] can be obtained from Fig. 2.10. To use the curve, start at a point on the temperature scale (15°C in the example). Proceed horizontally the correct relative humidity curve (50%). Move vertically to middle of the shaded area. From that point, move horizontally toward the right to the appropriate frequency curve (3 kc or 3 kHz in the example). Descend vertically to the K scale; make a mark on the K scale (2.4 in the example). Return to the temperature scale and move horizontally to the right to an intersection with the curve. From the intersection move upward to the (Log M) + 7 axis; make a point (0.06 in the example). Between the two points draw a straight line that intercepts the m_m scale (0.003 for the example). Calculate α_m from Eq. (2.110) giving 0.012 dB/m.

The total absorption is the sum of m_c and m_m, i.e.,

$$m_t = m_c + m_m \tag{2.112}$$

The value of m_t can be used in Eq. (2.109) to calculate the intensity ratio. Measured values of m_t appear in Fig. 2.11. One can obtain m_t either by using Fig. 2.11 or Fig. 2.10 and Eq. 2.111. Values of μ and k are given in Secs. 2.3, 2.4, and 2.7 for various gases.

The order of magnitude of the various mechanisms leading to absorption are summarized in Table 2.49. Equation 2.111 correctly predicts trends with changing frequency for viscosity and heat conduction for Table 2.49. When the gas is compressed by the sound wave, the temperature increases. A region of heated gas radiates; the loss of energy causes attenuation. Figure 2.10 applies to the row in Table 2.49 labeled "Internal energy of molecules." Examination of Fig. 2.10 shows

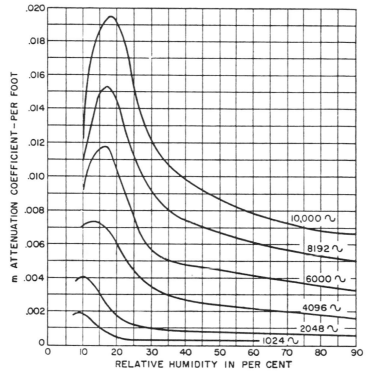

FIGURE 2.11 Measured values of intensity attenuation coefficient m_t as a function of relative humidity and frequency at constant temperature of 20 C. (Reproduced from Knudsen and Harris, 1950 with permission of Acoustical Society of America.)

the correct trends for m_m as frequency is increased; however, the maximum frequency in Fig. 2.10 is 80 kHz.

Dissipation due to diffusion occurs in gas mixtures. When the gas is compressed, concentration gradients occur. Molecules with larger diffusion coefficients will diffuse from the compressed region at a faster rate. The differential diffusion on un-mixing of the mixture results in dissipation of energy.

Evans and Bass (1972) present extension tables of absorption and velocity of sound in stationary air. The absorption in dB/m, α, is given as a function of frequency and relative humidity.

TABLE 2.49 **Order of Magnitude of the Various Mechanisms Causing Absorption [Wood (1966)]**

	Value of m is per/m at frequency		
Mechanism	6000 Hz	800,000 Hz	Infinite
Viscosity	3.6 E−4	6.4	infinite
Heat conduction	4 E−4	0.7	6 E+6
Heat radiation	1.5 E−6	negligible	negligible
Internal energy of molecules	0.0012	10	20
Diffusion (gas mixtures)	3E−5	0.5	infinite

Notation 3.6 E−4, which is similar to computer printouts, is 3.6×10^{-4}.

2.12 SURFACE TENSION
Peter E. Liley

TABLE 2.50   Surface Tension (N/m) of Liquids

							Temperature (K)									
	250	260	270	280	290	300	310	320	330	340	350	360	370	380	390	400
Acetone	0.0291	0.0279	0.0266	0.0253	0.0240	0.0228	0.0214	0.0201	0.0187	0.0174	0.0162	0.0150	0.0139	0.0128	0.0117	0.0106
Ammonia	0.0317	0.0294	0.0271	0.0248	0.0226	0.0203	0.0181	0.0159	0.0138	0.0117	0.0099	0.0080	0.0059	0.0040	0.0021	0.0003
Benzene	—	—	0.0320	0.0306	0.0292	0.0278	0.0265	0.0252	0.0239	0.0227	0.0215	0.0203	0.0191	0.0179	0.0167	0.0155
Butane	0.0177	0.0165	0.0153	0.0141	0.0129	0.0116	0.0104	0.0092	0.0080	0.0069	0.0059	0.0049	0.0040	0.0031	0.0023	0.0016
CO_2	0.0092	0.0071	0.0051	0.0032	0.0016	0.0003	—									
Chlorine	0.0244	0.0228	0.0213	0.0198	0.0183	0.0168	0.0153	0.0138	0.0123	0.0108	0.0094	0.0080	0.0066	0.0052	0.0044	0.0037
Ethane	0.0059	0.0047	0.0035	0.0024	0.0013	0.0005	—									
Ethanol	0.0271	0.0261	0.0251	0.0242	0.0232	0.0223	0.0214	0.0205	0.0196	0.0186	0.0177	0.0167	0.0158	0.0148	0.0137	0.0126
Ethylene	0.0032	0.0019	0.0009	0.0002	—											
Heptane	0.0244	0.0234	0.0224	0.0214	0.0205	0.0195	0.0185	0.0176	0.0166	0.0156	0.0147	0.0137	0.0127	0.0118	0.0109	0.0099
Hexane	0.0229	0.0218	0.0207	0.0197	0.0186	0.0175	0.0165	0.0154	0.0145	0.0135	0.0125	0.0115	0.0106	0.0096	0.0086	0.0076
Mercury	0.474	0.472	0.470	0.468	0.466	0.464	0.462	0.460	0.458	0.456	0.454	0.452	0.450	0.448	0.446	0.444
Methanol	—	—	—	—	—	0.0223	0.0214	0.0205	0.0196	0.0187	0.0178	0.0168	0.0159	0.0149	0.0139	0.0128
Octane	0.0251	0.0243	0.0234	0.0225	0.0216	0.0207	0.0197	0.0188	0.0179	0.0170	0.0161	0.0152	0.0143	0.0135	0.0127	0.0120
Propane	0.0132	0.0118	0.0104	0.0091	0.0079	0.0067	0.0056	0.0046	0.0037	0.0028	0.0018	0.0009	0.0000	—		
Propylene	0.0133	0.0120	0.0106	0.0091	0.0078	0.0065	0.0053	0.0042	0.0032	0.0023	0.0014	0.0006	—			
R12	0.0148	0.0135	0.0122	0.0109	0.0096	0.0083	0.0070	0.0058	0.0047	0.0036	0.0027	0.0018	0.0010	0.0003		
R13	0.0057	0.0042	0.0029	0.0018	0.0009	0.0003	—									
Toluene	0.0342	0.0327	0.0312	0.0298	0.0285	0.0272	0.0260	0.0249	0.0236	0.0225	0.0214	0.0203	0.0193	0.0183	0.0173	0.0163
Water	—	—	—	0.0746	0.0732	0.0716	0.0699	0.0684	0.0666	0.0650	0.0636	0.0614	0.0595	0.0575	0.0555	0.0535

2.13 PROPERTIES OF SELECTED ELEMENTS

Peter E. Lilley

TABLE 2.51 Physical Constants of Elements, Critical Conditions, and Melting and Vaporization Data

Name	Symbol	Formula Weight	T_m (K)	Δh_m (kJ/kg)	T_b (k)	Δh_v (kJ/kg)	T_c (K)	P_c (bar)	V_c (m³/kg)	Z_c (—)
Argon	Ar	39.948	83.8	29.6	87.3	160.8	150.9	49.0	0.00187	0.292
Bromine	Br_2	159.808	266.0	66.2	332.4	187.7	584.2	103.4	0.00079	0.270
Cesium	Cs	132.905	301.7	16.4	947.0	495.0	2020.0	154.0	0.00230	0.280
Chlorine	Cl_2	70.906	172.1	30.7	238.7	288.0	417.0	77.1	0.00175	0.277
Deuterium	D_2	4.028	18.7	48.9	23.7	304.0	38.3	16.7	0.00143	0.301
Fluorine	F_2	37.997	53.5	13.4	85.1	172.0	144.4	52.2	0.00174	0.288
Gallium	Ga	69.72	303.0	80.1	2500.0	3690.0	7125.0	4150.0	—	—
Helium[3]	He³	3.016	—	6.2	3.2	8.5	3.3	1.2	0.0229	0.307
Helium[4]	He⁴	4.003	—	2.1	4.2	20.4	5.2	2.3	0.0144	0.302
Hydrogen	H_2	2.016	14.0	58.0	20.4	454.0	33.3	13.0	0.0323	0.305
Krypton	Kr	83.80	115.8	19.6	119.8	107.6	209.4	55.0	0.00109	0.288
Mercury	Hg	200.59	234.4	11.4	630.1	295.6	1740.0	1505.0	—	—
Neon	Ne	20.179	24.6	16.3	27.1	85.8	44.4	26.5	0.00207	0.301
Nitrogen	N_2	28.013	63.1	25.7	77.3	197.6	126.2	34.0	0.00318	0.287
Oxygen	O_2	31.999	54.4	13.9	90.1	212.9	154.7	50.4	0.00229	0.288
Ozone	O_3	47.998	80.6	—	161.3	—	261.1	55.0	—	—
Radon	Rn	222.0	202.2	14.3	211.0	81.5	377.0	65.5	0.00625	0.293
Tritium	T_2	6.034	20.6	—	25.0	230.9	40.4	18.5	0.00094	0.311
Xenon	Xe	131.30	161.3	17.3	165.0	95.7	289.7	58.2	0.00091	0.285

TABLE 2.52   Melting, Vaporization, Critical Temperature, and Molecular Weight of Elements

Symbol	Formula Weight	T_m (K)	Δh_{fus} (kJ $\cdot$ kg^{-1})	T_b (K)	T_c (K)
Ac	227.028	1323	63	3475	—
Al	26.9815	933.5	398	2750	7850
Sb	121.75	903.9	163	1905	5700
Ar	39.948	83.	30	87.2	151
As	74.9216	885.	—	—	2100
Ba	137.33	1002	55.8	—	4450
Be	9.01218	1560	1355	2750	6200
Bi	208.980	544.6	54.0	1838	4450
B	10.81	2320	1933	4000	3300
Br	159.808	266	66.0	332	584
Cd	112.41	594	55.1	1040	2690
Ca	40.08	1112	213.1	1763	4300
C	12.011	3810		4275	7200
Ce	140.12	1072	390	—	9750
Cs	132.905	301.8	16.4	951	2015
Cl$_2$	70.906	172	180.7	239	417
Cr	51.996	2133	325.6	2950	5500
Co	58.9332	1766	274.7	3185	6300
Cu	63.546	1357	206.8	2845	8280
Dy	162.50	1670	68.1	2855	6925
Er	167.26	1795	119.1	3135	7250
Eu	151.96	1092	60.6	1850	4350
F$_2$	37.997	53.5	13.4	85.0	144
Gd	157.25	1585	63.8	3540	8670
Ga	69.72	303	080.1	2500	7125
Ge	72.59	1211	508.9	3110	8900
Au	196.967	1337	62.8	3130	7250
Hf	178.49	2485	134.8	4885	10400
He	4.00260	3.5	2.1	4.22	5.2
Ho	164.930	1744	73.8	2968	7575
H$_2$	2.0159	14.0	—	20.4	—
In	114.82	430	28.5	2346	6150
I$_2$	253.809	387	125.0	457	785
Ir	192.22	2718	13.7	4740	7800
Fe	55.847	1811	247.3	3136	8500
Kr	83.80	115.8	19.6	119.8	209.4
La	138.906	1194	44.6	3715	10500
Pb	207.2	601	23.2	2025	5500
Li	6.941	454	432.2	1607	3700
Lu	174.967	1937	106.6	3668	—
Mg	24.305	922	368.4	1364	3850
Mn	54.9380	1518	219.3	2334	4325
Hg	200.59	234.6	11.4	630	1720
Mo	95.94	2892	290.0	4900	1450
Nd	144.24	1290	49.6	3341	7900
Ne	20.179	24.5	16.4	27.1	44.5
Np	237.048	910	—	4160	12000

TABLE 2.52 (*Continued*)

Symbol	Formula Weight	T_m (K)	Δh_{vap} (kJ $\cdot$ kg^{-1})	T_b (K)	T_c (K)
Ni	58.70	1728	297.6	3190	8000
Nb	92.9064	2740	283.7	5020	12500
N$_2$	28.013	63.2	25.7	77.3	126.2
Os	190.2	3310	150.0	5300	12700
O$_2$	31.9988	54.4	13.8	90.2	154.8
Pd	106.4	1826	165.0	3240	7700
P	30.9738	317	—	553	995
Pt	195.09	2045	101	4100	10700
Pu	244	913	11.7	3505	10500
K	39.0983	336.4	60.1	1032	2210
Pr	140.908	1205	49	3785	8900
Pm	145	1353	—	2730	—
Pa	231	1500	64.8	4300	—
Ra	226.025	973	—	1900	
Rn	222	202	12.3	211	377
R	186.207	3453	177.8	5920	18900
Rh	102.906	2236	209.4	3980	7000
Rb	85.4678	312.6	26.4	964	2070
Ru	101.07	2525	256.3	4430	9600
Sm	150.4	1345	57.3	2064	5050
Sc	44.9559	1813	313.6	3550	6410
Se	78.96	494	66.2	958	1810
Si	28.0855	1684	1802	3540	5160
Ag	107.868	1234	104.8	2435	640
Na	22.9898	371	113.1	1155	2500
Sr	87.62	1043	1042	1650	4275
S	32.06	388	53.4	718	1210
Ta	180.948	3252	173.5	5640	16500
Tc	98	2447	232	4550	11500
Te	127.60	723	137.1	1261	2330
Tb	158.925	1631	67.9	3500	8470
Tl	204.37	577	20.1	1745	4550
Th	232.038	2028	69.4	5067	14400
Tm	168.934	1819	99.6	2220	6450
Sn	118.69	505	58.9	2890	7700
Ti	47.90	1943	323.6	3565	5850
W	183.85	3660	192.5	5890	15500
U	238.029	1406	35.8	4422	12500
Y	50.9415	2191	410.7	3680	11300
Xe	131.30	161.3	17.5	164.9	290
Yb	173.04	1098	44.2	1467	4080
Y	88.9059	1775	128.2	3610	8950
Zn	65.38	692.7	113.0	1182	—
Zr	91.22	2125	185.3	4681	10500

T_m = normal melting point. Δh_{fus} = enthalpy of fusion. T_b = normal boiling point, Δh_{vap} = enthalpy of vaporization, T_c = critical temperature, P_c = critical pressure, and T_{tr} = phase transition temperature.

TABLE 2.53 Refractive Index of Gases

Molecule	Symbol	Refractive Index
Air	—	1.000292
Nitrogen	N_2	1.000297
Oxygen	O_2	1.000271
Carbon dioxide	CO_2	1.000451
Helium	He	1.000036
Steam	H_2O	1.000254
Sulfur dioxide	SO_2	1.000686
Benzene	C_6H_6	1.001750

Values adjusted to 0°C and 760 Torr.
Refractive index at the sodium D-line at 589.3 nm.

TABLE 2.54 Refractive Index of Air as a Function of Wavelength

Wavelength, λ nanometers, nm	Refractive Index
486	1.0002951
546	1.0002936
579	1.0002930
656	1.0002919
671	1.0002918
6709	1.0003881
8678	1.0002888

Values are for a temperature of 273 K and pressure of one atmosphere.

2.14 REFRACTIVE INDICES
Allen E. Fuhs

Several experimental techniques require knowledge of the refractive index of the fluid. See Chaps. 1 and 15 as well as Sec. 23.12. The definition of refractive index is given in Sec. 23.12.

For values of refractive index of water and seawater, see Sec. 2.2.

Refractive index, n, is a function of wavelength. Values of n for several different gases are given in Table 2.53 at a fixed wavelength, $\lambda = 589.3$ nm. Table 2.54 provides the values of $n(\lambda)$ for air. In the visible region, n decreases slightly as λ increases.

2.15 ELECTRICAL CONDUCTIVITY
Allen E. Fuhs

Electrical conductivity becomes important for several problems in fluid dynamics. As discussed in *Magnetofluidmechanics* Sec. 13.8, the ability to extract power using magnetohydrodynamics or to accelerate a conductor depends critically on adequate electrical conductivity. For a coal-combustion-generated plasma, the range of conductivities σ, is 1–10 S/m. A unit for electrical conductivity is Siemens per meter or in the older literature, mho/m.

Electrical conductivity, σ, is important in seawater. See Sec. 2.2 for values of σ

TABLE 2.55 Electrical Conductivity of Liquids and a Few Metals for Comparison

Material	Symbol	Electrical Conductivity Siemens/meter or mho/m	Temperature °C
Mercury	Hg	1.044×10^6	20
Sodium	Na	1.03×10^7	100
Potassium	K	7.69×10^6	62
Silver	Ag	6.2×10^7	20
Copper	Cu	5.8×10^7	20
Eutectic[1]	NaK	5.4×10^6	100
Eutectic[2]	NaK	2.2×10^6	100

[1]56.5% K by weight.
[2]78.0% K by weight.

for seawater. The penetration of microwaves into seawater depends on the magnitude of σ. Salinity can be inferred from a measurement of σ.

Liquid metal nuclear reactors may use sodium, potassium, or the eutectic NaK as the coolant. Pumping by electromagnetic means as well as flow measurement using various electromagnetic techniques is now utilized. Table 2.55 presents values of electrical conductivity for several liquids.

During reentry, the air is transformed to a plasma by aerodynamic heating. Electrical conductivity of the air-plasma is an important feature of reentry. [Martin (1966), Cox and Crabtree (1965), Fuhs (1965), and Loh (1968)].

2.16 LAMINAR DIFFUSION COEFFICIENTS
Joseph A. Schetz

This section contains laminar binary diffusion coefficient information for a few selected pairs of fluids obtained from the computer code developed by Svehla and McBride (1973). This is an excellent, convenient source of information for a wide range of fluids for the working professional.

TABLE 2.56 Binary Diffusion Coefficients, cm^2/s

Pair	300K	400K	500K	600K	700K	800K	900K	1000K
CO_2–CO_2	0.110	0.190	0.287	0.399	0.524	0.658	0.803	0.958
CO_2–H_2	0.553	0.932	1.384	1.899	2.477	3.107	4.094	4.892
CO_2–H_2O	0.108	0.211	0.347	0.517	0.717	0.942	1.194	1.469
CO_2–N_2	0.174	0.293	0.436	0.600	0.782	0.982	1.034	1.233
CO_2–O_2	0.161	0.277	0.418	0.580	0.762	0.965	1.050	1.256
H_2–H_2	1.466	2.386	3.482	4.760	6.150	7.696	9.390	11.237
H_2–H_2O	0.886	1.502	2.237	3.080	4.024	5.057	6.179	7.386
H_2–N_2	0.622	1.013	1.473	1.994	2.576	3.217	3.912	4.660
H_2–O_2	0.914	1.490	2.170	2.944	3.800	4.745	5.772	6.876
H_2O–H_2O	0.088	0.202	0.373	0.603	0.891	1.235	1.636	2.086
H_2O–N_2	0.258	0.420	0.610	0.826	1.067	1.331	1.618	1.926
H_2O–O_2	0.258	0.442	0.665	0.922	1.210	1.527	1.871	2.241
N_2–N_2	0.204	0.335	0.490	0.665	0.861	1.075	1.309	1.561
N_2–O_2	0.206	0.340	0.498	0.679	0.884	1.113	1.368	1.643
O_2–O_2	0.206	0.341	0.502	0.686	0.894	1.126	1.381	1.656

2.17   UNITS, CONSTANTS, AND CONVERSION TABLES
Susan Fuhs

TABLE 2.57   The International System of Units (SI)

A. Basic Units		
Length	meter	m
Mass	kilogram	kg
Time	second	s
Electric current	ampere	A
Thermodynamic temperature	kelvin	K
Luminous intensity	candela	cd

B. Derived Units		
Quantity	Units	Symbol(s)
Acceleration	meter per second squared	m/s^2
Activity (of radioactive source)	1 per second	s^{-1}
Angular acceleration	radian per second squared	rad/s^{-1}
Angular velocity	radian per second	rad/s
Area	square meter	m^2
Density (mass density)	kilogram per cubic meter	kg/m^3
Dynamic viscosity	Newton-second per sq meter	$N \cdot s/m^2$
Electric capacitance	farad	F $(A \cdot s/V)$
Electric charge	coulomb	C $(A \cdot s)$
Electric field strength	volt per meter	V/m
Electric resistance	ohm	Ω (V/A)
Entropy	joule per kelvin	J/K
Force	Newton	N $(kg \cdot m/s^2)$
Frequency	hertz	Hz (s^{-1})
Illumination	lux	lx (lm/m^2)
Inductance	henry	H $(V \cdot s/A)$
Kinematic viscosity	sq meter per second	m^2/s
Luminance	candela per sq meter	cd/m^2
Luminous flux	lumen	lm
Magnetomotive force	ampere	A
Magnetic field strength	ampere per meter	A/m
Magnetic flux	weber	Wb $(V \cdot s)$
Magnetic flux density	tesla	T (Wb/m^2)
Plane angle	radian	rad
Power	watt	W (J/s)
Pressure	newton per sq meter (pascal)	N/m^2 (Pa)
Radiant intensity	watt per steradian	W/sr
Solid angle	steradian	sr
Specific heat	joule per kilogram kelvin	J/kg K
Thermal conductivity	watt per meter kelvin	W/m K
Velocity	meter per second	m/s
Volume	cubic meter	m^3
Voltage, potential difference, electromotive force	volt	V (W/A)
Wave number	1 per meter	m^{-1}
Work, energy, quantity of heat	joule	J $(N \cdot m)$

TABLE 2.57 (*Continued*)

C. Prefix Names for Multiples and Fractions of Units[a,b]

Prefix	Symbol	Multiplicative Factor	Decimal equivalent
tera	T	10^{+12}	1,000,000,000,000
giga	G	10^{+9}	1,000,000,000
mega	M	10^{+6}	1,000,000
kilo	k	10^{+3}	1,000
hecto[c]	h	10^{+2}	100
deka[c]	da	10	10
deci[c]	d	10^{-1}	0.1
centi[c]	c	10^{-2}	0.01
milli	m	10^{-3}	0.001
micro[d]	μ	10^{-6}	0.000 001
nano	n	10^{-9}	0.000 000 001
pico	p	10^{-12}	0.000 000 000 001
femto	f	10^{-15}	0.000 000 000 000 001
atto	a	10^{-18}	0.000 000 000 000 000 001

[a]Use only one prefix for a unit (1000 kg $-$ 1 Mg *not* 1 k kg).

[b]Prefixes used should differ from a unit in steps of 10^3.

[c]Use only if recommended prefixes[b] are inconvenient.

[d]The symbol μ is no longer used for the length unit micron (10^{-6} m), but is used only as a prefix. The old 1 micron is now 1 μm (as used, e.g., for infrared wavelengths).

D. Definitions of Basic Units

Meter (m): 1,650,763.73 wavelengths of $2p_{10}-d_5$ transition in ^{86}Kr.

Second (s): the duration of 9,192,631,770 periods of the radiation corresponding to the transition between the two hyperfine levels of the fundamental state of ^{133}Cs. The "ephemeris" second is 1/31,556,925.9747 of the tropical year 1900.

Kilogram (kg): mass of the international kilogram (Sèvres, France), a cylinder of platinum–iridium alloy.

Kelvin (K): unit of thermodynamic temperature; 273.16 K = triple point of water.

Liter (1): 0.001 m^3 = 1000 cm^3.

Mole (mol): amount of matter having the same number of formula units as atoms in 0.012 kg of ^{12}C.

Unified atomic mass unit (*u*): 1/12 the mass of an atom of ^{12}C = 1.66043×10^{-24} g.

Standard free fall acceleration (g_n): 9.80665 ms^{-2} = 980.665 cm s^{-2}.

Normal atmospheric pressure (atm): 101,325 N m^{-2} = 1,013,250 dyn cm^{-2}.

Thermochemical calorie (cal): 4.1840 J = 4.1840×10^7 erg.

International Steam Table calorie (cal$_{IT}$): 4.1868 J = 4.1868×10^7 erg.

Inch (in.): 0.0254 m = 2.54 cm.

Pound (avdp): 0.45359237 kg = 453.59237 g.

E. Other Units

curie	3.70×10^{10} disintegrations per s (dps) = 2.22×10^{12} dpm
roentgen	1 esu charge per cm^3 (0.001293 g) dry air at STP; 1.61×10^{12} ion pairs/g air; charge equivalent to 2×10^9 electrons/cm^3 air
rad	100 erg/g
rep	93 erg/g absorbed in tissue
1 barn	10^{-24} cm^2
96,500 coulombs	6.02×10^{23} electrons
1 gamma	10^{-6} g (μg)
1 lambda (λ)	10^{-6} 1 (10^{-3} ml)

Adapted from Gordon and Ford (1972).

TABLE 2.58 Fundamental Physical Constants

The numbers in parentheses are the standard deviation uncertainties in the last digits of the quoted value, computed on the basis of internal consistency.

Quantity	Symbol	Value	Error, ppm	Units SI	Units cgs
Velocity of light	c	2.9979250(10)	0.33	10^8 m sec^{-1}	10^{10} cm sec^{-1}
Fine-structure constant, $2\pi e^2/hc$	α	7.29351(11)	1.5	10^{-3}	10^{-3}
	α^{-1}	137.03602(21)	1.5		
Electron charge	e	1.6021917(70)	4.4	10^{-19} C	10^{-20} emu
		4.803250(21)	4.4		10^{-10} esu
Planck's constant	h	6.626196(50)	7.6	10^{-34} J $\cdot$ sec	10^{-27} erg $\cdot$ sec
	$\hbar = h/2\pi$	1.0545919(80)	7.6	10^{-34} J $\cdot$ sec	10^{-27} erg $\cdot$ sec
Avogadro's number	N	6.022169(40)	6.6	10^{26} kmole^{-1}	10^{23} mole^{-1}
Atomic mass unit	amu	1.660531(11)	6.6	10^{-27} kg	10^{-24} g
Electron rest mass	m_e	9.109558(54)	6.0	10^{-31} kg	10^{-28} g
	m_e^*	5.485930(34)	6.2	10^{-4} amu	10^{-4} amu
Proton rest mass	M_p	1.672614(11)	6.6	10^{-27} kg	10^{-24} g
	M_p^*	1.00727661(8)	0.08	amu	amu
Neutron rest mass	M_n	1.674920(11)	6.6	10^{-27} kg	10^{-24} g
	M_n^*	1.00866520(10)	0.10	amu	amu
Ratio of proton mass to electron mass	M_p/m_e	1836.109(11)	6.2		
Electron charge to mass ratio	e/m_e	1.7588028(54)	3.1	10^{11} C kg^{-1}	10^7 emu g^{-1}
		5.272759(16)	3.1		10^{17} esu g^{-1}
Magnetic flux quantum, $[c]^{-1}(hc/2e)$	Φ_0	2.0678538(69)	3.3	10^{-15} T $\cdot$ m^2	10^{-7} G $\cdot$ cm^2
	h/e	4.135708(14)	3.3	10^{-15} J $\cdot$ sec C^{-1}	10^{-7} erg sec emu^{-1}
		1.3795234(46)	3.3		10^{-17} erg sec esu^{-1}
Quantum of circulation	$h/2m$,	3.636947(11)	3.1	10^{-4} J $\cdot$ sec kg^{-1}	erg sec g^{-1}
	h/m_e	7.273894(22)	3.1	10^{-4} J $\cdot$ sec kg^{-1}	erg sec g^{-1}
Faraday constant, N_e	F	9.648670(54)	5.5	10^7 C kmole^{-1}	10^3 emu mole^{-1}
		2.892599(16)	5.5		10^{14} esu mole^{-1}
Rydberg constant, $[\mu_0 c^2/4\pi]^2(m_e e^4/4\pi h^3 c)$	R_∞	1.09737312(11)	0.10	10^7 m^{-1}	10^5 cm^{-1}

Description	Symbol	Value			
Bohr radius, $[\mu_0 c^2/4\pi]^{-1}(h^2/m_e e^2) = \alpha/4\pi R_\infty$	a_0	5.2917715(81)	1.5	10^{-11} m	10^{-9} cm
Classical electron radius, $[\mu_0 c^2/4\pi](e^2/m_e c^2) = \alpha^3/4\pi R_\infty$	r_0	2.817939(13)	4.6	10^{-15} m	10^{-13} cm
Electron magnetic moment in Bohr magnetons	μ_e/μ_B	1.0011596389(31)	0.0031		
Bohr magneton, $[c](eh/2m_e c)$	μ_B	9.274096(65)	7.0	10^{-24} J T^{-1}	10^{-21} erg G^{-1}
Electron magnetic moment	μ_e	9.284851(65)	7.0	10^{-24} J T^{-1}	10^{-21} erg G^{-1}
Gyromagnetic ratio of protons in H_2O	γ_p	2.67512708(82)	31	10^8 rad sec^{-1} · T^{-1}	10^4 rad sec^{-1} · G^{-1}
	$\gamma_p/2n$	4.257597(13)	3.1	10^7 Hz T^{-1}	10^3 Hz G^{-1}
γ_p corrected for diamagnetism of H_2O	γ_p	2.6751965(82)	3.1	10^8 rad sec^{-1} · T^{-1}	10^4 rad sec^{-1} · G^{-1}
	$\gamma_p/2n$	4.257707(13)	3.1	10^7 Hz T^{-1}	10^3 Hz G^{-1}
Magnetic moment of protons in H_2O in Bohr magnetons	μ_p/μ_B	1.52099312(10)	0.066	10^{-3}	10^{-3}
Proton magnetic moment in Bohr magnetons	μ_p/μ_B	1.52103264(46)	0.30	10^{-3}	10^{-3}
Proton magnetic moment	μ_p	1.4106203(99)	7.0	10^{-26} J T^{-1}	10^{-23} erg G^{-1}
Magnetic moment of protons in H_2O in nuclear magnetons	μ_p/μ_n	2.792709(17)	6.2		
μ_p/μ_n corrected for diamagnetism of H_2O	μ_p/μ_n	2.792782(17)	6.2		
Nuclear magneton, $[c](eh/2M_p c)$	μ_n	5.050951(50)	10	10^{-27} J T^{-1}	10^{-24} erg G^{-1}
Compton wavelength of the electron, $h/m_e c$	λ_C	2.4263096(74)	3.1	10^{-12} m	10^{-10} cm
	$\lambda_C/2\pi$	3.861592(12)	3.1	10^{-12} m	10^{-10} cm
Compton wavelength of the proton, $h/M_p c$	$\lambda_{C,p}$	1.3214409(90)	6.8	10^{-15} m	10^{-13} cm
	$\lambda_{C,p}/2\pi$	2.103139(14)	6.8	10^{-16} m	10^{-14} cm
Compton wavelength of the neutron, $h/M_n c$	$\lambda_{C,n}$	1.3196217(90)	6.8	10^{-15} m	10^{-13} cm
	$\lambda_{C,n}/2\pi$	2.100243(14)	6.8	10^{-16} m	10^{-14} cm
Gas constant	R_0	8.31434(35)	42	10^2 J kmole^{-1} · K^{-1}	10^7 erg mole^{-1} · K^{-1}
Boltzmann's constant, R_0/N	k	1.380622(59)	43	10^{-23} J K^{-1}	10^{-16} erg K^{-1}
Stefan–Boltzman constant, $\pi^2 k^4/60\hbar^3 c^2$	σ	5.66961(96)	170	10^{-8} W m^{-2} K^4	10^{-5} erg sec^{-1} · cm^{-2} · K^{-4}
First radiation constant, $8\pi hc$	c_1	4.992579(38)	7.6	10^{-24} J · m	10^{-15} erg · cm
Second radiation constant, hc/k	c_2	1.43833(61)	43	10^{-2} m · K	cm · K
Gravitational constant	G	6.6732(31)	460	10^{-11} N · m^2 kg^{-2}	10^{-8} dyn · cm^2 g^{-2}

TABLE 2.58 *(Continued)*

Quantity	Symbol	Value	Error, ppm	Units SI	cgs
kx-unit-to-angstrom conversion factor, $A = \lambda(\text{Å})/\lambda(kxu)$; $\lambda(CuK\alpha_1) \equiv 1.537400\,kxu$	Λ	1.0020764(53)	5.3		
Å*-to-angstrom conversion factor, $\Lambda = \lambda(\text{Å})/\lambda(\text{Å}^*)$; $\lambda(WK\alpha_1) \equiv 0.2090100\ \text{Å}^*$	Λ^*	1.0000197(56)	5.6		

[a]Note that the unified atomic mass scale $^{12}C \equiv 12$ has been used throughout, that amu = atomic mass unit, C = coulomb, G = gauss, Hz = hertz = cycles/sec, J = joule, K = kelvin (degrees kelvin), T = tesla (10^4 G), V = volt, and W = watt. In cases where formulas for constants are given (e.g., R_∞), the relations are written as the product of two factors. The second factor, in parentheses, is the expression to be used when all quantities are expressed in cgs units, with the electron charge in electrostatic units. The first factor, in brackets, is to be included only if all quantities are expressed in SI units. We remind the reader that with the exception of the auxiliary constants which have been taken to be exact, the uncertainties of these constants are correlated, and therefore the general law of error propagation must be used in calculating additional quantities requiring two or more of these constants. (From Taylor, Parker, and Langenberg, *Reviews of Modern Physics*, Vol. 41, pp. 375–496, July 1967. Reprinted by permission of the publisher, American Institute of Physics.)

TABLE 2.59 Values of Gas Constant

Value	Units
8.3166×10^7	erg/mole K
1.987	calorie/mole K
82.06	cm^3atm/mole K
0.08205	liter atm/mole K

$R = kN = $ (Boltzmann's constant)(Avogadro's number)

$R = (1.3806 \times 10^{-23} \text{J/K})(6.0222 \times 10^{26}/\text{kmole})$

$R = 8314.2$ J/K kmole

TABLE 2.60 Conversion Tables

To convert any physical quantity from one set of units, multiply by the appropriate table entry. For example, suppose that F is given as 10 poundals, but is desired in newtons. From this table, the result is:

$$F = 10 \times 1.386 \times 10^{-1} = 1.386 \; N$$

A. Conversion Factors for Quantities Having Dimensions of L (Length)

Multiply by table value to convert to these units →

Given a quantity in these units ↘	m	cm	mm	nm	Å	in	ft
1 m	1	10^2	10^3	10^9	10^{10}	39.37	3.281
1 cm	10^{-2}	1	10	10^7	10^3	0.3937	3.281×10^{-2}
1 mm	10^{-3}	10^{-1}	1	10^6	10^7	3.937×10^{-2}	3.281×10^{-3}
1 nm	10^{-9}	10^{-7}	10^{-6}	1	10	3.937×10^{-8}	3.281×10^{-9}
1 Å	10^{-10}	10^{-8}	10^{-7}	10^{-1}	1	3.937×10^{-9}	3.281×10^{-10}
1 in	0.0254	2.54	25.40	2.54×10^7	2.54×10^8	1	8.333×10^{-2}
1 ft	0.3048	30.48	304.8	3.048×10^8	3.048×10^9	12	1

Adapted from Gordon and Ford (1972).

B. Conversion Factors for Quantities Having Dimensions of L^2 (Area)

Multiply by table value to convert to these units →

Given a quantity in these units ↘	m^2	cm^2	mm^2	in^2	ft^2	mi^2
1 m^2	1	10^4	10^6	1.55×10^3	10.764	3.861×10^{-7}
1 cm^2	10^{-4}	1	10^2	0.155	1.076×10^{-3}	3.861×10^{-11}
1 mm^2	10^{-6}	10^{-2}	1	1.55×10^{-3}	1.076×10^{-5}	3.861×10^{-13}
1 in^2	6.452×10^{-4}	6.452	645.2	1	6.944×10^{-3}	2.49×10^{-10}
1 ft^2	9.29×10^{-2}	929.0	9.29×10^4	144	1	3.587×10^{-8}
1 mi^2	2.59×10^6	2.59×10^{10}	2.59×10^{12}	4.01×10^9	2.788×10^7	1
1 Acre	4.05×10^3	4.05×10^7	4.05×10^9	6.28×10^6	4.36×10^4	1.56×10^{-3}

Adapted from Gordon and Ford (1972).

TABLE 2.60 (*Continued*)

C. Conversion Factors for Quantities Having Dimensions of L^3 (Volume)

Given a quantity in these units ↘ / Multiply by table value to convert to these units →	m^3	cm^3	mm^3	in^3	ft^3	Liter
1 m^3	1	10^6	10^9	6.10×10^4	35.31	10^3
1 cm^3	10^{-6}	1	10^3	6.10×10^{-2}	3.53×10^{-5}	10^{-3}
1 mm^3	10^{-9}	10^{-3}	1	6.10×10^{-5}	3.53×10^{-8}	10^{-6}
1 in^3	1.639×10^{-5}	16.39	1.639×10^4	1	5.79×10^{-4}	1.639×10^{-2}
1 ft^3	2.832×10^{-2}	2.832×10^4	2.832×10^7	1728	1	28.316
1 liter	10^{-3}	10^3	10^6	61.02	3.53×10^{-2}	1
1 pint[a]	4.73×10^{-4}	473.18	4.73×10^5	28.88	1.67×10^{-2}	0.4731
1 quart[a]	9.46×10^{-4}	946.35	9.46×10^5	57.75	3.34×10^{-2}	0.9463
1 gallon[a]	3.78×10^{-3}	3.78×10^3	3.78×10^6	231	0.1334	3.785
1 ounce[a]	2.96×10^{-5}	29.57	2.96×10^4	1.805	1.04×10^{-3}	2.957×10^{-2}

[a]U.S., fluid. 1 British pint, quart, or gallon = 1.20 U.S. pints, quarts, or gallons; 1 British ounce = 0.9608 U.S. ounce.
Adapted from Gordon and Ford (1972).

D. Conversion Factors for Quantities Having Dimensions of M (Mass)

Given a quantity in these units ↘ / Multiply by table value to convert to these units →	g	kg	mg	Grain	Ounce[a]	Pound[a]
1 g	1	10^{-3}	10^3	15.432	3.527×10^{-2}	2.205×10^{-3}
1 kg	10^3	1	10^6	1.543×10^4	35.27	2.205
1 mg	10^{-3}	10^{-6}	1	1.543×10^2	3.527×10^{-5}	2.205×10^{-6}
1 grain	6.48×10^{-2}	6.48×10^{-5}	64.8	1	2.29×10^{-3}	1.43×10^{-4}
1 ounce[a]	28.3495	2.83×10^{-2}	2.83×10^4	437.5	1	6.25×10^{-2}
1 pound[a]	453.592	0.45359	4.53×10^5	7000	16	1
1 ton[b]	10^6	10^3	10^9	1.543×10^7	3.527×10^4	2.205×10^3

[a]Avoirdupois (avdp.).
[b]Metric. "Short" ton (U.S.) = 2000 pounds = 907.18 kg.
Adapted from Gordon and Ford (1972).

E. Conversion Factors for Quantities Having Dimensions of F or MLt^{-2}
(Force)

Given a quantity in these units ↘	Multiply by table value to convert to these units →			
	g cm sec^{-2} (dynes)	kg m sec^{-2} (newtons)	lb$_m$ ft sec^{-2} (poundals)	lb$_f$
g cm sec^{-2} (dynes)	1	10^{-5}	7.2330×10^{-5}	2.2481×10^{-6}
kg m sec^{-2} (newtons)	10^5	1	7.2330	2.2481×10^{-1}
lb$_m$ ft sec^{-2} (poundals)	1.3826×10^4	1.3826×10^{-1}	1	3.1081×10^{-2}
lb$_f$	4.4482×10^5	4.4482	32.1740	1

Reproduced from Bird, Stewart, and Lightfoot (1960).

F. Conversion Factors for Quantities Having Dimensions of F/L^2 or $ML^{-1}t^{-2}$
(Pressure, Momentum Flux)

Given a quantity in these units ↘	Multiply by table value to convert to these units →							
	g cm^{-1} sec^{-2} (dyne cm^{-2})	newtons m^{-2} (kg m^{-1} sec^{-2})	lb$_m$ ft^{-1} sec^{-2} (poundals ft^{-2})	lb$_f$ ft^{-2}	lb$_f$ in^{-2} (psia)[a]	Atmospheres[b] (atm)	mm Hg (torr)	in Hg
g cm^{-1} sec^{-2}	1	10^{-1}	6.7197×10^{-2}	2.0886×10^{-3}	1.4504×10^{-5}	9.8692×10^{-7}	7.5006×10^{-4}	2.9530×10^{-5}
Nm^{-2} (kg m^{-1} sec^{-2})	10	1	6.7197×10^{-1}	2.0886×10^{-2}	1.4504×10^{-4}	9.8692×10^{-6}	7.5006×10^{-3}	2.9530×10^{-4}
lb$_m$ ft^{-1} sec^{-2}	1.4882×10^1	1.4882	1	3.1081×10^{-2}	2.1584×10^{-4}	1.4687×10^{-5}	1.1162×10^{-2}	4.3945×10^{-4}
lb$_f$ ft^{-2}	4.7880×10^2	4.7880×10^1	32.1740	1	6.9444×10^{-3}	4.7254×10^{-4}	3.5913×10^{-1}	1.4139×10^{-2}
lb$_f$ in^{-2}	6.8947×10^4	6.8947×10^3	4.6330×10^3	144	1	6.8046×10^{-2}	5.1715×10^1	2.0360
Atmospheres	1.0133×10^6	1.0133×10^5	6.8087×10^4	2.1162×10^3	14.696	1	760	29.921
mm Hg	1.3332×10^3	1.3332×10^2	8.9588×10^1	2.7845	1.9337×10^{-2}	1.3158×10^{-3}	1	3.9370×10^{-2}
in Hg	3.3864×10^4	3.3864×10^3	2.2756×10^3	7.0727×10^1	4.9116×10^{-2}	3.3421×10^{-2}	25.400	1

[a] This unit is preferably abbreviated psia (pounds per square inch absolute) or psig (pounds per square inch gage). Gage pressure is absolute pressure minus the prevailing barometric pressure.

[b] 1 atm = 1.01325 bars.

Adapted from Bird, Stewart, and Lightfoot (1960).

TABLE 2.60 (*Continued*)

G. Conversion Factors for Quantities Having Dimensions of ML^2 (Moment of Inertia)

Given a quantity in these units ↘	Multiply by table value to convert to these units →				
	$g\ cm^2$	$kg\ m^2$	$lb_m\ in^2$	$lb_m\ ft^2$	$slug\ ft^2$
$g\ cm^2$	1	10^{-7}	3.4169×10^{-4}	2.37285×10^{-6}	7.37507×10^{-8}
$kg\ m^2$	10^7	1	3.4169×10^3	23.7285	0.737507
$lb_m\ in^2$	2.9266×10^3	2.9266×10^{-4}	1	6.944×10^{-3}	2.15841×10^{-4}
$lb_m\ ft^2$	4.21434×10^5	4.21434×10^{-2}	144	1	3.10811×10^{-2}
$slug\ ft^2$	1.3559×10^7	1.3559	4.63304×10^3	32.1739	1

Adapted from Eshbach and Souders (1975).

H. Conversion Factors for Quantities Having Dimensions of FL or ML^2t^{-2} (Work, Energy, Torque)

Given a quantity in these units ↘	Multiply by table value to convert to these units →							
	$g\ cm^2\ sec^{-2}$ (ergs)	$kg\ m^2\ sec^{-2}$ (absolute joules)	$lb_m\ ft^2\ sec^{-2}$ (ft-poundals)	$ft\ lb_f$	cal	BTU	hp-hr	kw-hr
$g\ cm^2\ sec^{-2}$	1	10^{-7}	2.3730×10^{-6}	7.3756×10^{-8}	2.3901×10^{-8}	9.4783×10^{-11}	3.7251×10^{-14}	2.7778×10^{-14}
$kg\ m^2\ sec^{-2}$	10^7	1	2.3730×10^1	7.3756×10^{-1}	2.3901×10^{-1}	9.4783×10^{-4}	3.7251×10^{-7}	2.7778×10^{-7}
$lb_m\ ft^2\ sec^{-2}$	4.2140×10^5	4.2140×10^{-2}	1	3.1081×10^{-2}	1.0072×10^{-2}	3.9942×10^{-5}	1.5698×10^{-8}	1.1706×10^{-8}
$ft\ lb_f$	1.3558×10^7	1.3558	32.1740	1	3.2405×10^{-1}	1.2851×10^{-3}	5.0505×10^{-7}	3.7662×10^{-7}
Thermochemical calories[a]	4.1840×10^7	4.1840	9.9287×10^1	3.0860	1	3.9657×10^{-3}	1.5586×10^{-6}	1.1622×10^{-6}
British thermal units	1.0550×10^{10}	1.0550×10^3	2.5036×10^4	778.16	2.5216×10^2	1	3.9301×10^{-4}	2.9307×10^{-4}
Horsepower-hours	2.6845×10^{13}	2.6845×10^6	6.3705×10^7	1.9800×10^6	6.4162×10^5	2.5445×10^3	1	7.4570×10^{-1}
Absolute kilowatt-hours	3.6000×10^{13}	3.6000×10^6	8.5429×10^7	2.6552×10^6	8.6042×10^5	3.4122×10^3	1.3410	1

[a]The unit, abbreviated cal, is used in chemical thermodynamic tables. To convert quantities expressed in International Steam Table calories (abbreviated I.T. cal) to this unit, multiply by 1.000654.

Reproduced from Bird, Stewart, and Lightfoot (1960).

I. Conversion Factors for Quantities Having Dimensions of FLt^{-1} or ML^2t^{-3}
(Power, Rate of Doing Work)

Given a quantity in these units ↘	Multiply by table value to convert to these units →					
	Btu min^{-1}	erg sec^{-1}	ft lb$_f$ min^{-1}	hp	cal min^{-1}	kw
Btu min^{-1}	1	1.758×10^8	7.782×10^2	2.358×10^{-2}	2.522×10^2	1.758×10^{-2}
erg sec^{-1}	5.687×10^{-9}	1	4.425×10^{-6}	1.341×10^{-10}	1.434×10^{-6}	10^{-10}
ft lb$_f$ min^{-1}	1.285×10^{-3}	2.260×10^5	1	3.030×10^{-5}	3.240×10^{-1}	2.260×10^{-5}
hp	42.41	7.457×10^9	3.300×10^4	1	1.069×10^4	7.457×10^{-1}
cal min^{-1}	3.966×10^{-3}	6.973×10^5	3.086	9.352×10^{-5}	1	6.973×10^{-5}
kw	5.687×10^1	10^{10}	4.425×10^4	1.341	1.434×10^4	1

Reproduced from Bird, Stewart, and Lightfoot (1960).

J. Conversion Factors for Quantities Having Dimensions[a] of $ML^{-1}t^{-1}$ of FtL^{-2} or Moles $L^{-1}t^{-1}$
(Viscosity, Density times Diffusivity, Concentration times Diffusivity)

Given a quantity in these units ↘	Multiply by table value to convert to these units →					
	g cm^{-1} sec^{-1} (poises)	Pa · s (N sec m^{-2})	lb$_m$ ft^{-1} sec^{-1}	lb$_f$ sec ft^{-2}	Centipoises	lb$_m$ ft^{-1} hr^{-1}
g cm^{-1} sec^{-1}	1	10^{-1}	6.7197×10^{-2}	2.0886×10^{-3}	10^2	2.4191×10^2
Pa · s (N sec m^{-2})	10	1	6.7197×10^{-1}	2.0886×10^{-2}	10^3	2.4191×10^3
lb$_m$ ft^{-1} sec^{-1}	1.4882×10^1	1.4882	1	3.1081×10^{-2}	1.4882×10^3	3600
lb$_f$ sec ft^{-2}	4.7880×10^2	4.7880×10^1	32.1740	1	4.7880×10^4	1.1583×10^5
Centipoises	10^{-2}	10^{-3}	6.7197×10^{-4}	2.0886×10^{-5}	1	2.4191
lb$_m$ ft^{-1} hr^{-1}	4.1338×10^{-3}	4.1338×10^{-4}	2.7778×10^{-4}	8.6336×10^{-6}	4.1338×10^{-1}	1

[a]When moles appear in the given and desired units, the conversion factor is the same as for the corresponding mass units.

Adapted from Bird, Stewart, and Lightfoot (1960).

TABLE 2.60 *(Continued)*

K. Conversion Factors for Quantities Having Dimensions of $MLt^{-3}T^{-1}$ or $Ft^{-1}T^{-1}$
(Thermal Conductivity)

Given a quantity in these units ↘ / Multiply by table value to convert to these units →	g cm sec⁻³ K⁻¹ (ergs sec⁻¹ cm⁻¹ K⁻¹)	kg m sec⁻³ K⁻¹ (watts m⁻¹ K⁻¹)	lb_m ft sec⁻³ R⁻¹	lb_f sec⁻¹ R⁻¹	cal sec⁻¹ cm⁻¹ K⁻¹	BTU hr⁻¹ ft⁻¹ R⁻¹
g cm sec⁻³ K⁻¹	1	10^{-5}	4.0183×10^{-5}	1.2489×10^{-6}	2.3901×10^{-8}	5.7780×10^{-6}
kg m sec⁻³ K⁻¹	10^5	1	4.0183	1.2489×10^{-1}	2.3901×10^{-3}	5.7780×10^{-1}
lb_m ft sec⁻³ R⁻¹	2.4886×10^4	2.4886×10^{-1}	1	3.1081×10^{-2}	5.9479×10^{-4}	1.4379×10^{-1}
lb_f sec⁻¹ R⁻¹	8.0068×10^5	8.0068	3.2174×10^1	1	1.9137×10^{-2}	4.6263
cal sec⁻¹ cm⁻¹ K⁻¹	4.1840×10^7	4.1840×10^2	1.6813×10^3	5.2256×10^1	1	2.4175×10^2
BTU hr⁻¹ ft⁻¹ R⁻¹	1.7307×10^5	1.7307	6.9546	2.1616×10^{-1}	4.1365×10^{-3}	1

Reproduced from Bird, Stewart, and Lightfoot (1960).

L. Conversion Factors for Quantities Having Dimensions of $L^2 t^{-1}$
(Momentum Diffusivity, Thermal Diffusivity, Molecular Diffusivity)

Given a quantity in these units ↘ / Multiply by table value to convert to these units →	cm² sec⁻¹	m² sec⁻¹	ft² hr⁻¹	Centistokes
cm² sec⁻¹	1	10^{-4}	3.8750	10^2
m² sec⁻¹	10^4	1	3.8750×10^4	10^6
ft² hr⁻¹	2.5807×10^{-1}	2.5807×10^{-5}	1	2.5807×10^1
Centistokes	10^{-2}	10^{-6}	3.8750×10^{-2}	1

Reproduced from Bird, Stewart, and Lightfoot (1960).

214

M. Conversion Factors for Quantities Having Dimensions of $Mt^{-3}T^{-1}$ or $FL^{-1}t^{-1}T^{-1}$
(Heat Transfer Coefficients)

Given a quantity in these units ↘ / Multiply by table value to convert to these units →	g sec⁻³ K⁻¹	kg sec⁻³ K⁻¹ (watts m⁻² K⁻¹)	lb$_m$ sec⁻³ R⁻¹	lb$_f$ ft⁻¹ sec⁻¹ R⁻¹	cal cm⁻² sec⁻¹ K⁻¹	Watts cm⁻² K⁻¹	BTU ft⁻² hr⁻¹ R⁻¹
g sec⁻³ K⁻¹	1	10^{-3}	1.2248×10^{-3}	3.8068×10^{-5}	2.3901×10^{-8}	10^{-7}	1.7611×10^{-4}
kg sec⁻³ K⁻¹	10^{3}	1	1.2248	3.8068×10^{-2}	2.3901×10^{-5}	10^{-4}	1.7611×10^{-1}
lb$_m$ sec⁻³ R⁻¹	8.1647×10^{2}	8.1647×10^{-1}	1	3.1081×10^{-2}	1.9514×10^{-5}	8.1647×10^{-5}	1.4379×10^{-1}
lb$_f$ ft⁻¹ sec⁻¹ R⁻¹	2.6269×10^{4}	2.6269×10^{1}	32.1740	1	6.2784×10^{-4}	2.6269×10^{-3}	4.6263
cal cm⁻² sec⁻¹ K⁻¹	4.1840×10^{7}	4.1840×10^{4}	5.1245×10^{4}	1.5928×10^{3}	1	4.1840	7.3686×10^{3}
Watts cm⁻² K⁻¹	10^{7}	10^{4}	1.2248×10^{4}	3.8068×10^{2}	2.3901×10^{-1}	1	1.7611×10^{3}
BTU ft⁻² hr⁻¹ R⁻¹	5.6782×10^{3}	5.6782	6.9546	2.1616×10^{-1}	1.3571×10^{-4}	5.6782×10^{-4}	1

Reproduced from Bird, Stewart, and Lightfoot (1960).

N. Conversion Factors for Quantities Having Dimensions[a] of $ML^{-2}t^{-1}$ or Moles $L^{-2}t^{-1}$ or $FL^{-3}t$
(Mass Transfer Coefficients k_{xi})

Given a quantity in these units ↘ / Multiply by table value to convert to these units →	g cm⁻² sec⁻¹	kg m⁻² sec⁻¹	lb$_m$ ft⁻² sec⁻¹	lb$_f$ ft⁻³ sec	lb$_m$ ft⁻² hr⁻¹
g cm⁻² sec⁻¹	1	10	2.0482	6.3659×10^{-2}	7.3734×10^{3}
kg m⁻² sec⁻¹	10^{-1}	1	2.0482×10^{-1}	6.3659×10^{-3}	7.3734×10^{2}
lb$_m$ ft⁻² sec⁻¹	4.8824×10^{-1}	4.8824	1	3.1081×10^{-2}	3600
lb$_f$ ft⁻³ sec	1.5709×10^{1}	1.5709×10^{2}	32.1740	1	1.1583×10^{5}
lb$_m$ ft⁻² hr⁻¹	1.3562×10^{-4}	1.3562×10^{-3}	2.7778×10^{-4}	8.6336×10^{-6}	1

[a]When moles appear in the given and desired units, the conversion factor is the same as for the corresponding mass units.

Reproduced from Bird, Stewart, and Lightfoot (1960).

TABLE 2.60 (*Continued*)

O. Conversion Factors for Quantities With Dimensions of Q
(Quantity of Electricity, Dielectric Flux)

Given a quantity in these units ↘ / Multiply by table value to convert to these units →	Abcoulomb	Ampere-hour	Coulomb	Faraday	Statcoulomb
abcoulomb	1	2.778×10^{-3}	10	1.036×10^{-4}	2.998×10^{10}
ampere–hour	360	1	3600	3.731×10^{-2}	1.080×10^{13}
coulomb	0.1	2.778×10^{-4}	1	1.036×10^{-5}	2.998×10^{9}
Faraday	9649	26.80	9.649×10^{4}	1	2.893×10^{14}
statcoulomb	3.335×10^{-11}	9.259×10^{-14}	3.335×10^{-10}	3.457×10^{-15}	1

Reproduced from Eshbach and Souders (1975).

REFERENCES

Anon., *Military Specification: Hydraulic Fluid, Fire-Resistant, Non-Neurotoxic*, MIL-H-19457C (SH), Apr. 1981.

Anon., *Military Specification: Hydraulic Fluid, Non-Petroleum Base, Aircraft*, MIL-H-84456B, June 1978.

Anon., *Military Specification: Hydraulic Fluid, Petroleum Base, For Machine Tools*, MIL-H-46001C, 11 March 1982 with Amendment 1, Nov. 1982.

Anon., *Military Specification: Hydraulic Fluid, Petroleum Base; Aircraft, Missile, and Ordnance*, MIL-H-5606E, Jan. 1978.

Anon., *U.S. Standard Atmosphere, 1976*, U.S. Government Printing Office, 1976.

Anon., *Gylcols*, Union Carbide Corp., Publication F-41515B, 1978.

ASTM, "Standard Specification for Aviation Turbine Fuels," ASTM D 1655-89, *1990 Annual Book of ASTM Standards*, Vol. 05.01, Philadelphia, PA, 1990a.

ASTM, "Standard Specification for Aviation Gasolines," ASTM D 910-89, *1990 Annual Book of ASTM Standards*, Vol. 05.01, Philadelphia, PA, 1990b.

ASTM, "Standard Specification for Automotive Gasolines," ASTM D 439-89, *1990 Annual Book of ASTM Standards*, Vol. 05.01, Philadelphia, PA, 1990c.

ASTM, "Standard Specification for Diesel Fuel Oils," ASTM D 975-89, *1990 Annual Book of ASTM Standards*, Vol. 05.01, Philadelphia, PA, 1990d.

ASTM, "Standard Specification for Fuel Oils," ASTM D 396-89, *1990 Annual Book of ASTM Standards*, Vol. 05.01, Philadelphia, PA, 1990e.

ASTM, "Standard Method for Distillation of Petroleum Products," ASTM D 86-82, *1990 Annual Book of ASTM Standards*, Vol. 05.01, Philadelphia, PA, 1990f.

ASTM, "Estimation of Solubility of Gases in Petroleum and Other Organic Liquids," ASTM D 3827-86, *1990 Annual Book of ASTM Standards*, Vol. 05.03, Philadelphia, PA, 1990g.

ASTM, "Standard Viscosity—Temperature Charts for Liquid Petroleum Products," ASTM D 341-89, *1990 Annual Book of ASTM Standards*, Vol. 05.01, Philadelphia, PA, 1990h.

Athey, S. W., *Acoustics Technology, A Survey*, NASA SP-5093, 1970.

Austin, R. W. and Halikas, G., "The Index of Refraction of Seawater," Visibility Laboratory, Scripps Institution of Oceanography, University of California at San Diego, *SIO Reference No. 76-1*, pp. 121, Jan. 1976.

"Aviation Fuels Handbook," Coordinating Research Council, CRC Report No. 530, 1983.

Barnett, H. C. and Hibbard, R. R., "Properties of Aircraft Fuels," NACA TN-3276, Aug. 1956.

Beranek, L. L., *Acoustic Measurements*, John Wiley & Sons, Inc., New York, 1949.

Bird, R. B., Stewart, W. E., and Lightfoot, E. N., *Transport Phenomena*, John Wiley & Sons, New York, 1960.

Broecker, W. S., *Chemical Oceanography*, Harcourt Brace Jovanovich Inc., New York, 1974.

Bromley, L. A., Desonssure, V. A., Clipp, J. C., and Wright, J. S., "Heat Capacities of Seawater Solutions at Salinities of 1 to 12‰ and temperatures of 2 to 80°C," *J. Chem. Eng. Data*, Vol. 12, pp. 202–206, 1967.

Chen, C.-T. and Millero, F. J., "Reevaluation of Wilson's Sound Speed Measurements for Pure Water," *J. Acous. Soc. Amer.*, Vol. 60, pp. 1270–1273, 1976.

Chen, C.-T. and Millero, F. J., "Speed of Sound in Seawater at High Pressures," *J. Acous. Soc. Amer.*, Vol. 62, No. 5, pp. 1129–1135. Nov. 1977.

Cohen, E. R. and Taylor, B. N., "The 1973 Least Squares Adjustment of the Fundamental Constants," *J. Phys. Chem. Ref. Data*, Vol. 2, No. 4, pp. 663–734, 1973.

Counsell, J. F., Lawrenson, I. J., and Lees, E. B., "The Thermodynamic Properties of Benzene and Toluene," National Physical Laboratory Report Chem., Vol. 52, pp. 13, 1976.

Cox, R. N. and Crabtree, L. F., *Elements of Hypersonic Aerodynamics*, Academic Press, New York, 1965.

Dickson, C. L., "Aviation Turbine Fuels, 1991," NIPPER-174 PPS-92/2, Mar. 1992.

Dickson, C. L., "Aviation Turbine Fuels, 1992," NIPPER-179 PPS-93/2, Mar. 1993.

Eshbach, O. W., and Souders, M., *Handbook of Engineering Fundamentals*, John Wiley & Sons, New York, 1975.

Evans, L. V. and Bass, H. E., "Tables of Absorption and Velocity of Sound in Still Air at 68F," *Rept. WR-72-2 AROD-8725*, Wylie Labs. (National Technical Information Service, U.S. Department of Commerce, Springfield VA, AD 738 576), 1972.

Fofonoff, N. P. and Millard, R. C., Jr., "Algorithms for Computation of Fundamental Properties of Seawater," *UNESCO Technical Papers in Marine Science*, No. 44, UNESCO, Paris, 1983.

Francois, R. E. and Garrison, G. R., "Sound Absorption Based on Ocean Measurements. Part II: Boric Acid Contribution and Equation for Total Absorption," *J. Acous. Soc. Amer.*, Vol. 72, No. 6, pp. 1879–1890, Dec. 1982.

Fuhs, A. E., *Instrumentation for High Speed Plasma Flow*, Gordon and Breach Science Publishers, New York, 1965.

General Electric Co., Aircraft Gas Turbine Development Dept., Cincinnati, OH, *Properties of Combustion Gases/Systems: CnH2n—Air. Vol. 1, Thermodynamic Properties*, Mc-Graw-Hill Book Co., New York, 1955.

Gordon, A. J. and Ford, R A., *The Chemist's Companion*, John Wiley & Sons, New York, 1972.

Gordon, H. R., Smith, R. C., and Zaneveld, J. R. V., "Introduction to Ocean Optics," *Society of Photo-Optical Instrumentation Engineers, Proceedings*, Vol. 208, pp. 14–55, 1979.

Gordon, S. and McBride, B. J., "Computer Program for Calculation of Complex Chemical Equilibrium Compositions, Rocket Performance, Incident and Reflected Shocks, and Chapman—Jouguet Detonations," *NASA SP-273, Interim Revision*, 1976.

Gordon, S. "Thermodynamic and Transport Combustion Properties of Hydrocarbons With Air. I—Properties in SI Units," *NASA TP 1906*, July 1982a.

Gordon, S., "Thermodynamic and Transport Combustion Properties of Hydrocarbons With Air. II—Compositions Corresponding to Kelvin Temperature Schedules in Part I," *NASA TP 1907*, July 1982b.

Hanley, H. J. M. and Ely, J. F., "The Viscosity and Thermal Conductivity Coefficients of Dilute Nitrogen and Oxygen," *J. Phys. Chem. Ref. Data*, Vol. 2, No. 4, pp. 735–755, 1973.

Hatton, R. E., *Introduction to Hydraulic Fluids*, Van Nostrand Reinhold, New York, 1962.

Holden, N. E., "Atomic Weights of the Elements, 1977," *Pure Appl. Chem.*, Vol. 51, pp. 405–433, 1979.

Holzbock, W. G., *Hydraulic Power and Equipment*, Industrial Press, New York, 1968.

Hood, D. and Pytkowicz, R. M., "Chemical Oceanography," *Marine Sciences*, Vol. 1, Smith, F. G. W. (Ed), CRC Press, Cleveland, OH, 1974.

Horne, R. A., *Marine Chemistry*, Arthur D. Little, Inc., Wiley–Interscience, New York, 1969.

Hougen, O. A., Watson, K. M., and Ragatz, R. A., *Chemical Process Principles, Part I*, 2nd ed., John Wiley & Sons, New York, 1954.

Humphreys, W. J., *Physics of the Air*, Dover Publications, New York, 1964.

JANAF Thermochemical Tables, Dow Chemical Co., Midland, MI, Dec. 31, 1960–Mar. 31, 1979.

Jerlov, N. G., *Marine Optics*, Elsevier Scientific Publishing Company, Amsterdam, Oxford, New York, 1976.

Juza, J., An *Equation of State for Water and Steam, Academia*, Prague, Czechoslovakia, 1966.

Keenan, J. H., Chao, J., and Kaye, J., *Gas Tables*, 2nd ed., John Wiley & Sons, New York, 1980.

Kesselman, P. H. and Blank, Yu. I., Paper B-11, *7th Int. Conf. on Properties of Steam*, Tokyo, Japan, 1968.

Klein, L. A. and Swift, C. T., "An Improved Model for the Dielectric Constant of Sea Water at Microwave Frequencies," *IEEE Trans. Antennas and Propagation*, Vol. AP-25, No. 1, pp. 104–111, Jan. 1977.

Knudsen V. O. and Harris, C. M., *Acoustical Designing in Architecture*, John Wiley & Sons, Inc., New York, 1950.

Lewis, G. N. and Randall, M., *Thermodynamics*, 2nd ed., John Wiley & Sons, New York, 1980.

Leyendekkers, J. V., "Prediction of Heat Capacities of Seawater and Other Multicomponent Solutions From the Tammann–Tait–Gibbs Model," *Marine Chemistry*, Vol. 9, pp. 25–35, 1980.

Loh, W. H. T. (Ed.), *Re-Entry and Planetary Entry Physics and Technology*, Springer–Verlag, New York, 1968.

Lovett, J. R., "Geographic Variation of Low-Frequency Sound Absorption in the Atlantic, Indian, and Pacific Oceans," *J. Acous. Soc. Amer.*, Vol. 67, pp. 338–340, 1980.

Mackenzie, K. V., "Discussion of Seawater Sound-Speed Determinations," *J. Acous. Soc. Amer.*, Vol. 70, No. 3, pp. 801–806, Sept. 1981.

Maitland, G. C. and Smith, E. B., "Critical Reassessment of Viscosities of 11 Common Gases," *J. Chem. Eng. Data*, Vol. 17, No. 2, pp. 150–156, 1972.

Martin, J. J., *Atmospheric Reentry*, Prentice-Hall Space Technology Series, Englewood Cliffs, NJ, 1966.

McNeil, G. T., "Underwater Photography Handbook," Published under the direction of the Commander, U.S. Naval Air Systems Command, NAVAIR 10-1-799, U.S. Government Printing Office: 1983-605-018/6665, Mar. 1983.

Mehu, A. and Johannin–Gilles, A., "Variation de l'indice de réfraction de l'eau de mer étalon de Copenhague et des dilutions en fonction de longeur d'onde, de la température et de la chlorinité," *Cahiers Océanographiques*, Vol. 20, No. 9, pp. 803–812, 1968.

MIL, "Specification, Turbine Fuel, Aviation, JP-4 and JP-5," MIL-T-5624N, Amend. 1, Feb. 1989a.

MIL, "Specification, Turbine Fuel, Aviation, JP-8," MIL-T-83133C-Amend. 1, Feb. 1989b.

"Specification, Turbine Fuel, Aviation, JP-4 and JP-5," MIL-T-5624N, Amend. 1, Feb. 1989a.

Millero, F. J. and Kubinski, T., "Speed of Sound in Seawater as a Function of Temperature and Salinity at 1 atm," *J. Acous. Soc. Amer.*, Vol. 57, No. 2, pp. 312–319, Feb. 1975.

Millero, F. J. and Leung, W. H., "The Thermodynamics of Seawater at One Atmosphere," *Amer. J. Sci.*, Vol. 276, pp. 1035–1077, 1976.

Millero, F. J. and Poisson, A., "International One-Atmosphere Equation of State of Seawater," *Deep Sea Research*, Vol. 28A, No. 6, pp. 625–629, 1981.

Millero, F. J. and Schreiber, D. R., "Use of the Ion-Pairing Model to Estimate Activity Coefficients of the Ionic Components of Natural Waters," *Amer. J. Sci.*, Vol. 282, pp. 1508–1540, 1982.

Millero, F. J. and Sohn, M. L., *Chemical Oceanography*, CRC Press, Boca Raton, FL, 1992.

Millero, F. J., Perron, G., and Desnoyers, J. E., "The Heat Capacity of Seawater Solutions From 5 to 35°C and 0.5 to 22‰ Chlorinity," *J. Geophys. Res.*, Vol. 78, pp. 4499–4507, 1973.

Millero, F. J., "Seawater as a Multicomponent Electrolytic Solution," *The Sea*, Vol. 5: *Marine Chemistry*, Goldberg, E. D. (Ed.), John Wiley & Sons, New York, 1974.

Millero, F. J., "The Thermodynamics of Seawater, Part II. Thermochemical Properties," *Ocean Science and Engineering*, Vol. 8, No. 1, pp. 1–40, 1983.

Millero, F. J., "The Thermodynamics of Seawater. Part I. The PVT Properties," *Ocean Science and Engineering*, Vol. 7, No. 4, pp. 403–406, 1982.

Morel, A., "Optical Properties of Pure Water and Pure Sea Water," *Optical Aspects of Oceanography*, Jerlov, N. G. and Steenman Nielsen, E. (Eds.), Academic Press, London and New York, 1974.

Morel, A., "Etude experimentale de la diffusion de la lumiére par l'eau, les solutions de chlorure de sódium et l'eau de mer optiquement pures," *Journal de Chimie Physique*, Vol. 10, pp. 1359–1366, 1966.

Padilla, A., "High Temperature Thermodynamic Properties of Sodium," Argonne National Laboratory Report ANL 8095, 1974.

Pangborn, J. and Gillis, J., "Alternative Fuels for Automotive Transportation a Feasibility Study, Vol. III, Appendices," EPA-460/3-74-012-C, July 1974.

Petroleum "Measurement Tables," ASTM D 1250-80, American Society for Testing Materials, Philadelphia, PA, 1980a.

Petzold, T. J., "Volume Scattering Functions for Selected Ocean Waters," *Scripps Institution of Oceanography Reference No. 72-28*, 1972.

Poferl, D. J. and Svehla, R. A., "Thermodynamic and Transport Properties of Air and Its Products of Combustion with ASTM-A-1 Fuel and Natural Gas at 20, 30, and 40 Atmospheres," *NASA TN D-7488*, 1973.

Poisson, A., "Conductivity/Salinity/Temperature Relationship of Diluted and Concentrated Standard Seawater," *IEEE J. Oceanic Eng.* vol. OE-5, No. 1, pp. 41–50, Jan. 1980. (Reprinted by UNESCO in "Background Papers and Supporting Data on the Practical Salinity Scale 1978," *UNESCO Technical Papers in Marine Science*, No. 37, Paris, 1982.)

Preisendorfer, R. W., *Hydrologic Optics, Volume I. Introduction*, U.S. Department of Commerce, National Oceanic and Atmospheric Administration, Environmental Research Laboratories, Pacific Marine Environmental Laboratory, Honolulu, Hawaii, 1976.

Quinby–Hunt, M. S. and Turekian, K. K. "Distribution of Elements in Seawater," *EOS*, Vol. 64, No. 14, 1983.

Rayleigh, Lord B., *The Theory of Sound*, Vols. I and II, Dover Publications, New York, 1945.

Reynard, K. W., *Engineering Science Data Unit*, London, England, Item 67031, 1967.

Ribe, R. L. and Howe, J. G., "An Empirical Equation Relating Sea Water Salinity, Temperature, Pressure, and Electrical Conductivity," *Marine Tech. Soc. J.*, Vol. 9, No. 9, pp. 3–13, Oct.–Nov. 1975.

Schmidt E., *Properties of Water and Steam in S.I. Units* Springer–Verlag, Berlin, 1969.

Shelton, E. M. and Dickson, C. L., "Aviation Fuels, 1983," NIPER-134 PPS-84/2, Apr. 1984.

Shelton, E. M., "Diesel Fuel Oils, 1981," DOE/BETC/PPS-81/5, 1981.

Shelton, E. M., "Motor Gasolines, Summer 1981," DOE/BETC/PPS-82/1, 1982a.

Shelton, E. M., "Motor Gasolines, Winter 1981–82," DOE/BETC/PPS-82/3, 1982b.

Smith, M., *Aviation Fuels*, Foulis Co., Henley-on-Thames, 1970.

So, B. Y. C. and Klaus, E. E., "Viscosity-Pressure Correlation of Liquids," *Trans. ASLE*, Vol. 23, p. 409, 1980.

Stanley, E. M., "The Refractive Index of Seawater as a Function of Temperature, Pressure, and Two Wavelengths," *Deep-Sea Research*, Vol. 18, pp. 833–840, 1971.

Svehla, R. A. and McBride, B. J., "FORTRAN IV Computer Program for Calculation of Thermodynamic and Transport Properties of Complex Chemical Systems," *NASA TN D-7056*, 1973.

Sverdrup, H. W., Johnson, M. W., and Fleming, R. H., *The Oceans*, Prentice-Hall, Inc., Englewood Cliffs, NJ, 1942.

Taylor, B. N., Parker, W. H., and Langenberg, D. N., "Determination of e/n Using Macroscopic Quantum Phase Coherence in Superconductors: Implications for Quantum Electrodynamics and the Fundamental Physical Constants," *Reviews of Modern Physics*, Vol. 41, p. 375, 1969.

Technical Data Book—Refining, American Petroleum Institute, Washington, D.C., 1970.

"Technical Data Book—Petroleum Refining," American Petroleum Institute, Division of Refining, Port City Press, Baltimore, MD, 1966.

Tilton, L. W. and Taylor, J. K., "Refractive Index and Dispersion of Distilled Water for Visible Radiation at Temperatures 0° to 60°C," *J. Research of the National Bureau of Standards* [U.S.], Vol. 20, pp. 419–477, 1938.

Touloukian, Y. S., Liley, P. E., and Saxena, S. C., *Thermophysical Properties of Matter, Vol. 3, Thermal Conductivity*, IFI/Plenum, 1979.

Touloukian, Y. S., Saxena, S. C., and Hestermans, P., *Thermophysical Properties of Matter, Vol. 11, Viscosity*, IFI/Plenum, 1975.

UNESCO, "Background Papers and Supporting Data on the International Equation of State of Seawater 1980," UNESCO Technical Papers in Marine Science, No. 38, 1981.

UNESCO, "Background Papers and Supporting Data on the Practical Salinity Scale 1978," UNESCO Technical Papers in Marine Science, No. 37, 1981.

Van Winkle, M., "Tag Flash vs. ASTM Distillation," *Petroleum Refiner*, Vol. 33, No. 11, pp. 171, 1954.

Vargaftik, N. B., *Tables of the Thermophysical Properties of Liquids and Gases*, Halstead Press, New York, 758 pp., 1975.

Vukalovich, M. P., Ivanov, A. I., Fokin, L. R., and Yakovlev, A. T., *Thermophysical Properties of Mercury*, Izd. Standartov, Moscow, U.S.S.R., 1971.

Williams, P. J. le B., "Biological and Chemical Aspects of Dissolved Organic Material in Seawater," *Chemical Oceanography*, Vol. 2, Riley, J. P. and Skirrow, G. (Eds.), Academic Press, New York, 1975.

Wilson, W. D., "Ultrasonic Measurement of the Speed of Sound in Distilled Water and in Seawater," *U.S. Naval Ordnance Laboratory Report No. 6746*, 1959.

Wood, A., *Acoustics*, Dover Publications, New York, 1966.

3 Non-Newtonian Liquids

R. BYRON BIRD

University of Wisconsin–Madison
Madison, WI

JOHN M. WIEST

Purdue University
West Lafayette, IN

CONTENTS

Handbook of Fluid Dynamics and Fluid Machinery, Edited by Joseph A. Schetz and Allen E. Fuhs
ISBN 0-471-12598-9 Copyright © 1996 John Wiley & Sons, Inc.

3.1 INTRODUCTION

Non-Newtonian liquids are those liquids that are not described by the Newtonian constitutive equation, in which the components of the stress tensor are linear in the velocity gradients. In this section, examples of some non-Newtonian liquids and also some illustrations of non-Newtonian flow phenomena are given [see also Bird, *et al.* (1987), Lodge (1964), Schowalter (1978), Barnes *et al.* (1989), Boger and Walters (1993), and Giesekus (1994)].

3.1.1 Conservation Equations

For incompressible liquids the laws of conservation of mass, momentum, and energy lead to the following set of equations [Bird *et al.* (1987) and Bird *et al.* (1960)].

Continuity:
$$(\nabla \cdot \bar{\mathbf{v}}) = 0 \tag{3.1}$$

Motion:
$$\rho \frac{D\bar{\mathbf{v}}}{Dt} = - [\nabla \cdot \bar{\bar{\pi}}] + \rho\bar{\mathbf{g}} \tag{3.2}$$

Energy:
$$\rho \frac{D\hat{U}}{Dt} = -(\nabla \cdot \bar{\mathbf{q}}) - (\bar{\bar{\pi}}{:}\nabla\bar{\mathbf{v}}) \tag{3.3}$$

Here ρ is the fluid density, considered throughout this chapter to be constant; $\bar{\mathbf{v}}$ is the fluid velocity, $\hat{U}$ is the internal energy per unit mass, and $\bar{\mathbf{g}}$ is the gravitational acceleration; D/Dt is the *substantial derivative* defined by $\partial/\partial t + (\bar{\mathbf{v}} \cdot \nabla)$. These conservation equations contain two *fluxes* with respect to the flow velocity $\bar{\mathbf{v}}$: the heat flux vector $\bar{\mathbf{q}}$, and the momentum flux tensor (or stress tensor) $\bar{\bar{\pi}}$. It is conventional to write the latter as

$$\bar{\bar{\pi}} = p\bar{\bar{\delta}} + \bar{\bar{\tau}} \tag{3.4}$$

in which p is the pressure (not uniquely determined for incompressible fluids) and $\bar{\bar{\tau}}$ is the *extra stress tensor*, which becomes zero at equilibrium; $\bar{\bar{\delta}}$ is the unit tensor, For incompressible fluids it is convenient to introduce a *modified pressure* as

$$\mathcal{P} = p + \rho g h \tag{3.5}$$

where h is the distance upward (i.e., in the direction opposite to $\bar{\mathbf{g}}$) from some arbitrary datum plane. Then the equations of motion and energy for incompressible fluids may be put into an alternative form:

Motion:
$$\rho \frac{D\bar{\mathbf{v}}}{Dt} = - [\nabla \cdot \bar{\bar{\tau}}] - \nabla \mathcal{P} \tag{3.6}$$

Energy:
$$\rho \frac{D\hat{U}}{Dt} = -(\nabla \cdot \bar{\mathbf{q}}) - (\bar{\bar{\tau}} : \nabla \bar{\mathbf{v}}) \tag{3.7}$$

Note that $-(\bar{\bar{\tau}} : \nabla \bar{\mathbf{v}})$ is the rate of conversion of mechanical to thermal energy per unit volume, which is very important in high-speed flows of materials with large viscosity (see Chaps. 1 and 4).

The conservation equations cannot in general be solved until expressions are given for the heat flux vector $\bar{\mathbf{q}}$ and the extra stress tensor $\bar{\bar{\tau}}$. For pure fluids it has been found that *Fourier's law* is generally applicable

$$\bar{\mathbf{q}} = -k\nabla T \tag{3.8}$$

in which k is the thermal conductivity, and T is the temperature. For fluids made up of small molecules, $\bar{\bar{\tau}}$ is given by *Newton's law*

$$\bar{\bar{\tau}} = -\mu \left(\nabla \bar{\mathbf{v}} + (\nabla \bar{\mathbf{v}})^{\dagger} \right) = -\mu \dot{\bar{\bar{\gamma}}} \tag{3.9}$$

Here μ is the viscosity, $(\nabla \bar{\mathbf{v}})^{\dagger}$ is the transpose of $(\nabla \bar{\mathbf{v}})$, and $\dot{\bar{\bar{\gamma}}}$ is the rate-of-strain tensor.

3.1.2 Non-Newtonian Fluids

This chapter deals with fluids for which Eq. (3.9) does not provide an adequate description of the stresses within the flowing fluid [Bird *et al.* (1987), Lodge (1964), Schowalter (1978), Eirich (1969), Astarita *et al.* (1980), Barnes *et al.* (1989), and Giesekus (1994)]. These *non-Newtonian fluids* include several broad categories of structurally complex materials: (i) polymeric liquids, (ii) inorganic glasses, (iii) soap solutions, (iv) liquid crystals, and (v) two-phase fluids. All of these materials contain some kind of substructures that interact with one another and are orientable or deformable. It is the behavior of these substructures that leads to marked deviations from Eq. (3.9). These non-Newtonian fluids exhibit phenomena that require a ''constitutive equation'' far more complicated than that for the Newtonian fluids made up of small molecules. It should be emphasized that non-Newtonian fluids do not exhibit just small quantitative deviations from Eq. (3.9), but rather striking qualitative deviations.

The five categories of non-Newtonian fluids mentioned above can be further subdivided. For example, polymeric fluids include biopolymers and synthetic polymers; they include undiluted polymers (polymer melts) and polymer solutions; they include straight-chain molecules and branched molecules; in solutions the solvent may be classed as a good solvent or a poor solvent. Because of this enormous diversity, it can be expected that a very wide variety of rheological behavior will be encountered. Similarly, under the category of two-phase fluids, there are suspensions of solid particles of many shapes in Newtonian liquids or in polymeric liquids, dispersions of one liquid phase in another liquid phase, dispersions of gas bubbles in a liquid phase, and solid powders with gas in the interstitial regions. Here again, very diverse rheological phenomena can be encountered, and many kinds of particle motion and particle–particle interactions can be observed.

Non-Newtonian fluids are commercially very important [Pearson (1983), Tadmor and Gogos (1979), Dealy and Wissbrun (1990), and Baird and Collias (1995)]. Polymers are used in the manufacture of plastics, in lubricants as additives, and in the food industry for modification of texture. Slurries, suspensions, pastes, and emulsions are encountered in a wide range of chemical industries. In spite of that, basic rheological data for many materials are lacking, many phenomena await discovery, many theories have yet to be checked by experiment, and sometimes existing theories are in conflict with one another. The fact is that non-Newtonian fluid dynamics is still very much an experimental field. Because of this, it must be emphasized that flow visualization and collection of basic experimental information are often to be preferred over blind reliance on theoretical results. It should be kept in mind that even for Newtonian fluids, for which a constitutive equation is known, analytical solutions of the equations of motion and continuity are available for a limited number of flows, and therefore flow visualization [Boger and Walters (1993)] and optical measurements [Janeschitz–Kriegl (1983)] are of great importance; for non-Newtonian fluids, for which only limited knowledge about constitutive equations exists, the collection of experimental data is even more important.

3.1.3   Examples of Non-Newtonian Behavior

To illustrate the qualitative differences between Newtonian and non-Newtonian fluids, Fig. 3.1 shows a number of phenomena that have been observed for polymeric liquids [Bird *et al.* (1987), Lodge (1964), and Boger and Walters (1993)].

(a)   A rotating rod is inserted into a beaker of liquid. For Newtonian liquids the centrifugal force causes the liquid to pile up in the neighborhood of the beaker wall, and the surface of the liquid is depressed in the vicinity of the rotating rod. For polymeric liquids, however, there are additional *normal stresses* present in the fluid that result in a lowering of the surface level near the wall and a rise in the surface level near the rotating rod. That is, one observes a *rod-climbing effect*.

(b)   A cylindrical container with inside diameter D is filled with a Newtonian fluid. When a rotating disk of diameter D is placed in contact with the fluid surface, it imparts a tangential motion to the liquid (the *primary flow*). At steady-state there is, in addition to the primary flow, a *secondary flow*—outward along the rotating disk, caused by the centrifugal force, and then downward along the cylinder wall, inward along the bottom, and upward near the cylinder axis. For polymeric fluids, however, there is a *reversal of the secondary flow in the disk-and-cylinder system*: the fluid moves inwards along the rotating disk, down along the axis and then upwards at the cylinder wall.

(c)   If a Newtonian fluid flows down an inclined semicylindrical trough, the liquid surface is flat (except near the edges where meniscus effects may be seen). Polymeric liquids, however, display a slightly *convex surface in tilted-trough flow*.

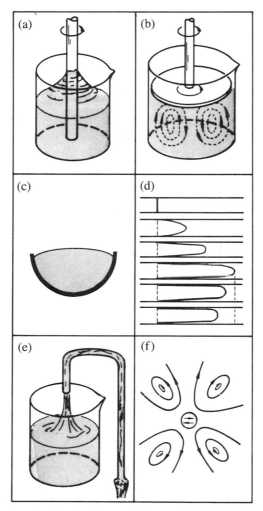

FIGURE 3.1 Experiments showing the non-Newtonian behavior of polymeric liquids: (a) rod-climbing effect near a rotating rod, (b) secondary flows in the disk-and-cylinder system, (c) convex surface in a tilted trough flow, (d) recoil in tube flow after the superposed pressure difference has been removed, (e) siphoning of a liquid when siphon is raised above the liquid level, and (f) reversal of the secondary flows in an acoustical streaming experiment.

(d) When the pumping of a Newtonian fluid through a tube is suddenly stopped, the fluid comes to rest almost immediately. But when one stops supplying a driving force for the flow of a polymeric liquid in a tube, the polymeric liquid begins to retreat and then gradually comes to rest. This *recoil* is symptomatic of a fluid that can *remember* its past kinematic experience; the fluid does not retreat all the way to its initial configuration, since as time proceeds it *forgets* its kinematic history. Polymeric fluids are often referred to as *fluids with fading memory*.

(e) When a Newtonian liquid is being siphoned, the siphon ceases to function when the end of the tube is removed from the liquid. For a polymeric liquid, however, the tube may be withdrawn from the liquid and the siphoning continues; there may be a gap of several centimeters between the liquid surface and the end of the tube. This effect is called the *tubeless siphon*.

(f) A long cylinder parallel to the *y*-axis is made to oscillate transversely in the *x*-direction. For a Newtonian liquid a secondary flow, called acoustical streaming, is produced. Fluid moves towards the cylinder from above and below and then moves away from the cylinder in the vicinity of the *xy*-plane. For polymeric fluids, however, there is a *reverse acoustical streaming* in which the fluid approaches the cylinder in the *xy*-plane and then moves away from it above and below the cylinder.

These are just a few of the many experiments [Bird *et al.* (1987) and Boger and Walters (1993)] that show conclusively that polymeric liquids are qualitatively different from liquids containing small molecules, and that Newton's law of viscosity (and therefore the Navier–Stokes equations of classical fluid dynamics) cannot be used to describe polymeric liquids. It is not well known to what extent other categories of non-Newtonian fluids—glasses, soap solutions, liquid crystals, and two-phase fluids—behave similarly to polymeric liquids in the experiments summarized in Fig. 3.1.

In addition to the experiments shown in Fig. 3.1, which all involve laminar flow, there is the *drag reduction* effect in turbulent flow. The addition of very small amounts of polymers to water produces a substantial reduction of the friction factor in the turbulent flow regime. Only 5–500 ppm by weight of polymer can lower the friction factor by 20–40 percent in some instances. The drag reduction phenomenon has been used to increase the range of fire-fighting systems and to decrease pumping costs for transport of petroleum in pipelines [Burger *et al.* (1980) and Burger *et al.* (1982)].

3.1.4 Development of Constitutive Equations

The central problem in non-Newtonian fluid dynamics is the establishment of useful expressions for the stress tensor to replace the Newtonian expression in Eq. 3.9 [Bird *et al.* (1987), Lodge (1964), Schowalter (1978), Lodge (1974), Astarita and Marrucci (1974), Larson (1988), Tanner (1985), and Giesekus (1994)]. The relation between the stress tensor and various kinematic tensors (strain tensors, rate of strain tensor, etc.) is called the *constitutive equation* or the *rheological equation of state*. There are three sources of information for obtaining the information needed to construct this key equation:

(a) *Rheometric measurements*: One may attempt to measure the stress tensor components in simple, carefully controlled flows; in such experiments, one can usually measure only one or two components of the stress tensor. The number of simple flows for which the data can be satisfactorily interpreted is rather small. Therefore, only fragmentary information can be obtained from experiments. Although obtaining this kind of experimental data is extremely important in the overall program of understanding rheological behavior and interpreting continuum and molecular theories, rheometry alone is not sufficient for generating constitutive equations.

(b) *Continuum mechanics:* This subject provides a very useful and important formalism that enables one to draw some conclusions about the form of the stress tensor. Also, continuum mechanics studies have led to the development of some general expansions (retarded motion expansion, memory integral expansion, etc.) and some general results for certain classes of flows (shear flows, shear-free flows, substantially stagnant motions, commutative motions, etc.) which have been of enormous help in systematizing our knowledge of liquid rheology. Continuum mechanics also suggests certain rules that have to be obeyed by empirical constitutive equations. Continuum mechanics, however, cannot provide the explicit functional forms that constitutive equations should take for various kinds of liquids nor can continuum mechanics relate the constants or functions in the general expansions to concentration, temperature, or structure of particles, polymers, or liquid crystals.

(c) *Structural theories*: For polymeric liquids nonequilibrium statistical mechanics is used to obtain relationships between macroscopically measurable quantities (viscosity, normal stresses, elongational viscosity, etc.) and the parameters in various kinds of mechanical models (bead-spring chains, bead-rod chains, ellipsoids, etc.) that are used to represent the macromolecules. There are several problems here: the inherent difficulty of the kinetic theories for macromolecules, and the virtual necessity of using some kind of oversimplified mechanical model to describe the internal motions of the macromolecule and its interactions with other macromolecules or with the solvent. For suspensions and emulsions one tries to obtain relations between the macroscopically measurable rheological properties and the mechanical properties of the materials making up the two phases as well as the mechanical properties of the interfaces, such as surface tension and surface viscosity.

No one of the above three approaches is entirely successful at the present time for obtaining a constitutive equation. Consequently in most engineering work, one makes use of empirical relations containing a small number of constants or functions which have simple physical significance and can be determined from rheological measurements. One chooses the empirical relations in such a way as to conform to the rules of continuum mechanics, and if possible one makes use of structural theories to suggest useful forms for the empirical relations. It is anticipated that in the future rheologists and fluid mechanicians will rely more and more on the structural theories as these become available.

3.2 RHEOMETRY AND MATERIAL FUNCTIONS

In order to characterize non-Newtonian liquids, one has to measure a number of rheological properties (the *material functions*). These properties are also necessary for the evaluation of empirical constitutive equations and molecular theories. The material functions are measured in bench-scale devices, and the science dealing with this activity is called "rheometry" [Bird *et al.* (1987), Walters (1973), Dealy (1982), and Barnes *et al.* (1989)].

3.2.1 Shear Flows and Elongational Flows

Much of the present knowledge about non-Newtonian liquids has been obtained from measurements of stress components in carefully controlled flows. The two most studied flows are *shear flows* for which

$$v_x = \dot{\gamma}(t)y; \qquad v_y = 0; \qquad v_z = 0 \tag{3.10}$$

where $\dot{\gamma}(t)$ is the *shear rate*, and *elongational flows* for which

$$v_x = -\tfrac{1}{2}\,\dot{\epsilon}(t)x; \qquad v_y = -\tfrac{1}{2}\,\dot{\epsilon}(t)y; \qquad v_z = \dot{\epsilon}(t)z \qquad (3.11)$$

where $\dot{\epsilon}(t)$ is the *elongation rate*. There is a general feeling among non-Newtonian fluid dynamicists that continuum or molecular theories that can describe the above flows might be successful for more complex flows.

It can be shown that for the two classes of flows just defined the stress tensor has the form

Shear Flows: $\quad \overline{\overline{\tau}} \underset{xyz}{=\!=\!=\!=} \begin{pmatrix} \tau_{xx} & \tau_{xy} & 0 \\ \tau_{yx} & \tau_{yy} & 0 \\ 0 & 0 & \tau_{zz} \end{pmatrix} \quad [\tau_{xy} = \tau_{yx}] \qquad (3.12)$

Elongational Flows: $\quad \overline{\overline{\tau}} \underset{xyz}{=\!=\!=\!=} \begin{pmatrix} \tau_{xx} & 0 & 0 \\ 0 & \tau_{yy} & 0 \\ 0 & 0 & \tau_{zz} \end{pmatrix} \quad [\tau_{xx} = \tau_{yy}] \qquad (3.13)$

The symbol $\underset{xyz}{=\!=\!=}$ means "is displayed in matrix form with order of indices being *xyz*." Since the discussion is restricted to incompressible flow, the diagonal components of the stress tensor are defined only to within an additive constant. Therefore there are only three determinable quantities in shear flows; by general agreement these are taken to be the *shear stress* τ_{yx}, the *first normal-stress difference* $\tau_{xx} - \tau_{yy}$, and the *second normal-stress difference* $\tau_{yy} - \tau_{zz}$. For elongational flows there is only one determinable quantity; this is taken to be the normal stress difference $\tau_{zz} - \tau_{xx}$. Many choices can be made for the time dependence of the shear rate $\dot{\gamma}(t)$ and the elongation rate $\dot{\epsilon}(t)$; for each of these choices various "material functions" are defined. Some examples of shear flows that have been studied are shown in Fig. 3.2.

3.2.2 Definitions of Material Functions

For *steady-state shear flow*, with $\dot{\gamma} = $ constant (see Fig. 3.2a), three material functions $\eta(\dot{\gamma})$, $\Psi_1(\dot{\gamma})$, and $\Psi_2(\dot{\gamma})$ are defined by [Bird *et al.* (1987) and Lodge (1964)]

$$\tau_{yx} = -\eta\dot{\gamma} \qquad (3.14)$$

$$\tau_{xx} - \tau_{yy} = -\Psi_1\dot{\gamma}^2 \qquad (3.15)$$

$$\tau_{yy} - \tau_{zz} = -\Psi_2\dot{\gamma}^2 \qquad (3.16)$$

The viscosity η and the first normal-stress coefficient Ψ_1 are positive, and the most reliable experimental data on the second normal stress coefficient Ψ_2 for polymeric liquids show that it is negative. The secondary flow in the disk-and-cylinder experiment is related to Ψ_1, and the negative Ψ_2 explains the convex surface in the tilted-trough flow. Experimental data on polymeric liquids indicate that the ratio Ψ_2/Ψ_1 varies between about -0.04 and -0.4. When the shear rate becomes vanishingly

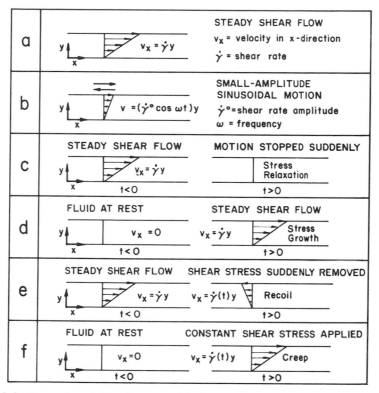

FIGURE 3.2 Examples of shear flows used in rheometric measurements. It should be noted that in all time-dependent experiments it is *assumed* that the viscosity of the liquid is sufficiently high that the flow is "homogeneous" throughout the gap (i.e., the velocity v_x is a linear function of the coordinate y).

small, the limiting values of these three material functions are designated by η_0, $\Psi_{1,0}$, and $\Psi_{2,0}$.

For *small-amplitude sinusoidal shearing motion* (see Fig. 3.2b) the shear rate varies according to $\dot{\gamma}(t) = \dot{\gamma}^0 \cos \omega t$, where ω is the frequency of oscillation and the amplitude of the displacement gradient $\dot{\gamma}^0/\omega$ is understood to be vanishingly small. For this flow eight frequency-dependent material functions may be defined by

$$\tau_{yx} = -\eta'\dot{\gamma}^0 \cos \omega t - \eta''\dot{\gamma}^0 \sin \omega t \tag{3.17}$$

$$\tau_{xx} - \tau_{yy} = -\Psi_1^d\dot{\gamma}^{02} - \Psi_1'\dot{\gamma}^{02} \cos 2\omega t$$
$$-\Psi_1''\dot{\gamma}^{02} \sin 2\omega t \tag{3.18}$$

$$\tau_{yy} - \tau_{zz} = -\Psi_2^d\dot{\gamma}^{02} - \Psi_2'\dot{\gamma}^{02} \cos 2\omega t$$
$$- \Psi_2''\dot{\gamma}^{02} \sin 2\omega t \tag{3.19}$$

The η's and Ψ's are all real, frequency-dependent functions. Continuum mechanics arguments suggest that $\Psi_1^d = \eta''/\omega$. The small amplitude oscillatory motion experiment has been the most popular time-dependent experiment and has been widely used for the study of the structure of polymeric liquids; in most experiments only the shear stress is measured. Note that the normal stress differences, which are far more difficult to monitor, oscillate with a frequency 2ω about a nonzero mean. [Note on notation: The quantities $\eta'(\omega)$ and $\eta''(\omega)$ are often combined to form a "complex viscosity" $\eta^*(\omega) = \eta'(\omega) - i\eta''(\omega)$. In addition, a "complex modulus" $G^*(\omega) = i\omega\eta^*(\omega) = G'(\omega) + iG''(\omega)$ is a widely used term. The quantity $\eta'(\omega)$ has a limiting value of η_0 at $\omega = 0$, and $2\eta''/\omega$ has the limiting value of $\Psi_{1,0}$ at $\omega = 0$.]

Next, *stress relaxation* after cessation of steady shear flow is discussed [see Fig. 3.2c]. For $t < 0$, the velocity gradient is $\dot{\gamma}$, and for $t > 0$ the gradient is zero. Then, for $t > 0$ one defines the material function as

$$\tau_{yx}(t) = -\eta^-(t, \dot{\gamma})\dot{\gamma} \tag{3.20}$$

$$\tau_{xx}(t) - \tau_{yy}(t) = -\Psi_1^-(t, \dot{\gamma})\dot{\gamma}^2 \tag{3.21}$$

$$\tau_{yy}(t) - \tau_{zz}(t) = -\Psi_2^-(t, \dot{\gamma})\dot{\gamma}^2 \tag{3.22}$$

For polymeric liquids the stresses relax monotonically to zero, and the relaxation process is faster as the shear rate $\dot{\gamma}$ is increased. The shear stress relaxes more rapidly than the first normal-stress difference.

For *stress growth* upon inception of steady shear flow with shear rate $\dot{\gamma}$ at time $t = 0$ [see Fig. 3.2d], one defines material functions for $t > 0$ as follows

$$\tau_{yx}(t) = -\eta^+(t, \dot{\gamma})\dot{\gamma} \tag{3.23}$$

$$\tau_{xx}(t) - \tau_{yy}(t) = -\Psi_1^+(t, \dot{\gamma})\dot{\gamma}^2 \tag{3.24}$$

$$\tau_{yy}(t) - \tau_{zz}(t) = -\Psi_2^+(t, \dot{\gamma})\dot{\gamma}^2 \tag{3.25}$$

For polymeric liquids it is known that the shear stress and the first normal-stress difference go through a maximum before settling down to the steady-state values; little is known about the behavior of the second normal-stress difference.

There are many other experiments that are performed in shear flow, such as the recoil and creep experiments shown in Fig. 3.2e,f. Also superposed flows have been studied; for example, one can superpose the motions in (a) and (b) of Fig. 3.2. In addition, experiments have been performed using a step-function strain, multiple step-function strains, and a ramp function velocity gradient (with $\dot{\gamma}(t) = at$). In each experiment additional material functions can be defined.

A wide variety of elongational flow measurements can also be made. For *steady-state elongational flow*, one elongates a sample of a material with a constant elongation rate $\dot{\epsilon}$. Then the elongational viscosity $\bar{\eta}$ is the material function defined by

$$\tau_{zz} - \tau_{xx} = -\bar{\eta}\dot{\epsilon} \tag{3.26}$$

The elongational viscosity in the zero-elongational-rate limit is $3\eta_0$. If $\dot{\epsilon}$ is negative, the experiment is called *biaxial stretching*. Additional material functions such as $\bar{\eta}^-$ and $\bar{\eta}^+$ in stress relaxation and stress growth experiments may be defined by analogy with Eqs. (3.20) and (3.23).

It should be emphasized that for Newtonian fluids all of these rheometric experiments lead just to the measurement of the Newtonian viscosity μ (a material constant). But for non-Newtonian fluids each experiment leads to a different set of material functions.

3.2.3 Data on Material Functions

In Figs. 3.3–3.7 are shown representative data for the viscosity, the first normal stress coefficient, and the components of the complex viscosity for a variety of non-Newtonian fluids. Several points should be noted: (i) the non-Newtonian viscosity as a function of shear rate can change by several orders of magnitude; this emphasizes the fact that for non-Newtonian materials the assumption of a constant viscosity can lead to erroneous results in engineering flow calculations; (ii) the first normal stress coefficient changes with shear rate even more dramatically than the viscosity; (iii) the curve for $\eta'(\omega)$ lies somewhat beneath the curve $\eta(\dot{\gamma})$, and a similar relationship exists for $2\eta''(\omega)/\omega$ and $\Psi_1(\dot{\gamma})$. For many polymeric liquids a useful

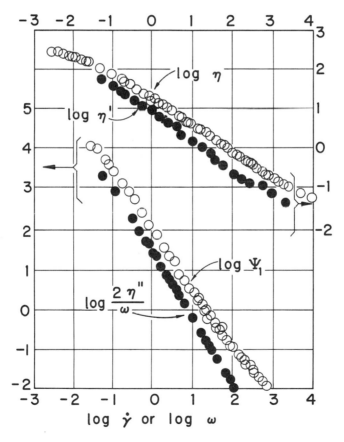

FIGURE 3.3 Material functions for a 1.5% polyacrylamide (Separan AP30) solution in a 50/50 mixture, by weight, of water and glycerin. The functions η, η', and η'' are given in Pa · s and Ψ_1 and Pa · s^2; both $\dot{\gamma}$ and $\dot{\epsilon}$ are in s^{-1}. (Replotted from J. D. Huppler, E. Ashare, and L. Holmes, *Trans. Soc. Rheol.*, **11**, pp. 159–179, 1967, The Society of Rheology Inc.)

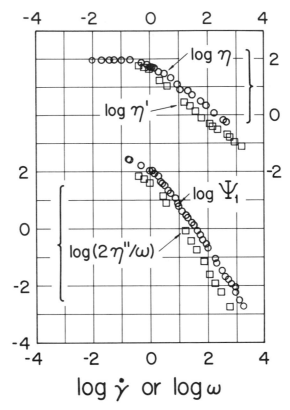

FIGURE 3.4 Material functions for a 7% solution of aluminum laurate in a mixture of decalin and m-cresol. The functions η, η', and η'' are given in Pa · s and Ψ_1 is in Pa · s^2; both $\dot{\gamma}$ and ω are in s^{-1}. (Replotted from J. D. Huppler, E. Ashare, and L. A. Holmes, *Trans. Soc. Rheol.*, Vol. 11, pp. 159–179, 1967, The Society of Rheology Inc.)

empirical relation, the *Cox–Merz rule*, states that $\eta(\dot{\gamma})$ and $|\eta^*(\omega)| = ([\eta'(\omega)]^2 + [\eta''(\omega)]^2)^{1/2}$ are one and the same curve; this rule is sometimes used to estimate the non-Newtonian viscosity curve from measurements of the complex viscosity.

Very few reliable data are available for second normal stress coefficients since the measurement techniques have not been standardized; one set of data for Ψ_2 as a function of $\dot{\gamma}$ is given in Fig. 3.8.

Stress relaxation and stress growth functions are shown in Figs. 3.9 and 3.10. Most data of this type have been taken with a cone-and-plate instrument. In Fig. 3.11a and 3.11b are shown some sample data on $\bar{\eta}(\dot{\epsilon})$ for uniaxial and biaxial extension; only very few data are available for this property, and few commercial instruments for measuring it have been developed.

3.2.4 Measurement of Material Functions

To measure a material function, it is necessary to have an apparatus with a carefully controlled flow pattern for which an expression can be obtained for the material function in terms of measurable quantities, without the use of a constitutive equation

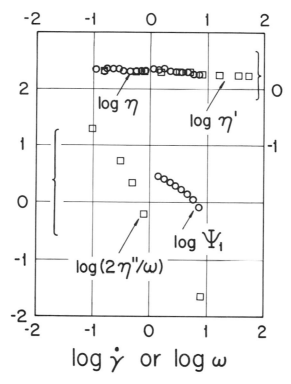

FIGURE 3.5 Material functions for a 0.03% solution of polyacrylamide (Separan MG500) in corn syrup. The functions η, η', and η'' are given in Pa $\cdot$ s, Ψ_1 in Pa $\cdot$ s^2, and $\dot{\gamma}$ and ω in s^{-1}. (Replotted from D. V. Boger, *Proc. of the von Kármán Inst. for Fluid Dynamics*, May 4–8, 1981.)

in the analysis [Walters (1973), Whorlow (1980), Tanner (1985), and Barnes *et al.* (1989)]. There are only a few geometrical arrangements for which this is possible. Even then there are many annoying factors which have to be considered in the equipment operation and data analysis: (i) At high speeds of operation viscous heating can produce erroneous results, since most rheological properties are very sensitive to temperature. (ii) End effects and edge effects have to be considered, inasmuch as the flow fields are usually not the same as the theoretical flow field in the vicinity of the system boundary. (iii) Solvent evaporation can produce temperature effects and also render the fluid sample nonhomogeneous. (iv) Maintenance of the geometrical arrangements is sometimes made difficult because of alignment problems, lack of sufficient rigidity of the instrument, and distortion of the containing surfaces because of normal stress effects. (v) Many non-Newtonian materials undergo constitutional changes because of bacterial growth, changes in pH, photodecomposition, shear degradation, and absorption of water or oxygen; therefore sample characterization, careful recording of all procedures used in preparing the fluid, and frequent monitoring of physical properties are all essential in taking rheometric data. All of the above points must be taken into account in the various rheometric systems considered below.

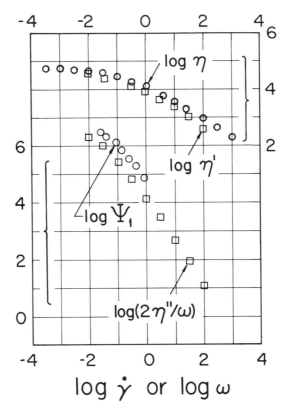

FIGURE 3.6 Material functions for a low density polyethylene melt (IUPAC sample A). The functions η, η', and η'' are given in Pa · s and Ψ_1 in Pa · s^2; both $\dot{\gamma}$ and ω are in s^{-1}. (Replotted from J. Meissner, *Pure and App. Chem.*, Vol. 42, pp. 551–612, 1975, International Union of Pure and Applied Chemistry.)

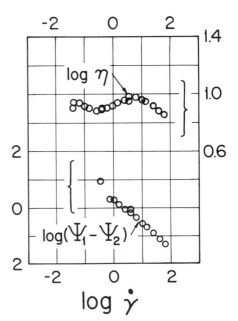

FIGURE 3.7 Material functions for a 50.1 volume percent suspension of 40–50 μm diameter polystyrene spheres in UCON oil 75-H-450 ($\eta = 0.193$ Pa · s). The function η is given in Pa · s, ($\Psi_1 - \Psi_2$) in Pa · s^2, and $\dot{\gamma}$ in s^{-1}. (Replotted from F. A. Gadala-Maria, Ph.D. Thesis, Stanford University, 1979.)

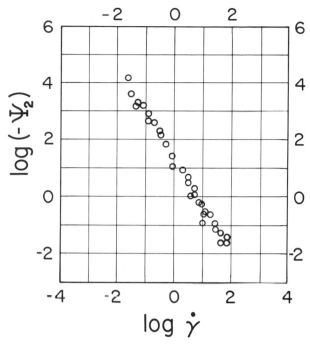

FIGURE 3.8 Dependence of the second normal stress coefficient on shear rate for a 2.5% solution of polyacrylamide in a 50/50 mixture of water and glycerin. The function Ψ_2 is given in Pa $\cdot$ s^2 and $\dot{\gamma}$ in s^{-1}. (Replotted from E. B. Christiansen and W. R. Leppard, *Trans. Soc. Rheol.*, Vol. 18, pp. 65–86, 1974, The Society of Rheology Inc.)

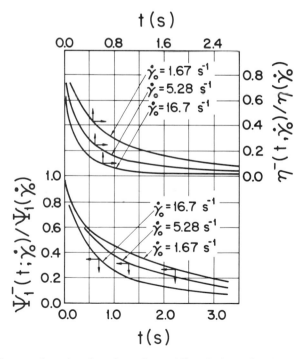

FIGURE 3.9 Stress relaxation functions for a 4% solution of polystyrene in Aroclor at 298 K. [From I. F. Macdonald, Ph.D. Thesis, University of Wisconsin, Madison (1968).]

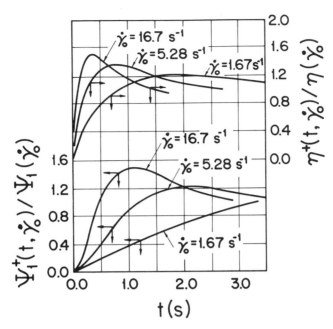

FIGURE 3.10 Stress growth functions for a 4% solution of polystyrene in Aroclor at 298 K. [From I. F. Macdonald, Ph.D. Thesis, University of Wisconsin, Madison (1968).]

Tube Viscometer. One of the oldest viscometers is the tube (or capillary) viscometer [Bird *et al.* (1987) and Walters (1973)]; here define the radius of the tube to be R and its length to be L (see Fig. 3.12a). One measures the volume rate of flow Q as a function of the difference between the modified pressures at the ends of the tubes, $\mathcal{P}_0 - \mathcal{P}_L$. It can then be shown that the shear rate at the wall, $\dot{\gamma}_R = -dv_z/dr|_{r=R}$, and the shear stress at the wall, τ_R, are

$$\dot{\gamma}_R = \frac{1}{\tau_R^2}\frac{d}{d\tau_R}\left(\frac{\tau_R^3 Q}{\pi R^3}\right) \tag{3.27}$$

$$\tau_R = (\mathcal{P}_0 - \mathcal{P}_L)R/2L \tag{3.28}$$

Then the viscosity at the wall η_R, is given by

$$\eta_R = \tau_R/\dot{\gamma}_R \tag{3.29}$$

It is then presumed that the viscosity-shear-rate relation obtained at the wall is valid throughout the entire cross section of the tube, so that the function $\eta_R(\dot{\gamma}_R)$ can then be replaced by $\eta(\dot{\gamma})$. If the pressure difference across the tube is very large, then the compressibility of the fluid may have to be taken into account; usually temperature rise due to viscous heating is a disturbing factor. Note that the shear rate varies with the radial coordinate r in tube flow.

Slit Instrument (Lodge Stressmeter). A slit viscometer offers several advantages over a tube viscometer: it is easier to clean, and temperature control is more effec-

(a)

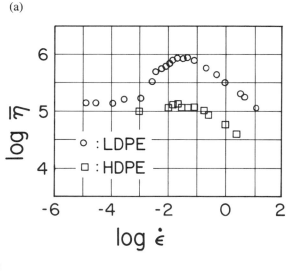

(b)

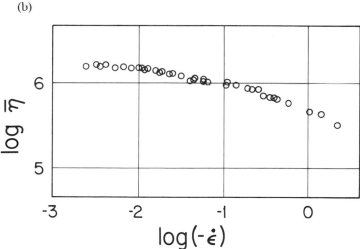

FIGURE 3.11 (a) Elongational viscosity for uniaxial stretching of low and high density polyethylene. The function $\bar{\eta}$ is given in Pa · s and the elongation rate in s^{-1}. [From H. Münstedt and H. M. Laun, *Rheol. Acta*, Vol. 20, pp. 211–221 (1981), Dietrich Steinkopff Verlag.] (b) Elongational viscosity for biaxial stretching of low density polyethylene deduced from flow birefringence data. The function $\bar{\eta}$ is given in Pa · s and $\dot{\epsilon}$ in s^{-1}. [From J. A. van Aken and H. Janeschitz-Kriegl, *Rheol. Acta*, Vol. 20, pp. 419–432 (1981), Dietrich Steinkopff Verlag.]

tive. In addition, by using a pair of pressure transducers at the bottom of transverse slits and another flush-mounted transducer (see Fig. 3.12b), it is possible to measure both $\eta(\dot{\gamma})$ and $\Psi_1(\dot{\gamma})$. The viscosity measurement is similar to that for tubes. The shear rate at the wall $\dot{\gamma}_B = -dv_x/dy|_{y=B}$ and the shear stress at the wall τ_B are obtained from

$$\dot{\gamma}_B = \frac{1}{\tau_B} \frac{d}{d\tau_B} \left(\frac{\tau_B^2 Q}{2WB^2} \right) \tag{3.30}$$

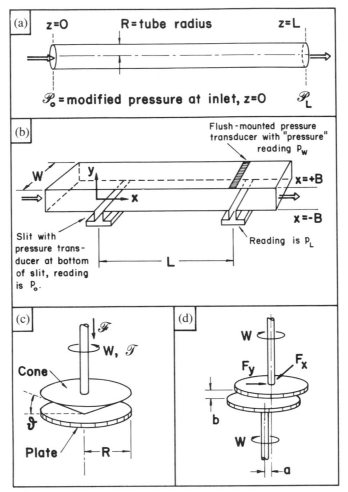

FIGURE 3.12 Rheometric devices: (a) tube viscometer, (b) Lodge stressmeter, (c) cone-plate rheometer, and (d) eccentric-disk rheometer.

$$\tau_B = \frac{(p_0 - p_L)B}{L} \tag{3.31}$$

Then the viscosity at the wall η_B is given by

$$\eta_B = \tau_B/\dot{\gamma}_B \tag{3.32}$$

The first normal-stress difference can be found by using the fact that the "pressure" p_w measured at the wall by a flush-mounted transducer is not the same as the pressure measured in a pressure tap p_L (both transducers at $x = L$). The difference $p_H = p_L - p_W$ is the "hole pressure error" and is a negative quantity. The primary normal-stress difference is obtained from

$$\tau_{xx} - \tau_{yy} = 2p_H \frac{d \ln (-p_H)}{d \ln \tau_B} \tag{3.33}$$

This formula has been refined by making a small inertial correction [Lodge (1987)].

Cone-and-Plate Instrument. A number of commercial instruments are now available that make use of a cone-and-plate arrangement (see Fig. 3.12c) [Walters (1973) and Whorlow (1980)]. Here one puts the fluid sample in the space between a circular plate and a cone of radius R, the angle ϑ between the cone and plate being quite small (usually less than $4°$). The cone is then caused to rotate with an angular velocity W, and the torque $\mathfrak{J}$ required to maintain the motion is measured; in addition one has to apply a normal thrust $\mathfrak{F}$ in order to prevent the cone from moving away from the plate because of the normal stresses. One can also measure $\pi_{\theta\theta}$ as a function of the radial coordinate r along the plate surface by means of flush-mounted pressure transducers. Then the material functions for steady-state shear flow are:

$$\eta(\dot{\gamma}) = \frac{3\mathfrak{J}}{2\pi R^3 \dot{\gamma}} \tag{3.34}$$

$$\Psi_1(\dot{\gamma}) = \frac{2\mathfrak{F}}{\pi R^2 \dot{\gamma}^2} \tag{3.35}$$

$$\Psi_1(\dot{\gamma}) + 2\Psi_2(\dot{\gamma}) = -\frac{1}{\dot{\gamma}^2} \frac{\partial \pi_{\theta\theta}}{\partial \ln r} \tag{3.36}$$

in which the shear rate is given by

$$\dot{\gamma} = W/\vartheta \tag{3.37}$$

It should be emphasized that the shear rate has the same value throughout the gap between the cone and plate. For steady-state flow measurements the cone-and-plate system is limited to low shear rates, since at high rates of rotation the sample may be thrown out of the gap by centrifugal force. Also both viscous heating and shear degradation can be serious problems for this instrument.

The instrument is rather versatile in that the motion of the cone can be programmed in such a way that a variety of transient experiments can be performed. The cone-and-plate system has been used to obtain data on complex viscosity, stress relaxation, stress growth, response to step function strain, etc., just by making W be the appropriate function of time.

Eccentric-Disk Rheometer. In this instrument the fluid sample is placed in the space between two circular disks (radius R) separated by a vertical distance b with the disk centers displaced by a distance a (see Fig. 3.12d) [Walters (1973) and Whorlow (1980)]. Both disks are made to rotate in the same direction with an angular velocity W. In the limit as the ratio a/b becomes very small, the components of the complex viscosity can be obtained from the forces F_x and F_y that have to be applied in the direction along the line of centers and in the direction normal to the line of centers of the two disks

$$\eta'(W) = \frac{F_x}{\pi R^2 (Wa/b)} \tag{3.38}$$

$$\eta''(W) = \frac{F_y}{\pi R^2 (Wa/b)} \tag{3.39}$$

It should be emphasized that although the flow pattern established in this instrument is a steady-state flow, when a/b becomes very small, the complex viscosity components can be measured.

3.3 GENERALIZED NEWTONIAN FLUID MODELS

These empirical models are the simplest to use and have been widely employed in industrial calculations [Bird *et al.* (1987) and Middleman (1977)]. They account for the variation of viscosity with shear rate which is the most important property in many fluid-flow and heat transfer problems.

3.3.1 Definition of the Generalized Newtonian Fluid

In many industrial processes engineers are concerned with steady-state shear flows. Frequently encountered flows of this type are:

  A. Flow between parallel planes
    i. Pressure-difference or gravity driven flow between two fixed parallel planes
    ii. Flow between two parallel planes one of which is moving
    iii. Combination of (i) and (ii) (plane Couette flow)
  B. Axial flow in a circular tube, resulting from a pressure difference or gravity
  C. Flow between coaxial cylinders
    i. Axial flow between two fixed cylinders resulting from a pressure difference or gravity
    ii. Axial flow in an annular region resulting from axial motion of the inner cylinder
    iii. Tangential flow resulting from the relative rotatory motion of the cylinders
    iv. Helical annular flow (a combination of (i) and (iii))
  D. Slow tangential flow between a pair of parallel circular disks, one of which is rotating
  E. Slow tangential flow between a circular disk and a cone, one of which is rotating

In each of these systems a flow field of type (a) of Fig. 3.2 prevails locally. For the calculation of torques, pressure drops, and volume rates of flow, the only rheological property of importance in these flows is the non-Newtonian viscosity. It is therefore not surprising that early non-Newtonian fluid dynamicists suggested modifying Newton's law of viscosity to allow the viscosity to depend on the shear rate.

To generalize the Newtonian fluid, one replaces the material constant μ of Eq. 3.9 by a variable viscosity η. The "non-Newtonian viscosity" η must depend on the rate-of-strain tensor $\dot{\bar{\bar{\gamma}}}$; the simplest hypothesis is that η should depend on the scalar invariants of $\dot{\bar{\bar{\gamma}}}$: I = tr $\dot{\bar{\bar{\gamma}}}$; II = tr $\dot{\bar{\bar{\gamma}}}^2$; III = tr $\dot{\bar{\bar{\gamma}}}^3$. For incompressible fluids I is identically zero; for shear flows, III vanishes. Hence the viscosity η must depend on II. It is somewhat more convenient to let η be a function of $\dot{\gamma} = [(1/2)\text{II}]^{1/2} = [(1/2)(\dot{\bar{\bar{\gamma}}}:\dot{\bar{\bar{\gamma}}})]^{1/2}$, which is the "magnitude of $\dot{\bar{\bar{\gamma}}}$;" the scalar quantity $\dot{\gamma}$ defined in this way is also called the "shear rate," and for the flow shown in Fig. 3.2a it is exactly the same as $\dot{\gamma}$ in the velocity distribution shown there. The *generalized Newtonian fluid* model is then given by the constitutive equation

$$\bar{\bar{\tau}} = -\eta(\dot{\gamma})\dot{\bar{\bar{\gamma}}} \tag{3.40}$$

This very simple empiricism has been widely used by engineers for over a half-century, and many empirical expressions have been suggested for $\eta(\dot{\gamma})$. It should be used only for shear flows that are time-independent. However, it has been successfully applied to some flows that deviate somewhat from simple shear flows and that are slowly varying in time.

3.3.2 Empirical Expressions for the Non-Newtonian Viscosity

Sample experimental viscosity data for non-Newtonian fluids are shown in Figs. 3.3–3.6. For analytical derivations, it is convenient to fit these curves to simple equations containing a small number of constants. Many expressions have been suggested; some of the more popular ones are listed here:

The Power Law Model. In many industrial problems the plateau region of the $\eta(\dot{\gamma})$ curve for small $\dot{\gamma}$ is of little importance; most applications in the plastics and lubrication areas involve high shear rates, and therefore it is the rapidly descending region of the curve that is crucial. Since that region is found to be linear on a plot of log η vs. log $\dot{\gamma}$, a particularly useful empiricism is

$$\eta = m\dot{\gamma}^{n-1} \tag{3.41}$$

in which m and n are parameters to be determined for each fluid; if one is dealing with nonisothermal flows the parameters m and n must be known as functions of the temperature. The quantity n is less than unity for most fluids. When $n = 1$ and $m = \mu$ the Newtonian liquid is recovered. The *power law* has been very popular because it is easy to use, and because many fluid dynamics and heat-transfer problems have been solved with it.

The Eyring–Powell, Sutterby, and Eyring Models. These two- and three-parameter models have been found to give better fits to experimental data than the power-law model. They are all of the general form

$$\eta = (\eta_0 - \eta_\infty)\{(\text{arcsinh } t_0\dot{\gamma})/(t_0\dot{\gamma})\}^\alpha + \eta_\infty \tag{3.42}$$

where η_0 is the zero-shear-rate viscosity, η_∞ is the infinite-shear-rate viscosity, t_0 is a characteristic time, and α is a parameter. The Eyring–Powell model is obtained by setting α equal to unity, the Sutterby model is obtained by setting η_∞ equal to zero, and the Eyring model is obtained by setting α equal to unity and η_∞ equal to zero. The Eyring model was the first $\eta(\dot{\gamma})$ expression obtained from a molecular theory.

The Carreau Model. The four-parameter model of Carreau includes a time constant λ, the zero-shear-rate viscosity η_0, and an infinite-shear-rate viscosity η_∞, as well as the parameter n which has the same physical meaning as the corresponding parameter in the power law model

$$\eta - \eta_\infty = (\eta_0 - \eta_\infty) \, [1 + (\lambda\dot{\gamma})^2]^{(n-1)/2} \tag{3.43}$$

Experimental data for many polymer solutions and polymer melts have been described quite well by this equation.

The Bingham Model. For concentrated suspensions and pastes that exhibit a yield stress, τ_0, the simplest two-constant equation is the Bingham model

$$\eta = \infty \qquad\qquad \tau \leq \tau_0 \tag{3.44a}$$

$$\eta = \mu_0 + (\tau_0/\dot{\gamma}) \qquad \tau \geq \tau_0 \tag{3.44b}$$

in which τ is the magnitude of the stress tensor defined by $\tau = [(1/2)(\bar{\bar{\tau}}:\bar{\bar{\tau}})]^{1/2}$. Empirical relations are available [Thomas (1961), (1963)] to relate the parameters μ_0 and τ_0 to quantities describing the suspension: volume fraction of solids λ, suspended particle diameter D_p, and the viscosity of the suspending medium μ_s

$$\tau_0 = 312.5 \, (\phi^3/D_p^2) \tag{3.45}$$

$$\mu_0 = \mu_s \exp\left[\left(\frac{5}{2} + \frac{14}{\sqrt{D_p}}\right)\phi\right] \tag{3.46}$$

Here D_p is given in μm and τ_0 in *Pa*.

3.3.3   Solution to Flow Problems for Generalized Newtonian Fluids

Many laminar flow problems have been solved for the generalized Newtonian fluid. The procedure is not unlike that used for Newtonian fluids [Bird *et al.* (1987), Tadmor and Gogos (1979), Middleman (1977), and Bird *et al.* (1982)].

The method is illustrated by obtaining the relation between the volume rate of flow Q and the modified pressure difference $\Delta\mathcal{P} = \mathcal{P}_0 - \mathcal{P}_L$ for the steady-state axial flow of an Eyring fluid [Eq (3.42)] in a circular tube of radius R and length L. One begins by postulating a solution of the form $v_z = v_z(r)$, $v_r = 0$, and $\mathcal{P} = \mathcal{P}(z)$; for the postulated velocity field, the only nonzero stress tensor components are $\tau_{rz} = \tau_{zr}$. Then the equation of continuity is automatically satisfied, and the equation of motion in Eq. (3.2), in cylindrical coordinates, becomes

$$0 = -\frac{d\mathcal{P}}{dz} - \frac{1}{r}\frac{d}{dr}(r\tau_{rz}) \tag{3.47}$$

An integration with respect to z gives

$$0 = +\frac{\Delta\mathcal{P}}{L} - \frac{1}{r}\frac{d}{dr}(r\tau_{rz}) \tag{3.48}$$

Then integration with respect to r gives

$$\tau_{rz} = \left(\frac{\Delta\mathcal{P} R}{2L}\right)\frac{r}{R} \equiv \tau_R \cdot \frac{r}{R} \tag{3.49}$$

where $\tau_R = \Delta\mathcal{P}R/2L$ is the shear stress at the wall. The constant of integration is set equal to zero since τ_{rz} must be finite at $r = 0$. The Eyring model of Eq. (3.42) is now inserted into the rz-component of Eq. (3.40), and use is made of the definition of the shear rate (a positive quantity) $\dot\gamma = -dv_z/dr$; this gives

$$\tau_{rz} = \frac{\eta_0}{t_0} \operatorname{arcsinh}\left(-t_0\frac{dv_z}{dr}\right) \tag{3.50}$$

When Eqs. (3.49) and (3.50) are combined and the resulting differential equation for $v_z(r)$ integrated, the following velocity distribution is obtained

$$v_z = \frac{\eta_0 R}{t_0^2\tau_R}\left(\cosh\frac{t_0\tau_R}{\eta_0} - \cosh\frac{t_0\tau_R}{\eta_0}\frac{r}{R}\right) \tag{3.51}$$

Then, integration of $v_z(r)$ over the tube cross section gives

$$Q = \int_0^{2\pi}\int_0^R v_z(r)\, r\, dr\, d\theta$$

$$= \frac{2\pi R^3}{t_0 X^3}\left[\left(1 + \frac{1}{2}X^2\right)\cosh X - X\sinh X - 1\right] \tag{3.52}$$

where $X = t_0\tau_R/\eta_0$. In the limit that $t_0 \to 0$, the Newtonian result $Q = \pi\Delta\mathcal{P}R^4/8\eta_0 L$ is obtained. The tube-flow entries in Table 3.1 for the power law and Bingham fluids may be obtained in a similar way.

As a second example, consider the helical flow of a power-law fluid in an annulus as shown in Fig. 3.13a. Only thin annuli are considered because curvature may then be neglected (see Fig. 3.13b). The following relations are postulated: $v_x = v_x(y)$, $v_z = v_z(x)$, $v_y = 0$, and $\mathcal{P} = \mathcal{P}(z)$. For these postulates the equation of continuity is satisifed, and the y- and z-components of motion become

y-component:
$$0 = -\frac{d}{dx}\tau_{xy} \tag{3.53a}$$

z-component:
$$0 = \frac{\mathcal{P}_0 - \mathcal{P}_L}{L} - \frac{d}{dx}\tau_{xz} \tag{3.53b}$$

where $\mathcal{P}_0 - \mathcal{P}_L$ is the pressure drop between $z = 0$ and $z = L$. The stress tensor components, according to Eqs. (3.40) and (3.41) are

$$\tau_{xy} = -\eta(\dot\gamma)\dot\gamma_{xy} = -m\dot\gamma^{n-1}\frac{dv_y}{dx} \tag{3.54a}$$

TABLE 3.1 Analytical Solutions for Power-Law[a] and Bingham Fluids[b,c,d]

Flow System	Power-Law Fluid with Parameters m and n (Eq. 3.41)	Bingham Fluid with Parameters μ_0 and τ_0 (Eq. 3.44)
Volume rate of flow Q in a thin slit of length L, width W, and thickness $2B$ ($B \ll W \ll L$), resulting from a modified pressure difference $\Delta\mathcal{P} = \mathcal{P}_0 - \mathcal{P}_L$; the magnitude of the wall shear stress is $\tau_B = \Delta\mathcal{P}B/L$.	$Q = \dfrac{2WB^2}{(1/n) + 2}\left(\dfrac{\tau_B}{m}\right)^{1/n}$	$Q = \dfrac{2WB^2\tau_B}{3\mu_0}\left[1 - \dfrac{3}{2}\left(\dfrac{\tau_0}{\tau_B}\right) + \dfrac{1}{2}\left(\dfrac{\tau_0}{\tau_B}\right)^3\right]$ when $\tau_B \geq \tau_0$
Volume rate of flow Q in a circular tube of length L and radius R ($R \ll L$), resulting from a modified pressure difference $\Delta\mathcal{P}$; the wall shear stress is $\tau_R = \Delta\mathcal{P}R/2L$	$Q = \dfrac{\pi R^3}{(1/n) + 3}\left(\dfrac{\tau_R}{m}\right)^{1/n}$	$Q = \dfrac{\pi R^3\tau_R}{4\mu_0}\left[1 - \dfrac{4}{3}\left(\dfrac{\tau_0}{\tau_R}\right) + \dfrac{1}{3}\left(\dfrac{\tau_0}{\tau_R}\right)^4\right]$ when $\tau_B \geq \tau_0$
Torque $\mathfrak{I}$ on the outer cylinder of an annulus (filled with liquid) required to maintain an angular velocity W of the outer cylinder; annulus is made up of cylinders of radii κR and R and length L ($\kappa < 1$)	$\mathfrak{I} = 2\pi(\kappa R)^2 L m \left(\dfrac{2W/n}{1 - \kappa^{2/n}}\right)^n$	$\mathfrak{I} = 4\pi(\kappa R)^2 L\mu_0 \left(\dfrac{W - (\tau_0/\mu_0)\ln\kappa}{1 - \kappa^2}\right)$ (This expression is valid only if $r_0 \geq R$, where $r_0 = \sqrt{\mathfrak{I}/2\pi\tau_0 L}$)

246

Volume rate of flow Q for axial flow in an annulus (inner radius κR, outer radius R) produced by axial motion of inner cylinder with velocity v_0

$$Q = \pi R^2 v_0 \left[\frac{(1 - \kappa^2 v_0)\,[1 + (\tau_0 R/\mu_0 v_0)\,(1 - \kappa)]}{2 \ln (1/\kappa)} - \frac{1}{3}\left(\frac{\tau_0 R}{\mu_0 v_0}\right)(1 - \kappa^3) - \kappa^2 \right]$$

This expression is valid only for $r_0 \geq R$, where r_0 is obtained from:

$$\frac{\tau_0 r_0}{\mu_0 v_0} \ln \frac{r_0}{\kappa R} = 1 + \frac{\tau_0 r_0}{\mu_0 v_0}\left(1 - \frac{\kappa R}{r_0}\right)$$

Volume rate of flow Q for axial flow in an annulus (radii κR and R) produced by an imposed modified-pressure difference, $\Delta\mathcal{P}$; length of annulus is L, and τ_R is an abbreviation for $\Delta\mathcal{P}R/2L$

$$Q = \frac{2\pi R^2 v_0}{1 - \kappa^{1 - (1/m)}} \left[\frac{1 - \kappa^2}{2} - \frac{1 - \kappa^{3 - (1/n)}}{3 - (1/n)} \right]$$

$$Q = \frac{\pi R^3}{(1/n) + 3}\left(\frac{\tau_R}{m}\right)^{1/2} \left[(1 - \lambda^2)^{1 + (1/n)} - \kappa^{1 - (1/m)}(\lambda^2 - \kappa^2)^{1 + (1/n)} \right]$$

λR is the location of the maximum in the velocity profile determined from:

$$\int_\kappa^\lambda \left(\frac{\lambda^2}{\xi^2} - \xi\right)^{1/n} d\xi = \int_\lambda^1 \left(\xi - \frac{\lambda^2}{\xi}\right)^{1/n} d\xi$$

$$Q = \frac{\pi R^3 \tau_R}{4\mu_0}\left[(1 - \kappa^4) - 2\lambda_+ - \left(\lambda_+ - \frac{\tau_0}{\tau_R}\right)(1 - \kappa^2) - \frac{4}{3}(1 + \kappa^3)\frac{\tau_0}{\tau_R} + \frac{1}{3}\left(2\lambda_+ - \frac{\tau_0}{\tau_R}\right)^3 \frac{\tau_0}{\tau_R} \right]$$

$\lambda_+ R$ and $\lambda_- R$ are the locations of the boundaries of the plug-flow region, given by:

$$\lambda_\pm - \frac{\lambda^2}{\lambda_\pm} = \pm\frac{\tau_0}{\tau_R}$$

Volume rate of flow Q for the radial flow between a pair of parallel disks resulting from a modified pressure difference $\Delta\mathcal{P} = \mathcal{P}_1 - \mathcal{P}_2$ between radii $r = R_1$ and $r = R_2$; disks are separated by a distance $2B^*$

$$Q = \frac{4\pi B^2}{(1/n) + 2}\left(\frac{\Delta\mathcal{P}B(1 - n)}{m(R_2^{1-n} - R_1^{1-n})} \right)^{1/n}$$

Q given as a function of $\Delta\mathcal{P}$ in tabular form[c]

TABLE 3.1 (*Continued*)

Flow System	Power-Law Fluid with Parameters m and n (Eq. 3.41)	Bingham Fluid with Parameters μ_0 and τ_0 (Eq. 3.44)
Fluid placed between two circular disks of radius R; plates are squeezed together so that the plates are instantaneously at $\pm h(t)$, and the instantaneous force required to maintain the motion is $F(t)$*	$F = \dfrac{(-dh/dt)^n}{h^{2n+1}} \left(\dfrac{2n+1}{2n} \right)^n \dfrac{\pi m R^{n+3}}{n+3}$	$F = \dfrac{3}{8} \dfrac{\pi \mu_0 R^4}{h^3} \left(-\dfrac{dh}{dt} \right) + \dfrac{\pi \tau_0 R^3}{2h} + \cdots$ (valid for fast squeezing)[d]

[a]Bird, R. B., Armstrong, R. C., and Hassager, O., ''Dynamics of Polymeric Liquids,'' John Wiley & Sons, New York (1987).

[b]Bird, R. B., Dai, G. C., and Yarusso, B. J., *Revs. in Chem. Engr.*, Vol. 1, 1–70 (1982).

[c]Dai, G. C. and Bird, R. B., *J. Non-Newtonian Fl. Mech.*, Vol. 8, 349–355 (1981).

[d]Covey, G. H. and Stanmore, B. R., *J. Non-Newtonian Fl. Mech.*, Vol. 8, 249–260 (1981).

*These are approximate results in which the ''lubrication approximation'' and the ''quasi-steady-state assumption'' have been used.

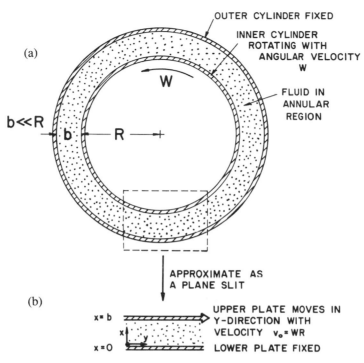

FIGURE 3.13 (a) Helical flow in an annulus produced by rotation of inner cylinder and an imposed pressure gradient in the axial direction. The length of the annulus is L. (b) Equivalent problem when curvature is neglected; the z-direction is perpendicular to the plane of the paper. A modified pressure drop $\Delta \mathcal{P} = \mathcal{P}_0 - \mathcal{P}_L$ acts in the z-direction.

$$\tau_{xz} = -\eta(\dot{\gamma})\dot{\gamma}_{xz} = -m\dot{\gamma}^{n-1}\frac{dv_z}{dx} \qquad (3.54b)$$

The shear rate $\dot{\gamma}$ is given by

$$\dot{\gamma} = \sqrt{\frac{1}{2}(\dot{\bar{\gamma}}:\dot{\bar{\gamma}})} = \sqrt{\left(\frac{dv_y}{dx}\right)^2 + \left(\frac{dv_z}{dx}\right)^2} \qquad (3.55)$$

When Eqs. (3.54) and (3.55) are substituted into the equations of motion, it is clear that these equations are coupled, both equations containing v_y and v_z. A first integration yields

$$\frac{d\bar{v}_y}{d\bar{x}} = C_1[C_1^2 + (C_2 - a\bar{x})^2]^{(1-n)/2n} \qquad (3.56a)$$

$$\frac{d\bar{v}_z}{d\bar{x}} = (C_2 - a\bar{x})[C_1^2 + (C_2 - a\bar{x})^2]^{(1-n)/2n} \qquad (3.56b)$$

Where C_1 and C_2 are constants of integration, and $\bar{v}_i = v_i/v_0$ $(i = y, z)$, $\bar{x} = x/b$, and $a = (\mathcal{P}_0 - \mathcal{P}_L)b^{n+1}/mLv_0^n$. To continue the solution numerical methods are needed for arbitrary n; for $n = 1/3$, however, Eqs. (3.56a) and (3.56b) can be integrated. The boundary condition that $\bar{v}_z = 0$ at $\bar{x} = 1$ leads to $C_2 = a/2$, and the boundary condition that $\bar{v}_y = 1$ at $\bar{x} = 1$ leads to $C_1^3 + (1/12)a^2C_1 - 1 = 0$, from which $C_1 = (12/a^2) - (12/a^2)^4 + \cdots$ for large values of a. Then, if just the first term in the expansion of C_1 is used, the velocity components are given by

$$\bar{v}_y \cong 3\bar{x} - 6\bar{x}^2 + 4\bar{x}^3 \tag{3.57a}$$

$$\bar{v}_z \cong (a^3/64)\,[1 - (1 - 2\bar{x})^4] \tag{3.57b}$$

and the volume flow rate in the z-direction, Q, is

$$Q = \frac{\pi R^2 bWa^3}{40}\left[1 + \frac{960}{a^6} + \cdots\right]$$

$$= \frac{\pi Rb^2}{40}\left(\frac{b\Delta\mathcal{P}}{mL}\right)^3\left[1 + 960\left(\frac{WR}{b}\right)^2\left(\frac{mL}{b\Delta\mathcal{P}}\right)^6 + \cdots\right] \tag{3.58}$$

The factor $(\pi Rb^2/40)(b\Delta\mathcal{P}/mL)^3$ is just the flow in a thin slit with no cylinder rotation (this may be obtained from the slit formula in Table 3.1, by replacing W by $2\pi R$, B by $b/2$, and n by $1/3$); the factor in [ ] is the flow enhancement resulting from the cylinder rotation. Note also that in the absence of a pressure-driven flow in the z-direction the velocity profile in the y-direction would just be linear: $\bar{v}_y = \bar{x}$; the distortion of the velocity profile $\bar{v}_y(\bar{x})$ is evident in Eq. (3.57a). The flow enhancement and the distortion of velocity profiles does not arise in the analogous Newtonian flow problem, where the y- and z-motions are entirely uncoupled. The coupling in the non-Newtonian flow problem occurs through the shear rate expression [Eq. (3.55)], so that flow in one direction alters the viscosity of the fluid for flow in another direction.

In Table 3.1 some analytical solutions of flow problems are given for the power-law fluid and for the Bingham fluid. These have been widely used for making rough flow calculations or for determining the fluid parameters (m, n or μ_0, τ_0) from flow measurements. A variational principle is also available for solving more difficult generalized Newtonian flow problems [Bird *et al.* (1981)].

Although the generalized Newtonian fluid was originally suggested as an empiricism, it has been shown to be the first term in the Criminale–Ericksen–Filbey (C.E.F.) equation for steady shear flows [see Eq. (3.115)].

3.3.4 Solution to Heat-Transfer Problems for Generalized Newtonian Fluids

Generalized Newtonian fluid models have been widely used in the solution of non-Newtonian heat-transfer problems [Bird *et al.* (1987), Metzner (1965), and Winter (1977)]. The equations of continuity, motion, and energy that have to be solved are

Continuity: $\qquad\qquad (\nabla \cdot \bar{v}) = 0 \tag{3.59}$

Motion: $\qquad\qquad \rho\frac{D\bar{v}}{Dt} = [\nabla \cdot \eta\dot{\bar{\gamma}}] - \nabla\mathcal{P} \tag{3.60}$

Energy:
$$\rho \hat{C}_p \frac{DT}{Dt} = (\nabla \cdot k\nabla T) + \frac{1}{2}\eta(\dot{\bar{\bar{\gamma}}}:\dot{\bar{\bar{\gamma}}}) \qquad (3.61)$$

The energy equation is obtained from Eq. (3.7) by making the assumption that the internal energy is a function of pressure and temperature only [Bird *et al.* (1960) and Bird and Wiest (1995)], and by using Fourier's law of heat conduction [Eq. (3.8)] for the heat flux vector $\bar{q}$. In order to use these equations one has to insert one of the empirical $\eta(\dot{\gamma})$ expressions from Sec. 3.3.2; in addition one has to introduce appropriate empiricisms for the dependence of the model parameters on temperature. For example, for the power law model [Eq. (3.41)] the relations

$$m(T) = m^0 e^{-A(T-T_0)/T_0} \qquad (3.62a)$$

$$n(T) = n^0 + \frac{B(T-T_0)}{T_0} \qquad (3.62b)$$

have been used to describe the change of m and n in the neighborhood of their values m^0 and n^0 at some reference temperature T_0; the constants A and B have to be determined from rheometric data. The assumption of constant thermal conductivity is usually not serious for most liquids; however, viscosity is very sensitive to temperature and the dependence of rheological constants must be taken into account in any realistic computation. Also in many industrial processes [Pearson (1983) and Tadmor and Gogos (1979)], such as polymer extrusion, wire coating, and lubrication, one deals with large velocity gradients and high viscosities; because of this, the viscous dissipation term $(1/2)\eta(\dot{\bar{\bar{\gamma}}}:\dot{\bar{\bar{\gamma}}})$ in the energy equation (Eq. (3.61)] cannot be neglected. Because of these complexities almost all realistic heat transfer problems have to be solved numerically; analytical results for constant rheological properties and no viscous heating (such as those in Tables 3.2 and 3.3) are of use only for checking computer programs in limiting cases and for making order-of-magnitude estimates.

The summaries of Nusselt numbers [Bird *et al.* (1987)] in Tables 3.2 and 3.3 give the standard results needed for estimating heat transfer coefficients in tubes and slits, for both the thermal entrance region and the thermally fully-developed region. These are all "local" Nusselt numbers defined by

Tubes: $\qquad$ Nu = hD/k $\qquad$ $D = 2R$ = tube diameter $\qquad$ (3.63a)

Slits: $\qquad$ Nu = $4hB/k$ $\qquad$ $2B$ = slit thickness $\qquad$ (3.63b)

and the heat transfer coefficient h is given by

$$q_w = h(T_b - T_w) \qquad (3.64)$$

where q_w and T_w are the heat flux through the wall and the wall temperature, and T_b is the local "bulk temperature" of the fluid defined by $T_b = \int v_z T dS / \int v_z dS$ where the integrals are taken over a cross section of the conduit. The derivations of the results in these tables are given in Bird *et al.* (1987).

As an example of a heat transfer problem in which viscous heating and the tem-

TABLE 3.2 Asymptotic Results for Nusselt Numbers (Tube Flow)*; Nu = hD/k

All values are local Nu numbers

		Constant Wall Temperature		Constant Wall Heat Flux
"Thermal Entrance Region" (small z)	Plug Flow	$Nu = \dfrac{1}{\sqrt{\pi}}\left(\dfrac{\langle v_z\rangle D^2}{\alpha z}\right)^{1/2}$	Plug Flow	$Nu = \dfrac{\sqrt{\pi}}{2}\left(\dfrac{\langle v_z\rangle D^2}{\alpha z}\right)^{1/2}$
	Laminar Non-Newtonian Flow	$Nu = \dfrac{2}{9^{1/3}\Gamma(\frac{4}{3})}\left[\dfrac{\langle v_z\rangle D^2}{\alpha z}\left(-\dfrac{1}{4}\dfrac{d\phi}{d\xi}\Big|_{\xi=1}\right)\right]^{1/3}$	Laminar Non-Newtonian Flow	$Nu = \dfrac{2\Gamma(\frac{2}{3})}{9^{1/3}}\left[\dfrac{\langle v_z\rangle D^2}{\alpha z}\left(-\dfrac{1}{4}\dfrac{d\phi}{d\xi}\Big|_{\xi=1}\right)\right]^{1/3}$
	Laminar Newtonian Flow	$Nu = \dfrac{2}{9^{1/3}\Gamma(\frac{4}{3})}\left(\dfrac{\langle v_z\rangle D^2}{\alpha z}\right)^{1/3}$	Laminar Newtonian Flow	$Nu = \dfrac{2\Gamma(\frac{2}{3})}{9^{1/3}}\left(\dfrac{\langle v_z\rangle D^2}{\alpha z}\right)^{1/3}$
"Thermally Fully-Developed Region" (large z)	Plug Flow	$Nu = 5.772$	Plug Flow	$Nu = 8$
	Laminar Non-Newtonian Flow	$Nu = \beta_1^2$, where β_1 is the *lowest* eigenvalue of $\dfrac{1}{\xi}\dfrac{d}{d\xi}\left(\xi\dfrac{dX_n}{d\xi}\right) + \beta_n^2\phi(\xi)X_n = 0;$ $\quad X_n'(0) = 0,\; X_n(1) = 0$	Laminar Non-Newtonian Flow	$Nu = \left[2\int_0^1 \dfrac{1}{\xi}\left[\int_0^\xi \xi'\phi(\xi')\,d\xi'\right]^2 d\xi\right]^{-1}$
	Laminar Newtonian Flow	$Nu = 3.657$	Laminar Newtonian Flow	$Nu = \dfrac{48}{11}$

NOTE: $\phi(\xi) = v_z/\langle v_z\rangle$ where $\xi = r/R$ and $R = D/2$; for Newtonian fluids $\langle v_z\rangle D^2/\alpha z = RePr(D/z)$ with $Re = D\langle v_z\rangle\rho/\mu$

*From Bird, R. B., Armstrong, R. C., and Hassager, O., "Dynamics of Polymeric Liquids," Vol. 1, *Fluid Mechanics*, John Wiley & Sons, New York, 1987.

TABLE 3.3 Asymptotic Results for Nusselt Numbers (Thin-Slit Flow)*, Nu $= 4Bh/k$

All values are local Nu numbers		Constant Wall Temperature	Constant Wall Heat Flux
"Thermal Entrance Region" (small z)	Plug Flow	$Nu = \dfrac{4}{\sqrt{\pi}} \left(\dfrac{\langle v_z \rangle B^2}{\alpha z} \right)^{1/2}$	$Nu = 2\sqrt{\pi} \left(\dfrac{\langle v_z \rangle B^2}{\alpha z} \right)^{1/2}$
	Laminar Non-Newtonian Flow	$Nu = \dfrac{4}{9^{1/3}\Gamma(\frac{4}{3})} \left[\dfrac{\langle v_z \rangle B^2}{\alpha z} \left(-\dfrac{d\phi}{d\sigma}\Big|_{\sigma=1} \right) \right]^{1/3}$	$Nu = \dfrac{4\Gamma(\frac{2}{3})}{9^{1/3}} \left[\dfrac{\langle v_z \rangle B^2}{\alpha z} \left(-\dfrac{d\phi}{d\sigma}\Big|_{\sigma=1} \right) \right]^{1/3}$
	Laminar Newtonian Flow	$Nu = \dfrac{4}{3^{1/3}\Gamma(\frac{4}{3})} \left(\dfrac{\langle v_z \rangle B^2}{\alpha z} \right)^{1/3}$	$Nu = \dfrac{4\Gamma(\frac{4}{3})}{3^{1/3}} \left(\dfrac{\langle v_z \rangle B^2}{\alpha z} \right)^{1/3}$
"Thermally Fully-Developed Region" (large z)	Plug Flow	$Nu = \pi^2$	$Nu = 12$
	Laminar Non-Newtonian Flow	$Nu = 4\beta_1^2$, where β_1 is the lowest eigenvalue of the equation: $\dfrac{d^2 X_n}{d\sigma^2} + \beta_n^2 \phi(\sigma) X_n = 0;\ X_n(\pm 1) = 0$	$Nu = \left[\dfrac{1}{4} \int_0^1 \left[\int_0^\sigma \phi(\sigma')\, d\sigma' \right]^2 d\sigma \right]^{-1}$
	Laminar Newtonian Flow	$Nu = 7.54$	$Nu = \dfrac{140}{17}$

NOTE: $\phi(\sigma) = v_z/\langle v_z \rangle$ where $\sigma = y/B$; for Newtonian fluids $\langle v_z \rangle D^2/\alpha z = 4\,Re\,Pr(B/z)$ with $Re = 4B\langle v_z \rangle \rho/\mu$

*From Bird, R. B., Armstrong, R. C., and Hassager, O., "Dynamics of Polymeric Liquids," Vol. 1, *Fluid Mechanics*, John Wiley & Sons, New York, 1987.

perature variation of the non-Newtonian viscosity are taken into account, the equations of motion and energy are solved for a power-law fluid in the region between two parallel planes, a fixed plane at $x = 0$ and a plane at $x = b$ that moves in the y-direction at speed v_0. The two planes are maintained at constant temperature T_0. Postulate that $v_y = v_y(x)$, $\mathcal{P} = $ constant, and $T = T(x)$. Then for the generalized Newtonian fluid the equations of motion and energy are

$$0 = \frac{d}{dx}\left(\eta \frac{dv_y}{dx}\right) \tag{3.65}$$

$$0 = \frac{d}{dx}\left(k \frac{dT}{dx}\right) + \eta \left(\frac{dv_y}{dx}\right)^2 \tag{3.66}$$

The non-Newtonian viscosity is taken to be the power law expression of Eq. (3.41), or $\eta = m(dv_y/dx)^{n-1}$. Let m vary according to Eq. (3.62a), but let n and k be constants. It is convenient to use the dimensionless variables $\Theta = (T - T_0)/T_0$, $\phi = v_y/v_0$, and $\xi = x/b$. Then the equation of motion (integrated once) and the energy equation become

$$e^{-A\Theta}(d\phi/d\xi)^n = C \tag{3.67}$$

$$d^2\Theta/d\xi^2 + \text{Br}\, e^{-A\Theta}(d\phi/d\xi)^{n+1} = 0 \tag{3.68}$$

in which $\text{Br} = m^0 v_0^{n+1}/b^{n+1} k T_0$ is the power-law fluid Brinkman number, which is a measure of the importance of viscous heating, and C is an integration constant. The exponentials in the above equations can be expanded so that

$$d\phi/d\xi = C^s(1 + a_1\Theta + a_2\Theta^2 + \cdots) \tag{3.69}$$

$$d^2\Theta/d\xi^2 = -\text{Br}\, C^{s+1}(1 + a_1\Theta + a_2\Theta^2 + \cdots) \tag{3.70}$$

where $s = 1/n$, and $a_j = (As)^j/j!$. Perturbation expansions are now sought of the form

$$\phi = \phi_0(\xi) + \text{Br}\, \phi_1(\xi) + \text{Br}^2\, \phi_2(\xi) + \cdots \tag{3.71}$$

$$\Theta = \Theta_0(\xi) + \text{Br}\, \Theta_1(\xi) + \text{Br}^2\, \Theta_2(\xi) + \cdots \tag{3.72}$$

It is also convenient to expand the constant of integration similarly ($C = C_0 + \text{Br}C_1 + \cdots$). Substituting these expansions into Eqs. (3.69) and (3.70) and collecting terms of the same order in Br leads to a set of ordinary differential equations that are solved so that the boundary conditions are satisifed. In this way one obtains

$$\phi = \xi - \frac{1}{12}\, \text{Br}\left(\frac{A}{n}\right)(\xi - 3\xi^2 + 2\xi^3) + \cdots \tag{3.73}$$

$$\Theta = \frac{1}{2}\, \text{Br}\, (\xi - \xi^2)$$

$$-\frac{1}{24}\, \text{Br}^2\left(\frac{A}{n}\right)(n\xi - (n+1)\xi^2 + 2\xi^3 - \xi^4) + \cdots \tag{3.74}$$

The first term in each equation is the result for constant m (that is $A = 0$); higher-order terms give the distortion of the velocity and temperature profiles resulting from the coupling of the motion and energy equations, which in turn results from the temperature dependence of the non-Newtonian viscosity.

3.4 LINEAR VISCOELASTIC FLUID MODELS

Linear viscoelastic behavior—and in particular the small-amplitude sinusoidal response experiment—has been and will continue to be of great importance in characterizing polymeric materials. Also an understanding of linear viscoelasticity is a *sine qua non* for the study of nonlinear viscoelastic behavior [Tschoegl (1989), Giesekus (1994), Bird *et al.* (1987), and Ferry (1980)].

3.4.1 Definition of Linear Viscoelastic Fluids

Many fluids of complex structure exhibit elastic effects, such as recoil. This suggests that the mechanical behavior of such fluids can be described by some combination of Newton's law for viscous liquids and Hooke's law for elastic solids (both written for incompressible materials)

$$\textit{Newton:} \qquad \bar{\bar{\tau}} = -\mu \dot{\bar{\gamma}} \qquad (\dot{\bar{\gamma}} = \nabla \bar{v} + (\nabla \bar{v})^\dagger) \qquad (3.75)$$

$$\textit{Hooke:} \qquad \bar{\bar{\tau}} = -G\gamma \qquad (\bar{\bar{\gamma}} = \nabla \bar{u} + (\nabla \bar{u})^\dagger) \qquad (3.76)$$

Here $\bar{\bar{\gamma}}$ is the infinitesimal strain tensor, and $\bar{u}$ is the displacement of a material particle from its equilibrium position at time t_0. It should be emphasized that Hooke's law in Eq. (3.76) is valid only for small displacement gradients.

Maxwell combined the above two equations thus

$$\textit{Maxwell:} \qquad \bar{\bar{\tau}} + \lambda_0 \frac{\partial}{\partial t} \bar{\bar{\tau}} = -\eta_0 \dot{\bar{\bar{\gamma}}} \qquad (3.77)$$

where $\eta_0 \equiv \mu$ and $\lambda_0 \equiv \mu/G$ is a time constant. For steady-state flows with $\partial \tau / \partial t = 0$ Eq. (3.77) simplifies to Eq. (3.75); when stresses are suddenly imposed on the fluid, the $\partial \tau / \partial t$ term is much larger than the τ term, so that the equation can be integrated with respect to time to recover Eq. (3.76). Thus the Maxwell equation contains both the notions of viscosity and elasticity; however it is valid only for infinitesimal displacement gradients.

Maxwell's equation can be generalized by including additional time derivatives of $\bar{\bar{\tau}}$ and/or $\dot{\bar{\bar{\gamma}}}$; the simplest generalization of this type is the Jeffreys model

$$\textit{Jeffreys:} \qquad \bar{\bar{\tau}} + \lambda_1 \frac{\partial}{\partial t} \bar{\bar{\tau}} = -\eta_0 \left(\dot{\bar{\bar{\gamma}}} + \lambda_2 \frac{\partial}{\partial t} \dot{\bar{\bar{\gamma}}} \right) \qquad (3.78)$$

where λ_1 and λ_2 are time constants.

Maxwell's equation can also be generalized by superposing a number of Maxwell equations thus

$$\text{Generalized Maxwell:} \begin{cases} \overline{\overline{\tau}} = \sum_{k=1}^{\infty} \overline{\overline{\tau}}_k & (3.79a) \\[2ex] \overline{\overline{\tau}}_k + \lambda_k \frac{\partial}{\partial t} \overline{\overline{\tau}}_k = -\eta_k \dot{\overline{\overline{\gamma}}} & (3.79b) \end{cases}$$

Such an expression for the stress tensor allows for a spectrum of relaxation times. Linear viscoelastic data can often be fit satisfactorily by letting

$$\eta_k = \eta_0 \lambda_k / \Sigma_k \lambda_k; \qquad \lambda_k = \lambda / k^\alpha \qquad (3.80a,b)$$

In this way two-times-infinity constants (the λ_k and η_k) are replaced by three constants (the zero-shear-rate viscosity η_0, a time constant λ, and a dimensionless parameter α).

The above linear viscoelastic constitutive equations are all written in differential form. By integration they can be put into two useful integral forms. This is illustrated for the generalized Maxwell model

$$\overline{\overline{\tau}}(t) = -\int_{-\infty}^{t} \underbrace{\left\{ \sum_{k=1}^{\infty} \frac{\eta_k}{\lambda_k} e^{-(t-t')/\lambda_k} \right\}}_{\text{"relaxation modulus"}} \dot{\overline{\overline{\gamma}}}(t') \, dt' \qquad (3.81)$$

$$\overline{\overline{\tau}}(t) = +\int_{-\infty}^{t} \underbrace{\left\{ \sum_{k=1}^{\infty} \frac{\eta_k}{\lambda_k^2} e^{-(t-t')\lambda_k} \right\}}_{\text{"memory function"}} \overline{\overline{\gamma}}(t, t') \, dt' \qquad (3.82)$$

To obtain the first of these, one integrates the first-order linear differential equation in Eq. (3.79b). To obtain the second, one performs an integration by parts and uses the relation $\overline{\overline{\gamma}}(t, t') = \int_t^{t'} \dot{\overline{\overline{\gamma}}}(t'') dt''$; note that for liquids the reference state for the strain tensor is chosen to be the configuration at the current time t. Equation (3.81) is reminiscent of Newton's law, with $\overline{\overline{\tau}}$ at time t depending on the velocity gradients that have been experienced over all past times t'; the velocity gradients in the most recent past receive the greatest emphasis. Equation (3.82) is reminiscent of Hooke's law, with the stress at the current time t depending on the displacement gradients for all past times t'. The quantities within the { } are called the *relaxation modulus* [in Eq. (3.81)] and the *memory function* [in Eq. (3.82)]; these sums of exponentials capture in mathematical form the notion of "fading memory"—the fact that the fluid "remembers" kinematic events of the recent past rather well but "forgets" kinematic changes of the more distant past.

Equations (3.81) and (3.82) suggest that the most general linear integral constitutive equation connecting stress to the kinematic tensors $\dot{\overline{\overline{\gamma}}}$ or $\overline{\overline{\gamma}}$ are for incompressible liquids

$$\text{General Linear Viscoelastic Fluid:} \begin{cases} \overline{\overline{\tau}}(t) = -\int_{-\infty}^{t} G(t - t') \dot{\overline{\overline{\gamma}}}(t') dt' & (3.83) \\[2ex] \overline{\overline{\tau}}(t) = +\int_{-\infty}^{t} M(t - t') \overline{\overline{\gamma}}(t, t') dt' & (3.84) \end{cases}$$

Here $G(t - t')$ is the relaxation modulus and $M(t - t') = \partial G(t - t')/\partial t'$ is the memory function. These equations have been used principally by chemists and physicists in the analysis of experiments with very small displacement gradients. The main aim of these experiments has been the elucidation of fluid structure by comparison of linear viscoelastic measurements with the predictions of the various molecular or structural kinetic theories.

3.4.2 Expressions for Material Functions in Terms of the Relaxation Modulus

Various material functions of Sec. 3.2.2 may now be expressed in terms of the relaxation modulus by inserting the appropriate flow pattern into Eq. (3.83) [Bird *et al.* (1987) and Ferry (1980)]. These are summarized in Table 3.4. In all these flows it is assumed that the flow field is homogeneous (i.e., $\nabla \overline{v}$ is independent of position).

The most important of these experiments for small-displacement gradient flows is that of sinusoidal displacements. Note that the complex viscosity $\eta^*(\omega) = \eta'(\omega) - i\eta''(\omega)$ and the relaxation modulus are interrelated thus

$$\eta^*(\omega) = \int_0^\infty G(s)e^{-i\omega s}\,ds \tag{3.85}$$

The real and imaginary parts of η^* are not independent of one another, but are related by

$$\eta''(\omega) = \frac{2\omega}{\pi}\int_0^\infty \frac{\eta'(\overline{\omega}) - \eta'(\omega)}{\omega^2 - \overline{\omega}^2}\,d\overline{\omega} \tag{3.86}$$

$$\eta'(\omega) - \eta'(\infty) = \frac{2}{\pi}\int_0^\infty \frac{\overline{\omega}\eta''(\overline{\omega}) - \omega\eta''(\omega)}{\overline{\omega}^2 - \omega^2}\,d\overline{\omega} \tag{3.87}$$

These are known as the *Kramers–Kronig relations*.

TABLE 3.4 Small-Displacement Material Functions in Terms of $G(s)$

Equation Giving the Material Function	Flow Field Specified by	Material Function	
3.14	$\dot{\gamma} = $ constant	$\eta(0) \equiv \eta_0 = \int_0^\infty G(s)\,ds$	(A)
3.17	$\dot{\gamma}(t) = \dot{\gamma}^\circ \cos \omega t$ ($\dot{\gamma}^\circ/\omega$ vanishingly small)	$\eta'(\omega) = \int_0^\infty G(s) \cos \omega s\,ds$	(B)
		$\eta''(\omega) = \int_0^\infty G(s) \sin \omega s\,ds$	(C)
3.18	$\dot{\gamma} = $ constant $\quad t < 0$ $\dot{\gamma} = 0 \quad\quad\quad t > 0$	$\eta^-(t, 0) = \int_t^\infty G(s)\,ds$	(D)
3.21	$\dot{\gamma} = 0 \quad\quad\quad t < 0$ $\dot{\gamma} = $ constant $\quad t > 0$	$\eta^+(t, 0) = \int_0^t G(s)\,ds$	(E)
3.24	$\dot{\epsilon} = $ constant	$\overline{\eta}(0) \equiv 3\eta_0 = 3\int_0^\infty G(s)\,ds$	(F)

3.4.3 Solution of Flow Problems using the General Linear Viscoelastic Fluid Model

As an example of the use of the linear viscoelastic model in conjunction with the conservation equations, the transmission of a shear wave through a viscoelastic fluid is discussed [Bird *et al.* (1987)]. Let the fluid be located in the half space $0 < y < \infty$. The xz-plane is made to oscillate in the x-direction so that at $y = 0$, $v_x = v_0 \cos \mu t = v_0 \, \mathrm{Re} \, \{e^{i\omega t}\}$, where v_0 is the (very small) velocity amplitude and ω is the frequency of oscillation. It is desired to find the velocity distribution $v_x(y, t)$ throughout the fluid after the initial transients have subsided.

Postulate that in the "oscillatory steady state"

$$v_x = \mathrm{Re} \, \{v_x^0(y) \, e^{i\omega t}\} \tag{3.88}$$

in which v_x^0 is the complex amplitude, and "Re$\{z\}$" means "real part of z." For this velocity distribution the linear viscoelastic constitutive equation gives

$$\tau_{yx} = -\int_{-\infty}^{t} G(t - t') \frac{\partial v_x(t')}{\partial y} \, dt'$$

$$= -\int_{-\infty}^{t} G(t - t') \, \mathrm{Re} \left\{ \frac{dv_x^0}{dy} \, e^{i\omega t'} \right\} dt'$$

$$= -\mathrm{Re} \left\{ \frac{dv_x^0}{dy} \int_0^{\infty} G(s) \, e^{i\omega(t-s)} \, ds \right\}$$

$$= -\mathrm{Re} \left\{ \frac{dv_x^0}{dy} \, e^{i\omega t} \, \eta^*(\omega) \right\} \tag{3.89}$$

where use has been made of Eq. (3.85).

The x-component of the equation of motion is

$$\rho \frac{\partial v_x}{\partial t} = -\frac{\partial}{\partial y} \tau_{yx} \tag{3.90}$$

Substitution of Eqs. (3.88) and (3.89) into this equation of motion gives

$$\frac{d^2 v_x^0}{dy^2} - \frac{\rho i \omega v_x^0}{\eta^*} = 0 \tag{3.91}$$

This has to be solved for the boundary conditions: $v_x^0 = v_0$ at $y = 0$, and $v_x^0 = 0$ at $y = \infty$. If $\alpha(\omega)$ and $\beta(\omega)$ are defined by

$$\frac{\rho i \omega}{\eta^*} = (\alpha + i\beta)^2 \tag{3.92}$$

then the solution to Eq. (3.92) is

$$v_x^0(y) = v_0 \, e^{-(\alpha + i\beta)y} \tag{3.93}$$

When this is inserted into Eq. (3.88) one obtains

$$v_x(y, t) = v_0 \, e^{-\alpha y} \cos (\omega t - \beta y) \qquad (3.94)$$

It is thus seen that the quantity $\alpha(\omega)$ describes the attenuation and $\beta(\omega)$ describes the phase shift; $\alpha(\omega)$ and $\beta(\omega)$ are given by

$$\alpha = \frac{\sqrt{(\rho\omega/2)(|\eta^*| - \eta'')}}{|\eta^*|} \qquad (3.95a)$$

$$\beta = \frac{\eta'}{|\eta^*|} \sqrt{\frac{(\rho\omega/2)}{|\eta^*| - \eta''}} \qquad (3.95b)$$

For the Maxwell model $\eta^* = \eta_0/(1 + i\lambda\omega)$, and hence,

$$\alpha = \sqrt{\rho\omega/2\eta_0} \, \sqrt{\sqrt{1 + \lambda_0^2\omega^2} - \lambda_0\omega}. \qquad (3.96)$$

As λ_0 goes from 0 to ∞, the attenuation factor α goes from $\sqrt{\rho\omega/2\eta_0}$ to 0; hence an increase in *elasticity* (as measured by the time constant λ_0) results in a decrease in the attenuation of the shear wave.

3.5 NONLINEAR VISCOELASTIC FLUID MODELS

There are many continuum mechanics approaches to the subject of nonlinear viscoelasticity [Bird *et al.* (1987), Lodge (1964), Sedov (1965), Schowalter (1978), Lodge (1974), Astarita and Marrucci (1974), Joseph (1990), Larson (1988), and Giesekus (1994)]. The one given here, [Oldroyd (1950)] based on a convected coordinate system, is the most widely accepted. The subject may also be studied by using a corotating coordinate frame that moves with the fluid particle [Bird *et al.* (1987) and Oldroyd (1984)]. Still another approach is the use of "body tensors" [Lodge (1974)]. Each approach has its advantages and disadvantages. In each approach the main idea is to establish a procedure for constructing constitutive equations that are *objective*, that is, free from any unwanted dependence on the instantaneous orientation of a fluid particle.

3.5.1 Convected Base Vectors and Reciprocal Base Vectors

To describe the kinematics of a moving fluid, it is useful to introduce the notion of a convected coordinate system $\hat{x}^1$, $\hat{x}^2$, $\hat{x}^3$, which is imprinted in the fluid and moves with it [Oldroyd (1950), (1958), (1984) and Bird *et al.* (1987)]. This coordinate system may be chosen so that, at the current time t, it coincides with a space-fixed Cartesian coordinate system x_1, x_2, x_3 (see Fig. 3.14). For times $t' \neq t$ the coordinate system is in general nonorthogonal, and the coordinates of a fluid particle in this system are designated by $\hat{x}^1$, $\hat{x}^2$, $\hat{x}^3$; the Cartesian coordinates of the fluid particle are x_1', x_2', x_3'. The relations

$$x_i' = x_i'(x_1, x_2, x_3, t, t') \qquad (3.97)$$

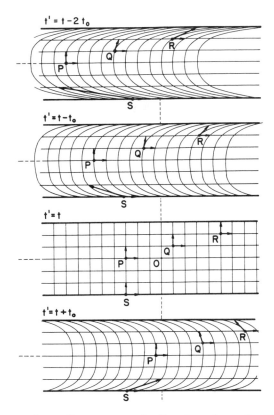

FIGURE 3.14 Convected coordinate system for flow in a plane slit. The fluid viscosity is given by $\eta = m\dot\gamma^{n-1}$ with $n = 1/3$. At $t' = t$ fluid particle R has coordinates $x_1 = 5$, $x_2 = 2$, $x_3 = 0$. At time $t' = t - 2t_0$ fluid particle R has cartesian coordinates $x_1' = 1$, $x_2' = 2$, $x_3' = 3$; it has convected coordinates $\hat{x}^1 = 5$, $\hat{x}^2 = 2$, $\hat{x}^3 = 0$ for all times t'. O is the origin of the cartesian coordinate system. The base vectors $\hat{\bar{\mathbf{g}}}_1$ and $\hat{\bar{\mathbf{g}}}_2$ are shown for fluid particles P, Q, R, S; $\hat{\bar{\mathbf{g}}}_3$ is of unit length perpendicular to the plane of the paper. At $t' = t$, the base vectors $\hat{\bar{\mathbf{g}}}_1$, $\hat{\bar{\mathbf{g}}}_2$, $\hat{\bar{\mathbf{g}}}_3$ coincide with the unit vectors $\bar{\boldsymbol{\delta}}_1$, $\bar{\boldsymbol{\delta}}_2$, $\bar{\boldsymbol{\delta}}_3$.

are called the *displacement functions*. They give the location at time t' of the fluid particle that is at x_1, x_2, x_3 at time t.

At each fluid particle a set of *convected base vectors* $\hat{\bar{\mathbf{g}}}_i(\bar{\mathbf{r}}, t, t')$ can be defined by the standard relation used in tensor calculus: $\hat{\bar{\mathbf{g}}}_i = (\partial/\partial \hat{x}^i)\bar{\mathbf{r}}' = (\partial/\partial x_i)(\bar{\boldsymbol{\delta}}_1 x_1' + \bar{\boldsymbol{\delta}}_2 x_2' + \bar{\boldsymbol{\delta}}_3 x_3')$; these are the base vectors at time t' associated with the fluid particle that is located at position $\bar{\mathbf{r}}$ at time t. That is, the "$\bar{\mathbf{r}}$, t" in the argument of $\hat{\bar{\mathbf{g}}}_i$ is a "fluid particle label." At time t the base vectors coincide with the unit vectors $\bar{\boldsymbol{\delta}}_i$. The base vector $\hat{\bar{\mathbf{g}}}_i(\bar{\mathbf{r}}, t, t')$ is tangent to the $\hat{x}^i$-coordinate curve. The set of $\hat{\bar{\mathbf{g}}}_i$ shows how a fluid element is oriented and distorted as it moves along in the flow field (see Fig. 3.14).

A set of *convected reciprocal base vectors* $\hat{\bar{\mathbf{g}}}^i(\bar{\mathbf{r}}, t, t')$ can also be defined by the standard relation $\hat{\bar{\mathbf{g}}}^i = (\partial/\partial \bar{\mathbf{r}}')\hat{x}^i = (\bar{\boldsymbol{\delta}}_1 \partial/\partial x_1' + \bar{\boldsymbol{\delta}}_2 \partial/\partial x_2' + \bar{\boldsymbol{\delta}}_3 \partial/\partial x_3')\hat{x}^i$. The reciprocal base vector $\hat{\bar{\mathbf{g}}}^i(\bar{\mathbf{r}}, t, t')$ is normal to the surface $\hat{x}^i = $ constant. For an incompressible fluid the base vectors and reciprocal base vectors are interrelated by $\hat{\bar{\mathbf{g}}}^i = [\hat{\bar{\mathbf{g}}}_j \times \hat{\bar{\mathbf{g}}}_k]$ where $ijk = 123, 231, 312$. In addition $(\hat{\bar{\mathbf{g}}}_i \cdot \hat{\bar{\mathbf{g}}}^j) = \delta_{ij}$, where δ_{ij} is the Kronecker delta. The dot products

$$\hat{g}_{ij} = (\hat{\bar{\mathbf{g}}}_i \cdot \hat{\bar{\mathbf{g}}}_j); \qquad \hat{g}^{ij} = (\hat{\bar{\mathbf{g}}}^i \cdot \hat{\bar{\mathbf{g}}}^j) \tag{3.98}$$

are of particular interest since they describe the deformation of a fluid element independently of its orientation in space; for example $\hat{g}_{22}$ is a measure of the stretching of the fluid along the $\hat{x}^2$-coordinate direction, and $\hat{g}_{12} = \hat{g}_{21}$ is a measure of the angle between the $\hat{x}^1$ and $\hat{x}^2$ coordinate curves.

The convected base vectors and reciprocal base vectors may be given in terms of their Cartesian components

$$\hat{\bar{\mathbf{g}}}_i(\mathbf{r}, t, t') = \Sigma_j \bar{\delta}_j \Delta_{ji}(\bar{\mathbf{r}}, t, t') \tag{3.99a}$$

$$\hat{\bar{\mathbf{g}}}^i(\bar{\mathbf{r}}, t, t') = \Sigma_j \, \mathrm{E}_{ij}(\bar{\mathbf{r}}, t, t')\bar{\delta}_j \tag{3.99b}$$

where the components of the tensors $\bar{\bar{\boldsymbol{\Delta}}}$ and $\bar{\bar{\mathbf{E}}}$ are given by

$$\Delta_{ij}(\bar{\mathbf{r}}, t, t') = \partial x'_i(x, t, t')/\partial x_j \tag{3.100a}$$

$$\mathrm{E}_{ij}(\bar{\mathbf{r}}, t, t') = \partial x_i(x', t, t')/\partial x'_j \tag{3.100b}$$

In the expression for Δ_{ij} one differentiates the displacement function in Eq. (3.97); to get E_{ij} one differentiates the inverse of Eq. (3.97) and then replaces the x'_i by x_i using the displacement function. Note further that $\bar{\bar{\mathbf{E}}} = \bar{\bar{\boldsymbol{\Delta}}}^{-1}$, and that at $t' = t$, both Δ_{ij} and E_{ij} become the Kronecker delta δ_{ij}.

As the fluid particle moves, the base vectors change with the time t' as follows [Bird *et al.* (1987)]

$$\frac{\partial}{\partial t'} \hat{\bar{\mathbf{g}}}_i = [\hat{\bar{\mathbf{g}}}_i \cdot \nabla \bar{\mathbf{v}}] \tag{3.101a}$$

$$\frac{\partial}{\partial t'} \hat{\bar{\mathbf{g}}}^i = -[\nabla \bar{\mathbf{v}} \cdot \hat{\bar{\mathbf{g}}}^i] \tag{3.101b}$$

Here the base vectors and the velocity gradient tensor $\nabla \bar{\mathbf{v}}$ all depend on the particle label $\bar{\mathbf{r}}$, t and the time t'. Eq. (3.101a,b) may be derived by considering the motion of a fluid line element during the flow. From Eqs. (3.99) and (3.101) the changes of Δ_{ij} and E_{ij} following a fluid particle are

$$\frac{\partial}{\partial t'} \Delta_{ij} = \Sigma_n \, (\nabla \bar{\mathbf{v}})^\dagger_{in} \Delta_{nj} \tag{3.102a}$$

$$\frac{\partial}{\partial t'} \mathrm{E}_{ij} = -\Sigma_n \mathrm{E}_{in}(\nabla \bar{\mathbf{v}})^\dagger_{nj} \tag{3.102b}$$

These equations enable one to calculate Δ_{ij} and E_{ij} from a knowledge of the flow field $\bar{\mathbf{v}}(\bar{\mathbf{r}}, t)$, whereas Eq. (3.100a,b) provide a way to get Δ_{ij} and E_{ij} from the particle trajectories.

3.5.2 Strain Tensors and Rate-of-Strain Tensors

The strain of a fluid element at time t' relative to the reference configuration at time t can be described by giving the difference between $(\hat{\mathbf{g}}_i \cdot \hat{\mathbf{g}}_j)$ at time t' and $(\hat{\mathbf{g}}_i \cdot \hat{\mathbf{g}}_j)$ at time t, or the analogous quantity involving the reciprocal base vectors [Bird et al. (1987), Lodge (1964), Sedov (1965), and Oldroyd (1950)]

$$[\hat{g}_{ij}(\bar{\mathbf{r}}, t, t') - \hat{g}_{ij}(\bar{\mathbf{r}}, t, t)] = \Sigma_k \Delta_{ki}(\mathbf{r}, t, t') \, \Delta_{kj}(\mathbf{r}, t, t') - \delta_{ij} \quad (3.103a)$$

$$-[\hat{g}^{ij}(\mathbf{r}, t, t') - \hat{g}^{ij}(\bar{\mathbf{r}}, t, t)] = \delta_{ij} - \Sigma_k E_{ik}(\bar{\mathbf{r}}, t, t') \, E_{jk}(\bar{\mathbf{r}}, t, t') \quad (3.103b)$$

These are the components of the *finite strain tensors*; designate the tensors corresponding to the right side of these equations by $\bar{\bar{\gamma}}^{[0]}(\bar{\mathbf{r}}, t, t')$ and $\bar{\bar{\gamma}}_{[0]}(\bar{\mathbf{r}}, t, t')$ respectively. They are closely related to the *Cauchy strain tensor* $\bar{\bar{\mathbf{C}}}$ and its inverse, the *Finger strain tensor* $\bar{\bar{\mathbf{B}}}$

$$\bar{\bar{\gamma}}^{[0]} = \bar{\bar{\mathbf{C}}} - \bar{\bar{\delta}} \quad (3.104a)$$

$$\bar{\bar{\gamma}}_{[0]} = \bar{\bar{\delta}} - \bar{\bar{\mathbf{B}}} \quad (3.104b)$$

In the limit of vanishingly small deformations, both finite strain tensors $\bar{\bar{\gamma}}^{[0]}(\bar{\mathbf{r}}, t, t')$ and $\bar{\bar{\gamma}}_{[0]}(\bar{\mathbf{r}}, t, t')$ reduce to the *infinitesimal strain tensor* $\bar{\bar{\gamma}}(\mathbf{r}, t, t')$ used in linear viscoelasticity [cf. Eq. (3.76)]; when $t' = t$, all three of these strain tensors vanish.

Successive time derivatives $(\partial/\partial t', \partial^2/\partial t'^2, \cdots)$ may be taken of Eq. (3.103a,b) to generate the *rate-of-strain tensors*, *second rate-of-strain tensors*, etc., at time t', by using Eq. (3.102a,b). In this way the entries in Table 3.5 are generated. The *nth rate-of-strain tensors* at time t, $\bar{\bar{\gamma}}^{(n)}$ and $\bar{\bar{\gamma}}_{(n)}$, appearing in the table are generated by "convected differentiation"

$$\bar{\bar{\gamma}}^{(n+1)} = \frac{D}{Dt} \bar{\bar{\gamma}}^{(n)} + \{(\nabla\mathbf{v}) \cdot \bar{\bar{\gamma}}^{(n)} + \bar{\bar{\gamma}}^{(n)} \cdot (\nabla\mathbf{v})^{\dagger}\} \quad (3.105a)$$

$$\bar{\bar{\gamma}}_{(n+1)} = \frac{D}{Dt} \bar{\bar{\gamma}}_{(n)} - \{(\nabla\bar{\mathbf{v}})^{\dagger} \cdot \bar{\bar{\gamma}}_{(n)} + \bar{\bar{\gamma}}_{(n)} \cdot (\nabla\bar{\mathbf{v}})\} \quad (3.105b)$$

The $\bar{\bar{\gamma}}^{(n)}$ are twice the *Oldroyd nth rate-of-strain tensors* and exactly the same as the *Rivlin–Ericksen tensors* commonly called $\bar{\bar{\mathbf{A}}}_n$. In Eq. (3.105a,b) both $\bar{\bar{\gamma}}^{(1)}$ and $\bar{\bar{\gamma}}_{(1)}$ are defined to be $\dot{\bar{\bar{\gamma}}} = \nabla\mathbf{v} + (\nabla\bar{\mathbf{v}})^{\dagger}$. In addition to the convected derivatives defined above, one frequently encounters the *Jaumann* (or *corotational*) *time derivative* [Oldroyd (1950) and Bird et al. (1977)] defined by

$$\frac{\mathfrak{D}}{\mathfrak{D}t} \dot{\bar{\bar{\gamma}}} = \frac{1}{2} (\bar{\bar{\gamma}}^{(2)} + \bar{\bar{\gamma}}_{(2)}) \quad (3.106)$$

This derivative describes the time rate of change of $\dot{\bar{\bar{\gamma}}}$ in a coordinate frame that is moving along with a fluid particle and rotating with the instantaneous local angular velocity.

TABLE 3.5 Kinematic Tensors: Strain Tensors and Their Time Derivatives

Kinematic Tensors at t' in Terms of $\hat{g}_{ij}$ or $\hat{g}^{ij}$ (Convected Metric Tensor Components)		Kinematic Tensors at t' in Terms of Δ_{ij} or E_{ij} (Fixed Cartesian Components)		Kinematic Tensors at t (Fixed Cartesian Components)
$[\hat{g}_{ij}(\mathbf{r}, t, t') - \hat{g}_{ij}(\mathbf{r}, t, t)]$	$=$	$\gamma_{ij}^{[0]} = \Sigma_k \Delta_{ki}\Delta_{kj} - \delta_{ij}$ $\xrightarrow{\ t'=t\ }$		0
$-[\hat{g}^{ij}(\mathbf{r}, t, t') - \hat{g}^{ij}(\mathbf{r}, t, t)]$	$=$	$\gamma_{[0]ij} = \delta_{ij} - \Sigma_k E_{ik}E_{jk}$		0
$\partial/\partial t' \downarrow$		$\partial/\partial t' \downarrow$		
$\partial \hat{g}_{ij}/\partial t'$	$=$	$\gamma_{ij}^{[1]} = \Sigma_m \Sigma_n \gamma_{mn}^{(1)}\Delta_{mi}\Delta_{nj}$ $\xrightarrow{\ t'=t\ }$		$\gamma_{ij}^{(1)} \equiv \dot{\gamma}_{ij} = \dfrac{\partial}{\partial x_i}v_j + \dfrac{\partial}{\partial x_j}v_i$
$-\partial \hat{g}^{ij}/\partial t'$	$=$	$\gamma_{[1]ij} = \Sigma_m \Sigma_n E_{im}E_{jn}\gamma_{(1)mn}$		$\gamma_{(1)ij} \equiv \dot{\gamma}_{ij} = \dfrac{\partial}{\partial x_i}v_j + \dfrac{\partial}{\partial x_j}v_i$
$\partial/\partial t' \downarrow$		$\partial/\partial t' \downarrow$		convected $\downarrow$ differentiation
$\partial^2 \hat{g}_{ij}/\partial t'^2$	$=$	$\gamma_{ij}^{[2]} = \Sigma_m \Sigma_n \gamma_{mn}^{(2)}\Delta_{mi}\Delta_{nj}$ $\xrightarrow{\ t'=t\ }$		$\gamma_{ij}^{(2)} = \dfrac{D}{Dt}\gamma_{ij}^{(1)} + \Sigma_p \left(\dfrac{\partial v_p}{\partial x_i}\gamma_{pj}^{(1)} + \gamma_{ip}^{(1)}\dfrac{\partial v_p}{\partial x_j}\right)$
$-\partial^2 \hat{g}^{ij}/\partial t'^2$	$=$	$\gamma_{[2]ij} = \Sigma_m \Sigma_n E_{im}E_{jn}\gamma_{(2)mn}$		$\gamma_{(2)ij} = \dfrac{D}{Dt}\gamma_{(1)ij} - \Sigma_p \left(\dfrac{\partial v_i}{\partial x_p}\gamma_{(1)pj} + \gamma_{(1)ip}\dfrac{\partial v_j}{\partial x_p}\right)$
etc. $\downarrow$		etc. $\downarrow$		etc. $\downarrow$

All quantities in the first column are functions of $\bar{\mathbf{r}}, t, t'$, as are all $\bar{\bar{\Delta}}, \bar{\bar{E}}, \bar{\bar{\gamma}}^{[n]}, \bar{\bar{\gamma}}_{[n]}$ in the second column. The $\bar{\bar{\gamma}}^{(n)}$ and $\bar{\bar{\gamma}}_{(n)}$ are functions of $\bar{\mathbf{r}}, t'$ in the second column and functions of $\bar{\mathbf{r}}, t$ in the third column.

3.5.3   The Stress Tensor and its Time Derivatives

The stress tensor at the fluid particle at time t' may be written in terms of the co-variant convected components $\hat{\tau}_{ij}$ or the contravariant convected components $\hat{\tau}^{ij}$ [Bird *et al.* (1987) and Oldroyd (1950)]

$$\bar{\bar{\tau}}(\bar{r}, t, t') = \sum_i \sum_j \hat{\bar{g}}^i(\bar{r}, t, t') \, \hat{\bar{g}}^j(\bar{r}, t, t') \, \hat{\tau}_{ij}(\bar{r}, t, t') \tag{3.107a}$$

$$\bar{\bar{\tau}}(\bar{r}, t, t') = \sum_i \sum_j \hat{\bar{g}}_i(\bar{r}, t, t') \, \hat{\bar{g}}_j(\bar{r}, t, t') \, \hat{\tau}^{ij}(\bar{r}, t, t') \tag{3.107b}$$

It may also be written in terms of fixed Cartesian components

$$\bar{\bar{\tau}}(\bar{r}, t, t') = \sum_i \sum_j \bar{\delta}_i \bar{\delta}_j \, \tau_{ij}(\bar{r}, t, t') \tag{3.108}$$

By using Eq. (3.99a,b), one can obtain the relations between the fixed and convected coordinates

$$\hat{\tau}_{ij}(\bar{r}, t, t') = \sum_m \sum_n \tau_{mn}(\bar{r}, t, t') \, \Delta_{mi}(\bar{r}, t, t') \, \Delta_{nj}(\bar{r}, t, t') \tag{3.109a}$$

$$\hat{\tau}^{ij}(\bar{r}, t, t') = \sum_m \sum_n E_{im}(\bar{r}, t, t') \, E_{jn}(\bar{r}, t, t') \, \tau_{mn}(\bar{r}, t, t') \tag{3.109b}$$

The right-hand sides of these equations are designated by $\bar{\bar{\tau}}^{[0]}(\bar{r}, t, t')$ and $\bar{\bar{\tau}}_{[0]}(\bar{r}, t, t')$ respectively. Both $\bar{\bar{\tau}}^{[0]}$ and $\bar{\bar{\tau}}_{[0]}$ are equal to $\bar{\bar{\tau}}(\bar{r}, t)$ at $t' = t$.

Time derivatives $(\partial/\partial t', \partial^2/\partial t'^2, \cdots)$ of $\bar{\bar{\tau}}^{[0]}$ and $\bar{\bar{\tau}}_{[0]}$ can then be formed as shown in Table 3.6. The quantities $\bar{\bar{\tau}}^{(n)}$ and $\bar{\bar{\tau}}_{(n)}$ which appear are obtained exactly as in Eq. (3.105a,b) with $\bar{\bar{\tau}}^{(0)}$ and $\bar{\bar{\tau}}_{(0)}$ both being equal to $\bar{\bar{\tau}}$.

3.5.4   Formulation of Constitutive Equations

Oldroyd (1950) established two postulates for formulating constitutive equations: (1) The relation between the stress and kinematic tensors for a fluid element should not depend on the kinematics of neighboring fluid elements (later researchers called this the "simple fluid" assumption); (2) The constitutive equation should not depend on the orientation of the fluid element in space (later called the assumption of "objectivity"). Therefore the constitutive equation can involve any relation among the convected components of the stress tensor $\hat{\tau}_{ij}$ and $\hat{\tau}^{ij}$ and the convected quantities $\hat{g}_{ij}$ and $\hat{g}^{ij}$, including derivatives or integrals of these quantities with respect to time. In terms of fixed components (which are more convenient to use, since the equations of continuity and motion are normally written in terms of fixed components), the constitutive equation can involve any of the quantities $\bar{\bar{\gamma}}^{[n]}, \bar{\bar{\gamma}}_{[n]}, \bar{\bar{\tau}}^{[n]}, \bar{\bar{\tau}}_{[n]}$ at time t' [indices in brackets], and/or the quantities $\bar{\bar{\gamma}}^{(n)}, \bar{\bar{\gamma}}_{(n)}, \bar{\bar{\tau}}^{(n)}, \bar{\bar{\tau}}_{(n)}$ at time $t' = t$ (indices in parentheses). Constitutive equations constructed in this way automatically satisfy the Oldroyd postulates given above. The [ ]-suffixed tensors depend on t and t' and normally appears only in the integrands of integrals over the variable t'.

Constitutive equations have been developed in several ways:

  a. By *empirical combinations* of the above [ ]- or ( )-suffixed tensors designed to fit available experimental data.

TABLE 3.6 The Stress Tensor and Its Time Derivatives

Convected Components at t'		Fixed Cartesian Components at t'		Fixed Cartesian Components at t
$\hat{\tau}_{ij}(\bar{\mathbf{r}}, t, t')$	$=$	$\tau_{ij}^{[0]} = \Sigma_m \Sigma_n \tau_{mn}^{(0)} \Delta_{mi} \Delta_{nj}$		$\tau_{ij}^{(0)} \equiv \tau_{ij}$
$\hat{\tau}^{ij}(\bar{\mathbf{r}}, t, t')$	$=$	$\tau_{[0]ij} = \Sigma_m \Sigma_n \mathrm{E}_{im}\mathrm{E}_{jn}\tau_{(0)mn}$	$t'=t$ $\uparrow$	$\tau_{(0)ij} \equiv \tau_{ij}$
$\partial/\partial t'$ $\downarrow$		$\partial/\partial t'$ $\downarrow$		convected $\downarrow$ differentiation
$\partial\hat{\tau}_{ij}/\partial t'$	$=$	$\tau_{ij}^{[1]} = \Sigma_m \Sigma_n \tau_{mn}^{(1)} \Delta_{mi} \Delta_{nj}$		$\tau_{ij}^{(1)} = \dfrac{D}{Dt}\tau_{ij} + \sum_p \left(\dfrac{\partial v_p}{\partial x_i}\tau_{pj} + \tau_{ip}\dfrac{\partial v_p}{\partial x_j}\right)$
$\partial\hat{\tau}^{ij}/\partial t'$	$=$	$\tau_{[1]ij} = \Sigma_m \Sigma_n \mathrm{E}_{im}\mathrm{E}_{jn}\tau_{(1)mn}$	$t'=t$ $\uparrow$	$\tau_{(1)ij} = \dfrac{D}{Dt}\tau_{ij} - \sum_p \left(\dfrac{\partial v_i}{\partial x_p}\tau_{pj} + \tau_{ip}\dfrac{\partial v_j}{\partial x_p}\right)$
$\partial/\partial t'$ $\downarrow$				convected $\downarrow$ differentiation
$\partial^2\hat{\tau}_{ij}/\partial t'^2$	$=$	$\tau_{ij}^{[2]} = \Sigma_m \Sigma_n \tau_{mn}^{(2)} \Delta_{mi} \Delta_{nj}$		$\tau_{ij}^{(2)} = \dfrac{D}{Dt}\tau_{ij}^{(1)} + \sum_p \left(\dfrac{\partial v_p}{\partial x_i}\tau_{pj}^{(1)} + \tau_{ip}^{(1)}\dfrac{\partial v_p}{\partial x_j}\right)$
$\partial^2\hat{\tau}^{ij}/\partial t'^2$	$=$	$\tau_{[2]ij} = \Sigma_m \Sigma_n \mathrm{E}_{im}\mathrm{E}_{jn}\tau_{(2)mn}$	$t'=t$ $\uparrow$	$\tau_{(2)ij} = \dfrac{D}{Dt}\tau_{(1)ij} - \sum_p \left(\dfrac{\partial v_i}{\partial x_p}\tau_{(1)pj} + \tau_{(1)ip}\dfrac{\partial v_j}{\partial x_p}\right)$
etc. $\downarrow$		etc. $\downarrow$		etc. $\downarrow$

b. By developing *ordered expansions* in some kinematic variable(s).

c. By obtaining expressions valid for some *special classes of flows*.

d. By using *molecular theory* (for polymeric liquids) or *structural theory* (for two-phase systems).

Constitutive equations are evaluated by comparing their predictions for material functions with experimental data. To facilitate this process, Table 3.7 summarizes a number of the most commonly occurring quantities for shear flows and elongational flows.

Several examples are now given of constitutive equations from each of the above categories:

Empirical Expressions. The *generalized Newtonian models* discussed in Sec. 3.3 fall into this category, since they can be written in terms of ( )-suffixed tensors: $\bar{\bar{\tau}}^{(0)} = -\eta(\dot{\gamma})\bar{\bar{\gamma}}^{(1)}$. Such models do not, of course, describe elastic effects.

The linear viscoelastic models of Sec. 3.4 are not objective since they involve ordinary time derivatives (as in Eq. 3.77) or non-[ ]-suffixed quantities in the time integrals (as in Eqs. (3.81) and (3.82)]. These models may be converted into objective models by replacing the time derivatives by the corresponding convected derivatives with ( )-suffixes in the differential constitutive equations or by using [ ]-suffixed quantities in the time integrals. In this way, *quasilinear viscoelastic models* may be generated. For example, from the linear Maxwell model of Eqs. (3.77), (3.81), or (3.82), a *quasilinear* (or *convected*) *Maxwell model* in three equivalent forms may be generated

$$\bar{\bar{\tau}}_{(0)} + \lambda_0 \bar{\bar{\tau}}_{(1)} = -\eta_0 \bar{\bar{\gamma}}_{(1)} \tag{3.110a}$$

$$\bar{\bar{\tau}}_{(0)} = -\int_{-\infty}^{t} \left\{ \frac{\eta_0}{\lambda_0} e^{-(t-t')/\lambda_0} \right\} \bar{\bar{\gamma}}_{[1]}(t, t') \, dt' \tag{3.110b}$$

$$\bar{\bar{\tau}}_{(0)} = +\int_{-\infty}^{t} \left\{ \frac{\eta_0}{\lambda_0^2} e^{-(t-t')/\lambda_0} \right\} \bar{\bar{\gamma}}_{[0]}(t, t') \, dt' \tag{3.110c}$$

The ( )-superscript tensors could have been used and the equation written as: $\bar{\bar{\tau}}^{(0)} + \lambda_0 \bar{\bar{\tau}}^{(1)} = -\eta_0 \bar{\bar{\gamma}}^{(1)}$. However, because $\bar{\bar{\tau}}_{(1)}$ and $\bar{\bar{\tau}}^{(1)}$ are not the same, this constitutive equation is not the same as Eq. (3.110a). Only comparisons with experimental data or molecular theories can determine which equation is preferable. Some real fluids behave qualitatively like Eq. (3.110), at least over a limited range of kinematics, whereas the corresponding ( )-superscripted model does not seem to correspond to any real fluids. Equation (3.110) has been used for making some viscoelastic fluid dynamics calculations. It cannot be recommended for general use, however, because it gives $\eta = $ constant, $\Psi_1 = $ constant, $\Psi_2 = 0$, $\eta^+(t, \dot{\gamma})$ monotonically increasing, and $\bar{\eta} \rightarrow \infty$ at a finite value of $\dot{\epsilon}$.

One can also construct a wide variety of *nonlinear viscoelastic models*. For example, one can add to the convected analogue of Eq. (3.78) (the Jeffreys model) an additional nonlinear term to obtain the 4-constant Oldroyd model [Bird *et al.* (1987) and Oldroyd (1958)]

$$\bar{\bar{\tau}}_{(0)} + \lambda_1 \bar{\bar{\tau}}_{(1)} + (\mu_0/2) (tr\, \bar{\bar{\tau}}_{(0)})\bar{\bar{\gamma}}_{(1)} = -\eta_0(\bar{\bar{\gamma}}_{(1)} + \lambda_2 \bar{\bar{\gamma}}_{(2)}) \tag{3.111}$$

TABLE 3.7 Continuum Mechanics Tensors

Velocity Distribution	Shear	Elongation
	$v_x = \dot\gamma(t)y,\ v_y = 0,\ v_z = 0$	$v_x = -\tfrac{1}{2}\dot\epsilon(t)x,\ v_y = -\tfrac{1}{2}\dot\epsilon(t)y,\ v_z = \dot\epsilon(t)z$
$\nabla\bar{\mathbf{v}}$	$\begin{pmatrix} 0 & 0 & 0 \\ 1 & 0 & 0 \\ 0 & 0 & 0 \end{pmatrix}\dot\gamma(t)$	$\begin{pmatrix} -1/2 & 0 & 0 \\ 0 & -1/2 & 0 \\ 0 & 0 & 1 \end{pmatrix}\dot\epsilon(t)$
$\bar{\bar{\Delta}}$	$\begin{pmatrix} 1 & \gamma & 0 \\ 0 & 1 & 0 \\ 0 & 0 & 1 \end{pmatrix}$	$\begin{pmatrix} \lambda_x^{-1} & 0 & 0 \\ 0 & \lambda_y^{-1} & 0 \\ 0 & 0 & \lambda_z^{-1} \end{pmatrix}$
$\bar{\bar{E}}$	$\begin{pmatrix} 1 & -\gamma & 0 \\ 0 & 1 & 0 \\ 0 & 0 & 1 \end{pmatrix}$	$\begin{pmatrix} \lambda_x & 0 & 0 \\ 0 & \lambda_y & 0 \\ 0 & 0 & \lambda_z \end{pmatrix}$
$\bar{\bar{\gamma}}^{[0]}$	$\begin{pmatrix} 0 & \gamma & 0 \\ \gamma & \gamma^2 & 0 \\ 0 & 0 & 0 \end{pmatrix}$	$\begin{pmatrix} \lambda_x^{-2}-1 & 0 & 0 \\ 0 & \lambda_y^{-2}-1 & 0 \\ 0 & 0 & \lambda_z^{-2}-1 \end{pmatrix}$
$\bar{\bar{\gamma}}_{[0]}$	$\begin{pmatrix} -\gamma^2 & \gamma & 0 \\ \gamma & 0 & 0 \\ 0 & 0 & 0 \end{pmatrix}$	$\begin{pmatrix} 1-\lambda_x^2 & 0 & 0 \\ 0 & 1-\lambda_y^2 & 0 \\ 0 & 0 & 1-\lambda_z^2 \end{pmatrix}$
$\bar{\bar{\gamma}}^{[1]}$	$\begin{pmatrix} 0 & 1 & 0 \\ 1 & 2\gamma & 0 \\ 0 & 0 & 0 \end{pmatrix}\dot\gamma(t')$	$\begin{pmatrix} -\lambda_x^{-2} & 0 & 0 \\ 0 & -\lambda_y^{-2} & 0 \\ 0 & 0 & 2\lambda_z^{-2} \end{pmatrix}\dot\epsilon(t')$

TABLE 3.7 (Continued)

Velocity Distribution	Shear	Elongation
	$v_x = \dot{\gamma}(t)y,\ v_y = 0,\ v_z = 0$	$v_x = -\tfrac{1}{2}\dot{\epsilon}(t)x,\ v_y = -\tfrac{1}{2}\dot{\epsilon}(t)y,\ v_z = \dot{\epsilon}(t)z$
$\bar{\bar{\gamma}}_{[1]}$	$\begin{pmatrix} -2\gamma & 1 & 0 \\ 1 & 0 & 0 \\ 0 & 0 & 0 \end{pmatrix}\dot{\gamma}(t')$	$\begin{pmatrix} -\lambda_x^2 & 0 & 0 \\ 0 & -\lambda_y^2 & 0 \\ 0 & 0 & 2\lambda_z^2 \end{pmatrix}\dot{\epsilon}(t')$
$\bar{\bar{\gamma}}^{(1)} = \bar{\bar{\gamma}}_{(1)}$	$\begin{pmatrix} 0 & 1 & 0 \\ 1 & 0 & 0 \\ 0 & 0 & 0 \end{pmatrix}\dot{\gamma}(t)$	$\begin{pmatrix} -1 & 0 & 0 \\ 0 & -1 & 0 \\ 0 & 0 & 2 \end{pmatrix}\dot{\epsilon}(t)$
$\bar{\bar{\gamma}}^{(2)}$	$\begin{pmatrix} 0 & 1 & 0 \\ 1 & 0 & 0 \\ 0 & 0 & 0 \end{pmatrix}\dfrac{\partial \dot{\gamma}}{\partial t} + \begin{pmatrix} 0 & 0 & 0 \\ 0 & 2 & 0 \\ 0 & 0 & 0 \end{pmatrix}\dot{\gamma}^2$	$\begin{pmatrix} -1 & 0 & 0 \\ 0 & -1 & 0 \\ 0 & 0 & 2 \end{pmatrix}\dfrac{\partial \dot{\epsilon}}{\partial t} + \begin{pmatrix} 1 & 0 & 0 \\ 0 & 1 & 0 \\ 0 & 0 & 4 \end{pmatrix}\dot{\epsilon}^2$
$\bar{\bar{\gamma}}_{(2)}$	$\begin{pmatrix} 0 & 1 & 0 \\ 1 & 0 & 0 \\ 0 & 0 & 0 \end{pmatrix}\dfrac{\partial \dot{\gamma}}{\partial t} - \begin{pmatrix} 2 & 0 & 0 \\ 0 & 0 & 0 \\ 0 & 0 & 0 \end{pmatrix}\dot{\gamma}^2$	$\begin{pmatrix} -1 & 0 & 0 \\ 0 & -1 & 0 \\ 0 & 0 & 2 \end{pmatrix}\dfrac{\partial \dot{\epsilon}}{\partial t} - \begin{pmatrix} 1 & 0 & 0 \\ 0 & 1 & 0 \\ 0 & 0 & 4 \end{pmatrix}\dot{\epsilon}^2$
$\bar{\bar{\tau}}^{(0)} = \bar{\bar{\tau}}_{(0)}$	$\begin{pmatrix} \tau_{xx} & \tau_{xy} & 0 \\ \tau_{yx} & \tau_{yy} & 0 \\ 0 & 0 & \tau_{zz} \end{pmatrix}$ with $[\tau_{xy} = \tau_{yx}]$	$\begin{pmatrix} \tau_{xx} & 0 & 0 \\ 0 & \tau_{yy} & 0 \\ 0 & 0 & \tau_{zz} \end{pmatrix}$ with $[\tau_{xx} = \tau_{yy}]$
$\bar{\bar{\tau}}^{(1)}$	$\dfrac{\partial}{\partial t}\begin{pmatrix} \tau_{xx} & \tau_{xy} & 0 \\ \tau_{yx} & \tau_{yy} & 0 \\ 0 & 0 & \tau_{zz} \end{pmatrix} + \begin{pmatrix} 0 & \tau_{xx} & 0 \\ \tau_{xx} & 2\tau_{xy} & 0 \\ 0 & 0 & 0 \end{pmatrix}\dot{\gamma}(t)$	$\dfrac{\partial}{\partial t}\begin{pmatrix} \tau_{xx} & 0 & 0 \\ 0 & \tau_{yy} & 0 \\ 0 & 0 & \tau_{zz} \end{pmatrix} + \begin{pmatrix} -\tau_{xx} & 0 & 0 \\ 0 & -\tau_{yy} & 0 \\ 0 & 0 & 2\tau_{zz} \end{pmatrix}\dot{\epsilon}(t)$

$$\bar{\bar{\tau}}_{(1)}$$

$$\frac{\partial}{\partial t}\begin{pmatrix} \tau_{xx} & \tau_{xy} & 0 \\ \tau_{yx} & \tau_{yy} & 0 \\ 0 & 0 & \tau_{zz}\end{pmatrix} - \begin{pmatrix} 2\tau_{yx} & \tau_{yy} & 0 \\ \tau_{yy} & 0 & 0 \\ 0 & 0 & 0\end{pmatrix}\dot{\gamma}(t)$$

Displacement Functions

$$x = x' - y'\gamma(t, t')$$

$$y = y'$$

$$z = z'$$

where $\quad \gamma(t, t') = \int_t^{t'} \dot{\gamma}(t'')\, dt''$

$$\frac{\partial}{\partial t}\begin{pmatrix} \tau_{xx} & 0 & 0 \\ 0 & \tau_{yy} & 0 \\ 0 & 0 & \tau_{zz}\end{pmatrix} - \begin{pmatrix} -\tau_{xx} & 0 & 0 \\ 0 & -\tau_{yy} & 0 \\ 0 & 0 & 2\tau_{zz}\end{pmatrix}\dot{\epsilon}(t)$$

$$x = \lambda_x x'; \quad \lambda_x = \exp[\epsilon(t, t')/2]$$

$$y = \lambda_y y'; \quad \lambda_y = \exp[\epsilon(t, t')/2]$$

$$z = \lambda_z z'; \quad \lambda_z = \exp[-\epsilon(t, t')]$$

where $\quad \epsilon(t, t') = \int_t^{t'} \dot{\epsilon}(t'')\, dt''$

This contains a zero-shear-rate viscosity, η_0, and three time constants λ_1, λ_2, and μ_0. This differential constitutive equation allows for variation of η and Ψ_1 with $\dot{\gamma}$.

Similarly, nonlinear integral constitutive equations can be invented. One single-integral equation that has enjoyed considerable success at describing the properties of many materials is the K-BKZ equation

$$\bar{\bar{\tau}}_{(0)} = \int_{-\infty}^{t} [(\partial V/\partial I_1)\bar{\bar{\gamma}}_{[0]}(t, t') + (\partial V/\partial I_2)\bar{\bar{\gamma}}^{[0]}(t, t')] \, dt' \tag{3.112}$$

In this equation, $V(t - t', I_1, I_2)$ is a material-dependent function, and I_1 and I_2 are the scalar invariants of the Finger strain for deformation of an incompressible material (i.e., $I_1 = tr \, \bar{\bar{\mathbf{B}}}$, $I_2 = tr \, \bar{\bar{\mathbf{B}}}^{-1}$). Most of the work that has been done with the K-BKZ equation has been with a factorized form that is obtained by assuming that the scalar function V may be written as the product of a time-dependent function and a strain-dependent function. The time dependent function is the linear viscoelastic memory function [$M(t - t')$ in Eq. (3.84)], and the strain dependent function, $W(I_1, I_2)$ is called the "potential function." When $W(I_1, I_2) = I_1$ the resulting constitutive equation is the convected analogue of Eq. (3.84), and it is known as Lodge's rubberlike liquid [Lodge (1964)].

Equations (3.110)–(3.112) are just a few of the many empirical constitutive relations that have been proposed [Larson (1988)]. A few others are shown in Table 3.8. Solving industrially important flow problems with these expressions for τ is extremely difficult, and numerical methods must be used [Tanner (1985) and Crochet *et al.* (1984)].

Ordered Expansions. The stress tensor can be expanded in various *memory integral expansions* [Bird *et al.* (1987), Lodge (1964), and Rivlin and Ericksen (1955)], one example of which is

$$\bar{\bar{\tau}}_{(0)} = -\int_{-\infty}^{t} G(t - t')\bar{\bar{\gamma}}_{[1]}(t, t') \, dt'$$

$$- (1/2) \int_{-\infty}^{t} \int_{-\infty}^{t} H(t - t', t - t'') \{\bar{\bar{\gamma}}_{[1]}(t, t') \cdot \bar{\bar{\gamma}}_{[1]}(t, t'')$$

$$+ \bar{\bar{\gamma}}_{[1]}(t, t'') \cdot \bar{\bar{\gamma}}_{[1]}(t, t')\} \, dt''dt' - \cdots \tag{3.113}$$

The first term of this expansion includes the general linear viscoelastic fluid in the form of Eq. (3.110b); the first term is also equivalent to the Lodge rubberlike liquid (i.e., Eq. (3.112 with $V = M(t - t')I_1$. The memory-integral expansion provides a systematic description of deviations from linear viscoelastic behavior, but it is not useful for solving fluid dynamics problems, since the higher order kernels $H(t - t', t - t'')$, etc., have never been determined experimentally.

If in Eq. (3.113) $\bar{\bar{\gamma}}_{[1]}(t, t')$ is expanded as a function of t' in a Taylor series about $t' = t$, then one gets the *retarded motion expansion*

$$\bar{\bar{\tau}}_{(0)} = -b_1\bar{\bar{\gamma}}_{(1)} + b_2\bar{\bar{\gamma}}_{(2)} - b_{11}\{\bar{\bar{\gamma}}_{(1)} \cdot \bar{\bar{\gamma}}_{(1)}\}$$

$$-b_3\bar{\bar{\gamma}}_{(3)} - b_{12}\{\bar{\bar{\gamma}}_{(1)} \cdot \bar{\bar{\gamma}}_{(2)} + \bar{\bar{\gamma}}_{(2)} \cdot \bar{\bar{\gamma}}_{(1)}\}$$

$$-b_{1:11}(\bar{\bar{\gamma}}_{(1)}:\bar{\bar{\gamma}}_{(1)}) \, \bar{\bar{\gamma}}_{(1)} + \cdots \tag{3.114}$$

TABLE 3.8 Examples of Constitutive Equations

Name	Reference	Constants or Functions	Constitutive Equation	Comments
			Differential Models	
(Upper) Convected Maxwell	a	η_0, λ_0	$\bar{\bar{\tau}}_{(0)} + \lambda_0 \bar{\bar{\tau}}_{(1)} = -\eta_0 \bar{\bar{\gamma}}_{(1)}$	$\eta, \Psi_1 = $ constant; $\Psi_2 = 0$; $\bar{\eta} \to \infty$ at $\dot{\epsilon} = (1/2\lambda)$ (A special case of Oldroyd's 4 constant model, Eq. (3.111)
White–Metzner	b	$\eta(\dot{\gamma}), G$	$\bar{\bar{\tau}}_{(0)} + (\eta/G)\bar{\bar{\tau}}_{(1)} = -\eta\bar{\bar{\gamma}}_{(1)}$	Can fit $\eta(\dot{\gamma})$ exactly; $\Psi_1 = 2\eta^2/G$; $\Psi_2 = 0$; $\bar{\eta} \to \infty$ at finite $\dot{\epsilon}$
Giesekus	c	$\eta_0, \lambda_0, \lambda_1$	$\bar{\bar{\tau}}_{(0)} + \lambda_0\bar{\bar{\tau}}_{(1)} - (\lambda_1/\eta_0)\{\bar{\bar{\tau}}_{(0)} \cdot \bar{\bar{\tau}}_{(0)}\} = -\eta_0\bar{\bar{\gamma}}_{(1)}$	η, Ψ_1, Ψ_2 depend on $\dot{\gamma}$; $\Psi_{2,0}/\Psi_{1,0} = -\lambda_1/2\lambda_0$; $\bar{\eta}$ is finite
Co	d	$\eta(\dot{\gamma}), \lambda(\dot{\gamma})$	$\bar{\bar{\tau}}_{(0)} + \lambda(\bar{\bar{\tau}}_{(1)} + \frac{1}{2}\{\bar{\bar{\gamma}}_{(1)} \cdot \bar{\bar{\tau}}_{(0)} + \bar{\bar{\tau}}_{(0)} \cdot \bar{\bar{\gamma}}_{(1)}\})$ $= -\eta\,(1 + \lambda^2\dot{\gamma}^2)\,\bar{\bar{\gamma}}_{(1)}$	Can fit $\eta(\dot{\gamma})$ exactly; $\Psi_1 = 2\lambda\eta$; $\Psi_2 = -(1/2)\Psi_1$; $\bar{\eta}$ is finite
Oldroyd 8-constant	e	$\eta_0, \lambda_1, \lambda_2, \mu_0,$ $\mu_1, \mu_2, \nu_1, \nu_2$	$\bar{\bar{\tau}}_{(0)} + \lambda_1\bar{\bar{\tau}}_{(1)} + \frac{1}{2}\mu_0(\text{tr }\bar{\bar{\tau}}_{(0)})$ $+ \frac{1}{2}\mu_1\{\bar{\bar{\gamma}}_{(1)} \cdot \bar{\bar{\tau}}_{(0)} + \bar{\bar{\tau}}_{(0)} \cdot \bar{\bar{\gamma}}_{(1)}\} + \frac{1}{2}\nu_1(\bar{\bar{\tau}}_{(0)}:\bar{\bar{\gamma}}_{(1)})\bar{\bar{\delta}}$ $= -\eta_0[\bar{\bar{\gamma}}_{(1)} + \lambda_2\bar{\bar{\gamma}}_{(2)} + \mu_2\{\bar{\bar{\gamma}}_{(1)} \cdot \bar{\bar{\gamma}}_{(1)}\}$ $+ \frac{1}{2}\nu_2(\bar{\bar{\gamma}}_{(1)}:\bar{\bar{\gamma}}_{(1)})\bar{\bar{\delta}}]$	Includes the convected Maxwell model, the convected Jeffreys model, the 4-constant Oldroyd model, and the 2nd order fluid

TABLE 3.8 (*Continued*)

Name	Reference	Constants or Functions	Constitutive Equation	Comments
Hand	f	α_i, β_i	$\begin{aligned}\bar{\bar{\tau}}_{(0)} &= \beta_0\bar{\bar{\delta}} + \beta_1\bar{\bar{\alpha}} + \beta_2\bar{\bar{\gamma}}_{(1)} + \beta_3\bar{\bar{\alpha}}^2 \\ &+ \beta_4\bar{\bar{\gamma}}_{(1)}^2 + \beta_5\{\bar{\bar{\alpha}}\cdot\bar{\bar{\gamma}}_{(1)} + \bar{\bar{\gamma}}_{(1)}\cdot\bar{\bar{\alpha}}\} \\ &+ \beta_6\{\bar{\bar{\alpha}}^2\cdot\bar{\bar{\gamma}}_{(1)} + \bar{\bar{\gamma}}_{(1)}\cdot\bar{\bar{\alpha}}^2\} \\ &+ \beta_7\{\bar{\bar{\alpha}}\cdot\bar{\bar{\gamma}}_{(1)}^2 + \bar{\bar{\gamma}}_{(1)}^2\cdot\bar{\bar{\alpha}}\} \\ &+ \beta_8\{\bar{\bar{\alpha}}^2\cdot\bar{\bar{\gamma}}_{(1)}^2 + \bar{\bar{\gamma}}_{(1)}^2\cdot\bar{\bar{\alpha}}^2\}\end{aligned}$ $\begin{aligned}\frac{\mathcal{D}}{\mathcal{D}t}\bar{\bar{\alpha}} &= \alpha_0\bar{\bar{\delta}} + \alpha_1\bar{\bar{\alpha}} + \alpha_2\bar{\bar{\gamma}}_{(1)} + \alpha_3\bar{\bar{\alpha}}^2 \\ &+ \alpha_4\bar{\bar{\gamma}}_{(1)}^2 + \alpha_5\{\bar{\bar{\alpha}}\cdot\bar{\bar{\gamma}}_{(1)} + \bar{\bar{\gamma}}_{(1)}\cdot\bar{\bar{\alpha}}\} \\ &+ \alpha_6\{\bar{\bar{\alpha}}^2\cdot\bar{\bar{\gamma}}_{(1)} + \bar{\bar{\gamma}}_{(1)}\cdot\bar{\bar{\alpha}}^2\} \\ &+ \alpha_7\{\bar{\bar{\alpha}}\cdot\bar{\bar{\gamma}}_{(1)}^2 + \bar{\bar{\gamma}}_{(1)}^2\cdot\bar{\bar{\alpha}}\} \\ &+ \alpha_8\{\bar{\bar{\alpha}}^2\cdot\bar{\bar{\gamma}}_{(1)}^2 + \bar{\bar{\gamma}}_{(1)}^2\cdot\bar{\bar{\alpha}}^2\}\end{aligned}$	$\mathcal{D}/\mathcal{D}t$ is defined in Eq. (3.106)
Oldroyd–Walters–Fredrickson	g	$G(t-t')$	$\bar{\bar{\tau}}_{(0)} = -\int_{-\infty}^{t} G\,\bar{\bar{\gamma}}_{[1]}(t, t')\,dt'$	This can be integrated by parts to give the Lodge elastic liquid. $G(t - t')$ is the relaxation modulus of linear viscoelasticity.
Lodge's rubberlike liquid	h	$M(t-t')$	$\bar{\bar{\tau}}_{(0)} = +\int_{-\infty}^{t} M\,\bar{\bar{\gamma}}_{[0]}(t, t')\,dt'$	Same as (upper) convected Maxwell model, if $M(t - t') = (\eta_0/\lambda_0^2)\exp\left(-(t - t')/\lambda_0\right)$

Wagner	i	$M(t - t')$, $h(I_1, I_2)$	$\bar{\bar{\tau}}_{(0)} = +\int_{-\infty}^{t} M h \, \bar{\bar{\gamma}}_{[0]}(t, t') \, dt'$	The "damping function" h depends on two invariants of the Finger strain tensor at time t'
Kaye–BKZ	j	$M_1(t - t', I_1, I_2)$; $M_2(t - t', I_1, I_2)$	$\bar{\bar{\tau}}_{(0)} = +\int_{-\infty}^{t} [M_1 \bar{\bar{\gamma}}_{[0]}(t, t') + M_2\{\bar{\bar{\gamma}}_{[0]}(t, t') \cdot \bar{\bar{\gamma}}_{[0]}(t, t')\}] \, dt'$	

a. Oldroyd, J. G., *Proc. Roy. Soc.*, Vol. A200, pp. 45–63, 1950.
b. White, J. L. and Metzner, A. B., *J. Appl. Polym. Sci.*, Vol. 7, pp. 1867–1889, 1963.
c. Giesekus, H., *J. Non-Newtonian Fl. Mech.*, Vol. 11, pp. 69–109, 1982.
d. Co, A. and Stewart, W. E., *AIChE Journal*, Vol. 28, pp. 644–655, 1982.
e. Oldroyd, J. G., *Proc. Roy. Soc.*, Vol. A245, pp. 278–297, 1958.
f. Hand, G. L., *J. Fluid Mech.*, Vol. 13, pp. 33–46, 1962.
g. Oldroyd, J. G. (1950, loc. cit.); Walters, K., *Quart J. Mech. Appl. Math.*, Vol. 13, pp. 444–461, 1960; Fredrickson, A. G., *Chem. Engr. Sci.*, Vol. 17, pp. 155–166, 1962.
h. Lodge, A. S., *Trans. Faraday Soc.*, Vol. 52, pp. 120–130, 1956.
i. Wagner, M. H., *Rheol. Acta*, Vol. 18, pp. 13–50, 1979.
j. Kaye, A., College of Aeronautics, Cranfield, Note No. 134, 1962; Bernstein, B., Kearsley, E. A., and Zapas, L. J., *Trans. Soc. Rheol.*, Vol. 7, pp. 391–410, 1963.

in which the b's are constants (they involve integrals over the kernel functions G, H, etc., in Eq. (3.113). If the series is truncated after the first term on the right side, the Newtonian fluid results; truncation after the b_{11}-terms gives the *second-order fluid*; truncation after the $b_{1:11}$ term gives the *third-order fluid*, etc. These lower-order fluids are useful only for flows with small velocity gradients and with small time variations.

Special Classes of Flows. There are several classes of flows for which the infinite series in Eq. (3.114) collapses and useful expressions are obtained. The most important of these is the class of *steady state shear flows* for which Eq. (3.114) simplifies to

$$\bar{\bar{\tau}}_{(0)} = -\eta\bar{\bar{\gamma}}_{(1)} + \tfrac{1}{2}\Psi_1\bar{\bar{\gamma}}_{(2)} - \Psi_2\{\bar{\bar{\gamma}}_{(1)} \cdot \bar{\bar{\gamma}}_{(1)}\} \qquad (3.115)$$

in which η, Ψ_1, and Ψ_2 are the "viscometric functions" which are defined in Eqs. (3.14)–(3.16) and which can be measured in several different kinds of rheometers. This equation, known as the *Criminale–Ericksen–Filbey (CEF) equation* [Bird *et al.* (1987)], is the most general equation available for steady-state shear flows and can be applied to any of the flow systems listed at the beginning of Sec. 3.3.1. If the Ψ_1 and Ψ_2 terms in Eq. (3.115) are omitted, the generalized Newtonian fluid model is recovered. Equation (3.115) is particularly useful for the analysis of the steady-state operation of shear-flow rheometers.

Molecular or Structural Theory Expressions. In this approach to deriving constitutive equations, one begins by proposing a model for the entities making up a fluid. For polymer solutions the polymer molecules are generally modeled as mechanical assemblages of "beads," "rods," and "springs" that can mimic the internal motions of the molecules; the solvent is generally regarded as a continuum that provides a hydrodynamic drag on the beads and also Brownian motion forces to simulate the thermal motion. For suspensions one specifies the concentration and shapes of the suspended particles and the viscosity of the suspending medium; for emulsions one specifies the viscosities of the two fluid phases as well as the interfacial tension. The object is then to derive a constitutive equation containing some constants or functions that depend on the microscopic structural parameters. In this way the rheological behavior of the fluid is related to the mechanical properties of the constituent elements (polymer molecules, suspended rigid particles, suspended deformable droplets, suspended compressible gas bubbles, etc.). Examples are given in Sec. 3.6.

3.5.5 Material Functions from Constitutive Equations

In order to show how to use Table 3.7, some of the material functions for the 4-constant Oldroyd model of Eq. (3.111) will be obtained. Table 3.7 is used to translate this equation into matrix form for *steady-state shear flow*

$$\begin{pmatrix} \tau_{xx} & \tau_{yx} & 0 \\ \tau_{yx} & \tau_{yy} & 0 \\ 0 & 0 & \tau_{zz} \end{pmatrix} - \lambda_1\dot{\gamma}\begin{pmatrix} 2\tau_{yx} & \tau_{yy} & 0 \\ \tau_{yy} & 0 & 0 \\ 0 & 0 & 0 \end{pmatrix} + \tfrac{1}{2}\mu_0(\tau_{xx} + \tau_{yy} + \tau_{zz})\begin{pmatrix} 1 & 0 & 0 \\ 0 & 1 & 0 \\ 0 & 0 & 1 \end{pmatrix}\dot{\gamma}$$

$$= -\eta_0\dot{\gamma}\begin{pmatrix} 1 & 0 & 0 \\ 0 & 1 & 0 \\ 0 & 0 & 1 \end{pmatrix} + \eta_0\lambda_2\dot{\gamma}^2\begin{pmatrix} 2 & 0 & 0 \\ 0 & 0 & 0 \\ 0 & 0 & 0 \end{pmatrix} \qquad (3.118)$$

From this, one can obtain expressions for the individual stress components

$$\tau_{yx} - \lambda_1\dot{\gamma}\tau_{yy} + \tfrac{1}{2}\mu_0(\tau_{xx} + \tau_{yy} + \tau_{zz})\dot{\gamma} = -\eta_0\dot{\gamma} \qquad (3.119a)$$

$$\tau_{xx} - 2\lambda_1\dot{\gamma}\tau_{yx} = 2\eta_0\lambda_2\dot{\gamma}^2 \qquad (3.119b)$$

$$\tau_{yy} = 0 \qquad (3.119c)$$

$$\tau_{zz} = 0 \qquad (3.119d)$$

By solving the first two equations simultaneously, one obtains

$$\tau_{yx} = -\eta_0\dot{\gamma}\left(\frac{1 + \lambda_2\mu_0\dot{\gamma}^2}{1 + \lambda_1\mu_0\dot{\gamma}^2}\right) \qquad (3.120a)$$

$$\tau_{xx} = -2\eta_0\lambda_1\dot{\gamma}^2\left(\frac{1 + \lambda_2\mu_0\dot{\gamma}^2}{1 + \lambda_1\mu_0\dot{\gamma}^2} - \frac{\lambda_2}{\lambda_1}\right) \qquad (3.120b)$$

When these results are compared with Eqs. (3.14)–(3.16) one gets for the material functions

$$\eta(\dot{\gamma}) = \eta_0\frac{1 + \lambda_2\mu_0\dot{\gamma}^2}{1 + \lambda_1\mu_0\dot{\gamma}^2} \qquad (3.121a)$$

$$\Psi_1(\dot{\gamma}) = 2\eta_0\frac{\lambda_1 - \lambda_2}{1 + \lambda_1\mu_0\dot{\gamma}^2} \qquad (3.121b)$$

$$\Psi_2(\dot{\gamma}) = 0 \qquad (3.121c)$$

Therefore one obtains shear-rate dependent η and Ψ_1 functions, which are qualitatively correct; they are not, however, of the proper functional form to fit experimental data over a wide range of $\dot{\gamma}$. Note that to describe a monotonically decreasing $\eta(\dot{\gamma})$, one has to require that $\lambda_2 < \lambda_1$.

For *small-amplitude oscillatory shearing motion* a similar treatment yields

$$\eta'(\omega) = \eta_0\frac{1 + \lambda_1\lambda_2\omega^2}{1 + \lambda_1^2\omega^2} \qquad (3.122a)$$

$$\eta''(\omega) = \eta_0\frac{(\lambda_1 - \lambda_2)\omega}{1 + \lambda_1^2\omega^2} \qquad (3.122b)$$

Here again for $\lambda_2 < \lambda_1$ the shapes of the predicted curves are qualitatively, but not quantitatively, correct. In order for the η curve to lie above the η' curve, and for Ψ_1 to lie above $2\eta''/\omega$ (as the data in Fig. 3.3 suggest), the inequality, $\lambda_1 > \mu_0$, is required in addition. For *steady-state elongational flow*, use of the right-hand column of Table 3.7 gives

$$\bar{\eta}(\dot{\epsilon}) = 3\eta_0 \frac{1 - \lambda_2\dot{\epsilon} + (3\mu_0 - 2\lambda_1)\lambda_2\dot{\epsilon}^2}{1 - \lambda_1\dot{\epsilon} + (3\mu_0 - 2\lambda_1)\lambda_1\dot{\epsilon}^2} \tag{3.123}$$

The requirement that $\lambda_2 < \lambda_1$ guarantees that the slope of $\bar{\eta}$ vs. $\dot{\epsilon}$ is positive at $\dot{\epsilon} = 0$, which is in qualitative agreement with experiment. However, $\bar{\eta}$ will go to ∞ for some finite value of $\dot{\epsilon}$, a result that is counterintuitive.

This example gives a glimpse of what one has to do to assess the merits of a proposed constitutive equation. No empirically proposed equation can be expected to describe all the rheological material functions for one fluid.

3.5.6   Solution of Flow Problems

In Table 3.9 a summary is given of some of the solutions of flow problems which are now available for nonlinear viscoelastic models [Bird *et al.* (1987), Schowalter (1978), Astarita and Marrucci (1974), and Tanner (1985)]. For the mathematical aspects of viscoelastic fluid mechanics, see the book by Joseph (1990). For stochastic techniques, the treatise of Öttinger (1995) should be consulted.

As an illustration of solving a fluid dynamics problem, the creeping flow of a second-order fluid into a very thin slit is considered (see Fig. 3.15); the volume flow rate through the slit is Q, and the width of the slit in the z-direction is W. For the Newtonian fluid, with viscosity μ, the solution to this problem is known

$$v_r = -(2q/\pi r) \cos^2\theta; \quad v_\theta = 0; \quad v_z = 0 \tag{3.124a,b,c}$$

where $q = Q/W$ and p_∞ is the pressure at $r = \infty$. For plane creeping flows it is known that any Newtonian velocity field with given boundary conditions is also a solution to the second-order fluid with the same boundary conditions (the *Giesekus–Tanner theorem*). Hence Eq. (3.124) gives the flow field for the second-order fluid. The pressure distribution, p, for the second-order fluid may be obtained from that for the Newtonian fluid, p_N, by

$$p = p_N - \frac{b_2}{b_1} \frac{D}{Dt} p_N + \left(b_{11} + \frac{1}{2} b_2\right) \gamma_{(1)}^2 \tag{3.125}$$

in which $\gamma_{(1)} = \sqrt{\frac{1}{2}(\gamma_{(1)} : \gamma_{(1)})}$ and in which the μ appearing in p_N is replaced by b_1 (the *Giesekus–Tanner–Pipkin theorem*). Hence for the second order flow into the slit (for which $(D/Dt)p_N$ is $v_r(\partial p_N/\partial r)$) is

$$p = p_\infty - b_1(2q/\pi r^2) \cos 2\theta$$

$$+ b_2 (8q^2/\pi^2 r^4) \cos^2\theta \cos 2\theta$$

$$+ (b_{11} + \tfrac{1}{2} b_2) \cos^2\theta \tag{3.126}$$

TABLE 3.9 Summary of Solutions to Non-Newtonian Fluid Dynamics Problems to 1986*

	a	b	c	d	e	f	g	h	i	j	k	l	m	n	p
Constitutive Equation → Flow system ↓	Power Law	Other Generalized Newtonian Fluid	Second-Order Fluid	Third-and higher order fluids	Convected Maxwell, White–Metzner, and Denn	Oldroyd	Lodge's rubberlike liquid	Goddard–Miller	Criminale–Eriksen–Filbey	Kaye–BKZ	Bird–Carreau	Wagner	Johnson–Segalman, Phan–Tien–Tanner	Tanner; Bird–Dotson–Johnson	Giesekus
1. Pulsatile flow in tubes		5	80	119		7	53								78.1
2. Flow in corrugated tubes	46.12				78.4	50 28									
3. Flow in curved tubes				7 47	46.13		107 110								
4. Flow in rotating tubes			117			42	52								
5. Flow in tubes of non-circular cross section	3 87 123 125	74	81	84 85					29						
6. Journal bearing	102 31	46	22 9.1	22	9.1	22	34		78.2						9.1

TABLE 3.9 *(Continued)*

Flow system	a Power Law	b Other Generalized Newtonian Fluid	c Second-Order Fluid	d Third-and higher order fluids	e Convected Maxwell, White-Metzner, and Denn	f Oldroyd	g Lodge's rubberlike liquid	h Goddard-Miller	i Criminale-Ericksen-Filbey	j Kaye-BKZ	k Bird-Carreau	l Wagner	m Johnson-Segalman, Phan-Tien-Tanner	n Tanner; Bird-Dotson-Johnson	p Giesekus
7. Secondary flow in disk-and-cylinder systems				40					77		64				
8. Secondary flow in cone-and-plate systems			38 41	120			75								
9. Flow near a rotating sphere or drop			108	37 122											
10. Flow near a translating sphere	93 112 121	45 93 127	14	37 118	116	68 111.1									
11. Flow in a converging or diverging tube	20 104	78 97 98	2	88	62	56									

278

TABLE 3.9 (*Continued*)

Flow system	a Power Law	b Other Generalized Newtonian Fluid	c Second-Order Fluid	d Third-and higher order fluids	e Convected Maxwell, White-Metzner, and Denn	f Oldroyd	g Lodge's rubberlike liquid	h Goddard-Miller	i Criminale-Ericksen-Filbey	j Kaye-BKZ	k Bird-Carreau	l Wagner	m Johnson-Segalman, Phan-Tien-Tanner	n Tanner; Bird-Dotson-Johnson	p Giesekus
22. Flow around a cylinder					116										
23. Unsteady motion of a sphere						60							96.1		
24. Centrifugal pump						113									
25. Exit from a tube to a tube		17			24								23.1 59.1		
26. Exit from a tube or slit; die swell	46.11		126		20.2 20.3 19.2 13.1 14.1	20.1				69.2			13.2 69.1		
27. Hole pressure error			59 106 115.1		46.2				44						

No.					
28.	Spinning a threadline	70	25	15	
			78.3		
29.	Withdrawal of flat plates from fluids	43			
		95			
		100			
30.	Flow near a re-entrant corner			19.1	
31.	Flow into a slit from a semi-infinite region		46.1		
32.	Eccentric annuli	58.1			

*For the period after 1986 the primary repository for non-Newtonian fluid dynamics problems is the *Journal of Non-Newtonian Fluid Mechanics*.

The numbers marked with an * after each reference refer to the entries in Table 3.9.

1. Acrivos, A., Shah, M. J., and Peterson, E. E., *AIChE J.*, Vol. 6, pp. 312–317, 1960, *15a.
2. Adams, E. B., Whitehead, J. C., and Bogue, D. C., *AIChE J.* Vol. 11, pp. 1026–1032, 1965, *11c.
3. Arai, T. and Toyoda, H., *Proc. Int. Congr. Rheol.*, 5th, Vol. 4, pp. 461–470, 1971, *5a.
4. Balmer, R. T., *Trans. Soc. Rheol.*, Vol. 16, pp. 277–293, 1972, *26h.
5. Barnes, H. A., Townsend, P., and Walters, K., *Nature*, Vol. 224, pp. 585–587, 1969, *1b.
6. Barnes, H. A., Townsend, P., and Walters, K., *Rheol. Acta*, Vol. 10, pp. 517–525, 1971, *1f.
7. Barnes, H. A. and Walters, K., *Proc. Roy. Soc.*, Vol. A314, pp. 85–109, 1969, *3d.
8. Beard, D. W., Davies, M. H., and Walters, K., *Proc. Camb. Phil. Soc.*, Vol. 60, pp. 667–674, 1964, *15c.
9.1. Beris, A. N., Armstrong, R. C., and Brown, R. A., *J. Non-Newtonian Fluid Mech.*, Vol. 16, pp. 141–172, 1984; Vol. 19, pp. 323–348, 1985; Vol. 22, pp. 129–167, 1987, *6c.f.p.
10. Berger, J. L. and Gogos, C. G., *Polymer Eng. Sci*, Vol. 13, pp. 102–112, 1973, *12a.
11. Bird, R. B., *AIChE J.*, Vol. 5, p. 565, 1959, *17a.
12. Bird, R. B., Stewart, W. E., and Lightfoot, E. N., *Transport Phenomena*, John Wiley & Sons, New York, 1960, *17a, *21a.
13. Bird, R. B. and Turian, R. M., *I.E.C. Fund.*, Vol. 3, p. 87, 1964, *19a.
13.1. Bush, M. B., Milthorpe, J. F., and Tanner, R. I., *J. Non-Newtonian Fluid Mech.*, Vol. 16, pp. 37–51, 1984, *26f.
13.2. Bush, M. B., Tanner, R. I., and Phan-Thien, N., *J. Non-Newtonian Fluid Mech.*, Vol. 18, pp. 143–162, 1985, *26m.
14. Caswell, B. and Schwarz, W. H., *J. Fluid Mech.*, Vol. 13, pp. 417–426, 1962, *10c.
14.1. Caswell B. and Viriyayuthakorn, M., *J. Non-Newtonian Fluid Mech.*, Vol. 12, pp. 13–30, 1983, *26f.

TABLE 3.9 *(Continued)*

15. Chang, H. and Lodge, A. S., *Rheol. Acta.*, Vol. 10, pp. 448–449, 1971, *28g.
16. Chan Man Fong, C. F. and Walters, K., *J de Mécanique*, Vol. 4, pp. 439–453, 1965, *16g.
17. Chen, S. S., Fan, L. T., and Hwang, C. L., *AIChE J.*, Vol. 16, pp. 293–299, 1970, *25b.
18. Christopher, R. H. and Middleman, S., *I.E.C. Fund.*, Vol. 4, pp. 422–426, 1965, *21a.
19. Co, A. and Bird, R. B., Rheology Research Center Report No. 31, University of Wisconsin, 1974, *13h.
19.1. Cochrane, T., Walters, K., and Webster, M. F., *J. Non-Newtonian Fluid Mech.*, Vol. 10, pp. 95–114, 1982, *30f.
20. Collins, M. and Schowalter, W. R., *AIChE J.*, Vol. 9, pp. 98–102, 1963, *11a, *15a.
20.1. Crochet, M. J. and Keunings, R., *J. Non-Newtonian Fluid Mech.*, Vol. 10, pp. 339–356, 1982, *26f.
20.2. Crochet, M. J. and Keunings, R., *J. Non-Newtonian Fluid Mech.*, Vol. 10, pp. 85–94, 1982, *26e.
20.3. Crochet, M. J. and Keunings, R., *J. Non-Newtonian Fluid Mech.*, Vol. 7, pp. 199–212, 1980, *26e.
21. Datta, J. M. and Walters, K., *Rheology of Lubricants* (T. C. Davenport, Ed.) John Wiley & Sons, New York, pp. 65–80, 1973, *6c,d,f.
23. Davies, M. H., *Z.A.M.P.*, Vol. 17, pp. 189–191, 1966, *15c.
23.1. Debbaut, B. and Crochet, M. J., *J. Non-Newtonian Fluid Mech.*, Vol. 20, pp. 173–185, 1986, *25m.
24. Denn, M. M., *Chem. Eng. Sci.*, Vol. 22, pp. 395–405, 1967, *25e.
25. Denn, M. M., *Stability of Reaction and Transport Processes*, Prentice-Hall, Englewood Cliffs, NJ, 1975, *28e.
26. Denn, M. M. and Porteous, K. C., *Chem. Eng. J.*, Vol. 2, pp. 280–285, 1971, *17e.
27. Denn, M. M., and Roisman, J. J., *AIChE J.*, Vol. 15, pp. 454–459, 1969, *16c.
28. Dodson, A. G., Townsend, P., and Walters, K., *Rheol. Acta*, Vol. 10, pp. 508–516, 1971, *2f.
29. Dodson, A. G., Townsend, P., and Walters, K., *Computers and Fluids*, Vol. 2, pp. 317–338, 1974, *5i.
30. Duda, J. L. and Vrentas, J. S., *Trans. Soc. Rheol.*, Vol. 17, pp. 89–108, 1973, *25b.
31. Ehrlich, R. and Slattery, J. C., *I.E.C. Fund.*, Vol. 7, pp. 239–246, 1968, *6a.
32. Etter, I. and Schowalter, W. R., *Trans. Soc. Rheol.*, Vol. 9, pp. 351–369, 1965, *1f.
33. Feinberg, M. R. and Schowalter, W. R., *I.E.C. Fund.*, Vol. 8, pp. 332–338, 1969, *16g.
34. Fix, G. J. and Paslay, P. R., *J. Appl. Mech.*, Vol. 34, pp. 579–782, 1967, *6h.
35. Fogler, H. S. and Goddard, J. D., *Phys. Fluids*, Vol. 13, pp. 1135–1141, 1970, *20h.
36. Fox, V. G., Erickson, L. E., and Fan, L. T., *AIChE J.*, Vol. 15, pp. 327–333, 1969, *15a.
37. Giesekus, H., *Rheol. Acta*, Vol. 3, pp. 59–71, 1963, *9d, *10d.
38. Giesekus, H., *Rheol. Acta*, Vol. 6, pp. 339–353, 1967, *8c.
39. Ginn, R. F. and Denn, M. M., *AIChE J.*, Vol. 15, pp. 450–454, 1969, *16c.
40. Griffiths, D. F., Jones, D. T., and Walters, K., *J. Fluid Mech.*, Vol. 36, pp. 161–175, 1969, *7d.
41. Griffiths, D. F. and Walters, K., *J. Fluid Mech.*, Vol. 42, pp. 379–399, 1970, *8c.
42. Gunn, R. W., Mena, B., and Walters, K., *Z.A.M.P.*, Vol. 25, pp. 591–606, 1974, *4f.
43. Gutfinger, C. and Tallmadge, J. A., *AIChE J.*, Vol. 11, pp. 403–413, 1965, *29a.
43.1. Hassager, O. and Bisgaard, C., *J. Non-Newtonian Fluid Mech.*, Vol. 12, pp. 153–164, 1983, *23e.

44. Higashitani, K. and Pritchard, W. G., *Trans. Soc. Rheol.*, Vol. 16, pp. 687–696, 1972, *27i.

45. Hopke, S. W. and Slattery, J. C., *AIChE J.*, Vol. 16, pp. 224–229, 1970, *10b.

46. Horowitz, H. H. and Steidler, F. E., *ALSE Trans.*, Vol. 3, pp. 124–133, 1960, *6b.

46.1. Hull, A. M., *J. Non-Newtonian Fluid Mech.*, Vol. 8, pp. 327–336, 1981, *31e.

46.11. Huynh, B. P., *J. Non-Newtonian Fluid Mech.*, Vol. 13, pp. 1–20, 1983, *26a.

46.12. Iemoto, Y., Nagata, M., and Yamamoto, F., *J. Non-Newtonian Fluid Mech.*, Vol. 19, pp. 161–184, 1985, *3a.

46.13. Iemoto, Y., Nagata, M., and Yamamoto, F., *J. Non-Newtonian Fluid Mech.*, Vol. 22, pp. 101–114, 1986, *3e.

46.2. Jackson, N. R. and Finlayson, B. A., *J. Non-Newtonian Fluid Mech.*, Vol. 10, pp. 71–84, 1982, *27e.

47. Jones, D. T., Ph.D. Thesis, University of Wales, 1967, *3d.

48. Jones, J. R., *J. de Mécanique*, Vol. 3, pp. 79–99, 1964, *18f.

49. Jones, J. R., *J. de Mécanique*, Vol. 4, pp. 121–132, 1965, *18f.

50. Jones, J. R., *J. de Mécanique*, Vol. 6, pp. 443–448, 1967, *2f.

51. Jones, J. R., and Jones, R. S., *J. de Mécanique*, Vol. 5, pp. 375–395, 1966, *18f.

52. Jones, J. R., and Lewis, M. K., *Rheol. Acta*, Vol. 7, pp. 307–316, 1968, *4g.

53. Jones, J. R., and Walters, T. S., *Rheol. Acta*, Vol. 6, pp. 240–245, 1967, *1g.

54. Joseph, D. D., Beavers, G. S., and Fosdick, R. L., *Arch. Rat. Mech.*, Vol. 49, pp. 381–401, 1973, *14c.

55. Joseph, D. D., and Fosdick, R. L., *Arch. Rat. Mech.*, Vol. 49, pp. 321–380, 1973, *14d.

56. Kaloni, P. N., *J. Phys. Soc.* (Japan), Vol. 20, pp. 132–138, 1965, *11f.

57. Karlsson, S. K. F., Sokolov, M., and R. I. Tanner, *Chem. Eng. Prog. Symp.*, Vol. 67, No. 111, pp. 11–20, 1971, *16j.

58. Kaye, A., *Rheol. Acta*, Vol. 12, pp. 207–211, 1973, *14c.

58.1. Kazakia, J. Y., and Rivlin, R. S., *J. Non-Newtonian Fluid Mech.*, Vol. 8, pp. 311–317, 1981, *32c.

59. Kearsley, E. A., *Trans. Soc. Rheol.*, Vol. 14, pp. 419–424, 1970, *27c.

59.1. Keunings, R. and Crochet, M. J., *J. Non-Newtonian Fluid Mech.*, Vol. 14, pp. 279–299, 1984, *25m.

60. King, M. J. and Waters, N. D., *J. Phys.*, Vol. D5, pp. 141–150, 1972, *23f.

61. Kobayashi, T. and Tomita, Y., *Bull. JSME*, Vol. 14, pp. 208–216, 1971, *25a,e.

62. Kobayashi, T. and Tomita, Y., *Bull. JSME*, Vol. 15, pp. 236–243, 1972, *11e.

63. Kramer, J. M., *Appl. Sci. Res.*, Vol. 30, pp. 1–16, 1974, *13g.

64. Kramer, J. M. and Johnson, M. W., *Trans. Soc. Rheol.*, Vol. 16, pp. 197–212, 1972, *7k.

65. Laurencena, B. R. and Williams, M. C., *Trans. Soc. Rheol.*, Vol. 18, pp. 331–355, 1974, *12a.

66. Lee, S. Y. and Ames, W. F., *AIChE J.*, Vol. 12, pp. 700–708, 1966, *15b.

67. Leider, P. J. and Bird, R. B., *I.E.C. Fund.*, Vol. 13, pp. 336–341, 1974, *13a.

68. Leslie, F. M. and Tanner, R. I., *Quart. J. Mech. Appl. Math.*, Vol. 14, pp. 36–48, 1961, *10f.

69. Lockett, F. J. and Rivlin, R. S., *J. de Mécanique*, Vol. 7, pp. 475–498, 1968, *16d.

69.1. Luo, X.-L. and Tanner, R. I., *J. Non-Newtonian Fluid Mech.*, Vol. 21, pp. 179–199, 1986, *26m.

69.2. Luo, X.-L. and Tanner, R. I., *J. Non-Newtonian Fluid Mech.*, Vol. 22, pp. 61–89, 1986, *26j.

70. Matovich, M. A. and Pearson, J. R. A., *I.E.C. Fund.*, Vol. 8, pp. 512–520, 1969, *28c.

TABLE 3.9 *(Continued)*

71. McIntire, L. V. and Schowalter, W. R., *AIChE J.*, Vol. 18, pp. 102–110, 1972, *16c.k.
72. Mitsuishi, N. and Aoyagi, Y., *Chem. Eng. Sci.*, Vol. 24, pp. 309–319, 1969, *5b.
73. Mitsuishi, N. and Aoyagi, Y., *J. Chem. Eng.*, (Japan), Vol. 6, pp. 402–408, 1973, *5b.
74. Mitsuishi, N., Aoyagi, Y., and Soeda, H., *Kagaku Kōgaku*, Vol. 36, pp. 186–192, 1972, *5b.
75. Mohan Rao, D. K., *Proc. Indian Acad. Sci*, Sect. A, Vol. 56, 198–205, 1962, *8g.
76. Na, T. Y. and Hansen, A. G., *Int. J. Nonlinear Mech.*, Vol. 2, pp. 261–273, 1967, *12b.
77. Nirschl, J. P. and Stewart, W. E., *J. Non-Newtonian Fluid Mech.*, Vol. 16, pp. 233–250, 1984, *7i.
77.1. O'Donovan, E. J., and Tanner, R. I., *J. Non-Newtonian Fluid Mech.*, Vol. 15, pp. 75–83, 1984, *13m.
78. Oka, S. and Murata, T., *Japan J. Appl. Phys.*, Vol. 8, pp. 5–8, 1969, *11b.
78.1. Phan-Thien, N. and Tanner, R. I., *J. Non-Newtonian Fluid Mech.*, Vol. 11, pp. 147–161, 1982, *1m.
78.2. Phan-Thien, N. and Tanner, R. I., *J. Non-Newtonian Fluid Mech.*, Vol. 9, pp. 107–117 1981 *6i.
78.3. Phan-Thien, N., and Caswell, B., *J. Non-Newtonian Fluid Mech.*, Vol. 21, pp. 225–234, 1986, *28e.
78.4. Phan-Thien, N. and Khan, M. M. K., *J. Non-Newtonian Fluid Mech.*, Vol. 24, pp. 203–220, 1987, *2e.
79. Piau, J. M. and Piau, M., *Can. Roy. Acad. Sci.*, Vol. A270, pp. 159–161, 1970, *12c.
80. Pipkin, A. C., *Phys. Fluids*, Vol. 7, pp. 1143–1146, 1964, *1c.
81. Pipkin, A. C., *Proc. Int. Congr. Rheol.*, 4th, Vol. 1, pp. 213–222, 1965, *5c.
82. Porteous, K. C. and Denn, M. M., *Trans. Soc. Rheol.*, Vol. 16, pp. 295–308, 1972, *16c.e.
83. Rajeswara, G. K. and Rathna, S. L., *Z.A.M.P.*, Vol. 13, pp. 43–57, 1962, *15c.
84. Rivlin, R. S., *Second-Order Effects in Elasticity, Plasticity, and Fluid Dynamics*, Reiner, M. and Abir, D. (Eds.), Macmillan, New York, pp. 668–677, 1964, *5d.
85. Rivlin, R. S. and Langlois, W. E., *Rend. Mat.*, Vol. 22, pp. 169–185, 1963, *5d.
86. Sadowski, T. J. and Bird, R. B., *Trans. Soc. Rheol.*, Vol. 9, pp. 243–250, 1965, *21b.
87. Schechter, R. S., *AIChE J.*, Vol. 7, pp. 445–448, 1961, *5a.
88. Schummer, P. von, *Rheol. Acta*, Vol. 6, pp. 192–200, 1967, *11d.
89. Schwarz, W. H. and Bruce, C., *Chem. Eng. Sci.*, Vol. 24, pp. 399–413, 1969, *12d.
90. Scott, J. R., *Trans. Inst. Rubber Ind.*, Vol. 7, pp. 169–186, 1931, *13a.
91. Scott, J. R., *Trans. Inst. Rubber Ind.*, Vol. 8, pp. 481–493, 1932, *13b.
92. Šesták, J. and Ambros, F., *Rheol. Acta*, Vol. 12, pp. 70–76, 1973, *19a.
93. Slattery, J. C., *AIChE J.*, Vol. 8, pp. 663–667, 1962, *10a.b.
94. Smith, M. M. and Rivlin, R. S., *J. de Mécanique*, Vol. 11, pp. 70–94, 1972, *16d.
95. Spiers, R. P., Subbaraman, C. V., and Wilkinson, W. L., *Chem. Eng. Sci.*, Vol. 30, pp. 379–395, 1975, *29a.
96. Street, J. R., *Trans. Soc. Rheol.*, Vol. 12, pp. 103–131, 1968, *20f.
96.1. Sugeng, F. and Tanner, R. I., *J. Non-Newtonian Fluid Mech.*, Vol. 20, pp. 281–292, 1986.
97. Sutterby, J. L., *Trans. Soc. Rheol.*, Vol. 9, pp. 227–242, 1965, *11b.

98. Sutterby, J. L., *AIChE J.*, Vol. 12, pp. 63–68, 1966, *11b.

99. Tadmor, Z. and Bird, R. B., *Polym. Eng. Sci.*, Vol. 14, pp. 124–136, 1974, *18i.

100. Tallmadge, J. A., *AIChE J.*, Vol. 16, pp. 929–930, 1970, *29a.

101. Tanner, R. I., *Z.A.M.P.*, Vol. 136, pp. 573–580, 1962, *17f.

102. Tanner, R. I., *Aust. J. Appl. Sci.*, Vol. 14, pp. 129–136, 1963, *6a.

103. Tanner, R. I., *ASLE Trans.*, Vol. 8, pp. 179–183, 1965, *13e.

104. Tanner, R. I., *I.E.C. Fund.*, Vol. 5, pp. 55–59, 1966, *11a.

105. Tanner, R. I., *J. Polym. Sci.*, Vol. A8, pp. 2067–2078, 1970, *26j.

106. Tanner, R. I., and Pipkin, A. C., *Trans. Soc. Rheol.*, Vol. 13, pp. 471–484, 1969, *27c.

107. Thomas, R. H. and Walters, K., *Proc. Roy. Soc.*, Vol. A274, pp. 371–385, 1963, *3a.

108. Thomas, R. H. and Walters, K., *Quart. J. Mech. Appl. Math.*, Vol. 17, pp. 39–53, 1964, *9c.

109. Thomas, R. H. and Walters, K., *J. Fluid Mech.*, Vol. 19, pp. 557–560, 1964, *16g.

110. Thomas, R. H. and Walters, K., *J. Fluid Mech.*, Vol. 21, pp. 173–182, 1965, *3g.

111. Thomas, R. H. and Walters, K., *Rheol. Acta*, Vol. 5, pp. 23–27, 1966, *23e.

111.1. Tiefenbruck, G. and Leal, L. G., *J. Non-Newtonian Fluid Mech.*, Vol. 10, pp. 115–155, 1982, *10f.

112. Tomita, Y., *Bull. JSME*, Vol. 2, pp. 469–474, 1959, *10a.

113. Tomita, Y., *Bull. JSME*, Vol. 4, pp. 77–86, 1961, *15a.

114. Tomita, Y. and Kato, H., *Nippon Kikai Gakkai Ronbun-shi (Daini-bu)*, Vol. 32, pp. 1399–1408, 1966, *24f.

115. Townsend, P., *Rheol. Acta*, Vol. 12, pp. 13–18, 1973, *1f.

115.1. Trogdon, S. A. and Joseph, D. D., *J. Non-Newtonian Fluid Mech.*, Vol. 10, pp. 185–213, 1982, *27c.

116. Ultman, J. S. and Denn, M. M., *Chem. Eng. J.*, Vol. 2, pp. 81–89, 1971, *10e, *22e.

117. Vidyanidhi, V. and Sithapathi, A., *J. Phys. Soc.*, (Japan), Vol. 29, pp. 215–219, 1970, *4c.

118. Wagner, M. G. and Slattery, J. C., *AIChE J.*, Vol. 17, pp. 1198–1207, 1971, *10d.

119. Walters, K. and Townsend, P., *Proc. Int. Congr. Rheol.*, 5th, Vol. 4, pp. 471–483, 1970, *1d,f.

120. Walters, K. and Waters, N. D., *Polymer Systems Deformation and Flow*, Wetton, R. E. and Whorlow, R. W. (Eds.), Macmillan, London, pp. 211–235, 1968, *8d.

121. Wasserman, M. L. and Slattery, J. C., *AIChE J.*, Vol. 10, pp. 383–388, 1964, *10a.

122. Waters, N. D. and King, M. J., *Quart. J. Mech. Appl. Math.*, Vol. 24, pp. 331–346, 1971, *9d.

123. Wheeler, J. A. and Wissler, E. H., *AIChE J.*, Vol. 11, pp. 207–212, 1965, *5a.

124. Yang, W. H. and Yeh, H. C., *AIChE J.*, Vol. 12, pp. 927–931, 1966, *20a,b.

125. Young, D. M. and Wheeler, M. F., *Nonlinear Problems in Engineering*, Ames, W. F. (Ed.), Academic Press, New York, pp. 220–246, 1964, *5a.

126. Zidan, M., *Rheol. Acta*, Vol. 8, pp. 89–123, 1969, *26c.

127. Ziegenhagen, A. J., Bird, R. B., and Johnson, M. W., Jr., *Trans. Soc. Rheol.*, Vol. 5, pp. 47–49, 1961, *10b.

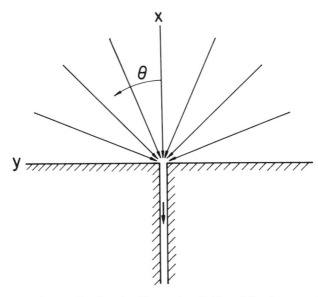

FIGURE 3.15 Flow into a slit. For the Newtonian fluid and for the second-order fluid the flow is radial. The volume rate of flow into the slit is Q; the width of the slit (in the z-direction) is W.

Note that at the solid surface $\theta = \pi/2$ the terms containing b_2 and b_{11} drop out. Next, to get the components of τ, the kinematic tensors must be evaluated

$$\nabla \bar{\mathbf{v}} \underset{r,\theta,z}{=} \begin{pmatrix} \partial v_r/\partial r & 0 & 0 \\ (1/r)(\partial v_r/\partial\theta) & v_r/r & 0 \\ 0 & 0 & 0 \end{pmatrix} = \begin{pmatrix} \cos^2\theta & 0 & 0 \\ \sin 2\theta & -\cos^2\theta & 0 \\ 0 & 0 & 0 \end{pmatrix} \frac{2q}{\pi r^2} \quad (3.127)$$

$$\bar{\bar{\gamma}}_{(1)} \underset{r,\theta,z}{=} \begin{pmatrix} 2\cos^2\theta & \sin 2\theta & 0 \\ \sin 2\theta & -2\cos^2\theta & 0 \\ 0 & 0 & 0 \end{pmatrix} \frac{2q}{\pi r^2} \quad (3.128)$$

$$\{\bar{\bar{\gamma}}_{(1)} \cdot \bar{\bar{\gamma}}_{(1)}\} \underset{r,\theta,z}{=} \begin{pmatrix} 1 & 0 & 0 \\ 0 & 1 & 0 \\ 0 & 0 & 0 \end{pmatrix} \frac{16q^2 \cos^2\theta}{\pi^2 r^4} \quad (3.129)$$

$$\bar{\bar{\gamma}}_{(2)} = v_r \frac{\partial}{\partial r} \bar{\bar{\gamma}}_{(1)} - \{(\nabla\bar{\mathbf{v}})^\dagger \cdot \bar{\bar{\gamma}}_{(1)} + \bar{\bar{\gamma}}_{(1)} \cdot (\nabla\bar{\mathbf{v}})\}$$

$$\underset{r,\theta,z}{=} \begin{pmatrix} -2\sin^2\theta & \sin 2\theta & 0 \\ \sin 2\theta & -2\cos^2\theta & 0 \\ 0 & 0 & 0 \end{pmatrix} \frac{16q^2 \cos^2\theta}{\pi^2 r^4} \quad (3.130)$$

When $\bar{\bar{\gamma}}_{(1)}$, $\bar{\bar{\gamma}}_{(2)}$, and $\{\bar{\bar{\gamma}}_{(1)} \cdot \bar{\bar{\gamma}}_{(1)}\}$ are substituted into the first three terms on the right side of Eq. (3.114), the expression for $\bar{\bar{\tau}}$ is obtained. From this expression one finds that at the surface $\theta = \pi/2$, the nomal stress $\tau_{\theta\theta}$ vanishes identically. Hence in a measurement of $\theta_{\theta\theta}$ at $\theta = \pi/2$ by means of flush-mounted pressure transducer, one cannot distinguish between a Newtonian fluid and a second-order fluid. The shear stress $\tau_{r\theta}$ also vanishes identically at $\theta = \pi/2$. These same conclusions are obtained for the convected Maxwell model of Eq. (3.110); the stress fields obtained for the latter have been compared with experimental birefringence data [Hull (1981)].

3.6 KINETIC THEORIES AND RHEOLOGICAL PROPERTIES

It should be evident from the discussions in Sec 3.5 that the empirical approach to constitutive equations and the ordered-expansion approach both leave much to be desired. The best constitutive equations in the future will probably be those based on or suggested by kinetic theories that take into account the structure of the non-Newtonian fluid. In this connection, stochastic methods can be particularly helpful for solving kinetic theory problems for realistic molecular models [Öttinger (1995)]. Also, if one wishes to know how to modify or control rheological properties by structural modifications, the kinetic theory approach is essential.

3.6.1 Molecular Theories for Polymeric Liquids

There are several kinds of problems that require an understanding of the molecular aspects of polymer rheology: (a) the determination of polymer orientation and stretching in polymer processing operations; (b) the interpretation of rheological material functions in terms of structure (e.g., concentration, molecular weight distribution, chain stiffness, chain branching); (c) the analysis of drag reduction in terms of polymer distortion and orientation; (d) the understanding of the role of polymer additives in lubrication [Bird *et al.* (1987b) and Yamakawa (1971)]. Three points must be kept in mind in connection with the kinetic theory of polymers: the enormous variety of polymeric liquids (chemical structure, polymer–solvent interactions, concentration), the enormous number of internal degrees of freedom in a polymer molecule, and the fact that most polymer samples are polydisperse (i.e., they contain polymer molecules with a distribution of molecular weights).

In the kinetic theory of polymers, one begins by specifying some kind of mechanical model for a polymer molecule; a sampling of these models is shown in Fig. 3.16. These models represent, to varying degrees of faithfulness, the orientability, stretching, and internal degrees of motion of linear polymer molecules. Because of the artificiality of these models, the parameters that appear in them—number of beads, spring constants, rod lengths, etc.—are very difficult to relate to the true polymer structure. Usually these parameters are left to be determined from rheological or rheo-optical data.

The kinetic theories may be divided into three concentration ranges as shown in Fig. 3.17. In the dilute-solution theories, one generally regards the solvent as a Newtonian continuum and then determines the effect of the motions of a single polymer molecule on the stress tensor for the polymer solution. In concentrated solutions or melts, two competing theories have been developed: (a) network theories, which regard the fluid as a weak network resulting from temporary physical junctions, and (b) mean field theories, in which the motion of a single molecule is described, but

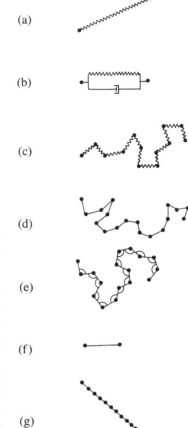

FIGURE 3.16 Structural models used to represent polymer molecules in kinetic theory: (a) the elastic dumbbell, consisting of two mass points ("beads") connected by a spring, (b) the elastic dumbbell with "internal viscosity," consisting of two mass points connected by a spring and dashpot in parallel, (c) the Rouse chain, a freely jointed bead-spring chain, (d) the Kramers chain, a freely jointed bead-rod chain, (e) the Kirkwood–Riseman chain, a bead-rod chain with fixed angles between rods, (f) the rigid dumbbell, consisting of two mass points connected by a rigid rod, (g) the multibead rod. Models (a) through (e) have been used to represent flexible chain polymers, whereas models (f) and (g) have been used to represent rigid rodlike polymers.

the constraints provided by the neighboring molecules are included in some way. For a survey of kinetic theories see the review by Bird and Öttinger (1992).

Dilute Polymer Solutions. The methods of kinetic theory are illustrated by formulating the basic equations for a dilute suspension of *elastic dumbbells* (model (a) of Fig. 3.16 [Bird *et al.* (1987b), Yamakawa (1971), and Doi and Edwards (1986)]. The configuration of the dumbbell may be specified by giving the position vectors of the two beads $\bar{\mathbf{r}}_1$ and $\bar{\mathbf{r}}_2$, or alternatively the position vector of the center of mass $\bar{\mathbf{r}}_c$ and the vector $\overline{\mathbf{Q}}$ going from bead "1" to bead "2." The number of dumbbells with configurations in the range $d\,\bar{\mathbf{r}}_1 d\,\bar{\mathbf{r}}_2$ about $\bar{\mathbf{r}}_1$, $\bar{\mathbf{r}}_2$ at time t is called $\Psi(\bar{\mathbf{r}}_1, \bar{\mathbf{r}}_2, t)$. This distribution function is assumed to be independent of $\bar{\mathbf{r}}_c$ so that one can write $\Psi(\bar{\mathbf{r}}_1, \bar{\mathbf{r}}_2, t) = n\psi(\overline{\mathbf{Q}}, t)$ where n is the number density of dumbbells. These distribution functions have to satisfy an equation of continuity in the configuration space

$$\frac{\partial \Psi}{\partial t} = -\sum_{\nu=1}^{2} \frac{\partial}{\partial \bar{\mathbf{r}}_\nu} \cdot (\llbracket \dot{\bar{\mathbf{r}}}_\nu \rrbracket \Psi) \tag{3.131}$$

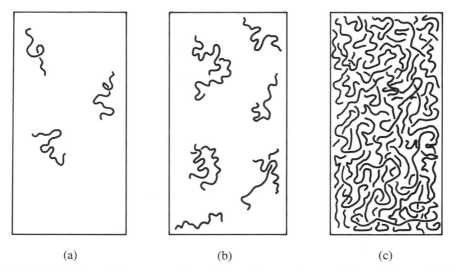

(a)   (b)   (c)

FIGURE 3.17 Pictorial representation of (a) a dilute polymer solution, (b) a semi-dilute polymer solution, and (c) a concentrated polymer solution or a polymer "melt."

$$\frac{\partial \psi}{\partial t} = -\frac{\partial}{\partial \overline{\mathbf{Q}}} \cdot (\llbracket \dot{\overline{\mathbf{Q}}} \rrbracket \psi) \tag{3.132}$$

where the double-square brackets indicate an average over the bead-velocity space. These equations cannot be used until expressions are available for the average velocities, and these are obtained by writing equations of motion for the beads (assuming that the accelerations are negligible). That is

$$0 = -\zeta[\llbracket \dot{\overline{\mathbf{r}}}_\nu \rrbracket - \overline{\overline{\kappa}} \cdot \overline{\mathbf{r}}_\nu] - kT \frac{\partial}{\partial \overline{\mathbf{r}}_\nu} \ln \psi + \overline{\mathbf{F}}_\nu \tag{3.133}$$

Here ζ is a friction coefficient, $\overline{\overline{\kappa}}$ is $(\nabla \mathbf{v})^\dagger$ for the solvent, kT is Boltzmann's constant times temperature, and $\overline{\mathbf{F}}_\nu$ is the force on bead ν through the connecting spring. When Eq. (3.133) for bead "1" is subtracted from the same equation for bead "2," the equation for internal motion is obtained; when this is solved for $\llbracket \dot{\overline{\mathbf{Q}}} \rrbracket$, the following result is obtained

$$\llbracket \dot{\overline{\mathbf{Q}}} \rrbracket = [\overline{\overline{\kappa}} \cdot \overline{\mathbf{Q}}] - \frac{2kT}{\zeta} \frac{\partial}{\partial \overline{\mathbf{Q}}} \ln \psi - \frac{2}{\zeta} \overline{\mathbf{F}}^{(c)} \tag{3.134}$$

where $\overline{\mathbf{F}}^{(c)} = \overline{\mathbf{F}}_1 = -\overline{\mathbf{F}}_2$ is the tension in the connecting spring. This expression for $\llbracket \dot{\overline{\mathbf{Q}}} \rrbracket$ is now substituted into Eq. (3.132) to give the *diffusion equation for* $\Psi(\overline{\mathbf{Q}}, t)$

$$\frac{\partial \psi}{\partial t} = -\left(\frac{\partial}{\partial \overline{\mathbf{Q}}} \cdot \left\{ [\overline{\kappa} \cdot \overline{\mathbf{Q}}] \psi - \frac{2kT}{\zeta} \frac{\partial}{\partial \overline{\mathbf{Q}}} \psi - \frac{2}{\zeta} \overline{\mathbf{F}}^{(c)} \psi \right\} \right) \tag{3.135}$$

Once the flow field (i.e., $\overline{\overline{\kappa}}$) and the spring force law (i.e., $\mathbf{F}^{(c)}$) are specified, then the distribution function $\psi(\mathbf{Q}, t)$ can in principle be obtained.

Next, an expression is needed for the stress tensor $\overline{\overline{\pi}}$, which is presumed to be the sum of a solvent contribution $\overline{\overline{\pi}}_s$ and a polymer contribution $\overline{\overline{\pi}}_p$. The polymer contribution is in turn the sum of two contributions

(a) the *connector* contribution, $\overline{\overline{\pi}}_p^{(c)} = -n\langle \overline{\mathbf{F}^{(c)}\mathbf{Q}} \rangle$, which accounts for the stress resulting from the tension in the springs; here $\langle B \rangle = \int B\psi d\mathbf{Q}$ is an average value, calculated with the configurational distribution function ψ.

(b) the *bead* contribution, $\overline{\overline{\pi}}_p^{(b)} = 2nkT\overline{\delta}$, which accounts for the stress associated with the momentum transfer of the beads; this is calculated by assuming that the bead velocity distribution is Maxwellian about the mean stream velocity at the center of mass of the dumbbell.

Therefore, the total stress tensor is

$$\overline{\overline{\pi}} = \overline{\overline{\pi}}_s - n\langle \overline{\mathbf{F}^{(c)}\mathbf{Q}} \rangle + 2nkT\overline{\overline{\delta}} \qquad (3.136)$$

At equilibrium, this equation becomes

$$p\overline{\overline{\delta}} = p_s\overline{\overline{\delta}} - nkT\overline{\overline{\delta}} + 2nkT\overline{\overline{\delta}} \qquad (3.137)$$

Subtracting these two equations gives the final expression for the *Kramers form for the stress tensor*

$$\overline{\overline{\tau}} = \overline{\overline{\tau}}_s - n\langle \overline{\mathbf{F}^{(c)}\mathbf{Q}} \rangle + nkT\overline{\overline{\delta}} \qquad (3.138)$$

in which $\overline{\overline{\tau}}_s = \eta_s\overline{\overline{\dot{\gamma}}}$ where η_s is the solvent viscosity. By multiplying Eq. (3.135) by $\overline{\mathbf{QQ}}$ and integrating over all configurations one obtains $\langle \overline{\mathbf{F}^{(c)}\mathbf{Q}} \rangle = kT\overline{\overline{\delta}} - (\zeta/4)\langle \overline{\mathbf{QQ}} \rangle_{(1)}$; then Eq. (3.138) may be written as

$$\overline{\overline{\tau}} = \overline{\overline{\tau}}_s + (n\zeta/4) \langle \overline{\mathbf{QQ}} \rangle_{(1)} \qquad (3.139)$$

which is the *Giesekus form for the stress tensor*. Equations (3.135) and (3.138) or (3.139) are the principal results for the kinetic theory of elastic dumbbells. Once $\psi(\mathbf{Q}, t)$ is known from the solution of the diffusion equation, the stress tensor can be obtained from Eq. (3.138). (A somewhat improved theory can be obtained by taking into account the *hydrodynamic interaction*, which is the perturbation of the solvent flow field at bead "1" due to the motion of bead "2" and vice versa. The elementary theory has been modified in many other ways, such as letting ζ depend on Q, replacing Stokes' law by a anisotropic drag force, including an *internal viscosity* [model (6) of Fig. 3.16], etc.)

Several examples of constitutive equations obtained from the above theory are now given:

Hookean Dumbbells. $\mathbf{F}^{(c)} = H\overline{\mathbf{Q}}$, with $H = $ constant [Bird *et al.* (1987) and Giesekus (1966)]: When this force expression is inserted into Eq. (3.138), the quantity $\langle \overline{\mathbf{QQ}} \rangle$ then appears; but $\langle \overline{\mathbf{QQ}} \rangle$ can then be eliminated between Eqs. (3.138) and (3.139) to give $\bar{\bar{\tau}} = \eta_s \dot{\bar{\bar{\gamma}}} + \bar{\bar{\tau}}_p$, with $\bar{\bar{\tau}}_p$ determined from

$$\bar{\bar{\tau}}_p + \lambda_H \bar{\bar{\tau}}_{p(1)} = -nkT\lambda_H \dot{\bar{\bar{\gamma}}} \tag{3.140}$$

where $\lambda_H = \zeta/4H$ is the time constant. This has the same form as the convected Maxwell model in Eq. (3.110). Note also that the constitutive equation for $\bar{\bar{\tau}}$ may be put into the form of Eq. (3.111) with

$$\eta_0 = \eta_s + nkT\lambda_H$$

$$\lambda_1 = \lambda_H$$

$$\lambda_2 = [\eta_s/(\eta_s + nkT\lambda_H)]\lambda_H$$

$$\mu_0 = 0 \tag{3.141}$$

Hookean Dumbbells. $\mathbf{F}^{(c)} = H\mathbf{Q}$ with $H = H(T)$: The above procedure leads to the constitutive equation for the polymer contribution τ_p

$$\bar{\bar{\tau}}_p + \frac{\zeta}{4H}\left[\bar{\bar{\tau}}_{p(1)} - \bar{\bar{\tau}}_p\left(\frac{d\ln H}{d\ln T}\right)\frac{D\ln T}{Dt} + nkT\bar{\bar{\delta}}\left(\frac{d\ln H}{d\ln T} - 1\right)\frac{D\ln T}{Dt}\right]$$

$$= -nkT\left(\frac{\zeta}{4H}\right)\dot{\bar{\bar{\gamma}}} \tag{3.142}$$

This emphasizes that in nonisothermal problems the temperature history actually enters the constitutive equation [Marrucci (1972)].

Hookean Dumbbells. $\overline{\mathbf{F}}^{(c)} = H\overline{\mathbf{Q}}$ with an effective velocity gradient [Bird (1979) and Tanner (1975)]. One can replace $\bar{\bar{\kappa}}$ in Eq. (3.135) by an "effective velocity gradient" tensor $\bar{\bar{\kappa}} - (1/2)\xi\dot{\bar{\bar{\gamma}}}$, where ξ is a small positive constant, sometimes called the *slip factor*. This accounts for the fact that the velocity gradient in the vicinity of the beads is slightly different from that imposed on the solution as a whole. When this is done, Eq. (3.140) is replaced by

$$\bar{\bar{\tau}}_p + \lambda_H(\bar{\bar{\tau}}_{p(1)} + \tfrac{1}{2}\xi\{\dot{\bar{\bar{\gamma}}} \cdot \bar{\bar{\tau}}_p + \bar{\bar{\tau}}_p \cdot \dot{\bar{\bar{\gamma}}}\}) = -(1 - \xi)nkT\lambda_H\dot{\bar{\bar{\gamma}}} \tag{3.143}$$

This result gives a shear-rate-dependent viscosity and has a number of other desirable features—it gives a negative second normal stress coefficient, which Eq. (3.140) does not—but it predicts an infinite elongational viscosity at some finite elongation rate. Equation (3.143) is a special case of the Oldroyd 8-constant model (Table 3.8).

Finitely Extensible Nonlinear Elastic (FENE) Dumbbells. $\overline{\mathbf{F}}^{(c)} = H\overline{\mathbf{Q}}/[1 - (Q/Q_0)^2]$, [Bird *et al.* (1980)]: If, in the force law, $(Q/Q_0)^2$ is replaced by $\langle(Q/Q_0)^2\rangle$ (the "Peterlin approximation") then the procedure outlined for Hookean dumbbells leads to the "Tanner equation"

$$Z\bar{\bar{\tau}}_p + \lambda_H[\bar{\bar{\tau}}_{p(1)} - (\bar{\bar{\tau}}_p - nkT\bar{\bar{\delta}})\,D\ln Z/Dt] = -nkT\lambda_H\dot{\bar{\bar{\gamma}}} \tag{3.144a}$$

$$Z = 1 + (3/b)\,(1 - \text{tr } \bar{\bar{\tau}}_p/3nkT) \tag{3.144b}$$

Here $b = HQ_0^2/kT$ is a dimensionless quantity, which goes to infinity in the limit of Hookean dumbbells, whereupon Eq. (3.140) is recovered. Although Eq. (3.144) is more complicated than Eq. (3.140), because of the appearance of terms nonlinear in the stresses, it has been shown that the model contains a number of promising features: viscosity that decreases with shear rate, elongational viscosity that remains finite, shear-stress growth curves with overshoot, and an elastic response that can change sign in superposed steady shear and sinusoidal flow; on the other hand this constitutive equation gives $\Psi_2 = 0$. Equation (3.144) may be regarded as a promising candidate for fluid dynamics calculations for polymer solutions.

This discussion of dumbbells shows that the molecular approach can lead to structural interpretation of some of the constitutive equations proposed in Sec. 3.5 and can also lead to new suggestions for forms of constitutive equations [such as Eq. (3.144)]. The above ideas have been extended to bead-spring chain models, which can describe the linear viscoelastic data much more faithfully because they have more internal degrees of freedom.

Models such as the rigid dumbbell, the Kramers bead-rod chain, and the Kirkwood-Riseman bead-rod chain require a much more careful formulation inasmuch as some internal degrees of freedom have been frozen out. A formal kinetic theory for such systems has been developed [Bird *et al.* (1987) and Curtiss *et al.* (1976)].

All the theories discussed above are for infinitely dilute solutions. They should therefore be appropriate for the study of drag reduction. It has been found that many rheological properties of moderately concentrated solutions are described qualitatively by, for example, the constitutive equation based on the FENE dumbbell model.

Table 3.10 provides a few sample constitutive equations for dilute polymer solutions.

Concentrated Polymer Solutions and Melts. All the *network theories* [Lodge (1964) and Bird *et al.* (1977)] picture the polymer as being a temporary network of Hookean springs (*strands*) which is continually undergoing changes as portions of the network break up and are rearranged. The number of strands of given orientation changes because of fluid motion, strand creation, and strand destruction according to the following differential equation

$$\frac{\partial f_{\text{in}}}{\partial t} = -\left(\frac{\partial}{\partial \mathbf{R}} \cdot [\overline{\overline{\kappa}} \cdot \mathbf{R}] f_{\text{in}}\right) + L_{\text{in}}(\mathbf{R}, t) - \frac{f_{\text{in}}}{\lambda_{\text{in}}(t)} \qquad (3.145)$$

Here the distribution function has the following meaning: $f_{\text{in}}(\overline{\mathbf{R}}, t)d\overline{\mathbf{R}}$ is the concentration at time t of strands of complexity i and composed of n equivalent random links (of length ℓ) with ensemble-averaged end-to-end vectors within the range $\overline{\mathbf{R}}$ to $\overline{\mathbf{R}} + d\overline{\mathbf{R}}$. The L_{in} are strand creation rates, and the λ_{in} are functions, with dimension of time, that describe the rate of loss of strands. The stress tensor is given by

$$\overline{\overline{\tau}} = -\Sigma_n \Sigma_i \, H_n \, \langle \overline{\mathbf{RR}} \rangle_{\text{in}} \qquad (3.146)$$

Here symbol $H_n = 3kT/n\ell^2$ is the effective Hookean spring constant of an n-link strand, so that $H_n\overline{\mathbf{R}}$ can be interpreted as a force in the strand. The $\langle \ \rangle_{\text{in}}$ brackets indicate an average value calculated with the f_{in}. Equations (3.145) and (3.146) can be manipulated to give a constitutive equation for the fluid that is represented as a network of temporary physical junctions

$$\bar{\bar{\tau}} = -\int_{-\infty}^{t} m(t, t') \, \bar{\bar{\gamma}}_{[0]}(t, t') \, dt' \tag{3.147a}$$

$$m(t, t') = \Sigma_n \Sigma_i \, H_n \bar{L}_{in}(t') \exp\left[-\int_{t'}^{t} dt''/\lambda_{in}(t'')\right] \tag{3.147b}$$

where $L_{in}(t) = (4\pi/3) \int_0^\infty R^4 L_{in}(R, t) dR$. If all the L_{in} and λ_{in} are taken to be constants, and if all strands have the same complexity, then the Lodge rubberlike-liquid is obtained (see Table 3.8). Other choices have been made for the loss and creation rates, and various alternatives have been made in the basic theory; this has led to a wide variety of constitutive equations. The interrelations of these various models have been carefully studied in Bird, *et al.* (1987).

The *mean field theories* are formally very similar to the dilute-solution theories. If the polymer chains are modeled as Kramers freely jointed bead-rod chains with N beads and N-1 rods of length a, the Curtiss-Bird constitutive equation is obtained [Curtiss and Bird (1981)]

$$\bar{\bar{\tau}} = NnkT \left[\frac{1}{3} \bar{\bar{\delta}} - \int_{-\infty}^{t} \mu(t - t') \, \bar{\bar{A}} \, dt' \right.$$

$$\left. - \epsilon \int_{-\infty}^{t} \nu(t - t') \, \bar{\bar{B}} \, dt' \right] \tag{3.148}$$

where ϵ is the *link tension coefficient*, which can vary between 0 and 1, and

$$\nu(t - t') = \frac{16}{\pi^2 \lambda} \sum_{\alpha, \text{odd}} \alpha^{-2} e^{-\pi^2 \alpha^2 (t - t')/\lambda} \tag{3.149}$$

and $\mu(t - t') = (\lambda/2) d\nu(t - t')/dt$. The time constant $\lambda = N^{3+\beta} \zeta a^2 / 2kT$ contains the friction coefficient ζ and the *chain constraint exponent* β, the latter having values about 0.3 to 0.5. The tensors $\bar{\bar{A}}$ and $\bar{\bar{B}}$ are weighted averages of the finite strain tensor $\gamma^{[0]}$, of Table 3.5

$$\bar{\bar{A}} = \frac{1}{4\pi} \int [1 + \bar{\bar{\gamma}}^{[0]} : \overline{uu}]^{-3/2} \, \overline{uu} \, d\bar{u} \tag{3.150}$$

$$\bar{\bar{B}} = \frac{1}{4\pi} \lambda \bar{\bar{\kappa}} : \int [1 + \bar{\bar{\gamma}}^{[0]} : \overline{uu}]^{-3/2} \, \overline{uuuu} \, d\bar{u} \tag{3.151}$$

Here $\bar{u}$ is a unit vector given by the polar angles θ and ϕ, and $d\bar{u} = \sin \theta \, d\theta \, d\phi$. Experimental data on material functions for monodisperse systems have been fit by determining ϵ, λ, and N from experimental data; values of ϵ determined from viscosity are in the range 0.3 to 0.5.

TABLE 3.10 A Few Sample Constitutive Equations for Polymeric Liquids, with Polymers Modeled as Bead-Rod-Spring Systems

η_s = viscosity of solvent

n = number of density of polymer molecules

kT = Boltzmann's constant × absolute temperature

ζ = friction coefficient for bead

Reference	System	Constitutive Equation (Equation number)	Constants or Functions in Equation	Comments
a	Dilute solution of elastic dumbbells with Hookean springs	Oldroyd's 4-constant model (Eq. 3.111)	$\eta_0 = \eta_s + nkT\lambda_H$ $\lambda_1 = \lambda_H$ $\lambda_2 = [\eta_s/(\eta_s + nkT\lambda_H)]\lambda_H$ $\mu_0 = 0$	$\lambda_H = \zeta/4H$ = time constant H = Hookean spring constant
b	Dilute solution of elastic dumbbells with Hookean springs	Oldroyd-Walters-Fredrickson model (Table 3.8)	$G(t - t') = 2\eta_s\delta(t - t')$ $\qquad + nkT\, e^{-(t-t')/\lambda_H}$	
c	Dilute solution of elastic dumbbells with nonlinear springs (FENE dumbbells)	Retarded-motion expansion (Eq. 3.114)	$b_1 = \eta_s + \dfrac{bnkT\lambda_H}{b + 5}$ $b_2 = \dfrac{b^2 nkT\lambda_H^2}{(b + 5)(b + 7)}$ $b_{11} = 0$ (constants available through third order)	$b = HQ_0^2/kT$ $\lambda_H = \zeta/4H$ H = spring constant Q_0 = maximum length of dumbbells

	Model	Equation	Constants
d	Dilute solution of rigid dumbbells Memory-integral expansion (Eq. 3.113)	$G(t - t') = 2\eta_s\delta(t - t')$ $\quad + nkT\left[\frac{4}{5}\lambda\delta(t - t')\right.$ $\quad \left. + \frac{3}{5}e^{-(t - t')/\lambda}\right]$ (Kernel functions available through third order)	$\lambda = \zeta L^2/12kT$ L = length of rigid dumbbell
b	Dilute solution of freely jointed bead-spring chain with Hookean springs (Rouse model) Oldroyd-Walters-Fredrickson model (Table 3.8)	$G(t - t') = 2\eta_s\delta(t - t')$ $\quad + nkT\sum_{j=1}^{N-1} e^{-(t - t')/\lambda_j}$	$\lambda_j = \dfrac{\zeta/2H}{4\sin^2\left(j\pi/2N\right)}$ N = number of beads in chain H = Hookean spring constants
e	Dilute solution of freely jointed bead-rod chain (Kramers model) Retarded-motion expansion (Eq. 3.113)	$b_1 = \eta_s + \dfrac{(N^2 - 1)n\zeta L^2}{36}$ $b_2 = \dfrac{(N^2 - 1)(10N^3 - 12N^2 + 35N - 12)}{32400\,N}$ $\qquad \cdot \dfrac{n\zeta^2 L^4}{kT}$ $b_{11} = 0$ (constants available through second order)	N = number of beads in chain (the expression for b_2 is approximate)

a. Giesekus, H., *Rheol. Acta*, Vol. 5, pp. 29–35, 1966.

b. Lodge, A. S. and Wu, Y., *Rheol. Acta.*, Vol. 10, pp. 539–553, 1971.

c. Armstrong, R. C., *J. Chem. Phys.*, Vol. 60, pp. 724–728, pp. 729–733, 1944; Christiansen, R. L. and Bird, R. B., *J. Non-Newtonian Fluid Mech.*, Vol. 3, pp. 161–177, 1977/1978.

d. Armstrong, R. C. and Bird, R. B., *J. Chem. Phys.*, Vol. 58, pp. 2715–2723, 1973.

e. Hassager, O., *J. Chem. Phys.*, Vol. 60, pp. 2111–2124, 1974.

3.6.2 Structural Theories for Suspensions

A suspension is defined as a dispersion of one phase within another. The basic goal of structural theories for suspensions [Schowalter (1978), Brenner (1972), (1974), Russel (1987), Kim and Kerrila (1991), and Russel *et al.* (1989)] is modeling the dispersed multi-phase system as an equivalent continuum with its constitutive equation containing constant (or functions) that depend on the properties of the constituent phases. This modeling is valid provided the length scales of the suspended particles are much smaller than those describing the motion of the suspension as a whole.

The approach taken for structural theories of suspensions is quite different from that taken for polymeric liquids. Whereas polymer theories have to resort to some mechanical model to describe the extremely complex behavior of the polymer molecules, the theories for suspensions involve description of the actual flow phenomena taking place within the dispersed phases. Hence, for suspensions, an attempt is made to handle the hydrodynamic interaction in a more rigorous fashion.

Suspensions are generally characterized by the nature of the phases involved, the sizes and shapes of the suspended particles, and the relative amounts of each phase. Emulsions are suspensions of one liquid in another and have properties notably different from those of suspensions of solid particles. At large particle concentrations, the mechanical properties of the solid phase can have a significant effect upon the bulk suspension properties [Goddard (1977)]. Suspensions of colloidal particles behave differently than do suspensions of larger particles because of the influence of Brownian motion and nonhydrodynamic forces in the former. Suspensions in which the volume fraction of the suspended phase is below 0.1% are classified as dilute, and the interaction between particles is ignored. Structural theories for more concentrated suspensions must take into account the particle–particle interactions and are, therefore, much more complex. However, most suspensions of practical interest are nondilute so these theories are important.

There are essentially two methods for formulating rheological equations of state for suspensions:

1. equating the viscous dissipation of energy in the hypothetical one-phase fluid to that in the two-phase system.
2. ensemble averaging the stress for the two phase flow.

The first of these is the classical method of Einstein (1911). It has the disadvantage of assuming at the outset that the suspension behaves as a Newtonian fluid. Hence, it gives only the zero-shear-rate viscosity of the suspension as a function of the volume fraction of the dispersed phase. The second method [Landau and Lifschitz (1959)] has the advantage of giving the entire expression for the stress tensor. It is generally applied by expressing the ensemble average as a volume average [Batchelor (1970)]. It has the disadvantage of requiring calculation of the actual stress tensor for the often extremely complex geometry.

Several of the prominent constitutive relations for suspensions are given in Table 3.11. The first two entries in this table were obtained by Einstein's energy dissipation method. Batchelor and Green (1972) have extended Einstein's solution to include a term of order ϕ^2. The equations of Simha, Frankel and Acrivos, and Goddard

TABLE 3.11 Constitutive Relations for Neutrally Buoyant Suspensions

μ_s = viscosity of Newtonian suspending fluid

ϕ = volume fraction dispersed phase

Reference	System	Constitutive Equation (Equation Number)	Constants in Equation	Definitions of Structural Constants
a	Dilute Rigid spheres	Newton (3.9)	$\mu = \mu_s\left(1 + \frac{5}{2}\phi\right)$	
b	Dilute Emulsion	Newton (3.9)	$\mu = \mu_s\left(1 + \dfrac{\mu_s + (5/2)\mu_1}{\mu_s + \mu_1}\,\phi\right)$	μ_1 = viscosity of suspended fluid
c	All Concentrations Rigid spheres	Newton (3.9)	$\mu = \mu_s(1 + \lambda(\phi)\phi)$	$\lambda(\phi)$ = structure-dependent function of concentration
d	Concentrated Rigid spheres	Newton (3.9)	$\mu = \dfrac{9}{8}\mu_s\left(\dfrac{(\phi/\phi_M)^{1/3}}{1 - (\phi/\phi_M)^{1/3}}\right)$	ϕ_M = maximum concentration possible
e	Concentrated Elastic Spheres	(Linear) Maxwell (3.77)	$\lambda_0 = \dfrac{1}{4}\dfrac{\mu_s}{G}\dfrac{N}{\epsilon}$ $\eta_0 = \dfrac{1}{4}\phi\mu_s\dfrac{N}{\epsilon}$	G = Young's modulus of solid $\dfrac{N}{\epsilon}$ = function of structure
f	Dilute Elastic Spheres	(Linear) Jeffreys (3.78)	$\eta_0 = \mu_s\left(1 + \dfrac{5}{2}\phi\right)$ $\lambda_1 = \dfrac{3}{2}\mu_s\left(1 + \dfrac{5}{2}\phi\right)\bigg/G$ $\lambda_2 = \dfrac{3}{2}\mu_s\left(1 - \dfrac{5}{2}\phi\right)\bigg/G$	G = Young's modulus of solid

TABLE 3.11 (*Continued*)

μ_s = viscosity of Newtonian suspending fluid

ϕ = volume fraction dispersed phase

Reference	System	Constitutive Equation (Equation Number)	Constants in Equation	Definitions of Structural Constants
g	Dilute Emulsion	(Linear) Jeffreys (3.78)	$\eta_0 = \mu_s\left(1 + \dfrac{\mu_s + (5/2)\mu_1}{\mu_s + \mu_1}\right)\phi$ $\lambda_1 = \Lambda\left(3\mu_s + 2\mu_1 + \dfrac{\mu_s(16\mu_s + 19\mu_1)\phi}{5(\mu_s + \mu_1)}\right)$ $\lambda_2 = \Lambda\left(3\mu_s + 2\mu_1 - \dfrac{3\mu_s(16\mu_s + 19\mu_1)\phi}{10(\mu_s + \mu_1)}\right)$	μ_1 = viscosity of suspended fluid σ = interfacial tension a = drop radius $\Lambda = \dfrac{a(16\mu_s + 19\mu_1)}{40(\mu_s + \mu_1)\sigma}$
h	Dilute Rigid Spheres includes the inertia of the fluid	C.E.F. (3.115)	$\eta = \mu_s\left[1 + \left(\dfrac{5}{2} + 1.34\dfrac{a^2\dot\gamma\rho}{\mu_s}\right)\phi\right]$ $\Psi_1 = a^2\phi\rho\left[0.287\sqrt{\dfrac{a^2\dot\gamma\rho}{\mu_s}} - \dfrac{4}{3}\right]$ $\Psi_2 = -a^2\phi\rho\left[0.252\sqrt{\dfrac{a^2\dot\gamma\rho}{\mu_s}} - \dfrac{2}{3}\right]$	a = sphere radius ρ = density of fluid
i	Dilute Colloidal (including electrostatic interactions)	G.N.F. (Bingham) (3.44)	$\mu_0 = \left(1 + \dfrac{5}{2}\phi + \dfrac{49}{10}\phi^2\right)\mu_s$ $\tau_0 = \dfrac{3\lvert V(2a + \delta)\rvert\,\phi^2}{\pi^2 a^3}$	V = electrostatic pair potential a = particle radius δ = minimum particle separation
j	Dilute Emulsion	Second Order Fluid (3.114)	$b_1 = \mu_s\left(1 + \dfrac{5\lambda + 2}{2(\lambda + 1)}\phi\right)$	$\lambda = \dfrac{\mu_1}{\mu_S}$

298

$$b_2 = \frac{\mu_s^2 a \phi}{\sigma} \frac{(19\lambda + 16)^2}{80(\lambda + 1)^2}$$

$$b_{11} = \frac{\mu_s^2 a \phi}{\sigma} \left(\frac{19\lambda + 16}{20\,(\lambda + 1)^2} \right)$$
$$\cdot \left[\frac{3(25\lambda^2 + 41\lambda + 4)}{28\,(\lambda + 1)} - \frac{19\lambda + 16}{4} \right]$$

$$\beta_0 = \frac{2}{3}\,\mu_s\,\phi(r - 1)^2$$

$$\beta_1 = -\frac{12}{5}\,D\mu_s\phi(r - 1)^2$$

$$\beta_2 = -\mu_s \left(1 + \frac{5}{2}\,\phi + \frac{26}{147}\,\phi(r - 1)^2 \right)$$

$$\beta_5 = -\frac{1}{7}\,\mu_s\phi(r - 1)^2$$

$$\beta_3 = \beta_4 = \beta_6 = \beta_7 = \beta_8 = 0$$

$$\alpha_1 = -6D \quad \alpha_2 = 3/2$$

$$\alpha_0 = \alpha_3 = \alpha_4 = \alpha_5 = \alpha_6 = \alpha_7 = \alpha_8 = 0$$

μ_1 = viscosity of suspended fluid

a = droplet radius

σ = interfacial tension

r = particle aspect ratio $= \dfrac{\text{major axis}}{\text{minor axis}}$

D = rotational Brownian Motion Coefficient for particles

k	Dilute Nearly spherical rigid particles (subject to Brownian motion)	Hand (Table 3.8)

a. Einstein, A., *Ann. Phsik*, Vol. 19, pp. 289–302, 1906; erratum Vol. 34, pp. 591–592, 1911.

b. Taylor, G. I., *Proc. Roy. Soc.* (London), Vol. A138, pp. 41–48, 1932.

c. Simha, R., *J. Research, Nat. Bus. Stds.*, Vol. 42, pp. 409–418, 1949.

d. Frankel N. A. and Acrivos, A., *Chem. Engr. Sci.*, Vol. 22, pp. 847–853, 1967; Graham, A. L., *Appl. Sci. Res.*, Vol. 37, pp. 275–286, 1981.

e. Goddard, J. D., *J. Non-Newtonian Fluid Mech.*, Vol. 2, pp. 169–189, 1977.

f. Fröhlich, H. and Sack, R., *Proc. Roy. Soc.*, Vol. A185, pp. 415–430, 1946.

g. Oldroyd, J. G., *Proc. Roy. Soc.* Vol. A218, pp. 122–132, 1953.

h. Lin, C.-J., Peery, J. H., and Schowalter, W. R., *J. Fluid Mech.*, Vol. 44, pp. 1–17, 1970.

i. Russel, W. B., *J. Rheol.*, Vol. 24, pp. 287–317, 1980.

j. Schowalter, W. R., Chaffey, C. E., and Brenner, H., *J. Coll. and Interfac. Sci.*, Vol. 24, pp. 258–269, 1967.

k. Leal, L. G. and Hinch, E. J., *J. Fluid Mech.*, Vol. 55, pp. 745–765, 1972.

were obtained by assuming that the hydrodynamic interaction between particles for the concentrated suspension can be represented by some type of cell around an individual particle. Although these theories are of great historical interest, they greatly oversimplify the interactions between particles.

All of the theories previously mentioned assume that the inertia of the particles and suspending fluid is negligible. Lin, Peery, and Schowalter included the inertia of the fluid in their calculations and obtained shear rate dependent viscosity and normal stress coefficients.

In considering suspensions of spheroidal particles subject to Brownian motion, Leal and Hinch obtained a simplified form of the Hand equation (see Table 3.8). Barthés–Biesel and Acrivos (1973) have also obtained simplified forms of the Hand equation for dilute suspensions of solid ellipsoids, elastic spheres, and liquid droplets.

By considering the electrostatic interactions between colloidal particles, Russel obtained a constitutive relation of the Bingham type. Many other expressions ranging from the theoretical to empirical have been obtained for the yield stresses exhibited by suspensions [Bird *et al.* (1982)].

It should be emphasized that the field of suspension rheology is still in its infancy and much more information is needed—particularly about the structure of these media—before unambiguous theories can be obtained.

ACKNOWLEDGMENTS

The authors wish to acknowledge financial support provided by National Science Foundation grant CPE-8104705, the Vilas Trust Fund of the University of Wisconsin, and a grant from the John D. MacArthur Foundation. In addition, we wish to point out that we have drawn generously from material in *Dynamics of Polymeric Liquids*, Vol. 1, *Fluid Mechanics* (by R. B. Bird, R. C. Armstrong, and O. Hassager) and Vol. 2, *Kinetic Theory* (by R. B. Bird, C. F. Curtis, R. C. Armstrong, and O. Hassager), John Wiley & Sons, New York, 1987. Thanks are also due to Dr. Barbara J. Yarusso, who prepared an earlier version of Table 3.9.

REFERENCES

Astarita, G. and Marrucci, G., *Principles of Non-Newtonian Fluid Mechanics*, McGraw-Hill, New York, 1974.

Astarita, G., Marrucci, G., and Nicolais, L. (Eds.), *Rheology*, Plenum Press, New York, 1980.

Baird, D. G. and Collias, D. I., *Polymer Processing: Principles and Design*, Butterworth–Heinemann, Boston, 1995.

Barnes, H. A., Hutton, J. F., and Walters, K., *An Introduction to Rheology*, Elsevier, Amsterdam, 1989.

Barthes-Biesel, D. and Acrivos, A., *Int. J. Multiphase Flow*, Vol. 1, pp. 1–24, 1973.

Batchelor, G. K., *J. Fluid Mech.*, Vol. 41, pp. 545–570, 1970.

Batchelor, G. K. and Green, J. T., *J. Fluid Mech.*, Vol. 56, pp. 401–427, 1972.

Bird, R. B., *J. Non-Newtonian Fluid Mech.*, Vol. 5, pp. 1–12, 1979.

Bird, R. B., Armstrong, R. C., and Hassager, O., *Dynamics of Polymeric Liquids, Vol. 1, Fluid Mechanics*, 1st ed., John Wiley & Sons, New York, 1977.

Bird, R. B., Armstrong, R. C., and Hasager, O., *Dynamics of Polymeric Liquids, Vol. 1, Fluid Mechanics*, 2nd ed., John Wiley & Sons, New York, 1987.

Bird R. B., Dai, G. C., and Yarusso, B. J., *Revs. in Chem. Engr.*, Vol. 1, pp. 1–70, 1982.

Bird, R. B., Curtiss, C. F., Armstrong, R. C., and Hassager, O., *Dynamics of Polymeric Liquids*, Vol. 2, *Kinetic Theory*, 2nd ed., John Wiley & Sons, New York, 1987b.

Bird, R. B., Dotson, P. J., and Johnson, N. L., *J. Non-Newtonian Fluid Mech.*, Vol. 7, pp. 213–235, 1980.

Bird, R. B. and Öttinger, H. C., *Ann. Rev. Phys. Chem.*, Vol. 43, pp. 371–406, 1992.

Bird, R. B., Stewart, W. E., and Lightfoot, E. N., *Transport Phenomena*, John Wiley & Sons, New York, 1960.

Bird, R. B. and Wiest, J. M., *Ann. Rev. Fluid Mech.*, Vol. 27, pp. 169–193, 1995.

Boger, D. V. and Walters, K., *Rheological Phenomena in Focus*, Elsevier, Amsterdam, 1993.

Brenner, H., *Int. J. Multiphase Flow*, Vol. 1, pp. 195–341, 1974.

Brenner, H., *Prog. Ht. and Mass Trf.*, Vol. 5, Schowalter W. R. (Ed.), Pergamon Press, New York, pp. 89–129, 1972.

Burger, E. D., Chorn, L. G., and Perkins, T. K., *J. Pet. Tech.*, pp. 377–386, Feb. 1982.

Burger, E. D., Chorn, L. G., and Perkins, T. K., *J. Rheol.*, Vol. 24, pp. 603–626, 1980.

Crochet, M. J., Davies, A. R., and Walters, K., *Numerical Simulation of Non-Newtonian Flow*, Elsevier, Amsterdam, 1984.

Curtiss, C. F. and Bird, R. B., *J. Chem. Phys.*, Vol. 74, pp. 2016–2025, pp. 2026–2033, 1981 [see also Bird, R. B., Saab, H. H., and Curtiss, C. F., *J. Phys. Chem.*, Vol. 86, pp. 1102–1106, 1982 and *J. Chem. Phys.*, Vol. 77, pp. 4747–4757, 1982; Saab, H. H., Bird, R. B., and Curtiss, C. F., *J. Chem. Phys.*, vol. 77, pp. 4758–4766, 1982; Fan, X. and Bird, R. B., *J. Non-Newtonian Fluid Mech.*, Vol. 15, pp. 341–373, 1984; Schieber, J. D., Curtiss, C. F., and Bird, R. B., *I.E.C. Fund.*, Vol. 25, pp. 471–475, 1986; Schieber, J. D., *J. Chem. Phys.*, Vol. 87, pp. 4917–4927 and 4928–4936, 1987].

Curtiss, C. F., Bird, R. B., and Hassager, O., *Adv. Chem. Phys.*, Vol. 35, pp. 31–117, 1976.

Dealy, J. M., *Rheometers for Molten Plastics: A Practical Guide to Testing and Property Measurement*, Van Nostrand Reinhold, New York, 1982.

Dealy, J. M. and Wissbrun, K. F., *Melt Rheology and Its Role in Plastics Processing*, Van Nostrand Reinhold, New York, 1990.

Doi, M. and Edwards, S. F., *The Theory of Polymer Dynamics*, Oxford University Press, 1986.

Einstein, A., *Ann. Phys.*, Vol. 19, pp. 289–302, 1906, erratum 34, pp. 591–592, 1911.

Eirich, F. R. (Ed.), *Rheology: Theory and Application*, Academic Press, New York, Vol. 1, 1956, Vol. 2, 1958, Vol. 3, 1960, Vol. 4, 1967, and Vol. 5, 1969.

Ferry, J. D., *Viscoelastic Properties of Polymers*, 3rd ed., John Wiley & Sons, New York, 1980.

Giesekus, H., *Rheol. Acta*, Vol. 5, pp. 29–35, 1966.

Giesekus, H., *Phänomenologische Rheologie: Eine Einführung*, Springer-Verlag, Berlin, 1994.

Goddard, J. D., *J. Non-Newtonian Fluid Mech.*, Vol. 2, pp. 169–189, 1977.

Hull, A. M., *J. Non-Newtonian Fluid Mech.*, Vol. 8, pp. 327–336, 1981.

Janeschitz-Kriegl, H., *Polymer Melt Rheology and Flow Birefringence*, Springer, New York, 1983.

Joseph, D. D., *Fluid Dynamics of Viscoelastic Fluids*, John Wiley & Sons, New York, 1990.

Kim, S. and Karrila, S. J., *Microhydrodynamics*, Butterworth–Heinemann, Boston, 1991.

Landau, L. D. and Lifshitz, E. M., *Fluid Mechanics*, Addison–Wesley, Reading, MA, pp. 76–79, 1959.

Larson, R. G., *Constitutive Equations for Polymer Melts and Solutions*, Butterworths, Boston, 1988.

Lodge, A. S., *Body Tensor Fields in Continuum Mechanics*, Academic Press, New York, 1974.

Lodge, A. S., *Elastic Liquids*, Academic Press, New York, 1964.

Lodge, A. S., SAE Paper No. 872043, 1987.

Marrucci, G., *Trans. Soc. Rheol.*, Vol. 16, pp. 321–330, 1972.

Metzner, A. B., *Adv. in Heat Transfer*, Vol. 2, pp. 357–397, 1965.

Middleman, S., *Fundamentals of Polymer Processing*, McGraw-Hill, New York, 1977.

Oldroyd, J. G., *Proc. Roy. Soc.*, Vol. A200, pp. 45–63, 1950 [see also Vol. A245, pp. 278–297, 1958, Vol. A283, pp. 115–133, 1965; *J. Non-Newtonian Fluid Mech.*, Vol. 14, pp. 9–46, 1984].

Öttinger, H. C., *Stochastic Processes in Polymer Fluids*, Springer–Verlag, Berlin, 1995.

Pearson, J. R. A., *Mechanics of Polymer Processing*, Hemisphere Publishing Co., New York, 1983.

Rivlin, R. S. and Ericksen, J. L., *J. Rat. Mech. Anal.*, Vol. 4, pp. 323–425, 1955 [see also Rivlin, R. S., Chap. 5 in *Research Frontiers in Fluid Dynamics*, Seeger, R. J. and Temple, G. (Eds.), John Wiley & Sons, New York, 1965].

Russel, W. B., *The Dynamics of Colloidal Systems*, University of Wisconsin Press, Madison, WI, 1987.

Russel, W. B., Saville, D. A., and Schowalter, W. R., *Colloidal Dispersions*, Cambridge University Press, 1989.

Schowalter, W. R., *Mechanics of Non-Newtonian Fluids*, Pergamon, New York, 1978.

Sedov, L. I., *Introduction to the Mechanics of a Continuous Medium*, Addison–Wesley, Reading, MA, 1965.

Tadmor, Z. and Gogos, C. G., *Principles of Polymer Processing*, John Wiley & Sons, New York, 1979.

Tanner, R. I., *Engineering Rheology*, Oxford Press, Oxford, 1985 (see also Crochet, M. J., Davies, A. R., and Walters, K., *Numerical Simulation of Non-Newtonian Flow*, Elsevier, Amsterdam, 1984).

Tanner, R. I., *Trans. Soc. Rheol.*, Vol. 19, pp. 37–65, 1975.

Thomas, D. G., *AIChE J.*, Vol. 7, pp. 431–437, 1961; Vol. 9, pp. 310–316, 1963.

Tschoegl, N. W., *The Phenomenological Theory of Linear Viscoelastic Behavior: An Introduction*, Springer–Verlag, Berlin, 1989.

Walters, K., *Rheometry*, Chapman and Hall, London, 1973 (see also Dealy, J. M., *Rheometers for Molten Plastics*, Van Nostrand Reinhold, New York, 1982).

Whorlow, R. W., *Rheological Techniques*, John Wiley & Sons, New York, 1980.

Winter, H. H., *Adv. in Heat Transfer*, Vol. 13, pp. 205–267, 1977.

Yamakawa, H., *Modern Theory of Polymer Solutions*, Harper & Row, New York, 1971 (see also Doi, M. and Edwards, S. F., *Theory of Polymer Dynamics*, Oxford Press, Oxford, 1986).

4 Flow of Fluids with Viscosity and Thermal Conductivity

HARRY A. DWYER
University of California
Davis, CA

MUJEEB R. MALIK and CHAU-LYAN CHANG
High Technology Corporation
Hampton, VA

JAMES C. WILLIAMS III
Auburn University
Auburn, AL

JOSEPH A. SCHETZ
Virginia Polytechnic Institute and State University
Blacksburg, VA

CONTENTS

Handbook of Fluid Dynamics and Fluid Machinery, Edited by Joseph A. Schetz and Allen E. Fuhs
ISBN 0-471-12598-9 Copyright © 1996 John Wiley & Sons, Inc.

4.1   INTRODUCTION
Harry A. Dwyer

4.1.1   Scope and Parameter Range

This chapter of the Handbook explains the behavior of Newtonian fluids (Sec. 4.14 introduces non-Newtonian fluids) flowing at high Reynolds number and Prandtl numbers within an order of magnitude of one. For someone who is not familiar with the richness of possibilities in fluid dynamics, this choice may seem somewhat narrow, but as is well known by those with some experience, this range of parameters covers a vast spectrum of flows of practical interest and of extreme difficulty. The primary reason for practical interest stems from the fact that the flow of air or water at the macroscopic level and at observable velocities almost always is associated with a high Reynolds number. If one wishes to study the flow of air or water at normal conditions and a low Reynolds number, it is usually necessary to introduce very small dimensions or extremely slow velocities. Therefore, since air and water flows are frequently associated with a high Reynolds number, it follows directly that this parameter range is of practical importance.

The variety of high Reynolds number flows is a direct result of the fact that a typical flow can be laminar, transitional, turbulent, or all three simultaneously, depending on the geometry and the particular value of the Reynolds number. At the present time, this variety is so vast that the majority of researchers in fluid dynamics restrict themselves to a limited range of Reynolds numbers, as well as limiting or eliminating the influence of other parameters such as Prandtl and Grashof numbers. In fact, it can be said that there are very few experts in more than one parameter range of Reynolds number. The majority of researchers spend their careers in a rather restricted range in order to make progress in our understanding of fluid dynamics.

The majority of the material to be presented will consist of attached boundary layer type flows, however some material on more complex flows will be presented. See Sec. 4.9 for separated flows. This range of flows represents a vast amount of material, and it is not possible to be exhaustive in the presentation. Also, research techniques and methods for transitional and turbulent flows are presently in the state of rapid change due to such influences as the digital computer and the use of lasers in experimental work (see Chap. 15). Therefore, the material presented will represent only some of the more widely used turbulence models and will not be the latest state-of-the-art research models. There is a large amount of material in this chapter which can be applied to practical flows and also form the basis for a physical understanding of the complex phenomena associated with fluid dynamics.

4.1.2 Some Important Dimensionless Groups

It is important when approaching any problem associated with fluid dynamics to know the parameter range, or values of Reynolds and Prandtl numbers, in which the problem is posed. Only when one knows these parameters is it possible to have an understanding of the relative importance of the various terms which appear in the equations of motion and energy. From this understanding, the various approximate forms of the equations can be understood and applied in a useful way. Therefore, it will be very useful to begin this chaper with a discussion of the basic equations and the very important parameters which arise from these equations.

The equations for a compressible, perfect gas can be written as

$$\frac{\partial \rho}{\partial t} + \frac{\partial}{\partial x_j} (\rho u_j) = 0 \qquad \text{(Continuity)} \qquad (4.1)$$

$$\frac{\partial}{\partial t}(\rho u_i) + \frac{\partial}{\partial x_j}(\rho u_i u_j) = -\frac{\partial p}{\partial x} + \frac{\partial \tau_{ij}}{\partial x_j} \qquad \text{(Momentum)} \tag{4.2}$$

$$\frac{\partial(\rho h)}{\partial t} + \frac{\partial}{\partial x_j}(\rho h u_j) = \frac{Dp}{Dt} + \tau_{ij}\frac{\partial u_i}{\partial x_j} - \frac{\partial}{\partial x_j}\left(k\frac{\partial T}{\partial x_j}\right) + \tau_{ij}\frac{\partial u_i}{\partial x_j} \tag{4.3}$$

$$\text{(Energy)}$$

where the stress tensor τ_{ij} is given by

$$\tau_{ij} = \lambda \delta_{ij}\frac{\partial u_i}{\partial x_j} + \mu\left(\frac{\partial u_i}{\partial x_j} + \frac{\partial u_j}{\partial x_i}\right) \tag{4.4}$$

here λ is the bulk viscosity μ the dynamic viscosity, k the thermal conductivity, and h the static enthalpy. With the introduction of the characteristic velocities and property values of the flow, the dimensionless parameters on which the solution depends will naturally appear. For example, if the following dimensionless quantities are introduced into the above equations: $u_i' = u_i/U_\infty$, $h' = h/c_\infty T_\infty$, $\mu' = \mu/\mu_\infty$, $p' = p/\rho_\infty V_\infty^2$, $k' = k/k_\infty$, $\rho' = \rho/\rho_\infty$, $x_i' = x_i/L$, $t' = t(U_\infty/L)$, and $T' = T/T_\infty$, these equations become, with the superscripts neglected,

$$\frac{\partial \rho}{\partial t} + \frac{\partial}{\partial x_j}(\rho u_j) = 0 \qquad \text{(Continuity)} \tag{4.5}$$

$$\frac{\partial}{\partial t}(\rho u_i) + \frac{\partial}{\partial x_j}(\rho u_i u_j) = -\frac{\partial p}{\partial x_i} + \frac{1}{\text{Re}}\frac{\partial \tau_{ij}}{\partial x_j} \qquad \text{(Momentum)} \tag{4.6}$$

$$\frac{\partial}{\partial t}(\rho h) + \frac{\partial}{\partial x_j}(\rho h u_j) = (\gamma - 1)M^2\frac{Dp}{Dt} + \frac{\gamma - 1}{1}\frac{M^2}{\text{Re}}\tau_{ij}\frac{\partial u_i}{\partial x_j}$$

$$+ \frac{1}{\text{Pr}R_e}\frac{\partial}{\partial x_j}\left(k\frac{\partial T}{\partial x_j}\right) \qquad \text{(Energy)} \tag{4.7}$$

where

$$\text{Re} = \frac{\rho_\infty U_\infty L}{\mu_\infty}, \ \text{Pr} = \frac{\mu_\infty c_{p\infty}}{k_\infty}, \ M = \frac{V}{\sqrt{\gamma R T_\infty}}, \ \text{and} \ \gamma = \frac{c_p}{c_v}$$

It can be seen from the above results that the three parameters which arise from the basic equations are the Reynolds number, Re, Prandtl number, Pr, and Mach number, M. The flow over a body of given geometry will be dramatically different depending on the magnitude of these parameters. Typically, as the Reynolds number takes on large values, thin regions of high gradients develop in the flow such as boundary layers; with large Mach number, shock waves appear. These regions of high gradient are the result of the basic time scales associated with convection, diffusion, and wave propagation having much different values and present serious problems for both numerical and analytical modeling.

Both large and small values of these parameters also present the opportunity for approximate modeling, since many terms sometimes can be eliminated from the basic equations. In fact, the majority of the work presented in this chapter results from making some type of approximation, and it will be a major objective of this work to explain, justify, and limit the approximation used in high Reynolds number flows.

Another factor which must be taken into account at this time is the influence of the digital computer on the presentation of material. It is not only necessary to present subject material on numerical methods, but the basic equations and their nature should be described in a manner that lends itself to a numerical solution with a digital computer. It will be assumed that many readers will have some type of computer program available to them, and that he or she is using this Handbook as a guide for physical insight. However, there are many results which are timeless and universal and have not been affected by the digital computer. These results still form a strong foundation for this Chapter.

4.1.3 Wall Characteristics and Boundary Conditions

When dealing with a problem in fluid dynamics, there is normally considerable information contained in the solution since a fluid is a continuum. However, in a majority of cases, one is primarily interested in certain properties or characteristics of the solution near the wall. These characteristics represent the interaction between the fluid and a wall; the ones of primary interest are usually wall shear stress, τ_w, and heat transfer, q_w. For a *Newtonian fluid*, the wall stress in a boundary layer is given in terms of the velocity gradient as

$$\tau_w = \mu_w \left. \frac{\partial u}{\partial y} \right|_w \tag{4.8}$$

where u is the velocity parallel to the wall and y is the coordinate normal to the wall.
The *local skin friction coefficient* is defined as

$$C_f = \frac{\tau_w}{\rho_\infty U_\infty^2 / 2} \tag{4.9}$$

and one can obtain an average coefficient along a given line L with the expression

$$\overline{C_f} = \frac{1}{\dfrac{1}{2} \rho_\infty U_\infty^2 L} \int_0^L \tau_w \, dx \tag{4.10}$$

For the determination of wall heat transfer, one applies *Fourier's law of conduction*, and a typical formulation is

$$q_w = -k_w \left. \frac{\partial T}{\partial y} \right|_w \tag{4.11}$$

where k_w is the thermal conductivity of the fluid at the wall location. For low speed flows, the heat flux is usually normalized with a convection term to form the *local Stanton number*

$$C_h = \frac{q_w}{\rho_\infty U_\infty c_p (T_w - T_\infty)} \tag{4.12}$$

or, in terms of the *Nusselt number*,

$$\mathrm{Nu} = \frac{q_w L}{k_w (T_w - T_\infty)} = C_h \frac{\rho_\infty U_\infty L}{\mu_\infty} \frac{\mu_\infty c_p}{k_w}$$

$$= C_h \, \mathrm{Re} \, \mathrm{Pr} \tag{4.13}$$

The above definitions are a few of the many currently used. Other wall characteristics will be explained and developed as presented. The Chapter proceeds to specific technical topics beginning with a discussion of laminar incompressible boundary layers.

4.2   LAMINAR INCOMPRESSIBLE BOUNDARY LAYERS
Harry A. Dwyer

4.2.1   Two-Dimensional Incompressible Flow

One of the most useful approximations which has been used for high Reynolds number flow has been the *boundary layer concept*. This concept has formed the basis of the majority of knowledge about turbulent wall flows, and the field is developed sufficiently that boundary layer analysis is currently used as a design tool. In fact, the results and development of the subject have been so useful and fruitful that it is necessary to have many sections on different aspects and applications of boundary layer theory.

A good starting point for the development of boundary layer theory is two-dimensional, incompressible, steady flow. There are many ways to arrive at these equations and it is advised for a reader without experience to consult the many excellent physical derivations based on physical approximations [Schlichting (1979), Rosenhead (1963), White (1983), and Schetz (1993)]. In this section, the approximation is obtained by taking the following limits of the parameters and coordinates which appear in Eqs. (4.5) to (4.7): $M \to 0$, $\mathrm{Pr} \to O(1)$ (order of magnitude one), $\mathrm{Re} \to \infty$, $y \to 0$ (coordinate normal to wall), and ρ = constant. The overall physical picture contained in these limits is that the inertia or flow terms dominate throughout the flow field except for a thin region near a solid body surface. In the thin region near the body surface, the no-slip condition on velocity (or applied wall temperature) results in large velocity and temperature gradients and, thus, important influences due to viscosity and thermal conductivity. However, since the flow must be parallel to the body surface, only shear stresses parallel to the wall are important. Also, for $M \to 0$ and $\mathrm{Pr} \cong 1$, the only significant term changing fluid particle enthalpy is thermal conduction normal to the wall. The resulting equations from these limits are

$$\frac{\partial(\rho u)}{\partial x} + \frac{\partial(\rho v)}{\partial y} = 0 \tag{4.14}$$

$$\frac{\partial}{\partial x}(\rho u u) + \frac{\partial}{\partial y}(\rho u v) = -\frac{\partial p}{\partial x} + \frac{\partial}{\partial y}\left(\mu \frac{\partial u}{\partial y}\right) \tag{4.15}$$

$$\frac{\partial}{\partial x}(\rho u h) + \frac{\partial}{\partial y}(\rho v h) = \frac{\partial}{\partial y}\left(k \frac{\partial T}{\partial y}\right) \tag{4.16}$$

where x is the local coordinate parallel to the solid body surface, y is the local coordinate normal to the body surface, and u and v are the respective local velocities in the x- and y-directions.

The above equations contain many simplifications, and it is useful to summarize them in a logical and precise fashion. The major simplifications in the boundary layer approximation are the following:

i) *Coordinate system*—Due to the extreme thinness of the boundary layer, all solid surface length scale variations are neglected, as for example, when applying solid body mechanics over short distances on the earth's surface.

ii) *Continuity equation*—No approximations are made, since this equation does not contain any of the characteristic parameters for which limits have been taken.

iii) *Momentum equation, x-direction*—All stress terms have been eliminated except for shear stress parallel to the wall, while the convective and pressure terms are unchanged. This implies that convection of momentum in the x-direction dominates diffusion of momentum in the x-direction.

iv) *Momentum equation, y-direction*—This equation is neglected completely on the assumption that a significant momentum cannot be generated normal to the wall surface in the boundary layer.

v) *Energy equation*—There are two major approximations in this equation; the first says that, at low Mach number, the mechanical energy change in the boundary layer is small compared with thermal energy change. The second approximation is that diffusion of heat is significant only normal to the wall.

vi) *Mathematical implications*—The full Navier–Stokes equations (see Chap. 1) are classified as being *elliptic* in nature, which implies that fluid particles interact simultaneously in all directions through stresses, diffusion, convection, and pressure interaction. The boundary layer equations are *parabolic*, since the second derivative terms in the x-direction are eliminated as well as the entire y-momentum equation. This change in mathematical nature converts the boundary layer equations into an initial value problem along the x coordinate and is the primary reason for the relatively easy success which both analytical and numerical methods have had in solving these equations.

vii) *Numerical method implications*—The initial value and parabolic nature of the boundary layer equations drastically reduce the computer storage requirements to solve the resulting two-dimensional problem. The equations have a strong analogy with unsteady, one-dimensional diffusion problems, and there exist many efficient numerical techniques for solving problems of this type.

The next step in the use of the boundary layer approximation is to know where and when the approximation can be applied. When applying the boundary layer approximation to a physical problem, the following information must be available or calculated from other sources: 1) the location of the start of boundary layer formation or the initial conditions, 2) an inviscid ($\mu = 0$) flow field solution which yields the pressure, velocity, and temperature at the edge of the boundary layer y_e, 3) the regions of the boundary layer where the flow is laminar, transitional, and turbulent, 4) the values of viscosity and thermal conductivity must be obtained from experiments or other theories which calculate these transport properties, and 5) the conditions at the body surface concerning mass transfer and temperature. Also, it should be emphasized that the boundary layer equations cannot be used directly in regions of separated flow, since the equations do not contain enough physical information to treat a flow of this complexity.

The above information is usually very difficult to obtain in many problems, and that is one of the principal reasons for continued research on direct solution of the Navier–Stokes equations rather than the boundary layer equations. However, in a large number of problems, this information can be obtained or approximated, and boundary theory has provided a useful and efficient tool for understanding a flow process.

4.2.2 Axisymmetric Boundary Layers

A useful extension of the two-dimensional, planar, boundary layer equations is for axisymmetric flow. For this situation, the flow will be rotationally symmetric about an axis of symmetry, z, and the body surface location is given by the local surface radius, r_o, which is a function of axial location. The boundary layer positions and velocities are formulated using the same notation x, y and u, v, however the governing equations are influenced by the three-dimensional aspects of the flow. A significant contribution in this area has been made by Mangler [see Schlichting (1979)], and the boundary layer equations of continuity and momentum which result are

$$\frac{\partial}{\partial x}(r_o u) + \frac{\partial(r_o v)}{\partial y} = 0 \qquad \text{(Continuity)} \qquad (4.17)$$

$$u\frac{\partial u}{\partial x} + v\frac{\partial u}{\partial y} = -\frac{1}{\rho}\frac{\partial p}{\partial x} + \nu\frac{\partial^2 u}{\partial y^2} \qquad \text{(x-Momentum)} \qquad (4.18)$$

Remember that r_o is an implicit function of x, since x itself is defined by r_o.

It is evident immediately that the momentum equation is unchanged (due to the fact that the boundary layer thickness is small compared to r_o), while the continuity equation explicitly contains r_o, which precludes a direct analog between two-dimensional and axisymmetric boundary layers. However, Mangler devised a coordinate transformation which considerably closes the gap between the two types of boundary layer flows. The Mangler transformation defines the new variables

$$x' = \frac{1}{L}\int_o^x r_o^2 \, dx, \quad y' = \frac{r_o y}{L}$$

$$u' = u, \quad v' = \frac{L}{r_o}\left(v + \frac{yu}{r_o}\frac{dr_o}{dx}\right) \qquad (4.19)$$

where L is a characteristic reference length for the axisymmetric body. Application of the chain rule of mathematics and use of the above definitions for u' and v' results in the following equations

$$\frac{\partial u'}{\partial x'} + \frac{\partial v'}{\partial y'} = 0 \tag{4.20}$$

$$u' \frac{\partial u'}{\partial x'} + v' \frac{\partial u'}{\partial y'} = -\frac{\partial p}{\partial x'} + \nu \frac{\partial^2 u'}{\partial y'^2} \tag{4.21}$$

which are formally identical to the two-dimensional, planar, boundary-layer equations.

A major difference between Eqs. (4.14) and (4.15) and Eqs. (4.20) and (4.21) is that the pressure field (obtained from the inviscid flow) is a greatly changed function of x' as compared to x. For semi-analytical studies this can be very cumbersome. However, for numerical solutions there is essentially no difference between the transformed axisymmetric and two-dimensional equations. It should be mentioned, however, that there are solutions of the boundary layer equations which are useful for both flow geometries, and the discussion moves onward to these flow geometries.

4.2.3 Some Useful Flow Solutions of the Boundary Layer Equations

In general, since the laminar boundary layer equations are nonlinear, coupled, and partial differential equations, there are only numerical or approximate solutions available for these equations. However, there are families of relatively simple flows which can reduce the boundary layer equations to forms which are amenable to analysis which does not involve the solution of a partial differential equation. These simple flows are the so-called *similarity flows* of which the most famous is the *Blasius solution*. The conditions of the validity of these solutions are still debated, but it cannot be debated that they have been very useful in a wide number of boundary layer studies. The most readily recognizable condition for their application is the lack of length scale in the geometry of the flow. The discussion of these solutions will begin with the simple flat plate case which is better known after its founder Blasius [see Rosenhead (1963)].

Boundary Layer Flow Over a Semi-Infinite Flat Plate The geometry of this flow consists of a thin, semi-infinite flat plate which only perturbs the flow significantly in a thin region near the surface. Flow at a location far away from the surface (at the outer edge of the boundary layer) will be denoted by the symbols U_e and V_e and will consist of $U_e =$ constant and $V_e = 0$. This is the inviscid flow field needed to begin the boundary layer solution. It will be assumed that a solution away from the leading edge, where $x = 0$ is desired and that the Reynolds number based on x is large $(10{,}000 \le \mathrm{Re}_x = U_e x/\nu)$. Under these conditions, the pressure gradient is zero (by Bernoulli's equation with $U_e =$ constant) and the equations take on the form

$$\frac{\partial u}{\partial x} + \frac{\partial v}{\partial y} = 0 \tag{4.22}$$

$$u \frac{\partial u}{\partial x} + v \frac{\partial u}{\partial y} = \nu \frac{\partial^2 u}{\partial y^2} \tag{4.23}$$

with the boundary conditions

$$y = 0 \qquad u = v = 0$$

and

$$y \to \infty \qquad u \to U_e \tag{4.24}$$

These equations will only accept one boundary condition for v, and it is not entirely clear where to apply the boundary condition $u = U_e$.

The momentum equation is analogous to the unsteady diffusion equation of heat transfer, which is

$$\frac{\partial T}{\partial t} = \frac{\partial}{\partial y} \left(\alpha \frac{\partial T}{\partial y} \right) \tag{4.25}$$

and which also accepts a reduction in the number of variables to a similar solution for semi-infinite geometries [Birkhoff (1961)]. Assuming similarlity, Blasius introduced the *similarity variable*

$$\eta = y \left(\frac{U_e}{\nu x} \right)^{1/2} \tag{4.26}$$

and defined the following *stream function*

$$\psi = (\nu x U_e)^{1/2} f(\eta) \tag{4.27}$$

which leads to the velocities

$$u = \frac{\partial \psi}{\partial y} = \frac{\partial \psi}{\partial \eta} \frac{\partial \eta}{\partial y} = U_e f'(\eta)$$

$$v = \frac{1}{2} \left(\frac{\nu U_e}{x} \right)^{1/2} (\eta f' - f) \tag{4.28}$$

where the prime denotes differentiation with respect to the variable η. The continuity and momentum equations reduce to

$$ff'' + 2f''' = 0 \qquad \text{(Blasius' Eq.)} \tag{4.29}$$

and the boundary conditions become

$$\eta = 0 \qquad f = f' = 0$$
$$\eta \to \infty \qquad f' = 1 \tag{4.30}$$

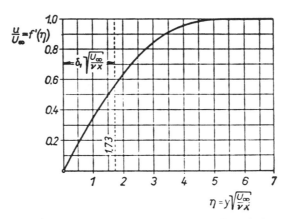

FIGURE 4.1 Primary velocity distribution predicted by the Blasius solution for the flat plate. (From Schlichting, 1979.)

The Blasius equation was difficult to solve at the time of its introduction but can now be solved relatively easily with a wide variety of numerical techniques a few of which can be found in the text of White (1993). For practial purposes, the edge of the boundary layer can be located in the vicinity of $\eta \cong 6.0$, and this location can be used quite accurately for numerical solutions.

The distributions of the primary flow velocity u and the normal velocity v are shown as functions of η in Figs. 4.1 and 4.2, and a detailed listing of the numerical values of f, f', and f'' is given in Table 4.1. From the numerical values given in the table, some useful surface quantities and boundary layer characteristics can be derived in a straightforward fashion. For example, the wall shear and local drag coefficient are given by

$$\tau_w(x) = \mu \left.\frac{\partial u}{\partial y}\right|_{y=0} = \mu \, U_e \left(\frac{U_e}{\nu x}\right)^{1/2} f''(0) = 0.332 \, \mu \, U_e \left(\frac{U_e}{\nu x}\right)^{1/2} \quad (4.31)$$

and

$$C_{fx} = \frac{\tau_w(x)}{\frac{1}{2} \rho \, U_e^2} = 0.664 \left(\frac{\nu}{U_e x}\right)^{1/2} = \frac{0.664}{(\mathrm{Re}_x)^{1/2}} \quad (4.32)$$

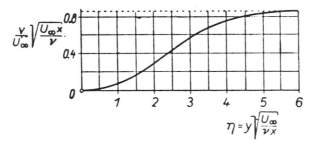

FIGURE 4.2 Normal velocity distribution for the flat plate. (From Schlichting, 1979.)

TABLE 4.1 Tabulated Values of the Blasius Function

$\eta = y \sqrt{\dfrac{U_\infty}{\nu x}}$	f	$f' = \dfrac{u}{U_\infty}$	f''
0	0	0	0.33206
0.2	0.00664	0.06641	0.33199
0.4	0.02656	0.13277	0.33147
0.6	0.05974	0.19894	0.33008
0.8	0.10611	0.26471	0.32739
1.0	0.16557	0.32979	0.32301
1.2	0.23795	0.39378	0.31659
1.4	0.32298	0.45627	0.30787
1.6	0.42032	0.51676	0.29667
1.8	0.52952	0.57477	0.28293
2.0	0.65003	0.62977	0.26675
2.2	0.78120	0.68132	0.24835
2.4	0.92230	0.72899	0.22809
2.6	1.07252	0.77246	0.20646
2.8	1.23099	0.81152	0.18401
3.0	1.39682	0.84605	0.16136
3.2	1.56911	0.87609	0.13913
3.4	1.74696	0.90177	0.11788
3.6	1.92954	0.92333	0.09809
3.8	2.11605	0.94112	0.08013
4.0	2.30576	0.95552	0.06424
4.2	2.49806	0.96696	0.05052
4.4	2.69238	0.97587	0.03897
4.6	2.88826	0.98269	0.02948
4.8	3.08534	0.98779	0.02187
5.0	3.28329	0.99155	0.01591
5.2	3.48189	0.99425	0.01334
5.4	3.68094	0.99616	0.00793
5.6	3.88031	0.99748	0.00543
5.8	4.07990	0.99838	0.00365
6.0	4.27964	0.99898	0.00240
6.2	4.47948	0.99937	0.00155
6.4	4.67938	0.99961	0.00098
6.6	4.87931	0.99977	0.00061
6.8	5.07928	0.99987	0.00037
7.0	5.27926	0.99992	0.00022
7.2	5.47925	0.99996	0.00013
7.4	5.67924	0.99998	0.00007
7.6	5.87924	0.99999	0.00004
7.8	6.07923	1.00000	0.00002
8.0	6.27923	1.00000	0.00001
8.2	6.47923	1.00000	0.00001
8.4	6.67923	1.00000	0.00000
8.6	6.87923	1.00000	0.00000
8.8	7.07923	1.00000	0.00000

(From Schlichting, 1979)

where $\mathrm{Re}_x = U_e x / \nu$. The overall or integrated drag coefficient for a flat plate is

$$\overline{C}_{f_L} = \frac{1}{L} \int_0^L C_{f_x} \, dx = \frac{1.33}{\sqrt{\mathrm{Re}_L}} \tag{4.33}$$

There are a wide variety of boundary layer characteristics which have been defined over the years. The most widely used are *boundary layer thickness*, δ, *displacement thickness*, δ^*, and *momentum thickness*, θ. The boundary layer thickness gives a measure of the region where the viscous terms are important and is usually defined as the location where the primary flow velocity approaches about 99% of the potential or inviscid flow velocity. Although the definition is not precise and δ is not a good measure of the extent of the boundary layer, it is frequently used in analytical, numerical, and experimental investigations.

The displacement thickness, δ^*, is a measure of the amount of mass displaced away from the wall by viscous terms compared to a potential flow solution. The decrease in mass is

$$\rho \delta^* U_e = \int_0^\infty \rho (U_e - u) \, dy \tag{4.34}$$

Thus

$$\delta^* = \int_0^\infty \left(1 - \frac{u}{U_e} \right) dy \tag{4.34a}$$

The definitions of δ^* is precise and does give a definite measure of the state of the boundary layer. Displacement thickness has played a key role in approximate theories which were more important before the advent of the digital computer.

The momentum thickness, θ, is used as a local measure of the momentum deficit in the boundary layer compared to a potential flow. The mathematical definition is

$$\rho U_e^2 \theta = \rho \int_0^\infty u (U_e - u) \, dy \tag{4.35}$$

or

$$\theta = \int_0^\infty \frac{u}{U_e} \left(1 - \frac{u}{U_e} \right) dy \tag{4.35a}$$

where a common mass flux has been used between the potential and boundary layer flow. As can be seen from the integral, θ is more sensitive to the shape of the velocity profile and provides a different measure of the state of boundary layer.

During the development of boundary layer theory, the use of δ^* and θ along with semi-empirical information about boundary layer structure provided a strong foundation for an approximate theory of the incompressible boundary layer for laminar and turbulent flow. A good account of these developments can be given in the books of Schlichting (1979), White (1993), and Schetz (1994). For most flows, it is not possible to develop simple expressions for the thicknesses δ^* and θ, however, for the flat plate case, simple expressions can be found by numerical integration of the values given in Table 4.1. These expressions are

$$\delta^* = 1.7208 \, \frac{\nu x^{1/2}}{U_e}$$

$$\theta = 0.664 \, \frac{\nu x^{1/2}}{U_e} \tag{4.36}$$

The thicknesses δ^* and θ can be used to obtain a measure of viscous scales at high Reynolds number.

At this point, it is useful to discuss the range of validity of the above solution. The Blasius solution is limited at both low and high Reynolds number. At low Reynolds number (approximately 10^4), the full Navier–Stokes equations must be employed. At high Reynolds number, the solution will be limited by transition to turbulent flow. The transition process for the flat plate flow is extremely sensitive to small disturbances in the inviscid flow and can be much different for external free-flight conditions compared to an internal fan driven flow. See, for example, Chap. 16 on fluid mechanics facilities. The discussion of transition is delayed to the next section.

General Similarity Solutions. More general similarity solutions have been found. They have value in providing information concerning the shape of the velocity profile with pressure gradients and in providing solutions at the stagnation point for both plane and axisymmetric geometries. The general family of similarity solutions can be approached by introducing the following transformations and definitions into the boundary layer equations

$$\xi = \frac{x}{L}, \quad \eta = \frac{y(R)^{1/2}}{Lg(x)} \tag{4.37}$$

where L is a characteristic length, $g(x)$ is a dimensionless scale factor, and R is $U_\infty L/\nu$.

In terms of a dimensionless stream function, $f(\xi, \eta)$, the continuity and momentum equations become

$$\frac{\partial^3 f}{\partial \eta^3} + \alpha f \frac{\partial^2 f}{\partial \eta^2} + \beta(1 - f'^2)$$

$$= \frac{U_e}{U_\infty} g^2 \left(\frac{\partial f}{\partial \eta} \frac{\partial^2 f}{\partial \eta \partial \xi} - \frac{\partial^2 f}{\partial \eta^2} \frac{\partial f}{\partial \xi} \right) \tag{4.38}$$

where

$$f(\xi, \eta) = \frac{\psi(x, y) \, (R)^{1/2}}{L U_e(x) \, g(x)},$$

$$\alpha = \frac{Lg}{U_\infty} \frac{d}{dx} (U_e g), \quad \beta = \frac{L}{U_\infty} g^2 \frac{dU_e}{dx} \tag{4.39}$$

and the boundary conditions for f are

$$f = \frac{\partial f}{\partial \eta} = 0 \qquad \text{at} \qquad \eta = 0$$

$$\frac{\partial f}{\partial \eta} \to 1 \qquad \text{at} \qquad \eta \to \infty \qquad (4.40)$$

Equation (4.38) is not self-similar and cannot be made self-similar by any exact deductions. However, upon neglecting the right hand side and assuming α and β are constant, a self-similar form is obtained. This form is

$$f''' + \alpha f f'' + \beta(1 - f'^2) = 0 \qquad (4.41)$$

where the prime again denotes differentiation with respect to η. The general conditions for the application of similar solutions are

$$U_e(x) = Cx^m \qquad (4.42)$$

which physically corresponds to potential flow over a wedge and which has the following values of α, β, g, and η.

$$\alpha = 1$$

$$\beta = \frac{2m}{m + 1}$$

$$g = \left(\frac{2}{m + 1} \frac{x}{L} \frac{U_\infty}{U_e} \right)^{1/2}$$

$$\eta = y \left(\frac{m + 1}{2} \frac{U_e}{\nu x} \right)^{1/2} \qquad (4.43)$$

The important special cases are the following three:

I. Blasius Flow

$\beta = 0 \qquad m = 0$

II. Plane Stagnation Flow

$\beta = 1 \qquad m = 1$

III. Axisymmetric Stagnation Flow

$\beta = 1/2 \qquad m = 1/3$

(*Note*: III applies after the Mangler transformation)

The Blasius solution needs no further discussion, however it is appropriate to comment on the stagnation flows. Both stagnation point equations and solutions are remarkable in that they can be made identical to the exact solutions to the Navier–

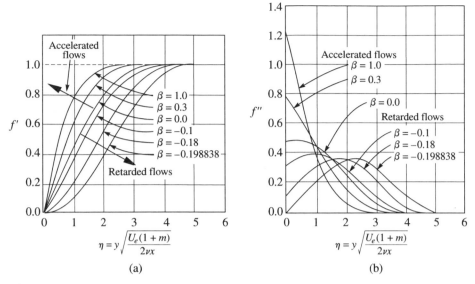

FIGURE 4.3 Velocity and shear functions for the self similar velocity profiles. (From White, 1993.)

Stokes equations for the stagnation point conditions. This fact implies that there is no need to correct a boundary layer solution at the stagnation point for low Reynolds number effects, and indeed this is the case for flows at high overall Reynolds number.

The velocity profiles for various values of m and β are shown in Fig. 4.3, while detailed values of the solution are given in Table 4.2. These velocity profiles have been useful for the development of approximate theories, and the literature of boundary layer theory has copious solutions based on these results. Values of m greater than zero correspond to flows with favorable pressure gradients ($\partial p/dx < 0$), while negative values correspond to unfavorable or positive pressure gradients.

A key point to be gained from these results is the very small pressure gradients needed for the condition of separation of the boundary layer. This result is borne out by full numerical solutions of the boundary layer equations, however turbulent flows have considerably more ability to flow against adverse pressure gradients. The *separation of boundary layers* is important for all high Reynolds number flows because of its dramatic influence on the potential flow and the process of transition to turbulence. Because of the strong interaction of the separation process with other processes in the flow, the discussion of separation is delayed to subsequent sections.

4.3 BOUNDARY-LAYER STABILITY AND TRANSITION
Mujeeb R. Malik and Chau-Lyan Chang

4.3.1 Introduction

Since Obsorne Reynolds' pipe flow experiment in 1883, the subject of laminar-turbulent transition has remained a challenge for researchers in the field of fluid

TABLE 4.2 Tabulated Values of Self-Similar Velocity Profiles. Solutions of the Equation of Similar Profiles, $f''' + \alpha f f'' + \beta(1 - f'^2) = 0$, $f(0) = f'(0) = 0$. $f'(\infty) = 1$

$\alpha =$	1	1	1	1	1	1	1	1	1	0	-1	-1
$\beta =$	-0.15	-0.18	-0.1988	-0.18	-0.15	0	0.5	1	2	1	4	1
$F''(0) =$	-0.132	-0.097	0	0.1285	0.2161	0.4696	0.928	1.233	1.687	1.1547	2.273	1.086
η	F'	F'	F'	F'	F'	F'	F'	F'	F'	F'	F'	F'
0	0	0	0	0	0	0	0	0	0	0	0	0
0.1			0.001	0.014	0.022	0.047	0.090	0.118	0.159	0.110	0.207	0.104
0.2			0.004	0.029	0.046	0.094	0.176	0.227	0.298	0.211	0.377	0.197
0.3			0.009	0.047	0.072	0.141	0.256	0.325	0.419	0.302	0.513	0.282
0.4		-0.024	0.016	0.066	0.098	0.188	0.331	0.414	0.522	0.384	0.620	0.357
0.5	-0.041		0.025	0.087	0.125	0.234	0.402	0.495	0.610	0.458	0.705	0.425
0.6			0.036	0.109	0.156	0.281	0.467	0.566	0.683	0.523	0.770	0.485
0.7		-0.020	0.049	0.134	0.187	0.327	0.528	0.630	0.745	0.582	0.821	0.538
0.8	-0.058		0.064	0.160	0.220	0.372	0.583	0.686	0.786	0.633	0.860	0.585
0.9			0.080	0.188	0.253	0.417	0.634	0.735	0.838	0.679	0.891	0.627
1.0			0.099	0.217	0.287	0.461	0.681	0.778	0.872	0.719	0.914	0.664
1.2	-0.050	+0.013	0.142	0.279	0.359	0.545	0.761	0.847	0.921	0.786	0.947	0.726
1.4			0.193	0.346	0.433	0.624	0.826	0.897	0.953	0.837	0.967	0.774
1.6	-0.019	0.076	0.250	0.417	0.507	0.697	0.876	0.932	0.973	0.876	0.979	0.813
1.8			0.313	0.490	0.581	0.761	0.914	0.957	0.985	0.906	0.986	0.843
2.0	+0.035	0.167	0.380	0.562	0.651	0.817	0.942	0.973	0.991	0.929	0.991	0.868
2.4	0.119	0.283	0.523	0.700	0.776	0.901	0.976	0.991	0.998	0.960	0.996	0.904
2.8	0.226	0.419	0.664	0.815	0.872	0.953	0.991	0.997	0.999	0.977	0.998	0.928
3.2	0.354	0.563	0.786	0.899	0.936	0.980	0.997	0.999	1.000	0.987	0.999	0.944
3.6	0.493	0.699	0.879	0.952	0.972	0.993	0.999	1.000		0.993	1.000	0.956
4.0	0.633	0.814	0.940	0.980	0.989	0.998	1.000			0.996		0.965
4.4	0.759	0.898	0.974	0.993	0.997	0.999				0.998		0.971
4.8	0.858	0.951	0.990	0.998	0.999	1.000				0.999		0.976
5.2	0.926	0.979	0.997	0.999	1.000					0.999		0.980
5.6	0.966	0.992	0.999	1.000						1.000		0.983
6.0	0.986	0.997	1.000									0.985
6.4	0.995	0.999										0.986
6.8	0.999	1.000										

(Reproduced from Rosenhead, 1963.)

319

mechanics. Reynolds took the view that transition is related to the instability of laminar flows and that once a critical Reynolds number is exceeded, the instability in the flow causes it to change to turbulent motion. Lord Rayleigh, in a series of papers dating from 1880 to 1913, studied the instability of inviscid flows and obtained many useful results, including the well-known inflectional instability criterion. This theory, however, did not explain the transition observed in Reynold's experiment and other flows involving solid boundaries. Both Taylor and Prandtl realized that viscosity may play an important role in hydrodynamic stability. Tollmien (1929) solved the viscous stability equation (Orr–Sommerfeld equation) for a flat plate boundary layer, assuming that the mean flow variation along the plate can be ignored. Under this assumption, Tollmien was able to find a neutral boundary within which infinitesimally small waves could grow exponentially. Schlichting computed the amplification factors for fixed frequency disturbances and correlated them with transition Reynolds numbers obtained in wind-tunnel experiments. He found logarithmic amplification factors to be 4 and 9 for transition in two different tunnels. The flat-plate experiment of Schubauer and Skramstad (1947) clearly showed that in wind tunnels with low turbulence levels, boundary-layer transition is preceded by the amplification of instability waves whose characteristics are in general agreement with the theory of Tollmien and Schlichting (TS).

The transition process is actually composed of several distinct physical processes (see Fig. 4.4). External disturbances in the form of free stream vorticity, sound, entropy spots, disturbances associated with particulates, surface roughness, and vi-

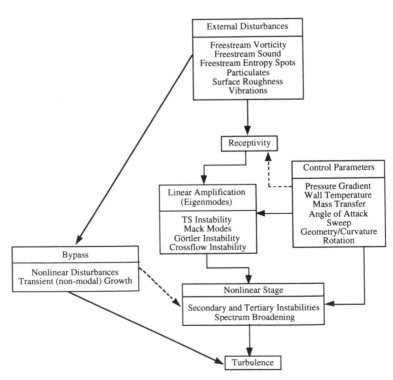

FIGURE 4.4 A simplified view of the laminar-turbulent transition process.

brations get internalized in the boundary layer through a process known as *receptivity*—a term first coined by Morkovin (1969). Thus, receptivity is the genesis of instability waves in the boundary layer and is an initial boundary value problem as opposed to an eigenvalue problem considered in hydrodynamic stability theory. The identification and complete definition of the initial disturbance field is a fundamental issue in the prediction of transition. In wind tunnels, the dominant disturbances, except for roughness and vibrations, may originate in the settling chamber and/or the tunnel walls depending upon the Mach number and flow conditions. The flow on a particular part of an aerospace vehicle in flight may see disturbances directly coming from the free stream or those emanating from other parts of the vehicle configuration itself. Thus, the disturbance field, which is internalized by the boundary layer through the process of receptivity, is generally not known. However, considerable progress has been made in understanding the receptivity process when the disturbance environment is given. The basic question is the identification of mechanisms responsible for removing any disparity of scales between external disturbance fields and boundary layer instability waves. One such mechanism is provided by mean flow and surface inhomogeneities [see Goldstein and Hultgren (1989)] which rescale the free-stream disturbances and provide forcing at the wavelengths appropriate for boundary-layer instabilities. The most significant contributions in the field of receptivity have been made by Goldstein (1983, 1985) and Ruban (1985). For a recent review, see Saric *et al.* (1994).

In general, the receptivity stage is a localized process and takes place in the *subcritical region* in which the internalized disturbances begin to damp until the *critical region* is reached. In cases of distributed receptivity, the process of internalization of disturbances into the boundary layer is most effective in the critical region. The resulting disturbance level in the critical region determines the subsequent behavior of the transition process. Beyond the critical region, both existing and newly forced disturbances, if any, are amplified. For low amplitudes, this amplification is described by linear stability theory, and the particular normal modes responsible for the amplification are a function of the particular mean flow under consideration. For boundary-layer flows, these instabilities include the viscous TS waves referred to above, Rayleigh (inflectional) waves (e.g., instability due to crossflow, adverse pressure gradient or first-mode disturbances at supersonic Mach numbers), Mack modes (for hypersonic flows) and, for curved streamlines, Görtler vortices. The linear amplification takes place at a slow scale involving long streamwise distances. This is the stage at which the disturbance amplification may be most efficiently modulated by parameters such as pressure gradient, wall heat/mass transfer. These control parameters may also affect the receptivity process and, to a lesser extent, the nonlinear stage discussed below.

When sufficient disturbance amplitude is achieved through linear amplification, nonlinearity sets in and, through secondary and teritary instabilities, the flow becomes *transitional*. Because the disturbance growth is fast in this case, this nonlinear region subtends a relatively small portion (spatially) of the entire transition process (perhaps, with the exception of high Mach number flows). On the contrary, for low initial disturbances, the linear process extends over a much longer distance along the body, and this very fact provides the rationale for use of linear stability theory, in the form of e^N method (see Sec. 4.3.6 below), for estimate of onset of transition and design of laminar-flow control (LFC) surfaces [Pfenninger (1977)]. However, it must by emphasized that a truly predictive method for transition should include the physical richness of both the receptivity and the various secondary and tertiary instability processes.

When high levels of initial disturbances are present, transition may take place through nonlinear mechanisms completely bypassing the linear regime. An example of such transition is the well-known problem of attachment-line contamination [Pfenninger (1965)], a problem which is central in the application of laminar flow control technology to swept-wing aircraft. Pipe flow is stable to infinitesimal disturbances, yet transition can occur at a Reynolds number of about 2000 based upon mean-flow velocity and pipe diameter [see Schlichting (1979)]. While this limit can be substantially increased by minimizing the disturbances in the entrance region of the pipe, the fact remains that linear stability theory fails to be a guide for transition in Reynolds' pipe flow. Similarly, the critical Reynolds number for plane Poiseuille flow is found to be about 5772 (based upon centerline velocity and channel half-height) [Orszag (1971)] while transition has been observed at Reynolds numbers as low as 1500 [Carlson et al. (1982)]. To date, two explanations of this phenomenon have been put forth: 1) there are finite-amplitude equilibrium states which become unstable to three-dimensional disturbances (the theory of secondary instability [Orszag and Patera (1983) and Herbert (1983), (1988)] and 2) transient or nonmodal linear growth [Trefethen et al. (1993)] which may be followed by nonlinear mechanisms. Landahl's (1980) algebraic instability concept is related to the second scenario. Each flow has a $Re_{\theta_{min}}$ [Morkovin (1985)] below which turbulent spots cannot be sustained, and in some flows this $Re_{\theta_{min}}$ is much below the critical Reynolds number for amplified eigenmodes. For specific design applications, it is important to know this minimum Reynolds number for sustenance of turbulence.

It is clear that the road to turbulence is not unique, but, for a class of external flows, the experience has shown that linear theory can be used as a guide for estimation of the onset of transition (see, however, the later section on the attachment-line instability). The theory is particularly useful for LFC studies, since it can be utilized to correlate transition location and parameterize for such effects as Mach number, pressure gradient, wall suction/blowing, heat transfer, body geometry, and sweep.

In what follows, a brief description of the theory of boundary-layer stability is given. The application of the theory to various instability mechanisms relevant in boundary-layer transition is also discussed. For additional material, the reader may refer to various IUTAM symposia on Laminar-Turbulent Transition [Eppler and Fasel (1980), Kozlov (1985), and Arnal and Michel (1990)], ICASE/NASA workshops on Instability and Transition [Dwoyer and Hussaini (1987), Hussaini and Voigt (1990), and Hussaini et al. (1992)] and AGARD proceedings on this subject [AGARD (1977), AGARD (1984), AGARD (1988), AGARD (1992), and AGARD (1993)]. For additional readings see Reshotko (1976), Smith (1979), Morkovin (1993), and Drazin and Reid (1981).

4.3.2 Mathematical Formulation

Let us consider a three-dimensional boundary-layer such as the one which develops over an infinite-swept wing as shown in Fig. 4.5. This will allow us to develop the theory in a more general framework which can be simplified when a specific appli-

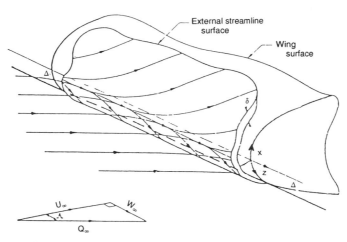

FIGURE 4.5 Flow near the leading edge of a swept wing.

cation is under consideration. Let the surface chordwise, wall-normal and spanwise coordinates be denoted by $x = X/l_0$, $y = Y/l_0$, and $z = Z/l_0$, respectively ($y = 0$ denotes the wall), where the length scale l_0 will be prescribed later. Note that the wall is curved only along the x-coordinate. Now, let the x, y, and z components of the velocity and pressure be given by

$$(u^\dagger, v^\dagger, w^\dagger) = U_s\{U(x, y) + u(x, y, z, t), V(x, y) + v(x, y, z, t),$$

$$W(x, y) + w(x, y, z, t)\} \tag{4.44}$$

$$p^\dagger - \rho U_s^2(P + p(x, y, z, t)) \tag{4.45}$$

where the superscript † represents a dimensional quantitity and U_s in an appropriate velocity scale. Here U, V, and W are mean-flow velocity components obtained by solving the boundary-layer equations, whereas u, v, w represent the perturbation velocity components in x, y, z directions, respectively. Similarly, P and p represent the mean and perturbation pressures. We assume that the Reynolds number is large and that the local radius of curvature $r^\dagger(x) = 1/k^\dagger(x)$ is much larger than the boundary-layer thickness, δ (i.e., $k^\dagger\delta \ll 1$). In this case, the relevant first-order incompressible boundary-layer equations can be written as

$$\frac{\partial \bar{u}}{\partial X} + \frac{\partial \bar{v}}{\partial Y} = 0 \tag{4.46}$$

$$\bar{u}\frac{\partial \bar{u}}{\partial X} + \bar{v}\frac{\partial \bar{u}}{\partial Y} = U_e\frac{dU_e}{dX} + \nu\frac{\partial^2 \bar{u}}{\partial Y^2} \tag{4.47}$$

$$\bar{u}\frac{\partial \bar{w}}{\partial X} + \bar{v}\frac{\partial \bar{w}}{\partial Y} = \nu\frac{\partial^2 \bar{w}}{\partial Y^2} \tag{4.48}$$

where $(\bar{u}, \bar{v}, \bar{w}) = U_s\{U, V, W\}$ and ν is the kinematic viscosity. The boundary conditions are

$$\bar{u}(0) = 0, \ \bar{v}(0) = \bar{v}_w, \ \bar{w}(0) = 0$$

$$\bar{u} \to U_e, \ \bar{w} \to W_e, \ Y \to \infty \tag{4.49}$$

where nonzero $\bar{v}_w$ indicates the possibility of a transpiring wall. Note that the free-stream spanwise velocity component, W_e, does not vary with x.

By considering the transformations

$$\eta = \sqrt{\frac{U_e}{\nu X}} \, Y,$$

$$f' = \frac{df}{d\eta} = \frac{\bar{u}}{U_e},$$

$$g = \frac{\bar{w}}{W_e}, \tag{4.50}$$

the above equations can be rewritten as

$$f''' + \frac{m+1}{2} ff'' + m(1 - f'^2) = X \left(f' \frac{\partial f'}{\partial X} - f'' \frac{\partial f}{\partial X} \right) \tag{4.51}$$

$$g'' + \frac{m+1}{2} fg' = Xf' \frac{\partial g}{\partial X} \tag{4.52}$$

where m is the pressure gradient parameter

$$m = \frac{X}{U_e} \frac{dU_e}{dX} \tag{4.53}$$

For self-similar flows ($U_e = cX^m$), the right-hand side drops out and Eqs. (4.51) and (4.52) reduce to

$$f''' + \frac{m+1}{2} ff'' + m(1 - f'^2) = 0 \tag{4.54}$$

$$g'' + \frac{m+1}{2} fg' = 0 \tag{4.55}$$

Equations (4.54) and (4.55) are subject to the boundary-conditions

$$f(0) = f_w, \ f'(0) = 0, \ g(0) = 0$$

$$f' \to 1, \ g \to 1 \text{ as } Y \to \infty \tag{4.56}$$

As a prerequisite to any stability analysis involving boundary layers, the above equations [Eqs. (4.51) and (4.52) for nonsimilar flows and Eqs. (4.54) and (4.55) for similar flows] must be solved very accurately. Various numerical methods exist which can be used to readily solve these equations [Cebeci and Smith (1974) and Schetz (1994)].

The nonlinear perturbation equations can be derived by substituting Eqs. (4.44) and (4.45) into the incompressible Navier–Stokes equations and subtracting the mean-flow equations. If $\kappa = \kappa^{\dagger} l_0$ and $|\partial k / \partial x| \ll 1$, the equations governing the perturbation quantities are

$$\frac{\partial u}{\partial t} + U \frac{\partial u}{\partial x} + u \frac{\partial U}{\partial x} + V \frac{\partial u}{\partial y} + v \frac{\partial U}{\partial y} + W \frac{\partial u}{\partial z} + \kappa(Uv + Vu) + \frac{\partial p}{\partial x} - \frac{1}{R} \nabla^2 u = F_1$$

$$(4.57)$$

$$\frac{\partial v}{\partial t} + U \frac{\partial v}{\partial x} + u \frac{\partial V}{\partial x} + V \frac{\partial v}{\partial y} + v \frac{\partial V}{\partial y} + W \frac{\partial v}{\partial z} - 2\kappa Uu + \frac{\partial p}{\partial y} - \frac{1}{R} \nabla^2 v = F_2$$

$$(4.58)$$

$$\frac{\partial w}{\partial t} + U \frac{\partial w}{\partial x} + V \frac{\partial w}{\partial y} + W \frac{\partial w}{\partial z} + \frac{\partial p}{\partial z} - \frac{1}{R} \nabla^2 w = F_3$$

$$(4.59)$$

$$\frac{\partial u}{\partial x} + \frac{\partial v}{\partial y} + \frac{\partial w}{\partial z} + \kappa v = 0$$

$$(4.60)$$

where F_1, F_2, and F_3 represent the nonlinear terms

$$F_1 = -u \frac{\partial u}{\partial x} - v \frac{\partial u}{\partial y} - w \frac{\partial u}{\partial z} - \kappa u v$$

$$F_2 = -u \frac{\partial v}{\partial x} - v \frac{\partial v}{\partial y} - w \frac{\partial v}{\partial z} + \kappa u^2$$

$$F_3 = -u \frac{\partial w}{\partial x} - v \frac{\partial w}{\partial y} - w \frac{\partial w}{\partial z} \qquad (4.61)$$

and $\nabla^2 = \partial^2/\partial x^2 + \partial^2/\partial y^2 + \partial^2/\partial z^2 + \kappa \, \partial/\partial y$. The boundary conditions are

$$u = v = w = 0 \quad \text{at} \quad y = 0$$

$$u \to 0, \, v \to V_m(x), \, w \to 0, \quad \text{as} \quad y \to \infty \qquad (4.62)$$

where V_m signifies a nonzero value in order to compensate for the change of displacement thickness associated with instability waves. In Eqs. (4.57)–(4.60), the Reynolds number R is defined as $R = U_s l_0/\nu$, where the proper length scale l_0 and the velocity scale U_s are problem dependent. Another important parameter which is a measure of the wall curvature is the Görtler number, $G = R\sqrt{|\kappa|}$, which is relevant for boundary-layer flow over concavely curved walls. In principle, the evolution of the disturbances is governed by Eqs. (4.57)–(4.60). Numerical solutions of these equations can be obtained by accurate discretization schemes as is done in direct numerical simulations (DNS) (see Sec. 21.5 for a discussion of DNS). We will discuss two alternatives to DNS in the following sections.

4.3.3 Linear Parabolized Stability Equations

In the linear Parabolized Stability Equations (PSE), the nonlinear terms, F_1, F_2, and F_3 are set to zero and the solution of Eqs. (4.57)–(4.60) is sought in the form

$$\phi(x, y, z, t) = \hat{\phi}(x, y)E + \text{complex conjugate} \tag{4.63}$$

where

$$\phi = \{u, v, w, p\}^T, \quad \hat{\phi}\{\hat{u}, \hat{v}, \hat{w}, \hat{p}\}^T$$

$$E = \exp\left[i\left(\int_{x_0}^x \alpha(\bar{x})\, d\bar{x} + \beta z - \omega t\right)\right]. \tag{4.64}$$

The assumption implicit in Eq. (4.63) and the PSE approximation, in general, is that the disturbance ϕ can be compartmented into an amplitude function, $\hat{\phi}$, and a wave part represented by the exponent in E. Thus,

$$\phi_x = (\hat{\phi}_x + i\alpha\hat{\phi})E, \quad \phi_{xx} = (-\alpha^2\hat{\phi} + 2i\alpha_x\hat{\phi} + \hat{\phi}_{xx})E, \tag{4.65}$$

$$\phi_y = \hat{\phi}_y E, \quad \phi_{yy} = \hat{\phi}_{yy}E, \quad \hat{\phi}_z = i\beta\hat{\phi}E, \quad \hat{\phi}_{zz} = -\beta^2\hat{\phi}E \tag{4.66}$$

where complex conjugates are implied so that the physical disturbance ϕ is real. In the PSE approximation, it is assumed that $\hat{\phi}$ is a slowly varying function of x so that $\hat{\phi}_{xx}$ can be set to zero. Under this assumption, Eqs. (4.57)–(4.60) and (4.63), (4.65), and (4.66) yield

$$\gamma_1\hat{u} + \hat{v}U_y + i\alpha\hat{p} - \gamma_2\hat{u} = -\{\hat{u}U_x + V\hat{u}_y\} - \left\{U\hat{u}_x + \hat{p}_x - 2\frac{i\alpha}{R}\hat{u}_x - \frac{i\alpha_x}{R}\hat{u}\right\}$$

$$- \kappa\left(U\hat{v} + V\hat{u} + \frac{1}{R}\frac{\partial\hat{u}}{\partial y}\right) \tag{4.67}$$

$$\gamma_1\hat{v} + \hat{p}_y - \gamma_2\hat{v} = -\{\hat{u}V_x + V\hat{v}_y + \hat{v}V_y\} - \left\{U\hat{v}_x + 2\frac{i\alpha}{R}\hat{v}_x - \frac{i\alpha_x}{R}\hat{v}\right\}$$

$$+ \kappa\left(2U\hat{u} + \frac{1}{R}\frac{\partial\hat{u}}{\partial y}\right) \tag{4.68}$$

$$\gamma_1 \hat{w} + \hat{v} W_y + i\beta \hat{p} - \gamma_2 \hat{w} = -\{\hat{u} W_x + V\hat{w}_y\} - \left\{ U\hat{w}_x + 2\frac{i\alpha}{R}\hat{w}_x - \frac{i\alpha_x}{R}\hat{w} \right\}$$

$$+ \frac{\kappa}{R}\frac{\partial \hat{w}}{\partial y} \tag{4.69}$$

$$i\alpha\hat{u} + \hat{v}_y + i\beta\hat{w} = -\hat{u}_x - \kappa\hat{v} \tag{4.70}$$

where

$$\gamma_1 = i(\alpha U + \beta W - w)$$

$$\gamma_2 = \frac{1}{R}\left[-\alpha^2 + \frac{\partial^2}{\partial y^2} - \beta^2 \right] \tag{4.71}$$

The boundary conditions are

$$\hat{u}(x, 0) = \hat{v}(x, 0) = \hat{w}(x, 0) = 0$$

$$\hat{u}(x, y) \to 0, \ \hat{v}(x, y) \to 0, \ \hat{w}(x, y) \to 0, \ y \to \infty \tag{4.72}$$

Note that for the stability problem, homogeneous boundary conditions are used. Nonhomogeneous boundary conditions would need to be employed for receptivity computations. The initial conditions at $x = x_0$ are

$$\hat{u}(x_0, y) = g_1(y), \ \hat{v}(x_0, y) = g_2(y), \ \hat{w}(x_0, y) = g_3(y) \tag{4.73}$$

Since Eqs. (4.67)–(4.70) are parabolic in the streamwise direction, no outflow boundary conditions are needed.

Note that the left-hand side of Eqs. (4.67)–(4.70) contain only y derivatives, while the right-hand side contains boundary layer growth terms and the x derivatives of the disturbance amplitude function. The only curvature term which can be shown to be of significance in boundary-layer stability is the $2\kappa U\hat{u}$ term in Eq. (4.68). If the right-hand side of these equations is dropped, then we obtain homogeneous ordinary differential equations leading to the well-known Orr–Sommerfeld problem to be discussed in Sec. 4.3.5.

Numerical solution of the parabolized stability equations requires discretization in both the x and y directions. We discretize the streamwise derivative by a backward Euler step and wall-normal derivatives by fourth-order accurate compact differences [Malik *et al.* (1982)]. Homogeneous boundary conditions at the wall and in the free stream are imposed. The initial conditions (g_1, g_2, g_3) are obtained by a local approximation to Eqs. (4.67)–(4.70) and by solving the associated eigenvalue problem. It can be shown that, just as for parabolized Navier–Stokes (PNS), the above system is ill-posed for an initial value problem due to upstream propagation of acoustic waves. However, stable marching solution can be obtained if one of the following two conditions are satisfied (for incompressible flow): $\Delta x > 1/|\alpha_r|$ or $\hat{p}_x = 0$. The first condition implies that about six steps per disturbance wavelength are allowed. This is not a severe limitation, and calculations for two-dimensional boundary layers show that PSE results with only three steps per wavelength agree quite well with accurate Navier–Stokes computations using 60 grid points per wavelength [see Joslin *et al.* (1992)]. For TS waves, if α is chosen such that most of the ellipticity in the problem is absorbed in $i\alpha\hat{p}$ term, then $\hat{p}_x$ can be dropped which allows the condition first to be relaxed. It is also possible to consider a global iteration scheme in which multiple marching sweeps are performed and elliptic terms such as $\hat{p}_x$ and even $\hat{\phi}_{xx}$ are completely retained in the solution. Such a scheme may be particularly useful for flows involving small separation bubbles.

A crucial step in PSE calculations is the determination of the complex wave number, α. There are various ways in which this can be accomplished. For a given initial guess, the new value of α can be determined in an iterative fashion from the following equation

$$\alpha_n = \alpha_{n-1} - \frac{i \int_0^\infty (\hat{u}^*, \hat{v}^*, \hat{w}^*) \frac{\partial}{\partial x} \begin{pmatrix} \hat{u} \\ \hat{v} \\ \hat{w} \end{pmatrix} dy}{\int_0^\infty (|\hat{u}|^2 + |\hat{v}|^2 + |\hat{w}|^2)\, dy} \tag{4.74}$$

where * denotes a complex conjugate, and α_{n-1}, α_n are the old and new values of α. Iterations to solve for α_n continue until the second term drops below a specified tolerance.

4.3.4 Nonlinear Parabolized Stability Equations

For finite-amplitude disturbances, the right-hand side of Eqs. (4.57)–(4.60) is not negligible, and one must resort to nonlinear PSE. Nonlinear parabolized stability equations can be derived from Eqs. (4.57)–(4.60) in a manner similar to that for linear PSE, except that F_1, F_2, F_3 are nonzero. In this case, Eq. (4.63) is replaced by the discrete Fourier series representation

$$\phi = \sum_{m=-M}^{M} \sum_{n=-N}^{N} \chi_{mn}(x, y) e^{i(n\beta z - m\omega t)} \tag{4.75}$$

where M and N represent one-half the number of modes kept in the analysis. Here, the frequency, ω, and wave number, β, are chosen such that the longest period and wave length are $2\pi/\omega$ and $2\pi/\beta$ in the temporal and spanwise domains, respectively. Substituting Eq. (4.75) into Eqs. (4.57)–(4.60), we obtain governing equations for χ_{mn} which are elliptic. As in Eq. (4.63), we decompose the disturbance into a fast varying wave-like part and a slowly varying amplitude function and write χ_{mn} as

$$\chi_{mn}(x, y) = A_{mn}(x)\Psi_{mn}(x, y)$$

$$A_{mn}(x) = e^{i\int_{x0}^{x} \alpha_{mn}(\bar{x})\, d\bar{x}} \tag{4.76}$$

where Ψ_{mn} is the amplitude function ($\hat{u}_{mn}$, $\hat{v}_{mn}$, etc.) for the Fourier mode ($m\omega$, $n\beta$) and α_{mn} is the associated streamwise (complex) wave number. The parabolized stability equations (PSE) for the amplitude function of a single Fourier mode (m, n) can be written after the same approximation used for linear PSE, as

$$L_0\Psi_{mn} + L_1\frac{\partial\Psi_{mn}}{\partial x} + L_2\frac{\partial\Psi_{mn}}{\partial y} + L_3\frac{\partial^2\Psi_{mn}}{\partial y^2} = F_{mn}/A_{mn} \qquad (4.77)$$

where L_0, L_1, L_2, L_3 are coefficient matrices and are given, for example, in Malik *et al.* (1994). The nonlinear forcing function, F_{mn}, is the Fourier component of the total forcing, $F = (F_1, F_2, F_3)$, and can be evaluated by the Fourier series expansion

$$F(x, y, z, t) = \sum_{m=-M}^{M} \sum_{n=-N}^{N} F_{mn}(x, y)e^{i(n\beta z - m\omega t)} \qquad (4.78)$$

The PSE, Eq. (4.77), can be used to study nonlinear interaction of various modes (e.g., crossflow/crossflow, Görtler/TS, etc.) or one can study the onset of transition to turbulence provided appropriate initial conditions are prescribed.

4.3.5 The Quasi–Parallel Approximation

In the quasi-parallel approximation, the effect of boundary-layer growth is assumed to be of a lower order and, hence, ignored. In other words, the mean flow is a function of the normal coordinate y only i.e., $U = U(y)$, $V = 0$, $W = W(y)$ and the wave form, Eq. (4.63), can be rewritten as

$$\phi(x, y, z, t) = \hat{\phi}(y)e^{i(\alpha x + \beta z - \omega t)} + \text{complex conjugate} \qquad (4.79)$$

If the curvature effect is also ignored, then the governing linear stability equations can be obtained by setting the right-hand side of Eqs. (4.67)–(4.70) to zero. The homogeneous equations along with the boundary conditions, Eq. (4.72), then constitute an eigenvalue problem described by the complex *dispersion relation*

$$\alpha = \alpha(\omega, \beta) \qquad (4.80)$$

which may be solved numerically [see, e.g., Malik (1990)]. In the temporal stability theory, the wavenumber α, β are real and ω is complex. In this case, disturbances of frequency ω_r (the real part of ω) grow in time if the imaginary part $\omega_i > 0$. In the spatial stability theory, the frequency ω is real but α, β are complex. Here, the real parts of α, β (i.e., α_r and β_r) represent the disturbance wave numbers and imaginary parts (α_i, β_i) the growth rates in the x and z directions, respectively. In the present case of two-dimensional or quasi-three-dimensional flow (infinite swept wing), β_i is set to zero and disturbances are assumed to grow in x when $-\alpha_i > 0$. The temporal and spatial growth rates are related through Gaster's (1962) group velocity transformation.

The homogeneous form of Eqs. (4.67)–(4.70) can be manipulated and combined to yield a single fourth-order equation

$$\frac{d^4\hat{v}}{dy^4} - 2(\alpha^2 + \beta^2)\frac{d^2\hat{v}}{dy^2} + (\alpha^2 + \beta^2)^2\hat{v} = iR\left[(\alpha U + \beta W - \omega)\right.$$
$$\left.\cdot\left\{\frac{d^2\hat{v}}{dy^2} - (\alpha^2 + \beta^2)\hat{v}\right\} - (\alpha U'' + \beta W'')\hat{v}\right] \qquad (4.81)$$

which is the well-known *Orr–Sommerfeld equation*. The associated boundary conditions are

$$\hat{v} = \frac{d\hat{v}}{dy} = 0 \quad \text{at} \quad y = 0$$

$$\hat{v}, \frac{d\hat{v}}{dy} \to 0 \quad \text{as} \quad y \to \infty \qquad (4.82)$$

In the limit of infinite Reynolds number, Eq. (4.81) reduces to the second-order equation

$$\frac{d^2\hat{v}}{dy^2} - \left[\alpha^2 + \beta^2 + \frac{\alpha U'' + \beta W''}{\alpha U + \beta W - \omega}\right]\hat{v} = 0 \qquad (4.83)$$

For a two-dimensional wave in a two-dimensional boundary layer, this equation can be rewritten as

$$\frac{d^2\hat{v}}{dy^2} - \left(\alpha^2 + \frac{U''}{(U - c)}\right)\hat{v} = 0 \qquad (4.84)$$

where $c = c_r + ic_i$ with c_r the disturbance phase speed and αc_i the growth rate for temporal stability analysis. Note that if $U = c_r$, the above equation is singular for a neutral disturbance. The transverse region near such a singularity is called the *critical layer*, and the effect of viscosity becomes important in this region. Using Eq. (4.84), Rayleigh showed that a necessary condition for instability is that the basic velocity profile should have an inflection point ($U'' = 0$). There are various flows where the basic flow profile contains an inflection point and is unstable to inviscid disturbances. However, the Blasius velocity profile is not inflected and is stable to inviscid disturbances, but, in this case, viscosity plays a destabilizing role and this instability is described by the Orr–Sommerfeld equation.

4.3.6 e^N Method

In a low-disturbance environment, the streamwise extent where linearly unstable waves grow is much larger than the final nonlinear breakdown. Therefore, the onset of transition (the location where mean boundary-layer characteristics, such as skin friction, begin to depart from the laminar boundary layer) can be estimated by correlating with linear disturbance amplification. This idea was first utilized by Smith and Gamberoni (1956) and Van Ingen (1956) who correlated the linear stability results with experimentally determined transition locations.

The e^N (or N-factor) method is an amplitude-ratio method where the amplitude A of a fixed frequency disturbance at some location x is related to the A_0 at the neutral point x_0 by

$$\frac{A}{A_0} = e^N \tag{4.85}$$

and the exponent is computed from the integral

$$N(\omega) = -\int_{x_0}^{x} \alpha_i \, dx \tag{4.86}$$

The procedure is to repeat the N-factor computation for different disturbance frequencies and determine the maximum N factor at the experimentally observed transition location. It has been found that the value $N \sim 10$ best describes the low-disturbance experimental data [Bushnell et al. (1988)]. However, it is clear that the value of N will depend upon factors such as free-stream disturbance level, surface finish, etc. To account for the effect of free-stream turbulence, Mack (1977) suggested the empirical correlation

$$N = -8.43 - \ln (Tu)^{2.4} \tag{4.87}$$

where Tu is the free-stream turbulence level. Hence, for example, if $Tu = .05\%$ (low disturbance) N will be 9.8 and if $Tu = 0.5\%$ (high disturbance) N will drop to 4.3.

For two-dimensional disturbances, the above N-factor computations involve repetitive calculations for various frequencies ω to find the maximum $N(\omega)$. For three-dimensional disturbances, these calculations have to also be repeated for various spanwise wave numbers to find maximum $N(\omega, \beta)$. For subsonic two-dimensional flows, $\beta = 0$ yields the highest N factors, while for supersonic two-dimensional flows, the highest N factors are associated with the oblique waves ($\beta \neq 0$). In three-dimensional flows, β will always be nonzero.

4.3.7 Tollmien–Schlichting Instability

A two-dimensional boundary-layer is subject to viscous instability first found by Tollmien and Schlichting which results in downstream traveling waves known as *TS waves*. Let us consider flow over a flat plate for which $m = 0$. In this case, for $g = 0$, the equations reduce to the Blasius equation

$$f''' + \tfrac{1}{2} ff'' = 0 \tag{4.88}$$

with

$$f(0) = 0, f'(0) = 0$$
$$f' \to 1, \eta \to \infty \tag{4.89}$$

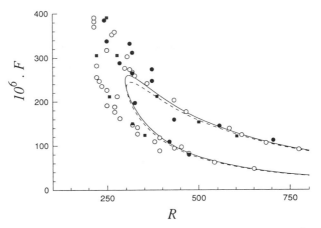

FIGURE 4.6 Neutral curve for flat-plate boundary layer: — — — quasi-parallel, —— non-parallel; experiment: $\bigcirc$ Ross *et al.* (1970), $\bullet$ Schubauer and Skramstad (1947), $\blacksquare$ Kachanov *et al.* (1977).

Here if $U_s = U_e$, $l_0 = \sqrt{\nu X_0/U_e}$, then the appropriate Reynolds number, R, can be represented as $R = \sqrt{U_e X_0/\nu} = \sqrt{\mathrm{Re}_x}$ where X_0 is the distance from the leading edge of the plate.

The solution of the stability equations for the Blasius mean flow yields the neutral boundary ($\alpha_i = \omega_i = 0$) in the F–R plane as shown in Fig. 4.6. Here, F is the nondimensional frequency

$$F = \omega_r/R = \frac{2\pi\nu S}{U_e^2} \tag{4.90}$$

where S is the disturbance frequency in Hertz.

Two curves are shown in the figure. The one labeled "quasi-parallel" was obtained by solving the Orr–Sommerfeld equation while the other labeled "nonparallel" was solved by using the method of multiple-scale [Nayfeh (1981)]. As shown by Gaster (1974), the effect of nonparallelism in this case is small [see Bertolotti *et al.* (1992) for further discussion of nonparallel effects]. This is also evidenced in Fig. 4.7 which shows the variation of growth rate of a fixed frequency disturbance with Reynolds number. Except at high frequencies, the effect of boundary layer growth is small. Here, the nonparallel results are obtained by solving the linear PSE's, in which case, as with any other method which accounts for nonparallel effect, the growth rate depends upon the variable chosen to measure the amplification of the disturbance, e.g.,

$$\sigma_u = -\alpha_i + \mathrm{Real}\left(\frac{1}{\hat{u}}\frac{\partial \hat{u}}{\partial x}\right)_{\max}$$

$$\sigma_v = -\alpha_i + \mathrm{Real}\left(\frac{1}{\hat{v}}\frac{\partial \hat{v}}{\partial x}\right)_{\max} \tag{4.91}$$

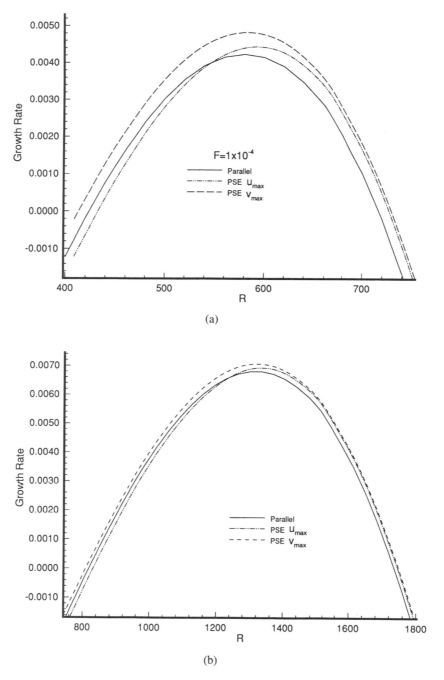

FIGURE 4.7 Linear growth of fixed frequency disturbances in flat-plate boundary layer: —— quasi-parallel, —··—··— PSE ($\hat{u}_{max}$), — — — PSE ($\hat{v}_{max}$). (a) $F = 1 \times 10^{-4}$, (b) $F = 0.3 \times 10^{-4}$.

where the subscript max indicates that the quantity is computed at the location where the amplitude function $\hat{u}$ or $\hat{v}$ is maximum.

The quasi-parallel N factors, computed by using Eqs. (4.81), (4.82), and (4.86), for the flat-plate boundary-layer, are given in Fig. 4.8. The first application by Smith and Gamberoni (1956) suggested that a value of $N = 9$ correlates the onset of transition, hence the method came to be known as e^9 method. If $N = 9$ is taken as the transition criterion, then transition in a flat plate boundary layer occurs at a Reynolds number of $Re_x = 3.1 \times 10^6$. Transition onset is influenced by various parameters such as wall suction, heat transfer, pressure-gradient and compressibility. Wall suction, heat transfer (cooling in air and heating in water), favorable pressure-gradient and compressibility all act to stabilize the flow and, hence increase the transition Reynolds number. Wall blowing, heat transfer (cooling in water, and heating in air) and adverse pressure gradients all tend to decrease the transition Reynolds number. In the case of adverse pressure gradients, boundary-layer profiles develop an inflection point and become inviscidly unstable [Wazzan *et al.* (1967)] and the critical Reynolds number decreases with the value of m (from about 520, based on displacement thickness, $m = 0$ to as low as 64 at $m = -0.091$ which corresponds to the Falkner–Skan profile at separation). Govindarajan and Narasimha (1993) show that the effect of nonparallelism is significant for flows with adverse-pressure gradients.

The reason why the above mean flow modifiers stabilize or destabilize the boundary layer is linked to the term proportional to the curvature of the boundary-layer profile ($\alpha U'' + \beta W''$) in Eq. (4.81). For simplicity, let us consider two-dimensional flow, in which case $W = W'' = 0$. We know from the inviscid stability equation, Eq. (4.84), that the bondary-layer profile becomes inviscidly unstable if it has an inflection point within the layer. The Blasius profile is not inflected but is unstable to viscous disturbances. However, if it is modified (for example by adverse pressure

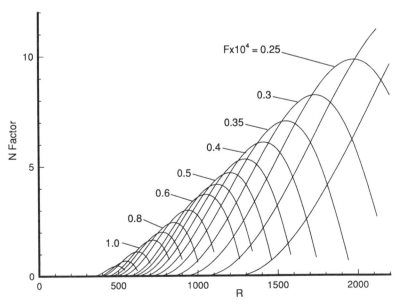

FIGURE 4.8 *N*-factor calculations for various fixed frequency TS waves in flat-plate boundary layer.

gradient) so that it tends to have an inflection point, it will become more unstable. In contrast, if it is modified so that it becomes fuller, it will be more stable. In order to explain this, let us rewrite the mean x-momentum equation in the neighborhood of the wall, assuming temperature-dependent dynamic viscosity $\mu = \mu(T)$,

$$\mu \bar{u}_{YY} = \rho \bar{v}_w \bar{u}_Y + P_X - \frac{d\mu}{dT} T_Y \bar{u}_Y \qquad (4.92)$$

Now for Blasius flow with no suction $\bar{u}_{YY}(0) = 0$ and $\bar{u}_{YY}$ is negative within the boundary layer. But, if $P_X < 0$ (favorable) or $\bar{v}_w < 0$ (suction) or $T_y > 0$ (cooling) with $d\mu/dT > 0$ (e.g., air) then $\bar{u}_{YY}(0)$ will become negative, i.e., $\bar{u}(Y)$ will become fuller and hence more stable. On the other hand, if $P_X > 0$ (adverse), $\bar{v}_w > 0$ (blowing) or $T_y < 0$ (heating) with $d\mu/dT > 0$ (e.g., air) then $\bar{u}_{YY}(0)$ will be positive but somewhere in the boundary layer $\bar{u}_{YY}$ will change sign, hence the boundary layer will be more unstable. For water $d\mu/dT < 0$, therefore, to stabilize the boundary layer, heating ($T_Y < 0$) must be applied.

Assuming that $N = 9$ correlates the transition location for a low-disturbance environment, Masad and Malik (1993) developed an analytical correlation to account for the effect of uniform wall suction, heat transfer and Mach number ($M_\infty \leq 0.8$). The correlation for transition Reynolds number $(\text{Re}_x)_{tr}$ is given by

$$(\text{Re}_x)_{tr} = [a(a_1 + M_\infty^{a_2} + a_3 M_\infty^{a_4})(b_1 + 10^8 |v_w|^{b_2} + b_3 10^8 |v_w|^{b_4})$$
$$\cdot (c_1 + (T_{ad}/T_w)^{c_2} + c_3(T_{ad}/T_w)^{c_4})]^2 \qquad (4.93)$$

where constants are given as: $a = 0.0183$, $a_1 = 4.1$, $a_2 = 3.5$, $a_3 = 1.41$, $a_4 = 1.83$, $b_1 = 1310$, $b_2 = 1.28$, $b_3 = 0.082$, $b_4 = 1.0$, $c_1 = 7.9$, $c_2 = 5.8$, $c_3 = 9.1$, and $c_4 = 5.9$. In Eq. (4.93), v_w is the uniform nondimensional suction velocity given by $v_w = \bar{v}_w/U_e$, T_w is the actual wall temperature, and T_{ad} is the adiabatic wall temperature. The correlation is valid for suction only and for mild heat transfer ($0.95 \leq T_w/T_{adb} \leq 1.05$) in air. According to the results of Masad and Malik (1993), the disturbance frequency which yields transition for $N = 9$, is given by

$$F \times 10^6 = \left(\frac{254 \times 10^6}{\text{Re}_{x_{tr}} - 431800}\right)^{0.719} \qquad (4.94)$$

The effect of surface roughness on transition is an important issue. For realistic sufaces, roughness could result from manufacturing, corrosion, inset impingement, and icing. Furthermore, the surfaces might include steps and gaps such as the steps at the joints between a wing and control surfaces. These surface inhomogeneities influence transition in two ways: 1) enhanced receptivity, and 2) mean-flow modification directly influencing the stability characteristics. An example of the latter is provided from the work of Masad and Iyer (1994) who considered the flow past a hump on an otherwise smooth flat plate (see Fig. 4.9). The corresponding pressure distribution is given in Fig. 4.10 which shows four regions of pressure gradients in the neighborhood of the hump-adverse, favorable, and a strong-adverse followed by a mild favorable region. Away from the hump on either side, the zero-pressure gradient of a smooth flat plate is recovered. Several parameters (such as the hump

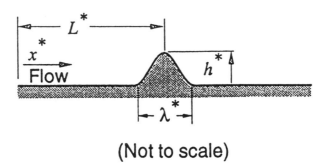

(Not to scale)

FIGURE 4.9   A two-dimensional hump on a flat plate.

height, length, location, and free-stream unit Reynolds number) govern the effect of a hump on transition, but the overall effect is destabilzing as shown in Fig. 4.11. The figure shows that as the hump height increases, the estimated transition location (based on e^9) moves continuously upstream until it reaches the separated flow region on the back side of the hump.

The above examples are provided to indicate how the results of the linear theory can be used to estimate the onset of transition. However, as pointed out earlier, the actual physical mechanism of boundary-layer transition are much more complex.

Figure 4.12 shows the results of nonlinear PSE calculations where a two-dimensional TS wave with frequency $F = 10^{-4}$ enters into parametric resonance with two oblique waves ($\beta/R = 3.6 \times 10^{-4}$) of the same frequency (fundamental secondary instability). In Fig. 4.12 the initial amplitude of the TS wave ($\bar{u}_{rms}$) is 0.3% and of the oblique waves 0.05%. The oblique waves [(1, 1) mode; $m = 1$, $n = 1$ in Eq. (4.75)] decay initially, but when the two-dimensional TS wave [(1, 0) mode] reaches

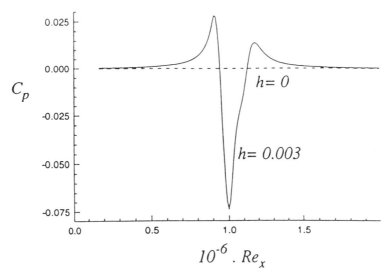

FIGURE 4.10   Variation of pressure coefficient (C_p) for a separating incompressible flow due to a hump with $\lambda = \lambda^*/L^* = 0.2$, $h = h^*/L^* = 0.003$, and $U_\infty L^*/\nu_\infty = 10^6$.

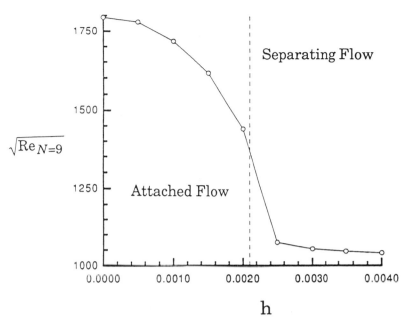

FIGURE 4.11 Effect of 2D roughness/hump on predicted transition Reynolds number using $N = 9$.

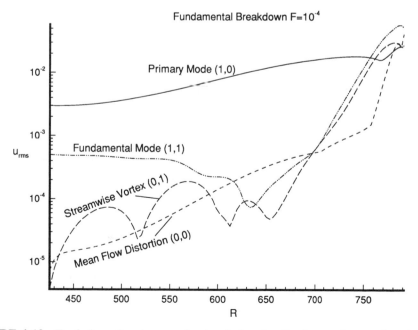

FIGURE 4.12 Evolution of various modes (u_{rms}) showing fundamental secondary instability in flat plate boundary layer. The initial amplitude of the 2D TS wave [(1, 0) mode] of frequency $F = 10^{-4}$ is 0.3% and that of oblique waves [(1, 1) mode] 0.05%.

an amplitude of about 1% they begin to grow much faster than the TS wave due to parametric resonance until they gain amplitude which is higher than the TS wave itself. Another important mode which grows is the streamwise vortex (0, 1) mode generated by the interaction of (1, 0) and (−1, 1) modes. The mean-flow distortion [(0, 0) mode] becomes significant only towards the end of the simulation. The distribution of the instantaneous streamwise velocity component in x, z plane is given in Fig. 4.13. Here, x and z have been scaled with $l_0 = \sqrt{\nu X_0 / U_e}$ where $X_0 = 54$ mm. The figure first shows straight vertical lines indicating the 2D, TS wave which is then modulated in the z direction due to the presence of oblique waves, and finally the aligned lambda vortex structure appears towards the end of the computational domain. This constitutes the first stage in the breakdown of the *orderly* two-dimensional TS waves which is followed by other stages where various harmonics come to play an increasingly important role as the flow heads for transition to turbulence. Direct numerical simulation of Navier–Stokes equations [see, e.g., Kleiser and Zang (1991)] is needed for these highly-nonlinear stages. Figures 4.14 and 4.15 show the results for subharmonic resonance where a sufficiently amplified TS (2, 0) mode causes rapid growth of oblique modes with half the frequency [(1, ±1) modes]. Streamwise vortices [(0, 2) mode or (0, 1) in Fig. 4.12] constitute an essential ingredient in all routes to transition. Figure 4.16 taken from Kegelman (1994) for flow over an ogive cylinder shows a striking resemblance to the computed results in Fig. 4.13, indicating that the transition in the experiment was caused by the fundamental resonance mechanism.

Figure 4.17 shows the computed variation of the normalized wall shear ($\partial u / \partial y$) given by both secondary instability mechanisms. Calculations were performed with the same initial amplitudes for both cases. It can be seen that nonlinear PSE predicts

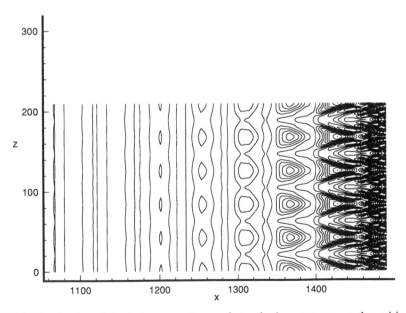

FIGURE 4.13   A plot of instantaneous streamwise velocity contours at the critical layer height for the fundamental secondary instability case of Fig. 4.12.

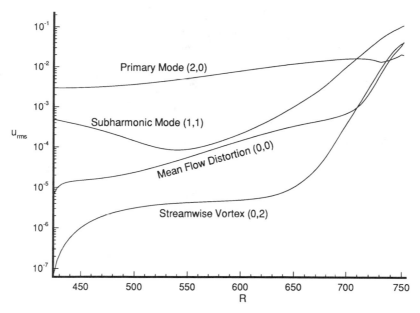

FIGURE 4.14 Evolution of various modes (u_{rms}) showing subharmonic secondary instability in flat-plate boundary layer. The initial amplitude of the 2D TS wave [(2, 0) mode] of frequency $F = 10^{-4}$ is 0.3% and that of oblique waves [(1, 1) mode] of frequency $F = 0.5 \times 10^{-4}$ is 0.05%.

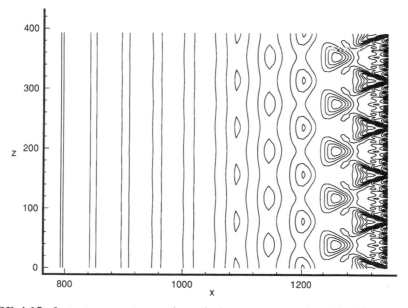

FIGURE 4.15 Instantaneous streamwise velocity contours at the critical layer height for the subharmonic resonance case of Fig. 4.14.

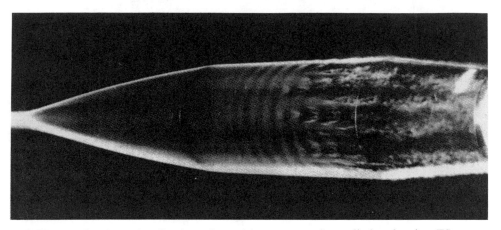

FIGURE 4.16 Flow visualization of transition on an ogive cylinder showing TS waves followed by aligned lambda vortices indicating fundamental secondary instability similar to the one in Fig. 4.13. (Photo courtesy of Kegelman of NASA Langley.)

the onset of transition which weakly depends upon the mechanism responsible for transition. In this example, the initial amplitude of the disturbance modes was arbitrarily assigned. However, if the initial amplitudes are coupled with external disturbances through the receptivity mechanisms, then PSE is capable of predicting transition from first principles.

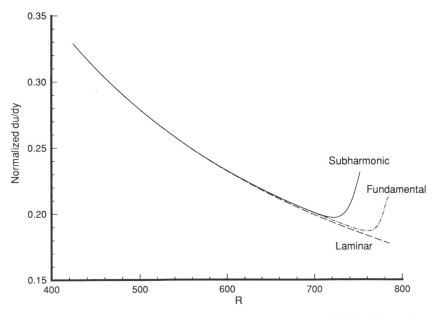

FIGURE 4.17 Comparison of computed wall shear (normalized $\partial u/\partial y$) distribution associated with the breakdown mechanisms of Figs. 4.12 and 4.14.

4.3.8 Crossflow Instability

In swept-wing flows, the chordwise pressure gradient near the leading edge causes inviscid streamlines to be curved in the planes parallel to the wing surface (see Fig. 4.5). Associated with this streamline curvature is a pressure gradient which acts in a direction normal to the streamlines and introduces a secondary flow within the boundary layer. This secondary flow, known as *crossflow*, is subject to inviscid instability due to the presence of an inflection point [Gregory, Stuart, and Walker (1955)] and is the main cause of transition in swept-wing flows. For a recent review on the subject, see Reed and Saric (1989). Crossflow instability often results in the formation of stationary corotating vortices commonly called *crossflow vortices*. This phenomenon is observed in swept-wing boundary layers as well as in other geometries such as rotating disks and cones [Gray (1952), Owen and Randall (1952), Poll (1985), and Kobayashi *et al.* (1980)]. Traveling crossflow disturbances are also possible, and the role of traveling vs. stationary disturbances is of interest in swept-wing boundary-layer transition. Stationary crossflow disturbances are directly generated due to surface roughness [Wilkinson and Malik (1985)]. On the other hand, traveling disturbances are generated due to the interaction of free-stream disturbances (e.g., acoustic waves) and surface inhomogeneities [Crouch (1993), (1994), and Choudhari (1994)] and therefore have lower initial amplitude.

Here, we discuss the crossflow instability for a model 3D boundary layer, but the qualitative features are applicable to flow near the leading edge of a swept wing and to any other three-dimensional boundary layer. The boundary layer we consider is that constructed by introducing a spanwise velocity component (W_e along z axis) to the classical Hiemenz problem ($m = 1$ in $U_e = cX^m$). If we use W_e as the velocity scale and $l_0 = \sqrt{\nu/c}$ (where ν is the kinematic velocity) as a typical length scale, then

$$U = (x/\bar{R})f'(y)$$

$$V = -(1/\bar{R})f(y)$$

$$W = g(y) \tag{4.95}$$

where f and g are governed by the ordinary differential equations

$$f''' + ff'' + (1 - f'^2) = 0 \tag{4.96}$$

$$g'' + fg' = 0 \tag{4.97}$$

which can be obtained from Eqs. (4.54) and (4.55) by setting $m = 1$.

The Reynolds number $\bar{R}$ in Eq. (4.95) is given as $\bar{R} = W_e l/\nu$. A Reynolds based upon chordwise velocity components can also be defined as $R = U_e l_0/\nu = X/l_0$. The local angle of the inviscid streamline, with respect to x axis, is given as

$$\theta = \arctan(W_e/U_e) = \arctan(\bar{R}/R) \tag{4.98}$$

Figure 4.18 shows the mean velocity profiles in directions tangential and across the inviscid stream at an R of 500 and $\bar{R} = 500$. The velocity profiles U_t and U_c are

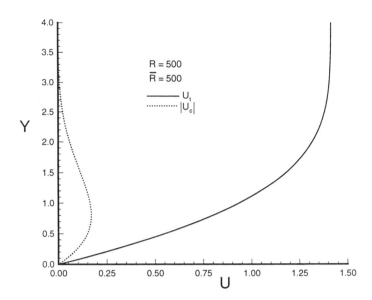

FIGURE 4.18   Streamwise U_t (——) and crossflow U_c ($\cdots\cdots$) velocity profiles in the swept Hiemenz flow. $R = \overline{R} = 500$.

defined as

$$U_t = \overline{u} \cos \theta + \overline{w} \sin \theta$$

$$U_c = \overline{u} \sin \theta - \overline{w} \cos \theta \tag{4.99}$$

where the streamline angle θ is defined in Eq. (4.98). These velocity profiles have been scaled with the spanwise inviscid velocity, W_e.

Crossflow instability is associated with the inflectional velocity profile U_c which, for swept wings, is positive towards the center of curvature of the streamline. The flow becomes unstable when crossflow Reynolds number $R_{cf} \gtrsim 40$ where R_{cf} is defined by

$$R_{cf} = \frac{\overline{U}_c \delta_{.1}}{\nu} \tag{4.100}$$

where $\overline{U}_c$ is the maximum value of the crossflow velocity, U_c, and $\delta_{.1}$ is the thickness where the crossflow velocity has dropped to 10% of $\overline{U}_c$. In swept-wing flows, transition usually occurs where R_{cf} becomes of $O(200)$.

Quasi-parallel calculations are shown for stationary as well as traveling disturbances with frequency $F = 0.75 \times 10^{-4}$ (where $F = 2\pi\nu S/W_e^2$, S being the frequency in Hertz) in Fig. 4.19. It is clear that traveling disturbances amplify more than the stationary disturbances according to linear theory. However, stationary disturbances are found to dominate when experiments are performed in low-disturbance wind tunnels. This is due to the lower initial amplitude of traveling modes because of the difference in the receptivity mechanisms of stationary and traveling disturbances. In the figure, there are two curves associated with $F = 0.75 \times 10^{-4}$, one

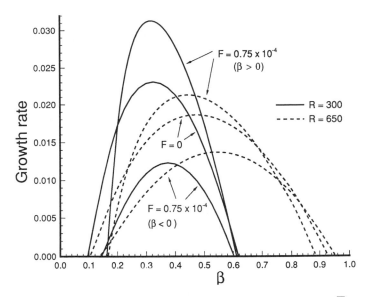

FIGURE 4.19 Quasi-parallel growth rate for positive and negative β at $\bar{R} = 500$. ——, $R = 300$; – – –, $R = 650$.

with positive β and another with negative β, the latter with smaller growth rates. The stationary vortex and $\beta > 0$ traveling disturbance has peak growth rate at $\beta \approx 0.35$ as shown in the figure for $R = 300$. In the downstream, the peak shifts to higher wavenumbers and lies, for example, at $\beta \approx 0.45$, $R = 650$. The two families of unstable traveling disturbances are further shown in Fig. 4.20, where the growth rate of the most amplified (among various wave orientations) disturbance is plotted

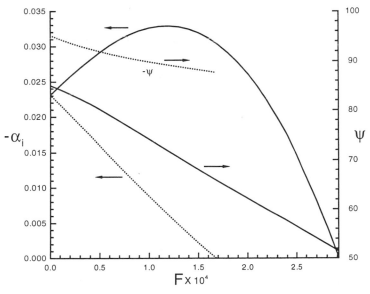

FIGURE 4.20 Growth rates (most amplified) and orientation (ψ) of the two families of unstable modes at $\bar{R} = 500$, $R = 300$.

as a function of frequency. The family with high growth rates has its wave vector oriented at positive angles with respect to the inviscid flow streamline (angles measured from the convex side), while the family with lower growth rates has its wave vector oriented at negative angles. The relative sense of the two modes depends upon the direction of the crossflow, with the more amplified mode always oriented opposite to the crossflow direction. In both cases, the direction of the group velocity lies at small angles to the inviscid streamline direction. Thus, the disturbance energy propagates downstream for both modes as also noted by Mack (1985).

We now compare the quasi-parallel growth rate results with those obtained by solving linear PSE equations, Eqs. (4.67)–(4.70). Figure 4.21(a) shows the results for $\bar{R} = 250$ for stationary vortices while Fig. 4.21(b) shows the results for a frequency of $F = 0.75 \times 10^{-4}$. In case of PSE, different growth rate results are obtained for the $\hat{u}$, $\hat{v}$ and $\hat{w}$ components of velocity as pointed out earlier. At low R (between 200 and 400), there is considerable difference between these growth rates with $\hat{u}$ growth higher than $\hat{v}$ and $\hat{w}$, but the latter two approach the same value at higher Reynolds numbers. The quasi-parallel growth rate is, in general, close to the growth rate based upon the $\hat{w}$ component, except at lower Reynolds numbers where it lies somewhere in between the three growth rates. The growth rate can also be defined based upon the total disturbance energy which accounts for all the velocity components, and growth rates based upon this definition suggests that the nonparallel effect is usually destabilizing. The effect of concave surface curvature on crossflow instability is destabilizing [Zurigat and Malik (1994)] while the effect of convex curvature is stabilizing [Masad and Malik (1994)].

The evolution of the maximum (in y) disturbance amplitude for the stationary disturbance for $\bar{R} = 500$, $\beta = 0.4$ is given in Fig. 4.22 using nonlinear PSE's. The initial amplitude of $\max(\hat{u}^2 + \hat{w}^2)^{1/2}$ is assigned to be $0.001 \, W_e$. Amplitudes of all the velocity components are given. Initially, the spanwise velocity, $\hat{w}$, is higher than the chordwise velocity, $\hat{u}$. Later, the magnitude of the two switches as the inviscid streamline angle decreases (note that $\theta = 45°$ when $R = 500$). The magnitude of the normal velocity, $\hat{v}$, is much lower than $\hat{u}$ and $\hat{w}$ at all R. It is clear that the nonlinear N factors ($\ln A/A_0$) at $R = 650$ are 7 and 5 for $\hat{u}$ and $\hat{w}$, respectively. The corresponding value from quasi-parallel linear analysis is about 9. Thus, as the crossflow disturbances gain larger amplitudes ($\sim 5\%$), the growth rate decreases from that given by the linear theory, and these disturbances tend to reach a quasi-equilibrium when they gain amplitudes of about 30%.

At such large amplitudes, the crossflow disturbances roll up into corotating vortices which appear to be half-mushroom as shown in Fig. 4.23. The figure represents a plot of the u velocity in the y, z plane. The high-speed fluid moves down near $z \approx 12$ resulting in fuller velocity profiles. In contrast, the low-speed fluid moves up near $z \approx 5$ resulting in a thicker boundary layer. Associated with these profiles are regions of high- and low-wall shear which would cause these vortices to leave an imprint on a solid wall in a flow visualization experiment. The footprint of crossflow instability leaves the impression of streaks as depicted in Fig. 4.24 for a swept wing taken from Dagenhart *et al.* (1989). The jagged imprint shows the boundary between laminar and turbulent flow. The reason why this boundary is so nonuniform is linked to the nonuniform distribution of small surface imperfections near the wing leading edge which directly feed into the generation of stationary crossflow vortices. Hence,

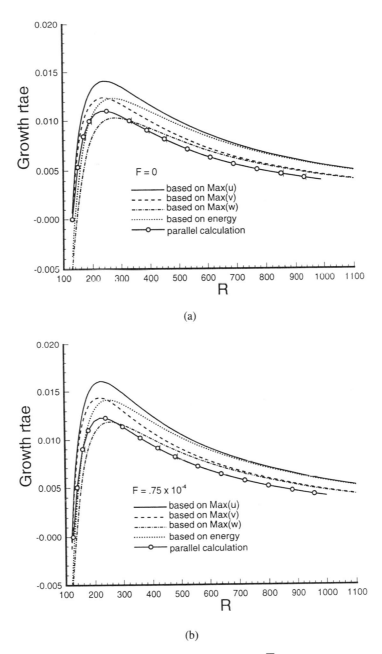

(a)

(b)

FIGURE 4.21 Disturbance growth rate for $\beta = 0.4$ and $\overline{R} = 250$. (a) stationary vortex, (b) traveling disturbance with frequency $F = 0.75 \times 10^{-4}$, ——, based on max $(\hat{u})$, – – –, based on max $(\hat{v})$, –·–·–, based on max $(\hat{w})$, ·····, based on energy, — ○ —, parallel calculation.

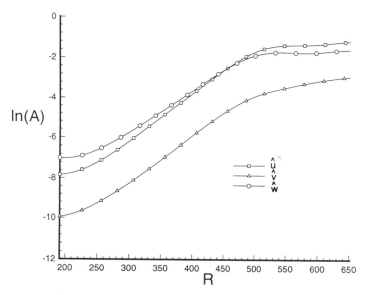

FIGURE 4.22 Nonlinear evolution of maximum (over y) amplitude of the disturbance velocity components $\hat{u}$ (— □ —), $\hat{v}$ (— △ —), $\hat{w}$ (— ○ —) for stationary disturbances with initial amplitude of 0.1%.

the nonuniform distribution of initial amplitude of crossflow vortices leads to breakdown at different locations along the wing.

How a stationary disturbance leads to turbulence is a question of considerable interest. This question has been addressed in the experiments of Kohama (1984), Kohama *et al.* (1991), Kohama and Motegl (1994), and theoretical studies of Balachander *et al.* (1992) and Malik *et al.* (1994). It has been found that these vortices become unstable to high-frequency, secondary disturbances which are caused by the highly inflectional nature of the velocity profiles in the flange region of the half-mushroom shown in Fig. 4.23.

4.3.9 Görtler Instability

Two-dimensional boundary-layer flow over a concavely curved wall is subject to *Görtler instability* due to the action of centrifugal force, and this results in the for-

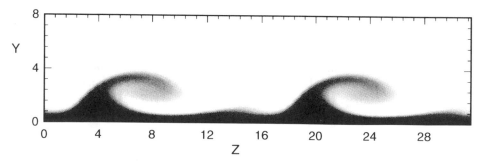

FIGURE 4.23 The structure of the crossflow vortices as viewed from upstream.

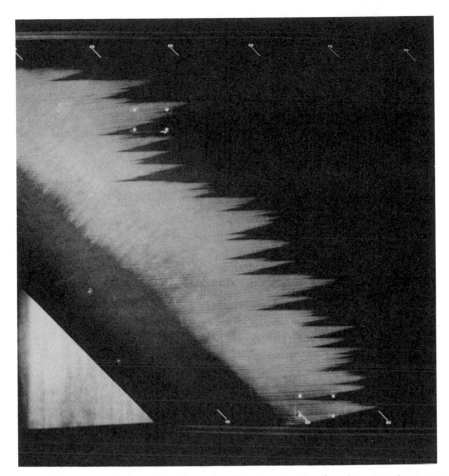

FIGURE 4.24 Naphthalene surface flow visualization showing imprint of stationary cross-flow vortices on a swept wing, $R_c = 3.27 \times 10^6$ (from Dagenhart *et al.*, 1989). Flow is from left to right.

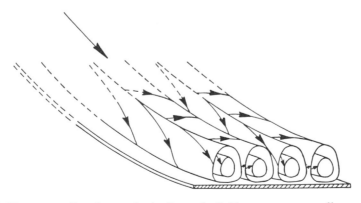

FIGURE 4.25 Vortex disturbances in the flow of a fluid on a concave wall, axes of vortices parallel to principal flow direction. (From Görtler, 1954.)

mation of steady counter-rotating streamwise vortices (see Fig. 4.25) as first predicted by Görtler (1940). *Görtler vortices* play a dominant role in boundary-layer transition in many aerodynamic flows such as on turbine blades and supersonic nozzle walls [e.g., Beckwith *et al.* (1985)]. Due to their technological importance, Görtler vortices have been the subject of a number of investigations [for recent reviews, see Floryan (1991)]. Görtler vortices could be generated at surface inhomogeneities [Denier *et al.* (1991)] or due to spanwise variations in the oncoming stream.

For zero pressure gradient flow over a curved wall with small curvature, the first-order mean flow is governed by the Blasius equation, Eq. (4.88). In the limit $R \to \infty$, and $\kappa \to 0$ with G held fixed, and by rescaling the dependent and independent variables $[V = 0(1/R)U, (v, w) = O(1/R)u, y = O(1/R)x]$, the parabolic equations derived by Hall (1983, 1988) can be recovered from Eqs. (4.67)–(4.70) for $W = 0$. For linear disturbances, Floryan and Saric (1983) considered a further local approximation whereby terms such as $\partial \hat{u}/\partial x$, $\partial \hat{v}/\partial x$, etc., in Eqs. (4.67)–(4.70) are set to $\sigma \hat{u}$, $\sigma \hat{v}$, etc., where σ is the growth rate. With the scalings used in Floryan and Saric (1983), $\sigma = -R\alpha_i$. While criticism can be leveled against the local analysis [Hall (1983)], the approach does provide information which is useful for engineering purposes. Using this approach, the Görtler instability results can be summarized in Fig. 4.26, with a cautionary remark that the results progressively become meaningless as the wave number β and Görtler number G decrease.

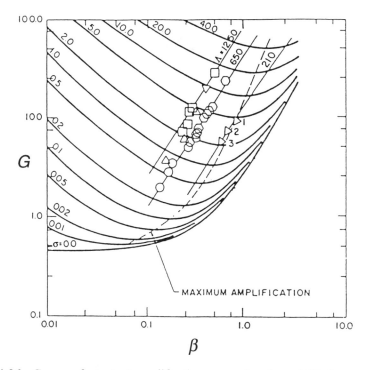

FIGURE 4.26   Curves of constant amplification σ as a function of Görtler number G and spanwise wavenumber β for the Blasius boundary layer. Symbols show experimental points and lines of constant Λ signify constant physical wavelengths. (From Floryan and Saric, 1983.)

Smith (1955) was the first one to use the N-factor method for Görtler instability and found that an N of around 10 would correlate with the experimentally observed transition location. Liepmann (1945) suggested that transition may occur if a Görtler number based upon momentum thickness ($G_\theta = \mathrm{Re}_\theta \sqrt{\theta|k^\dagger|}$) reaches a value of about 9. For a given flow, while the value of the Görtler number indicates the possibility of weak or strong Görtler instability, the onset of transition can be better correlated with an N factor. It has been found that crossflow tends to stabilize the Görtler vortex flow [Hall (1985)], however a sufficiently high crossflow leads to the crossflow instability discussed above.

Just like crossflow vortices, Görtler vortices are steady, and the question how they breakdown to turbulent motion is a problem of fundamental interest in fluid mechanics. This problem has been studied experimentally by Bippes (1978), Aihara and Koyama (1981), and Swearingen and Blackwelder (1987), and theoretically by Liu and Domaradzki (1993), Hall and Horseman (1991), and Malik and Li (1993). We will discuss the phenomenon by presenting computational results for the conditions of the experiment of Swearingen and Blackwelder (1987). The experiment was performed on a concave surface with a constant radius of curvature. The streamwise range of interest in the theoretical analysis lies approximately between Reynolds number, $\mathrm{Re}_x = 3.3 \times 10^4$ to 4×10^5, and the Görtler number G_θ ranges from 1.3 to 8.3. The Blasius equations are solved to obtain the basic flow (U, V) for the vortex stability analysis. The calculation is started at $X_0 = 10$ cm for the disturbance wavelength $\lambda_z = 1.8$ cm and initial amplitude of $\hat{u} = 0.0187\, U_e$ estimated from the experimental data. The particular Görtler vortex chosen here is close to the most amplified Görtler vortex according to linear theory. The experiment produced *mushroom-like* structures for the streamwise velocity due to the pumping action of the counter-rotating vortices. The contours of the computed streamwise velocity [using Eq. (4.77)] at various downstream locations are shown in Fig. 4.27. In the early stages of the development, the amplitude of the u perturbation is small, and the velocity contours show a wavy spanwise structure. As the Görtler vortices gather strength at relatively large distances downstream the same *mushroom* structures observed in the experiment are clearly seen (compare with half-mushrooms in Fig. 4.23 for crossflow vortex). The regions in the neighborhood of the centerlines of the *mushrooms* are referred to as *peak* regions where the streamwise velocity is relatively low, and the regions between the *mushrooms* are referred to as *valley* regions where the streamwise velocity is relatively high. The skin friction in the valley region increases, while it decreases in the peak region. The trend changes as the Görtler vortices break down en route to turbulence. The streamwise velocity profiles at the peak and the valley are shown in Fig. 4.28. The high shear layer region up in the peak plane will become subject to a particular mode of secondary instability. The spanwise variation of the streamwise velocity component at fixed y is given in Fig. 4.29. Again, the inflected profiles will be subject to another mode of secondary instability discussed below.

There are two types of secondary instability mode—odd and even (with respect to the Görtler vortex), see Fig. 4.30. The *odd mode* leads to the well-known *sinuous mode* of breakdown and is associated with the inflectional profiles shown in Fig. 4.29. The *even mode* leads to a *horse-shoe type vortex* structure and is associated with the inflectional profiles shown in Fig. 4.28. This helps explain experimental observations that Görtler vortices breakdown sometimes to sinuous motion [Swear-

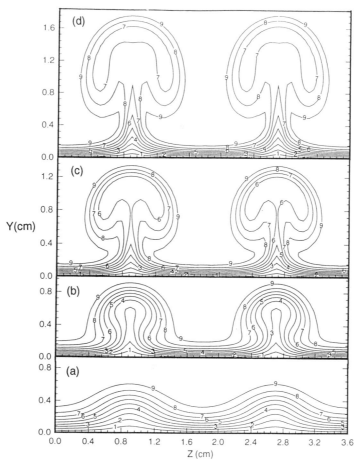

FIGURE 4.27 Variation of streamwise velocity in the Y, Z plane due to the presence of Görtler vortices at (a) $X = 60.1$, (b) 79.8, (c) 100.4, and (d) 120.1 cm. Two spanwise wavelengths are shown and contours range from $u/u_e = 0.1$ to 0.9 in increments of 0.1.

ingen and Blackwelder (1987)] and sometimes by developing a horse-shoe vortex structure [Aihara and Koyama (1981)].

4.3.10 Transition in the Attachment-Line Boundary Layer

The attachment-line boundary layer on a swept wing is a special class of three-dimensional boundary layers which is subjected to viscous, TS-like disturbances propagating along the attachment-line. For a large leading-edge radius, the mean flow is appropriately modeled by the swept-Hiemenz flow discussed in Sec. 4.3.8. The stability of this boundary layer was studied by Hall *et al* (1984) who considered disturbances of the form

$$\{u, v, w\} = \{x\hat{u}(y), \hat{v}(y), \hat{w}(y)\}e^{i(\beta z - \omega t)} + \text{complex conjugate} \quad (4.101)$$

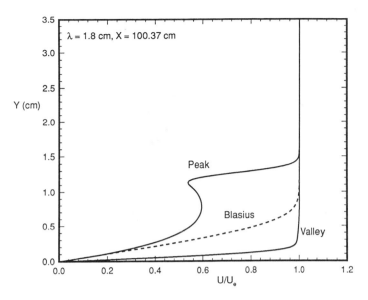

FIGURE 4.28 Streamwise velocity profiles in the peak and the valley plane of the Görtler vortex.

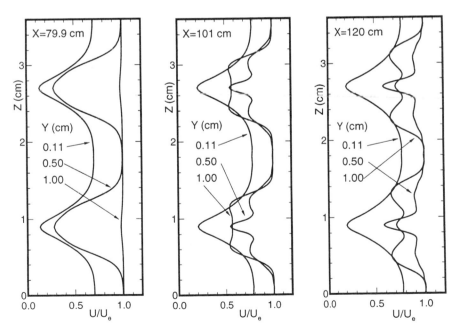

FIGURE 4.29 Spanwise variation of streamwise velocity at fixed Y and three streamwise locations.

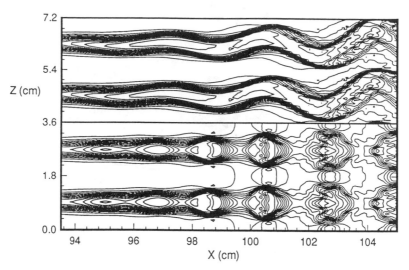

FIGURE 4.30   Instantaneous streamwise velocity in the X–Z plane at $Y = 1.08$ cm. Upper half—sinuous mode (odd), lower half—varicose mode (even). (From Lin and Malik, 1994.)

Note that an unstable disturbance of real frequency, ω, grows in the spanwise direction (i.e., $-\beta_i > 0$). Stability equations for the attachment-line boundary layer can be derived by substituting Eq. (4.101) in the linearized version of Eqs. (4.57)–(4.60). The results indicate that the boundary layer is stable to small disturbances provided $\bar{R}$ ($\equiv W_e l/\nu$) is greater than 583.1 (momentum thickness Reynolds number $Re_\theta > 235$). Lin and Malik (1994) considered a more general class of disturbances and found that the special class of disturbances, Eq. (4.101), is indeed the most dominant.

The N-factor results for this boundary layer ($N = -\int_{z_0}^{z}\beta_i dz$) are shown in Fig. 4.31 where the momentum thickness Reynolds number at transition, $Re_{\theta_{tr}}$, is plotted against ξ/θ where $\xi = z - z_0$ is the distance along the attachment line. Comparison with the experiments show that e^{10} correlates the data fairly well. In any case, the results show that the flow is stable to small disturbances provided $Re_\theta < 235$. However, some experiments show that turbulence can be sustained at Re_θ as low as about 100 [Pfenninger (1965), Poll (1979)]. This situation is similar to the case of plane Poiseuille flow where transition occurs at subcritical Reynolds numbers. This problem is of great practical significance to swept-wing design, since, if the attachment-line boundary layer is turbulent, the flow downstream on the entire wing is expected to be turbulent. Therefore, if the wing boundary layer is to be laminarized, special attention needs to be paid to the disturbances propagating from the turbulent fuselage boundary layer. This is generally accomplished by providing local flow diverters in the fuselage-wing junction region which prevents the fuselage boundary layer from contaminating the attachment-line boundary layer [Gaster (1965)].

4.3.11   The Effect of Mach Number

The stability of compressible boundary layers has been studied by many authors [see, e.g., Lees and Lin (1946), Lees and Reshotko (1962), Mack (1969), (1984),

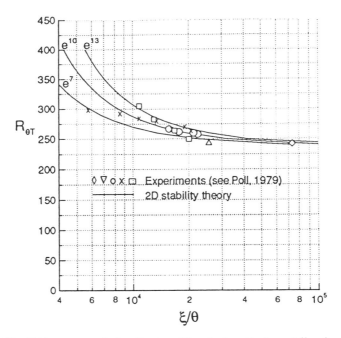

FIGURE 4.31 *N*-Factor correlation for transition in the attachment-line boundary layer.

Gapanov (1981), Bertolotti and Herbert (1991), Chang *et al.* (1991), and Chang and Malik (1994)]. There are two important modes of instability present in a compressible, flat plate boundary layer. The first mode is an extension to high speeds of the Tollmien-Schlichting instability present in incompressible flows, though for supersonic Mach numbers it differs in one aspect in that it is most amplified when oblique. This mode represents viscous instability at low Mach numbers, but the inviscid nature of the instability begins to dominate when the Mach number increases, since compressible flat plate boundary layer profiles contain a generalized inflection point [i.e., $d/dy(\rho\, du/dy) = 0$ at some point in the boundary layer]. This mode may be stabilized by wall cooling, suction, and favorable pressure gradient. The second mode is the result of an inviscid instability which is present due to a region of supersonic mean flow relative to the disturbance phase velocity. In fact, when this supersonic relative flow is present, there exist an infinite number of modes in the boundary layer. However, the first mode belonging to this family, commonly known as *Mack's second mode*, is most relevant, since it is known to have the highest growth rate. The second mode becomes important at Mach numbers above about 4 and has growth rates much higher than the first mode. The existence of both the first and second modes was established by the experiments of Kendall (1975), Demetriades (1974), and Stetson *et al.* (1983). The second mode is different in character with respect to the first mode; it is most amplified when two dimensional and is destabilized with wall cooling [Mack (1969)]. However, this mode can be stabilized with wall suction and favorable pressure gradient [Malik (1989)].

The effects of Mach number on the stability of a two-dimensional boundary layer are shown in Fig. 4.32 where the growth rate of the most amplified mode is plotted as a function of a Mach number. As pointed out earlier, compressibility tends to

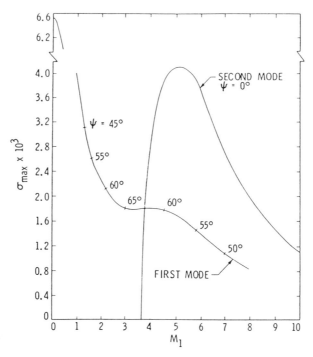

FIGURE 4.32   Effect of Mach number on the maximum spatial amplification rate of first- and second-mode waves at $R = 1500$. Insulated wall, wind-tunnel temperatures. (From Mack 1984.)

stabilize the flow. Hence, the growth rate decreases with the Mach number. Above Mach 1, the most amplified wave angle is in the range of 45–65° for the first mode waves. At a Mach number above about 4, the second mode disturbance is amplified at this Reynolds number of $R = 1500$. The effect of the Mach number on crossflow instability has been studied by Balakumar and Reed (1991) and on the Görtler instability by Spall and Malik (1989).

4.4   TURBULENCE AND TURBULENCE MODELING
Harry A. Dwyer

As has been mentioned previously, numerous flows of practical interest occur at a Reynolds number where the flow is turbulent. In general, turulence appears in many different costumes, and it has not yet been fully understood. For problems with additional influences such as electric and magnetic forces or significant gravitational forces, the present state of knowledge on the structure of turbulence is poor. In fact, it can be said that turbulence is an unsolved problem of classical physics, and there does not seem to be a major advance on the horizon. However, for certain flows, extensive experimental information is available which has been used to model turbulence processes semi-empirically. These semi-empirical models have high visibility when used with the digital computer, and some turbulence calculations are applied routinely to design engineering systems. One of the areas of turbulent flow which is relatively well developed is the turbulent boundary layer as applied to external aerodynamic flows.

This Section describes some important models of turbulent transport which have been used successfully, as well as the limitations and pitfalls using these models. The discussion is not complete and focuses on those results which have received some time-stable recognition over the years. At the present time, a vast research effort in many countries seeks improvements of turbulence modeling. Many of these efforts will be useful for specialized systems, but the majority will be replaced by new results in the future. Such is the nature of turbulence research at the present time, with many new models appearing which will only have limited usefulness. The above criticism can also be applied to some of the results presented herein, however the results are presented mainly to indicate the direction of research efforts and to help the reader understand the current literature.

4.4.1 Nature of Turbulent Flow

For most problems involving flow over external bodies, there are three distinct regions of flow which must be modeled. These three regions are laminar flow, transitional flow, and fully developed turbulent flow. In general, a sharp boundary does not exist between these different flows, and the size and nature of these different flow regions will depend on the following factors: 1) the values of the parameters in the problem such as Reynolds and Prandtl number, 2) the geometry of the boundaries of the flow system, and 3) the magnitude of the free stream turbulence or the disturbances in the free stream which always exist in a practical system. The last factor is particularly revealing about turbulence. It should always be remembered that turbulence is an inherently unsteady and three-dimensional process which arises due to unsteady and three-dimensional disturbances. However, for many flows such as turbulent boundary layers at high Reynolds number, the three-dimensional and time-dependent influences can be statistically averaged to yield a useful, steady, and, if applicable, two-dimensional description of the flow. This will be the case for the majority of the results presented, but this Section will discuss some of the structural aspects of turbulent flow.

A relatively simple example that reveals many aspects of these processes is flow over a flat plate which is shown in Fig. 4.33. Near the leading edge or attachment point of the inviscid flow, a laminar boundary layer develops. The laminar boundary layer responds to any disturbances in the flow but does not amplify the disturbances until a critical Reynolds number, Re_{crit}, is reached. The flow then begins to extract energy from the mean flow and deposit this energy into a growing three-dimensional instability. The amount of time and spatial distance required for the instability process to become fully developed depends strongly on the nature of the disturbances in the flow. However, the final fully developed turbulence process is usually independent of the disturbances in the free stream.

For internal flows, ample disturbances are usually available to be amplified due to the compressors or pumps used to generate the flow. However, for free flight conditions, the disturbances in the flow can be quite small; consequently the transition region to fully developed flow can be substantially extended. The influence of freestream turbulence can be seen in Fig. 4.34, where it is seen that the location of the transition zone can vary by a factor of 30. Another factor which has a strong

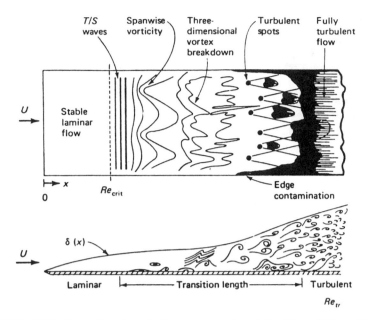

FIGURE 4.33   Transition to turbulence over a flat plate. (From White, 1993.)

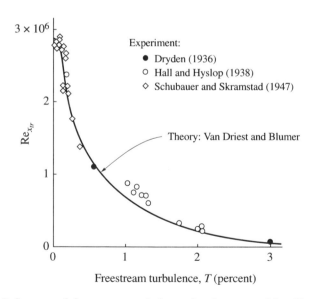

FIGURE 4.34   Influence of freestream turbulence level on transition Reynolds number. (From White, 1993.) The symbol T is

$$T = \frac{q}{U} \times 100 = \frac{[(\overline{u'^2} + \overline{v'^2} + \overline{w'^2})/3]^{1/2}}{\overline{U}} \times 100$$

influence on transition on a flat plate is the wall roughness. In cases where the wall roughness becomes large, it can be considered as a *trip* to force the boundary layer to proceed quickly from laminar to fully developed turbulent flow. In many applications a *trip* is used since transition needs to be avoided, or a turbulent boundary layer is desired to delay a separation process.

For more complex geometries, the transition process is strongly influenced by the pressure gradients in the flow. Laminar boundary layers with adverse pressure gradients are extremely unstable.

For most practical flows, it is necessary to use experience, semi-empirical models, or experimental information to predict the location of fully developed turbulent flow. At the present time, there is modeling capability for both the laminar and turbulent portions of a flow, but less is available for the transition region. However, in some problems, especially with *trips* or adverse pressure gradients, the transition region is relatively small and can be neglected.

4.4.2 Description of Turbulent Flow

Almost all descriptions of a fully developed turbulent flow involve a statistical measure of the velocity field to some degree. For any given turbulent flow, there is so much time-dependent and three-dimensional information in the flow that its average properties and some measure of variations about the average are all that are usually utilized. One of the simplest approaches is to divide the velocity components into mean and fluctuating parts as

$$u = U + u' \tag{4.102}$$

where

$$U = \frac{1}{\tau} \int_{t_0}^{t_0 + \tau} u \, dt \tag{4.103}$$

(It is difficult to relate this time average to an ensemble average, however, for the purposes of this chapter, ensemble and time averages are assumed to be equivalent.) The standard measure of the variation about the mean is the mean square of u' and is given by

$$\overline{u'^2} = \frac{1}{\tau} \int_{t_0}^{t_0 + \tau} u'^2 \, dt \tag{4.104}$$

and the level of turbulence is defined by

$$T = \frac{1}{U} \left[\frac{(\overline{u'^2} + \overline{v'^2} + \overline{w'^2})}{3} \right]^{1/2} \tag{4.105}$$

The distribution of these quantities in a fully turbulent flat plate boundary layer is shown in Fig. 4.35.

It is well known that the above two measures of a velocity field, U and $\overline{u'^2}$, are not a complete description of the flow. Many other quantities are needed if one is

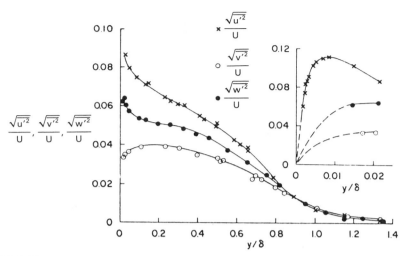

FIGURE 4.35   Turbulence intensity distributions in a flat plate boundary layer. (Reprinted with permission, Cebeci, T. and Smith, A. M. O., *Analysis of Turbulent Boundary Layers*, 1st ed., Academic Press, New York, 1974.)

to model a turbulent flow. Some of these other statistical quantities are: 1) Reynolds stresses or velocity correlation functions, $\overline{u_i' u_j'}$, 2) time scales of the turbulent transport processes, 3) length scales of turbulent transport processes, 4) distribution of turbulent kinetic energy, $k = (\overline{u_1'^2} + \overline{u_2'^2} + \overline{u_3'^2})/2$, among the turbulent length scales, and 5) similar quantities for the transport of heat and mass in turbulent flows. The amount of required information depends to a large extent on the complexity of the model being employed. In many flows, simple algebraic effective or *eddy viscosity* laws will be employed, while for others detailed partial differential equations will be solved for turbulent quantities. Some of the widely used models will be developed in the following sections.

4.4.3   Turbulent Momentum Transport

The equations of the turbulent flow of an incompressible fluid can be obtained by substituting the mean, U_i, are fluctuating, u_i', velocity components into the continuity and momentum equations and averaging these equations. For steady flows, this process yields

$$\frac{\partial}{\partial x_j} (U_j) = 0 \tag{4.106}$$

$$\rho \frac{\partial}{\partial x_j} (U_i U_j + \overline{u_i' u_j'}) = -\frac{\partial P}{\partial x_i} + \frac{\partial}{\partial x_j} (\overline{\tau}_{ij}) \tag{4.107}$$

The major difference in Eq. (4.107) compared to a laminar flow is the appearance of the *Reynolds stress* terms, $\overline{u_i' u_j'}$ which are responsible for momentum transport by turbulent motion. It has been one of the major goals of turbulence research to model these terms for shear flows of practical interest. The most straightforward and difficult approach is to develop transport equations for the Reynolds stresses. This can be accomplished by multiplying the momentum equations by u'_j and subtracting the mean equation, Eq. (4.107), to form the following equation for the *Reynolds stresses*

$$\rho \left[\frac{\partial}{\partial x_k} (U_k \overline{u'_i u'_j} + \overline{u'_k u'_i u'_j}) + \overline{u'_j u'_k} \frac{\partial U_i}{\partial x_k} + \overline{u'_i u'_k} \frac{\partial U_j}{\partial x_k} \right]$$

$$= -\frac{\partial}{\partial x_i} (\overline{p' u'_j}) + \frac{\partial}{\partial x_j} \overline{p' u'_i} + \overline{p' \left(\frac{\partial u'_j}{\partial x_i} + \frac{\partial u'_i}{\partial x_j} \right)}$$

$$+ \mu \left(\frac{\partial^2 \overline{u'_i u'_j}}{\partial x_l \, \partial x_l} - 2 \overline{\left(\frac{\partial u'_i}{\partial x_l} \right) \left(\frac{\partial u'_j}{\partial x_l} \right)} \right) \tag{4.108}$$

The above equations for the Reynolds stresses are exact but not closed. This lack of closure is the <u>direct</u> result of the appearance of unknown statistical correlations such as $\overline{u'_k u'_k u'}$, $\partial(\overline{p' u'_j})/\partial x_i$, etc., which require an endless succession of equations. Over the past fifty years, there has been extensive research attempting to model the unknown terms in these equations, but the majority of the successful methods have exploited experimental results. Also, the most widely used models are based on simplifications of the full Reynolds stress equations. Therefore, in the present discussion, the models to be presented are based on simplifications of the full equations.

At this time, the majority of successful models use the Reynolds stress equation in contracted form. This contraction $(i = j)$ results in the equation for turbulent kinetic energy, k, and is

$$U_l \frac{\partial k}{\partial x_l} + \overline{u'_i u'_l} \frac{\partial U_l}{\partial x_l} = -\frac{\partial}{\partial x_l} \overline{\left(\frac{p}{\rho} + k \right) u'_l} + \nu \frac{\partial^2 k}{\partial x_l \partial x_l} - \nu \overline{\frac{\partial u'_i \partial u'_i}{\partial x_l \, \partial x_l}} \tag{4.109}$$

The most popular model at the present time, is the k-ϵ model, which uses the turbulent kinetic energy equation together with the equation for the dissipation of energy, ϵ. Other popular models are k-ω (vorticity) and k-l (length scale) all of which are quite similar. The nature of all these models is to develop semi-empirical equations for a length scale and a time scale of turbulence. To a large extent, all models indicated above have had some pragmatic success, but all lack a substantial amount of theoretical justification. Also, for many applications the more complex models have not been more successful than the simpler eddy viscosity or mixing length models.

4.4.4 Models for Turbulent Stresses

The majority of the turbulence calculations make extensive use of the digital computer. In these calculations, the Reynolds stresses must be modeled independently of the mean flow equations which can be written as

$$\frac{\partial U_j}{\partial x_j} = 0 \tag{4.110}$$

$$\rho \frac{\partial}{\partial x_j} (U_j U_i) = -\frac{\partial P}{\partial x_j} + \frac{\partial}{\partial x_j} (\overline{\tau_{ij}} + \overline{R_{ij}}) \tag{4.111}$$

where R_{ij} is the Reynolds stress tensor defined by $\overline{R}_{ij} = -\rho\,\overline{u_i' u_j'}$. In this section, both the mixing length and k-ϵ models will be discussed. Both models are based on the *eddy viscosity* concept developed by Boussinesq. In this approximation, the Reynolds stress tensor is represented as an effective kinematic viscosity, ν_t,

$$\frac{\overline{R}_{ij}}{\rho} = \nu_t\left(\frac{\partial U_i}{\partial x_j} + \frac{\partial U_j}{\partial x_i}\right) - \frac{k}{3}\delta_{ij} \tag{4.112}$$

where ν_t must be supplied by independent modeling assumptions. The mean momentum equation, Eq. (4.111), can then be written as

$$\frac{\partial}{\partial x_j}(U_i U_j) = -\frac{\partial P}{\rho \partial x_i} + \frac{\partial}{\partial x_j}\left[(\nu + \nu_t)\left(\frac{\partial U_i}{\partial x_j} + \frac{\partial U_j}{\partial x_i}\right) - \frac{k}{3}\delta_{ij}\right] \tag{4.113}$$

It is obvious that the influence of turbulence has been relegated to a *pseudo-laminar* format.

Although the eddy viscosity concept can be easily questioned theoretically, it is the basis for the most widely used digital computer models. More experimental results are being incorporated into its semi-empirical structure, thus it seems to be firmly established.

Mixing Length Models. The mixing length model of turbulence attempts to represent the eddy viscosity in terms of a length scale, l_t, and time scale of turbulence, τ_t. The time scale is chosen proportional to the mean flow gradients

$$\tau_t \propto \frac{1}{\left|\dfrac{\partial U_j}{\partial x_i} + \dfrac{\partial U_i}{\partial x_j}\right|} \tag{4.114}$$

and the length scale is determined from experimental results. The eddy viscosity is then given by

$$\nu_t = \frac{l_t^2}{\tau_\tau} = l_t^2\left|\frac{\partial U_i}{\partial x_j} + \frac{\partial U_j}{\partial x_i}\right| \tag{4.115}$$

and the Reynolds stresses become

$$\frac{\overline{R}_{ij}}{\rho} = -\frac{k}{3}\delta_{ij} + l_t^2\left(\frac{\partial U_i}{\partial x_j} + \frac{\partial U_j}{\partial x_i}\right)^2 \tag{4.116}$$

This is generally modified for the particular flow under consideration; the most widely employed case is the two-dimensional boundary layer.

The two-dimensional turbulent boundary layer equations are

$$\frac{\partial u}{\partial x} + \frac{\partial v}{\partial y} = 0$$

$$u \frac{\partial u}{\partial x} + v \frac{\partial u}{\partial y} = -\frac{1}{\rho}\frac{\partial p}{\partial x} + v \frac{\partial^2 u}{\partial y^2} - \frac{\partial}{\partial y}(\overline{u'v'}) \qquad (4.117)$$

where the averaging notation for mean quantities has been neglected and the normal stress, u'^2, has been assumed small. Substitution of the mixing length equation, Eq. (4.116), yields

$$u \frac{\partial u}{\partial x} + v \frac{\partial u}{\partial y} = -\frac{1}{\rho}\frac{\partial p}{\partial x} + v \frac{\partial^2 u}{\partial y^2} + \frac{\partial}{\partial y}\left(l_i^2 \left[\frac{\partial u}{\partial y}\right]^2 \right) \qquad (4.118)$$

which introduces a significant nonlinearity into the equation. To proceed further with the model, recourse to experimental data is required.

Most of the models on turbulent wall shear flows have been based on fully developed flow over a flat plate. The mean velocity profile for a turbulent flat plate flow can be correlated in terms of the wall stress, τ_w. The velocity profile normalized with the *friction velocity*, $v^* = (\tau_w/\rho)^{1/2}$, is given in Fig. 4.36, where $u^+ = u/v^*$ and $y^+ = yv^*/v$. The data in Fig. 4.36 shows four distinct regions which must be described by any mixing length model. The inner region consists of a viscous, *laminar sublayer* dominated by laminar viscosity, which is then followed by a *transitional region* where instabilities grow rapidly. After this transitional layer, a fully developed turbulent wall region exists. Finally, we see a turbulent outer region which is decoupled to some degree from the wall. A popular treatment of the viscous,

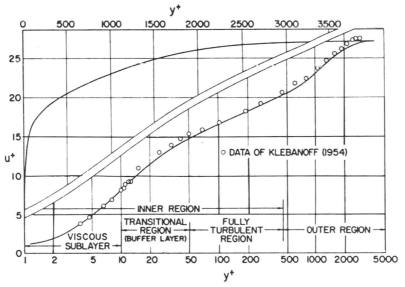

FIGURE 4.36 Mean velocity distribution in a zero pressure gradient boundary layer. (Reprinted with permission, Cebeci, T. and Smith, A. M. O., *Analysis of Turbulent Boundary Layers*, 1st ed., Academic Press, New York, 1974.)

transitional, and inner turbulent layers has been developed by Cebeci and Smither (1974), where the turbulent length scale is given by

$$l_t = 0.4y \, [1 - \exp \, (-y^+/26)] \tag{4.119}$$

and the outer region is represented as

$$l_t = 0.09\delta \tag{4.120}$$

or by the empirical formula

$$\nu_t = 0.0168 \, U_e \, \delta^* \, \gamma \tag{4.121}$$

where γ is an *intermittency correction* of the outer flow and δ^* is the local displacement thickness.

The intermittency factor, γ, is an attempt to take into account that the outer part of the boundary layer contains the nonregular boundary between rotational and nonrotational fluid. The empirical equation used by Cebeci and Smith (1974) is

$$\gamma = [1 + 5.5 \, (y/\delta)^6]^{-1} \tag{4.122}$$

For wall flows, this intermittency correction seems to play a minor role and is neglected in many treatments. However, for free turbulent flows such as jets and wakes, it may play a critical role.

The simple mixing length model described above may seem to be very empirical, however the model has been used successfully. It is considered inferior to the more complicated k-ϵ model by many investigators, but it continues to perform well. Also, because of their conceptual simplicity and ease of application, mixing length models are preferred in many practical cases. Descriptions of the result of applying these models to boundary layer type flows will be given in other Sections.

k-ϵ Model Equations. The k-ϵ model of turbulent shear flows was developed by the research team at the Imperial College of Science and Technology in London and is widely used. A good discussion of its development can be found in the paper by Launder and Spalding (1972). The model assumes that the turbulent viscosity, $\mu_t = \rho\nu_t$, is related to the turbulent kinetic energy, k, and dissipation rate of turbulent kinetic energy per unit volume, ϵ. That is,

$$\mu_t = C_\mu \, \rho k^2/\epsilon \tag{4.123}$$

where

$$\epsilon = \mu \left(\frac{\overline{\partial u_i'}}{\partial x_k} \right)^2 \tag{4.124}$$

and C_μ is an empirical constant determined from experiments. This model is equivalent to a specification of a length scale

$$l_t = C_D \frac{k^{3/2}}{\epsilon} \tag{4.125}$$

and a time scale

$$\tau_t = C_\tau \frac{k}{\epsilon} \tag{4.126}$$

where both k and ϵ are determined by semi-empirical turbulent transport equations. The equations for k and ϵ are modeled as

$$\frac{Dk}{Dt} = \frac{1}{\rho} \frac{\partial}{\partial x_l} \left[\frac{\mu_t}{\sigma_k} \frac{\partial k}{\partial x_l} \right] + \frac{\mu_t}{\rho} \left(\frac{\partial U_i}{\partial x_l} + \frac{\partial U_l}{\partial x_i} \right) \frac{\partial U_i}{\partial x_l} - \epsilon \tag{4.127}$$

$$\frac{D\epsilon}{Dt} = \frac{1}{\rho} \frac{\partial}{\partial x_l} \left[\frac{\mu_t}{\sigma_\epsilon} \frac{\partial \epsilon}{\partial x_l} \right] + \frac{C_1 \mu_t}{\rho} \frac{\epsilon}{k} \left(\frac{\partial U_i}{\partial x_l} + \frac{\partial U_l}{\partial x_i} \right) \frac{\partial U_i}{\partial x_l} - \frac{C_2 \epsilon^2}{k} \tag{4.128}$$

where the constants are chosen to be

$$C_\mu = 0.09, \ C_1 = 1.44, \ C_2 = 1.92, \ \sigma_k = 1.0, \ \text{and} \ \sigma_\epsilon = 1.3.$$

The equation for k is derived in a straightforward fashion from Eq. (4.109) but the ϵ equation has empirical roots. In fact, the lack of physical interpretation of the ϵ equation may lead to its downfall as more complicated flows are attempted. Also, the model as currently formulated does not predict accurately the transition and viscous layers near the wall. In order to treat these regions, wall functions or additional low Reynolds number modeling must be added to both the k and ϵ equations. Both of these near-wall, corrections continue to undergo extensive study and revision.

The major success of the k-ϵ model has been in treating more complex flows other than boundary layers, such as separated flows and diffusion flames. For attached boundary layers, the k-ϵ model is on a par with mixing length models but has not been used as extensively. A survey of models of complex turbulent flows [Kline *et al.* (1982)] has rated the k-ϵ model as the most widely applicable of the existing models, but with considerable deficiencies.

Other Turbulence Models. There exists a wide variety of other turbulence models which are simpler than and more complex than the two highlighted here. The reader can refer to texts such as Schetz (1994) for further discussion and details.

4.4.5 Averaging Procedures for Compressible Flows

A major problem area of turbulence modeling is the appearance of density fluctuations for flows with variable density caused by compressible flow or chemically reacting flow (combustion). The decomposition of the density, $\rho(x, t)$, into its mean and fluctuation components yields $\rho = \bar{\rho} + \rho'$, and substitution into the continuity and momentum equations yields, upon averaging,

$$\frac{\partial \bar{\rho}}{\partial t} + \frac{\partial}{\partial x_j} (\bar{\rho} \, U_j + \overline{\rho' u'}) = 0 \tag{4.129}$$

$$\frac{\partial}{\partial t} (\bar{\rho} \, U_i + \overline{\rho' u_i'}) + \frac{\partial}{\partial x_j} (\bar{\rho} \, U_i U_j) = -\frac{\partial P}{\partial x_i}$$

$$+ \frac{\partial}{\partial x_j} (\bar{\tau}_{ij} - U_j \overline{\rho' u_i'} - U_i \overline{\rho' u_j'} - \overline{\rho' u_i' u_j'} - \bar{\rho} \, \overline{u_i' u_j'}) \tag{4.130}$$

Little knowledge is available about the structure of the time averaged product of density times velocity fluctuations in turbulent shear flows, thus many investigators have neglected the fluctuating density-times-velocity terms. If the density fluctuations are neglected, the resulting equations have variable mean density but additional terms due to density fluctuations are not included and hence, cannot influence turbulent transport.

An alternate approach to variable-density turbulent flows is the use of *mass-weighted averaging procedures* which were developed by Van Driest (1951) and Favre (1969). A mass-weighted average for any quantitity $\phi(x_i, t)$ is defined as

$$\tilde{\phi} = \frac{\overline{\rho \phi}}{\bar{\rho}} \tag{4.131}$$

while the fluctuating component is given by the decomposition into mass-weighted average and fluctuating parts

$$\phi(x_i, t) = \tilde{\phi}(x_i, t) + \phi'' \tag{4.132}$$

An immediate advantage of this procedure is the much simpler forms for the mean continuity and momentum equations as shown by Eqs. (4.133) and (4.134)

$$\frac{\partial \bar{\rho}}{\partial t} + \frac{\partial}{\partial x_j} (\bar{\rho} \, \tilde{u}_j) = 0 \tag{4.133}$$

$$\frac{\partial}{\partial t} (\bar{\rho} \, \tilde{u}_i) + \frac{\partial}{\partial x_j} (\bar{\rho} \, \tilde{u}_i \, \tilde{u}_j) = -\frac{\partial P}{\partial x_i} + \frac{\partial}{\partial x_j} (\bar{\tau}_{ij} - \overline{\rho u_i'' u_j''}) \tag{4.134}$$

The form of these equations is almost identical to those of incompressible flow, however the terms have a different meaning.

Because of their simpler form, the mass-weighted description of the equations of turbulent flow have formed the starting point for the majority of turbulent modeling efforts which have been attempted for variable-density flows. A large percentage of these efforts have simply assumed that the modeling relations of incompressible flow can be used directly for variable density flows. Although these assumptions cannot be rigorously justified, some success has been achieved [Cebeci and Smith (1974) and Schetz (1994)]. Another potential difficulty for mass-weighting methods is that most of the instruments used for the measurement of statistical quantities in turbulent flow employ volume averaging techniques and do not measure mass-averaged variables. However, as advances are made in instrumentation, it may be possible to measure the appropriate quantities directly. Until that time, there will probably continue to be considerable speculation on the merits of mass-weighting and the turbulent modeling assumptions which should be employed.

4.5 TURBULENT BOUNDARY LAYERS
Harry A. Dwyer

4.5.1 Turbulent Boundary Layer Equations

In this Section, some features of incompressible, turbulent boundary layer flow are presented. At the present time, many methods are used to calculate the properties of turbulent boundary layers and the methods do not differ substantially in their accuracy. This presentation features those results and techniques which lend themselves to the numerical approach.

The equations for boundary layer flow can be derived independently of whether the flow is laminar or turbulent, with the only stipulation being that the flow be steady. Taking the same limits as were employed for laminar flow and time averaging, the incompressible, two-dimensional, turbulent boundary equations are

$$\frac{\partial U}{\partial x} + \frac{\partial V}{\partial y} = 0 \tag{4.135}$$

$$\frac{\partial}{\partial x}(\rho UU) + \frac{\partial}{\partial y}(\rho UV) = -\frac{\partial P}{\partial x} + \frac{\partial}{\partial y}\left(\mu \frac{\partial U}{\partial y} - \rho \overline{u'v'}\right) \tag{4.136}$$

where x and y are the local boundary layer coordinates and U and V the respective mean velocities.

The major difficulty, of course, with these equations is the appearance of the additional Reynolds stresses, $-\rho\overline{u'v'}$, which can have a value many orders of magnitude larger than the laminar stress in some parts of the boundary layer. In order to proceed further, almost all techniques of modeling of turbulent flow make extensive use of the experimental data on turbulent boundary layers. Therefore, it is necessary to understand thoroughly the behavior of turbulent boundary layers from the point of view of experimental data.

4.5.2 Turbulent Boundary Layer on a Flat Plate

Most of our knowledge on turbulent flow has been obtained from flow over a flat plate. Some details of the mean velocity profile have already been shown in Fig. 4.36, and another presentation of similar data is shown in Fig. 4.37. The basic idea behind a presentation in terms of u^+ and y^+ is that the mean velocity profile acquires a *universal* form when normalized in terms of the wall property v^* which is essentially the wall shear stress. As can be seen from Fig. 4.37, this is indeed true for $y^+ < 1000$, and the velocity profile can be described with the following logarithmic function in the fully turbulent region

$$u^+ = 2.44 \ln y^+ + 4.9 \tag{4.137}$$

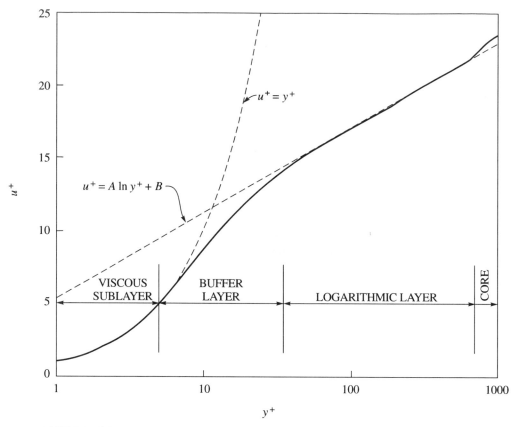

FIGURE 4.37 Universal velocity profile for a turbulent flat plate boundary layer.

The major difficulty with Eq. (4.137) is that the wall shear stress must be known, and the velocity profile for y^+ greater than 1000 is not well described. In order to obtain the wall stress, recourse is made to measurements, and for flow over a smooth plate, the skin friction can be obtained from the empirical formula proposed by Schultz–Grunow [see Rubesin and Inouye (1973)]

$$\frac{C_f}{2} = \frac{0.185}{(\log_{10} \text{Re}_x)^{2.584}} \tag{4.138}$$

which is valid over the fully developed turbulent range.

For the outer boundary layer, Coles (1956) has developed the concept of the *Law of the Wake* which effectively correlates the mean velocity profile in terms of a velocity defect $(u_e^+ - u^+)$ and the boundary layer thickness, δ. The velocity defect is shown in Fig. 4.38. A combined formula for the velocity profile in the inner turbulent region and the outer *wake* region has been developed by Coles (1956) as

$$u^+ = 2.43 \ln y^+ + 5.0 + 1.3 \, w \, (y/\delta) \tag{4.139}$$

where the function $w \, (y/\delta)$ can be accurately represented by $w = 1 - \cos (\pi y/\delta)$.

The above empirical laws can be quite useful, but they do not easily lend them-

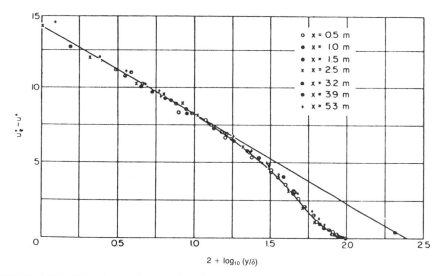

FIGURE 4.38 Velocity defect profile for a turbulent flat plate boundary layer. (From Rubesin and Inouye, 1973 with permission.)

selves to an extension to more complex flows and also do not give much information on the turbulence quantities in the boundary layer. In order to describe more complex flows, a *turbulence model*, along with the numerical solution of the boundary layer equations, is usually required. A good example of this type of approach can be found in the text by Cebeci and Smith (1974), and this reference is used extensively here.

The numerical solution of the boundary layer equations is considerably easier than the Navier–Stokes equations since the equations are parabolic, partial differential equations and can be solved by a marching procedure. There are many techniques available and most of these methods are variations on the Crank–Nicolson scheme [see Blottner (1979)]. See Chaps. 19 and 20 for more discussion of numerical methods for viscous flows. The major complication caused by turbulent flow is the validity of the turbulence model and the iteration required for the calculation of the turbulent stresses. The most popular models are the k-ϵ and mixing length, with the mixing length models being more highly developed for external boundary layers.

A typical comparison of a numerical calculation of the turbulent boundary layer for flow over a flat plate is shown in Fig. 4.39 for the skin friction, C_f; mean velocity, u; velocity defect, $u_e^+ - u^+$; and shear stress. As can be seen, the results are quite good and agree well with the experimental data. The mixing length formula used in the inner region was Eq. (4.119), and the eddy viscosity in the outer region was modeled as Eq. (4.121). It is perhaps not surprising that the agreement is good, since the constants have been chosen to agree with the turbulent flat plate data. However, it will be seen that the same constants play a key role in more complex problems and are useful for a wide class of problems.

4.5.3 General Turbulent Boundary Layer Flows

The real strong point of a numerical calculation, if it works, is the ability to calculate a wide range of flows of practical interest. The majority of turbulent boundary layer flows have both favorable and unfavorable pressure gradients as for example in flow

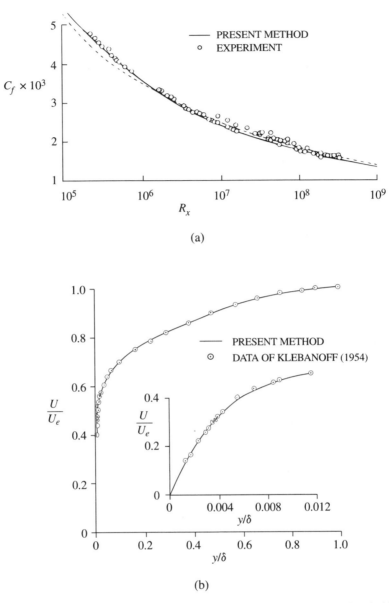

FIGURE 4.39   Numerical predictions for a flat plate boundary layer. (a) local skin-friction coefficient, (b) mean velocity profile, (c) velocity defect profile, and (d) turbulent shear stress distribution. (Reprinted with permission, Cebeci, T. and Smith, A. M. O., *Analysis of Turbulent Boundary Layers*, 1st ed., Academic Press, New York, 1974.)

over an airfoil. Also, there are influences such as wall suction and blowing and the flow over axisymmetric bodies. Turbulent boundary layer calculations have been performed with various models for a wide variety of such flows, and it can now be said that the models are a useful design tool. However, as has been the case for the majority of turbulent flow models, additional empirical models are introduced to account for the added complexity.

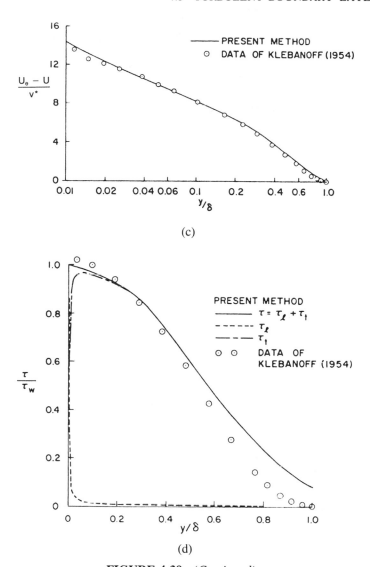

FIGURE 4.39 (*Continued*)

Flows With Pressure Gradients. The calculation of the mean velocity profiles, skin friction, and boundary layer momentum thickness is shown for four flows in Figs. 4.40 through 4.43. These flows represent, in order, the effects of favorable pressure gradient, adverse pressure gradient, *boundary layer relaxation*, and flow over a large airfoil. The calculations were performed with a modification for the influence of the pressure gradient on the mixing length. This modification influences the transition region between the laminar sublayer and the fully turbulent wall region. Cebeci changes the mixing length to be

$$l_t = 0.4y \left[1 - \exp \left(-\frac{y^+}{26} N \right) \right] \tag{4.140}$$

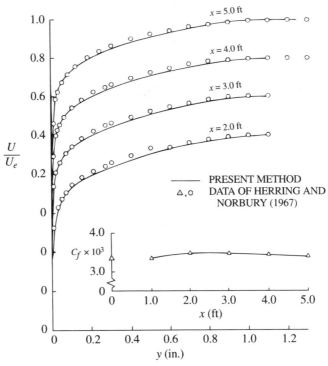

FIGURE 4.40   Prediction of turbulent velocity profiles in a boundary layer with favorable pressure gradients. (Reprinted with permission, Cebeci, T. and Smith, A. M. O., *Analysis of Turbulent Boundary Layers*, 1st ed., Academic Press, New York, 1974.)

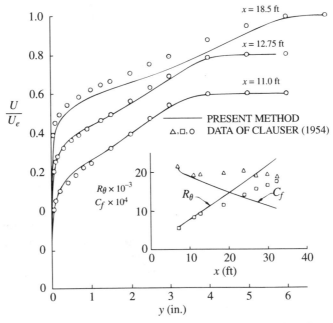

FIGURE 4.41   Prediction of turbulent velocity profiles in a boundary layer with adverse pressure gradients. (Reprinted with permission, Cebeci, T. and Smith, A. M. O., *Analysis of Turbulent Boundary Layers*, 1st ed., Academic Press, New York, 1974.)

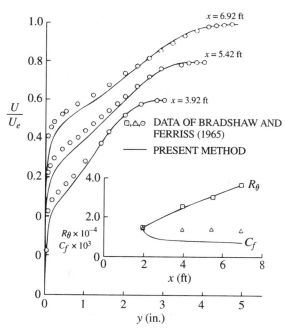

FIGURE 4.42 Prediction of mean velocity profiles in a turbulent boundary relaxing from an adverse pressure gradient. (Reprinted with permission, Cebeci, T. and Smith, A. M. O., *Analysis of Turbulent Boundary Layers*, 1st ed., Academic Press, New York, 1974.)

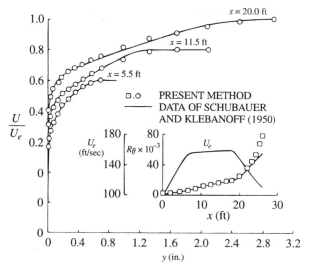

FIGURE 4.43 Prediction of mean velocity profiles over an airfoil. (Reprinted with permission, Cebeci, T. and Smith, A. M. O., *Analysis of Turbulent Boundary Layers*, 1st ed., Academic Press, New York, 1974.)

where N is a function of the local applied pressure given by

$$N = [1 - 11.8 \, p^+]^{1/2}$$

$$p^+ = \frac{\nu_e \, U_e}{(v^*)^3} \frac{dU_e}{dx} \tag{4.141}$$

Essentially, the factor N moves the transition layer closer or further from the wall depending on whether a favorable or adverse pressure gradient is applied.

A survey of the results in Figs. 4.40 through 4.43 shows that the numerical calculations are good in regions of favorable pressure gradient and leave something to be desired in the adverse pressure gradient regions. The most sensitive parameter is the wall shear, and the velocity profile does not reflect changes very close to the wall. In general, it is difficult to pinpoint the specific reason for the failure of the mixing length model, other than lack of a proper physical description of turbulence. It should be pointed out that all turbulent flow models, including k-ϵ and the advanced Reynolds stress models, have difficulty predicting external flow separation. Therefore, the flow of turbulent boundary layers with adverse pressure gradients is and will continue to be one of the major research areas. Other comparisons can be found in Schetz (1994). In general the more complex models do not give superior predictions for attached, external flows.

Axisymmetric Flows. An important variation of the previous analysis is the application to axisymmetric, turbulent boundary layer flow. The axisymmetric geometry can be a significant influence on the turbulence model when the boundary layer thickness becomes comparable to the transverse body curvature. Under these circumstances, account must be made for the increasing area associated with boundary layer growth, and the local radius must appear in the turbulence model. Cebeci has developed the following correction for the mixing length in this circumstance

$$l_t = 0.4 \, r_0 \ln \left(\frac{r}{r_0} \right) \left[1 - \exp \left(- \frac{r_o^+}{26} \ln \left(\frac{r}{r_0} \right) \right) \right] \tag{4.142}$$

where r_0 is the local body radius.

Typical calculations where this effect is important are shown in Fig. 4.44. The geometries of these flows all contain a rear axisymmetric section where the boundary layer is thick and r_0 approaches zero. As can be seen from the figure, the predictions agree well with measurements. It should also be noted that this experiment and calculation contained a trip to trigger turbulent flow, and it seems to have been simulated well by the calculation technique.

Wall Roughness. The influence of wall roughness on the skin friction coefficient in turbulent flow can be large due to the intrusion of the roughness elements into high momentum regions of the boundary layer. The only systematic studies have been confined to flat plate flows and the texts by Schlichting (1979) and Schetz (1994) are good references. In general, there is considerable difficulty defining the many types of roughness which can occur in practical flows, but all methods of analysis use the concept of roughness Reynolds number Re_k

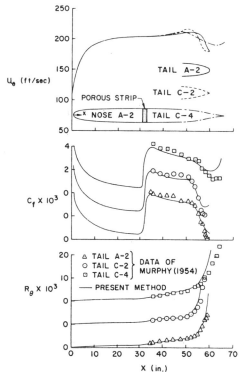

FIGURE 4.44 Prediction of turbulent mean velocity profiles in a thick axisymmetric boundary layer. (Reprinted with permission, Cebeci, T. and Smith, A. M. O., *Analysis of Turbulent Boundary Layers*, 1st ed., Academic Press, New York, 1974.)

$$\mathrm{Re}_k = \frac{v^* k_s}{\nu} \tag{4.143}$$

where k_s is the characteristic length scale for the wall roughness.

In terms of Re_k, the regions of wall roughness can be classified as follows:

Aerodynamically Smooth	$0 \le \mathrm{Re}_k \le 5$
Transitional Roughness	$5 \le \mathrm{Re}_k \le 70$
Fully Rough	$70 \le \mathrm{Re}_k$

For the aerodynamically smooth case, the turbulence model need not be modified for the influence of roughness, whereas an influence exists in the other roughness regions. For the fully rough case, the laminar sublayer and transition regions effectively disappear, and the influence of roughness can be modeled as an increase of the mixing length in the fully turbulent layer. The turbulent viscosity concept can be applied to the rough wall, and Kayes and Crawford (1980) have suggested that the mixing length becomes

$$l_t = 0.4 \, (y + \delta y_0) \tag{4.144}$$

where

$$\delta y_0 = 0.031 \frac{\nu_w}{v^*} (\text{Re}_k - 43) \tag{4.145}$$

In the outer part of the flow, changes are not necessary, since this region is effectively decoupled from the small scales near the wall caused by roughness. The final case of transitional roughness must be handled in an empirical way based on experimental information. As Re_k increases from 5 to 70, the transition region should be systematically reduced until it disappears at $\text{Re}_k = 70$. For further information, the reader should refer to the work of Kayes and Crawford (1980).

4.6 INFLUENCE OF HEAT TRANSFER AND COMPRESSIBILITY ON BOUNDARY LAYER FLOW
Harry A. Dwyer

4.6.1 Laminar Flows

Material is now presented for both laminar and turbulent flows with influence of heat transfer and compressibility. It is natural that heat transfer and compressibility be treated together, since one of the major influences of compressible flow is to cause temperature changes and thus heat flow. In some ways, the subject of heat transfer can be more tractable than fluid mechanics, since the energy equation is linear with temperature for constant-property flows. In other ways it can be much more complex when the fluid flow and heat transfer interact strongly to create a family of complex flows. In the following Section, examples of both types of behavior are discussed.

The starting point for the majority of the results will be the boundary layer energy equation, which can be written in terms of static enthalpy, h, or total enthalpy, $H = h + u^2/2$. The total enthalpy equation is

$$\rho u \frac{\partial H}{\partial x} + \rho v \frac{\partial H}{\partial x} = \frac{\partial}{\partial y} \left[\frac{\mu}{\text{Pr}} \frac{\partial H}{\partial y} + \mu \left(1 - \frac{1}{\text{Pr}} \right) \frac{\partial (u^2/2)}{\partial y} \right] \tag{4.146}$$

Note that Eq. (4.146) takes on a simpler form for $\text{Pr} = 1$. For low speed flows, $M \ll 1$, the velocity term can be neglected with respect to the static enthalpy, h, if the Prandtl number is nearly unity. Under these conditions and the additional constraint of constant properties, the boundary layer energy equation becomes

$$\rho u \frac{\partial T}{\partial x} + \rho v \frac{\partial T}{\partial y} = \frac{\mu}{\text{Pr}} \frac{\partial^2 T}{\partial y^2} \tag{4.147}$$

which is a linear equation in T. For laminar flow over a flat plate, the velocity field is given by the Blasius solution of Sec. 4.3, and the temperature field also becomes self-similar for constant temperature boundary conditions. With the following boundary conditions and dimensionless variable

$$T_N = \frac{T - T_e}{T_w - T_e}$$

$$x = 0 \qquad y > 0 \qquad T = T_e$$

$$x > 0 \qquad y = 0 \qquad T = T_w$$

$$y \to \infty \qquad T = T_e \qquad \text{(4.148)}$$

the energy equation, Eq. (4.147), takes on the self similar form

$$T_N'' + \tfrac{1}{2} \operatorname{Pr} f T_N' = 0$$

$$\eta = 0 \qquad T_N = 1$$

$$\eta \to \infty \qquad T_N \to 0 \qquad \text{(4.149)}$$

The term f is the Blasius stream function. The solution of Eq. (4.149) is shown in Fig. 4.45 where the Blasius similarity parameter, η, has been modified by a factor of $\operatorname{Pr}^{1/3}$. The close agreement of the curves leads to the following approximate relationship for the thickness of the *thermal boundary layer*, δ_{th}

$$\frac{\delta_{\text{th}}}{x} = \frac{5 \operatorname{Pr}^{-1/3}}{(\operatorname{Re}_x)^{1/2}} = \frac{\delta}{x} \operatorname{Pr}^{-1/3} \qquad \text{(4.150)}$$

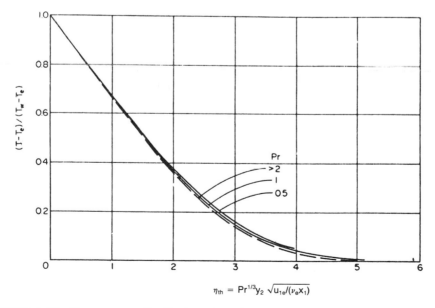

FIGURE 4.45 Temperature distributions in a laminar flat-plate boundary layer. (From Rubesin and Inouye, 1973.)

Equation (4.150) indicates that fluids with Prandtl numbers less than one have relatively thick thermal boundary layers, while fluids with higher Prandtl numbers have a thermal boundary layer which is thinner than the momentum boundary layer. Another useful approximation can be made for the local Stanton number, and Pohlhausen has found that

$$C_h = \frac{0.332 \ \text{Pr}^{-2/3}}{(\text{Re}_x)^{1/2}} \tag{4.151}$$

or, in terms of *Reynolds analogy*,

$$C_h = \frac{C_f}{2} \text{Pr}^{-2/3} \tag{4.152}$$

There are many other constant-property solutions available in the literature for other self-similar flows [Schlichting (1979), Schetz (1994)] including high speed flows. However, a more general approach is to present the variable-property solutions and then to recover the constant-property solution as a special case. The development of variable-property solutions of the laminar boundary layer equations owes a great deal to the coordinate transformation attributed to Howard and Dorodnitsyn [Rubesin and Inouye (1973)]. This transformation involves both x and y and is

$$\xi = U_e \, \mu_r \, \rho_r \, x$$

$$\eta = \sqrt{\frac{U_e}{\nu_r \, x}} \int_0^y \frac{\rho}{\rho_r} \, dy \tag{4.153}$$

where the subscript r refers to a reference condition for the properties. The transformation is a combination of a generalized Blasius transformation and a transformation for removing density variations as well as limiting the influence of viscosity variations.

Under an assumption of self similarlity, the full compressible boundary layer equations become, for uniform flow over a flat plate

$$(C_r f'')' + \tfrac{1}{2} f f'' = 0 \tag{4.154}$$

$$\left(\frac{C_r \, g'}{\text{Pr}}\right)' + \frac{1}{2} f g' + \frac{U_e^2}{H_e}\left[C_r \left(1 - \frac{1}{\text{Pr}}\right) f' f'' \right] = 0 \tag{4.155}$$

where

$$f' = \frac{u}{U_e}, \qquad g = \frac{H}{H_e}, \qquad C_r = \frac{\mu \rho}{\mu_r \rho_r} \tag{4.156}$$

with the boundary conditions

$$\eta = 0 \quad f = f' = 0 \quad g = h_w/H_e \text{ (constant)}$$

$$\eta \to \infty \quad f' = 1 \quad g \to 1 \tag{4.157}$$

It should be clear that the influences of density and viscosity have been confined to the term C_r, and that for an ideal gas behavior with a linear viscosity-temperature relationship, C_r will be constant. The above form of the compressible boundary layer equations leads to a large family of useful solutions which have been well documented by Rubesin and Inouye (1973). A few of the solutions are presented here.

The special case of Prandtl number of unity yields a particularly simple relationship between the velocity field and the total enthalpy ratio, g, which is

$$g = \left(1 - \frac{h_w}{H_e}\right) f' + \frac{h_w}{H_e} \tag{4.158}$$

In terms of static enthalpy, h, the above equation can be rewritten as

$$\frac{h}{h_e} = 1 + \frac{(h_w - H_e)}{h_e}(1 - f') + \frac{U_e^2}{2h_e}(1 - f'^2) \tag{4.159}$$

which gives a direct relationship between temperature and the velocity field f'. The *Reynolds analogy* takes the particularly simple form

$$C_h = \frac{C_f}{2} \tag{4.160}$$

The simplicity of the above relationship is somewhat misleading in that one must still solve the momentum equation for the velocity field, and this requires a statement concerning the viscosity of the compressible fluid in question. The most straightforward case occurs when the viscosity obeys the equation

$$\frac{\mu}{\mu_e} = \frac{T}{T_e} \tag{4.161}$$

and C_r has the value of unity. From the momentum equation, Eq. (4.155), it is obvious that the velocity field can be obtained from the incompressible Blasius solution under these restrictions. These results, which show the usefulness of the coordinate transformations, can serve as the basis for a very rapid approximate calculation of boundary layer structure.

A realistic representation of the viscosity temperature relationship for many gases is *Sutherland's Law*

$$\frac{\mu\rho}{\mu_e\rho_e} = \sqrt{\frac{T}{T_e}} \left(\frac{1 + \theta_{\text{ref}}}{T/T_e + \theta_{\text{ref}}}\right) \tag{4.162}$$

where θ_{ref} is a constant depending on the gas and is given in Table 4.3.

The momentum and energy equations have been solved by Crocco (1956) for a

TABLE 4.3 Variation of Sutherland Constant for Gases

Gas	Pr $T = 230$ K	Sutherland constant T_c K	$\theta_{ref} = T_c/T_e$ T_e 218 K	T_e 300 K	T_e 3000 K
H_2	0.717	90	0.413	0.300	0.030
CO	0.765	104	0.477	0.347	0.035
N_2	0.739	112	0.514	0.373	0.037
Air	0.725	116	0.532	0.387	0.039
O_2	0.731	131	0.601	0.437	0.044
CO_2	0.805	266	1.220	0.887	0.089
H_2O	1.08	673	3.09	2.24	0.224

(From Rubesin and Inouye, 1973.)

wide variety of conditions, and a good discussion can be found in the article by Rubesin and Inouye (1973). Particularly useful results for air with Pr = 0.75 and $\theta_{ref} = 0.505$ have been calculated by Van Driest and are shown in Figs. 4.46–4.48 for various wall conditions. The major influence of compressibility in these results is the increased boundary layer thickness and the corresponding decrease in skin friction coefficient.

4.6.2 Computer Solutions of Variable Temperature Flows

In the early development of compressible boundary layer theory, the workers in the field had great difficulty with the viscosity laws presented in laminar flows by real fluids. These viscosity variations are not nearly as complex as turbulent flows, and the digital computer can readily solve this type of problem if the physics is properly developed. Therefore, it is not surprising that digital computer solutions dominate

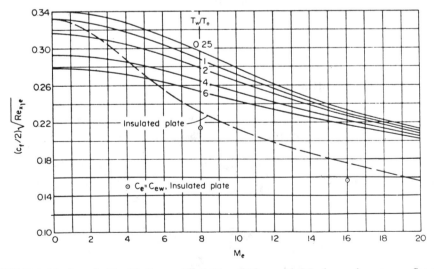

FIGURE 4.46 Local skin friction coefficient variation with Mach number over a flat plate; Pr = 0.75, $\theta_{ref} = 0.505$. (From Rubesin and Inouye, 1973.)

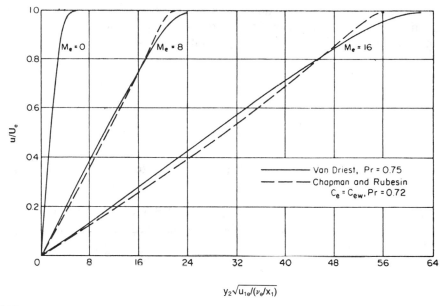

FIGURE 4.47 Velocity profiles in a laminar boundary layer on a flat plate, $\theta_{\text{ref}} = 0.505$; insulated surface. (From Rubesin and Inouye, 1973.)

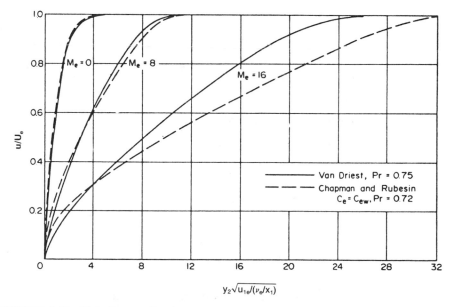

FIGURE 4.48 Velocity profiles in a laminar boundary layer on a flat plate, $\theta_{\text{ref}} = 0.505$; $h_w/h_e = 1/4$. (From Rubesin and Inouye, 1973.)

the modern approach to compressible flows with heat transfer, whether these flows are laminar or turbulent. However, the solution of compressible boundary layers can be considerably simplified by use of a generalization of the *Howard–Dorodnitsyn transformation* which is known as the *Levi–Lees transformation* [Blottner (1979)]. The new independent variables are

$$\eta(x, y) = U_e r_o^j \sqrt{\frac{K}{2\xi}} \int_0^y \rho \, dy$$

$$\xi(x) = K \int_0^x (\rho\mu)_r \, U_e r_o^{2j} \, dx \tag{4.163}$$

where K is a constant, $j = 0$ for two-dimensional flows, $j = 1$ for axisymmetric flows, and u_e is the local velocity at the boundary layer edge. For the following new dependent variables

$$F = u/U_e$$

$$\theta = h/h_e$$

$$V = 2\xi \, \frac{(F \, \partial\eta/\partial x + \rho v \, r_o^j/\sqrt{2\xi/K})}{[K \, (\rho\mu)_r \, U_e r_o^{2j}]} \tag{4.164}$$

the governing equations become

$$2\xi \, \frac{\partial F}{\partial \xi} + \frac{\partial V}{\partial \eta} + F = 0 \qquad \text{(Continuity)} \tag{4.165}$$

$$2\xi \, \frac{\partial F}{\partial \xi} + V \, \frac{\partial F}{\partial \eta} + \beta \, (F^2 - \theta) = \frac{\partial}{\partial \eta} \left(l \, \frac{\partial F}{\partial \eta} \right) \qquad \text{(Momentum)} \tag{4.166}$$

$$2\xi \, \frac{\partial \theta}{\partial \xi} + V \, \frac{\partial \theta}{\partial \eta} - \alpha l \left(\frac{\partial F}{\partial \eta} \right)^2 = \frac{1}{\Pr} \frac{\partial}{\partial \eta} \left(l \, \frac{\partial \theta}{\partial \eta} \right) + \frac{2\xi\theta F}{h_e} \frac{dh_e}{d\xi} \qquad \text{(Energy)} \tag{4.167}$$

where the following notation has been used for laminar flows

$$l = \frac{\rho\mu}{(\rho\mu)_r}$$

$$\alpha = \frac{U_e^2}{c_p T_e}$$

$$\beta = \left(\frac{2\xi}{U_e} \right) \frac{dU_e}{d\xi} \tag{4.168}$$

Equations (4.163)–(4.168) serve as the basis for most modern studies of boundary layers, whether the boundary layers are laminar, turbulent, incompressible, or compressible. The transport properties can be obtained from experimental data, and the major limitation is the use of the boundary layer approximation itself. All of the solutions presented previously can be easily obtained by a wide variety of numerical techniques which have been described by Blottner (1979). For the calculation of turbulent flows, the central factor becomes the model used for turbulent momentum and heat transfer. As was the case for momentum transfer, the most highly developed models for boundary layers are the mixing length models. It is true that more complex models offer the hope of calculating complex flows, but the mixing length models still play a key role in attached flows.

The basic equations given in Eqs. (4.165) to (4.167) remain unchanged in the mixing length formulation except the transport coefficient term, l/Pr, becomes

$$\frac{l}{\mathrm{Pr}} \rightarrow \frac{\rho\mu}{(\rho\mu)_r}\left(\frac{1}{\mathrm{Pr}} + \frac{\nu_t}{\nu\,\mathrm{Pr}_t}\right) \qquad (4.169)$$

where Pr_t is a turbulent Prandtl number, and ν_t is the eddy viscosity for compressible turbulent transport. A general formula given by Cebeci for the turbulent length scale l_t is

$$l_t = 0.4y\left[1 - \exp\left\{-\frac{y^+}{26}N\frac{\nu_w}{\nu}\left(\frac{\rho_w}{\rho}\right)^{1/2}\right\}\right] \qquad (4.170)$$

where

$$N = \left[1 - 11.8\left(\frac{\mu_w}{\mu_e}\right)\left(\frac{\rho_e}{\rho_w}\right)^2 p^+\right]^{1/2} \qquad (4.171)$$

and

$$y^+ = \frac{y\sqrt{\dfrac{\tau_w}{\rho_w}}}{\nu_w} \qquad (4.172)$$

and

$$\nu_t = l_t^2\left|\frac{\partial U}{\partial y}\right| \qquad (4.173)$$

One approach to introducing the *turbulent Prandtl number* is to define a heat transfer length scale correction, l_h, such that

$$\alpha_t = \frac{\nu_t}{\mathrm{Pr}_t} = l_t l_h\left|\frac{\partial U}{\partial y}\right| \qquad (4.174)$$

where Cebeci has chosen

$$l_h = 0.44y \left[1 - \exp \left\{ - \frac{y^+}{B^+} N \frac{\nu_w}{\nu} \left(\frac{\rho_w}{\rho} \right) \right\} \right] \qquad (4.175)$$

The coefficient B^+ is a strong function of the molecular Prandtl number and is given by

$$B^+ = B^{++}/\mathrm{Pr}^{1/2} \qquad (4.176)$$

and the function B^{++} is shown in Fig. 4.49. The complexity of these results is a direct result of the multiple length scales introduced by the molecular Prandtl number in the transition region between laminar and turbulent flow near the wall. For the outer part of the boundary layer, a correction is not needed, and the turbulent Prandtl number is essentially unity, or a little less. Also, for the important case of air, the coefficient B^+ has the value of 35.

Typical results for this model for step and ramp type wall temperature distributions are shown in Fig. 4.50 for incompressible flow over a flat plate. The agreement with experimental data is good; other similar results can be found in the text by Cebeci and Smith (1974). In the calculation of compressible turbulent boundary layers, computer programs based on models of this type are routinely used in design studies for the aerospace industry. The computer programs have reproduced experimental results for the skin friction and Stanton number for flat plate flows as shown in Figs. 4.51 and 4.52, as well as flows over complex bodies before separation. The situation of predicting compressible turbulent separated flows is not well understood at this time, and there is a need for more experimental results. One problem that has

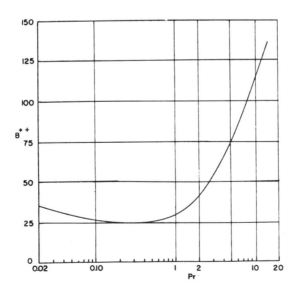

FIGURE 4.49 Variation of B^{++} with molecular Prandtl number Pr. (Reprinted with permission, Cebeci, T. and Smith, A. M. O., *Analysis of Turbulent Boundary Layers*, 1st ed., Academic Press, New York, 1974.)

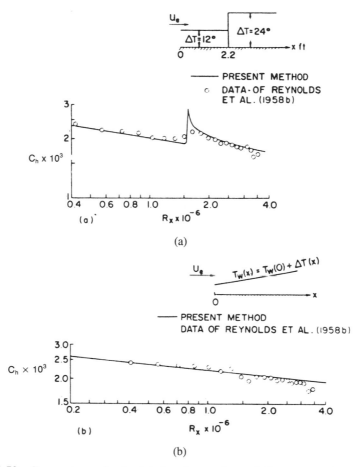

FIGURE 4.50 Comparison of calculated and experimental Stanton numbers for the flat-plate boundary layer measured by Reynolds *et al.* (a) double-step wall temperature, (b) ramp temperature distribution. (Reprinted with permission, Cebeci, T. and Smith, A. M. O., *Analysis of Turbulent Boundary Layers*, 1st ed., Academic Press, New York, 1974.)

not proven as difficult as expected was the influence of turbulent density fluctuations on turbulent transport. The corrections applied for wall turbulent flows seem to work well as just demonstrated. However, the same is not true in free turbulent flows where the role of density fluctuations on turbulent transport can be large [Schetz (1994)].

4.7 INFLUENCE OF INJECTION AND SUCTION ON BOUNDARY LAYER FLOW

Harry A. Dwyer

Both laminar and turbulent boundary layer flows are very sensitive to relatively small amounts of mass addition or suction at the wall. This sensitivity, which usually extends to all the boundary layer characteristics such as heat transfer, skin friction, and the separation point location, is caused by the influence that suction or blowing

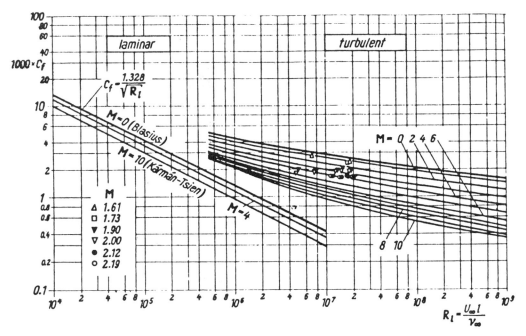

FIGURE 4.51 Coefficient of total skin friction for an adiabatic flat plate at zero incidence for laminar and turbulent boundary layer. Theoretical curves for turbulent flow $\gamma = 1.4$, $\omega = 0.76$, Pr = 1. (From Schlichting, 1979.)

have on the small component of the boundary layer velocity normal to the wall. The normal velocity component plays a key role in the transfer of primary velocity momentum through the term $v\,\partial u/\partial y$, and an increase in v increases the transport of u momentum away from the wall, thus lowering wall gradients. A decrease in v brings fluid particles inward from the outer parts of the boundary layer and thus increases

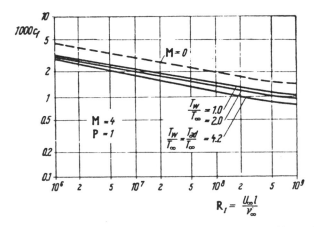

FIGURE 4.52 Skin friction coefficient for a flat plate at zero incidence in turbulent flow with heat transfer as a function of Reynolds number for different values of the temperature ratio T_w/T_∞. The curve for $T_w/T_\infty = T_{aw}/T_\infty = 4.2$ corresponds to the case with zero heat transfer; M = 4, Pr = 1. (From Schlichting, 1979.)

wall gradients. Due to these significant influences, the use of mass transfer at the wall has often been considered and used to control the structure of the boundary layer flow near the wall. The following is a small list of the applications that suction and injection have had on boundary layer flows:

Suction:

i) Delay the separation process and thus allow pressure recovery.

ii) Delay the process of transition to turbulence.

Injection:

i) Inject a cool gas near the wall to accomplish film cooling.

ii) Decrease the wall heat transfer by reducing the gradients near the wall.

Due to the high performance characteristics of some aircraft power plants, this area of research has been very active, and there have been many numerical investigations performed on turbine blade injection systems. As has been the case for other types of boundary layer flows, numerical investigations now dominate because of the ability to handle a wide variety of boundary conditions, as well as gas species. Also, the major unsolved research question is the influence of wall mass transport on the turbulent transport processes.

Laminar Flows. A major distinction must be made in all injection systems depending on whether the injected gas has a different chemical composition than the flowing gas. If the chemical composition is different, a species transport equation must be solved simultaneously to include the influence of mass diffusion in the boundary layer (see Sec. 4.13). The case of the same fluid used for mass transfer as for the main flow can be analyzed in a simple fashion for flow over a flat plate. The flow will be self-similar only if a length scale is not introduced at the wall, and this is the situation for the improbable case where the surface velocity varies as $x^{-1/2}$, and the boundary condition itself becomes self-similar.

The major influence of mass transfer is to modify the boundary condition at the wall, and for uniform injection, the boundary conditions for Eq. (4.154) become

$$\eta = 0 \quad f' = 0 \quad f(0) = -\frac{2\rho_w v_w}{\rho_e U_e}\sqrt{\frac{\rho_e U_e x}{\mu_e}}$$

$$\eta \to \infty \quad f' = 1 \tag{4.177}$$

The velocity profiles for various values of $f(0)$ are shown in Fig. 4.53; a dramatic thickening of the boundary layer is clearly evident. The influence on wall skin friction coefficient and Stanton number is shown in Fig. 4.54 and the corresponding decrease in these wall characteristics with the boundary layer thickening are exhibited clearly.

With wall suction, the only change in the analysis is the change in sign of the wall boundary condition, and an extensive series of calculations have been conducted by Schlichting (1979) for both flat plate and stagnation point flows. The results have been calculated for incompressible flow, but the results can be extended

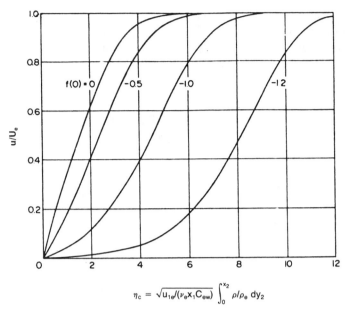

$$\eta_c = \sqrt{u_{1e}/(\nu_e x_1 C_{ew})} \int_0^{x_2} \rho/\rho_e \, dy_2$$

FIGURE 4.53 Velocity profiles in a laminar boundary layer with surface injection. (From Rubesin and Inouye, 1973.)

to compressible flow for $C_r = 1$. The influence of wall suction is shown in Fig. 4.55, where the increased skin friction coefficient is given as a function of wall suction velocity.

Turbulent Flow. In the case of turbulent flow with wall mass transfer, the flow is best analyzed by numerical methods, and the most critical facet again becomes the

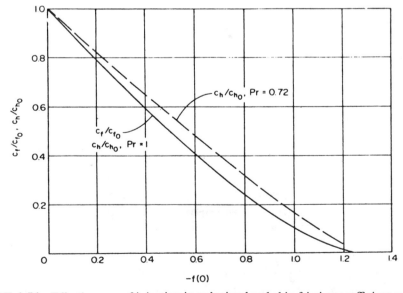

FIGURE 4.54 Effectiveness of injection in reducing local skin friction coefficient and Stanton number on a flat plate with a laminar boundary layer. (From Rubesin and Inouye, 1973.)

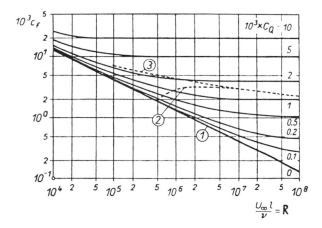

FIGURE 4.55 Drag coefficients for the flat plate at zero incidence with uniform suction c_Q = $(-v_w)/U_\infty$ = volume coefficient of suction. Curves (1), (2), and (3) refer to no suction; (1) laminar, (2) transition from laminar to turbulent, (3) fully turbulent. (From Schlichting, 1979.)

model for turbulent transport. Cebeci has suggested a variation of the mixing length in the transition wall region, and this extension of Eq. (4.170) is

$$l_t = 0.4y \left[1 - \exp\left\{ -\frac{y^+}{26} N \frac{\nu_w}{\nu} \left(\frac{\rho_w}{\rho}\right)^{1/2} \right\} \right] \qquad (4.178)$$

where

$$N^2 = \frac{\mu}{\mu_w} \left(\frac{\rho_e}{\rho_w}\right) \frac{p^+}{u_w^+} \left[1 - \exp 11.8 \frac{\mu_w}{\mu} v_w^+ \right]$$

$$+ \exp\left[11.8 \frac{v_w}{\mu} v_w^+ \right] \qquad (4.179)$$

and

$$v_w^+ = \frac{v_w}{v^*} = v_w \left(\frac{\rho_w}{\tau_w}\right)^{1/2} \qquad (4.180)$$

The expression given in Eq. (4.178) is representative of the complexity that mixing length models have become for two-dimensional boundary layer flows. Other suggestions for turbulence models can be found in Schetz (1994).

The results of calculations with the mixing length model in Eq. (4.178) have been good considering that only the effective location of the transition layer has been changed. Shown in Fig. 4.56 is the calculation of the velocity profiles for a step input of mass injection which is then discontinued. Another calculation by Cebeci for the prediction of the Stanton number for both suction and blowing is shown in Fig. 4.57. Similar results have been obtained when the mixing length models are extended to compressible turbulent flow; an example of these results are shown in Fig. 4.58.

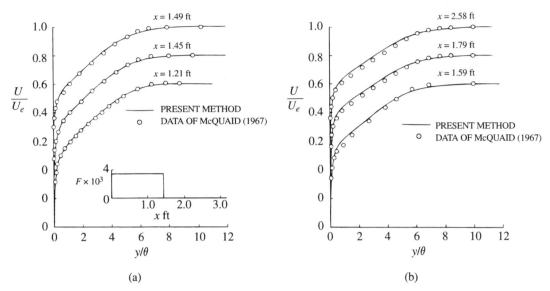

FIGURE 4.56   Comparison of calculated and measured velocity profiles for the discontinuous-injection flow. The calculations with blowing were started at $x = 0.958$ ft. $F = 0.0034$. (a) Profiles upstream of discontinuity, (b) profiles downstream. (Reprinted with permission, Cebeci, T. and Smith, A. M. O., *Analysis of Turbulent Boundary Layers*, 1st ed., Academic Press, New York, 1974.)

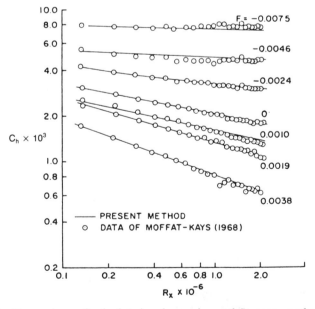

FIGURE 4.57   Comparison of calculated and experimental Stanton numbers for a turbulent boundary layer with suction and blowing. (Reprinted with permission, Cebeci, T. and Smith, A. M. O., *Analysis of Turbulent Boundary Layers*, 1st ed., Academic Press, New York, 1974.)

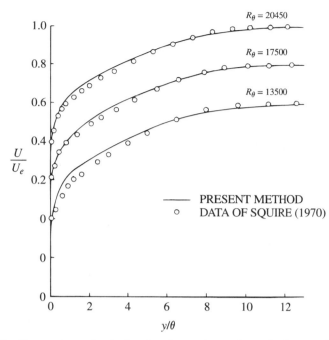

FIGURE 4.58   Comparison of calculated and experimental results for an adiabatic flat plate flow with mass transfer. $M_e = 1.8$; $F = 0.0013$. (Reprinted with permission, Cebeci, T. and Smith, A. M. O., *Analysis of Turbulent Boundary Layers*, 1st ed., Academic Press, New York, 1974.)

4.8   ASPECTS OF TIME-DEPENDENT AND THREE-DIMENSIONAL BOUNDARY LAYER FLOWS
Harry A. Dwyer

4.8.1   Introduction

The addition of time and/or three-dimensional aspects to a high Reynolds number flow opens a very rich array of flow possibilities. From the experimental, analytical, and computational points of view, these influences present a large collection of phenomena. In order to comprehend these phenomena, it will be very important that three-dimensional graphical presentations be developed and that the data or computational results be placed in a movie format, because there are serious limitations to the comprehension of results with more than two independent variables.

In this article, the results will be restricted to boundary layer flows because of their relative simplicity compared to more complex separated flows. However, even with this restriction, the most basic flow fields require a full numerical solution and a substantial computational effort and facility. The starting point for a derivation of the equations is again the Navier–Stokes equations, and the boundary layer equations can be obtained by taking the simultaneous limit of Reynolds number approaching infinity and y, the coordinate normal to a no-slip surface, approaching zero [Rosenhead (1963)]. The results for unsteady incompressible, laminar flow are

$$\frac{\partial u}{\partial x} + \frac{\partial v}{\partial y} + \frac{\partial w}{\partial z} = 0 \qquad (4.181)$$

$$\frac{\partial u}{\partial t} + u \frac{\partial u}{\partial x} + v \frac{\partial u}{\partial y} + w \frac{\partial u}{\partial z} = -\frac{1}{\rho} \frac{\partial p_e}{\partial x} + \nu \frac{\partial^2 u}{\partial y^2} \qquad (4.182)$$

$$\frac{\partial w}{\partial t} + u \frac{\partial w}{\partial x} + v \frac{\partial w}{\partial y} + w \frac{\partial w}{\partial z} = -\frac{1}{\rho} \frac{\partial p_e}{\partial z} + \nu \frac{\partial^2 w}{\partial y^2} \qquad (4.183)$$

where x and z are two Cartesian coordinates locally parallel to a body or wall surface, u, v, and w are the velocity components in the x, y, and z directions, respectively, p_e is the pressure, etc.

The mathematical nature of the equation set, Eqs. (4.181) to (4.183), is parabolic, with boundary conditions needed for u and w along y and initial conditions associated with t, x, and z. The continuity equation determines v from a local initial-value problem associated with the no-slip surface, and the pressure must be determined from other sources (either experimental results or inviscid flow results). A more physical way of describing the mathematical assumption would be to say that convection occurs only in the x and z directions, while the combined processes of diffusion and convection occur along the y coodinate. Convection by itself defines an initial-value problem, while diffusion defines a boundary-value problem, and of course, time defines an initial-value problem.

Because of the parabolic nature of the three-dimensional boundary-layer equations, initial conditions play a significant role in the flow develpoment and the history of most flows. In general, the two types of initial conditions for the spatial coordinates x and z correspond to either a sharp leading edge or a stagnation point, while for time, t, a wide variety of conditions appear in practice (the impulsively started flow being of general interest). From the numerical viewpoint these various initial conditions present serious scaling problems, and there is a great advantage to be gained using coordinate transformations. Coordinate transformations can remove leading-edge and time-dependent singularities, as well as generally smooth the variation of all variables in the transformed plane, and the small amount of time spent in transforming the equations is usually well worth the effort.

A transformation has been developed [Dwyer and Sherman (1981)] that incorporates many features of other investigators. This transformation is

$$\xi = x, \quad \eta = y \left(\frac{1 + U_e t/x}{\nu t} \right)^{1/2}, \quad \zeta = z, \quad \tau = t \qquad (4.184)$$

(It is assumed for the discussion of initial conditions that the leading edge or stagnation point is located at $x = 0$ and U_e is the inviscid-flow velocity.) In terms of these new independent variables, Eqs. (4.181)–(4.183) become

$$\xi \frac{\partial f'}{\partial t} + \xi \frac{\partial f'}{\partial \eta} \frac{\partial \eta}{\partial x} + \frac{\partial v'}{\partial \eta} + \beta_x f' + \frac{\xi}{U_e} \frac{\partial w}{\partial z} = 0 \qquad (4.185)$$

$$\frac{\xi}{U_e} \frac{\partial f'}{\partial t} + \xi f' \frac{\partial f'}{\partial \xi} + \left[\frac{\xi}{U_e} \frac{\partial \eta}{\partial t} + \xi f' \frac{\partial \eta}{\partial x} + v' \right] \frac{\partial f'}{\partial \eta}$$

$$= (1 - f')\beta_\tau + (1 - f'^2)\beta_x + \epsilon \frac{\partial^2 f'}{\partial \eta^2} \qquad (4.186)$$

$$\frac{\xi}{U_e} \frac{\partial \psi}{\partial t} + \xi f' \frac{\partial \psi}{\partial \xi} + \left[\frac{\xi}{U_e} \frac{\partial \eta}{\partial t} + \xi f' \frac{\partial \eta}{\partial x} + v' \right] \frac{\partial \psi}{\partial \eta}$$

$$= (1 - \psi)\psi_t + (1 - f'\psi)\psi_x + \xi \frac{\partial W_e/\partial z}{U_e} (1 - \psi^2) + \epsilon \frac{\partial^2 \psi}{\partial \eta^2} \quad (4.187)$$

where the following definitions have been employed

$$f' = \frac{u}{U_e}, \ \psi = \frac{\partial w/\partial z}{\partial W_e/\partial z}, \ \beta_x = \frac{x}{U_e} \frac{\partial U_e}{\partial x}, \ \beta_t = \frac{x}{U_e^2} \frac{\partial U_e}{\partial t},$$

$$\epsilon = \frac{\xi + U_e t}{U_e t}, \ \psi_x = \frac{x}{\partial W_e/\partial z} \frac{\partial^2 W_e}{\partial x \, \partial z}, \ \psi_1 = \frac{x}{U_e \partial W_e/\partial z} \frac{\partial^2 W_e}{\partial z \, \partial t},$$

$$v' = \frac{v\xi}{U_e} \left(\frac{1 + U_e t/x}{vt} \right)^{1/2} \quad (4.188)$$

It should be pointed out that Eq. (4.187) represents the z derivative of Eq. (4.183); this form is chosen to obtain stagnation-point solutions. An equation for w is useful at a sharp leading edge and has a form very similar to Eq. (4.186) [Dwyer (1968)].

The use of this transformation for stagnation-point and sharp leading edge flows will now be illustrated. The stagnation-point solution is obtained from the conditions

$$\frac{\partial U_e}{\partial x} = A, \quad \frac{\partial W_e}{\partial z} = B, \quad C = A/B \quad \text{as} \quad \begin{cases} \xi \to 0 \\ U_e \to 0 \\ W_e \to 0 \end{cases} \quad (4.189)$$

and Eqs. (4.185) to (4.187) become

$$\frac{\eta}{2} \frac{\partial f'}{\partial \eta} + \frac{\partial v'}{\partial \eta} + f' + \frac{\partial w/\partial z}{A} = 0 \quad (4.190)$$

$$\frac{1}{A} \frac{\partial f'}{\partial t} + \left[-\frac{f'\eta}{2} + v' \right] \frac{\partial f'}{\partial \eta} = (1 - f'^2) + \frac{\partial^2 f'}{\partial \eta^2} \quad (4.191)$$

$$\frac{1}{A} \frac{\partial \psi}{\partial t} + \left[-\frac{f'\eta}{2} + v' \right] \frac{\partial \psi}{\partial \eta} = \frac{\partial W_e/\partial z}{A} (1 - \psi^2) + \frac{\partial^2 \psi}{\partial \eta^2} \quad (4.192)$$

The above system of equations must be solved simultaneously with Eqs. (4.181)–(4.183) for time-dependent problems or once at the start of a calculation for a steady-state problem.

For a sharp leading edge where U_e and W_e are both nonzero, it can be shown that Eqs. (4.185)–(4.187) reduce to the well-known Blasius equation [Dwyer (1968)], therefore tabulated values of u, w, and v can be utilized to obtain initial conditions in the transformed plane. However, it should be pointed out that a singularity exists in the physical-plane solution and that the flow cannot be calculated accurately near $x = 0$ unless the singularity is removed by a coordinate transformation.

A singularity of a different type also exists near $t = 0$ for any three-dimensional boundary layer that is impulsively started. This singularity is again removed by the transformation Eq. (4.184), and it can be shown [Dwyer and Sherman (1981)] that

$$\frac{\partial \eta}{\partial x} \to 0, \quad \frac{\partial \eta}{\partial t} \to -\frac{\eta}{2t}, \quad \text{at } x \neq 0 \qquad (4.193)$$

The equations for u and w have the form

$$\frac{\eta}{2}\frac{\partial f'}{\partial \eta} + \frac{\partial^2 f'}{\partial \eta^2} = 0 \qquad (4.194)$$

Again, tabulated values can be found in standard references [Rosenhead (1963)]. These same solutions are also useful along lines of symmetry, and it can be shown that Eqs. (4.190)–(4.192) reduce to the form of Eq. (4.194).

As mentioned previously, boundary conditions are needed at the effective boundaries of the η coordinate for the variables u and w. For the wall location, $\eta = 0$, the conditions are the conventional no-slip ones, unless there is surface mass transfer, which is not considered here. At the outer edge of the boundary layer, or the inner edge of the inviscid flow, both the locations of the boundary-layer edge and the values of U_e and W_e must be determined. For laminar boundary layers without heat or mass transfer, the effective edge of the boundary-layer flow is usually located $\eta \leq 6.0$. (The edge of the boundary layer is defined to be the point where the variables u and w have returned to 99% of their inviscid-flow values.) This value is essentially the same as that for two-dimensional flow, and the three-dimensional nature of the problem does not seem to influence the location of the boundary-layer edge.

The actual values of U_e and W_e and the pressure gradients within the boundary layer cannot be obtained from boundary-layer theory. In most cases, these values are obtained from inviscid-flow analysis or experimental measurements. Although the boundary conditions rarely cause difficulties with numerical simulations, the boundary conditions can be a problem in boundary-layer theory. This is due to the influence of separation. For most nonstreamlined bodies, the influence of the downstream separated-flow region has a major effect on the inviscid flow over the boundary layer, and inviscid analyses that do not take account of this fact result in large errors. Unfortunately for boundary-layer theory, the only remedy is a complete solution of the Navier–Stokes equations which effectively negates the need for boundary-layer solutions.

4.8.2   Calculation of Some Typical Flows

The first example to be presented is that of the impulsive motion of a flat plate through an incompressible fluid at velocity U_e. This problem has been investigated previously in detail by Stewartson, who considered the problem in the two-dimensional coordinate system

$$\tau = \frac{U_e t}{x}, \quad \xi = \frac{y}{(\nu t)^{1/2}} \qquad (4.195)$$

In this coordinate system, the equations of motion are reduced to two independent variables. Stewartson showed that for $\tau \le 1$ the flow is an *impulsive Rayleigh flow* past an infinite plate, and that in the region $1 \le \tau \le \infty$ the flow transitions to a completely Blasius form as τ becomes large. Stewartson also gave a lucid physical description of how the leading edge influences reach the region $\tau = 1$ by convection in the outer region of the boundary layer. [This description is basically the same as the *region of influence principle* given in Dwyer (1981) for three dimensional boundary layers]. A further investigation by Stewartson studied the behavior of the flow in the $\tau = 1$ region with the linearized form of the boundary layer equations (*Oseen Equations*). It was found that an analytic matching of the solutions for $\tau < 1.0$ and $\tau > 1.0$ could not be completed at $\tau = 1.0$. With the full equations, convection tends to diffuse leading edge influences as the $\tau = 1$ region is approached, and only the infinitesimally small influences at the edge of the boundary layer reach the $\tau = 1$ flow region. Therefore, the $\tau = 1$ problem is exaggerated by the artificial convection in the Oseen problem. This point is verified by the numerical results.

Shown in Figs. 4.59 and 4.60 are results of the numerical computation of the boundary layer flow over an impulsively started semi-infinite flat plate. Figure 4.59 gives a non-dimensional plot of the wall shearing stress as a function of τ. It is seen from the data that the parameter τ does correlate the flow variables. For $\tau \le 1.0$ and almost to $\tau \le 2.0$, the flow has a strong Rayleigh character, while for $\tau \ge 3.5$ the behavior becomes almost the same as a Blasius flow. The region $\tau = 1$ does not seem to be unusual, and as can be seen from the data, is hardly influenced by the leading edge. The convergence to the Blasius flow appears to be a little slower than would be expected from Stewartson's work. The development of the velocity profiles as a function of τ are shown in Fig. 4.60. Again, notice that the variables η and τ effectively correlate the data for all x, y, and t.

As proof of the general usefulness of the flat plate flow to illustrate clearly important features of different physical problems, results are now presented for a rotating flat plate airfoil. A general airfoil geometry and coordinate system is shown in Fig. 4.61. It is evident that the influences of Coriolis and centrifugal forces play a significant role.

The results of the numerical solution of the boundary-layer equations for flow over a rotating flat plate are shown in Fig. 4.62, along with an extended perturbation

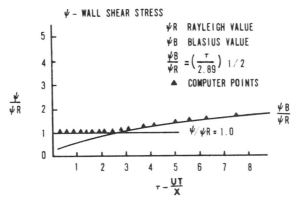

FIGURE 4.59 Wall shear stress as a function of time and space. (From Dwyer, 1968.)

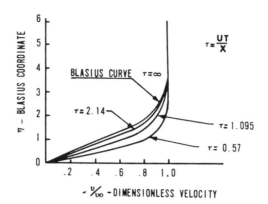

FIGURE 4.60   Velocity profile as a function of time and space. (From Dwyer, 1968.)

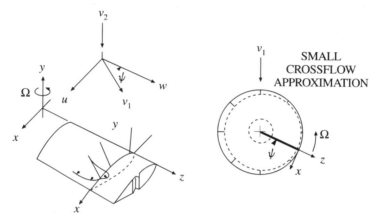

FIGURE 4.61   Coordinates in the rotating system. (From McCroskey and Dwyer, 1971.)

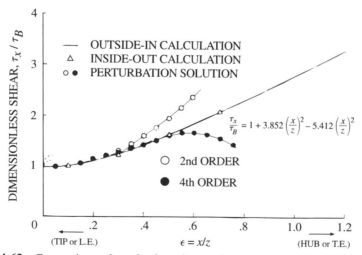

FIGURE 4.62   Comparison of results for primary shear stress on a steady rotating flat plate blade. (From McCroskey and Dywer, 1971.)

analysis. The numerical calculations were performed in two different ways. The first method started far away from the hub ($z/c = 6.0$) and proceeded inward, while the second method started at $z/c = 1.0$ and marched outward. Both methods gave results within 1/2% of each other, thus indicating that the iterative scheme for starting the calculation worked well. These results also indicate that it is possible to integrate against the flow for some problems and obtain unique results. In general, however, numerical calculations of flows with both negative and positive crossflow should be analyzed carefully for nonuniqueness.

Figure 4.62 shows several other important results. The first is that the numerical results give a direct confirmation of the predictions of the perturbation analysis, both with regard to the magnitude of the wall shear and with regard to the assumed parameters. For small values of $\epsilon = x/z$ (<0.3), the perturbation results carried out to second order $[(x/z)^2]$ are adequate to describe the flow, while the fourth-order results $[(x/z)^4]$ are valid for $x/z < 0.5$. Also, it is important to note that the character of the flow is completely determined by the single parameter, x/z. The reason for this correlation with x/z is that the pressure gradients and Coriolis forces are functions only of x/z in the transformed plane for the flat-plate problem. (They are not functions of x/z alone in the physical plane and that is why the local shear stress is divided by the local Blasius value.)

Another important result in Fig. 4.62 is that the shear stress can become much larger than its two-dimensional value in the region of the hub, although for the range of interest for helicopters, $x/z < 0.1$, the effects of rotation are very small. The reason for the high shear in the hub region is twofold: 1) an increased crossflow in the bottom regions of the boundary layer, and 2) an induced crossflow pressure gradient from the term $W_e(\partial W_e/\partial x)$. In the vicinity of the hub, the induced pressure gradient term can become very large, and its influence can dominate the entire problem, as will be seen later.

Additional insight into the flat-plate flow may be obtained by studying the velocity profiles and crossflow velocity derivatives. A comparison of the velocity profiles between the numerical and perturbation solutions shows that the velocity profiles are identical for small x/z, however, when x/z increases to 0.7, there is substantial deviation. Similarly, it was observed that the crossflow derivative was large for large values of x/z near the hub. It is the large crossflow derivative which the perturbation analysis does not calculate well, and this derivative is partially responsible for the deviation between the numerical and perturbation solutions. It should be pointed out that the perturbation method does give correct qualitative trends.

As is well known, the flow over a flat plate is a drastic simplification of the flow over an airfoil. Now, results of laminar boundary-layer calculations over a finite thickness rotating blade are presented. The main purpose of these calculations will be to investigate the importance of crossflow effects compared to the primary flow pressure gradient. Only numerical results and some limited related experimental data are presented, since the perturbation analysis does not readily lend itself to finite thickness aerofoils.

The flow over a rotating NACA 0012 aerofoil section is representative of contemporary helicopter blades (the calculation is somewhat unrealistic because of the assumption of laminar flow). For the inviscid flow, the constant circulation solution of Sears was used for an angle of attack of 0°, and the two-dimensional potential function required for this flow was obtained using the method of Theodorsen and

Garrick (1933). It should also be mentioned that the calculations over the aerofoil proceed only to the point where the primary flow velocity reverses itself (*flow reversal line*). Because of the fact that the crossflow velocity is small for this problem, the flow reversal line will be essentially identical to the *limiting surface streamline* [McCroskey and Dwyer (1971)]. (The limiting surface streamline divides upstream and downstream fluid at the bottom of the boundary layer.)

Figure 4.63 shows the results of the calculations for the flow reversal line as defined previously. Two important features are exhibited: 1) the flow reversal moves aft as z decreases, and 2) a large rotationally induced pressure gradient exists near the hub. For the NACA 0012 airfoil section, the limiting streamline is behind and very close to the flow reversal line. It is also apparent, that at a distance of about six chord lengths from the hub, the flow reversal line begins to deviate substantially from its two-dimensional value, and near the hub no separation will occur at all. The important influence of the crossflow-induced pressure gradient on separation can be seen by observing the line in Fig. 4.63 labeled "Locus of $\partial p/\partial x = 0$." This line separates the regions of favorable pressure gradients from regions of adverse gradients. For large z, the pressure gradient is favorable near the leading edge of the rotor and then becomes adverse past the point of maximum thickness, as in two-dimensional airfoil theory. However, for small z (i.e., close to the hub), the pressure gradient is favorable over the entire blade. The favorable pressure gradient is a direct result of the crossflow-induced pressure gradient, $W_e(\partial W_e/\partial x)$.

The above results show clearly that three-dimensional and/or time dependent boundary layers can be calculated, however there is still one very substantial factor that has hindered the practical application of these techniques. This factor is a turbulence model which accounts for the influence of time dependence and three-dimensionality. At the present time, there are relatively few models or data available to build models for the numerical solution of turbulent flows of the above type. See Schetz (1994) for a discussion of models for 3-D boundary-layer flows.

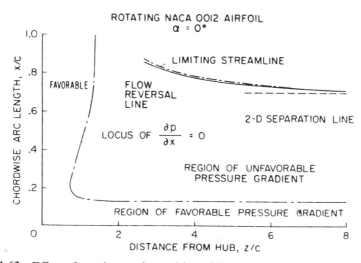

FIGURE 4.63   Effect of rotation on the position of the limiting streamline and flow reversal point for a NACA 0012 airfoil. (From McCroskey and Dwyer, 1971.)

4.9 SEPARATED FLOWS
James C. Williams, III

4.9.1 Introduction

Separation of the boundary layer from a solid wall is the controlling, if not dominant, feature of many practical fluid flows. For flow in passages such as diffusers or nozzles, or in components of rotating machinery, such as compressors or turbines, separation frequently alters the flow field and hence the performance of the device. For flow over bluff bodies, it is the location of separation which determines the pressure drag of the body. For airfoils and blades at high angles of attack, the conditions at separation dictate the circulation about the airfoil or blade and hence the lift and moment.

It is generally perceived that separation has a deleterious effect on the performance of fluid mechanical devices, as in the examples noted above. There are cases in which separation is promoted and controlled in order to make a fluid device more effective. In certain fluidics devices, for example, separation is designed into the device so that the flow can be altered in a desired fashion.

Whether separation limits or enhances the performance of a fluid handling device, it is important to understand the physical nature of separation so that it can be predicted and controlled. Unfortunately, however, this requirement is not easily fulfilled. In spite of its profound practical importance, separation is one of the least well understood phenomena in fluid mechanics. At high Reynolds numbers, separation marks the division between two very different types of fluid flow. Upstream of separation the effects of viscosity are limited to a thin layer next to the bounding surface; downstream of separation the effects of viscosity influence an extensive region of the flow field. Beyond separation there may be a large recirculating flow (wakes) and extensive interactions between the viscous and inviscid portions of the flow.

One might think that with all the developments in boundary layer theory in the past 90-odd years and with all the experimental investigations conducted in the past century separation would be completely understood by now. Unfortunately this is not the case, for separation, by its very nature, is difficult to inspect closely. For two-dimensional laminar or turbulent separation the neighborhood of separation is characterized by small velocities and small shear. Direct measurement of the velocity field or the surface shear distribution is generally difficult, and the results obtained are often inaccurate. From the analytical point of view, the problem is also extremely difficult. In general, separation represents a point beyond which the usual simplifications associated with approximate theories, such as boundary layer theory, break down, and it is necessary to either return to the full Navier–Stokes equations to describe the flow field or to seek new methods of simplification for the problem. In short, it must be concluded that present knowledge of separation is far from complete. At this point in time, it is only possible to present a review of our basic knowledge of the phenomenon of separation rather than a detailed catalog of accurate prediction methods.

One might hope that there would be at least some simple correlations which would be useful in predicting the location of separation. Unfortunately, except in cases where the location of separation can be forced, such as at shock impingement or compression corners in supersonic flow, such correlations do not exist. Separation on a continuous surface defines the end of a region where the flow is parabolic (history dependent) in nature. Beyond this point, the flow field characteristics depend not only on the local flow properties, but also on the entire flow field upstream of separation. Thus, any correlation based on local flow conditions cannot hope to succeed.

In recent years there have been several reviews of separation [Brown and Stewartson (1969), Chang (1976), Williams (1977), and Simpson (1981)]. These provide significant insight into the phenomenon of separation. In the present section an attempt is made to encapsulate the knowledge of separation reviewed in earlier publications and to add to this summary some of the more recent work on this phenomenon.

Attention is limited to flows at high Reynolds numbers. Most practical fluid flows occur at high Reynolds numbers so that this limitation is not a serious one. The discussion will begin with consideration of two-dimensional steady separation, for this case is most clearly defined and well understood. This is followed by a discussion of three-dimensional steady separation and two-dimensional unsteady separation. The differences and similarities between laminar and turbulent flows will be noted. The last section is devoted to two special cases in which separation is forced by discontinuities on the bounding surfaces, rearward-facing steps and cavity-type flows.

4.9.2 Boundary Layer Separation From Continuous Surfaces

Two-Dimensional Separation

Two-Dimensional Separation Criteria: Two-dimensional laminar separation was first described by Prandtl who used the term *separation* to describe the phenomenon in which a thin layer of viscous fluid was ejected from the surface into the surrounding fluid. The boundary layer passing over a continuous surface develops under the influence of two forces: 1) the pressure gradient, which is imposed at all levels of the shear layer, and 2) the wall shear, which is imposed at the lowest layer in the fluid flow but whose influence is transmitted to the upper layers by means of viscosity and turbulence. Prandtl argued that under the influence of a sustained adverse pressure gradient ($dp/dx > 0$) those particles near the wall will have their momentum depleted to the point where their forward motion must cease, and the main fluid will be reversed as flow comes in under the separated layer. Separation in the two-dimensional, steady case is directly related to flow reversal near the wall and hence to a condition of vanishing wall shear. This concept of separation (for two-dimensional, steady flows) proven to be so accurate that it has been universally accepted since its inception.

There are flows in which the recirculating flow downstream of separation is relatively thin, of the same order as the thickness of the boundary layer upstream of separation. The cases of a thin separation bubble trapped within the boundary layer on an airfoil and of separation very near the trailing edge of a thin airfoil are typical examples of such flows. Such situations lead to speculation that the boundary layer equations might yield acceptable solutions for the flow field both upstream and downstream of separation, since the basic boundary layer assumption of a thin viscous layer is not violated in such flows. There are two problems which must be overcome if the boundary layer equations are to be used in the analysis of such flows. First, some way must be found to eliminate or avoid the singularity which occurs at separation whenever the boundary layer equations are solved for a prescribed pressure gradient. Second, some way must be found to account for the displacement effects resulting from the thickened boundary layer near and beyond separation.

Even when separation does not occur on the body, the presence of the boundary layer, in effect, alters the shape of the body by an amount proportional to the boundary layer displacement thickness. As a result, the pressure distribution on the body is different from that determined by potential flow theory. This is particularly true when separation occurs since one of the major effects of separation is a thickening of the viscous layer. Figure 4.64 from Thwaites (1960) depicts the effects of the boundary layer alone, of trailing edge separation and of a separation bubble on the pressure distribution of an airfoil. One notes the essentially constant pressure in the separated region. In effect, every flow in which separation occurs is an *interacting flow*. The boundary layer development alters the potential flow pressure field, and the pressure field influences (*interacts* with) the boundary layer development. This interaction occurs whether the wake-like region downstream of separation is small or not. It is particularly influential when the wake region is large.

It is possible now to obtain accurate numerical solutions of the boundary layer equations, solutions which determine the detailed structure of the boundary layer, by modern numerical techniques. One finds, however, in applying such methods to flows with a prescribed adverse pressure gradient, that as the point of zero wall shear is approached it becomes impossible to complete the calculation. This difficulty is easily traceable to the existence of a singularity in the solution of the boundary layer equations at separation. The structure of this singularity was investigated by Goldstein (1948). In recent years, a number of numerical investigations have confirmed the features of the separation singularity so that its structure is now well understood. The problem, however, is not understanding the singularity, but instead, finding some means of avoiding it in detailed calculations.

The way to avoiding the singularity appears to lie in the pressure gradient. Gold-

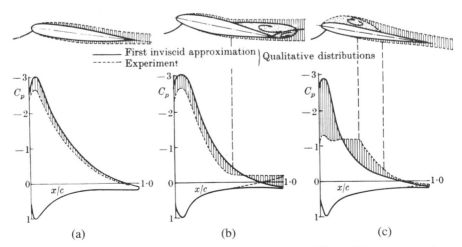

FIGURE 4.64 Viscous flow over airfoils. (a) Unseparated flow, (b) rear separation, and (c) leading-edge separation and long bubble. (Reprinted with permission of Oxford University Press, Oxford, *Incompressible Aerodynamics*, by Thwaites, Bryan, 1960.)

stein apparently recognized this fact and noted that the possibility exists "that a singularity will always occur except for certain very special pressure variations in the neighborhood of separation and that experimentally, whatever we may do, the pressure variations near separation will always be such that no singularity will occur." It has already been noted that the thickened boundary layer in the vicinity of separation is expected to alter the pressure variation on the body. Could it be that the pressure due to the displacement effect is altered in such a way as to eliminate the separation singularity? If so, then the problem of integrating the boundary layer equations through separation is truly an interaction problem, and it should be possible to integrate the boundary layer equations through separation using an inverse technique in which some property other than the pressure distribution is prescribed. This idea was proven correct by a number of investigators [Catherall and Mangler (1966), Klineberg and Steger (1974), and Horton (1974)] who prescribe a regular distribution of either the displacement thickness or the wall shear. In either case, the pressure distribution on the body is obtained as a result of the calculation. The validity of such inverse techniques was convincingly demonstrated by Carter (1973) who used inverse techniques to calculate Briley's (1971) Navier–Stokes solution for the flow through an incompressible separation bubble.

These results indicate that, as a practical matter, the problem of the separation singularity has been resolved. Regular solutions to the two-dimensional, steady laminar boundary layer equations may be obtained for flow passing through separation if, instead of prescribing the pressure gradient, one prescribes either the wall shear stress distribution or the displacement thickness distribution.

Accounting for the interaction, when the interaction is weak, appears to offer no insurmountable difficulties. Carter and Wornom (1975), for example, have calculated flow fields with embedded laminar separation bubbles using a technique in which the separation singularity is avoided by using a displacement thickness distribution obtained from consideration of the interaction between the boundary layer and the inviscid flow.

Figure 4.65 presents the results obtained by Carter and Wornom for the case of a flat plate with a depression in the surface. The parameter t (legend, Figure 4.65) is a measure of the depth of the depression. The actual body shape is presented in Figure 4.65(a) together with the *displacement body* obtained by adding the boundary layer displacement thickness which results from the interaction calculation. Also shown in Fig. 4.65(a) are the pressure coefficient variations for the body with and without interactions. Clearly, the interaction serves to mitigate the strong adverse pressure gradient caused by the depression in the body. Figure 4.65(b) compares the boundary layer displacement thickness for a flat plate (the body without a depression), the body with no interaction and for the body with interaction. Figure 4.65(c) compares the skin friction for three cases: 1) the body without a depression (flat plate), 2) the body with the depression but with no interaction, and 3) the interaction calculation. The noninteraction calculation shows the typical singularity at separation where the slope of the skin friction line is infinite. The interaction calculation has a large but finite gradient in skin friction at separation, and the calculation proceeds smoothly through the region of negative shear.

When the interaction resulting from the separation is strong, as in the case of bluff bodies, the problem is a great deal more difficult. Here, the problem cannot be handled within the framework of boundary layer theory, and it appears that it re-

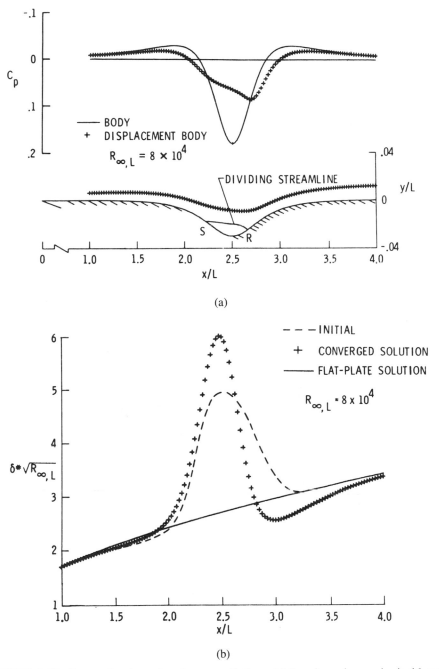

FIGURE 4.65 Interacting boundary layer prediction. (a) Laminar viscous-inviscid results; $t = -0.03$, (b) initial and final displacement thickness distribution; $t = -0.03$, and (c) laminar interaction skin friction; $t = -0.03$. (Reprinted with permission of NASA Scientific and Technical Information Facility, ''Aerodynamic Analysis Requiring Advanced Computers,'' Part I, NASA SP 347 by Carter, J. E. and Wornom, S. F., 1975.)

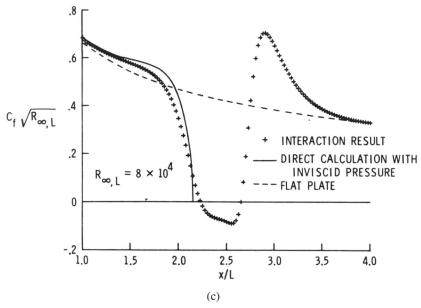

(c)

FIGURE 4.65   (*Continued*)

quires extensive analysis to match the upstream boundary layer flow with the extensive region of separated flow downstream. For many external flows with large separated regions, such as airfoils at high angles of attack [e.g., Fig. 4.64(b)] or bluff bodies, the pressure on the body surface beyond separation appears to be constant so that there is a region of constant pressure flow in the wake and full pressure recovery occurs in the wake downstream of the body. Sychev (1972) and Messiter (1975) have suggested that the theory of asymptotic expansions might be combined with free streamline theory in order to provide a solution to the problem in the case of flow past a circular cylinder.

For internal flows such as found in two-dimensional diffusers, there seem to be four flow regimes, depending upon the wall divergence angle. These regimes, for the case of straight, two-dimensional diffusers, are shown in Figs. 4.66 and 4.67. In the first of these regimes, there is no separation (*stall*) from either wall [Fig. 4.67(b)] and the pressure recovery in the diffuser increases with an increase in the wall divergence angle (Fig. 4.66). In the second, the *large transitory stall* region, there are pulsating separations on one or the other of the diffuser walls. A positive pressure gradient exists on the wall [Fig. 4.67(c)]. The peak pressure recovery occurs in this region (Fig. 4.66). At still larger wall divergence angles, the flow is fully stalled with the flow separating from one wall near the throat, while the flow remains attached to the other wall [Fig. 4.67(d)]. Here, the flow is much like the free streamline flow separation found in external flows, and complete pressure gradient relief occurs (Fig. 4.66). Finally, at large wall divergence angles, the flow is separated at both walls [Fig. 4.67(e)] and has a jet-like character. In this case, the pressure recovery is quite small (Fig. 4.66).

Laminar Separation: The singularity at separation and the interaction between the separating boundary and the essentially inviscid external stream are important fea-

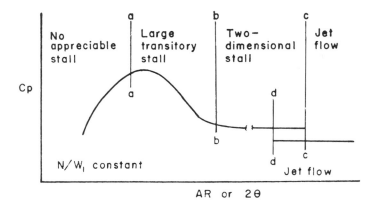

FIGURE 4.66   Relationship between flow regimes and a pressure recovery curve. (Reprinted with permission from Reneau, L. R., Johnson, J. P., and Kline, S. J., *J. Basic Eng.*, American Society of Mechanical Engineers, 1967.)

tures of many separating flows. Research work in this area continues and will continue for years to come. Nevertheless, it is important in many practical engineering applications to determine, at least approximately, the location of separation on the body or in an internal flow. If the interaction between the separated flow and the external flow is weak, then it is possible to make a reasonably accurate prediction of the separation point by straightforward integration of the boundary layer equations.

Table 4.4, presents an interesting comparison of 12 methods of predicting laminar separation. The table presents the pressure coefficient at separation $C_{ps} = (p_s - p_0)/(\rho U_0^2/2)$ where p_0 and U_0 are the minimum pressure and maximum velocity, respectively, of the outer flow and p_s and U_s are the pressure and velocity, respectively, of the outer flow at separation. Methods A through E are momentum integral methods employing either a prescribed velocity profile (A, B, and C) or correlations of integral functions (D and E). Methods F and G employ an energy integral combined with the momentum integral equation, while Method H employs a system of moments of the momentum equation. Methods I through L are approximate methods which divide the boundary layer into two parts and result in simple expressions for conditions at separation. The results of these approximate methods are compared with results obtained by numerical integration of the boundary layer equation (*accurate method*).

It is interesting to note that the Pohlhausen method [see Schetz (1994)], which is quite commonly used, gives the poorest results when compared with the accurate solutions. One has to be impressed with the accuracy (at least in comparison to the accurate numerical solution) of the particularly simple results of Curle and Skan (Method L) in which the position of separation is determined from the simple expression

$$[x^2 C_p (dC_p/dx)^2]_s = 0.0104 \tag{4.196}$$

Special note is made of this result not only because of its accuracy and simplicity, but also because there is a turbulent boundary layer counterpart to be discussed later.

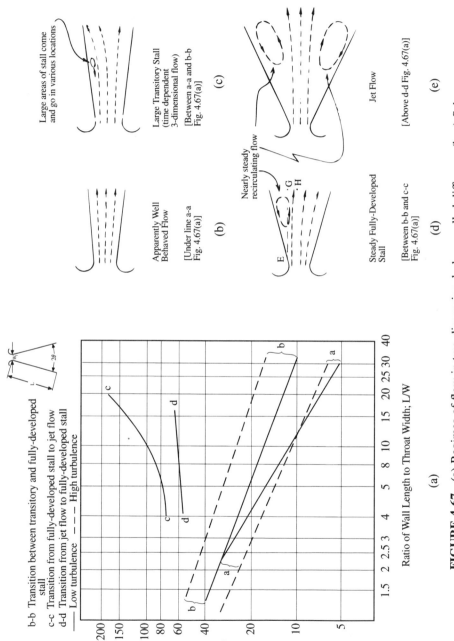

FIGURE 4.67 (a) Regimes of flow in two-dimensional plane-walled diffusers, (b–e) Schematic diagrams of regimes of diffuser flow. (Reprinted with permission from Kline, S. J., *Basic Eng.*, American Society of Mechanical Engineers, 1959.)

TABLE 4.4 Pressure Coefficients at Separation Esimated by Various Methods. (From Gadd, Jones and Watson, 1963.)

Case Method	1	2	3	4	5	6	7
Accurate	0.082	0.226	0.142	0.089	0.056	0.056	0.062
A	· ·	0.305	0.192	0.118	0.067	· ·	· ·
B	0.072	0.230	0.140	0.080	0.043	· ·	· ·
C	0.049	0.204	0.119	0.067	0.035	· ·	· ·
D	0.063	0.217	0.133	0.074	0.040	0.041	0.46
E	0.069	0.231	0.138	0.080	0.047	0.047	0.051
F	· ·	0.225	0.142	0.088	0.057	· ·	· ·
G	· ·	0.215	· ·	· ·	· ·	· ·	· ·
H	· ·	0.236	0.141	0.080	0.044	0.066	· ·
I	0.069	0.194	· ·	· ·	· ·	· ·	· ·
J	0.067	0.194	0.135	0.071	0.051	· ·	· ·
K	0.074	0.204	0.125	0.078	0.050	0.048	· ·
L	0.088	0.227	0.141	0.088	0.055	0.062	0.068

KEY

Case	Mainstream flow
1	Hartree–Schubauer ellipse Hartree (1939b)
2	$U = U_0(1 - x/l)$ Howarth (1938)
3	$U = U_0(1 - x^2/l^2)$ Tani (1940)
4	$U = U_0(1 - x^4/l^4)$ Tani (1940)
5	$U = U_0(1 - x^8/l^8)$ Tani (1940)
6	$U = \frac{3}{2}\sqrt{3}\, U_0(x/l - x^3/l^3)$ Curie (1958a)
7	$U = U_0 \sin(x/l)$ Terril (1960)

Method	
A	Pohlhausen (1921)
B	Timman (1949)
C	Walz (1941)
D	Thwaites (1940b)
E	Curle and Skan (Thwaites modification) (1957)
F	Tani (1954)
G	Truckenbrodt (1952a,b)
H	Loitsianskii (1949)
I	Karman and Millikan (1934)
J	Doenhoff (1938)
K	Stratford (1954) (equation (219); $K = 0.0076$)
L	Curle and Skan (Stratford modification) (1957) and Curle (1960) ($K = 0.0104$)

As a result of the development of high speed, large capacity computers, and sophisticated numerical techniques, the direct numerical integration of the nonlinear laminar boundary layer equations has become rather routine [See Blottner (1970) and Schetz (1994)]. There are a number of these numerical techniques which are suitable for obtaining solutions up to the vicinity of separation. Since this technique provides an accurate prediction of the internal structure of the boundary layer, one would expect difficulty as separation is approached, and indeed that is the case. The difficulty manifests itself in an increase in the number of iterations required for convergence at each station, as separation is approached, and finally, the impossibility of obtaining a converged solution in a reasonable (but large) number of iterations at one station. Fortunately, it appears that the region influenced by the singularity is small, hence the integration can be carried quite closely to separation. The actual separation point is then determined by extrapolation of the wall shear to zero.

Turbulent Separation: Most practical fluid machinery is designed for and operates at conditions which are likely to promote turbulent rather than laminar flow. This is certainly expected, since it is well known that the turbulent boundary layer can sustain a much larger pressure rise without separation than can a laminar boundary layer and, in general, larger pressure rises without separation translate directly into better performance. Figure 4.68 shows, for example, the pressure rise across a shock wave required to separate a laminar or a turbulent boundary layer. At higher Reynolds numbers, the pressure rise required to separate a turbulent boundary layer is ten or more times that required to separate a laminar boundary layer. In practice then, the engineer generally desires to design machinery to operate with a turbulent boundary layer and as close to separation as possible without actually encountering it.

While the basic symptom of two-dimensional, steady separation is the same in both laminar and turbulent flow, there is a major difference in the physical details of the two flows as well as in our ability to predict them. It must be remembered that turbulent flow, and hence turbulent separation, is basically an unsteady phenomenon. While it is convenient and expedient to consider turbulence as an unsteady perturbation of a statistically steady mean flow, one cannot escape the consequences of this unsteadiness. There is some evidence that for turbulent flow, the separation is actually a region of unsteady and perhaps even three-dimensional flow. A major question is: how should the *point of zero shear* be defined for essentially time-dependent turbulent separation? Should one consider separation to be the farthest upstream point at which the wall shear is either zero or positive, or should separation be defined in terms of the percentage of time the flow is directed downstream (or upstream) next to the wall? This is a question which will ultimately have to be answered. At present, however, for engineering purposes, it is sufficient to consider separation to be that point at which the mean velocity profile has zero shear.

Even when it comes to the actual calculation of turbulent separation, the picture is not completely clear. There are those who feel that there are additional terms which should be retained in the x moment equation in order to predict adequately the turbulent boundary layer leading to separation. Here again, this is a question which must be deferred until we have a better understanding of turbulence and turbulent separation.

The development of prediction techniques for turbulent boundary layer develop-

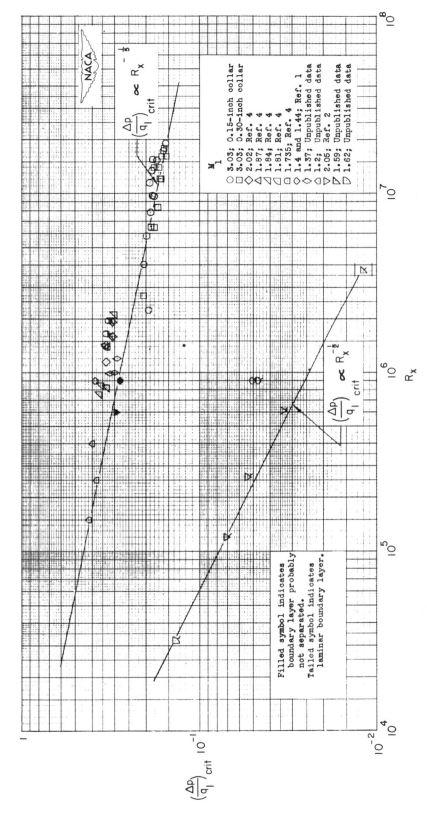

FIGURE 4.68 Variation with Reynolds number of critical pressure coefficient across shock waves which cause separation of the boundary layer. (From Donaldson, C. DuP. and Lange, R. H., 1952.)

407

ment to separation is complicated as compared to the laminar case, by the addition of the Reynolds stress to the problem. There are two general approaches to overcoming the problem posed by the Reynolds stresses. In the first of these, the need for a detailed knowledge of the local Reynolds stress is eliminated by integrating the boundary layer equations across the boundary layer. This approach leads to a momentum integral equation which relates, in a single equation, two boundary layer thicknesses (momentum and displacement) and the wall shear. Now, however, one must generate in some way additional relations between the thickness, the wall shear and the pressure gradient. The relationships employed amount to global assumptions regarding the Reynolds stresses.

These prediction methods which employ an integral of the momentum equations are known as *integral methods*. The required additional relations come from empirical correlations of the parameters involved, empirical correlations of velocity profiles or by generating additional integral relationships between certain integral quantities (e.g., the dissipation integrals, entrainment). In this latter, one can evaluate the integral quantities which appear in the new integral relations either by making an assumption regarding the local Reynolds stress distribution or by making assumptions, based on empirical observations, regarding the relationships between those new integral quantities and the mean velocity field.

In the second of the general approaches to overcoming the problem posed by the lack of knowledge regarding the Reynolds stresses, one seeks solutions to the differential equation for the mean flow (the momentum equation) employing models for the local distribution of the Reynolds stresses. In the simplest of these models, the Reynolds stresses are related to the mean velocity field by very simple algebraic relationships. In the more complex models, the Reynolds stresses are related to turbulent field properties (such as the mean turbulent energy and/or the mean turbulence scale) which in turn are determined from certain *model equations* for these quantities. These differential equation methods are the subject of considerable interest at the moment, and hence, are undergoing considerable development. See Sec. 4.4 and Schetz (1994) for further discussions of turbulence models.

In either the integral methods or the differential equation methods, the lack of complete knowledge of the turbulence process leads to a certain degree of empiricism. There is always the concern that the empirical constants, determined from some set of data, may not be appropriate for other practical cases.

The problem of major concern here, however, is separation. Assuming one can carry out the necessary calculations required in the prediction techniques, is it possible to predict separation? Many existing prediction techniques are not capable of being carried out to the point where the wall shearing stress is nominally zero. In some cases, the very inability to carry the calculation further is taken as the criterion for separation. For cases in which the external pressure gradient is specified *a priori*, one would expect that the separation singularity would cause a breakdown of the numerical method so that a failure of the method to achieve convergence might provide a reasonable prediction of separation. On the other hand, some engineers prefer the more indirect criteria in which the boundary layer shape factor ($H \equiv \delta^*/\theta$) increases rapidly as separation is approached. Limiting values of the shape factor of 1.8 to 4.05 have been proposed, but this wide range is misleading, since the shape factor generally increases very rapidly near separation so that the predicted location of separation is not very sensitive to the exact value of the shape factor chosen.

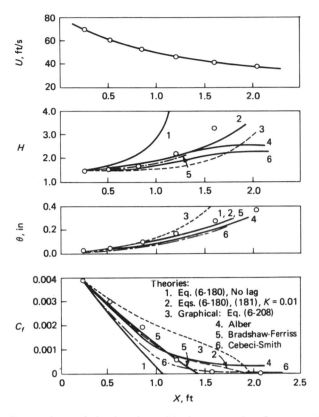

FIGURE 4.69 Comparison of six theories with the separating-flow experiment of Moses (1964). (Reprinted with permission of McGraw-Hill, New York, *Viscous Fluid Flow*, by White, Frank M., 1974.)

An interesting comparison of the results for the prediction of scparation by several methods has been made by White (1993) and is shown in Fig. 4.69. White compares the computational results of six theories with the experimental results for a separating flow obtained by Moses (1964). The theories are numbered 1 through 6. Theories 1 and 2 combine a von Karman momentum integral equation with a skin friction coefficient correlation and a shape factor correlation. The method numbered 3 is an integral method due to White (1993) which uses so-called *inner variables*. The fourth method again uses the von Karman momentum integral plus a skin Friction-Law of the Wake correlation and a mechanical energy relation. Methods 5 and 6 are boundary layer differential equation methods as opposed to integral methods. Method 5 uses an algebraic expression for the eddy viscosity to obtain closure while Method 6 solves the momentum equation plus a turbulent energy equation and employs dissipation turbulent diffusion scales. In general, the results obtained are reasonable considering there is no accounting for the interactive nature of the flow. Still, Method 4 does not even predict separation and the other methods are not terribly accurate.

In general, one must conclude that at present it is only possible to be moderately successful in predicting two-dimensional turbulent boundary layer separation. Two recent developments indicate that it may not be long before there is considerable improvement in this situation. First, there is a growing body of experimental data on two-dimensional turbulent boundary layer separation becoming available. Two examples of such data are the studies of Simpson *et al.* (1974) and Chu and Yang (1975). Second, it is gratifying to see that Pletcher (1978) has calculated the turbulent boundary layer development for a flow for which experimental data is available well into the separation region. Pletcher carried out these calculations employing a differential equation approach with several turbulence models. All the models employed the same basic turbulence model for the inner portion of the boundary layer, a simple mixing length model with damping. The models used however, differed in the manner in which the turbulence was modeled in the outer portion of the boundary layer. The various models employed included: 1) two simple algebraic models, 2) a one-equation model in which the turbulent kinetic energy is obtained from a transport equation, 3) a one-equation model for the characteristic mixing length scale, and finally, 4) a model which includes both a transport equation for turbulent kinetic energy and a transport equation for the mixing length scale. Pletcher uses an inverse technique in making the calculations, prescribing the displacement thickness and determining the external velocity. Figure 4.70 shows Pletcher's results for the skin friction. He finds that only one of the models proposed actually predicts, with a fair degree of accuracy, the skin friction distribution and the velocity profiles

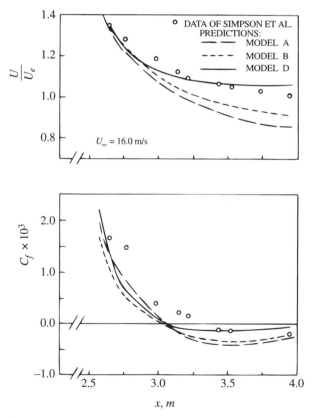

FIGURE 4.70 Comparison of skin friction coefficients and edge velocities predicted by inverse calculation procedure and measured by Simpson *et al.* (Reprinted with permission from Pletcher, R. H., *J. Fluid Eng.*, American Society of Mechanical Engineers, 1978.)

both before and after separation. The model which gives these results (Pletcher's Model D) somewhat surprisingly combines the algebraic inner mixing length model with a simplified transport equation for the length scale in the outer portion of the flow. These results lead one to believe that before long it will be possible to make accurate predictions of the two-dimensional turbulent boundary layer separation point on practical bodies.

The previous discussion considers in some detail methods for predicting turbulent boundary layer separation which are based on integration of the boundary layer equations. It is often necessary or desirable to use less accurate but simpler methods to predict separation, i.e. methods which are generally based on simple analysis plus a correlation. There are a number of these. One useful method is that of Stratford (1955) which is based on a two layer analysis of the boundary layer development. This analysis results in the particularly simple separation criteria

$$\left[C_p \left(x \frac{dC_p}{dx} \right)^{1/2} \right]_s = 0.39 \, (10^{-6} \, \text{Re})^{1/10} \tag{4.197}$$

One will note the similarity of this equation, for turbulent separation, with Eq. (4.196), for laminar separation. Such a comparison also provides an indication of the larger pressure rise to separation which can be obtained in turbulent flow compared to that obtainable in laminar flow.

Three-Dimensional Separation

Three-Dimensional Separation Criteria: In practical flow situations, the flow is three-dimensional rather than two-dimensional, hence prediction of three-dimensional separation is of considerable practical importance. Unfortunately, however, our knowledge of three-dimensional separation is more limited than that of two-dimensional separation. Various investigators have proposed that three-dimensional separation is: 1) a line along which some component of the skin friction vanishes, 2) a line along which the solution to the boundary layer equations is singular, 3) an envelope of limiting streamlines, or 4) a line which divides flow which has come from different regions. None of these definitions appear to be universally valid, but it now appears that each may have some element of validity.

The idea that a three-dimensional separation is a line on which some component of the skin friction vanishes certainly cannot be an acceptable definition. There is always some component of shear which vanishes regardless of the proximity to separation. The classical example of this occurs on the cone at angle of attack in a uniform flow. On such a cone, there is a line along which the azimuthal shear vanishes, called the cross-flow separation line. The vanishing of the azimuthal shear, however, is the only characteristic of this line. There is no local thickening of the boundary layer or ejection of boundary fluid into the inviscid flow. Indeed, if the angle of attack is sufficiently large, true separation clearly occurs beyond *crossflow separation* and near the leeward line of symmetry.

Any definition of three-dimensional separation based on a line along which solutions to the boundary layer equations are singular is of limited usefulness, since singularities are a feature of the mathematics and not the physics of the flow. Only in the case where the singularity can be tied to some physical characteristic of the flow can such an analysis be of some value. The results of Williams (1975) appear to indicate that the line along which the boundary layer solution is singular is also an envelope of limiting streamlines. The idea that separation is characterized by an envelope of the limiting streamlines seems to have some validity and forms the basis for certain flow visualization techniques.

The concept of separation as a line dividing flows which come from different regions [Eichelbrenner (1973), Moore (1956), and Stewartson (1964)] appears to have some validity in certain cases and also forms the basis for some flow visualization techniques. Wang (1972) has argued that there is a new type of separation which he terms *open separation* which does not divide the body into two distinct regions—one accessible to streamlines entering a zone of attachment and one which is not accessible to these streamlines.

A simple and elegant demonstration of the nature of three-dimensional separation has been given by Lighthill [Rosenhead (1963)] who poses the question: "How parallel to the surface are the neighborhood streamlines?" If the streamline close to the body surface closely parallels the surface, the boundary layer is thin and separation cannot occur. If these streamlines diverge from the surface, then separation is to be expected. Let τ_x and τ_y be the shearing stresses in the x and y directions, respectively, where x and y are the coordinates on the body surface, and z is the coordinate normal to the surface. If two streamlines are a distance n apart and the height of a rectangular stream tube is h (see Fig. 4.71), then the mass flow $\dot{m} = \rho n h u_{\text{ave}}$, through the stream tube must be constant. The average velocity, u_{ave}, is related to the wall shear by

$$\tau_w = \sqrt{\tau_x^2 + \tau_y^2} = \mu \frac{u_{\text{ave}}}{(h/2)} \tag{4.198}$$

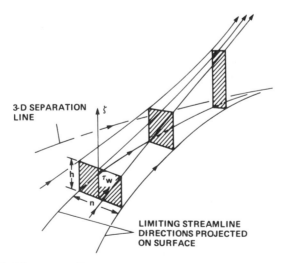

FIGURE 4.71   Limiting streamlines near a three-dimensional separation line. (Reprinted with permission of Annual Reviews Inc., Tobak, Murray and Peake, David J., "Topology of Three-Dimensional Flows," *Ann. Rev. Fluid Mech.*, 1982.)

so that

$$\dot{m} = \frac{\rho n h^2}{2\mu} \sqrt{\tau_x^2 + \tau_y^2} \qquad (4.199)$$

There are two mechanisms by which separation can occur. Either τ_x and τ_y must approach zero simultaneously, in which case h must become large (unbounded) or n may approach zero and again h must become large (unbounded). In either case, as h becomes large, the thin layer which was near the wall is projected out away from the wall and separation occurs. Using the concept of limiting streamlines, Maskell (1955) reached the same conclusion, i.e. that there are two mechanisms for three-dimensional separation. Limiting streamlines are those streamlines closest to the body surface. There projections on the body surface have the local slope (on the body surface)

$$\frac{dy}{dx} = \frac{\tau_y}{\tau_x} \qquad (4.200)$$

The projections of the limiting streamlines on the surface coincide with skin friction lines. These limiting streamlines are often called *surface streamlines* although, clearly, there are no streamlines on the body surface. The points at which τ_x and τ_y vanish simultaneously are singular points, hence Maskell defined this type of separation as *singular separation*. The other case in which the distance between limiting streamlines approaches zero is defined by Maskell as *orinary separation*. Figure 4.72 shows examples of ordinary and singular separation. One can easily see how the separated fluid along a line of ordinary separation can roll up into a vortex sheet.

Maskell shows how the limiting streamlines can be used to construct a *skeleton structure of the viscous region* and shows a number of examples of such skeleton

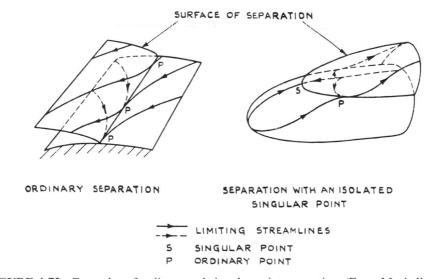

FIGURE 4.72 Examples of ordinary and singular point separation. (From Maskell, 1955.)

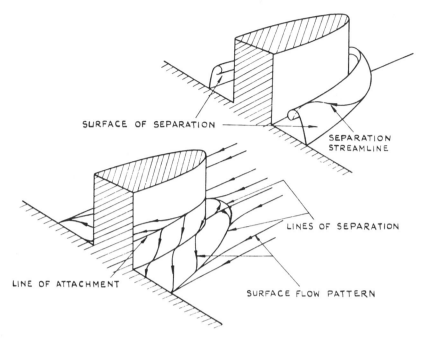

FIGURE 4.73   Formation of a separation vortex at the junction of two bodies. (From Maskell, 1955.)

structures. One of the most interesting of these is the formation of a free vortex at the intersection of two bodies (e.g., wing-fuselage). This flow is shown graphically in Fig. 4.73. Other examples of the complex nature of three-dimensional separation may be found in Smith (1975) and Peake, Rainbird, and Arraghje (1972).

Separation is viewed here again as the termination of a thin viscous layer—the boundary layer. Since the boundary layer equations are parabolic in nature, an accurate prediction of separation requires the prediction of the entire viscous layer upstream of separation. In this respect, the integration of the three-dimensional laminar or turbulent boundary layer equations to separation presents some special problems. First, there is a problem of just what coordinate system should be used. In some cases, the correct choice for the coordinate system is obvious, while in others several choices may be available, and the manner in which the calculation proceeds may depend upon the choice of the coordination system. Second, and more importantly, while the momentum equations for the three-dimensional boundary layer are diffusive in the direction normal to the body, they are *wave-like* in planes parallel to the body. In these planes, the direction of propagation of information is along the local streamline. Thus, at any point on the body, there is a *zone of influence* and a *zone of dependence* in the shape of curvilinear wedges, one opening in the downstream direction and one opening in the upstream direction. The zones of influence and dependence are bounded by the streamlines of maximum and minimum angles passing through the normal to the body at the point in question as shown in Fig. 4.74. Any computational mesh set up for numerical integration of the boundary layer equations must account for these zones so that the influence principle of Ritz is not violated [Wang (1974)].

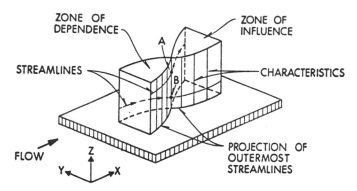

FIGURE 4.74 Zones of dependence and influence in three-dimensional boundary layers. (Reprinted with permission of Ballistic Research Laboratory, Kitchens, Jr., C. W., Sedney, S., and Gerber, H., "The Role of the Zone of Dependence Concept in Three-Dimensional Boundary Layer Calculations," USA Ballistic Research Laboratories Rep. 1821, 1975.)

Laminar Three-Dimensional Separation: As in the two-dimensional case, the advent of fast, large-capacity computers together with sophisticated numerical analysis techniques have made possible the rapid solution of the 3-D, laminar boundary layer equations, hopefully up to separation. The boundary layer development on a prolate spheroid at angle of attack has been calculated by several authors [Wang (1974), Geissler (1974), and Cebeci *et al.* (1981)]. The results of these investigations appear to agree for small angle of attack or for the windward side separation at large angle of attack and indicate that the separation line is indeed an envelope of limiting streamlines. On the leeward side of the body at high angle of attack, however the situation is far from clear. Wang (1974) terminates his calculations when the circumferential shear became vanishingly small and argued that since the circumferential shear drops rapidly while the meridional shear changes little, the limiting streamlines turn rapidly to become parallel to the meridians, a situation corresponding to an envelope of limiting streamlines or separation. Geissler (1974) calculated a similar flow employing a coordinate system of inviscid streamlines and constant potential lines (Fig. 4.75). He terminates his calculation on the line where his numerical stability criteria is violated and takes this line as his separation line. His *separation line* corresponds roughly to Wang's and is open at its forward end, giving credence to Wang's concept of separation. On the other hand, Cebeci *et al.* (1981) investigating separation on the same type body find that, as one moves circumferentially around the body from the windward side, the separation line moves forward but on the leeward side turns and moves rearward. The calculations cannot be completed near the leeward line of symmetry, since they encounter a region where the calculation would require information from streamlines which have passed over the region of separation.

The experimental study of Ham and Patel (1979) seems to verify the concept of *open separation* on a prolate spheroid at large angle of attack (see Fig. 4.76). This work also indicates the rather complex nature of the separation structure on this simple body. Clearly, in this case, there is extensive interaction between the separated boundary layer and the external stream, an interaction which will need to be modeled in improved calculations.

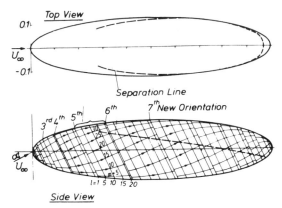

FIGURE 4.75 Streamline net and separation line for an ellipsoid of revolution of axis ratio $a/b = 4$ at $= 15°$ angle of attack, $l =$ number of equipotential line, $m =$ number of streamline. (Reprinted with permission of AIAA, Geissler, W., "Three-Dimensional Laminar Boundary Layer Over a Body of Revolution at Incidence and With Separation," *AIAA Journal*, 1974.)

Turbulent Three-Dimensional Separation: While practical examples of three-dimensional, turbulent boundary layer separation abound in engineering, the major problems in predicting such flows are the problem of interaction and the problem of developing appropriate turbulence models for both the unseparated and separated regions. A number of investigations have extended various turbulence models from two-dimensional flow [Rotta (1977), Rastagi and Rodi (1978), and Cebeci (1974)].

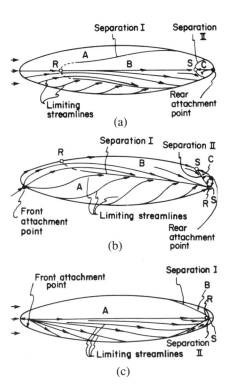

FIGURE 4.76 Wang's concept of *open separation*. (a) Top view, (b) side view, and (c) bottom view. (Reprinted with permission of AIAA, Wang, K. C., "Separating Patterns of Boundary Layer Over an Inclined Body of Revolution," *AIAA Journal*, 1972.)

Most of these models are truly only validated well upstream of separation where the interaction with the external stream is weak, and they must be used with care near separation. Simpson (1981) reports that, in general, the mixing length model does not work well for vortex separation flows.

Some fairly successful turbulent separation predictions have been made by treating the three-dimensional separated flow field as the asymptotic flow field (at long times) using a time-dependent computational approach. Here, the time-dependent, Reynolds-averaged Navier–Stokes equations are solved using an eddy-viscosity model and an explicit finite difference method. The existence of a separated region changes the problem to one which is elliptic in nature, and downstream boundary conditions are necessary. An example of this type of calculation is given in Horstman and Hung (1979).

Unsteady Two-Dimensional Separation

Separation Criterion: It has been known for some time that the two-dimensional, steady separation criterion, i.e. the vanishing of the wall shear, is not a valid separation criterion for two-dimensional unsteady flow. It is now generally believed that the Moore–Rott–Sears (MRS) criterion of vanishing shear and velocity at some point within the boundary layer, *in a coordinate system moving with separation*, provides an accurate model for separation. Figure 4.77 shows the velocity distribution at separation according to this model. The reviews of Williams (1977), Telionis (1979), and Shen (1978) provide insight into the history of the development of this model, its verification—at least for upstream moving separation—and the relationship be-

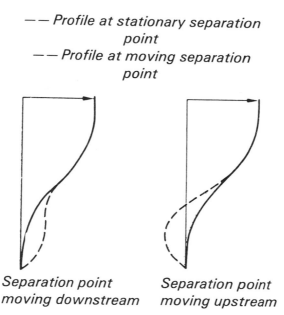

$—\!—$ Profile at stationary separation
point
$—\!—$ Profile at moving separation
point

Separation point
moving downstream

Separation point
moving upstream

FIGURE 4.77 Comparison of velocity profiles at separation for stationary flow and when separation is in motion. [Reprinted with permission of Springer–Verlag, Moore, F. K., "On the Separation of the Unsteady Laminar Boundary Layer," Boundary Layer Research (IUTAM Symposium, Freiburg/BR 1957), 1958.]

tween unsteady separation over fixed walls and unsteady separation over moving walls. It suffices here to note that the analytical results obtained by a number of investigators tend to substantiate the MRS at least for upstream moving separation. For the case of downstream moving separation, the model remains to be verified. The analytical studies involving solutions of the boundary layer equations which tend to verify this model strongly indicate the existence of a singularity in the solution to the laminar boundary layer equations at the MRS point.

Intuitively, one might expect that in the case of oscillating external streams the MRS model should also be valid. At present, however, this expectation remains to be verified completely.

The MRS model was developed from consideration of laminar boundary layers. There is, however, no reason to expect that the model should not also be valid for unsteady, turbulent boundary layer separations.

Laminar Separation: In view of the fact that the unsteady separation criterion is not well established, it is not surprising to find that there are, at present, no general techniques available for predicting unsteady separation. There have been a number of investigations of unsteady separation but these have all been rather specialized in nature, seeking to prove (or disprove) the MRS criterion or investigating the associated singularity. Taneda (1977) has presented the results of an extensive visual study of separation associated with the bodies in a variety of unsteady motions. Although much of this work is at low to moderate Reynolds number, the results presented show quite complex patterns of separation and separation vortices.

Intuitively, one expects that there is a qualitative relationship between separation of unsteady flow past a stationary wall and unsteady separation on a moving wall. Indeed Williams and Johnson (1974) have shown that, for certain flows, there is a qualitative relationship between these two types of flows and have used the relationship to verify the MRS model analytically. The experimental results of Kormoilas and Telionis (1980) tend to substantiate the MRS model for steady flow past both upstream and downstream moving walls. Two analytical studies of steady separation on upstream moving walls have yielded contradictory results [Tsahalis and Telionis (1977) and Inone (1981)].

The case in which the external flow is oscillating is of particular interest because of the importance of oscillating separation in helicopter and compressor aerodynamics. Despard and Miller (1971) conducted a basic study of separation resulting from an oscillating external stream. They found that the first indications of wake formation occurred where the velocity gradient at the wall is less than or equal to zero throughout the entire cycle of oscillation and that "This point may logically be defined as the *separation point*, marking the beginning of boundary layer separation." It should be noted that this definition of separation is not consistent with the MRS separation criterion. On the other hand, the results show that a thin layer of reverse flow may be embedded in the lower portion of an attached viscous layer, consistent with the MRS model. Despard and Miller also point out that for an oscillating external stream, the separation point occurs upstream of the steady separation point and that increasing the frequency of oscillation tends to move the separation point downstream towards the steady separation point.

Turbulent Separation: It has been noted earlier that all turbulent flows are inherently unsteady. The unsteadiness of concern here is that of the highly repeatable and organized unsteady flow. Two types of unsteady flow are of current interest. The first of these occurs when a periodic external flow is imposed on a turbulent boundary layer as in the case of dynamic stall on helicopter blades [McCroskey (1977)]. The second occurs when a turbulent flow interacts with an inviscid region of flow to set up a quasi-periodic motion as is observed in transitory stall in a diffuser [Kline (1959)]. Extension of boundary layer prediction techniques involves modifying turbulence models to allow for unsteadiness—a process which has not met with a great deal of proven success.

Prediction techniques employing the time-dependent, Reynolds-averaged Navier-Stokes equations with a quasi-steady mixing length turbulence model show some promise. The quasi-steady approximation appears to be reasonable as long as the period of the unsteadiness is long as compared to the turbulence time scales. When the energy-containing turbulence frequencies and the organized unsteadiness frequency are comparable, the quasi-steady approximation fails to provide accurate results [Simpson (1981)].

4.9.3 Separation from Corners

The behavior of shear layers which have separated from sharp corners is important in a number of practical engineering configurations. For these flows, the location of separation is fixed, and it is the post-separation behavior which is of interest. The actual physics of such flows is quite complex, but in recent years, there has been a significant improvement in the understanding of these flows. In the present section, we consider two such flows: 1) the flow field downstream of the backward facing step or in a sudden expansion of an internal channel, and 2) flows past cavities in which self-sustaining oscillations are generated. The first of these is perhaps the simplest of the possible separation-reattachment flows, hence it provides basic information for many related, but more complex, flows. The second provides some interesting examples of flows where instabilities play an important role in the flow physics.

Separation From and Reattachment Beyond Rearward Facing Steps. Although the flow downstream of a rearward facing step is the simplest reattachment flow, embodying a single separated zone with a fixed straight separation line and nearly parallel flow and little influence upstream of the separation point, the reattaching flow field is still very complex.

The boundary layer separates at the sharp corner of the step, forming a free shear layer as illustrated in Fig. 4.78. If the upstream boundary layer is laminar, transition begins shortly after separation, unless the Reynolds number is quite low. In the first half of the separated flow region, the separated shear layer behaves very much like a normal plane mixing layer (see Sec. 4.10) except that the flow on the low speed side of the shear layer is highly turbulent. In the reattachment zone, the shear layer curves sharply towards and impinges upon the wall. As a result of a strong adverse pressure gradient, a portion of the fluid in the reattaching shear layer is deflected upstream into the recirculating flow, and the flow in this region is very unsteady. The Reynolds and normal stresses decay rapidly in this region, but there are quite large scale turbulent structures passing through the reattachment zone.

The Reynolds stresses continue to decay downstream of reattachment, while a new sub-boundary layer grows up through the reattachment shear layer. The outer

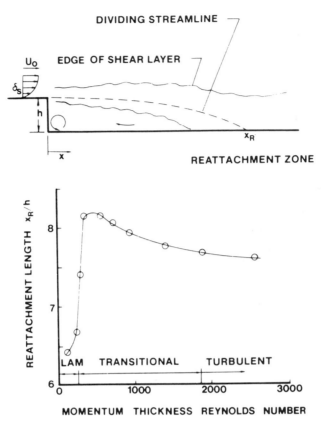

FIGURE 4.78 Flow field for a backwards facing step and variation of reattachment length for this flow field. (Reprinted with permission of AIAA, Eaton, J. K. and Johnson, J. P., "A Review of Research on Subsonic Turbulent Flow Reattachment," *AIAA Journal*, 1981.)

portion of the reattached shear layer continues to display the characteristics of a free shear layer for a considerable distance downstream [Bradshaw and Wong (1972) and Smythe (1979)]. In the recirculating zone trapped behind the step and beneath the shear layer, the back flow velocities may reach 20% of the free stream velocity, and the negative skin friction is relatively large [Eaton and Johnson (1981)].

Eaton and Johnson (1981) have presented an extensive review of some 23 sets of data on the flow field downstream of backward facing steps. Their review indicates the following effects of system parameters on the reattachment:

i) The initial state of the boundary layer has a major effect on the reattachment length (Fig. 4.78). When the separating boundary layer is laminar, the reattachment length is on the order of 6 to 6.5 step-heights. For transitional boundary layers, the reattachment increases to roughly 8 step-heights and for fully developed turbulent separation, it is roughly 7.5 step-heights (see Fig. 4.78).

ii) The effects of initial boundary layer thickness are not well defined. In general, the reattachment length decreases with increased boundary layer thickness.

iii) When the separation occurs at a rearward-facing step in a sudden expansion, there is an overall streamwise adverse pressure gradient acting on the flow field. This pressure gradient is controlled by the system geometry. The effect of increasing the geometric expansion ratio (the overall pressure increase) is to increase the reattachment length. For an increase in expansion ratio from just over 1.0 to approximately 1.8, the reattachment length increases from approximately 6.0 to approximately 7.5 times the step height.

iv) For aspect ratio (channel width to step height) in excess of 10, there is a negligible effect of aspect ratio. For smaller aspect ratios, the reattachment length decreases if the boundary layer is turbulent at separation and increases if the boundary layer is laminar at separation.

v) High levels of freestream turbulence apparently decrease the reattachment length.

Flows Past Cavities. Geometrically, a cavity in a body is but a simple extension of rearward facing step, but physically, the flow past a step and the flow over a cavity are very different. For many cases of fluid flow past a cavity, the flow develops oscillations which, rather than being damped out, become self-sustaining. Such flows are found in a number of engineering applications. Rockwell and Naudascher (1978) catalog such fluid oscillations as *fluid dynamic, fluid resonant,* and *fluid elastic*.

Fluid dynamic oscillations arise as a result of an inherent instability in flows past cavities. Such oscillations occur if the ratio of the cavity length to the acoustic wave length is very small and provided no free surface wave effects are present. Here, the primary mechanism for excitation of fluid-dynamic oscillations is the selective amplification of unstable disturbances in the cavity shear layer. These flows also involve an upstream propagation (*feedback*) of disturbances emanating from the vicinity of the downstream cavity edge. The selective amplification of disturbances is demonstrated by the organized character of pressure fluctuations measured on the downstream wall of the cavity even when the separation is turbulent. The organized oscillation Strouhal number increases with cavity length to depth ratio (L/W) for L/W less than approximately two but appears to be independent of L/W for ratios greater than about two. This behavior is predictable with existing theories [Rockwell and Naudascher (1978)].

When the boundary layer at separation is laminar, the cavity depth, W, apparently has an insignificant effect on the oscillations, except when the cavity depth and the boundary layer thickness are of the same order of magnitude. In addition, there is a minimum cavity length below which cavity oscillations do not occur.

For fluid-resonant oscillations, there is a strong coupling between the fluid oscillations and the resonant wave effects in the cavity. Here, the acoustic wave length is of the same order or smaller than the characteristic length. If the length to depth ratio, L/W, is sufficiently large, longitudinal standing waves may exist, and the cavity is termed shallow. If L/W is sufficiently small, transverse waves may be present and the cavity is termed *deep*.

For shallow cavities, up to five oscillation frequencies may coexist at a given Mach number in contrast to two coexisting frequencies for fluid-dynamic oscillations. However, one of the modes of oscillation seems to predominate. For a given mode, the Strouhal number for the oscillation decreases slightly with increasing Mach number ahead of the cavity. The variation of the Strouhal number with Mach number can be predicted quite successfully by one of several semi-empirical methods, except at very low Mach number.

For deep cavities, the Strouhal number for the oscillations is dependent upon the depth-to-width ratio, W/L, of the cavity. As W/L increases, the Strouhal number increases rapidly at first, but beyond W/L of three, the rate of increase decreases. There may be several modes of oscillation coexisting in this type of flow, but this has not been firmly established. The variation of Strouhal number with depth-to-width ratio may be predicted theoretically [Rockwell and Naudascher (1978)].

Fluid resonant oscillations may also be found in a variety of practical cavity shapes other than rectangular.

Fluid-elastic oscillations occur when one or more of the cavity walls undergoes an elastic displacement large enough to exert a feedback on the shear layer perturbations. The perturbed shear layer flow is then enhanced by the resonance type of process. These flows are quite complex, and there appears to be only limited data available on this mode of fluid oscillation.

Finally, it should be noted that in all three of the foregoing cases, the amplitude of the cavity oscillations may be attenuated by appropriate modifications of the cavity geometry [Rockwell and Naudascher (1978)].

4.10 SHEAR LAYERS
Joseph A. Schetz

Here, and in the next two sections, we shall be concerned with flows where there is no surface in the region of interest. The three representatives of this class of flows are shear layers, jets and wakes, as shown in Fig. 4.79. Attention is focused on the flow downstream of the body producing the wake or the walls that initially bound the jet or shear layer. Thus, for turbulent cases one is only concerned with the equivalent of the outer region of a wall-bounded turbulent layer.

4.10.1 The Laminar Shear Layer

The simplest case of a free mixing flow is that formed by two parallel streams with different velocities that are initially separated by a thin splitter plate [see Fig. 4.79(a)]. An exact solution can be developed for the laminar, incompressible case when the boundary layers on both sides of the splitter plate can be neglected. The initial profile for the mixing layer at the end of the plate ($x = 0$) is then simply a discontinuous jump at $y = 0$ from the velocity above the plate, U_1, to that below the plate, U_2. The boundary layer equations admit to a *similarity solution* (see Sec. 4.2) for this case with

$$\frac{u(x, y)}{U_1} = f'(\eta(x, y))$$

$$\psi(x, y) = U_1 \, b(x) \, f(\eta) \tag{4.201}$$

$$\eta = \frac{y}{b(x)}$$

where $b(x)$ is a length which scales the width of the viscous region. Substituting into the boundary layer equations gives

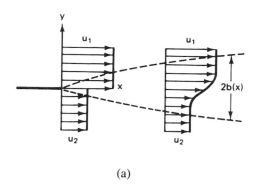

(a)

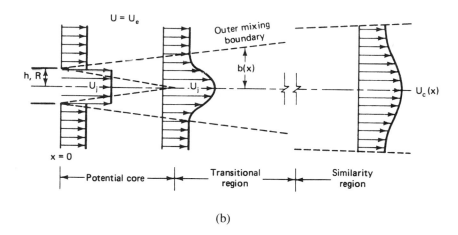

(b)

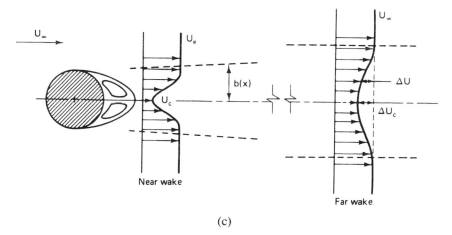

(c)

FIGURE 4.79   Schematic illustrations of free shear flows. (a) Simple shear layer, (b) typical jet flow, and (c) typical wake flow.

$$f''' + \frac{U_1 b}{\nu} \frac{db}{dx} ff'' = 0 \qquad (4.202)$$

For a similarity solution to hold, the coefficient on the second term must be a constant which can be conveniently taken as equal to $\frac{1}{2}$. This leads to

$$b(x) = \left(\frac{\nu x}{U_1}\right)^{1/2}, \qquad \eta = \frac{y}{x}\left(\frac{U_1 x}{\nu}\right)^{1/2} \qquad (4.203)$$

and

$$f''' + \tfrac{1}{2} ff'' = 0 \qquad (4.204)$$

This is essentially the same equation as for the Blasius solution in Sec. 4.2, but the boundary conditions are different here.

$$f(0) = 0$$
$$\lim._{\eta \to \infty} f'(\eta) = U_1 \qquad (4.205)$$
$$\lim._{\eta \to -\infty} f'(\eta) = U_2$$

A numerical solution is again required, and some typical results from Lock (1951) are shown in Fig. 4.80. It can be seen that the width of the mixing zone grows as $x^{1/2}$.

The case of initial boundary layer(s) on the splitter for $U_2 = 0$ has been treated by Kubota and Dewey (1964) with an approximate analysis based on the integral momentum equation approach for wall boundary layers. The profiles develop downstream as shown in Fig. 4.81(a).

The analysis of Kubota and Dewey (1964) was extended to variable density cases by the use of compressibility transformations. Some profile results at Mach 3.0 and 8.0 are shown in Figs. 4.81(b)–(e). Note, however, the vertical coordinate is based on the transformed values.

4.10.2 The Turbulent Shear Layer

For all turbulent flows, the mathematical specification of the flow is not complete with the usual equations of motion, and a turbulent transport model must also be given (see Sec. 4.4). For the simple shear layer problem, this is usually expressed in terms of a *spreading parameter* denoted as σ, which reflects the rate at which the shear layer grows laterally with axial distance.

The dependence of the width growth on axial distance, $b(x)$, can be developed from a simple physical argument. We postulate that the rate of growth of the turbulent region is directly proportional to the value of the transverse velocity fluctuations, i.e.,

$$\frac{Db}{Dt} \sim v' \qquad (4.206)$$

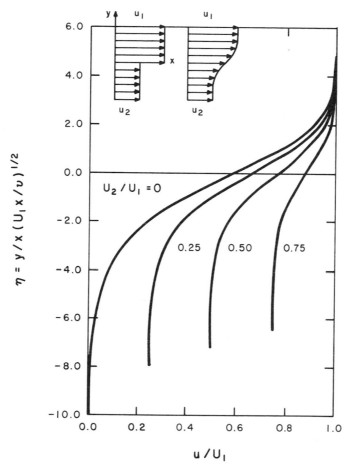

FIGURE 4.80   Velocity profiles for the laminar mixing layer with various values of U_2/U_1. (Reprinted with permission from Lock, R. C., "The Velocity Distribution in the Laminar Boundary Layer between Parallel Streams," *Quart. J. Mech.*, Vol. 4, pp. 42–43, 1951.)

where $D(\ )/Dt$ is the *convective derivative*. Introducing the mixing length, ℓ_t

$$v' \sim \ell_t \frac{\partial U}{\partial y} \tag{4.207}$$

and approximating $\partial U/\partial y$ by $(U_1 - U_2)/b$

$$v' \sim \frac{\ell_t}{b}(U_1 - U_2) \tag{4.208}$$

Next, we may say

$$\frac{Db}{Dt} \approx \left(\frac{U_1 + U_2}{2}\right)\frac{db}{dx} \tag{4.209}$$

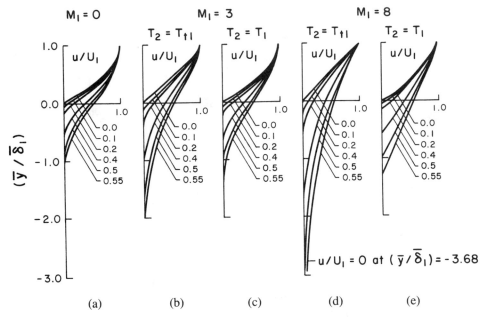

FIGURE 4.81 Effect of an initial boundary layer and Mach number on velocity profiles in a laminar mixing layer. (From Kubota and Dewey, 1964.)

so

$$\frac{db}{dx} \approx \text{const.} \left(\frac{\ell_t}{b}\right) = \text{const.} \tag{4.210}$$

since (ℓ_t/b) is generally taken as a constant which leads to $b \sim x$.

Beginning with the case of negligible initial boundary layers, we can again postulate that the velocity profiles will be similar as in laminar flows if the turbulent eddy viscosity, μ_t, may be assumed constant across the layer. The similarity variable, η, is now written in terms of the spreading parameter, σ as

$$\eta \equiv \frac{\sigma y}{x} \tag{4.211}$$

Since $b \sim x$, we write

$$\psi \sim x$$
$$= x \left(\frac{U_1 + U_2}{2}\right) f(\eta) \tag{4.212}$$

then, since $U = \partial\psi/\partial y$

$$U = \left(\frac{U_1 + U_2}{2}\right) \sigma f'(\eta) \tag{4.213}$$

Görtler (1942) used these relations with the Prandtl (1942) eddy viscosity model

$$\mu_t \sim \rho b(U_1 - U_2) \sim \rho x(U_1 - U_2) \tag{4.214}$$

Prandtl simply reasoned that $\mu_t \sim density \; x \; velocity \; x \; length$ and then took the length scale as the width and the velocity scale as the velocity difference across the layer. This reduces the equations of motion to

$$f''' + 2\sigma^2 ff'' = 0 \tag{4.215}$$

The solutions for various values of σ were compared to the experiments of Reichardt (1942) with $U_2 = 0$, and the value $\sigma_0 = 13.5$ was selected. The results are shown in Fig. 4.82. Other workers have proposed $9.0 \leq \sigma_0 \leq 13.5$; a value of $\sigma_0 = 11$ is now generally accepted based on more recent data. The influence of the parameter U_2/U_1 on σ has also been studied by several groups. A survey by Birch and Eggers (1973) suggests

$$\frac{\sigma_0}{\sigma} = \frac{U_1 - U_2}{U_1 + U_2} \tag{4.216}$$

as being in the best agreement with data. The analysis of the initial boundary layer effects for laminar flows in Kubota and Dewey (1964) was extended to turbulent flows with $U_2 = 0$ by Alber and Lees (1968).

There has been some work aimed at predicting the development of turbulent shear layers with advanced turbulence models. The 1981 Stanford conference [see Kline *et al.* (1982)] had the low-speed simple shear layer as a test case. Few predictors treated it, but a prediction with a $K\epsilon$ model (see Sec. 4.4) from Rodi is shown compared to several sets of data in Fig. 4.83. Calculations for the simple shear layer were also made with the Donaldson Reynolds stress model. The results are presented in Fig. 4.84, where the fair agreement can be seen.

The effect of variable density on the value of the spreading parameter has been

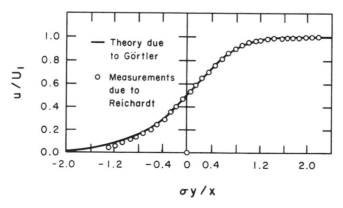

FIGURE 4.82 Comparison of analysis from Görtler (1942), and experiment from Reichardt (1951), for velocity profiles in a turbulent mixing layer. Analysis assumes $\sigma_0 = 13.5$.

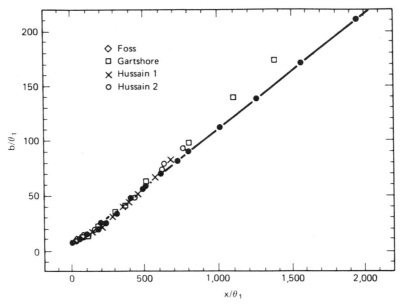

FIGURE 4.83 Comparison of prediction by Rodi, using a $K\epsilon$ model, with experiment for the low-speed turbulent simple shear layer. (From Kline *et al.*, 1982.)

the subject of considerable interest. Studies by Alber and Lees (1968) suggest

$$\frac{(\sigma_\theta)_{\text{var.}\rho}}{(\sigma_\theta)_{\text{const.}\rho}} = \left(\frac{\rho_1}{\rho_t}\right)^2 \tag{4.217}$$

The data collection in Fig. 4.85 illustrates the scope of the problem. The data indicate a reduction in mixing rate at high Mach number (a higher σ means slower

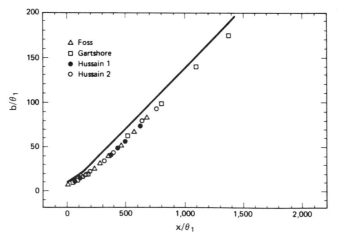

FIGURE 4.84 Comparison of prediction from the Donaldson Reynolds stress model with experiment for the low-speed turbulent simple shear layer. (From Kline *et al.*, 1982.)

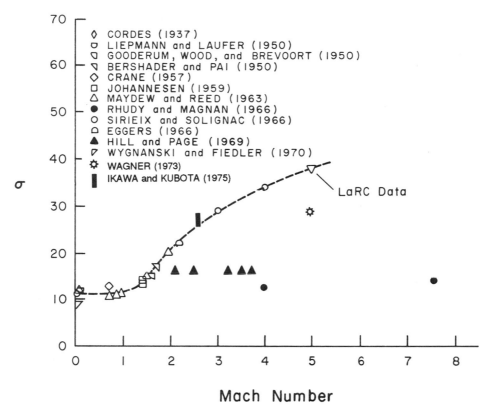

FIGURE 4.85 Correlation of the variation of σ with Mach number. (From Birch and Eggers, 1973; additional data from Wagner, 1973 and Ikawa and Kubota, 1975.)

spreading) if the experiments are free of the influences of the initial boundary layers on the splitter plate. This is generally viewed as an effect of *compressibility*, not just density variation, and it is now possible to estimate with reasonable certainty when such effects might be expected. Following Morkovin and Bradshaw in Bradshaw (1977), one criteria is that a representative density fluctuation level

$$\frac{\rho'}{\bar{\rho}} \approx (\gamma - 1)M^2 \frac{\Delta U}{U} \tag{4.218}$$

should be larger than $\approx 1/10$. Another criteria can be found from Brown and Roshko (1974), Bogdanoff (1983), and Papamoschou and Roshko (1986) as the conditions when a *convective Mach number, M_{c1},* approaches unity where

$$M_{c1} \equiv \frac{U_1 - u_{\text{conv}}}{a_1} \tag{4.219}$$

and u_{conv} is the convective speed of large-scale turbulent structures. Those results have been presented as a width growth rate

$$\frac{d\delta}{dx} = C_3 f(M_{c1}) \frac{(1 - U_2/U_1)(1 + \sqrt{\rho_2/\rho_1})}{\left(1 + \sqrt{\frac{\rho_2 U_2^2}{\rho_1 U_1^2}}\right)} \tag{4.220}$$

with $C_3 \approx 0.17$ and $f(M_{c1})$ a function equal to unity up to about $M_{c1} = 0.5$ and then decreasing to a value of about 0.3 by $M_{c1} \approx 1.0$ to mimic the data shown in Fig. 4.86.

The $K\epsilon$ model will not predict the effects of compressibility as demonstrated in Fig. 4.87. In order to reflect that effect, it is necessary to introduce a "correction" to the $K\epsilon$ model. Dash et al. (1975) developed an ad hoc correction factor $K(M_\tau)$ to the eddy viscosity relation. Here, $M_\tau = (K_{max})^{1/2}/a$, and a is the local speed of sound. The correction factor was simply determined to reproduce the variation of σ found experimentally. The result is given in Fig. 4.88, and the behavior of the new model called $K\epsilon2$, cc (the cc denotes compressibility corrected) is shown in Fig. 4.87.

Direct numerical simulations (see Sec. 21.5) have been made for some simple free shear flows. The matter is a little easier for such cases than for wall-bounded flows, but these calculations are still much too expensive for routine engineering use. To illustrate the kind of results that are achievable, we present some calculations by Greenough et al. (1989) in Fig. 4.89. These calculations use the inviscid equations of motion and start with a perturbation to a simple shear layer in the range where linear stability theory predicts instability. The case with $M_1 = 1.5$ and $M_{c1} = 0.75$ in Fig. 4.89 has an initial time rate of change of the turbulent kinetic energy [see Fig. 4.89(a)] that agrees closely with the linear stability theory. It levels off at

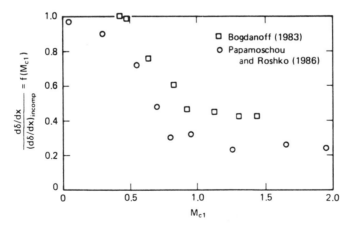

FIGURE 4.86 The effect of convective Mach number on the spreading rate in the turbulent simple shear layer. (Reprinted with permission from AIAA Paper 86-0162, "Observations of Supersonic Free Shear Layers," by Papamoschou, D. and Roshko, A. A., 1986, also from *AIAA J.*, Vol. 21, No. 6, "Compressibility Effects in Turbulent Shear Layers," by Bogdanoff, D. W., 1983.)

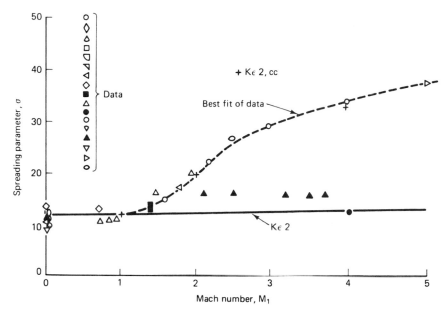

FIGURE 4.87 Comparison of predictions, using $K\epsilon$ models, with experiment for the spreading parameter for the supersonic turbulent simple shear layer. (Reprinted with permission from *AIAA J.*, Vol. 23, No. 4, "Analysis of Turbulent Underexpanded Jets, Part I: Parabolized Navier–Stokes Model SCIPVIS," by Dash, S. M., Wolf, D. E., and Seiner, J. M., 1975.)

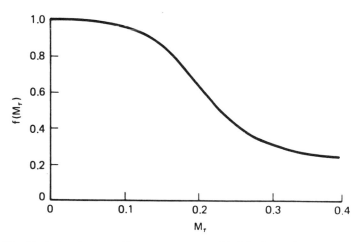

FIGURE 4.88 Variation in spreading parameter σ with Mach number and compressibility correction factor for $K\epsilon 2$ turbulence model. (Reprinted with permission from *AIAA J.*, Vol. 23, No. 4, "Analysis of Turbulent Underexpanded Jets, Part I: Parabolized Navier–Stokes Model SCIPVIS," by Dash, S. M., Wolf, D. E., and Seiner, J. M., 1975)

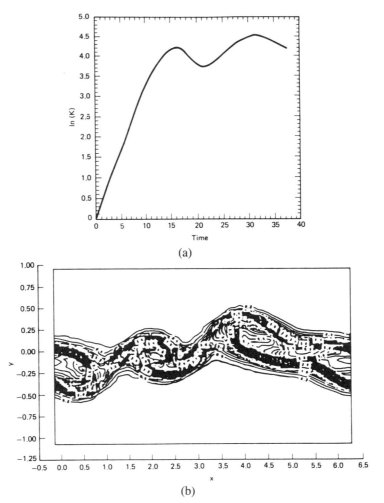

(a)

(b)

FIGURE 4.89 Direct numerical simulation of the unsteady development of a supersonic ($M_1 = 1.5$ $M_{c1} = 0.75$) simple shear layer. (a) Evolution of turbulent kinetic energy and (b) vorticity field at $t = 10.5$. (Reprinted with permission from AIAA Paper 89-0372, "The Effects of Walls on a Compressible Mixing Layer," by Greenough, J., Riley, J., Soertrisno, M., and Eberhardt, D., 1989.)

a dimensionless time of about 8.0. The vorticity field in Fig. 4.89(b) shows not only the fundamental mode which was excited, but also the third harmonic.

4.11   JETS

Joseph A. Schetz

4.11.1   Laminar Jets

The simplest jet flows [see Fig. 4.79(b)] are laminar, constant-density cases where the surrounding fluid is at rest. (Jets where buoyancy forces are important, termed *plumes*, are discussed in Sec. 13.3.) These flows allow *similarity solutions* (see Sec.

4.2). One imagines the jet to be from a point or line source, so actual similarity is only achieved for $x/D \gg 1$ in real cases. The planar, two-dimensional geometry will be treated here as an example. Since the surrounding flow is at rest and at constant static pressure, we may assume that the static pressure is constant in the mixing region. To the lowest order, the total axial momentum flux must remain constant, since there is no net shear force acting on the flow

$$J \equiv \int_{-\infty}^{\infty} \rho u^2 \, dy = \text{constant} \qquad (4.221)$$

The similarity condition for velocity profiles is

$$\frac{u(x, y)}{u_c(x)} = f\left(\frac{y}{b(x)}\right) \qquad (4.222)$$

where u_c is the centerline velocity and $b(x)$ is a suitably defined width scale. Assume that $b(x)$ varies as, x, to some power, and thus take

$$\psi \sim x^p f\left(\frac{y}{x^q}\right) \qquad (4.223)$$

The exponents, p and q, can be determined by satisfying two conditions: 1) the flux of streamwise momentum remains constant, as stated above, and 2) the inertia and viscous terms in the equations of motion are of the same order of magnitude. This results in $p = 1/3$ and $q = 2/3$. Using

$$\eta = \frac{1}{3\nu^{1/2}} \frac{y}{x^{2/3}}, \qquad \psi = \nu^{1/2} x^{1/3} f(\eta)$$

$$u = \frac{1}{3x^{1/3}} f'(\eta) \qquad (4.224)$$

$$v = -\frac{1}{3} \nu^{1/2} x^{-2/3} (f - 2\eta f')$$

The equations of motion reduce to

$$f''' + ff'' + (f')^2 = 0 \qquad (4.225)$$

The boundary conditions are

$$\lim_{\eta \to \infty} f'(\eta) = f(0) = f''(0) = 0 \qquad (4.226)$$

and Schlichting (1933) developed a closed form solution to this non-linear equation to give

$$u = \tfrac{2}{3} a^2 x^{-1/3} (1 - \tanh^2(a\eta)) \qquad (4.227)$$

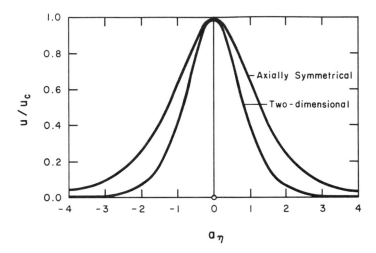

FIGURE 4.90   Velocity distribution in circular and planar laminar free jets. (Reprinted with permission of McGraw-Hill, Inc., *Boundary Layer Theory*, 6E, by Schlichting, H., 1968.)

where

$$a = \left(\frac{9J}{16\rho\nu^{1/2}} \right)^{1/3}$$ (4.228)

Velocity profiles for this and the corresponding axisymmetric case are shown in Fig. 4.90. Also, $b \sim x^{2/3}$ and $u_c \sim x^{-1/3}$ for the planar jet and $b \sim x$ and $u_c \sim x^{-1}$ for the axisymmetric jet.

In recent times, careful analysis [Schneider (1985)] has shown that the momentum flux does not remain constant if there are any walls in the flowfield. Thus, if the jet issues from a hole in a wall, the axial momentum flux in the jet actually decreases slowly with distance from the orifice.

There are many practical cases where a fluid of one phase is injected into surroundings of a different phase, e.g., a liquid jet injected into a gas. Obviously, such problems are much more complex than single-phase flows as a result of the added effects of surface tension, among other things. Practical treatment of these cases is largely empirical, with analyses of only the simplest cases available. For purposes of illustration, consider a very simple problem that does have practical application— a liquid, laminar jet injected into a gaseous medium at rest. The interaction between aerodynamic forces and surface tension forces on the interface produces an instability such that small background disturbances can grow, leading to fracture of the liquid column into droplets. Rayleigh (1945) linearized the equations of motion for small disturbances and determined which wavelengths of the disturbances were stable and unstable and the unstable wavelength with the greatest growth rate

$$\lambda_{\max} = 4.51d$$ (4.229)

More details of this kind of analysis can be found in Chap. 6.

4.11.2 Turbulent Jets

This section begins with a brief presentation of data for a round, turbulent jet flow. This then forms the basis for the discussion of turbulence modeling. A more general treatment of the subject can be found in Schetz (1980).

In Fig. 4.91, we show the variation of the centerline velocity in terms of $U_e/[U_c(x/d) - U_e]$ and a characteristic width $r_{1/2}$ defined by the relation

$$U(r_{1/2}, x) = \frac{U_c(x) + U_e}{2} \tag{4.230}$$

with the axial distance. This latter quantity is frequently called the *half-radius*, and it can be more easily determined accurately than some ill-defined total width where the mixing zone merges asymptotically into the external stream. First, it can be seen that the influence of the parameter m ($\equiv U_e/U_j$) on the flow field is quite profound. Second, the total distance to the final decay where $U_c \to U_e$ is very long when measured in terms of jet diameters. Nondimensional radial velocity profiles are given in Fig. 4.92 for various x/d. These data indicate that the profiles are apparently *similar* for $x/d \geq 40$, where the term *similar* means that the profiles, expressed in terms of coordinates such as those in Fig. 4.92, remain unchanged with x/d. That will prove useful for analysis later.

Some data for the axial turbulence intensity are given in Fig. 4.93. Note that the average fluctuations are a substantial fraction of the mean flow velocity difference. The characteristic shape of the transverse profiles can easily be seen. The maximum generally occurs somewhat off the centerline, since the mean velocity gradient, and hence the turbulence production, is greater there. Observe that the profiles of this variable have not attained a similarity condition by $x/d \approx 200$. The data of Gibson (1963) for tests with $U_e = 0$ showed that the relative magnitudes of the axial, transverse, and peripheral turbulence intensities were about the same.

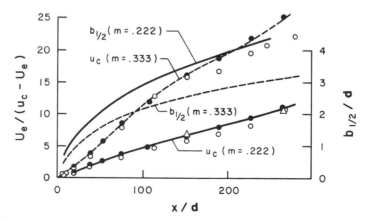

FIGURE 4.91 Variation of centerline velocity and half-radius for an axisymmetric jet: ○ △ hot wire; + Pitot tube. (Reprinted with permission of Cambridge University Press, "An Experimental Investigation of an Axisymmetric Jet in a Co-Flowing Air Stream," by Antonia, R. A. and Bilger, R. W., *J. Fluid Mech.*, Vol. 61, pp. 805–822, 1973.)

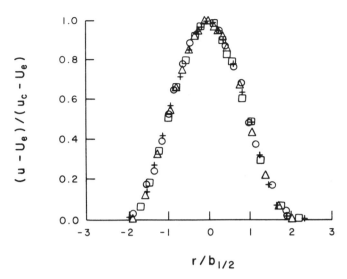

FIGURE 4.92 Nondimensional radial profiles across an axisymmetric jet: $m = 0.222$, x/D = 38 (+), 76 ($\triangle$), 152 ($\square$), 266 ($\bigcirc$). (Reprinted with permission of Cambridge University Press, "An Experimental Investigation of an Axisymmetric Jet in a Co-Flowing Air Stream," by Antonia, R. A. and Bilger, R. W., *J. Fluid Mech.*, Vol. 61, pp. 805–822, 1973.)

The simplest cases with density variations are produced with fluid injection at a temperature different than the main flow. Experiments of that type in the axisymmetric geometry were reported by Landis and Shapiro (1951). Mean flow results in terms of the axial variation of the centerline velocity and temperature are shown here in Fig. 4.94 for a case with $T_e/T_j = 0.77$ and $U_e/U_j \equiv m = 0.50$. The density ratio

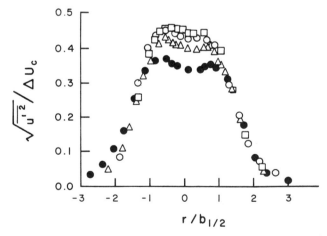

FIGURE 4.93 Radial profiles of axial turbulence intensity in axisymmetric jets x/D = 38 (+), 76 ($\triangle$), 152 ($\square$), 266 ($\bigcirc$); $m = 0.333$. (Reprinted with permission of Cambridge University Press, "An Experimental Investigation of an Axisymmetric Jet in a Co-Flowing Air Stream," by Antonia, R. A. and Bilger, R. W., *J. Fluid Mech.*, Vol. 61, pp. 805–822, 1973.)

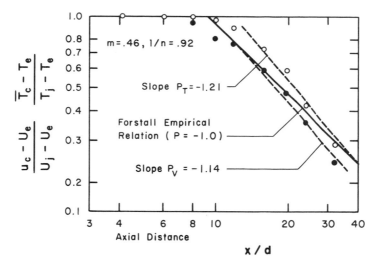

FIGURE 4.94 Streamwise variation of centerline velocity and temperature in a heated, axisymmetric jet. (Reprinted with permission from Landis, F. and Shapiro, A. H., "The Turbulent Mixing of Co-Axial Gas Jets," Heat Transfer and Fluid Mechanics Inst., Reprints and Papers, Stanford University Press, 1951.)

$\rho_e/\rho_j \equiv n$ for such cases is related to the temperature ratio as $T_e/T_j = \rho_j/\rho_e = 1/n$. It can be seen that the power decay law exponent P_i (i.e., $\Delta \overline{T} \sim x^{P_T}$, $\Delta U \sim x^{P_V}$, etc.) is slightly greater for this case than that for the nearly constant-density flow studied by Forstall and Shapiro (1950).

The axial variations of the half-widths for the velocity and temperature profiles are shown in Fig. 4.95. From these data and that in Fig. 4.96, we can see the important result that nondimensionalized temperature profiles are always wider or

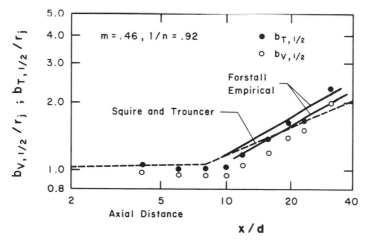

FIGURE 4.95 Variation of half-radii based on velocity and temperature in a heated, axisymmetric jet. (Reprinted with permission from Landis, F. and Shapiro, A. H., "The Turbulent Mixing of Co-Axial Gas Jets," Heat Transfer and Fluid Mechanics Inst., Reprints and Papers, Stanford University Press, 1951.)

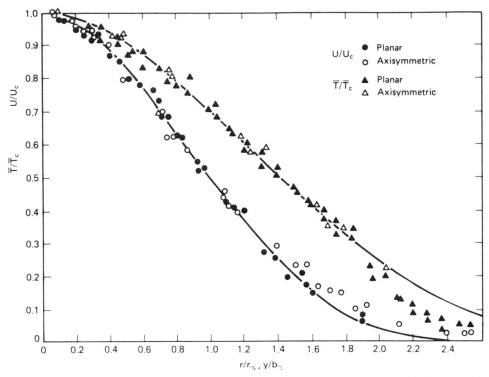

FIGURE 4.96 Transverse velocity and temperature profiles for planar and axisymmetric jets with $m = 0$. (From Ginevskii, 1966.)

fuller than the corresponding velocity profiles, which indicates that the transverse transport of thermal energy by turbulence is more rapid than that for momentum. This means that a *turbulent thermal conductivity*, k_t, defined by

$$-k_t \frac{\partial \overline{T}}{\partial y} = \rho c_p \overline{v'T'} = q_t \tag{4.231}$$

when combined with $\nu_t \equiv \mu_t/\rho$, to give a *turbulent Prandtl number* will correspond to values of $\mathrm{Pr}_t < 1$.

$$\mathrm{Pr}_t = \frac{\nu_t}{k_t/\rho c_p} \tag{4.232}$$

For the axisymmetric case of Corrsin and Uberoi (1949), this quantity was determined to be approximately 0.7, while planar jet data indicate a value of $\mathrm{Pr}_t \approx 0.5$. No satisfactory explanation has been found why geometry should affect Pr_t.

Many important practical applications involve injection of one fluid into surroundings of a different fluid. In this section, the discussion is restricted to cases of one phase, either gas or liquid, and binary mixtures. A discussion of more general cases can be found in Schetz (1980). In Fig. 4.97, we have the influence of density and velocity ratios on the centerline velocity decay. The axial variation of the centerline composition of injectant is presented in Fig. 4.98.

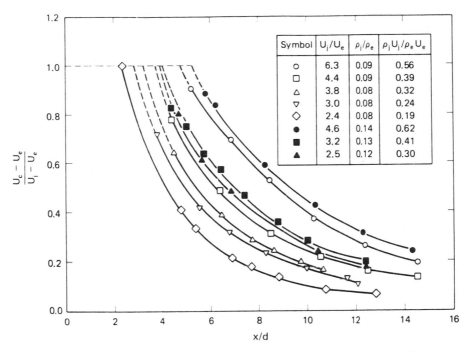

FIGURE 4.97 Streamwise variation in centerline velocity for axisymmetric H_2 jets into air. (From Chriss, 1968.)

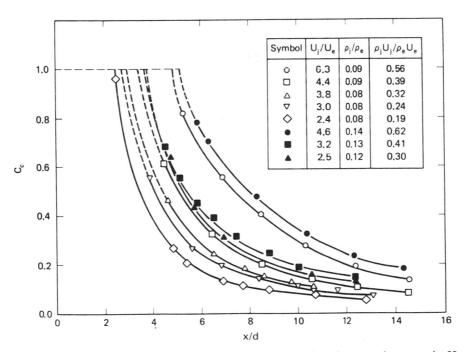

FIGURE 4.98 Streamwise decay of centerline concentration for an axisymmetric H_2 jet into air. (From Chriss, 1968.)

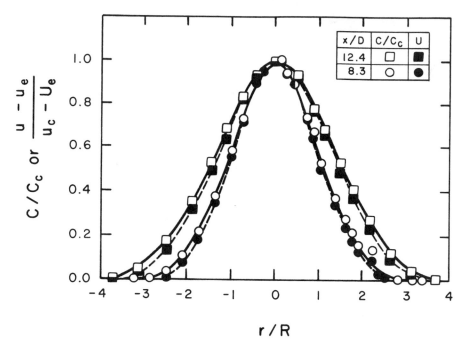

FIGURE 4.99 Radial profiles of velocity and concentration for an axisymmetric H_2 jet into air. (From Chriss, 1968.)

As for density changes produced by temperature variations alone, the half-width of concentration profiles is larger than for velocity, and the profiles themselves are consequently *fuller*, as shown in Fig. 4.99. Indeed, nondimensional temperature and concentration profiles have been found to be virtually identical.

Introducing a *turbulent mass diffusion coefficient*, D_t, through

$$-\rho D_t \frac{\partial C_i}{\partial y} = \overline{\rho v' c_i'} \tag{4.233}$$

we may express this quantity in relation to turbulent momentum or heat transfer through a *turbulent Schmidt number*

$$Sc_t = \frac{\nu_t}{D_t} \tag{4.234}$$

or a *turbulent Lewis number*

$$Le_t = \frac{D_t}{k_t/\rho c_p} = \frac{Pr_t}{Sc_t} \tag{4.235}$$

Most workers agree that for the axisymmetric case $Pr_t \approx Sc_t \approx 0.7$, which leads to $Le_t \approx 1.0$. For planar flows $Pr_t \approx Sc_t \approx 0.5$ and again $Le_t \approx 1.0$.

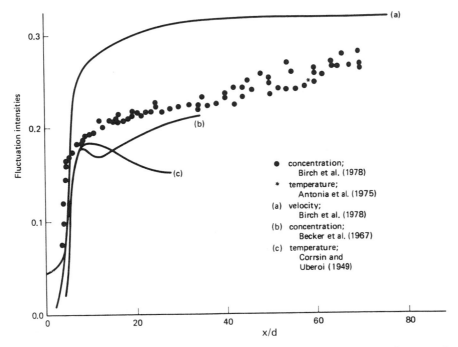

FIGURE 4.100 Streamwise variation of centerline fluctuation intensities for an axisymmetric jet with $m = 0$. (Reprinted with permission of Cambridge University Press, "The Nozzle Fluid Concentration Field of the Round Turbulent Free Jet," by Becker, H. A., Hottel, H. C., and Williams, G. C., *J. Fluid Mech.*, Vol. 30, pp. 285–303, 1967; also from "The Turbulent Concentration Field of a Methane Jet," by Birch, A. D., Brown, D. R., Dodson, M. S., and Thomas, J. R., *J. Fluid Mech.*, Vol. 88, pp. 431–449, 1978.)

The accurate measurement of quantities involving concentration fluctuations is generally conceded to be a difficult task, and various methods have been used (see Secs. 15.11 to 15.12). The axial variation of the intensity of the concentration fluctuations is shown in Fig. 4.100, together with corresponding results for the velocity and temperature fields.

The simplest analyses for turbulent jet flows are based on the observed similarity of profiles, as was displayed in Figs. 4.92, 4.96, and 4.99. Taking the case of a constant-density, round jet with $U_e = 0$ as an easy example, one need only assert that the integral of the streamwise momentum remains constant at its initial value

$$J = \rho \int U^2 \, dA = \text{constant} = \rho_j U_j^2 A_j \tag{4.236}$$

and that the width of the mixing zone $b(x) \sim x$ from experiment. Since the profile may be taken as *similar* for $x/D \gg 1$, we can write

$$\frac{U(x, r)}{U(x, 0)} = f\left(\frac{r}{b}\right) \tag{4.237}$$

Substituting into Eq. (4.236)

$$U(x, 0) = \text{constant} \left(\frac{1}{b}\right) \sqrt{\frac{J}{\rho}} \qquad (4.238)$$

where the *constant* depends on the particular *shape* of the profile [i.e., the form of $f(r/b)$]. Finally,

$$U_c(x) = U(x, 0) = \text{constant} \left(\frac{1}{x}\right) \sqrt{\frac{J}{\rho}} \qquad (4.238a)$$

since b is proportional to x. The corresponding result for a planar jet with $U_e = 0$ is $U_c \sim 1/x^{1/2}$. The analysis can and has been carried further to produce complete solutions for the velocity profiles within an unknown constant by Tollmien (1926) and Görtler (1942). The unknown constant must be determined by experiment, and then good agreement with the shape of experimental profiles is obtained, as shown in Fig. 4.101. These solutions are not, however, easily extended to cases with $U_e \neq 0$.

Again as for the laminar case, more recent analysis [Schneider (1985)] shows that the total momentum flux does not really stay constant in these flows. Schneider (1985) derived

$$\frac{dJ}{dx} = -\frac{C}{2} J x^{-1} \qquad (4.239)$$

with $C \approx 0.013$ for axisymmetric jets. Using these results, the axial velocity in the jet decays slightly more rapidly than predicted by the classical analysis and the induced velocities and induced pressure are slightly smaller than they are for constant

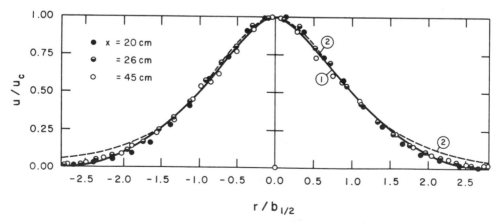

FIGURE 4.101   The velocity distribution in a circular turbulent free jet. Data from Forthman (1934), and analysis based on mixing length (1) from Tollmein (1976), and eddy viscosity (2) from Görtler (1942).

J. For planar jets, the value of *C* can be deduced from the measured entrainment rate as $C \approx 0.085$.

The availability of digital computers that permit *numerically exact* solutions of the equations of motion incorporating complex turbulence models has rendered the older approximate methods for more general cases than that just discussed obsolete. Once it has been assumed that the equations of motion can be solved numerically, the discussion reduces, within the limits of mean flow turbulence models first, to a choice of an eddy viscosity or mixing-length model.

Consider the widely-used Prandtl model, based on the half-width $b_{1/2}(x)$ and ΔU_e

$$\nu_t = 0.037b_{1/2}|U_{max} - U_{min}| \tag{4.240}$$

and also an extended version of the Clauser model originally developed for the outer region of a turbulent boundary layer (see Sec. 4.5),

$$\nu_t = \frac{\mu_t}{\rho} = CU_e\delta* \tag{4.241}$$

where $C \approx 0.018$, and $\delta*$ must be interpreted to be based on the absolute value of $[1 - (U/U_e)]$. For simple assumed profile shapes, one finds that the two expressions agree in form and 0.036 versus 0.037 as the proportionality constant. We present in Figs. 4.102 and 4.103 some comparisons of experimental data and predictions based on Eq. (4.241) for a constant-density, planar jet flow from Schetz (1971), where the good agreement can be seen.

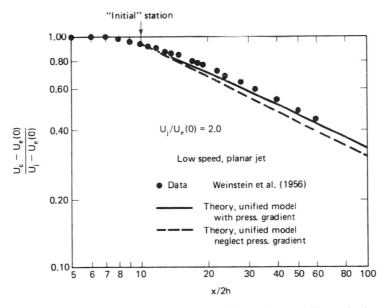

FIGURE 4.102 Comparison of prediction with experiment for centerline velocity variation of a planar jet. (Reprinted with permission of Elsevier Science Ltd., "Some Studies of the Turbulent Wake Problem," by Schetz, J. A., *Astronautica Acta*, Vol. 16, pp. 107–117, 1971.)

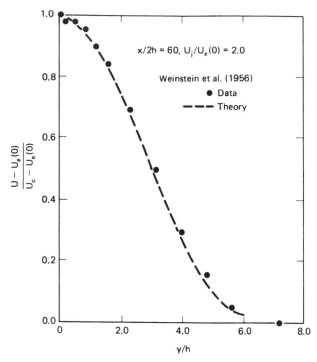

FIGURE 4.103   Comparison of prediction with experiment for velocity profile in a planar jet. (Reprinted with permission of Elsevier Science Ltd., "Some Studies of the Turbulent Wake Problem," by Schetz, J. A., *Astronautica Acta*, Vol. 16, pp. 107–117, 1971.)

Schetz (1968) chose to further extend the Clauser model to varying density cases as,

$$\nu_t = 0.018 U_e \Delta^* = 0.018 U_e \int_{-\infty}^{\infty} \left| 1 - \frac{\overline{\rho} U}{\rho_e U_e} \right| dy \qquad (4.241a)$$

and still further to the axisymmetric geometry as

$$\mu_t = \frac{C_3 \rho_e U_e \pi (\delta_r^*)^2}{R} \qquad (4.241b)$$

where the R is needed to make the dimensions correct. The proportionality constant C_3 had to be determined by a one-time comparison between theory and experiment, as is done with all eddy viscosity models, and the experiments of Forstall and Shapiro (1950) were employed. The results are shown in Fig. 4.104; the value of $C_3 \pi$ determined is 0.018. This model gives excellent qualitative as well as quantitative agreement with the data. Schlichting's extension of the Prandtl model to axisymmetric jets produces poor predictions for the rate of centerline velocity decay.

To consider variable density cases, it is necessary to add energy and/or species conservation equations to the system. Heat transfer is, in general, simply modeled using an eddy viscosity or mixing-length model and a constant value of the *turbulent*

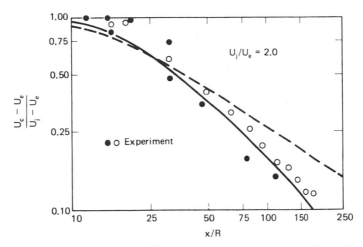

FIGURE 4.104 Comparison of prediction with experiment for centerline velocity variation of an axisymmetric jet. Dashed line: analysis with Schlichting axisymmetric version of Prandtl model; solid line: analysis with Eq. (4.241b). (From Schetz, 1968; data from Forstall and Shapiro, 1950.)

Prandtl number [see Eq. (4.232)]. Mass transfer is correspondingly modeled using a constant value of the *turbulent* Schmidt or Lewis number [see Eqs. (4.234) and (4.235)].

The two basic TKE models in use for wall-bounded flows—the Prandtl energy method and the Bradshaw model, are described in Schetz (1994). For free mixing flows, they are used in the same forms, except it is not necessary to modify ℓ for wall effects. Launder *et al.* (1973) have used the Prandtl energy method, and Harsha (1973) extended the Bradshaw TKE approach to free shear flows. Some comparisons of predictions with experiment for the case of $U_j/U_\infty = 4.0$ from Forstall and Shapiro (1950) selected for the 1972 NASA Conference on Free Turbulent Shear Flows [see NASA SP-321 (1973)] are of interest. In Fig. 4.105, the centerline velocity predictions based on the eddy viscosity model from Schetz (1971), the mixing-length model, the TKE model from Launder *et al.* (1973), and the extended Bradshaw TKE model from Harsha (1973) are shown, compared with experiment. The predictions are all in reasonable agreement with the data in the far field.

For test cases with very large density differences across the mixing layer, we again use the H_2-air mixing cases of Chriss (1968). Figure 4.106 shows the centerline variation of concentration, comparing some of the same models as for Fig. 4.105 with the data. Again in this case, it is hard to detect any advantage of the TKE models over the simpler mean flow models when one looks at the whole range of x/D covered.

The basic material on $K\epsilon$ models was developed in Sec. 4.5, but again, it is not necessary here to treat a special wall-dominated region. Launder *et al.* (1973) presented calculations for the 1972 NASA conference, and the final modeled equation for ϵ was taken as (note: $j \equiv 0$ for planar flows and $j \equiv 1$ for axisymmetric flows)

$$\rho\left(U\frac{\partial\epsilon}{\partial x} + V\frac{\partial\epsilon}{\partial y}\right) = \frac{1}{y^j}\frac{\partial}{\partial y}\left(y^j\frac{\mu_t}{\sigma_\epsilon}\frac{\partial\epsilon}{\partial y}\right) + C_{\epsilon 1}\frac{\rho\epsilon}{K}\nu_t\left(\frac{\partial U}{\partial y}\right)^2 - C_{\epsilon 2}\frac{\rho\epsilon^2}{K} \quad (4.242)$$

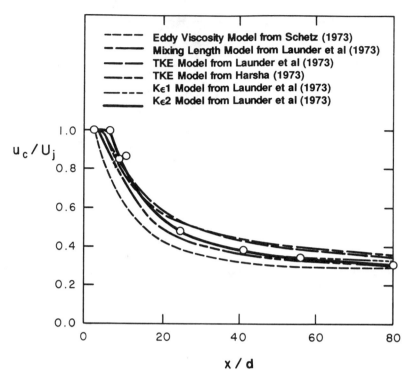

FIGURE 4.105 Comparison of predictions for velocity with mean-flow, one-equation, and two-equation models with the axisymmetric jet experiment. [From NASA SP-321 (1973).]

with $\sigma_\epsilon = 1.3$, $C_{\epsilon 1} = 1.43$, and $C_{\epsilon 2} = 1.92$ for planar flows. For axisymmetric flows,

$$C_{\epsilon 2} = 1.92 - 0.0667F \tag{4.243}$$

where

$$F = \left[\frac{b}{2\Delta U_c} \left(\frac{dU_c}{dx} - \left| \frac{dU_c}{dx} \right| \right) \right]^{1/5} \tag{4.244}$$

The TKE equation is retained with $\sigma_K = 1.0$ and

$$\nu_t = \frac{C_\mu K^2}{\epsilon} = \frac{(0.09 - 0.04F)K^2}{\epsilon} \tag{4.245}$$

This model is called $K\epsilon 1$. For weak shear flows (i.e., for cases where the velocity defect or excess, ΔU_c, is a small fraction of U_e), an extended version called $K\epsilon 2$ was presented. This model used the same equations, but now with $C_{\epsilon 1} = 1.40$, $C_{\epsilon 2} = 1.94$, and

$$C_\mu = 0.09g(\overline{\mathcal{P}/\epsilon}), \qquad C_\mu = 0.09g(\overline{\mathcal{P}/\epsilon}) - 0.0534F \tag{4.246}$$

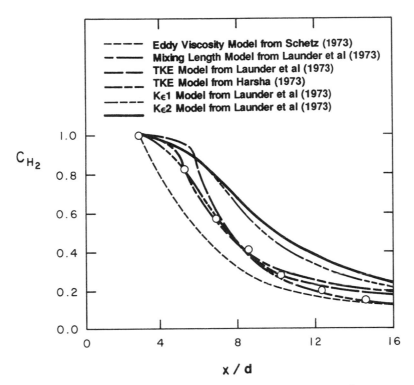

FIGURE 4.106 Comparison of predictions for composition with mean-flow, one-equation, and two-equation models with the axisymmetric H_2 jet into air experiment. [From NASA SP-321 (1973).]

for planar and axisymmetric flows, respectively, and where

$$\overline{\mathcal{P}/\epsilon} = \frac{\int_0^b \tau_t \left(\frac{\mathcal{P}}{\epsilon}\right) y^j \, dy}{\int_0^b \tau_t \, y^j \, dy} \tag{4.247}$$

$\mathcal{P}$ is production of K, and $g(\overline{\mathcal{P}/\epsilon})$ is given as a graph shown here in Fig. 4.107.

Consider the low-speed, axisymmetric jet case of Forstall and Shapiro (1950) which was discussed before; comparison of predictions and data are given in Fig. 4.105. It can be seen that the $K\epsilon 1$ and $K\epsilon 2$ models perform very well. In axisymmetric H_2-air cases, the situation is more confused. For the tests of Eggers (1971), the TKE and $K\epsilon 1$ and $K\epsilon 2$ models perform essentially equally. However, for the data of Chriss (1968), the relative performance of both $K\epsilon$ models comes out poorer, as shown in Fig. 4.106.

Rodi has conducted a thorough study of $K\epsilon$ models. Some results for planar and axisymmetric jets are shown in Fig. 4.108. Plotting the results in the manner shown exaggerates any differences between prediction and measurement. The "1 Eqn. model" mentioned on the figures is the Prandtl energy method.

The predictions of the Hanjalic–Launder Reynolds stress model [Launder *et al.*

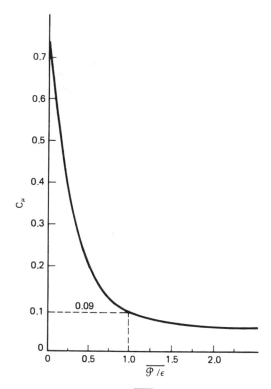

FIGURE 4.107   Empirical function $C_\mu = f(\overline{\mathcal{P}/\epsilon})$ for the $K\epsilon 2$ model. (From Launder *et al.*, 1973.)

(1973)] are compared with data for a planar jet with $U_e = 0$ in Fig. 4.109. From that comparison and only the one other given by Launder *et al.* (1971), no clear advantage over the two-equation models is apparent. The Daly and Harlow (1970) Reynolds stress model has been applied by Rodi to the planar jet with $U_e = 0$. The mean flow was predicted well, but the individual normal stresses are predicted poorly. The shear stress and turbulence energy are reported to have been predicted accurately.

4.12   WAKES
Joseph A. Schetz

4.12.1   Laminar Wakes

The analysis of wakes [see Fig. 4.79(c)] is simple only in the far wake where the velocity difference across the wake has decayed to a small value, since this permits a linearization of the equations of motion. Consider the planar, constant-density, constant-property case as an example. More general cases can be treated by numerical methods.

If $\Delta u = U_\infty - u_c \ll 1$, the boundary layer momentum equation can be approximated as

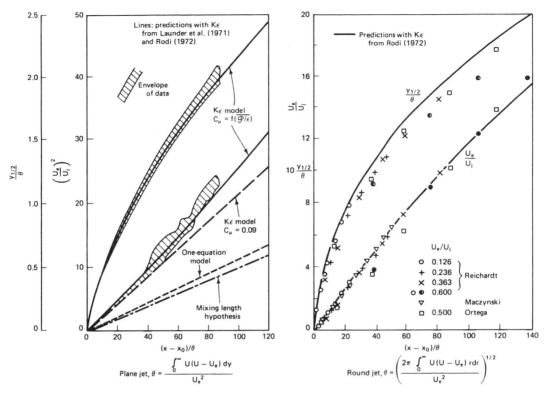

FIGURE 4.108 Development of maximum excess velocity and half-width for turbulent jets. (Reprinted with permission of IADR-AIRH, "Turbulence Models and Their Application in Hydraulics," by Rodi, W., 1980.)

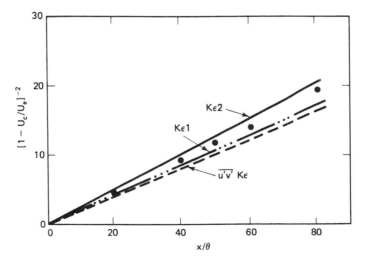

FIGURE 4.109 Comparison of predictions from two-equation and Reynolds stress models with the planar jet experiment of Everitt and Robins (1978). (From Launder *et al.*, 1973.)

$$U_\infty \frac{\partial u}{\partial x} = \nu \frac{\partial^2 u}{\partial y^2} \tag{4.248}$$

With boundary conditions of $\lim_{y \to \infty} u(x, y) = U_\infty$ and symmetry on the axis, the solution is

$$u(x, y) = U_\infty - \frac{C}{\sqrt{x}} U_\infty \exp\left(-\frac{U_\infty y^2}{4\nu x}\right) \tag{4.249}$$

The remaining constant C is to be determined to match the momentum defect in the wake with the drag of the body. If the body is a general, two-dimensional (planar) cylinder, we may write the drag, D, for a body length L perpendicular to the plane of the flow as

$$D = \rho L \int_{-\infty}^{\infty} u(U_\infty - u) \, dy \tag{4.250}$$

Noting, the definition of Δu and the fact $u \approx U_\infty$ (since $\Delta u \ll U_\infty$), this can be approximated as

$$D \approx \rho U_\infty L \int_{-\infty}^{\infty} (\Delta u) \, dy \tag{4.250a}$$

With this and the definition of a drag coefficient as $C_D = D/(1/2\rho U_\infty^2 L d)$, where d is a characteristic thickness of the cylinder (e.g., the diameter for a circular cylinder), we can set drag as expressed by the two relations [Eq. (4.250a) and the definition of C_D] equal to determine C and then obtain

$$\frac{U_\infty - u(x, y)}{U_\infty} = 2^j C_D \left(\frac{\text{Re}}{16\pi} \frac{d}{x}\right)^{(j+1)/2} \exp\left(-\frac{U_\infty y^2}{4\nu x}\right) \tag{4.251}$$

where $\text{Re} = U_\infty d/\nu$. The axisymmetric result obtained by a parallel analysis is included here for $j = 1$; the planar result corresponds to $j = 0$.

4.12.2 Vortex Streets

For *bluff* bodies at moderate Reynolds numbers, a new, unsteady phenomena appears even with a steady external flow and a stationary body. Consider a circular cylinder as an example. The laminar boundary layer separates slightly upstream of the 90° and 270° points on the top bottom of the cylinder forming a pair of vortices. These vortices are very unstable, and above some Reynolds number, they break away from the body. Moreover, they are unstable to asymmetrical disturbances, so they develop and move down behind the body in the neat alternating pattern called a *Karman Vortex Street* shown in Fig. 4.110. This pattern exists for $60 \leq \text{Re}_d \leq 5000$ for circular cylinders. At higher Reynolds numbers, the regular pattern rapidly breaks down into turbulence. The frequency at which the vortices are shed is im-

FIGURE 4.110 Visualization of vortex shedding behind a circular cylinder. (From Corning, 1982.)

portant for studying vibrations induced on bodies in a flow and also noise production. The experimental results of Roshko (1954) are shown in Fig. 4.111 in terms of the dimensionless frequency, the Strouhal number, $S = nd/U_\infty$.

4.12.3 Turbulent Wakes

For the turbulent wake behind a body, the profiles only become *similar* at a distance of many characteristic body dimensions downstream where the velocity defect $\Delta U \equiv (U_\infty - U)$ has become small compared to U_∞. This situation is like that for the laminar far wake discussed just above, so Eq. (4.250a) holds here also. The value of the integral will be proportional to $(\Delta U_c)(b(x))$ with the proportionality factor depending on the details of the profile shape $\Delta U(y)$. Here, ΔU_c is the centerline value of ΔU. With this and the definition of the drag coefficient, C_D, we can set drag as expressed by the two relations equal and obtain

$$\frac{\Delta U_c}{U_\infty} \sim \frac{C_D d}{2b} \tag{4.252}$$

The growth of the width $b(x)$ is again determined based on the assumption that the time rate of increase of the width is proportional to the transverse velocity fluctuation v'; that is,

$$\frac{Db}{Dt} \sim v' \tag{4.253}$$

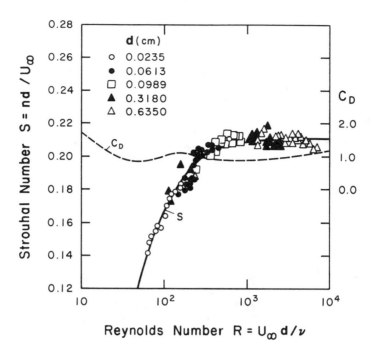

FIGURE 4.111 The Strouhal number for vortex shedding behind a circular cylinder vs. Reynolds number. (From Roshko, 1954.)

with

$$\frac{Db}{Dt} \approx U_\infty \frac{db}{dx} \tag{4.254}$$

and

$$v' \sim \ell_t \frac{\partial U}{\partial y} \approx \ell_t \frac{\Delta U_c}{b} \tag{4.255}$$

Taking, as usual for a flow not restricted by a rigid wall, ℓ_t as proportional to the width of the layer (i.e., $\ell_t = K_2 b$), one obtains

$$\frac{db}{dx} \sim K_2 \frac{\Delta U_c}{U_\infty} \tag{4.256}$$

Now, substitute for ΔU_c from above, and the result is

$$2b \frac{db}{dx} \sim K_2 C_D d \tag{4.257}$$

Integration gives

$$b \sim (K_2 x C_D d)^{1/2} \tag{4.258}$$

Using this in Eq. (4.252) yields

$$\frac{\Delta U_c}{U_\infty} \sim \left(\frac{C_D d}{K_2 x}\right)^{1/2} \tag{4.259}$$

Comparison with experiment gives the values of the proportionality constants and leads to

$$b = 0.57(x C_D d)^{1/2} \tag{4.258a}$$

and

$$\frac{\Delta U_c}{U_\infty} = 0.98 \left(\frac{C_D d}{x}\right)^{1/2} \tag{4.259a}$$

The corresponding results for a circular wake are

$$b \sim (K_2 C_D A x)^{1/3} \tag{4.260}$$

and

$$\frac{\Delta U_c}{U_\infty} \sim \left(\frac{C_D A}{K_2^2 x^2}\right)^{1/3} \tag{4.261}$$

where A is the projected area of the body.

Modern treatments of more general turbulent wake problems utilize numerical solutions of the equations of motion incorporating a turbulence model. At the mean flow level of turbulence models, it might seem that the treatments used for jets described in Sec. 4.11.2 could be applied directly here. However, eddy viscosity models that have been successfully applied to jet or wake cases often do not perform well when applied to the other cases. The discrepancy is primarily centered on the value of the proportionality constant, and the difference is not simply academic. In Sec. 4.11.2, it was shown that Eq. (4.241) is capable of good predictions of planar jet mixing flows. What happens if one uses it for a planar wake case? Figure 4.112 shows such a comparison for the wake behind a circular cylinder using the data of Schlichting (1930). Good agreement with experiment can be achieved only with an increase in the value of the proportionality constant significantly above that successfully used for jet problems.

In Schetz (1971) the connection between the observed turbulence field in various flow problems and the proportionality constant in eddy viscosity models was investigated. It was shown that, rather than being a constant, the numerical factor in Eq. (4.241) should be proportional to $u_c'^2/(\Delta U_c)^2$. Some data for this turbulence/mean flow information are presented in Fig. 4.113. The fact that this quantity is more or less constant for a given flow problem, except in the near-wake or potential core, is comforting, since the numerical factor is presumed to be a constant for a given flow. Comparing the results for the planar jet with $U_e \neq 0$ with those for the wake behind a circular cylinder in Fig. 4.113, we see that the increase of the factor by a ratio of

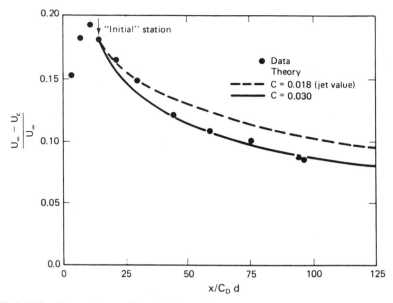

FIGURE 4.112 Comparison of prediction with experiment for centerline velocity in the wake behind a circular cylinder. [Reprinted with permission of Elsevier Science Ltd., "Some Studies of the Turbulent Wake Problem," by Schetz, J. A., *Astronautica Acta*, Vol. 16, pp. 107–117, 1971; data from Schlichting, 1930; analysis with Eq. (4.241), using different constants.]

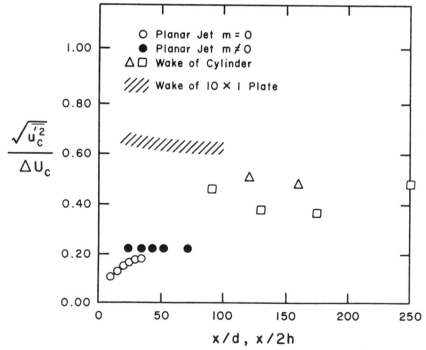

FIGURE 4.113 Turbulence intensity in various planar free mixing flows. (Reprinted with permission of Elsevier Science Ltd., "Some Studies of the Turbulent Wake Problem," by Schetz, J. A., *Astronautica Acta*, Vol. 16, pp. 107–117, 1971.)

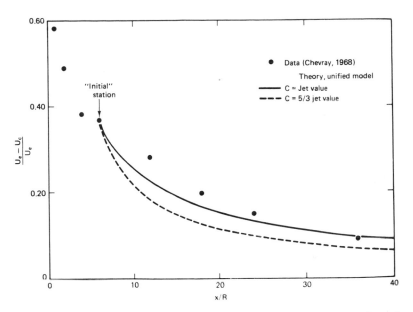

FIGURE 4.114   Centerline velocity distribution for a low-speed axisymmetric wake. (From Schetz, 1971; data from Chevray, 1968.) (Reprinted with permission of Elsevier Science Ltd., "Some Studies of the Turbulent Wake Problem," by Schetz, J. A., *Astronautica Acta*, Vol. 16, pp. 107–117, 1971; also data from Chevray, *ASME Journal of Basic Engineering*, Vol. 90, pp. 275–284, 1968.)

5/3 required in going from a jet to a wake behind a circular cylinder as found for the case in Fig. 4.112, is roughly predicted by this representation. For wakes behind streamlined bodies, $\overline{u_c'^2}/(\Delta U_c)^2$ is quite close to the values found in jets, so the appropriate proportionality constant is correspondingly the same as for jets. This is demonstrated in Fig. 4.114 which shows comparisons of predictions and data for the wake behind a slender, axisymmetric body. No increase in the *jet* value of the proportionality constant is necessary for this *wake* case. In summary, it isn't that all

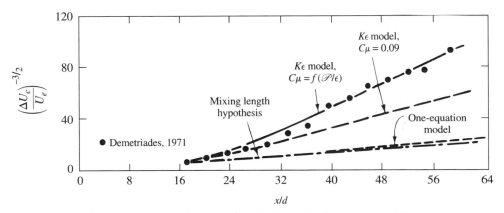

FIGURE 4.115   Development of maximum velocity defect in a compressible, axisymmetric wake. (Reprinted with permission of IAHR-AIRH, "Turbulence Models and Their Application in Hydraulics," by Rodi, W., 1980.)

jets are different than all wakes, but rather that wakes behind bodies of different shapes have different turbulence levels in relation to the mean flow velocity difference.

Rodi's studies with the $K\epsilon$ model also included wakes, and some results are illustrated in Fig. 4.115. In this flow, the $K\epsilon 2$ model (see Sec. 4.11.2) performs much better than the mixing length, TKE or $K\epsilon 1$ model.

Calculations for high-speed wakes, and thus significant density variations, using eddy viscosity models including Eq. (4.241) can be found in Schetz (1971). Up to about Mach 3.0, reasonable agreement with measurement can be achieved with such models. Even very high-speed wakes seldom have density variations as large as H_2-air jets, for example, but turbulence modeling still is often a problem. In particular, the compressibility effects discussed in Sec. 4.10.2 can be very important as the Mach number increases.

4.13 MASS TRANSFER IN BOUNDARY LAYERS
Joseph A. Schetz

4.13.1 Introduction

For the flow of a fluid consisting of a mixture of two (or more) different fluids with a nonuniform composition, there is a process of diffusion akin to viscous shear and heat transfer. If the local concentration of a species is described by a *mass fraction*, c_i, defined as the mass of species i in a volume divided by the total mass of all fluids in that volume, the application of the principle of conservation of mass of a species i requires only a slight generalization of the development for the conservation of the total mass in Sec. 1.2. The net mass flow of species i between that in the left side and out the right side of a differential control volume is

$$\frac{\partial}{\partial x} (\rho u c_i) \, dx \, dy \tag{4.262}$$

For the flows in the bottom and out the top, we must now allow for a mass flux of species i by diffusion as well as by convection. Denote the diffusive flux as $\dot{m}_i$. The net mass flow of species i in the y direction is then

$$\frac{\partial}{\partial y} (\rho v c_i + \dot{m}_i) \, dy \, dx \tag{4.263}$$

The diffusive flux in the x direction is neglected with the boundary layer assumption in the same way as for shear and heat transfer.

Any accumulation of mass of species i in the volume must appear as an unsteady term

$$\frac{\partial}{\partial t} (\rho c_i) \, dx \, dy \tag{4.264}$$

Applying conservation of mass of species i and assuming no production or consumption of species by chemical reactions, one obtains after using the continuity equation (for the total mass),

$$\rho \left[\frac{\partial c_i}{\partial t} + u \frac{\partial c_i}{\partial x} + v \frac{\partial c_i}{\partial y} \right] = -\frac{\partial \dot{m}_i}{\partial y} \qquad (4.265)$$

This holds for laminar or turbulent flow. The diffusive flux must still be modeled.

4.13.2 Laminar Diffusion

The diffusive flux can be written in terms of a *diffusive velocity*, v_i, as

$$\dot{m}_i = \rho c_i v_i \qquad (4.266)$$

where

$$v_i = v_{ic} + v_{iT} + v_{ip} \qquad (4.267)$$

indicating contributions from concentration, temperature, and pressure gradients. Here, we will neglect the effects of temperature and pressure gradients compared to concentration gradients, and we can write from Hirschfelder *et al.* (1954)

$$v_i = \frac{n^2}{\rho n_i} \sum_j m_j D_{ij} \frac{\partial}{\partial y} \left(\frac{n_j}{n} \right) \qquad (4.268)$$

where n_i is the number of particles of species i in a volume, m_i is the mass of such particles and D_{ij} are *Binary diffusion coefficients* for pairs of fluids. One must sum over all pairs. For many problems of engineering interest, there are only two fluids, or the various fluids can be placed into two groups by molecular weight. Then, we have so-called *binary diffusion*, and we shall restrict our coverage here to that simpler case. The treatment of unrestricted problems is covered in Hirschfelder *et al.* (1954). Assuming a binary mixture, *Fick's Law* gives

$$\dot{m}_i = \rho c_i v_i = -\rho D_{ij} \frac{\partial c_i}{\partial y} \qquad (4.269)$$

Thus, Eq. (4.265) becomes

$$\rho \left[\frac{\partial c_i}{\partial t} + u \frac{\partial c_i}{\partial x} + v \frac{\partial c_i}{\partial y} \right] = \frac{\partial}{\partial y} \left(\rho D_{ij} \frac{\partial c_i}{\partial y} \right) \qquad (4.270)$$

Mass transfer can contribute to energy transfer even if the fluid is at constant temperature. This is a consequence of the fact that different species generally have different specific heats (e.g., c_p). For a laminar flow, this results in an additional term in the energy equation expressed as a function of stagnation enthalpy, which is the form commonly used for mass transfer problems, as

$$-\frac{\partial}{\partial y} \left[\left(\frac{1}{\text{Le}} - 1 \right) \rho D_{ij} \sum_i h_i \frac{\partial c_i}{\partial y} \right] \qquad (4.271)$$

where

$$h_i = \int c_{pi}\, dT + h_i^\circ \tag{4.272}$$

Here, h_i° is the heat of formation of species i at the reference temperature for enthalpy.

The general-purpose computer code of Svehla and McBride (1973) used for the diffusion coefficient data in Sec. 2.16 will also calculate properties of mixtures of Newtonian fluids. One should not suppose that a property (e.g., viscosity) of a mixture is a linear function of the concentration between the property values of each single constituent. For mixtures of only two constituents, the following approximate rules may be used. For the viscosity of gases, the equation of Buddenberg and Wilke (1949) can be applied

$$\mu = \frac{X_1^2}{(X_1^2/\mu_1) + 1.385(X_1 X_2 RT/pW_1 D_{12})}$$
$$+ \frac{X_2^2}{(X_2^2/\mu_2) + 1.385(X_2 X_1 RT/pW_2 D_{21})} \tag{4.273}$$

where X_i is the *mole fraction* of species i. For the thermal conductivity of gases one can use

$$k = \frac{1}{2}\left[(k_1 X_1 + k_2 X_2) + \frac{k_1 k_2}{(X_1\sqrt{k_2} + X_2\sqrt{k_1})^2}\right] \tag{4.274}$$

There are no corresponding simple expressions for liquid mixtures, so one must refer to published data.

The diffusion coefficient is combined with other properties to form dimensionless groupings in two ways. The first is the *Schmidt number*,

$$Sc = \frac{\mu}{\rho D_{ij}} \tag{4.275}$$

The second is the *Lewis number*,

$$Le = \frac{\rho c_p D_{ij}}{k} \tag{4.276}$$

Obviously, they are related by

$$Pr = Le \cdot Sc \tag{4.277}$$

The Schmidt number is the ratio of coefficients for diffusion of momentum to diffusion of mass, and the Lewis number gives the ratio for mass transfer to heat transfer.

Under assumptions of constant density and properties, only rather restricted cases of foreign fluid injection can be treated. More general cases are discussed below. Here, we must assume that not only are the properties independent of temperature and pressure, but also that the properties of the two fluids are essentially equal. Taking this all together, the complete system of equations to be used in such a case for a steady flow over a flat plate is

$$\frac{\partial u}{\partial x} + \frac{\partial v}{\partial y} = 0$$

$$u \frac{\partial u}{\partial x} + v \frac{\partial u}{\partial y} = \nu \frac{\partial^2 u}{\partial y^2}$$

$$u \frac{\partial T}{\partial x} + v \frac{\partial T}{\partial y} = \frac{\nu}{\mathrm{Pr}} \frac{\partial^2 T}{\partial y^2} \qquad (4.278)$$

$$u \frac{\partial c_i}{\partial x} + v \frac{\partial c_i}{\partial y} = \frac{\nu}{\mathrm{Sc}} \frac{\partial^2 c_i}{\partial y^2}$$

Actually, only one species equation, say for c_1, is needed, since we know that $c_1 + c_2 \equiv 1.0$.

A problem governed by Eq. (278) with $v_w \neq 0$ was treated by Hartnett and Eckert (1957). In a case where $v_w \sim x^{-1/2}$, this system of equations admits to a similar solution, and that property was used. The results are shown in Fig. 4.116(a), (b), and (c) for velocity, temperature, and concentration profiles in the boundary layer and film coefficients for heat and mass transfer. The profiles can be seen to be greatly influenced by either suction $(v_w/U_e)\,(\mathrm{Re}_x)^{1/2} < 0$ or injection $(v_w/U_e)\,(\mathrm{Re}_x)^{1/2} > 0$. For injection, the wall friction and surface heat and mass transfer are dramatically reduced. Indeed, for values of the injection parameter greater than 0.62, the surface transfer rates all go to zero, and the boundary layer is said to be *blown off*. With the constant-density, constant-property assumptions of this analysis, the velocity profiles in Fig. 4.116(a) also apply if there is no temperature difference and thus no heat transfer and only a single specie and thus no mass transfer.

The transverse mass flux of the air (species 2) at any point is composed of a part due to a *diffusive velocity* and a part due to the *convective velocity*, v. Thus,

$$\rho_2 v_2 = -\rho D_{12} \frac{\partial c_2}{\partial y} + \rho c_2 v \qquad (4.279)$$

or

$$v_2 = -\frac{D_{12}}{c_2} \frac{\partial c_2}{\partial y} + v \qquad (4.280)$$

Since no air penetrates into the wall, $v_2(x, 0) \equiv 0$, and

$$v_w = \frac{D_{12}}{c_{2w}} \frac{\partial c_2}{\partial y}\bigg|_w \qquad (4.281)$$

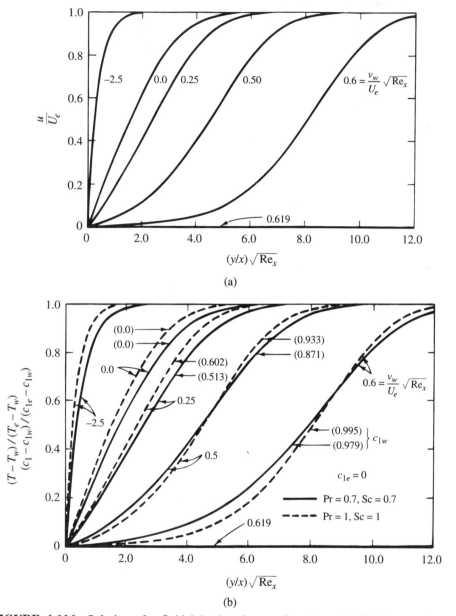

FIGURE 4.116 Solutions for fluid injection from a flat plate. (a) Velocity profiles, (b) temperature and mass fraction profiles, and (c) nondimensional heat and mass transfer rates. (From Hartnett and Eckert, 1957.)

Using $c_1 + c_2 \equiv 1.0$, this can be rewritten

$$v_w = -\frac{D_{12}}{1 - c_{1w}} \left.\frac{\partial c_1}{\partial y}\right|_w \tag{4.282}$$

Therefore, one is not free to prescribe both v_w and c_{1w}.

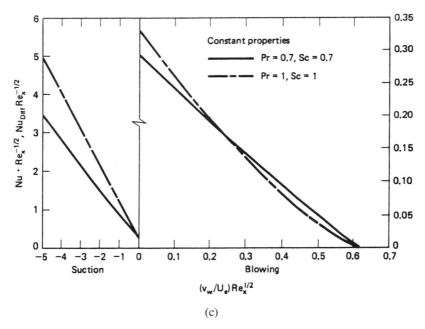

(c)

FIGURE 4.116   (*Continued*)

The results can also be expressed in terms of a *diffusion film coefficient*, h_D, defined by

$$\dot{m}_{iw} = -\rho D_{12} \left.\frac{\partial c_1}{\partial y}\right|_{y=0} \equiv \rho h_D (c_{1w} - c_{1e}) \qquad (4.283)$$

or as a *Nusselt number for diffusion*

$$\mathrm{Nu}_{\mathrm{Diff}} = \frac{h_D x}{D_{12}} = 0.332 \, \mathrm{Re}_x^{1/2} \mathrm{Sc}^{1/3} \qquad (4.284)$$

Note the close relationship to the corresponding result for a constant-temperature plate case. Data for mass transfer from a flat plate in low-speed flow are scarce. In Fig. 4.117, the measurements of Christian and Kezios (1959) for naphthalene sublimation (at a low rate, so $v_w \approx 0$) into air flowing axially over cylinders with $(\delta/R) \ll 1$ (so the effects of axisymmetry compared to a planar case are $\approx 2\%$) are compared to Eq. (4.284).

Exact solutions for injection of a light gas (H_2) through a porous wall into a high-speed airstream without restrictive assumptions on the physical properties and with variable density have been published by Eckert *et al.* (1958). Such a case is of practical interest for cooling a surface in a high-speed flow. The solutions, however, were again restricted to a *similar* distribution of injection (i.e., $v_w \sim x^{-1/2}$), and that is unrealistic from a practical viewpoint. Nonetheless, some useful conclusions can be drawn from the results.

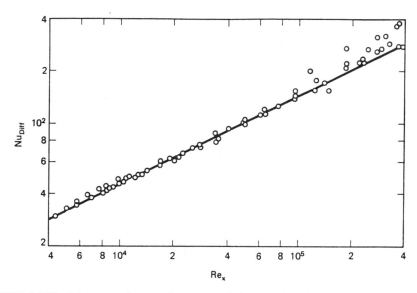

FIGURE 4.117   Mass transfer rate from a napthalene surface in an airstream compared to analysis. (Reprinted with permission of American Institute of Chemical Engineers, "Sublimation from Sharp-Edged Cylinders in Axisymmetric Flow, Including Influence of Surface Curvature," by Christian, W. J. and Kezios, S. P., *AIChE J.*, Vol. 5, pp. 61–68, 1959.)

Results corresponding to an air-only, flat plate case treated earlier by Van Driest (1952) without injection are shown in Fig. 4.118 for various values of c_{1w} (H_2 concentration at the wall) corresponding to various dimensionless injection rates, f_w, through a generalized form of Eq. (4.282).

$$\rho_w v_w = \frac{-(\rho D_{12})_w}{1 - c_{1w}} \left.\frac{\partial c_1}{\partial y}\right|_w \tag{4.285}$$

where

$$v_w = -\frac{1}{2} \left(\frac{\rho_w}{\rho_e}\right) \left(\frac{U_e \nu_e}{x}\right)^{1/2} f_w \tag{4.286}$$

The solutions for $c_{1w} = 0$ (i.e., no H_2 injection) differ a little from those of Van Driest because of slightly different assumed values of the properties of pure air. The peaks in the static temperature profiles are reduced sharply by H_2 injection even in low amounts ($c_{1w} \approx 0.1$). The wall temperature gradients, and thus the heat transfer, are also reduced substantially.

Results for skin friction and heat transfer rate in terms of Nusselt number are given in Fig. 4.119. Large reductions in both quantities are produced by H_2 injection. Figure 4.119(b) also illustrates how T_{aw} is strongly reduced by H_2 injection at $M_e = 12$. These results show that light gas injection is very effective at reducing heat transfer and static temperature peaks in high-speed flows. Heavier gases are not as effective per unit mass flow, but substantial benefits are still achievable. The use

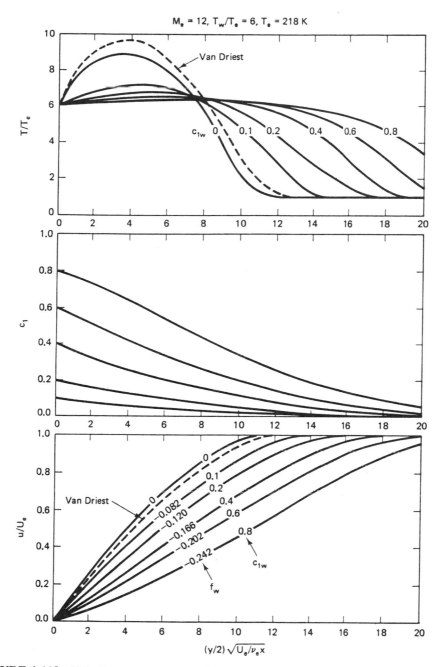

FIGURE 4.118 Velocity, temperature, and mass fraction profiles for H_2 injection into high-speed airflow over a flat plate. (Reprinted with permission from AIAA, "Mass Transfer Cooling of a Laminar Boundary by Injection of a Light-Weight Foreign Gas," by Eckert, E. R. G., Schneider, P. J., Hayday, A. A., and Larson, R. M., *Jet Propulsion*, Vol. 28, No. 1, 1958.)

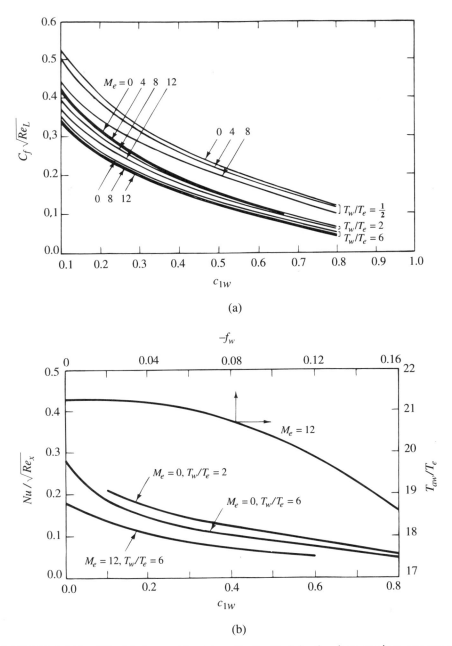

FIGURE 4.119 Skin friction and heat transfer for H_2 injection into an airstream over a flat plate. (a) Skin friction coefficient and (b) dimensionless heat transfer rate. (Reprinted with permission from AIAA, ''Mass Transfer Cooling of a Laminar Boundary by Injection of a Light-Weight Foreign Gas,'' by Eckert, E. R. G., Schneider, P. J., Hayday, A. A., and Larson, R. M., *Jet Propulsion*, Vol. 28, No. 1, 1958.)

of heavy gas "injection" by allowing teflon to sublime thus has found considerable use, because the system is simple requiring no valves, meters, controls, etc.

4.13.3 Turbulent Diffusion

Most of the consideration of turbulent flows with surface mass transfer has been in terms of the analogy between heat and mass transfer. Within that framework, one presumes a direct correspondence between a film coefficient for heat, h, and for diffusion, h_D, usually presented in dimensionless terms as Nusselt numbers, Nu, and, Nu_{Diff}, or Stanton numbers, C_h, and $C_{h,\text{Diff}}$. The subscripts D and Diff denote flows with diffusion (i.e., mass transfer). Further, the role of the Prandtl number, Pr, in heat transfer correlations is taken by the Schmidt number, Sc, for mass transfer. The correspondence between mass transfer data and a transformed heat transfer law ($C_h \rightarrow C_{h,\text{Diff}}$ with Pr $\rightarrow$ Sc), where the heat transfer law is for flow over a solid surface, is only obtained for very low surface mass transfer rates. The basic analogy between heat and mass transfer still holds for nonnegligible surface mass transfer, but, in such cases, one must begin with a heat transfer law for flow over a porous surface with injection or suction. The influence of injection or suction on heat transfer is substantial as was shown earlier for laminar flow (see Sec. 4.13.2) and now in Fig. 4.120 for turbulent flow. This kind of information has been generalized considerably by relating both the ratio of skin friction coefficients without and with injection or suction and the ratio of the Stanton numbers without and with injection or suction through two parameters: $B_f = (\rho_w v_w / \rho_e U_e)/(C_f/2)$, $B_h = (\rho_w v_w / \rho_e U_e)/C_h$. The simple correlation formula

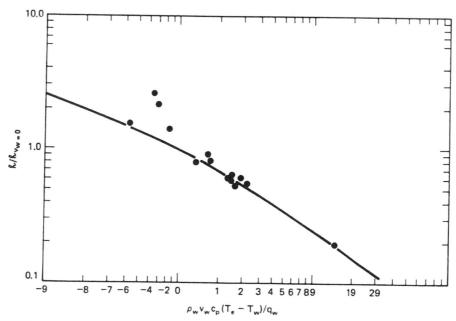

FIGURE 4.120 Influence of injection or suction on the film coefficient for flow over a flat plate. (From Mickley *et al.*, 1954.)

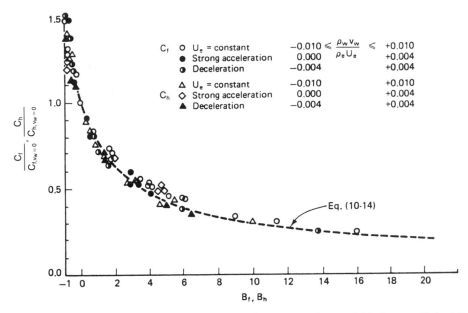

FIGURE 4.121 The effect of transpiration on Stanton number and friction coefficient for a wide range of flows. (Reprinted with permission of Academic Press, Inc., "The Behavior of Transpired Turbulent Boundary Layers," by Kays, W. W. and Moffat, R. J., in *Studies in Convection*, Vol. 1, pp. 223–319, 1975.)

$$\frac{C_h}{C_{h,\,v_w=0}} = \frac{C_f}{C_{f,\,v_w=0}} = \left[\frac{\ln(1 + B_{f,h})}{B_{f,h}}\right]^{5/4} (1 + B_{f,h})^{1/4} \qquad (4.287)$$

works very well over a considerable range of injection or suction rates and different pressure gradients as illustrated in Fig. 4.121. Here, Re_θ for the two flows must match. This information is useful not only for mass transfer work but also for momentum and/or heat transfer in single specie flows.

The study of turbulent flows with variable composition has not progressed to the same point as for the mean velocity and temperature fields. Thus, there are not well-developed equivalents of the *Law of the Wall* and *Defect Law* for species profiles.

4.14 NON-NEWTONIAN BOUNDARY LAYER FLOWS
Joseph A. Schetz

The same tools that have developed for boundary-layer flows of Newtonian fluids have been applied to non-Newtonian fluid flows (see Chap. 3) in those cases where simple forms of the stress-strain relation are appropriate.

4.14.1 Laminar Flows

The momentum integral method [Schetz (1994)] has been successfully applied to the laminar flow of some non-Newtonian fluids modeled as power law or Bingham plastic fluids (see Chap. 3), for example. Consider first the case of a *power law fluid* over a flat plate. The biggest change that must be made is in the relationship of the wall shear to the assumed velocity profile. Now, we must see

$$\tau_w = \mu_{PL} \left(\frac{\partial u}{\partial y} \right)_{y=0}^p \tag{4.288}$$

It is common to employ a cubic velocity profile [see Acrivos *et al.* (1960)]. Substituting into the integral momentum equation for a flat plate case, gives

$$\frac{39}{280} \rho U_e^2 \frac{d\delta}{dx} = \mu_{PL} \left(\frac{3}{2} \right)^p U_e^p \left(\frac{1}{\delta} \right)^p \tag{4.289}$$

This can be integrated using $\delta(0) = 0$ to yield

$$\frac{\delta}{x} = \left[\frac{280(p+1)}{39} \left(\frac{3}{2} \right)^p \right]^{1/(p+1)} \left(\frac{x^p U_e^{2-p} \rho}{\mu_{PL}} \right)^{-1/(p+1)} \tag{4.290}$$

For a Newtonian fluid $p = 1$ and $\mu_{PL} = \mu$, and the numerical factor for the same assumed velocity profile is $(280/13)^{1/2}$. Where a Newtonian fluid result has $\mathrm{Re}^{1/2}$, we now find the last term in Eq. (4.290). This is apparently a *Reynolds number for power law fluids* which can be denoted as Re_{PL}, i.e.,

$$\mathrm{Re}_{PL} \equiv \frac{x^p U_e^{2-p} \rho}{\mu_{PL}} \tag{4.291}$$

Our primary interest is usually in the wall shear which becomes

$$C_f(x) \equiv \frac{\tau_w}{1/2\rho U_e^2} = 2 \left[\left(\frac{3}{2} \right) \frac{39}{280(p+1)} \right]^{p/(p+1)} \mathrm{Re}_{PL}^{-1/(p+1)} \tag{4.292}$$

For flow at the same edge velocity, a power law fluid will generally have a lower wall shear and a smaller boundary layer thickness than a Newtonian fluid with the same density and $\mu_{PL} = \mu$. This has led to some interest in these fluids for drag reduction in water flows by adding a small amount of a power law fluid in solution.

In the case of a *Bingham plastic fluid* (see Chap. 3), the integral momentum equation itself must also be modified. This is because $\tau_\delta \neq 0$ as $(\partial u/\partial y)_{y=\delta} = 0$ for such a fluid, which has a *yield stress* at zero velocity gradient. Thus, the integral momentum equation for flow over a flat plate must be rewritten as

$$\frac{d\theta}{dx} = \frac{\tau_w - \tau_\delta}{\rho U_e^2} \tag{4.293}$$

Using the stress model for a Bingham plastic fluid, we can say

$$\tau_w - \tau_\delta = \left[\tau_o + \mu_{BP} \left(\frac{\partial u}{\partial y} \right)_{y=0} \right] - \left[\tau_o + \mu_{BP} \left(\frac{\partial u}{\partial y} \right)_{y=\delta} \right] = \mu_{BP} \left(\frac{\partial u}{\partial y} \right)_{y=0} \tag{4.294}$$

so the effect appears to cancel out. That is not quite true as will be clear below. The solution for the boundary layer thickness with a cubic profile assumption is, however, the same as for a Newtonian fluid [Schetz (1994)].

$$\frac{\delta}{x} = 4.64 \ \text{Re}_{\text{BP}}^{-1/2} \tag{4.295}$$

where

$$\text{Re}_{\text{BP}} = \frac{\rho U_e x}{\mu_{\text{BP}}} \tag{4.296}$$

The yield stress does enter into the expression for the wall shear

$$C_f(x) = \frac{\tau_w}{1/2 \rho U_e^2} = \frac{\tau_o}{1/2 \rho U_e^2} + \frac{0.646}{\text{Re}_{\text{BP}}^{1/2}} \tag{4.297}$$

Similar solutions (see Sec. 4.2) are also possible under restricted conditions for some non-Newtonian fluid cases. The case of a power law fluid flowing over a flat plate has been treated by Acrivos *et al.* (1960). The similarity variable used is

$$\eta_{\text{PL}} \equiv y \left[\frac{\rho U_e^{2-p}}{\mu_{\text{PL}} x} \right]^{1/(p+1)} \tag{4.298}$$

and the velocity is expressed as

$$\frac{u}{U_e} = f'(\eta_{\text{PL}}) \tag{4.299}$$

Note the close relationship to the Blasius, Newtonian case (see Sec. 4.2) if $p = 1$ and $\mu_{\text{PL}} = \mu$. With all this, the boundary layer equations for a power law fluid become

$$p(p+1)f''' + (f'')^{2-p} f = 0 \tag{4.300}$$

Some velocity profiles are plotted in Fig. 4.122 for typical values of p including $p = 1$ which is the Newtonian case. Clearly the result of $p < 1$ is fuller profiles and thinner boundary layers. The wall shear turns out to be lower. These results are in qualitative agreement with the approximate solutions presented above. The exact result for the skin friction is written as

$$C_f = 2c(p) \ \text{Re}_{\text{PL}}^{-1/(p+1)} \tag{4.301}$$

The variation of $c(p)$ with p is given in Table 4.5 along with the corresponding results from the integral momentum solution in Eq. (4.292).

Since the momentum equation contains the shear stress appearing only as $\partial \tau / \partial y$, substituting the shear model for a Bingham plastic results in the same equation as for a Newtonian fluid with μ_{BP} written in the place of the usual μ. Thus, the Blasius

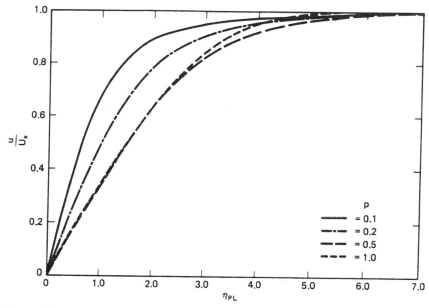

FIGURE 4.122 Velocity profiles for power law fluid over a flat plate. (Reprinted with permission of American Institute of Chemical Engineers, "Momentum and Heat Transfer in Laminar Boundary-Layer Flows of Non-Newtonian Fluids Past External Surfaces," by Acrivos, A., Shah, M. J., and Petersen, E. E., *AIChE J.*, Vol. 6, pp. 312–317, 1960.)

solution given in Sec. 4.2 can be used for a Bingham plastic case. The important difference between a Bingham plastic fluid case and a Newtonian fluid case emerges in the wall shear stress or skin friction results where an extra term due to the yield stress, τ_0, arises.

$$C_f = \frac{\tau_0}{1/2\rho U_e^2} + \frac{0.664}{\sqrt{\text{Re}_x}} \tag{4.302}$$

4.14.2 Turbulent Flows

Most non-Newtonian fluids of practical concern have such a high effective viscosity that they are usually in a laminar state for conditions of interest. One important

TABLE 4.5   Solutions for Coefficient in Skin Friction Relation (Eq. 4.301) for Power Law Fluid over a Flat Plate

p	$c(p)$, Exact	$c(p)$, Eq. (4.292)
0.05	1.017	0.926
0.1	0.969	0.860
0.2	0.873	0.747
0.3	0.733	0.655
0.5	0.576	0.518
1.0	0.332	0.323

exception is the addition of a small amount of a long-chain-molecule polymer to a water flow under high Reynolds number conditions. The amounts of the additive are so small that the viscosity of the fluid is about the same as pure water, but substantial drag reduction can occur. Much of the work in this area has been for flow in pipes [see Hoyt (1989), for example].

Skelland (1966) has extended the simple integral analysis for turbulent flow over a flat plate [Schetz (1994)] to non-Newtonian fluids of the power law type. The usual 1/7th expression for the velocity profile is extended to

$$\frac{U}{U_e} = \left(\frac{y}{\delta}\right)^{\beta p/(2 - \beta(2 - p))} \tag{4.303}$$

Here, $\beta(p)$ is a function of the power law exponent p given as a graph. It is actually the exponent on the effective Reynolds number in a skin friction law corresponding to that usually used for Newtonian fluids. For the Newtonian case, $p = 1$ and $\beta = 0.25$, so the exponent in Eq. (4.303) becomes 1/7. Using an extension of the usual skin friction law to power law fluids and proceeding as for Newtonian fluids gives the following result for the growth of the boundary layer along the plate

$$\frac{\delta}{x} = \left[\frac{(\beta p + 1)\phi_1(\alpha,\beta)}{\phi_2(\beta)}\left(8^{p-1}\left(\frac{3p+1}{4p}\right)^p\right)^\beta\left(\frac{\rho U_e^{2-p}x^p}{\mu_{PL}}\right)^{-\beta}\right]^{1/(\beta p + 1)}$$

$$\phi_1(\alpha, \beta) = \frac{\alpha(0.817)^{2 - \beta(2 - p)}}{2^{\beta p + 1}}$$

$$\phi_2(\beta) = \frac{2 - \beta(2 - p)}{2(1 - \beta + \beta p)} - \frac{2 - \beta(2 - p)}{2 - 2\beta + 3\beta p} \tag{4.304}$$

where $\alpha(p)$ is the coefficient in the skin friction law which is also provided as a graph.

There have been some efforts to apply more elaborate turbulence models to non-Newtonian fluid flows. For example, simple extensions of the $K\epsilon$ model have been proposed, but the database upon which such models can be constructed and against which they can be tested is very sparse.

REFERENCES

Acrivos, A., Shah, M. J., and Petersen, E. E., "Momentum and Heat Transfer in Laminar Boundary-Layer Flows of Non-Newtonian Fluids Past External Surfaces," *AIChE J.*, Vol. 6, pp. 312–317, 1960.

AGARD, *Laminar-Turbulent Transition*, AGARD Conference Proceedings No. 224, 1977.

AGARD, *Special Course on Stability and Transition of Laminar Flow*, AGARD Conference Proceedings No. 709, 1984.

AGARD, *Fluid Dynamics of Three-Dimensional Turbulent Shear Flows and Transition*, AGARD Conference Proceedings No. 438, 1988.

AGARD, *Special Course on Skin Friction Drag Reduction*, AGARD Report 786, 1992.

AGARD, *Progress in Transition Modelling*, AGARD Report 793, 1993.

Aihara, Y. and Koyama, H., "Secondary Instability of Görtler Vortices: Formation of Periodic Three-Dimensional Coherent Structures," *Trans. Japan Soc. Aero. Space Sci.*, Vol. 24, 1981.

Alber, I. and Lees, L., "Integral Theory for Supersonic Turbulent Base Flows," *AIAA J.*, Vol. 6, pp. 1343–1351, 1968.

Antonia, R. A. and Bilger, R. W., "An Experimental Investigation of an Axisymmetric Jet in a Co-Flowing Air Stream," *J. Fluid Mech.*, Vol. 61, pp. 805–822, 1973.

Arnal, D. and Michel, R. (Eds.), *Laminar-Turbulent Transition*, Springer-Verlag, New York, 1990.

Balachandar, S., Streett, C. L., and Malik, M. R., "Secondary Instability in a Rotating Disk Flow," *J. Fluid Mech.*, Vol. 242, 1992.

Balakumar, P. and Reed, H. L., "Stability of Three-Dimensional Supersonic Boundary Layers," *Physics of Fluids A*, Vol. 3, No. 4, 1991.

Becker, H. A., Hottel, H. C., and Williams, G. C., "The Nozzle Fluid Concentration Field of the Round Turbulent Free Jet," *J. Fluid Mech.*, Vol. 30, pp. 285–303, 1967.

Beckwith, I. E., Malik, M. R., Chen, F.-J., and Bushnell, D. M., "Effects of Nozzle Design Parameters on the Extent of Quiet Test Flow at Mach 3.5," *Laminar-Turbulent Transition*, Kozlov, V. V. (Ed.), Springer-Verlag, New York, 1985.

Bertolotti, F. P. and Herbert, T., "Analysis of the Linear Stability of Compressible Boundary Layers Using the PSE," *J. Theor. Comput. Fluid Dyn.*, Vol. 3, 1991.

Bertolotti, F. P., Herbert, T., and Spalart, P. R., "Linear and Nonlinear Stability of the Blasius Boundary Layer," *J. Fluid Mech.*, Vol. 242, 1992.

Bippes, H., "Experimental Study of the Laminar-Turbulent Transition of a Concave Wall in a Parallel Flow," NASA TM-75243, 1978.

Birch, A. D., Brown, D. R., Dodson, M. G., and Thomas, J. R., "The Turbulent Concentration Field of a Methane Jet," *J. Fluid Mech.*, Vol. 88, pp. 431–449, 1978.

Birch, S. F. and Eggers, J. M., "A Critical Review of the Experimental Data for Developed Free Turbulent Shear Layers," *Free Turbulent Shear Flows*, NASA SP-321, 1973.

Birkhoff, G., *Hydrodynamics: A Study in Logic, Fact, and Similitude*, 2nd ed., Princeton University Press, 1961. (Reprinted by Dover, New York, 1960.)

Blottner, F. G., "Finite Difference Methods of Solution of the Boundary Layer Equations," *AIAA J.*, Vol. 8, pp. 193–205, 1970.

Blottner, F. G., "Introduction to Computational Techniques for Boundary Layers," Sandia Laboratories Report SAND 79-0893, Sandia National Laboratories, Albuquerque, NM, 1979.

Bogdanoff, D. W., "Compressibility Effects in Turbulent Shear Layers," *AIAA J.*, Vol. 21, pp. 926–927, 1983.

Bradshaw, P. and Wong, F. Y. F., "The Reattachment and Relaxation of a Turbulent Shear Flow," *J. Fluid Mech.*, Vol. 52, pp. 113–135, 1972.

Bradshaw, P., "Compressible Turbulent Shear Layers," *Annual Review of Fluid Mech.*, Annual Reviews Inc., Palo Alto, CA, 1977.

Briley, W. R., "A Numerical Study of Laminar Separation Bubbles Using the Navier–Stokes Equations," *J. Fluid Mech.*, Vol. 47, pp. 713–736, 1971.

Brown, G. L. and Roshko, A. A., "On Density Effects and Large Structure in Turbulent Shear Layers," *J. Fluid Mech.*, Vol. 64, pp. 775–816, 1974.

Brown, S. N. and Stewartson, K., "Laminar Separation," *Ann. Rev. Fluid Mech.*, Vol. 1, pp. 45–72, 1969.

Buddenberg, J. W. and Wilke, C. R., "Calculation of Gas Mixture Viscosities," *Ind. Eng. Chem.*, Vol. 41, pp. 1345–1347, 1949.

Bushnell, D. M., Malik, M. R., and Harvey, W. D., "Transition Prediction in External Flows Via Linear Stability Theory," *Symposium Transsonicum III*, Zierep, J. and Ortel, H. (Eds.), Springer-Verlag, Berlin, 1988.

Carlson, D. R., Widnall, S. E., and Peeters, M. T., "A Flow-Visualization Study of Transition in a Plane Poiseuille Flow," *J. Fluid Mech.*, Vol. 121, pp. 487–505, 1982.

Carter, J. E., "Inverse Solutions for Laminar Boundary-Layer Flows with Separation and Attachment," NASA TR-R447, 1973.

Carter, J. E. and Wornom, S. F., "Solutions for Incompressible Separated Boundary Layers Including Viscous-Inviscid Interaction," *Aerodynamic Analyses Requiring Advanced Computers, Part I*, NASA SP 347, pp. 125–150, 1975.

Catherall, D. and K. W. Mangler, "The Integration of Two-Dimensional Laminar Boundary Layer Equations Past a Point of Vanishing Skin Friction," *J. Fluid Mech.*, Vol. 26, pp. 163–192, 1966.

Cebeci, T. and Smith, A. M. O., *Analysis of Turbulent Boundary Layers*, 1st ed., Academic Press, New York, 1974.

Cebeci, T., "Calculation of Three-Dimensional Boundary Layers I, Swept Infinite Cylinders and Small Cross Flow," *AIAA J.*, Vol. 12, pp. 779–786, 1974.

Cebeci, T., Kattab, A. K., and Stewartson, K., "Three-Dimensional Laminar Boundary Layers of the Ok of Accessibility," *J. Fluid Mech.*, Vol. 107, pp. 57–87, 1981.

Chang, C.-L. and Malik, M. R., "Oblique-Mode Breakdown and Secondary Instability in Supersonic Boundary Layers," *J. Fluid Mech.*, Vol. 273, 1994.

Chang, C.-L., Malik, M. R., Erlebacher, G., and Hussaini, M. Y., "Compressible Stability of Growing Boundary Layers Using Parabolized Stability Equations," AIAA 91-1636, 1991.

Chang, P. K., *Control of Flow Separation*, Hemisphere, Washington, D.C., 1976.

Chevray, R., "The Turbulent Wake of a Body of Revolution," *J. Basic Eng.*, Vol. 90, pp. 275–284, 1968.

Choudhari, M. M., "Roughness-Induced Generation of Crossflow Vortices in Three-Dimensional Boundary Layers," *Theoret. Comput. Fluid Dynamics*, Vol. 6, 1994.

Chriss, D. E., "Experimental Study of Turbulent Mixing of Subsonic Axisymmetric Gas Streams," Arnold Engineering Development Center, AEDC-TR-68-133, Aug. 1968.

Christian, W. J. and Kezios, S. P., "Sublimation from Sharp-Edged Cylinders in Axisymmetric Flow, Including Influence of Surface Curvature," *AIChE J.*, Vol. 5, pp. 61–68, 1959.

Chu, J. and Young, A. D., "Measurement of Separating Two-Dimensional Turbulent Boundary Layers," AGARD Conference Proceedings, No. 168, *Flow Separation*, pp. 13-1–13-12, 1975.

Coles, D., "The Law of the Wake in the Turbulent Boundary Layer," *J. Fluid Mech.*, Vol. 1, p. 191, 1956.

"Computation of Turbulent Boundary Layers—1968," *AFOSR-IFP—Stanford Conference—Proceedings*, Kline, S. J., Morkovin, M. V., Sovran, G., Cochrell, D. J. (Eds.), Stanford University, Stanford, CA, 1968.

Corning, J. J., "Vortex Flows," 8th Quick-Göthert Lecture Series, Tullahoma, TN, 1982.

Corrsin, S. and Uberoi, M., "Further Experiments on the Flow and Heat Transfer in a Heated Turbulent Air Jet," NACA TN 1895, 1949.

Crocco, L. and Cohen, C. B., "Compressible Laminar Boundary Layer with Heat Transfer and Pressure Gradient, Fifty Years of Boundary Layer Research," NACA Rep. No. 1294, 1956.

Crouch, J. D., "Receptivity of Three-Dimensional Boundary Layers," AIAA 93-0074, 1993.

Crouch, J. D., "Theoretical Studies on the Receptivity of Boundary Layers," AIAA 94-2224, 1994.

Dagenhart, J. R., Saric, W. S., Mousseaux, M. C., and Stack, J. P., "Crossflow Vortex Instability and Transition on a 45-Degree Swept Wing," AIAA No. 89-1892, 1989.

Daly, B. J. and Harlow, F. H., "Transport Equations in Turbulence," *Phys. Fluids*, Vol. 13, p. 2634, 1970.

Dash, S. M., Weilerstein, G., and Vaglio-Laurin, R., "Compressibility Effects in Free Turbulent Shear Flows," AFOSR-TR-75-1436, 1975.

Demetriades, A., "Computation of Numerical Data on the Mean Flow from Compressible Turbulent Wake Experiments," Publ. No. U-4970, Aero. Div., Philco–Ford Corp., 1971.

Demetriades, A., "Laminar Boundary Layer Stability Measurements at Mach 7 Including Wall Temperature Effects," AIAA 74-535, 1974.

Denier, J. P., Hall, P., and Seddougni, S. O., "On the Receptivity Problem for Görtler Vortices: Vortex Motion Introduced by Wall Roughness," *Phil. Trans. Roy. Soc. London*, Series A335, pp. 51–85, 1991.

Despard, R. A. and Miller, J. A., "Separation in Oscillating Boundary Layer Flows," *J. Fluid Mech.*, Vol. 47, pp. 21–31, 1971.

Donaldson, C. Du P. and Lange, R. H., "Study of Pressure Rise Across Shock Waves Required to Separate Laminar and Turbulent Boundary Layers," NACA Tech. Note. 2770, 1952.

Donaldson, C. Du P., "A Progress Report on an Attempt to Construct an Invariant Model of Turbulent Shear Flows," AGARD CP-93, 1971.

Drazin, P. G. and Reid, W. H., *Hydrodynamic Stability*, Cambridge University Press, Cambridge, UK, 1981.

Dwoyer, D. L. and Hussaini, M. Y. (Eds.), *Stability of Time Dependent and Spatially Varying Flows*, Springer-Verlag, New York, 1987.

Dwyer, H. A. and Sherman, R., "Some Characteristics of Unsteady Two- and Three-Dimensional Reversed Boundary Layer Flows," *Numerical and Physical Aspects of Aerodynamics Flows*, Cebeci, T. (Ed.), Springer-Verlag, New York, 1981, pp. 313–325.

Dwyer, H. A., "Calculation of Unsteady Leading-Edge Boundary Layers," *AIAA J.*, Vol. 6, No. 12, pp. 2447–2448, 1968.

Dwyer, H. A., "Solution of a Three-Dimensional Boundary Layer Flow with Separation," *AIAA J.*, Vol. 6, No. 7, pp. 1336–1342, 1968.

Dwyer, H. A., "Some Aspects of Three-Dimensional Laminar Boundary Layers," *Ann. Rev. Fluid Mech.*, Vol. 13, pp. 217–229, 1981.

Eaton, J. K. and Johnson, J. P., "A Review of Research on Subsonic Turbulent Flow Reattachment," *AIAA J.*, Vol. 19, pp. 1093–1100, 1981.

Eckert, E. R. G., Schneider, P. J., Hayday, A. A., and Larson, R. M., "Mass-Transfer Cooling of a Laminar Boundary by Injection of a Light-Weight Foreign Gas," *Jet Propul.*, Vol. 3, pp. 34–39, 1958.

Eggers, J. M., "Turbulent Mixing of Coaxial Compressible Hydrogen-Air Jets," NASA TN D-6487, 1971.

Eichelbrenner, E. A., "Three-dimensional Boundary Layers," *Ann. Rev. Fluid Mech.*, Vol. 5, pp. 339–360, 1973.

Eppler, R. and Fasel, H. (Eds.), *Laminar-Turbulent Transition*, Springer-Verlag, New York, 1980.

Everitt, K. W. and Robins, A. G., "The Development and Structure of Turbulent Plane Jets," *J. Fluid Mech.*, pp. 563–583, 1978.

Favre, A. "Statistical Equations of Turbulent Gases," *Problems of Hydrodynamics and Continuum Mechanics, Sedov 60th Birthday Volume*, SIAM, 1969.

Floryan, J. M. and Saric, W. S., "Stability of Görtler Vortices in Boundary Layers," *AIAA J.*, Vol. 20, No. 3, 1983.

Floryan, J. M., "On the Görtler Vortex Instability of Boundary Layers," *Prog. Aerospace Sci.*, Vol. 28, 1991.

Forstall, W., Jr. and Shapiro, A. H., "Momentum and Mass Transfer in Coaxial Gas Jets," *J. Appl. Mech.*, Vol. 72, pp. 339–408, 1950.

Förthman, E., "Über Turbulente Strahlansbreitung," *Ing-Arch.*, Vol. 5, p. 42, 1934 also NACA TM 789, 1936.

Gadd, G. E., Jones, C. W., and Watson, E. J., *Laminar Boundary Layers*, Rosenhead, L. (Ed.), Oxford, London, p. 331, 1963.

Gapanov, S. A., "The Influence of Flow Non-Parallelism on Disturbances Development in the Supersonic Boundary Layer," *Proc. of the Eighth Canadian Cong. of Appl. Mech.*, 1981.

Gaster, M., "A Simple Device for Preventing Turbulent Contamination on Swept Leading Edges," *J. Royal Aero. Soc.*, Vol. 69, 1965.

Gaster, M., "On the Effect of Boundary Layer Growth in Flow Stability," *J. Fluid Mech.*, Vol. 66, 1974.

Geissler, W., "Three-Dimensional Laminar Boundary Layer over a Body of Revolution at Incidence and with Separation," *AIAA J.*, Vol. 12, pp. 1743–1745, 1974.

Gibson, M. M., "Spectra of Turbulence in a Round Jet," *J. Fluid Mech.*, Vol. 15, pp. 161–173, 1963.

Ginevskii, A. S., "Turbulent Nonisothermal Jets of a Compressible Gas of Variable Composition," *Promyshlennaya Aerodinamika*, No. 27, pp. 31–54, 1966.

Goldstein, M. E., "The Evolution of Tollmien–Schlichting Waves Near a Leading Edge," *J. Fluid Mech.*, Vol. 127, 1983.

Goldstein, M. E., "Scattering of Acoustic Waves into Tollmien–Schlichting Waves by Small Streamwise Variations in Surface Geometry," *J. Fluid Mech.*, Vol. 154, 1985.

Goldstein, M. E. and Hultgren, L. S., "Boundary-Layer Receptivity to Long-Wave Free-Stream Disturbances," *Ann. Rev. Fluid Mech.*, Vol. 21, 1989.

Goldstein, S., "On Laminar Boundary Layer Flow Near a Position of Separation," *Quarterly J. Mech. and Appl. Math.*, Vol. 1, pp. 43–69, 1948.

Görtler, H., "Berechnung von Aufgaben der freien Turbulenz auf Grund eines neuen Näherungsansatzes," *Z. Angew. Math. Mech.*, Vol. 22, pp. 244–254, 1942.

Görtler, H., "On the Three-Dimensional Instability of Laminar Boundary Layers on Concave Walls," NACA, Tech. Mem. 1375, June 1954. (also Ges. d. Wiss. Göttingen, Nachr. a.d. Math., Bd. 2, Nr. 1, 1940).

Govindarajan, R. and Narasimha, R., "Stability of Spatially Developing Boundary Layers in Pressure Gradients," *Proc. Int. Symp. Aero. Fluid Sci.*, Institute of Fluid Science, Tohoku University, Sendai, Japan, 1993.

Gray, W. E., "The Nature of the Boundary Layer Flow at the Nose of a Swept Wing," Royal Aero. Est., Tech. Memo. (Aero) 256, 1952.

Greenough, J., Riley, J., Soertrisno, M., and Eberhardt, D., "The Effects of Walls on a Compressible Mixing Layer," AIAA 89-0372, 1989.

Gregory, N., Stuart, J. T., and Walker, W. S., "On the Stability of Three-Dimensional Boundary Layers with Application to the Flow Due to a Rotating Disk," *Phil. Trans. Roy. Soc. A*, Vol. 248, 1955.

Hall, P. and Horseman, N. J., "The Linear Inviscid Secondary Instability of Longitudinal Vortex Structures in Boundary Layers" *J. Fluid Mech.*, Vol. 232, 1991.

Hall, P., "The Görtler Vortex Instability Mechanism in Three-Dimensional Boundary Layers," *Proc. Roy. Soc. Lond. A*, Vol. 399, 1985.

Hall, P., "The Linear Development of Görtler Vortices in Growing Boundary Layer," *J. Fluid Mech.*, Vol. 30, 1983.

Hall, P., "The Nonlinear Development of Görtler Vortices in Growing Boundary Layers," *J. Fluid Mech.*, Vol. 193, 1988.

Hall, P., Malik, M. R., and Poll, D. I. A., "On the Stability of an Infinite Swept Attachment Line Boundary Layer," *Proc. Roy. Soc. Lond. A*, Vol. 395, 1984.

Han, T. and Patel, V. C., "Flow Separation on a Spheroid at Incidence," *J. Fluid Mech.*, Vol. 92, pp. 643–657, 1979.

Hanjalic, K. and Launder, B. E., "A Reynolds Stress Model of Turbulence and Its Application to Asymmetric Shear Flows," *J. Fluid Mech.*, Vol. 52, p. 609, 1972.

Harsha, P. T., "Prediction of Free Turbulent Mixing Using a Turbulent Kinetic Energy Method," *Free Turbulent Shear Flows*, NASA SP-321, 1973.

Hartnett, J. P. and Eckert, E. R. G., "Mass Transfer Cooling in a Laminar Boundary Layer with Constant Fluid Properties," *Trans. ASME*, Vol. 79, pp. 247–254, 1957.

Herbert, T., "Secondary Instability of Boundary Layers," *Ann. Rev. Fluid Mech.*, Vol. 20, 1988.

Herbert, T., "Secondary Instability of Plane Channel Flow to Subharmonic Three-Dimensional Disturbances," *Phys. Fluids*, Vol. 26, No. 4, 1983.

Hirschfelder, J. O., Curtis, C. F., Bird, R. B., and Spotz, E. L., *The Molecular Theory of Gases and Liquids*, John Wiley & Sons, New York, 1954.

Horstman, C. C. and Hung, C. M., "Computation of Three-Dimensional Turbulent Separated Flows at Supersonic Speeds," *AIAA J.*, Vol. 17, pp. 1155–1156, 1979.

Horton, H. P., "Separating Laminar Boundary Layers with Prescribed Wall Shear," *AIAA J.*, Vol. 12, pp. 1772–1774, 1974.

Hoyt, J. W., "Drag Reduction by Polymers and Surfactants," in *Viscous Drag Reduction in Boundary Layers*, Bushnell, D. M. and Hefner, J. (Eds.), AIAA, New York, 1989.

Hussaini, M. Y. and Voigt, R. G. (Eds.), *Instability and Transition*, Vols. I and II, Springer-Verlag, New York, 1990.

Hussaini, M. Y., Kumar, A., and Streett, C. L. (Eds.), *Instability, Transition, and Turbulence*, Springer-Verlag, New York, 1992.

Ikawa, H. and Kubota, T., "Investigation of Supersonic Turbulent Mixing Layer with Zero Pressure Gradient," *AIAA J.*, Vol. 13, pp. 566–572, 1975.

Inone, O., "MRS Criterion for Flow Separation over Moving Walls," *AIAA J.*, Vol. 19, pp. 1108–1111, 1981.

Joslin, R. D., Streett, C. L., and Chang, C.-L., "3-D Incompressible Spatial Direct Numerical Simulation Code Validation Study—A Comparison with Linear Stability and Parabolic Stability Equation Theories for Boundary-Layer Transition on a Flat Plate," NASA TP-3205, 1992.

Kachanov, Yu. S., Kozlov, V. V., and Levenchko, V. Ya, "Nonlinear Development of a Wave in a Boundary Layer," *Fluid Des.*, Vol. 3, 1977.

Kays, W. M. and Crawford, M. L., *Convective Heat and Mass Transfer*, 2nd ed., McGraw-Hill, New York, 1980.

Kays, W. M. and Moffat, R. J., "The Behavior of Transpired Turbulent Boundary Layers," in *Studies in Convection*, Vol. 1, Academic Press, London, pp. 223–319, 1975.

Kegelman, J. T., Personal Communication, 1994.

Kendall, J. M., "Wind Tunnel Experiments Relating to Supersonic and Hypersonic Boundary-Layer Transition," *AIAA J.*, Vol. 13, No. 3, 1975.

Kitchens, C. W., Jr., Sedney, R., and Gerber, H., "The Role of the Zone of Dependence Concept in Three-Dimensional Boundary Layer Calculations," Report 1821, USA Ballistic Research Laboratories, Aberdeen Proving Ground, MD, 1975.

Kleiser, L. and Zang, T. A., "Numerical Simulation of Transition in Wall-Bounded Shear Flows," *Ann. Rev. Fluid Mech.*, Vol. 23, 1991.

Kline, S. J., "On the Nature of Stall," *J. Basic Eng.*, Vol. 81, pp. 305–320, 1959.

Kline, S. J., Cantwell, B. J., and Lilley, G. M., *The 1980–81 AFOSR-HTTM—Stanford Conference on Complex Turbulent Flows: Comparison of Computation and Experiment*, Stanford Univ., Stanford, CA, 1982.

Kline, S. J., Morkovin, M. V., Sovran, G., Cochrell, D. J. (Eds.), "Computation of Turbulent Boundary Layers—1968," *AFOSR-IFP—Stanford Conference—Proceedings*, Stanford University, Stanford, CA, 1968.

Klineberg, J. M. and Steger, J. L., "On Laminar Boundary Layer Separation," AIAA 74-94, 1974.

Kobayashi, R., Kohama, Y., and Takamadate, Ch., "Spiral Vortices in Boundary Layer Transition Regime on a Rotating Disk," *Acta Mech.*, Vol. 35, 1980.

Kohama, Y. and Motegl, D., "Traveling Disturbances Appearing in Boundary Layer Transition in a Yawed Cylinder," *Exp. Thermal and Fluid Sci.*, Vol. 8, 1994.

Kohama, Y., "Study on Boundary-Layer Transition of a Rotating Disk," *Acta Mechanica*, Vol. 50, 1984.

Kohama, Y., Saric, W. S., and Hoos, J. A., "A High-Frequency, Secondary Instability of Crossflow Vortices that Leads to Transition," *Proc. of The Royal Aero. Soc. Conf. Boundary Layer Transition and Control*, Cambridge, UK, 1991.

Koromilas, C. A. and Telionis, D. P., "Unsteady Separation—An Experimental Study," *J. Fluid Mech.*, Vol. 97, pp. 347–384, 1980.

Kozlov, V. V. (Ed.), *Laminar-Turbulent Transition*, Springer-Verlag, New York, 1985.

Kubota, T. and Dewey, C. F., Jr., "Momentum Integral Methods for the Laminar Free Shear Layer," *AIAA J.*, Vol. 2, pp. 625–629, 1964.

Landahl, M. T., "A Note on Algebraic Instability of Inviscid Parallel Shear Flows," *J. Fluid Mech.*, Vol. 98, pp. 243, 1980.

Landis, F. and Shapiro, A. H., "The Turbulent Mixing of Co-Axial Gas Jets," Heat Transfer and Fluid Mechanics Inst., Reprints and Papers, Stanford University Press, Stanford, CA, 1951.

Launder, B. E. and Spalding, D. B., *Mathematical Models of Turbulence*, Academic Press, New York, 1972.

Launder, B., Morse, A., Rodi, W., and Spalding, D. B., "Prediction of Free Shear Flows—A Comparison of the Performance of Six Turbulence Models," *Free Turbulent Shear Flows*, NASA SP-321, 1973.

Lees, L. and Lin, C. C., "Investigation of the Stability of the Laminar Boundary Layer in a Compressible Fluid," NACA TN 1115, 1946.

Lees, L. and Reshotko, E., "Stability of the Compressible Laminar Boundary Layer," *J. Fluid Mech.*, Vol. 12, Part 4, April 1962.

Liepmann, H. W., "Investigations of Boundary Layer Transition on Concave Walls," NACA Wartime Report, W-87, 1945.

Lighthill, M. J., *Laminar Boundary Layers*, Rosenhead, L. (Ed.), Oxford, London, pp. 1–113, 1963.

Lin, R.-S. and Malik, M. R., "The Stability of Incompressible Attachment-Line Boundary Layers—A 2D-Eigenvalue Approach," AIAA 94-2372, 1994.

Liu, W. and Domaradzki, J. A., "Direct Numerical Simulation of Transition to Turbulence in Görtler Flow," *J. Fluid Mech.*, Vol. 246, 1993.

Lock, R. C., "The Velocity Distribution in the Laminar Boundary Layer between Parallel Streams," *Quart. J. Mech.*, Vol. 4, pp. 42–63, 1951.

Mack, L. M., "Boundary Layer Linear Stability Theory," AGARD Report No. 709, 1984.

Mack, L. M., "Boundary Layer Stability Theory," Rept. 900-277 Rev. A, Jet Propulsion Lab., Pasadena, CA, 1969.

Mack, L. M., "The Wave Pattern Produced by a Point Source on a Rotating Disk," AIAA 85-0490, 1985.

Mack, L. M., "Transition Prediction and Linear Stability Theory," AGARD CP-224, May 1977.

Malik, M. R. and Li, F., "Secondary Instability of Görtler and Crossflow Vortices," *Proc. Int. Symp. Aero. and Fluid Science*, Institute of Fluid Science, Tohoku University, Sendai, Japan, 1993.

Malik, M. R., "Numerical Methods for Hypersonic Boundary Layer Stability," *J. Comp. Physics*, Vol. 86, No. 2, 1990.

Malik, M. R., "Prediction and Control of Transition in Supersonic and Hypersonic Boundary Layers," *AIAA J.*, Vol. 27, No. 11, 1989.

Malik, M. R., Chuang, S., and Hussaini, M. Y., "Accurate Numerical Solution of Compressible Stability Equations," *ZAMP*, Vol. 33, 1982.

Malik, M. R., Li, F., and Chang, C.-L., "Crossflow Disturbances in Three-Dimensional Boundary Layers: Nonlinear Development, Wave Interaction and Secondary Instability," *J. Fluid Mech.*, Vol. 268, 1994.

Masad, J. A. and Iyer, V., "Transition Prediction and Control in Subsonic Flow Over Hump," *Physics of Fluids*, Vol. 6, No. 1, 1994.

Masad, J. A. and Malik, M. R., "Effects of Body Curvature and Nonparallelism on the Stability of Flow Over a Swept Cylinder," *Physics of Fluids*, Vol. 6, No. 7, 1994.

Masad, J. A. and Malik, M. R., "Transition Correlation in Subsonic Flow Over a Flat Plate," *AIAA J.*, Vol. 31, 1993.

Maskell, E. C., "Flow Separation in Three Dimensions," RAE Report, Aero. 2565, Royal Aircraft Establishment, England, 1955.

McCroskey, W. J. and Dwyer, H. A., "Crossflow and Unsteady Boundary-Layer Effects on Rotating Blades," *AIAA J.*, Vol. 9, No. 8, pp. 498–506, 1971.

McCroskey, W. J., "Some Current Research in Unsteady Fluid Dynamics," *J. Fluids Eng.*, Vol. 99, pp. 8–37, 1977.

Messiter, A. F., "Laminar-separation—A Local Asymptotic Flow Description for Constant Pressure Downstream," AGARD Conference Proceedings, No. 160—Flow Separation, pp. 4-1-4-10, 1975.

Mickley, H. S., Ross, R. C., Squyers, A. L., and Stewart, W. E., "Heat, Mass and Momentum Transfer for Flow over a Flat Plate with Blowing and Suction," NACA TN 3208, 1954.

Moore, F. K., "On the Separation of the Unsteady Laminar Boundary Layer," *Boundary Layer Research*, Görtler, H. G. (Ed.), pp. 296–310, Springer, Berlin, 1957.

Moore, F. K., "Three-Dimensional Boundary Layer Theory," *Adv. in Appl. Mech.*, Vol. 4, pp. 159–228, 1956.

Morkovin, M. V., "Bypass Transition to Turbulence and Research Desiderata," *Transition in Turbines*, NASA Conference Publication 2386, 1985.

Morkovin, M. V., "Critical Evaluation of Transition from Laminar to Turbulent Shear Layers with Emphasis on Hypersonically Traveling Bodies," AFFDL-TR-68-149, Wright-Patterson Air Force Base, Ohio, Air Force Flight Dynamics Laboratory, March 1969.

Morkovin, M., "Bypass-Transition Research: Issues and Philosophy," *Instabilities and Turbulence in Engineering Flows*, Ashpis, D. E., Gatski, T. B., and Hirsh, R. (Eds.), Kluwer Academic, The Netherlands, 1993.

Moses, H. L., "The Behavior of Turbulent Boundary Layers in Adverse Pressure Gradients," Ph.D. thesis, Massachusetts Institute of Technology, Cambridge, MA (also MIT Gas Turbine Lab. Report 73), 1964.

Nayfeh, A. H., *Introduction to Perturbation Techniques*, John Wiley & Sons, New York, 1981.

Obremski, H. J., Morkovin, M. V., and Landahl, M., "A Portfolio of Stability Characteristics of Incompressible Boundary Layers," AGARDograph No. 134, Paris.

Orszag, S. A. and Patera, A. T., "Secondary Instability of Wall-Bounded Shear Flows," *J. Fluid Mech.*, Vol. 128, 1983.

Orszag, S. A., "Accurate Solution of the Orr-Sommerfeld Stability Equation," *J. Fluid Mech.*, Vol. 50, 1971.

Owen, P. R. and Randall, D. G., "Boundary Layer Transition on a Swept Back Wing," Royal Aircraft Est., Tech. Memo. (Aero) 256, 1952.

Papamoschou, D. and Roshko, A. A., "Observations of Supersonic Free Shear Layers," AIAA 86-0162, 1986.

Peake, D. J., Rainbird, W. J., and Atragligi, E. G., "Three-Dimensional Flow Separation on Aircraft and Missiles," *AIAA J.*, Vol. 10, pp. 567–580, 1972.

Pfenninger, W., "Flow Phenomena at the Leading Edge of Swept Wings. Recent Developments in Boundary Layer Research—Part IV," AGARDograph 97, 1965.

Pfenninger, W., "Laminar Flow Control—Laminarization," AGARD Report No. 654, 1977.

Pletcher, R. H., "Prediction of Incompressible Turbulent Separating Flow," *J. Fluids Eng.*, Vol. 100, p. 427, 1978.

Poll, D. I. A., "Some Observations of the Transition Process on the Windward Face of a Long Yawed Cylinder," *J. Fluid Mech.*, Vol. 150, 1985.

Poll, D. I. A., "Three-Dimensional Boundary Layer Transition via the Mechanisms of 'Attachment Line Contamination' and 'Cross Flow Instability,'" *Laminar-Turbulent Transition*, Eppler, R. and Fasel, H. (Eds.), Springer-Verlag, New York, 1980.

Prandtl, L., "Bermerkungen zur Theorie der freien Turbulenz," *Z. Angew. Math. Mech.*, Vol. 22, pp. 241–243, 1942.

Rastogi, A. K. and Rodi, W., "Calculation of General Three-Dimensional Turbulent Boundary Layers," *AIAA J.*, Vol. 16, p. 151, 1978.

Rayleigh, Lord, *Theory of Sound*, Vol. II, 2nd ed., Dover, New York, 1945.

Reed, H. L. and Saric, W. S., "Stability of Three-Dimensional Boundary Layers," *Ann. Rev. Fluid Mech.*, Vol. 21, 1989.

Reichardt, H., "Gesetzmassigkeitender freien Turbulenz," *VDI-Forschungsh.*, No. 414 (1942), 2nd ed., 1951.

Reneau, L. R., Johnson, J. P., and Kline, S. J., "Performance and Design of Straight Two-Dimensional Diffuses," *J. Basic Eng.*, Vol. 98, No. 3, pp. 141–150, 1967.

Reshotko, E., "Boundary Layer Stability and Transition," *Ann. Rev. Fluid Mech.*, Vol. 8, 1976.

Reshotko, E., "Stability Theory as a Guide to the Evaluation of Transition Data," *AIAA J.*, Vol. 7, No. 6, 1969.

Rockwell, D. and Naudascher, E., "Review—Self-Sustaining Oscillations of Flow Past Cavities," *J. Fluids Eng.*, Vol. 100, pp. 152–165, 1978.

Rodi, W., *Turbulence Models and Their Application in Hydraulics*, Int. Assoc. Hydraulics Res., Delft, 1980.

Rosenhead, L. (Ed.), *Laminar Boundary Layers*, Oxford University Press, London, 1963.

Roshko, A., "On the Development of Turbulent Wakes from Vortex Streets," NACA Rept. 1191, 1954.

Ross, J. A., Barnes, F. H., Burns, J. G., and Ross, M. A. S., "The Flat Plate Boundary Layer. Part 3. Comparison of Theory with Experiment," *J. Fluid Mech.*, Vol. 43, 1970.

Rotta, J., "A Family of Turbulence Models for Three-Dimensional Thin Shear Layers," *Symposium on Turbulent Shear Flows*, Vol. 1, pp. 10-27–10-34, 1977.

Ruban, A. I., "On Tollmien–Schlichting Wave Generation by Sound," *Laminar-Turbulent Transition*, Kozlov, V. V. (Ed.), Springer-Verlag, New York, 1985.

Rubesin, M. W. and Inouye, M., "Forced Convection-External Flows," *Handbook of Heat Transfer*, Rohsenow, W. M. and Hartnett, J. P. (Eds.), McGraw-Hill, New York, 1973.

Saric, W. S., Reed, H. L., and Kerschen, E. J., "Leading Edge Receptivity to Sound: Experiments, DNS, and Theory," AIAA 94-2222, 1994.

Schetz, J. A., "Free Turbulent Mixing in a Co-Flowing Stream," *Free Turbulent Shear Flows*, NASA SP-321, 1973.

Schetz, J. A., "Some Studies of the Turbulent Wake Problem," *Astronaut. Acta*, Vol. 16, pp. 107–117, 1971.

Schetz, J. A., "Turbulent Mixing of a Jet in a Coflowing Stream," *AIAA J.*, Vol. 6, pp. 2008–2010, 1968.

Schetz, J. A., *Boundary Layer Analysis*, Prentice Hall, Englewood Cliffs, NJ, 1994.

Schetz, J. A., *Injection and Mixing in Turbulent Flow*, AIAA, New York, 1980.

Schlichting, H., "Über das ebene Windschattenproblem," Diss., Göttingen, 1930; *Ing.-Arch.*, Vol. 1, pp. 533–571, 1930.

Schlichting, H., "Laminare Strahlausbreitung," *Z. Angew. Math. Mech.*, Vol. 13, pp. 260–263, 1933.

Schlichting, H., *Boundary Layer Theory*, 7th ed. (translated by Kestin, J.), McGraw-Hill, New York, 1979.

Schneider, W., "Decay of Momentum Flux in Submerged Jets," *J. Fluid Mech.*, Vol. 154, pp. 91–110, 1985.

Schubauer, G. B. and Skramstad, H. K., "Laminar Boundary-Layer Oscillations and Transition on a Flat Plate," *J. Res. Natl. Bur. Stand.*, Vol. 38, 1947.

Shen, S. F., "Unsteady Separation According to the Boundary Layer Equation," *Adv. Appl. Mech.*, Academic Press, Vol. 18, pp. 177–220, 1978.

Simpson, R. L., "A Review of Some Phenomenon of Turbulent Flow," *J. Fluid Eng.*, Vol. 103, pp. 520–533, 1981.

Simpson, R. L., Strickland, J. H., and Barr, P. W., "Laser and Hot Film Anemometer Measurements in a Separating Turbulent Boundary Layer," Tech. Rep. WT-3, Southern Methodist University, Dallas, TX, 1974.

Skelland, A. H. P., "Momentum, Heat, and Mass Transfer in Turbulent Non-Newtonian Boundary Layers," *AIChE J.*, Vol. 12, pp. 69–75, 1966.

Smith, A. M. O. and Gamberoni, N., "Transition, Pressure Gradient, and Stability Theory," Douglas Aircraft Company, Inc., Report ES 26388, 1956.

Smith, A. M. O., "On the Growth of Taylor–Görtler Vortices Along Highly Concave Walls," *Quarterly Applied Math.*, Vol. XIII, No. 3, 1955.

Smith, F. T., "On the Non-Parallel Flow Stability of the Blasius Boundary Layer," *Proc. Roy. Soc. Lond. A.*, Vol. 366, 1979.

Smith, J. H. B., "A Review of Separation in Steady, Three-Dimensional Flow," AGARD Conference Proceedings, No. 168, *Flow Separation*, pp. 31-1–31-17, 1975.

Smythe, R., "Turbulent Flow over a Plane Symmetric Sudden Expansion," *J. Fluids Eng.*, Vol. 103, pp. 348–353, 1979.

Spall, R. E. and Malik, M. R., "Goertler Vortices in Supersonic and Hypersonic Boundary Layers," *Phys. Fluids A*, Vol. 1, No. 11, 1989.

Stetson, K. F., Thompson, E. R., Donaldson, J. C., and Siler, L. G., "Laminar Boundary Layer Stability Experiments on a Cone at Mach 8. Part 1: Sharp Cone," AIAA-83-1761, 1983.

Stewartson, K., *The Theory of Laminar Boundary Layers in Compressible Fluids*, Oxford University Press, Oxford, 1964.

Stratford, B. S., "The Prediction of the Turbulent Boundary Layer," *J. Fluid Mech.*, Vol. 5, pp. 1–16, 1955.

Svehla, R. A. and McBride, B. J., "FORTRAN IV Computer Program for Calculation of Thermodynamic and Transport Properties of Complex Chemical Systems," NASA TN D-7056, 1973.

Swearingen, J. G. and Blackwelder, R. F., "The Growth and Breakdown of Streamwise Vortices in the Presence of a Wall," *J. Fluid Mech.*, Vol. 182, 1987.

Sychev, V. V., "Laminar Separation," *Mekh. Fluid in Gaza*, No. 3, pp. 47–59, 1972.

Taneda, S., "Visual Study of Unsteady Separated Flows Around Bodies," *Progress in Aero. Sci.*, Pergamon Press, Vol. 17, pp. 287–348, 1977.

Telionis, D. P., "Unsteady Boundary Layers Separated and Attached," *J. Fluids Eng.*, Vol. 101, pp. 29–43, 1979.

Theodorsen, T. and Garrick, I. E., "General Potential Theory of Arbitrary Wing Section," NACA TR452, 1933.

Thwaites, B., *Incompressible Aerodynamics*, Oxford Press, Oxford, 1960.

Tobak, M. and Peake, D. J., "Topology of Three-Dimensional Separated Flows," *Ann. Rev. Fluid Mech.*, Vol. 14, pp. 61–85, 1982.

Tollmien, W., "Berechnung turbulenter Ausbreitungsvorgänge," *Z. Angew. Math. Mech.*, Vol. 6, pp. 468–478, 1926.

Tollmien, W., "Uber die Entstehung der Turbulenz," *Nachr. Ges. Wiss. Göttingen. Math.-Phys. Klasse,* 1929.

Trefethern, L. N., Trefethern, A. E., Reddy, S. C., and Driscoll, T. A., "Hydrodynamic Stability without Eigenvalues," *Science*, Vol. 261, 1993.

Tsahalis, D. T. and Telionis, D. P., "Laminar Boundary Layer Separation from an Upstream Moving Wall," *AIAA J.*, Vol. 15, pp. 561–566, 1977.

Van Driest, E. R., "Turbulent Boundary Layer in Compressible Fluids," *J. Aero. Sci.*, Vol. 18, p. 145, 1951.

Van Driest, E. R., "Investigation of Laminar Boundary Layer Compressible Fluids Using the Crocco Method," NACA TN 2597, 1952.

van Ingen, J. L., "A Suggested Semi-Empirical Method for the Calculation of the Boundary Layer Transition Region," University of Techn., Dept. of Aero. Eng., Rept. UTH-74, Delft, 1956.

Wagner, R. D., "Mean Flow and Turbulence Measurements in a Mach 5 Shear Layer," NASA TN D-7366, 1973.

Wang, K. C., "Boundary Layer over a Blunt Body at High Incidence with an Open Type Separation," *Proc. Roy. Soc.* (London), Series A. Vol. 340, pp. 33–35, 1974.

Wang, K. C., "Separating Patterns of Boundary Layer over an Inclined Body of Revolution," *AIAA J.*, Vol. 10, pp. 1044–1050, 1972.

Wazzan, A. R., Okamura, T. T., and Smith, A. M. O., "Stability of Laminar Boundary Layers at Separation," *Phys. Fluids*, Vol. 10, 1967.

Weinstein, A. S., Osterle, J. F., and Forstall, W., "Momentum Diffusion from a Slot Jet into a Moving Secondary," *J. Appl. Mech.*, pp. 437–443, 1956.

White, F. M., *Viscous Fluid Flow*, 2nd ed. McGraw-Hill, New York, 1993.

Wilkinson, S. P. and Malik, M. R., "Stability Experiments in the Flow over a Rotating Disk," *AIAA J.*, Vol. 23, 1985.

Williams, J. C. III, "Semi-Similar Solutions to the Three-Dimensional Boundary Layer," *Appl. Sci. Res.*, Vol. 31, pp. 161–186, 1975.

Williams, J. C. III and Johnson, W. D., "Note on Unsteady Boundary Layer Separation," *AIAA J.*, Vol. 12, pp. 1427–1429, 1974.

Williams, J. C. III, "Incompressible Boundary Layer Separation," *Ann. Rev. Fluid Mech.*, Vol. 9, pp. 113–144, 1977.

Zurigat, Y. H. and Malik, M. R., "Effect of Crossflow on Görtler Instability," presented at ICASE LaRC Workshop on Transition, Turbulence, and Combustion, 1993.

5 Internal Flows

DR. BHARATAN R. PATEL
Fluent Incorporated
Lebanon, NH

CONTENTS

5.1 INTRODUCTION

Internal flows are flows bounded by stationary or moving walls. At one end of the spectrum of internal flows lie the laminar fully developed flows in constant area ducts which can be analyzed with relative ease. At the other end lie developing turbulent flows in curved and rotating ducts such as the blade passages of a centrifugal compressor, where basic knowledge is far from complete. In between lie the myriad of commonplace, yet challenging, internal flow situations where the practicing fluids engineer must rely on a blend of analysis and empiricism to get on with the task at hand. Understandably therefore, this section begins with the simpler internal flows such as flows in constant area ducts (Sec. 5.2) and builds on the under-

Handbook of Fluid Dynamics and Fluid Machinery, Edited by Joseph A. Schetz and Allen E. Fuhs
ISBN 0-471-12598-9 Copyright © 1996 John Wiley & Sons, Inc.

standing and information available to describe more complex internal flows. Section 5.3 deals with the effects of curvature and rotation on the wall shear layers. Ducts with varying cross-sectional area are discussed in Sec. 5.4. Sustained wall curvature gives rise to secondary flows, and these are discussed in Sec. 5.5. Finally, Sec. 5.6 describes separated internal flows where, in addition to the shear layers at the duct walls, *free shear layers* are generated in the core flow due to separation or injection of a secondary fluid stream. Given the rather wide ranging scope of this chapter, the discussion is limited to incompressible flows and physical geometries that are representative of a given class of internal flows.

5.2   FLOWS IN STRAIGHT CONSTANT AREA DUCTS

Consider a straight long duct of constant area as shown in Fig. 5.1. The flow enters the duct from a large plenum such that the velocity profile at the inlet is uniform. In reality, a close approximation to this ideal case can be obtained by providing a well designed convergent nozzle at the inlet. The fluid near the wall is retarded due to viscous forces, and a boundary layer develops along the duct walls. Near the entrance of the duct, the boundary layer is thin compared to the height of the duct, and the development of the boundary layer in this region can be viewed in the same way as that for a boundary layer developing on a plate in an infinite flow. As the boundary layer grows, a greater fraction of the flow is retarded by the viscous effects and develops a velocity deficit with respect to the inviscid core. From mass conservation this implies that the velocity in the inviscid core increases. Since the total pressure in the inviscid core is a constant and equal to the stagnation pressure in the plenum, the increase in the core velocity results in a drop in the static pressure in the flow direction. In this zone where the boundary layer affects the core flow, the development of the boundary layer is no longer similar to that on a flat plate in an infinite flow. Therefore, flow calculation in this zone must account for the interactions between the core flow and the boundary layer. Note that due to the fact that an inviscid core still exists in this zone, coupled potential flow/boundary layer computational methods can still be used.

At some point along the duct axis the boundary layer will have grown to the centerline of the duct and the inviscid core is nonexistent. The boundary layers from opposite walls will then start to interact with each other. Finally, at some point downstream, the flow becomes *fully developed*, that is, the velocity profile and pressure gradient along the duct axis remain constant from this point onwards. Note that for turbulent flows, fully developed flow implies that not only the mean velocity profile, but the profiles of the various statistical turbulence properties must also attain equilibrium values that remain unchanged in the downstream direction. For both laminar and turbulent flows, fully developed flow is attained considerably downstream of the point where the boundary layers from opposite walls start to interact with each other.

5.2.1   Developing Duct Flows

The flow in the portion of the duct before fully developed flow is attained is called developing flow. This portion of the flow has also been referred to in the literature as *inlet* or *entrance flow*.

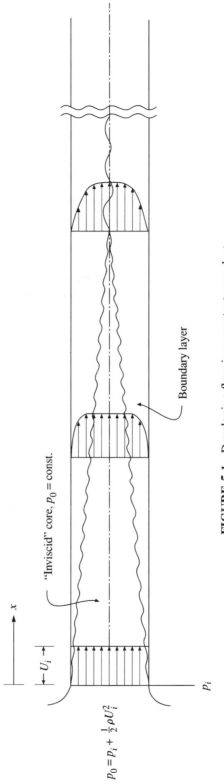

FIGURE 5.1 Developing flow in a constant area duct.

485

Laminar Developing Flow. The developing incompressible laminar flow in the entrance region of a duct has been the subject of many investigations in the past. An exhaustive review of these investigations is given by Ward–Smith (1980) and Fargie and Martin (1971). There are several experimental investigations of note on developing flows in ducts. Shapiro *et al.* (1954) measured friction factors in the entrance region of laminar flows in smooth tubes. More recently, Fargie and Martin (1971) also presented data on developing flows in tubes. Mohanty and Asthana (1977) measured the developing flow in a pipe at Reynolds numbers of 1875, 2500, and 3250. They present velocity profiles and pressure drop data. Experimental data for circular pipes at low Reynolds number (<300) are presented by Atkinson *et al.* (1969). Beavers *et al.* (1970) performed a careful set of experiments for developing flows in rectangular ducts. They covered a wide range of duct aspect ratios, and the data for pressure drop were obtained for Reynolds numbers of 500–3000. Data on velocity profiles are not presented, however.

Entrance Length. The length of the duct from the inlet to the point where the flow reaches the fully developed velocity profile is generally known as the *entrance length*. In theory, the fully developed profile is reached in an asymptotic manner. Therefore, for practical applications, the flow is considered fully developed when the velocity profile is 99% of the fully developed profile. Based on the analytical and experimental investigations available to date, the best estimate of the entrance length x_e for laminar flow in circular pipes is

$$x_e/D = 0.6 + 0.056 \, \text{Re} \qquad (5.1)$$

where D is the pipe diameter, and Re is the Reynolds number based on the pipe diameter. It should be noted that the above correlation is valid for the case where the inlet velocity profile is uniform, which might not be the case in many practical applications. Sparrow and Anderson (1977) show that the effect of ducting upstream of the pipe inlet can be quite significant at very low Reynolds numbers (<100), and the entrance length can be reduced by greater than a factor of two from that given by the above equation.

The entrance length for rectangular ducts can be estimated from the following expressions

$$x_e/D_h = A + 0.03 \, \text{Re, for aspect ratio} <2 \qquad (5.2)$$

and,

$$x_e/D_h = A + 0.015 \, \text{Re, for aspect ratio} \geq 2. \qquad (5.3)$$

In the above expressions, D_h is the hydraulic diameter and Re is the Reynolds number based on the hydraulic diameter. The constant A is the value of the entrance length in the limit where the Reynolds number goes to zero, that is, for creeping flow. The value of A for a two-dimensional channel (infinite aspect ratio) is 0.25. Corresponding values for ducts of finite aspect ratios are not available.

Pressure Drop. To obtain insight into the pressure drop in developing flows, consider the momentum balance between the duct inlet and a plane which is a distance

x from the inlet (the duct is assumed to be circular)

$$\pi R^2 \, (p_i - p_x) = \int_0^R \rho \, (U_x^2 - U_i^2) \, 2\pi r \, dr + \int_0^x \tau_w \, 2\pi R \, dx \qquad (5.4)$$

where p is the static pressure, U, the velocity, ρ, the density, R, the radius of the pipe, and τ_w, the shear stress at the pipe wall. The subscripts i and x refer to the inlet and the exit planes, respectively. The first term on the right-hand side represents the pressure drop required to accelerate the flow as the viscosity effects at the wall cause a *blockage*. The second term on the right-hand side is the pressure drop caused by the shear stress at the wall. Equation (5.4) can be recast into the following form

$$\frac{p_i - p_x}{\frac{1}{2}\rho U_i^2} = 4f \frac{x}{D} + K \qquad (5.5)$$

where, f is the *Fanning friction coefficient* defined as $(\tau_d)/(\frac{1}{2}\rho U_i^2)$, τ_d is the shear stress in the fully developed flow regime, and K is the excess pressure drop due to the developing flow given by

$$K = \frac{4}{\rho U_i^2} \int_0^1 (U_x^2 - U_i^2) \left(\frac{r}{R}\right) d\left(\frac{r}{R}\right) + \frac{8}{\rho U_i^2} \int_0^{x/D} (\tau_w - \tau_d) \, d\left(\frac{x}{D}\right) \qquad (5.6)$$

Again, the first term in the right-hand side of Eq. (5.6) represents the pressure drop associated with accelerating the core flow; the second term on the right-hand side is due to the fact the wall shear stress in the developing flow is higher than that for a fully developed flow. Shah (1978) has correlated extensive pressure drop data for laminar developing flows in circular, rectangular, annular, and triangular ducts. Based on the pressure drop correlations presented by Shah (1978), the following correlation for K is obtained

$$K(x) = \frac{K(\infty) + 13.74 \, (C/(x^+)^{3/2}) - 4f \, \mathrm{Re} \, (C/(x^+))}{(1 + (C/(x^+)^2))} \qquad (5.7)$$

where $x^+ = (x/D_h)\mathrm{Re}$, D_h is the hydraulic diameter, f is the Fanning friction coefficient, and C and $K(\infty)$ are empirically derived coefficients whose values are listed in Table 5.1.

Turbulent Developing Flow. The developing turbulent flow in constant area ducts is far more complex than that for the laminar case discussed in the previous section. If it is assumed that the flow to the duct is from a quiescent plenum through a smooth and well designed nozzle, the initial boundary layer will be laminar and will transition to turbulent some distance downstream of the inlet. Available data on transition in developing pipe flow have been crossplotted and analyzed by Klein (1981) in his comprehensive review of developing turbulent flows in pipes. For naturally developing flows, it is found that the lower and upper transitional Reynolds numbers (based on length from inlet) are 6×10^5 and 1.85×10^6. That is, for a flow Rey-

TABLE 5.1 $K(\infty)$, $f(\text{Re})$, and C to be used in Eq. (5.7) for Rectangular Triangular, and Concentric Annular Ducts, after Shah (1978)

	$K(\infty)$	$f(\text{Re})$	C
α^*	Rectangular Ducts		
1.00	1.43	14.227	0.00029
0.50	1.28	15.548	0.00021
0.20	0.931	19.071	0.000076
0.00	0.674	24.000	0.000029
ϕ	Equilateral Triangular Duct		
60°	1.69	13.333	0.00053
R_i/R_o	Concentric Annular Ducts		
0	1.25	16.000	0.000212
0.05	0.830	21.567	0.000050
0.10	0.784	22.343	0.000043
0.50	0.688	23.813	0.000032
0.75	0.678	22.967	0.000030
1.00	0.674	24.000	0.000029

α^* = short side/long side
ϕ = apex angle

nolds number Re (based on the pipe diameter) of 1×10^6, a fully turbulent boundary layer will be established in 1.85 pipe diameters downstream of the inlet. It should be kept in mind that the above values for the transitional Reynolds numbers are for a carefully designed inlet to the pipe, and therefore, for most practical applications, transition will occur in shorter distances.

Virtually all analytical and experimental investigations of developing turbulent flow in a pipe address the case where the boundary layer is turbulent from the pipe inlet. The majority of the analytical investigations assume that the developing boundary layers in the pipe have a velocity distribution that is similar to that in the fully developed pipe flow. Some form of empirical relation is used for the shear stress at the wall. A brief survey of such analytical investigations is presented by Wang and Tullis (1974). Comparisons with experimental data show that such analytical methods are unable to satisfactorily predict the entire flow development and miss some of the key physical features of the developing flow.

A careful experimental investigation of developing turbulent flow in a smooth pipe was performed by Barbin and Jones (1963). They found that fully developed turbulent flow was not achieved in 45 diameters and that the centerline velocity was in fact higher than for the fully developed condition. However, the values of the wall shear and, hence, the static pressure gradient achieved the fully developed values in about 15 diameters. The velocity profiles in the developing inlet boundary layers were not similar and differed from the fully developed profiles. The boundary layers reached the centerline of the pipe at approximately 28 diameters, and the velocity profile at that point was significantly different from the fully developed profile. The centerline velocity was larger by about 4% than that for the fully developed profile, whereas the velocity at radial locations closer to the wall was lower by as much as 10%.

This behavior of developing turbulent flow has been observed by other investigators as well. Klein (1981) in his survey of developing turbulent pipe flows found that in experiments where care was taken to prevent disturbances at the pipe inlet, the centerline velocity always shows a peak at approximately 40 diameters and then gradually decays to the fully developed value. This behavior of the developing turbulent flow is due to the interactions between the boundary layers from opposite sides of the pipe as they merge. Simple computational methods mentioned earlier would not be able to predict the complex interactions at the merger of the boundary layers. Bradshaw, Dean, and McEligot (1973) successfully predicted the boundary layer interactions described above by assuming that the turbulent fields of the two boundary layers can be directly superimposed. That is, the shear stress gradient in the overlap region between the two boundary layers is equal to the sum of the shear stress gradients for the individual boundary layers as if the boundary layers were developing in the absence of each other.

The typical centerline velocity development discussed above is not always observed. Klein (1981) in his analysis of developing pipe flow cites several experimental investigations where the centerline velocity shows a monotonic increase to the fully developed value. Klein's explanation is that such a monotonic behavior of the centerline velocity is observed if significant flow perturbations are present at the inlet of the pipe. He further observed that the experimental data analyzed showed a larger than expected amount of scatter indicating that the developing turbulent flow is remarkably sensitive to disturbances at the inlet, and it is very difficult to obtain truly axisymmetric flow in the inlet region of pipes. Patel (1974) found that it took as many as 140 pipe diameters before the fully developed turbulent flow was independent of inlet conditions.

Several investigations have been performed on developing turbulent flow in rectangular ducts [see Gessner et al. (1979) and Melling and Whitelaw (1976)]. The main difference between developing flow in circular and rectangular ducts is the development of secondary flows in the corners. Such secondary flows will be discussed later in this chapter. As with circular ducts, developing turbulent flow in rectangular ducts is very sensitive to inlet conditions, and the effect of inlet disturbances can be observed at a considerable distance downstream of the inlet [Melling and Whitelaw (1976)].

Entrance Length. In contrast to the case of laminar developing flows, there is no clear consensus on the *entrance length* for developing turbulent flow in ducts. If one defines the entrance length as the distance over which the wall shear stress and, hence, the pressure gradient reaches the fully developed flow value, then the entrance length is about 15 pipe diameters. On the other hand, if the entrance length is defined as the length required so that the fully developed profiles of the mean and turbulence parameters are independent of the pipe inlet conditions, then the entrance length can exceed 140 pipe diameters.

Pressure Drop. Following the definition of Eq. (5.5), the excess pressure loss in the developing flow region is the sum of the pressure drop assuming a fully developed flow over the entry length and a constant K that represents the additional pressure drop due to the flow acceleration and the excess shear stress in the entrance region. Note that the entrance length used here is the one required to establish the fully developed wall shear which is approximately $L/D = 15$. The value of K is nearly 0.04 for a smooth pipe with a well rounded inlet and Re $> 10^5$. The value of K for a *fully rough* pipe is two to three times higher than that for a smooth pipe.

It is interesting to note that the value of K for turbulent developing flows is an order of magnitude lower than that for developing laminar flows. The reason for this difference becomes evident when the two terms in Eq. (5.6) that make up K are examined. The first term on the right-hand side of Eq. (5.6) represents the pressure drop due to acceleration of the flow to the fully developed profile. The turbulent fully developed flow profile is much flatter than that for the laminar case. This means that the pressure drop to accelerate the flow to the fully developed profile will be significantly lower for turbulent flow than for the laminar case. For example, assuming a 1/8th power law for the fully developed turbulent velocity profile, the value of the acceleration term on the right-hand side of Eq. (5.6) is 0.03. The corresponding value of this term for a fully developed laminar flow in a pipe is 0.67. The second term on the right-hand side of Eq. (5.6), which represents the pressure drop due to the excess wall shear in a developing flow, is also lower than that for laminar developing flow; the difference between the fully developed value of the wall shear stress and the actual wall shear stress in developing turbulent flow is smaller than that for the laminar case.

5.2.2 Fully Developed Flows

Fully developed flow marks the end of the *entrance* or developing flow region in a constant area duct. In the fully developed state, the cross stream profiles of the mean velocity and turbulence quantities do not vary along the duct axis; the pressure gradient (pressure drop per unit length) is constant. Further, the cross stream profiles and the pressure gradients are independent of the inlet conditions such as the inlet geometry or flow disturbances.

As seen from the previous section, except at very low Reynolds numbers, a substantial length of duct from the inlet is required to achieve fully developed flow. In practical applications, therefore, fully developed flow is seldom encountered except in long straight pipes and ducts such as oil and gas pipelines or shell and tube heat exchangers with long straight tubes. Nonetheless, fully developed flows are one of the most extensively studied class of flows. This is not surprising given that the invariance of the flow properties with distance along the duct axis leads to considerable simplification of the governing equations. Exact solutions for most commonly encountered duct cross section shapes can be derived for fully developed laminar flows, and semi-empirical solutions are available for their turbulent counterparts. The following subsections cover fully developed laminar and turbulent flows in straight ducts of constant cross section.

Fully Developed Laminar Flows. Consider a straight duct of arbitrary but constant cross section. The duct axis and the flow direction are along the x axis, and the flow is incompressible and laminar. Consider a point in the duct, where the flow is fully developed, that is, the profiles of the u, v, and w velocities are no longer a function of the location in the streamwise direction x. Stated another way, all derivatives of the velocities with respect to x are zero (i.e., $\partial/\partial x = 0$).

A further simplification can be made by using the fact that in fully developed laminar flows $v = w = 0$. The resulting equations for fully developed laminar flow are

$$\frac{\partial^2 u}{\partial y^2} + \frac{\partial^2 u}{\partial z^2} = \frac{1}{\mu}\frac{\partial p}{\partial x} \tag{5.8}$$

$$\partial p/\partial y = 0 \tag{5.9}$$

$$\partial p/\partial z = 0 \tag{5.10}$$

From Eqs. (5.9) and (5.10) it can be concluded that the pressure p is not a function of the cross stream directions y and z. That is, the pressure is constant across any given cross section of the duct. Further, since the left hand side of Eq. (5.8) is a function of y and z only, and the right-hand side a function of x only, both the right-hand side and left-hand side must be equal to a constant and, therefore, $dp/dx =$ constant.

This conclusion can also be derived by performing a simple momentum balance. Consider the momentum balance between two planes a distance Δx apart and perpendicular to the flow direction x. The profiles of the axial velocity u at both planes are identical in fully developed flow, therefore the net change in momentum is zero. Hence, the force due to the pressure difference between the two cross sections is exactly balanced by the force due to the shear stress τ_w at the wall

$$(p_1 - p_2) A + \Delta x \int_P \tau_w \, ds = 0 \tag{5.11}$$

where p_1 and p_2 are the respective pressures at the upstream and downstream planes, A is the cross-sectional area, and P is the perimeter. Since the axial velocity profiles are identical at the two sections,

$$\int_P \tau_w \, ds = \text{constant} \equiv \tau_m P \tag{5.12}$$

where τ_m is the average shear stress around the perimeter. Therefore,

$$\frac{p_1 - p_2}{\Delta x} = \frac{dp}{dx} = \tau_m \frac{P}{A} = \text{constant} \tag{5.13}$$

The governing equation for steady, laminar, incompressible, fully developed flows in straight ducts of arbitrary cross section is

$$\nabla^2 u = \frac{1}{\mu}\frac{dp}{dx} \tag{5.14}$$

Using a characteristic length L and a characteristic velocity $(L^2/\mu)\,(dp/dx)$, Eq. (5.14) in dimensionless form is

$$\nabla^2 u^* = -1 \tag{5.15}$$

The boundary conditions are that the velocity is zero at the duct walls (the nonslip condition for a viscous fluid). Equation (5.15) is the well-known *Poisson equation*, and solution techniques for this equation can be found in standard text books on applied mathematics.

The fact that led to the simplifications of the nonlinear Navier–Stokes equations to the linear Poisson equation was that the cross stream velocity components are zero in fully developed laminar flows. As will be seen later, this is not the case in general for fully developed turbulent flows in ducts with arbitrary cross sections. In fact, in turbulent flows, cross stream components are zero only for very few cross section shapes such as axisymmetric (circular) ducts. Given this situation, it is not entirely satisfactory to simply accept the fact that cross stream velocities are zero for fully developed laminar flows. Unfortunately, the Navier–Stokes equations do not shed any light on the behavior of the cross stream velocity components v and w. The presence of cross stream velocity components implies the presence of the vorticity component along the flow direction. Let us therefore examine the vorticity transport equation for the x component of the vorticity

$$u \frac{\partial \Omega_x}{\partial x} + v \frac{\partial \Omega_x}{\partial y} + w \frac{\partial \Omega_x}{\partial z}$$

$$= \Omega_x \frac{\partial u}{\partial x} + \Omega_y \frac{\partial u}{\partial y} + \Omega_z \frac{\partial u}{\partial z} + v \left(\frac{\partial^2 \Omega_x}{\partial x^2} + \frac{\partial^2 \Omega_x}{\partial y^2} + \frac{\partial^2 \Omega_x}{\partial z^2} \right) \quad (5.16)$$

This equation is obtained by taking the curl of the Navier–Stokes equation. In the above equation, Ω_x, Ω_y, and Ω_z are the components of vorticity in the x, y, and z directions, respectively, defined as

$$\Omega_x = \frac{1}{2} \left(\frac{\partial w}{\partial y} - \frac{\partial v}{\partial z} \right), \ \Omega_y = \frac{1}{2} \left(\frac{\partial u}{\partial z} - \frac{\partial w}{\partial x} \right), \ \Omega_z = \frac{1}{2} \left(\frac{\partial v}{\partial x} - \frac{\partial u}{\partial y} \right) \quad (5.17)$$

Using the property of fully developed flows that $\partial/\partial x = 0$, Eqs. (5.16) and (5.17) become

$$v \frac{\partial \Omega_x}{\partial y} + w \frac{\partial \Omega_x}{\partial z} = \Omega_y \frac{\partial u}{\partial y} + \Omega_z \frac{\partial u}{\partial z} + v \left(\frac{\partial^2 \Omega_x}{\partial y^2} + \frac{\partial^2 \Omega_x}{\partial z^2} \right) \quad (5.18)$$

and

$$\Omega_x = \frac{1}{2} \left(\frac{\partial w}{\partial y} - \frac{\partial v}{\partial z} \right)$$

$$\Omega_y = \frac{1}{2} (\partial u/\partial z) \quad (5.19)$$

$$\Omega_z = -\frac{1}{2} (\partial u/\partial y)$$

Using the relations for Ω_y and Ω_z from Eq. (5.19), it is seen that the first two terms on the right-hand side of Eq. (5.18) cancel out giving

$$v \frac{\partial \Omega_x}{\partial y} + w \frac{\partial \Omega_x}{\partial z} = v \left(\frac{\partial^2 \Omega_x}{\partial y^2} + \frac{\partial^2 \Omega_x}{\partial z^2} \right) \tag{5.18a}$$

The terms on the left-hand side represent the convection of vorticity, whereas the terms on the right-hand side represent the dissipation of vorticity due to viscosity. Note that there is no source term that creates vorticity on the right-hand side. This means that any streamwise vorticity Ω_x, created by a disturbance at the inlet for example, is continually dissipated by viscosity and eventually Ω_x becomes zero in the fully developed flow region. It is then straightforward to show that $\Omega_x = 0$ implies $v = w = 0$.

Having demonstrated that the cross stream velocity components are zero in fully developed flow, we can confidently proceed to derive solutions to Eq. (5.14) for fully developed laminar flow in ducts. The circular pipe is by far the most common duct shape encountered in practice. Writing Eq. (5.14) in cylindrical coordinates (r, θ, x), making use of the symmetry of a circular cross section, and setting $\partial / \partial \theta = 0$

$$\frac{1}{r} \frac{\partial}{\partial r} \left(r \frac{\partial u}{\partial r} \right) = \frac{1}{\mu} \frac{dp}{dx} \tag{5.20}$$

i.e., $u = u(r)$.

Integrating the above equation once

$$\frac{du}{dr} = \frac{1}{\mu} \frac{dp}{dx} \frac{r}{2} + \frac{B}{r} \tag{5.21}$$

Integrating once more

$$u = \frac{1}{4\mu} \frac{dp}{dx} r^2 + B \ln r + C \tag{5.22}$$

The above equation is the exact solution for fully developed laminar flow in a circular pipe. The only task remaining is to evaluate the constants B and C using the boundary conditions that the velocity be finite at the pipe center ($r = 0$), and zero at the pipe walls ($r = R$). The first boundary condition requires that $B = 0$. The second boundary condition gives

$$C = -\frac{1}{4\mu} \frac{dp}{dx} R^2 \tag{5.23}$$

Substituting the values of the constants B and C into Eqs. (5.21) and (5.22)

$$\frac{du}{dr} = \frac{1}{2\mu} \frac{dp}{dx} r \tag{5.24}$$

and

$$u = -\frac{1}{4\mu} \frac{dp}{dx} R^2 \left[1 - \left(\frac{r}{R} \right)^2 \right] \tag{5.25}$$

The preceding equations are the solutions for steady, incompressible, fully-developed, laminar flow in circular pipes. The velocity profile is parabolic with zero slope at the centerline. The next step is to relate the solution to parameters such as the flow rate through the pipe and the Darcy or Fanning friction factors.

The flow rate through the pipe is obtained by

$$Q = \int_0^R u \, 2\pi r \, dr = -\frac{\pi}{8\mu} \frac{dp}{dx} R^4 \tag{5.26}$$

The average velocity through the pipe is

$$u_m = \frac{Q}{\pi R^2} = -\frac{1}{8\mu} \frac{dp}{dx} R^2 = \frac{u_c}{2} \tag{5.27}$$

where u_c is the maximum velocity at the centerline. Also,

$$\frac{dp}{dx} = -\frac{8\mu}{R^2} u_m \tag{5.28}$$

This equation relates the average velocity to the pressure gradient and shows that the pressure drop in laminar flow is directly proportional to the mean velocity. The shear stress at the wall can be obtained from the velocity gradient in Eq. (5.24) evaluated at the pipe wall

$$\tau_w = -\mu \left. \frac{du}{dr} \right|_{r=R} = -\frac{dp}{dx} \frac{R}{2} \tag{5.29}$$

or, since the pressure gradient is a constant,

$$(p_1 - p_2) = 4\tau_w \, (L/D) \tag{5.30}$$

where p_1 and p_2 are pressures at two points a distance L apart along the axis of the pipe. Traditionally, the shear stress has been defined in terms of the *Darcy* or *Fanning friction factors*

$$\text{Darcy friction factor: } \lambda = 8\tau_w/\rho u_m^2$$

$$\text{Fanning friction factor: } f = \tau_w/\tfrac{1}{2}\rho u_m^2 \tag{5.31}$$

and $\lambda = 4f$.

Note that the Fanning friction factor is also known as the *skin friction* factor. The pressure drop can be written in terms of these friction factors

$$(p_1 - p_2) = \lambda \frac{L}{D} \left(\frac{1}{2} \rho u_m^2 \right) = 4f \frac{L}{D} \left(\frac{1}{2} \rho u_m^2 \right) \tag{5.32}$$

Using Eqs. (5.28) and (5.29), the friction factors for a circular pipe can be expressed in terms of the Reynolds number based on the pipe diameter as follows

$$\lambda = 64/\text{Re} \tag{5.33}$$

$$f = 16/\text{Re}$$

Note that Eq. (5.20) also describes the fully developed flow in the annulus between concentric cylinders. Therefore, the expression for the velocity given in Eq. (5.24) is also valid for this case except that the constants of integration must now be evaluated using the boundary conditions: $u = 0$ at $r = R_i$, $u = 0$ at $r = R_o$; R_i and R_0 are the inner and outer radii of the annulus, respectively. Using these boundary conditions, the velocity profile for a concentric annulus is given by

$$u = -\frac{1}{4\mu}\frac{dp}{dx}\left[(R_i^2 - r^2) + \frac{n^2 - 1}{\ln n}R_i^2 \ln \frac{r}{R_i}\right] \tag{5.34}$$

where $n = R_o/R_i$.

The average velocity through the annulus is obtained by integrating Eq. (5.34) across the annulus and dividing by the annulus cross section area giving

$$u_m = -\frac{(R_o - R_i)^2}{8\mu}\frac{dp}{dx}\left[\frac{n^2 + 1}{(n-1)^2} - \frac{(n-1)}{(n+1)\ln n}\right] \tag{5.35}$$

and the pressure gradient can be expressed in terms of the mean velocity as

$$\frac{dp}{dx} = -\frac{8\mu \; \phi(n)}{(R_o - R_i)^2}u_m \tag{5.36}$$

where

$$\phi(n) = \left[\frac{n^2 + 1}{(n-1)^2} - \frac{(n+1)}{(n-1)\ln n}\right]^{-1} \tag{5.36a}$$

Note that the expression for the pressure gradient is identical to that for the circular pipe except that the characteristic "radius" for the annulus is $((R_o - R_i)/(\sqrt{\phi(n)}))$. In fact, for fully developed laminar flow in all ducts, the expression for the pressure drop will always have the form

$$dp/dx = \text{constant} \times \mu u_m/D_m^2 \tag{5.37}$$

where D_m is a characteristic cross-sectional dimension. Therefore, the friction factors λ and f will have the form

$$\lambda = \text{constant}/\text{Re}; \quad f = \text{constant}/\text{Re} \tag{5.38}$$

where the Reynolds number is based on the mean velocity and the characteristic diameter D_m. The above observations lead to the concept of the *hydraulic diameter*. This concept is based on the assumption that the friction factor for a duct with an arbitrary cross section is the same as that for a circular pipe of appropriate equivalent diameter. The task now is to determine a relationship between the properties of the duct cross section and the equivalent or hydraulic diameter. As derived earlier in this section, a momentum balance yields

$$dp/dx = \tau_m \, P/A \qquad (5.39)$$

where τ_m is the average shear stress along the perimeter P of the duct. Comparing the above expression with that for a circular pipe given in Eq. (5.29) and assuming that the average shear stress in the duct is identical to that for a circular pipe of radius R_h, we obtain

$$R_h = 2A/P = 2 \times \text{cross-sectional area/wetted perimeter} \qquad (5.40)$$

or, the hydraulic diameter is

$$D_h = 2R_h = 4 \times \text{cross-sectional area/wetted perimeter} \qquad (5.41)$$

The friction factor for the duct is then given by

$$\lambda = 64/\text{Re}_h \quad \text{or} \quad f = 16/\text{Re}_h \qquad (5.42)$$

where Re_h is the Reynolds number based on the hydraulic diameter. Note that in these manipulations it was assumed that the average shear stress along the perimeter of the duct was equal to that of a circular duct of diameter D_h. Let us test the validity of this assumption by comparing the friction factors for the concentric annulus derived using the hydraulic diameter hypothesis and the exact solution presented earlier.

The hydraulic diameter for the annulus is

$$D_h = 4\pi(R_o^2 - R_i^2)/2\pi(R_o + R_i) = 2(R_o - R_i) \qquad (5.41\text{a})$$

Therefore, according to the hydraulic radius hypothesis, the Darcy friction factor is

$$\lambda_h = 64/\text{Re}_h \qquad (5.43)$$

Using the exact solution for the pressure gradient given by Eq. (5.36) and the relationship between the pressure gradient and the friction factor given in Eq. (5.32)

$$\lambda = 64\phi(n)/\text{Re}_h \qquad (5.44)$$

Comparing Eqs. (5.43) and (5.44) it is seen that the λ_h differs from the exact λ by the function $\phi(n)$. Figure 5.2 shows $\phi(n)$ as a function of the ratio of the inner to outer radii of the annulus. As expected, when R_i goes to zero, i.e., when the annulus degenerates to a circular pipe, $\phi(n)$ goes to unity. However, $\phi(n)$ quickly attains a value of 1.5 at R_i/R_o of approximately 0.5. The friction factor λ_h is now 1.5 times lower than the exact λ. One would conclude from this that the hydraulic diameter concept does not work well. In fact, similar comparisons for elliptical and rectangular ducts shown in Figs. 5.3 and 5.4 confirm that for laminar fully developed flows, using friction factors based on the hydraulic radius concept can result in significant errors in the prediction of the pressure drop. However, as will be seen in the next section, the use of the hydraulic radius leads to rather good predictions of

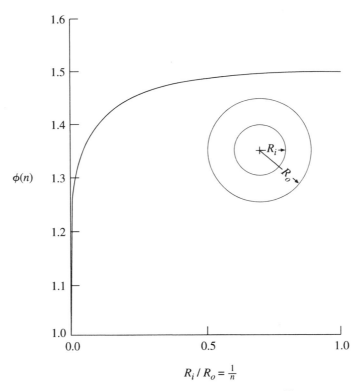

FIGURE 5.2 $\phi(n)$ for concentric annulli.

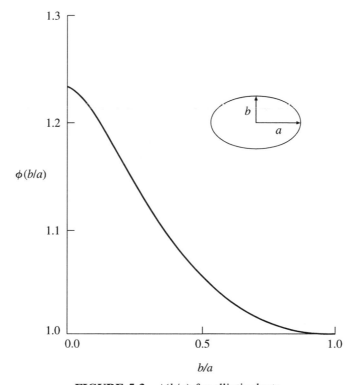

FIGURE 5.3 $\phi(b/a)$ for elliptic ducts.

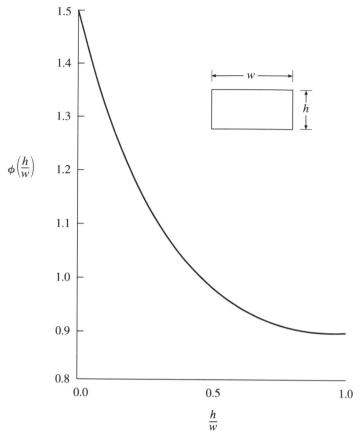

FIGURE 5.4 $\phi(h/w)$ for rectangular ducts.

the pressure drops for fully developed turbulent flows. Solutions of fully developed laminar incompressible flows in other duct cross section shapes are given in Berker (1963).

Fully Developed Turbulent Flows. Fully developed turbulent flows are more commonly encountered in actual practice than their laminar counterparts discussed in the previous subsection. Unfortunately, the analysis of fully developed turbulent flows is more complex, since, in general, the cross stream velocity components are not zero as was the case in laminar fully developed flows. These cross stream velocities or secondary flows occur due to variations in the turbulent stresses across the cross section. Consider Eq. (5.16), the transport equation for the vorticity in the streamwise direction x. Replacing the velocity and vorticity components by the sum of the mean and fluctuating components and time averaging the resulting equation, the transport equation for the mean streamwise vorticity for fully developed (i.e., $\partial/\partial x = 0$) turbulent flow is obtained

$$V \frac{\partial \Omega_x}{\partial y} + W \frac{\partial \Omega_x}{\partial z} = \frac{\partial^2}{\partial y dx}(\overline{w'^2} - \overline{v'^2}) - \left(\frac{\partial^2}{\partial z^2} - \frac{\partial^2}{\partial y^2}\right)\overline{v'w'} + \nu\left(\frac{\partial^2 \Omega_x}{\partial y^2} + \frac{\partial^2 \Omega_x}{\partial z^2}\right)$$

$$(5.45)$$

where, u', v' and w' are the fluctuating components of the velocities. Comparing with Eq. (5.19a) for laminar flow, it is seen that a *source* term appears on the right-hand side of the above equation due to the turbulence. It is this term that is responsible for the generation and sustenance of cross stream vorticity or secondary flows in fully developed turbulent flows. The only exceptions are fully developed turbulent flow in axisymmetric ducts and two-dimensional channels where the symmetry makes the source term in Eq. (5.45) go to zero. Therefore, accurate calculation of the fully developed turbulent flow in a duct with an arbitrary cross section requires the solution of the three-dimensional Navier–Stokes equations with sophisticated turbulence closure models. Fortunately, these secondary flows make the shear stress at the wall more uniform around the perimeter of the duct, therefore the hydraulic radius approach is quite successful in fully developed turbulent flows.

As mentioned earlier, due to its symmetry, the fully developed turbulent flow in a circular pipe is devoid of cross stream flow velocities, hence $V = W = 0$. Further, the axial velocity U is not a function of the streamwise location. Using these facts, the equations for fully developed flow in a circular pipe are

$$\frac{1}{r}\frac{d}{dr}r\left(\mu\frac{dU}{dr} - \rho\,\overline{u'v'}\right) = \frac{\partial p}{\partial x} = \text{constant} \tag{5.46}$$

$$\frac{1}{r}\frac{d}{dr}(r\,\rho\overline{v'^2}) - \rho\,\frac{\overline{w'^2}}{r} = \frac{\partial p}{\partial r} \tag{5.47}$$

$$\frac{d(\overline{v'w'})}{dr} + \frac{\overline{v'w'}}{r} = 0 \tag{5.48}$$

From Eq. (5.46) it can be deduced that the sum of the viscous stress $\mu(dU/dr)$ and turbulent stress $-\rho\overline{u'v'}$ varies linearly with pipe radius r and vanishes at the pipe center. The pressure varies across the pipe cross section as given by Eq. (5.47). However, the magnitude of this cross stream pressure variation is small (of the order $\rho\overline{v'^2}$) and is therefore negligible. Equation (5.48) implies that $\overline{v'w'} = 0$. Integration of Eq. (5.46) to obtain the mean velocity distribution requires the variation of $\overline{u'v'}$ as a function of r, which is obtained empirically given the current state-of-the-art in turbulence modeling.

Figure 5.5 shows a schematic of the velocity distribution for a fully developed turbulent flow in a smooth pipe. From a zero value at the wall, the velocity rises very steeply and almost linearly. This region is known as the *laminar* or *viscous sublayer*. Viscous effects dominate in this region, and, to the first order, the mean velocity profile is therefore linear. The mean velocity is related to the shear stress at the wall as follows

$$\tau_w = \mu(U/y) \tag{5.49}$$

where y is the distance from the wall. The above relation is generally cast into the dimensionless form

$$U^+ = y^+ \tag{5.50}$$

where $U^+ = U/v^*$ and $y^+ = y/y^*$. The length and velocity scales y^* and v^* are defined as

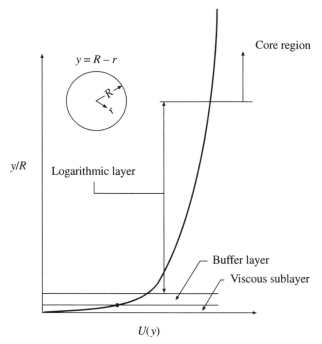

FIGURE 5.5   Mean turbulent velocity variation in a smooth pipe.

$$v^* = \sqrt{\tau_w/\rho} \text{ and } y^* = \mu/\sqrt{\tau_w\rho} = \nu/v^* \qquad (5.51)$$

These length and velocity scales follow from simple dimensional arguments based on the fact that the velocity in the viscous sublayer depends on the wall shear, fluid properties, and distance from the wall.

Figure 5.6 shows the variation of the turbulence intensities and the turbulent shear stress in the near wall region. These are based on the measurements of Laufer (1950) for a Reynolds number of 5×10^5. Defining the viscous sublayer as the region where the turbulent shear stress is less than 10% of the wall shear stress, the viscous sublayer extends out to $y^+ = 5$ (or $y/R = 0.0005$ at Re $= 5 \times 10^5$). The viscous sublayer is indeed very thin. Substituting this value of y^+ into Eq. (5.50), the velocity at the edge of the viscous sublayer $U^+ = 5$ (or $U/U_c = 18\%$).

Progressing outwards from the viscous sublayer ($y^+ > 5$), Fig. 5.6 shows that the turbulence intensities and the turbulent shear stress rise rapidly. The axial turbulence component u' peaks at $y^+ = 15$ and then settles down to a nearly constant value for $y^+ > 35$. The other two turbulence components and the shear stress rise monotonically and become nearly constant for $y^+ > 35$. The turbulent shear stress reaches 90% of the wall shear stress at $y^+ = 35$ which implies that the contribution due to molecular viscosity is very small beyond this point. This region between $y^+ = 5$ and 35 where the transition from the viscous to the fully turbulent region occurs, is called the *buffer layer*. The buffer layer extends to $y/R = 0.0035$ at a Reynolds number of 5×10^5. The velocity distribution in the buffer layer is generally derived from dimensional analysis and matching with the velocity profiles in the viscous sublayer and the *logarithmic layer* that extends outwards from the buffer

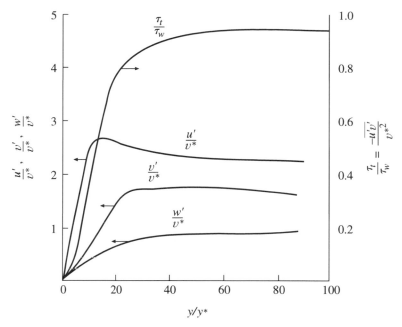

FIGURE 5.6 Variation of turbulence intensities and shear stress near a smooth pipe wall (after Laufer, 1954).

layer. The details of the velocity in the buffer layer are not important in the determination of the frictional pressure drop. Therefore, the reader is referred to White (1992) and Reynolds (1974) for the various formulations that have been proposed for the buffer layer.

Moving outward from the edge of the buffer layer, the fully turbulent region is reached. As discussed earlier, the turbulent shear stress reaches a value very close to the wall shear stress at the edge of the buffer layer. If one plots the velocity variation in the pipe where the dimensionless velocity u^+ is plotted as a function of the logarithm of the dimensionless distance from the wall y^+, it is evident that beyond the buffer layer ($y^+ > 35$), the velocity assumes a logarithmic profile of the form (see Fig. 1.39)

$$u^+ = A \ln (y^+) + B \qquad (5.52)$$

where A and B are constants. The portion of the pipe flow that is represented by the above expression is called the *logarithmic layer*. Strictly speaking, the logarithmic layer extends to $y/R = 0.19$ i.e., 19% of the pipe radius. However, it is often assumed that the logarithmic velocity profile extends all the way across the pipe cross section. As shown later, this is a fairly good assumption.

The constants in Eq. (5.52) can be given the following physical interpretation: the constant B represents the *slip* imparted by the viscous sublayer and the buffer layer. The constant A is related to the eddy viscosity. From the definition of the eddy viscosity ν_t where

$$\nu_t = \frac{\tau_w}{\rho(dU/dy)} \qquad (5.53)$$

Differentiating Eq. (5.52) and substituting the resulting value of dU/dy into the above expression gives

$$\nu_t/\nu = y^+/A \tag{5.54}$$

The constant A therefore relates the eddy viscosity to the distance to the wall. The value of these constants are obtained through empirical fits of experimental data. Reynolds (1974) lists the range of values for these constants based on the available experimental data. The most probable *universal* values for these constants are $A = 2.5$ and $B = 5.5$. Substituting these values into Eq. (5.52), the velocity in the *logarithmic layer* is given as

$$u^+ = 2.5 \ln (y^+) + 5.5 \tag{5.55}$$

The above equation is a very good approximation to the velocity profile to about 19% of the pipe radius from the pipe wall. The *core region* starts at this point and extends to the center of the pipe.

At the edge of the logarithmic layer (and the start of the *core region*), the velocity reaches approximately 85% of the centerline velocity. In the core region, therefore, the total variation in the velocity is only about 15%, and the velocity profile with radius is virtually flat. From dimensional analysis, the form of the velocity variation in the core region has the form

$$(U_c - U)/v^* = f(y/R) \tag{5.56}$$

Requiring that the above velocity variation match with that at the edge of the logarithmic region

$$(U_c - U)/v^* = 2.5 \ln (y/R) \tag{5.57}$$

This equation gives the velocity variation in the core region of the fully developed flow in a smooth pipe. It should be noted that this expression is not a unique formulation for the core flow [see Reynolds (1974) and Schlichting (1979)]. However, given that the velocity variation in the core is small (approximately 15% of the core velocity), the question of selecting the most accurate expression is a moot one. In fact, the commonly made assumption is that the logarithmic variation given in Eq. (5.56) is also valid for the core flow. Setting $y = R$ in Eq. (5.56) the approximate centerline velocity U_{ca} is given by

$$u_{ca}^+ = 2.5 \log (R^+) + 5.5 \tag{5.58}$$

where $u_{ca}^+ = U_{ca}/v^*$ and $R^+ = R/y^*$. Comparison of the centerline velocities predicted using this expression and actual data show that the difference is of the order of a few percent. Therefore, using the logarithmic profile given by Eq. (5.56) in the core region is reasonable.

The average velocity through the pipe is obtained by integrating the velocity profile across the pipe

$$U_m = \frac{1}{\pi R^2} \int_0^R 2\pi U r \, dr \tag{5.59}$$

where $r = R - y$.

Using the characteristic length y^* and velocity v^*, Eq. (5.60) can be written in the dimensionless form

$$U_m^+ = \frac{2}{R^+} \int_0^{R^+} u^+ \, dy^+ - \frac{2}{(R^+)^2} \int_0^{R^+} u^+ y^+ \, dy^+ \tag{5.60}$$

Assuming that the logarithmic profile is valid across the entire pipe cross section,

$$u_m^+ = A \ln (R^+) + B - \frac{3A}{2} \tag{5.61}$$

where A and B are the constants in Eq. (5.55). Further,

$$u_m^+ = U_m/v^* = \sqrt{\rho U_m^2/\tau_w} = \sqrt{8/\lambda}$$

$$R^+ = R/y^* = \frac{U_m R}{\nu} \sqrt{\tau_w/\rho U_m^2} = \text{Re} \sqrt{\lambda/32} \tag{5.62}$$

where λ is the Darcy friction factor, and Re is the Reynolds number based on the pipe diameter. Substituting the above expressions into Eq. (5.61), an implicit relationship for λ is obtained

$$1/\sqrt{\lambda} = (A/\sqrt{8}) \ln (\text{Re}\sqrt{\lambda}) - 1.14A + 0.35B \tag{5.63}$$

A slight adjustment in the values of the constants A and B is needed to match the data (as a compensation for assuming that the log profile extends across the entire pipe cross section). The commonly cited values for these constants when used in Eq. (5.63) are $A = 2.46$ and $B = 5.67$. Using these values

$$1/\sqrt{\lambda} = 0.87 \ln (\text{Re}\sqrt{\lambda}) - 0.8 \tag{5.64}$$

Alternately, using logarithms with base 10,

$$1/\sqrt{\lambda} = 2 \log (\text{Re}\sqrt{\lambda}) - 0.8 \tag{5.65}$$

This expression predicts the friction factor to within 2% for $\text{Re} > 10^4$.

Power Law Formulae. The logarithmic profiles and the resulting expression for the friction factor are somewhat cumbersome to use. For example, the determination of the friction factor from Eq. (5.64) requires an iterative procedure. Therefore, turbulent velocity profiles and friction factor formulae in smooth pipes are often presented in the power law form

$$U/U_c = (y/R)^{1/n}$$

$$\lambda = C + D \, \text{Re}^{-m} \tag{5.66}$$

where n, C, D, and m are empirically derived constants over specific ranges of the Reynolds number. The most commonly used value of n is 7 which gives a fairly good fit for the velocity profile for $10^4 < \text{Re} < 10^6$. The so called 1/7th power law is therefore

$$U/U_c = (y/R)^{1/7} \tag{5.67}$$

Similarly, commonly used expressions for the friction factor are,

$$\lambda = 0.316 \, \text{Re}^{-1/4} \qquad \text{Re} < 10^5 \text{ (Blasius formula)}$$

$$\lambda = 0.184 \, \text{Re}^{-1/5} \qquad 10^5 < \text{Re} < 10^6 \tag{5.68}$$

$$\lambda = 0.0156 + 0.5 \, \text{Re}^{-0.32} \qquad 10^6 < \text{Re} < 10^7.$$

Of note is a formula provided by White (1992) which is accurate to about $\pm 3\%$

$$\lambda = 1.02 \, (\log \text{Re})^{-2.5} \text{ for } 10^4 < \text{Re} < 10^8 \tag{5.69}$$

Effect of Wall Roughness. So far, we have considered fully developed turbulent flow in smooth walled pipes. Surface roughness significantly affects turbulent flow in ducts, in contrast to laminar flow where it has no effect on the wall shear and hence the pressure drop. Roughness is traditionally characterized in terms of a length k which is related to the average height of the roughness elements. Tests with different types of roughness elements such as sand grains, hemispheres etc., show that characterization of roughness in terms of a single length scale is inadequate. Nonetheless, given that the interactions between surface roughness and the turbulent duct flow are not yet fully understood, the use of a single length scale and empiricism is the most expedient option at the present time.

The experiments of Nikuradse [see Schlichting (1979)] form the basis of the empirical correlations for fully developed turbulent flows in pipes. Nikuradse used uniformly graded sand glued to the pipe wall to obtain various degrees of roughness. The sieve size for the uniformly graded sand was used as the characteristic roughness length scale k. The effect of roughness was found to fall into three regimes. For $k/y^* < 5$, roughness has no effect, i.e., the pipe can be considered as smooth walled. For $k/y^* > 60$, the flow characteristics are independent of the Reynolds number and only depend on the ratio of k to pipe diameter; this regime is known as the *fully rough* regime. For $5 < k/y^* < 60$, flow characteristics depend on both the Reynolds number and the roughness k; this is the *transitional roughness* regime.

In the fully rough regime, the velocity profile over the entire cross section of the pipe can be approximated by

$$U/v^* = A \ln (y/k) + B_s \tag{5.70}$$

This is the familiar logarithmic profile similar to that in a smooth pipe except that the roughness scale k has replaced the *viscous length scale* y^*. The value of the constant A is approximately 2.5 which is the same as that for the smooth pipe. The value of the constant B_s is approximately 8.5 (for sand grain roughness) and is therefore significantly higher than its smooth wall counterpart. Unfortunately, in the fully rough region, the constant B_s is a function of the type of roughness and therefore has different values for roughness other than sand grain type roughness. The common practice is to classify other types of roughness in terms of *equivalent sand grain roughness*. Equivalent sand grain roughness for common pipe surface finishes are given in Sec. 25.1.

Using the logarithmic profile given by Eq. (5.70) and following the procedure used earlier for the smooth pipe, the mean velocity through the pipe for the fully rough regime is given by

$$U_m/v^* = 2.5 \ln (R/k) + 4.75 \tag{5.71}$$

and the friction factor is given by

$$1/\sqrt{\lambda} = 2 \log (D/k) + 1.1 \tag{5.72}$$

where D is the pipe diameter.

Colebrook (1939) combined the expressions for the friction factors for smooth and rough pipes to obtain the following composite formula

$$1/\sqrt{\lambda} = 1.74 - 2.0 \log \left(\frac{2k}{D} + \frac{18.7}{\text{Re}\sqrt{\lambda}} \right) \tag{5.73}$$

The above expression covers all three regimes of fully developed turbulent pipe flow—smooth, transitional roughness, and fully rough regimes. Moody developed a graph, the Moody diagram, for friction factor versus Reynolds number for laminar and turbulent pipe flow. The Moody diagram is discussed in Sec. 25.1 and is often used in practice to determine the friction factors for fully developed pipe flows.

Fully Developed Turbulent Flow in Noncircular Ducts. As discussed earlier in this section, secondary flows occur in fully developed noncircular ducts due to the variations in the wall shear stresses around the periphery of the duct. The only exception is fully developed turbulent flow in two-dimensional channels. In practice, such flows are encountered in the center portions of rectangular ducts of high aspect ratio. The most comprehensive measurements of fully developed turbulent channel flow were made by Hussain and Reynolds (1975). The flow characteristics of two-dimensional fully developed channel flow are very similar to those for the circular pipe discussed above. The velocity profile can again be approximated by the logarithmic profile

$$U/v^* = A \ln (y/y^*) + B \tag{5.74}$$

Based on a survey of available data, Dean (1978) suggests the following values for the constants

$$A = 2.44 \text{ and } B = 5.17 \tag{5.75}$$

Note that the above values are very close to those for the circular pipe, and, in fact, using the circular pipe values instead of the above values is almost as good. Also, based on Dean's survey, the following formula for calculating friction factors is suggested

$$1/\sqrt{\lambda} = 1.99 \log{(\text{Re}\sqrt{\lambda})} - 0.56 \qquad (5.76)$$

The Reynolds number is based on the total height of the channel, i.e., the spacing between the upper and lower walls. A power law expression for the friction factor valid for $6 \times 10^3 < \text{Re} < 6 \times 10^5$ is

$$\lambda = 0.292 \, \text{Re}^{-1/4} \qquad (5.77)$$

In general, fully developed turbulent flows in rectangular ducts give rise to a secondary flow that sweeps the core flow towards the corners. The secondary flow is driven by the difference in the wall shear stress, and hence the pressure, at the corners of the rectangular duct and the sidewalls. The magnitude of the cross stream velocities is less than a few percent of the axial flow velocity. Nonetheless, this secondary flow makes the shear stress around the periphery of the duct more uniform than for laminar fully developed flow in a similar duct. The secondary flow also becomes significant to local wall heat and mass transfer processes. Detailed measurements for fully developed flows in rectangular ducts are given by Melling and Whitelaw (1976).

The pressure drop for fully developed turbulent flow in ducts of arbitrary cross sections is calculated using the hydraulic diameter approach. The hydraulic diameter was introduced in the previous section on fully developed laminar flows and is defined as $D_h = 4 \times$ cross-sectional area/wetted perimeter. The pressure drop is then calculated for an equivalent pipe with a diameter equal to the hydraulic diameter. This approach works reasonably well for fully developed turbulent flows, and the calculated pressure drop is generally accurate to better than 20%. The accuracy improves as the aspect ratio (ratio of the maximum to minimum cross section dimensions) approaches unity. This is in contrast to the fully developed laminar flow situation where the pressure drop predictions using the hydraulic diameter approach can have errors as much as 50%. The relative success of the hydraulic diameter concept for fully developed turbulent flows is directly attributed to the secondary flows that tend to make the shear stress along the duct periphery more uniform.

Jones and Lueng (1981) and Jones (1976) have proposed improving the accuracy of the pressure drop prediction using the hydraulic diameter approach by introducing an additional geometric function, ϕ, similar to that discussed in the previous section on fully developed laminar flows. Using this geometric function in conjunction with the hydraulic diameter, they predict pressure drop data for rectangular and annular ducts to within 5%.

5.3 EFFECT OF CURVATURE AND ROTATION

Ducts with curved walls are encountered in many engineering applications where either a change in flow direction and/or a gradual change in the flow area is required. Common examples are pipe elbows, diffusers and turbomachine blade passages.

Further, most fluid machines have rotating flow passages, and the flow through these is subjected to the effects of rotation. The blade passage in a centrifugal compressor is an example of a curved duct with rotation. Wall curvature and rotation impose additional acceleration fields on the flow which have significant effects on the viscous layers at the duct walls and lead to the formation of secondary flows in the duct. This section describes the effects of wall curvature and rotation on thin shear layers before the onset of secondary flows.

Consider the boundary layer developing on the walls of a curved and rotating duct as shown in Fig. 5.7. The local boundary layer thickness is δ, and the local wall radius of curvature is R. Further, the duct is rotating at a constant angular velocity Ω with the sign convention that Ω is positive when the rotation is counterclockwise as shown in Fig. 5.7. The Navier–Stokes equations for such a flow written in the generalized curvilinear coordinate system [Van Dyke (1969) and Halleen (1967)] will have several additional terms compared to those for a plane flow. For example, the additional terms for the accelerations (the centrifugal and Coriolis accelerations) imposed along the y axis on the boundary layer by the mean flow due to curvature and rotation are

$$A_{cy} = -U^2/(R + y), \quad A_{wy} = 2\Omega U \qquad (5.78)$$

The corresponding accelerations in the x-direction are

$$A_{cx} = A_{cy}(V/U), \quad A_{wx} = -A_{wy}(V/U) \qquad (5.79)$$

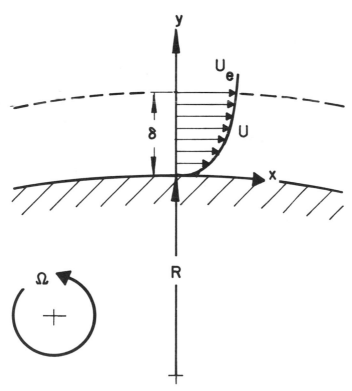

FIGURE 5.7 Coordinates for a curved and rotating wall.

All the additional terms appearing in the Navier-Stokes equations due to curvature are of the order δ/R. Similarly, those due to rotation are of the order $(\Omega\delta/U)$. Therefore, curvature and rotation have first order effects on the flow. This is certainly the case for laminar flows. For example, the skin friction for a laminar boundary layer on a curved surface will be of the form [Van Dyke (1969)]

$$f = f_0 (1 + K\delta/R) \tag{5.80}$$

where, f_0 is the skin friction for a plane boundary layer, and K is a constant of order unity. For example, K has a value of about -1.24 for a convex wall laminar boundary layer which implies that the skin friction decreases on a convex wall. The converse is true for a concave wall laminar boundary layer for a small amount of curvature. However, concave curvature has a significant effect on the stability of the boundary layer and can lead to the formation of Goertler vortices and/or earlier transition to turbulence. Further details on the effect of wall curvature on the stability of boundary layers are provided in reviews by Schlichting (1979) and Tani (1969).

The situation is considerably more complicated for turbulent flows in curved and/or rotating ducts. Bradshaw (1973) shows that additional terms due to curvature in the mean flow equations for turbulent flows exhibit the same first order dependence as for laminar flow. However, experiments indicate that the effect of curvature is almost ten times greater than the first order effects expected. That is, the value of the factor K in Eq. (5.80) is of the order of ten in turbulent flow as opposed to unity for laminar flows. Turbulence is strongly suppressed on a convex surface (or the *suction surface* in a rotating passage) and increased on a concave surface (or the *pressure surface* in a rotating passage).

A qualitative, albeit inexact, explanation for the effect of curvature and rotation on turbulence initially forwarded by Prandtl and expanded upon by Bradshaw (1969, 1973) is as follows: consider the case of a turbulent boundary layer on a convex wall as shown in Fig. 5.7. For now assume that the wall is not rotating ($\Omega = 0$). The pressure gradient normal to the wall due to curvature is $\rho(U^2/R)$ [see Eq. (5.78)]. Due to the fact that the flow is retarded in the boundary layer ($dU/dy > 0$), if a fluid element in the boundary layer is displaced radially outward, it will have a lower angular momentum than the surrounding fluid at the new radial location. The angular momentum deficit will give rise to a body force which will force the displaced fluid element back towards its original location in the boundary layer. That is, radial displacements caused by turbulence are counteracted (or damped) by the radial pressure gradient. The net effect is to reduce turbulence. The converse situation occurs on a concave wall. Similar arguments can be used for rotating ducts where the pressure gradient normal to the mean flow direction is caused by the Coriolis acceleration.

The above qualitative explanation is analogous to the one used in examining the stability of buoyancy driven flows. Extending this analogy, the stability of turbulent flow with curvature can be related to a *flux Richardson number* Ri_c

$$\mathrm{Ri}_c = S(1 + S) \tag{5.81}$$

where

$$S = 2(U/R)\Big/\left(\frac{\partial U}{\partial y} - \frac{U}{R}\right) \tag{5.81a}$$

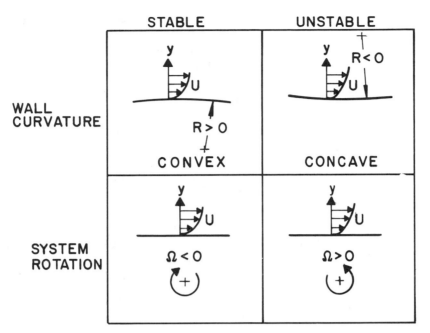

FIGURE 5.8 Stable and unstable curvature/rotation configurations.

Similarly, the *flux Richardson number for rotation* Ri_w is

$$Ri_w = S(1 + S) \tag{5.82}$$

where

$$S = -2\Omega/(\partial U/\partial y). \tag{5.82a}$$

The flow is stabilized, i.e., turbulence damped, when Ri_c or $Ri_w > 0$ and amplified when Ri_c or $Ri_w < 0$; Ri_c or $Ri_w = 0$ denotes neutral stability. Figure 5.8 shows the curvature and rotation configurations that stabilize or de-stabilize turbulent flow. Note that although the Richardson number provides information about the stability of the flow, the quantitative effect of curvature and rotation on the turbulence has to be derived from experimental data at the present time. Empirical formulae relating the Richardson number to various turbulence parameters have been proposed and used in flow computations. A brief discussion of the experimental data on the effects of curvature and rotation is presented before proceeding to these empirical formulae.

5.3.1 Effect of Curvature on Turbulent Flows

As discussed above, for turbulent flow over a convex wall, $Ri_c > 0$ and the flow is stabilized. The opposite is true for turbulent flow over a concave wall. The neutral stability condition of $Ri_c = 0$ is satisfied when $(\partial U/\partial y) + (U/R) = 0$, that is, in the free stream (or inviscid) portion of the flow field. Bradshaw (1973) has surveyed the work prior to 1973; the more recent experiments in this area are briefly reviewed by Adams and Johnston (1983). The data of So and Mellor (1973) are representative of strong curvature ($\delta/R \cong 0.1$) whereas those of Meroney and Bradshaw (1975) and Shivaprasad and Ramaprian (1978) are for mild curvature ($\delta/R \cong 0.01$). All the data show similar overall trends. At the convex wall, the turbulent intensity and shear stress drop rapidly and reach equilibrium levels that are considerably lower than those for a plane wall boundary layer. The suppression of turbulence is most dramatic in the outer layers of the boundary layer as shown in Fig. 5.9, and for δ/R of 0.1, the shear stress drops to zero well within the boundary layer. The consequence is that although the logarithmic velocity profile remains valid near the wall, the region over which it remains valid is considerably reduced. The net result of convex curvature is that the wall shear stress is reduced. About a 10% decrease in wall shear was observed for mild curvature ($\delta/R \cong .01$), whereas about a 20% decrease occurred for strong curvature ($\delta/R \cong 0.1$).

On the concave wall, the effects of curvature are to increase the turbulent intensities and shear stress. The logarithmic velocity region grows, and wall shear stress increases significantly in the outer layer of the boundary layer. Again, about a 10% increase in wall shear stress was observed for mild curvature ($\delta/R \cong .01$). Concave curvature may cause the growth of parallel sets of longitudinal vortices [Meroney and Bradshaw (1975)] similar to the Goertler vortices in laminar flow.

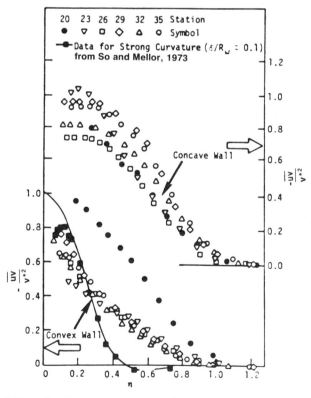

FIGURE 5.9 Effect of wall curvature on turbulent shear stress. (Reprinted with permission of ASME, ''Turbulence Measurement in Boundary Layers Along Mildly Curved Surfaces,'' by Shivaprasad, B. G. and Ramaprian, B. R., *J. Fluids Eng.*, Vol. 100, No. 1, pp. 37–46, 1978.)

The most significant aspect of the data on wall curvature effects is that even small amounts of curvature produce large effects on the turbulence properties of the boundary layer. Further, the available data suggest that the effect is nonlinear with respect to the parameter δ/R. Finally, when curvature is removed, the boundary layer recovers rather slowly. That is, the turbulence properties relax to the plane wall boundary layer profiles at significant distances downstream of the point where curvature is removed. The implication of these observations is that simple turbulence models will not correctly predict curvature effects, and empirical modifications are needed.

There are several models that are currently used in the computation of turbulent flows. A detailed discussion of these models and their appropriateness in terms of predicting the effects of curvature is beyond the scope of this section. However, empirical modifications to incorporate the effects of curvature in two of the most commonly used turbulence models, the mixing length model and the k-ϵ model, will be discussed briefly. The mixing length model is the simplest model for the computation of turbulent boundary layers. The turbulent shear stresses are written as (see Sec. 4.5)

$$-\overline{u'v'} = \nu_t \, \partial U/\partial y \tag{5.83}$$

where ν_t is the eddy viscosity given by

$$\nu_t = \ell_t^2 \, |\partial U/\partial y| \tag{5.84}$$

The length scale ℓ_t is known as the mixing length. A prescription of the mixing length across the boundary layer is needed to proceed further. In most current procedures, the boundary layer is divided into at least two layers, the wall layer and the outer layer [Bradshaw (1978)]. In the wall layer the mixing length is taken to be a function of the distance from the wall y. Note that the log law implies that $\ell_t = \kappa y$, where κ is the von Karman constant (inverse of the constant A in the log law given in the previous section). In the outer region of the boundary layer, the mixing length is presumed to be proportional to the boundary layer thickness, i.e., $\ell_t = \alpha\delta$, where α is a constant with a value of around 0.08, and δ is the local boundary layer thickness.

Based on the data on the effects of curvature, Bradshaw (1969) suggested the following modification to the mixing length due to curvature

$$\ell_{to}/\ell_t = (1 + \beta \, \mathrm{Ri}_c) \tag{5.85}$$

where ℓ_{to} is the mixing length for $\mathrm{Ri}_c = 0$, i.e., for the zero curvature case, and β is an empirical constant that has a value about 7 for $\mathrm{Ri}_c > 0$ (stable flows) and a value of about 4 for $\mathrm{Ri}_c < 0$ (unstable flows). Note that the above expression would result in a negative mixing length for $\mathrm{Ri}_c < -0.25$. In such cases, the following formulation known as *Keyps formula* is suggested

$$\ell_{to}/\ell_t = 1 - 2 \, \mathrm{Ri}_c \tag{5.86}$$

Most mixing length formulations use the above empirical formulae or some variation thereof [Adams and Johnston (1983)] to account for the effects of curvature.

Another popular model for turbulence is the k-ϵ model (see Sec. 4.5). In this model, the eddy viscosity is related to the local turbulent kinetic energy k and dissipation ϵ which are obtained from the solution of two additional equations for the transport of k and ϵ. Details of the k-ϵ model are given by Launder and Spaulding (1974). Launder *et al.* (1977) included the effect of curvature by modifying the decay term in the transport equation for ϵ by a multiplicative factor $(1 - C_c \, \text{Ri}_c)$. The constant C_c has a value of approximately 0.2, and Ri_c is a modified Richardson number

$$\text{Ri}_c = 2(U/R) \frac{\partial(UR)}{\partial y} (k/\epsilon)^2 \tag{5.87}$$

Here, the denominator used in the definition of the Richardson number given earlier has been replaced by a turbulent time scale $(k/\epsilon)^2$. Another two-equation closure model that includes the effects of curvature is that of Wilcox and Chambers (1977).

5.3.2   Effect of Rotation on Turbulent Shear Layers

The effects of rotation on turbulent flow discussed in this section are derived from the work of Johnston *et al.* (1973). Other experimental data on this subject may be found in Moore (1973). Consider a two-dimensional rotating channel shown in Fig. 5.10. The mean flow is along the x-axis, and the duct is rotating about an axis normal to the x–y plane. As discussed earlier, the cross stream Coriolis acceleration will be $2U\Omega$ in the positive y direction for a counterclockwise rotation direction. This will give rise to a negative pressure gradient in the y direction with the pressure decreasing from the trailing side wall of the duct (the *pressure* surface) to the leading side wall (the *suction* surface, see Fig. 5.10). The relative magnitude of this Coriolis acceleration with respect to other terms in the momentum equations is given by the rotational number Ro (inverse of the Rossby number)

$$\text{Ro} = 2 \, \delta\Omega/U \tag{5.88}$$

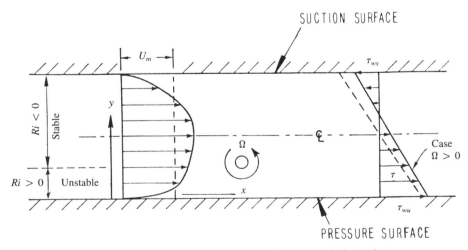

FIGURE 5.10   A rotating two-dimensional channel.

where δ is the boundary layer thickness. For a channel with fully developed flow, Ro is defined as

$$\text{Ro} = h\Omega/U \qquad (5.89)$$

where h is the channel height. The parameter Ro is analogous to the curvature parameter δ/R, and significant effects on turbulent flow have been observed for Ro $\geq$ 0.01.

The stability of the flow in a rotating duct is governed by the Richardson number Ri_w introduced earlier. Stabilization occurs when $\text{Ri}_w > 0$ and *vice versa*. $\text{Ri}_w = 0$ defines the neutral stability condition. Therefore, for a viscous turbulent flow in the channel shown in Fig. 5.10, the flow is stabilized on the suction surface and destabilized on the pressure surface. Neutral stability is found in the region of the flow that satisfies the condition $(\partial U/\partial y) - 2\Omega = 0$, that is, the absolute mean vorticity is zero. This condition is satisfied in the free stream (or inviscid) flow regions in the duct.

In the experiments by Johnston *et al.* (1973), the flow through the channel at zero rotation was close to the fully developed turbulent state. With rotation, the turbulence was strongly suppressed on the suction surface, and the wall shear stress decreased. This effect increases as Ro increases. In fact, rotation can cause the turbulent boundary layer on the suction surface to relaminarize. Figure 5.11 shows the

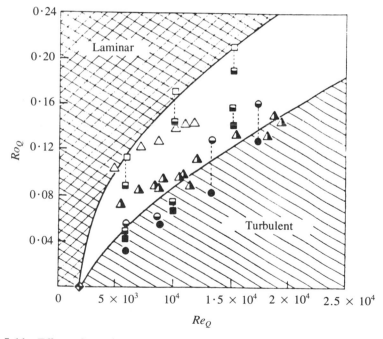

FIGURE 5.11 Effect of rotation on the flow regimes at the suction surface of a rotating two-dimensional channel. (Reprinted with permission of Cambridge University Press, "Effects of spanwise rotation on the structure of 2-D fully developed turbulent channel flow," by Johnson, J. P., Halleen, R. M., and Lezius, D. K., *J. Fluid Mechan.*, Vol. 56, No. 3, pp. 533–558, Dec. 1972.)

laminar, transitional and turbulent regimes as a function of the rotational number and the Reynolds number obtained by Johnston *et al.* (1973). On the pressure surface where $Ri_w < 0$, rotation increases the turbulence levels, and the wall shear stress is increased. However, the wall shear stress and turbulence intensity level off beyond a certain value of Ro. This is probably due to the formation of roll cells with axes parallel to the mean flow direction, similar to Goertler vortices observed in laminar and turbulent flows over concave walls.

Johnston (1973) has proposed the following relationship between wall shear stress τ_w and the rotation number Ro

$$\tau_w / \tau_{wo} = 1 \pm 1.75 \, R_o \tag{5.90}$$

where, τ_{wo} is the wall shear stress at zero rotation. The negative sign applies to the duct wall where $Ri > 0$, i.e., the suction surface. The positive sign applies to the unstable wall where $Ri < 0$. The above expression is valid as long as the flow remains turbulent at the wall on the suction surface (see Fig. 5.11). Further, for the pressure surface, the expression is only valid up to values of Ro < 0.05–0.15, after which the shear stress levels off as discussed above.

As seen from the above discussion, the effects of rotation and curvature are analogous. It is not surprising therefore that the methods discussed in the previous subsection for predicting the effects of curvature are directly applicable to flows with rotation. See Johnston and Eide (1976) and Howard *et al.* (1980) for further information on turbulence models for predicting the effects of rotation.

5.4 CONVERGING AND DIVERGING DUCTS

In this section, straight ducts with changing cross-sectional area will be discussed. Consider a duct with a gradually varying cross-sectional area and with one-dimensional incompressible mean flow along the axis of the duct. Mass conservation implies

$$d(UA) = 0 \tag{5.91}$$

or

$$dU/U = -dA/A \tag{5.91a}$$

This implies that the velocity U increases with decreasing area A and *vice versa*. At any point in the duct, the total pressure is defined as

$$P_0 = p + \tfrac{1}{2}\rho U^2 \tag{5.92}$$

where p is the static pressure. For small wall curvature, p will be nearly constant across a given cross section of the duct, and the above equation can be differentiated to give the change along the flow axis x

$$dP_0 = dp + \rho U \, dU \tag{5.93}$$

or combining with Eq. (5.91a) and rearranging,

$$dp = \left(\rho \, \frac{U^2}{A} \right) dA + dP_0 \qquad (5.94)$$

For inviscid flow with no energy transfer, P_0 is constant throughout the flow, therefore

$$dp = \left(\rho \, \frac{U^2}{A} \right) dA \qquad (5.95)$$

This equation states that a variation in duct cross-sectional area gives rise to a pressure gradient along the axis of the duct. Converging ducts ($dA/dx < 0$) produce a negative pressure gradient ($dp/dx < 0$), i.e., the pressure drops along the duct axis. In diverging ducts ($dA/dx > 0$) the pressure increases along the flow axis ($dp/dx > 0$). These pressure gradients have a profound effect on the wall shear layers especially in the case of diverging ducts.

Another interesting result can be deduced from Eq. (5.93). Assume that the velocity is nonuniform at the entrance of the duct. Note that this implies that the total pressure is also nonuniform for a uniform static pressure at the inlet. Eq. (5.93) can be written along a streamline

$$dU = -[dp - dP_0]/(\rho U) \qquad (5.96)$$

Again, as before $dP_0 = 0$, hence, for a positive pressure gradient ($dp > 0$), the velocity decrease is inversely proportional to the magnitude of the velocity along the streamline. That is, the lower velocity regions will be decelerated more than the higher velocity regions. Thus, ducts with increasing area and hence positive pressure gradients will cause an initially nonuniform velocity profile to become more nonuniform. On the other hand, for a given negative pressure gradient along the duct axis such as for a converging duct, the lower velocity regions are accelerated more than the higher velocity regions causing the flow velocity profile across the duct to become more uniform.

Given the rather disparate behavior of flow in converging and diverging ducts, the two types of ducts will be discussed separately in the following subsections. The discussions in these subsections will be restricted to incompressible flows.

5.4.1 Converging Ducts—Nozzles

In converging ducts, the cross-sectional area of the duct decreases along the flow axis. As discussed above, for incompressible or subsonic flows, this area convergence leads to an increase in the flow velocity and a corresponding decrease in the flow static pressure along the axis of the duct. Converging ducts are commonly known as *nozzles* and will be referred to as such in the remainder of this section. Typical applications of nozzles in the role of converting pressure to flow kinetic energy are in turbine blade passages and turbojets. The flow acceleration through a nozzle causes inlet velocity profile distortions to be smoothed as shown in the introductory remarks to this section. The exit flow velocity profile from well designed nozzles is very uniform, hence nozzles are commonly used as flow metering devices. Nozzles also reduce turbulence levels. That is, turbulence intensities at the exit of

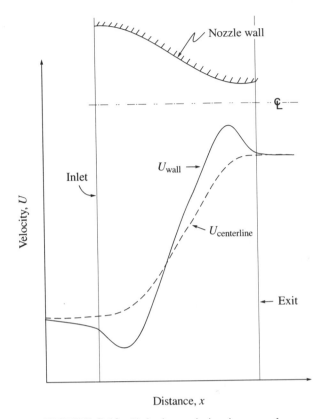

Velocity, U

Distance, x

FIGURE 5.12   Velocity variation in a nozzle.

the nozzle are generally lower than those at the nozzle inlet. Therefore, nozzles are used upstream of wind tunnel test sections to produce a uniform and low turbulence flow at the test section.

Figure 5.12 shows a typical velocity variation along the nozzle axis and wall. The *wall velocity* is the velocity deduced from a static pressure tap at the wall and the core flow stagnation pressure. Therefore, the wall velocity shown is the velocity at the outer edge of the wall boundary layer. The velocity along the nozzle centerline generally shows a monotonic increase. The wall velocity initially decreases, starting at some point upstream of the nozzle, and reaches a minimum downstream of the nozzle inlet. This behavior is caused by the curvature of the streamlines near the wall as the flow anticipates the converging area. The more abrupt the curvature, the further upstream this effect extends. After reaching a minimum just inside the nozzle inlet, the flow at the wall also accelerates but at a faster rate than the centerline velocity. The wall velocity exceeds the centerline velocity at some point inside the nozzle and reaches a maximum just upstream of the nozzle exit. The wall velocity then decreases and is slightly higher than the centerline velocity at the nozzle exit. The exact location and magnitude of the wall velocity undershoot at the nozzle inlet, and overshoot at the nozzle exit, depends on the nozzle wall contour. A very gradual change in the wall curvature at the inlet and exit attenuates the wall velocity over- and under-shoots, whereas abrupt wall curvature changes amplify them. These wall velocity excursions cause regions of adverse pressure gradients that affect the growth of the wall boundary layer and can lead to local separation. Note that due to the overall favorable pressure gradient through the nozzle, the boundary layers will reattach and large separated zones are rare. Nonetheless, the extent of these adverse pressure gradients should be minimized if the objective is to design a nozzle with a very uniform exit velocity profile. Morel (1975, 1977) has developed design charts for nozzle shapes comprised of two matched cubics based on an inviscid treatment of the core flow coupled with criteria for the acceptable size of the velocity over- and under-shoots.

Due to the overall favorable pressure gradient along the flow direction, the wall boundary layers are thin and attached for all nozzle shapes that do not have a sharp variation in the local wall curvature. Therefore, the common approach to nozzle design is to simply select a shape that looks smooth. Such an approach normally works, and the exit mean velocity profile will be uniform to less than 10%. Typically, the nonuniformity in the exit velocity profile for a poorly designed nozzle is in the form of velocity overshoots near the wall. Axisymmetric nozzle shapes that are known to produce exit velocity profiles with nonuniformities of 2% or less [Hussain and Ramjee (1976)], are the *Batchelor–Shaw* (BS) *nozzle*, the *cubic equation nozzle* and the *ASME low and high beta nozzles*. The BS nozzle contour is given by

$$1/A^2 = 1/A_i^2 + (1/A_e^2 - 1/A_i^2)\left(\frac{x}{L} - \frac{1}{2\pi}\sin\left(2\pi\frac{x}{L}\right)\right) \qquad (5.97)$$

where A is the cross-sectional area at axial location x, A_i and A_e are the inlet and exit areas respectively, and L is the nozzle length. The cubic equation (CE) nozzle has a wall contour described by a cubic equation. By requiring that the cubic match the inlet and exit radii and that the wall slope be zero at the nozzle inlet and exit, the following expression results

$$R = D_i/2 - (3/2)(D_i - D_e)(x/L)^2 + (D_i - D_e)(x/L)^3 \qquad (5.98)$$

where, R is the nozzle radius at axial location x, and D_i and D_e are the inlet and exit diameters, respectively. Straight sections of lengths R_e and $R_e/2$ are added at the nozzle inlet and exit, respectively (R_e is the exit radius). For both these nozzle shapes, the nozzle length L is an independent parameter. Hussain and Ramjee (1976) used a length L equal to $2D_e$. The ASME nozzle shapes are specified in the ASME Fluid Meters Handbook. The wall contour, starting at the nozzle inlet, is a quadrant of an ellipse followed by a length of constant diameter. Hussain and Ramjee (1976) tested the above four nozzle shapes at low subsonic Mach numbers. The contraction ratio for all four nozzles was 11, where the contraction ratio is $(D_i/D_e)^2$. They found that the CE nozzle had the thinnest exit boundary layer thickness ($\delta = 0.02\, R_e$). However, the CE nozzle had a 2% velocity overshoot near the wall ($.02 < y/R_e < 0.1$). The other three nozzle shapes had a larger boundary layer thickness ($\delta = 0.03\, R_e$), however the mean velocity profiles were more uniform than that for the CE nozzle. The boundary layer profile at the nozzle exit exhibited the characteristics of a laminar boundary layer. The inlet boundary layer to the nozzle was a fully developed turbulent boundary layer. Therefore, laminarization of the boundary layer occurred inside the nozzle. This is not surprising given the effects of flow acceleration and wall curvature; see Sec. 5.3. The flow turbulence intensity in the core flow at the nozzle exit was constant and had a value of 0.58% compared to 7.5% at the nozzle inlet.

The effects of nozzle shape and contraction ratio were studied experimentally by Hussain and Ramjee (1976) and Ramjee and Hussain (1976), respectively. A qualitative understanding of the effect of a converging duct on the core flow turbulence can be obtained by viewing the turbulence as being caused by a set of vortex filaments and considering the distortions of these vortex filaments as they pass through the nozzle. The vortex filaments aligned with the nozzle axis are the primary contributors to the velocity fluctuations in the cross stream direction (v' and w' components). As the flow accelerates through the nozzle, these vortex filaments are stretched and from Kelvin's Theorem it can be shown that v' and w' increase as the square root of the contraction ratio. The vortex filaments aligned normal to the cross stream direction produce the velocity fluctuations parallel to the free stream velocity, u'. The length of these filaments decreases as they pass through the nozzle, i.e., the filaments are compressed. Again, Kelvin's Circulation Theorem requires that u' decrease in proportion to the contraction ratio. Therefore, a nozzle is expected to amplify the lateral turbulent fluctuations and attenuate the axial fluctuations. Note that the turbulent intensities, defined as the ratio of the *rms* values of the turbulent velocity fluctuations to the mean velocity, decrease for all three components, since the mean velocity increases in proportion to the contraction ratio.

Hussain and Ramjee (1976) found that in the initial portions of the nozzle, the axial turbulence component was attenuated rapidly. This is in agreement with the trend predicted by linear theories. However, after this initial decrease, the axial turbulence component increased, which contradicts the continuous attenuation predicted by linear theories. The lateral turbulence increased monotonically through the nozzle as predicted by linear theories, however the magnitude of this increase was found to be lower than the predictions by a factor of two. The other salient observation made by Hussain and Ramjee (1976) was that nozzle shape (for a fixed contraction ratio) did not affect the final levels of the turbulence in the core flow at the nozzle exit. Nozzle shape did affect the turbulence levels in the boundary layer region at the nozzle exit. The cubic equation nozzle had the lowest turbulence level in this region.

In their investigations of the effect of contraction ratio on turbulence, Ramjee and Hussain (1976) found that the turbulence intensities at the nozzle exit for a given contraction ratio were independent of the initial turbulence levels. The axial turbulence intensity, u'/U, decreases rapidly with increasing contraction ratio and reaches a minimum for a contraction ratio of 45 after which it increases slowly. The lateral turbulence intensities, v'/U, and w'/U, decrease monotonically with contraction ratio, and the decrease after a contraction ratio of 45 is small.

In closing, it should be remarked that due to the overall favorable pressure gradient along the nozzle, the boundary layers are thin and for most applications, the core flow field can be treated as inviscid. Accurate boundary layer predictions, and hence predictions of local wall shear and heat transfer, however, require that the effects of the additional strains imposed by the bulk flow acceleration and the wall curvature be considered. Most nozzle boundary layer calculations ignore these.

5.4.2 Diverging Ducts—Diffusers

As discussed in the introduction to this section, diverging ducts cause the velocity to decrease, *diffuse*, along the duct axis producing a rise in, *recovery* of, the static pressure. Hence, diverging ducts are commonly known as *diffusers*. Applications for diffusers are numerous. Diffusers are used downstream of wind tunnel test sections to reduce the net power consumption. Diffusers are also used downstream of turbines and compressors. In turbines, the total work extracted is proportional to the pressure drop across the turbine. For a given pressure drop, say between a boiler and a condenser, the use of a diffuser downstream of the turbine reduces the turbine back pressure and allows for a larger power output. In centrifugal pumps and compressors, the rotor primarily imparts kinetic energy to the flow. A downstream diffuser is therefore crucial in converting this kinetic energy to pressure.

Diffusers come in many shapes: planar, conical, annular, axi-radial, radial, etc. Although differing in geometrical shape, the flow phenomena encountered in virtually all types of diffusers are qualitatively similar. Given the importance of diffusers in many internal flow applications, the diffuser literature is vast. However, most of the literature tends to be design oriented and, therefore, concerned only with the overall performance of a particular geometry. Detailed measurements of diffuser internal flow fields are few, and virtually all fundamental investigations have been on the planar diffuser. Therefore, in this section, attention will be focused on the planar diffuser. It is the intent of this section to convey an understanding of diffuser flows (insofar as limited data allow this). A more design-oriented view and empirical design procedures currently practiced are given in Sec. 25.2 of this handbook.

A planar diffuser geometry is shown in Fig. 5.13. Two-dimensional mean flow can be assumed if the aspect ratio, b/W_1, is greater than 4 to 5. The divergence angle is 2θ, the axial length is N, and the following relationship can be derived from geometrical considerations

$$AR = W_2/W_1 = 1 + (2N \tan \theta)/W_1 \tag{5.99}$$

where, AR is the area ratio and W_2 the exit width. Therefore, the geometry of a straight-walled, planar diffuser can be described by two independent parameters, any two of 2θ, N/W_1, and AR.

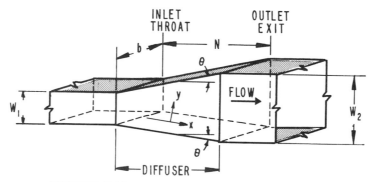

FIGURE 5.13 Two-dimensional diffuser geometry.

In most applications, the inlet duct leading to the diffuser is short, and the mean flow Reynolds numbers are in the turbulent range. The inlet flow is therefore composed of an *inviscid core* with thin turbulent wall boundary layers. Laminar boundary layers are unable to sustain significant pressure gradients and, therefore, are not desirable in a diffuser. Further, for typical applications with high inlet mean flow Reynolds numbers, even if the inlet boundary layer was laminar, it would separate right at the inlet, transition to turbulent flow and reattach at a short distance downstream as a turbulent boundary layer. In most applications, the exit of the diffuser is either connected to a *tailpipe* as shown in Fig. 5.13 or to a plenum. The exit condition has a significant effect on the overall performance of the diffuser and, therefore, is an important consideration in the overall diffuser design.

Figure 5.14 from Reneau *et al.* (1967) shows the flow regimes as a function of the geometric parameters 2θ and N/W_1. This flow regime map is best understood if one starts from the bottom of the map and progresses upwards (increasing 2θ). This is analogous to setting up a diffuser with a fixed wall length and then successively opening up the divergence angle of the diffuser. At small values of 2θ, the boundary layers remain attached (on the average) along the entire length of the walls, and the flow is generally steady. This regime is therefore known as the regime of *no appreciable stall*. Figure 5.15, also from Reneau *et al.*, (1967) shows the pressure recovery coefficient C_p as a function of 2θ. The pressure recovery coefficient is defined as

$$C_p = (p_2 - p_1)/(\tfrac{1}{2}\rho U_1^2) \tag{5.100}$$

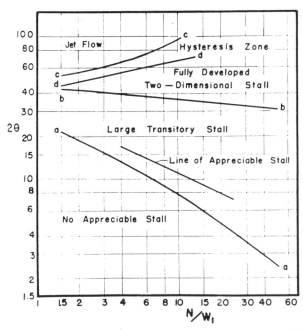

FIGURE 5.14 Diffuser Flow Regimes. (Reprinted with permission of ASME, "Performance and design of straight, two-dimensional diffusers," by Reneau, L. R., Johnston, J. P., and Kline, S. J., *J. Basic Eng.*, Vol. 89, No. 1, 141, 1967.)

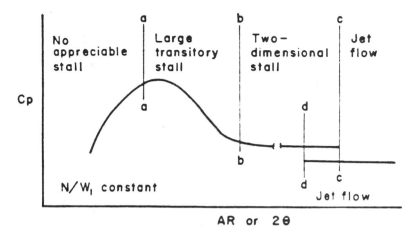

FIGURE 5.15 Relationship between flow regimes and a pressure recovery curve. (Reprinted with permission of ASME, ''Performance and design of straight, two-dimensional diffusers,'' by Reneau, L. R., Johnston, J. P., and Kline, S. J., *J. Basic Eng.*, Vol. 89, No. 1, p. 141, 1967.)

where, p_1 and p_2 are the static pressures at the diffuser inlet and exit, respectively, and U_1 is the mean flow velocity at the inlet. It is seen that in the *no appreciable stall* regime, C_p continues to rise with increasing 2θ. That is, the pressure gradient increases with increasing 2θ. At some value of 2θ, the boundary layer at one of the diverging walls will start to separate, i.e., *stall*, at the corner near the exit of the diffuser. Line a–a in Fig. 5.14 represents this condition and is known as the line of *first stall*.

As the divergence angle is increased beyond line a–a, the boundary layer separates at one of the walls, and a stall region with reverse flow is formed. The stall region varies in size with time and can switch from one wall to the other. Hence, this regime above line a–a and below line b–b is known as the *transitory stall regime*. In this transitory stall regime, the entire flow field becomes unsteady in response to the time varying stall cell size and location. Ashjaee and Johnston (1980) observed that in the region just slightly above line a–a, the stall region tends to completely washout followed by a slow build-up. At values of 2θ closer to line c–c, the stall region is always present, and only its size and location varies with time.

Data on flow unsteadiness in the transitory stall region are presented by Smith and Layne (1979) and Smith and Kline (1974). The maximum unsteadiness occurs at a value of 2θ about 20°; peak excursions in C_p attain $\pm35\%$, and peak inlet velocity fluctuations of up to $\pm12\%$ are observed. For short diffusers ($N/W_1 \leq 6$) the magnitude of the unsteadiness increases with diffuser length for a fixed value of 2θ. For long diffusers ($N/W_1 > 6$) the opposite trend is observed implying, therefore, that for a given 2θ, the unsteadiness peaks at some diffuser length. However, sufficient data are not available to definitely substantiate this. Available data show that the typical time scales are of the order 100 to 1000 N/U_1. This implies that the stall cell oscillation time scale is much larger than the through-flow time scale. The stall build-up portion of the stall oscillation cycle is observed to be considerably larger than the stall washout (total or partial) portion.

The variation of C_p with 2θ in the transitory stall regime is shown schematically in Fig. 5.15. The peak value of C_p lies in the transitory stall regime close to the a–a line. However, C_p starts dropping rapidly as 2θ is increased further and then becomes nearly constant as the *two-dimensional stall* regime is reached. The two-dimensional stall regime lies above the transitory stall regime. In this regime, a stable stall zone is found along one of the diffuser walls, and the main flow adheres to the other wall. Small perturbations at the diffuser exit can make the main flow switch from one wall to another. The value of C_p is virtually constant with 2θ in this regime. As 2θ is increased, the flow eventually transitions to the *jet flow* regime at line c–c. In the jet flow regime, the main flow forms a central jet in the diffuser, and stall regions are present on both walls. If 2θ is decreased from the jet flow regime, the transition to the two-dimensional stall regime occurs at line d–d. The region between lines c–c and d–d, therefore, represents a bi-stable region where either the jet flow regime or the two-dimensional stall regime can exist. The C_p in the jet flow regime is lower than that in the two-dimensional stall regime as shown in Fig. 5.15.

The above discussion was for the case where thin boundary layers were assumed at the inlet to the diffuser. However, Reneau *et al.* (1967) found that the regime boundaries are remarkably insensitive to inlet boundary layer thickness. The line a–a moves downwards by about 1 or 2 degrees while the other boundaries stay unchanged. Inlet Reynolds number has no effect on line a–a, but, in some instances, it can change the location of line b–b significantly. Similarly, there is some evidence that high levels of inlet turbulence (greater than 7–10%) can shift line b–b upwards. Finally, aspect ratio has little or no effect on the regime boundaries. The flow regime map is therefore universal for planar diffusers. There are no such detailed maps for other types of diffusers. However, it is expected that these flow regimes will exist in other types of diffusers as well. In fact, Sovran and Klomp (1967) in their correlation of diffuser performance found that optimum pressure recovery lines on an area ratio versus diffuser length map for planar, annular, and conical diffusers were remarkably similar.

The effect of various flow parameters on diffuser performance is described briefly in the following. The inlet Mach number has little or no effect on diffuser performance, except when the inlet chokes. Of course, if sonic velocity is reached at the diffuser inlet, the traditional subsonic diffuser with diverging walls will act as a supersonic nozzle and accelerate the flow instead of diffusing it. Reynolds number, in the range 10^5 to 10^6 (based on the inlet flow parameters), has a weak effect on C_p. Boundary layer thickness at the inlet is the most important flow parameter in terms of diffuser performance. Sovran and Klomp (1967) show that *simple area blockage* due to the boundary layer is a sufficient parameter to correlate diffuser performance. They define blockage as the ratio of the area occupied by the displacement thickness of the boundary layer to the inlet cross-sectional area. The pressure recovery C_p decreases almost linearly with blockage. Finally, inlet swirl is the only other flow parameter that has a substantial effect on C_p, however the effects of swirl are too geometry-specific to allow generalizations. Therefore, the effects of swirl will be left to Sec. 25.2 of this handbook where the actual design of diffusers is considered.

From the discussions of the flow regimes, it is clear that flow in diffusers is quite complex. However, steady turbulent boundary layer calculations have been developed by Bardina *et al.* (1981) that allow the prediction of C_p for two-dimensional planar diffusers for all except the jet flow regime. The flow in the diffuser is assumed to be composed of a inviscid core-flow zone and turbulent boundary layer zones. A one-dimensional flow model is used for the core flow. In the boundary layer zone, the integral momentum equation and an entrainment equation are used. A modified Coles boundary layer profile, capable of representing separated, reattaching, and attached zones, is used. Correlations are used for flow separation and for boundary layer behavior near separation. The core flow equation and boundary layer equations are solved simultaneously. It should be noted that even though this approach represents an important step in the prediction of diffuser performance, it does not properly account for the flow physics in the transitory stall region where the flow is far from steady or two-dimensional. Diffusers are challenging internal flow elements indeed.

5.5 SECONDARY FLOWS

In most ducts there is a dominant flow direction aligned with the duct axis. However, flow velocity components, *secondary flows*, in the cross stream plane can develop. Development of secondary flows requires the presence of cross stream stress gradients in the flow field. Cross stream normal stresses, i.e., pressure gradients, can occur due to turning of the mean flow or rotation of the duct about an axis perpendicular to the mean flow direction. In addition, cross stream gradients can occur due to unbalanced turbulent normal stresses in non-axisymmetric ducts. Secondary flows due to the cross stream pressure gradients in the mean flow are known as *secondary flows of the first kind*. The secondary flows in pipe bends are examples of such secondary flows. The secondary flows due to cross stream gradients of the turbulent stresses are called *secondary flows of the second kind*. An example of such secondary flows is the corner vortices in fully developed turbulent flow through a straight rectangular duct. Secondary flows of the second kind were discussed in Sec. 5.2 on fully developed flows in straight ducts. This section will focus on secondary flows of the first kind. Further, the discussion in this section will concentrate on secondary flows in stationary curved ducts such as pipe bends.

Consider a curved duct of circular cross section shown in Fig. 5.16. The duct radius is a, and the radius of curvature is R. The cross stream pressure gradient due to a mean flow velocity U through the duct is proportional to U^2/R, with lower

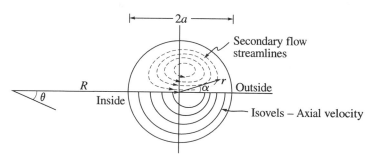

FIGURE 5.16 Toroidal coordinate system and flow field at small Dean numbers.

pressure at the inside of the bend. The shear layers at the duct walls do not have sufficient momentum to counteract this pressure gradient, therefore the shear layer at the outside of the bend migrates to the inside of the bend as shown schematically in Fig. 5.16. At the innermost point of the bend ($\alpha = \pi$), the shear layers from the opposite halves meet and flow towards the outside of the bend setting up two (or in some cases four) counter-rotating vortices. The effect of these secondary flows is to push the core flow towards the outside of the bend, and low momentum fluid accumulates on the inside of the bend. The accumulation of low velocity fluid at the inner wall creates a *blockage*, therefore conservation of mass requires that the axial velocity of the core flow increase causing a pressure drop in the streamwise direction. Also, as the core flow is pushed towards the outside of the bend, the shear layer there is thinned out causing a higher wall shear stress. Both the higher shear stress in the outer wall region and the acceleration of the core flow are the reasons for increased pressure losses in a bend.

Due to the rather widespread use of curved ducts such as bends in piping systems and turbomachinery blade passages, they have been the subject of numerous investigations. Excellent reviews of flow through curved ducts are given by Ward-Smith (1980) and Berger *et al.* (1983).

5.5.1   Fully Developed Secondary Flows—Flow Through Coiled Ducts

Fully developed flow is attained when the flow velocities (and hence the secondary flows) are independent of the axial position in the duct. Such a state is reached in constant radius coils. The terms in the flow equations containing the derivatives with respect to axial direction can be dropped. The governing equations (for the toroidal coordinate system shown in Fig. 5.16) remain nonlinear, and analytical solutions cannot be obtained. However, approximate solutions can be obtained for certain limiting cases. The parameter that governs the relative importance of the viscous, inertia, and curvature terms is the *Dean number*, named after Dean (1927) who obtained approximate solutions for the case of small curvature. The Dean number as originally defined by Dean is

$$K = \frac{2U_c^2 a^3}{\nu^2 R} \tag{5.101}$$

where U_c is the velocity at the center of the pipe. This form is inconvenient in that U_c is not known *a priori* and is a part of the solution itself. A more convenient alternative form used in reporting experimental data is

$$D_u = \text{Re } (a/R)^{1/2} \tag{5.102}$$

where the Reynolds number Re is based on the mean velocity through the pipe and the pipe diameter. A third form of the Dean number favored by some analysts is

$$D_p = \frac{Ga^2}{\mu} \left(\frac{2a^3}{\nu^2 R}\right)^{1/2} \tag{5.103}$$

where G is the pressure gradient at the center of the pipe. Note that for fully developed flow, G is a constant. The various Dean numbers can be related to each other for the case when the pipe curvature is very small and the fully developed flow is very close to that in a straight pipe

$$D_p = 4K^{1/2} = 4\sqrt{2}D_u \qquad (5.104)$$

The above relationship holds for the *small Dean number limit* where $K \leq 576$, $D_p \leq 96$ and $D_u \leq 17$. For higher values of the Dean number, there is no simple relationship between the three forms. This author prefers D_u over the other two formulations simply because it can be calculated in terms of the mean flow velocity which is both a primary design parameter and a primary experimental parameter. In all the above forms, the Dean number represents the ratio of the product of the inertia and centrifugal forces to the viscous forces.

For small Dean number, $D_u \leq 17$ (i.e., small curvature and small Reynolds number), Dean (1928) obtained a solution using a perturbation technique. The Fanning friction factor f_c for flow through the curved pipe obtained by Dean was

$$f_s/f_c = \left[1 - 0.03508 \left(\frac{D_u}{16.97}\right)^4 + 0.01195 \left(\frac{D_u}{16.97}\right)^6 + O\left(\frac{D_u}{16.97}\right)^8\right] \qquad (5.105)$$

where $f_s = 16/Re$, the Fanning friction factor for fully developed flow in a straight circular pipe (Poiseuille flow). Note that the above expression represents a perturbation of Poiseuille flow, when $R \rightarrow \infty$, $D_u \rightarrow 0$ and hence $f_c = f_s$. The secondary flow at small Dean number is a small perturbation on the laminar through flow. The secondary flow streamlines and axial velocity contours are shown in Fig. 5.16. Two helical vortices occur with their centers located at $r = 0.43$, $\alpha = \pm 90°$. The core flow is shifted slightly towards the outside of the bend. The location of the maximum velocity occurs at $r = 8.4 \times 10^4 \, D_u$.

As the Dean number is increased, the maximum velocity location moves further towards the outside of the bend, and the secondary flow becomes stronger. Several investigators including Barua (1963), Mori and Nakayama (1965), and Ito (1969), have obtained solutions in the limit of large Dean number, $D_u > 200$. They assume that the flow can be represented as an inviscid core flow with the secondary flow in the form of thin boundary layers along the walls and uniform flow through the core from the inside of the bend to the outside. These analyses show that f_c is proportional to $D_u^{1/2}$. Collins and Dennis (1975) performed numerical integration of the governing equations for $17 < D_u < 400$. Their numerical solutions show that for large D_u, f_c takes the following form

$$f_c/f_s = 0.1028 D_u^{1/2} \left[1 + 3.70 D_u^{-1/2} + O(D_u)^{-1}\right] \qquad (5.106)$$

White (1929) derived the following empirical correlation based on experimental data

$$f_c/f_s = \{1 - [1 - (11.6/D_u)^{0.45}]^{2.222}\}^{-1} \qquad (5.107)$$

for $17 < D_u < 1000$. Ward–Smith (1980) compared the predictions from the above expression and Eq. (5.106) against data and recommends the use of Eq. (5.106) for $D_u > 1000$. Equation (5.107) is more accurate in the range $17 < D_u < 1000$.

Strictly speaking, these expressions are only valid for circular coils with very small helix angles. Based on experimental data of Farrugia (1967) cited by Ward–Smith (1980), for radius ratios (a/R) between 17 to 30, varying the helix angle from 2° to 43° reduced the friction factor. The maximum deviation from values predicted by Eq. (5.107) was 12%. Further, at small helix angles, Farrugia's data showed that Eq. (5.107) was adequate (within 3%) for radius ratios down to 5. A further reduction in radius ratio produced a significant increase in the friction factor over that given by Eq. (5.107).

The above discussion relates to laminar flow in coiled ducts of circular cross section. Cheng *et al.* (1976) have analyzed laminar flow through rectangular cross section coils over a wide range of Dean numbers. Fully developed laminar flow in elliptical cross section coils has been investigated by Cuming (1952) and Ito (1951); flow through annuli have been analyzed by Kapur *et al.* (1964) and Topakoglu (1967). Collins and Dennis (1976) and Smith (1976) have analyzed flow through triangular cross section coils.

The effect of curvature is to delay transition of laminar flow to turbulent flow. Ito (1959) has proposed the following expression for the critical Reynolds number Re_{crit}, at which transition occurs in curved circular pipes

$$Re_{crit} = 2 \times 10^4 (a/R)^{0.32} \tag{5.108}$$

This expression is applicable for radius ratios in the range $10 < R/a < 860$. Note that at the largest radius ratio, the critical Reynolds number is 2300 which is close to that for a straight pipe. For a radius ratio of 10, the critical Reynolds number is 9600. Ward–Smith (1980) compared predictions from the above expression with available data and found good agreement except for data where flow disturbances were avoided. These low disturbance data generally indicate transition at Reynolds numbers higher than those predicted by Eq. (5.108). For most engineering applications, this expression should suffice.

Turbulent flow in circular pipe coils has been investigated experimentally by Ito (1959). He also correlated all previous data. The empirical expression derived by Ito is

$$f_c (R_c/a)^{1/2} = 0.00725 + 0.076[Re \ (a/R)^2]^{-0.25} \tag{5.109}$$

for $0.034 < Re \ (a/R)^2 < 300$. There are no data for coils of noncircular cross section. However, it is expected that the use of the hydraulic radius should prove adequate for most applications.

5.5.2 Secondary Flows in Bends

Circular arc bends are commonplace in ducting systems. Fully developed secondary flow is seldom encountered in bends except possibly for very large radius bends, therefore the correlations for fully developed flow presented in the previous subsection are generally inapplicable. Consider a circular arc bend with straight constant area ducts at the inlet and exit of the bend. These inlet and exit sections are referred to as *tangents*. The pressure distribution along the bend and tangents is shown in Fig. 5.17. The bend affects both the upstream and downstream flow in the tangents. The influence of the bend is felt at $x_i/D = 2$ to 5 from the physical entrance to the

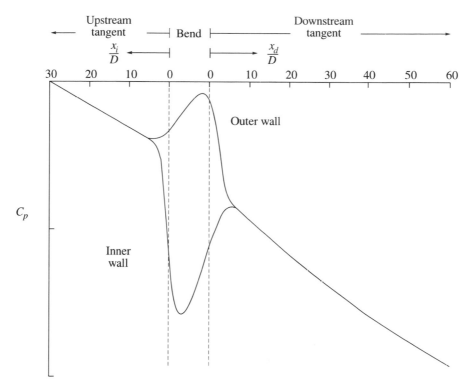

FIGURE 5.17 Pressure distributions in a bend.

bend. The downstream influence on the other hand extends much farther (up to $x_d/D = 50$). It can be surmised that the shape and length of the tangents have significant effects on the flow in the bend. The pressure distribution in the bend shows that on the inner surface, the streamwise pressure gradient is initially favorable, followed by an unfavorable pressure gradient. The situation is reversed on the outer wall. The development of the flow through the bend is significantly affected by these streamwise pressure gradients which are a function of the radius ratio, shape of the bend, and boundary conditions imposed by the inlet and outlet tangents. Other parameters affecting the development of the secondary flows in the bend are inlet boundary layer thickness and the Reynolds number. Due to the complexity of the flow through bends, detailed measurements of the secondary flow are difficult. Therefore, most experimental investigations have focused on measuring the pressure drops and pressure distributions along the bend walls. Ito (1960) and Ward–Smith (1980) present a comprehensive set of such data. These data are primarily used to deduce the pressure loss through bends and will be discussed further in Chap. 25 of this handbook.

Measurement of the streamwise total pressure and velocities have been obtained by Ward–Smith (1968), Rowe (1970), Mori *et al.* (1971). Measurements of the secondary flow velocity components using the laser doppler anemometer have been reported by Agrawal *et al.* (1978) for laminar flow in a circular cross section bend, and by Humphrey *et al.* (1977, 1981) and Taylor *et al.* (1982) for laminar and turbulent flows in 90° bends of square cross section. The picture that emerges from these detailed measurements shows a complex interaction between the core flow and the viscous wall layers. For developing and fully developed laminar flow at the inlet to the bend, the thick viscous layers cause a large secondary flow to develop which accumulates the low velocity fluid at the inner wall of the bend pushing the core to the outside of the bend. The secondary flow velocities can reach 60% of the mean throughflow velocity, and in the bend, a pair of helical vortices form. Figure 5.18a shows the U/U_m contours measured by Taylor *et al.* (1982) for a developing laminar flow at the inlet; U_m is the mean velocity through the bend. The shift of the core flow towards the outside is clearly evident as the flow progresses through the bend. In the straight duct following the bend, a second set of vortices is formed as the flow distortions caused in the bend even out. For a fully developed laminar flow at the entry of a square cross section bend with a radius ratio of 2.3, Humphrey *et al.* (1977) found a separation region on the outside wall near the bend inlet. This separation is caused by the initial adverse pressure gradient at the outer wall near the inlet, however a separated region was not observed for a developing laminar profile at the inlet to the same bend.

In the case of the fully developed turbulent flow at the inlet to the same 90°, 2.3 radius ratio bend, Taylor *et al.* (1982) find that the core remains near the inside wall of the bend out to about the 60° plane and then rapidly moves to the outer wall. This is seen in the contours of U/U_m measured by Taylor *et al.* (1982) shown in Fig. 5.18b. The reason forwarded is that due to the thinner inlet boundary layers, the strength of the secondary flows is less. Therefore, the core flow is dominated by the streamwise pressure gradients, and the initial favorable pressure gradient at the inner wall causes the core flow to stay near the inner wall. The core flow then shifts to the outer wall later in the bend as it encounters the adverse pressure gradient at the inner wall and the increasing effect of the secondary flow. The magnitude of the secondary flow velocities is about 40% of the mean through flow velocity. The secondary flow is in the form of a pair of helical vortices in the bend and in the straight exit tangent. Taylor *et al.* (1982) also compared these data with data for a 90° bend with a radius ratio of 7 and an identical square cross section. They found that the secondary flow velocities in the radius ratio 7 bend were half of those in the 2.3 radius ratio bend. That is, the secondary flows decrease with increasing radius ratio. This is expected, given that the cross stream pressure gradient that drives the secondary flow is reduced as the radius ratio is increased. However, reduced radius ratio also decreases the streamwise pressure gradients. Hence, the core flow continually moved towards the outside of the bend. Recall that for the 2.3 radius ratio bend, the core flow initially moved to the inside of the bend. The preceding discussion illustrates the complex interactions that occur in bends.

Due to the complex nature of the secondary flow development in bends, analytical investigations are limited. The pressure distribution through the bend has been calculated by some investigators assuming the flow through the bend to be inviscid. As expected, such predictions are not accurate and do not give any information about the secondary flow or the pressure losses incurred. Another class of analyses examines the transport of inlet cross stream vorticity through the bend and calculates the resulting streamwise vorticity, and hence the strength of the secondary flow at the bend exit. This approach has been used to calculate the effect of secondary flows at the exit of turbomachine blades; a review of this approach is given by Horlock

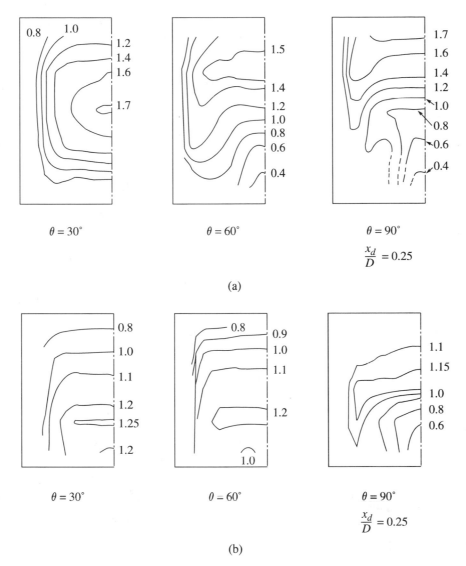

FIGURE 5.18 Isovelocity (U/U_m) contours in a bend with a square a duct. (a) Developing laminar flow at the inlet and (b) fully developed turbulent flow at the inlet. (Reprinted with permission of ASME, ''Curved Ducts with Strong Secondary Motion: Velocity Measurements of Developing Laminar and Turbulent Flow,'' by Taylor, A. M. K. P., Whitelaw, J. H., and Yianneskis, M., *J. Fluids Eng.*, Vol. 104, No. 3, pp. 350–359, Sept. 1982.)

and Lakshminarayana (1973). Briefly, the streamwise component of the vorticity transport equation written in the streamline coordinates (s, n, b) for incompressible laminar flow is

$$U \frac{\partial \xi}{\partial s} + \frac{U\eta}{R} = \xi \frac{\partial U}{\partial s} + \eta \frac{\partial U}{\partial n} + \zeta \frac{\partial U}{\partial b} + \nu \nabla^2 \xi \qquad (5.110)$$

where ξ, η, and ζ are the vorticity components in the s, n, and b directions, respectively; see Horlock and Lakshminarayana for a more rigorous form of this equation. Note that s is the direction along the stream lines, n is the normal to the streamline and is aligned with the radius of curvature of the duct, and b is orthogonal to s and n. Recognizing that $\eta = \partial U/\partial b$ and $\zeta = -(\partial U/\partial n + U/R)$, the above equation can be written as

$$\frac{\partial}{\partial s}(\xi/U) \cong \frac{-2\eta}{UR} + \frac{\nu\nabla^2\xi}{U^2} \tag{5.111}$$

Similarly, the vorticity transport in the n direction is given by

$$\frac{\partial}{\partial s}(U\eta) = \frac{U\zeta}{\tau} + \nu\nabla^2\eta \tag{5.112}$$

where τ is the radius of torsion, i.e., the radius of curvature of the streamline in the b direction. If it is assumed that the flow is inviscid and that $\tau = \infty$, that is, the stream-surfaces (*Bernoulli surfaces*) are flat as they traverse the bend, Eq. (5.112) shows that η is constant along a stream line. Finally, for a constant bend radius R and constant flow velocity U through the passage, Eq. (5.111) can be integrated yielding

$$\xi_2 - \xi_1 = -2\eta\theta \tag{5.113}$$

where θ is the total turning angle through the bend and subscripts 1 and 2 refer to the inlet and exit planes, respectively. At the inlet, η is equal to dU/db, therefore an inlet velocity profile with sidewall boundary layers will generate streamwise vorticity, i.e., secondary flow at the bend outlet. The secondary flow velocities V and W at the bend exit (or any other plane in the bend) can be calculated from the secondary vorticity ξ_2 as follows. The bend cross section is assumed to be rectangular for convenience (therefore, n and b are rectilinear coordinates), and it is further assumed that at the exit, $(\partial V/\partial n)$, $(\partial W/\partial b) \gg (\partial U/\partial s)$. Therefore, the continuity equation can be written as

$$(\partial V/\partial n) + (\partial W/\partial b) = 0 \tag{5.114}$$

and a *secondary stream function* ψ can be defined such that $V = -(\partial\psi/\partial b)$ and $W = (\partial\psi/\partial n)$. Therefore, from the definition of ξ and Eq. (5.113)

$$(\partial^2\psi/\partial n^2) + (\partial^2\psi/\partial b^2) = -\xi_2 = 2\eta\theta - \xi_1 \tag{5.115}$$

This equation is a form of the Poisson equation, and solution procedures are straightforward. Most of the investigations using the above approach have been for inviscid flows and give solutions to Eq. (5.111) with varying degrees of refinements; see Horlock and Lakshminarayana (1973). It should be noted that this approach is only valid for small amounts of turning and when the secondary flows are a small perturbation superimposed on the main flow, the assumption of undistorted Bernoulli surfaces implies that the secondary flow does not displace the core flow. As discussed earlier, in most bends encountered in practice, strong secondary flows develop, and the surfaces of constant pressures are severely distorted. Therefore, the usefulness of the above approach is limited.

In recent years, numerical techniques for solving the Navier–Stokes equations have become available [see Patankar (1980)]. Several investigators have used these to calculate the laminar flow through bends with varying degrees of success [Pratap and Spaulding (1975), Humphrey *et al.* (1977), and Buggeln *et al.* (1980), to cite but a few]. In the simulation of the flow through the rectangular cross section bend they tested, Humphrey *et al.* (1977) found that a rather large number of grids points are needed to obtain adequate solution accuracy. This implies large computational times and hence cost. Nonetheless, these approaches offer the only alternative to complete reliance on empirical data.

5.6 SEPARATED INTERNAL FLOWS

Separated internal flows are defined as flows where shear layers occur in the main body of the flow. This situation is most commonly encountered when the flow in a duct separates from the wall such as in a stalled diffuser or at a sudden expansion. A *free* shear layer (see Sec. 4.10) occurs between the core flow and the separated region. Another example is the case where a secondary fluid stream is introduced into the duct with a velocity different than that of the primary flow in the duct. In this case, the shear layer is formed between the two streams with dissimilar velocities as in the case of ducted coaxial jets, e.g., an *ejector*. Separated internal flows are one of the most complex flows and are far from totally understood. Among the numerous duct geometries with separated internal flows, two of the more extensively investigated are the sudden expansion and ducted coaxial jets. This section will, therefore, focus on these two geometries as a means of providing more general insight on separated internal flows.

5.6.1 Sudden Expansions

Consider the duct shown in Fig. 5.19 where the duct area abruptly increases. The incoming wall shear layer separates at point A, and develops as a free shear layer

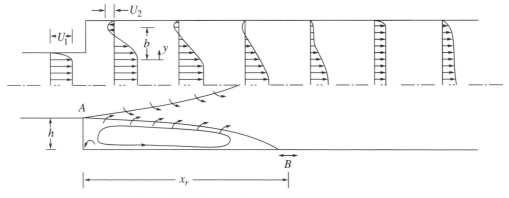

FIGURE 5.19 Sudden expansion duct flow.

downstream of the sudden expansion. The free shear layer entrains fluid from the separated zone and the core flow as it grows, and, if the duct is sufficiently long, the free shear layer reattaches to the duct wall at some downstream location B. The flow continues to decelerate downstream of the reattachment point until finally, fully developed flow is attained. Due to flow entrainment by the shear layer from the separated zone, continuity requires the formation of a primary recirculation zone as shown in Fig. 5.19 with the flow going in the upstream direction near the wall. The velocities in this primary recirculation zone are 10 to 20% of the inlet velocity. In addition to the primary recirculation eddy in the separated zone, a smaller secondary vortex system will occur in the corner. The size of the separated zone, i.e., the distance x_r from the sudden expansion to the reattachment point B, is determined by the entrainment rate and growth of the shear layer. These properties of the shear layer depend on whether the shear layer is laminar or turbulent.

First, consider the asymmetric two-dimensional sudden expansion where the area expansion is due to a step at only one wall of a two-dimensional channel. Goldstein *et al.* (1970) performed experiments on backward facing two-dimensional steps and found that for the case where the shear layer was laminar at separation and reattachment, the following expression correlated the observed reattachment length

$$x_r/h = 2.13 + 0.021 \, \text{Re}_h \qquad (5.116)$$

where $\text{Re}_h = U_1 h/\nu$, and h is the step height.

Other investigations of a backward facing step in a two-dimensional channel were performed by Armaly *et al.* (1983). In these experiments, the expansion ratio was 1.94, that is, the step height h was nearly equal to the width of the inlet channel W_1, and a fully developed flow was present at the entrance of the sudden expansion. In the laminar shear layer case, the variation of the reattachment length with Reynolds number is not quite linear, however the data are in remarkable agreement with Eq. (5.116). Armaly *et al.* (1983) contend that this agreement is accidental and that their numerical studies show that in laminar flows, the reattachment length is also a function of area ratio and the characteristics of the velocity profile at the sudden expansion. Nonetheless, the above correlation should give reasonable estimates of the laminar reattachment length for asymmetric two-dimensional expansions with area ratios less than 2.

The situation is far more complex for the symmetric planar sudden expansion (with steps of equal heights at both walls of a two-dimensional planar channel). One would expect that for such geometries, separated regions of equal size would form at both walls, i.e., the flow pattern would be symmetric about the axis of the duct. Durst *et al.* (1974), in their experimental investigations of laminar flow in a symmetric planar expansion of area ratio 3, found that at a Reynolds number (based on the peak upstream velocity and the step height) of 56, the flow was symmetric about the duct axis and the separated zones at both walls were similar with similar reattachment lengths. However, at higher Reynolds numbers, the flow became asymmetric about the duct axis, with the separation zone on one wall being considerably larger. Also, additional separation regions were observed further downstream. The Reynolds number at which the flow becomes asymmetric appears to be a function of the area ratio of the sudden expansion; symmetric flow is maintained for higher Reynolds numbers as the area ratio (and hence step size) decreases. There are insufficient data at the present time to provide any correlations for the laminar reattachment lengths for symmetric planar sudden expansions.

Laminar flow in axisymmetric sudden expansions was studied by Macagno and Hung (1967). The separated region was found to be nearly axisymmetric. It is speculated that this is due to the fact that the separated region for axisymmetric sudden expansions is a single annular region. Further, Macagno and Hung (1967) found that the reattachment length was a linear function of the Reynolds number for the 4:1 sudden expansion investigated. The following correlation for the reattachment length was found

$$x_r/h = 0.6 + 0.17 \, \mathrm{Re}_h \tag{5.117}$$

Note that the above correlation implies that the reattachment length is a linear function of the Reynolds number similar to that for backward facing planar steps; see Eq. (5.116). However, the multiplicative coefficient for the Reynolds number for the axisymmetric case is eight times greater than that for the planar case. When using the above correlation, it should be kept in mind that it is derived from investigations of a 4:1 expansion only, and its validity at other area ratios has not been demonstrated.

Data on the transition of shear layers from laminar to turbulent flow in internal flows are scarce. Based on the investigations of Goldstein *et al.* (1970) on backward facing planar steps, transition is governed by two criteria. The shear layer remains laminar only if $\delta^*/h > 0.4$ and $\mathrm{Re}_h < 520$, where δ^* is the displacement thickness of the boundary layer at the inlet to the sudden expansion. If either of these conditions is not met, the shear layer transitions to turbulent flow before reattachment. It should be noted that in the investigations of Goldstein *et al.* (1970), the flow to the inlet of the sudden expansion was not fully developed in that the flow had an inviscid core and a laminar boundary layer at the inlet to the sudden expansion. Measurements of Armaly *et al.* (1983) with a fully developed flow at the entrance to an asymmetric planar sudden expansion show that the shear layer is laminar for $\mathrm{Re}_h < 570$. For $570 < \mathrm{Re}_h < 3170$ the flow is transitional, and the fully turbulent regime is reached when $\mathrm{Re}_h > 3170$. Based on these limited data, the criterion for shear layer transition from laminar to turbulent appears to be that $\mathrm{Re}_h > 520$ to 570.

Once the shear layer becomes fully turbulent, properties such as growth rate and hence the reattachment length become independent of Reynolds number. In turbulent shear layers, the entrainment rate varies with time due to the unsteady nature of turbulent processes. Therefore, the size of the separated zone, and hence the reattachment length, varies with time about some mean value. The values of reattachment lengths discussed in the following are the time averaged values. For asymmetric planar sudden expansions (i.e., step at only one wall of a planar channel), Kim *et al.* (1980) present data showing that the reattachment length was 7 step heights (i.e., $x_r/h = 7$) for area ratios of 1.3–2. The reattachment length decreases somewhat to $x_r/h = 5.5$ to 6, for area ratios 1.2 and less (small step heights). Abbott and Kline (1962) measured reattachment lengths for symmetric planar expansions and found that for area ratios less than 1.5, the separated zones at both walls of the channel were approximately equal in size ($x_r/h \sim 8$), and the flow was symmetric about the centerline of the channel. At larger area ratios, the separated zones were different in size with $x_r/h < 7$ for the smaller separated zone and $x_r/h > 7$ for the larger one. The reattachment lengths for both zones increased with increasing area ratios, the longer reattachment length growing at a faster rate. At an area ratio of 4, the reattachment lengths for the two zones differed by a factor of 4. The flow was therefore highly asymmetric about the channel center line.

Although no concrete data are available on the symmetry of separated zones in axisymmetric sudden expansions, it is expected that significant circumferential asymmetries will not occur due to the fact that the entire separated zone is a single connected region. Available data [Lipstein (1962), Chaturvedi (1963), Moon and Rudinger (1977), Kangovi and Page (1979)] indicate that the reattachment length is independent of the Reynolds number and $6 < x_r/h < 9$. Note, therefore, that the reattachment length for the turbulent shear layer case is similar for both planar and axisymmetric sudden expansions. The reattachment length is a measure of the size of the separated zone and hence the growth of the shear layer. It can, therefore, be surmised that the characteristics of the shear layer are reasonably geometry independent. If it is assumed that the growth of the shear layer is similar to that for a free jet (see Sec. 4.10), the width of the shear layer b grows linearly with distance x from the jet origin. That is,

$$b = cx, \qquad (5.118)$$

where for free jets, the constant c has a value between 0.2 and 0.3 (Abramovich (1963)). If it is further assumed that the centerline of the shear layer remains parallel to the duct axis, reattachment will occur when $b/2 = h$. Using a value of $c = 0.25$ and assuming that the origin of the shear layer coincides with the inlet to the sudden expansion, the reattachment length predicted by Eq. (5.118) is 8 step heights, which is in good agreement with the observed reattachment lengths.

The mean velocity profiles and turbulence profiles in sudden expansions have been measured by several investigators [Kim *et al.* (1980), Chaturvedi (1963), Davies and Snell (1977), Moss *et al.* (1977)]. The variation of the mean velocity and turbulence quantities across the shear layer are similar to those in the mixing region of a free jet. The mean velocity profile across the shear layer can be approximated by the following expression suggested by Abramovich (1963) for free shear layers

$$(U - U_2)/(U_1 - U_2) = (1 - \eta^{3/2})^2 \qquad (5.119)$$

where, U is the velocity at some location η, U_1, and U_2 are the velocities in the core and separated regions, and $\eta = y/b$; see Fig. 5.19. As stated earlier, the velocity near the wall in the separated region is negative and about 10 to 20% of the core flow velocity. The turbulence intensities (u'/U_1, etc.) and the turbulence shear stress ($\overline{u'v'}/U_1^2$) peak in the shear layer. Figure 5.20 shows the variation of the peak longitudinal turbulence intensity u'/U_1 and the turbulent shear stress $\overline{u'v'}/U_1^2$ as a function of the axial distance x. Note that the fluctuating components u', v', etc., have been normalized by the inlet velocity U_1 instead of the local velocity, U. The data band shown includes data for asymmetric planar and axisymmetric sudden expansions. The initial development of the shear layer is dependent on the properties of the incoming boundary layer at the separation point and is, therefore, test facility dependent. The data band partially reflects this and also the fact that the measurements near the upstream separation point are extremely difficult and prone to sig-

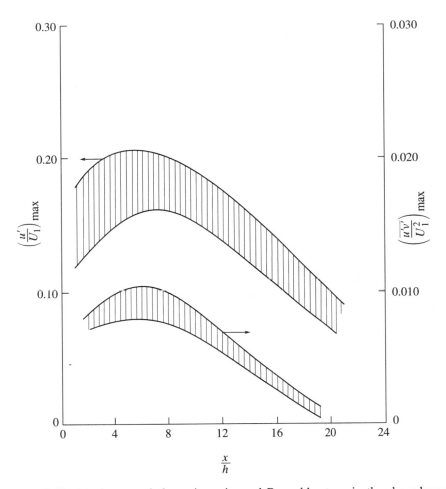

FIGURE 5.20 Maximum turbulence intensity and Reynolds stress in the shear layer in a sudden expansion.

nificant uncertainties. The general trend, however, is that the turbulence intensity increases as the shear layer develops downstream of the separation point and reaches a maximum value of between 16 and 20% near the reattachment point. The turbulent shear stress shows a similar but less pronounced trend. The turbulent shear stress $\overline{u'v'}/U_1^2$ reaches a maximum value of between 0.008 and 0.010 near the reattachment point.

For free shear layers, based on Prandtl's theory, the turbulent shear stress is given by

$$-\overline{u'v'} = v_t \, dU/dy \qquad (5.120)$$

The eddy diffusivity v_t is given by

$$v_t = kb \, (U_1 - U_2) \qquad (5.121)$$

where k is an empirical constant. If the mean velocity profile in Eq. (5.119) is used, a value of k of 0.008 produces the shear stress levels shown in Fig. 5.20. The mixing length l_t is related to the constant k and the shear layer width b by $l_t = b\sqrt{k}$, hence $l_t = 0.09\,b$.

Teyssandier and Wilson (1974) used the relationships given by Eq. (5.118)–(5.121) to analyze the flow through sudden expansions up to the reattachment point with reasonable success. The region around the reattachment point and downstream is difficult to analyze, since the relationships for the shear layer given above are no longer valid. After the shear layer reattaches, its characteristics change from those of a *free shear layer* to those of a *wall shear layer*. This process is complex, and Kim *et al.* (1980) present empirical formulations for the mixing length in this transition zone. It should also be mentioned that two-equation turbulence models such as the k-ϵ models do reasonably well in prediction of the flow through the sudden expansion including the region downstream of the reattachment point.

The pressure rise across a sudden expansion is traditionally obtained by a *Borda–Carnot analysis* where a momentum balance is performed between a plane at the sudden expansion and a downstream plane where the velocity is assumed to be uniform. Shear at the walls is neglected, and it is assumed that the static pressure is uniform across the two planes. The following expression results for the pressure rise across the sudden expansion normalized by the inlet dynamic pressure

$$C_{p_{B-C}} = (p_1 - p_2)/(\tfrac{1}{2}\rho\,U_1^2) = (2/AR)\,(1 - 1/AR) \tag{5.122}$$

where AR is the area ratio. This expression gives surprisingly good predictions of the actual pressure rise for turbulent flow through asymmetric planar and axisymmetric sudden expansions. Figure 5.21 shows the pressure rise along the axis of a symmetric sudden expansion constructed from the data of Ackeret (1967). The actual pressure rise is normalized by the pressure rise predicted by Eq. (5.122). It is seen that the net pressure rise across the sudden expansion is within 5% of that

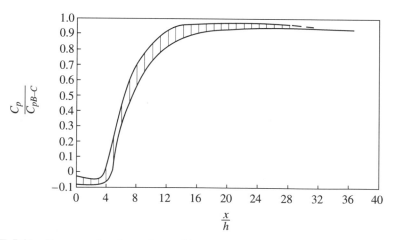

FIGURE 5.21   Pressure rise through a sudden expansion in a duct. [Based on data in "Aspects of Internal Flows," by Ackeret, Jakob, *Fluid Mechanics of Internal Flows*, Sovran, G. (Ed.), pp. 1–26, 1967 with permission of Elsevier Science Publishing Co., Inc.]

predicted by the Borda–Carnot analysis. Also note that a substantial portion of the pressure rise occurs after the reattachment zone, and the maximum pressure recovery occurs about 20 step heights downstream of the plane of the sudden expansion. In contrast, Eq. (5.122) will not do well in laminar flows where the wall shear is relatively large.

5.6.2 Confined Coaxial Jets

The confined coaxial jet geometry is shown in Fig. 5.22. The jet inlet velocity is U_1, and the velocity of the outer stream is U_2. The pressure along the centerline increases for $U_1/U_2 > 2$, therefore this geometry is normally referred to as a *jet pump* or *ejector*. In this section, the discussion will be limited to the geometry where the diameter of the mixing tube is constant and the flow is incompressible. For this case, the maximum pressure occurs at x/R_2 between 10 and 14 where R_2 is the radius of the mixing tube. The flow in a confined coaxial jet geometry can be divided into three regions as shown in Fig. 5.22. In Region 1, potential cores are present in both the jet and the outer stream, with a shear layer between the two, and a shear layer is present at the wall of the mixing tube. In Region 2, the jet shear layer has grown to the point where a potential core does not exist in the jet. A potential core still exists in the outer stream, but it is decreasing due to the growth of the jet shear layer and the wall shear layer. Finally, in Region 3, the jet and wall shear layers have merged and the flow proceeds to develop into a fully developed pipe flow. In all practical applications, the flow is turbulent, therefore the following discussion will be limited to turbulent ducted coaxial jets.

In Region 1, the development of the shear layer between the jet and the annulus flow is similar to that between unconstrained coflowing streams. Razinsky and Brighton (1971) show that the mean velocity profile in the jet shear layer is well represented by Eq. (5.119). The width of the shear layer, b, increases in the downstream direction. Although there are no detailed measurements available, the growth of the jet shear layer can be estimated by

$$b = c'x, \tag{5.123}$$

where

$$c' = c\,(1 - U_2/U_1)/(1 + U_2/U_1). \tag{5.123a}$$

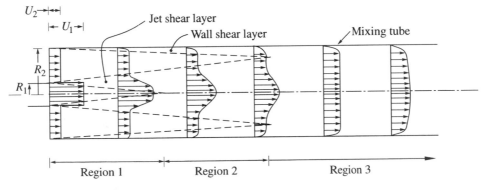

FIGURE 5.22 Coaxial jets in a duct.

The empirical constant c is the same as the one in Eq. (5.118) and has a value between 0.2 and 0.3. The effect of the annulus flow, therefore, is to reduce the rate of spread of the jet shear layer compared to that for a free jet. The variation of the turbulent parameters across the shear layer is similar to that for shear layers in the sudden expansion discussed in the previous subsection, with the maximum values of the turbulence intensities and shear stress occurring at approximately $b/2$. The magnitude of the turbulence quantities increases with downstream distance, and typical values of the u'/U_1 and $\overline{u'v'}/U_1^2$ are 0.2 and 0.01 at the end of Region 1. The value of the empirical constant k in the expression for eddy viscosity given by Eq. (5.121) is around 0.007. Therefore the mixing length $l_t \cong 0.084\,b$. The developing jet shear layer is, therefore, quite similar to the shear layer in a sudden expansion described in the previous subsection. In Region 1, the wall shear layer at the mixing tube wall is similar to a developing turbulent boundary layer.

In Region 2, the jet shear layer now extends from the duct centerline to the annulus flow region which still has a potential core. The mean velocity distribution across the jet shear layer in this region is also approximated well by Eq. (5.119). The turbulence shear stress in the jet shear layer $\overline{u'v'}/U_1^2$ peaks in this region. Hedges and Hill (1974) used a slightly higher value of mixing length in this region than in Region 1; $l_t = 0.09b$ was used which results in a value of 0.008 for the constant k in Eq. (5.121). The development of the mixing tube wall layer in this region depends on the velocity ratio U_1/U_2. For values of $U_1/U_2 > 2$, an adverse pressure gradient is present in this region, and that can cause the wall shear layer to separate. In any case, the wall shear layer grows rapidly in this zone until it merges with the jet shear layer at the end of Region 2.

In Region 3, the entire width of the mixing tube is now a shear flow, and the flow begins its adjustment to the fully developed state. This zone is akin to the region downstream of the reattachment point in a sudden expansion, and the turbulent processes that occur are not well understood. Hedges and Hill (1974) used two separate relations for the mixing length in this region. Near the wall, the mixing length was $l_t = 0.41\,(R_2 - r)$ where R_2 is the radius of the mixing tube. In the core region, $l_t = 0.28\,R_2$ giving a value of $k = 0.078$ which is an order of magnitude greater than that in Regions 2 and 3. Oosthuizen and Wu (1977) used the mixing length models proposed by Hedges and Hill (1974) and obtained satisfactory agreement with the mean velocity data from their experiments on ducted coaxial jets. In fact, the predictions based on the empirical mixing length formulation were somewhat better than those using two equation models such as the $k - l$ and $k - \epsilon$ models. The implication of the large change in the mixing length required to match experimental data is that the turbulence characteristics in Region 3 are significantly different than those in *free* or *wall shear layers*.

REFERENCES

Abbott, D. E. and Kline, S. J., "Experimental Investigation of Subsonic Turbulent Flow over Single and Double Backward Facing Steps," *J. Basic Engrg., Trans. ASME*, Vol. 84, p. 317, 1962.

Abramovich, G. N., *The Turbulent Theory of Jets*, MIT Press, Cambridge, MA, 1963.

Ackeret, J., "Aspects of Internal Flows," *Fluid Mechanics of Internal Flows*, Sovran, G. (Ed.), Elsevier, New York, 1967.

Adams, E. W. and Johnston, J. P., A Mixing-Length Model for Prediction of Convex Curvature Effects on Turbulent Boundary Layers,'' ASME Paper No. 83-GT-80, 1983.

Agrawal, Y., Talbot, L., and Gong, K., ''Laser Anemometer Study of Flow Development in Curved Circular Pipes,'' *J. Fluid Mech.*, Vol. 85, p. 497, 1978.

Armaly, B. F., Durst, F., Periera, J. C. F., and Schonung, B., ''Experimental and Theoretical Investigation of Backward-Facing Step Flow,'' *J. Fluid Mech.*, Vol. 127, p. 473, 1983.

Ashjaee, J. and Johnston, J. P., ''Straight Walled, Two-Dimensional Diffusers—Transitory Stall and Peak Pressure Recovery,'' *J. Fluids Engrg., Trans. ASME*, Vol. 102, p. 275, 1980.

Atkinson, B., Brocklebank, M. P., Card, C. C. H., and Smith, J. M., ''Low Reynolds Number Developing Flows,'' *AIChE J.*, Vol. 15, p. 548, 1969.

Barbin, A. R. and Jones, J. B., ''Turbulent Flow in the Inlet Region of a Smooth Pipe,'' *J. Basic Engrg., Trans. ASME*, D, Vol. 85, p. 29, 1963.

Bardina, J., *et al.*, ''A Prediction Method for Planar Diffusers,'' *J. Fluids Engrg., Trans. ASME*, Vol. 103, p. 315, 1981.

Barua, S. N., ''On Secondary Flow in Stationary Curved Pipes,'' *Q. J. Mech. Appl. Math.*, Vol. 16, p. 61, 1963.

Beavers, G. S., Sparrow, E. M., and Magnuson, R. A., ''Experiments on Hydrodynamically Developing Flow in Rectangular Ducts of Arbitrary Aspect Ratio,'' *Int. J. Heat Mass Transfer*, Vol. 13, p. 689, 1970.

Berger, S. A., Talbot, L., and Yao, L-S., ''Flow in Curved Pipes,'' *Ann. Rev. of Fluid Mech.*, Vol. 15, p. 46, 1983.

Berker, R., *Handbuch der Physik*, Vol. VIII, Springer–Verlag, Berlin, 1963.

Bradshaw, P. (Ed.), *TURBULENCE: Topics in Applied Physics*, Springer–Verlag, New York, 1978.

Bradshaw, P., ''Effect of Streamline Curvature on Turbulent Flow,'' AGARDograph No. 169, 1973.

Bradshaw, P., ''The Analogy Between Streamline Curvature and Buoyancy in Turbulent Shear Flow,'' *J. Fluid Mech.*, Vol. 36, p. 177, 1969.

Bradshaw, P., Dean, R. B., and McEligot, D. M., ''Calculation of Interacting Turbulent Shear Layers: Duct Flow,'' *J. Fluid Mech.*, Vol. 95, p. 214, 1973.

Buggeln, R. C., Briley, W. R., and McDonald, H., ''Computation of Laminar and Turbulent Flow in Curved Ducts, Channels, and Pipes using the Navier–Stokes Equations,'' Science Res. Assoc. Rept. R80-920006-F, 1980.

Chambers, T. L. and Wilcox, D. C., ''Critical Examination of Two Equation Turbulence Closure Models for Boundary Layers,'' *AIAA J.*, Vol. 15, p. 821, 1977.

Chaturvedi, M. C., ''Flow Characteristics of Axisymmetric Expansions,'' *ASCE J. Hydraulic Div.*, Vol. 89, p. 61, 1963.

Cheng, K. C., Lin, R-C., and Ou, J-W., ''Fully Developed Laminar Flow in Curved Rectangular Channels,'' *J. Fluids Engrg., Trans. ASME*, Vol. 98, p. 41, 1976.

Colebrook, C. F., ''Turbulent Flow in Pipes, with Particular Reference to the Transition Between the Smooth and Rough Pipe Laws,'' *J. Inst. Civ. Eng.*, Vol. 11, p. 133, 1939.

Collins, W. M. and Dennis, S. C. R., ''The Steady Motion of a Viscous Fluid in a Curved Tube,'' *Q. J. Mech. Appl. Math.*, Vol. 28, p. 133, 1975.

Collins, W. M. and Dennis, S. C. R., ''Viscous Eddies Near a 90° and a 45° Corner in Flow Through a Curved Tube of Triangular Cross Section,'' *J. Fluid Mech.*, Vol. 76, p. 417, 1976.

Cuming, H. G., "The Secondary Flow in Curved Pipes," *ARC Reports and Memoranda 2880*, 1952.

Davies, T. W. and Snell, D. J., "Turbulent Flow over a Two-Dimensional Step and its Dependence upon Upstream Flow Conditions," *Proc. Symp. Turbulent Shear Flows*, p. 13.29, 1977.

Dean, R. B., "Reynolds Number Dependence of Skin Friction and Other Bulk Flow Variables in Two-Dimensional Rectangular Duct Flow," *J. Fluids Eng., Trans. ASME*, Vol. 100, p. 215, 1978.

Dean, W. R., "Note on the Motion of Fluid in a Curved Pipe," *Phil. Mag.*, Vol. 4, p. 208, 1927.

Dean, W. R., "The Stream-Line Motion of Fluid in a Curved Pipe," *Phil. Mag.*, Vol. 4, p. 673, 1928.

Durst, F., Melling, A., and Whitelaw, J. H., "Low Reynolds Number Flow over a Plane Symmetric Sudden Expansion," *J. Fluid Mech.*, Vol. 64, p. 111, 1974.

Fargie, D. and Martin, B. W., "Developing Laminar Flow in a Pipe of Circular Cross Section," *Proc. Roy. Soc.*, Vol. 321, p. 461, 1971.

Farrugia, M., "Characteristics of Fluid Flow in Helical Tubes," Ph.D. Thesis, Univ. of London, 1967.

Gessner, F. B., Po, J. K., and Emery, A. F., "Measurements of Developing Turbulent Flow in a Square Duct," *Turbulent Shear Flows I*, New York, Springer–Verlag, 1979.

Goldstein, R. J., Eriksen, V. L., Olson, R. M., and Eckert, E. R. G., "Laminar Separation, Reattachment, and Transition of the Flow over a Downstream-facing Step," *J. Basic Engrg., Trans. ASME*, Vol. 92, 1970.

Halleen, R. M., Ph.D. Dissertation, Mech. Eng. Dept., Stanford University, 1967.

Hedges, K. R. and Hill, P. G., "Compressible Flow Ejectors Parts I & II," *J. Fluids Engrg., Trans. ASME*, Vol. 96, p. 272, 1974.

Horlock, J. H. and Lakshminarayana, B., "Secondary Flows: Theory, Experiment, and Application in Turbomachinery Aerodynamics," *Ann. Rev. Fluid Mech.*, Vol. 5, p. 247, 1973.

Howard, J. H. G., Patankar, S. V., and Bordynuik, R. M., "Flow Prediction in Rotating Ducts Using Coriolis Modified Turbulence Models," *J. Fluids Engrg., Trans. ASME*, Vol. 102, p. 456, 1980.

Humphrey, J. A. C., Taylor, A. M. K. P., and Whitelaw, J. H., "Laminar Flow in a Square Duct of Strong Curvature," *J. Fluid Mech.*, Vol. 83, p. 509, 1977.

Humphrey, J. A. C., Whitelaw, J. H., and Yee, G., "Turbulent Flow in a Square Duct with Strong Curvature," *J. Fluid Mech.*, Vol. 103, p. 443, 1981.

Hussain, A. K. M. F. and Ramjee, V., "Effects of Axisymmetric Contraction Shape on Incompressible Turbulent Flow," *J. Fluids Engrg., Trans. ASME*, Vol. 98, p. 58, 1976.

Hussain, A. K. M. F. and Reynolds, W. C., "Measurements in Fully Developed Turbulent Channel Flow," *J. Fluids Engrg., Trans. ASME*, Vol. 97, p. 568, 1975.

Ito, H., "Friction Factors for Turbulent Flow in Curved Pipes," *J. Basic Eng., Trans. ASME*, Vol. 81, p. 123, 1959.

Ito, H., "Laminar Flow in Curved Pipes," *Z. Angew. Math. Mech.*, Vol. 49, p. 653, 1969.

Ito, H., "Pressure Losses in Smooth Pipe Bends," *J. Basic Eng., Trans. ASME*, Vol. 82, No. 1, p. 131, 1960.

Ito, H., "Theory on Laminar Flows Through Curved Pipes of Elliptic and Rectangular Cross-Sections," *Rep. Inst. High Speed Mech.*, Vol. 1, 1951.

Johnston, J. P. and Eide, S. A., "Turbulent Boundary Layers on Centrifugal Compressor Blades: Prediction of the Effects of Surface Curvature and Rotation," *J. Fluids Engrg., Trans. ASME*, Vol. 98, p. 374, 1976.

Johnston, J. P., "The Suppression of Shear Layer Turbulence in Rotating Systems," *J. Fluids Engrg., Trans. ASME*, Vol. 95, p. 229, 1973.

Johnston, J. P., Halleen, R. M., and Lezius, D. K., "Effects of Spanwise Rotation on the Structure of Two-Dimensional Fully Developed Turbulent Channel Flow," *J. Fluid Mech.*, Vol. 56, p. 533, 1973.

Jones, O. C. and Leung, J. C. M., "An Improvement in the Calculation of Turbulent Friction in Smooth Annuli," *J. Fluids Engrg., Trans. ASME*, Vol. 103, p. 615, 1981.

Jones, O. C., "An Improvement in the Calculation of Turbulent Friction in Rectangular Ducts," *J. Fluids Engrg., Trans. ASME*, Vol. 98, p. 173, 1976.

Kangovi, S. and Page, R. H., "Subsonic Turbulent Flow Past a Downstream Facing Annular Step," *J. Fluids Engrg., Trans. ASME*, Vol. 101, p. 230, 1979.

Kapur, J. N., Tyagi, V. P., and Srivastawa, R. C., "Streamline Flow Through a Curved Annulus," *Appl. Sci. Res.*, Vol. 14, p. 253, 1964.

Kim, J., Kline, S. J., and Johnston, J. P., "Investigation of a Reattaching Turbulent Shear Layer: Flow over a Backward-Facing Step," *J. Fluids Engrg., Trans. ASME*, Vol. 102, p. 302, 1980.

Klein, A., "Review: Turbulent Developing Pipe Flow," *J. Fluids Engrg., Trans. ASME*, Vol. 103, p. 243, 1981.

Laufer, J., "The Structure of Turbulence in Fully Developed Pipe Flow," NACA Report 1174, *U. S. Nat. Adv. Com. Aero.*, 1954.

Launder, B. E. and Spaulding, D. B., "The Numerical Computation of Turbulent Flows," *Comp. Method. Appl. Mech. and Eng.*, Vol. 3, p. 269, 1974.

Launder, B. E., Priddin, C. H., and Sharma, B. I., "The Calculation of Turbulent Boundary Layers on Spinning and Curved Surfaces," *J. Fluids Engrg., Trans. ASME*, Vol. 99, p. 231, 1977.

Lipstein, N. J., "Low Velocity Sudden Expansion Pipe Flow," *ASHRAE J.*, Vol. 4, p. 43, 1962.

Macagno, E. O. and Hung, T. K., "Computational and Experimental Study of a Captive Annular Eddy," *J. Fluid Mech.*, Vol. 28, p. 43, 1967.

Melling, A. and Whitelaw, J. H., "Turbulent Flow in a Rectangular Duct," *J. Fluid Mech.*, Vol. 78, p. 289, 1976.

Meroney, R. N. and Bradshaw, P., "Turbulent Boundary-layer Growth over Longitudinally Curved Surfaces," *AIAA J.*, Vol. 13, p. 1448, 1975.

Mohanty, A. K. and Asthana, S. B. L., "Laminar Flow in the Entrance Region of a Smooth Pipe," *J. Fluid Mech.*, Vol. 90, p. 433, 1978.

Moon, L. F. and Rudinger, G., "Velocity Distribution in an Abruptly Expanding Circular Duct," *J. Fluids Engrg., Trans. ASME*, Vol. 99, p. 226, 1977.

Moore, J., "A Wake and an Eddy in a Rotating Radial Flow Passage," *J. Engrg. for Power, Trans. ASME*, Vol. 95, p. 205, 1973.

Morel, T., "Comprehensive Design of Axisymmetric Wind Tunnel Contractions," *J. Fluids Engrg., Trans. ASME*, Vol. 97, p. 225, 1975.

Morel, T., "Design of Two-Dimensional Wind Tunnel Contraction," *J. Fluids Engrg., Trans. ASME*, Vol. 99, p. 371, 1977.

Mori, Y. and Nakayama, W., "Study of Forced Convective Heat Transfer in Curved Pipes," *Int. J. Heat Mass Trans.*, Vol. 10, p. 681, 1967.

Mori, Y., Uchida, Y., and Ukon, T., "Forced Convective Heat Transfer in a Curved Channel with a Square Cross Section," *Int. J. Heat Mass Trans.*, Vol. 14, p. 1787, 1971.

Moss, W. D., Baker, S., and Bradbury, J. S., "Measurements of Mean Velocity and Reynolds Stresses in Some Regions of Recirculating Flows," *Proc. Symp. Turbulent Shear Flows*, p. 13.1, 1977.

Oosthuizen, P. H. and Wu, M. C., "Experimental and Numerical Study of Constant Diameter Ducted Jet Mixing," *Proc. Symp. Turbulent Shear Flows*, p. 10.1, 1977.

Patankar, S. V., *Numerical Heat Transfer and Fluid Flow*, Hemisphere, Washington, D.C., 1980.

Patel, R. P., "A Note on Fully Developed Turbulent Flow Down a Circular Pipe," *Aeronautical J.*, Vol. 78, p. 93, 1974.

Pratap, V. S. and Spaulding, D. B., "Numerical Computations of the Flow in Curved Ducts," *Aero. Quart.*, Vol. 26, p. 219, 1975.

Ramjee, V. and Hussain, A. K. M. F., "Influence of the Axisymmetric Contraction Ratio on Free-Stream Turbulence," *J. Fluids Engrg., Trans. ASME*, Vol. 98, p. 506, 1976.

Razinsky, E. and Brighton, J. A., "Confined Jet Mixing for Nonseparating Conditions," *J. Basic Engrg., Trans. ASME*, Vol. 94, p. 333, 1971.

Reneau, L. R., Johnston, J. P., and Kline, S. J., "Performance and Design of Straight, Two-Dimensional Diffusers," *J. Basic Engrg., Trans. ASME*, Vol. 89, No. 1, p. 141, 1967.

Reynolds, A. J., "Turbulent Flows in Engineering," John Wiley & Sons, New York, 1974.

Rowe, M., "Measurements and Computations of Flow in Pipe Bends," *J. Fluid Mech.*, Vol. 43, p. 771, 1970.

Schlichting, H., *Boundary-Layer Theory*, 7th ed., McGraw-Hill, New York, 1979.

Shah, R. K., "A Correlation for Laminar Hydrodynamic Entry Length Solutions for Circular and Noncircular Ducts," *J. Fluids Engrg., Trans. ASME*, Vol. 100, p. 177, 1978.

Shapiro, A. H., Siegel, R., and Kline, S. J., "Friction Factor in the Laminar Entry Region of a Round Tube," *Proc. 2nd U.S. National Congress of Applied Mechanics*, ASME, p. 733, 1954.

Shivaprasad, B. G. and Ramaprian, B. R., "Turbulence Measurements in Boundary-Layers Along Mildly Curved Surfaces," *J. Fluids Engrg., Trans. ASME*, Vol. 100, p. 37, 1978.

Smith, C. R. and Kline, S. J., "An Experimental Investigation of the Transitory Stall Regime in Two-Dimensional Diffusers," *J. Fluids Engrg., Trans. ASME*, Vol. 96, p. 11, 1974.

Smith, C. R. and Layne, J. L., "An Experimental Investigation of Flow Unsteadiness Generated by Transitory Stall in Plane-Wall Diffusers," *J. Fluids Engrg., Trans. ASME*, Vol. 101, p. 181, 1979.

Smith, F. T., "Steady Motion Within a Curved Pipe," *Proc. Roy. Soc. London, Series A*, Vol. 347, p. 345, 1976.

So, R. M. C. and Mellor, G. I., "Experiment on Convex Curvature Effects in Turbulent Boundary Layers," *J. Fluid Mech.*, Vol. 60, p. 43, 1973.

Sovran, G. and Klomp, E. D., "Experimentally Determined Optimum Geometries for Rectilinear Diffusers with Rectangular, Conical or Annular Cross Section," *Fluid Mech. of Int. Flow*, G. Sovran (Ed.), Elsevier, New York. 1967.

Sparrow, E. M. and Anderson, C. E., "Effect of Upstream Flow Processes on Hydrodynamic Development in a Duct," *J. Fluids Engrg., Trans. ASME*, Vol. 99, p. 556, 1977.

Tani, I., "Boundary Layer Transition," *Ann. Rev. Fluid Mech.*, Vol. 1, p. 169, 1969.

Taylor, A. M. K. P., Whitelaw, J. H., and Yianneskis, M., "Curved Ducts with Strong Secondary Motion: Velocity Measurements of Developing Laminar and Turbulent Flow," *J. Fluids Eng., Trans. ASME*, Vol. 104, p. 350, 1982.

Teyssandier, R. G. and Wilson, M. P., "An Analysis of Flow Through Sudden Enlargements in Pipes," *J. Fluid Mech.*, Vol. 64, p. 85, 1974.

Topakoglu, H. C., "Steady Laminar Flows of a Incompressible Viscous Fluid in Curved Pipes," *J. Math. Mech.*, Vol. 16, p. 1321, 1967.

Van Dyke, M., "Higher Order Boundary Layer Theory," *Ann. Rev. Fluid Mech.*, Vol. 1, p. 265, 1969.

Wang, J.-S. and Tullis, J. P., "Turbulent Flow in the Entry Region of a Rough Pipe," *J. Fluids Engrg., Trans. ASME*, Vol. 96, p. 62, 1974.

Ward-Smith, A. J., *Internal Fluid Flow*, Clarendon Press, England, 1980.

Ward–Smith, A. J., "Some Aspects of Fluid Flow in Ducts," Ph.D. Thesis, Univ. of Oxford, 1968.

White, C. M., "Streamline Flow Through Curved Pipes," *Proc. R. Soc. London, Ser. A*, Vol. 23, p. 645, 1929.

White, F. M., *Viscous Fluid Flow*, 2nd ed., McGraw-Hill, New York, 1992.

Wilcox, D. C. and Chambers, T. L., "Streamline Curvature Effects on Turbulent Boundary Layers," *AIAA J.*, Vol. 15, p. 574, 1977.

6 Waves in Fluids

PHILIP G. SAFFMAN
California Institute of Technology
Pasadena, CA

HENRY C. YUEN
TRW Space and Technology Group
Redondo Beach, CA

CONTENTS

6.1 INTRODUCTION

The most fundamental waves in fluid are sound waves, which can exist in the absence of external force fields. Waves involve a balance between a restoring force and the inertia of the system. Sound waves propagate independently of external forces as the restoring force balancng the fluids inertia arises from the fluids own compressibility. Sound propagation is consequently isotropic, since the compressibility properties of fluids are independent of direction. The study of sound waves forms the special discipline of acoustics which is discussed in Sec. 23.2.

The present article describes waves produced by external restoring forces, primarily gravity but also surface tension. Other external forces that may be important include electromagnetic fields and Coriolis effects due to rotation of the system; see also Sec. 13.2, Rotating Flows. Most wave motions produced by external restoring forces are anisotropic. In an inhomogeneous situation, important features are often confined to interfacial regions. The most well-known example is waves on a horizontal water surface.

From a mathematical point of view, two main classes of waves can be distin-

Handbook of Fluid Dynamics and Fluid Machinery, Edited by Joseph A. Schetz and Allen E. Fuhs
ISBN 0-471-12598-9 Copyright © 1996 John Wiley & Sons, Inc.

guished as characterized by their behavior. The first is known as *hyperbolic waves*, which typically obey an equation of the form

$$\phi_{tt} = C^2 \nabla^2 \phi \qquad (6.1)$$

commonly known as the *wave equation*. This is the equation for small amplitude sound waves in a homogeneous medium, and it also arises in elasticity, electromagnetism and other fields. A feature of hyperbolic waves is that the speed of propagation C is essentially independent of the wavelength.

The second class of waves is known as *dispersive waves* and is characterized by the property that simple harmonic waves can propagate in the system with wave speed that depends on wavelength. A disturbance therefore tends to disperse, since different Fourier components travel at different speeds. An important property of dispersive waves is that the energy tends to move at a speed called the group velocity which is different from the speed of the wave itself.

The effect of nonlinearity is the dependence of wave properties on the amplitude of the wave. In recent years, research in the understanding of wave properties has been concentrated on this aspect, and many important and unexpected phenomena have been discovered, such as the modulational instability of the uniform wave train, existence of three-dimensional steady wave patterns, and chaotic time evolution [see the review by Yuen and Lake (1982)]. Nonlinear wave motion poses serious mathematical difficulties, since the principle of linear superposition no longer applies and the problems are less amenable to the techniques of classical analysis. The advent of high speed computers and rapid development in numerical methods have led to significant progress in the study of nonlinear wave motion.

6.2 KINEMATIC WAVES

The simplest mathematical example of hyperbolic waves is provided by the class of kinematic waves which occurs in a system in which there is a density ρ and a flux of mass Q obeying a conservation relation

$$\rho_t + Q_x = 0 \qquad (6.2)$$

and a functional relationship $Q = Q(\rho)$. Physical situations which are described by these equations to a first approximation include flood waves in long rivers, glacier dynamics, and chemically reacting flows.

Substitution of the expression for the mass flux Q into Eq. (6.2) yields

$$\rho_t + C(\rho)\rho_x = 0 \qquad (6.3)$$

which is a first order hyperbolic wave equation where $C(\rho) = dQ/d\rho$. For small perturbations about a uniform state, ρ can be written in the form

$$\rho = \rho_0 + \rho', \, \rho' \ll \rho_0 \qquad (6.4)$$

where (ρ_0) is a constant. Equation (6.3) then reduces to

$$\rho_t + C_0 \rho'_x = 0 \qquad (6.5)$$

where $C_0 = C(\rho_0)$ is a constant. This equation has solutions of the form

$$\rho' = f(x - C_0 t) \qquad (6.6)$$

where f is an arbitrary function determined by initial or boundary conditions. This solution describes a wave which propagates without change of shape at a speed C_0 which is independent of the waveform.

When the amplitude is not small so that the approximation Eq. (6.4) cannot be made, the solution to Eq. (6.2) can be written exactly in an implicit form

$$\rho = f[x - C(\rho)t] \qquad (6.7)$$

where again f is arbitrary and determined by initial or boundary conditions. This solution shows that a constant value of ρ propagates with a constant speed $C(\rho)$. These trajectories or paths in the x-t space are called *characteristics*. In the small amplitude approximation, these trajectories are straight lines parallel to each other. For arbitrary amplitudes, different values of ρ travel at different speeds and may lead to a crossing of characteristics and the failure of the simple solution Eq. (6.7). When this occurs, a discontinuity or *shock* is introduced into the solution (see Chap. 8). The speed of the shock is determined by the need to satisfy appropriate conservation laws of the system.

Applications of kinematic wave theory to flood waves, traffic flow, chromatography, glacier motion, chemical processes, sedimentation and gas dynamics can be found in Whitham (1974).

6.3 LINEAR DISPERSIVE WAVES

The most important example of dispersive waves is water waves. The water is assumed to be inviscid and incompressible, and the motion irrotational. The governing equations are Laplace's equation inside the fluid, and the dynamic condition of constant pressure and the kinematic condition that the interface remains as a material surface applied on the free surface. The interface shape must be determined. Laplace's equation is linear, but the boundary conditions are nonlinear.

A gravity wave is one in which gravity is the main restoring force, whereas a capillary wave has surface tension as the main restoring force. For clean water, waves having wavelengths substantially greater than 1.7 cm can be considered gravity waves, since surface tension effects are negligible except for regions of very large curvature, and waves having wavelengths much shorter than 1.7 cm are capillary waves. Waves of intermediate wavelengths respond to both gravity and surface tension and are known as gravity-capillary waves or *ripples*. In situations where the wavelength is small compared with the depth of the water, the waves are deep-water waves, whereas shallow-water waves denote the opposite case.

For small disturbances on water otherwise at rest, the free surface and the conditions to be applied on it can be linearized. For deep-water, gravity waves, the boundary conditions reduce to

$$\phi_t + g\eta = 0$$

$$\eta_t - \phi_z = 0 \quad \text{on } z = 0 \tag{6.8}$$

where ϕ is the velocity potential, η is the free surface elevation measured from $z = 0$, g is gravitational acceleration, and z is the vertical coordinate pointing upwards. Elimination of η by differentiating both equations yields an equation for ϕ

$$\phi_{tt} + g\phi_z = 0 \quad \text{on } z = 0. \tag{6.9}$$

The velocity potential of ϕ satisfies the Laplace equation inside the fluid, and a set of fundamental solutions satisfying the linearized boundary conditions, Eq. (6.8), is

$$\phi(\bar{\mathbf{x}}, z, t) = \frac{\omega a}{|\bar{\mathbf{k}}|} \exp(-|\bar{\mathbf{k}}|z) \sin(\bar{\mathbf{k}} \cdot \bar{\mathbf{x}} - \omega t) \tag{6.10}$$

$$\eta(\bar{\mathbf{x}}, t) = a \cos(\bar{\mathbf{k}} \cdot \bar{\mathbf{x}} - \omega t) \tag{6.11}$$

where $\bar{\mathbf{k}}$ is the wave vector, ω is the wave frequency, and a is the wave amplitude. This represents a periodic wave with the crest propagating at the phase speed

$$\mathbf{C} = \frac{\omega}{|\bar{\mathbf{k}}|^2} \bar{\mathbf{k}}. \tag{6.12}$$

Substitution into the boundary conditions eliminates the wave amplitude and yields the expression relating the wave frequency ω and wave vector $\bar{\mathbf{k}}$

$$\omega^2 = g|\bar{\mathbf{k}}|. \tag{6.13}$$

This is known as the *dispersion relation*, and it is the most important characterization of the wave system. Note that the wave speed, applying Eq. (6.13), is

$$\mathbf{C} = \sqrt{\frac{g}{|\bar{\mathbf{k}}|}} \frac{\bar{\mathbf{k}}}{|\bar{\mathbf{k}}|}. \tag{6.14}$$

is a function of $\bar{\mathbf{k}}$, indicating that waves with different values of $\bar{\mathbf{k}}$ (and hence different wavelengths) travel at different speeds, a property which characterizes dispersive systems.

For two-dimensional waves which are independent of the transverse direction, the dispersion relation and the phase speed take the forms

$$\omega^2 = gk$$

$$C = \sqrt{\frac{g}{k}} \tag{6.15}$$

and the wavelength λ is given by $2\pi/k$. For waves on finite depth, the corresponding expressions are

$$\omega^2 = gk \tanh kh$$

$$C = \sqrt{\frac{g}{k} \tanh kh}. \tag{6.16}$$

When surface tension effects are included, the expressions become

$$\omega^2 = gk \tanh kh + \sigma k^3$$

$$C = \sqrt{\frac{g}{k} \tanh kh + \sigma k} \tag{6.17}$$

The energy density per unit area of surface associated with a deep-water wave is $\frac{1}{2}\rho g a^2$ where ρ is the density of the water. The rate of transmission of energy is $\frac{1}{4}\rho g a^2 C$. This implies that the speed of energy propagation is $\frac{1}{2}C$. Independent arguments establish that the rate of energy propagation is given by the *group velocity*, defined as

$$\frac{\partial \omega}{\partial k} = \frac{1}{2}\sqrt{\frac{g}{k}} \tag{6.18}$$

which is equal to one-half of the phase velocity in deep water. The group velocity is a fundamental concept in dispersive wave systems, and it is also a consequence of the conservation of wave crests. A detailed discussion of group velocity can be found in Whitham (1974) and Lamb (1932).

Sinusoidal waves on the water surface are attenuated through three main processes of energy dissipation. There is bottom friction, which is the most important when the water depth is less than one wave length, so that there is significant horizontal motion near the bottom induced by the waves. The energy dissipation takes place in a thin boundary layer on the solid bottom. The proportional energy lost per wave period is given by

$$\left(\frac{\nu}{2\omega}\right)^{1/2} \frac{4\pi k}{\sinh 2kh} \tag{6.19}$$

where h is the depth of water. Waves on deep water produce no motion and no frictional dissipation near the bottom. Their attenuation is caused by internal dissipation associated with water particle motion which is confined substantially to within one wavelength from the surface. The proportional loss of energy per wave period through this mechanism is

$$\frac{8\pi\nu k^2}{\omega} \tag{6.20}$$

where ν is the kinematic viscosity. There is also surface dissipation associated with nonuniformity of the surface tension and departures from thermodynamic equilibrium. Surface dissipation may be especially important when surface contaminants are present.

The longer gravity waves with wave lengths greater than one meter on deep water are very lightly damped by the action of viscosity, but short capillary waves of millimeter wave lengths are damped within several wave periods.

Specific applications to wave patterns generated by a local disturbance and a moving disturbance, such as a ship, can be found in Lamb (1932) and Lighthill (1978).

6.4 NONLINEAR DISPERSIVE WAVES

The infinitesimal or small amplitude solution Eqs. (6.10) and (6.11) is the simplest example of a wave train. In a more general context, the wave amplitude, the wave frequency, and the wave length vary in space and time. Provided that the rate of variation is slow compared with the scales associated with an individual wave, a complex wave envelope $A = a \exp{(i\theta)}$ can be introduced where A is a slowly varying function of space and time. The time derivative of θ gives the modulation to the wave frequency, and the spatial derivatives of θ give the modulations to the wave vector. For the case of unidirectional deep water waves, the complex amplitude satisfies the nonlinear Schrödinger equation

$$i \left(A_t + \frac{\omega_0}{2k_0} A_x \right) - \frac{\omega_0}{8k_0^2} A_{xx} - \frac{1}{2} \omega_0 k_0^2 |A|^2 A = 0. \qquad (6.21)$$

The free surface η is related to A by the expression

$$\eta = A \exp{(i\theta)} + A^* \exp{(-i\theta)}. \qquad (6.22)$$

This equation has been solved both analytically and numerically. Its solutions indicate that a weakly nonlinear wave train is unstable to modulational perturbations with wave numbers K lying within the range

$$0 < K < 2\sqrt{2}k_0^2 a_0 \qquad (6.23)$$

where k_0 and a_0 are the wave number and amplitude of the unperturbed wave train. The most unstable disturbance has wave number $K = 2k_0^2 a_0$ with a growth rate of $\frac{1}{2}\omega_0 k_0^2 a_0^2$, where ω_0 is the frequency of the unperturbed wave train. This instability is a result of nonlinear effects, since both the range and growth rate of the instability approach zero when the wave steepness $k_0 a_0$ of the unperturbed surface approaches zero, which is the limit for infinitesimal waves.

The modulation on an unstable wave train grows to a maximum and then subsides, and this modulation-demodulation cycle would repeat indefinitely in time in the absence of attenuation. Slow attenuation would lengthen the cycle period and reduce the maximum modulation magnitude, and the instability eventually becomes insignificant with the decrease of wave steepness.

Equation (6.21) also predicts that a localized wave packet would evolve into a definite number of *envelope solitons*, which are packets with a fixed shape. These envelope solitons can survive interactions or collisions with other packets without change of form [Yuen and Lake (1974)].

The above properties are associated with two-dimensional wave trains, where the motion is confined in the vertical and one horizontal direction. In general, the motion is three-dimensional being dependent also on the transverse direction.

The uniform wave train is unstable to three-dimensional disturbances, but, in the weakly nonlinear regime, the most unstable disturbances are two-dimensional in the sense that they are aligned with the unperturbed wave train. However, the envelope soliton, which is not unstable to two-dimensional disturbances, is unstable to three-dimensional cross-wave disturbances of long wave length.

For deep-water gravity waves, there is a limiting ratio of wave height to wave length, namely 0.142, beyond which smooth steady waves no longer exist. When the waves reach a steepness of about 60% of this limiting steepness, a second type of instability becomes dominant. This instability is fully three-dimensional in nature, in the sense that the most unstable disturbance has a transverse as well as a longitudinal structure. The most unstable disturbance propagates with the same speed as the unperturbed wave train, causes the wave train to develop a modulation which enhances every other crest, and has a transverse length scale of approximately 60% of the unperturbed wave length.

The uniform wave train is not the only admissible steady solution, since there exists an infinite class of three-dimensional solutions which are also steady. It has been experimentally observed that a uniform wave train of large amplitude which experiences the three-dimensional co-propagating instability may evolve into a non-uniform, three-dimensional steady wave train through a bifurcation mechanism.

A more general situation involves a field of waves. If the wave field is infinitesimal, then its properties can be obtained by linearly superposing independent wave trains. When the wave field is nonlinear, superposition cannot be used since the various component wave trains interact with each other.

A simple nonlinear wave field can be characterized by a characteristic wave steepness and a characteristic width of the wave spectrum. When the characteristic wave steepness approaches zero, the wave field is considered linear or infinitesimal. When the spectral width approaches zero, the wave field becomes a wave train with a single frequency or wave length. The behavior of wave spectra depends on the ratio S, the characteristic wave steepness to the spectral width. For large S, which corresponds to a steep, narrow-banded wave field, the behavior is nonlinear and resembles that of a wave train. In fact, an initially homogeneous wave field under these conditions will develop modulational structures. For small S, which corresponds to a small amplitude, broadbanded wave field, the behavior is near-linear, and an initially homogeneous wave field will not develop modulational structures.

On a longer time scale, a weaker interaction which acts to transfer energy among the different components in the wave field can become effective. The time scale for this mechanism to take effect is proportional to the fourth power of the characteristic wave steepness, whereas the modulational structures develop on a time scale proportional to the square of the steepness. A survey of some results is given by Yuen and Lake (1982).

6.5   INTERNAL WAVES

Internal waves can be produced by a variety of mechanisms. One of the principal types of internal gravity waves is produced by the action of gravity on variations of density within the fluid when the fluid is stably stratified. Waves propagate in such systems due to the balance of the gravitational restoring force and the total fluid inertia.

The basic quantity which characterizes these waves is the *Brunt–Vaisala frequency*, defined as

$$N(z) = \left[\frac{-g \dfrac{d\rho_0}{dz}}{\rho_0(z)} \right]^{1/2} \tag{6.24}$$

where ρ_0 is the density distribution. The infinitesimal waves are anisotropic. In the case of constant Brunt–Vaisala frequency, the simple harmonic wave whose crest lies on the plane $kx + \ell y + mz = \omega t$ has a frequency given by

$$\omega^2 = \frac{N^2(k^2 + \ell^2)}{k^2 + \ell^2 + m^2}. \tag{6.25}$$

The particle velocity is parallel to the surfaces of constant phase, and the wave energy flux is perpendicular to the direction of propagation. The Brunt–Vaisala frequency, N, is the maximum allowable frequency of these waves. These internal waves are widely encountered in the study of the atmosphere and the ocean.

Furher complications ensue when the fluid, in addition to being stratified, is in a state of relative motion. The wave motions even under the approximations of the linear theory are complex and may become unstable. In this case, a second parameter of importance, which measures the relative effects of velocity shear and stratification, is the Richardson number. This number is the ratio of the Brunt–Vaisala frequency to the rate of the shear. Instability may occur if this number is too small.

The term *inertial wave* is used to describe wave motions that occur in the fluid which is in uniform solid body rotation. In this case, the dispersion relation for plane waves is

$$\omega^2 = \frac{\Omega_0^2 m^2}{k^2 + \ell^2 + m^2} \tag{6.26}$$

where Ω_0 is the frequency of the solid body rotation. Qualitative properties of these waves are similar to those of the internal waves due to stratification in the case of constant Brunt–Vaisala frequency. See Sec. 13.2.

REFERENCES

Lamb, Sir Horace, *Hydrodynamics*, 6th ed., Dover, New York, 1932.

Lighthill, Sir James, *Waves in Fluids*, Cambridge University Press, London, 1978.

Whitham, G. B., *Linear and Nonlinear Waves*, John Wiley & Sons, New York, 1974.

Yuen, H. C. and Lake, B. M., "Nonlinear Deep Water Waves: Theory and Experiment," *Phys. Fluids*, Vol. 18, pp. 956–960, 1974.

Yuen, H. C. and Lake, B. M., "Nonlinear Dynamics of Deep Water Gravity Waves," *Adv. Appl. Mech.*, Vol. 22, pp. 67–229, 1982.

7 Thermodynamics

RICHARD S. TANKIN
Northwestern University
Evanston, IL

CONTENTS

7.1 INTRODUCTION

Thermodynamics is a science that deals with the bulk properties of a system—it makes no hypotheses about the structure of matter. It is mainly concerned with the transformation of heat into mechanical work. Restrictions pertaining to such transformations, which were obtained through experience, are embodied in two general laws.

In thermodynamics, one separates the region for consideration from its surroundings by an envelope which may be real or imaginary. The region under consideration is called a *system*, and the envelope is called a boundary or control surface. The boundary may move or be stationary; heat may or may not flow across the boundary. This idea of separating the system from its surroundings and analyzing the system, with the effects of the surrounding incorporated into appropriate boundary conditions, is a fundamental aspect of engineering analysis. For example, this technique

Handbook of Fluid Dynamics and Fluid Machinery, Edited by Joseph A. Schetz and Allen E. Fuhs
ISBN 0-471-12598-9 Copyright © 1996 John Wiley & Sons, Inc.

is employed in rigid body mechanics where the *free body diagram* is analogous to the system (or control surface) with its appropriate boundary conditions.

Thermodynamics is a science that can only analyze states that are in *equilibrium*. To be in equilibrium there must be: (a) no unbalanced forces within the system, nor between the system and its surroundings, (b) no temperature gradients within the system, and (c) no chemical reactions nor mass diffusion occurring within the system. A system will reach a state of equilibrium if it is isolated from its surroundings, that is, no exchange of energy or mass between the system and the surroundings. When equilibrium is achieved, the thermodynamic properties, such as, pressure, temperature, and specific volume, etc., will not change with time. Thermodynamics can be used to analyze many problems where the system is initially in a state of equilibrium and proceeds in a nonequilibrium manner to a final state of equilibrium. For example, one might have a rigid, thermally conducting tank that is divided into two separate specified regions by a membrane. A gas may exist on one side of the membrane at a specified pressure with a temperature the same as the surroundings; on the other side of the membrane is a vacuum. The membrane is ruptured, and the gas rapidly fills the entire volume of the rigid tank. In the process of filling the tank, shock waves, expansion waves, reflections, etc., occur within the rigid tank. These phenomenona cannot be analyzed using thermodynamics, since the system proceeds to its final equilibrium state through a sequence of nonequilibrium states where there are variations in temperature, pressure, and specific volume both spatially and temporally. However due to viscosity, the gas eventually comes to a state of equilibrium, and this equilibrium state can be computed by thermodynamic principles.

Some definitions of terms that will be used throughout this section are the following:

1) Q (kJ) is the *heat transfer* between the system and its surroundings. By convention, Q is positive when heat flows into the system from the surroundings; Q is negative when heat flows from the system to the surroundings. The specific heat is represented by q (kJ/kg).

2) W (kJ) is *work*, which is any form of energy except heat that is transferred across the boundary separating the system from the surroundings. By convention, W is positive when the system does work on the surrounding; W is negative when the surroundings does work on the system. Specific work is represented by w (kJ/kg).

3) E (kJ) is *internal energy*. It is the energy within the system but does not include potential or kinetic energy. Specific internal energy is represented by e (kJ/kg).

The *state* of a system is defined by thermodynamic properties such as pressure, volume, and temperature. When a system undergoes a change in state, a *process* is said to occur. Typical processes, which may occur depend on the restrictions imposed. A few examples are the following:

1) An *isothermal process* is one in which the temperature of the system remains constant.
2) An *isochoric process* is one in which the volume of the system remains constant.

3) An *isobaric process* is one in which the pressure within the system remains constant.

4) An *adiabatic process* is one in which heat is not transferred between the system and its surroundings.

5) An *isolated process* is one in which neither work nor heat transfer occurs between the system and surroundings.

7.2 EQUATION OF STATE

The system under consideration may be classified as homogeneous or heterogeneous. A system whose properties are the same in all parts of the system is *homogeneous*. Even though a system is not in complete chemical equilibrium, it may often be appropriately assumed to be homogeneous. For example, in combustion, the chemical changes may occur so rapidly when compared to characteristic flow times that the system may be assumed to be homogeneous. A system composed of two or more phases is called a *heterogeneous* system. The number of independent variables consistent with equilibrium between the various phases is given by

$$F = C - P + 2 \qquad (7.1)$$

where F = number of independent variables
   C = number of chemical species in the system
   P = number of phases

This is the *phase rule* without chemical reactions.

In a homogeneous system composed of a pure substance (simple homogeneous) only two thermodynamic properties may be independent; the remaining thermodynamic properties are dependent. Let us consider a specified mass (kmoles) of gas in a container of specified volume (m^3) and at specified pressure (kPa). Since p, and v [specific volume (m^3/kmole)] are specified, the temperature T(K) is likewise specified. This implies there is a relation between p, v, and T. The equation relating these variables is called an *equation of state*.

A particularly simple equation of state is

$$pv = RT \qquad (7.2)$$

where R is a constant equal to 8.31434 kJ/kmole K. Equation (7.2) is the equation of state for an *ideal gas* (often the term *perfect gas* is used) where the interaction between the gas molecules is negligible. This equation of state will apply to gases when the density approaches small values (high temperature and low pressure); air closely approximates an ideal gas at temperatures above 300 K (room temperature) and pressures up to 10 MPa (approximately 100 atmospheres).

At high densities, the gas or vapors may deviate greatly from ideal gas behavior. Consider a more extreme case in which a pure substance changes phase. Figure 7.1 is a graphical representation of the equation of state where liquid phase, vapor phase, and mixed phases may exist. The dome-shaped region is where mixed phases, liquid and vapor, co-exist. This is called the *saturated* region. It is seen that in this region

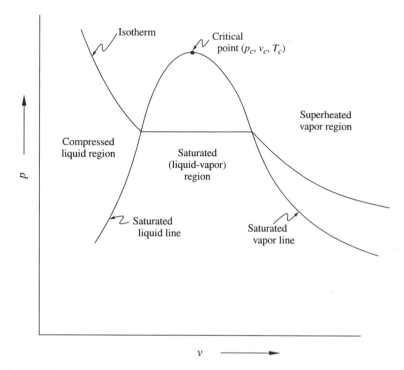

FIGURE 7.1 Schematic drawing of p-v diagram showing liquid-vapor phases.

p and T cannot be specified independently. To the left of the dome region, the liquid phase exists, which is frequently called the compressed liquid region. To the right of the dome, the vapor phase exists and is called the *superheated* vapor region. The highest pressure at which distinct liquid phase and vapor phases can co-exist is called the *critical pressure* (p_c). Associated with this *critical point* are critical specific volume (v_c) and critical temperature (T_c). The saturated-liquid line separates the compressed liquid from the saturated region; the saturated-vapor line separates the saturated region from the superheated vapor region. Similar regimes can be defined for solid-liquid phases and solid-vapor phases.

In the saturated region, where temperature and pressure are not independent, an additional thermodynamic property is needed. The quantity introduced is the mass fraction of the vapor which is called the *quality* of the vapor–liquid mixture and is generally represented by "x". Thus,

$$v = v_f + x(v_g - v_f) = v_f + xv_{fg} \tag{7.3}$$

where the subscript f refers to saturated liquid, and the subscript g refers to saturated vapor. Similar expressions exist for other thermodynamic properties (e, h, s, a, and g) which will be defined later. Values of p, T, v (or v_f, v_g) as well as other thermodynamic properties are listed in tables or presented graphically for many substances that are frequently used in engineering systems—steam, ammonia, etc. (see Chap. 2 for fluid properties).

An equation of state which corrects for *real gas effects*—attractive forces between molecules—is the van der Waals equation of state.

$$\left(p + \frac{a}{v^2}\right)(v - b) = RT \tag{7.4}$$

where the constants a and b depend on the gas. These constants can be expressed in terms of the critical temperature (T_c) and critical specific volume (v_c); that is,

$$b = \frac{v_c}{3}, a = \frac{9v_c T_c}{8R} \tag{7.5}$$

The van der Waals equation of state agrees qualitatively with experimental data in the compressed liquid region, the saturated region (with suitable interpretation concerning stability), and the superheat region. However, quantitative agreement between van der Waals equation and experiments is only approximate.

There are two laws which are the foundation of thermodynamics. The *first law* is simply the conservation of energy and introduces the definition of internal energy. The *second law* excludes the possibility of certain processes even though the process may obey the first law of thermodynamics; the second law introduces the concept of entropy.

7.3   FIRST LAW OF THERMODYNAMICS

If a system undergoes a displacement (change in volume), then work may be performed on the surroundings. The amount of work is equal to the product of the external force acting on the system and the component of the displacement parallel to the force. This work is external mechanical work and occurs between the system and its surroundings. By convention (used in most engineering thermodynamics texts), work done by the system on its surroundings is positive; if work is done on the system, it is negative.

To apply the laws of thermodynamics, the system must be in equilibrium; that is, the system must be in thermal equilibrium, chemical equilibrium, and mechanical equilibrium. For the present, consider only mechanical equilibrium where unbalanced forces are not acting on any part of the system. One cannot, strictly speaking, describe thermodynamically a process in which a system undergoes a displacement, since the displacement is due to unbalanced forces. However, if the system is at all times near a state of equilibrium, the unbalanced force is infinitesimal. Such a process can be described thermodynamically; the work is performed quasi-statically.

Let us assume a cylinder is fitted with a movable piston. The fluid inside the cylinder is the system. The piston is in an equilibrium position where the force due to the pressure of the system balances the force of the surroundings on the piston. If the pressure within the cylinder increases and displacement occurs, the system does work on the surroundings which is displacement of the surroundings and overcoming friction between the piston and the cylinder. If this process is performed quasi-statically, the infinitesimal work performed is

$$dW = pA \, dx = p \, dV \tag{7.6}$$

where dx = infinitesimal displacement
   A = cross section of the piston
   dV = infinitesimal volume change of the cylinder

If the work is not done quasi-statically, the work will be equal to the work done in displacing the surroundings and overcoming friction; but the work will not equal $p \, dV$. Such a process cannot be represented on a p-V diagram.

   Under normal conditions, quasi-static work depends on how the work is done; that is, it depends on the path followed. However, as a result of many experiments, it was observed that adiabatic work is independent of path. This particular work is defined as change in internal energy of the system and is a state property. For a system that proceeds adiabatically from equilibrium state 1 to equilibrium state 2

$$W_{1 \to 2}^{\text{ad}} = (E_1 - E_2) \tag{7.7}$$

where E is the internal energy of the system and is a function of state. For nonadiabatic processes, the first law of thermodynamics for a closed system can be written as

$$dQ = dE + dW \tag{7.8}$$

where dQ is the heat added to the system. Both Q and W depend on the process.

   Often it is inconvenient to consider a fixed mass of the working fluid as in a closed system. This is particularly true when the working fluid is in motion such as the flow of a fluid through turbomachinery, nozzles, etc. In these cases, it is usually simpler to define a volume, fixed in space, through which the fluid is flowing. Such a volume is called a *control volume*, and the system is called an *open system* as distinct from a closed system. The first law of thermodynamics using the control volume concept can be expressed as

$$\dot{Q}_{cv} - \int_{cs} \left(e + \frac{V^2}{2} + gZ + pv \right) \rho V \cdot dA$$

$$- \frac{d}{dt} \int_{cv} \rho \left(+ \frac{V^2}{2} + gZ \right) dV - \dot{W}_{cv} = 0 \tag{7.9}$$

The first term in Eq. (7.9) is the heat added per unit time to the control volume; the second term is the flux of the various forms of energy (state quantities) plus the flow work terms into and out of the control volume; the third term is the change of energy within the control volume per unit of time; and the last term is the power output associated with shaft effects (pumps, turbines, etc.) and electrical effects. The flow work terms are usually combined with the internal energy terms to form another state function called *enthalpy*.

$$h = e + pv \tag{7.10}$$

In many applications, the thermodynamic state within the control volume may be changing with time, however, the state of the masses entering and leaving the control volume may remain constant with time. Then Eq. (7.9) can be written as

$$Q_{cv} + \sum_i m_i \left(h_i + \frac{V_i^2}{2} + gZ_i \right) - \sum_e m_e \left(h_e + \frac{V_e^2}{2} + gZ_e \right)$$
$$+ m_1 \left(e_1 + \frac{V_1^2}{2} + gZ_1 \right) - m_2 \left(e_2 + \frac{V_2^2}{2} + gZ_2 \right)$$
$$- W_{cv} = 0 \tag{7.11}$$

Subscript i refers to inlet and subscript e refers to exit. Subscripts 1 and 2 refer to two different instances of time. For steady state, steady flow processes

$$\sum_i m_i = \sum_e m_e \tag{7.12}$$

and

$$m_1 \left(e_1 + \frac{V_1^2}{2} + gZ_1 \right) = m_2 \left(e_2 + \frac{V_2^2}{2} + gZ_2 \right) \tag{7.13}$$

Thus, Eq. (7.11) becomes

$$Q_{cv} + \sum_i m_i \left(h_i + \frac{V_i^2}{2} + gZ_i \right) - \sum_e m_e \left(h_e + \frac{V_e^2}{2} + gZ_e \right)$$
$$- W_{cv} = 0 \tag{7.14}$$

For cases where only one flow stream enters and leaves the control volume, Eq. (7.14) is further reduced to

$$q_{cv} + \left(h_i + \frac{V_i^2}{2} + gZ_i \right) - \left(h_e + \frac{V_e^2}{2} + gZ_e \right) - w_{cv} = 0 \tag{7.15}$$

7.4 REVERSIBLE AND IRREVERSIBLE PROCESSES

As its name implies, a *reversible* process is one in which the process can be exactly reversed. That is, after the conclusion of the process, the system *and* surroundings can be returned to their initial states without producing any changes in any mechanical devices or reservoirs. To be reversible, a process must be performed quasi-statically without dissipative effects.

Let us consider a quasi-static process with dissipative effects. A piston of weight W and cross sectional area A confines a gas in a cylinder. By supplying heat slowly to the gas in the cylinder, the piston will rise. The direction of the frictional force (F_f) is opposite to the direction of displacement. During upward displacement, the frictional force will be acting down. The forces on the piston are shown schematically in Fig. 7.2.

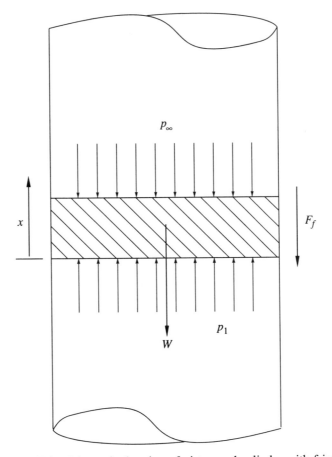

FIGURE 7.2 Schematic drawing of piston and cylinder with friction.

Since this process is performed quasi-statically, the pressure in the cylinder, which remains constant during expansion can be expressed as

$$p_1 = \frac{W + F_f}{A} + p_\infty \tag{7.16}$$

By removing heat slowly, the piston will move downward. The pressure in the cylinder during this process is constant and less than during expansion,

$$p_2 = \frac{W - F_f}{A} + p_\infty \tag{7.17}$$

When the heat removed during contraction equals the heat supplied during expansion, the process is completed. The final state of the gas is not the same as its original state. Such a process is quasi-static but not reversible. A reversible process implies the process is quasi-static; the inverse is not true.

All natural processes are irreversible and involve one or more of the following features: the system is not in mechanical, thermal or chemical equilibrium, and dissipative effects are always present. Therefore, one may question the importance of assuming a reversible process. The assumption of reversibility allows one to compute work knowing the thermodynamic properties alone. Many natural processes which are essentially adiabatic can be approximated by a reversible adiabatic process. Such an approximation defines the path of the process; there is a relation between the thermodynamic functions such as p and V, p and T, etc. In addition, the concept of a reversible process is important in discussing the second law of thermodynamics.

7.5 THE SECOND LAW OF THERMODYNAMICS

The second law of thermodynamics involves inequalities. From experience it is known that certain processes will not occur, and the second law is a statement of the impossibility of a process occurring. For example, two thermal reservoirs at temperatures T_1 and T_2 are brought into contact. Heat will not flow from the low temperature reservoir to the high temperature reservoir. Although such a process is possible based on the first law of thermodynamics, it is excluded by the second law of thermodynamics.

The second law of thermodynamics may be stated in different ways, all of which are equivalent. Clausius' statement is as follows: "No process is possible whose sole result is the transfer of heat from a cooler body to a hotter body." Rather than attempting to interpret all processes in terms of Clausius' statement, it would be desirable if this law could be stated in terms of a well defined thermodynamic property. This leads to the concept of *entropy* and the following proposition: "The entropy change of the system plus its surroundings can not be negative. It is positive if the process is irreversible (natural) and zero if the process is reversible."

If we proceed reversibly from thermodynamic state A to thermodynamic state B, entropy is defined as

$$S_B - S_A = \int_A^B \left(\frac{dQ}{T} \right)_{\text{rev}} \tag{7.18}$$

Since entropy is a thermodynamic property, it is independent of path; entropy depends on the state of the system. For an ideal gas, specific entropy changes can be expressed as

$$s_B - s_A = \int_{T_A}^{T_B} c_p \frac{dT}{T} - R \ln p_B/p_A \tag{7.19}$$

or

$$s_B - s_A = \int_{T_A}^{T_B} c_v \frac{dT}{T} + R \ln v_B/v_A \tag{7.20}$$

Specific volume, i.e., volume/mass, is v whereas volume is V.

Compute the entropy change for the following irreversible process. An insulated, rigid vessel has an evacuated compartment (volume V_1) separated from a second compartment (volume V_2) by a membrane. The gas in the second compartment is at temperature T_2 and pressure p_2. The membrane is ruptured, and the gas fills the entire volume ($V_1 + V_2$). Since there is no heat flow and no work, there is no change in the internal energy. For an ideal gas this implies no change in temperature. Thus, from Eq. (7.19)

$$\Delta s = R \ln \left(\frac{v_1 + v_2}{v_2} \right) \qquad (7.21)$$

Let us now reexamine the same problem but proceed reversibly from initial to final state. The membrane is replaced with an insulated, frictionless piston which moves quasi-statically, and a heat reservoir at temperature T_2 is attached to the vessel. (Note: It is not stated how this quasi-static process is to be achieved. It may be with a series of membranes containing gas at prescribed pressures replacing the evacuated chamber or connecting the piston to a work reservoir.) The gas is maintained at constant temperature T_2 by the heat reservoir. Thus,

$$\Delta s = \frac{1}{T_2} \int (dq)_{rev} \qquad (7.22)$$

The specific work done by the gas during this quasi-static, isothermal expansion is

$$w = RT_2 \ln \left(\frac{v_2 + v_1}{v_2} \right) \qquad (7.23)$$

which equals the heat added to the system from the heat reservoir. Thus, the entropy change of the system is again

$$\Delta s = R \ln \left(\frac{v_2 + v_1}{v_2} \right) \qquad (7.24)$$

7.6 HELMHOLTZ FUNCTION AND GIBBS FUNCTION

In addition to internal energy, enthalpy, and entropy other thermodynamics functions can be defined which are convenient to use for certain processes. One is the *Helmholtz function* (or *free energy*), and another is the *Gibbs function*. For a reversible process

$$de = T \, ds - p \, dv \qquad (7.25)$$

and

$$dh = T \, ds + v \, dp \qquad (7.26)$$

The Helmholtz function, a, is defined as

$$a = e - Ts \qquad (7.27)$$

Thus,

$$da = -s\,dT - p\,dv \qquad (7.28)$$

Note that a has the same units as e, that is, kJ/kg. For an ideal gas, the specific Helmholtz function is

$$a = \int C_v\,dT - T \int \frac{C_p\,dT}{T} + RT \ln p + \text{constant} \qquad (7.29)$$

For a system that undergoes a reversible process at constant temperature, the change in the Helmholtz function is equal to the work done on the system.

For some processes, the temperature and pressure remain constant; this is particularly true for many problems with change in phase. A thermodynamic function can be defined which is constant for such processes—the Gibbs function,

$$g = h - Ts \qquad (7.30)$$

Thus,

$$dg = -s\,dT + v\,dp \qquad (7.31)$$

For an ideal gas, the specific Gibbs function is

$$g = \int c_p\,dT - t \int \frac{c_p\,dT}{T} + RT \ln p + \text{constant} \qquad (7.32)$$

For a mixture of phases of a pure substance in equilibrium, the Gibbs function in one phase equals the Gibbs function in the other phase. For a liquid–vapor system, $g_f = g_g$.

This in turn leads to the Clayperon equation

$$\ln p = \frac{h_{fg}}{R}\left(-\frac{1}{T}\right) + \text{constant} \qquad (7.33)$$

which describes the pressure–temperature relation along the boiling curve. In a similar manner, this may be extended to phase equilibria of a pure substance as shown in Fig. 7.3. There are other thermodynamic functions that have been defined; however e, h, a, and g are the best known and most widely used set. Table 7.1 is a summary table for some of the processes and functions mentioned.

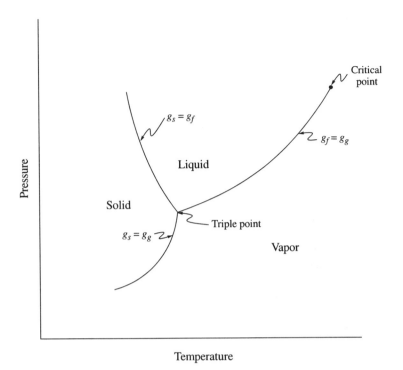

FIGURE 7.3 Schematic drawing showing change of phase relations.

7.7 ADIABATIC REVERSIBLE PROCESS

An adiabatic process is important because it closely approximates many problems in fluid mechanics. Adiabatic approximations are reasonable under the following conditions: (1) the system is well insulated; thus, the heat loss (or gain) compared to other terms in the energy equation is negligible, and (2) the process may occur rapidly so that heat losses (or gain) again are negligible. If, in addition, the energy dissipated is small, it is reasonable to assume an adiabatic reversible process. When an adiabatic reversible process is assumed, then

$$de = -p \, dv \tag{7.34}$$

For an ideal gas with constant specific heat

$$c_v \ln T = -R \ln v + \text{constant} \tag{7.35}$$

or

$$Tv^{\gamma - 1} = \text{constant} \tag{7.36}$$

where $\gamma = (c_p/c_v)$.

TABLE 7.1 Summary Table for Various Thermodynamic Processes for an Ideal Gas With Constant Specific Heats

Process → Function ↓	const V	const p	const T	const s	$pv^n = \text{const}$
$\displaystyle\int_{v_1}^{v_2} p\,dv$	0	$P(v_2 - v_1)$	$RT\ln v_2/v_1$	$\dfrac{p_2v_2 - p_1v_1}{1 - \gamma}$	$\dfrac{p_2v_2 - p_1v_1}{1 - n}$
$e_2 - e_1$	$c_v(T_2 - T_1)$	$c_v(T_2 - T_1)$	0	$c_v(T_2 - T_1)$	$c_v(T_2 - T_1)$
$q_{1\to 2}$	$c_v(T_2 - T_1)$	$c_p(T_2 - T_1)$	$RT\ln v_2/v_1$	0	$c_v(T_2 - T_1) +$ $\left(\dfrac{p_2v_2 - p_1v_1}{1 - n}\right)$
$h_2 - h_1$	$c_p(T_2 - T_1)$	$c_p(T_2 - T_1)$	0	$c_p(T_2 - T_1)$	$c_p(T_2 - T_1)$
$s_2 - s_1$	$c_v\ln T_2/T_1$	$c_p\ln T_2/T_1$	$R\ln v_2/v_1$	0	$C_v\ln T_2/T_1 + R\ln v_2/v_1$

Also

$$pv^{\gamma} = \text{constant} \tag{7.37}$$

and

$$pT^{\frac{\gamma-1}{\gamma}} = \text{constant} \tag{7.38}$$

This concept can often be applied to an adiabatic process in which the overall process is irreversible. Consider the system shown in Fig. 7.4. The gas in tank A (volume V_1) is initially at pressure p_A and temperature T_A. The valve is opened, and the gas flows from tank A into cylinder B, raising the piston. The mass of the piston is such that p_B ($p_B < p_A$) is required to raise the piston. Flow continues until the pressure in tank A equals p_B. Assume the overall process is performed adiabatically. To solve for the final state of the gas in tank A and cylinder B, some reasonable assumptions are needed. Irreversibility is primarily due to the flow through the valve and piping connecting the tank to the cylinder. Therefore, it will be assumed that the gas which remains in tank A has undergone a reversible adiabatic process. For an ideal gas

$$p_A T_A^{\frac{\gamma-1}{\gamma}} = p_B T_{Af}^{\frac{\gamma-1}{\gamma}} \tag{7.39}$$

where T_{Af} is the final temperature of gas remaining in the tank. Knowing p_A, T_A, and p_B, T_{Af} can be computed. From the equation of state, the mass that remains in the tank can be determined.

$$m_{Af} = \frac{p_B V_A}{R T_{Af}} \tag{7.40}$$

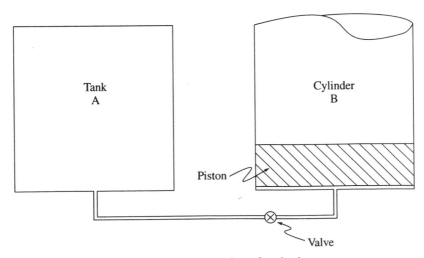

FIGURE 7.4   Schematic drawing of tank-piston system.

From the first law of thermodynamics

$$0 = m_{Af}c_v T_{Af} + m_{Bf}c_v T_{Bf} - m_A c_v T_A + p_B V_{Bf} \tag{7.41}$$

Thus

$$T_{Bf} = \frac{c_v(m_A T_A - m_{Af}T_{Af})}{m_{Bf}(c_v + R)} \tag{7.42}$$

The final temperature in cylinder B (T_{Bf}) is greater than the final temperature in tank A (T_{Af}).

7.8 EQUILIBRIUM CONDITIONS

For an isolated system—a system with rigid, adiabatic walls—any change from one equilibrium state to another equilibrium state with removal of constraints implies $\Delta s > 0$. This is a statement of the second law of thermodynamics applied to an isolated system. Such a process is called spontaneous and is irreversible.

The entropy of an isolated system will, if permitted by the constraints, increase; that is $s \rightarrow s_{max}$. Experience indicates that the internal energy will, if permitted by the constraints, decrease; that is $e \rightarrow e_{min}$. Hence, any infinitesimal virtual change in these quantities is equal to zero for an isolated system in equilibrium. [Virtual change means all conceivable variations in the thermodynamic property (s or e) subject to the constraints of the system.] That is, $\delta s = \delta e = 0$.

The following relations exist for an equilibrium condition:

for a given s and v, e is a minimum

for a given s and p, h is a minimum

for a given T and v, a is a minimum

for a given T and p, g is a minimum

Usually T is a more convenient independent variable than s, thus the last two relations are more frequently used. However, this is not to imply that any one relation is more fundamental than the other.

7.9 NONREACTING MIXTURES

Working fluids often are composed of mixtures of gases and vapors. The composition of the mixture may be described by specifying either the mass or number of moles of each constituent. Two commonly used parameters are

$$\text{mass fraction} = x_A = \frac{m_A}{\sum\limits_{i} m_i} \tag{7.43}$$

and

$$\text{mole fraction} = y_A = \frac{n_A}{\sum\limits_i n_i} \tag{7.44}$$

The vapors and gaseous constituents are treated as a mixture of ideal gases.

The assumption of treating the constituents as ideal gases implies that each component of the mixture exists independent of the other components. Thus, each component will occupy the total volume of the system (V) and be at the mixture temperature (T). This leads to *Dalton's law*

$$\sum p_i = p \tag{7.45}$$

where p_i is the partial pressure of the ith component.

For a mixture, with constant specific heats, changes from state 1 to state 2 are given by the following expressions

$$E_2 - E_1 = \sum_i E_{i2} - \sum_i E_{i1} = \sum_i m_i c_{vi}(T_2 - T_1) \tag{7.46}$$

$$H_2 - H_1 = \sum_i H_{i2} - \sum_i H_{i1} = \sum_i m_i c_{pi}(T_2 - T_1) \tag{7.47}$$

$$S_2 - S_1 = \sum_i S_{i2} - \sum_i S_{i1} = \sum_i m_i \left[c_{pi} \ln T_2/T_1 - R_i \ln \frac{p_{i2}}{p_{i1}} \right] \tag{7.48}$$

$$G_2 - G_1 = \sum_i G_{i2} - \sum_i G_{i1} = \sum_i m_i \left[c_{pi}(T_2 - T_1) \right.$$
$$\left. - T_2(c_{pi} \ln T_2 - R_i \ln p_{i2}) + T_1(c_{pi} \ln T_1 - R_i \ln p_{i1}) \right] \tag{7.49}$$

For air–water vapor mixtures, the composition is usually given by specially defined terms, humidity ratio or relative humidity. Humidity ratio is the ratio of the mass of the water vapor to the mass of the air in the mixture

$$\omega = \frac{m_v}{m_a} = 0.622 \frac{p_v}{p_a} \tag{7.50}$$

Relative humidity is defined as the ratio of the partial pressure of the water vapor to the saturation pressure at the mixture temperature

$$\phi = \frac{p_v}{p_g} \tag{7.51}$$

Thus

$$\phi = \frac{\omega}{0.622} \frac{p_a}{p_g} \tag{7.52}$$

7.10 CHEMICAL REACTIONS

Other sections of the Handbook may be also of interest; see Sec. 2.7 on Properties of Combustion Products, Sec. 23.11 on Fluid Dynamics of Combustion and Chap. 22 on Computational Methods for Chemically Reacting Flow. For chemical reacting systems, in which chemical equilibrium is assumed, the difficulty is generally in determining the composition of the products of reaction. For example, consider a chemical reaction between the reactants

$$H_2 + 1/2\ O_2 + 2.5\ N_2 \qquad (7.53)$$

First, one must list the possible products, which in this case could be H_2O, N_2, NO, H_2, H, O, and OH. Thus

$$H_2 + 1/2\ O_2 + 2.5\ N_2 = (\alpha)H_2O + (\beta)NO + (\gamma)OH$$
$$+ (\omega)N_2 + (r)H_2 + (s)O_2 + (t)H \qquad (7.54)$$

From conservation of atoms, 3 equations are obtained (H, N, O); however there are 7 unknowns (α, β, γ, ω, r, s, t). The remaining equations, which are needed to determine the composition, are obtained from the equilibrium relations that exist between the products. For example, the following 4 reactions may occur between the products

$$OH + H_2 \rightleftharpoons H_2O$$
$$2\ O \rightleftharpoons O_2$$
$$NO \rightleftharpoons 1/2\ O_2 + 1/2\ N_2$$
$$H_2 + 1/2\ O_2 \rightleftharpoons H_2O \qquad (7.55)$$

For each of these reactions, there is an *equilibrium constant* which is a function of temperature. Let us now see how this equilibrium constant is evaluated.

Any equilibrium reaction, such as those in Eq. (7.55) proceeding at a constant temperature and pressure will be such that the Gibbs function of the mixture will be a minimum value at the equilibrium composition. The molar chemical potential of a constituent (μ_i) is equal to its molar Gibbs function evaluated at the mixture temperature and its partial pressure. That is

$$\mu_i = g_i(T, P_i) = g_i(T, p) + RT \ln (p_i/p) \qquad (7.56)$$

Consider the following chemical reaction

$$\nu_1 C_1 + \nu_2 C_2 \rightleftharpoons \nu_3 C_3 + \nu_4 C_4 \qquad (7.57)$$

where $\nu_1 \ldots$ are the stoichiometric coefficients and $C_1 \ldots$ are the chemical constituents. The condition for chemical equilibrium is

$$\nu_1 \mu_1 + \nu_2 \mu_2 = \nu_3 \mu_3 + \nu_4 \mu_4 \qquad (7.58)$$

Thus,

$$\nu_1 g_1(t, p) + \nu_2 g_2(T, p) - \nu_3 g_3(T, p) - \nu_4 g_4(T, p) = RT \ln \frac{(p_{3/p})^{\nu_3} (p_{4/p})^{\nu_4}}{(p_{1/p})^{\nu_1} (p_{2/p})^{\nu_2}} \quad (7.59)$$

The Gibbs functions can be written in terms of T and a selected pressure p_0. Then

$$\frac{(p_{3/p})^{\nu_3} (p_{4/p})^{\nu_4}}{(p_{1/p})^{\nu_1} (p_{2/p})^{\nu_2}} \left(\frac{p}{p_0}\right)^{\nu_3 + \nu_4 - \nu_1 - \nu_2} = K(T) \quad (7.60)$$

where

$$K(T) = \exp\left[\frac{\nu_1 g_1(T, p_0) + \nu_2 g_2(T, P_0) - \nu_3 g_3(T, p_0) - \nu_4 g_4(T, p_0)}{RT}\right] \quad (7.61)$$

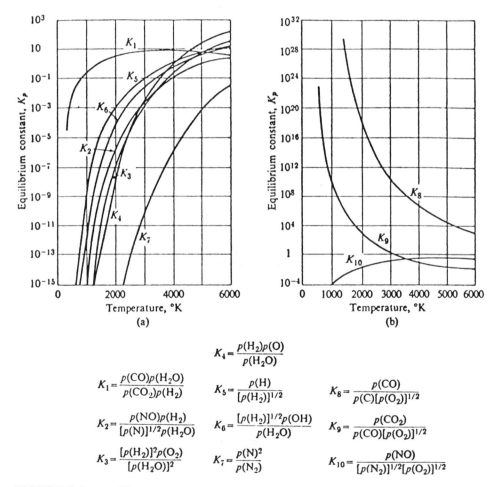

$$K_4 = \frac{p(H_2)p(O)}{p(H_2O)}$$

$$K_1 = \frac{p(CO)p(H_2O)}{p(CO_2)p(H_2)} \qquad K_5 = \frac{p(H)}{[p(H_2)]^{1/2}} \qquad K_8 = \frac{p(CO)}{p(C)[p(O_2)]^{1/2}}$$

$$K_2 = \frac{p(NO)p(H_2)}{[p(N)]^{1/2}p(H_2O)} \qquad K_6 = \frac{[p(H_2)]^{1/2}p(OH)}{p(H_2O)} \qquad K_9 = \frac{p(CO_2)}{p(CO)[p(O_2)]^{1/2}}$$

$$K_3 = \frac{[p(H_2)]^2 p(O_2)}{[p(H_2O)]^2} \qquad K_7 = \frac{p(N)^2}{p(N_2)} \qquad K_{10} = \frac{p(NO)}{[p(N_2)]^{1/2}[p(O_2)]^{1/2}}$$

FIGURE 7.5 Equilibrium constants in terms of partial pressures. Pressure in atmospheres. (Reprinted with permission of Addison–Wesley Publishing Co., *Mechanics and Thermodynamics of Propulsion*, by Hill, Philip G. and Peterson, Carl R. © 1965.)

Equation (7.62) is known as the *law of mass action*. $K(T)$ has been calculated for specific chemical reactions such as those listed in Eq. (7.55) and tabulated in tables or graphs such as Fig. 7.5. The calculation of the composition of the products is complicated by the fact that the equilibrium relations require the simultaneous solution of a set of linear and nonlinear algebraic equations.

Once the composition is determined, an energy balance can be obtained from the enthalpy of formation of compounds from their elements as well as the enthalpy change in raising the products from their initial temperature to the final temperature. The values of the enthalpy of formation for several common compounds are given in Table 7.2, and the enthalpy change in the products due to temperature rise can be calculated from expressions listed in Table 7.3.

7.11 DISSOCIATION AND IONIZATION

At high temperatures—of the order of several thousand degrees—diatomic molecules dissociate into neutral atoms (partially discussed in Sec. 7.10). Since dissociation of a molecule requires considerable energy, this process has an appreciable effect on the thermodynamics of the system. At very high temperatures—about ten thousand degrees and above—the atoms violently collide with each other, and valency electrons (outermost) are knocked off. When such an ionized atom and electron collide, the ion may capture the electron. These processes of dissociation and ionization are analogous to chemical reactions discussed in Sec. 7.10.

Consider nitrogen, for which the equilibrium chemical reaction for dissociation is

$$N_2 \rightleftharpoons 2\,N \tag{7.62}$$

Let x_1 and x_2 represent the mass fractions of N and N_2, respectively, present in the mixture. Subscript 1 denotes N which is *monatomic*, and subscript 2 indicates N_2 which is *diatomic*. Consequently, $(x_1 + x_2 = 1)$. The degree of dissociation (α is commonly used symbol) is defined as the mass fraction of nitrogen atoms present; that is, $x_1 = \alpha$ and $x_2 = (1 - \alpha)$. Assuming an ideal gas, and using Dalton's law of partial pressures, one obtains

$$p_1 v = x_1 R_1 T = \alpha R_1 T$$
$$p_2 v = x_2 R_2 T = (1 - \alpha) R_2 T/2 \tag{7.63}$$

From the law of mass action, Eq. (7.62)

$$\frac{p_1^2}{p_2} = \left(\frac{4\alpha^2}{1 - \alpha}\right)(p) = K(T) \tag{7.64}$$

The symbol $K_p(T)$ is used occasionally; the ordinate of Fig. 7.5 has K_p. The equilibrium degree of dissociation (α) is determined by the temperature and pressure of the gas. $K(T)$ is the dissociative equilibrium constant. This constant depends only

TABLE 7.2* Enthalpy of Formation, Gibbs Function of Formation and Absolute Entropy of Various Substances of 25°C, 0.1 MPa Pressure

Substance	Formula	M	State	h_f kJ/kmol	g_f kJ/kmol	s kJ/kmol K
Carbon monoxide[a]	CO	28.011	gas	−110 529	−137 150	197.653
Carbon dioxide[a]	CO_2	44.011	gas	−393 522	−394 374	213.795
Water[a,b]	H_2O	18.015	gas	−241 827	−228 583	188.833
Water[b]	H_2O	18.015	liq.	−285 838	−237 178	70.049
Methane[a]	CH_4	16.043	gas	−74 873	−50 751	186.256
Acetylene[a]	C_2H_2	26.038	gas	+226 731	+209 234	200.958
Ethene[a]	C_2H_4	28.054	gas	+52 283	+68 207	219.548
Ethane[c]	C_2H_6	30.070	gas	−84 667	−32 777	229.602
Propane[c]	C_2H_8	44.097	gas	−103 847	−23 316	270.019
Butane[c]	C_4H_{10}	58.124	gas	−126 148	−16 914	310.227
Octane[c]	C_8H_{10}	114.23	gas	−208 447	+16 859	466.835
Octane[c]	C_8H_{10}	114.23	liq.	−249 952	+6 940	360.896
Carbon[a] (graphite)	C	12.011	solid	0	0	5.795

*From van Wylen and Sonntag, *Fundamentals of Classical Thermodynamics*, John Wiley & Sons, New York, 1978.
[a]From JANAF *Thermochemical Data*, The Dow Chemical Company, Thermal Laboratory, Midland, MI.
[b]From *Circular 500*, National Bureau of Standards.
[c]From F. D. Rossini et al., *API Research Project 44.*

TABLE 7.3*   Constant-Pressure Specific Heats of Various Ideal Gases[a]

$$C_p = kJ/kmol\ K$$

$$\theta = T(Kelvin)/100$$

Gas		Range K	Max. Error %
N_2	$C_p = 39.060 - 512.79\theta^{-3.8} + 1072.7\theta^{-2} - 820.40\theta^{-3}$	300–3500	0.43
O_2	$C_p = 37.432 + 0.020102\theta^{1.5} - 178.57\theta^{-3.5} + 236.88\theta^{-2}$	300–3500	0.30
H_2	$C_p = 56.505 - 702.74\theta^{-0.75} + 1165.0\theta^{-1} - 560.70\theta^{-1.5}$	300–3500	0.60
CO	$C_p = 69.145 - 0.70463\theta^{0.75} - 200.77\theta^{-0.5} + 176.76\theta^{-0.75}$	300–3500	0.42
OH	$C_p = 81.546 - 59.350\theta^{0.25} + 17.329\theta^{0.75} - 4.2660\theta$	300–3500	0.43
NO	$C_p = 59.283 - 1.7096\theta^{0.5} - 70.613\theta^{-0.5} + 74.889\theta^{-1.5}$	300–3500	0.34
H_2O	$C_p = 143.05 - 183.54\theta^{0.25} + 82.751\theta^{0.5} - 3.6989\theta$	300–3500	0.43
CO_2	$C_p = -3.7357 + 30.529\theta^{0.5} - 4.1034\theta + 0.024198\theta^2$	300–3500	0.19
NO_2	$C_p = 46.045 + 216.10\theta^{-0.5} - 363.66\theta^{-0.76} + 232.550\theta^{-2}$	300–3500	0.26
CH_4	$C_p = -672.87 + 439.74\theta^{0.25} - 24.875\theta^{0.76} + 323.88\theta^{-0.5}$	300–2000	0.15
C_2H_4	$C_p = -95.395 + 123.15\theta^{0.5} - 35.641\theta^{0.72} + 182.77\theta^{-2}$	300–2000	0.07
C_2H_6	$C_p = 6.895 + 17.26\theta - 0.6402\theta^2 + 0.00728\theta^3$	300–1500	0.83
C_2H_8	$C_p = -4.042 + 30.46\theta - 1.571\theta^2 + 0.03171\theta^3$	300–1500	0.40
C_4H_{30}	$C_p = 3.954 + 37.12\theta - 1.833\theta^2 + 0.03498\theta^3$	300–1500	0.54

*From Van Wylen and Sonntag, *Fundamentals of Classical Thermodynamics*, John Wiley & Sons, New York, 1978.

[a]From Scott, T. C. and Sonntag, R. E., Univ. of Michigan, unpublished, 1971, except C_7H_6, C_2H_8, C_4H_{30} from Kobe, K. A., Petroleum Refiner 28, No. 2, 113, 1949.

upon temperature, molecular constants, and atomic constants. It can be derived from statistical mechanics by expressing the Hemholtz function in terms of partition functions (translational, rotational, vibrational, and electronic). The equilibrium composition of the mixture corresponds to a minimum of the Helmholtz function. Assuming the molecular vibrations are fully excited and the electronic partition functions contain only ground states, the following expression is obtained

$$\frac{\alpha^2}{1 - \alpha} = \frac{M_1^{3/2}\nu g_{01}^2}{16\pi^{1/2}I_2(kT)^{1/2}q_{02}n_2^0} e^{-U/kT} \tag{7.65}$$

where M_1 = mass of N atom
k = Boltzman's constant (8.31 kJ/(K kmole)
g_0 = statistical weight for the electronic ground state
n_2° = the original number of molecules per unit volume of gas
ν = frequency of vibrational excitation
I_2 = linear moment of inertia of N_2 molecules
U = dissociation energy (9.3 × 10⁵ kJ/k mole for N_2)

The equation for a single ionized nitrogen atom can be represented as

$$N \rightleftharpoons N^+ + e^- \tag{7.66}$$

Assuming that this mixture of neutral atoms, positive ions, and electrons behaves as a mixture of ideal gases, the following expression is obtained

$$\frac{\alpha^2}{1 - \alpha} = \frac{(2\pi m)^{3/2}(kT)^{5/2}}{ph^3} \exp\left(-I/kT\right) \tag{7.67}$$

where m = mass of an electron, kg
k = Boltzmann's constant, 1.38026×10^{-23} J/K
T = temperature, K
I = ionization potential, kJ
p = pressure, N/m^2
h = Planck's constant, 6.62377×10^{-34} U/s
α = mass fraction of ions, i.e. $N^+/(N + N^+)$

Equation (7.67) is known as the *Saha Equation*. This equation was obtained in a similar manner to the one for dissociation, Eq. (7.65).

REFERENCES

Epstein, P. S., *Textbook of Thermodynamics*, John Wiley & Sons, New York, 1937.

Fermi, E., *Thermodynamics*, Dover, New York, 1956.

Reiss, H., *Methods of Thermodynamics*, Blaisdell, Massachusetts, 1965.

Van Wylen, G. J. and Sonntag, R. E., *Fundamentals of Classical Thermodynamics*, John Wiley & Sons, New York, 1978.

Zeldovich, V. B. and Raizer, Y. P., *Physics of Shock Waves and High-Temperature Hydrodynamic Phenomena*, Academic Press, New York, 1966.

Zemansky, M. W., Abbott, M. M., and Van Hess, H. C., *Basic Engineering Thermodynamics*, McGraw-Hill, New York, 1975.

8 Compressible Flow

GEORGE EMANUEL
University of Oklahoma
Norman, OK

CONTENTS

8.1 INTRODUCTION

The basic elements of compressible gas dynamics are largely a product of the first half of the 20th century. Interest in supersonic flow was spurred by the development of the supersonic wind tunnel, which preceded applications such as high-speed missiles or aircraft. The principal nondimensional parameters are the Mach number M, which is the ratio of the flow speed V to the speed of sound a, and the ratio of specific heats γ. Generally speaking, gas dynamics refers to inviscid, nonconducting flows where the Mach number exceeds about 0.4. Above this Mach number, compressibility effects become significant. In this chapter, the inviscid, nonconducting

Handbook of Fluid Dynamics and Fluid Machinery, Edited by Joseph A. Schetz and Allen E. Fuhs
ISBN 0-471-12598-9 Copyright © 1996 John Wiley & Sons, Inc.

restriction applies to most topics, but not all. In particular Rayleigh flow and Fanno flow, which respectively involve heat addition and wall shear, are discussed in Sec. 8.3. Another assumption that is made, but which is not basic, is that of a thermally and calorically perfect gas. This approximation provides a major simplification in all aspects of the subject, e.g., γ is a constant. Under most conditions of interest ($M \lesssim 5$), air is well represented by this approximation.

In contrast to incompressible flow, there is a strong coupling not only between the equations of motion, but also with thermodynamics. With the perfect gas approximation, the connection is by means of thermal and caloric state equations, both of which are simple. The solution for the flows in this chapter utilize these state equations.

There are many useful compressible flow references that the reader can consult, such as Shapiro (1953), Liepman and Roshko (1957), Anderson (1982), and Emanuel (1986). One aspect of this chapter is the consistent formulation of the equations in a form suitable for computer solution. A number of frequently occurring special Mach number functions are systematically employed; they are defined at the end of this paragraph. Since tables of the Mach number functions are easily generated, they are not included. This type of table, however, can be found in some of the above references.

Mach Number Functions

$$W = 1 + \gamma M^2$$

$$X = 1 + \frac{\gamma - 1}{2} M^2$$

$$Y = \frac{2\gamma}{\gamma + 1} M^2 - \frac{\gamma - 1}{\gamma + 1}$$

$$Z = M^2 - 1$$

$$\mu = \sin^{-1}\left(\frac{1}{M}\right)$$

$$\nu = \left(\frac{\gamma + 1}{\gamma - 1}\right)^{1/2} \tan^{-1}\left\{\left[\left(\frac{\gamma - 1}{\gamma + 1}\right) Z(M)\right]^{1/2}\right\} - \tan^{-1}\left[Z(M)\right]^{1/2}$$

8.2 ISENTROPIC FLOW

In an isentropic process, a particle of fluid undergoes a reversible adiabatic change, where viscous effects are negligible. The actual changes experienced by the bulk of the flow field are often well approximated by this process. This is the case in acoustics, internal flows such as for nozzles, and external flows such as around an airfoil. Wall boundary layers and the flow inside shock waves are excluded here [see Sec. 4.6 for Compressible Boundary Layers].

In this brief section, an isentropic process is assumed for a thermally and calorically perfect gas. The thermal and entropy equations of state are taken to be

$$p = \rho RT \tag{8.1}$$

$$\frac{s - s_0}{R} = \ln\left[\left(\frac{p_0}{p}\right)\left(\frac{T}{T_0}\right)^{\gamma/(\gamma - 1)}\right] \tag{8.2}$$

where the stagnation condition (where $V = 0$) is denoted by an 0 subscript, and p, ρ, R, and T are, respectively, the pressure, density, gas constant, and absolute temperature. Relations for the specific heats at constant volume and constant pressure

$$c_v = \frac{1}{\gamma - 1} R, \; c_p = \frac{\gamma}{\gamma - 1} R \tag{8.3}$$

are utilized. For an isentropic process, set $s = s_0$ to obtain

$$\frac{p}{p_0} = \left(\frac{\rho}{\rho_0}\right)^{\gamma} = \left(\frac{T}{T_0}\right)^{\gamma/(\gamma - 1)} = \left(\frac{h}{h_0}\right)^{\gamma/(\gamma - 1)} = \left(\frac{a}{a_0}\right)^{2\gamma/(\gamma - 1)} \tag{8.4}$$

for the more important thermodynamic variables, where h is enthalpy.

For a steady, adiabatic flow, the energy equation is

$$h + \tfrac{1}{2} V^2 = h_0 \tag{8.5}$$

where the stagnation enthalpy h_0 is a constant. By introducing

$$h = \frac{\gamma}{\gamma - 1} RT, \tag{8.6a}$$

$$M = \frac{V}{a}, \tag{8.6b}$$

$$a = (\gamma RT)^{1/2} \tag{8.6c}$$

into Eq. (8.5), we have

$$\frac{T}{T_0} = \frac{1}{1 + (\gamma - 1)M^2/2} = X^{-1} \tag{8.7a}$$

and from Eqs. (8.4)

$$\frac{p}{p_0} = X^{-\gamma/(\gamma - 1)}, \tag{8.7b}$$

$$\frac{\rho}{\rho_0} = X^{-1/(\gamma - 1)} \tag{8.7c}$$

Equations (8.7) are the isentropic relations for an actual process. They also hold for a fictitious point process whereby a fixed fluid particle is isentropically brought to rest at its stagnation state.

8.3 STEADY ONE-DIMENSIONAL FLOW

8.3.1 Formulation

Unsteady, quasi-one-dimensional flow is first considered wherein the cross-sectional area A of a streamtube may change with time or with the axial position x. Heat transfer into or out of the streamtube and viscous shear on the streamtube's surface are both allowed, but (gravitational) body forces are neglected. For consistency with the one-dimensional approximation, these effects must be averaged across the streamtube. Although referred to as a streamtube, the area A need not be small. However, its rate of change, dA/dx, cannot be too rapid if the one-dimensional approximation is to be valid.

Under the above assumptions, the conservation equations for mass, momentum, and energy can be written as

$$\frac{D(\rho A)}{Dt} + \rho A \frac{\partial V}{\partial x} = 0, \tag{8.8a}$$

$$\rho \frac{DV}{Dt} = -\frac{\partial p}{\partial x} - \frac{c}{A}\tau, \tag{8.8b}$$

$$\frac{Dh_0}{Dt} = \frac{1}{\rho}\frac{\partial p}{\partial t} + \dot{q} \tag{8.8c}$$

where $c(x, t)$ is the streamtube's perimeter, τ is the shear stress on the wall, and the substantial derivative, which follows a fluid particle, is

$$\frac{D}{Dt} = \frac{\partial}{\partial t} + V\frac{\partial}{\partial x} \tag{8.9}$$

The stagnation enthalpy is given by Eq. (8.5) but is not a constant, and $\tau c/A$ is the shear stress per unit volume. The hydraulic diameter d and coefficient of skin friction c_f are defined by

$$d = \frac{4A}{c}, \tag{8.10}$$

$$c_f = \frac{2\tau}{\rho V^2} \tag{8.11}$$

and reduce the shear term to

$$\frac{c}{A}\tau = 2\frac{c_f}{d}\rho V^2 \tag{8.12}$$

The parameter $\dot{q}(x, t)$ is the heat transferred by conduction, radiation, etc., per unit mass and per unit time and is positive when the surroundings supply the heat. For a complete system of equations, Eqs. (8.8) are supplemented with thermal, Eq. (8.1), and caloric, Eq. (8.6a), equations of state.

For a steady flow, Eqs. (8.8) become

$$\rho V A = \dot{m},$$

(8.13a)

$$V dV + \frac{1}{\rho} dp = -\frac{1}{2} V^2 \left(4 c_f \frac{dx}{d} \right),$$

(8.13b)

$$dh_0 = \dot{q} \frac{dx}{V} = dq$$

(8.13c)

where $\dot{m}$ is the constant mass flow rate. Equation (8.13a) is logarithmically differentiated, h_0 is replaced by Eq. (8.5), h and p are eliminated by the state equations, and V is removed in favor of M. In a nondimensional matrix form, Eqs. (8.13) become

$$\frac{d\rho}{\rho} + \frac{1}{2} \frac{dT}{T} + \frac{1}{2} \frac{dM^2}{M^2} = -\frac{dA}{A}$$

(8.14a)

$$\frac{d\rho}{\rho} + \left(1 + \frac{\gamma}{2} M^2 \right) \frac{dT}{T} + \frac{\gamma}{2} M^2 \frac{dM^2}{M^2} = -\frac{\gamma}{2} M^2 \left(4 c_f \frac{dx}{d} \right)$$

(8.14b)

$$\frac{d\rho}{\rho} - \frac{1}{\gamma - 1} \frac{dT}{T} = -\frac{dq}{RT} - \frac{\gamma}{2} M^2 \left(4 c_f \frac{dx}{d} \right)$$

(8.14c)

and can be solved for $d\rho/\rho$, dT/T, and dM^2/M^2. For dM^2/M^2 the result is

$$Z \frac{dM^2}{M^2} = 2X \frac{dA}{A} - \frac{\gamma - 1}{\gamma} W \frac{dq}{RT} - \gamma M^2 X \left(4 c_f \frac{dx}{d} \right)$$

(8.15)

where W, X, and Z, are defined at the end of the first section. It is convenient to eliminate dq/RT in favor of dT_0/T_0. For this, Eq. (8.7a) and

$$h_0 = c_p T_0 = \frac{\gamma R}{\gamma - 1} T_0$$

(8.16)

are used to obtain

$$\frac{dM^2}{M^2} = \frac{2X}{Z} \frac{dA}{A} - \frac{WX}{Z} \frac{dT_0}{T_0} - \frac{\gamma M^2 X}{Z} \left(4 c_f \frac{dx}{d} \right)$$

(8.17)

8.3.2 General Results

Equations similar to (8.17) are obtained for the other derivatives by the same procedure. Each derivative is proportional to the dA/A, dT_0/T_0, and $4 c_f dx/d$ terms with the coefficients depending only on γ and M. This influence coefficient approach is summarized in Table 8.1, where a comparison with Eq. (8.17) illustrates how the table is used.

The solution of the equations in Table 8.1 requires a knowledge of the variation

TABLE 8.1   Influence Coefficients for a Thermally and Calorically Perfect Gas

	$\dfrac{dA}{A}$	$\dfrac{dT_0}{T_0}$	$4c_f\dfrac{dx}{d}$
$\dfrac{dM^2}{M^2}$	$\dfrac{2X}{Z}$	$-\dfrac{WX}{Z}$	$-\dfrac{\gamma M^2 X}{Z}$
$\dfrac{dp}{p}$	$-\dfrac{\gamma M^2}{Z}$	$\dfrac{\gamma M^2 X}{Z}$	$\dfrac{\gamma M^2(W-M^2)}{2Z}$
$\dfrac{dT}{T}$	$-\dfrac{(\gamma-1)M^2}{Z}$	$-\dfrac{(1-\gamma M^2)X}{Z}$	$\dfrac{\gamma(\gamma-1)M^4}{2Z}$
$\dfrac{d\rho}{\rho}$	$-\dfrac{M^2}{Z}$	$\dfrac{X}{Z}$	$\dfrac{\gamma M^2}{2Z}$
$\dfrac{dp_0}{p_0}$	0	$-\dfrac{\gamma M^2}{2}$	$-\dfrac{\gamma M^2}{2}$
$\dfrac{dV}{V}$	$\dfrac{1}{Z}$	$-\dfrac{X}{Z}$	$-\dfrac{\gamma M^2}{2Z}$
$\dfrac{ds}{R}$	0	$\dfrac{\gamma X}{\gamma-1}$	$\dfrac{\gamma M^2}{2}$

of dA/A, dT_0/T_0, and c_f with x or with M. There is no general closed-form solution to these equations. However, three cases of importance can be solved in closed form. They are isentropic flow with area change (Sec. 8.3.3), Rayleigh flow for a change in T_0 (Sec. 8.3.4), and Fanno flow caused by friction (Sec. 8.3.5).

Table 8.2 contains these results, where the asterisk denotes an $M = 1$ reference condition, which is process dependent. Thus, the $M = 1$ condition for an area change is not the same state as when only T_0 changes. This choice of a reference state avoids the singularities that would occur had a stagnation reference state been chosen.

Consider two transverse sections of a streamtube whose axis is a straight line. By definition, no flow crosses the lateral surface of the streamtube. The net force F_{net} produced by the stream is determined by conservation of momentum, and is (see Sec. 8.3.5)

$$F_{\text{net}} = (pA + \rho AV^2)_2 - (pA + \rho AV^2)_1 \tag{8.18}$$

where the 1 and 2 subscripts denote the upstream and downstream transverse sections, respectively. This relation includes the pressure and viscous forces and is not limited with regard to heat transfer, or to irreversible processes, such as shock waves, that may occur interior to the length of streamtube under consideration. It does require appropriate averages at stations 1 and 2 for p, ρ, and V. A positive value for F_{net} means a net force on the surroundings in a direction opposite to the fluid's motion. Equation (8.18) can be written as

$$F_{\text{net}} = F_2 - F_1 \tag{8.19}$$

TABLE 8.2 Equations for Simple Flows of a Thermally and Calorically Perfect Gas with c_f Constant

	Area Change Only	Change in T_0 Only	Friction Only
A/A^*	$\dfrac{1}{M}\left(\dfrac{2X}{\gamma+1}\right)^{(\gamma+1)/2(\gamma-1)}$	1	1
T_0/T_0^*	1	$\dfrac{2(\gamma+1)M^2 X}{W^2}$	1
$4c_f\dfrac{L_m}{d}$	0	0	$-\dfrac{Z}{\gamma M^2}+\dfrac{\gamma+1}{2\gamma}\ln\left[\dfrac{(\gamma+1)M^2}{2X}\right]$
p/p^*	$\left(\dfrac{\gamma+1}{2X}\right)^{\gamma/(\gamma-1)}$	$\dfrac{\gamma+1}{W}$	$\dfrac{1}{M}\left(\dfrac{\gamma+1}{2X}\right)^{1/2}$
T/T^*	$\dfrac{\gamma+1}{2X}$	$\dfrac{(\gamma+1)^2 M^2}{W^2}$	$\dfrac{\gamma+1}{2X}$
ρ/ρ^*	$\left(\dfrac{\gamma+1}{2X}\right)^{1/(\gamma-1)}$	$\dfrac{W}{(\gamma+1)M^2}$	$\dfrac{1}{M}\left(\dfrac{2X}{\gamma+1}\right)^{1/2}$
p_0/p_0^*	1	$\dfrac{\gamma+1}{W}\left(\dfrac{2X}{\gamma+1}\right)^{\gamma/(\gamma-1)}$	$\dfrac{1}{M}\left(\dfrac{2X}{\gamma+1}\right)^{(\gamma+1)/2(\gamma-1)}$
V/V^*	$M\left(\dfrac{\gamma+1}{2X}\right)^{1/2}$	$\dfrac{(\gamma+1)M^2}{W}$	$M\left(\dfrac{\gamma+1}{2X}\right)^{1/2}$
$(s-s^*)/R$	0	$-\dfrac{\gamma}{\gamma-1}\ln\left[M^2\left(\dfrac{\gamma+1}{W}\right)^{(\gamma+1)/\gamma}\right]$	$-\dfrac{\gamma}{\gamma-1}\ln\left[M^2\left(\dfrac{\gamma+1}{2X}\right)^{(\gamma+1)/2\gamma}\right]$

where F is the impulse function defined by

$$F = pA + \rho A V^2 = pA + \dot{m}V = pAW \qquad (8.20a)$$

or as

$$\frac{F}{F^*} = \frac{W}{M[2(\gamma + 1)X]^{1/2}} \qquad (8.20b)$$

8.3.3   Isentropic Flow with Area Change

When only the area changes, Table 8.2 yields

$$\frac{A}{A^*} = \frac{1}{M}\left(\frac{2X}{\gamma + 1}\right)^{(\gamma + 1)/2(\gamma - 1)}, \quad \frac{T_0}{T_0^*} = 1, \ldots \qquad (8.21)$$

The variation with Mach number of p/p_0, T/T_0, and ρ/ρ_0 is monotonically decreasing. On the other hand, A/A^* has a minimum of unity when $M = 1$, and the streamtube's cross-sectional area is double valued with respect to M. The section with the minimum area is referred to as a throat. As shown in Fig. 8.1, there are several different types of solutions as a consequence of the double valued area. Curve a denotes a purely subsonic solution where the maximum Mach number, which is less than unity, occurs at the throat. Curve b denotes a solution where $M = 1$ at the throat, a condition referred to as choking. Downstream of the throat, subsonic (b_1) and supersonic (b_2) solutions are possible. Curves c and d have a similar interpretation. Nozzle flow corresponds to a subsonic Mach number upstream of the throat (curves a and b), while curves c and d correspond to a diffuser flow. In practice, curves a and b, b_2 are often closely approximated. On the other hand, curves d and c-c_1 are not obtainable owing to the presence of boundary-layer separation and shock waves.

For curve a, $1 > M(x^*)$, $A(x^*) > A^*$, and the throat area exceeds the fictitious reference area A^*. The two areas, $A(x^*)$ and A^*, coincide only when the flow is

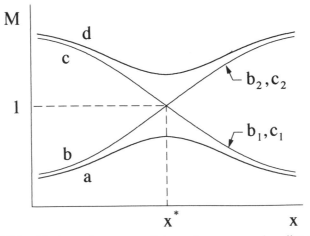

FIGURE 8.1   Mach number versus distance for a converging-diverging duct.

choked and $M(x^*) = 1$. By means of Eq. (8.13a) and Table 8.2, the constant mass flow rate $\dot{m}$ is given by

$$\dot{m} = MX^{-(\gamma + 1)/2(\gamma - 1)} (\gamma/RT_0)^{1/2} p_0 A \tag{8.22}$$

When the flow is choked, this becomes

$$\dot{m} = \left(\frac{\gamma + 1}{2}\right)^{-(\gamma + 1)/2(\gamma - 1)} (\gamma/RT_0)^{1/2} p_0 A^* \tag{8.23}$$

This discussion continues in Sec. 8.6 after the shock and Prandtl–Meyer sections, since these topics are necessary for a more complete understanding of nozzle flow.

8.3.4 Rayleigh Flow

A constant area duct without friction is assumed. Heat can be removed or added to the gas by heat exchange through the duct wall, by radiative heat transfer, by combustion, or by evaporation or condensation. Although the inviscid assumption may appear to be unrealistic, Rayleigh flow is nevertheless useful for the analysis of jet engine combustors and flowing gaseous lasers. In these devices, the heat addition process dominates viscous effects.

The speed V is eliminated from the mass and momentum equations, to obtain for average conditions across the duct

$$p + \left(\frac{\dot{m}}{A}\right)^2 \frac{1}{\rho} = \frac{F}{A} \tag{8.24}$$

Since only thermodynamic variables occur in Eq. (8.24), it is a curve, called the Rayleigh line, on a Mollier diagram (see Fig. 8.2). The location where the maximum entropy occurs is denoted by an asterisk, and $M = 1$ at this point. On the upper branch the flow is subsonic, while on the lower branch it is supersonic.

The foregoing discussion is applicable to a gas with any state equations. For a thermally and calorically perfect gas, however, the equations reduce to those in the T_0 column of Table 8.2. In this case, Eqs. (8.1), (8.6a), and

$$\frac{p}{p_r} = \left(\frac{\rho}{\rho_r}\right)^{\gamma} e^{(\gamma - 1)(s - s_r)/R} \tag{8.25}$$

apply, where the r subscript denoted an arbitrary reference state. Hence, the p and ρ functions for Eq. (8.24) are

$$\frac{p}{p_r} = \left(\frac{h}{h_r}\right)^{\gamma/(\gamma - 1)} e^{-(s - s_r)/R} \tag{8.26}$$

$$\frac{\rho}{\rho_r} = \left(\frac{h}{h_r}\right)^{1/(\gamma - 1)} e^{-(s - s_r)/R} \tag{8.27}$$

In Figs. 8.2–8.4, the Rayleigh and Fanno curves are for a perfect gas with $\gamma = 1.4$.

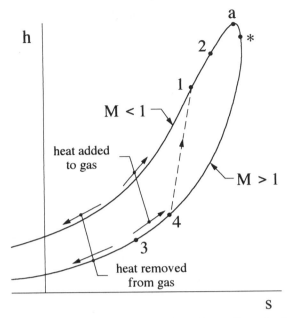

FIGURE 8.2 Rayleigh line on a Mollier diagram for a thermally and calorically perfect gas with $\gamma = 1.4$.

According to the Second Law of Thermodynamics

$$ds \geq \frac{\delta q}{T} \qquad (8.28)$$

where δq is the actual heat added to the system. Consequently, heat addition (removal) to the gas increases (decreases) s and the state of the system moves along a Rayleigh line as described in Fig. 8.2. Consider a duct where subsonic inlet conditions are fixed at state 1 and, as a result of heat addition, outlet conditions are at state 2. As the heat addition is increased, state 2 moves to the right. At state a, h, and T have maximum values. Additional heating beyond state a then cools the gas, because more energy goes into kinetic energy than is being added. Thermal choking occurs when outlet conditions become sonic at the asterisk state. With fixed inlet conditions at state 1, additional heat cannot be added, since this would violate the second law. If more heat is added to a choked flow, inlet conditions must change. In this circumstance, $\dot{m}/A$ and possibly F/A in Eq. (8.24) change and a different Rayleigh line is generated. The manner of change is described in the examples at the end of this section. Cooling a subsonic flow merely requires considering state 2 as the inlet and state 1 as the outlet.

Heat addition on the supersonic branch requires consideration of a normal shock wave. The discussion here, and in the Fanno flow section, thus utilizes material in Sec. 8.4.1.

With state 3 as the fixed supersonic inlet condition (Fig. 8.2), heat addition will result in state 4 as an outlet condition. Further heat addition causes the flow to choke when state 4 becomes sonic. Cooling the flow simply reverses the inlet and outlet states.

A shock wave may occur when the flow is supersonic. As will be evident in Sec. 8.4.1, Eq. (8.24) still holds. Hence, the upstream and downstream states associated with a normal shock both fall on the Rayleigh line, as shown in Fig. 8.2. Inlet conditions are at state 3 and heat addition is presumed. The normal shock occurs between states 4 and 1, where

$$M_4 > 1, 1 > M_1, s_1 > s_4 \tag{8.29}$$

The energy equation for Rayleigh flow is

$$q = h_{02} - h_{01} = \frac{\gamma R}{\gamma - 1} (T_{02} - T_{01}) \tag{8.30}$$

where subscripts 1 and 2 denote inlet and outlet stations, respectively. With the T_0/T_0^* relation in Table 8.2, the maximum amount of heat q_{m1} that can be added to a flow, with an inlet Mach number M_1, is

$$\frac{q_{m1}}{RT_0^*} = \frac{\gamma}{\gamma - 1} \left(\frac{Z_1}{W_1} \right)^2 \tag{8.31}$$

where the exit Mach number is sonic. Equation (8.30) thus becomes

$$\frac{q}{RT_0^*} = \frac{q_{m1}}{RT_0^*} - \frac{q_{m2}}{RT_0^*} \tag{8.32}$$

where T_0^* is constant on a given Rayleigh line.

In Fig. 8.2, let the heat added between states 1 and 2 be q_a, while that added between states 3 and 4 be q_b, with

$$q = q_a + q_b \tag{8.33}$$

By means of Eqs. (8.31) and (8.32), and the normal shock equations, one can show [see Emanuel (1986)] that the exit Mach number M_2 is independent of q_a/q, under the proviso that M_3 and q are fixed. The fraction of heat added to the flow upstream or downstream of the shock does not determine the shock location inside the duct. If a shock-free supersonic Rayleigh flow is choked, then inserting a shock wave in the flow will not unchoke it. On the other hand, if additional heat is added to a choked flow, inlet conditions must change. Since the inlet conditions are supersonic, the adjustment occurs via a shock system that starts upstream of the inlet station. This type of adjustment has been observed in supersonic chemical laser flows [Emanuel (1982a)].

A Rayleigh flow analysis requires uniform inlet and outlet conditions, or inlet and outlet conditions that can be averaged. In subsonic duct flow, there often is adequate flow time for suitable average outlet conditions to be established. Rayleigh theory does not require a known heat addition $q(x)$ profile, i.e., the computation is of a ''black-box'' type. The theory is of less value for shock-free supersonic flow, in part because such duct flows are difficlt to maintain over any appreciable distance. Lateral heat transfer is also slow in this type of flow. An oblique shock system, of course, would enhance the lateral energy transfer. Nevertheless, the analysis is applicable to supersonic heat addition as occurs in chemical lasers, gas-dynamic lasers, laser-induced chemistry, and laser-isotope separation. In the latter two technologies, a cold gas at low pressure is generally required for the enrichment or separation process. These conditions are achieved by expanding the gas through a converging-diverging nozzle. Subsequently, laser energy is absorbed by the gas in a supersonic flow [Emanuel *et al.* (1981)].

As an example, consider nitrogen at 202.7 kPa (2 atm) pressure and 500 K which is fed to a straight duct through a converging-only adiabatic nozzle. At the exit of the duct, the flow is choked and the ambient pressure p_a equals the exit pressure, which is 101.35 kPa (1 atm). We are to determine M_2, q, $\dot{m}/A$, and F/A, where A is the duct's cross-sectional area.

Since the gas is N_2, γ, and R are

$$\gamma = 1.4, R = 296 \text{ (N} - \text{m)/(kg} - \text{K)} = 296 \text{ J/(kg} - \text{K)} \tag{8.34a}$$

Station 1 denotes the entrance to the nozzle, and stations 2 and 3 denote, respectively, the inlet and outlet sections of the constant area duct. Stagnation conditions are assumed for state 1, hence,

$$p_{01} = p_{02} = 202.6 \text{ kPa}, T_{01} = T_{02} = 500 \text{ K}$$

$$M_2 < 1, M_3 = 1, p_3 = 101.35 \text{ kPa} \tag{8.34b}$$

The known pressure ratio p_3/p_{01} is used to determine M_2

$$\frac{p_3}{p_{01}} = \frac{p_3}{p^*} \frac{p^*}{p_2} \frac{p_2}{p_{01}} = \frac{W_2}{\gamma + 1} X_2^{-\gamma/(\gamma - 1)} \tag{8.35}$$

where $p_3 = p^*$, $W_2 = W(M_2)$, $X_2 = X(M_2)$, and p_2/p^* comes from the T_0 column in Table 8.2, while p_2/p_{01} is given by Eq. (8.7b). Equation (8.35) is iteratively solved for M_2 with the result that $M_2 = 0.656$. Equation (8.22) is used with M_2 to yield $(\dot{m}/A) = 319 \text{ kg/m}^2$-s. Equation (8.22) cannot be used downstream of station 2 since the flow is not isentropic inside the duct. Equation (8.18), written as

$$\frac{F}{A} = p_{01} \frac{p_2}{p_{01}} W_2 \tag{8.36}$$

is utilized with the result $F/A = 0.243 \text{ MPa}$. Both stations 2 and 3 have this value and the net thrust is zero. Table 8.2 is again used to determine

$$T_{02}^* = \frac{W_2^2 T_{02}}{2(\gamma + 1)M_2^2 X_2} = 572 \text{ K} \tag{8.37a}$$

Finally, Eqs. (8.31) and (8.32) result in

$$\frac{q_{m2}}{RT_{02}^*} = \frac{\gamma}{\gamma - 1} \left(\frac{Z_2}{W_2}\right)^2 = 0.443, \frac{q_{m3}}{RT_{02}^*} = 0$$

$$q = RT^*_{02} \frac{q_{m2}}{RT^*_{02}} = 7.51 \times 10^4 \text{ J/kg} \tag{8.37b}$$

Suppose the reservoir pressure p_{01} is increased above 2 atm with a corresponding increase in $\rho_{01}T_{01}$. This results in an increased pressure level in the duct and increased values for $\dot{m}/A$ and F/A. By utilizing nondimensional variables, h/h_{01} and $(s - s_{01})/R$, and nondimensional parameters, $\dot{m}/A(p_{01}\rho_{01})^{1/2}$ and F/Ap_{01}, we observe that the nondimensional Rayleigh curve is unaltered by this pressure increase. Hence, the above Mach numbers, M_2 and M_3, and the heat addition q/RT^*_0 are also unaltered. However, the pressure p_3, which still equals p^*, is determined by Eq. (8.35), but now exceeds the ambient pressure. The flow coming out of the duct initially will go through a supersonic expansion. This mode of pressure adjustment is discussed in Sec. 8.6.

On the other hand, if p_{01} is reduced below 2 atm, the exit pressure p_3 will still equal the 1 atm ambient pressure. Both M_2 and M_3 are altered, with $M_3 < 1$ (the duct is not choked) and $\dot{m}/A$ and F/A decrease. With the above value for q, the Mach numbers are found by a procedure similar to the one in the following example.

The heat addition is now doubled, $q' = 2q$, where a prime is used to distinguish this case from the preceding one. With this one change, we determine M'_2 and M'_3.

If there is only a moderate increase in q, it is reasonable to assume that conditions at location 1 are unaltered. (This assumption must be dropped if the increase in q is sufficiently large.) Next assume $M'_3 < 1$, which is shortly verified. As a consequence, $p'_3 = p_a$, but p'_3 is not equal to $p^{*'}$, since $M'_3 < 1$. Equation (8.35) now has the form

$$W'_3 = \frac{p_{01}}{p_a} \frac{W'_2}{(X'_2)^{\gamma/(\gamma-1)}} \tag{8.38a}$$

where $W'_i = W(M'_i)$, and both M'_2 and M'_3 are unknowns. Two additional relations stem from Eqs. (8.31), (8.32), and (8.37a):

$$T^{*'}_{02} = \frac{T'_{02}(W'_2)^2}{2(\gamma+1)(M'_2)^2 X'_2} \tag{8.38b}$$

$$\frac{q'}{RT^{*'}_{02}} = \frac{\gamma}{\gamma-1}\left[\left(\frac{Z'_2}{W'_2}\right)^2 - \left(\frac{Z'_3}{W'_3}\right)^2\right] \tag{8.38c}$$

where $T'_{02} = T_{02}$. Both $T^{*'}_{02}$ and M'^2_3 are eliminated from Eqs. (8.38), thus yielding a single equation for M'^2_2, which is iteratively solved with the result that $M'_2 = 0.562$. The remaining Mach number is obtained from Eq. (8.38a), which yields $M'_3 = 0.974$. Since M'_2 is less than M_2, $\dot{m}'/A$ and F'/A are less than $\dot{m}/A$ and F/A, respectively.

Both the q and q' solutions are clarified by sketching them on a perfect gas Mollier diagram. As shown in Fig. 8.3, the flow expands isentropically from state 1 to state 2 or 2'. From state 2, the flow moves along the Rayleigh line until it chokes at state 3, where the static pressure equals p_a, the ambient pressure. Increasing the heat addition alters $\dot{m}/A$ and F/A in Eq. (8.24); hence, the flow proceeds on a different Rayleigh line to a subsonic exit condition, where $p'_3 = p_a$. The earlier discussion in

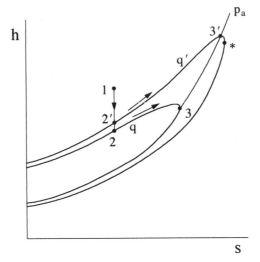

FIGURE 8.3   Shift in the Rayleigh line as a result of increasing the heat addition of a choked flow.

this section tended to emphasize fixed duct inlet conditions, which here are not fixed. As these examples show, increasing the heat addition unchokes the flow, contrary to expectation. Thus unchoking is accompanied by a reduction in the mass flow rate and impulse function.

If q is decreased instead of being increased with state 1 and p_a held fixed, the flow again adjusts to a new Rayleigh line, which is interior to the q curve in Fig. 8.3, but now $M_3' = 1$ and $p_3' > p_a$. The pressure adjusts to the ambient by means of a supersonic expansion at the exit of the duct.

8.3.5   Fanno Flow

A constant area, adiabatic duct flow with wall friction is considered. Eliminate V from Eqs. (8.5) and (8.13a) to obtain

$$h + \frac{1}{2}\left(\frac{\dot{m}}{A}\right)^2 \frac{1}{\rho^2} = h_0 \tag{8.39}$$

which is the equation for a Fanno line. Since h_0 is constant, this equation is a curve on a Mollier diagram, see Fig. 8.4. The maximum entropy point corresponds to $M = 1$, while the upper (lower) branch is subsonic (supersonic).

Equation (8.27) is used in conjunction with Eq. (8.39) to obtain the h, s Fanno line for a perfect gas. From Table 8.1, the change in entropy is given by

$$\frac{ds}{R} = 2\gamma M^2 c_f \frac{dx}{d} \tag{8.40}$$

For x increasing in the flow direction, the right side of Eq. (8.40) is positive. According to the Second Law of Thermodynamics, $ds > 0$, for an irreversible adiabatic

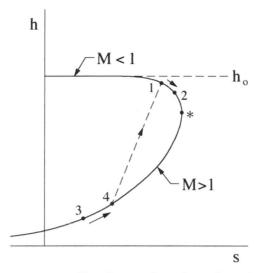

FIGURE 8.4 Fanno line on a Mollier diagram for a thermally and calorically perfect gas with $\gamma = 1.4$.

process, and flow conditions always move to the right along the Fanno curve with increasing x. With subsonic inlet conditions, state 1, exit conditions are at state 2, due to wall friction. From Eq. (8.5), $h_0 - h_2$ equals $V_2^2/2$, and friction actually increases the average cross-sectional speed when the flow is subsonic. By increasing the duct length, with state 1 fixed, state 2 moves to the right, until the flow chokes and $M_2 = 1$. A further increase in duct length causes inlet conditions to change; namely, $\dot{m}/A$ decreases and the flow shifts to a different Fanno line.

A similar discussion applies when the inlet, state 3, and outlet, state 4, are on the supersonic branch. A normal shock wave is permissible as sketched in Fig. 8.4, where states 3 and 2 represent the duct inlet and exit conditions, respectively. Choking occurs when state 2 becomes sonic. Unlike Rayleigh flow, the shock location does have an effect on the overall drag. Suppose the flow in Fig. 8.4 is choked with $M_2 = 1$. If the duct is further lengthened, the normal shock wave moves upstream, and state 4 approaches state 3. As it moves upstream, the shock becomes stronger because M_4 increases. Any further duct increase, once states 3 and 4 coincide, changes inlet conditions and the Fanno line. The method for determining shock location is outlined in the second example at the end of this section.

Equation (8.17), with $dA = dT_0 = 0$, yields the appropriate form for conservation of momentum

$$\frac{Z}{\gamma M^4 X} dM^2 = -4c_f \frac{dx}{d} \tag{8.41}$$

If the local skin friction coefficient c_f is assumed constant, the integrated result shown in Table 8.2 is obtained. The parameter L_m is the distance from station x to a usually fictitious location x^*, where the flow would be choked. Hence, for states 1 and 2 in Fig. 8.4

$$x_2 - x_1 = (x^* - x_1) - (x^* - x_2) = L_{m1} - L_{m2} \tag{8.42a}$$

and

$$4c_f \frac{x_2 - x_1}{d} = \left(4c_f \frac{L_m}{d}\right)_1 - \left(4c_f \frac{L_m}{d}\right)_2 \qquad (8.42b)$$

This equation relates the skin friction term on the left side with the Mach number function in Table 8.2.

Inasmuch as the result in Table 8.2 for $4c_f L_m/d$ is valid only when c_f is constant, i.e., independent of the Mach number, it is important to establish when this assumption holds. For fully developed, subsonic, laminar duct flow c_f is given by

$$c_f = \frac{16}{\mathrm{Re}}, \; \mathrm{Re} < 2300 \qquad (8.43)$$

where the Reynolds number is

$$\mathrm{Re} = \frac{\rho V d}{\mu} \qquad (8.44)$$

and μ is the shear viscosity. When the Reynolds number exceeds 2300, the flow is transitional or turbulent, and

$$c_f = c_f(\mathrm{Re}, e/d) \qquad (8.45)$$

where e is the wall roughness. Whether laminar or turbulent, subsonic experiments show little dependence of c_f on Mach number. Equations (8.43) and (8.45) lead to the familiar Moody (see Sec. 1.4.7) diagram. In the upper right-hand part of the diagram, where

$$\left(10^2 \frac{d}{e} \frac{1}{\mathrm{Re}}\right)^2 < c_f \qquad (8.46)$$

the fully developed flow is turbulent, and c_f is constant.

As noted earlier, shock-free supersonic flow is difficult to maintain in a duct for any appreciable distance. Viscous effects are thus relegated to a thin boundary layer, and the outlet flow may be far from uniform. The shock waves that frequently occur are a system of mostly oblique shocks distributed over several duct diameters, and not a single, normal shock as assumed in the theory. Between shocks, regions of separated flow adjacent to the wall are usually present. There is no known theory for c_f in the supersonic case, other than to assume it is constant at an empirically determined average value. It is worth noting that the difficulty in a supersonic flow is the *a priori* inability to estimate c_f, not the presence of a shock system in lieu of a normal shock. Under the assumption of a constant c_f, the Fanno theory equations are an exact solution of the conservation equations between any two uniform flow stations. In this case, a normal shock is a useful one-dimensional representation of the various frictional dissipative mechanisms that are present in the duct. An analogous difficulty occurs for an entrance flow; namely, the variation of c_f with distance from the entrance.

It is worth extending the theory in the laminar case, where Eq. (8.43) provides a nonconstant skin friction coefficient. The viscosity μ is assumed to be proportional to the temperature

$$\mu = \frac{T}{T_0} \mu(T_0) = \frac{\mu_0}{X} \tag{8.47}$$

and the Reynolds number can be written as

$$\text{Re} = \frac{4\dot{m}}{\pi \, d\mu} = \frac{4\dot{m}}{\pi \, d\mu_0} X = \text{Re}_0 X \tag{8.48}$$

where $\text{Re}_0 = \rho V d/\mu_0$, which is constant. With Eqs. (8.43) and (8.48), Eq. (8.41) becomes

$$\frac{Z}{\gamma M^4} \, dM^2 = - \frac{64}{\text{Re}_0} \frac{dx}{d} \tag{8.49}$$

which integrates to

$$4c_{f0} \frac{L_m}{d} = \frac{1}{\gamma} \left(\ln M^2 - \frac{Z}{M^2} \right) \tag{8.50}$$

where $c_{f0} = 16/\text{Re}_0$.

Shapiro (1953) can be consulted for the treatment of isothermal flow in a duct with friction.

The drag D on a section of duct is given by

$$D = \int_{x_1}^{x_2} \tau c \, dx = \frac{1}{2} \left(\frac{\gamma + 1}{2} \right)^{1/2} \dot{m} V* \int_{x_1}^{x_2} \frac{M}{X^{1/2}} \left(4c_f \frac{dx}{d} \right) \tag{8.51a}$$

Equation (8.41) yields

$$D = - \frac{1}{2\gamma} \left(\frac{\gamma + 1}{2} \right)^{1/2} \dot{m} V* \int_{M_1}^{M_2} \frac{Z \, dM^2}{M^3 X^{3/2}} \tag{8.51b}$$

The integral can be performed and, after simplification, results in

$$D = F_1 - F_2 \tag{8.51c}$$

where the impulse function F is given by Eqs. (8.20). Equation (8.51c) is quite general, holding for subsonic or supersonic flow (with or without shock waves and boundary-layer separation), and for any skin friction distribution.

As an illustration, we use the first example in the Rayleigh section (Sec. 8.3.4) that is associated with Eqs. (8.34), except for M_2 and M_3. We are to determine M_2, M_3, and $\dot{m}/A$ when the insulated straight section of the duct is 0.1 m in diameter, is 50 m long, and has a skin friction coefficient of 7×10^{-3}.

With the assumption that M_3 is subsonic, set $p_3 = p_a = 101.35$ kPa. As a consequence, we have

$$\frac{p_3}{p_{01}} = \frac{p_3}{p^*}\frac{p^*}{p_2}\frac{p_2}{p_{01}} = \frac{M_2}{M_3}\left(\frac{X_2}{X_3}\right)^{1/2} X_2^{-\gamma/(\gamma-1)} = \frac{1}{2} \qquad (8.52a)$$

A second relation is provided by Eq. (8.42b), namely,

$$4c_f\frac{x_3 - x_2}{d} = \left(4c_f\frac{L_m}{d}\right)_2 - \left(4c_f\frac{L_m}{d}\right)_3 = 14 \qquad (8.52b)$$

Equations (8.52) are iteratively solved for M_2 and M_3, and result in $M_2 = 0.186$ and $M_3 = 0.360$. Equation (8.22) with M_2 known then yields $(\dot{m}/A) = 114$ kg/m^2 − s. With $\mu = 2.57 \times 10^{-5}$ N − s/m^2 (Pa − s) for N_2 at 500 K, the Reynolds number is $Re_2 = (4\dot{m}/\pi d\mu) = 4.42 \times 10^5$. According to Eq. (8.46) the assumption of a constant turbulent value for c_f is appropriate, providing e/d exceeds 2.7×10^{-3}.

As a second example, consider a diatomic gas flowing in a duct fed by a supersonic nozzle. There is a normal shock located at x_s, see Fig. 8.5. Known conditions consist of

$$L = 2.5 \text{ m}, d = 0.1 \text{ m}, c_f = 10^{-3}$$

$$p_{01} = 385.13 \text{ kPa}, p_a = 101.35 \text{ kPa}, M_1 = 3, M_4 = 0.6 \qquad (8.53)$$

and we are to determine M_2, M_3, and x_s.

Since $M_4 < 1$ and $p_4 = p_a$, the overall pressure ratio is

$$\frac{p_{01}}{p_4} = \frac{p_{01}}{p_1}\frac{p_1}{p^*}\frac{p^*}{p_2}\frac{p_2}{p_3}\frac{p_3}{p^*}\frac{p^*}{p_4} = 3.8 \qquad (8.54a)$$

The four p^* pressures are the same, since there is but one Fanno line, and for a normal shock (Sec. 8.4.1)

$$\frac{p_3}{p_2} = Y_2 \qquad (8.55)$$

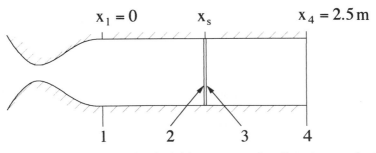

FIGURE 8.5 Duct with friction that is fed by a converging-diverging nozzle. There is a normal shock wave at station 2.

Equation (8.54a) thus becomes

$$\left(\frac{p_{01}}{p_4}\right)^2 = \left(\frac{M_2 M_4}{M_1 M_3}\right)^2 \frac{X_1^{(\gamma+1)/(\gamma-1)} X_2 X_4}{Y_2^2 X_3} \tag{8.54b}$$

where the unknowns are M_2 and M_3. A second relation for these unknown results from adding the following equations

$$4c_f \frac{x_s}{d} = \left(4c_f \frac{L_m}{d}\right)_1 - \left(4c_f \frac{L_m}{d}\right)_2 \tag{8.56a}$$

$$4c_f \frac{x_4 - x_s}{d} = \left(4c_f \frac{L_m}{d}\right)_3 - \left(4c_f \frac{L_m}{d}\right)_4 \tag{8.56b}$$

thereby eliminating x_s. The iterative solution of these equations yields $M_2 = 2.69$ and $M_3 = 0.589$. With this value for M_2, Eq. (8.56a) then yields $x_s = 1.32$ m for the shock location.

8.4 SHOCK WAVES

When the velocity exceeds the speed of sound ($M > 1$), adjustments in the flow often take place through abrupt discontinuous surfaces called shock waves. Flows with the most important type of shock waves are examined here: steady, normal shocks; unsteady, normal shocks; steady, oblique planar shocks; and steady, conical shocks.

8.4.1 Steady Normal Shock Waves

Conditions upstream and downstream of a steady, normal shock are denoted by subscripts 1 and 2, respectively. Conservation equations relate states 1 and 2 according to

$$\dot{m}_1 = \dot{m}_2, \ F_1 = F_2, \ h_{01} = h_{02} \tag{8.57}$$

The first and last of these relations means the shock process conserves mass flow rate and is adiabatic. The impulse relation stems from a momentum balance and results in an adiabatic process that is not isentropic. When written out, these relations yield

$$\rho_1 V_1 = \rho_2 V_2, \ p_1 + \rho_1 V_1^2 = p_2 + \rho_2 V_2^2, \ h_1 + \tfrac{1}{2} V_1^2 = h_2 + \tfrac{1}{2} V_2^2 \tag{8.58}$$

We assume a thermally and calorically perfect gas and introduce the Mach number into Eq. (8.58). After simplification, the steady shock results in Table 8.3 are obtained. All subscript 1 quantities refer to M_1, which is called the shock Mach number, and is then denoted as M_s. (In a laboratory coordinate system, where a shock of speed V_1 is moving into a quiescent gas which has a sound speed a_1, the

TABLE 8.3 Equations for Steady and Unsteady Normal Shock Waves for a Thermally and Calorically Perfect Gas

Steady Shock	Unsteady Shock
$M_2 = \left(\dfrac{2}{\gamma+1}\dfrac{X_1}{Y_1}\right)^{1/2}$	$M_2' = M_2\left(\dfrac{\gamma+1}{2}\dfrac{M_1 M_s'}{X_1} - 1\right)$
$\dfrac{V_2}{V_1} = \dfrac{2}{\gamma+1}\dfrac{X_1}{M_1^2}$	$\dfrac{V_2'}{V_1'} = \dfrac{\dfrac{V_2}{V_1}\left(\dfrac{\gamma+1}{2}\dfrac{M_1 M_s'}{X_1} - 1\right)}{1 - \dfrac{M_s'}{M_1}}$
$\dfrac{p_2}{p_1} = Y_1$	$\dfrac{p_2'}{p_1'} = \dfrac{p_2}{p_1}$
$\dfrac{T_2}{T_1} = \dfrac{2}{\gamma+1}\dfrac{X_1 Y_1}{M_1^2}$	$\dfrac{T_2'}{T_1'} = \dfrac{T_2}{T_1}$
$\dfrac{\rho_2}{\rho_1} = \dfrac{V_1}{V_2}$	$\dfrac{\rho_2'}{\rho_1'} = \dfrac{\rho_2}{\rho_1}$
$\dfrac{p_{02}}{p_{01}} = \dfrac{\left[\left(\dfrac{\gamma+1}{2}\right)\dfrac{M_1^2}{X_1}\right]^{\gamma/(\gamma-1)}}{Y_1^{1/(\gamma-1)}}$	$\dfrac{p_{02}'}{p_{01}'} = Y_1\left(\dfrac{X_2'}{X_1'}\right)^{\gamma/(\gamma-1)}$
$\dfrac{T_{02}}{T_{01}} = 1$	$\dfrac{T_{02}'}{T_{01}'} = \dfrac{T_2}{T_1}\dfrac{X_2'}{X_1'}$
$\dfrac{s_2 - s_1}{R} = \ln\left(\dfrac{p_{01}}{p_{02}}\right)$	$\dfrac{s_2' - s_1'}{R} = \dfrac{s_2 - s_1}{R}$

Mach number associated with the shock is $M_s = M_1$.) Other ratios are easily determined, e.g.,

$$\frac{\rho_{02}}{\rho_{01}} = \frac{p_{02}}{p_{01}}, \frac{p_{02}}{p_1} = \frac{p_{02}}{p_{01}}\frac{p_{01}}{p_1} \tag{8.59}$$

where the p_{02}/p_1 relation is the Rayleigh Pitot formula. Another well-known result is the Rankine-Hugoniot equation that relates p_2/p_1 to ρ_2/ρ_1. This equation is obtained by eliminating M_1^2 from the p and ρ equations in Table 8.3.

8.4.2 Unsteady Normal Shock Waves

In many situations, such as in explosions, the shock wave is moving. If it travels at a constant speed, a simple transformation reduces the moving case to the steady one. Even when the shock wave is accelerating or decelerating, this approach is valid at any one instant, since the shock thickness is usually quite negligible [Shapiro (1953)].

A prime is used to distinguish the unsteady case from when the shock is fixed. As is evident in Fig. 8.6, the gas ahead of the shock has a speed V_1', while the shock

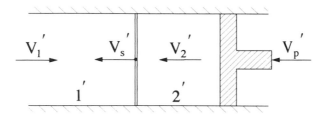

FIGURE 8.6 Schematic of a normal shock wave driven by a constant speed piston.

speed is V_s'. A rightward (leftward) facing velocity hereafter is considered positive (negative). Thus, V_s' and V_2' are viewed as negative. In many situations, the speed V_1' is zero. For a reflected shock wave, however, as occurs in shock tube flow, this is not the case. The steady and unsteady flows are connected by the transformation

$$V_1 = V_1' - V_s', \tag{8.60a}$$

$$V_2 = V_2' - V_s' \tag{8.60b}$$

Generally, conditions upstream of the moving shock, such as V_1', and the shock speed are known, whereas conditions downstream are unknown and are to be determined. It is convenient to introduce the positively defined Mach numbers

$$M_s' = -\frac{V_s'}{a_1'}, \; M_1' = \frac{V_1'}{a_1'}, \; M_2' = -\frac{V_2'}{a_2'} \tag{8.61}$$

Transformation (8.60) leaves all thermodynamic static quantities, on both sides of the shock, unaltered. As a consequence,

$$a_1' = a_1, \; a_2' = a_2, \; s_1' = s_1, \; \dots \tag{8.62}$$

Equations (8.60) and (8.61) can be shown to yield

$$M_2' = M_2 \left(\frac{\gamma + 1}{2} \frac{M_1 M_s'}{X_1} - 1 \right), \tag{8.63a}$$

$$M_1 = M_1' + M_s' \tag{8.63b}$$

A knowledge of M_1' and M_s' determines M_1, which in turn determines M_2 by the first relation on the left side of Table 8.3. With M_1 and M_2 known, Eq. (8.63a) then yields M_2'. Other unsteady parameters are then provided by the right column of Table 8.3.

An important difference between a steady shock and a moving one is that T_{02}'/T_{01}' is not unity for a moving shock. Several other points need to be kept in mind. For instance, it is wrong to write $Y(M_1')$ for p_2'/p_1' in place of $Y(M_1)$, which does equal p_2/p_1. On the other hand, the isentropic relations, as point functions (see the last paragraph in Sec. 8.2), are correct for both steady and unsteady flows; consequently,

$$\frac{p_2}{p_{02}} = [X(M_2)]^{-\gamma/(\gamma-1)}, \frac{p_2'}{p_{02}'} = [X(M_2')]^{-\gamma/(\gamma-1)}, \dots \tag{8.64}$$

As an illustration, consider a piston with a speed V_p' of -200 m/s moving into quiescent air, as shown in Fig. 8.6. We are to determine V_s', M_s', p_2'/p_1', p_{02}'/p_{01}', T_2'/T_1', T_{02}'/T_{01}', and compare the last four quantities with their steady-flow counterparts.

Given parameters include

$$\gamma = 1.4, R = 287 \text{ J/kg} - \text{K}, T_1' = 300 \text{ K} \tag{8.65}$$

and help establish

$$a_1 = 347 \text{ m/s}, V_1' = 0, M_1' = 0, V_2' = V_p' = -200 \text{ m/s} \tag{8.66}$$

A relation for V_p' is obtained by dividing Eq. (8.60b) by a_1 and using Table 8.3, to obtain

$$\frac{V_p'}{a_1} = -\frac{2}{\gamma+1}\frac{Z_1}{M_1} \tag{8.67}$$

When solved for M_1, this equation yields $M_1 = 1.40$, and Eqs. (8.63b) and (8.66) result in $M_s' = M_1 = 1.40$. The use of Table 8.3 for M_2 and Eqs. (8.63a) and (8.66), provide $M_2' = 0.514$. Equations (8.60) result in

$$V_s' = -V_1 = \frac{V_p'}{1 - \frac{V_2}{V_1}} = \frac{V_p'}{1 - \frac{2}{\gamma+1}\frac{X_1}{M_1^2}} = -487 \text{ m/s} \tag{8.68}$$

The pressure and temperature ratios are found using Table 8.3 and can be summarized as follows:

	$\frac{p_2}{p_1}$	$\frac{T_2}{T_1}$	$\frac{p_{02}}{p_{01}}$	$\frac{T_{02}}{T_{01}}$
stationary shock	2.13	1.26	0.957	1
moving shock	2.13	1.26	2.55	1.32

For a steady shock, the stagnation pressure ratio p_{02}/p_{01} is less than one. The ratio here is not much less than unity, since $M_1 - 1$ is only 0.4. For a moving shock wave, the ratios p_{02}'/p_{01}' and T_{02}'/T_{01}' are greater than unity as a result of the momentum imparted by the piston to the gas between the piston and the shock.

8.4.3 Oblique Planar Shock Waves

Oblique shock waves are analyzed by applying a uniform velocity transformation that is parallel to the shock, as shown in Fig. 8.7. The components of velocity

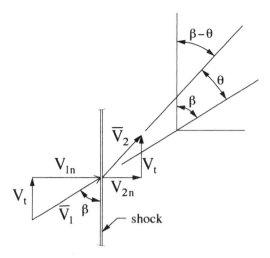

FIGURE 8.7 Schematic for various speeds and angles for an oblique shock wave.

normal and tangential to the shock are denoted with n and t subscripts, respectively. The normal component must be supersonic for a shock wave to exist, i.e.,

$$M_{1n} = \frac{V_{1n}}{a_1} > 1 \tag{8.69}$$

The component V_t is vectorially added to both sides of the shock and results in $\overline{\mathbf{V}}_1$ and $\overline{\mathbf{V}}_2$. Since $V_{1n} > V_{2n}$, $\overline{\mathbf{V}}_2$ is turned toward the shock wave by θ, which is the angle between $\overline{\mathbf{V}}_1$ and $\overline{\mathbf{V}}_2$. A second angle of importance is β, which denotes the angle between $\overline{\mathbf{V}}_1$ and the shock wave. Figure 8.8 indicates how the foregoing pattern applies to the flow over a wall with an abrupt turn. The shock turns the flow by an angle θ so that it is parallel to the downstream wall. In this case, two uniform flow regions are separated by the shock; it is the instrument whereby the incoming supersonic flow abruptly learns of the increase in wall slope.

Table 8.3 provides

$$\frac{V_{2n}}{V_{1n}} = \frac{2}{\gamma + 1} \frac{X_{1n}}{M_{1n}^2} \tag{8.70}$$

where M_{1n} is given by

$$M_{1n} = \frac{V_{1n}}{a_1} = \frac{V_1}{a_1} \sin \beta = M_1 \sin \beta \tag{8.71}$$

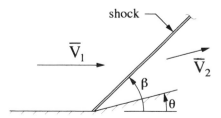

FIGURE 8.8 Oblique shock wave caused by a change in wall slope.

and from Fig. 8.7

$$\frac{V_{1n}}{V_t} = \tan \beta, \quad \frac{V_{2n}}{V_t} = \tan (\beta - \theta) \tag{8.72}$$

Combining the foregoing, yields

$$\tan \theta = \cot \beta \, \frac{Z(M_1 \sin \beta)}{1 + (\gamma + \cos 2\beta)M_1^2/2} \tag{8.73}$$

This is a basic relation for an oblique shock that provides θ as a function of M_1 and β, and produces a family of M_1 curves, as shown in Fig. 8.9. For given γ and M_1 values, θ is double valued as β varies over its range

$$\mu(M_1) \leq \beta \leq \pi/2 \tag{8.74}$$

where μ is the Mach angle, the angle of propagation of a very weak disturbance,

$$\mu(M) = \sin^{-1} \left(\frac{1}{M} \right) \tag{8.75}$$

Equation (8.75) is obtained by noting that as $\theta \rightarrow 0$ in Eq. (8.73), Z also goes to zero. To the left of the $M_2 = 1$ curve in Fig. 8.9, the flow behind the shock is supersonic, to the right it is subsonic. Slightly to the right of the curve is a curve, labeled θ_m, that passes through the peak of the individual curves. The equation for this line is given by

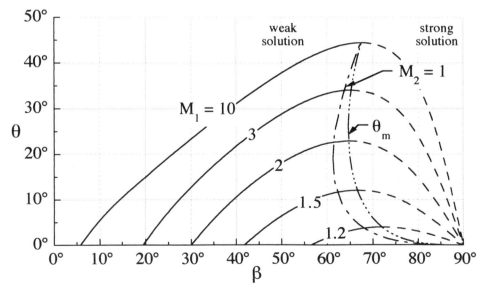

FIGURE 8.9 Wall turn angle as a function of upstream Mach number M_1 and shock wave angle when $\gamma = 1.4$.

$$\sin^2 \beta_m = \frac{1}{4\gamma M_1^2} ((\gamma + 1)M_1^2 - 4 + \{(\gamma + 1)$$

$$\cdot [(\gamma + 1)M_1^4 + 8(\gamma - 1)M_1^2 + 16]\}^{1/2}) \qquad (8.76)$$

in conjunction with Eq. (8.73). To the left of the peak is the weak solution, while to the right is the strong solution. The reason for this terminology is simply that as β increases with M_1 fixed, the pressure ratio p_2/p_1 increases.

When the wall turn angle θ, for a given M_1 value, exceeds its maximum permissible value θ_m, an attached shock cannot occur. A curved, detached shock system is established, which is located upstream of where the attached shock would have been. When $\theta < \theta_m$, the weak solution is the one experimentally observed. In supersonic flow about a bluff body, however, where the shock wave is detached, the strong solution prevails in the vicinity of the nose of the body. As indicated in Salas and Morgan (1982) and Emanuel (1982b), the occurrence of the strong versus weak solution depends on a variety of factors, including downstream pressure conditions and whether M_2 is subsonic or supersonic.

A planar or conical shock in a uniform flow has a constant entropy (referred to as homentropic) downstream of the shock. A detached shock, however, has longitudinal curvature and its strength varies along it. In this case, the downstream flow is isentropic (constant entropy along individual streamlines) but not homentropic.

Equation (8.71) relates M_1 and M_{1n}. The corresponding relation downstream of the shock is

$$M_{2n} = M_2 \sin (\beta - \theta) \qquad (8.77)$$

These two relations enable the steady shock column of Table 8.3 to be used for oblique shock waves by replacing M_1 and M_2 with M_{1n} and M_{2n}, respectively. [Equation (8.73) is a notable exception to this otherwise general substitution procedure.] Thus, M_2 and X_1 in the table are replaced with $M_2 \sin (\beta - \theta)$ and $X_{1n} = X(M_1 \sin \beta)$.

More often than not, β is preferred as a function of θ and M_1. This inversion is analytically possible, since Eq. (8.73) is a cubic when written in terms of $\tan \beta$. The exact result is provided by the convenient sequence of equations

$$\lambda = \left[Z_1^2 - 3X_1 \left(1 + \frac{\gamma + 1}{2} M_1^2 \right) \tan^2 \theta \right]^{1/2} \qquad (8.78a)$$

$$\chi = \frac{Z_1^3 - 9X_1 \left(X_1 + \frac{\gamma + 1}{4} M_1^4 \right) \tan^2 \theta}{\lambda^3} \qquad (8.78b)$$

$$\tan \beta = \frac{Z_1 + 2\lambda \cos [(4\pi\delta + \cos^{-1} \chi)/3]}{3X_1 \tan \theta} \qquad (8.78c)$$

where $\delta = 0$ or 1 for the strong or weak shock solutions, respectively. For an attached shock, $|\chi| \leq 1$, and $\cos^{-1} \chi$ is in radians.

As an illustration, consider air at $M_1 = 3$ impinging on a 20° wedge at a 5° angle of attack in a wind tunnel, see Fig. 8.10. We are to determine p_2/p_1, p_3/p_1, M_4, and

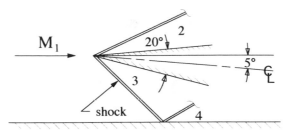

FIGURE 8.10   Schematic for a 20° wedge at a 5° angle of attack in a wind tunnel. Only the lower tunnel wall is shown.

p_4/p_1, where region 4 results from the incident oblique shock reflecting from the lower tunnel wall.

Since $\theta_2 = 5°$ and $\theta_3 = 15°$, Eqs. (8.78) yield for the weak solution angles $\beta_2 = 23.1°$ and $\beta_3 = 32.2°$. Equation (8.71) then yields $M_{1n2} = M_1 \sin \beta_2 = 1.18$ and $M_{1n3} = M_1 \sin \beta_3 = 1.60$. Two of the desired pressure ratios are $(p_2/p_1) = Y_{1n2} = 1.45$ and $(p_3/p_1) = Y_{1n3} = 2.82$. It should be noted that the wedge is treated as two separate ramps with no communication between regions 2 and 3. This is appropriate providing both shock waves are attached to the tip of the wedge, as is the case here. (If one shock is detached, then the other shock is detached, and there is only a single curved bow shock.) The normal component of the region 3 Mach number associated with the upstream shock and M_3 are given by

$$M_{3n} = \left(\frac{2}{\gamma + 1} \frac{X_{1n3}}{Y_{1n3}} \right)^{1/2} = 0.648 \tag{8.79a}$$

$$M_3 = \frac{M_{3n}}{\sin (\beta_3 - \theta_3)} = 2.25 \tag{8.79b}$$

With $\theta_4 = 15°$ and M_3 known, the weak solution shock angle is $\beta_4 = 40.4°$ (see Fig. 8.11). Consequently, the normal component of the region 3 Mach number, associated with the reflected shock, is

$$M_{3n4} = M_3 \sin \beta_4 = 1.46 \tag{8.80a}$$

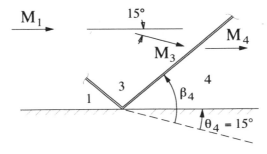

FIGURE 8.11   Reflected shock region for the flow shown in Fig. 8.10.

and the final pressure ratio is given by

$$\frac{p_4}{p_1} = \frac{p_3}{p_1}\frac{p_4}{p_3} = \frac{p_3}{p_1} Y_{3n4} = 6.55 \tag{8.80b}$$

The normal component of the Mach number in region 4 and M_4 are

$$M_{4n} = \left(\frac{2}{\gamma + 1}\frac{X_{3n4}}{Y_{3n4}}\right)^{1/2} = 0.716 \tag{8.81a}$$

$$M_4 = \frac{M_{4n}}{\sin (\beta_4 - \theta_4)} = 1.67 \tag{8.81b}$$

Each time a β value is determined, the weak solution is chosen. With an increasing angle of attack, θ_3 increases and M_4 decreases. When M_4 is supersonic but close to unity, a transition to the so-called Mach reflection pattern occurs [Emanuel (1986)].

8.4.4 Conical Flow

Supersonic flow about a cone at zero angle of attack frequently occurs, particularly at the nose of missiles and in supersonic inlets. If the semi-vertex angle θ_b is not too large, the flow will appear as in Fig. 8.12, with a conical shock wave attached to the apex of the body. A streamline is turned θ_2 degrees by the shock, where $0 < \theta_2 < \theta_b$. With increasing distance the streamlines gradually become parallel to the wall.

The conical shock wave is of uniform strength, and the streamline turn angle θ_2 at the shock is given by Eq. (8.73). The Mach number just downstream of the shock, M_2, is provided Eq. (8.77), where M_{2n} is determined by the first equation on the left in Table 8.3 with M_1 replaced by the normal component, $M_1 \sin \beta$. Thus, γ, M_1, and β are sufficient to establish conditions at location 2.

Downstream of the shock wave the flow is homentropic and irrotational. The inviscid tangency condition for the velocity holds at the wall.

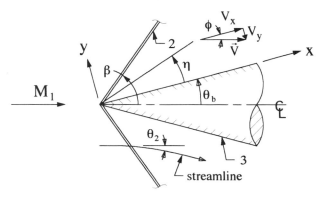

FIGURE 8.12 Schematic for supersonic flow about a cone of half-angle θ_b at zero angle of attack.

As in the oblique shock case, there is no characteristic length scale. Consequently, the solution for the flow between the shock and the body depends only on the angular coordinate η. Thus $\overline{V}$ and its angle relative to the surface of the body ϕ (see Fig. 8.12) are constant on a conical surface with a semi-vertex angle $\eta + \theta_b$. The orientation of the velocity vector is restricted to the range

$$\theta_2 - \theta_b \le \phi \le 0 \tag{8.82}$$

In terms of the x, y coordinate system in Fig. 8.12, the equations of motion are

$$\frac{\partial(\rho V_x)}{\partial x} + \frac{\partial(\rho V_y)}{\partial y} + \sigma\rho \left(\frac{V_x \sin\theta_b + V_y \cos\theta_b}{x \sin\theta_b + y \cos\theta_b} \right) = 0 \tag{8.83a}$$

$$V_x \frac{\partial V_x}{\partial x} + V_y \frac{\partial V_x}{\partial y} + \frac{1}{\rho}\frac{\partial p}{\partial x} = 0 \tag{8.83b}$$

$$V_x \frac{\partial V_y}{\partial x} + V_y \frac{\partial V_y}{\partial y} + \frac{1}{\rho}\frac{\partial p}{\partial y} = 0 \tag{8.83c}$$

$$V_x \frac{\partial p}{\partial x} + V_y \frac{\partial p}{\partial y} = \gamma \frac{p}{\rho}\left(V_x \frac{\partial \rho}{\partial x} + V_y \frac{\partial \rho}{\partial y} \right) \tag{8.83d}$$

where $\sigma = 0$, 1 for two-dimensional and axisymmetric flow, respectively. In view of the foregoing discussion, new dependent and independent (similarity) variables are introduced

$$\xi = x, \quad \eta = \tan^{-1}(y/x) \tag{8.84}$$

$$p = p_3 P(\eta) \tag{8.85a}$$

$$\rho = \rho_3 R(\eta) \tag{8.85b}$$

$$V_x = V_{x3} Q(\eta) \cos\phi \tag{8.85c}$$

$$V_y = V_{x3} Q(\eta) \sin\phi \tag{8.85d}$$

On the cone, where $\eta = 0$,

$$P(0) = R(0) = Q(0) = 1 \tag{8.86}$$

and the tangency condition is automatically satisifed. It is also convenient to introduce the Mach number

$$M^2 = \frac{V^2}{\gamma(p/\rho)} = M_3^2 \frac{RQ^2}{P} \tag{8.87}$$

After some algebra, the foregoing equations can be shown [Emanuel (1967)] to yield two integrals of the motion and two, first-order, ordinary, differential equations

$$P = \left[\frac{X(M_3)}{X(M)} \right]^{\gamma/(\gamma - 1)} \tag{8.88a}$$

$$Q = \frac{M}{M_3} \left[\frac{X(M_3)}{X(M)} \right]^{1/2} \tag{8.88b}$$

$$\frac{d\phi}{d\eta} = \sigma \frac{\sin(\phi + \theta_b)}{\sin(\eta + \theta_b)} \frac{\cos(\phi - \eta)}{Z(M \sin(\phi - \eta))} \tag{8.89a}$$

$$\frac{dM}{d\eta} = \sigma \frac{\sin(\phi + \theta_b)}{\sin(\eta + \theta_b)} \frac{MX(M) \sin(\phi - \eta)}{Z(M \sin(\phi - \eta))} \tag{8.89b}$$

where the entity $M \sin(\phi - \eta)$ is the argument of the Z function. The dependent variables in Eq. (8.89) are ϕ and M and can be found by a numerical integration, starting from the body with

$$\phi(0) = 0, \; M(0) = M_3. \tag{8.90}$$

It is worth noting that the shock angle β and the wall values ρ_3, p_3, and V_{x3} do not enter into Eqs. (8.88)–(8.90). The only required parameters are σ, γ, θ_b, and M_3.

For the two-dimensional case where $\sigma = 0$, ϕ, and M retain their wall values everywhere downstream of the shock, and the solution in Sec. 8.4.3 is recovered.

In the $\sigma = 1$ case, integration of Eqs. (8.89) terminates when [Emanuel (1967)]

$$\frac{Z(M \sin(\phi - \eta))}{X(M \sin(\phi - \eta))} \frac{\sin(\eta + \theta_b)}{\sin(\phi + \theta_b)} \cos(\phi - \eta) + 1 = 0 \tag{8.91}$$

is satisfied, where this relation stems directly from the shock wave discussion earlier in this section. When Eq. (8.91) is satisfied

$$M_2 = M, \; \theta_2 = \phi_2 + \theta_b, \; \beta_2 = \eta_2 + \theta_b \tag{8.92}$$

and the freestream Mach number and pressure are given by [Emanuel (1967)]

$$M_1^2 = \left[\sin^2(\eta_2 + \theta_b) - \frac{\gamma + 1}{2} \frac{\sin(\phi_2 + \theta_b) \sin(\eta_2 + \theta_b)}{\cos(\phi_2 - \eta_2)} \right]^{-1} \tag{8.93}$$

$$\frac{p_1}{p_3} = \frac{P(\eta_2)}{Y(M_1 \sin(\eta_2 + \theta_b))} \tag{8.94}$$

If M_1^2 is negative, then a smaller value for M_3 must be used. If the shock is not attached, then Eq. (8.91) is not satisfied for any η in the range

$$0 < \eta < \frac{\pi}{2} - \theta_b \tag{8.95}$$

and a larger value for M_3 is necessary.

Usually both M_1 and θ_b are prescribed, thereby resulting in a two-point boundary value problem. In the foregoing approach, θ_b is fixed as the simpler of the two choices, thus requiring an iteration on M_3 until M_1 is matched. Once M_1 is matched, matching p_1 is simple, since p_3 only appears in Eqs. (8.85) and (8.94).

There is considerable similarity between the two-dimensional and axisymmetric solutions. For instance, for a given value of θ_b there are different values of M_1 for an attached shock wave. When the shock is attached, there are two possibilities, the weak and strong solutions. Generally, only the weak solution is expected to occur because the flow field downstream of the shock is everywhere subsonic for the strong solution. However, this expectation is subject to the nature of the boundary conditions, [Salas and Morgan (1982)] especially the downstream pressure condition when some or all of the conical flow field is subsonic.

Except for the streamline that wets the body, a particle of fluid experiences an isentropic compression as it travels downstream from the shock wave. It is therefore possible to have a sonic condition on a conical surface that is situated between the body and the shock, where the flow is supersonic (subsonic) upstream (downstream) of the sonic surface. Such flow fields have been experimentally observed and demonstrate the ability of a supersonic flow to negotiate a smooth, shock free, transition to subsonic flow.

Various figures relating to the weak solution for $\gamma = 1.405$ can be found in Ames Research Staff (1953). Approximate relations for the conical flow field are available [Rasmussen (1967)], and can be used in place of Eqs. (89) or for establishing initial estimates for boundary conditions.

8.5 PRANDTL–MEYER FLOW

Section 8.4.3 analyzes a two-dimensional flow that is disturbed by a sudden wall turn, as seen in Fig. 8.8. However, when the turn is clockwise, the flow must expand so that no characteristic length appears and the wave nature of the flow is preserved. This adjustment occurs by means of a centered Prandtl–Meyer expansion fan, as sketched in Fig. 8.13.

Because there is no characteristic length, the same equations that describe the flow in Sec. 8.4.4 apply, namely, Eqs. (8.88) and (8.89). To counter the $\sigma = 0$ in

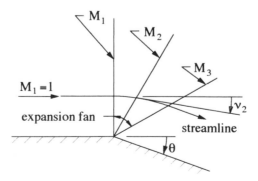

FIGURE 8.13   Prandtl–Meyer expansion caused by a wall turn of angle θ.

Eqs. (8.89), they are made indeterminate by setting the Z term in the denominator equal to zero, which becomes

$$\tan (\phi - \eta) = [Z(M)]^{-1/2} \tag{8.96}$$

The indeterminacy is removed by dividing Eq. (8.89b) by (8.89a), to obtain

$$\frac{dM}{d\phi} = MX(M) \tan (\phi - \eta) \tag{8.97}$$

Eliminate the tangent term from the above two equations and rearrange, to yield

$$\int_0^v d\phi = \int_1^M \frac{[Z(M)]^{1/2}}{MX(M)} dM \tag{8.98a}$$

which integrates to

$$\nu(M) = \left(\frac{\gamma + 1}{\gamma - 1}\right)^{1/2} \tan^{-1} \left\{\left[\left(\frac{\gamma - 1}{\gamma + 1}\right) Z(M)\right]^{1/2}\right\} - \tan^{-1} [Z(M)]^{1/2} \tag{8.98b}$$

The Prandtl–Meyer function $\nu(M)$ is in radians, although it is customary to give results in degrees. It varies monotonically from $\nu(1) = 0$ to its final value $\nu(\infty)$, which equals $130.5°$ when $\gamma = 1.4$.

Recall that ϕ is the slope of a streamline (see Fig. 8.12). The physical interpretation of ν can thus be deduced from Fig. 8.13, where $M_1 = 1$ and $\nu(M_1) = 0$. Let the Mach number be M_2 along a ray within the fan, where $M_2 > M_1$. An arbitrary streamline then has a slope $\nu(M_2)$, where the M_2 ray crosses the streamline. If the trailing edge of the fan has a Mach number M_3, then the streamline slope $\nu(M_3)$ equals θ, with $\theta \geq 0$, and the flow is parallel to the downstream wall.

As is evident from Eqs. (8.88), the flow in the expansion is isentropic. Equation (8.88a), for example, is just Eq. (8.7b), while Eq. (8.88b) is $V = aM \sim MX^{-1/2}$. As a consequence, the use of ν is not restricted to a centered expansion. Consider the $\gamma = 1.4$ flow in Fig. 8.14, where $M_1 = 2$ and M_2 is unknown. The relation governing the Mach number change is

$$\nu(M_2) - \nu(M_1) = \theta_2 - \theta_1 \tag{8.99}$$

and with $\theta_2 - \theta_1 = 20°$ and $\nu(2) = 26.38°$, we have $\nu(M_2) = 20 + 26.38 = 46.38°$, thus yielding $M_2 = 2.83$. The same result is obtained for a wall with a sharp $20°$ expansive turn.

The orientation of the constant Mach number lines (on which other parameters such as $V, p, \phi, \ldots$, are also constant) is determined by the Mach angle μ. Figure 8.15 shows how ν and μ are applied to a sharp turn. The notation LE and TE refer to the leading and trailing edges of the fan, respectively. Upstream and downstream of the fan are two uniform, supersonic flow regions, and M_2 is determined by Eq. (8.99). The lines labeled with a 1 or a 2 are called Mach lines or characteristics. By geometry,

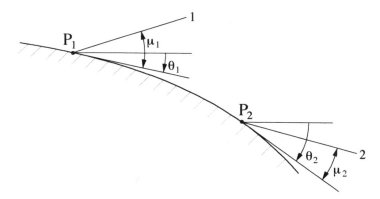

FIGURE 8.14   Prandtl–Meyer expansion for a gradual expansive wall turn. The angles θ_1 and θ_2 are measured from an arbitrary common reference, while the Mach angles μ_1 and μ_2 are measured relative to a wall tangent. Lines 1 and 2 are Mach lines.

$$\lambda + \mu_2 = \mu_1 + \theta \qquad (8.100)$$

which determines the magnitude of the fan angle λ.

For a gradual turn, Fig. 8.14, Mach lines are shown starting from points P_1 and P_2 on the wall. The mach angle μ_i $(i = 1, 2)$ is measured from the tangent to the wall, whereas θ_i is measured from an arbitrary reference. To determine M_2, use Eq. (8.99) with $\theta_2 - \theta_1 > 0$. The M_2 value holds along the Mach line from P_2, whose orientation is determined by $\mu(M_2)$.

As the angle θ in Fig. 8.15 increases, $\nu(M_2)$ and M_2 increase. For a sufficiently large θ, M_2 becomes infinite, and (in radians)

$$\nu(\infty) = \left(\frac{\gamma + 1}{\gamma - 1}\right)^{1/2} - 1, \ \mu(\infty) = 0 \qquad (8.101)$$

For a still larger value of θ, there is a vacuum between the trailing edge of the expansion and the wall, as sketched in Fig. 8.16. On the trailing edge, we not only

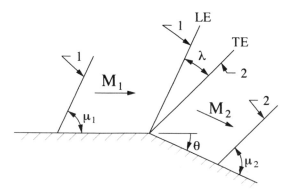

FIGURE 8.15   Prandtl–Meyer expansion caused by a sharp turn with a supersonic upstream Mach number M_1.

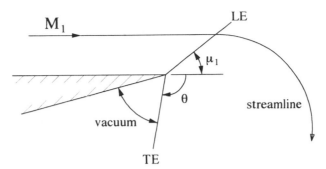

FIGURE 8.16 Prandtl–Meyer expansion into a vacuum.

have $M = \infty$, but also $p = T = \rho = 0$. This type of flow with a vacuum occurs when a vernier rocket on an orbiting satellite is fired.

Prandtl–Meyer flow also applies close to the wall for a gradual compressive turn that is not too large [Emanuel (1986)]. The Mach lines that emanate from the wall converge to gradually form the oblique shock of Sec. 8.4.3. However, close to the wall, the flow is compressed isentropically; hence, $1 \leq M_2 < M_1$, which means that $\theta_2 - \theta_1$ is negative in Eq. (8.99). Although strictly valid only for planar flow, Prandtl–Meyer flow also holds near a turn in the wall for an axisymmetric body.

8.6 NOZZLE FLOW

Section 8.3.3 considers isentropic flow with an area change. Shock waves can occur inside a nozzle, however, and the flow is then no longer isentropic. The nature of the flow in the jet that emanates from the nozzle is also examined.

The analysis in Sec. 8.3.3 is renewed by considering the effect of a change in the ambient back pressure, $\hat{p}_a$, on the flow in a converging-diverging nozzle. Figure 8.17 sketches the nozzle and several traces for the pressure. In the discussion, conditions in the plenum or reservoir are held fixed. For a sufficiently high ambient pressure, there is subsonic flow everywhere in the nozzle as indicated by curve a in Figs. 8.1 and 8.17. The pressure in the jet at the exit plane, p_2, equals $\hat{p}_a$, and the flow is isentropic; consequently, Eqs. (8.21) and (8.22) hold. With a reduced ambient pressure, $\hat{p}_a = p_b$, curve b in Fig. 8.17 and curve b, b_1 in Fig. 8.1 are obtained. The flow rate through the nozzle has achieved its maximum value, given by Eq. (8.23), since $M_1 = 1$ and $A_1 = A^*$, and the flow is now choked. Any further decrease in $\hat{p}_a$ leaves $\dot{m}$ and the flow upstream of the throat unaltered, and $M_1 = 1$. The condition when M_1 first becomes sonic is called the first critical point.

When $\hat{p}_a = p_c$, there is a supersonic region downstream of the throat followed by a normal shock (curve c in Fig. 8.17). Downstream of the shock, the flow is subsonic. The presence of an internal shock is necessary if the pressure in the jet, p_2, is to match the ambient pressure $\hat{p}_a = p_c$. This description is convenient for modeling purposes but differs from an actual flow. The presence of an internal shock wave frequently causes portions of the boundary layer to separate from the wall, resulting in an oblique shock system that is more complex than a simple normal shock.

As $\hat{p}_a$ decreases, the shock moves downstream and becomes stronger. When p_2

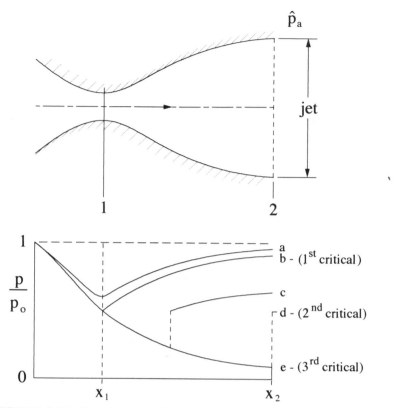

FIGURE 8.17 Pressure traces for flow in a converging-diverging nozzle.

$= p_d = \hat{p}_a$, the shock is at the nozzle exit plane. Here, p_d is the static pressure just downstream of the shock; the pressure just upstream of the shock is p_e. This condition is referred to as the second critical point.

For any ambient pressure $\hat{p}_a$ below p_d, the nozzle exit pressure p_2 is constant at p_e. The third critical point occurs when $\hat{p}_a = p_e$. In general, whenever $\hat{p}_a < p_d$ holds, the jet downstream of the nozzle adjusts to the ambient pressure by means of shock waves or expansion fans as sketched in Fig. 8.18. The only exception is at the third critical point when no adjustment is needed.

Figure 8.18 covers the entire range of possibilities. Starting from the top diagram, the flow is everywhere subsonic, and the nozzle is basically a venturi. Between the first and second critical points, there is a shock wave inside the nozzle. For $\hat{p}_a \leq p_d$, the flow inside the nozzle is isentropic and unchanging. Between the second and third critical points, the external jet adjusts to the ambient pressure by means of a shock system. The flow is called *overexpanded*, since $p_2 = p_e < \hat{p}_a$, i.e., the nozzle has provided too large an expansion relative to the ambient pressure. Different shock configurations are possible in the jet when the nozzle flow is overexpanded. These configurations key to $\hat{p}_a/p_0$. Below the third critical point, the flow is termed *under-expanded* with the pressure adjustment occurring by means of expansion fans. These fans reflect off the edge of the jet as compression waves thereby forming downstream shock waves.

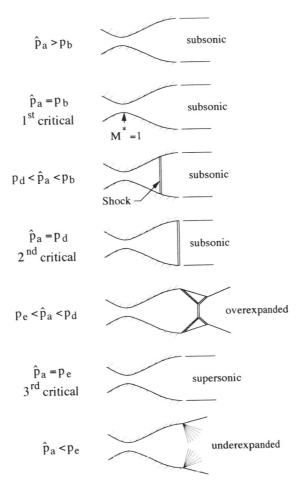

FIGURE 8.18 Schematic illustrating possible inviscid flows in a converging-diverging nozzle.

Critical point conditions are determined in terms of the pressure ratios p_b/p_0, p_d/p_0, and p_e/p_0. The first and third critical points are for isentropic nozzle flow. One can show that

$$\left(\frac{p}{p_0}\right)^{2/\gamma} - \left(\frac{p}{p_0}\right)^{(\gamma+1)/\gamma} = \frac{\gamma-1}{2}\left(\frac{2}{\gamma+1}\right)^{(\gamma+1)/(\gamma-1)}\left(\frac{A_1}{A_2}\right)^2 \quad (8.102)$$

where p is either p_b or p_e, and A_2/A_1 is the area ratio of the nozzle. This equation has two solutions, where the smaller (larger) value for p/p_0 is p_e/p_0 (p_b/b_0). The second critical point is given by

$$\frac{p_d}{p_0} = \frac{p_e}{p_0}\frac{p_d}{p_e} = \frac{p_e}{p_0}Y(M_2) \quad (8.103)$$

where M_2 is determined by Eq. (8.21). Thus, all three points depend only on γ and A_2/A_1.

As an example, nitrogen with stagnation conditions $p_0 = 111.5$ kPa and $T_0 = 400$ K is flowing in a nozzle with an area ratio $(A_4/A_1) = 5$, where A_4 is now the exit area. If a normal shock is present inside the nozzle, the upstream and downstream states are respectively denoted with subscripts 2 and 3. With an ambient pressure of 1 atm (101.35 kPa), we are to determine: (i) the flow regime, (ii) the location of an internal shock, if there is one, and (iii) $\dot{m}/A_4$, M_i, and p_i ($i = 1, \ldots, 4$).

With $\gamma = 1.4$ and $p = p_b$, Eq. (8.102)

$$\left(\frac{p_b}{p_0}\right)^{1.429} - \left(\frac{p_b}{p_0}\right)^{1.714} = 2.68 \times 10^{-3} \tag{8.104}$$

yields $(p_b/p_0) = 0.9905$. With a large area ratio, p_b/p_0 is expected to be close to unity. Since $(\hat{p}_a/p_0) = 0.909$, the flow is below the first critical point. Because the two pressure ratios are close, we assume the flow is above the second critical point with a shock wave inside the nozzle. This assumption is verified if the subsequent solution with an internal shock satisfies $p_4 = \hat{p}_a$. We also know that $M_1 = 1$, and Eq. (8.7b) yields

$$p_1 = p_0 X_1^{-\gamma/(\gamma-1)} = 58.90 \text{ kPa} \tag{8.105}$$

The mass flow rate is obtained from Eq. (8.23) as

$$\frac{\dot{m}}{A_4} = \left(\frac{\gamma+1}{2}\right)^{-(\gamma+1)/2(\gamma-1)} \left(\frac{\gamma}{RT_0}\right)^{1/2} p_0 \frac{A_1}{A_4} = 4.44 \times 10^{-2} \text{ kg/s} - \text{m}^2 \tag{8.106}$$

Since $\dot{m}/A$ and T_0 are constant across a steady normal shock, Eq. (8.23) yields

$$A_2^* p_{02} = A_3^* p_{03} \tag{8.107}$$

We also have

$$A_2^* = A_1^* = A_1, \ A_3^* = A_4^*, \ p_{02} = p_0, \ p_{03} = p_{04} \tag{8.108}$$

Hence, Eq. (8.107) becomes $A_1^* p_0 = A_4^* p_{04}$, which is rearranged as

$$\frac{A_4}{A_4^*} \frac{p_4}{p_{04}} = \frac{A_4}{A_1} \frac{\hat{p}_a}{p_0} \tag{8.109}$$

With Eqs. (8.7b) and (8.21), this becomes

$$M_4^2 X_4 = \left(\frac{2}{\gamma+1}\right)^{(\gamma+1)/(\gamma-1)} \left(\frac{A_1}{A_4}\right)^2 \left(\frac{p_0}{\hat{p}_a}\right)^2 \tag{8.110}$$

which is a quadratic equation for M_4^2, and where the right-hand side is known. Its solution yields $M_4 = 0.121$. This result provides

$$p_{03} = p_{04} = \hat{p}_a X_4^{\gamma/(\gamma-1)} = 102.4 \text{ kPa} \tag{8.111}$$

From Table 8.3, we have

$$\frac{p_{03}}{p_{02}} = \frac{p_{03}}{p_0} = \left[\left(\frac{\gamma + 1}{2}\right)\frac{M_2^2}{X_2}\right]^{\gamma/(\gamma - 1)}\frac{1}{Y_2^{1/(\gamma - 1)}} \qquad (8.112)$$

which results in $M_2 = 1.53$. Utilizing Table 8.3 again, we obtain $M_3 = 0.690$, and $p_2 = 29.09$ kPa, $p_3 = 74.59$ kPa, and $(A_2/A_1) = 1.20$. Here, p_2 is provided by Eq. (8.7b), and A_2/A_1 provides the location of the shockwave, which is close to the throat.

8.7 METHOD-OF-CHARACTERISTICS

The method-of-characteristics is the traditional tool for the computation of inviscid supersonic flows (see Sec. 20.5 for numerical methods). Before the computer era, hand calculations were performed with the method-of-characteristics to design the wall configuration of the supersonic section of a wind tunnel nozzle in order to provide uniform flow at the start of the test section. One-dimensional analysis is quite inadequate for this task. For brevity, the presentation is limited to steady ir-rotational, two-dimensional or axisymmetric, supersonic flow. The method itself is not so restricted, being applicable to three-dimensional, rotational, steady, super-sonic flow, and to unsteady, one-dimensional flow over the full Mach number range.

The equations of motion reduce the two partial differential equations when the speed V and the streamline angle ϕ are used to [Liepman and Roshko (1957) and Emanuel (1986)]

$$\frac{Z}{V}\frac{\partial V}{\partial s} - \frac{\partial \phi}{\partial n} = \sigma\frac{\sin\phi}{r}, \qquad (8.113a)$$

$$\frac{1}{V}\frac{\partial V}{\partial n} - \frac{\partial \phi}{\partial s} = 0 \qquad (8.113b)$$

where s and n are natural coordinates, and r is the transverse distance from the symmetry axis to the point of interest on a streamline. The coordinate s is along the streamline, while n, which is orthogonal to s, is directed toward the center of cur-vature of the streamline. The σ term in Eq. (8.113a) accounts for axisymmetric flow and corresponds to the σ term in Eq. (8.83a). The second equation is the condition of irrotationality.

It is convenient to replace V with the Prandtl–Meyer functions ν by means of $d\nu = Z^{1/2}(dV/V)$

$$\frac{\partial \nu}{\partial s} - Z^{-1/2}\frac{\partial \phi}{\partial n} = \sigma Z^{-1/2}\frac{\sin\phi}{r}, \quad Z^{-1/2}\frac{\partial \nu}{\partial n} - \frac{\partial \phi}{\partial s} = 0 \qquad (8.114)$$

where both dependent variables, ν and ϕ, are angles. By adding and subtracting Eq. (8.114), we have

$$\frac{\partial R}{\partial s} + Z^{-1/2}\frac{\partial R}{\partial n} = \sigma Z^{-1/2}\frac{\sin\phi}{r}, \quad \frac{\partial Q}{\partial s} - Z^{-1/2}\frac{\partial Q}{\partial n} = \sigma Z^{-1/2}\frac{\sin\phi}{r} \qquad (8.115)$$

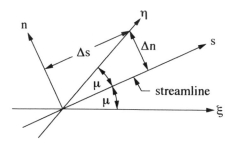

FIGURE 8.19   Natural (s, n) and characteristic (ξ, η) coordinate systems.

where the *Riemann invariants*, R and Q, are given by

$$R = v - \phi, \quad Q = v + \phi \tag{8.116}$$

In a two-dimensional or axisymmetric flow, the Mach lines form a nonorthogonal coordinate system η, ξ as indicated in Fig. 8.19. By means of the figure, one can show that the transformation from n,s coordinates to characteristic coordinates η, ξ is given by

$$\frac{\partial}{\partial s} = \frac{M}{2Z^{1/2}} \left(\frac{\partial}{\partial \eta} + \frac{\partial}{\partial \xi} \right), \quad \frac{\partial}{\partial n} = \frac{M}{2} \left(\frac{\partial}{\partial \eta} - \frac{\partial}{\partial \xi} \right) \tag{8.117}$$

Equation (8.115) simplifies to

$$\frac{\partial R}{\partial \eta} = \sigma \frac{\sin \phi}{Mr}, \quad \frac{\partial Q}{\partial \xi} = \sigma \frac{\sin \phi}{Mr} \tag{8.118}$$

where ϕ and v are given by the inverse of Eq. (8.116)

$$\phi = \tfrac{1}{2}(Q - R), \quad v = \tfrac{1}{2}(Q + R) \tag{8.119}$$

Because only one partial derivative appears in each of Eq. (8.118), they can be treated as ordinary derivatives. That is, $\partial R/\partial \eta$ becomes $dR/d\eta$ along each of the ξ = constant lines. The use of the characteristic η, ξ mesh is responsible for the partial differential equations reducing to ordinary differential equations.

In the two-dimensional case, $\sigma = 0$ and Eq. (8.118) integrates to

$$v - \phi = R = \text{constant on a } \eta\text{-characteristic} \tag{8.120a}$$

$$v + \phi = Q = \text{constant on a } \xi\text{-characteristic} \tag{8.120b}$$

In the axisymmetric case, $\sigma = 1$, Eq. (8.118) must be numerically integrated, as discussed in Liepman and Roshko (1957).

The differential Eqs. (8.118), or their solution, are called the *compatibility equations*. As is evident from Fig. 8.19, the characteristic mesh provided by the η and ξ coordinates depends on the Mach number. In general, therefore, it is necessary to solve the compatibility equations at the same time the mesh is determined. Furthermore, mesh points on walls, symmetry lines (such as a centerline), free boundaries (such as the edge of a free jet where the pressure is prescribed), contact surfaces or slipstreams, and shock waves all require special consideration. Special treatment of the centerline in an axisymmetric flow where $r = 0$, is also required, since the right sides of Eqs. (8.118) are indeterminate.

More details and helpful worked examples can be found in Shapiro (1953).

REFERENCES

Ames Research Staff, "Equations, Tables, and Charts for Compressible Flow," Report 1135, 1953.

Anderson, J. D., Jr., *Modern Compressible Flow*, McGraw-Hill, New York, 1982.

Emanuel, G., "Blowing from a Porous Cone or Wedge when the Contact Surface is Straight," *AIAA J.*, Vol. 5, p. 534, 1967.

Emanuel, G., Cline, M. C., and Witte, K. H., "Laser-Induced Disturbance with Application to a Low Reynolds Number Flow," *AIAA J.*, Vol. 19, p. 226, 1981.

Emanuel, G., "Choking analysis for a CW, HF or DF Chemical Laser," *AIAA J.*, Vol. 20, p. 1401, 1982a.

Emanuel, G., "Near-Field Analysis of a Compressive Supersonic Ramp," *Phys. Fluids*, Vol. 25, p. 1127, 1982b.

Emanuel, G., *Gasdynamics: Theory and Applications*, AIAA Education Series, Washington, D.C., 1986.

Liepmann, H. W. and Roshko, A., *Elements of Gasdynamics*, John Wiley & Sons, New York, 1957.

Rasmussen, M. L., "On Hypersonic Flow Past an Unyawed Cone," *AIAA J.*, Vol. 5, p. 1945, 1967.

Salas, M. D. and Morgan, B. D., "On the Instability of Shock Waves Attached to Wedges and Cones," AIAA-82-0288, 1982.

Shapiro, A. H., *The Dynamics and Thermodynamics of Compressible Fluid Flow*, Vols. I and II, Ronald Press, New York, 1953.

9 Transonic Flow

H. YOSHIHARA (Retired)
Boeing Company
Seattle, WA

CONTENTS

9.1 INTRODUCTION

A steady transonic flow is defined as a flow in which subsonic and supersonic regions co-exist. Such flows play an important role in aeronautical applications ranging from the cruise optimization of commercial and military aircraft to the design of efficient rotating machinery in turbofan engines. As a result, there has been an extensive effort in the development of computational tools and experimental techniques for the transonic regime.

The fluid dynamics of either subsonic or supersonic flow and their respective mathematical theory and computational methods are well-founded and applied in a routine fashion. Here subsonic flow is characterized by an analytic *smoothness* brought about by the ever-present tendency of flow perturbations to attenuate. In contrast, supersonic flow is characterized by flow *unsmoothnesses* as Mach waves (characteristics), zones of silence, and shock waves (see Chap. 8).

It is clear that if the boundary between the subsonic and supersonic regions, namely the sonic surface, were known in advance, the transonic problem would be greatly simplified since the above classical procedures could then be directly applied. Unfortunately, this is not the case, and one has to treat a difficult nonlinear subsonic/supersonic boundary value problem, the solution in most cases requiring shock waves.

With shock waves arising adjacent to the configuration surface, significant viscous effects can arise through the shock/boundary layer interactions. Such interac-

Handbook of Fluid Dynamics and Fluid Machinery, Edited by Joseph A. Schetz and Allen E. Fuhs
ISBN 0-471-12598-9 Copyright © 1996 John Wiley & Sons, Inc.

tions usually greatly alter the location of the shocks relative to their inviscid flow location, resulting in a significant change in the forces and moments on the configuration. Further, the pressure distribution on an airfoil or wing is known to be sensitive to small changes in the configuration shape. As a result, the viscous displacement effects along the configuration surface and the effects of the downstream viscous wake can additionally be significant.

Despite the complexities sketched above, there has been significant progress, particularly during the past decade, in the development of computational tools and test techniques which are being applied widely in the aeronautical industry. It will be the objective here and in Sec. 16.3 to discuss this progress.

In Sec. 9.2 the flow over an airfoil will be sketched for a fixed angle of attack and for various free stream Mach numbers from a high subsonic value to a supersonic value. Here, the flows for the inviscid and viscous cases are sketched in juxtaposition together with their pressure distributions. These sketches will serve to identify important flow features as well as to show the nature and consequences of the important viscous interactions. The flow over a typical swept wing configuration characterizing a jet transport is next sketched, again for a fixed angle of attack and various Mach numbers in the high subsonic regime.

In Sec. 9.3 an example of the use of the Transonic Navier–Stokes code from NASA Ames Research Center is given for the case of the General Dynamics F-16A aircraft to illustrate the type of calculations possible. Test/theory comparison of the pressure distribution on various parts of the aircraft is given.

9.2 TRANSONIC FLOW PATTERNS OVER AIRFOILS AND WINGS

In the present section, the flow over an airfoil at a fixed angle of attack and for varying transonic free stream Mach numbers is sketched. Here, the inviscid and viscous cases are shown side-by-side together with their pressure distributions. Important flow features are pointed out, in particular the significant effects of viscosity. The shock wave and separation patterns for a typical swept wing are then sketched for various high subsonic Mach numbers.

The shock configuration and its interaction with the boundary layer play an important role in determining the lift, pitching moment, and drag on the configuration. The entropy rise through the shocks changes the surface pressure creating a pressure drag. As the free stream Mach number is increased with the lift held invariant, the drag abruptly increases due to a sudden strengthening of the shock. The Mach number at which this drag divergence occurs is of practical importance, since it establishes the cruise Mach number at which an aircraft achieves maximum range.

In Figs. 9.1, 9.2, and 9.3, respectively, the flow patterns are sketched for a typical aft-cambered supercritical airfoil at (a) the design cruise Mach number, (b) a higher subsonic Mach number where buffet is present, and (c) a moderate supersonic Mach number.

At the cruise point (Fig. 9.1), the terminating shock wave in the inviscid case is located at the trailing edge. At lower Mach numbers, the shock forms further upstream and drifts downstream with increases in Mach number. When the shock reaches approximately the 60% chord station, just upstream of the large aft upper surface curvature, a small further increase in Mach number will cause the shock to

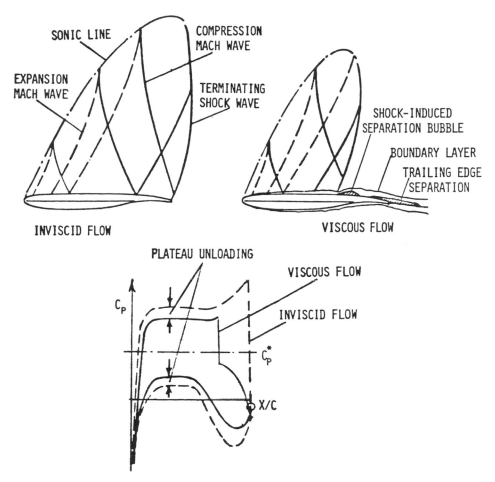

FIGURE 9.1 Inviscid/viscous flow comparison-cruise.

slide all the way down the aft curvature to the trailing edge. This behavior in the inviscid case is typical of aft-cambered airfoils and is due to the fact that as the shock moves onto the aft camber region, the high curvature region is exposed to supersonic flow. The flow then rapidly expands generating increased Mach numbers upstream of the shock. The strengthened shock then displaces downstream, magnifying the above effects until the shock reaches the trailing edge.

Viscous effects are seen in Fig. 9.1 to have an important effect. As the shock starts to displace onto the aft camber region, the strengthened shock, representing a severe adverse pressure gradient, causes an abrupt thickening of the boundary layer. The wedging displacement effect then impedes and finally halts the downstream movement of the shock.

The differences of the boundary layer displacement thicknesses on the aft upper and lower surfaces will reduce the effective aft camber and accordingly cause the decrease of the plateau loading shown in the pressure distributions. The effects of the near-wake will also affect the pressure distributions but in a less significant manner.

For thick airfoils and for thinner airfoils at moderate to high lift, an important compound viscous interaction (Type B) can arise identified by Pearcey, Osborne, and Haines (1968). Such a case arises when the boundary layer encounters two successive adverse pressure gradients as the shock pressure rise and the trailing edge pressure recovery. After encountering a sufficiently strong shock pressure rise, the boundary layer thickens and the velocity profile loses its customary turbulent fullness. If the boundary layer then encounters the second adverse pressure gradient in this vulnerable state, it is less able to remain unseparated. Usually, the result is a sudden severe separation which extends from the shock to a point downstream of the trailing edge.

A case with such a severe separation is shown in Fig. 9.2 where the airfoil is in buffet. Here, buffet on the stationary airfoil is characterized by severe broad-band fluctuations of the lift and moment caused by the unsteadiness of the severely separated boundary layer. Such unsteadiness produces both fluctuations of the aft pressures as well as the shock locations. Buffet onset occurs when the shock-induced separation no longer reattaches on the airfoil, but attaches at some point downstream

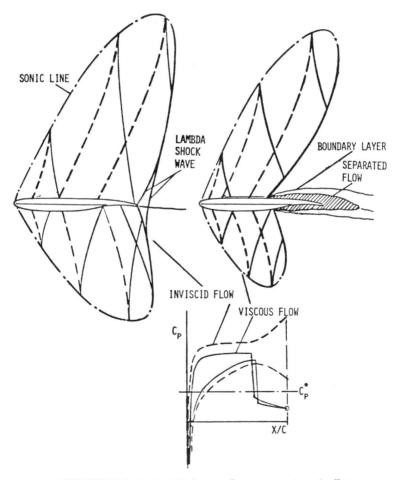

FIGURE 9.2 Inviscid/viscous flow comparison-buffet.

in the wake. It is characterized by a sudden drop in the trailing edge pressure as the free stream Mach number is increased at a fixed angle of attack, for example. Buffet must not be confused with flutter where the unsteady aerodynamic driving force is produced by the motion of the airfoil or wing itself. Section 23.4 discusses Aeroelasticity and Hydroelasticity.

It will be appropriate to comment here on the role of the Kutta condition at the airfoil trailing edge in establishing the circulation and hence the lift on the airfoil. Circulation is in general a global feature of the flow determined by the relative apportionment of the free stream into the portions passing above or below the airfoil. Fortunately, airfoils have sharp trailing edges, and this apportionment can be established by postulating the flow to stream smoothly off the airfoil trailing edge. Such a model avoids infinite trailing edge velocities in the inviscid flow and is consistent with experimental observations. As the trailing edge is approached along the upper and lower surfaces of the airfoil and along the rear stagnation streamline, the pressure must then tend to a common value which, in the case of a finite trailing edge angle, must be the stagnation pressure. That is, the Kutta condition must be satisfied. The reader is referred to Sec. 23.3, Lifting Surfaces.

In the case of the high subsonic flow of Fig. 9.2 with the termination shocks at the trailing edge, the Kutta condition no longer plays a role in establishing the circulation. Here, the match of the pressure and the flow direction on the downstream side of the trailing edge is achieved, independent of the circulation on the airfoil, by the adjustment of the obliqueness of the two trailing edge shocks. In this case, the circulation is determined by the relative mass flux impedance (choking) of the space above or below the airfoil measured by the "throat" geometry at the shoulder of the airfoil.

At lower free stream Mach numbers than represented in Fig. 9.2 where the shock waves are located upstream of the trailing edge, one must also expect a reduced role of the Kutta condition and a dominant role of the choking at the airfoil "throat" in establishing the circulation. That is, a relaxation of the Kutta condition in such a flow, for example, would affect the pressures near the trailing edge but not distort the overall circulation.

Let us consider next the case with supersonic free stream Mach numbers. As the Mach number just exceeds the sonic value, a weak near-normal bow shock appears far upstream. Downstream of such a shock, the flow is locally a near-uniform subsonic flow so that the overall flow would correspond closely to that for a high subsonic Mach number. With an increase in Mach number, the bow shock approaches the airfoil as shown in Fig. 9.3. Here, the subsonic region becomes embedded in the overall supersonic flow. A limiting Mach wave then arises as shown in the above figure which forms the downstream limit of the flow which influences the embedded subsonic region. With the shock at the trailing edge, the viscous effects are small, so that the sketch for the pressure distributions has been omitted.

Before the transonic flow over a swept wing is described, the case of a constant-chord, yawed wing of infinite span is considered (Fig. 9.4). In an inviscid flow, the pressure distribution along the chord normal to the wing leading edge is invariant along the span. It corresponds to the planar flow over the airfoil section normal to the leading edge with a free stream velocity V_n, the component of the free stream velocity normal to the leading edge. Since transonic difficulties such as drag divergence are tied to the Mach number based on V_n, it follows that yawing the wing has

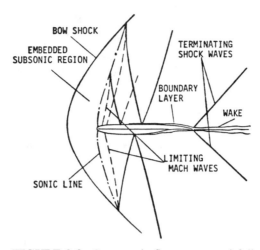

FIGURE 9.3 Supersonic flow over an airfoil.

increased the drag divergence Mach number and hence the cruise Mach number, since they will now be based on V. Such a deceptively simple idea (sweepback theory) was proposed independently by A. Busemann of Germany and R. T. Jones of the United States during World War II as a means to improve the transonic performance of aircraft by sweeping the wings.

In the case of a swept wing, the above yawed wing flow is distorted by the symmetry plane at the upstream end, a wing tip at the downstream end, and wing taper. The consequence of the three dimensionalization on the degradation of the transonic

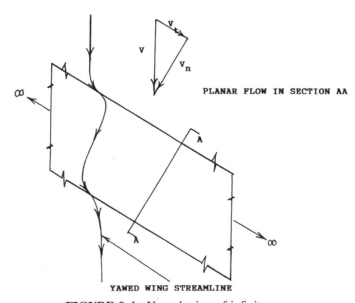

FIGURE 9.4 Yawed wing of infinite span.

performance is due primarily to the symmetry plane which constrains the flow to stream along the symmetry plane rather than along the yawed wing streamline shown in Fig. 9.4. The latter is constructed by vectorally adding the planar surface velocity to the constant tangential free stream component, V_t. The consequence is that a family of compression waves is generated at the symmetry plane which, at a sufficiently high free stream Mach number, coalesce outboard to form the rear shock as shown in Fig. 9.5(a). As the Mach number is increased further, the wing apex region becomes embedded in an extensive supersonic region. A forward shock is then formed [Fig. 9.5(b)] which turns the inward-directed oncoming flow to the direction of the symmetry plane. The forward shock is a weak shock being inclined in the plane of the wing at approximately the Mach angle. Both the inboard and forward shocks are three dimensional shocks having no counterparts in planar flow.

With still further increases in Mach number [Fig. 9.5(c)], the rear shock strength-

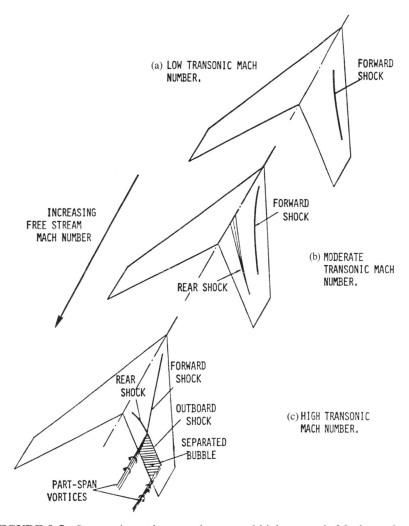

FIGURE 9.5 Swept wing at low, moderate, and high transonic Mach number.

ens and extends in the inboard direction, and the foward shock assumes a greater sweep eventually intersecting the rear shock. The shock outboard of the point of intersection is labeled the outboard shock and is usually the strongest of the shocks. Shock-induced separation thus first appears downstream of this shock. The separated zone usually then spreads tipward and eventually inward with increase in Mach number. With the increase in the severity of the shock-induced separation, the outboard shock is displaced upstream, leading to an unsweeping and a further strengthening of the outboard shock. This worsening shock-induced separation contaminates the trailing edge separation usually present resulting in the compound (Type B) interaction described earlier for the planar case.

For a given lift, the appearance of an extensive separation in the outboard regions of the wing will shift the loading on the wing to the inboard more upstream portion of wing generating a nose-up pitching moment. Such a moment will then increase the angle of attack, further worsening the outboard separation. This sequence of events will lead to a longitudinal instability known as *pitchup*. Pitchup is further greatly aggravated by the interference of the wing flow on the horizontal tail, which in the usual case, has a downward lift to trim the aircraft. The shift of the wing loading will increase the inboard wake downwash. This then increases the negative lift on the tail, thereby contributing to the pitchup. Pitchup is an inherent undesirable feature of swept wings which must be eliminated in the operating range of the aircraft in an acceptable design.

The pattern of three-dimensional shocks may differ for a given swept wing, since it depends on the wing parameters such as the sweep, taper ratio, and aspect ratio and the airfoil section as well as on the free stream Mach number. More details of swept wing flows are given, for example, in Rogers and Hall (1960) and Taylor (1975).

It was seen earlier that the interference of the wing and the symmetry plane or fuselage had a negative effect on the flow. To counter this, a standard design procedure is to contour the fuselage in the juncture region with the wing. The contouring of the fuselage is guided by an empirically derived *area rule* developed by R. Whitcomb, formerly of NASA–Langley. Earlier, W. Hayes of Princeton in his Cal Tech thesis developed a drag formulae based on the slender body theory. He showed that the wave drag for a slender wing/fuselage was equal to that of an axisymmetric body whose axial area distribution was generated from the area intercepted by a sequence of cuts by planes normal to the wing/fuselage axis. Clearly, a modern swept-wing commercial transport is not even remotely a slender body, and Hayes' rule applied, not to high subsonic flows, but to supersonic flow. But, Whitcomb demonstrated in the wind tunnel that Hayes' area rule was a viable design rule for commercial transports in the high subsonic speed range. The equivalent area distribution of a swept-wing transport would show an abrupt bump where the cutting planes intercepted the wing. The Whitcomb tests showed that, to achieve reduced drag, the maximum area of the equivalent body must be reduced and the area distribution made as "smooth" as possible. Thus, in military aircraft, a *coke bottling* of the fuselage was incorporated. In a high subsonic commercial transport the coke bottling reduces the passenger seating and, instead, large wing fillets were added to smoothen the area distribution. In the Boeing 747, the large fuselage hump just aft of the cockpit served the same purpose.

9.3 EXAMPLE FLOWFIELD COMPUTATION: SOLUTIONS FOR THE F-16A

The complexity of transonic flows as described in the previous sections demands the use of advanced computer aided methods to obtain flowfield solutions. Here, we discuss a representative case treated by such methods. The details of these techniques are discussed in Chaps. 18–21.

To illustrate the use of the Transonic Navier–Stokes code developed at NASA Ames Research Center [Kaynak, Holst, and Cantwell (1986)], results for the complete General Dynamics F-16A aircraft from Flores and Chaderjian (1988) are described next.

For these calculations, the Baldwin/Lomax algebraic turbulence model was used. A total of 27 grid blocks was employed with the mesh in each block generated by the elliptic method of Sorenson and Steger (1983). It was a major achievement to develop a code that kept track of the patching of so many blocks in a reasonably cost-effective fashion. The free stream Mach number for the test results was 0.9, the angle of attack was 6°, and the Reynolds number based on the mean chord was 4.5×10^6. In the calculations, however, the free stream Mach number was arbitrarily increased to 0.902 to obtain a match of the rear shock location. Such a practice is almost routine for computational fluid dynamicists who frequently also alter the angle of attack to achieve improved test/theory agreement. The rationale is that such alterations compensate for the wind tunnel wall interference effects. It perhaps is more straightforward to simply use the Mach number and angle of attack of the test.

In Fig. 9.6, test/theory comparisons of the pressure distributions are given on various parts of the configuration. The ordinate in the plots is the pressure coefficient. It is seen that the general agreement is good. An important exception, however, is in the disagreement in the leading edge suction peaks caused by the extreme sharpness of the leading edge. Details are lacking to determine whether a more refined mesh might improve the agreement here. It is also to be noticed that the forward shock assumes a significant thickness. Such thickening arises when the swept shock is not aligned with the coordinate lines, and the mesh spacing in the lateral direction is very much larger than the streamwise mesh spacing, which is usually the case. The shock then assumes a thickness characteristic of the lateral mesh, thereby spreading out over many more streamwise meshes.

An important result of the above F-16A calculations was the determination of the flow uniformity entering the inlet and the adequacy of the boundary layer plough to divert the fuselage boundary layer flow from the inlet.

The F-16A calculation required 5000 iterations to reduce the residual error by three orders of magnitude and required 25 hours on the Cray XMP (1 CPU). On the Cray YMP C-90 (1 CPU) it would accordingly require about 6 hours, a more reasonable turnaround for design purposes.

The matter of solution validation can be put into perspective in the above F-16A calculations. The residual errors have been reduced by three orders of magnitude. With so many grid blocks, it is an impractical task to incorporate a multigrid scheme to reduce the residual errors further. The mesh undoubtedly can be refined in local critical areas as about the wing leading edge and about the forward shock to reduce

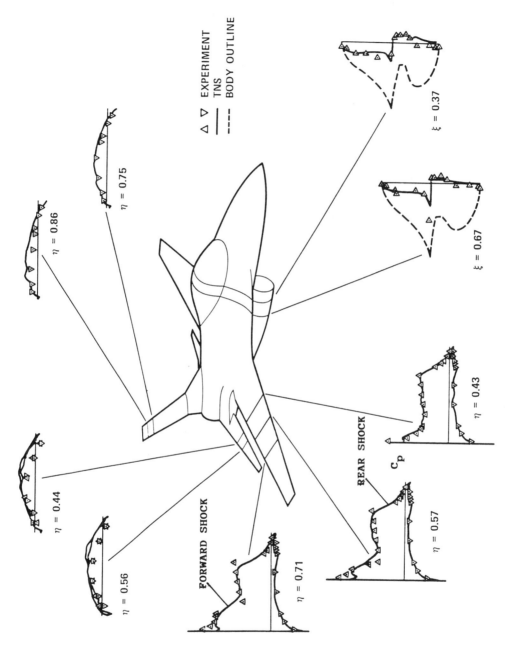

FIGURE 9.6 Test/theory comparisons on F-16A at Mach 0.9. (From Flores and Chaderjian, 1988.)

the truncation errors in these regions. Further iterations need be carried out only in those blocks containing these refinements. It is not impossible to carry out these improvements, but it would be difficult to justify the additional programming efforts and computer costs for a demonstration calculation. It clearly would be difficult to validate a turbulence model under these circumstances.

Despite these uncertainties, the F-16A results can be used to reveal undesirable flow features such as excessively strong shocks and separations, and to determine the performance of the inlet and rear empennage. Accurate determination of the drag, however, is still well out of reach, even if the prediction of the leading edge suction forces could be improved.

Overall, the above F-16A results are an outstanding accomplishment. The Applied Computational Fluids Branch at NASA–Ames has pioneered the calculation of complete aircraft configurations, and more recently completed results for the YAV-8B, an advanced Harrier STOVL configuration, in ground effect [Smith *et al.* (1991)] and the F-18 at large angle of attack where the vortices off the strake will play an important role [Rizk and Gee (1991)].

9.4 SUMMARY

The fluid dynamics of transonic flow is now reasonably well understood. There exist, however, problems of substantial magnitude both in the computational as well as experimental aspects of the problem. On the computational side, the outstanding problem is turbulence modeling. For planar flows over airfoils, the simple algebraic model has yielded promising results, though not yet adequate to predict drag. Separated flows are important to applications. No existing turbulence models have been successful in treating such flows, and none appears on the horizon. Despite these shortcomings, computations can still play a valuable role by identifying potential problems as premature shock strengthening, appearance of separation, as well as adverse interference problems as the local bucketing of the span loading on the wing by the nacelle-pylons. In addition, computations with even relatively crude turbulence models are useful in developing configuration improvements to remove the above shortcomings.

On the experimental side, two problems that need to be resolved are wind tunnel wall interference and Reynolds number simulation. Satisfactory resolution of these problems is elusive. In the case of commercial aircraft development, there is extensive correlation of wind tunnel and flight measurements from past related aircraft developments to calibrate the wind tunnel results. More details on these matters are provided in Sec. 16.3.

REFERENCES

Flores, J. and Chaderjian, N. M., "The Numerical Simulation of Transonic Separated Flow about the Complete F-16A," AIAA Paper 88-2506, 1988.

Kaynak, U., Holst, T., and Cantwell, B. J., "Computation of Transonic Separated Wing Flows using a Euler/Navier–Stokes Zonal Approach," NASA TM 88311, 1986.

Pearcey, H., Osborne, J., and Haines, A. B., "The Interaction between Local Effects at the Shock and Rear Separation," AGARD CP No. 35, 1968.

Rizk, Y. M. and Gee, K., "Numerical Prediction of the Unsteady Flowfield Around the F-18 Aircraft at Large Incidence," AIAA Paper 91-0020, 1991.

Rogers, E. and Hall, I. M., "An Introduction to the Flow about Plane Swept-Back Wings at Transonic Speeds," *J. Royal Aero. Soc.*, Vol. 64, No. 596, 1960.

Smith, M., Chawla, K., and Van Dalsem, W., "Numerical Simulation of a Complete Stovl Aircraft in Ground Effect," AIAA Paper 91-3293, 1991.

Sorenson, R. L. and Steger, J. L., "Grid Generation in Three Dimensions by Poisson Equations with Control of Cell Size and Skewness at Boundary Surfaces," *Advances in Grid Generation-FED*, Vol. 5, ASME, 1983.

Taylor, C. R. (Ed.), "Aircraft Stalling and Buffetting," AGARD LS-74, 1975.

10 Hypersonic Flow

JOHN D. ANDERSON, JR.
University of Maryland
College Park, MD

CONTENTS

10.1 HYPERSONIC FLOW—WHAT IS IT?

There is a conventional *rule of thumb* that defines hypersonic aerodynamics as those flows where the Mach number, M, is greater than 5. However, this is no more than just a rule of thumb; when a flow is accelerated from $M = 4.99$ to $M = 5.01$, there is no "clash of thunder" and the flow does not "instantly turn from green to red." Rather, hypersonic flow is best defined as that regime where certain physical flow phenomena become progressively more important as the Mach number is increased to higher values. In some cases, one or more of these phenomena may become important above Mach 3, whereas in other cases they may not be compelling until

Handbook of Fluid Dynamics and Fluid Machinery, Edited by Joseph A. Schetz and Allen E. Fuhs
ISBN 0-471-12598-9 Copyright © 1996 John Wiley & Sons, Inc.

Mach 7 or higher. The purpose of this section is to briefly describe these physical phenomena.

10.1.1 Thin Shock Layers

For a given flow deflection angle, the density increase across a shock wave becomes progressively larger as the Mach number is increased. At higher density, the mass flow behind the shock can more easily "squeeze through" smaller areas. For flow over a hypersonic body, this means that the distance between the body and the shock wave can be small. The flowfield between the shock wave and the body is defined as the *shock layer*, and for hypersonic speeds this shock layer can be quite thin. For example, consider the Mach 36 flow of a calorically perfect gas with a ratio of specific heats, $\gamma = c_p/c_v = 1.4$, over a wedge of 15° half angle. From standard oblique shock theory the shock wave angle will be only 18°, as shown in Fig. 10.1. If high-temperature, chemically reacting effects are included, the shock wave angle will be even smaller. Clearly, this shock layer is thin. It is a basic characteristic of hypersonic flows that shock waves lie close to the body and that the shock layer is thin. In turn, this can create some physical complications, such as the merging of the shock wave itself with a thick, viscous boundary layer growing from the body surface—a problem which becomes important at low Reynolds numbers. However, at high Reynolds numbers, where the shock layer is essentially inviscid, its thinness can be used to theoretical advantage, leading to a general analytical approach called *thin shock-layer theory*. In the extreme, a thin shock layer approaches the fluid dynamic model postulated by Isaac Newton in 1687; such *Newtonian theory* is simple and straightforward, and is frequently used in hypersonic aerodynamics for approximate calculations (to be discussed in Sec. 10.2).

10.1.2 Entropy Layer

Consider the wedge shown in Fig. 10.1 except now with a blunt nose as shown in Fig. 10.2. At hypersonic Mach numbers, the shock layer over the blunt nose is also very thin, with a small shock detachment distance, d. In the nose region, the shock wave is highly curved. The entropy of the flow increases across a shock wave, and the stronger the shock, the larger the entropy increase. A streamline passing through the strong, nearly normal portion of the curved shock near the centerline of the flow will experience a larger entropy increase than a neighboring streamline which passes through a weaker portion of the shock further away from the centerline. Hence, there are strong entropy gradients generated in the nose region; this *entropy layer* flows

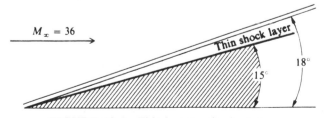

FIGURE 10.1 Thin hypersonic shock layer.

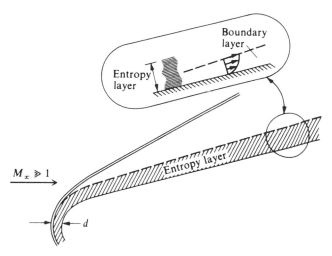

FIGURE 10.2 The entropy layer.

downstream and essentially wets the body for large distances from the nose as shown in Fig. 10.2. The boundary layer along the surface grows inside this entropy layer and is affected by it. Since the entropy layer is also a region of strong vorticity, this interaction is sometimes called a *vorticity interaction*. The entropy layer causes analytical problems when we wish to perform a standard boundary layer calculation on the surface, because there is a question as to what the proper conditions should be at the outer edge of the boundary layer.

10.1.3 Viscous Interaction

Consider a boundary layer on a flat plate in a hypersonic flow as shown in Fig. 10.3. A high velocity, hypersonic flow contains a large amount of kinetic energy; when this flow is slowed by viscous effects within the boundary layer, the lost kinetic energy is transformed (in part) into internal energy of the gas. This is called *viscous dissipation*. In turn, the temperature increases within the boundary layer; a typical temperature profile within the boundary layer is also sketched in Fig. 10.3. The characteristics of hypersonic boundary layers are dominated by such temperature

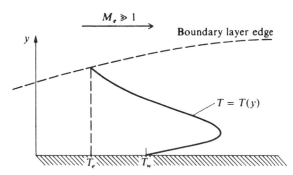

FIGURE 10.3 Temperature profile in a hypersonic boundary layer.

increases. For example, the viscosity coefficient increases with temperature, and this by itself will make the boundary layer thicker. In addition, because the pressure, p, is constant in the normal direction through a boundary layer, the increase in temperature T results in a decrease in density, ρ, through the equation of state $\rho = p/RT$, where R is the specific gas constant. In order to pass the required mass flow through the boundary layer at reduced density, the boundary layer thickness must be larger. Both of these phenomena combine to make hypersonic boundary layers grow more rapidly than at slower speeds. Indeed, the flat plate compressible laminar boundary layer thickness δ grows essentially as $\delta \propto M_\infty^2/\sqrt{Re_x}$, where M_∞ is the freestream Mach number, and Re_x is the local Reynolds number. Clearly, since δ varies as the square of M_∞, it can become inordinately large at hypersonic speeds.

The thick boundary layer in hypersonic flow can exert a major displacement effect on the inviscid flow outside the boundary layer, causing a given body shape to appear much thicker than it really is. Due to the extreme thickness of the boundary layer flow, the outer inviscid flow is greatly changed; the changes in the inviscid flow in turn feed back to affect the growth of the boundary layer. This major interaction between the boundary layer and the outer inviscid flow is called *viscous interaction*. Viscous interactions can have important effects on the surface pressure distribution, hence lift, drag, and stability on hypersonic vehicles. Moreover, skin friction and heat transfer are increased by viscous interaction. For example, Fig. 10.4 illustrates the viscous interaction on a sharp, right-circular cone at zero degrees of angle of attack. Here, the pressure distribution on the cone surface, p, is given as a function of distance from the tip. These are experimental results obtained from Anderson (1962). If there were no viscous interaction, the inviscid surface pressure would be constant, equal to p_c (indicated by the horizontal dashed line in Fig. 10.4). However, due to the viscous interaction, the pressure near the nose is considerably greater; the surface pressure distribution decays further downstream, ultimately approaching the inviscid value far downstream.

The boundary layer on a hypersonic vehicle can become so thick that it essentially

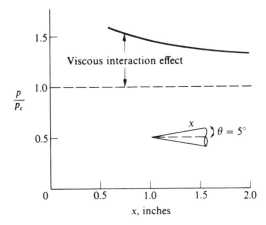

FIGURE 10.4 Viscous interaction effect. Induced pressure on a sharp cone at $M_\infty = 11$ and $Re = 1.88 \times 10^5$ per foot. (From Anderson, 1962.)

merges with the shock wave, a *merged shock layer*. When this happens, the shock layer must be treated as fully viscous, and the conventional boundary layer analysis must be completely abandoned.

10.1.4 High Temperature Flows

As discussed previously, the kinetic energy of a high-speed, hypersonic flow is dissipated by the influence of friction within a boundary layer. The extreme viscous dissipation that occurs within hypersonic boundary layers can create very high temperatures—high enough to excite vibrational energy internally within molecules, and to cause dissociation and even ionization within the gas. If the surface of a hypersonic vehicle is protected by an ablative heat shield, the products of ablation are also present in the boundary layer giving rise to complex hydrocarbon chemical reactions. On both accounts, we see that the surface of a hypersonic vehicle can be wetted by a *chemically reacting boundary layer*.

The boundary layer is not the only region of high-temperature flow over a hypersonic vehicle. Consider the nose region of a blunt body, as sketched in Fig. 10.5. The bow shock wave is normal, or nearly normal, in the nose region, and the gas temperature behind this strong shock wave can be enormous at hypersonic speeds. For example, Fig. 10.6 is a plot of temperature behind a normal shock wave as a function of free-stream velocity, for a vehicle flying at a standard altitude of 52 km; this figure is taken from Anderson (1990). Two curves are shown: (1) the upper curve, which assumes a calorically perfect nonreacting gas with the ratio of specific

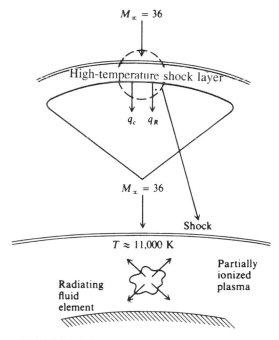

FIGURE 10.5 High-temperature shock layer.

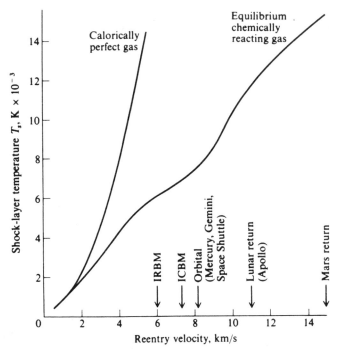

FIGURE 10.6 Temperature behind a normal shock wave as a function of free-stream velocity at a standard altitude at 52 km. (From Anderson, 1990.)

heats $\gamma = 1.4$, gives an unrealistically high value of temperature; and (2) the lower curve, which assumes an equilibrium chemically reacting gas, is usually closer to the actual situation. This figure illustrates two important points:

i) By any account, the temperature in the nose region of a hypersonic vehicle can be extremely high, for example, reaching approximately 11,000 K at a Mach number of 36 (Apollo reentry).

ii) The proper inclusion of chemically reacting effects is vital to the calculation of an accurate shock-layer temperature; the assumption that γ is constant and equal to 1.4 is no longer valid.

So we see that, for a hypersonic flow, not only can the boundary layer be chemically reacting, but the entire shock layer can be dominated by *chemically reacting flow*.

For a moment, let us examine the physical nature of a high-temperature gas. In introductory studies of thermodynamics and compressible flow, the gas is assumed to have constant specific heats, hence the ratio $\gamma = c_p/c_v$, is also constant. This leads to some ideal results for pressure, density, temperature, and Mach number variations in a flow. However, when the gas temperature is increased to high values, the gas behaves in a *nonideal* fashion. Specifically:

i) The vibrational energy of the molecules becomes excited, and this causes the specific heats, c_p and c_v, to become functions of temperature. In turn, the ratio of specific heats, $\gamma = c_p/c_v$, also becomes a function of temperature. For air, this effect becomes important above a temperature of 800 K.

ii) As the gas temperature is increased further, chemical reactions can occur. For an equilibrium chemically reacting gas, c_p and c_v are functions of both temperature and pressure, and hence $\gamma = f(T, p)$. For air at 1 atm pressure, O_2 dissociation ($O_2 \rightarrow 2O$) begins at about 2000 K, and molecular oxygen is essentially totally dissociated at 4000 K. At this temperature, N_2 dissociation ($N_2 \rightarrow 2N$) begins and is essentially totally dissociated at 9000 K. Above a temperature of 9000 K, ions are formed ($N \rightarrow N^+ + e^-$, and $O \rightarrow O^+ + e^-$), and the gas becomes a partially ionized *plasma*.

All of these phenomena are called *high temperature effects*. (They are frequently referred to in the aerodynamic literature as *real gas effects*, but there are good technical reasons to discourage the use of that label.) If the vibrational excitation and chemical reactions take place very rapidly in comparison to the time it takes for a fluid element to move through the flowfield, we have vibrational and chemical *equilibrium flow*. If the opposite is true, we have *nonequilibrium flow*, which is considerably more difficult to analyze.

High temperature chemically reacting flows can have an influence on lift, drag, and moments on a hypersonic vehicle. For example, such effects have been found to be important for estimating the amount of body flap deflection necessary to trim the space shuttle during high-speed reentry. However, by far the most dominant aspect of high temperatures in hypersonics is the resultant high heat-transfer rates to the surface. Aerodynamic heating dominates the design of such a vehicle, or a wind tunnel to test the vehicle. This aerodynamic heating takes the form of heat transfer from the hot boundary layer to the cooler surface, called *convective heating*, denoted by q_c in Fig. 10.5. Moreover, if the shock-layer temperature is high enough, the thermal radiation emitted by the gas itself can become important, giving rise to a radiative flux to the surface, called *radiative heating*, denoted by q_R in Fig. 10.5. (In the winter, when you warm yourself beside a roaring fireplace, the warmth you feel is not hot air blowing out of the fireplace, but rather radiation from the flame itself. Imagine how *warm* you would feel standing next to the gas behind a strong shock wave at Mach 36, where the temperature is 11,000 K—about *twice* the surface temperature of the sun.) For example, for Apollo reentry, radiative heat transfer was more than 30% of the total heating. For a space probe entering the atmosphere of Jupiter, the radiative heating will be more than 95% of the total heating.

Another consequence of high-temperature flow over hypersonic vehicles is the *communications blackout* experienced at certain altitudes and velocities during atmospheric entry, where it is impossible to transmit radio waves either to or from the vehicle. This is caused by ionization in the chemically reacting flow, producing free electrons which absorb radio-frequency radiation. Therefore, the accurate prediction of electron density within the flowfield is important.

10.1.5 Interim Summary

In summary, it is too simplistic to define hypersonic flow as *flow where the Mach number is greater than five*. Instead, hypersonic flow denotes that part of the Mach number spectrum where the Mach number is high enough such that the flow is dominated by certain physical effects which are not important at lower Mach numbers; these physical effects have been itemized above.

It is also important to note that hypersonic flow theory is nonlinear. This is in contrast to subsonic and supersonic flows, where various linearized theories abound. The equations of fluid flow, when applied at high Mach numbers, cannot be linearized even for the case of small disturbances in the flow (such as the flow over slender bodies at small angles of attack). There is an established *hypersonic small disturbance theory*, but it is a nonlinear theory. *Hypersonic flow is inherently nonlinear*; intuitively, this should be no surprise in light of the interesting and complex physical phenomena associated with hypersonic flow as described earlier.

Various aspects of hypersonic flow are covered in several basic texts, such as Anderson (1989), Cox and Crabtree (1965), Chernyi (1961), and Hayes and Probstein (1966). Much of this chapter is drawn from material in Anderson (1989), which should be consulted for more details.

In the rest of this chapter, we will highlight several of the most important (and in some cases classic) theories for the prediction of hypersonic flows. We will also discuss pertinent physical aspects in order to promote a better understanding of the nature of hypersonic flows.

10.2 NEWTONIAN THEORY

Three centuries ago, Isaac Newton established a fluid dynamic theory which later was used to derive a law for the force on an inclined plane in a moving fluid. This law indicated that the force varies as the square of the sine of the deflection angle, the famous Newtonian *sine-squared law*. Experimental investigations carried out by d'Alembert more than a half-century later indicated that Newton's sine-squared law was not very accurate; indeed, the preponderance of fluid dynamic experience up to the present day confirms this finding. The exception to this is the modern world of hypersonic aerodynamics. Ironically, Newtonian theory, developed 300 years ago for the application to low speed fluid dynamics, has direct application to the prediction of pressure distributions on hypersonic bodies.

In Propositions 34 and 35 of his *Principia*, first published in 1687, Newton modeled a fluid flow as a stream of particles in rectilinear motion, much like a stream of pellets from a shotgun blast which, when striking a surface, would lose all their momentum normal to the surface but would move tangentially to the surface without loss of tangential momentum. This picture is illustrated in Fig. 10.7, which shows a stream with velocity, V_∞, impacting on a surface of area, A, inclined at the angle,

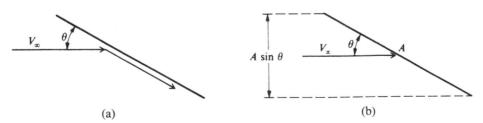

FIGURE 10.7 Schematic for Newtonian impact theory.

θ, to the free stream. From this figure, we see that: 1) change in normal velocity = $V_\infty \sin \theta$, 2) mass flux incident on an area A of the surface = $\rho_\infty V_\infty A \sin \theta$, and 3) time rate of change of momentum of this mass flux = $(\rho_\infty V_\infty A \sin \theta)(V_\infty \sin \theta)$ = $\rho_\infty V_\infty^2 A \sin^2 \theta$. From Newton's second law, the time rate of change of momentum is equal to the force, F, exerted on the surface

$$F = \rho_\infty V_\infty^2 A \sin^2 \theta \tag{10.1}$$

or

$$\frac{F}{A} = \rho_\infty V_\infty^2 \sin^2 \theta \tag{10.1a}$$

The force, F, in Eq. (10.1) requires some interpretation. Newton assumed the stream of particles to be rectilinear, i.e., he assumed that the individual particles do not interact with each other and have no random motion. Due to this lack of random motion, F, in Eq. (10.1) is a force associated only with the directed linear motion of the particles. On the other hand, modern science recognizes that the static pressure of a gas or liquid is due to the purely *random* motion of the particles—motion not included in Newtonian theory. Hence, in Eq. (10.1a), F/A, which has the dimensions of pressure, must be interpreted as the *pressure difference* above the free stream static pressure, namely $F/A = p - p_\infty$, where p is the surface pressure, and p_∞ is the free-stream static pressure. Hence, from Eq. (10.1)

$$p - p_\infty = \rho_\infty V_\infty^2 \sin^2 \theta \tag{10.2}$$

or

$$\frac{p - p_\infty}{\frac{1}{2} \rho_\infty V_\infty^2} = 2 \sin^2 \theta \tag{10.2a}$$

or

$$\boxed{C_p = 2 \sin^2 \theta} \tag{10.2b}$$

Equation (10.2b) is the famous *Newtonian sine-squared law* for the pressure coefficient.

What does the Newtonian pressure coefficient have to do with hypersonic flow? To answer this question, recall Fig. 10.1 which illustrated the shock wave and thin shock layer on a 15° wedge at Mach 36. An elaboration of this picture is given in Fig. 10.8 which shows the streamline pattern for the same Mach 36 flow over the same wedge. Here, upstream of the shock wave, we see straight, parallel streamlines in the horizontal free stream direction; downstream of the shock wave, the streamlines are also straight but parallel to the wedge surface inclined at a 15° angle. Now imagine that you examine Fig. 10.8 from a distance, say from across the room. Because the shock wave lies so close to the surface at hypersonic speeds, Fig. 10.8 "looks" as if the incoming flow is directly impinging on the wedge surface, and

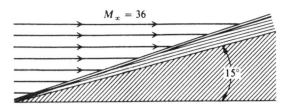

FIGURE 10.8   Streamlines in the thin hypersonic shock layer.

then is running parallel to the surface downstream—*precisely the picture Newton drew in 1687.* Therefore, the geometric picture of hypersonic flowfields has some characteristics which closely approximate Newtonian flow; Newton's model had to wait for more than two-and-a-half centuries before it came into its own. By this reasoning, Eq. (10.2b) should approximate the surface pressure coefficient in hypersonic flow. Indeed, it has been used extensively for this purpose since the early 1950's.

In applying Eq. (10.2) to hypersonic bodies, θ is taken as the local deflection angle, i.e., the angle between the tangent to the surface and the free stream. Clearly, Newtonian theory is a *local surface inclination method* where C_p depends only on the local surface deflection angle; it does not depend on any aspect of the surrounding flowfield. To be specific, consider Fig. 10.9(a) which shows an arbitrarily shaped two-dimensional body. Assume that we wish to estimate the pressure at point P on the body surface. Draw a line tangent to the body at point P; the angle between this line and the free stream is denoted by θ. Hence, from Newtonian theory, the value of C_p at this point is given by $C_p = 2 \sin^2 \theta$. Now consider a three-dimensional body such as sketched in Fig. 10.9(b). We wish to estimate the pressure at an arbitrary point P on this body. Draw a unit normal vector $\bar{\mathbf{n}}$ to the surface at point P. Consider the free stream velocity as a vector $\overline{\mathbf{V}}_\infty$. Then by definition of the vector dot product, and using a trigonometric identity, we obtain

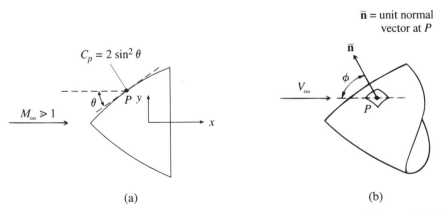

FIGURE 10.9   Geometry for Newtonian applications (a) two-dimensional flow and (b) three-dimensional flow.

$$\overline{\mathbf{V}}_\infty \cdot \overline{\mathbf{n}} = |V_\infty| \cos \phi = |V_\infty| \sin \left(\frac{\pi}{2} - \phi\right) \tag{10.3}$$

where ϕ is the angle between $\overline{\mathbf{n}}$ and $\overline{\mathbf{V}}_\infty$. The vectors $\overline{\mathbf{n}}$ and $\overline{\mathbf{V}}_\infty$ define a plane, and in that plane the angle $\theta = \pi/2 - \phi$ is the angle between a tangent to the surface and the free-stream direction. Thus, from Eq. (10.3)

$$\overline{\mathbf{V}}_\infty \cdot \overline{\mathbf{n}} = |V_\infty| \sin \theta$$

or

$$\sin \theta = \frac{\overline{\mathbf{V}}_\infty}{|\overline{\mathbf{V}}_\infty|} \cdot \overline{\mathbf{n}} \tag{10.4}$$

The Newtonian pressure coefficient at point P on the three-dimensional body is then $C_p = 2 \sin^2 \theta$, where θ is given by Eq. (10.4).

In the Newtonian model of fluid flow, the particles in the free stream impact only on the frontal area of the body; they cannot curl around the body and impact on the back surface. Hence, for that portion of a body which is in the *shadow* of the incident flow, such as the shaded region sketched in Fig. 10.10, no impact pressure is felt. Hence, over this shadow region it is consistent to assume that $p = p_\infty$, and therefore $C_p = 0$, as indicated in Fig. 10.10.

It is instructive to examine Newtonian theory applied to a flat plate, as shown in Fig. 10.11. Here, a two-dimensional flat plate with chord length, c, is at an angle of attack, α, to the free-stream. Since we are not including friction, and because surface pressure always acts normal to the surface, the resultant aerodynamic force is perpendicular to the plate. In turn, N is resolved into lift and drag, denoted by L and D, respectively, as shown in Fig. 10.11. Defining the normal force coefficient as $C_N = N/q_\infty S$ where $S = (c)(1)$, we can readily calculate C_N by integrating the pressure coefficients over the lower and upper surfaces. We obtain

$$C_N = 2 \sin^2 \alpha \tag{10.5}$$

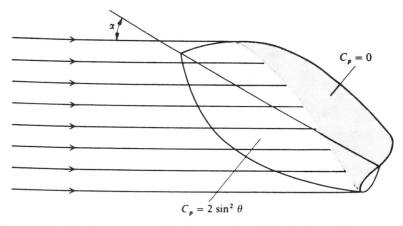

$$C_p = 2 \sin^2 \theta$$

FIGURE 10.10 Shadow region on the leeward side of a body from Newtonian theory.

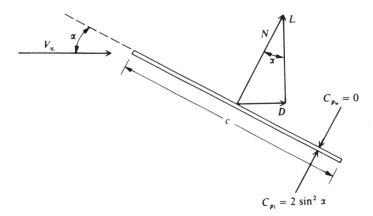

FIGURE 10.11   Flat plate at angle of attack. Illustration of aerodynamic forces.

From the geometry of Fig. 10.11, we see that the lift and drag coefficients, defined as $C_L = L/q_\infty S$ and $C_D = D/q_\infty S$, respectively, where $S = (c)(1)$, are given by

$$C_L = C_N \cos \alpha \qquad (10.6)$$

and

$$C_D = C_N \sin \alpha \qquad (10.7)$$

Substituting Eq. (10.5) into Eqs. (10.6) and (10.7), we obtain

$$C_L = 2 \sin^2 \alpha \cos \alpha \qquad (10.8)$$

and

$$C_D = 2 \sin^3 \alpha \qquad (10.9)$$

Finally, from the geometry of Fig. 10.11, the *lift-to-drag ratio* is given by

$$\frac{L}{D} = \cos \alpha \qquad (10.10)$$

These results are shown in Fig. 10.12, where, L/D, C_L, and C_D are plotted versus angle of attack α. From this figure, note the following aspects:

i) The value of L/D increases monotonically as α is decreased. Indeed, $L/D \rightarrow \infty$ as $\alpha \rightarrow 0$. However, this is misleading; when skin friction is added to this picture, D becomes finite at $\alpha = 0$, and then $L/D \rightarrow 0$ as $\alpha \rightarrow 0$.

ii) The lift curve peaks at about $\alpha \approx 55°$. (To be exact, it can be shown from Newtonian theory that the maximum C_L occurs at $\alpha = 54.7°$.) It is interesting to note that $\alpha \approx 55°$ for maximum lift is fairly realistic; the maximum lift coefficient for many practical hypersonic vehicles occurs at angles of attack in this neighborhood.

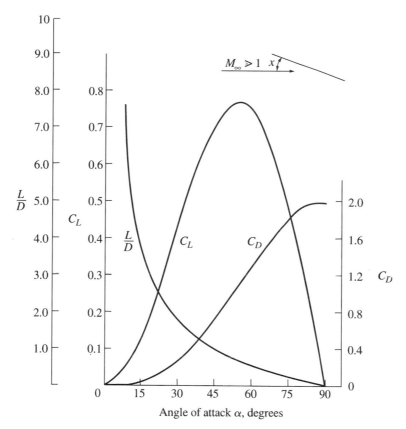

FIGURE 10.12 Newtonian results for a flat plate.

iii) Examine the lift curve at low angle of attack, say in the range of α from 0–15°. Note that the variation of C_L with α is very *nonlinear*. This is in direct contrast to the familiar result for subsonic and supersonic flow, where for thin bodies at small α, the lift curve is a linear function of α. (Recall, for example, that the theoretical lift slope from incompressible thin airfoil theory is 2π per radian.) Hence, the nonlinear lift curve shown in Fig. 10.12 is a graphic demonstration of the nonlinear nature of hypersonic flow.

Consider two other basic aerodynamic bodies, the circular cylinder of infinite span and the sphere. Newtonian theory can be applied to estimate the hypersonic drag coefficients for these shapes. The results are for a circular cylinder of infinite span: $C_D = 4/3$ ($C_D = D/q_\infty S$ and $S = 2R$, where R = radius of cylinder). For a sphere $C_D = 1$ ($C_D = D/q_\infty S$ and $S = \pi R^2$, where R = radius of sphere).

To predict the pressure distribution over blunt bodies, Eq. (10.2b) can be improved by replacing the coefficient 2 with the pressure coefficient at the stagnation point. This leads to the *modified Newtonian law*

$$C_p = C_{p_{max}} \sin^2 \theta \qquad\qquad (10.11)$$

where $C_{p_{\max}}$ is the maximum value of the pressure coefficient evaluated at a stagnation point behind a normal shock wave, i.e.,

$$C_{p_{\max}} = \frac{p_{O_2} - p_\infty}{\frac{1}{2}\rho_\infty V_\infty^2} \tag{10.12}$$

where p_{O_2} is the total pressure behind a normal shock wave at the free stream Mach number. From exact normal shock wave theory,

$$\frac{p_{O_2}}{p_\infty} = \left[\frac{(\gamma + 1)^2 M_\infty^2}{4\gamma M_\infty^2 - 2(\gamma - 1)}\right]^{\gamma/(\gamma - 1)} \left[\frac{1 - \gamma + 2\gamma M_\infty^2}{\gamma + 1}\right] \tag{10.13}$$

Noting that $1/2\,\rho_\infty V_\infty^2 = (\gamma/2)p_\infty M_\infty^2$, Eq. (10.12) becomes

$$C_{p_{\max}} = \frac{2}{\gamma M_\infty^2}\left[\frac{p_{O_2}}{p_\infty} - 1\right] \tag{10.14}$$

Combining Eqs. (10.13) and (10.14), we obtain

$$C_{p_{\max}} = \frac{2}{\gamma M_\infty^2}\left\{\left[\frac{(\gamma + 1)^2 M_\infty^2}{4\gamma M_\infty^2 - 2(\gamma - 1)}\right]^{\gamma/(\gamma - 1)} \left[\frac{1 - \gamma + 2\gamma M_\infty^2}{\gamma + 1}\right] - 1\right\} \tag{10.15}$$

For the prediction of pressure distributions over blunt-nosed bodies, modified Newtonian, Eq. (10.11), is considerably more accurate than the straight Newtonian, Eq. (10.2b). This is illustrated in Fig. 10.13 which shows the pressure distribution over a paraboloid at Mach 8. The solid line is an exact finite difference solution of the blunt body flowfield; the solid symbols are the modified Newtonian results from

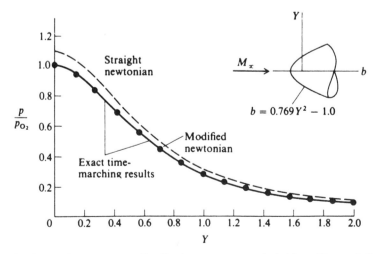

FIGURE 10.13 Surface pressure distribution over a paraboloid at $M_\infty = 8.0$; p_{O_2} is the total pressure behind a normal shock wave at $M_\infty = 8.0$.

Eqs. (10.11) and (10.15). Note the excellent agreement, particularly over the forward portion of the nose. The dashed line is the straight Newtonian result from Eq. (10.2); it lies 9% above the exact result.

10.3   OTHER SIMPLE HYPERSONIC PRESSURE DISTRIBUTION THEORIES

There are two other, well-established, simplified theories that yield reasonable engineering results for surface pressure distributions on hypersonic bodies, namely the tangent-wedge/cone method, and the shock-expansion method. Both will be discussed here.

The *tangent-wedge method* works as follows. Consider the two-dimensional body shown as the hatched area in Fig. 10.14. Assume that the nose of the body is pointed and that the local surface inclination angle θ at all points along the surface is less than the maximum deflection angle for the free stream Mach number. Consider point i on the surface of the body, where we wish to calculate the pressure. The local deflection angle at point i is θ_i. Imagine a line drawn tangent to the body at point i; this line makes an angle θ_i with respect to the free stream and can be imagined as the surface of an equivalent wedge with a half angle of θ_i, as shown by the dashed line in Fig. 10.14. The tangent-wedge approximation assumes that the pressure at point i is the same as the surface pressure on the equivalent wedge at the free stream Mach number, M_∞, that is, p_i is obtained directly from the exact oblique relations for a deflection angle of θ_i and a Mach number of M_∞.

The *tangent-cone method* for application to axisymmetric bodies is analogous to the tangent-wedge method and is illustrated in Fig. 10.15. Consider point i on the body; a line drawn tangent to this point makes the angle θ_i with respect to the free stream. Shown as the dashed line in Fig. 10.15, this tangent line can be imagined as the surface of an equivalent cone, with a semiangle of θ_i. The tangent-cone ap-

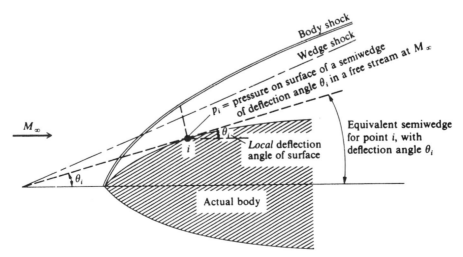

FIGURE 10.14   Illustration of the tangent-wedge method.

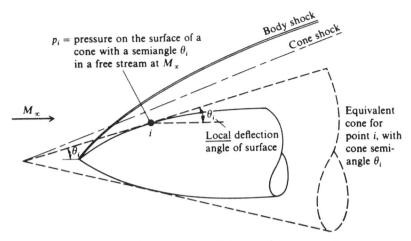

FIGURE 10.15 Illustration of the tangent-cone method.

proximation assumes that the pressure at point i is the same as the surface pressure on the equivalent cone at a Mach number of M_∞.

Both the tangent-wedge and tangent-cone methods are very straightforward. However, they are approximate methods, not based on any theoretical grounds. We cannot *derive* these methods from a model of the flow to which basic mechanical principles are applied, in contrast to the theoretical basis for Newtonian flow. Nevertheless, the tangent-wedge and tangent-cone methods frequently yield reasonable results at hypersonic speeds. Several, physically based reasons for this are given in Anderson (1989).

The *shock-expansion method* is carried out as follows. Consider the hypersonic flow over a sharp-nosed, two-dimensional body with an attached shock wave at the nose, as sketched in Fig. 10.16. The deflection angle at the nose is θ_n. The essence of the shock-expansion theory is as follows. First, assume the nose is a wedge with semiangle θ_n, and calculate M_n and p_n behind the oblique shock at the nose by means of exact oblique-shock theory. Second, assume a local Prandtl–Meyer expansion along the surface downstream of the nose. We wish to calculate the pressure at point i, p_i. To do this, we must first obtain the local Mach number at point i, M_i. This is obtained from the Prandtl–Meyer function, assuming an expansion through the deflection angle $\Delta\theta = \theta_n - \theta_i$

$$
-\Delta\theta = \sqrt{\frac{\gamma + 1}{\gamma + 1}} \left[\tan^{-1} \sqrt{\frac{\gamma - 1}{\gamma + 1}(M_n^2 - 1)} - \tan^{-1}\sqrt{\frac{\gamma - 1}{\gamma + 1}(M_i^2 - 1)} \right]
$$

$$
- [\tan^{-1}\sqrt{M_n^2 - 1} - \tan^{-1}\sqrt{M_i^2 - 1}] \tag{10.16}
$$

In Eq. (10.16), M_i is the only unknown; M_n is known from the first step above and $\Delta\theta = \theta_n - \theta_i$ is a known geometric quantity. Of course, for air with $\gamma = 1.4$, tables for the Prandtl–Meyer function abound [see Anderson (1990)], and in such a case the tables would be used to calculate M_i rather than attempting to solve Eq. (10.16) implicitly for M_i. Third, calculate p_i from the isentropic flow relation

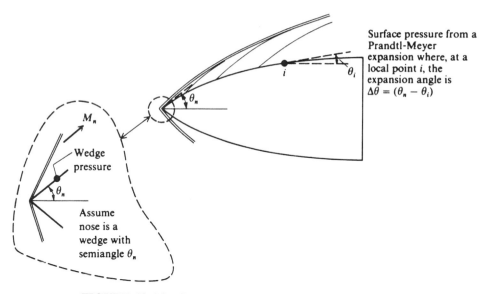

FIGURE 10.16   Illustration of the shock-expansion method.

$$\frac{p_i}{p_n} = \left\{ \frac{1 + [(\gamma - 1)/2]M_n^2}{1 + [(\gamma - 1)/2]M_i^2} \right\}^{\gamma/(\gamma - 1)}$$ (10.17)

[again, for air with $\gamma = 1.4$, the isentropic flow tables, such as found in Anderson (1990), can be used to obtain p_i in a more convenient manner].

The shock–expansion method can also be applied to bodies of revolution. The method is essentially the same as shown in Fig. 10.16, except now θ_n is assumed to be the semiangle of a cone, and M_n and p_n at the nose are obtained from the exact Taylor–Maccoll cone results. Then, the Prandtl–Meyer expansion relations are applied locally downstream of the nose. This implies that the flow downstream of the nose is locally two-dimensional, which assumes that the divergence of streamlines in planes tangential to the surface is much smaller than the divergence of streamlines in planes normal to the surface. For bodies of revolution at 0° angle of attack, this condition is usually met.

10.4   MACH NUMBER INDEPENDENCE

At high Mach numbers, certain aerodynamic quantities such as pressure coefficient, lift and wave–drag coefficients, and flowfield structure (such as shock wave shapes and Mach wave patterns) become essentially independent of Mach number. This is the essence of the *Mach number independence principle* in hypersonic flow. This principle can be derived from the governing continuity, momentum, and energy equations of fluid flow, specialized to high, hypersonic Mach numbers [see Anderson (1989) and Cox and Crabtree (1965) for more details].

Mach number independence is demonstrated by the data shown in Fig. 10.17. In Fig. 10.17, the measured drag coefficients for spheres and for a large-angle cone-

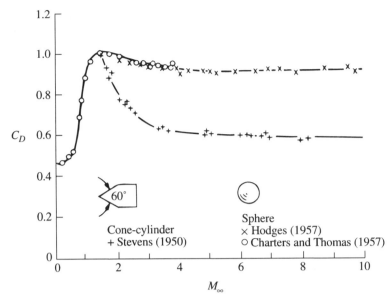

FIGURE 10.17 Drag coefficient for a sphere and a cone-cylinder from ballistic range measurements; an illustration of Mach number independence. (From Cox and Crabtree, 1965.)

cylinder are plotted versus Mach number, cutting across the subsonic, supersonic, and hypersonic regimes. Note the large drag rise in the subsonic regime associated with the drag–divergence phenomena near Mach 1 and the decrease in C_D in the supersonic regime beyond Mach 1. Both of these variations are expected and well understood [see Anderson (1990)]. For our purposes in the present section, note in particular the variation of C_D in the hypersonic regime; for both the sphere and cone-cylinder, C_D approaches a plateau and becomes relatively independent of Mach number as M_∞ becomes large.

10.5 HYPERSONIC SIMILARITY

For the special case of *slender* hypersonic bodies at small angle of attack, there exists an approximate similarity rule at hypersonic speeds. This rule can be derived by assuming small perturbations in the flowfield and reducing the governing inviscid flow equations to a simpler, albeit still nonlinear, system called the *hypersonic small disturbance equations*. See Anderson (1990) for a derivation and presentation of these equations. Here, we will simply state and discuss the resulting principle of hypersonic similarity.

We define the *hypersonic similarity parameter*, K, as $K \equiv M_\infty \tau$ where τ is the body *slenderness ratio*, defined by the ratio of the base of the body to its length. (The smaller is τ, the more slender is the body.) For slender bodies (small τ) at hypersonic speeds, the principle of hypersonic similarity states that

$$\frac{C_p}{\tau^2} = f(K, \gamma) \tag{10.18}$$

For two-dimensional shapes

$$\frac{C_L}{\tau^2} = f_2\left(\gamma, K, \frac{\alpha}{\tau}\right) \tag{10.19}$$

$$\frac{C_D}{\tau^3} = f_3\left(\gamma, K, \frac{\alpha}{\tau}\right) \tag{10.20}$$

For three-dimensional bodies

$$\frac{C_L}{\tau} = F_1(\gamma, K, \alpha/\tau) \tag{10.21}$$

$$\frac{C_D}{\tau^2} = F_2(\gamma, K, \alpha/\tau) \tag{10.22}$$

In Eqs. (10.18)–(10.22), C_p is the pressure coefficient, C_L and C_D are the lift and drag coefficients, respectively [for a two-dimensional shape based on planform area (per unit span) and for a three-dimensional body based on base area]. The principle of hypersonic similarity states that affinely related bodies with the same values of γ, $M_\infty\tau$, and α/τ will have: 1) the same values of C_L/τ^2 and C_D/τ^3 for two-dimensional flows, when referenced to planform area, and 2) the same values of C_L/τ and C_D/τ^2 for three-dimensional flows when referenced to base area. The validity of the hypersonic similarity principle is verified by the results shown in Figs. 10.18 and 10.19, obtained from the work of Neice and Ehret (1951). Consider first Fig. 10.18(a), which shows the variation of C_p/τ^2 as a function of distance downstream of the nose of a slender ogive-cylinder (as a function of $X = x/l$, expressed in percent of nose length). Two sets of data are presented, each for a different M_∞ and τ, but such that the product $K \equiv M_\infty\tau$ is the same value, namely 0.5. The data are exact calculations made by the *Method of Characteristics*. Hypersonic similarity states that the two sets of data should be identical, which is clearly the case shown in Fig. 10.18(a).

A similar comparison is made in Fig. 10.18(b), except for a higher value of the hypersonic similarity parameter, namely $K = 2.0$. The conclusion is the same; the data for two different values of M_∞ and τ, but with the same K, are identical. An interesting sideline is also shown in Fig. 10.18(b). Two different Method of Characteristics calculations are made—one assuming irrotational flow (the solid line), and the other treating rotational flow (the dashed line). There are substantial differences in implementing the Method of Characteristics for these two cases. In reality, the flow over the ogive-cylinder is rotational because of the slightly curved shock wave over the nose. The effect of rotationality is to increase the value of C_p, as shown in Fig. 10.18(b). This effect is noticeable for the high value of $K = 2$ in Fig. 10.18(b). However, Neice and Ehret (1951) state that no significant differences between the rotational and irrotational calculations resulted for the low value of $K = 0.5$ in Fig. 10.18(a), which is why only one curve is shown. One can conclude from this comparison the almost intuitive fact that the effects of rotationality become more important as M_∞, τ, or both are progressively increased. However, the main reason for bringing up the matter of rotationality is to state that the principle of hypersonic similarity holds for *both* irrotational and rotational flows. This is clearly demon-

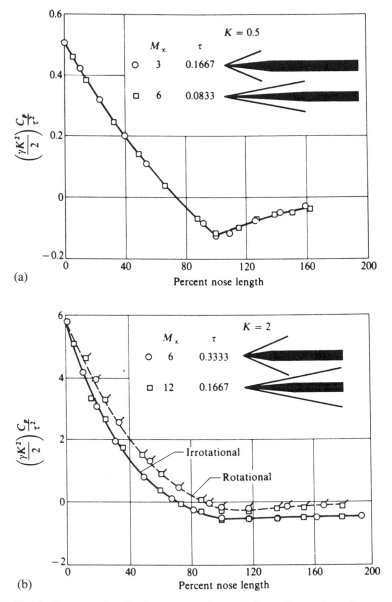

FIGURE 10.18 Pressure distributions over ogive cylinders; illustration of hypersonic similarity. (a) $K = 0.5$ and (b) $K = 2.0$. (Reprinted with permission, Neice, S. E. and Ehret, D. M., "Similarity Laws for Slender Bodies of Revolution in Hypersonic Flows," *J. Aero. Sci.*, Vol. 18, No. 8, pp. 527–530, 568, Copyright © 1951.)

strated in Fig. 10.18(b), where the data calculated for irrotational flow for two different values of M_∞ and τ (but the same K) fall on the same curve, and the data calculated for rotational flow for the two different values of M_∞ and τ (but the same K) also fall on the same curve (but a different curve than the irrotational results).

Figure 10.18(a) and (b) contain results at zero angle of attack. For the case of bodies at angle of attack, α/τ is an additional similarity parameter. This, as well as

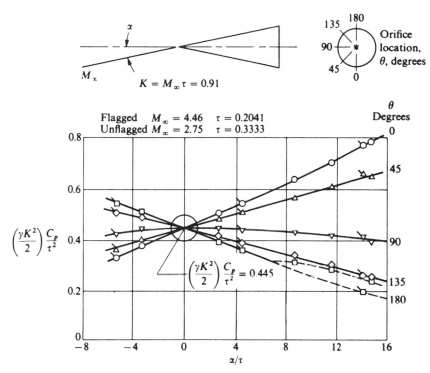

FIGURE 10.19 Cone pressure at angle of attack correlated by hypersonic similarity. (Reprinted with permission, Neice, S. E. and Ehret, D. M., "Similarity Laws for Slender Bodies of Revolution in Hypersonic Flows," *J. Aero. Sci.*, Vol. 18, No. 8, pp. 527–530, 568, Copyright © 1951.)

the general principle of hypersonic similarity, is experimentally verified by the wind tunnel data shown in Fig. 10.19. Neice and Ehret (1951) reported some experimental pressure distributions over two sharp, right-circular cones at various angles of attack obtained in the NACA Ames 10 by 14 in. supersonic wind tunnel. The free stream Mach numbers were 4.46 and 2.75, and the cones had different slenderness ratios such that $K = 0.91$ for both cases. Since the flow was conical, the values of C_p on the surface were constant along a given ray from the nose, but, because of the angle of attack, C_p varied from one ray to another around the cone as a function of angular location. Note in Fig. 10.19 that the data along any given ray for the two different values of M_∞ and τ (but both such that $K = 0.91$) fall on the same curve when plotted versus α/τ. Hence, the data in Fig. 10.19 is a direct experimental verification of hypersonic similarity for bodies at angle of attack. (Note that at $\alpha = 0$, all the curves pass through the value of C_p predicted from exact cone theory.)

10.6 HYPERSONIC BOUNDARY LAYERS AND AERODYNAMIC HEATING

In the next two sections, we will deal with hypersonic *viscous* flows. For a calorically perfect gas (constant specific heats), the hypersonic boundary layer is governed by the same equations as those for classical compressible boundary layer flow. See

Anderson (1989) for a development of these equations and how they are applied to hypersonic flows. Some of the practical results from this theory are discussed below.

A familiar result for the local skin friction coefficient on a flat plate in incompressible laminar flow is (see Chap. 4)

$$C_f \text{ (incompressible)} = \frac{0.664}{\sqrt{\text{Re}_x}} \tag{10.23}$$

where $C_f = \tau_w / 1/2 \rho_e U_e^2$, ρ_e and U_e are the density and velocity at the edge of the boundary layer, τ_w is the local shear stress, and Re_x is the local Reynolds number defined as $\text{Re}_x = \rho_e U_e x / \mu_e$. For *compressible flow*, the above result is modified as

$$C_f \text{ (compressible)} = \frac{F(M_e, \text{ Pr}, \gamma, T_w/T_e)}{\sqrt{\text{Re}_x}} \tag{10.24}$$

where the constant 0.664 in Eq. (10.23) is replaced by a function of Mach number, M_e, Prandtl number, $\text{Pr} = \mu_e c_p / k_e$, ratio of specific heats, $\gamma = c_p/c_v$, and the wall-to-freestream temperature ratio. This function, F, in Eq. (10.24) must be found by a numerical solution of the boundary layer flow equations. In the same vein, defining a heat transfer coefficient as the Stanton number C_h, as

$$C_h = \frac{q_w}{\rho_e U_e (h_{aw} - h_w)} \tag{10.25}$$

where q_w is the local heat transfer rate, h_{aw} is the *adiabatic wall enthalpy* (that wall enthalpy for *no* heat transfer), and h_w is the wall enthalpy, we find for laminar compressible flow over a flat plate,

$$C_h \text{ (compressible)} = \frac{G(M_e, \text{ Pr}, \gamma, T_w/T_e)}{\sqrt{\text{Re}_x}} \tag{10.26}$$

where the function G is also found from a numerical solution of the compressible boundary layer equations.

Computed variations of F and G with Mach numbers are given in Figs. 10.20 and 10.21, respectively. Note that the effects of increasing Mach number are to reduce both C_f and C_h, well into the hypersonic range. This decrease in C_f and C_h goes hand-in-hand with the fact that the boundary layer thickness increases as Mach number increases. The results in Figs. 10.20 and 10.21 were calculated by Van Driest (1952), a classic in the field.

Typical velocity profiles through hypersonic laminar boundary layers on an insulated flat plate, also calculated by Van Driest, are shown in Fig. 10.22. Note that at a given x station at a given Re_x, the boundary layer thickness increases markedly as M_e is increased to hypersonic values. *This clearly demonstrates one of the most important aspects of hypersonic boundary layers, namely, that the boundary layer thickness becomes large at large Mach numbers.* Indeed, the laminar boundary layer thickness varies approximately as M_e^2. Figure 10.23 illustrates the temperature profiles for the same case as Fig. 10.22. Note the obvious physical trend that, *as M_e*

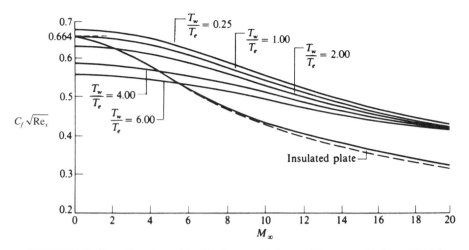

FIGURE 10.20   Flat plate skin friction coefficients. (From van Driest, 1952.)

increases to large hypersonic values, the temperatures increase markedly. Also note in Fig. 10.23 that at the wall ($y = 0$), $(\partial T/\partial y)_w = 0$, as it should be for an insulated surface ($q_w = 0$). Figures 10.24 and 10.25 also contain results by Van Driest, but now for the case of heat transfer to the wall. Such a case is called a *cold wall* case, because $T_w < T_{aw}$. (The opposite case would be a *hot wall*, where heat is transferred from the wall into the flow; in this case, $T_w > T_{aw}$.) For the results shown in Figs. 10.24 and 10.25, $T_w/T_e = 0.25$ and Pr $= 0.75 =$ constant. Figure 10.24 shows velocity profiles for various different values of M_e, again demonstrating the rapid growth in boundary layer thickness with increasing M_e. In addition, the effect of a cold wall on the boundary layer thickness can be seen by comparing Figs. 10.22 and 10.24. For example, consider the case of $M_e = 20$ in both figures. For the insulated wall at Mach 20 (Fig. 10.22), the boundary layer thickness reaches out beyond a value of $(y/x)\sqrt{\text{Re}_x} = 60$, whereas for the cold wall at Mach 20 (Fig.

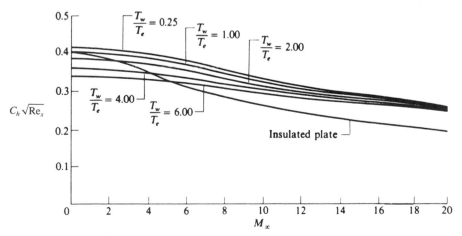

FIGURE 10.21   Flat plate Stanton numbers. (From van Driest, 1952.)

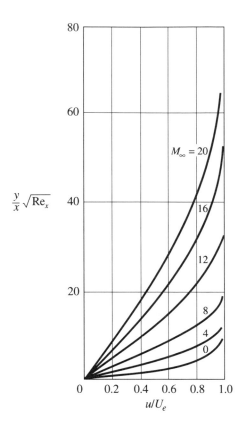

FIGURE 10.22 Velocity profiles in a compressible laminar boundary layer over an insulated flat plate. (From van Driest, 1952.)

10.24), the boundary layer thickness is slightly above $(y/x)\sqrt{\text{Re}_x} = 30$. *This illustrates the general fact that the effect of a cold wall is to reduce the boundary layer thickness.* This trend is easily explainable on a physical basis when we examine Fig. 10.25, which illustrates the temperature profiles through the boundary layer for the cold wall case. Comparing Figs. 10.23 and 10.25, we note that, as expected, the temperature levels in the cold wall case are considerably lower than in the insulated case. In turn, because the pressure is the same in both cases, we have from the equation of state $p = \rho RT$, that the *density in the cold wall case is much higher*. If the density is higher, the mass flow within the boundary layer can be accommodated within a smaller boundary layer thickness, hence the effect of a cold wall is to *thin* the boundary layer. Also note in Fig. 10.25 that, starting at the outer edge of the boundary layer and going toward the wall, the temperature first increases, reaches a peak somewhere within the boundary layer, and then decreases to its prescribed cold wall value T_w. The peak temperature inside the boundary layer is an indication of the amount of viscous dissipation occurring within the boundary layer. Figure 10.25 clearly demonstrates the rapidly growing effect of this viscous dissipation as M_e increases—yet another basic aspect of hypersonic boundary layers.

The physical results shown in Figs. 10.22–10.25 are so important that we summarize them below:

i) Boundary thickness δ increases rapidly with M_e.

ii) Temperature inside the boundary layer increases rapidly with M_e due to viscous dissipation.

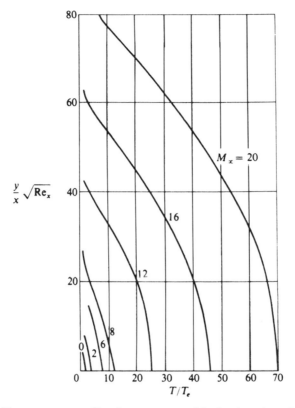

FIGURE 10.23 Temperature profiles in a compressible laminar boundary layer over an insulated flat plate. (From van Driest, 1952.)

iii) Cooling the wall reduces δ.

iv) Both C_f and C_h decrease as M_e increases.

v) Both C_f and C_h increase as the wall is cooled.

Let us consider some further aspects of aerodynamic heating at hypersonic speeds. Return to the definition of C_h given by Eq. (10.25). Note from this definition that aerodynamic heating to the surface is given by

$$q_w = \rho_e U_e C_h (h_{aw} - h_w) \tag{10.27}$$

This equation is important, because it emphasizes that the *driving potential* for aerodynamic heating to the surface is the enthalpy difference $(h_{aw} - h_w)$. We will find this to be the case for virtually all cases in aerodynamic heating to high speed vehicles. In turn, the calculation of the adiabatic wall enthalpy, h_{aw}, is an important consideration before Eq. (10.27) can be used to obtain q_w. In most engineering related calculations, the value of h_{aw} (and of $T_{aw} = h_{aw}/c_p$) is expressed in terms of the *recovery factor*, r, defined as

$$h_{aw} = h_e + r \frac{U_e^2}{2} \tag{10.28}$$

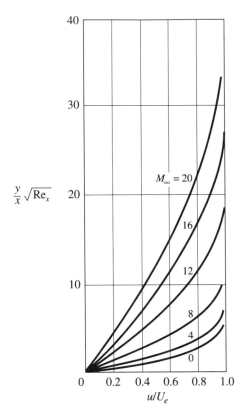

FIGURE 10.24 Velocity profiles in a laminar, compressible boundary layer over a cold flat plate. (From van Driest, 1952.)

At the outer edge of the boundary layer, we have

$$h_o = h_e - \frac{U_e^2}{2} \tag{10.29}$$

where h_o is the total enthalpy in the inviscid flow outside the boundary layer. Substituting Eq. (10.29) into (10.28) we have

$$h_{aw} = h_e + r(h_o - h_e)$$

$$r = \frac{h_{aw} - h_e}{h_o - h_e} \tag{10.30}$$

For a calorically perfect gas, where $h = c_p T$, Eq. (10.30) can be written as

$$r = \frac{T_{aw} - T_e}{T_o - T_e} \tag{10.31}$$

For incompressible flow, the value of r is related to the Prandtl number as $r = \sqrt{\text{Pr}}$. Exact results for r for compressible flow are shown in Fig. 10.26, obtained from Ref. 12, and are compared with the $\sqrt{\text{Pr}} = \sqrt{0.715} = 0.845$. Note that r decreases as M_e increases through the hypersonic regime. However, also note that

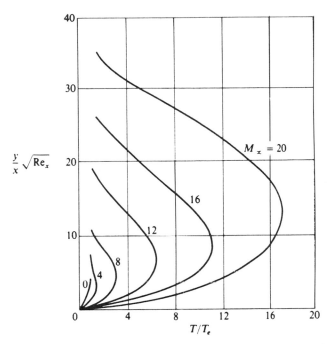

FIGURE 10.25 Temperature profiles in a laminar, compressible boundary layer over a cold flat plate. (From van Driest, 1952.)

the ordinate is an expanded scale, showing that r decreases by only 2.4 percent from $M_e = 0$ to 16. Hence, for all practical purposes, we can assume for laminar hypersonic flow over a flat plate that

$$r = \sqrt{Pr} \qquad (10.32)$$

With Eqs. (10.32) and (10.30), we can readily estimate h_{aw} for use in Eq. (10.27). To complete an engineering analysis of q_w using Eq. (10.27), we must obtain an

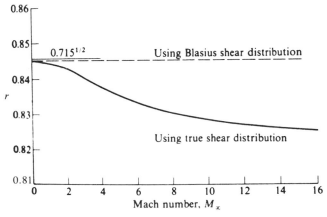

FIGURE 10.26 Comparison of exact and approximate recovery factor for laminar flow over a flat plate.

estimate of C_h. We can estimate C_h using Reynolds analogy. The general, exact value for Reynolds analogy, which links C_f and C_h, is given by

$$\frac{C_h}{C_f} \text{ (compressible)} = \frac{G}{F} = f\left(M_e, \text{Pr}, \gamma, \frac{T_w}{T_e}\right) \tag{10.33}$$

Results for this ratio are given in Fig. 10.27, obtained from Van Driest (1952). Note that the ratio C_f/C_h decreases as M_e increases across the hypersonic regime. However, again note that the ordinate is an expanded scale, and C_f/C_h decreases by only 2% from $M_e = 0$ to 16. Thus, the incompressible result is a reasonable approximation at hypersonic speeds, namely

$$\frac{C_h}{C_f} = \frac{1}{2}\,\text{Pr}^{-2/3} \tag{10.34}$$

Consider now the heat transfer to a stagnation point. For the stagnation heating on a two-dimensional cylinder oriented perpendicular to the flow, we have

$$\text{Cylinder: } q_w = 0.57\,\text{Pr}^{-0.6}\,(\rho_e\mu_e)^{1/2}\sqrt{\frac{dU_e}{dx}}\,(h_{aw} - h_w) \tag{10.35}$$

where ρ_e and μ_e are the density and viscosity coefficients at the edge of the stagnation point boundary layer, and dU_e/dx is the inviscid velocity gradient at the stagnation point. The corresponding expression for an axisymmetric stagnation point, such as a sphere, is

$$\text{Sphere: } q_w = 0.763\,\text{Pr}^{-0.6}\,(\rho_e\mu_e)^{1/2}\sqrt{\frac{dU_e}{dx}}\,(h_{aw} - h_w) \tag{10.36}$$

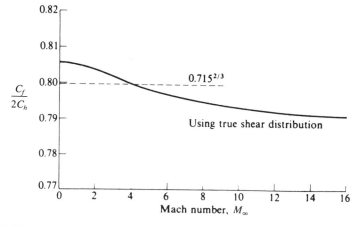

FIGURE 10.27 Comparison of exact and approximate Reynolds analogy factor for laminar flow over a flat plate. (From van Driest, 1952.)

Compare Eq. (10.35) for the two-dimensional cylinder with Eq. (10.36) for the axisymmetric sphere. The equations are the same except for the leading coefficient, which is higher for the sphere. Everything else being the same, this demonstrates that stagnation point heating to a sphere is larger than to a two-dimensional cylinder. Why? The answer lies in a basic difference between two- and three-dimensional flows. In a two-dimensional flow, the gas has only two directions to move when it encounters a body—up or down. In contrast, in an axisymmetric flow, the gas has three directions to move—up, down, and sideways, hence the flow is somewhat *relieved*, i.e., in comparing two- and three-dimensional flows over bodies with the same longitudinal section (such as a cylinder and a sphere), there is a well-known three-dimensional *relieving effect* for the three-dimensional flows. As a consequence of this relieving effect, the boundary layer thickness, δ, at the stagnation point is smaller for the sphere than for the cylinder. In turn, the temperature gradient at the wall, $(\partial T/\partial y)_w$, which is $O(T_e/\delta)$, is larger for the sphere. Since $q_w = -k(\partial T/\partial y)_w$, then q_w is larger for the sphere. This confirms the comparison between Eqs. (10.35) and (10.36).

The above results for aerodynamic heating to a stagnation point have a stunning impact on hypersonic vehicle design—namely, they impose the requirement for the vehicle to have a blunt, rather than a sharp nose. To see this more clearly, we note that at hypersonic speeds, the velocity gradient at the stagnation point can be calculated from Newtonian theory as

$$\frac{dU_e}{dx} = \frac{1}{R}\sqrt{\frac{2(p_e - p_\infty)}{\rho_e}} \tag{10.37}$$

where R is the nose radius, and p_e and p_∞ are the edge and freestream values, respectively, of the pressure. See Anderson (1989) for a derivation of Eq. (10.37).

Examine Eqs. (10.35) and (10.36) in light of Eq. (10.37), and we see that

$$q_w \propto \frac{1}{\sqrt{R}} \tag{10.38}$$

This states that *stagnation-point heating varies inversely with the square root of the nose radius, hence, to reduce the heating, increase the nose radius*. This is the reason why the nose and leading edge regions of hypersonic vehicles are blunt, otherwise the severe aerothermal conditions in the stagnation region would quickly melt a sharp leading edge. Indeed, for earth entry bodies, such as the Mercury and Apollo space vehicles, the only viable design to overcome aerodynamic heating is a very blunt body.

An approximate method for estimating aerodynamic heating and skin friction in hypersonic flow is the *reference temperature (or reference enthalpy) method*. This method is based on the simple idea of utilizing the formulas obtained from incompressible flow theory, wherein the thermodynamic and transport properties in these formulas are evaluated at some reference temperature indicative of the temperature somewhere inside the boundary layer. This idea was first advanced by Rubesin and Johnson (1949), and was modified by Eckert (1956) to include a reference enthalpy. In this fashion, in some sense the classical incompressible formulas were "corrected'' for compressibility effects. Reference temperature (or reference enthalpy) methods have enjoyed frequent application in engineering-oriented hypersonic analyses, because of their simplicity. For this reason, we briefly describe the approach here.

Consider the *incompressible* laminar flow over a flat plate. The local skin friction and heat transfer coefficients, obtained from classical theory are, respectively

$$C_f = \frac{0.664}{\sqrt{\mathrm{Re}_x}} \tag{10.39}$$

$$C_h = \frac{0.332}{\sqrt{\mathrm{Re}_x}} \mathrm{Pr}^{-2/3} \tag{10.40}$$

where Re and Pr are based on properties at the edge of the boundary layer, that is $\mathrm{Re}_x = \rho_e U_e x / \mu_e$ and $\mathrm{Pr} = \mu_e c_{p_e} / k_e$.

Now consider the *compressible* laminar flow over a flat plate. In the reference temperature method, the compressible local skin friction and heat transfer coefficient are given by expressions analogous to Eqs. (10.39) and (10.40)

$$C_f^* = \frac{0.664}{\sqrt{\mathrm{Re}_x^*}} \tag{10.41}$$

$$C_h^* = \frac{0.332}{\sqrt{\mathrm{Re}_x^*}} \mathrm{Pr}^{*-2/3} \tag{10.42}$$

where C_f^*, C_h^*, Re_x^*, and Pr* are evaluated at a reference temperature T^*. That is,

$$\mathrm{Re}_x^* = \frac{\rho^* U_e x}{\mu^*} \tag{10.43}$$

$$\mathrm{Pr} = \frac{\mu^* c_p^*}{k^*} \tag{10.44}$$

$$C_f^* = \tau_w / \tfrac{1}{2} \rho^* U_e^2 \tag{10.45}$$

$$C_h^* = q_w / \rho^* U_e (h_{aw} - h_w) \tag{10.46}$$

where ρ^*, μ^*, c_p^*, and k^* are evaluated for the reference temperature T^*. From Eqs. (10.24) and (10.26) we know that, for compressible flow, C_f and C_h depend on M_e and T_w/T_e. Hence T^* must be a function of M_e and T_w/T_e. From the above references and White (1992), this function is

$$\frac{T^*}{T_e} = 1 + 0.032 M_e^2 + 0.58 \left(\frac{T_w}{T_e} - 1 \right) \tag{10.47}$$

Return to Fig. 10.20, where the solid curves give the exact solutions for compressible laminar flow over a flat plate. The approximate results obtained from the reference temperature method using Eq. (10.41) where T^* is given by Eq. (10.47) are shown as dashed curves in Fig. 10.20. For most of the curves, the reference temperature method falls directly on the exact results, and hence no distinction can be made between the two sets of results; only for the insulated plate is there some discernible difference, and that is small.

To apply the above results to cones, simply multiply the right-hand sides of Eqs. (10.41) and (10.42) by the *Mangler factor*, $\sqrt{3}$. It makes sense that, everything else being equal, the skin friction and heat transfer to the cone should be higher than the flat plate. For the cone, there is a three-dimensional relieving effect which makes the boundary layer thinner. This, in turn, results in larger velocity and temperature gradients throughout the boundary layer including at the wall and, hence, yields a higher skin friction and heat transfer than in the two-dimensional boundary layer over a flat plate. Also, the idea of the reference temperature method has been carried over to general three-dimensional flows simply by defining R_x^* as the running length Reynolds number along a streamline (where now x denotes distance along the streamline).

For turbulent flow over a flat plate, a reasonable incompressible result is [see White (1992)]

$$C_f = \frac{0.0592}{(\mathrm{Re}_x)^{0.2}} \tag{10.48}$$

Carrying over the reference temperature concept to the turbulent case, the compressible turbulent flat plate skin friction coefficient can be approximated as

$$C_f^* = \frac{0.0592}{(\mathrm{Re}_x^*)^{0.2}} \tag{10.49}$$

where Re_x^* is evaluated at the reference temperature given by Eq. (10.47). The turbulent flat plate heat transfer can be estimated from a form of Reynolds analogy, written as

$$C_h = \frac{C_f}{2s} \tag{10.50}$$

where s is defined as the *Reynolds analogy factor*.

The matter of transition from laminar to turbulent flow at hypersonic speeds is important, and has been the subject of extensive investigation. Although transition has been *well-observed*, it certainly is not *well-understood*, even to the present day. Turbulence, and transition to turbulence, is one of the unsolved problems in basic physics. Our only recourse in aerodynamics is to treat these problems in an approximate, engineering sense, depending always on as large a dose of empirical data as we can find to swallow. This situation is particularly severe at hypersonic speeds, where transition seems to exhibit some peculiar anomalies in comparison to our experience at lower speeds. For simplicity, first consider the picture of transition, as sketched in Fig. 10.28 for flow over a flat surface. The flow starts out at the leading edge as laminar; this laminar flow is highly stable, and any disturbances are

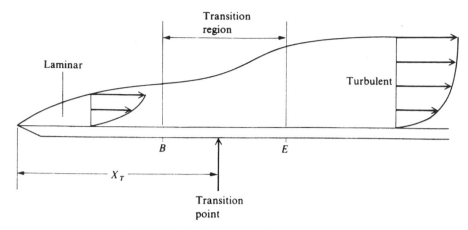

FIGURE 10.28 Schematic of transition.

not amplified. However, at some location downstream, the laminar flow becomes unstable, and any disturbances (say from the free stream or from the surface such as surface roughness) are now amplified. This point is labeled B in Fig. 10.28, for the beginning of transition. As the amplification of disturbances continues in this unstable flow, transition to turbulence takes place, finally becoming fully turbulent at point E in Fig. 10.28, where point E is the end of transition. The region between points B and E is called the *transition region*. Since our knowledge of transition is so imprecise, including our knowledge of the extent of the transition region, engineering analyses frequently assume that transition takes place at a point, labeled the transition point in Fig. 10.28. For purposes of analysis, the flow is assumed laminar upstream of the transition point, and fully turbulent downstream. The location of the transition point is given by x_T in Fig. 10.28, and we define a *transition Reynolds number* as

$$\text{Re}_T = \frac{\rho_e U_e x_T}{\mu_e} \tag{10.51}$$

For the accurate prediction of skin friction and aerodynamic heating to a body, knowledge of the transition Reynolds number is critical, To date, no theory exists for the accurate prediction of Re_T; any knowledge concerning its value for a given situation must be obtained from experimental data. If the desired application is outside the existing data base, then an estimate of Re_T is essentially guesswork.

Given this situation, in the present section we can only discuss some guidelines for transition at hypersonic speeds. Many of our remarks will be influenced by a survey by Stetson (1987). Indeed, Stetson begins by the flat statement that "there is no transition theory," although our data base at hypersonic speeds is sufficient to establish some general trends based on experiment. The hypersonic transition Reynolds number can be expressed functionally as

$$\text{Re}_T = f\left(M_e,\ \theta_c,\ T_w,\ \dot{m},\ \alpha,\ k_R,\ E,\ \frac{\partial p}{\partial x},\ R_N,\ \text{Re}_\infty/\text{ft},\ \frac{x}{R_N},\ V,\ C,\ \frac{\partial w}{\partial z},\ T_0,\ d^*,\ \tau,\ Z \right)$$

where:

M_e is the Mach number at the edge of the boundary layer,

θ_c is a characteristic defining the shape of the body (for a cone, θ_c would be the cone angle),

T_w is the wall temperature,

$\dot{m}$ is mass addition or removal at the surface,

α is the angle of attack,

k_R is a parameter expressing the roughness of the surface,

E is a general term characterizing the *environment* (such as free stream turbulence or acoustic disturbances propagating from the nozzle boundary layer in a wind tunnel),

$\partial p/\partial x$ is the local pressure gradient,

R_N is the radius of a blunt nose tip,

$\mathrm{Re}_\infty/\mathrm{ft}$ is the Reynolds number per foot,

x/R_N is the location of the boundary layer while it is immersed in the entropy layer generated by the nose (effects of the entropy layer can be felt more than a hundred nose radii downstream of the tip),

V is an index of the vibration of the body,

C is the body curvature,

$\partial w/\partial z$ is the cross flow velocity gradient,

T_0 is the stagnation temperature,

$d*$ is a characteristic dimension of the body,

τ is a chemical reaction time, and

Z is an index of the magnitude of chemical reactions taking place in the boundary layer.

One look at this list, and the reader is justified in becoming frustrated. Clearly, the transition Reynolds number is an elusive quantity, and it is no surprise that our knowledge of it is so imprecise. However, the situation is not hopeless; for any given situation, Re_T will be dominated by only a few of the parameters listed above, and the others will be secondary.

In the hypersonic regime, the effect of Mach number on transition is of primary importance, because at high Mach numbers the transition Reynolds number tends to be large, with corresponding large regions of laminar flow existing over hypersonic flight vehicles. Specifically, the Mach number at the edge of the boundary layer, M_e, has a strong influence on the stability of the laminar boundary layer and through this on Re_T. Boundary layer stability theory shows that stability of the laminar boundary layer is generally enhanced by an increasing Mach number, hence Re_T is increased with increased M_e, especially above $M_e = 4$. This is dramatically shown in Fig. 10.29, obtained from Stetson (1987). Here we see a plot of Re_T versus M_e for sharp cones in both wind tunnels and free flight. Clearly, above Mach 4, Re_T increases rapidly with M_e. A virtual rule of thumb places the transition Reynolds number for incompressible flow over a flat plate near 5×10^5; in contrast, at high hypersonic Mach numbers, Re_T can be on the order of 10^8. This effect of Mach number on transition is extremely beneficial. Since skin friction and aerodynamic heating are considerably smaller for laminar in comparison to turbulent flows, the relatively large region of laminar flow that can occur over a body at hypersonic speeds is a very advantageous design feature.

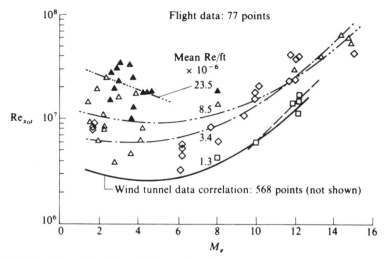

FIGURE 10.29   Transition Reynolds number data on sharp cones from wind tunnels and free flight. (From Stetson, 1987.)

10.7   HYPERSONIC VISCOUS INTERACTION

The laminar boundary layer thickness grows essentially as the square of the Mach number. Hence, hypersonic boundary layers can be orders of magnitude thicker than low speed boundary layers at the same Reynolds number. In turn, this thick hypersonic boundary layer displaces the outer inviscid flow, changing the nature of the inviscid flow. This is called *hypersonic viscous interaction*. For example, inviscid flow over a flat plate is sketched in Fig. 10.30(a); the streamlines are straight and parallel, and the pressure on the surface is constant, as sketched above the streamlines. In contrast, for hypersonic viscous flow with a thick boundary layer, the inviscid streamlines are displaced upward, creating a shock wave at the leading edge as sketched in Fig. 10.30(b). Moreover, the pressure varies over the surface of the flat plate, as sketched above the flow picture in Fig. 10.30(b). This is the source of the viscous interaction. The increased pressure (hence increased density) tends to make the boundary layer thinner than would be expected (although δ is still large on a relative scale), hence the velocity and temperature gradients at the wall are increased. In turn, the skin friction and heat transfer is increased over their values that would exist if a constant pressure equal to p_∞ were assumed. In the viscous interaction, the pressure increase (and the resulting C_f and C_η increases) become more severe closer to the leading edge. The magnitude of the viscous interaction increases as Mach number is increased and Reynolds number is decreased. Therefore, viscous interaction effects are important for slender hypersonic vehicles flying at high Mach numbers and high altitudes.

There are two general classes of hypersonic viscous interactions, *weak* and *strong interaction*. They are defined as follows. Consider the sketch shown in Fig. 10.31, which illustrates the hypersonic viscous flow over a flat plate. Two regions of viscous interaction are illustrated here—the strong interaction region immediately downstream of the leading edge, and the weak interaction region further down-

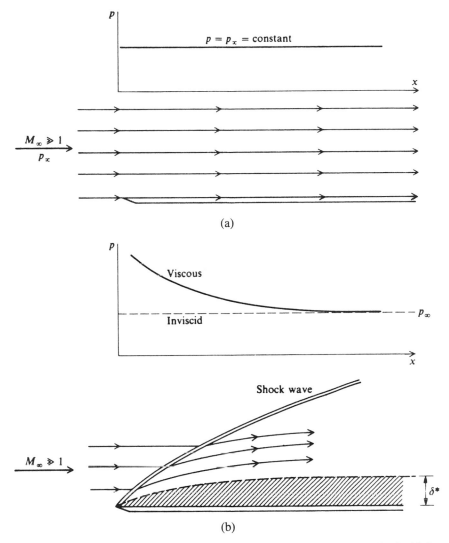

FIGURE 10.30 Illustration of pressure distributions over a flat plate. (a) Inviscid flow and (b) viscous flow.

stream. By definition, the *strong interaction* region is one where the following physical effects occur:

i) In the leading edge region, the rate of growth of the boundary layer displacement thickness is large, that is, $d\delta^*/dx$ is large.

ii) Hence, the incoming freestream *sees* an effecive body with rapidly growing thickness; the inviscid streamlines are deflected upward, into the incoming flow, and a shock wave is consequently generated at the leading edge of the flat plate, i.e., the inviscid flow is *strongly affected* by the rapid boundary layer growth.

iii) In turn, the substantial changes in the outer inviscid flow feed back to the boundary layer, affecting its growth and properties.

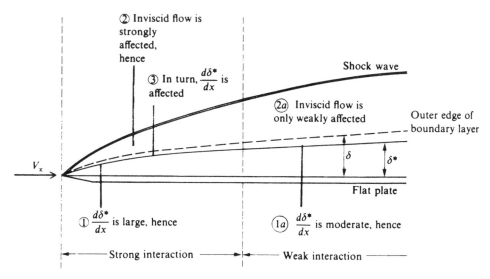

FIGURE 10.31 Illustration of strong and weak viscous interactions.

This mutual interaction process, where the boundary layer substantially affects the inviscid flow, which in turn substantially affects the boundary layer, is called a *strong viscous interaction*, as sketched in Fig. 10.31.

In contrast, further downstream a region of weak interactions is eventually encountered. By definition, the *weak interaction* region is one where the following physical effects occur:

i) The rate of growth of the boundary layer is moderate, that is, $d\delta*/dx$ is reasonably small.

ii) In turn, the outer inviscid flow is only weakly affected.

iii) As a result, the changes in the inviscid flow result in a negligible feedback on the boundary layer, and this is ignored.

Therefore, as indicated in Fig. 10.31, the region of flow where the feedback effect is ignored is called a *weak viscous interaction*.

The similarity parameter that governs laminar viscous interactions, both strong and weak, is *chi-bar*, defined as

$$\overline{\chi} = \frac{M_\infty^3}{\sqrt{\text{Re}}} \sqrt{C} \qquad (10.52)$$

where

$$C = \frac{\rho_w \mu_w}{\rho_e \mu_e} \qquad (10.53)$$

The values of $\overline{\chi}$ can be used to ascertain whether an interaction region is strong or weak; large values of $\overline{\chi}$ correspond to the strong interaction region, and small values of $\overline{\chi}$ denote a weak interaction region.

The induced pressures on the surface of a flat plate due to the hypersonic viscous interaction are given as follows, from Hayes and Probstein (1959). These results are for air with a constant ratio of specific heats equal to 1.4.

For an *insulated flat plate*,

$$\text{Strong interaction} \quad \frac{p}{p_\infty} = 0.514\bar{\chi} + 0.759 \tag{10.54}$$

$$\text{Weak interaction} \quad \frac{p}{p_\infty} = 1 + 0.31\bar{\chi} + 0.05\bar{\chi}^2 \tag{10.55}$$

For a *cold wall case*, where $T_w \ll T_{aw}$,

$$\text{Strong interaction} \quad \frac{p}{p_\infty} = 1 + 0.15\bar{\chi} \tag{10.56}$$

$$\text{Weak interaction} \quad \frac{p}{p_\infty} = 1 + 0.078\bar{\chi} \tag{10.57}$$

Note that a cold wall mitigates to some extent the magnitude of the viscous interaction. This makes sense, because for a cold wall the density in the boundary layer will be higher, hence the boundary layer thickness will be smaller, thus diminishing the root cause of the viscous interaction in the first place.

Note that in Eqs. (10.54)–(10.57) the induced pressure ratios depend only on the hypersonic viscous interaction parameter, $\bar{\chi}$.

This viscous interaction effect on pressure coefficient, as well as on lift and drag coefficients, is governed by a *modified viscous interaction parameter*, $\bar{V}$, which varies as M_∞ (not as M_∞^3). The modified parameter is defined as

$$\bar{V} = M_\infty \sqrt{\frac{C}{\text{Re}}} \tag{10.58}$$

where

$$C_p = f_1(\bar{V})$$
$$C_L = f_2(\bar{V})$$
$$C_D = f_3(\bar{V}) \tag{10.59}$$

See Anderson (1989) for more details.

10.8 HIGH TEMPERATURE EFFECTS

Hypersonic flows frequently involve high temperatures. This is easily seen when an initially hypersonic flow with a lot of kinetic energy is slowed to a much lower velocity, for example in the nose region of a blunt body, where the flow kinetic energy is small. *Question:* Where has all the kinetic energy gone? *Answer:* For the most part, into the internal energy of the gas, hence increasing the temperature of the gas. The temperatures can be high enough that the assumption made previously in this chapter, namely that of a calorically perfect gas (constant specific heats) does not hold. Various thermo/chemical effects prevail at high temperatures, such as excitation of the molecular vibrational energy, chemical dissociation, and ionization. In comparison to a calorically perfect gas, a high-temperature, chemically reacting flow is characterized by the following aspects:

i) The thermodynamic properties (e, h, p, T, ρ, s, etc.) are completely different.

ii) The transport properties (μ and k) are completely different. Moreover, the additional transport mechanism of diffusion becomes important, with the associated diffusion coefficients, D_{ij} (see Sec. 4.13).

iii) High heat transfer rates are usually a dominant aspect of any high temperature application.

iv) The ratio of specific heats, $\gamma = c_p/c_v$, is a variable. In fact, for the analysis of high-temperature flows, γ loses the importance it has for the classical constant γ flows.

v) In view of the above, virtually all analyses of high temperature gas flows require some type of numerical rather than closed form, solutions (see Chap. 22).

vi) If the temperature is high enough to cause ionization, the gas becomes a partially ionized plasma, which has a finite electrical conductivity. In turn, if the flow is in the presence of an exterior electric or magnetic field, then electromagnetic body forces act on the fluid elements. This is the purview of an area called *magnetohydrodynamics*, MHD (see Sec. 13.8).

vii) If the gas temperature is high enough, there will be nonadiabatic effects due to radiation to or from the gas.

From the applications discussed in this chapter, it is clear that we are frequently concerned with air as the working gas. At what temperatures do chemically reacting effects become important in the air? An answer is given in Fig. 10.32, which illustrated the ranges of dissociation and ionization in air at a pressure of 1 atm. Let us go through the following *thought experiment.* Imagine that we take the air in the room around us and progressively increase the temperature, holding the pressure constant at 1 atm. At about a temperature of 800 K, the vibrational energy of the molecules becomes significant (as noted on the right of Fig. 10.32). This is not a chemical reaction, but it does have some impact on the properties of the gas. When the temperature reaches about 2500 K, the dissociation of O_2 begins. At 4000 K, the O_2 dissociation is essentially complete; most of the oxygen is in the form of atomic oxygen, O. Moreover, by an interesting quirk of nature, 4000 K is the temperature at which N_2 begins to dissociate, as shown in Fig. 10.32. When the temperature reaches 9000 K, most of the N_2 has dissociated. Coincidentally, this is the temperature at which both oxygen and nitrogen ionization occurs, and, above 9000 K, we have a partially ionized plasma consisting mainly of O, O^+, N, N^+, and electrons. Not shown in Fig. 10.32 (because it would become too cluttered) is a region of mild ionization that occurs around 4000 to 6000 K; here, small amounts of NO are formed, some of which ionize to form NO^+ and free electrons. In terms of the overall chemical composition of the gas, these are small concentrations, how-

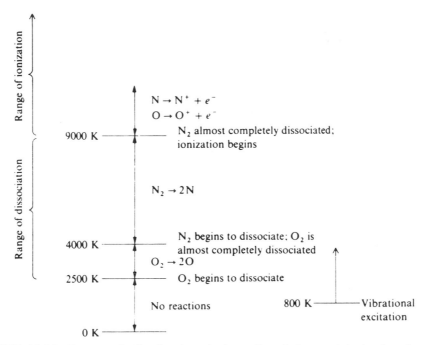

FIGURE 10.32 Ranges of vibrational excitation, dissociation, and ionization for air at 1-atm pressure.

ever the electron number density due to NO ionization can be sufficient to cause the communications blackout discussed in Sec. 10.14. Reflecting upon Fig. 10.32, it is very useful to fix in your mind the *onset temperatures*; 800 K for vibrational excitation, 2500 K for O_2 dissociation, 4000 K for N_2 dissociation, and 9000 K for ionization. With the exception of vibrational excitation, which is not affected by pressure, if the air pressure is lowered, these onset temperatures decrease; conversely, if the air pressure is increased these onset temperatures are raised.

The information in Fig. 10.32 leads directly to the velocity-altitude map shown in Fig. 10.33. In Fig. 10.33, we show the flight paths of lifting entry vehicles with different values of the lift parameter, $m/C_L S$. Superimposed on this velocity-altitude map are the flight regions associated with various chemical effects in air. The 10 and 90% labels at the top of Fig. 10.33 denote the effective beginning and end of various regions where these effects are important. Imagine that we start in the lower left corner and mentally rideup the flight path in reverse. As the velocity becomes larger, vibrational excitation is first encountered in the flowfield, at about $V = 1$ km/s. At the higher velocity of about 2.5 km/s, the vibrational mode is essentially fully excited, and oxygen dissociation begins. This effect covers the shaded region labeled *oxygen dissociation*. The O_2 dissociation is essentially complete at about 5 km/s, where N_2 dissociation commences. This effect covers the shaded region labeled *nitrogen dissociation*. Finally, above 10 km/s, the N_2 dissociation is complete, and ionization begins. It is most interesting that regions of various dissociations and ionization are so separate on the velocity-altitude map, with very little overlap. This is, of course, consistent with the physical data shown in Fig. 10.32. In a sense, this is a situation where nature is helping to simplify things for us.

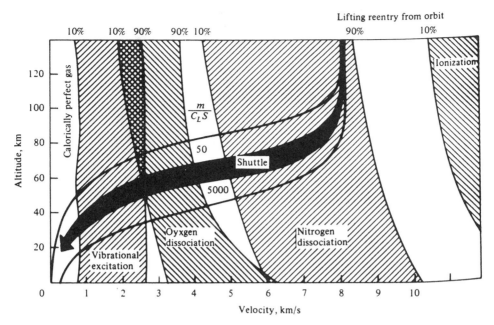

FIGURE 10.33 Velocity-altitude map showing where various chemical reactions are important in the blunt-nosed region of a hypersonic vehicle.

Finally, we can make the following general observation from Fig. 10.33. The entry flight paths slash across major sections of the velocity-altitude map where chemical reactions and vibrational excitation are important. Indeed, the vast majority of any given flight path is in such regions. From this, we can clearly understand why high-temperature effects are so important to entry body flows.

In hypersonic flow, high temperatures are found in those regions where the flow kinetic energy has been greatly decreased, hence the internal energy is greatly increased. For example, a hypersonic free stream at $M_\infty = 20$ has a kinetic energy that is 112 times larger than its internal energy. However, when this flow enters a boundary layer, it is slowed by the effects of friction. In such a case, the kinetic energy decreases rapidly and is converted in part into internal energy, which zooms in value. Since the gas temperature is proportional to internal energy, it also increases rapidly. Thus, hypersonic boundary layers are high-temperature regions of the flow, due to viscous dissipation of the flow kinetic energy. Another region of high temperature flow is the shock layer behind a strong bow shock wave. In this case, the flow velocity discontinuously decreases as it passes through the shock wave. Once again, the lost kinetic energy reappears as an increase in internal energy and, hence, an increase in temperature behind the shock wave. Therefore, that portion of the shock layer behind a strong bow shock wave on a blunt-nosed body is a region of high temperature flow.

As an example of the nature of hypersonic chemically reacting flow, consider a typical variation of chemical species in the flowfield around a blunt-nosed body, as shown in Fig. 10.34. The body shape, shock-wave shape, and two streamlines labeled A and B are shown in Fig. 10.34(a) for $V_\infty = 23,000$ ft/s at an altitude of 250,000 ft. The nose radius of the body is about 0.5 ft. In Fig. 10.34(b), the variation in concentration of atomic oxygen and atomic nitrogen along streamlines A

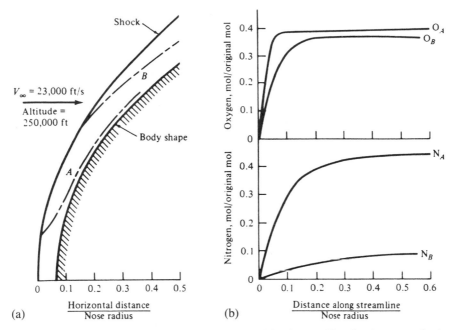

FIGURE 10.34 Hypersonic flow over a blunt-nosed body. (a) The shock wave, the body, and the shape of two streamlines labeled *A* and *B*. (b) Variation of concentrations of atomic oxygen and atomic nitrogen along the two streamlines in (a). Concentrations are given on the ordinates as moles of nitrogen or oxygen per original mole of air upstream of the shock wave. (Reprinted with permission, Hall, J. G., Eschenroeder, A. A., and Marrone, P. V., "Blunt-Nose Inviscid Airflows with Coupled Nonequilibrium Processes," *J. Aero. Sci.*, Vol. 29, No. 9, pp. 1038–1051, Copyright © 1962.)

and *B* is shown as a function of distance *s* along the streamlines. Note that dissociation occurs rapidly behind the shock wave and that large amounts of oxygen and nitrogen atoms are formed in the shock layer. These results are taken from Hall *et al.* (1962). More detailed information on the calculation of flows with chemical reactions can be found in Chap. 22.

10.9 SUMMARY

We return to our original discussion on the definition of hypersonic flow as that region of flow at the high Mach number end of the flight spectrum where certain physical phenomena become important which were not so important at lower Mach numbers. This entire chapter has been devoted to discussing these physical phenomena. Clearly, the reader can now appreciate that hypersonic flow stands by itself as an identifiable and important aspect within the general field of fluid dynamics.

REFERENCES

Anderson, J. D., Jr., "Hypersonic Viscous Flow Over Cones at Nominal Mach 11 in Air," ARL Report 62-387, Aerospace Research Labs., Wright–Patterson Air Force Base, Ohio, 1962.

Anderson, J. D., Jr., *Hypersonic and High Temperature Gasdynamics*, McGraw-Hill, New York, 1989.

Anderson, J. D., Jr., *Modern Compressible Flow: With Historical Perspective*, 2nd ed., McGraw-Hill, New York, 1990.

Charters, A. C. and Thomas, R. N., "The Aerodynamic Performance of Small Spheres from Subsonic to High Supersonic Velocities," *J. Aero. Sci.*, Vol. 12, pp. 468–476, 1945.

Chernyi, G. G., *Introduction to Hypersonic Flow* (translated from Russian by Probstein, R. F.), Academic Press, New York, 1961.

Cox, R. N. and Crabtree, L. F., *Elements of Hypersonic Aerodynamics*, Academic Press, New York, 1965.

Eckert, E. R. G., "Engineering Relations for Heat Transfer and Friction in High-Velocity Laminar and Turbulent Boundary-Layer Flow Over Surfaces with Constant Pressure and Temperature," *Trans. ASME*, Vol. 78, No. 6, p. 1273, 1956.

Hall, J. G., Eschenroeder, A. A., and Marrone, P. V., "Blunt-Nose Inviscid Airflows with Coupled Nonequilibrium Processes," *J. Aero. Sci.*, Vol. 29, No. 9, pp. 1038–1051, 1962.

Hayes, W. D. and Probstein, R. F., *Hypersonic Flow Theory*, Academic Press, New York, 1959.

Hayes, W. D. and Probstein, R. F., *Hypersonic Flow Theory,* Vol. 1: Inviscid Flows, Academic Press, New York, 1966.

Hodges, A. J., "The Drag Coefficient of Very High Velocity Spheres," *J. Aero. Sci.*, Vol. 24, pp. 755–758, 1957.

Neice, S. E. and Ehret, D. M., "Similarity Laws for Slender Bodies of Revolution in Hypersonic Flows," *J. Aero. Sci.*, Vol. 18, No. 8, pp. 527–530, 568, 1951.

Rubesin, M. W. and Johnson, H. A., "A Critical Review of Skin-Friction and Heat Transfer Solutions of the Laminar Boundary Layer of a Flat Plate," *Trans. ASME*, Vol. 71, No. 4, pp. 383–388, 1949.

Stetson, K. F., "On Predicting Hypersonic Boundary Layer Transition," AFWAL-TM-84-160-FIMG, Air Force Wright Aeronautical Laboratories, Wright–Patterson Air Force Base, Ohio, 1987.

Stevens, V. I., "Hypersonic Research Facilities at the Ames Aeronautical Laboratory," *J. Appl. Phys.*, Vol. 21, pp. 1150–1155, 1950.

Van Driest, E. R., "Investigation of Laminar Boundary Layers in Compressible Fluids Using the Crocco Method," NACA TN 2597, 1952.

White, F. M., *Viscous Fluid Flow*, 2nd ed., McGraw-Hill, New York, 1992.

11 Rarefied Gas Dynamics

J. LEITH POTTER
Vanderbilt University
Nashville, TN

CONTENTS

11.1 INTRODUCTION

Rarefied gas dynamics is a term identifying fluid flow phenomena wherein the molecular nature of the gas medium is of significance and continuum fluid concepts may not be admissible. A broader and closely related category of flow processes including free-molecular flows at one extreme and viscous but continuum flows at the other extremity is usually identified as the *transitional flow regime*. The terms rarefied and transitional often are used interchangeably. Results obtained by applying continuum fluid theory usually do not merge smoothly with free-molecular flow theory as the limit of low Reynolds number or high Knudsen number is approached, even though experimental data such as drag and heat transfer coefficients vary smoothly from continuum to free-molecular values.

Handbook of Fluid Dynamics and Fluid Machinery, Edited by Joseph A. Schetz and Allen E. Fuhs
ISBN 0-471-12598-9 Copyright © 1996 John Wiley & Sons, Inc.

In the continuum flow regime, the Reynolds and Mach numbers are the most generally useful dimensionless simulation parameters. However, noncontinuum flows most often are described in terms of a Knudsen number,

$$Kn = \lambda/L \qquad (11.1)$$

where λ = mean free path of gas molecules and L = body dimension most characteristic of the flow being considered. It is obvious that most problems will involve several classes of gas molecules having different mean free paths, and no single body dimension that adequately characterizes the flow. Therefore, it is impossible to assign values to an overall Knudsen number that defines the boundaries of continuum, transitional, and free-molecular flow regimes. As a crude approximation, $Kn < O\,(0.1)$ may be said to suggest continuum, viscous flow and $Kn > O\,(10)$ implies nearly free-molecular flow. To attain collisionless, free-molecular flow requires $Kn > O\,(100)$. By using the *billiard ball* molecular model, an effective mean free path may be defined, and when Knudsen and Reynolds numbers (Re) are based on equal values of L,

$$Kn = 1.277\sqrt{\gamma}\ M/Re, \qquad (11.2)$$

where M = Mach number and γ = ratio of specific heats. The numerical coefficient in Eq. (11.2) varies, depending on the method of calculation of the mean free path. The numerical coefficient given is appropriate when the mean free path, λ_∞, of an undisturbed, uniform body of elastic-sphere molecules is considered. If significant interference between undisturbed, ambient (freestream) molecules and the molecules impacted and reflected by a solid body are considered, e.g., when a spacecraft moves through a low-density planetary atmosphere, then the distance, λ_w, a reflected molecule moves before colliding with a freestream molecule is a more relevant parameter than the freestream mean free path, λ_∞. For example, when the Mach number is greater than approximately six, the mean free path of molecules reflected upstream from a blunt body and interacting with freestream molecules is

$$\lambda_w \approx \lambda_\infty \sqrt{2}/[1 + M_\infty \sqrt{8\gamma T_\infty/(9\pi T_w)}] \qquad (11.3)$$

where T represents temperature, subscript ∞ identifies freestream conditions, and w identifies wall condition. Again an elastic-sphere or billiard ball molecule is assumed.

Bird (1986) has shown that local Knudsen number, rather than overall Knudsen number, is the parameter of importance in predicting rarefied-flow effects. He defines a Knudsen number, based on local density, ρ, and density gradient, i.e.,

$$Kn = (\lambda/\rho)\ d\rho/dx \qquad (11.4)$$

This clearly is a more accurate guide to departure from continuum flow. It does, however, require detailed flow-field information to implement. Bird (1986) also includes comments on the limits of continuum theory for differently shaped bodies.

The purpose of this Chapter is to provide the handbook user with an introduction to the concepts of rarefied flow and some reference sources pertaining to problems involving rarefied gases. Considering that most applications will concern transitional flows, the presentation of kinetic theory is minimal. The derivation of an expression for the mean free path is chosen as a means for introducing the reader to the concepts and analysis of rarefied flows. More extensive coverage may be found in the texts by Loeb (1961), Present (1958), Patterson (1956), (1971), Shidlovskiy (1967), Kogan (1969), and Bird (1976). Earlier handbook publications are those of Schaaf (1963), and Schaaf and Chambré (1961). A recent review by Muntz (1989) covers may aspects of this general subject.

An unusual feature of the rarefied flow research literature is the concentration of many of the research reports in one convenient set of books. This results from the publication of the proceedings of each Rarefied Gas Dynamics Symposium which has been held every two years beginning in 1958. Throughout this article, the Symposium proceedings will be referred to as the RGDS volumes. All published to date are listed in Table 11.1. Readers interested in topics not represented herein, such as separation of gaseous species, ionospheric phenomena, and condensation, should consult the RGDS volumes for relevant sources of information.

11.2 MOLECULAR MODELS

The kinetic theory of gases is based on the concept of matter being composed of molecules, the molecules being subject to the law of classical mechanics, and that statistical methods may be applied to molecular motion when a large enough sample is considered. The latter hypothesis requires that the number of molecules involved in a sample of gas be great enough so that the average behavior is representative of the whole. However, that does not prevent the assumption that gas molecules are separated from one another by distances much in excess of molecular size. The justification for these assumptions at normal pressures and temperatures is evident from the published data on gases. One finds that a fluid volume of 1 cm^3 at 1 atm and 273 K contains on the order of 10^{19} molecules, and the mean free paths are of the order of 10^2 to 10^3 times the collision diameters of the molecules.

Molecular collisions evidently are elastic because molecular motion inside a container does not subside. Thus, the motion must be incessant and the molecules must preserve their mean speeds and kinetic energies. This equality of kinetic energies before and after a collision allows the simplified modeling of a molecule as a sphere with perfectly elastic properties. More realistic and sophisticated models are described in texts and papers on kinetic theory, but many useful results can be obtained with the *billiard ball* or *hard sphere* concept. The internal structure of atoms and molecules is a factor when internal degrees of freedom have to be taken into account.

Intermolecular force is repulsive at short range and attractive at longer range. The repulsion at short range prevents two molecules from occupying the same space. Resistance to merging of molecules is evidenced by the high resistance to compression of liquids or solids. The existence of liquid states, for example, suggests that molecules attract at long range. The models usually adopted for approximating intermolecular force fields are: (1) rigid elastic spheres interacting only upon contact, (2) weakly attracting rigid elastic spheres, (3) point centers of inverse power repulsion, and (4) combinations of inverse power attraction and repulsion.

The forces on a molecule are random, so after a time and many collisions a

TABLE 11.1 Rarefied Gas Dynamics
Proceedings of the International Symposia on Rarefied Gas Dynamics

Symposium Number/ Place	Date of Publication	Technical Editor	Publisher
1/Nice, France	1960	M. Devienne	Pergamon Press, Paris
2/Berkeley, Cal.	1961	L. Talbot	Academic Press, New York
3/Paris, France	1963	J. A. Laurmann	Academic Press, New York
4/Toronto, Canada	1965	J. H. deLeeuw	Academic Press, New York
5/Oxford, England	1967	C. L. Brundin	Academic Press, New York
6/Cambridge, Mass.	1969	L. Trilling & H. Y. Wachman	Academic Press, New York
7/Pisa, Italy	1971	D. Dini	Edizioni Tecnico- Scientifiche, Pisa
8/Palo Alto, Cal.	1974	K. Karamcheti	Academic Press, New York
9/Gottingen, Ger.	1974	M. Becker & M. Fiebig	DFVLR Press, Porz-.Wahn, West Germany
10/Aspen, Colo.	1977	J. L. Potter	American Inst. of Aero. & Astro. (AIAA), Washington, D.C.
11/Cannes, France	1979	R. Campargue	Commissariate A L'Energie Atomique, Paris
12/Charlottesville, Va.	1981	S. S. Fisher	American Inst. of Aero. & Astro. (AIAA), Washington, D.C.
13/Novosibirsk, USSR	1982	O. M. Belotserkovskii M. N. Kogan S. S. Kutateladze A. K. Rebrov	Plenum Press, New York
14/Tsukuba, Japan	1984	H. Oguchi	Univ. of Tokyo Press, Tokyo
15/Grado, Italy	1986	V. Boffi & C. Cercignani	B. G. Teubner, Stuttgart
16/Pasadena, Cal.	1989	E. P. Muntz D. Weaver D. Campbell	American Inst. of Aero. & Astro. (AIAA), Washington, D.C.
17/Aachen, Germany	1991	A. E. Beylich	VCH Publishers Weinheim, Germany, New York
18/Vancouver, Canada	1994	B. D. Shizgal D. P. Weaver	American Inst. of Aero. & Astro. (AIAA), Washington, D.C.

molecule must "forget" its prior condition. This characteristic relaxation time will depend inversely on the number of collisions, and it may depend on the types of molecules encountered. The characteristic time will also vary with the effect being considered. For example, the number of collisions or the time to change the vibrational energy may be different than that required to change the rotational or translational energy. Relaxation processes become very important when nonequilibrium thermo-chemical-kinetic processes are considered; some lasers and the flow fields of reentry vehicles are excellent examples.

In summary, it is hypothesized that:

(1) All gases are composed of molecules, including polyatomic molecules, atoms, ions, electrons, etc.

(2) The molecules are separated by distances great enough that molecules interact only weakly except when they *collide*, i.e., molecular dimensions are small in comparison to mean free path.

(3) Ternary (three body) collisions are infrequent compared to binary (two body) collisions.

(4) Molecular chaos prevails so that the behavior of any two molecules before they collide is statistically unrelated.

(5) Perfect gas laws prevail.

(6) Classical Newtonian mechanics may be applied.

11.3 MEAN FREE PATH

The concept of a *mean free path*, λ, relates to the hard sphere model of a molecule. It is defined as the mean distance traveled by a molecule between two successive collisions. Let v_r be the *relative speed* of a reference molecule in a collection of hard sphere molecules, σ be the *equivalent* collision cross section, and n be the number of molecules per unit volume, then it follows that the reference molecule will sweep through a volume $v_r \sigma$ in unit time. This volume will be occupied by $n\sigma v_r$ other molecules, and there will be $n\sigma v_r$ collisions. Therefore, the time between collisions will be $1/(n\sigma v_r)$. The mean distance covered in unit time is $\bar{v}$, the *mean molecular speed*. Therefore, the mean free path will be

$$\lambda = \bar{v}/(n\sigma\bar{v}_r), \tag{11.5}$$

where $\bar{v}_r$ represents the average *relative molecular speed*, i.e., the average distance between points in velocity space. The ratio, $\bar{v}/\bar{v}_r$, will not differ greatly from unity, and sometimes it is taken as unity in introductory treatments. Later, it is shown to be $1/\sqrt{2}$ for a Maxwellian speed distribution. Knowledge about the distribution of molecular speeds is necessary before $\bar{v}$ and $\bar{v}_r$ can be evaluated.

It should be noted that *collision cross section* is not the same as the cross section of a molecule. Hard-sphere molecules of equal diameter, d, will come in contact if their centers are a distance d apart. Thus, the collision cross section is πd^2.

The mean time between collisions may be expressed as

$$t_c = \lambda/\bar{v}_r \tag{11.6a}$$

and the collision frequency will be

$$1/t_c = \sigma n\bar{v}_r. \tag{11.6b}$$

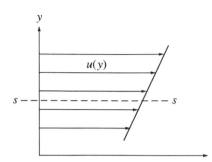

FIGURE 11.1 Two-dimensional shear layer.

Equation (11.6a) suggests that the relaxation time, t_r, may be expressed as

$$t_r = \alpha_r t_c = \alpha_r/(\sigma n \bar{v}_r) \tag{11.6c}$$

where α_r = a proportionality factor which, in reality, may vary by orders of magnitude for different types of molecules and relaxation processes.

The relation between mean free path, gas viscosity, and density can be found by considering the transfer of momentum across a plane, as illustrated in Fig. 11.1, a two-dimensional shear layer with mass velocity u in the direction of the x axis. It can be seen that each molecule, of mass m, having experienced its last collision at a distance λ from the surface $s - s$ arrives there with a momentum in the x direction of $m(u_s + \lambda \, \partial u/\partial y)$ if it comes from above and $m(u_s - \lambda \, \partial u/\partial y)$ if it comes from below. The average number of molecules crossing the plane $s - s$ per unit area per unit time is proportional to the number density, n, times the average random thermal speed of the molecules, $\bar{v}$. Thus, the net transport of momentum upward, in the direction of the gradient of u, is proportional to

$$n\bar{v}[m(u_s - \lambda \, \partial u/\partial y) - m(u_s + \lambda \, \partial u/\partial y)] \propto n\bar{v}m\lambda \, \partial u/\partial y, \tag{11.7}$$

where $\propto$ means *proportional to*. Shear is created in a gas by the molecular transport of momentum in the direction opposite to a velocity gradient. The conventional Newtonian shear stress is expressed as

$$\tau = \mu \, \partial u/\partial y, \tag{11.8}$$

where the coefficient μ is the dynamic viscosity.

Combining Eq. (11.7) and (11.8), it is seen that

$$\lambda \propto \mu/(n\bar{v}m) \propto \mu/(\rho\bar{v}), \tag{11.9}$$

11.4 DISTRIBUTION FUNCTIONS

If $\bar{v}$ is a molecular velocity vector with components, v_x, v_y, and v_z, it can be represented in velocity space by a point having the coordinates (v_x, v_y, v_z) as shown in Fig. 11.2.

For a gas at rest, with no external force fields and no flow of matter or energy,

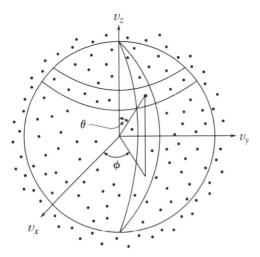

FIGURE 11.2 Molecular velocity space.

there is no preferred location or direction of motion of molecules. A velocity distribution function, $F(\overline{v})$, may be defined such that the number of molecules with velocities in the interval between v_x and $v_x \pm dv_x/2$, v_y and $v_y \pm dv_y/2$, v_z and $v_z \pm dv_z/2$ is $F(\overline{v})\ dv_x\ dv_y\ dv_z$.

Referring to Fig. 11.2, the element of volume in velocity space is $v\ d\theta(v \sin \theta)(d\phi)\ dv$, where θ and ϕ are the polar and azimuthal angles associated with spherical coordinates. Defining a solid angle element as $d\omega = \sin \theta\ d\theta\ d\phi$, then $v^2F(\overline{v})\ d\omega\ dv$ is the fraction of molecules with speeds and directions in the infinitesimal interval about $\overline{v}$. The number of molecules with speeds in this range is simply obtained by integrating over all possible velocity directions, which is seen in Fig. 11.2 to be a solid angle of 4π. That procedure yields the *Maxwellian distribution function* for molecular speeds,

$$f(v)\ dv = \int_0^{4\pi} v^2F(\overline{v})\ dv\ d\omega. \qquad (11.10)$$

Determining $f(v)$ from mechanical and statistical considerations is fully discussed in texts on kinetic theory. A simplified argument can be based on the independence of each velocity component and isotropy of the velocity distribution. Under those circumstances, the distribution is expected to be a normal or Gaussian one, i.e., a function of the form

$$F(\mathbf{v}) = A \exp (-\beta v^2), \qquad (11.11)$$

where A and β are to be determined. This expression was obtained by Maxwell. The basic assumption is not altogether evident, as recognized by Maxwell, and other more rigorous derivations were given by Boltzmann and by Gibbs, cf. Present (1958).

By definition, the distribution function must satisfy the relation

$$\int_{-\infty}^{\infty} f(v)\ dv = 1. \qquad (11.12)$$

Therefore,

$$\int_{-\infty}^{\infty} \int_{-\infty}^{\infty} \int_{-\infty}^{\infty} A \exp \left[-\beta(v_x^2 + v_y^2 + v_z^2) \, dv_x \, dv_y \, dv_z = A(\pi/\beta)^{3/2} = 1 \quad (11.13) \right.$$

or

$$A = (\beta/\pi)^{3/2}. \tag{11.14}$$

An approach for finding β is to consider the pressure exerted by a confined gas on the walls of a container. Assuming the walls perfectly smooth and the gas composed of billiard-ball molecules, elastic collisions can be assumed. The angle of reflection of a molecule will equal the angle of incidence. The normal component of molecular velocity will be exactly reversed upon impact with a wall, and the tangential component will not be altered. From Newton's second law, the pressure on the container wall is equal to the momentum imparted per unit area per unit time by the gas molecules. It can be shown [cf. Present (1958)] that the pressure exerted by all molecules striking the wall is

$$p = (1/3)nm \int_0^\infty v^2 f(v) \, dv = (1/3)nm\overline{v^2}, \tag{11.15}$$

where $\overline{v^2}$ is the mean-squared speed of the molecules.

Let the number of moles of gas in the container be

$$\nu = nV/N_0, \tag{11.16}$$

where

n = molecular number density,
V = volume of gas,
N_0 = number of molecules per mole.

Then from Eqs. (11.15 and 11.16),

$$pV = (2/3)(nV/N_0)(N_0 m \overline{v^2}/2), \tag{11.17}$$

where the kinetic energy of translation of the molecules in one mole is represented by the term $(N_0 m\overline{v^2}/2)$. The perfect gas law, confirmed by experimental observations is

$$pV = \nu \Re T, \tag{11.18}$$

where $\Re$ = universal gas constant per mole.

Comparing Eqs. (11.17) and (11.18), it is seen that

$$T = (2/3\Re)(N_0 m \overline{v^2}/2). \tag{11.19}$$

Boltzmann's constant, k, is

$$k = \Re/N_0. \tag{11.20}$$

Thus, Eq. (11.19) can be written as

$$T = (1/3k)m\overline{v^2}. \tag{11.21}$$

Now

$$\overline{v_x^2} = (\beta/\pi)^{3/2} \int_{-\infty}^{\infty} \int_{-\infty}^{\infty} \int_{-\infty}^{\infty} \cdot v_x^2 \exp\left[-\beta(v_x^2 + v_y^2 + v_z^2)\right] dv_x\, dv_y\, dv_z$$

$$= (\beta/\pi)^{1/2} \int_{-\infty}^{\infty} v_x^2 \exp\left(-\beta v_x^2\right) dv_x$$

$$= -(\beta/\pi)^{1/2}(d/d\beta) \int_{-\infty}^{\infty} \exp\left(-\beta v_x^2\right) dv_x = 1/2\beta \tag{11.22}$$

and a similar result would be obtained for $\overline{v_y^2}$ or $\overline{v_z^2}$. Furthermore,

$$\overline{v^2} = \overline{v_x^2} + \overline{v_y^2} + \overline{v_z^2} \tag{11.23}$$

and, with equally probable molecular motion in any coordinate direction, i.e., no mass velocity,

$$\overline{v_x^2} = \overline{v_y^2} = \overline{v_z^2}. \tag{11.24}$$

Thus, by combining Eqs. (11.21) through (11.24), we obtain

$$\beta = m/(2kT). \tag{11.25}$$

The substitution of Eqs. (11.12) and (11.25) into Eq. (11.11) gives

$$F(\mathbf{v}) = (m/2\pi kT)^{3/2} \exp\left[(-m/2kT)v^2\right]. \tag{11.26}$$

Introducing the specific gas constant, R, where

$$R = \Re/\text{molecular weight} = k/m,$$

Eq. (11.26) becomes

$$F(\mathbf{v}) = (1/2\pi RT)^{3/2} \exp\left(-v^2/2RT\right). \tag{11.27}$$

The speed distribution function, $f(v)\, dv$, is obtained by integrating the velocity distribution over all possible directions. This may be thought of as a process of counting the number of representative points in velocity space (Fig. 11.2) within a spherical shell of thickness dv and radius v. The volume of such a shell is $4\pi v^2\, dv$.

By substitution into Eq. (11.27), the number of points (or molecules) is then

$$f(v) \, dv = (1/2\pi RT)^{3/2} 4\pi v^2 \exp\left(-v^2/2RT\right) dv. \qquad (11.28)$$

The mean molecular speed may be calculated from Eq. (11.28) as follows

$$\bar{v} = \int_0^\infty v f(v) \, dv = 4\pi (1/2\pi RT)^{3/2} \int_0^\infty v^3 \exp\left(-v^2/2RT\right) dv = (8RT/\pi)^{1/2}$$

$$(11.29)$$

The mean squared speed is

$$\overline{v^2} = 3RT. \qquad (11.30)$$

The most probable speed can be shown to be $\sqrt{2RT}$.

Recalling from Eq. (11.5) that a mean relative velocity is also of interest, it is possible to calculate that quantity using the Maxwellian distribution function. When the mean value of the relative velocity of two unlike molecules in a binary mixture of gases is calculated [cf. Present (1958)], it is found that $\bar{v}_r = (v_a^2 + v_b^2)^{1/2}$. Then, if the molecules are all alike,

$$\bar{v}_r = \bar{v}\sqrt{2}, \qquad (11.31)$$

and Eq. (11.5) can be written

$$\lambda = 1/(\sqrt{2}n\sigma). \qquad (11.32)$$

Although Eqs. (11.5) and (11.9) demonstrate that a relation between σ and μ exists, it also depends on the details of the molecules. It can be shown [cf. Patterson (1956)] that, for smooth elastic-sphere molecules,

$$\mu = 5m(\overline{v^2}/3)^{1/2}/(16d^2\sqrt{\pi}), \qquad (11.33)$$

where d = molecular "diameter."

With Eqs. (11.29), (11.30), (11.32), and (11.33) it is seen that

$$\sigma = (5m\sqrt{\pi}/16\mu)(\overline{v^2}/3)^{1/2} \qquad (11.34)$$

and

$$\lambda = \frac{16\mu\sqrt{3}}{5mn\sqrt{2\pi\overline{v^2}}}. \qquad (11.35)$$

But, the mass density or mass per unit volume, $\rho = nm$, and the mean squared speed, $\overline{v^2} = 3RT$. After substituting into Eq. (11.35),

$$\lambda = 16\mu/(5\rho\sqrt{2\pi RT}) = 1.277\mu/(\rho\sqrt{RT}). \qquad (11.36)$$

Present (1958) shows that an elementary analysis of the problem leads to a proportionality factor of 3 in Eq. (11.9). That is,

$$\lambda = 3\mu/(\rho\bar{v}). \qquad (11.37)$$

Combining Eqs. (11.29) and (11.37) gives a result differing from Eq. (11.36) only by a slight change in the constant. Mean free path concepts simplify molecular behavior. Variations in the numerical coefficients result when different approaches to the analysis or different molecular interactions are specified. For example, if one starts with an expression for the mean free path of molecules of a given speed and averages this mean free path for a distribution of velocities, Tait's free path is obtained, cf. Loeb (1961). Maxwell's free path is obtained by dividing the mean speed, $\bar{v}$, by the mean number of collisions per unit of time, averaged over all molecules. All such free paths will depend on $1/(n\sigma)$, but the numerical factors will differ.

Bird (1983) has proposed an alternative to the hard-sphere mean free path. His *variable hard sphere* concept leads to

$$\lambda = (2\mu/15)(7 - 2\omega)(5 - 2\omega)/[\rho\sqrt{2\pi RT}] \qquad (11.38)$$

where ω is the exponent in the relation $\mu_1/\mu_2 = (T_1/T_2)^\omega$.

The most relevant mean free path or Knudsen number for flow problems involving interaction of gases and solid bodies is not always clear, as stated earlier. It should also be noted that molecules in flows with a mass velocity or Mach number, M_∞, will travel farther between collisions in the flow direction than they would under static conditions. Thus, if the gas motion is relative to a fixed coordinate system, the mean free path in the flow direction is approximately

$$\lambda \approx \lambda_\infty M_\infty \quad \text{when } M_\infty \gg 1, \qquad (11.39)$$

where λ_∞ is the mean free path defined in Eq. (11.36) using freestream properties. Kogan (1969) has discussed this and points out the implications when data from wind tunnels and flight tests are compared.

11.5 THE BOLTZMANN EQUATION

Averaged molecular properties represent the macroscopic state and motion of a gas. Statistical concepts are used to express the dynamic properties of finite volumes of gases, and this requires the use of distribution functions. These functions are used in formulating the Boltzman equation.

In a stationary Cartesian coordinate system (x, y, z), let the molecular velocity components be denoted by v_x, v_y, v_z, and consider the elementary volume $dx\,dy\,dz$ at time t. The number of molecules in this volume with velocity components in the range v_x to $v_x + dv_x$, v_y to $v_y + dv_y$, and v_z to $v_z + dv_z$ may be expressed as $dN = F(x, y, z, t, v_x, v_y, v_z)\,dx\,dy\,dz\,dv_x\,dv_y\,dv_z$. The function F is the distribution function of molecular velocities. Taking m as the mass of a molecule, a mass density distribution function, $f = mF$, is defined. If dN is integrated over all values of the velocity components, the molecular number density $n(x, y, z, t)$ per unit volume at time t is found, i.e., $n = \int_{-\infty}^{\infty}\int_{-\infty}^{\infty}\int_{-\infty}^{\infty} F \, dv_x \, dv_y \, dv_z$. By use of distribution functions, it is possible to write equations for the average value of any quantity, g, which depends on molecular velocity, position, and time, e.g.,

$$\bar{g} = \int_{-\infty}^{\infty}\int_{-\infty}^{\infty}\int_{-\infty}^{\infty} gF \, dv_x \, dv_y \, dv_z \bigg/ \int_{-\infty}^{\infty}\int_{-\infty}^{\infty}\int_{-\infty}^{\infty} F \, dv_x \, dv_y \, dv_z$$

$$= (1/n) \int_{-\infty}^{\infty}\int_{-\infty}^{\infty}\int_{-\infty}^{\infty} gF \, dv_x \, dv_y \, dv_z. \qquad (11.40)$$

If X, Y, Z represent the components of external force per unit mass acting on each gas molecule, Boltzmann's equation is

$$\partial F/\partial t + v_x(\partial F/\partial x) + v_y(\partial F/\partial y) + v_z(\partial F/\partial z) + X(\partial F/\partial v_x)$$

$$+ Y(\partial F/\partial v_y) + Z(\partial F/\partial v_z) = \Delta_{coll}F. \qquad (11.41)$$

The right-hand term represents the rate of change of the distribution function F, at a point due to intermolecular collisions. It is expressed as a *collision integral*. When the gas is in a uniform, steady state, so that F is independent of position and time, and there are no external forces, a dynamic quantity associated with a molecule will depend only on velocity. Changes within an element of volume will be caused only by transport of molecules into and out of that volume and changes owing to molecular collisions inside the volume.

A first step beyond the simplification, $\Delta_{coll}F^{(0)} = 0$, is taken by using the Maxwellian distribution function, i.e.,

$$F^{(0)} = n_0(2\pi RT_0)^{-3/2} \exp[-(\bar{v} - v)^2/2RT_0], \qquad (11.42)$$

where

n_0 = number density = $n_0(x, y, z, t)$,
v = macroscopic velocity = $v(x, y, z, t)$, and
T_0 = temperature = $T_0(x, y, z, t)$.

The latter function is referred to as a *local Maxwellian distribution function*. Note that v could represent a mass velocity, e.g., the freestream flow or the molecules reemitted after impacting a wall.

Boltzmann's integrodifferential equation for the velocity distribution function is the basis for the theory of transport phenomena. Methods of successive approximations are used in seeking a solution. Much effort has been devoted to finding better approximate solutions to the equation, see e.g., Ikenberry and Truesdell (1956) or Cercignani (1969), (1975). Overviews of various efforts toward solution are given by Schaaf and Chambré (1961), Shidlovskiy (1967), Muntz (1989), and Kogan (1969), (1973).

11.6 GAS-SURFACE INTERACTION

Before considering the prediction of forces on bodies or other problems involving solid surfaces exposed to rarefied gas flow, it is necessary to decide how gas-surface interaction will be modeled. The transfer of momentum or energy between gas and solid is influenced by the characteristics of both approaching and reemitted molecules. Molecules with a distribution of velocities approach the surface, molecule are reemitted from the surface with a different velocity distribution, and a variety of classes of collisions occur. For example, collisions occur between freestream and reemitted molecules, and reemitted molecules may be deflected back for a second impact on the surface. Under conditions of extreme rarefaction, in the free-molecular flow regime, only the impact of freestream molecules on the surface and the subsequent reemission are considered. Even that simplification does not eliminate the uncertainty concerning the nature of the reflection in a specific engineering application.

The gross features of gas molecular impact on a solid surface are indicated in Fig. 11.3. Diffuse reflection is represented by the spherical distribution, where the reemitted molecular flux is proportional to the cosine of the angle between the incident flow and the surface normal. All reemitted molecules may be at the surface temperature, or they may have a different temperature lying between that of the incident molecules and the surface temperature.

Laboratory data, as presented in the review papers of Hurlbut (1967), (1986), and (1989) show that the density distributions of the reflected molecules generally form lobes with their axes tilted toward the direction of the specular reflection. Increasing the temperature of the incident beam causes the reflected maxima to shift toward the surface, and the distribution lobe maxima shift away from the surface as surface temperature increases. Narrowing of the density distribution lobes occurs as gas temperature increases. Again, increasing surface temperature causes an opposite effect, namely, increasing width of the local pattern.

When describing energy exchange between the incoming gas and the surface, an energy accommodation coefficient may be defined as

$$\alpha = (E_i - E_r)/(E_i - E_w), \tag{11.43}$$

where

E_i = energy flux incident on wall,
E_r = energy flux reemited from wall,

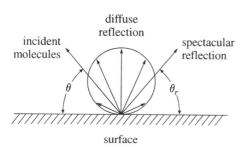

surface

FIGURE 11.3 Gas-surface interactions.

E_w = energy flux that would be reemitted if molecules were reemitted with a Maxwellian distribution corresponding to wall temperature.

For perfect accommodation, $\alpha = 1$. When $\alpha = 0$, there is no energy exchange. This concept implies that all energies associated with the molecular degrees of freedom are accommodated to the same degree. That may not be the case for all energy components, but is a common assumption. Numerous measurements of α for various surfaces and gases have been made, but the wide range of data serves to emphasize the critical influences of the various factors involved, and, at this time, there is no agreement on the value of α to use in a practical problem.

To gain some additional flexibility, two coefficients are used to specify the tangential and normal momentum transfer

$$\tau' = (\tau_i - \tau_r)/(\tau_i - \tau_w) \tag{11.44}$$

and

$$\sigma' = (P_i - P_r)/(P_i - P_w), \tag{11.45}$$

where

τ_i = tangential momentum flux incident,
τ_r = tangential momentum flux reflected,
τ_w = tangential momentum flux reflected with a Maxwellian distribution at wall temperature
P = normal momentum flux, and the subscripts have the same meanings as for the τ's.

In the case of specular reflection with no energy exchange, $\alpha = \tau' = \sigma' = 0$, and for diffuse reflection with full accommodation, $\alpha = \tau' = \sigma' = 1$. Sometimes Eq. (11.45) is written $\sigma' = (P_i + P_r)/(P_i + P_w)$, reflecting the reversal of momentum after impact. One must be careful to avoid error due to differences in definition. Equations (11.43) through (11.45) represent the traditional approach to energy and momentum transfer. Although it has been clear for a long time that this approach does not adequately model the gas-surface interaction revealed by modern experimental techniques, it is widely used because no other model has been generally accepted.

The values to be assigned to these coefficients remain in doubt despite much research. However, there is considerable evidence that intermediate values are more appropriate than either of the limits. It is indicated by experimental data that the accommodation coefficients may depend on surface material, cleanliness, roughness, temperature, and also gas temperature, velocity, species, and incidence angle. For more extensive discussion of this topic, the reviews by Hurlbut (1967), (1986), and (1989) and other papers on this subject in the latest RGDS volumes should be read.

Nocilla (1963) has proposed a new model for gas-surface interaction, and Hurlbut and Sherman (1968) have adapted Nocilla's model for predicting forces on bodies in free-molecular flow. Cercignani and Lampis (1971) also have proposed a new model. Nevertheless, the exploitation of the more recent interaction models is inhibited by the variability of conditions met in practice and the difficulty of measuring the needed data under conditions of most interest, e.g., spaceflight.

11.7 FORCES AND HEAT TRANSFER IN FREE-MOLECULAR FLOW

Classical free-molecular-flow heat transfer, pressure, and skin friction on an element of surface area are given by Schaaf and Chambré (1961). Some of their results are repeated here with minor modifications. [Errors which were made in certain force coefficients given in that publication have been corrected by Kemp (1979).] A Maxwellian velocity distribution is assumed for the freestream flow. The following expressions for Stanton number and recovery factor apply to a flat plate or an element of area at angle of attack, θ, (Fig. 11.4), with windward and leeward surfaces in perfect thermal contact, and with A equal to the total surface area of both sides. Results for the front surface, if it is insulated from the rear and A is the area of only the front side, may be obtained by replacing the erf $(S_\infty \sin \theta)$ term by $[1 + \mathrm{erf}\,(S_\infty \cdot \sin \theta)]$ in Eqs. (11.46) and (11.47). This modification also applies for the rear surface if it is insulated from the front, its area is A, and θ is replaced by $-\theta$. Figure 11.4 shows the geometry. Other quantities are

$$
\begin{aligned}
T_a &= \text{adiabatic recovery temperature} \\
T_\infty &= \text{static temperature of freestream flow} \\
T_0 &= \text{total temperature of freestream flow} \\
Q &= \text{heat transfer rate} \\
r &= \text{recovery factor} \\
S_\infty &= \text{molecular speed ratio} = U_\infty/(2RT_\infty)^{1/2} = M_\infty(\gamma/2)^{1/2} \\
S_t &= \text{Stanton number}
\end{aligned}
$$

$U_\infty, \rho_\infty, c_p = $ freestream velocity, density, and specific heat at constant pressure, respectively.

$$
\begin{aligned}
S_t &= Q/[\rho_\infty c_p A U_\infty (T_a - T_w)] \\
&= \frac{\alpha(\gamma + 1)}{4\gamma S_\infty \sqrt{\pi}} [\exp - (S_\infty \sin \theta)^2 + \sqrt{\pi}\, S_\infty \sin \theta \, \mathrm{erf}\,(S_\infty \sin \theta)], \quad (11.46)
\end{aligned}
$$

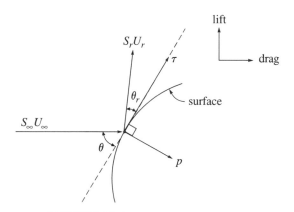

FIGURE 11.4 Aerodynamic forces.

$$r = (T_a - T_\infty)/(T_0 - T_\infty)$$

$$= \frac{\gamma}{S_\infty^2(\gamma + 1)} \{2S_\infty^2 + 1 - 1/[1 + \sqrt{\pi} (S_\infty \sin \theta)$$

$$\cdot \operatorname{erf} (S_\infty \sin \theta) \exp (S_\infty \sin \theta)^2]\} \tag{11.47}$$

where $T_0 = T_\infty[1 + (\gamma - 1)S_\infty^2/\gamma]$.

The pressure, p, and skin friction, τ, are

$$p = \frac{\rho_\infty U_\infty^2}{2S_\infty^2} \{[(2 - \sigma')\pi^{-1/2}S_\infty \sin \theta + (\sigma'/2)\sqrt{T_w/T_\infty}] \exp - (S_\infty \sin \theta)^2$$

$$+ ((2 - \sigma') [(S_\infty \sin \theta)^2 + 0.5] + (\sigma'/2)\sqrt{\pi T_w/T_\infty} \, S_\infty \sin \theta)$$

$$\cdot [1 + \operatorname{erf} (S_\infty \sin \theta)]\} \tag{11.48}$$

$$\tau = \frac{\tau'\rho_\infty U_\infty^2 \cos \theta}{2S_\infty \sqrt{\pi}} \{\exp - (S_\infty \sin \theta)^2 + \sqrt{\pi} (S_\infty \sin \theta) [1 + \operatorname{erf} (S_\infty \sin \theta)]\}$$

$$\tag{11.49}$$

Tables of the error function, which appears frequently in these equations, may be found in standard references. Because $\operatorname{erf} (x) \approx 1$ when $x > 2$ it is often satisfactory to simply use that approximation.

When the speed ratio, S_∞, is large, the wall temperature moderate, and $(S_\infty \sin \theta) \gg 1$, the following are useful approximations to Eqs. (11.46), (11.48), and (11.49)

$$Q = (\alpha/2) \rho_\infty U_\infty^3 A \sin \theta \tag{11.50}$$

$$p = (2 - \sigma') \rho_\infty U_\infty^2 \sin^2 \theta \tag{11.51}$$

$$\tau = \tau'\rho_\infty U_\infty^2 \sin \theta \cos \theta \tag{11.52}$$

11.8   HURLBUT–SHERMAN DEVELOPMENT OF THE NOCILLA MODEL

Hurlbut and Sherman (1968) have derived the aerodynamic coefficients for free-molecular flow on a basis consistent with Nocilla's gas-surface interaction model. The following expressions are directly from their paper. In order to retain their nomenclature, it is necessary to now let σ have the meaning given by Eq. (11.56). It is not a molecular collision cross section in the present usage.

Referring to Fig. 11.4, U_∞ and S_∞ represent freestream velocity and molecular speed ratio, and U_r and S_r represent the velocity and molecular speed ratio of the reemitted molecules, i.e., $S_r = U_r/(2RT_r)^{1/2}$, and it is assumed that U_∞ and U_r lie in the same plane. Hurlbut and Sherman obtain the following expressions for drag and lift coefficients on surface elements of unit area:

$$C_D = \frac{\chi(\sigma)}{\pi^{1/2} S_\infty} \left\{ B^{1/2} \left[\left(\sigma_r + \frac{\pi^{1/2}(1 + \text{erf } \sigma_r)}{2\chi(\sigma_r)} \right) \sin \theta - S_r \cos \theta_r \cos \theta \right] \right.$$
$$\left. + \left(1 + \frac{\pi^{1/2}\sigma(1 + \text{erf } \sigma)}{2S_\infty^2 \chi(\sigma)} \right) \right\} \tag{11.53}$$

and

$$C_L = \frac{\chi(\sigma)}{\pi^{1/2} S_\infty} \left\{ B^{1/2} \left[\left(\sigma_r + \frac{\pi^{1/2}(1 + \text{erf } \sigma_r)}{2\chi(\sigma_r)} \right) \cos \theta + S_r \cos \theta_r \sin \theta \right] \right.$$
$$\left. + \frac{1 + \text{erf } \sigma}{4S_\infty^2 \chi(\sigma)} \pi^{1/2}\sigma \cos \theta \right\}, \tag{11.54}$$

where

$$\chi(\sigma) = \exp(-\sigma^2) + \pi^{1/2} \sigma(1 + \text{erf } \sigma), \tag{11.55}$$

$$\sigma = S_\infty \sin \theta = (U_\infty \sin \theta)/(2RT_\infty)^{1/2}, \tag{11.56}$$

$$\sigma_r = S_r \sin \theta_r, \text{ and}$$

$$B = \left\{ (1 - \alpha_2) \left(1 + \frac{\gamma + 1}{2S_\infty^2(\gamma - 1)} + \frac{\pi^{1/2}\sigma(1 + \text{erf } \sigma)}{2S_\infty^2 \chi(\sigma)} \right) \right.$$
$$\left. + \frac{\alpha_2 RT_w(\gamma + 1)}{U_\infty^2 (\gamma - 1)} \right\}$$
$$\div \left\{ S_r^2 + \frac{\gamma + 1}{2(\gamma - 1)} + \frac{\pi^{1/2}\sigma_r(1 + \text{erf } \sigma_r)}{2\chi(\sigma_r)} \right\}. \tag{11.57}$$

The partial-accommodation coefficient, α_2, is defined as

$$\alpha_2 = [(\epsilon_i(\theta_i) - \epsilon_r]/[\epsilon_i(\theta_i) - \epsilon_w], \tag{11.58}$$

where

$\epsilon_i(\theta_i)$ = mean energy of particles of the incident flow,
ϵ_r = mean energy per particle taken over the entire reflected distribution of velocities, and
ϵ_w = mean energy of gas particles scattered in Maxwellian reemission from a wall at temperature T_w.

Hurlbut and Sherman suggest simple empirical guides for evaluating α_2, θ_r, and S_r, but specific values are not available, viz.,

1. The departure from diffuse scattering is determined by the magnitude of S_r.
2. The velocity U_r is directed at an angle numerically equal to θ.

3. The value of α_2 varies linearly from a maximum at normal incidence to a minimum at glancing incidence.

4. The value of S_r varies linearly from a minimum at normal incidence to a maximum at glancing incidence, but it never exceeds S_∞.

Upon making the assumptions described above, it was shown that

$$C_D = \frac{\chi(\sigma)}{\pi^{1/2} S_\infty} \left\{ B^{1/2} \left(S_r(2 \sin^2 \theta - 1) + \frac{\pi^{1/2}(1 + \text{erf } \sigma_r) \sin \theta}{2\chi(\sigma_r)} \right) \right.$$

$$\left. + 1 + \frac{\pi^{1/2}\sigma(1 + \text{erf } \sigma)}{2 S_\infty^2 \chi(\sigma)} \right\} \tag{11.59}$$

and

$$C_L = \frac{\chi(\sigma)}{\pi^{1/2} S_\infty} \left\{ \cos \theta \, B^{1/2} \left(\sigma_r + \frac{\pi^{1/2}(1 + \text{erf } \sigma_r)}{2 \, \chi(\sigma_r)} + S_r \sin \theta \right) \right.$$

$$\left. + \frac{\pi^{1/2}\sigma(1 + \text{erf } \sigma) \cos \theta}{4 \, S_\infty^2 \chi(\sigma)} \right\}. \tag{11.60}$$

Equations (11.59) and (11.60) give the results for flat plates if θ is treated as the angle of attack. By inserting the appropriate reference area, coefficients for wedges and cones may be calculated. Integrations are necessary to obtain values for spheres, cylinders, and other shapes with varying θ. The portions of bodies not "seen" by the oncoming freestream molecules are ignored in calculating forces when $S_\infty > 2$. When S_∞ is lower, the problem of accounting for leeside force components greatly complicates the problem.

Results for various conditions are presented in the referenced paper. Hurlbut (1994) has discussed the relations between the coefficients of Eqs. (11.43) through (11.45) and the reemitted molecular velocity distributions.

11.9 FLOW THROUGH ORIFICES AND TUBES

When two containers of gas at the same temperature are separated by a wall with an orifice, and if mean free paths are great in comparison with the orifice but much smaller than the container dimensions, then the mass flux is [Shidlovskiy (1967)]

$$\dot{m} = A(p_1 - p_2)/\sqrt{2\pi RT}, \tag{11.61}$$

where p_1, p_2 = the pressures, A = orifice area, and T = gas temperature.

In cases where there is an equilibrium gas on both sides of the wall containing the orifice and the mean free paths are very much greater than the orifice size, effusion of the gases will proceed independently in both directions, i.e., the components of a gas mixture effuse independently under collisionless or free-molecule flow conditions. Because mean molecular speed is inversely proportional to molecular mass, the amounts of different species in a gas mixture may be reduced or enhanced by the effusion process. The rates may be expressed as

$$\dot{m}_i = \rho_i (R_i T_i / 2\pi)^{1/2} A \tag{11.62}$$

where the subscript i designates a particular gas.

For tubes with free-molecular flow, an empirical formula due to Clausing is [Shidlovskiy (1967)]

$$\dot{m} = \{(20 + 8z/a)a^2(p_1 - p_2)\}\sqrt{\pi/(2RT)}/[20 + 19(z/a) + 3(z/a)^2], \tag{11.63}$$

where z and a are tube length and radius, respectively. Needless to say, severe lag-time problems may arise in experimental pressure measurements involving low-density flow through tubes or orifices. Temperature gradients induce molecular flows and affect pressures in these cases also.

Analyses of rarefied internal flows appear in the RGDS volumes and other fluid dynamics literature. Limited space prohibits the inclusion of additional material here.

11.10   TRANSITIONAL FLOW

Many of the applications involving rarefied flow lie in the *transitional flow* regime where neither continuum, thin boundary layer, nor collisionless, free-molecular flow analytical methods are adequate. Historically, attempts to devise means for engineering calculations in this area have been based on extensions of the Navier–Stokes equations to include slip effects, extensions of noncontinuum theory to include first-collision effects, or correlations of experimental data to bridge the transitional flow regime. In more recent years, the availability of increasingly powerful computers has made numerical simulation of transitional flows feasible, and the Direct Simulation Monte Carlo (DSMC) method has now become the dominant tool for predictions of rarefied flows when the cost of a computationally intensive solution is justified. In cases where quicker, less precise results for preliminary study are wanted, the bridging correlation approaches remain useful. Most of this work has concerned aerospace vehicles at hypersonic Mach numbers.

Discussions of theoretical approaches for analysis of transitional flows may be found in Bird (1976), Schaaf and Chambré (1961), and Kogan (1969). Slip and near-free-molecular flows are discussed. In addition to the moment and model-equation approaches to solution at the Boltzmann equation, Bird discusses the DSMC method. Various investigators have reported experimental measurements of aerodynamic forces, particularly drag, and heat transfer rates for simple bodies in transitional flows, cf., Sherman (1969), Potter (1967), and newer data appear in the RGDS volumes published since those reviews. Principal sources, in English, are the RGDS volumes, *The Physics of Fluids*, the *AIAA Journal*, the *Journal of Spacecraft and Rockets*, and the *Journal of Fluid Mechanics*.

The choice of correlation parameter for data on aerodynamic forces has been the subject of much discussion. For bluff shapes in continuum flow, a parameter such as a stagnation region Reynolds number, Re_0, is an effective correlating parameter. It may be defined using either the viscosity corresponding to the temperature immediately downstream of the bow shock wave or the stagnation temperature, T_0. If the latter is used, then,

$$\text{Re}_0 = \rho_\infty U_\infty r / \mu_0 = \text{Re}_\infty \, (T_\infty / T_0)^\omega \tag{11.64}$$

where ω is the exponent relating viscosity and temperature, and r is nose radius. For air or nitrogen, the exponent is approximately 0.78 when temperatures are at levels encountered in flight and near 0.90 at the low temperatures of unheated or modestly heated hypersonic wind tunnels. Owing to the difficulty and uncertainty of calculating T_0 in truly hypervelocity and frequently nonequilibrium thermochemical kinetic flow conditions, it is desirable to avoid T_0 as a parameter. Therefore, an enthalpy ratio, with $H_0 = U_\infty^2 / 2$, is sometimes used in Eq. (11.64) instead of temperature ratio, for hypersonic flows. Then, an alternate definition is

$$\text{Re}_0 = \text{Re}_\infty \, (2 H_\infty / U_\infty^2)^\omega \tag{11.65}$$

The parameter Re_0 is only appropriate for bluff bodies, i.e., blunt nosed bodies of low fineness ratio where the forces are largely determined by stagnation flow conditions. It also is limited in application to closely similar bodies for which a single length is characteristic, e.g., the nose radius.

There is an approach to a blunt-body similarity parameter involving the use of a *reference temperature* between T_0 and T_w to characterize the shock layer and boundary layer. These regions merge when Re_0 based on nose radius is below ≈ 40, forming a fully viscous shock layer, and an intermediate characteristic temperature seems appropriate. Sometimes $(T_0 + T_w)/2$ is used. Cheng and Chang (1963), Whitfield (1973), and Potter (1986) have used different parameters of that type.

When correlating data or trying to predict aerodynamic forces on bodies traversing the transitional-flow regime it is often convenient to estimate the coefficient for the higher Reynolds number, hypersonic Mach number conditions using Newtonian flow theory. Therefore, it is interesting to compare the pressure on a surface as calculated by free-molecular and Newtonian flow theories. In the former case, assuming $T_w = T_\infty, S_\infty \gg 1$, and high incidence angle,

$$p = \rho_\infty U_\infty^2 (2 - \sigma') \sin^2 \theta. \tag{11.66}$$

For Newtonian flow, it is assumed that the incoming molecules deliver all of their normal momentum to the wall, and the result is

$$p = \rho_\infty U_\infty^2 \sin^2 \theta \tag{11.67}$$

Under the *cold-wall*, high-speed conditions assumed, the difference in pressure is seen to depend on σ'. For fully diffuse reemission, $\sigma' = 1$ and the two pressures are equal. Note, however, that it does not follow that lift or drag coefficients are necessarily equal. The tangential momentum is a factor in establishing the free-molecular forces, while Newtonian flow theory disregards the influence of tangential momentum and wall temperature. Measured forces on various lifting bodies in transitional, hypersonic flow were compared with Newtonian and free-molecular theory by Boylan and Potter (1967).

More slender bodies present the major difficulty in finding an effective correlating parameter. Kogan has shown that the drag coefficients of cones in near-free-molecular flows may be greater or less than the free-molecular value, C_{Dfm}, depending on wall temperature ratio and incidence angle, θ. When $(M_\infty \sin \theta) \gg 1$, C_D does not exceed C_{Dfm}, because reflected molecules shield the body from incoming freestream molecules. However, when $(M_\infty \sin \theta) \ll 1$, the shielding is diminished, and molecules may strike the surface simply because of their random thermal velocities. Then, the C_D may exceed C_{Dfm}. Experimental evidence of this overshoot of C_D is not well established. However, DSMC calculations for slender bodies support its existence.

An effort to find an aerodynamic similarity parameter for high-speed, blunt-nosed bodies, including slender shapes of varying configurations and lifting bodies, has been described in Potter (1986), Kotov *et al.* (1985), and Potter (1989). A reference enthalpy ratio and a modified characteristic length are used to achieve a bridging correlation of a wide variety of shapes. Unfortunately, bridging methods, as well as DSMC numerical methods, suffer the uncertainty created by current lack of knowledge of the correct gas/surface interaction model. Lacking better information, fully accommodated, diffuse interaction generally is assumed.

11.11 VELOCITY SLIP AND TEMPERATURE JUMP

At the interface of a rarefied gas and a solid surface, the no-slip condition of continuum boundary layer theory is not valid. As indicated in Fig. 11.5, the slip velocity characterizes a layer of the gas adjacent to the solid surface. By a simple approach for monatomic gases, it can be shown that the slip or interface velocity, u_s, is [cf. Schaaf and Chambré (1961)]

$$u_s = (2 - \bar{\sigma})(\lambda/\bar{\sigma})(\partial u/\partial y)_w \tag{11.68}$$

The same source also gives results credited to Kennard, namely

$$u_s = (2 - \bar{\sigma})(\lambda/\bar{\sigma})(\partial u/\partial y)_w + (3\mu/4\rho T)(\partial T/\partial x)_w \tag{11.69}$$

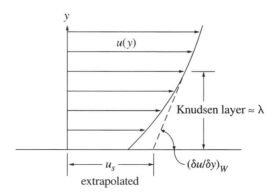

FIGURE 11.5 Slip velocity at a gas-solid interface.

and

$$T_s - T_w = [(2 - \alpha)/\alpha][2\gamma/(\gamma + 1)](\lambda/P_r)(\partial T/\partial y)_w, \qquad (11.70)$$

where

$\bar{\sigma}$ = the fraction of diffusely reflected molecules,
α = the thermal accommodation coefficient,
P_r = Prandtl number.

Subscript w indicates conditions at the wall, Subscript s indicates conditions in the gas adjacent to the wall, and x, y = coordinates along and normal to the wall, respectively. See Fig. 11.5. Kogan (1973) has reviewed more recent analyses. With the slip conditions in the form

$$u_s = \frac{A\mu}{n} \left(\frac{2}{mkT}\right)^{1/2} \partial u/\partial y + \frac{2B\mu}{nm} \partial \ln T/\partial x \qquad (11.71)$$

and

$$T_s - T_w = \frac{C\lambda}{2kn} \left(\frac{\pi m}{2kT}\right)^{1/2} \partial T/\partial y, \qquad (11.72)$$

he summarizes the work of numerous investigators. Rather than present all of these results individually, we disregard differences in basic approach and simply note that the reported values of A vary from 1.012 to 1.103, B varies from 0.3292 to 0.4456, and C varies from 0.8155 to 1.173.

It is noteworthy that the final terms in Eqs. (11.69) and (11.71) show that a temperature gradient along a surface induces a flow in a rarefied gas in the direction of increasing temperature. This phenomenon of *thermal creep* causes problems in transitional-flow pressure measurement when temperature is not equal throughout the tubing and transducer system, cf. Potter, *et al.* (1965), Kinslow and Arney (1967), Kinslow and Potter (1971), Guy and Winebarger (1967), and Potter and Blanchard (1991). Other phenomena associated with the gas flow induced by a temperature gradient are the *radiometric* and *thermophoresis* effects. Both arise because of the force of a flow from cooler to warmer regions, i.e., in the direction of the temperature gradient. Thermophoresis refers to the corresponding movement of small particles influenced by the temperature gradient. Talbot (1981) has reviewed that subject. Negative thermophoresis occurs when the particle has thermal conductivity much greater than the gas, or the particle has uniform temperature [see Sone and Aoki (1981)].

Other analyses of the velocity and temperature slip are to be found in the RGDS volumes and references cited therein, e.g., Cercignani (1977) and Loyalka (1975). Numerical studies, based on various computational techniques, also have been reported, e.g., Klemm and Giddens (1977) and Bird (1977).

11.12 DIRECT SIMULATION MONTE CARLO METHOD

Direct solution of the Boltzmann equation for the velocity distribution function is not the only means of solving transitional flow problems. The molecular nature of the gas suggests simulation by tracking the positions and velocities of thousands of representative molecules, making use of the capabilities of modern computers. By this approach the problem is formulated without having to use the distribution function. A reduced number of sample molecules are tracked, each representing millions of physical molecules. The trajectories of the molecules are followed throughout a sequence of time intervals, starting from a specified initial state. Each time a particular molecule moves, the possibility of collision with all other molecules has to be taken into account. Therefore, the computing process escalates in size with the number of molecules. Macroscopic fluid properties and the result of impacts on solid surfaces are obtained by averaging the contributions of individual molecules as they pass through computational cells or hit surfaces. A computational scheme based on probabilistic procedures reduces the effort to a feasible level. This is the Direct Simulation Monte Carlo (DSMC) method described by Bird (1976), (1977), (1978), and (1981), Moss (1986), and Nanbu (1986) and others. One of the most recent papers, which includes both a discussion of the computational technique and results pertaining to the NASA Space Shuttle Orbiter, has been presented by Rault (1993). Wilmoth *et al.* (1993) have shown how zonal decoupling enables combining Navier–Stokes based computations for higher-density forebody flows and DSMC computations for low-density wake flows. Many DSMC solutions for transition-regime flow problems have appeared in the latest RGDS volumes and technical journals.

REFERENCES

Bird, G. A., "Low Density Aerothermodynamics," *Thermophysical Aspects of Re-Entry Flows, Progress in Astronautics and Aeronautics*, Vol. 103, Moss, J. N. and Scott, C. D. (Eds.), AIAA, Washington, D.C., pp. 3–24, 1986.

Bird, G. A., "Definition of Mean Free Path for Real Gases," *Phys. Fluids*, Vol. 26, No. 11, pp. 3222–3223, 1983.

Bird, G. A., "Monte Carlo Simulations in an Engineering Context," *Rarefied Gas Dynamics: Progress in Astronautics and Aeronautics*, Vol. 74, Fisher, S. S. (Ed.), AIAA, New York, pp. 239–255, 1981.

Bird, G. A., "Monte Carlo Simulation of Gas Flows," *Annual Review of Fluid Mechanics*, Vol. 10, Van Dyke, M., Wehausen, J. V., and Lumley, J. L. (Eds.), Annual Reviews, Inc., Palo Alto, CA, pp. 11–31, 1978.

Bird, G. A., "Direct Simulation of the Incompressible Kramers Problem," *Rarefield Gas Dynamics: Progress in Astronautics and Aeronautics*, Vol. 51, Potter, J. L. (Ed.), AIAA, New York, pp. 323–333, 1977.

Bird, G. A., *Molecular Gas Dynamics*, Clarendon Press, Oxford, 1976.

Boylan, D. E. and Potter, J. L., "Aerodynamics of Typical Lifting Bodies under Conditions Simulating Very High Altitudes," *AIAA Journal*, Vol. 5, No. 2, pp. 226–232, 1967.

Cercignani, C., "Knudsen Layers: Some Problems and a Solution Technique," *Rarefied Gas Dynamics: Progress in Astronautics and Aeronautics*, Vol. 51, Potter, J. L. (Ed.), AIAA, New York, pp. 795–807, 1977.

Cercignani, C., *Theory of Application of the Boltzmann Equation*, Elsevier, New York, 1975.

Cercignani, C. and Lampis, M., "Kinetic Models for Gas-Surface Interaction," *Transport Theory and Statistical Physics*, Vol. 1, No. 2, pp. 101–114, 1971.

Cercignani, C., *Mathematical Methods in Kinetic Theory*, Plenum, New York, 1969.

Cheng, H. K. and Chang, A. L., "Stagnation Region in Rarefied, High Mach Number Flow," *AIAA Journal*, Vol. 1, No. 1, pp. 231–233, 1963.

Guy, R. W. and Winebarger, R. M., "Effect of Orifice Size and Heat-Transfer Rate on Measured Static Pressures in a Low-Density Arc-Heated Wind Tunnel," NASA TN D-3829, February 1967.

Hurlbut, F. C., "Two Contrasting Modes for the Description of Wall-Gas Interactions," *Rarefied Gas Dynamics: Progress in Astronautics and Aeronautics*, Vol. 158, Shizgal, B. D. and Weaver, D. P. (Eds.), AIAA, Washington, D.C., pp. 494–506, 1994.

Hurlbut, F. C., "Particle Surface Interactions in the Orbital Context," *Rarefied Gas Dynamics: Progress in Astronautics and Aeronautics*, Vol. 116, Muntz, E. P., Weaver, D., and Campbell, D. (Eds.), AIAA, Washington, D.C., pp. 419–450, 1989.

Hurlbut, F. C., "Gas–Surface Scatter Models for Satellite Applications," *Thermophysical Aspects of Re-Entry Flows, Progress in Astronautics and Aeronautics*, Vol. 103, Moss, J. N. and Scott, C. D. (Eds.), AIAA, Washington, D.C., pp. 97–119, 1986.

Hurlbut, F. C., "Current Developments in the Study of Gas–Surface Interactions," *Rarefied Gas Dynamics*, Vol. 1, Brundin, C. L. (Ed.), Academic Press, New York, pp. 1–34, 1967.

Hurlbut, F. C. and Sherman, F. S., "Application of the Nocilla Wall Reflection Model to Free-Molecule Kinetic Theory," *Phys. Fluids*, Vol. 11, No. 3, pp. 486–496, 1968.

Ikenberry, E. and Truesdell, C., "On the Pressures and the Flux of Energy in a Gas According to Maxwell's Kinetic Theory I," *J. Rational Mech. Anal.*, Vol. 5, No. 1, pp. 1–128, 1956.

Kemp, N. H., "Comment on Hypersonic Free Molecular Heating on Micron Size Particulate," *AIAA Journal*, Vol. 17, No. 4, pp. 445–446, 1979.

Kinslow, M. and Potter, J. L., "Reevaluation of Parameters Relative to the Orifice Effect," *Rarefied Gas Dynamics*, Vol. 1, Dini, D. (Ed.), Edizioni Tecnico-Scientifiche, Pisa, pp. 399–408, 1971.

Kinslow, M. and Arney, G. D., "Thermo-Molecular Pressure Effects in Tubes and at Orifices," AGARDograph No. 119, Advisory Group for Aeronaut. Res. and Dev., 1967.

Klemm, F. J. and Giddens, D. P., "A Numerical Solution to the Problem of Low-Density Hypersonic Wedge Flow," *Rarefied Gas Dynamics: Progress in Astronautics and Aeronautics*, Vol. 51, Potter, J. L. (Ed.), AIAA, New York, pp. 313–322, 1977.

Kogan, M. N., "Molecular Gas Dynamics," *Ann. Rev. Fluid Mech.*, Vol. 5, Van Dyke, M., Vincenti, W. G., and Wehausen, J. V. (Eds.), Annual Reviews, Inc., Palo Alto, CA, pp. 383–404, 1973.

Kogan, M. N., "Recent Developments in the Kinetic Theory of Gases," *Rarefied Gas Dynamics*, Vol. 1, Trilling, L. and Wachman, H. Y. (Eds.), Academic Press, New York, pp. 1–39, 1969.

Kogan, M. N., *Rarefied Gas Dynamics* (English trans., Trilling, L., Ed.), Plenum Press, New York, 1969.

Kotov, V. M., Lychkin, E. N., Reshetin, A. J., and Schelkonogov, A. N., "An Approximate Method of Aerodynamic Calculation of Complex Shape Bodies in a Transitional

Region,'' *Rarefied Gas Dynamics*, Vol. 1, Belotserkovskii, O. M., Kogan, M. N., Kutateladze, S. S., and Rebrov, A. K. (Eds.), Plenum Press, New York, pp. 487–494, 1985.

Loeb, L. B., *The Kinetic Theory of Gases*, 3rd ed., Dover, New York, 1961.

Loyalka, S. K., ''Velocity Profile in the Knudsen Layer for the Kramer's Problem,'' *Phys. Fluids*, Vol. 18, No. 12, pp. 1666–1669, 1975.

Moss, J. N., ''Direct Simulation of Hypersonic Transitional Flow,'' *Rarefied Gas Dynamics*, Boffi, V. and Cercignani, C. (Eds.), B. G. Tuebner, Stuttgart, pp. 384–399, 1986.

Muntz, E. P., ''Rarefied Gas Dynamics,'' *Annual Review of Fluid Mechanics*, Vol. 21, Lumley, J. L., Van Dyke, M., and Reed, H. L. (Eds.), Annual Reviews, Inc., Palo Alto, CA, 1989.

Nanbu, K., ''Theoretical Basis of the Direct Simulation Monte Carlo Method,'' *Rarefied Gas Dynamics*, Boffi, V. and Cercignani, C. (Eds.), B. G. Tuebner, Stuttgart, 1986.

Nocilla, S., ''The Surface Re-Emission Law in Free Molecular Flow,'' *Rarefied Gas Dynamics*, Laurmann, J. (Ed.), Academic Press, New York, pp. 327–346, 1963.

Patterson, G. N., *Introduction to the Kinetic Theory of Gas Flows*, Univ. of Toronto Press, Toronto, 1971.

Patterson, G. N., *Molecular Flow of Gases*, John Wiley & Sons, New York, 1956.

Potter, J. L., ''The Transitional Rarefied Flow Regime,'' *Rarefied Gas Dynamics*, Vol. 2, Brundin, C. L. (Ed.), Academic Press, New York, pp. 881–937, 1967.

Potter, J. L., ''Procedures for Estimating Aerodynamics of Three-Dimensional Bodies in Transitional Flow,'' *Rarefied Gas Dynamics: Progress in Astronautics and Aeronautics*, Vol. 118, Muntz, E. P., Weaver, D., and Campbell, D. (Eds.), AIAA, Washington, D.C., pp. 484–488, 1989.

Potter, J. L., ''Transitional, Hypervelocity Aerodynamic Simulation and Scaling,'' *Thermophysical Aspects of Re-Entry Flows, Progress in Astronautics and Aeronautics*, Vol. 103, Moss, J. N. and Scott, C. D. (Eds.), AIAA, Washington, D.C., pp. 79–96, 1986.

Potter, J. L. and Blanchard, R., ''Thermomolecular Effect on Pressure Measurements with Orifices in Transitional Flow,'' *Rarefied Gas Dynamics*, Beylich, A. E. (Ed.), VCH Publishers, Weinheim, Germany and New York, pp. 1459–1465, 1991.

Potter, J. L., Kinslow, M., and Boylan, D. E., ''An Influence of the Orifice on Measured Pressures in Rarefied Flow,'' *Rarefied Gas Dynamics*, Vol. 2, deLeeuw, J. H. (Ed.), Academic Press, New York, pp. 175–194, 1965.

Present, R. D., *Kinetic Theory of Gases*, McGraw-Hill, New York, 1958.

Rault, D. F. G., ''Aerodynamics of Shuttle Orbiter at High Altitudes,'' AIAA Paper 93-2815, 1993.

Schaaf, S. A., ''Mechanics of Rarefied Gases,'' *Encyclopedia of Physics*, Vol. VIII/2, Flugge, S. and Truesdell, C. (Eds.), Springer–Verlag, Berlin, 1963.

Schaaf, S. A. and Chambré, P. L., *Flow of Rarefied Gases*, Princeton University Press, Princeton, NJ, 1961.

Sherman, F. S., ''The Transition from Continuum to Molecular Flow,'' *Ann. Rev. Fluid Mech.*, Vol. 1, Sears, W. R. and Van Dyke, M. (Eds.), Annual Reviews, Inc., Palo Alto, CA, pp. 317–340, 1969.

Shidlovskiy, V. P., *Introduction to the Dynamics of Rarefied Gases* (English transl., Laurmann, J. A., Ed.), American Elsevier, New York, 1967.

Sone, Y. and Aoki, K., ''Negative Thermophoresis: Thermal Stress Slip Flow Around a Spherical Particle in a Rarefied Gas,'' *Rarefied Gas Dynamics: Progress in Astronautics and Aeronautics*, Vol. 74, Fisher, S. S. (Ed.), AIAA, New York, pp. 489–503, 1981.

Talbot, L., "Thermophoresis—A Review," *Rarefied Gas Dynamics: Progress in Astronautics and Aeronautics*, Vol. 74, Part 1, Fisher, S. S. (Ed.), AIAA, New York, pp. 467–488, 1981.

Whitfield, D. L., "Mean Free Path of Emitted Molecules and Correlation of Sphere Drag Data," *AIAA Journal*, Vol. 11, No. 12, pp. 1666–1669, 1973.

Wilmoth, R. G., Mitcheltree, R. A., Moss, J. N., and V. K. Dogra, "Zonally-Decoupled DSMC Solutions of Hypersonic Blunt Body Wake Flows," AIAA Paper 93-2808, 1993.

12 Unsteady Flows

TURGUT SARPKAYA
Naval Postgraduate School
Monterey, CA

CONTENTS

12.1 NATURE OF UNSTEADY MOTION

12.1.1 Unsteadiness, Diffusion, and Memory

The unsteady motion is of great interest in the solution of many applied technical problems in fluid mechanics, such as the motion of bodies through fluids, fluid motion in or about bodies, free surface flow phenomena, and the motion of fluids inside machinery. In the motion of all viscous fluids (here assumed to be Newtonian and incompressible) vorticity, $\overline{\omega} = \Delta \times \overline{q} = \epsilon_{ijk}q_{k,j}$ where $\overline{q}$ is the velocity, plays a

Handbook of Fluid Dynamics and Fluid Machinery, Edited by Joseph A. Schetz and Allen E. Fuhs
ISBN 0-471-12598-9 Copyright © 1996 John Wiley & Sons, Inc.

fundamental role. It is no exaggeration to state that the knowledge of the distribution of vorticity at any instant throughout the flow field is equivalent to the understanding of the fluid motion. Vorticity acquires an added importance in time-dependent flows, because the *memory* of fluid motion resides in its vorticity. The flow of an inviscid fluid instantly responds to the prevailing conditions. It cannot acquire or diffuse vorticity. Neither the past history nor the future state of its motion has any role to play in the determination of its instantaneous character. The awful correctness and unfailing promptness of the inviscid flow make researchers appreciate even more the challenges offered by the consequences of viscosity. Only viscous fluids can acquire and diffuse vorticity and, thereby, remember the past and anticipate the future.

It is natural that unsteadiness be intimately related to vorticity, or better, to the diffusion of vorticity. In fact, if diffusion has sufficient time to adjust to the unsteady conditions imposed on the flow, it may be said that the motion is a juxtaposition of steady states or a slowly varying unsteady flow. In this case, the effects of unsteadiness are negligible. If the diffusion cannot adjust to the conditions imposed on the flow, each succeeding state will be increasingly affected not only by the prevailing conditions but also by the past history of the motion. How far back in time and/or space the flow will remember depends on the nature of unsteadiness, on the nature of diffusion (molecular or turbulent), and on the state of flow (e.g., separated, unseparated, and transitional). For example, the upstream history has a profound effect on a boundary layer developing under a decelerating outer flow, probably due to the occurrence of separation and flow reversal near the wall. On the other hand, a boundary layer developing under an accelerating outer flow quickly forgets its early history.

The implication of the few available results is that flows with turbulent boundary layers respond in a more nearly instantaneous fashion than those with laminar boundary layers to changes in the characteristics of the ambient flow because the turbulent diffusion is much larger than the molecular diffusion, i.e., one expects much faster adjustment to the time-dependent conditions imposed on the flow. There are of course many transient cases where it is not possible to decompose the measured velocity into a conditionally-averaged velocity (at a discrete frequency which lies outside the spectrum of turbulence), a mean velocity, and the random fluctuations (the so-called *triple decomposition*). In turbulent flow, even strong unsteadiness may not directly affect the flow; it may, however, significantly change the character of the flow if the unsteadiness leads to transition or to flow separation (assuming none existed before), or to the significant excursion of the separation point(s) if one had already existed*. In other words, turbulence may weaken the memory of flow, but the imposed unsteadiness may completely change its character. There are a number of other turbulent flow situations where the unsteadiness may indirectly but significantly change the flow characteristics and flow/structure interactions. Some of these will be taken up later. Suffice it to note that complexities of unsteady nonequilibrium turbulence with or without separation under a variety of time-dependent flow conditions (periodic flows, transverse or bidirectional dynamic response of bodies) are far from understood. Even in major review articles, the subject is dealt with surprising brevity. For example, Bradshaw (1994), in his review of turbulence, noted only briefly

---

*Here it is assumed that the transition and separation point excursions are not due to unknowable facility-related constraints (*unsteady facility interference coupling*).

that "the effect of unsteadiness can be understood in the same way as that of pressure gradient—of course, unsteadiness is usually forced by a streamwise pressure gradient" and "if the pressure gradient is strong enough to cause separation (however defined), the internal layer is carried into the outer part of the flow and the "slip velocity" concept breaks down, as it would in steady separation."

12.1.2 Classification of Incompressible Time-Dependent Flows

There are an infinite number of time-dependent flows, and one can describe only a few that are caused by some well-defined time-dependence of the ambient flow and/or the fluid/structure interaction (e.g., change of freestream velocity; large-amplitude, time-dependent oscillations of the mean flow; the impulsive start from rest; change of angle of attack; motion of helicopter blades, dynamic response of a cable, or even turning of this page or closing of this book!). Nevertheless, it is useful to distinguish between several types of time-dependent flows. These may be ideal (inviscid), laminar, transitional, turbulent, and compressible or incompressible. We shall confine ourselves to a few basic, but practically significant, incompressible and mostly separated flows. These may be broadly classified as *effectively-inviscid unsteady flows, unseparated unsteady laminar flows, and separated, unsteady, laminar/turbulent flows*. The discussion of stability and turbulence will be excluded since they are not necessarily a consequence of externally imposed transient or periodic changes.

The unsteady flows may also be classified as slowly varying flows, flows with significant but manageable time-dependence, and flows with significant as well as arbitrary time-dependence. Defining x_o to be a reference length, U_o to be a reference velocity, and $(dU/dt)_o$ to be a reference acceleration, we note that both $x_o(dU/dt)_o/U_o^2$ and $(x_o^2/\nu)(1/U_o)(dU/dt)_o$ must be considered in assessing the significance of the local acceleration to the convective acceleration and to the rate of diffusion (one of the two parameters may be replaced by the Reynolds number $U_o x_o/\nu$). One may also state that a flow is unsteady if it cannot be approximated by or reduced to a juxtaposition of steady states, i.e. to flows with negligible or no history effects. In general the types of unsteady flows investigated (and some understood) are determined by the tractability of the flow through the use of the available or evolving means, one's ability to carry out theoretical analysis and physical and numerical experiments, the real or perceived needs for the investigation, and by the practical requirements of emerging technologies.

Temple (1953) reviewed the progress that had been made until 1953 in the theory of unsteady flows about wings and slender bodies. The aerodynamic aspect of unsteady flows has been reviewed by Jones (1962), Garrick (1966), Belotserkovskii (1977), and McCroskey (1982), among others. The phenomenon of periodic vortex shedding from a symmetrical bluff body has been the subject of extensive review [Rosenhead (1953), Marris (1964), Wille (1966), and, more recently by Oertel (1990)]. The solution of separation-free time-dependent laminar flows has enjoyed particular attention [Stewartson (1960), Stuart (1963), and Rott (1964)] due to its relative conceptual and mathematical simplicity. The hydrodynamics of unsteady flows, in particular those set in motion impulsively from rest, is most aptly described by Sedov (1965). The foregoing list of general references is by no means complete and there are, to be sure, a great many other text books [e.g., Schlichting (1979),

Pironneau, *et al.* (1992)], and review papers [Sarpkaya (1992)] dealing directly or peripherally with the unsteady motion of viscous or inviscid, separated, and cavitating flows.

In what follows, we will discuss inviscid flows, a few classical solutions of unsteady laminar flows, definitions of unsteady separation, and then a series of theoretically as well as practically important time-dependent flows. These are the types of flows which are of extreme importance in the scientific and engineering disciplines, and, they go far beyond merely being experimental and computational curiosities.

12.2 UNSTEADY FLOW OF INVISCID FLUIDS

12.2.1 Added Mass

The concept of *added mass* has several applications besides those of the problem of acceleration from rest. As examples of such applications, the following may be cited: the correction of the measured drag coefficient in wind tunnels with diverging walls, the estimation of the aerodynamic forces in cases of buffeting and flutter, the calculation of the periods of heaving, pitching, and rolling motions of a ship, the estimation of the initial acceleration of a fluid in the emptying of a reservoir [Sarpkaya (1962)], and, in general, in the calculation of forces and accelerations wherever the kinetic energy imparted to the fluid by the body is not negligible relative to the kinetic energy of the body itself.

The rigor of potential flow theory may be applied directly to the prediction of the characteristics of a class of fluid flows in which either boundary-layer separation does not occur or the stream surface constitutes an interface between a liquid and a gas (treated through the use of the *free-streamline theory* [see e.g., Birkhoff and Zarantonello (1957)]. For real fluids, the applicability of potential flow results is subject to the usual qualifications due to the effects of viscosity and separation, however small the viscosity may be. These will be discussed later.

The unseparated, unsteady, potential flows, most of which result from the unidirectional or periodic, linear and/or angular acceleration of bodies in an infinite or bounded fluid medium, give rise to an induced- or virtual-tensor inertia which must be added to the real mass or the real mass-moment-of-inertia of the body. In general, the virtual tensor inertia depends on the shape of the body, the type of the body motion, the nature of the fluid medium, the orientation of the body with respect to the direction of motion, and free surface or body-proximity effects.

The added mass and the added mass-moment of inertia may be determined either through the evaluation of the kinetic energy of the fluid in terms of the velocity potential or through the calculation of the *drift volume* from the analysis of the trajectories of the fluid particles*.

The kinetic energy of the flow may be written as

$$2T = \rho \iint |\text{grad } \phi|^2 \, ds = -\rho \iint \phi \, \frac{\partial \phi}{\partial n} \, ds \qquad (12.1)$$

---

*Darwin (1953) has shown that, for an unbounded fluid, the added mass really represents a mass of fluid (drift volume) entrained by the body during its motion.

for which

$$2T = \lambda_{ik} U_i U_k \tag{12.2}$$

where

$$\lambda_{ik} = \rho \int\int \phi_i n_k \, ds \tag{12.3}$$

The coefficients of λ_{ik} are symmetric and are called the coefficients of *virtual mass*. Because of symmetry, there are only 21 (not 36) distinct components of λ_{ik}. In an inviscid cross flow, the added mass coefficient* is $C_a = 1$ for a circular cylinder and $C_a = 1/2$ for a sphere. Once the parameters cited above (the body shape, and the type and direction of motion) are specified, the added mass coefficient remains invariant even if, for example, the amplitude of an imposed translatory or sinusoidal motion changes. The values of C_a have been evaluated for a large number of two- and three-dimensional bodies and extensive tables may be found in many texts [e.g., Sarpkaya and Isaacson (1981)].

There are powerful mathematical and numerical techniques available [Dryden *et al.* (1956), Segel (1960), Sarpkaya (1962), and Birkhoff and Zarantonello (1957)] for the evaluation of λ_{ik} for both two- and three-dimensional motions (except those for pure rotation) provided that the irrotational motion in the infinite fluid outside the body is known in terms of the internal singularities or surface singularity distributions (vortex element and panel methods) which cause it. Comprehensive reviews of the vortex methods are given by Sarpkaya (1989, 1994a). For quite arbitrary shapes, for which the evaluation of a suitable potential function may be quite difficult, and for bodies undergoing either impulsive or periodic motions in the proximity of a free surface, the evaluation of the virtual tensor inertia presents considerable, if not insurmountable, difficulties. Fortunately, however, these difficulties do not stem from a lack of understanding of the concept but rather from the lengthy computations involved in the approximations to the body using distributed singularities and their images. Cases of this type, such as the evaluation of the fluid free-surface-proximity effect on a body accelerating from rest, may be treated by a combination of theoretical and experimental methods [Waugh and Ellis (1969) and Sarpkaya (1960)]. It should be noted, in passing, that the effect of compressibility and the proximity of rigid walls is such as to increase the added mass, whereas, in the case of liquids, the effect of the free surface proximity is to decrease it. As a matter of fact, an infinite extent of fluid yields the maximum value of the added mass. The minimum value is reached when the object is moving just below a free surface. The experimental determination of the added mass for linear translation is often carried out through the use of forced oscillations in a viscous fluid with relative amplitudes (A/D) as small as possible so as to make the motion mimic inviscid flow. Even then the effects of separation cannot be avoided for some body shapes, as discussed in more detail later.

---

*The added mass coefficient C_a is often defined as the ratio of the added mass to the displaced mass of the body. However, there are exceptions for bodies with zero displaced mass, e.g., a thin disk or a plate moving normal to its plane. Then, a suitable effective displaced mass defined by the characteristic lengths of the body is used.

The differences between the ideal case of a body vibrating in an inviscid fluid and the nonideal case of a body vibrating in a viscous fluid demonstrate as powerfully as any other comparison in fluid dynamics the complexities resulting from the separation of flow and the creation, convection, and diffusion of vorticity.

12.3 UNSTEADY UNSEPARATED LAMINAR FLOWS

12.3.1 Introduction

Unseparated, unsteady laminar flows have been more thoroughly studied than any other class of flows partly because of their amenability to mathematical treatment and partly because many of them serve as the point of departure towards the understanding of stability and transition to turbulence. The solutions of the impulsive and periodic motions of a plate or cylinder, oscillatory fluid motion in or about a tube, unsteadily rotating bodies, just to name a few, have become classics of this family of problems. One can hardly improve on the treatments of the subject by Stewartson (1960), Stuart (1963), Rott (1964), Schlichting (1979), and Meksyn (1961). The present treatment is a brief review of some of these solutions with the objective of going beyond the classical solutions, towards real world problems.

12.3.2 Rayleigh's Problem

The analysis of flow over an infinite flat plate set in motion impulsively from rest in its own plane with velocity U_o shows that $u(y, t)$ is given by

$$u/U_o = \text{erfc } \eta \quad \text{with } \eta = y/(2\sqrt{\nu t}) \qquad (12.4)$$

where

$$\text{erfc } \eta = \frac{2}{\sqrt{\pi}} \int_{\pi}^{\eta} \exp(-\eta^2)\, d\eta = 1 - \text{erf } \eta \qquad (12.5)$$

in which erf is the *error function* and erfc is the *complementary error function*. Equation (12.4) shows that u/U_o tends to unity for a given value of η and the boundary layer thickness δ, defined as the value of η at which $u/U_o = 0.01$, reduces to $\delta = 4\sqrt{\nu t}$. The solution for a cylinder set in *rotational* motion impulsively from rest is given by Mallick (1957).

12.3.3 Stokes' Second Problem

Stokes provided the first theoretical solution of the motion of an unbounded fluid over an infinite flat plate executing sinusoidal oscillations with $u = U_o \sin \omega t$ along its plane (known as Stokes' second problem). Since then, considerable theoretical and experimental research have been carried out on *oscillatory flow* (with *zero mean*), and pulsatile flow (with nonzero mean) in pipes and over flat plates, including progressive water waves over rigid or mobile bottoms, through the use of velocity and turbulence measurements, flow visualization, and force measurements [for *pulsatile flow*, see e.g., Lighthill (1954), Sarpkaya (1966a), Mankbadi and Liu (1992), and for *oscillatory flow*, see, e.g., Li (1954), Spalart and Baldwin (1989), Eckmann and Grotberg (1991, 1994), and Sarpkaya (1993)]. The reasons most frequently invoked for these studies range from a search for physical insight as to how the transition process, bursting events, and turbulence production mechanisms are affected by flow periodicity to the potential application of the resulting understanding to the design of equipment and improvement of processes subjected to transient or pulsatile flow conditions (e.g., heat and mass transfer increases over that in a corresponding steady flow, respiratory ventilators, turbomachinery and fluid control systems, wave-height damping, and sediment movement by water waves, and transition to turbulence in the large arteries).

Stokes' classical solution which may be found in many textbooks is given by

$$u/U_o = \sin\left(\omega t - \frac{k}{\sqrt{2}} y\right) \exp\left(-\frac{k}{\sqrt{2}} y\right) \tag{12.6}$$

where $k = \sqrt{\omega/\nu}$. Equation (12.6) represents a strongly damped wave as shown in Fig. 12.1. This flow is stable for Re_δ (= $U_o \delta/\nu$, with $k\delta = \sqrt{2}$) less than about 420, above which the oscillatory motion gives rise to quasi-coherent structures [Sarpkaya (1993)]. Numerous variations to Stokes second problem (e.g., a second plate, at rest or in motion, above the first; uniform, impulsive, or arbitrary motions of the plate with uniform suction) have been solved [see, e.g., Rott (1964), Stuart (1963), and Panton (1984)].

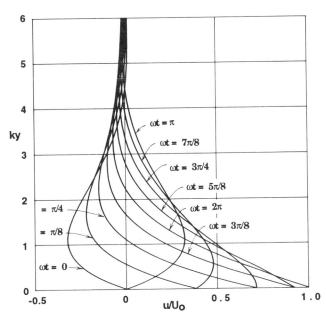

FIGURE 12.1 Velocity distribution in the neighborhood of an oscillating wall.

12.3.4 Oscillatory Flow along a Cylinder

An interesting variation to Stokes second problem, dealing with the effects of transverse curvature on oscillatory flow along the exterior of a circular cylinder, is presented by Casarella and Laura (1969) and by Chew and Liu (1989). In this case, the equation of motion in cylindrical coordinates reduces to

$$\frac{\partial u}{\partial t} = \nu\left(\frac{\partial^2 u}{\partial r^2} + \frac{1}{r}\frac{\partial u}{\partial r}\right) \tag{12.7}$$

With the usual boundary conditions: $r = a$, $u = U_o \cos \omega t$ and $r = \infty$, $u = 0$, the solution may be written as

$$\frac{u(r, t)}{U_o} = \frac{ker_o(ka)ker_o(kr) + kei_o(ka)kei_o(kr)}{ker_o^2(ka) + kei_o^2(ka)} \cos \omega t$$

$$+ \frac{kei_o(ka)ker_o(kr) - ker_o(ka)kei_o(kr)}{ker_o^2(ka) + kei_o^2(ka)} \tag{12.8}$$

in which *ker* and *kei* (and their cousins ber and bei and her and hei) are known as *Thomson functions* and are related to Bessel functions with complex arguments [see, e.g., Gradshteyn and Ryzhik (1965)].

As expected from the flat plate case, the velocity profiles have the form of damped harmonic oscillations. In the limit, as $a \to \infty$, Eq. (12.8) reduces to Eq. (12.6). In fact, the velocity profiles of the two cases are virtually the same for *ka* larger than about 5. The impulsive motion of an infinite cylinder parallel to its axis is discussed by Stuart (1963).

12.3.5 Oscillating Flow about Cylinders and Spheres

The theory of sinusoidally oscillating flow about bodies or the sinusoidal oscillation of a body in a viscous fluid otherwise at rest has long been of special interest to fluid dynamicists. The most celebrated case among these is the sinusoidal oscillation of a cylinder normal to its axis. Assuming that ν/fD^2 and U_o/fD are sufficiently small and that one is interested only in the calculation of the first order oscillatory motion together with the dominant part of the resulting second-order steady motion, one can use the boundary layer simplification given by

$$\frac{\partial u}{\partial t} - \nu\frac{\partial^2 u}{\partial y^2} - \frac{\partial U}{\partial t} = U\frac{\partial U}{\partial x} - u\frac{\partial u}{\partial x} - v\frac{\partial u}{\partial y} \tag{12.9}$$

where the terms on the right-hand are assumed to be small for the case under consideration (of order U_o/fD). Schlichting (1932), among others, has shown that a steady motion (streaming) is generated by the Reynolds stress associated with the oscillatory part of the flow within the boundary layer [Figs. 12.2(a) and (b)]. The solution is known to be valid for cases where the Reynolds numbers are small enough and the boundary layer simplification is a valid approximation. For relatively large

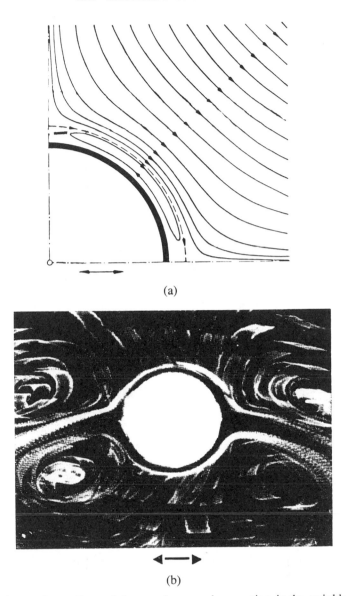

(a)

(b)

FIGURE 12.2 (a) Streamlines of the steady secondary motion in the neighborhood of an oscillating circular cylinder; (b) Secondary flow in the neighborhood of an oscillating circular cylinder. (Reprinted with permission from McGraw-Hill, *Boundary Layer Theory*, 7th ed., by Schlichting, H., and Kestin, J., 1979.)

Reynolds numbers (e.g., $\text{Re} = U_o^2/f\nu$ larger than about 1800), there exists a second, outer boundary layer within which the nonlinear inertia terms cannot be neglected.

The prediction of forces (shear and pressure) requires that the boundary layer assumption be abandoned and, instead, a linearized form of the Navier–Stokes equations be used. In one of his many monumental contributions, Stokes (1851) presented the solutions for both a cylinder and a sphere assuming that the amplitude of oscillations is small and, the flow about the bodies is laminar, unseparated, and stable. For a circular cylinder, the force F per unit length may be reduced to

$$F = (\pi \rho a^2) U_o \omega (C_a \sin \omega t - C' \cos \omega t) \tag{12.10}$$

in which C_a is the added mass coefficient, expressed in terms of the displaced mass $M' = (\pi \rho a^2)$, and C' is a damping coefficient, since the second part of the equation, namely $-C'M'U_o \omega \cos \omega t$, is out of phase with the acceleration and opposes the motion of the body. Both of these coefficients depend on $\beta = fD^2/\nu$ since the motion is controlled by viscous forces as well as the pressure forces. In inviscid fluids, only the latter is present and $C_a = 1$ for a circular cylinder and $C_a = 0.5$ for a sphere.

The coefficients C_a and C' decrease with increasing β. In fact, for a cylinder, for large values of β, one has

$$C_a = 1 + 4(\pi\beta)^{-1/2} \quad \text{and} \quad C' = 4[(\pi\beta)^{-1/2} + (\pi\beta)^{-1}] \tag{12.11}$$

For a sphere in sinusoidal motion, Stokes (1851) found

$$C_a = \tfrac{1}{2}[1 + 9(\pi\beta)^{-1}] \quad \text{and} \quad C' = 9(\pi\beta)^{-1} + \tfrac{9}{2}(\pi\beta)^{-1/2} \tag{12.12}$$

Once again, C' is the coefficient of $-M'\omega U_m \cos \omega t$. Wang (1968) extended Stokes analysis to $O[(\pi\beta)^{-3/2}]$ using the method of inner and outer expansions. His solution, valid for $A/D \ll 1$ and $\beta \gg 1$, for stable, unseparated laminar flow oscillating sinusoidally about a cylinder, may be reduced to

$$C_a = 1 + 4(\pi\beta)^{-1/2} + (\pi\beta)^{-3/2} \quad \text{and} \quad C' = 4(\pi\beta)^{-1/2} + 4(\pi\beta)^{-1} - (\pi\beta)^{-3/2}$$

$$\tag{12.13}$$

which may be compared with Eq. (12.11). The results differ only in the last terms and yield virtually identical results in the range of their validity. The discussion of these results for engineering applications will be taken up later.

There are, to be sure, a number of other unseparated unsteady laminar flow solutions dealing with porous flat plates, flow in pipes, stagnation flows, and heat transfer and temperature fluctuations. The extensive discussion of these may be found in classical reference books [e.g., Rosenhead (1963)].

12.3.6 A Decaying Vortex

Where there is continuous matter (air, liquids, gases, plasma of hot ionized gas) or interfaces of continuous matter (e.g., air/ocean interface) or discontinuous accumulations of all forms and states of aggregation of matter, vortices may be generated through the direct or indirect action of one or more forces of nature (density, temperature and pressure gradients, friction, gravity, and electrodynamic forces). Simply by turning this page over, one can impart rotation to air particles, generate new vorticity, circulation, numerous vortices, oppositely-signed vorticity, vortex-induced vibration, *Aeolian* tones, and a hierarchy of smaller scales of vorticity over which the larger structures distort, interact, convect and diffuse. Vortices are nature's favorite carriers of vorticity. Their beauty attracts attention; their power demands respect.

Vortices come in many shapes (circular, asymmetrical, spiral, columnar, disklike, ring like, etc.) and may appear in many forms: Quantized vortices, turbulent eddies, vortex rings, trailing vortices, a street of staggered vortices behind bodies, dust whirls, whirlpools, waterspouts, tornadoes (much feared funnel-shaped vortices of great power), hurricanes, ocean circulations, convection cells, intense circulations of the atmosphere (e.g., the great red spot of Jupiter), and spiral galaxies. Almost all three-dimensional vortices (e.g., tornadoes, hurricanes, and trailing vortices) have axial velocities near their core. Vortices may vary in size from as small as one angstrom, as in quantized vortices in Helium II, to as large as many light years, as in galaxies and *black holes*.

In a viscous fluid, diffusion could be used to create a nearly irrotational vortex motion (i.e., by rotating a single cylinder whose length is large compared with its radius) where circulation is approximately zero about all closed curves except those enclosing the origin. The establishment of a potential vortex in this manner requires infinite time, infinite energy, and the continual application of a shear stress at all radial distances by an amount equal to twice the dynamic viscosity times the angular velocity of the cylinder. This motion is not steady and the exact solution is not reducible from one based on the steady state solution of viscous flow between rotating concentric cylinders.

The growth of the core of a viscous vortex constitutes a fundamental solution of the time-dependent Navier–Stokes equations. The two dimensional version of the vorticity equation with the now scalar vorticity $\omega = \partial(rv)/r\,\partial r$ reduces to

$$\frac{\partial \omega}{\partial t} = \nu\left(\frac{\partial^2 \omega}{\partial r^2} + \frac{1}{r}\frac{\partial \omega}{\partial r}\right) \tag{12.14}$$

whose exact solution for an initially singular vortex-type motion is given by Oseen (1912) as

$$v = \frac{\Gamma_o}{2\pi}\left(1 - \exp\left(-\frac{r^2}{4\nu t}\right)\right) \tag{12.15}$$

which is valid only for *a single viscous vortex in an unbounded incompressible domain* and $\sqrt{2\nu t} = \sigma$ is the standard deviation of the vorticity distribution. The radius at which the tangential velocity reaches a maximum is $r_m = 2.24\sqrt{\nu t} = 1.584\,\sigma$. Obviously, *not even a single vortex whose entire vorticity is confined to an invariant finite core* (e.g., a Rankine vortex) is *an exact solution of the Navier–Stokes equations*. A single Oseen vortex (which has an infinite support in an unbounded domain) is an exact solution. However, the velocity field of a multi-Oseen-vortex system is not an exact solution, because the nonlinearity of the Navier–Stokes equations does not permit the superposition, without deformation, of a finite number of vortex fields. Interestingly enough, a linear sum of Oseen vortices *is* an exact solution of the linear *diffusion equation*, $(\partial \omega/\partial t = \nu \nabla^2 \omega)$, and it is often used to simulate viscous effects through the use of vortex element methods [see e.g., Sarpkaya (1989), (1994a)] as long as the convection and diffusion are treated separately.

12.4   UNSTEADY SEPARATED FLOWS

12.4.1   Unsteady Separation

The requirement for separation point specification constitutes the weakest link in the analysis of many steady and unsteady flows about *bluff bodies*. Even in *steady* two-dimensional flow, separation points can be predicted only approximately for laminar flows and hardly at all for turbulent flows. In unsteady flow, mobile separation points (when they are not fixed by sharp edges), may undergo large excursions. This experimental fact renders the treatment of boundary layers on bluff bodies subjected to periodic wake return extremely difficult, particularly when the state of the boundary layer changes during a given cycle. As noted by Simpson (1989), "Imposed and self-induced unsteadiness strongly influence detached flows. Detachment from sharply diverging walls leads to large transitory stall in diffusers and flapping shear layers around bluff bodies. Large-scale motions produced in such cases do not contribute much to turbulent shear stresses but change the mean flow and produce low-frequency pressure fluctuations. This behavior makes such flows difficult to calculate."

A definition of separation that is meaningful for all kinds of *unsteady flows* has not yet been established. It is now a well-known fact that the use of vanishing surface shear stress or flow reversal as the criterion for separation is unacceptable. In unsteady flow, the surface shear can change sign with flow reversal, but without *breakaway*. Conversely, the boundary layer concept may break down before any flow reversal is encountered. The many criteria that have been proposed require the use of laminar boundary-layer equations (some at an infinite Reynolds number!) and *a priori* knowledge of certain flow features, such as the freestream velocity, which must be identified either experimentally or theoretically. The so-called *M-R-S criterion* (the proposition that the unsteady separation is the simultaneous vanishing of the shear and the velocity at a point within the boundary layer, and away from the solid surface, in a coordinate convected with the separation velocity) requires *a priori* knowledge of the speed of separation [Moore (1957), Rott (1956), and Sears (1956)]. The outer flow is usually unknown and affected in an unknown manner by the boundary layer itself.

In calculations of laminar boundary layers, using the boundary layer approximation without interactive effects, it has been proposed that if at some location the calculated flow is found to be on the point of separation then the corresponding real boundary layer will actually separate at or near that location. The fact of the matter is that the boundary-layer assumption contains the very seeds of singularity which lead to the abrupt demise of the latter at a finite time (e.g., the displacement thickness shown in Fig. 12.3). The singularity structures (the spontaneous separation of the viscous layer) seem to describe only the nascent state of the separating shear layer, within an asymptotically short time, *without the interactive effects*, i.e., within the limitations of the classical boundary-layer formulation [e.g., Van Dommelen and Cowley (1990)].

Attempts to interpret the failure of an approximate formulation as a precursor to an instability or to a new state of bifurcation are not uncommon in fluid mechanics. For example, the failure of the quasi-cylindrical approximation in swirling flows approaching a critical state is proposed to correspond to vortex breakdown [Hall (1972), see also Sarpkaya (1971)]. On the other hand, *the solutions to the full Na-*

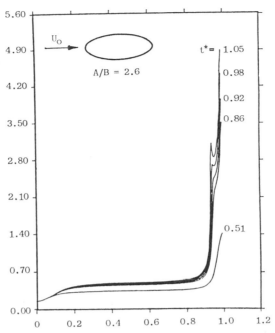

FIGURE 12.3 Development of spikes in the displacement thickness of a Carafoli airfoil at various times. (From Williams and Chen, 1990.)

vier–Stokes equations do not exhibit a singular behavior at separation. Taneda (1977, 1980) has concluded on the basis of his extensive flow visualization experiments that (i) the shedding of fluid particles from the wall is the most meaningful definition of separation for most time-dependent flows, (ii) this definition coincides with the Prandtl criterion in the case of steady two-dimensional flow over a fixed wall, and (iii) flow separation can be detected only by observing the integrated *streaksheet.* Although the streamline pattern changes according to the reference frame, the streakline pattern is invariant. One must keep in mind the fact that it is not always easy to determine accurately the streakline separation point from flow visualization. Also, it is not yet clear how to predict the motion of the streakline separation from a point undergoing large excursions on a body immersed in a time-dependent flow. Thus, not only the genesis but also the tracking at large times, of the unsteady separation remains unresolved, particularly for impulsively-started or harmonically-oscillating turbulent flow about bluff bodies. Beyond the time of spontaneous separation, the evolution of the large-scale unsteady recirculating structures is not yet calculable, except at low Reynolds numbers through the use of space-discretization methods and at unknown but relatively large Reynolds numbers through the use of Lagrangian or mixed Lagrangian–Eulerian vortex–element methods [for comprehensive reviews, see Sarpkaya (1989) and (1994a)].

12.4.2 Impulsively/Nonimpulsively Started Flows

The rapid acceleration or deceleration of flow about a body, from one steady state to another, *in times considerably smaller than the diffusion time,* results in a linear superposition of the initial fluid motion with the imposed irrotational motion. Subsequently, the two motions and their velocity fields interact nonlinearly and give rise to another state. In the case of flows subjected to varying accelerations and decelerations, with no steady-state to speak of, the motion of the separation points is indeed very complex. However, one can glean some useful information from two idealized states. If the flow is accelerated from a steady state, the existing boundary layer is washed downstream and a new boundary layer is created under the influence of the imposed flow. If the flow is decelerated, the existing boundary layer is washed upstream and the new boundary layer interacts with the existing wake returning to the body. This gives rise to one of the many complex problems in connection with the interaction of wakes and vortices with another body (e.g., a vortex over a helicopter blade, gusts about a building, wakes of micro-chips over other chips, and vortices over a Delta wing).

Among the numerous theoretical, numerical, and experimental investigations, impulsively-started steady flow about a circular cylinder has occupied a prominent place. The primary reasons for this are: it exhibits mobile separation points, it requires a relatively simple initial external flow, it is of intrinsic interest towards the understanding of the evolution of large-scale, unsteady, recirculating structures superimposed on a background of turbulence, it has practical importance in various aerodynamic applications (e.g., the impulsive flow analogy, flow about missiles, dynamic stall), it helps to understand the reasons and the consequences of the breakdown of the boundary-layer theory, and it provides the most fundamental case for the comparison and validation of numerical methods and codes.

For a circular cylinder of radius R in a stream of velocity U_o, the classical potential flow is given by $U(x) = 2U_\infty \sin (x/R)$ where x denotes the arc measured from the upstream stagnation point. The earliest time at which the absolute value of dU/dx reaches its maximum is $t^* = U_\infty t/R = S/R = 0.351$ at $\theta = 180°$ (the downstream stagnation point) [see e.g., Schlichting (1979)]. The point of zero wall shear then moves towards the front stagnation point along the cylinder. The *instantaneous pattern* of streamlines shows a relatively thin layer of reverse flow embedded in the boundary layer. At $t^* = 1.3$ to 1.4 (depending on the method of calculation) and in the vicinity of $\theta = 115°$, the viscous layer separates and the point of separation rapidly moves to $\theta = 104.5°$, provided that the interaction between the outer flow and the separated region is ignored. In reality, the flow upstream of the separation point is strongly affected by the occurrence and motion of the separation point and by the circumstances leading to the formation, growth, and motion of the wake. The definitions of unsteady separation on an impulsively-started cylinder depend on the imperfect nature of the boundary-layer equations, do not take into account the mutual interaction of the inner and outer flows, and do not enable one to make predictions on the evolution of large-scale structures beyond the time of spontaneous separation. Among the various numerical solutions of the evolution of large-scale structures in the wake of a circular cylinder, the works of Lecointe and Piquet (1984), Ta Phuoc Loc and Bouard (1985), Rumsey (1988), Wang (1989), and Chang and Chern (1991a) are the more recent and more accurate examples.

All numerical calculations using finite difference, finite element, or vortex–element methods [e.g., Chang and Chern (1991a), (1991b), Sarpkaya (1975, 1989), Sarpkaya and Shoaff (1979), Smith and Stansby (1988)] have assumed an *impulsively*-started flow. Experimental investigations of impulsively-started flow around circular and rectangular cylinders have been carried out by Sarpkaya (1966b, 1978a), Bouard and Coutanceau (1980), Sarpkaya and Kline (1982), Nagata *et al.* (1985), and by Sarpkaya and Ihrig (1986). The experimental efforts to generate impulsive or uniformly-accelerated flow at high Reynolds numbers are hampered by the generation of compression and rarefaction waves and regions of intense cavitation (in liquids). Because of this, at least one acceleration parameter such as $A_p = D(dU/dt)_o/U_o^2$ and $(D^2/\nu)(1/U_o)(dU/dt)_o$ must be added to the list of the parameters governing the phenomenon for the simplest case of flow accelerating from rest, with $(dU/dt)_o = $ constant, and then reaching a constant velocity of U_o at the end of a prescribed acceleration period.

Figures 12.4(a) and (b) show representative drag and lift coefficients obtained by Sarpkaya (1978a) for a nearly impulsively-started flow at Re $= 20,600 - 22,100$. A drag overshoot occurs at $S/R = 4$ in Fig. 12.4(a) where S is the actual displacement of the ambient flow. Depending on the unknown initial disturbances, the lift force exhibits wide variations from experiment to experiment as exemplified by Fig. 12.4(b). The period of vortex shedding decreases with increasing S/R. The Strouhal number increases from an initial value of about 0.17 to 0.205 within the range of S/R values shown in Fig. 12.4b. The important facts to be noted in any run are that (i) at the time of drag overshoot, C_L is negligibly small (indicating that the vortex pair is still nearly symmetrical, see Fig. 12.4(c)), (ii) the first reversal in the lift force occurs near $S/R = 12 - 14$, and (iii) in most cases, the period of lift oscillations decreases with increasing S/R.

The occurrence of a local maximum drag coefficient requires further discussion. The growth of the symmetric or nearly symmetric vortices (depending on the existing natural disturbances) is accompanied by other events. The so-called α-phenom-

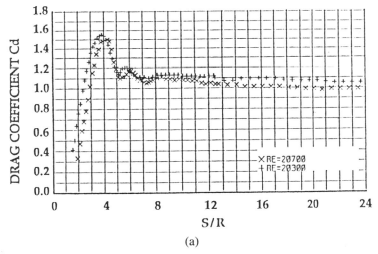

(a)

FIGURE 12.4 (a) Drag coefficient versus relative displacement (S/R) for $A_p = D(dU/dt)/V^2 = 1$. (b) Lift coefficient versus relative displacement for $A_p = 1$ and Re $= 20,600$ to 22,100 (note the increase in the vortex shedding frequency). (c) Impulsively-started flow about a circular cylinder. The contours of (from Koumoutsakos, 1993): (i) computed equivorticity lines, (ii) computed streamlines, and (iii) experimental streaklines. (From Bouard and Coutanceau, 1980.) (Fig. 12.4c(iii) is from *J. of Fluid Mech.*)

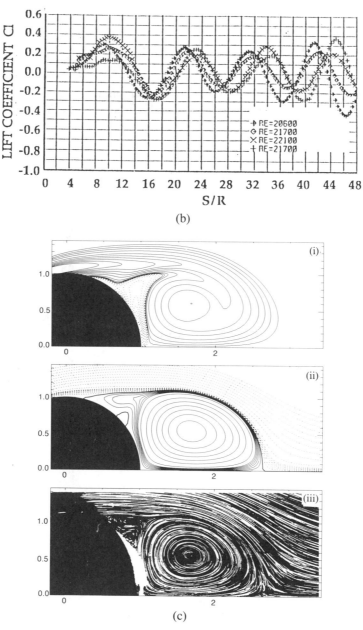

(b)

(c)

Figure 12.4 (*Continued*)

enon [Bouard and Coutanceau (1980)] disappears between $S/R \approx 4.8$ and 6.2, according to the calculations of Rumsey (1988) at Re = 1200 (the range of S/R values over which the said phenomenon disappears depends on Re). The fact of the matter is that the pressure distribution in the vicinity of the secondary vortices changes accordingly. Finally, during the early stages of the motion the vorticity flux is expected to be considerably larger than the cross-wake transfer of oppositely-signed vorticity (in fact, this is one of the reasons for the rapid accumulation of vorticity).

However, as the wake size grows and the vortices begin to exhibit asymmetry, as well as three-dimensionality, the cross-wake vorticity transfer is expected to increase significantly. It is surmised that it is the combination of the foregoing events that lead to the rapid rise and fall of the drag coefficient at or near $S/R \approx 4.5$.

The experiments with nonimpulsively accelerated flow past circular cylinders have shown that [Sarpkaya (1991)] the relative distance S/R at which the local maximum drag coefficient occurs does not depend on the acceleration parameter A_p for values larger than about 0.27 (corresponding to an almost impulsively-started flow beyond the period of acceleration) and agrees reasonably well with the previous numerical predictions [Ta Phuoc Loc (1980), Sarpkaya and Shoaff (1979), and Thoman and Szewczyk (1969)], based on an impulsive start. For accelerations which end near $S/R \approx 5$ (e.g., for $A_p = 0.2$), a smaller drag maximum occurs. The numerical simulation of nonimpulsively started flows might shed considerable light on the evolution of the primary and secondary vortices, migration of the separation points, and the time-variation of the lift and drag forces and the Strouhal number.

12.4.3 Periodic Flow About Bluff Bodies

Flow Topology. The effects of viscosity (creation and diffusion of vorticity) and separation place upper and lower limits on the values of the oscillation parameter $\beta = fD^2/\nu$ and the relative amplitude of the oscillation in the determination of the added mass and damping through the use of experimental and/or theoretical methods, employing unseparated flows with linear or angular oscillations. Then, the very meaning of the added mass becomes ambiguous. One is now confronted with the much broader question of resistance in time-dependent flows and with the nearly hopeless task of devising mathematical or semi-empirical means to decompose the instantaneous force into conceptually convenient components in terms of β and $K - U_o/fD$ or in terms of K and $\mathrm{Re} = U_o D/\nu$ (note that $\beta - \mathrm{Re}/K$). For very small values of β, the added mass and damping coefficients increase rapidly, as noted in connection with Eqs. (12.11) and (12.12). The parameter K (sometimes written as KC) is known as the Keulegan–Carpenter number, because of their seminal contribution to oscillating flows [Keulegan and Carpenter (1958)], even though U_o/fD was introduced much earlier by Schlichting (1932) as the ratio of the convective and local accelerations in connection with the discussion of periodic boundary layers.

In steady flow, the position of the separation points is nearly stationary, except for small excursions about ± 3 degrees (on a circular cylinder). Furthermore, the interference between the vortices and the body is confined mostly to the aft-body region. There is no obvious relation between the measured lift and drag coefficients and the excursion of the separation points. For example, while the separation angle remains nearly constant at about 80 degrees in the range of Reynolds numbers from about 2000 to 20,000, the drag coefficient increases from about 0.8 to 1.2 and the mean peak value of the lift coefficient varies anywhere from zero to 0.6 [see, e.g., Sarpkaya and Isaacson (1981)].

For oscillating flows, the net effect of the shed vortices is twofold. Firstly, their return to the body dramatically affects the boundary layer, outer flow, pressure distribution, and the generation and survival rate of the new vorticity. Secondly, they not only give rise to secondary separation points (during the early stages of the flow reversal) but also strongly affect the motion of the primary separation points. These effects are further compounded by the diffusion and decay of vortices, by the three-dimensional nature of the vortices (which gives rise to cycle-to-cycle variations and numerous flow modes, etc.), and by the fact that the boundary layer over a cylinder may change from a fully laminar to a partially or fully turbulent state, and the flow regime may range from subcritical to post-supercritical over a given cycle. The mobile separation points (when they are not fixed by sharp edges), undergo large excursions (as much as 120° during a given cycle of oscillation over a circular cylinder). This experimental fact [Sarpkaya and Butterworth (1992)] renders the treatment of boundary layers on bluff bodies subjected to periodic wake return extremely difficult, particularly when the state of the boundary layer changes during a given cycle. There are, at present, only a handful of applications of finite-difference and vortex-element methods to their analysis [e.g., Wang and Dalton (1991a, 1991b), Justesen (1991) and Sarpkaya and Putzig (1992)], (see Fig. 12.5). The use of DNS (Direct Numerical Simulation) and LES (Large Eddy Simulation) techniques are not likely to provide anything more than occasional samples for comparison with experiments at relatively low Reynolds numbers, at great expense even for two-dimensional flows. The most reliable recourse, at least at present, seems to be carefully conducted laboratory experiments. In this effort, flow visualization can serve as an excellent tool, not merely as an aid to understanding the physics of the flow but also as a means to acquire data and to guide the numerical calculations.

Even though every possible combination of the Keulegan–Carpenter number, K, Reynolds number, Re, and the relative roughness, k_s/D, represents an interesting state of oscillating flow about a circular cylinder, the approximate range of $8 < K < 13$ gives rise to a most interesting and precarious phenomenon known as the *transverse vortex street*. The case of $K = 11.5$ is chosen here to illustrate this intriguing phenomenon and the number of separation points which can occur simultaneously. Figure 12.6 shows the excursion of the separation points over two cycles of flow oscillation. It is clear that the numerical prediction of the excursion of multiple separation points in time-dependent flows and the development of methods to predict the instantaneous pressure distribution and the force acting on the body present extremely complex challenging problems in any flow regime.

Morison's Equation. The experimental studies of Morison *et al.* (1950) on forces on piles due to the action of progressive waves have provided a useful and somewhat

FIGURE 12.5 Oscillating flow about a circular cylinder for $K = 4$ and $\beta = 200$. (From Sarpkaya and Putzig, 1992.)

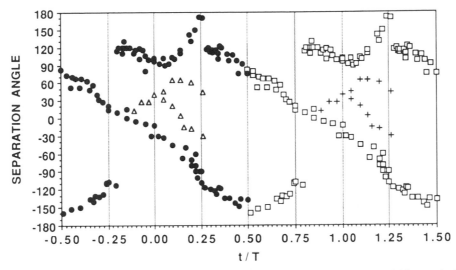

FIGURE 12.6 Separation angle versus time for $K = 11.5$ and Re $= 37,000$ (symbols x and + denote secondary separations). (From Sarpkaya and Butterworth, 1992.)

heuristic approximation. The forces are divided into two parts, one due to the drag, as in the case of flow at constant velocity, and the other due to acceleration or deceleration of the fluid. This concept necessitates the introduction of a drag coefficient C_d and inertia coefficient C_m in the expression for the in-line force. Morison's equation does not deal with the transverse force or lift force. If F is the force per unit length experienced by a cylinder, then one has

$$F = \frac{1}{2} \rho C_d D |U| U + \rho C_m \frac{\pi D^2}{4} \frac{dU}{dt} \qquad (12.16)$$

where U and dU/dt represent the undisturbed velocity and the acceleration of the fluid, respectively.

If one were to consider the superposition of mean plus oscillatory flow (e.g., waves and currents), then the above equation may be generalized, with some poetic license, to [Sarpkaya and Isaacson (1981)],

$$F(t) = \frac{1}{2} \rho C_d^u A_p |\{(U_o + U(t))\}| \{(U_o + U(t))\} + \rho C_m^u \; \forall \; \frac{dU(t)}{dt} \qquad (12.17)$$

where U_o represents the steady component of the fluid velocity; $U(t)$, the time-dependent fluid oscillations; $\forall$, the volume of the body; C_d^u, the Fourier-averaged drag coefficient and C_m^u, the Fourier-averaged inertia coefficient. The Fourier averages of the drag and added-mass coefficients over a period of T may be calculated by multiplying both sides of Eq. (12.17) once with $U(t)$ and once with dU/dt to yield

$$C_d^u = \frac{2 \int_0^T F(t) U(t) dt}{\rho A_p \int_0^T |\{U_o + U(t)\}| \{(U_o + U(t))\} U(t) dt} \qquad (12.18)$$

and

$$C_m^u = \frac{\displaystyle\int_0^T F(t)\,\frac{dU(t)}{dt}\,dt}{\rho\forall\displaystyle\int_0^T [dU(t)/dt]^2\,dt} \tag{12.19}$$

which may be evaluated readily provided that sufficiently reliable data are available for $F(t)$, U_o, $U(t)$, and $dU(t)/dt$.

In Stokes sphere problem where the Reynolds number is very small, drag is proportional to the first power of velocity. In Morison's equation, drag is proportional to the square of the velocity, since the flow is separated and the drag is primarily due to pressure rather than the skin friction. Clearly, Morison's equation is an heuristic extension to separated time-dependent flows of the solution obtained by Stokes. It is also clear that the validity of the equation and the limits of its application will have to be determined experimentally. The time-averaged force coefficients (C_d^u and C_m^u) are functions of a suitable Keulegan–Carpenter number, K, a Reynolds number, Re, (or $\beta = \text{Re}/K$), relative roughness, k_s/D, a parameter involving U_o (e.g., $U_o T/D$ or $V_r = U_o/[U(t)]_{max}$) [Sarpkaya and Storm (1985)], the proximity of other bodies, and the three-dimensional nature of the flow. In general, they show some scatter in the intermediate range of the relative amplitudes. For small values of A/D or K, where the inertial force is dominant, and for large values of A/D or K, where the drag forces are dominant, Morison's equation represents the measured force (with experimentally determined coefficients) with sufficient accuracy, as expected from the nature of the equation. In the intermediate range of A/D values, however, where both the inertial and drag forces are important, Morison's equation does require additional terms for a better representation of the measured force. This is due to the fact that one must now deal with a complex problem of vorticity-body interaction where the flow is neither like an unseparated laminar flow nor like a unidirectional steady flow. A detailed discussion of Morison's equation and comprehensive experimental data are given by Sarpkaya (1963, 1977a, 1986).

12.4.4 Bluff Bodies in Unsteady Motion

Unsteady aero/hydrodynamic phenomena are of far-reaching consequence in air and sea operations. They limit the speed, range, stability, and quiet operation of bodies and give rise to some of the most complex fluid-structure interaction problems. For example, the steady and unsteady characteristics of ship and submarine wakes in the region of the stern, and especially with regard to the effect of the wake and rolled-up body vortices on propulsor unsteadiness, and the consequences for noise and hull vibrations are important problems of great concern. Equally important are the unsteady motion of Delta-wing aircraft, the interaction of the trailing vortices with the control surfaces, wing-rock, vortex breakdown, and, in general, the dynamic interaction of vortices or quasi-coherent vorticity with bodies or fluid interfaces.

When an axisymmetric body moves at a sufficiently large angle of attack, the boundary layer vorticity lifts off the surface and sheets of vorticity are convected downstream while rolling up into streamwise vortices on the leeward side of the body (see Fig. 12.7). This is quite similar to the vortex-sheet roll up over sharp-

FIGURE 12.7 Streamwise vortices generated by three dimensional separation. (From Chang and Purtell, 1986.)

edged lifting surfaces (e.g., a Delta-wing) with the added complexities that the separation lines for an axisymmetric body are not fixed or known *a priori* and that the *afterbody* (part of the body downstream of the separation lines), protruding into the wake as it does, plays a significantly different roll in the coiling of the vortices relative to that over a delta wing. This separation and roll-up phenomenon, known as the cross-plane separation, occurs on almost all maneuvering bodies at sufficiently large angles of attack. The roll-up of the vortices and their subsequent evolution are extremely important for the proper design of a submerged body [Sarpkaya and Kline (1982)] and more importantly for the determination of the interaction of the vortices with the stern and the propulsion system. It is this interaction that determines the maneuverability and control of a submerged body and the vibration and noise transmitted to the body by the propulsion system. For example, the abrupt cavity-separation gives rise to violent vibrations on a hydrofoil. When a body is subjected to a turn of small radius, the local angle of attack varies significantly along the length of the body and will, in general, even reverse slightly near the bow.

A satisfactory numerical model would enable one to calculate the complete topological structure of three-dimensional separated flows, i.e., the development of the turbulent boundary layer, transition and separation lines, and the evolution of the roll-up of vortex sheets and their subsequent trajectory. The problems noted above are sufficiently difficult even for a simple axisymmetric body moving at a constant angle of attack. Considerable work has been done on unsteady laminar boundary layer on impulsively-started prolate spheroids and in particular on the boundary layer along the symmetry plane (Wang, 1972).

There are a number of two- and three-dimensional potential flow models (panel and vortex-element methods). These methods suffer from the necessity to specify the separation lines [Thrasher (1982), Chang and Purtell (1986), and Xu and Wang (1988)]. Numerous approximations have been introduced to calculate the shed vorticity, and none could be considered satisfactory or contributing to the understanding of the physics of the phenomenon. In more sophisticated potential flow models, the external surface (a source distribution), the camber surfaces of lifting components (e.g., a bilinear doublet distribution) and the wake surfaces (spanwise linear doublet distribution) are replaced by a large number of panels. The cross-flow separation lines on the body are either specified arbitrarily (on the basis of past experience with bluff bodies) or approximately determined from flow visualization experiments, at a much smaller Reynolds number [e.g., Fig. 12.8, reproduced from Conway and Mackay (1990)].

There are a number of Navier–Stokes computations of steady separated vortical flows past prolate spheroids at incidence [Newsome and Kandil (1987), Shirayama and Kuwahara (1987), and Wong *et al.* (1989)]. These use the unsteady incompres-

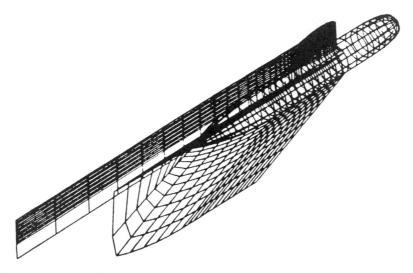

FIGURE 12.8 Hull-sail configuration and vortex wakes. (From Conway and MacKay, 1990.) (Courtesy of *AIAA*.)

sible thin-layer Navier–Stokes equations and the unsteady *compressible* thin-layer Navier–Stokes equations. In general, the two sets of equations are solved using a pseudo-time stepping scheme on a curvilinear grid. The Baldwin–Lomax algebraic eddy viscosity model is used to model the turbulent flow. Even for steady ambient flow, the predictions are not quite satisfactory as far as the skin friction coefficient and the details of the primary and secondary vortices are concerned. Furthermore, one needs better transition and turbulence models than the over-simplified approach of switching on a turbulence model at a prescribed position [Wong *et al.* (1989)]. A schematic representation of the pattern of vorticity lines near a spheroid at incidence is shown in Figs. 12.9(a–c), and particle paths and surface streamlines on a longer spheroid at 30° incidence are shown in Fig. 12.9d [reproduced from Shirayama and Kuwahara (1987)].

The problems to be dealt with become even more difficult, both numerically and experimentally, when the body is subjected to a turn. Experiments on turning models are conducted through the use of a rotating arm apparatus. The motion of the body makes the measurement of detailed velocity and pressure distribution difficult. Clearly, a stationary model in a water or wind tunnel would be preferable, but the curving shear flow required for a turn is difficult to achieve (even though it has been tried). One approximate approach is the transformation of *the straight body in a curving flow to an appropriate curved body in a straight flow*. In two dimensions, this merely involves a logarithmic conformal transformation. In three-dimensional flow, this approach has been further approximated so that the said transformation is applied to each plane of the body parallel to the plane of motion [Chang and Purtell (1986)].

As far as the numerical work is concerned [Cousteix (1986)], the existing methods require information on the location of separation lines either from boundary layer calculations or from approximate separation criteria [e.g., Stratford's (1959) turbulent separation criteria for 2-D flows], or experiments.

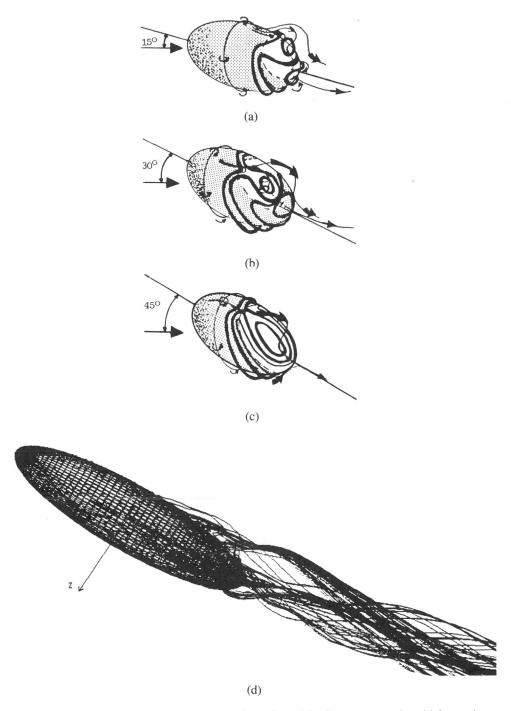

(a)

(b)

(c)

(d)

FIGURE 12.9 (a–c) Schematic representation of vorticity lines near a spheroid for various angles of incidence. (From Shirayama and Kuwahara, 1987.) (d) Particle paths and surface streamlines for Re = 10,000 on a spheroid at 30 deg. incidence. (Courtesy of *AIAA*.)

Chang and Purtell (1986) developed a numerical method for computing flow separation from a long, three-dimensional body, based on the potential flow approach. By considering the similarities of the local separating cross-flow to a two-dimensional separation, they have derived a *Kutta-like condition* [see, e.g., Crighton (1985)] and incorporated it into the method of calculation of the separated vortex sheet. A boundary layer model was used to predict the locations of the surface shear convergence lines. Computations were compared with experiments for an ogive-cylinder and prolate spheroid at angle of attack. As far as the turning body is concerned, they have developed a method to transform the body and the flow field into a curved body in a straight flow.

The submerged bodies encountered in the oceans are not simple axisymmetric bodies (autonomous underwater vehicles, submarines, offshore structures). For example, the asymmetry of a submarine (say because of a sail), introduces reactions in the vertical plane in response to the imposed motion, such as a turn, in the horizontal plane. These reactions take the form of a downward force on the hull aft of the sail and, in consequence, give rise to a *bow-up* pitching moment. The analysis of such motions and the resulting nonaxisymmetric flows offer many challenges which are not only timely from a practical point of view but also extremely important for increasing the physics content of our understanding of the hydrodynamic phenomena.

12.4.5   Fluid-Structure Interaction

Hydrodynamic Damping.   In fluid–structure interaction, *damping* is used to account for our inability to solve the complete fluid–structure interaction problem. Strictly speaking, there would be no need for the *fluid damping* if one were able to express all fluid forces resulting from the fluid–structure interaction, whether the excitation is imparted by the fluid motion or by outside sources. The prediction of time-dependent flows (unsteady ambient flow and/or body response) is far from developed in the range of parameters of practical significance.

Often, a distinction is made between bodies subjected to vortex-shedding excitation and those which undergo simple harmonic motion at very small amplitudes and high frequencies in water, presumably due to excitation that comes from outside, i.e., not caused by the flow itself. The latter is called *hydrodynamic damping* or *viscous damping*, meaning the decrease of the amplitude of an oscillation by forces in anti-phase with velocity. There are two reasons for such a distinction: The classical solution of Stokes [see also, Wang (1968)], valid only for $K \ll 1$ and $\beta \gg 1$, has shown that the oscillatory boundary layer gives rise to skin friction and normal pressure and, hence, to a damping force, in anti-phase with the velocity. The significant term of this force per unit length is given by $4\pi^3 \rho D^2 f_n^2 A/(\pi\beta)^{1/2}$, where $\beta = f_n D^2/\nu$ and A is the amplitude of oscillation. This force may be approximated by a drag coefficient [Sarpkaya and Isaacson (1981)] as $C_d = 3\pi^3/[2K(\pi\beta)^{1/2}] = \text{Constant}/K\beta^{1/2}$ or by a reduced damping coefficient $\delta_r = (4/3\pi)KC_d$ in which $K = U_o/f_n D = 2\pi A/D$. Here, the definition of a conventional drag coefficient does not imply velocity-squared damping since Stokes' damping force is proportional to velocity. However, as K increases, the effects of separation increase the contribution of the normal pressure, and the resistance changes from inertia with linear drag to inertia with both linear and quadratic drag, and eventually, to inertia with quadratic drag.

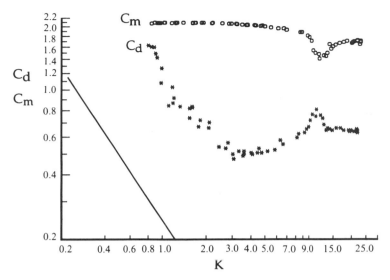

FIGURE 12.10 Drag and inertia coefficients versus Keulegan–Carpenter number. Experiment: ○, C_m, *, C_d; theory: — (all for $\beta = 11,240$).

Figure 12.10 shows sample data for a smooth cylinder over a sufficiently large range of K values. A detailed discussion of this and data for rough and smooth cylinders are given in Sarpkaya (1986a, 1986b). Data for larger values of β are given by Anaturk *et al.* (1992) for smooth as well as rough cylinders.

The second reason for the said distinction between the vortex-shedding excitation and the hydrodynamic damping is that the unseparated flow about a circular cylinder does not give rise to oscillatory forces in any direction, thus it cannot excite the body. It can only decrease the amplitude of a previously imparted (*plucking*) or externally-imposed oscillation (e.g., guyed towers, excited in the wave zone, may transmit their excitation downward into calmer regions). Having made this distinction, one must emphasize that any externally imposed current or wave on a body oscillating in a fluid otherwise at rest may change not only the topology of the flow but also the shear, the pressure distribution, hence the in-line and transverse pressure gradients acting on the body. The force-transfer coefficients obtained from the motion of a freely damping cylinder and those obtained from the forced oscillatory flow ($U = U_o \sin \omega t$) about a cylinder at very small amplitudes of oscillation are not identical.

Pluck tests may yield sufficiently accurate reduced damping for $K \ll 1$ only, provided that the starting value of the vibration has $K \ll 1$ also. The damping coefficients so obtained may be of some use in the calculation of the heave and pitch motions of the tendons of Tension Leg Platforms if one assumes that all other flow effects, such as waves and currents, are absent and the *springing* is taking place in a quiescent fluid, with very small amplitudes and periods. Otherwise, history effects and flow separation due to fluid motion in other directions may yield completely different apparent damping. Most industrial applications are beyond pluck tests where one must be concerned with the integrated effect of the various sources of energy dissipation.

Flow-Induced Vibrations. The nonlinear interaction between two shear layers, emanating from the sides of a sufficiently bluff body, influenced by feedback from the wake and the body shape, lead to vortex shedding and even excite the body under the right structural and fluid-mechanical conditions. That the vortex shedding could lead to structural excitation is neither new nor surprising. The smaller the structural damping of industrially-significant structures, the more prevalent the vortex excited oscillations and their undesirable consequences. What is surprising is that the excitation does not occur at $f_{so} = f_n$ only, but over a wider range of velocities (here f_{so} is the frequency of vortex shedding for the stationary cylinder, and f_n is the frequency of the transverse oscillations of the cylinder in water.) When the fluid–structure interaction is viewed in its totality, not just in terms of the transverse excitation, one becomes aware of the complex nonlinear interaction between the three-dimensional fluid motion and the response of the structure. For long rigid or flexible structures (e.g., a cable) the phenomenon is further complicated by the fact that the structure tends to respond at a variety of frequencies over its entire length. This, in turn, gives rise to additional and omni-directional fluid forces whose prediction is at best approximate. For an engineering guide to flow-induced vibrations, see [Naudascher and Rockwell (1994)].

When there is no synchronization (*lock-in*), the driving fluid force and the structure oscillate at their own frequencies. In field tests, a locked-in condition or a standing wave profile may not occur on long wires towed in the ocean [Alexander (1981)]. There is no total spanwise correlation along very long structures placed in the ocean environment partly due to the fact that instabilities prevent such coherence even in well-controlled laboratory experiments and partly due to the fact that larger amplitude disturbances and omni-directional waves and currents surely prevent anything other than short coherence lengths. Only relatively short test cylinders or cables result in well-separated modal frequencies, reducing the effects of modal interaction, and enabling single-mode lock-in to be studied in some detail [Iwan and Jones (1987)].

In the lock-in range *the frequency of the driving force in a given direction locks onto the natural frequency of the structure in that direction*. We shall not use the existing, often-repeated, definition that lock-in is *the capture of the vortex shedding frequency by the vibration frequency over a range of flow speeds*. If the driving force frequency in the transverse (in-line) direction brackets the natural frequency of the structure in that direction, one obtains lock-in in the transverse (in-line) direction, provided that the amplitude of the oscillating component of the force is sufficiently large. If the structure's natural frequency in the in-line direction, f_{nx}, is twice that of in the transverse direction, f_{ny}, in a sub-critical flow, the driving forces lock onto the natural frequency of their respective directions and the axis of the body traces an 8-shaped path. Clearly, the vortex shedding, the character and timing of vortices, and the amplification of the driving force are inextricably related. Also, excitation without lock-in is common but lock-in without excitation is impossible.

The natural frequency of the structure, f_n, does not remain constant over the excitation range, because of the change of added mass with A/D [Sarpkaya (1978b, 1979)]. Furthermore, in the range of velocities where f_{so} remains locked to f_n, the increase in flow velocity is partially compensated by the apparent increase in the diameter of the body in keeping $f_n D/V$ nearly constant. The correlation length along the body and the vortex strength tend to increase with oscillation, leading to larger exciting forces and oscillation amplitudes. For A/D larger than about 0.5, the forcing decreases due to changes in vortex shedding and A/D reaches a self-limiting value. Furthermore, the oscillations alter the phase, sequence, and pattern of vortices in the wake. The different synchronization regions have been photographically identified by Zdravkovich (1982) and by Williamson and Roshko (1988), but no force measurements were made. It appears that the highly nonlinear interaction between the shear layers is profoundly affected by the body motion, in an unknown and perhaps unknowable manner, leading to a self-limiting amplitude. The natural or artificial hindrance of the coherence of the driving force through various means may hinder the occurrence of the locked response, but not necessarily the nonlocked excitation. For additional reviews of vortex-induced oscillations see, e.g., Sarpkaya (1979), Griffin and Hall (1991), and Pantazopoulos (1994).

The fluid forces imposed on the structure are a function of the relative amplitude of response, mass parameter, structural damping, Reynolds number, length-to-diameter ratio, relative roughness, cable tension, ambient turbulence (intensity and integral length scale), parameters characterizing the wave, current, shear field, and wind, and proximity effects, just to name a few. The selection of the most important parameters is a case specific issue and must be done judiciously by the designer. Their specification is of extreme importance to minimize unwanted dependence on empiricism in dealing with the fluid/structure interaction. It is often assumed that the lock-in region is amplitude and, to a lesser extent, Reynolds number dependent. Even though this appears to be true in subcritical flows, cylinders subjected to or allowed to perform nonconstrained biharmonic oscillations in the *critical regime* may exhibit significantly different behavior due to the dramatic changes in vortex shedding and in the resulting forces. As far as the mass parameter is concerned, often the two independent dimensionless numbers $m/\rho D^2$ and δ are routinely combined into $m\delta/\rho D^2$. It is assumed that $m\delta/\rho D^2$ is a universal parameter to embody the effects of both the mass and damping even though the inadvisability of such a procedure has already been seriously questioned [Sarpkaya (1978b, 1979)].

A force model has to include information on the correlation of forces along the cable. Unfortunately, further information on such correlation has to come from tests. Sarpkaya (1977b, 1978b) was the first to measure the forces acting on a transversely oscillating rigid cylinder in a uniform water flow. Until then, the data of Feng (1968) were used to calibrate empirical models. It consisted of measurements of the amplitude and frequency response of a flexibly mounted cylinder in a wind tunnel as a function of flow velocity and structural damping.

Sarpkaya (1977b, 1978b) expressed the transverse (lift) force as,

$$C_L = C_{mh} \sin \omega t - C_{dh} \cos \omega t \qquad (12.20)$$

in which C_{mh} and C_{dh} are the inertia (in-phase) and drag (out-of-phase) coefficients, respectively, when the lift force is normalized by $0.5\rho DLV^2$. Equation (12.20) may be obtained either from a linearized form of Morison's equation or from a straightforward assumption that the lift force may be approximated by a sine wave. This represents a considerable simplification not only for the ocean environment but also for the laboratory environment. The transverse as well as the in-line force for more complex oscillations (e.g., *beating* phenomena, oscillations with irregular amplitudes) cannot be represented by a two-coefficient model. There will always be considerable uncertainty stemming from the ambient flow environment. This, in fact, is one of the fundamental reasons for the difficulty of dealing with fluid/structure interaction in a nonlaboratory environment.

Sarpkaya's (1977b) force data have been used with considerable success, [Iwan and Botelho (1985), Patrikalakis and Chryssostomidis (1985), Iwan and Jones (1987)] for the prediction of vortex-induced oscillations. During the past ten years, three additional force measurements with oscillating cylinders were reported [Staubli (1983), Moe and Wu (1990), Gopalkrishnan (1992)]. Staubli (1983), measured the fluid forces acting on a transversely oscillating circular cylinder in a wind tunnel. Then, he predicted the vibrations of a freely oscillating cylinder using the results of his measurements. He has shown that hysteresis effects, which are observed in experiments with elastically-mounted cylinders of certain damping and mass ratios, are caused by the nonlinear relation between the fluid force and the amplitude of oscillation. Staubli's data cannot be compared here with those obtained by others since its decomposition is based on a different assumption.

Figures 12.11(a) and (b) show the coefficients C_{mh} and C_{dh} for $A/D = 0.50$ for three sets of data [Sarpkaya (1978b), Moe and Wu (1990), and Gopalkrishnan (1992)], as a function of $V_r St = (U/f_n D) (f_{so} D/U) = f_{so}/f_n$ where f_{so} is the frequency of vortex shedding for the stationary cylinder and f_n is the frequency of the transverse oscillations of the cylinder in water. All three experiments were carried out in subcritical water flows. However, L/D ratios, Reynolds numbers, Strouhal numbers, and the experimental conditions were different. It is because of this reason that the coefficients were plotted with respect to $V_r St$ rather than with respect to V_r alone in order to account for the variations in Strouhal number (and Reynolds number) among the three experiments. The agreement between the three sets of data is rather remarkable considering the fact that the phenomenon dealt with is the crossflow vibration of a cylinder and that the gradients of the force coefficients in the neighborhood of $f_{so}/f_n = 1$ is very large, but not discontinuous. This agreement suggests that a suitable average of the data shown in each figure (e.g., a least squares fit curve) may be used with confidence to predict the lock-in response of similar or suitably idealized structures. Figure 12.11b shows that the drag coefficient or the normalized out-of-phase component of the total instantaneous transverse force becomes negative primarily in the neighborhood of $f_{so}/f_n = 1$. Outside this range, the drag is mostly positive, thus in the opposite direction to the motion of the cylinder. When the drag force is in phase with the motion of the cylinder, it helps to magnify the oscillations rather than damp them out. For this reason, the range in which C_{dh} is negative is sometimes referred to as the *negative damping region*. The fact of the matter is that this is not damping in the proper use of the word but rather an energy transfer from the fluid to the cylinder via the mechanism of excitation or synchronization. The maximum absolute value of C_{dh} in the synchronization range increases with A/D up to about $A/D = 0.5$ and then decreases. For A/D larger than about unity, the said component does not require energy extraction from the cylinder at each cycle. Thus, the oscillations are self-limiting. Throughout the entire A/D range, the relative in-line drag coefficient $C_d(A/D)/C_d(0)$ increases almost linearly with the parameter $W_r = (1 + 2A/D)/(V_r St)$, according to the data compiled by Griffin (1985a, 1995b) from various sources. The data discussed above confirm this finding.

In spite of the satisfactory agreement noted among the three sets of data, the fact remains that the test cylinders were constrained to remain in the transverse plane.

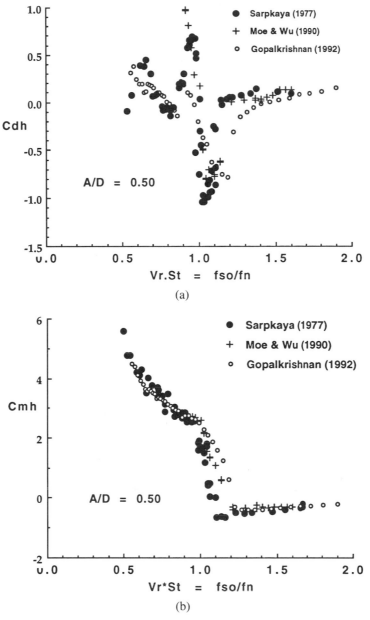

FIGURE 12.11   (a) Comparison of drag coefficients for three sets of data as a function of f_{so}/f_n for $A/D = 0.50$. (b) Comparison of inertia coefficients for three sets of data as a function of f_{so}/f_n for $A/D = 0.50$.

Only Moe and Wu (1990) carried out additional, but limited, tests with cylinders not constrained in the in-line direction (through the use of suitable springs). They have found that large self-excited cross-flow motions occur for a wider range of V_r values if the cylinder is not restrained in the in-line direction. Furthermore and interestingly enough, the random variations in the lift force are smaller for the in-line-unrestrained cases than for the in-line restrained cases. Clearly, the data obtained with in-line-restrained rigid cylinders may be of limited use under certain circumstances, particularly when the natural frequency of the cylinder in the in-line direction is made or becomes increasingly larger than that in the transverse direction. Thus, the exploration of the biharmonic motion of a cylinder elastically-supported in both directions becomes an issue even more important [Sarpkaya (1994b)] than that of an in-line restrained cylinder subjected to amplitude-modulated beating motions. Gopalkrishnan *et al.* (1992) subjected rigid cylinders, restrained in the in-line direction, to amplitude modulated (beating) motions and expressed the transverse force in terms of an *equivalent lift coefficient*. It is clear that flow-induced oscillations and hydrodynamic damping are not only industrially-significant time-dependent motions but also one of the chief outstanding difficulties of the subject.

12.5   CONCLUSIONS

The subject of unsteady flows is indeed very rich and offers many challenges and opportunities in novel experimentation and numerical analysis. The choice of topics treated herein is not unique and is based primarily on this writer's forty years of experience with unsteady flows.

The nature of unsteady motion, unsteady flow of inviscid fluids, some classical unsteady unseparated flows, and a number of representative unsteady separated flows (unsteady separation; characteristics of impulsively- and nonimpulsively-started flows; periodic flow about bluff bodies; bluff bodies in unsteady motion; and flow-induced oscillations) have been discussed. It appears that the prediction of the evolution of the wake, after the spontaneous separation of the viscous layer, is possible only for relatively small Reynolds numbers. For Reynolds numbers of any practical importance, the use of approximate turbulence models and experiments still appear to be the only recourse. A practical need exists for developing more general techniques capable of predicting the behavior of both laminar and turbulent unsteady boundary layers.

As in a great many steady flow problems, where the potential flow analysis provides many approximate solutions and a strong impetus for the need for the exact or approximate solution of the Navier–Stokes equations, it is especially challenging in unsteady flows to try to approximate real wake behavior using models involving potential theory. One may expect that some of these approximations will provide realistic solutions (in the sense of matching the observed characteristics) and will help to make it possible to introduce, eventually, the effects of roughness, viscosity, gravity, turbulence, past history, etc. The comparison between the theoretical results and the experimental observations may not be all that one wishes, but the objective is not always the achievement of a favorable comparison but rather the careful selection of those found to be reasonably accurate from among many cases tried. The potential flow model, i.e., the simulation of the shear layers by ideal fluid vortices, has been employed by many investigators with varying degrees of success. The problem in connection with the use of a potential flow model is essentially that of the formulation of a mechanism which will account for the generation of steady and unsteady vorticity in the boundary layers of a bluff body, its kinematic and dynamic convection downstream and redistribution and diffusion in the wake, and finally the resulting primary and secondary feed-back on the rate of shedding of organized vorticity and on the pressure and velocity fields near the body.

As far as the unsteady maneuvers of large bodies are concerned, the subject is at best in the hands of panel methods and numerous approximations, requiring the specification of (mobile) separation points, among other things. It appears that in spite of the enormous progress made in experimentation and computational fluid dynamics, the question of where the separation lines are remains unresolved and challenging as ever.

The fluid/structure interaction offers exceedingly complex and industrially-significant three-dimensional time-dependent flow problems. Experiments as well as direct numerical simulations and large-eddy simulation techniques may shed some light on such fluid/interaction problems.

In view of the foregoing, it seems appropriate to close this chapter with a thoughtful reminder by Sir Geoffrey Ingram Taylor (1974): ''Though the fundamental laws of the mechanics of the simplest fluids, which possess Newtonian viscosity, are known and understood, to apply them to give a complete description of any industrially significant process is often far beyond our power.''

REFERENCES

Alexander, C. M., ''The Complex Vibrations and Implied Drag of a Long Oceanographic Wire in Cross-flow,'' *Ocean Eng.*, Vol. 8, No. 4, pp. 379–406, 1981.

Anaturk, A. R., Tromans, P. S., van Hazendonk, H. C., Sluis, C. M., and Otter, A., ''Drag Force on Cylinders Oscillating at Small Amplitude: A New Model,'' *J. Offshore Mech. and Arctic Eng.*, Vol. 114, pp. 91–103, 1992.

Batchelor, G. K., ''Axial Flow in Trailing Line Vortices,'' *J. Fluid Mech.*, Vol. 20, pp. 645–658, 1964.

Belotserkovskii, S. M., ''Study of the Unsteady Aerodynamics of Lifting Surfaces Using the Computer,'' *Ann. Rev. Fluid Mech.*, Vol. 9, pp. 469–494, 1977.

Birkhoff, G. and Zarantonello, E. H., *Jets, Wakes, and Cavities*, Academic Press, New York, 1957.

Bouard, R. and Coutanceau, M., ''The Early Stage of Development of the Wake Behind an Impulsively Started Cylinder for $40 < Re < 10^4$,'' *J. Fluid Mech.*, Vol. 101, pp. 583–607, 1980.

Bradshaw, P., ''Turbulence: The Chief Outstanding Difficulty of Our Subject,'' *Exper. in Fluids*, Vol. 16, pp. 203–216, 1994.

Cassarella, M. J. and Laura, P. A., ''Drag on an Oscillating Rod with Longitudinal and Torsional Motion,'' *J. Hydronautics*, Vol. 3, No. 4, pp. 180–183, 1969.

Chang, C.-C. and Chern, R.-L., ''A Numerical Study of Flow Around an Impulsively Started Circular Cylinder by a Deterministic Vortex Method,'' *J. Fluid Mech.*, Vol. 233, pp. 243–263, 1991a.

Chang, C.-C. and Chern, R.-L., ''Vortex Shedding from an Impulsively Started Rotating and Translating Circular Cylinder,'' *J. Fluid Mech.*, Vol. 233, pp. 265–298, 1991b.

Chang, M.-S. and Purtell, L. P., ''Three-Dimensional Flow Separation and the Effect of Appendages,'' *Proc. 16th Symp. Naval Hydrodynamics*, National Academic Press, pp. 352–370, 1986.

Chew, Y. T. and Liu, C. Y., ''Effects of Transverse Curvature on Oscillatory flow along a circular cylinder,'' *AIAA J.*, Vol. 27, pp. 1137–1139, 1989.

Conway, J. and Mackay, M., "Prediction of the Effects of Body Separation Vortices on Submarine Configurations Using the CANAERO Panel Method," AIAA 90-0302, 1990.

Cousteix, J., "Three Dimensional and Unsteady Boundary-Layer Computations," *Ann. Rev. Fluid Mech.*, Vol. 18, pp. 173–196, 1986.

Crighton, D. G., "The Kutta Condition in Unsteady Flow," *Ann. Rev. Fluid Mech.*, Vol. 17, pp. 411–444, 1985.

Darwin, Sir Charles, "Notes on Hydrodynamics," *Proc. Camb. Phil. Soc.*, Vol. 49, pp. 342–354, 1953.

Dryden, H. L., Murnaghan, F. D., and Bateman, H., *Hydrodynamics*, Dover Publishing, New York, 1956.

Eckmann, D. M. and Grotberg, J. B., "Experiments on Transition to Turbulence in Oscillatory Pipe Flow," *J. Fluid Mech.*, Vol. 222, pp. 329–350, 1991.

Feng, C. C., "The Measurement of Vortex-Induced Effects in Flow Past Stationary and Oscillating Circular and D-Section Cylinders," M.Sc. Thesis, Univ. of British Columbia, Vancouver, Canada, 1968.

Garrick, I. E., "Unsteady Aerodynamics of Potential Flows," *Appl. Mech. Surveys*, Abramson, H. N. *et al.* (Eds.), Spartan Books, Washington, D.C., pp. 965–970, 1966.

Gopalkrishnan, R., "Vortex-Induced Forces on Oscillating Bluff Cylinders," Ph.D. Thesis, MIT, Cambridge, MA, 1992.

Gopalkrishnan, R., Grosenbaugh, M. A., and Triantafyllou, M. S., "Amplitude-Modulated Cylinders in Constant Flow: Fundamental Experiments to Predict Response in Shear Flow," *Proceedings of Bluff-Body/Fluid and Hydraulic Machine Interactions*, Vol. 245, No. G00727-1992, pp. 87–101, 1992.

Gradshteyn, I. S. and Ryzhik, I. M. (1965), *Table of Integrals, Series, and Products*, Academic Press, New York, 1965.

Griffin, O. M., "Vortex-Induced Vibrations of Marine Cables and Structures," NRL Memorandum Report 5600, Naval Res. Lab., Washington, D.C., 1985.

Griffin, O. M., "Vortex Shedding from Bluff Bodies in a Shear Flow: A Review," *J. Fluid Mech.*, Vol. 107, pp. 298–306, 1985.

Griffin, O. M. and Hall, M. S., "Review-Vortex Shedding Lock-on and Flow Control in Bluff Body Wakes," *J. Fluids Eng.*, Vol. 113, pp. 526–537, 1991.

Grotberg, J. B., "Pulmonary Flow and Transport Phenomena," *Ann. Rev. Fluid Mech.*, Vol. 26, pp. 529–571, 1994.

Hall, M. G., "Vortex Breakdown," *Ann. Rev. Fluid Mech.*, Vol. 4, pp. 195–218, 1972.

Iwan, W. D. and Botelho, D. L. R., "Vortex Induced Oscillation of Structures in Water," *J. Waterways, ASCE*, Vol. 111, No. WW2, pp. 289–303, 1985.

Iwan, W. D. and Jones, N. P., "On the Vortex-Induced Oscillation of Long Structural Elements," *J. Energy Resources Tech.*, Vol. 109, pp. 161–167, 1987.

Jones, W. P., "Research on Unsteady Flow," *J. Aerospace Sci.*, Vol. 29, No. 3, pp. 249–263, 1962.

Justesen, P., "A Numerical Study of Oscillating Flow Around a Circular Cylinder," *J. Fluid Mech.*, Vol. 222, pp. 157–196, 1991.

Keulegan, G. H. and Carpenter, L. H., "Forces on Cylinders and Plates in an Oscillating Fluid," *J. of Research, National Bureau of Standards*, Vol. 60, pp. 423–440, 1958.

Lecointe, Y. and Piquet, J., "On the Use of Several Compact Methods for the Study of Unsteady Incompressible Viscous Flow Round a Circular Cylinder," *Computers and Fluids*, Vol. 12, No. 4, pp. 255–280, 1984.

Li, H., "Stability of Oscillatory Laminar Flow Along a Wall," Beach Erosion Board, U.S. Army Corps Eng. Tech. Memo 47, 1954.

Lighthill, M. J., "The Response of Laminar Skin Friction and Heat Transfer to Fluctuations in the Stream Velocity," *Proc. Roy. Soc. (London)*, A, Vol. 224, pp. 1–23, 1954.

Mallick, D. D., "Nonuniform Rotation of an Infinite Circular Cylinder in an Infinite Viscous Fluid," *Z. Angew. Math. Mech.*, Vol. 37, 385–392, 1957.

Mankbadi, R. R. and Liu, T. C., "Near-wall Response in Turbulent Shear Flows Subjected to Imposed Unsteadiness," *J. Fluid Mech.*, Vol. 238, pp. 55–71, 1992.

Marris, A. W., "A Review on Vortex Streets, Periodic Wakes and Induced Vibration Phenomena," *J. Basic Eng.*, Vol. 86, Series D. pp. 185–196, 1964.

Meksyn, D., *New Methods in Laminar Boundary-Layer Theory*, Pergamon Press, Oxford, 1961.

McCroskey, W. J., "Unsteady Airfoils," *Ann. Rev. Fluid Mech.*, Vol. 14, pp. 285–311, 1982.

Moe, G. and Wu, Z.-J., "The Lift Force on a Cylinder Vibrating in a Current," *J. Offshore Mech. Arctic Eng.*, Vol. 112, pp. 297–303, 1990.

Moore, F. K., "On the Separation of the Unsteady Laminar Boundary Layer," IUTAM-Symposium (Berlin), *Boundary Layers*, Freiburg, pp. 296–311, 1957.

Morison, J. R., O'Brien, M. P., Johnson, J. W., and Schaaf, S. A., "The Forces Exerted by Surface Waves on Piles," *Petroleum Trans., AIME*, Vol. 189, pp. 149–157, 1950.

Nagata, H., Funada, H., and Matsui, T., "Unsteady Flows in the Vortex Region Behind a Circular Cylinder Started Impulsively, 2nd Report, Velocity Fields and Circulations," *J. Japanese Soc. Mech. Engrs.*, Vol. 28, No. 245, pp. 2608–2616, 1985.

Naudascher, E. and Rockwell, D., *Flow-Induced Vibrations, An Engineering Guide*, A. A. Balkema Publishers, Rotterdam, The Netherlands, 1994.

Newsom, R. W. and Kandil, O. A., "Vortex Flow Aerodynamics-Physical Aspects and Numerical Simulation," AIAA 87-0205, 1987.

Oertel, H. Jr., "Wakes Behind Blunt Bodies," *Ann. Rev. Fluid Mech.*, Vol. 22, pp. 539–564, 1990.

Oseen, C. W., "Uber Wirbelbewegung in Einer Reibenden Flussigkeit," *Ark. J. Mat. As trom. Fys.*, Vol. 7, pp. 14–21, 1912.

Pantazopoulos, M. S., "Vortex-Induced Vibration Parameters: Critical Review," *Proc. Offshore Mech. Arctic Eng.*, Vol. 1, pp. 199–254, 1994.

Panton, R. L., *Incompressible Flow*, John Wiley & Sons, New York, 1984.

Patrikalakis, N. M. and Chryssostomidis, C., "Vortex-Induced Response of a Flexible Cylinder in a Constant Current," *J. Energy Resources Tech.*, Vol. 107, pp. 244–249, 1985.

Pironneau, O., Rodi, W., Ryhming, I. L., Savill, A. M., and Truong, T. V. (Eds.), *Numerical Simulation of Unsteady Flows and Transition to Turbulence*, Cambridge Univ. Press, Cambridge, England, 1992.

Rosenhead, L., "Vortex Systems in Wakes," *Advances in Appl. Mech.*, von Mises, R. and Kármán, Th. von (Eds.), Academic Press, New York, Vol. III, pp. 185–193, 1953.

Rosenhead, L. (Ed.), *Laminar Boundary Layers*, Oxford University Press, Oxford, 1963.

Rott, N., "Diffraction of a Weak Shock with Vortex Generation," *J. Fluid Mech.*, Vol. 1, pp. 111–128, 1956.

Rott, N., "Theory of Time-Dependent Laminar Flows," *Theory of Laminar Flows*, Moore, F. K. (Ed.), Princeton University Press, Princeton, NJ, pp. 395–438, 1964.

Rumsey, C. L., "Details of the Computed Flow-field Over a Circular Cylinder at Reynolds Number 1200," *J. Fluids Eng.*, Trans. ASME, Vol. 110, pp. 446–452, 1988.

Sarpkaya, T., "Added Mass of Lenses and Parallel Plates," *J. Eng. Mech. Div. ASCE*, Vol. 86, No. EM3, pp. 141–152, 1960.

Sarpkaya, T., "Unsteady Flow of Fluids in Closed Systems," *J. Eng. Mech. Div. ASCE*, Vol. 88, pp. 1–15, 1962.

Sarpkaya, T., "Lift Drag, and Added-Mass Coefficients for a Circular Cylinder Immersed in a Time-Dependent Flow," *J. Appl. Mech.*, Vol. 30, No. 1, Trans. ASME, Vol. 85, Series E, pp. 13–15, 1963.

Sarpkaya, T., "Experimental Determination of the Critical Reynolds Number for Pulsating Poiseuille Flow," *J. Basic Eng.*, Vol. 88, pp. 589–598, 1966a.

Sarpkaya, T., "Separated Flow About Lifting Bodies and Impulsive Flow About Cylinders," *AIAA J.*, Vol. 3, No. 3, pp. 414–420, 1966b.

Sarpkaya, T., "On Stationary and Traveling Vortex Breakdowns," *J. Fluid Mech.*, Vol. 45, pp. 545–559, 1971.

Sarpkaya, T., "An Inviscid Model of Two-Dimensional Vortex Shedding for Transient and Asymptotically Steady Separated Flow Over an Inclined Flat Plate," *J. Fluid Mech.*, Vol. 68, pp. 109–128, 1975.

Sarpkaya, T., "In-Line and Transverse Forces on Cylinders in Oscillatory Flow at High Reynolds Numbers," *J. Ship Res.*, Vol. 21, No. 4, pp. 200–216, 1977a.

Sarpkaya, T., "Transverse Oscillations of a Circular Cylinder in Uniform Flow, Part I," Report No. NPS-69SL77071, Naval Postgraduate School, Monterey, CA, 1977b.

Sarpkaya, T., "Impulsive Flow About a Circular Cylinder," Naval Postgraduate School Technical Report NO: NPS-69SL-78-008, Monterey, CA, 1978a.

Sarpkaya, T., "Fluid Forces on Oscillating Cylinders," *J. Waterways, ASCE*, Vol. 104, No. WW4, pp. 275–290, 1978b.

Sarpkaya, T., "Vortex-Induced Oscillations—A Selective Review," *J. Appl. Mech.*, Vol. 46, pp. 241–258, 1979.

Sarpkaya, T., "Force on a Circular Cylinder in Viscous Oscillating Flow at Low Keulegan–Carpenter Numbers," *J. Fluid Mech.*, Vol. 165, pp. 61–71, 1986.

Sarpkaya, T., "Computational Methods with Vortices—The 1988 Freeman Scholar Lecture," *J. Fluids Eng.*, Vol. 111, No. 1, pp. 5–52, 1989.

Sarpkaya, T., "Nonimpulsively Started Steady Flow about a Circular Cylinder," *AIAA J.*, Vol. 29, No. 8, pp. 1283–1289, 1991.

Sarpkaya, T., "Brief Reviews of Some Time-Dependent Flows," *J. Fluids Eng.*, Vol. 114, No. 3, pp. 283–298, 1992.

Sarpkaya, T., "Coherent Structures in Oscillatory Boundary Layers," *J. Fluid Mech.*, Vol. 253, pp. 105–140, 1993.

Sarpkaya, T., "Vortex Element Methods for Flow Simulation," *Advances in Appl. Mech.*, Wu, T. Y. and Hutchinson, J. (Eds.), Vol. 31, pp. 113–247, Academic Press, New York, 1994a.

Sarpkaya, T., "Hydrodynamic Damping and Flow-Induced Vibrations: Reflections and New Results," in *Proc. Active/Passive Control of Flow-Induced Noise and Vib.*, ASME WAM-94, 1994b.

Sarpkaya, T. and Butterworth, W., "Separation Points on a Cylinder in Oscillating Flow," *J. Offshore Mech. Arctic Eng.*, Vol. 114, pp. 28–36, 1992.

Sarpkaya, T. and Ihrig, C. J., "Impulsively Started Steady Flow About Rectangular Prisms: Experiments and Discrete Vortex Analysis," *J. Fluids Engng.*, Vol. 108, pp. 47–54, 1986.

Sarpkaya, T. and Isaacson, M., *Mechanics of Wave Forces on Offshore Structures*, Van Nostrand Reinhold, New York, 1981.

Sarpkaya, T. and Kline, H. K., "Impulsively-Started Flow About Four Types of Bluff Body," *J. Fluids Eng.*, Vol. 104, pp. 207–213, 1982.

Sarpkaya, T. and Putzig, C., "Vortex Trajectories Around a Circular Cylinder in Oscillatory Plus Mean Flows," *Proc. 11th Int. Conf. Offshore Mech. Artic Eng.*, Vol. 1, pp. 69–77, 1992.

Sarpkaya, T. and Shoaff, R. L., "Inviscid Model of Two-Dimensional Vortex Shedding by a Circular Cylinder," *AIAA J.*, Vol. 17, No. 11, pp. 1193–1200, 1979.

Sarpkaya, T. and Storm, M., "In-line Force on a Cylinder Translating in Oscillating Flow," *Appl. Ocean Res.*, Vol. 7, No. 4, pp. 188–196, 1985.

Schlichting, H., "Berechnung ebener Periodischer Grenzschicht-strömungen," *Phys. Z.*, Vol. 33, pp. 327–335, 1932.

Schlichting, H., *Boundary-Layer Theory* (7th ed.), McGraw-Hill, New York, 1979.

Sears, W. R., "Some Recent Developments in Airfoil Theory," *J. Aerospace Sci.*, Vol. 23, pp. 490–499, 1956.

Sedov, L. I., *Two-Dimensional Problems in Hydrodynamics and Aerodynamics*, translated from the Russian, Chu, C. K. *et al.* (Eds.), Interscience Publishers, New York, 1965.

Segel, L. A., "A Uniformly Valid Asymptotic Expansion of the Solution to an Unsteady Boundary Layer Problem," *J. Math. Phys.*, Vol. 39, pp. 189–197, 1960.

Shirayama, S. and Kuwahara, K., "Patterns of Three-Dimensional Boundary Layer Separation," AIAA 87-0461, 1987.

Simpson, R. L., "Turbulent Boundary-Layer Separation," *Ann. Rev. Fluid Mech.*, Vol. 21, pp. 205–234, 1989.

Smith, P. A. and Stansby, P. K., "Impulsively Started Flow around a Circular Cylinder by the Vortex Method," *J. Fluid Mech.*, Vol. 194, pp. 45–77, 1988.

Spalart, P. R. and Baldwin, B. S., "Direct Simulation of a Turbulent Oscillating Boundary Layer," *Turbulent Shear Flows*, Vol. 6, pp. 417–440, Springer-Verlag, 1989.

Staubli, T., "Calculation of the Vibration of an Elastically Mounted Cylinder Using Experimental Data From Forced Oscillation," *J. Fluids Eng.*, Vol. 105, pp. 225–229, 1983.

Stewartson, K., "The Theory of Unsteady Laminar Boundary Layers," *Adv. Appl. Mech.* Dryden, H. L. *et al.* (Eds.), Academic Press, New York, pp. 1–37, 1960.

Stokes, G. G., "On the Effect of the Internal Friction of Fluids on the Motion of Pendulums," *Trans. Camb. Phil. Soc.*, Vol. 9, pp. 8–106, 1851.

Stratford, B. S., "The Prediction of Separation of the Turbulent Boundary Layer," *J. Fluid Mech.*, Vol. 5, pp. 1–16, 1959.

Stuart, J. T., "Unsteady Boundary Layers," *Laminar Boundary Layers*, Rosenhead, L. (Ed.), Oxford University Press, Oxford, pp. 349–406, 1963.

Taneda, S., "Visual Study of Unsteady Separated Flows around Bodies," *Prog. Aerospace Sci.*, Vol. 17, pp. 287–348, 1977.

Taneda, S., "Definition of Separation," *Reports of Res. Inst. Appl. Mech.*, Kyushu University, Vol. 28, No. 89, pp. 73–81, 1980.

Ta Phuoc Loc and Bouard, R., "Numerical Solution of the Early Stage of the Unsteady Viscous Flow around a Circular Cylinder; a Comparison with Experimental Visualization and Measurements," *J. Fluid Mech.*, Vol. 116, pp. 93–117, 1985.

Taylor, G. I. (Sir), "The Interaction Between Experiment and Theory in Fluid Mechanics," *Ann. Rev. Fluid Mech.*, Vol. 6, pp. 1–16, 1974.

Thoman, D. C. and Szewczyk, A. A., "Time-Dependent Viscous Flow Over a Cylinder," *Phys. Fluids*, Vol. 12, Supplement II, pp. 76–86, 1969.

Thrasher, D. F. "Application of the Vortex-Lattice Concept to Flows with Smooth-Surface Separation," *Proc. ONR Symp. Naval Hydrodynamics*, Ann Arbor, MI (see also DTNSRDC-85/041, 1985), 1982.

Temple, G., "Unsteady Motion," in *Modern Developments in Fluid Dynamics*, Howarth, L. (Ed.), Vol. 1, Oxford University Press, pp. 325–371, 1953.

Van Dommelen, L. L. and Cowley, S. J., "On the Lagrangian Description of Unsteady Boundary-Layer Separation. Part 1. General Theory," *J. Fluid Mech.*, Vol. 210, pp. 593–626, 1990.

Wang, C.-Y., "On High-Frequency Oscillating Viscous Flows," *J. Fluid Mech.*, Vol. 32, pp. 55–68, 1968.

Wang, K. C., "Separation Patterns on Boundary Layer Over an Inclined Body of Revolution," *AIAA J.*, Vol. 10, pp. 1044–1050, 1972.

Wang, X. and Dalton, C., "Numerical Solutions for Impulsively Started and Decelerated Viscous Flow Past a Circular Cylinder," *Int. J. Numerical Methods in Fluids*, Vol. 12, pp. 383–400, 1991a.

Wang, X. and Dalton, C., "Oscillating Flow Past a Rigid Circular Cylinder: A Finite-Difference Calculation," *J. Fluids Eng.*, Vol. 113, pp. 377–383, 1991b.

Waugh, J. G. and Ellis, A. T., "Fluid Free-Surface Proximity Effect on a Sphere Vertically Accelerated from Rest," AIAA 69-42, 1969.

Wille, R., "On Unsteady Flows and Transient Motions," *Prog. in Aeronautical Sci.*, Küchemann, D. *et al.* (Eds.), Pergamon Press, London, Vol. 7, pp. 195–207, 1966.

Williams, J. C. III and Chen, C.-G., "Unsteady Separation on an Impulsively Set into Motion Carafoli Airfoil," *Pro. Int. Symp. Nonsteady Fluid Dynamics*, Miller, J. A. and Telionis, D. P. (Eds.), FED-Vol. 92, pp. 291–301, 1990.

Williamson, C. H. K. and Roshko, A., "Vortex Formation in the Wake of an Oscillating Cylinder," *J. Fluids and Structures*, Vol. 2, pp. 355–381, 1988.

Wong, T. C., Kandil, O. A., and Liu, C. H., "Navier–Stokes Computations of Separated Vortical Flows Past Prolate Spheroid at Incidence," AIAA 89-0553, 1989.

Xu, W. C. and Wang, K. C., "Unsteady Laminar Boundary Layer on Impulsively Started Prolate Spheroid," *J. Fluid Mech.*, Vol. 195, pp. 413–435, 1988.

Zdravkovich, M. M., "Modification of Vortex Shedding in the Synchronization Range," *J. Fluids. Eng.*, Vol. 104, pp. 513–517, 1982.

13 Complex Flows

PHILIP G. SAFFMAN
California Institute of Technology
Pasadena, CA

FRITZ H. BARK
The Royal Institute of Technology
Stockholm, Sweden

THOMAS B. MORROW
Southwest Research Institute
San Antonio, TX

JOSEPH A. SCHETZ
Virginia Polytechnic Institute and State University
Blacksburg, VA

AYODEJI O. DEMUREN
Old Dominion University
Norfolk, VA

TAKAKAGE ARAI
Muroran Institute of Technology
Muroran, Japan

TAKAO INAMURA
Tohoku University
Sendai, Japan

ROBERT A. GREENKORN
Purdue University
West Lafayette, IN

DONALD O. ROCKWELL
Lehigh University
Bethlehem, PA

CARL H. GIBSON
University of California—San Diego
La Jolla, CA

Handbook of Fluid Dynamics and Fluid Machinery, Edited by Joseph A. Schetz and Allen E. Fuhs
ISBN 0-471-12598-9 Copyright © 1996 John Wiley & Sons, Inc.

SERGE T. DEMETRIADES
C. D. MAXWELL
STD Research Incorporated
Arcadia, CA

CONTENTS

13.1 VORTEX FLOWS
Philip G. Saffman

13.1.1 Introduction

The velocity of a fluid can be described by a vector function of space $\bar{\mathbf{x}}$ and time t, written as $\bar{\mathbf{u}}(\bar{\mathbf{x}}, t)$. The vector curl of the velocity

$$\bar{\boldsymbol{\omega}}(\bar{\mathbf{x}}, t) = \nabla \times \bar{\mathbf{u}} \tag{13.1}$$

is called the *vorticity*. In an incompressible fluid, the relation between velocity and vorticity is analogous to that between a magnetic field and current density. Vorticity can be interpreted physically as an *angular momentum density*; a spherical fluid particle, instantaneously frozen without loss of angular momentum, would rotate with angular velocity $\frac{1}{2}\bar{\boldsymbol{\omega}}$.

All flows of real fluids contain some vorticity. Boundary layers, for example, can be interpreted as vortex sheets of finite thickness. The term *vortex flow* refers to motions in which the vorticity is essentially confined to finite regions, usually called *vortices* in which the fluid is said to be in rotational motion, such that the principal mechanism of the flow is the movement and interaction of the vortex or vortices. In a fluid of uniform density, a fluid particle loses or acquires vorticity only by viscous diffusion or the action of nonconservative forces. Consequently, vortex flows are primarily, although not exclusively, of interest for motions in which viscous and compressible effects are of minor importance.

Examples of vortex motion are vortex rings (smoke rings), which can be visualized by carefully dropping dye into a beaker of water, the bathtub vortex, the *von Karman vortex street* in the wake of buff bodies at intermediate Reynolds numbers (see Fig. 4.110), the single row of vortices formed at the interface between two streams of different velocity, the trailing vortices behind an aircraft, often visible as vapor trails, the leading edge vortices produced by delta wings at high angles of attack, the wakes of propellers, *dust devils* and tornados, and the quantized vortices of superfluid Helium II. Propellers are discussed in Sec. 27.2.

The mechanics of vortex formation are in most cases not well understood. Vortices cannot be created in perfectly inviscid fluid, and their existence is due to subtle effects of viscosity, usually intimately related to the phenomena of boundary layer separation, or the application of nonconservative forces. Experimental observations are difficult, the flows are often unsteady, and nonintrusive measuring devices are desirable because vortex flows may be easily disturbed. The Laser Doppler Anemometer has been a useful tool. Flow visualization techniques are commonly employed, but may suffer from uncertainties in interpretation because of doubts about the extent to which vorticity is marked by dye or smoke. Flow separation is discussed in Sec. 4.9. Instrumentation for vorticity measurement is presented in Sec. 15.10, and test facilities are described in Chap. 16.

13.1.2 Laws of Vortex Motion

Three basic laws of vortex motion for incompressible fluids of negligible viscosity were formulated by Helmholtz [Serrin (1959), Batchelor (1967), Saffman and Baker (1979), Lugt (1983), and Saffman (1992)] and can be stated as follows (but note that alternative equivalent forms are often given):

    I. Fluid particles originally free of vorticity remain free of vorticity.
    II. Fluid particles on a vortex line remain on a vortex line, so that vortex lines move with the fluid.
    III. The strength of the vorticity is proportional to the length of the vortex line.

These laws are fundamental to the dynamics of fluid motions at large Reynolds numbers. The third law is particularly important for understanding the physics of turbulence and rotating fluids, the amplification of vorticity by the stretching of vortex lines being of the utmost importance. The laws can be derived from the *Helmholtz evolution equation* for the vorticity vector

$$\frac{\partial \overline{\omega}}{\partial t} + \overline{\mathbf{u}} \cdot \nabla \overline{\omega} = \overline{\omega} \cdot \nabla \overline{\mathbf{u}} \tag{13.2}$$

Alternative approaches utilize the important *Kelvin circulation theorem*, which states that the material circulation, Γ, defined as the line integral of the velocity field around a closed curve moving with the fluid,

$$\Gamma = \oint \overline{\mathbf{u}} \cdot \mathbf{d\overline{s}} \tag{13.3}$$

is constant when viscous effects are negligible and only conservative forces are acting.

    The Helmholtz laws and Kelvin theorem also hold with a minor modification to the third law if the fluid is compressible, provided the density is a function only of the pressure (i.e., the fluid is *barotropic*). The modification is that the vorticity is proportional to the product of the length of the line and the density.

13.1.3 Invariants of Vortex Motion

A *hydrodynamic impulse*, $\overline{\mathbf{I}}$, and *angular impulse*, $\overline{\mathbf{A}}$, can be associated with a vortex flow. These are defined in a fluid of uniform density ρ by

$$\bar{\mathbf{I}} = \tfrac{1}{2}\rho \int \bar{\mathbf{r}} \times \overline{\boldsymbol{\omega}} \, dV, \quad \overline{\mathbf{A}} = -\tfrac{1}{2}\rho \int r^2 \overline{\boldsymbol{\omega}} \, dV \qquad (13.4)$$

They can be interpreted physically as the impulse and moment of impulse of the body forces that must be applied to generate the motion in a fluid initially at rest. Confusion often occurs between the impulse and the momentum of the fluid. The latter is given by a conditionally convergent integral in an unbounded fluid and is zero if the fluid is confined within a rigid container. In two-dimensional flow, the impulse is given by Eq. (13.4) without the factor $\tfrac{1}{2}$, because of contributions over surfaces at infinity. The sums of $\bar{\mathbf{I}}$ and $\overline{\mathbf{A}}$ over all the vortices in an unbounded fluid are invariants of the motion of an inviscid fluid. (In fact, $\bar{\mathbf{I}}$ is constant even if viscosity is acting.) The kinetic energy is also an invariant if the boundaries are at rest; expressions for the kinetic energy as an integral over the vorticity exist but are somewhat complicated.

13.1.4 Vortex Filaments

Various idealizations prove useful in the study of vortex motions. The *vortex filament* is a vortex in the shape of a tube of small cross section. When the tube forms a circular ring, it models the vortex ring. The Helmholtz laws imply that the filament is always composed of the same fluid particles and its strength is constant, this being defined as the flux of vorticity through the cross section or, equivalently, the circulation around a circuit threaded by the filament. The limit of a vortex filament of finite strength and zero cross section is called a *line vortex* (not to be confused with a vortex line which is simply a curve in the fluid tangent to the local vorticity vector). The straight or rectilinear line vortex is of particular importance for modelling two-dimensional flows, where the absence of stretching in the transverse direction makes vorticity a conserved quantity. The Helmholtz laws imply that a rectilinear line vortex moves with the fluid unless external forces act on the core. The vortex is said to be *bound* if the external forces keep it at rest. Line vortices are a less useful concept in three dimensions, because the velocity field produced by a curved line vortex is infinite on the vortex, giving rise to an infinite self-induced velocity. To employ curved line vortices, it is necessary to specify a core cross section, upon which the results will depend, although only weakly.

Another common idealization is the *vortex sheet*, which is the limit when the vorticity is compressed into a surface of zero thickness. The velocity normal to the sheet is continuous. The tangential component is discontinuous, and the magnitude of the jump is the *strength* of the sheet. The flow past a wing or lifting body can be described by bound vortices in the wing together with a trailing vortex sheet downstream, which rolls up into a pair of oppositely rotating vortex tubes.

The velocity field induced at a point P by a vortex at Q can be obtained by integrating Eq. (13.1) and is

$$\overline{\mathbf{u}}_p = \frac{1}{4\pi} \int \frac{\overline{\boldsymbol{\omega}}_Q \times \bar{\mathbf{r}}_{PQ}}{r_{PQ}^3} \, dV_Q \qquad (13.5)$$

where $\bar{\mathbf{r}}_{PQ} = \bar{\mathbf{r}}_P - \bar{\mathbf{r}}_Q$. The contribution from a line vortex of strength κ is given by the *Biot–Savart law*

$$\overline{\mathbf{u}}_P = \frac{\kappa}{4\pi} \oint \frac{d\overline{\mathbf{s}}_Q \times \overline{\mathbf{r}}_{PQ}}{r_{PQ}^3} = \frac{\kappa}{4\pi} \nabla\Omega \tag{13.6}$$

where Ω is the solid angle subtended at P by the vortex. For a rectilinear line vortex at the origin, the (two-dimensional) velocity field takes the particularly simple form

$$u = \frac{-\kappa y}{2\pi r^2}, \quad v = \frac{\kappa x}{2\pi r^2} \tag{13.7}$$

The velocity given by Eq. (13.6) for a curved line vortex is divergent when the point P lies on the vortex, unless the radius of curvature R at the point is infinite. When $P \rightarrow Q$, this expression has the limit

$$\overline{\mathbf{u}}_P = \frac{\kappa}{2\pi} \frac{\overline{\mathbf{t}} \times \overline{\mathbf{r}}_{PQ}}{r_{PQ}^2} + \frac{\kappa \overline{\mathbf{b}}}{4\pi R} \log \frac{R}{r_{PQ}} \tag{13.8}$$

where $\overline{\mathbf{t}}$ and $\overline{\mathbf{b}}$ are the tangent and binormal to the line at Q. The first term describes the circulatory motion about the vortex. The singularity in the second term is removed when the vortex is given a finite cross section, and this term causes a self-induced translation of a curved filament in the direction of its binormal at a rate depending logarithmically on the core radius.

An axisymmetric rectilinear vortex filament of strength κ and finite cross section in unbounded fluid has the velocity field

$$u_r = 0, \quad u_\theta = V(r), \quad u_z = W(r) \tag{13.9}$$

where V and W are arbitrary functions satisfying $rV \rightarrow \kappa/2\pi$ and $W \rightarrow 0$ as $r \rightarrow \infty$. The streamlines are helices about the axis, circles when $W = 0$. This flow field is a good approximation to many cases of practical interest. The case most commonly considered is a uniform core of radius, a, with uniform vorticity and axial velocity

$$V = \frac{\kappa r}{2\pi a^2}, \quad W = W_0 \ (r < a), \quad V = \frac{\kappa}{2\pi r}, \quad W = 0 \ (r > a) \tag{13.10}$$

The case with $W_0 = 0$ is called the *Rankine vortex*. The hollow vortex in which the core is stagnant is also of interest. Exact solutions exist for nonaxisymmetric cores; these are not steady and rotate with some angular velocity. For the case (Kirchhoff) of an elliptical cross section with semiaxes a and b, the angular velocity is $\kappa/\pi(a + b)^2$. Numerical results exist for triangular and square-like shapes.

The diffusion of vorticity by viscosity in a rectilinear filament is described by

$$\frac{\partial \omega}{\partial t} = \nu \nabla^2 \omega \tag{13.11}$$

where ν is the kinematic viscosity. The solution (Oseen–Hamel)

$$\omega = \frac{\kappa}{4\pi\nu t} \exp\left[\frac{-r^2}{4\nu t}\right], \quad u_\theta = \frac{\kappa}{2\pi r}\left\{1 - \exp\left[\frac{-r^2}{4\nu t}\right]\right\} \tag{13.12}$$

describes the viscous decay of a rectilinear line vortex. An important question about which little is known is the diffusion of a vortex filament with a turbulent core, as vortices in applications are often turbulent. There are, however, reasons to believe that overshoots of circulation may occur, if the turbulence causes the vorticity to spread faster than it does as a result of molecular diffusion.

Rectilinear filaments of finite core size disintegrate if exposed to too large an external straining field. For the case of a uniform vortex with vorticity ω_0, it will be torn apart if the external rate of strain exceeds $0.15\omega_0$. Two rectilinear vortices amalgamate if placed too close together. These properties have generally to be studied by numerical means, but analytical results exist for special cases.

13.1.5 Vortex Rings

Kelvin's theory of vortex atoms stimulated 19th-century study of vortices and of vortex rings. In terms of cylindrical polar coordinates, (r, θ, z), the vorticity and swirl velocity in the core of a vortex ring are such that ω_θ/r and ru_θ are constant on streamlines moving with the vortex in steady inviscid flow. For thin rings of circulation Γ with core radius, a, small compared with the ring radius, R, the following formulas give the velocity V, kinetic energy E, and hydrodynamic impulse P of the ring

$$V = \frac{\kappa}{4\pi R}\left[\log\frac{8R}{a} - \frac{1}{2} + \frac{2\pi}{\kappa^2}\int (u_r^2 + u_z^2 - 2u_\theta^2)\,dS\right] \tag{13.13}$$

$$E = \frac{1}{2}\rho\kappa^2 R\left[\log\frac{8R}{a} - 2 + \frac{2\pi}{\kappa^2}\int (u_r^2 + u_z^2 + u_\theta^2)\,dS\right] \tag{13.14}$$

$$P = \rho\pi\kappa R^2 \tag{13.15}$$

The integral is over the cross section of the core. For uniform core without swirl, ω_θ/r is constant, $u_\theta = 0$, the value of the integral is $\kappa^2/8\pi$ leading to Kelvin's expression for the velocity

$$V = \frac{\kappa}{4\pi R}\left[\log\frac{8R}{a} - \frac{1}{4}\right] \tag{13.16}$$

According as $V > \kappa/2R$ or $< \kappa/2R$, the irrotational fluid carried along with the ring is ring shaped or oval. The relative error in these expressions is $O[(a^2/R^2)\log(R/a)]$.

For uniform cores without swirl, the thin ring is one extreme of a family of fat vortex rings. The other extreme is the *Hill spherical vortex*, which is an exact solution of the inviscid incompressible equations with the vortex core a sphere of radius a. For this flow

$$V = \frac{2\kappa}{5a}, \quad E = \frac{8}{35}\pi\rho a\kappa^2, \quad P = \frac{4}{5}\pi\rho a^2\kappa \tag{13.17}$$

It can be generalized to include *swirl*.

The steady motion of thin rings is a special case of the motion of curved vortex filaments. The velocity of a point P on the filament axis is given to first approximation by

$$\frac{d}{dt}\,\overline{\mathbf{r}}_P = \overline{\mathbf{U}}_{\text{ext}} + \frac{\kappa}{4\pi} \int_{[\delta]} \frac{d\overline{\mathbf{s}}_Q \times \overline{\mathbf{r}}_{PQ}}{r_{PQ}^3} \tag{13.18}$$

where $\overline{\mathbf{U}}_{\text{ext}}$ is the external irrotational velocity field, produced by other vortices or bodies, and $[\delta]$ denotes that a length 2δ centered on P is omitted from the Biot–Savart integral. The *cutoff length* δ is given by

$$\delta = \frac{1}{2}\,a\,\exp\left[\frac{1}{2} - \frac{2\pi a^2\overline{v^2}}{\kappa^2} + \frac{4\pi a^2\overline{w^2}}{\kappa^2}\right] \tag{13.19}$$

where a is the core radius and $\overline{v^2}$ and $\overline{w^2}$ are the average values in the core of the squares of the swirl velocity about the axis and axial flow parallel to the core. If $\overline{w^2} \gg \overline{v^2}$, the equation fails, and the vortex should be regarded as a weakly swirling jet. Filaments with significant axial velocity may undergo an *instability* or transition or *bifurcation* characterized by large changes in core radius and reverse axial flow, known as *vortex breakdown* or *bursting*, whose explanation is controversial.

The stability and periods of vibration of thin vortex rings have been studied. Kelvin argued that the uniform core without swirl was stable to axisymmetric disturbances, but left open the question of stability to three-dimensional disturbances. There is now both experimental and theoretical evidence of instability to twisting motions in which the cross section remains circular but the circular axis of the core is deformed.

There are a number of fascinating experimental studies of vortex ring interactions (see Chap. 17). However, vortex rings produced in the laboratory tend to be fattish and subject to turbulence and viscosity, and definitive results are sparse. The validity of the thin ring analysis is moot.

13.1.6 Two-Dimensional Vortex Motion

There are extensive studies of vortex motion in two dimensions. A popular approach is to replace continuous vorticity distributions and vortex sheets by large numbers of rectilinear line (point) vortices. The motion of N vortices is given by the integration of $2N$ ordinary differential equations and lends itself to high-speed computing. In this way, calculations of the evolution of two-dimensional vortices are possible. To overcome numerical difficulties, it is the practice to replace the velocity field of the individual point vortex by that of a circular vortex with some internal structure. Several variants are used. The results are probably qualitatively accurate, but the quantitative errors are unknown and caution should be employed. The methods are particularly suspicious for the roll up of vortex sheets, for which there is evidence that the formulation is ill-posed.

The stability of equilibrium configurations of point vortices, with the vortices in uniform translation or rigid body rotation, has been studied thoroughly for the case of two-dimensional disturbances. Rows and streets are generally unstable. Effects of finite core size on the stability of these configurations is of interest, and there is evidence that they may have a stabilizing effect, particularly for the Karman vortex street. The stability to three-dimensional disturbances is obscure; in this case, effects of *finite core size* and *core structure* are crucial. For example, an isolated vortex filament without axial velocity is stable to infinitesimal disturbances. If the core is uniform, it is unstable in an external flow to disturbances with axial wavelength comparable to core size, but hollow vortices do not have this instability.

13.2 ROTATING FLOWS
Fritz H. Bark

13.2.1 Scope

Rotating flows have, in many respects, properties that are markedly different from those of nonrotating flows. This section will, in brief form, describe some of these properties. However, the class of rotating flows is tremendously large, and restrictions are necessary. Only homogeneous Newtonian fluids will be considered, and the discussion will, with a few exceptions, be limited to flows that are strongly influenced by solid boundaries. Such flows occur quite frequently in industrial applications, e.g., vortex chambers, rotating tanks, and nozzles for fluid atomization. Other cases of interest are rotating devices for separation of suspensions (centrifuges, decanters, and cyclones), where the effects of the suspended matter on the flow sometimes are small and the fluid can, as a first approximation, be considered as homogeneous and Newtonian.

The dynamics of vortices is the subject of Sec. 13.1 of this handbook and will therefore not be dealt with here. Specific problems in turbomachinery are discussed in Chap. 27 and are also omitted.

The restriction to a homogeneous fluid means that rotating stratified fluids, which mainly occur in geophysics but also in some industrial applications, are not included. The interested reader is referred to the books by Holton (1972) and Pedlosky (1979).

Due to lack of space, many important phenomena, even within the restrictions outlined above, have been omitted in this section. The standard reference on rotating flows is the monograph by Greenspan (1990). For application-oriented material, the reader is refered to engineering archives such as the *Journal of Applied Mechanics* and the *Journal of Fluids Engineering*. Recent results of fundamental character can be found in the *Journal of Fluid Mechanics* and *Physics of Fluids*.

The disposition of the present article is the following: The equations of motion in a rotating frame of reference are briefly discussed in Sec. 13.2.2. In Sec. 13.2.3, some general properties of rotating flows are illustrated by two exact solutions of these equations. Small *Rossy number* flows (steady and oscillating inviscid flows, boundary layers) are considered in Sec. 13.2.4. Some properties of flows at finite Rossby numbers are given in Sec. 13.2.5. Stability of rotating boundary layers are considered in Sec. 13.2.6. Section 13.2.7 deals with rotating turbulent flows. The present article focuses on physical aspects of rotating flows. Issues related to numerical computation are not taken up.

13.2.2 Equations of Motion in a Rotating Frame

In many applications of rotating flows, e.g., centrifuges for separation, the motion of the fluid deviates only slightly from the rigid rotation of the container. In such cases, it is advantageous to use a coordinate system which rotates with the same constant angular velocity $\overline{\Omega}$ as the container. If a typical order of magnitude of the velocity in the rotating system is V, the following scales are convenient for the formulation of a nondimensional problem: Ω^{-1}, time, H, length, V, velocity, and $\rho H \Omega V$, pressure. Here $\Omega = |\overline{\Omega}|$, H is a typical dimension of the container (frequently taken as its height) and ρ is the constant density of the fluid. In terms of the nondimensional velocity and reduced* pressure fields, $\overline{u}$ and p respectively, the equations for conservation of mass and momentum are

$$\nabla \cdot \overline{u} = 0 \tag{13.20}$$

$$\frac{\partial \overline{u}}{\partial t} + \epsilon \overline{u} \cdot \nabla \overline{u} + 2\overline{k} \times \overline{u} = -\nabla p - E \nabla \times \nabla \times \overline{u} \tag{13.21}$$

where $\overline{k} = \overline{\Omega}/\Omega$. The nondimensional parameters ϵ and E are defined by

$$\epsilon = \frac{\nabla}{H\Omega}, \quad \text{Rossby number}$$

$$E = \frac{\nu}{H^2\Omega}, \quad \text{Ekman number} \tag{13.22}$$

where ν is the constant kinematic viscosity of the fluid. The Rossby number is a nondimensional measure of the deviation from rigid rotation or, equivalently, an estimate of the ratio between the convective acceleration of the flow in the rotating coordinate system and the *Coriolis force*. The *Ekman number* can be interpreted as an inverse Reynolds number based on the velocity, $H\Omega$, or as the ratio between viscous forces associated with fluid motions on the length scale, H, and the Coriolis force. In the present section, attention will be restricted to cases where the rate of rotation is large and where the deviation from rigid rotation is at most moderate so that $E \ll 1$ and $\epsilon \lesssim 1$. Proper boundary and initial conditions for Eqs. (13.20) and (13.21) are discussed in Sec. 1.4 of this handbook.

The difference between the law of conservation of momentum in a rotating coordinate system, Eq. (13.21) and its counterpart in an inertial coordinate system is the presence of the *Coriolis force* $-2\overline{k} \times \overline{u}$. This force appears because the coordinate system is accelerating and is thus of a fictitious nature. Nevertheless, when considering phenomena in a rotating frame, it is in many cases convenient to think about the Coriolis force as a real force. An important property of this fictitious (and nonconservative) force is its inability to perform work and thereby change the energy of fluid elements. It should be pointed out that, for frictionless motion, the circulation as measured in the rotating coordinate system is not a conserved quantity.

Henceforth, the axis of rotation is taken to coincide with the z-axis in a Cartesian or a cylindrical coordinate system, i.e., $\overline{k} = \overline{e}_z$ is what follows.

13.2.3   Simple Exact Solutions

General effects caused by rapid rotation on a moving fluid can be illustrated by exact solutions of Eq. (13.21) for idealized boundary conditions. For the solutions dis-

---

*The dimensional pressure Π_* is related to the dimensional reduced pressure p_* by the relation $p_* = \Pi_* - (\rho/2) |\overline{\Omega} \times \overline{x}_*|^2$ where $\overline{x}_*$ is the dimensional radius vector.

cussed in this section, the nonlinear term in Eq. (13.24) vanishes. Taking $\partial/\partial t = \partial/\partial x = \partial/\partial y = 0$ except for $\partial p/\partial x = -G = $ const. and prescribing $\bar{\mathbf{u}} = (u, v, w) = 0$ at $z = \pm 1$, one finds the *rotation-modified plane Poiseuille flow*

$$u - iv = \frac{iG}{2}\left[1 - \frac{\cosh (\vartheta z/E^{1/2})}{\cosh (\vartheta/E^{1/2})} \right], \quad \vartheta = 1 - i \qquad (13.23)$$

$$w = 0 \qquad (13.24)$$

The velocity profiles $u(z)$ and $v(z)$ are shown in Fig. 13.1(a) and (b). For large values of E, the nonrotating case is recovered. For small values of E, Eq. (13.22) has the following asymptotic behavior

$$u - iv \sim \frac{iG}{2}, \quad 1 \pm z = 0(1) \qquad (13.25)$$

$$u - iv \sim \frac{iG}{2}\{1 - \exp [-\vartheta(1 \pm z)/E^{1/2}]\}, \quad 1 \pm z = 0(E^{1/2}) \qquad (13.26)$$

For small values of E, the flow splits into an inviscid part, Eq. (13.25) and two boundary layers, Eq. (13.26). In the inviscid part, the Coriolis force balances the pressure gradient, i.e., the velocity is perpendicular to the pressure gradient. Such flows are called *geostrophic flows*. The boundary layers, whose thickness is $0(E^{1/2})$, are called *Ekman layers* [Ekman (1905)]. The oscillatory character of the velocity profiles in the Ekman layers should be noted.

For any value of the Reynolds number based on the velocity relative to the rotating boundaries, e.g., Re $= \epsilon E^{-1}$, the flow will have a boundary layer structure provided that the rotation is sufficiently strong, i.e., E is sufficiently small. This surprising effect is caused by the tendency of spinning vertical line elements of fluid

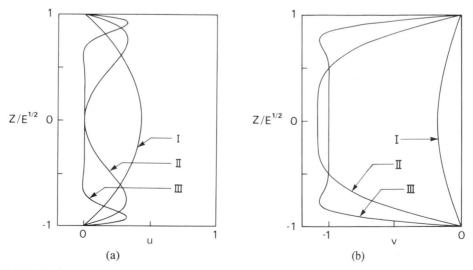

FIGURE 13.1 Velocity profiles u and v in rotation-modified plane Poiseuille flow. $G = 2$, I: $E = 2$, II: $E = 0.1$, III: $E = 0.01$. (a) u-velocity profiles and (b) v-velocity profiles.

to resist tilting in a way similar to a gyroscope, thereby preventing *viscous diffusion* away from the boundaries. Alternatively, one may say that the vortex lines associated with the rigid rotation have a stiffening effect on the fluid. Effects of viscosity will thus be felt only in the Ekman layers.

For $G = 0$ and $\overline{\mathbf{u}} = \pm\overline{\mathbf{e}}_x$ at $z = \pm 1$, one has the *rotation-modified Couette flow*, whose asymptotic behavior is

$$u - iv \sim 0, \qquad 1 \pm z = 0(1) \tag{13.27}$$

$$u - iv \sim \pm\exp\left[-\vartheta(1 \pm z)/E^{1/2}\right], \qquad 1 \pm z = 0(E^{1/2}) \tag{13.28}$$

The corresponding velocity profiles are shown in Fig. 13.2(a) and (b). In this case, there is no geostrophic flow; all motion occurs in the Ekman layers. The shear stress on the plates is shown in Fig. 13.2(c).

These simplified examples can be used to infer trends in much more complicated laminar flows occurring in applications. Consider, for example, two parallel circular plates that are rotating with the same angular velocity. For a given radial volume flux between the plates, the required radial pressure gradient will be $\sim \Omega^{1/2}$. The reason is that the radial transport will take place in boundary layers of thickness $\sim \Omega^{-1/2}$, even if the Rossby number is finite. Similarly, if one of the plates is rotating and the other is fixed, the torque is $\sim \Omega^{3/2}$ because the velocity is $\sim \Omega$ and the velocity gradient is $\sim \Omega^{1/2}$.

13.2.4 Flows at Very Small Rossby Numbers

This class of flows is often labelled *linear flows* and is described by solutions of the equations of motion with $\epsilon = 0$. An important question is when such solutions are approximations of solutions of the full nonlinear equations. In the general case, the requirement is $\epsilon \ll E^{1/2}$, which is quite restrictive. Less restrictive conditions can be derived in certain special cases [Smith (1976)]. Furthermore, linear theory is often useful for rough estimates in cases where nonlinear effects are finite [Bennetts and Hocking (1973)].

Even if the motion of the fluid relative to rotating boundaries is very slow, regions of inviscid flow and boundary layer flow do usually appear [see Smith (1976) and Bennetts and Hocking (1973)]. Linear flows can therefore often be computed approximately by using singular perturbation techniques [Greenspan (1990) and Kevorkian and Cole (1981)].

Steady Inviscid Flows. These kinds of flows are called geostrophic flows. One finds from Eq. (13.21) that

$$2\overline{\mathbf{e}}_z \times \overline{\mathbf{u}} = -\nabla p \tag{13.29}$$

$$\frac{\partial}{\partial z}[\overline{\mathbf{u}}, p] = 0 \tag{13.30}$$

The first equation says that the pressure is a stream function. The latter equation is the *Taylor–Proudman theorem* [Taylor (1923)], which means that the velocity and pressure fields are constant along lines that are parallel with the axis of rotation. A

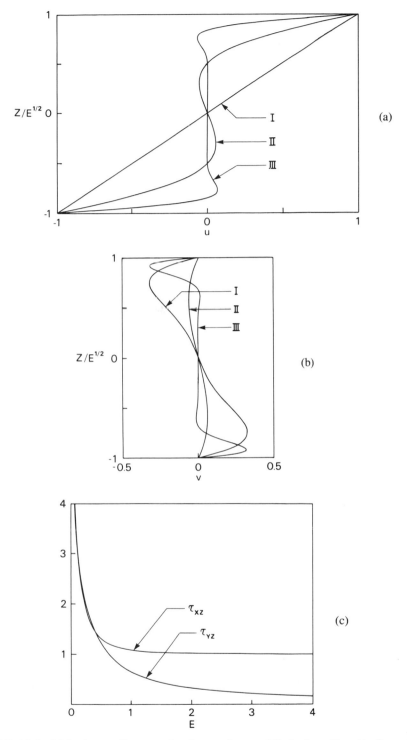

FIGURE 13.2 Velocity profiles u and v in rotation-modified plane Couette flow. I: $E = 2$, II: $E = 0.1$, III: $E = 0.01$. (a) u-velocity profiles, (b) v-velocity profiles, and (c) non-dimensional shear stress on the lower plate in the rotation-modified plane Couette flow defined in the text as function of E. For $E \to \infty$, $(\tau_{xz}, \tau_{yz}) = (1, 0)$. For $E \to 0$, $(\tau_{xz}, \tau_{yz}) = E^{-1/2}(1, 1)$.

consequence of the Taylor–Proudman theorem is that a vertical line element of fluid between the top and bottom of the container cannot change its height when moving around in the container. Therefore, the fluid has to move along a *closed* path on which the vertical distance between the top and bottom of the container is constant. Such paths are called *geostrophic curves*. The number of geostrophic curves may be zero or infinite depending on the shape of the container (see Fig. 13.3).

To determine a geostrophic flow, one must consider the Ekman layers on the top and bottom of the container, which are *controlling* the geostrophic flow. This is in contrast to the Blasius boundary layer on a flat plate, for example. In that case, the effect of the boundary layer on the outer inviscid flow is weak. For geostrophic flow between two horizontal flat plates that are rotating with the same angular velocity, consideration [Greenspan (1990)] of the Ekman layers gives that

$$\frac{\partial^2 p}{\partial x^2} + \frac{\partial^2 p}{\partial y^2} = 0 \qquad (13.31)$$

i.e., the function $p(x, y)$ is harmonic and the geostrophic velocity field is a two-dimensional potential flow. More general cases with, for example, curved boundaries in relative motion are complicated to compute, and the reader is referred to Greenspan (1990).

A consequence of the Taylor–Proudman theorem is shown in Fig. 13.4. The flow takes place between two infinitely large parallel plates. An obstacle is placed on one of the plates, and the geostrophic flow far from the obstacle is uniform. The fluid cannot move over the obstacle. Above the obstacle, there is a column of stagnant fluid, which affects the potential flow outside in the same way as a solid cylinder.

Oscillating Inviscid Flows. Wave motions are possible in a homogeneous incompressible fluid in a rotating container. Such waves are called *inertial waves*.

The simplest example is plane waves in an unbounded rotating fluid. The velocity

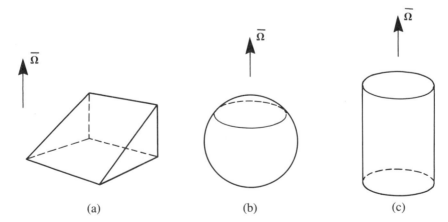

(a)   (b)   (c)

FIGURE 13.3 Geostrophic curves in containers. (a) Container with no geostrophic curves. (b) The geostrophic curves are defined by a constant latitude angle. (c) Any curve on the lid is a geostrophic curve.

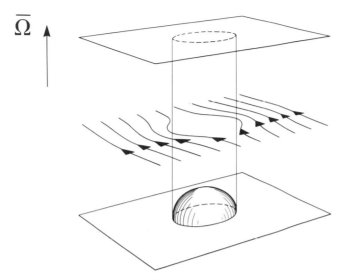

FIGURE 13.4 A Taylor column.

field for such a wave is written

$$\overline{u} = \overline{Q} \exp [i(\overline{\kappa} \cdot \overline{x} - \omega t)] \qquad (13.32)$$

where $\overline{Q}$ is a constant complex vector, $\overline{\kappa}$ is the wave number vector and ω is the frequency. Equations (13.20) and (13.32) give

$$\overline{\kappa} \cdot \overline{Q} = 0 \qquad (13.33)$$

i.e., the waves are transverse. The dispersion relation (see Chap. 6) is

$$\omega = \pm \frac{2\kappa_z}{\kappa} \qquad (13.34)$$

where $\kappa = |\overline{\kappa}|$. There is, consequently, an upper bound on the absolute value of the frequency

$$|\omega| \leq 2 \qquad (13.35)$$

A wave maker, whose frequency is larger than twice the rate of rotation will thus generate a local nonwavy response in the fluid. Equation (13.34) says that *plane inertial waves* are *anisotropic* and that the frequency depends only on the direction of the wave number vector. A peculiar property of plane inertial waves is that *phase velocity*

$$\overline{c} = \frac{\omega\overline{\kappa}}{\kappa^2} \qquad (13.36)$$

is perpendicular to the *group velocity*

$$\overline{\mathbf{c}}_g = \frac{2\overline{\mathbf{e}}_z}{\kappa} - \overline{\mathbf{c}} \tag{13.37}$$

A wave packet will thus propagate perpendicular to the motion of the wave crests.

Inertial waves have several properties in common with internal waves in a stratified fluid (see Sec. 6.3). For example, the wave pattern radiated from a body that is oscillating with a constant frequency (<2) is, at some distance from the body, confined to the neighborhood of a conical surface.

In vibration problems, oscillations interacting with the walls of the container are of interest. In such cases, the representation by Eq. (13.32) is replaced by

$$\overline{\mathbf{u}} = \overline{\mathbf{Q}}(\overline{\mathbf{x}}) \exp\left[i\omega t\right] \tag{13.38}$$

and the linear, inviscid version of Eqs. (13.20) and (13.21) leads to an eigenvalue problem for ω (the *Poincaré problem*). The result, Eq. (13.35), with $\leq$ replaced by $<$ holds also for oscillations in containers of finite size. There is an infinite number of eigenvalues in this interval, Eq. (13.39) below. This means that if the *eigenfrequencies* of the rotating container are >2, no resonance with the fluid can take place. On the other hand, the fluid can resonate with any oscillation of the container having a frequency <2.

In contrast to the case of geostrophic flow, the effect of Ekman layers on oscillating fluid motions is weak and manifests itself as exponential decay on the (usually long) time scale $E^{-1/2}\Omega^{-1}$.

The nature of the motion of a rotating oscillating fluid depends crucially on whether the geometry of the container is such that geostrophic flow is possible in the container. If this is the case, the Taylor–Proudman theorem, Eq. (13.30), does not hold, i.e., the motion has a vertical structure. However, if there are no geostrophic curves, two classes of waves occur. For the first class, Eq. (13.30) does not hold, whereas for the second class it does. The second class of waves are the so called *Rossby waves* occurring in geophysical fluid dynamics [Pedlosky (1979)].

Eigenvalues are known for a few geometrically simple containers [Greenspan (1990)]. For a cylindrical container, whose boundaries are defined by $z = 0, 1$, and $r = a$, one has

$$\omega_{mnk} = 2(1 + \xi_{mnk}/n^2\pi^2a^2)^{-1/2} \tag{13.39}$$

where k and n are the azimuthal and axial wave numbers, respectively ($k = 0, \pm1$, $\dots, n = \pm1, \pm2, \dots$). ξ_{mnk} is the mth positive root of

$$\xi \frac{d}{d\xi} J_{|k|}(\xi) + k(1 + \xi^2/n^2\pi^2a^2)^{1/2}J_{|k|}(\xi) = 0 \tag{13.40}$$

Viscous Boundary and Free Shear Layers. In the vicinity of solid boundaries, viscous effects are of importance, and the inviscid solutions described in the previous sections have to be replaced by boundary layer solutions in order to fulfil the no slip condition. If the boundary is not vertical, an Ekman layer of thickness $E^{1/2}$ will appear. Simple examples of Ekman layers have already been given, Eqs. (13.26) and (13.27). In a slightly more general case with the boundary at $z = 0$ but with a geostrophic flow whose horizontal velocity components u_g and v_g depend on x and y, one finds the following velocity field in the Ekman layer

$$u - iv = (u_g - iv_g)\{1 - \exp[-(1 - i)z/E^{1/2}]\}$$

$$w = \frac{E^{1/2}}{2}\left(\frac{\partial u_g}{\partial y} - \frac{\partial v_g}{\partial x}\right) \text{Im}[(1 + i)\{1 - \exp[-(1 - i)z/E^{1/2}]\}] \quad (13.41)$$

Unless the vorticity of the geostrophic flow is zero, the small axial velocity component will be nonzero at the edge of the boundary layer. This means that the Ekman layer can induce a weak axial velocity outside the layer. This effect is often referred to as the *Ekman suction mechanism* or *centrifugal pumping*. In cases where Ekman suction prevails, the geostrophic velocity field will thus be of the form

$$\bar{\mathbf{u}}_g = \left(u_g,\, v_g,\, \frac{E^{1/2}}{2}\left[\frac{\partial u_g}{\partial y} - \frac{\partial v_g}{\partial x}\right]\right) \quad (13.42)$$

The Ekman layer is then said to be *divergent*. An example of a divergent Ekman layer is shown in Fig. 13.5. The pressure field in the Ekman layer is, within an error $0(E^{1/2})$, the same as that outside the layer.

Linear boundary layers at vertical walls are called *Stewartson layers* [Stewartson (1957)]. There are two types of Stewartson layers, one of thickness $E^{1/3}$ and the other of thickness $E^{1/4}$. The structure of Stewartson layers is complicated. Some overall properties are illustrated by the following example. The flow is bounded by solid walls at $z = \pm1$ and $x = 0$. The geostrophic flow in the vicinity of the vertical wall is assumed to be given: $\bar{\mathbf{u}}_g(0, y) = f(y)\bar{\mathbf{e}}_y$. In this case, both types of Stewartson layers are present, the $E^{1/3}$ layer being nested within the $E^{1/4}$ layer. The *sandwich structure* of the flow is shown schematically in Fig. 13.6. In the $E^{1/4}$ layer, the velocity and pressure fields are of the following orders of magnitude

$$u = 0(E^{1/2}), \quad v = 0(1), \quad w = 0(E^{1/4}), \quad p = 0(1) \quad (13.43)$$

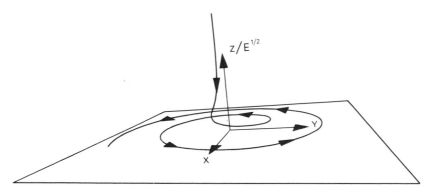

FIGURE 13.5 A divergent Ekman layer. The fluid far above the plate and the plate itself are rotating around the z-axis. The plate is rotating slightly faster than the fluid. The streakline is computed in a coordinate system that rotates with the fluid far above the plate.

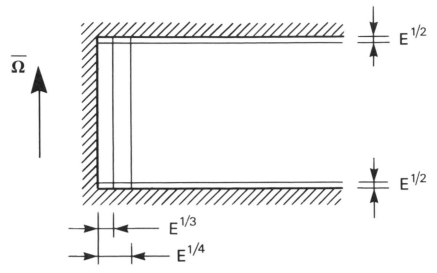

FIGURE 13.6 Sandwich structure of Stewartson layers at a vertical wall. Note that each of the Stewartson layers has its "own" Ekman layers at the horizontal walls.

The v component corrects $\bar{u}_g$ to zero at the wall, but u and w are nonzero at the wall. These velocity components are corrected by the inner $E^{1/3}$ layer. The orders of magnitude of the velocity and pressure fields in this layer are

$$u = 0(E^{1/2}), \quad v = 0(E^{1/6}), \quad w = 0(E^{1/6}), \quad p = 0(1) \tag{13.44}$$

At nonvertical walls, the velocity fields in both types of Stewartson layers are corrected by local Ekman layers. In cases where a boundary is almost vertical, e.g., near the equator of a sphere, other types of boundary layer flows appear [see, e.g., Stewartson (1966)].

Stewartson layers may also appear as vertical free shear layers. One example is the Taylor column shown in Fig. 13.4. At the vertical interface between the stagnant fluid in the column above the obstacle and the inviscid potential flow outside, the discontinuity in the velocity field is removed by an $E^{1/4}$ layer. There will also be an $E^{1/3}$ layer to correct shear stresses and the remaining velocity components in the $E^{1/4}$ layer. Free shear layers also appear if there is a discontinuity in the motion of a nonvertical boundary. Two examples are shown in Figs. 13.7 and 13.8. These graphs show the motion between two infinitely large and parallel split plates. The plates are at* $z = \pm 1$ and are both split along the line $x = 0$. In $x < 0$, both plates are stationary in the rotating coordinate system, whereas in $x > 0$, both plates are moving with constant velocities in the y-direction. These velocities are the same in Fig. 13.7. There is no geostrophic flow in $x < 0$. In $x > 0$, the geostrophic velocity field is uniform and the fluid moves with the plates. Between these regions, there is an $E^{1/4}$ layer that smoothly connects the geostrophic flow with the stagnant region. If Fig. 13.8, the plates are in $x > 0$ moving with the same velocity but in opposite

*The unit length is $(81\nu/\Omega)^{1/2}$.

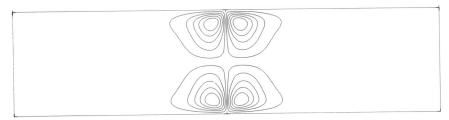

FIGURE 13.7 A Stewartson $E^{1/4}$ layer. The graph shows stream-lines in the y-z–plane. The notation and the boundary conditions are explained in the text.

directions. In this case, there is no geostrophic flow anywhere. The flow consists of an $E^{1/3}$ layer and, in $x > 0$, a rapidly rotating Couette flow that is distorted into two nondivergent Ekman layers. The net volume fluxes in the Ekman layers are in opposite directions and are connected by the axial flow in the $E^{1/3}$ layer. At the edges of this layer there are regions with closed streamlines.

13.2.5 Flows at Finite Rossby Numbers

For small but finite values of the Rossby number, the linear flows described in the previous sections become distorted by the inertia of the moving fluid. The Taylor column in Fig. 13.4 will still appear as an essentially straight cylinder but will be tilted in the downstream direction [Hide and Ibbetson (1968)]. The angle of tilt is approximately proportional to the Rossby number. For large Rossby numbers, only weak effects of rotation are felt. Instead of a column, a wake will appear downstream of the obstacle.

Rotating inviscid sink flows have unusual properties for finite Rossby numbers. A point sink at the center of the flat bottom cap of a slender vertical cylinder provides an example. The container is rotating around its axis of symmetry. If the motion far from the sink is taken as a uniform axial flow (in the rotating coordinate system), an exact solution of the nonlinear equations of motion was found by Long (1956). This solution, which agrees qualitatively with observations [Long (1956)], has essentially the same character as in the nonrotating case. However, the solution ceases to exist for $\epsilon < 0.267$. In this parameter range, theory [Pao and Shih (1973)] and experiments [Shih and Pao (1971)] show that the flow has a completely different

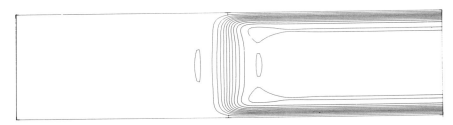

FIGURE 13.8 A Stewartson $E^{1/3}$ layer connecting the flow in two Ekman layers. The graph shows streamlines in the y-z–plane. The notation and the boundary conditions are explained in the text.

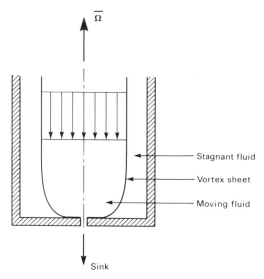

FIGURE 13.9   Blocked sink flow in a rotating cylinder.

character. The flow is partially blocked, and a region of stagnant fluid appears; see Fig. 13.9. The moving fluid is separated from the stagnant fluid by a vortex sheet. Due to the stiffness of vortex lines, a reasonably strong rotation will consequently cause a selective withdrawal of fluid by the sink. Other examples of similar phenomena are discussed by Yih (1965).

There are a few exact similarity solutions of the Navier–Stokes equations that describe flows with strong effects of rotation. Some of these solutions are of practical interest. von Karman (1921) considered an infinite flat plate in an unbounded incompressible viscous fluid. The plate is rotating with the constant angular velocity Ω. At large distances from the plate, the fluid is assumed to be in uniform axial motion toward the plate (in a coordinate system at rest). This part of the velocity field cannot be prescribed but will emerge as a result of the centrifugal pumping by the flow near the plate. The similarity solution is of the form

$$u_* = r_*\Omega F(\zeta), \quad v_* = r_*\Omega G(\zeta), \quad w_* = (\nu\Omega)^{1/2}H(\zeta)$$

$$p_* = \rho\nu\Omega \left[I - \tfrac{1}{2} H^2 - 2F \right] \tag{13.45}$$

where

$$\zeta = \left(\frac{\Omega}{\nu}\right)^{1/2} z_*, \quad F = -\frac{H'}{2}, \quad I = \text{const.} \tag{13.46}$$

and star subscripts denote dimensional quantities. The functions F, G, and H are shown in Fig. 13.10. In this flow, the diffusion of vorticity away from the plate is balanced by advection of vorticity toward the plate. It should be noted that the boundary layer thickness is independent of r_* and of the same order of magnitude as that of the linear Ekman layer.

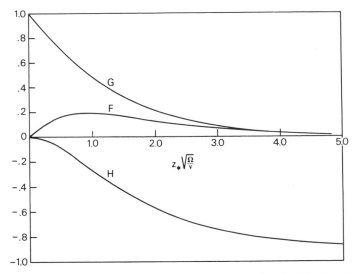

FIGURE 13.10   von Karman's solution for a rotating infinite flat plate.

For a circular disc of finite radius, R, von Karman's solution describes the flow very well except for the immediate neighborhood of the edge, provided that the boundary layer thickness is small compared to R and the flow is laminar. (The condition for stability is given in the next section.) In terms of the rotational Reynolds number Re_r, which is based on R and the tip speed of the plate, the *torque coefficient* C_M for a disc that is wetted on both sides is

$$C_M = 3.87 \, \mathrm{Re}_r^{1/2} \tag{13.47}$$

which agrees well with experiments. The radial volume flux Q at the periphery on each side of the disc is

$$Q = 2.78 R^3 \Omega \, \mathrm{Re}_r^{1/2} \tag{13.48}$$

The reversed problem with a stationary plate and a fluid velocity field far from the plate that is a rigid rotation with a uniform axial velocity away from the plate was considered by Bödewadt (1940). Apart from a slightly modified expression for the pressure, the form of the similarity solution is again Eq. (13.45). The solution for the velocity field is shown in Fig. 13.11. In this case, the diffusion of vorticity is balanced by the tendency of rotating vertical fluid elements to resist tilting. Due to nonlinear effects, the *Bödewadt velocity field* is *not* equal to a rigid rotation minus the von Karman velocity field.

For the finite disc, the applicability of Bödewadt's solution is restricted to $r_* \lesssim R/2$. Because $u_* < 0$ near the disc, effects of the edge are felt for larger radii. This part of the flow can be computed by using an expansion scheme given by Stewartson (1958) [see also Rogers and Lance (1964)]. Alternatively, the flow over the whole disc can be computed quite accurately by approximative momentum integral methods [Rott and Lewellen (1966) and Lewellen (1970)].

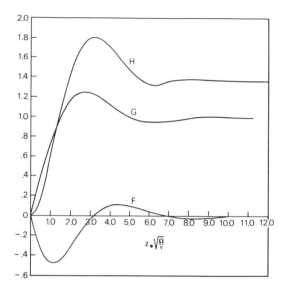

FIGURE 13.11 Bödewadt's solution for an infinite flat plate at rest below a fluid that is rotating rigidly at large distance from the plate.

Similarity solutions* of the type of Eq. (13.45) can be found for any angular velocity, Ω_p, of the plate and, Ω_f, of the fluid at large distances from the plate [Lentini and Keller (1980)]. Solutions can also be found for the flow between infinite parallel plates that are rotating around the same axis but with different angular velocities [Holodnick *et al.* (1977)]. Rotating flows in cavities can partly be described by the latter type of solutions. Consider a closed cavity made up by two parallel discs of radius, R, and a cylindrical wall at the periphery. The distance between the discs is H. One of the discs is rotating, and the other is fixed. If $E \ll 1$, there will be a boundary layer of thickness $E^{1/2}$ on each disc. Some distance from the boundary layer on the cylindrical wall, the inviscid flow, which is mainly a swirling motion, is controlled by the boundary layers on the discs. For $r_* \lesssim 0.5\,R$ the flow is approximately of the similarity form discussed above, and the inviscid part is essentially a rigid rotation with $v_* \approx 0.3 r_* \Omega$. Figure 13.12 shows a comparison between experimental data by Bien and Penner (1970) and numerical results by Pearson (1965). An approximate solution for the whole region (apart from the vertical boundary layer), which compares favorably with observations, can be computed with momentum integral methods [Lewellen (1970)]. If there is a strong net throughflow between the discs, the motion attains a complex structure with recirculating eddies. Such flows have been computed with Galerkin methods by Adams and Szeri (1982). Finite-difference methods have been successfully applied to complicated cavity flows by Chew (1984).

13.2.6 Stability

A thorough review of the general theory of hydrodynamic stability is given in the monograph by Drazin and Reid (1981) (see also Sec. 4.3). Instability criteria for

---

*Unfortunately, in some cases there is more than one similarity solution of the form of Eq. (13.45). The practical use of such solutions thus requires some care.

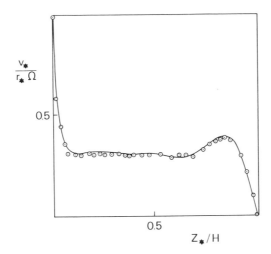

FIGURE 13.12 Swirl velocity distribution between a rotating and stationary disc in a cavity. ⊙: Experimental data from Bien and Penner (1970). −: Theoretical prediction from Pearson. $r_z/R = 0, 52, E = 10^{-3}$.

different types of flows are given in Sec. 13.7 of this handbook. The subtle relation between stability and transition are discussed in Secs. 1.4 and 4.3.

For inviscid rotating flows, e.g., geostrophic flows, the following instability criterion due to Rayleigh (1917) is useful: If the velocity field in a cylindrical coordiante system is of the form $\bar{\mathbf{u}} = (0, v(r), 0)$, the flow is unstable if

$$\frac{d}{dr}(r^2 v^2) < 0 \tag{13.49}$$

somewhere in the flow. There are several observations supporting this criterion [see, e.g., Drazin and Reid (1981) and Fjortoft (1950)].

Fjortoft (1950) has shown that an inviscid rectilinear shear flow may be unstable if the velocity profile has an inflection point that corresponds to a maximum in the vorticity distribution. At moderately large Reynolds numbers (typical values are given below), viscous forces are unable to stabilize such flows. Instabilities of this type can also appear in rotating flows with more complicated velocity fields if one (or several) of the velocity components has such an inflection point.

The stability properties of the Ekman layer are well known, and there is good agreement between theory and experiment [see Lilly (1966), Tatro and Mollo–Christensen (1967), and Caldwell and van Atta (1970)]. The onset of instability is determined by the local Rossby number $\epsilon = |\bar{\mathbf{u}}_g|/H\Omega$, where $\bar{\mathbf{u}}_g$ is the geostrophic flow outside the layer at the position under consideration, and by the local Reynolds number for the flow in the boundary layer, $\mathrm{Re} = \epsilon E^{-1/2}$. There are two *modes of instability*, which are called class A and class B waves respectively. These waves appear in the following parameter ranges

$$\mathrm{Re} > 56.3 + 58.4\,\epsilon \text{ class A waves}$$
$$\tag{13.50}$$
$$\mathrm{Re} > 124.5 + 3.66\,\epsilon \text{ class B waves}$$

Even if class A waves are linearly unstable for Re > 56.3, a neutrally stable weakly nonlinear form of these waves is observed for Re < 124.5. Fully developed turbulence has been observed for Re > 150 [Caldwell and van Atta (1970)].

Stability of the flow between narrowly-spaced, corotating parallel discs with a net radial volume flux has been investigated both experimentally [Pater *et al.* (1974)] and theoretically [Gusev and Bark (1980)]. For small Rossby numbers and moderately small Ekman numbers (based on the distance between the discs), this flow can approximately be described as two interacting nondivergent Ekman layers. Such flows are less stable than the single Ekman layer. The critical Reynolds number* has a minimum value Re $= 31$ for $E = 0.025$. For larger values of E, the critical Reynolds number increases rapidly with E.

The Stewartson $E^{1/4}$ layer in the form of an axisymmetric free shear layer has been subject to both experimental [Hide and Titman (1967)] and theoretical [Hashimoto (1979)] studies. In this flow, the instability appears as a periodic pattern of axial vortices in the vertical boundary layer. In terms of the azimuthal wave number, k, for the instability wave and the jump in angular velocity, $\Delta\Omega$, of the geostrophic flow across the layer, the theoretical criterion for instability is

$$k \frac{\Delta\Omega}{\Omega} > 38E^{1/2} \tag{13.51}$$

which agrees reasonably well with experiments.

For the Stewartson $E^{1/3}$ layer, the criterion for instability has been suggested by Gans (1975) as $\epsilon > 82.9E^{2/3}$, where the Rossby number is defined as the local magnitude of the axial velocity in the layer divided by the local azimuthal velocity of the unperturbed rigid rotation at the layer. There is experimental evidence in favor of this criterion [Danielson and Lundgren (1977)].

To this author's knowledge, no investigations of transition to turbulence in Stewartson layers have been carried out.

The stability of the von Karman boundary layer described in the previous section depends on the local rotational Reynolds number Re_r^l. This flow becomes unstable [Chan and Head (1967)] for $Re_r^l \approx 1.8 \cdot 10^5$ and attains fully developed turbulence for $Re_r^l > 2.9 \cdot 10^5$.

13.2.7 Rotating Turbulent Flows

In a rotating fluid in a container, approximately two-dimensional (*quasi-geostrophic*) turbulence can occur outside the boundary layers at the walls. Necessary conditions are that the Ekman and Rossby numbers, the latter being based on the r.m.s. value, are small. In such turbulence, which decays in time, the fluctuating fields vary much more weakly in the vertical direction than in the horizontal directions. A characteristic property of nearly two-dimensional turbulence is that, in wave number space, there is significant *anti-cascading* of energy toward small wave numbers, i.e., eddies are growing in size while their intensity decreases. This unusual kind of turbulence has been reviewed by Lesieur (1983).

In the turbulent counterpart of the *von Karman layer*, Coriolis effects are rela-

---

*The Reynolds number is here defined as Re $= (1/2\rho)|(\partial p_*/\partial r_*)|\Omega^{-3/2}\nu^{-1/2}$.

tively weak and the swirl velocity distribution is quite similar to the streamwise mean velocity distribution in a turbulent boundary layer on a flat plate with no pressure gradient in the free stream. Although very little experimental data are available [Chan and Head (1967) and Erian and Tong (1971)], there is some evidence that there are logarithmic and wake regions of essentially the same kind as in the nonrotating case. Results from velocity measurements are shown in Fig. 13.13. Although the maximum of the nondimensional radial velocity is larger in the laminar case (see Figs. 13.10 and 13.13), the radial volume flux, when scaled on $R^3\Omega$, expelled from the edge of the disc is larger in the turbulent case due to the increased boundary layer thickness. A semi-empirical formula for this flux is [von Karman (1921)]

$$Q = 0.22R^3\Omega \, \text{Re}_r^{-1/5} \qquad (13.52)$$

A simple and accurate estimate of the torque coefficient is

$$C_M = 0.15 \, \text{Re}_r^{-1/5} \qquad (13.53)$$

In contrast to the laminar case, the boundary layer thickness varies with r_*. This variation can be estimated as [Schlichting (1968)]

$$\delta = 0.53r_* \, \text{Re}^{-1/5} \qquad (13.54)$$

This type of boundary layer can be successfully computed by using simple eddy viscosity models designed for nonrotating turbulent flows [Cebeci and Abbot (1975)] (see Fig. 13.13).

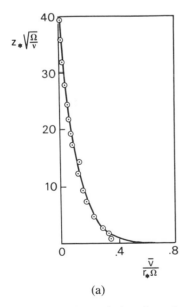

(a)

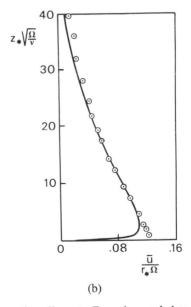

(b)

FIGURE 13.13 Turbulent boundary layer on a rotating disc. ⊙: Experimental data from Erian and Tong (1971). —: Theoretical prediction from Cebeci and Abbot (1975). $\text{Re}_r^l = 0.99 \cdot 10^6$. (a) v—velocity profiles and (b) u—velocity profiles.

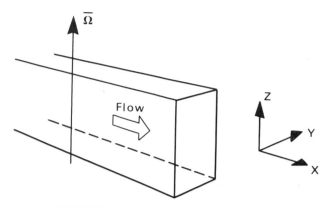

FIGURE 13.14   Sketch of rotating duct.

In flows between shrouded parallel discs, one disc being fixed and the other rotating, the turbulent boundary layers on the discs are separated by an almost rigidly rotating core. The angular velocity of this core is approximately 0.5Ω, which may be compared with the value 0.3Ω for $r_* \lesssim 0.5R$ in the laminar case. The physical reason for this difference is that the two boundary layers have essentially the same properties in the turbulent case, whereas this is not so if the flow is laminar. Eddy viscosity models can predict flows in closed cavities quite accurately [Cooper and Reshotko (1975)]. Flows with a net radial volume flux, which are complicated, can be predicted reasonably well by using momentum integral methods [Senodo and Hayami (1976)].

Turbulent duct flows are strongly affected by the Coriolis force. Figure 13.14 shows a configuration investigated by, among others, Johnston *et al.* (1972). In the major part of the cross section of the duct, the mean flow is a parallel shear flow in the x-direction and depends on y only. The mean velocity field is asymmetric with respect to the centerline as shown in Fig. 13.15. The magnitude of the mean velocity gradient is significantly different at the two walls. It can be shown [Johnston *et al.* (1972)] that the Coriolis force has a destabilizing effect on a shear flow of this type if $(\partial\bar{u}/\partial y_*) > 2\Omega$. For $(\partial\bar{u}/\partial y_*) < 2\Omega$, the flow is stabilized. Thus the flow is *unstably stratified* in some region adjacent to the wall at $y = -1$, which leads to enhanced mixing and a steep velocity profile. Near the wall at $y = 1$, the situation is reversed. Due to this effect, channel flows with both laminar and turbulent motion can be realized. Attempts to compute turbulent duct flows with strong Coriolis effects by using empirical turbulence models have so far been unsuccessful [Howard *et al.* (1980)].

13.3 FLOWS WITH BODY FORCES
Thomas B. Morrow

13.3.1 Buoyancy Driven Fluid Flows

Density differences between neighboring regions of a fluid can interact with body force fields to produce a wide variety of buoyant flow behavior in atmospheric,

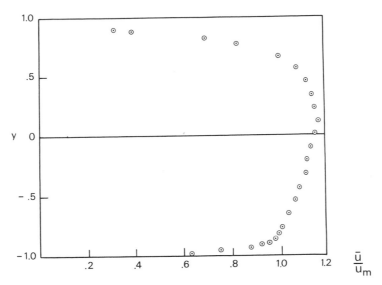

FIGURE 13.15 Asymmetric mean velocity profile in a rotating turbulent channel flow. Turbulence is suppressed in the upper parts of the graph but enhanced in the lower. (Data from Johnston *et al.*, 1972.)

industrial, and domestic environments. There are two basic geometrical configurations that give rise to buoyant flow behavior. The first, stable, configuration occurs when the density gradient is normal to the body force (usually the gravity force) vector. Buoyancy driven motion results immediately in this case. The second, unstable, configuration occurs when the density gradient is parallel to, but in the opposite direction from, the body force vector. The denser fluid resides above the lighter fluid in an unstable equilibrium until a critical value of density gradient is reached when a buoyancy driven flow begins spontaneously.

The buoyancy driven flow that results in either configuration can be laminar or turbulent, depending upon the physical scale of the flow geometry and the strength of the buoyant driving force. Atmospheric flows are of such large size that the dispersion of heated or moist effluent is usually accomplished by turbulent diffusion. Plumes from smokestacks and cooling towers are examples of turbulent, buoyancy driven flows. However, laminar and transitional buoyant flows are commonly encountered in smaller scale industrial and domestic geometries. A good example is the buoyancy driven flow in an insulated cavity filled with gas or liquid that can occur in flat plate solar collectors and double glazed windows.

Although buoyancy driven flow and heat transfer is a diverse and complex topic, a series of excellent articles have been written that provided survey coverage of this field [Ostrach (1972), (1982a), (1982b), Gebhart (1973), (1979), Chen and Rodi (1980), List (1982), Ede (1967), Cess (1973), and Catton (1978)]. The purpose of this section is to highlight the principal features of buoyancy driven flows, and to guide the reader to other sources where additional information may be found. Stratified flow is discussed in Sec. 13.7.

Free Convection. Free convection flows are distinguished from forced convection flows by the following characteristics:

1. The momentum and temperature fields are invariably coupled and must be considered together.
2. The flow field arises in response to a set of thermal (or concentration) conditions imposed on the flow boundaries.
3. The flow is relatively weak. Velocities are small with momentum and viscous effects of the same order.
4. Little is known *a priori* about the resulting flow field. This is especially true of internal free convection flows.

The first characteristic is the most significant. It requires that the equations of energy and momentum be solved simultaneously, rather than independently as for forced convection flows.

Governing Equations: A simplified set of the equations of motion for free convection flow are presented by Ostrach (1982a). These dimensional equations express the conservation of mass, momentum, energy, and chemical species.

$$\partial U_j / \partial X_j = 0 \tag{13.55}$$

$$\rho U_j \frac{\partial U_i}{\partial X_j} = \mu \frac{\partial^2 U_i}{\partial X_j \partial X_j} + (\rho_\infty - \rho)g_i - \frac{\partial(p - p_\infty)}{\partial X_i} \tag{13.56}$$

$$\rho C_p U_j \frac{\partial T}{\partial X_j} = k \frac{\partial^2 T}{\partial X_j \partial X_j} \tag{13.57}$$

$$U_j \frac{\partial C}{\partial X_j} = D \frac{\partial^2 C}{\partial X_j \partial X_j} \tag{13.58}$$

In deriving these equations, density variations were considered only in the buoyancy term in Eq. (13.56), and the temperature and concentration dependence of other fluid properties were neglected. Also, compressive work, viscous dissipation, *thermal diffusion (Soret)*, and *diffusion thermal (DuFour)* effects were neglected. The equations are coupled through the dependence of fluid density on concentration and temperature. The $(\rho_\infty - \rho)$ term is usually expressed as a sum of linear variations in C and T, and Eq. (13.56) as

$$U_j \frac{\partial U_i}{\partial X_j} = \nu \frac{\partial^2 U_i}{\partial X_j \partial X_j} + \beta g_i(T - T_\infty) + \bar{\beta} g_i(C - C_\infty) - \frac{1}{\rho} \frac{\partial(p - p_\infty)}{\partial X_i} \tag{13.59}$$

where β and $\bar{\beta}$ are the volumetric expansion coefficient and the concentration densification coefficient, respectively.

Dimensionless Parameters: It is instructive to nondimensionalize the free convection equations of motion and to inspect the dimensionless parameters that appear as coefficients. Following Ostrach (1982a) the variables are normalized by

$$u_i = U_i/U_R, \quad x_i = X_i/L, \quad \theta = (T - T_\infty)/(T_\omega - T_\infty),$$

$$\phi = (C - C_\infty)/(C_\omega - C_\infty), \quad p = p/\rho U_R^2. \tag{13.60}$$

Here U_R denotes a reference velocity, L is a characteristic length, and subscripts ω and ∞ represent two different reference values. The nondimensional equations of momentum, energy, and chemical species become

$$u_i \frac{\partial u_i}{\partial x_j} = \frac{\nu}{U_R L} \frac{\partial^2 u_i}{\partial x_j \partial x_j} + g_i \frac{L\beta \Delta T}{U_R^2} (\theta + N\phi) - \frac{\partial (p - p_\infty)}{\partial x_i} \quad (13.61)$$

$$u_j \frac{\partial \theta}{\partial x_j} = \frac{\alpha}{U_R L} \frac{\partial^2 \theta}{\partial x_j \partial x_j} \quad (13.62)$$

$$u_j \frac{\partial \phi}{\partial x_j} = \frac{D}{U_R L} \frac{\partial^2 \phi}{\partial x_j \partial x_j} \quad (13.63)$$

The dimensionless parameters in these equations are

$$\nu / U_R L = \text{Re}^{-1}$$

$$\alpha / U_R L = (\text{Pr} \cdot \text{Re})^{-1}$$

$$D / U_R L = (\text{Sc} \cdot \text{Re})^{-1}$$

$$g_i L\beta \Delta T / U_R^2 = \text{Gr}/\text{Re}^2 \quad (13.64)$$

The *Grashof number*, Gr, can be interpreted as the ratio of buoyant forces to viscous forces. It has the same importance in free convection heat transfer correlations as the *Reynolds number*, Re, has in forced convection. The *Rayleigh number*

$$\text{Ra} = \text{Gr} \cdot \text{Pr} \quad (13.65)$$

is also used to correlate heat transfer for fluids over a wide range of Prandtl number, Pr. Ostrach (1982a) shows that a consideration of the magnitude of the dimensionless parameters in Eqs. (13.61) to (13.63) can be used to estimate values of the fluid velocity in free convection flow.

External Convection Flows. Buoyancy driven flows are particularly important in environmental studies involving air and water pollution. In order to assess the impact of an emission source, it is necessary to predict both the trajectory and the diffusion of the plume or jet at several distances away from the source. In practice, most emissions of a buoyant effluent are accompanied by a discharge momentum flux. Therefore, the behavior of buoyant jets is discussed in the next three sections, which includes both pure (nonbuoyant) jets and plumes (without momentum).

Vertical Buoyant Jets: It is possible to assume that both the velocity and the buoyancy (temperature and/or concentration) profiles of a buoyant jet issued vertically upwards into quiescent surroundings become self-similar at a distance of from 10 to 20 diameters above the discharge point. The profile shapes are often assumed to be Gaussian for *turbulent buoyant jets* as,

$$U/U_{C_L} = \exp(-k_u \eta^2) \quad (13.66)$$

$$C^* = \frac{T - T_a}{T_{C_L} - T_a} \left(\text{or} \frac{C - C_a}{C_{C_L} - C_a} \right) = \exp(-k_c \eta^2) \quad (13.67)$$

Here, the subscripts C_L and a denote the value on the jet centerline and the ambient value away from the jet. k_u and k_c are profile constants. η is a dimensionless distance perpendicular to the jet axis defined as,

$$\eta = \frac{y}{x + x_0} \tag{13.68}$$

where x is the distance measured back to the source along the jet axis, and x_0 is the distance from the source to the virtual origin where the jet would emanate as a point or line source. Jets are discussed in Sec. 4.11.

The values of U_{C_L} and C^* in Eqs. (13.66) and (13.67) vary along the jet axis. Chen and Rodi (1980) describe three separate regions of development for turbulent buoyant jets. The first is a nonbuoyant region close to the emission source, where the jet momentum affects the diffusion process. The second is an intermediate region, where both momentum and buoyancy effects are important. The third is a plume region far from the source, where only the buoyancy flux is important. The locations of the boundaries of these three regions for a particular buoyant jet depend upon the value of the *densimetric Froude number*,

$$\mathrm{Fr} = \frac{\rho_o U_o^2}{gD(\rho_a - \rho_o)} \tag{13.69}$$

Here, D is the jet diameter at discharge and the subscript o denotes values at the jet discharge location. This Froude number could also be written as $\mathrm{Re}^2/\mathrm{Gr}$. It represents the ratio of inertia force to buoyancy force. Low values of Fr (high Gr and low Re) correspond to plumes that are buoyancy dominated. High values of Fr (low Gr and high Re) correspond to jets that are momentum dominated.

For plane vertical, turbulent buoyant jets, Chen and Rodi (1980) recommended the following equations for U_{C_L} and C^*

$$\underline{\text{Nonbuoyant Region, }} \mathrm{Fr}^{-2/3} \left(\frac{\rho_o}{\rho_a}\right)^{-1/3} \left(\frac{x}{D}\right) < 0.5$$

$$\frac{U_{C_L}}{U_o} = 2.4 \left(\frac{\rho_o}{\rho_a}\right)^{-1/2} \left(\frac{x}{D}\right)^{-1/2} \tag{13.70}$$

$$C^* = 2 \left(\frac{\rho_o}{\rho_a}\right)^{-1/2} \left(\frac{x}{D}\right)^{-1/2} \tag{13.71}$$

$$\underline{\text{Intermediate Region, }} 0.5 \leq \mathrm{Fr}^{-2/3} \left(\frac{\rho_o}{\rho_a}\right)^{-1/3} \left(\frac{x}{D}\right) \leq 5$$

$$\frac{U_{C_L}}{U_o} = 2.85 \, \mathrm{Fr}^{-1/3} \left(\frac{\rho_o}{\rho_a}\right)^{5/12} \left(\frac{x}{D}\right)^{-1/4} \tag{13.72}$$

$$C^* = 1.65 \, \mathrm{Fr}^{1/2} \left(\frac{\rho_o}{\rho_a}\right)^{-1/4} \left(\frac{x}{D}\right)^{-3/4} \tag{13.73}$$

Plume Region, $5 < \mathrm{Fr}^{-2/3} \left(\dfrac{\rho_o}{\rho_a}\right)^{-1/3} \left(\dfrac{x}{D}\right)$

$$\frac{U_{CL}}{U_o} = 1.9 \, \mathrm{Fr}^{-1/3} \left(\frac{\rho_o}{\rho_a}\right)^{1/3} \tag{13.74}$$

$$C^* = 2.4 \, \mathrm{Fr}^{1/3} \left(\frac{\rho_o}{\rho_a}\right)^{-1/3} \left(\frac{x}{D}\right)^{-1} \tag{13.75}$$

For round, vertical turbulent buoyant jets, Chen and Rodi (1980) recommended the following equations for U_{CL} and C^*

Nonbuoyant Region, $\mathrm{Fr}^{-1/2} \left(\dfrac{\rho_o}{\rho_a}\right)^{-1/4} \left(\dfrac{x}{D}\right) < 0.5$

$$\frac{U_{CL}}{U_o} = 6.2 \left(\frac{\rho_o}{\rho_a}\right)^{1/2} \left(\frac{x}{D}\right)^{-1} \tag{13.76}$$

$$C^* = 5 \left(\frac{\rho_o}{\rho_a}\right)^{-1/2} \left(\frac{x}{D}\right)^{-1} \tag{13.77}$$

Intermediate Region, $0.5 \leqq \mathrm{Fr}^{-1/2} \left(\dfrac{\rho_o}{\rho_a}\right)^{-1/4} \left(\dfrac{x}{D}\right) \leqq 5$

$$\frac{U_{Cl}}{U_o} = 7.26 \, \mathrm{Fr}^{-1/10} \left(\frac{\rho_o}{\rho_a}\right)^{9/20} \left(\frac{x}{D}\right)^{-4/5} \tag{13.78}$$

$$C^* = 0.44 \, \mathrm{Fr}^{1/8} \left(\frac{\rho_o}{\rho_a}\right)^{-7/16} \left(\frac{x}{D}\right)^{-5/4} \tag{13.79}$$

Plume Region, $5 < \mathrm{Fr}^{-1/2} \left(\dfrac{\rho_o}{\rho_a}\right)^{-1/4} \left(\dfrac{x}{D}\right)$

$$\frac{U_{CL}}{U_o} = 3.5 \, \mathrm{Fr}^{-1/3} \left(\frac{\rho_o}{\rho_a}\right)^{1/3} \left(\frac{x}{D}\right)^{-1/3} \tag{13.80}$$

$$C^* = 9.35 \, \mathrm{Fr}^{1/3} \left(\frac{\rho_o}{\rho_a}\right)^{-1/3} \left(\frac{x}{D}\right)^{-5/3} \tag{13.81}$$

Recommended values for the Gaussian profile constants in Eq. (13.66) are

Plane Jet $k_u = 62$, $k_c = 35$ and

Round Jet $k_u = 94$, $K_c = 57$

The preceding equations can be used to estimate buoyant jet diffusion in quiescent surroundings when the jet is turbulent and the turbulent Prandtl or Schmidt number controls the relative spreading rates of velocity and temperature (or concentration) profiles.

Pure jets are normally turbulent for values of the jet Reynolds number above 30–50 for a plane jet and 300–2000 for a round jet, depending on the nozzle exit conditions [Chen and Rodi (1980)]. For laminar buoyant jets, the transition to turbulence also depends upon jet Reynolds number. However, pure buoyant plumes may remain laminar for some distance above the heat source. The transition to turbulence depends upon local Grashof number for plumes. Plane plumes will be fully turbulent at values of local Grashof number above 5–8×10^9, and round plumes above 0.9–1.8×10^{10} [Chen and Rodi (1980)].

Chen and Rodi (1980) describe similarity models for both plane and round *laminar thermal plumes*. For plane plumes,

$$U_{CL} = \frac{\nu}{D} (G_{D,Q})^{2/5} \left(\frac{x}{D}\right)^{1/5} B_u \,(\text{Pr}) \qquad (13.82)$$

$$T_{CL} - T_a = \frac{Q_o}{C_p \rho \nu} (G_{D,Q})^{-1/5} \left(\frac{x}{D}\right)^{3/5} B_c \,(\text{Pr}) \qquad (13.83)$$

Here, Q_o is the source heat flux and $G_{D,Q}$ is the plume Grashof number.

$$G_{D,Q} = \frac{g\beta D^3}{\nu^2} \left(\frac{Q}{\nu \, C_p \rho_o D^j}\right) \qquad (13.84)$$

where $j = 1$ for a round jet and $j = 0$ for a plane jet.

The values of the *plume parameters* B_u and B_c are functions of the Prandtl number. For the Prandtl number range of 1–10, Chen and Rodi (1980) recommend values of

$$\begin{aligned} B_u &= 0.85 \\ B_c &= 0.38 \, \text{Pr}^{1/2} \end{aligned} \qquad (13.85)$$

The width of a plane laminar plume can be characterized by the velocity half width as

$$y_{0.5u} = 2.5D \left(\frac{x}{D}\right)^{2/5} (G_{D,Q})^{-1/5} \qquad (13.86)$$

The similarity equations for round, laminar plumes are

$$U_{CL} = \frac{\nu}{D} (G_{D,Q})^{1/2} B_u \,(\text{Pr}) \qquad (13.87)$$

$$T_{CL} - T_a = \frac{Q_o}{C_p \rho \nu D} \left(\frac{x}{D}\right)^{-1} B_c \,(\text{Pr}) \qquad (13.88)$$

and

$$y_{0.5u} = \alpha D(G_{D,Q})^{-1/4} \left(\frac{x}{D}\right)^{+1/2}$$

(13.89)

Chen and Rodi state that insufficient data are available to determine the values of the plume parameters for round laminar plumes.

Vertical Buoyant Jets in Crossflow: The prediction of the *trajectory* and *dilution* of vertical buoyant jets emitted into a crossflow is of particular interest in air and water pollution studies. Briggs (1975) reviews several models that have been developed to estimate the rise of plumes and buoyant jets in the open air downwind of emission sources. List (1982) also reviews the behavior of turbulent jets and buoyant plumes. List (1979) provides example calculations to illustrate the use of buoyant jet rise and dilution equations in water pollution studies.

List (1979) shows that the equations that govern the trajectory and dilution for a round, turbulent buoyant jet in a crossflow can be deduced by applying similarity theory to the conservation equations. Both the trajectory and the dilution depend upon the relative magnitudes of two length scales that are functions of the jet momentum, M, the jet buoyancy, B, and the crossflow velocity, U. The length scales are

$$Z_M = M^{1/2}/U$$

(13.90)

$$Z_B = B/U^3$$

(13.91)

where

$$M = QU_o, \quad B = g\frac{(\rho_a - \rho_o)}{\rho_o}Q, \quad \text{and} \quad Q = \frac{\pi D^2}{4}U_o$$

Wright (1977) showed that four different regimes of behavior were possible for buoyant jets, and he developed equations to predict trajectory and dilution in each regime. As List (1982) describes, there are two asymptotic cases in which the buoyant jet passes sequentially through three regimes. In the first case, $Z_M > Z_B$ and jet momentum has the dominant effect on trajectory and dilution. As the jet rises from its source, it passes through three regimes described by List as *vertical jet*, *bent jet*, and *bent plume*. Wright called these same regimes *momentum dominated nearfield*, *momentum dominated far field*, and *buoyancy dominated far field*.

In the second asymptotic case, $Z_B > Z_M$ and buoyancy has the dominant effect on trajectory and dilution. In this case, the buoyant jet passes through the *vertical jet*, *vertical plume*, and *bent plume* regimes. Wright called the vertical plume regime by the name *buoyancy dominated near field*.

In each regime, the following equations may be used to estimate z, the vertical rise height of the plume centerline above the source, and $\theta = (\rho_a - \rho)/\rho_o$, the dimensionless density difference. The *plume dilution*, S, can be calculated as

$$S = 4B/g\theta\pi D^2 U_o$$

(13.92)

Vertical Jet Regime, $D < z < Z_M$

$$\frac{z}{Z_M} = 1.8\text{--}2.3 \left(\frac{x}{Z_M}\right)^{1/2} \tag{13.93}$$

$$\frac{Mg\theta}{BU} = 2.9 \left(\frac{z}{Z_M}\right)^{-1} \tag{13.94}$$

Bent Jet Regime, $Z_M < z < \left(\frac{Z_M}{Z_B}\right)^{1/3} Z_M$

$$\frac{z}{Z_M} = 1.6\text{--}2.1 \left(\frac{x}{Z_M}\right)^{1/3} \tag{13.95}$$

$$\frac{Mg\theta}{BU} = 7.1 \left(\frac{z}{Z_M}\right)^{-2} \tag{13.96}$$

Vertical Plume Regime, $D < z < Z_B$

$$\frac{z}{Z_B} = 1.35\text{--}1.8 \left(\frac{x}{Z_B}\right)^{3/4} \tag{13.97}$$

$$\left(\frac{Z_B}{Z_M}\right)^2 \cdot \frac{Mg\theta}{BU} = 4.6 \left(\frac{z}{Z_B}\right)^{-5/3} \tag{13.98}$$

Bent Plume Regime, $Z_B < z$

$$\frac{z}{Z_B} = (0.85\text{--}1.4) \left(\frac{Z_M}{Z_B}\right)^{1/6} \left(\frac{x}{Z_B}\right)^{2/3} \tag{13.99}$$

$$\left(\frac{Z_B}{Z_M}\right)^2 \cdot \frac{Mg\theta}{BU} = 3.3 \left(\frac{z}{Z_B}\right)^{-2} \tag{13.100}$$

The constants in Eqs. (13.90) to (13.100) have been taken from List (1979) and Wright (1977). Two values appear in the trajectory equations. The lower value was determined from photographs, while the upper value was determined from tracer concentration measurements [Wright (1977)]. List shows that the values of the coefficients in the trajectory equations agree with data from other experiments. In particular, the coefficient values in Eqs. (13.95) and (13.99) agree well with the values of 1.8–2.1 for a bent jet and 0.85–1.3 for a bent plume recommended by Briggs (1975).

Wright's values for the coefficients in the dilution equations have been used. It should be noted that List (1979) recommended a single value of 2.4 for the dilution equations in all four regimes.

Vertical Buoyant Jets in Stratified Ambients: Ambient density stratification can occur naturally both in the atmosphere and in the oceans. When the ambient density of the environment is stably stratified, with denser fluid lying beneath lighter fluid, buoyant jets may reach a terminal rise height. The reason for this behavior is that the dense fluid entrained initially in the lower region of the jet will result in a negative buoyancy flux when the jet rises into a region of lower density. If the ambient density is stably stratified, it is often necessary to predict the elevation and the dilution of a turbulent vertical buoyant jet at its terminal rise height. Additional discussion on stratified flows is given in Sec. 13.7.

For linear ambient density stratification and negligible crossflow velocity, buoyant jet behavior is governed by two dimensionless parameters. These are the jet *Richardson number*, $\mathrm{Ri} = QB^{1/2}/M^{5/4}$, and a *plume stratification parameter*, $S = (MN/B)^2$. Here, Q is the volumetric flowrate, and N is the *Brunt–Väisälä frequency* for linear density stratification

$$N^2 = -(g/\rho)\,\frac{\partial \rho}{\partial z} \tag{13.101}$$

List (1979) applied *similarity theory* to develop equations that predict terminal rise height and dilution for the two asymptotic cases of buoyant jet behavior, $S \ll 1$ (pure plume) and $S \gg 1$ (pure jet). These equations are given here in their dimensional form for *terminal rise height*, h_T, and *volume flux*, μ_T. The dimensionless dilution ratio may be calculated as μ_T/Q. The values for the constants are taken from List (1982).

For a round buoyant jet

$$
\begin{array}{ll}
S \ll 1 & S \gg 1 \\[4pt]
h_T = 3.8(B/N^3)^{1/4} & h_T = 3.8(M/N^2)^{1/4} \\[4pt]
\mu_T = 0.84(B^3/N^5)^{1/4} & \mu_T = 0.67(M^3/N^2)^{1/4}
\end{array}
\tag{13.102}
$$

and, for a plane buoyant jet

$$
\begin{array}{ll}
S \ll 1 & S \gg 1 \\[4pt]
h_T = 2.5(B/N^3)^{1/3} & h_T = 3.6(M/N^2)^{1/3} \\[4pt]
\mu_T = 0.86(B^2/N^3)^{1/3} & \mu_T = 0.70(M^2/N)^{1/3}
\end{array}
\tag{13.103}
$$

Vertical Heated Surfaces: Another important type of external convective flow is the flow along a vertical heated or cooled surface. The surface heat flux alters the temperature and density of the fluid near the wall and produces natural convection boundary layers of momentum and energy along the surface. A general review of work in this area has been made by Ede (1967).

In the laminar flow regime, similarity solutions to the boundary layer equations of mass, momentum and energy have been obtained for certain wall boundary conditions. Cess (1973) presents results for *Nusselt number* as a function of Prandtl number and Grashof number for boundary conditions of constant wall temperature,

TABLE 13.1   Nusselt Number Results for Isothermal Wall (Pr $\leq$ 1)

Pr	Nu/(Pr^{2}Gr)$^{1/4}$
0	0.800
0.01	0.760
0.03	0.740
0.09	0.688
0.5	0.590
0.72	0.561
1.0	0.535

constant heat flux, and a temperature difference that varies as a power of the downstream distance.

Tables 13.1 and 13.2 present the results of computed solutions for natural convection along an isothermal surface for low and high values of Prandtl number. Following Cess (1973), the results are presented for low and high Prandtl numbers separately in dimensionless forms consistent with the asymptotic solutions for Pr = 0 and Pr = ∞, respectively. The Nusselt number and Grashof number are defined as

$$\text{Nu} = \frac{q_w L}{k(T_w - T_\infty)} \tag{13.104}$$

$$\text{Gr} = \frac{g\beta(T_w - T_\infty)L^3}{\nu^2} \tag{13.105}$$

where L is the length of the vertical plate. Results for local Nusselt number as a function of the local Grashof number and Prandtl number may be obtained by multiplying the coefficients in Tables 13.1 and 13.2 by 3/4 and replacing the plate length L with the local dimension X in Eqs. (13.104) and (13.105).

These laminar flow solutions to the boundary layer equations are valid for a range of local Rayleigh numbers between about 10^4 and 10^9, when transition to turbulent flow is assumed to occur.

In the turbulent flow regime, results for the Nusselt number have been obtained from analytical computations and as correlations of experimental data. Analytical

TABLE 13.2   Nusselt Number Results for Isothermal Wall (Pr $\geq$ 1)

Pr	Nu/(Pr Gr)$^{1/4}$
1	0.535
2	0.568
5	0.601
10	0.619
100	0.653
1000	0.665
∞	0.670

solutions to the momentum integral equations for turbulent, natural convection have been reviewed by Ede (1967). The experimental data for natural convection of gases along an isothermal wall in the laminar, transitional, and turbulent flow regimes has been reviewed by Clausing (1982). Equations (13.106) through (13.108) are the correlations proposed by Clausing that agree well with previous data as well as his own data taken under carefully controlled conditions. The effect of variable properties is contained in the function $f(\mathrm{Ra}, T_w/T_\infty)$ when the properties are evaluated at the film temperature, $T_f = (T_w + T_\infty)/2$.

In the laminar flow regime,

$$\mathrm{Nu} = 0.52\ \mathrm{Ra}^{1/4} \quad \text{for} \quad \mathrm{Ra} < 3.8 \times 10^8 \tag{13.106}$$

In the turbulent flow regime,

$$\mathrm{Nu} = 0.082\ \mathrm{Ra}^{1/3} f \quad \text{for} \quad \mathrm{Ra} > 1.6 \times 10^9 \tag{13.107}$$

where $f = -0.9 + 2.4(T_w/T_\infty) - 0.5(T_w/T_\infty)^2$ for $1 \leq T_w/T_\infty < 2.6$.

In the transitional flow regime,

$$\mathrm{Nu} = 1.6\ \mathrm{Ra}^{0.193} f_{tr} \quad \text{for} \quad 3.8 \times 10^8 \leq \mathrm{Ra} \leq 1.6 \times 10^9 \tag{13.108}$$

where

$$f_{tr} = (f - 1)\left\{\frac{\mathrm{Ra}^{1/3} - \mathrm{Ra}_l^{1/3}}{\mathrm{Ra}_t^{1/3} - \mathrm{Ra}_l^{1/3}}\right\} + 1 \tag{13.109}$$

In the last equation, f is evaluated using the expression for f in Eq. (13.107), and Ra_l and Ra_t represent the Rayleigh number limits for laminar and turbulent flow specified in Eq. (13.108).

Internal Convection Flows. Numerical and experimental investigations of natural convection in enclosures have been reviewed by Ostrach (1972) and Catton (1978). Internal natural convection flow fields are generally more complex than external flows due to the coupling that occurs at large Rayleigh number between the boundary layers near the walls and the flow in the core region. Also, for large Rayleigh number, the character of the velocity and temperature distribution in the core flow depends strongly upon the thermal boundary conditions. As a result, geometrical parameters like aspect ratio, and the angle of inclination between the heated surfaces and the body force direction appear in the correlations of results.

Ostrach (1972) and Catton (1978) describe the types of laminar and turbulent flow regimes that arise in natural convection flow in enclosures. The following sections present natural convection heat transfer correlations for two-dimensional vertical and rectangular enclosures. Ostrach (1972) and Catton (1978) should be consulted for flow regime information and correlations for tilted rectangular enclosures and other enclosure geometries.

Vertical Rectangular Enclosures: For vertical rectangular enclosures, heat is transferred through the vertical walls of the enclosure which are parallel to the direction of the gravitational force. For this geometry, Catton (1978) recommends two correlations of heat transfer data for Nusselt number as a function of Rayleigh number, Prandtl number and aspect ratio. Here, the aspect ratio, A, is defined as the extent, d, of the heated or cooled surfaces divided by the extent, L, of the unheated surfaces. For a vertical rectangular enclosure, the heated and cooled walls are the vertical walls, and the unheated walls are the horizontal walls. Vertical enclosures with aspect ratios >1 are tall and slender. The correlations below cover a range of aspect ratios, A, from 1 to 10. Reliable correlations are unavailable for aspect ratios <1.

$$\text{Nu} = 0.22 A^{-1/4} \left(\frac{\text{Pr}}{0.2 + \text{Pr}} \text{Ra} \right)^{0.28} \tag{13.110}$$

for $2 < A < 10$, $\text{Pr} < 10^5$, and $\text{Ra} < 10^{10}$.

$$\text{Nu} = 0.18 \left(\frac{\text{Pr}}{0.2 + \text{Pr}} \text{Ra} \right)^{0.29} \tag{13.111}$$

for $1 < A < 2$, $10^{-3} < \text{Pr} < 10^5$, and $10^3 < \text{RaPr}/(0.2 + \text{Pr})$.

Horizontal Rectangular Enclosures: A horizontal rectangular enclosure is equivalent to a vertical rectangular enclosure that has been inclined by an angle of $90°$, so that the heated and cooled walls are horizontal, and perpendicular to the direction of the gravitational force. When the heated wall lies above the fluid and the cooled wall below, a stable density stratification is produced, and natural convection flow is suppressed. When the heated wall is beneath the fluid and the cooled wall above, a natural convection flow will be produced when the Rayleigh number exceeds a lower critical limit. For very large aspect ratios (enclosures that are much wider than high), the *critical Rayleigh number* has a value of 1705. However, the critical Rayleigh number can be increased by several orders of magnitude if the aspect ratio is reduced below 1 [see Catton (1978)].

Catton (1978) recommends the following correlation for heat transfer in horizontal rectangular enclosures. This correlation has a power-integral form.

$$\text{Nu} = 1 + 2 \sum_{j=1}^{M} \min \left[1, \left(\frac{j}{5} \right) \right]^{1/5} \left(1 - \frac{\text{Ra}_j}{\text{Ra}} \right) U(\text{Ra}_j - \text{Ra}) \tag{13.112}$$

where $U(\text{Ra}_j - \text{Ra})$ is a unit step function. Ra_j is defined as

$$\text{Ra}_j = \frac{(a_j^2 + b_j^2)^3}{a_j^2}. \tag{13.113}$$

Also, $b_j = j\pi + 0.85$, $a_j^2 = \max (a^2, b_j^2/2)$.

The parameter a in these equations is a horizontal wave number. Catton (1978) recommends the following equation for calculating values of a for horizontal rectangular enclosures.

$$a = \pi \left(\frac{4L^2}{d^2} + \frac{L^2}{w^2} \right)^{1/2} \tag{13.114}$$

for perfectly conducting side walls. For adiabatic side walls multiply this value of *a* by 0.75.

Catton (1978) shows that these equations give satisfactory estimates of the heat transfer for horizontal rectangular enclosures for Rayleigh numbers in the range of 10^3 to 10^7.

13.3.2 Near-Zero Gravity Flows

Problems such as those associated with liquid propellent management in space vehicles, for example, have made it necessary to improve the understanding of fluid flow behavior in near-zero gravity environments. On earth, liquid and gas phases in static equilibrium are separated into distinct regions of space by the earth's normal gravitational force. In orbital space flight, the resultant *g* force is reduced to the level of surface capillary forces, and problems can arise with the separation and management of the liquid and vapor phases. Stark, Bradshaw *et al.* (1974) and Stark, Blatt *et al.* (1974) contain summaries of technical papers and reports concerning low-*g* fluid behavior and low-*g* fluid management systems.

The dynamic sloshing motion of liquid contained in a low-*g* environment is discussed in Sec. 23.13 of this handbook. Abramson (1966) provides a review of the dynamic behavior of liquids in moving containers as related to low gravity space vehicle technology.

The potential for developing materials processing techniques in space for improving the morphology of single crystals has prompted the need for better understanding of the low gravity convective flows. Buoyancy driven fluid flows caused by temperature and/or concentration gradients are important in crystal growing technology [Ostrach (1983)]. Crystal imperfections are often attributed to uncontrolled convective flows occurring within the process container during crystal growth. The low gravity environment of space appears to offer the potential for growing higher quality crystals by minimizing the buoyancy driven natural convection. However, natural convection can still occur in response to the low level steady and transient accelerations in an orbiting space vehicle [Ostrach (1982a)].

Near-Earth Orbit. A space vehicle orbiting the earth is often said to be in a *zero-gravity environment*. However, the term zero-gravity is a misnomer. Actually, the earth's gravitational force exerted on a mass in a space vehicle in near-earth orbit is only slightly less than it would be on the earth's surface. An additional body force due to centripetal acceleration is applied to the mass in near-earth orbit that balances the earth's gravitational body force. In this way, the resultant *g* force on an orbiting mass can be near-zero.

Micro-g Environment: The near weightless condition experienced by a mass in near-earth orbit can be upset by many different types of low-level disturbances. Naumann (1981) classifies these disturbances into four categories.

(1) Steady, low-level accelerations arising from aerodynamic drag, Keplerian effects, and slow vehicle rotations. These accelerations are on the order of $10^{-5}g$, and they are maintained relatively constant for periods of tens of minutes or longer.

(2) Compensated transient accelerations arising from crew activity or internal mechanical vibration. These are short term impulses that are compensated by

equal and opposite impulses, and they do not change the momentum of the space vehicle. Although the impulses may be as high as $10^{-2}g$, they are short in duration and random in nature.

(3) Uncompensated transient accelerations arising from external forces such as thruster firings for attitude control and orbital maneuvers. Each thruster firing for attitude control can produce an acceleration pulse with a magnitude of $10^{-4}g$, but the pulse duration is short, or the order of 100 milliseconds.

(4) Rotational flows that are induced in a contained fluid by vehicle rotation. These flows are introduced each time there is a change in angular momentum. The flow decays by Ekman damping with typical damping times ranging from seconds to tens of seconds (see Sec. 13.2).

Ostrach (1982a) describes a number of fluid flows that have been identified as being possible under low-gravity conditions. Naumann (1981) notes that different low-gravity experiments have varying degrees of susceptibility to each type of disturbance; he ranks the relative importance of each type of low-level disturbance for the material processing experiments under consideration for the Space Shuttle.

Gravitational Effects on Fluids: The main effects of the force of gravity on fluids are [Ostrach (1982a)],

(1) The need for a container or a means for levitating and confining the fluid.
(2) The development of a hydrostatic pressure distribution in the fluid.
(3) The sedimentation of freely suspended particles with densities different from the fluid density.
(4) The generation of natural convective flows driven by density gradients within the fields.

Each of these effects is proportional to the magnitude of the gravity force and can be reduced in proportion to the reduction in gravity. However, when the gravity force is reduced, other forces that are gravity independent, such as surface tension, may become significant and alter the types of natural convection flows that result.

Free Convection. Natural convection flows driven by gradients in temperature, solution concentration and surface tension may occur in fluids contained in a low-g environment. Both the unstable and the stable modes of convection are possible, depending upon the relative orientation of the density gradient to the direction of acceleration. Due to the complexity of natural convection and its sensitivity to configuration and boundary conditions, information is needed about: (1) the magnitude and direction of accelerations, (2) the geometrical configuration of the container, (3) the imposed boundary conditions, and (4) the material properties of the fluid to quantitatively predict the behavior of natural convection in low-g surroundings [Ostrach (1982a)].

Buoyancy Driven Convection: Ostrach (1981) has described the types of natural convection fluid flows that are anticipated under reduced gravity conditions. Unfortunately, there is little experimental data available on reduced gravity convection [Ostrach (1982a)]. Therefore, most of the information available is based upon order of magnitude estimates from dimensional analysis, and computational fluid flow studies.

Ostrach (1981) has estimated that Grashof numbers of the order of 1 for gases, 10 for liquids and 100 for liquid metals could be attained in a $10^{-6}g$ environment in containers with a characteristic dimension of 10 cm, a temperature level of 293 K, and a temperature difference of 10 K. For a reduced gravity level of $10^{-5}g$, convective velocities of the order of 0.13 mm/s for liquids and liquid metals and 0.5 mm/s for gases could be attained.

Transient accelerations of the type described in Sec. 13.3.2 could also produce natural convection flows under certain conditions. Ostrach (1981) has described the results of analytical and computer solutions for the flow in rectangles and cylinders heated from the sides under various types of transient acceleration conditions. These results showed that when the imposed accelerations were random, but had a zero time mean (like a sinusoid), that no significant fluid motion was produced. However, when the accelerations had a nonzero mean, a time mean flow equivalent to a natural convection flow was produced. Drop tower experiments have confirmed that fluid motion is not detectable for imposed accelerations with a zero mean.

Ostrach (1982a) described the results of a more comprehensive investigation of buoyant motions produced by random accelerations imposed normal to the direction of the temperature gradient. This investigation showed that the predominant type of fluid motion produced was a unicellular oscillatory fluid motion. For fully confined fluids, oscillatory flows in which fluid elements oscillate in a relatively small region would be produced by random accelerations in a reduced gravity environment.

Surface Tension Gradient Convection: Ostrach (1982a) has also considered the possibility for free convection flows produced by surface tension gradients under reduced gravity conditions. If the surface tension between two fluid phases varies from point to point, or if the interface between the phases has a finite curvature that is different from the equilibrium curvature, then fluid motion is possible. For reduced gravity conditions, the surface tension gradient flows are likely to be important, particularly for crystal growth processes where gradients in both temperature and concentration exist simultaneously.

Ostrach has derived the dimensionless parameters that govern surface tension gradient flows, and presented equations for predicting the characteristic flow velocity. These equations are given below for a thermo-capillary flow where the gradient in surface tension is related to a temperature gradient.

For low values of Reynolds number, Re < 1, inertial effects are negligible, and a viscous-type of flow would occur. For this case, the characteristic velocity is,

$$U_R = \left[\frac{|\partial \sigma / \partial T| \ \Delta T D}{\mu L} \right] \tag{13.115}$$

where μ is the absolute viscosity and D and L are different characteristic lengths.

When the combination of $Re(D/L)^2$ is $\gg 1$, a boundary layer type of flow would occur. For this case, the characteristic velocity is,

$$U_R = \left[\frac{(\partial \sigma / \partial T)^2 \Delta T^2 \nu}{\mu^2 L} \right]^{1/3} \tag{13.116}$$

These types of flows could be produced by a temperature gradient along a two-phase interface and could have an effect on transport processes in a reduced gravity environment. In addition, the possibility exists for capillary flows to be produced in response to both temperature and concentration gradients.

The review by Ostrach (1982a) should be consulted for a description of the results of studies of surface tension driven flow performed under normal gravity conditions.

13.4 JETS IN A CROSS FLOW

13.4.1 Observations of Single-Phase Flows
Joseph A. Schetz

Here, we are concerned with jet injection at large angles to a moving mainstream, which of course, produces strongly three-dimensional flowfields. This class of flows is often encountered in practice; smokestacks, some fuel injection systems, V/STOL aircraft, and sewage and cooling water outfalls can be cited as a few representative examples. Further, one or both fluid streams may be supersonic. Cases where buoyancy forces are important are not discussed here; the interested reader can refer to Turner (1969) and McGuirk and Rodi (1977).

The character of the flow for a low-speed case with 90° injection can be seen in Figs. 13.16 and 13.17 from Abramovich (1960). The *kidney-shaped* nature of the jet as it is deflected and distorted by the cross stream is noteworthy. The simplest quantity of engineering interest is the gross penetration of the injected fluid into the

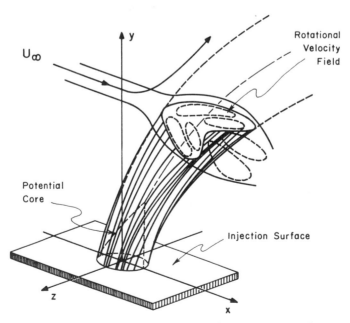

FIGURE 13.16 Schematic of transverse injection of a jet into a crossflow. (Reprinted with permission, Abramovich, G. N., *The Theory of Turbulent Jets*, MIT Press, Cambridge, MA, 1960.)

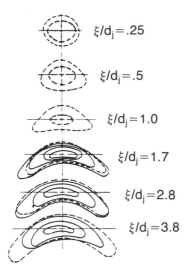

FIGURE 13.17 Pressure contours in cross sections of a transverse jet with $U_j/U_\infty = 2.2$. Solid and dashed lines correspond to constant total and static pressures, respectively. (Reprinted with permission, Abramovich, G. N., *The Theory of Turbulent Jets*, MIT Press, Cambridge, MA, 1960.)

mainstream. The trajectory of the center of the jet plume as a function of velocity ratio $R = U_j/U_\infty$ for 90° injection can be correlated as in Fig. 13.18. Here, h is the vertical penetration, and ξ is the arc length along the jet centerline trajectory. The next quantities of interest are the growth of the width of the mixing zone in the plane containing the trajectory, $\Delta\zeta$, and in the direction perpendicular to that plane, Δz,

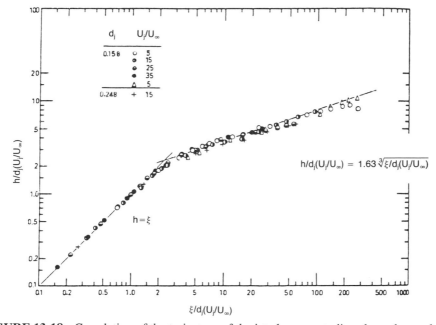

FIGURE 13.18 Correlation of the trajectory of the jet plume centerline along the arc length of normal (90°) injection. (Reprinted with permission, Pratte, B. and Baines, W., "Profiles of the Round Turbulent Jet in a Cross Flow," *J. Hydraulics Div.*, ASCE, pp. 56–63, Nov. 1967.)

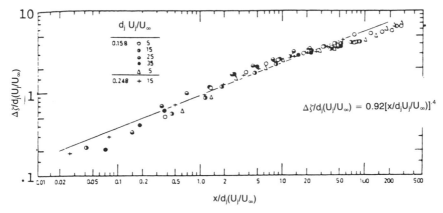

FIGURE 13.19   Correlation of the spreading of the plume in the plane of the trajectory for normal (90°) injection. (Reprinted with permission, Pratte, B. and Baines, W., "Profiles of the Round Turbulent Jet in a Cross Flow," *J. Hydraulics Div.*, ASCE, pp. 56–63, Nov. 1967.)

and some results are reproduced as Figs. 13.19 and 13.20. Some data for cases with injection at angles other than 90° is given in Campbell and Schetz (1973a).

Turning now to some details of the flow, we show the decay of the maximum velocity in the jet cross section in Fig. 13.21. Free jet results are also shown for comparison. It is not surprising that the transverse jet mixes faster, probably as a direct result of the vortices induced in the jet plume illustrated in Fig. 13.16. Velocity profiles across the plume are shown in Fig. 13.22, where a near-similarity condition can be observed.

The variation of the axial turbulence intensity along the jet centerline for three velocity ratios is presented in Fig. 13.23. Also shown for comparison are some

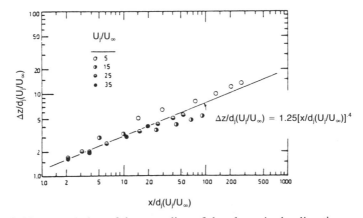

FIGURE 13.20   Correlation of the spreading of the plume in the direction perpendicular to the plane of the trajectory for normal (90°) injection. (Reprinted with permission, Pratte, B. and Baines, W., "Profiles of the Round Turbulent Jet in a Cross Flow," *J. Hydraulics Div.*, ASCE, pp. 56–63, Nov. 1967.)

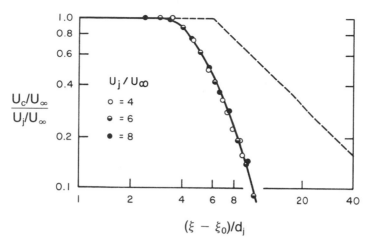

FIGURE 13.21 Variation of the centerline velocity along the jet trajectory. ----- free jet data. (Reprinted with permission, Keffer, J. F. and Baines, W. D., "The Round Turbulent Jet in a Cross Wind," *J. Fluid Mech.*, Vol. 15, pp. 481–496, 1963.)

coaxial jet results from Corrsin and Uberoi (1949). The turbulence is higher in the transverse jet case, which is consistent with the more rapid mixing noted earlier.

Some of the applications of jets in a cross-flow involve dual-jet arrangements, either in-line or side-by-side. Figure 13.24 shows some measurements of the flow-field above the surface for a dual jet case made with a yawhead probe. The inter-

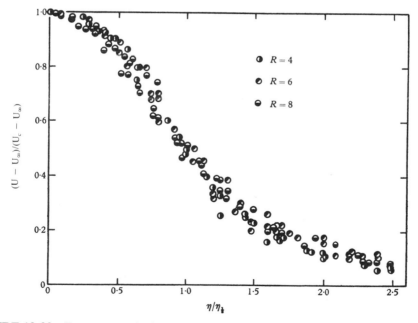

FIGURE 13.22 Transverse velocity profiles across the jet. η is a coordinate out of the plane of the trajectory and along the maximum velocity line. (Reprinted with permission, Keffer, J. F. and Baines, W. D., "The Round Turbulent Jet in a Cross-Wind," *J. Fluid Mech.*, Vol. 15, pp. 481–496, 1963.)

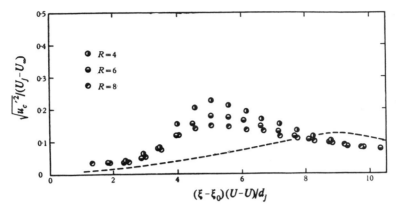

FIGURE 13.23   Turbulence intensity along the centerline of a transverse jet. ------ free jet data. (Reprinted with permission, Keffer, J. F. and Baines, W. D., "The Round Turbulent Jet in a Cross-Wind," *J. Fluid Mech.*, Vol. 15, pp. 481–496, 1963.)

section region with two jets at $R = 6.5$ is displayed. One can observe that the rear jet is *sheltered* strongly by the front jet; the trajectory of the rear jet is nearly vertical until the intersection.

One of the uses of transverse jets is to vertical and short take-off and landing (V/STOL) aircraft, where the pressure field induced on adjacent surfaces is of par-

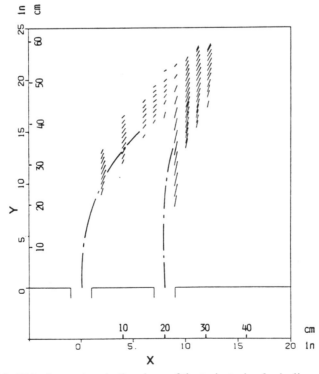

FIGURE 13.24   Velocity vectors in the plane of the trajectories for in-line, normal jets with $R = 6.5$ and a spacing of 4 diameters, —--— integral analysis prediction. (Reprinted with permission, Schetz, J. A., Jakubowski, A. K., and Aoyagi, K., "Jet Trajectories and Surface Pressures Induced on a Body of Revolution with Various Dual Jet Configurations," *J. Aircraft*, Vol. 20, No. 11, pp. 975–982, 1983.)

ticular importance. Reviews of the early work can be found in Garner (1967), and a more up-to-date tabulation on the available information is contained in Perkins and Mendenhall (1981). The jet generally induces negative (with respect to the free-stream) pressures on the nearby surfaces, and this results in a net loss of lift on the body viewed as a whole. The longitudinal variation of the surface pressures is also important, since that determines the resulting pitching movement. We will present data for single jets in comparison with that for dual jets, where the mutual interference as a function of center-to-center spacing is the issue. Another important item is the behavior of a jet (or jets) injected from a body of revolution as opposed to the large flat plates usually considered. This is of obvious importance for VTOL aircraft with lifting jets in the fuselage. One can anticipate substantial transverse pressure *relief* around a cylindrical body. The most comprehensive studies of these and other effects are in Schetz, Jakubowski, and Aoyagi (1983) and (1984).

Longitudinal pressure distributions at selected lateral distances are plotted in Fig. 13.25 for 90° injectors at $R = 6$ with a spacing of two diameters and the jets aligned one behind the other. The results for both jets are shown as circles, and those for a single jet only are stars. Looking at the single jet results first, the expected pattern of negative ΔC_p's [$\Delta C_p \equiv (C_{p\,\text{jet on}} - C_{p\,\text{jet off}})$] is evident. The pressure falls in the immediate vicinity of the jet, due largely to entrainment of external fluid into the jet. The dual jet results show that the influence of the rear jet is less than that of the

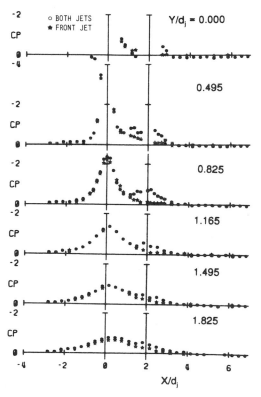

FIGURE 13.25 Surface pressure distributions, ΔC_p, for in-line, normal jets on a flat plate with $R = 6.0$ and a spacing of 2 diameters. (Reproduced with permission, Schetz, J. A., Jakubowski, A. K., and Aoyagi, K., "Surface Pressures on a Flat Plate with Dual Jet Configurations," *J. Aircraft*, Vol. 21, No. 7, pp. 484–490, 1984.)

front jet at this close spacing. On the other hand, the presence of the rear jet seems to strengthen the influence of the front jet slightly.

The effects of the important parameter R can be seen in the isobar plots in Fig. 13.26(a) and (b). As R increases, the area of the surface influenced by the jet increases. This increase is mostly in terms of the areas with small to moderate negative values of ΔC_p (e.g., $0 \geq \Delta C_p \geq -1.0$). Since the area of influence increases with R, the total normal force also increases with R, but the increase is slow, and the normal force normalized with the thrust of the jets actually decreases with increasing R. The effective center of force moves forward with increasing R. Estimates indicate

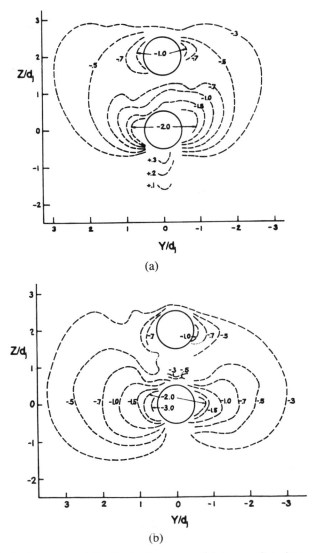

(a)

(b)

FIGURE 13.26 Isobar maps, ΔC_p, for in-line, normal jets on a flat plate with a spacing of 2 diameters. (a) $R = 4.0$ and (b) $R = 8.0$. (Reproduced with permission, Schetz, J. A., Jakubowski, A. K., and Aoyagi, K., "Surface Pressures on a Flat Plate with Dual Jet Configurations," *J. Aircraft*, Vol. 21, No. 7, pp. 484–490, 1984.)

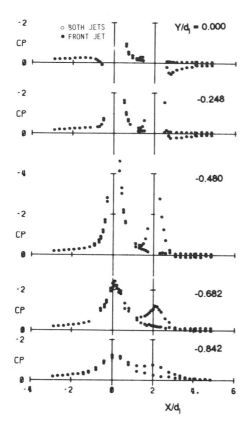

FIGURE 13.27 Surface pressure distributions, ΔC_p, for in-line, normal jets on a body of revolution with $R = 7.7$ and a spacing of 2 diameters. (Reproduced with permission, Schetz, J. A., Jakubowski, A. K., and Aoyagi, K., "Jet Trajectories and Surface Pressures Induced on a Body of Revolution with Various Dual Jet Configurations," *J. Aircraft*, Vol. 20, No. 11, pp. 975–982, 1983.)

that this center coincides with the center of the front jet at about $R - 10$. Lastly, the shape of the interaction region changes with increasing R. At low R, the isobars show asymmetrical lobes displaced in the downstream direction.

Surface pressure distributions on a body of revolution with $D_{jet}/D_{body} \approx 1/2$ obtained for $R = 7.7$, 90° injection through nozzles with exits contoured to the body surface at a spacing of 2.0 diameters are shown in Fig. 13.27. The pattern is qualitatively similar to that found on flat plates (see, e.g., Fig. 13.25 for $R = 6$). A more detailed comparison permits some interesting conclusions to be made. First, one finds that there are no positive ΔC_p values behind the second jet on the flat plate. Second, the peak values on the body of revolution are somewhat higher very near the nozzles. Third, by comparing values at $y/d_j = 0.8$ it can be seen that the peak values decay much faster with lateral distance on the body of revolution.

Transverse injection into supersonic streams is of engineering interest in several applications, including thrust vector control, fuel injection, and thermal protection. Since such a jet presents an obstruction to the main supersonic flow, an *interaction shock* is generally produced, and the whole flow differs from low-speed cases in important ways. The interaction shock produces a strong adverse pressure gradient that separates the boundary layer ahead of the jet. The turning of the external flow up over the separation zone produces an oblique shock that intersects the interaction shock. The case of a sonic, underexpanded transverse jet was considered in Schetz, Hawkins, and Lehman (1967), and the resulting flowfield is shown in Fig. 13.28. This very complex flow was related to the simpler situation of an underexpanded jet

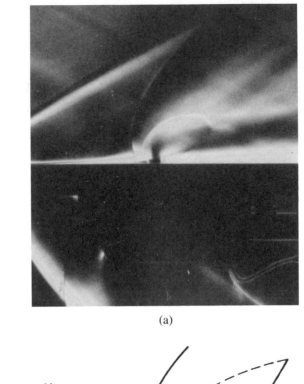

(a)

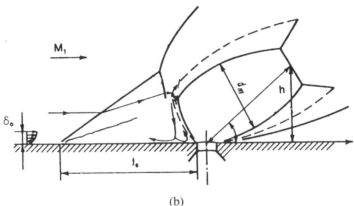

(b)

FIGURE 13.28 Flowfield for underexpanded, normal (90°), sonic injection into a supersonic crossflow. (a) Schlieren photo. (Reproduced with permission, Schetz, J. A., Hawkins, P. F., and Lehman, H., "The Structure of Highly Underexpanded Transverse Jets in a Supersonic Stream," *AIAA J.*, Vol. 5, No. 5, pp. 882–884, 1967.) (b) Schematic.

into quiescent surroundings. This relationship was accomplished through the notion of an *effective back pressure*, p_{eb}. It was then possible to use the experimental and theoretical results for the quiescent surroundings case to correlate data for the height (penetration) of the Mach disk in the transverse jet flow. This is shown in Fig. 13.29 using $p_{eb} = 0.8p_2$, where p_2 is the static pressure behind a normal shock in the main flow. Similar results for supersonic, underexpanded transverse jets are given in Schetz, Weinraub, and Mahaffey (1968). The effect of the shape of injectors with the same cross-sectional area on the height of the Mach disk was studied in Orth, Schetz, and Billig (1969) and found to be unimportant up to an aspect ratio of 4.0.

In addition to the initial penetration of the transverse jet as given here in terms

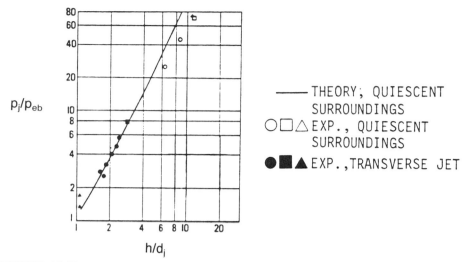

FIGURE 13.29 Correlation of the penetration height to the center of the Mach disk for underexpanded, normal (90°), sonic injection into a supersonic crossflow. (Reproduced with permission, Schetz, J. A., Hawkins, P. F., and Lehman, H., "The Structure of Highly Underexpanded Transverse Jets in a Supersonic Stream," *AIAA J.*, Vol. 5, No. 5, pp. 882–884, 1967.)

of the height of the Mach disk, the subsequent trajectory of the jet and mixing along that trajectory are also of interest. Some results for underexpanded, sonic, H_2 injection with $M_1 = 2.7$ from Orth, Schetz, and Billig (1969) are plotted in Fig. 13.30.

The shape of the interaction shock is often important; it determines the losses in fuel injection problems and the surface pressure for attitude control applications. Some results are shown in Fig. 13.31. The results of a simple analysis based upon an equivalent solid body obstruction with a hemisphere nose of radius h are also shown, along with the earlier *blast-wave* model.

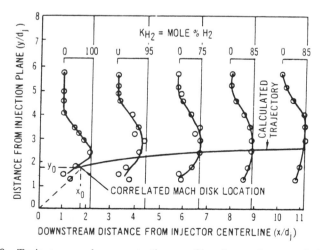

FIGURE 13.30 Trajectory and concentration profiles for underexpanded, normal (90°) sonic injection of H_2 into a supersonic air crossflow. ——— prediction of integral analysis. (From Orth *et al.*, 1969.)

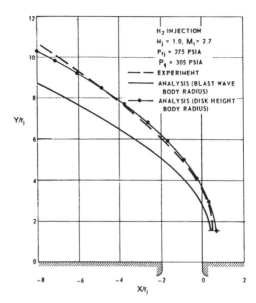

FIGURE 13.31 Comparison of predictions and experiment for the interaction shock shape for sonic H_2 injection into a supersonic air crossflow. (Reproduced with permission, Schetz, J. A., "Interaction Shock Shape for Transverse Injection," *J. Spacecraft and Rockets*, Vol. 7, No. 2, pp. 143–149, 1970.)

The size and shape of the separation zone on the surface and the pressure distribution in that zone has been the subject of numerous studies. Some results of the comprehensive tests of Zubkov and Glagolev (1979) are presented here to illustrate the information available. The notation used is given in Fig. 13.32. First, the influences of Mach number, M_1, pressure ratio ($n = p_j/p_1$) and boundary layer thickness on the length of the separation zone ahead of the jet are correlated in Fig. 13.33.

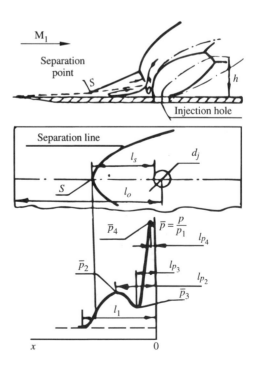

FIGURE 13.32 Notation for jet interaction flowfield.

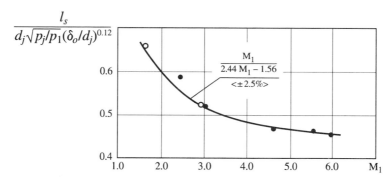

FIGURE 13.33 Correlation of the upstream length of the separation zone as a function of crossflow Mach number, pressure ratio, and boundary layer thickness. (Reprinted with permission, Zubkov, A. and Glagolev, A., "The Effect of Boundary Layer Thickness and Transverse Curvature of the Surface on the Geometry and Forces Produced by Injection of a Jet into Supersonic Flow over that Surface," *Fluid Mech.—Soviet Res.*, Vol. 8, No. 1, 1979.)

The effect of changing the injection angle is presented in Fig. 13.34. The important effect of injection from a body of revolution as opposed to a flat surface is shown in Fig. 13.35, where R_0 is the transverse radius of curvature. Finally, the wall pressure levels as they vary with some of the parameters are plotted in Fig. 13.36. The integrated effect of the size and shape and magnitude of the pressure variation manifests itself in the form of a total normal force, and Fig. 13.37 shows some typical results. The total force measured is considerably greater than the thrust of the jet, and this is often expressed as an *amplification factor* that is of the order of 3–5.

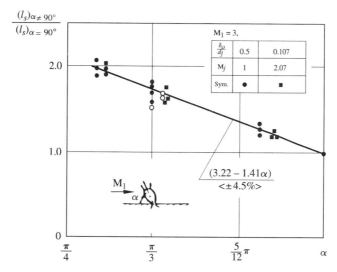

FIGURE 13.34 Correlation of the effect of injection angle on the upstream length of the separation zone. (Reprinted with permission, Zubkov, A. and Glagolev, A., "The Effect of Boundary layer Thickness and Transverse Curvature of the Surface on the Geometry and Forces Produced by Injection of a Jet into Supersonic Flow over that Surface," *Fluid Mech.—Soviet Res.*, Vol. 8, No. 1, 1979.)

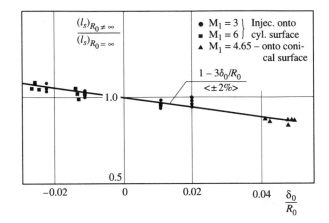

FIGURE 13.35 Correlation of the effect of transverse surface curvature on the upstream length of the separation zone. (Reprinted with permission, Zubkov, A. and Glagolev, A., "The Effect of Boundary Layer Thickness and Transverse Curvature of the Surface on the Geometry and Forces Produced by Injection of a Jet into Supersonic Flow over that Surface," *Fluid Mech.—Soviet Res.*, Vol. 8, No. 1, 1979.)

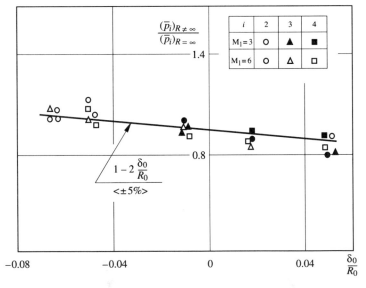

FIGURE 13.36 Correlation of the effect of boundary layer thickness and transverse surface curvature on the pressures in the interaction zone. (Reprinted with permission, Zubkov, A. and Glagolev, A., "The Effect of Boundary Layer Thickness and Transverse Curvature of the Surface on the Geometry and Forces Produced by Injection of a Jet into Supersonic Flow over that Surface," *Fluid Mech.—Soviet Res.*, Vol. 8, No. 1, 1979.)

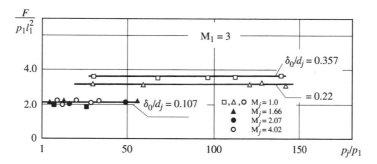

FIGURE 13.37 Correlation of the total normal force of the interaction as a function of crossflow Mach number, pressure ratio, and boundary layer thickness. (Reproduced with permission, Zubkov, A. and Glagolev, A., "The Effect of Boundary Layer Thickness and Transverse Curvature of the Surface on the Geometry and Forces Produced by Injection of a Jet into Supersonic Flow over that Surface," *Fluid Mech.—Soviet Res.*, Vol. 8, No. 1, 1979.)

13.4.2 Modeling of Jets in Cross Flow
Ayodeji O. Demuren

Introduction. In the earliest studies of jets in cross flow, empirical models were developed to correlate experimental data obtained under various idealized conditions. Such models are reviewed in detail by Abramovich (1960), Rajaratnam (1976), and Schetz (1980). They mostly give the jet trajectory and center-line decay rates. The earliest approach based on the actual solution of conservation equations belongs to the class of integral models. These are derived either by the application of conservation principles to a finite control volume or by the use of profile assumptions to simplify the partial differential equations which describe conservation laws. These models offer more flexibility than empirical models. Jet trajectories, decay rates, growth rates, and even cross-sectional shapes have been predicted. However, empirical input is usually required in the form of entrainment rates or a drag coefficient model. Further, it may be difficult to prescribe cross-sectional profiles in complex situations. Numerical models attempt to solve some form of the full partial differential equations, which represent the conservation laws, by using finite-difference, finite-volume or finite-element methods. Little empirical input is required, hence they have the potential for the widest range of applicability. However, there may be problems with inadequate grid resolution, imprecise boundary conditions and deficiencies in the turbulence model used for closure of the mean-flow equations. Recently, several models based on perturbation methods have been proposed. These are mostly of scientific interest, since drastic assumptions such as inviscid flow, negligible jet distortion, small deflection, etc., may be required for the perturbation analysis. Thus, they are used mainly to study the flow physics in limited regimes, either in the near-field or in the far-field.

In practical engineering applications, jets in cross flow are found in both confined and unconfined environments. Examples of confined jets in cross flow include: 1) Vertical and Short Take-Off and Landing (V/STOL) aircraft in transition from hover to forward flight, in which case, the jets from its engines impinge on the ground surface, 2) internal cooling of turbine blades by air jets impinging on the leading edge, and 3) dilution air jets in combustion chambers of gas-turbine engines, where the jets are injected radially into the chamber, through discrete holes along its circumference, in order to stabilize the combustion process near the head, and to dilute the hot combustion products near the end.

Practical examples of jets in unconfined or semi-infinite cross flow are more numerous. These include: 1) flow situations resulting from the action of cross winds on effluents from cooling towers, chimney stacks, or flames from petrochemical plants, 2) discharge of sewage or waste heat into rivers or oceans, 3) film cooling of turbine blades, 4) the use of air curtains to prevent cold air from entering open spaces in industrial buildings, and 5) thermal plumes rising into cross winds in the atmosphere.

The configuration of a jet in cross flow and the notation used in this section is illustrated in Fig. 13.38. The axis of the jet is usually defined as the locus of the maximum velocity or total pressure. The jet trajectory is referred to this line, as opposed to the center-line of the jet, which is mid-way between the inner and outer boundaries of the jet, usually determined from flow visualization. The main parameter which characterizes a jet in cross flow is the jet-to-cross-flow velocity ratio, R $(= U_j/U_0)$, or the momentum flux ratio $\bar{q}$ $[= (\rho_j/\rho_\infty)R^2]$. In confined jets, the normalized wall distance H/D may also be important if it is not very large. In multiple jets, the normalized spacing S/D will be a factor.

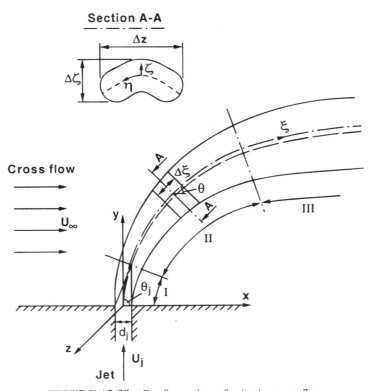

FIGURE 13.38   Configuration of a jet in cross flow.

As shown in Fig. 13.38, the jet in a cross flow has three main regions: the potential core zone (I), the zone (II) of maximum deflection, and the far-field zone (III). The potential core, in the central part of zone I remains relatively unaffected by the cross flow though its length is reduced in comparison to that of a jet in stagnant surroundings. Thus, for a turbulent jet it reduces from ($\sim 6D$) to $6.2De^{-3.3/R}$ [Fan (1967) and Pratte and Baines (1967)] or $6.4/(1 + 4.6/R)$ [Kamotani and Greber (1972)]. The two relations deviate at low R, where the potential core length is strongly influenced by actual exit flow conditions. In zone II, the jet experiences the most deflection. The pressure gradient across the jet is maximum as well as the entrainment rate. This is the most difficult zone to analyze accurately. In the far-field zone III, the jet axis approaches the crossflow direction asymptotically, and the flow field is nearly *self-similar*. All models can predict this region fairly well, given the correct boundary conditions at the end of zone II.

Four broad classes of models, namely, empirical, integral, perturbation, and numerical, are now described. Emphasis will be given to the last two, where most of the recent advances have been made. The first two methods were reviewed extensively in monographs by Abramovich (1963), Rajaratnam (1976), and Schetz (1980) and in a review article by Demuren (1985a).

Empirical Models. Empirical models present the simplest means of predicting the global properties of jets in cross flow. They depend largely on the correlation of experimental data, and the accuracy of the predictions may depend on the closeness of the conditions of the particular problem of interest to those in the data base used for the correlation. Due to their low cost and ease of use, empirical models are most useful for first-order estimates and as qualitative checks for results produced by other methods.

The most common parameter given by empirical models is the jet trajectory. For a single circular turbulent jet injected normally into a cross flow, the trajectory has the form

$$\frac{y}{d_j} = a\bar{q}^b \left(\frac{x}{d_j}\right)^c \tag{13.117}$$

where, in the range of $\bar{q}$ between 2 and 2,000, a has a value between 0.7 and 1.3, b has a value between 0.36 and 0.52 and c takes a value between 0.28 and 0.40, depending on experimental conditions. The values $a = 0.85$, $b = 0.47$ and $c = 0.36$ appear to be a good compromise for the intermediate range of $\bar{q}$. This equation should also be valid for confined jets, up to the point of contact with the walls, and for multiple jets with medium to large spacing ratios. Equation (13.117) with $b = 0.36$, and $c = 0.28$ also gives the physical boundaries of jets in cross flow [Pratte and Baines (1967)], with $a = 1.35$ and 2.63 for the inner and outer boundaries, respectively, and $a = 2.05$ for the center line. For plane jets in confined cross flow, Kamotani and Greber (1974) found that Eq. (13.117) can be used with $a = 2.0 (1 - e^{-H/D})$, $b = 0.28$, and $c = 0.50$. Equations for other parameters such as entrainment rates, velocity profiles, temperature trajectories, etc. can be found in Demuren (1985a).

Integral Models. Integral models were the first elaborate calculation procedures applied to predict the behavior of jets in cross flow. In these models, integral equations are derived either by considering the balance of forces and momentum changes over an elementary control volume of the jet, or by integrating in two spatial directions, the three-dimensional, partial differential equations governing the jet flow. In either case, a set of ordinary differential equations is obtained which can be solved analytically or numerically. Empirical input is required to prescribe pressure drag, entrainment rates, and spread rates. The first approach is easier to understand and to implement and is, therefore, more popular. On the other hand, the latter approach involves more extensive mathematical manipulation, but it is more transparent to the assumptions made and affords more flexibility in dealing with complex boundary conditions and trajectories.

Integral models flourished between the late 1960's, when more flexibility was required than could be obtained with empirical models and the early 1980's, when the rapid growth in computer hardware and software made elaborate numerical computations of three-dimensional flows feasible. Many of the earlier models are reviewed by Rajaratnam (1976). In these models, there was an assumption of the constancy of the momentum in either the initial jet direction, the cross flow direction or the axial direction, and the jet was bent over by a prescribed pressure drag force, or entrainment of ambient fluid. None of these models could predict correctly the jet trajectory over a range of R [Demuren (1985a)]. Thus, they offer no advantage over much simpler empirical models.

More refined integral models consider the effects of both the pressure drag and the entrainment of cross flow ambient fluid on the jet. A typical model is that proposed by Fan (1967) for buoyant jets in cross flow. The cross section of the jet was assumed circular with radius $\sqrt{2}\, b$, and the excess velocity profile was assumed to be *Gaussian*, i.e., $V - U_\infty \cos \theta = (V_{max} - U_\infty \cos \theta)e^{-\eta^2/b^2}$. The resulting set of ordinary differential equations can be written as:

Continuity

$$\frac{d}{d\xi}(\rho A V) = C\rho_\infty U_e \tag{13.118}$$

x-momentum

$$\frac{d}{d\xi}(\rho A \lambda_v V^2 \cos \theta) = C\rho_\infty U_e U_\infty + 0.5 C_D \Delta z \rho_\infty U_\infty^2 \sin^3 \theta \tag{13.119}$$

y-momentum

$$\frac{d}{d\xi}(\rho A \lambda_v V^2 \sin \theta) = -A(\rho - \rho_\infty)g + 0.5 C_D \Delta z \rho_\infty U_\infty^2 \sin^2 \theta \cos \theta \tag{13.120}$$

Scalar (Temperature/Concentration) ϕ

$$\frac{d}{d\xi}(\rho A \lambda_\phi V \phi) = 0 \tag{13.121}$$

where A is the cross-sectional area of the jet, C its circumference, ρ is the density of the jet fluid, g the acceleration due to gravity, and λ_v and λ_ϕ are, respectively, momentum or scalar flux coefficients which depend on the assumed velocity and scalar profiles. U_e is the entrainment velocity. Fan proposed that it should be proportional to the velocity vector difference between the jet and the cross flow, but this was found to be unreliable. Abraham (1971) proposed an entrainment model with two parts as

$$U_e = E_{\text{mom}}(V_{\text{max}} - U_\infty \cos \theta) + E_{\text{th}}U_\infty \sin \theta \cos \theta \qquad (13.122)$$

where the coefficients E_{mom} and E_{th} have the values 0.057 and 0.50, respectively. The first part of Eq. (13.122) represented the entrainment of a momentum jet in a nearly stagnant ambient fluid and the second part the entrainment into (momentum-free) thermals under similar conditions. Cos θ was introduced artificially into the second part to prevent it from contributing to entrainment when the jet was nearly perpendicular to the cross flow. With this entrainment model, the drag coefficient C_D was given a value of 0.3. Equations (13.118) through (13.122) were then integrated numerically. With this model, Abraham (1971) was able to obtain quite good agreement with experimental data of jet trajectory and axial concentration decay. Typical results are shown in Fig. 13.39. For a nonbuoyant jet, $E_{\text{th}} = 0$ and $\rho = \rho_0$ so that the buoyancy term in Eq. (13.120) is also zero.

Equations (13.118) through (13.122) may also be applied to predict plane jets in cross flow by substituting the appropriate expressions for the area A and the circumference C. The entrainment coefficients E_{mom} and E_{th} and the drag coefficient C_D must then be calibrated with plane jet data.

In order to extend the range of applicability of integral models to include more flow physics and to be able to deal with more practical situations, such as multiple jets in varied arrangements, highly nonuniform cross flow, etc., more elaborate models have been proposed by Campbell and Schetz (1973), Isaac and Schetz (1982), Makihata and Miyai (1983), among others. All these models were derived based on the control volume approach and are applicable mainly to jets with plane trajectories. Hirst (1972) and Schatzmann (1979) developed models based on the integration in the cross plane of the jet of the three-dimensional partial differential equations by making the assumption of axisymmetry and that profiles of the excess velocity are Gaussian. These latter models could be applied to situations with three-dimensional jet trajectories.

Although integal models allow economical prediction of several flow properties in comparison to full-blown numerical models, they have been criticized for the need to assume the shape of the jet cross section and profile functions, some of which may not be realistic for the whole evolution of the jet in cross flow, especially in the zone of maximum deflection. However, it appears that in spite of the apparent oversimplification, integral models can be made to perform well in some cases with proper calibration. Adler and Baron (1979) proposed a quasi-three-dimensional model which did not assume the cross-sectional shape or similarity profiles for the velocity, but these were computed along with other flow variables. The characteristic kidney-shaped cross-section of the jet was computed successfully by considering the evolution of vortices distributed along the boundaries of the jet in a Lagrangian manner, and the cross section was allowed to grow at a rate which was an average between the growth rates of free jets and vortex pairs. Similarly, velocity profiles were allowed to change in the zone of maximum deflection, culminating in self-similar profiles only in the far-field zone (III). The model gives quite good

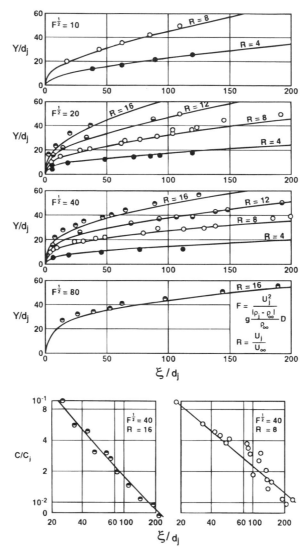

FIGURE 13.39   Prediction of jet trajectories and concentration decay; symbols—experimental data (from Fan, 1967), curves—calculations. (From Abraham, 1971.)

prediction of the three-dimensional flow fields of jets in cross flow studied experimentally by Kamotani and Greber (1972).

Perturbation Models.   If in the jet in cross flow problem a small parameter is defined, perturbation methods can be used to solve the governing equations. Most applications of perturbation models have been to study the flow of strong jets in a weak cross flow. In the initial stage, the flow can be considered to be a small perturbation from that of a free jet in stagnant surroundings, and the jet stiffness λ ($=1/R$) can be used as the small parameter. This places a severe restriction on the range of applicability of such models. However, they have the advantage of not being too dependent on empirical calibration as integral methods are, and they are computationally much cheaper than numerical methods. The goal is to predict the main features of jets at high R (> 10), including trajectory, cross-sectional shape, velocity field, vorticity field, mixing, etc., with minimal empirical input. It was believed [Needham et al. (1988), (1990)] that the jet distortion and deflection could be obtained by inviscid analyses based on the evolution of vortex filaments around the jet as it exits from a circular pipe or orifice. This approach was based on the original work of Chen (1942) which was also the basis of fairly successful computations with the integral model of Adler and Baron (1979).

Chen's model approximates the near-field as two regions of irrotational flow, the jet flow and the external cross flow, separated by a vortex sheet. The three-dimensional vortex sheet can then be approximated by a two-dimensional vortex sheet, which originates from the pipe or orifice exit and evolves in time in the axial direction. Needham et al. (1988), (1990) applied a three-dimensional model, with perturbation expansions for the potential flow within and outside of the jet. The distortion of the jet could be predicted reasonably well, but contrary to earlier studies, no jet deflection was obtained if the jet issued normally into the cross flow. Surprisingly, with a component of the cross flow in the direction of the jet, some deflection was obtained. This discrepancy was explained by Coelho and Hunt (1989) who showed that the two-dimensional, time-evolving vortex sheet model was a poor approximation for the fully three-dimensional vortex sheet model.

The two-dimensional vortex sheet equation can be written as

$$\frac{\partial \gamma}{\partial t} + \frac{\partial}{\partial s}(U_s \gamma) = 0 \tag{13.123}$$

where γ is the vortex strength and U_s is the average speed of the flow across the layer, using the nomenclature of Fig. 13.40(a). If $\gamma(\theta, t)$, $U_s(\theta, t)$, and $R(\theta, t)$ are approximated as Taylor series expansions with respect to t, derivatives of γ, U_s, and R of any order with respect to t, at $t = 0$, can be evaluated. This gives the shape of the vortex sheet or the jet boundary as

$$R(\theta, t) = R_0 - \left[\frac{1}{2}\frac{U_\infty^2}{R_0}\cos 2\theta\right]t^2 + \left[\frac{1}{6}\frac{U_\infty^3}{R_0^2}\{3\cos 3\theta - \cos \theta\}\right]t^3 + O(t^4) \tag{13.124}$$

If it is assumed that elements of the vortex-sheet travel at half-speed, such that $t = (2y/U_j)$, and U_j and R_0 are used for normalization, then Eq. (13.124) becomes

$$R(\theta, y) = 1 - [2y^2 \cos 2\theta]\lambda^2 + \left[\frac{4}{3}y^3\{3\cos 3\theta - \cos \theta\}\right]\lambda^3 + O(\lambda^4) \tag{13.125}$$

However, in the presence of the cross flow, the flow in the jet pipe is distorted [Andreopoulos (1983)], so that U_j is not uniform, and Eq. (13.125) may not be a good approximation for Eq. (13.124). The actual nonuniformity of the jet exit flow

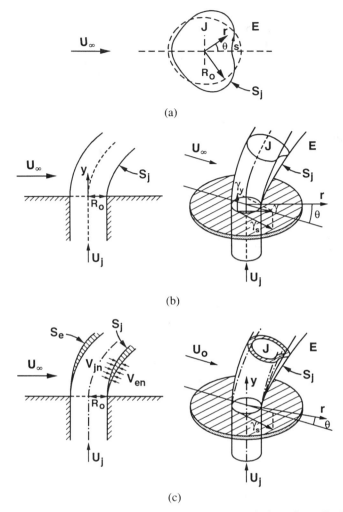

FIGURE 13.40 Nomenclature for perturbation model description (from Coelho and Hunt, 1989); (a) 2D vortex sheet model, (b) 3D vortex sheet model, and (c) entraining 3D vortex sheet model.

can be calculated using a fully three-dimensional vortex-sheet model. In this case, the longitudinal and transverse components of vorticity must be considered. As shown in Fig. 13.40(b), these can be approximated by the vertical and azimuthal components γ_y and γ_s, respectively. The general expression for the strength of the vortex sheet can be written in vector form as

$$\frac{\partial}{\partial t}\overline{\gamma} + (\overline{U}_v \cdot \nabla)\overline{\gamma} = (\overline{\gamma} \cdot \nabla)\overline{U}_v - \overline{\gamma}(\nabla \cdot \overline{U}_r) \qquad (13.126)$$

where

$$\overline{\gamma} = \{\gamma_s, \gamma_y\}, \ \overline{U}_v = \{U_s, U_y\} \text{ and } \nabla = \left\{\frac{\partial}{\partial s}, \frac{\partial}{\partial y}\right\} \qquad (13.127)$$

The vertical component of Eq. (13.126) is

$$U_y \frac{\partial \gamma_y}{\partial y} + \frac{\partial}{\partial s} (U_s \gamma_y) = \frac{\partial U_y}{\partial s} \gamma_s \qquad (13.126a)$$

which contains a source term, in contrast to Eq. (13.123). This source term expresses the rate at which fluid elements rotate as they travel up the vortex sheet. Thus, the vertical vortex strength may be strongly influenced by the azimuthal vortex strength and variations in the azimuthal velocity. The solution of the three-dimensional vortex-sheet problem, in terms of the potential flow inside the jet, and the external potential flow gives the shape of the vortex sheet to third order as

$$R(\theta, y) = 1 - \lambda^2 \left[y^2 - 2C_2 y - \sum_{n=1}^{\infty} A_n J_3(\sigma_n) e^{-\sigma_n y} - 1 \right] \cos 2\theta + O(\lambda^4)$$

$$(13.128)$$

where C_2 is a constant, and A_n are coefficients given by the boundary conditions. $J_3(\ )$ are Bessel functions of third order, and σ_n are zeros of $J_2(\ )$. Solutions for the velocity field (in terms of the velocity potential) and the pressure field are also given in terms of Bessel functions. Comparison of Eqs. (13.125) and (13.128) shows that the $O(\lambda^3)$ term in the former, which produces the deviation from symmetry and thus the jet deflection is absent in the latter. Therefore, Coelho and Hunt (1989) concluded that a three-dimensional inviscid vortex-sheet model could not produce jet deflection. Although, the two-dimensional model of Chen (1942) could produce a deflection, this was an incorrect approximation of the three-dimensional flow. However, by introducing a vertical component to the cross flow, the symmetry in Eq. (13.128) may be broken and a second smaller parameter is introduced into the perturbation expansion, as in the works of Needham *et al.* (1988), (1990). Then, jet deflection would occur.

Coelho and Hunt (1989) postulated that viscous or turbulent entrainment was necessary for jet deflection. They proposed an entraining vortex-sheet model [see Fig. 13.40(c) for nomenclature]. The entrainment velocity from the external cross flow is given by

$$V_{ce} = e(\gamma_s^2 + \gamma_y^2)^{1/2} + O(e^2) \qquad (13.129)$$

where e is the entrainment coefficient, which must be prescribed empirically, and it now becomes a second small parameter for the perturbation expansion. The mixing layer within the vortex-sheet entrains fluid from both the jet and the external cross flow. To the leading order, the entrainment rates are assumed proportional, so that

$$V_{je} = ec(\gamma_s^2 + \gamma_y^2)^{1/2} + O(e^2) \qquad (13.130)$$

where c is a constant of $O\ (1)$. The equation for the conservation of mass within the mixing layer is

$$\frac{\partial}{\partial \theta} [U_\theta (R_e - R_j)] + \frac{\partial}{\partial y} \left[U_y \left(\frac{R_e^2 - R_j^2}{2} \right) \right] = e(\gamma_s^2 + \gamma_y^2)^{1/2}$$

$$\cdot \left\{ \left[R_e^2 + \left(\frac{\partial R_e}{\partial \theta} \right)^2 \right]^{1/2} + c \left[R_j^2 + \left(\frac{\partial R_j}{\partial \theta} \right)^2 \right]^{1/2} \right\} + O(e^2) \qquad (13.131)$$

Equation (13.131) must be solved along with the potential flow equations for the jet flow and the external flow. The solution yields for the mean radius

$$R(\theta, y) = 1 + ye + 1 \; [(1 + c)y^2 \cos \theta]e\lambda - [Z(y) \cos 2\theta]\lambda^2 + O(e^2, \lambda^3, e\lambda^2)$$

(13.132)

where $Z(y) = y^2 - 2C_{2y} - \Sigma_{n=1}^{\infty} A_n J_3(\sigma_n) [e^{-\sigma_n y} - 1]$. Comparison of Eqs. (13.128) and (13.132) shows that the deviation from symmetry is now of $O(e\lambda)$, so that jet deflection would occur, as one would normally expect.

Higuera and Martinez (1993) have proposed a mixed perturbation/numerical model which does not use the vortex-sheet concept but solves the parabolized Navier–Stokes (PNS) equations in the distortion region of jets in weak cross flow ($R > 15$). The model is applicable to laminar flow or a turbulent flow in which the assumption of a constant eddy viscosity would be appropriate. The weak cross flow is necessary, so that there is only mild curvature in the distortion zone II, enabling the governing equations to be parabolized. Furthermore, there should be little deviation of the jet flow in the development zone I from that of a free jet, so that the flow field at the end of zone I and the beginning of zone II can be prescribed from Landau's self-similar solution for a point source of momentum [Batchelor (1967)]. A perturbation method is used to solve the PNS equations for small y, with y as the small parameter. For intermediate values of y, a parabolic numerical method is used. However, computations must be stopped once the distortion becomes too large for the assumptions of negligible axial diffusion and pressure gradient to be valid.

For small $y = [O(4y_w)]$, where $y_w = [(\sqrt{3}/16)RD]$, the deflection of the jet, β, will be small. The continuity and momentum equations, in dimensionless variables, can then be written for the axial velocity component, v and cross stream velocity vector $\overline{V} = \{u, w\}$ as

$$\nabla \cdot \overline{V} + \frac{\partial v}{\partial y} = 0$$

(13.133)

$$v \frac{\partial v}{\partial y} + (\overline{V} \cdot \nabla)v = \nabla^2 v + \beta^2 \left(-\frac{\partial p}{\partial y} + \frac{\partial^2 v}{\partial y^2} \right)$$

(13.134)

$$v \frac{\partial \overline{V}}{\partial y} + (\overline{V} \cdot \nabla)\overline{V} = -\nabla p + \nabla^2 \overline{V} + \beta^2 \frac{\partial^2 \overline{V}}{\partial y^2}$$

(13.135)

where $\nabla = \{\partial/\partial x, \partial/\partial z\}$ and $\nabla^2 = \{\partial^2/\partial x^2, \partial^2/\partial z^2\}$. The requirement that y_w should be beyond the development zone I indicates how large R must be for the analysis to be valid. For example for turbulent flow with a development length $\approx 6D$, $R > 15$. Therefore $\beta \to 0$, and terms in β^2 can be neglected. If the pressure gradient term is also eliminated by combining the divergence of Eq. (13.135), with the continuity equation, the governing equations become

$$v \frac{\partial v}{\partial y} + (\overline{V} \cdot \nabla)v = \nabla^2 v$$

(13.136)

$$v \frac{\partial \Omega}{\partial y} + (\overline{V} \cdot \nabla)\Omega = \Omega \frac{\partial v}{\partial y} + \frac{\partial u}{\partial y}\frac{\partial v}{\partial z} - \frac{\partial w}{\partial y}\frac{\partial v}{\partial x} + \nabla^2 \Omega$$

(13.137)

$$\Omega = \frac{\partial w}{\partial x} - \frac{\partial u}{\partial z} \tag{13.138}$$

where Ω is the vertical component of vorticity. Equations (13.136)–(13.138) are parabolic in y, so they can be solved by marching in zone II.

The initial conditions are derived from Landau's self-similar profiles as

$$(yu, yw, yv, y^2\Omega) \rightarrow (V_{rs} \cos \theta, V_{rs} \sin \theta, v_s, 0) \text{ as } y \rightarrow 0 \tag{13.139}$$

where,

$$V_{rs} = \frac{4\eta(1 - \eta^2)}{(1 + \eta^2)^2}, \; v_s = \frac{8}{(1 + \eta^2)^2}, \; \eta = \frac{r}{y}$$

The boundary conditions as $x \rightarrow \infty$ are

$$v = w = \Omega = 0; \; u \rightarrow 1 \tag{13.140}$$

the near-field solution has the form

$$(v, v_r, v_\theta, y\Omega) = \frac{1}{y} (v^{-1}, v_r^{(-1)}, 0, 0) + (v^{(0)}, v_r^{(0)}, v_\theta^{(0)}, \Omega^{(0)})$$

$$+ y(v^{(1)}, V_r^{(1)}, V_\theta^{(1)}, \Omega^{(1)}) + \cdots \tag{13.141}$$

The functions on the right-hand side of Eq. (13.141) depend only on η and θ, whereas Eqs. (13.136)–(13.138) are functions of η, θ, and y, Hence, by substituting Eq. (13.141) into Eqs. (13.136)–(13.138) and collecting terms of like order in y, solutions of different order can be found. Of course, terms of order (-1) will reproduce the initial conditions. This approach is really quite restrictive. The several requirements of large R, small y, constant turbulent eddy viscosity and low Reynolds number exclude it from consideration as a realistic tool for practical computations of jets in cross flow.

In general, perturbation models are not yet sufficiently mature to become more than curious tools of analysis. A redeeming factor is that it is especially in those high R flows for which they are valid, that most numerical models are least accurate. In these flows, there are substantial regions with high shear and rates of strain, where standard discretization schemes and turbulence models may become inadequate.

Numerical Models. Numerical models have the most potential for wide generality and can, in principle, be applied to the whole range of jet in cross flow situations: confined or unconfined, low, medium or high R, single or multiple jets, impinging on a wall or on other jets, swirling, homogeneous or heterogeneous cross flow, compressible or incompressible, etc. The analysis starts from the general conservation laws stated in partial differential equation form, which are the Navier–Stokes equations for the velocity field, and corresponding energy or species equations for the temperature or concentration fields, respectively. These equations, which describe unsteady, three-dimensional flow cannot be solved directly in practical applications for turbulent flows. In incompressible fluid flow, time-averaged forms, and in compressible fluid flow, density-weighted, time-averaged (or Favre-averaged) forms of the equations are solved. The process of time-averaging introduces a closure problem due to nonlinear correlation between fluctuating velocity and/or temperature/concentration fields. *Turbulence models* (see Secs. 4.4 and 4.5) are required to determine these correlations, thereby affecting closure of the system of equations. Most numerical models applied to the jet in cross problem use the eddy viscosity concept. In its simplest form, the turbulent eddy viscosity is prescribed as a constant, whereas more sophisticated models solve partial differential equations for turbulent quantities, from which the eddy viscosity distribution can then be obtained. Experimental studies by Andreopoulos and Rodi (1984) show that there are significant regions of the jet and cross flow interactions in which the eddy viscosity concept is invalid. Demuren (1992) proposed a numerical model in which the eddy viscosity concept is not invoked, but partial differential equations are solved to determine distributions of the turbulent correlations directly. In most numerical models, the computational domain encompasses the whole region in which the influence of the jet is felt or, if necessary, the whole field of the jet and cross flow. No assumptions are required as to the evolution of the jet within the flow domain, but this is obtained as a result of the computations. It is only necessary to prescribe boundary conditions at the chosen computational boundaries. The two major issues in the application of numerical models to jets in cross flow are the accuracy of the basic numerical method and the accuracy of the turbulence model.

The time-averaged, three-dimensional, steady-state mean flow equations can be written in Cartesian tensor notation as

Continuity

$$\frac{\partial}{\partial x_l}\,(\rho U_l) = 0 \tag{13.142}$$

Momentum

$$\frac{\partial}{\partial x_l}\,(\rho U_i U_l) = -\frac{\partial}{\partial x_i}\,p + \frac{\partial}{\partial x_l}\left[-\rho\overline{u_i u_l} + \mu\left(\frac{\partial U_i}{\partial x_l} + \frac{\partial U_l}{\partial x_i}\right)\right] \tag{13.143}$$

Scalar

$$\frac{\partial}{\partial x_l}\,(\rho U_l \phi) = S_\phi + \frac{\partial}{\partial x_l}\left[-\rho\overline{u_l \vartheta} + \frac{\mu}{\sigma}\frac{\partial \phi}{\partial x_l}\right] \tag{13.144}$$

with $i = 1, 2, 3$ and $l = 1, 2, 3$ representing properties in the lateral, vertical, and longitudinal directions, respectively. The equations are expanded with Einstein's summation rule for repeated indices. Here, x_i are the Cartesian coordinates, U_i the Cartesian velocity components, ϕ may represent any scalar such as the temperature or species concentration, and $-\rho\overline{u_i u_l}$ and $-\rho\overline{u_l \vartheta}$ represent the Reynolds stresses and the turbulent scalar fluxes, respectively. Distributions of these quantities are obtained from the turbulence model. Also, μ is the molecular viscosity and σ the corresponding Prandtl or Schmidt number. S_ϕ is the source term for the temperature or concentration equation.

Turbulence Models. The task of the turbulence model is to provide distributions for the Reynolds stresses and the scalar fluxes so that the mean flow Eqs. (13.142) to (13.144) can be closed. In the Boussinesq eddy viscosity concept, the Reynolds stresses are calculated from

$$-\rho \overline{u_i u_l} = \mu_t \left(\frac{\partial U_i}{\partial x_l} + \frac{\partial U_l}{\partial x_i} \right) - 2/3 \rho k \delta_{il} \tag{13.145}$$

The corresponding eddy diffusivity concept gives

$$-\rho \overline{u_l \vartheta} = \frac{\mu_t}{\sigma_\phi} \frac{\partial \phi}{\partial x_l} \tag{13.146}$$

where μ_t is the turbulent eddy viscosity, σ_ϕ is the turbulent Prandtl or Schmidt number, k is the turbulent kinetic energy (per unit mass), and ϕ_{il} is the Kronecker delta which is equal to unity when $i = l$ and zero otherwise.

The most common method for calculating the distribution of μ_t is through the k–ϵ turbulence model [Launder and Spalding (1974)]. This gives

$$\mu_t = c_\mu \rho \frac{k^2}{\epsilon} \tag{13.147}$$

The distributions of k and ϵ are then obtained from solution of transport equations which can be written in Cartesian tensor form as

$$\frac{\partial}{\partial x_l} (\rho U_l k) = \frac{\partial}{\partial x_l} \left(\frac{\mu_t}{\sigma_k} \frac{\partial k}{\partial x_l} \right) + \rho P_k - \rho \epsilon \tag{13.148}$$

$$\frac{\partial}{\partial x_l} (\rho U_l \epsilon) = \frac{\partial}{\partial x_l} \left(\frac{\mu_t}{\sigma_\epsilon} \frac{\partial \epsilon}{\partial x_l} \right) + c_{\epsilon_1} \rho P_k \frac{\epsilon}{k} - c_{\epsilon_2} \rho \frac{\epsilon^2}{k} \tag{13.149}$$

where ϵ is the rate of dissipation of k and P_k is the rate of production of k through the interaction of the Reynolds stresses with the mean flow. It is given by

$$P_k = -\overline{u_m u_l} \frac{\partial U_l}{\partial x_m} \tag{13.150}$$

The empirical coefficients which appear in Eqs. (13.145)–(13.149) are given the standard values $c_\mu = 0.09$, $c_{\epsilon_1} = 1.44$, $c_{\epsilon_2} = 1.92$, $\sigma_\phi = 0.9$, $\sigma_k = 1.0$, and $\sigma_\epsilon = 1.3$.

Simpler eddy viscosity relations have been utilized with reasonable success in some studies. Chien and Schetz (1975) prescribed a constant value for μ_t, proportional to the jet velocity excess and the jet diameter. Then in a subsequent study, Oh and Schetz (1990), calculated μ_t from a relation which takes into consideration the complex shape of the jet and the relative magnitude of the axial turbulence intensity to the velocity excess. Thus

$$\mu_t = 0.037 f \rho b_{1/2} \Delta U_c \tag{13.151}$$

where $b_{1/2}$ is the characteristic half width of the jet, ΔU_c the centerline velocity excess, and f $(=u'^2/\Delta U_c^2)$ takes a value of 2 in the potential core and $\{1 + \exp[-1.134(\xi - \xi_0)]\}$ in the main jet region. For computations at high R, Sykes *et al.* (1986) proposed a one-equation model which solves the k equation such as Eq. (13.148), but calculates the Reynolds stresses from

$$-\rho\overline{u_i u_l} = \rho k^{1/2} \Lambda \left(\frac{\partial U_i}{\partial x_l} + \frac{\partial U_l}{\partial x_i} \right) \tag{13.152}$$

where Λ is a length scale given by

$$\Lambda = 0.088D + 0.0088r \tag{13.153}$$

D is the jet diameter and r is the distance from the center. In spite of the rather crude length scale assumption, computed results, of the mean flow (for $R = 2$) agreed reasonably with experimental data. However, turbulent kinetic energy levels were grossly overpredicted, especially in the wake region.

Demuren (1992), (1994) and Alvarez and Jones (1993) have used various Reynolds stress models (RSM) to investigate the effect of the turbulence model on computations of jets in cross flow. Model computations in which the eddy viscosity concept is not invoked, but partial differential equations are solved for the Reynolds stresses, are compared to those using the k–ϵ model and to experimental data. The Reynolds stress equations can be written in Cartesian tensor notation as

$$\frac{\partial}{\partial x_l} (U_l \overline{u_i u_j}) = D_{ij} + P_{ij} + \pi_{ij} - \epsilon_{ij} \tag{13.154}$$

where D_{ij} is the turbulent diffusion, P_{ij} is the production, π_{ij} is the pressure-strain correlation, and ϵ_{ij} the dissipation rate. The production term $P_{ij} = -\overline{u_i u_l}\,(\partial U_j/\partial x_l)$ $- \overline{u_j u_l}\,(\partial U_i/\partial x_l)$. The dissipation rate is assumed to be locally isotropic, so that ϵ_{ij} $= 2/3\delta_{ij}\epsilon$. D_{ij} and π_{ij} contain higher-order correlations, so they must be approximated for closure at this level. In Demuren (1992), these terms are modeled after proposals of Daly and Harlow (1970) (denoted by DH) and Launder, Reece, and Rodi (1975) (denoted by LRR), respectively. In Demuren (1994) and Alvarez and Jones (1993), additional models for D_{ij} and π_{ij} are considered, including those proposed by Mellor and Herring (1973) (denoted by MH) and Speziale, Sarkar, and Gatski (1991) (denoted by SSG), respectively. The latter combination of models was found to give the best overall predictions of developed turbulent plane channel flow [Demuren and Sarkar (1993)]. The DH and MH diffusion models can be written, respectively, as

$$D_{ij} = Cs_1 \frac{\partial}{\partial x_k} \left(\frac{k}{\epsilon} \overline{u_k u_l} \frac{\partial \overline{u_i u_j}}{\partial x_l} \right) \tag{13.155}$$

$$D_{ij} = Cs_2 \frac{\partial}{\partial x_k} \left[\frac{k^2}{\epsilon} \left(\frac{\partial \overline{u_i u_j}}{\partial x_k} + \frac{\partial \overline{u_i u_k}}{\partial x_j} + \frac{\partial \overline{u_j u_k}}{\partial x_i} \right) \right] \tag{13.156}$$

with $C_{s1} = 0.22$ and $C_{s2} = 0.072$. The pressure-strain models can be written in the general form

$$\pi_{ij} = \alpha_0 \epsilon b_{ij} + \alpha_1 \epsilon \, (b_{ik} b_{kj} - \tfrac{1}{3} \Pi \delta_{ij}) + \alpha_2 k S_{ij}$$

$$+ \, \alpha_3 P_k b_{ij} + k \, \{ \alpha_4 \, (b_{ik} S_{jk} + b_{jk} S_{ik} - \tfrac{2}{3} \delta_{ij} b_{kl} S_{kl})$$

$$+ \, \alpha_5 (b_{ik} W_{jk} + b_{jk} W_{ik}) \} \tag{13.157}$$

where $b_{ij} \equiv [(\overline{u_i u_j}/2k) - \tfrac{1}{3} \delta_{ij}]$ is the Reynolds stress anisotropy tensor, $S_{ij} \equiv \tfrac{1}{2} [(\partial U_i / \partial x_j) + (\partial U_j / \partial x_i)$ is the rate of strain tensor, $W_{ij} \equiv \tfrac{1}{2} [\partial U_i / \partial x_j) - (\partial U_j / \partial x_i)]$ is the rotational tensor, and $\Pi = b_{lk} b_{kl}$ is the second invariant of b_{ij}. For the LRR model, $\alpha_0 = -3 + f_w$, $\alpha_1 = \alpha_3 = 0$, $\alpha_2 = 0.8$, $\alpha_4 = 1.745$, and $\alpha_5 = 1.309 - 0.24 f_w$. For the SSG model, $\alpha_0 = -3.4$, $\alpha_1 = 4.2$, $\alpha_2 = 0.8 - 1.3 \Pi^{1/2}$, $\alpha_3 = -1.8$, $\alpha_4 = 1.25$, and $\alpha_5 = 0.40$. Also, f_w is a wall proximity function which has a value of unity near a wall and zero in a turbulent flow free from walls. The correct rate of decay away from walls is a subject of controversy [Demuren and Rodi (1984)]. It is also difficult to specify in complex flows with curved walls or multiple walls. The absence of such a term makes the SSG pressure-strain model rather attractive for application to complex flows.

A turbulence modeling approach which is intermediate between the k–ϵ model and the full Reynolds stress model was utilized by Baker *et al.* (1987) to calculate the near field of jets in cross flow at high R. This is the so-called algebraic Reynolds stress model, which is derived by dropping the convection and diffusion terms in Eq. (13.154). Thus, implicit algebraic equations are obtained which can be solved simultaneously for the Reynolds stresses. Baker *et al.* (1987) used further simplifications of these equations to obtain explicit expressions for the Reynolds stresses. These expressions contain k and ϵ as unknowns, so Eqs. (13.148) and (13.149) must still be solved before the Reynolds stresses can be calculated. This approach falls within the general class of nonlinear k–ϵ models reviewed by Speziale (1991).

The boundary conditions applied to the equations depend on the particular problem. The various types of boundaries which may exist in these flow situations are: inflow, outflow, wall, symmetry planes, and the free stream. At inflow boundaries, the values of the dependent variables are prescribed or deduced from experimental data. At outflow planes, boundary conditions such as zero traction force or zero normal gradient are usually prescribed. It is normal to prescribe no-slip conditions along walls, but in order to bridge the flow between the fully turbulent region and the viscous sublayer near the wall, the wall-function method [Launder and Spalding (1974)], which is based on the assumption of local equilibrium, is usually employed to prescribe variable values along the first set of grid nodes nearest to the wall. Kim and Benson (1992) did not use this approach. Rather a two-layer model was used to integrate the equations all the way down to the wall. Sykes *et al.* (1986) and Oh and Schetz (1990) avoided this problem entirely by using slip conditions at the wall. Along symmetry planes, zero normal velocity and zero normal gradients for other variables are usually prescribed. Additionally, Reynolds shear stresses with a component in the normal direction will also be zero. Along the free stream, known variable values or zero surface stresses are prescribed. A major uncertainty exists as to the proper boundary condition at the jet exit plane. Experimental data by Andreopoulus (1983) had indicated that exit conditions are highly modified by the cross flow, especially at low R. Demuren (1983) found that by specifying constant local pressure at the jet exit, axial velocity profiles, similar to those observed experimentally could be simulated. However, in-plane velocity profiles had to be prescribed empirically. Kim and Benson (1992) overcame the uncertainty by placing the inflow boundary one diameter into the pipe from which the jet flow originates, and fully developed pipe flow conditions were prescribed at the recessed boundary. A different treatment would be required if the jet exited through an orifice rather than a pipe.

Equations (13.142)–(13.157) form closed sets which can be solved by a finite-difference, finite volume or finite-element methods to yield the mean flow and turbulence fields. By far the most popular approach is a combination of finite difference and finite volume methods. These are different manifestations and extensions of numerical techniques originally proposed by Chorin (1968) and Patankar and Spalding (1972). Notable exceptions are the finite-element computations of Baker *et al.* (1987) and Oh and Schetz (1990).

Current numerical models can, in principle, be used to predict most flows of jets in cross flow which occur in practice. Both laminar and turbulent flows can be computed, so long as the flow is not dominated by rapid distortion or coherent structures. Instabilities develop in jet flows for Reynolds numbers greater than $O(10)$, so that in most practical situations the flow will be turbulent or transitional. The former can be handled by current models with proper choice of the turbulence model. In the latter, large scale coherent structures may play a dominant role [Andreopoulos (1985)], and methods based on Reynolds averaging may be inadequate. Direct calculation of the unsteady flow will generally be required, but this cannot be done for any pracical Reynolds number, so a large eddy simulation (LES) appears to be a viable option. Such approaches are reviewed in Galperin and Orszag (1993), but much research work is still required before they can become reliable predictive tools in applications of interest here. In addition, current turbulence models cannot predict higher-order statistics such as the Reynolds stresses [Speziale (1994)] in flows with rapid distortion (ratio of turbulent-to-mean-flow time scales greater than 50) or in flows with high compressibility (free shear flows with turbulent Mach number of order 1). Of course, the question of the importance of such higher-order statistics in the present flows has been raised. Some studies indicate that they are less important at high R, where pressure effects dominate, than at low to medium R.

Earlier model computations of the fully elliptic type such as by Patankar *et al.* (1977) used grids ($15 \times 15 \times 10$ in the x, y, z directions) which are too coarse for the results to be considered reliable. Although correct trajectories were predicted, velocity fields deviated qualitatively and quantitatively from experimental observations. Finer grid computations have been reported, but grid independence could not be demonstrated conclusively in any of these. Multigrid methods [Demuren (1992)] allow systematic studies of grid dependence, since, on each finer grid level, there are twice as many points in each direction than on the coarser one (i.e., eight times as many total grid nodes in three dimensions). Although up to 2.4 million nodes (i.e., $256 \times 96 \times 96$ on the fifth grid level) were used, Claus and Vanka (1992) could still not demonstrate grid independence of the computed velocity and turbulence fields. Figure 13.41 shows comparisons of vertical profiles of the streamwise velocity component computed on the three finest grids. It is obvious that grid-convergence has not been achieved, especially in the near-field. Similar comparisons are shown for the turbulence level in Fig. 13.42. Far-field results appear closer to grid-convergence. With this type of grid refinement, better estimates of the true results can be obtained by using Richardson extrapolation techniques [Demuren and Wilson (1994)]. The question of grid resolution cannot be completely separated from

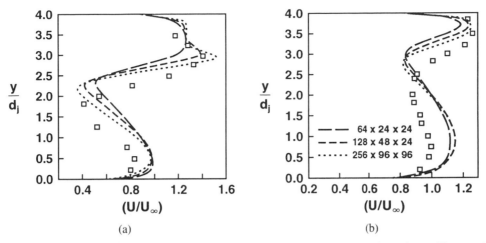

FIGURE 13.41 Grid dependency test: comparison between computations from Claus and Vanka (1992) and experimental data from Khan *et al.* (1982) of streamwise velocity in the center plane; (a) $x/d_j = 4$ and (b) $x/d_j = 8$.

that of the formal order of accuracy of the numerical scheme. Most studies of the Patankar and Spalding (1972) type have used the hybrid upwind/central difference scheme to approximate convection terms. This is monotonic and conservative, but it is known to be highly diffusive [Demuren (1985b)]. Studies in which higher-order differences, such as the quadratic-upstream-weighted (QUICK) scheme were utilized, e.g., Barata *et al.* (1991), showed that similar results, as with lower-order schemes could be obtained on coarser grids. However, higher-order schemes tend to suffer from lack of boundedness in regions with high gradients.

The uncertainty in the specification of boundary conditions for the jet hole exit

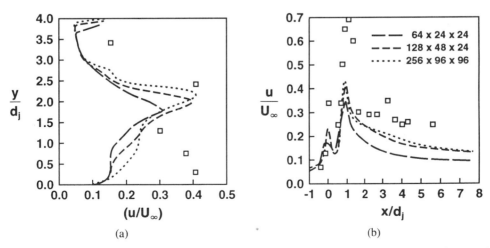

FIGURE 13.42 Grid dependency test: comparison between computations from Claus and Vanka (1992) and experimental data from Crabb *et al.* (1981) of turbulence intensity in the center plane; (a) $x/d_j = 2$ and (b) $y/d_j = 1.35$.

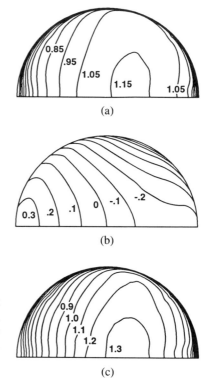

FIGURE 13.43 Contour plots of the flow field at the jet exit plane (from Kim and Benson, 1992); (a) axial velocity ($U/\overline{U}$), (b) static pressure ($p/\frac{1}{2}\rho\overline{U}^2$), and (c) total pressure [$(p + \frac{1}{2}\rho U^2)/\frac{1}{2}\rho\overline{U}^2$].

has been discussed above. Figure 13.43 shows contours of the jet velocity, static pressure and total pressure at the exit plane computed by Kim and Benson (1992). None of these is uniform, which is a compelling reason for including the jet pipe hole in the calculation domain. Other uncertainties involve the specification of inflow and near-wall conditions for turbulent quantities.

In many computational studies, the inadequacy of the turbulence model has been blamed for the lack of agreement between computed results and experimental data. Demuren (1992) tried to isolate the effect of the turbulence model by performing computations on the same grid with the k–ϵ model and the Reynolds stress (LRR–DH) model. The results are compared to experimental data of Atkinson et al. (1982) for opposed jets in cross flow in Figs. 13.44 and 13.45 for the mean flow and Reynolds stress fields, respectively. For the mean flow, there is little to choose between both model predictions, but the Reynolds stress model clearly gives better predictions of Reynolds stress profiles. From these results, it may be concluded that the mean flow was not strongly influenced by the turbulence field. Thus, if the interest is solely in the mean flow field, the k–ϵ turbulence model, or an even simpler model, would be adequate. But, if the turbulence field is required, e.g., to predict mixing, then a Reynolds stress model would give much better results, but at additional computational cost. The multigrid technique enables the additional cost to be minimized by ensuring a nearly grid-independent convergence rate. Thus, for a 3-level multigrid scheme, with (42 × 34 × 82) points on the finest grid, convergence of the Reynolds stress model computations could be obtained in less than 100 fine grid iterations.

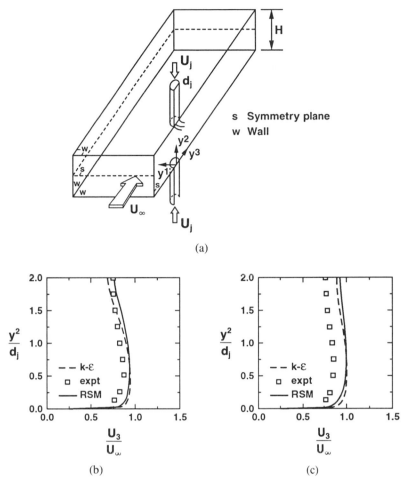

FIGURE 13.44 Effect of turbulence model: comparison between computations from De-muren (1992) and experimental data form Atkinson *et al.* (1982) of streamwise velocity in the center plane; (a) flow configuration—opposed jets in cross flow, (b) $x/d_j = 8$, and (c) $x/d_j = 12$.

Concluding Remarks. In the study of turbulent jets in cross flow, empirical models offer quick and simple methods for obtaining first-order estimates and a qualitative picture of the jet trajectory, its extent, and decay rates of its axial velocity and temperature. The main requirement for reasonable predictions is the use of corre-lation equations or curves derived from experimental data bases with similar char-acteristics as the problem of interest.

Integral models contain in simplified forms mathematical representations of the basic conservation laws, thus they can be applied much more widely than empirical models. Several physical phenomena which occur in the flow are modeled with re-lations which are more or less empirical. Combinations of these have been used successfully in integral models, so long as they are properly calibrated. One criticism of integral models is that they provide little insight into flow physics, since the same effect could be achieved in different ways. All integral models are computationally

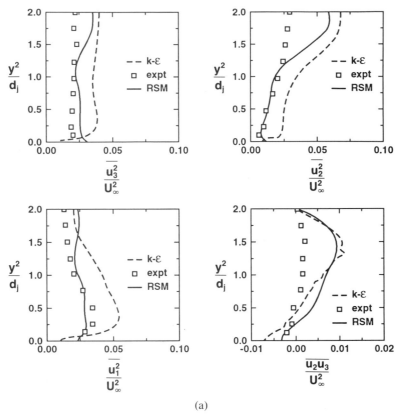

FIGURE 13.45 Effect of turbulence model: comparison between computations from Demuren (1992) and experimental data from Atkinson *et al.* (1982) of Reynolds stresses in the center plane; (a) $x/d_j = 8$ and (b) $x/d_j = 12$.

cheap to use. The basic models are conceptually simple, but more sophisticated models have been devised which enable more complex jet cross flow interactions to be analyzed.

Perturbation models do not require much empirical input, but they are mostly restricted to the near-field or far-field where small parameters required for expansions can be defined. They enable order-of-magnitude studies of the effects of various parameters, thus they are useful tools for the investigation of flow physics. Beyond that, they have limited practical utility.

Numerical models offer the best choice as practical predictive tools over a wide range of jets in cross flow applications. They require the least assumptions and empirical input. They are, however, the most computationally intensive. Quite complex jet–jet, jet-cross flow interactions can be analyzed. Depending on specific requirements, the choice of turbulence model may or may not be important. Although, several rather complex flow situations have been computed with a measure of quantitative accuracy, some questions remain as to the effects of grid resolution, turbulence model, and boundary conditions on the overall accuracy of the computed results. Reliability and computational accuracy are expected to improve with further developments in numerical techniques and turbulence models. These are clearly the

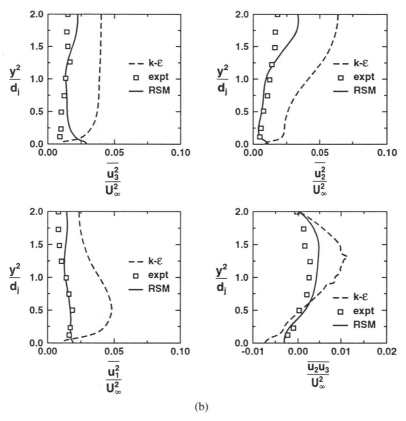

(b)

FIGURE 13.45 *(Continued)*

approaches of choice for the computation of practical jets in cross flow situations. Computer codes are available commercially for this purpose.

13.4.3 Two-Phase Jets in Cross Flow
Takakage Arai and Takao Inamura

Particle-Laden Gas Jets. The situation of particle-laden fluid jets injected transverse to a fluid stream is of interest because of the possible inertial effects of the particles. One might anticipate an appreciable *separation* of the fluid and particle streams, at least for large particles. The range of the variables that has been studied is very limited. In Salzman and Schwartz (1978), gas jets loaded with small particles (15-μ silicate) were injected at 90° to low-speed airstreams. The primary result is the trajectory of the center of mass of solid phase under various conditions. The results are correlated reasonably well by the equation

$$\frac{y}{d_j} = 1.92 \left(\left(\frac{\rho_j U_j^2}{\rho_\infty U_\infty^2} \right)^{1/2} \frac{x}{d_j} \right)^{1/3} \tag{13.158}$$

Liquid Jets into Supersonic Crossflow. For a liquid jet transverse to the gas stream, it happens that much of the work has been with supersonic gas streams. The interaction of the liquid column with the supersonic stream is highly unsteady, and only high-speed ($\sim 20{,}000$ pictures/s) motion pictures are adequate to display the features of the flow. Some stop-action (10^{-6} s) stills are shown in Fig. 13.46 from Kush and Schetz (1973) to show how the flowfield develops for various ranges of the parameter $\bar{q} = \rho_j U_j^2 / \rho_\infty U_\infty^2$. The most obvious phenomenon is the presence of waves on the jet column. The jet column fractures in the trough of these waves leading to large clumps of liquid. These clumps then break down into drops. This is all illustrated in Fig. 13.47. There are also extra shock waves produced by the waves on the jet, and the bow shock is unsteady. The wave processes on the jet are clearly very important for the processes of breakup and atomization.

Different workers have developed correlation formulas for the gross penetration of the liquid plume into the cross flow including the effects of various parameters. An overall correlation was developed in Baranovsky and Schetz (1980).

FIGURE 13.46 Spark photographs of transverse liquid jet injection into $M_\infty = 2.4$ air. (a) High $\bar{q}$, (b) moderate $\bar{q}$, and (c) low $\bar{q}$. (Reprinted with permission, Kush, E. A., Jr. and Schetz, J. A., "Liquid Jet Injection into a Supersonic Flow," *AIAA J.*, Vol. 11, No. 9, pp. 1223–1224, Copyright © 1973.)

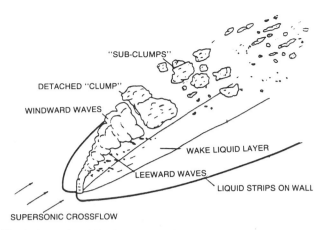

"SUB-CLUMPS"

DETACHED "CLUMP"

WINDWARD WAVES

WAKE LIQUID LAYER

LEEWARD WAVES

LIQUID STRIPS ON WALL

SUPERSONIC CROSSFLOW

FIGURE 13.47 Schematic of liquid jet in a supersonic cross flow. (Reprinted with permission, Kush, E. A., Jr. and Schetz, J. A., "Liquid Jet Injection into a Supersonic Flow," *AIAA J.*, Vol. 11, No. 9, pp. 1223–1224, Copyright © 1973.)

$$\frac{h}{d_j} = 1.32 \sqrt{\bar{q}} \left(\frac{d_{eq}}{d_f}\right)^2 \left(\frac{d_f}{d_s}\right)^{0.46} \ln\left(1 + 6\frac{x}{d_j}\right) \sin\left(\frac{2\theta}{3}\right) \qquad (13.159)$$

where h is the top of the liquid plume, d_f and d_s are the front and side dimensions of noncircular injectors, d_{eq} is the equivalent diameter, and θ is the injection angle.

The width or lateral spreading of the plume is also of interest in practical applications. The available correlation was suggested in Joshi and Schetz (1975)

$$\frac{w}{d_{eq}} = 11.2 \left(\bar{q}C_d + \frac{4p_{eb}}{5\gamma M_\infty p_\infty}\right)^{0.19} \qquad (13.160)$$

where p_{eb} is the effective back pressure behind the interaction shock. It is approximately 80% of the pressure downstream of a normal shock with upstream Mach number M_∞, γ is the specific heat ratio and C_d is the discharge coefficient of the injector.

The behavior of a linear array of jets is also useful for some applications, and results for such a configuration are presented in Arai and Schetz (1994). Figure 13.48 shows the progressively increasing penetration for each jet in an array of twelve jets spaced 13 diameters apart aligned with the main flow at $M_\infty = 2.4$ and $\bar{q} = 13$. Each jet shelters the one behind it, permitting greater penetration. The first jet has the penetration one would expect for a single jet, and the last has about a fivefold increase.

As mentioned above, the penetration of the jet plume depends strongly on the parameter $\bar{q}$. The idea of increasing $\bar{q}$ by adding a small amount of gas into the liquid jet (*bubbling jet*) was shown in Avrashkov and Baranovsky (1990) and Arai and Schetz (1992). If the liquid mass flow rate is constant, the ratio of momentum $\bar{q}_j$ to $\bar{q}_{j(\kappa=0)}$ is given by

$$\frac{\bar{q}_j}{\bar{q}_{j(\kappa=0)}} = (1 + \kappa)\left(1 + \frac{\kappa\rho_w}{\rho_g}\right) \qquad (13.161)$$

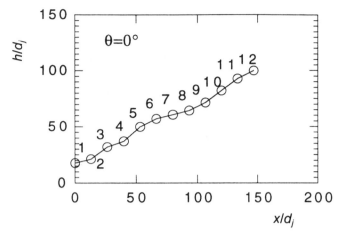

FIGURE 13.48   Penetration for liquid jets in a linear array. (Reprinted with permission, Arai, T. and Schetz, J. A., "Injection of Bubbling Liquid Jets from Multiple Injectors into a Supersonic Air Stream," *J. Propulsion and Power*, Vol. 10, No. 3, pp. 382–386, Copyright © 1994.)

where κ is the ratio of gas mass flow rate to liquid mass flow rate, ρ_g and ρ_w are the density of liquid and the gas, respectively. Figure 13.49 shows the effect of gas concentration κ on the penetration of the jet plume for the case of a constant liquid mass flow rate. The penetration was measured directly from the photographs at the position of $x/d_j = 30$. The penetration increases as gas concentration increases, and the penetration ratio gets over 2.0 under the condition of $\kappa > 0.01$. The similarity law, $h/d_j \propto \sqrt{q}$, was approximately verified also for the bubbling jet. On the linear array of bubbling jets, it was shown in Arai and Schetz (1994) that the width of the jet plume increased by using the bubbling jet.

Some measurements of the variation of d_{32} (the volume to surface average drop size) along and across the spray plume for water injection into at $M_\infty = 3.0$ and $\bar{q}$

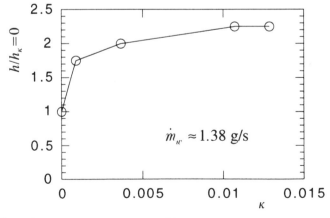

FIGURE 13.49   Effect of κ on penetration of jet plume for constant liquid mass flow rate of $\dot{m}_w \approx 1.38$ g/s. [Reprinted with permission, Arai, T. and Schetz, J. A., "Penetration and Mixing of a Transverse Bubbling Jet into a Supersonic Flow," *Proc. 2nd Nat. (Japan) Symp. RAM/SCRAM JET*, Sendai, Japan, pp. 131–136, Copyright © 1992.]

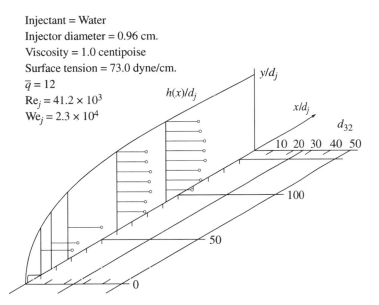

Injectant = Water
Injector diameter = 0.96 cm.
Viscosity = 1.0 centipoise
Surface tension = 73.0 dyne/cm.
$\bar{q} = 12$
$Re_j = 41.2 \times 10^3$
$We_j = 2.3 \times 10^4$

FIGURE 13.50   Drop sizes in $\bar{q} = 12$ water jet. (Reprinted with permission, Nejad, A. S. and Schetz, J. A., "Effects of Properties and Location in the Plume on Mean Droplet Diameter for Injection in a Supersonic Stream," *AIAA J.*, Vol. 21, No. 7, pp. 956–961, Copyright © 1983.)

$= 12$ are presented in Fig. 13.50 from Nejad and Schetz (1983). High values of $\bar{q}$ lead to smaller drops and more uniform atomization. The effects of reducing the surface tension of the liquid result in smaller drops and a more uniform spray. Increasing the viscosity to a very high level changes the break-up process, so that long ligaments are produced as the clumps are broken down. At every location in the plume, the drop size distribution varies significantly with time. Some typical results are shown in Fig. 13.51 from Less and Schetz (1986) for a case of $M_\infty = 3.0$ and $\bar{q} = 12$ from a $d_j = 0.51$ mm port. Each plot has three sequential drop size distributions—the solid line is recorded first followed by the dotted line and then the dashed line.

Liquid Jets into Subsonic Crossflow.   A penetration correlation has been developed in Schetz and Padhye (1977) for $M_\infty = 0.5 \sim 0.8$ as

$$\frac{h}{d_f} = 13.5 \sqrt{\bar{q}} \left(\frac{d_{eq}}{d_f}\right)^2 \left(\frac{d_f}{d_s}\right)^{0.46} \tag{13.162}$$

where the penetration is measured at $x/d_j = 30$. Comparing to the supersonice result in Eq. (13.159) at $x/d_j = 30$, the subsonic cases have greater penetration by a factor of about two. Under $U_j = 100$ m/s, the conditions in Inamura *et al.* (1991), the jet penetration was given by

$$\frac{h}{d_j} = (1.18 + 240d)(\bar{q})^{0.36} \ln \left\{ 1 + (1.56 + 480d) \frac{x}{d_j} \right\} \tag{13.163}$$

The correlation for the local liquid mass flow rate and width have been also developed in Oda *et al.* (1992) for under $U_j = 150$ m/s conditions as

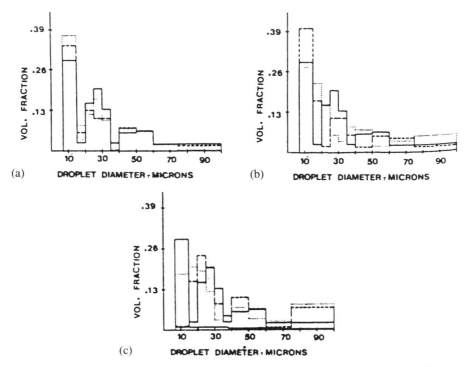

FIGURE 13.51 Instantaneous drop size distributions in a liquid spray. (Reprinted with permission, Less, D. and Schetz, J. A., "Quantitative Study of Time Dependent Character of Spray Plumes," *AIAA J.*, Vol. 24, No. 12, pp. 1974–1986, Copyright © 1986.)

$$\frac{\dot{m}_l(x,\,y,\,z)}{\dot{m}_l(x,\,0,\,z)} = \exp\left(-2.58\left(\frac{y}{w}\right)^2\right) \tag{13.164}$$

and

$$\frac{w}{d_j} = 1.42(\overline{q})^{0.17}\ln\left(1 + 2.8\left(\frac{x}{d}\right)\right) \tag{13.165}$$

It should be noted that all the above correlation equations were obtained under conditions where the incoming boundary layer was very thin.

Slurry Jets. A slurry fuel including metal particles is of interest because of its high density and high volumetric heating value, contributing to compact fuel storage and high specific thrust [see Northam (1985)]. Furthermore, Yatsuyanagi (1989) reported that a slurry fuel including aluminum particles stabilizes the oscillatory combustion in a rocket engine combustor.

Schetz and his colleagues have investigated the disintegration phenomena of liquid and slurry jets with 3–50 wt% loading (5 μm silicon dioxide) traversing a supersonic airstream [Less and Schetz (1983) and Thomas and Schetz (1985)]. They reported that as the loading of the solid particles is increased the fluctuations in the jet decrease and the slurry jet becomes stable [Less and Schetz (1983)]. Figure 13.52 shows nanoflash photographs for various loadings. From the photographs, it was clarified that the penetration of the jet decreases as the loading is increased with a

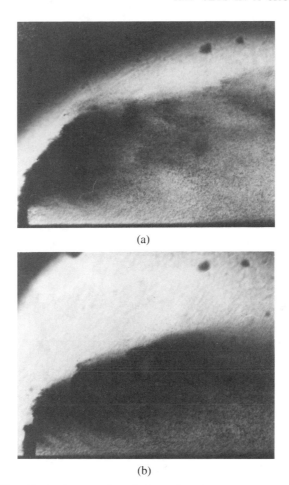

(a)

(b)

FIGURE 13.52 Nanoflash photographs of slurry jets for various loading. (a) loading $=$ 16%, $\bar{q} = 4.2$ and (b) loading $= 52\%$, $\bar{q} = 4.2$. (Reprinted with permission, Less, D. M. and Schetz, J. A., "Penetration and Breakup of Slurry Jets in a Supersonic Stream," *AIAA J.*, Vol. 21, No. 7, pp. 1045–1046, Copyright © 1983.)

constant momentum flux ratio. They attributed this to the combined factors of particle/liquid interactions and the decrease of the jet velocity. They observed that water was sheared away from the particles, and the slurry particles were separated into agglomerated particles and water particles. This phase separation results in a decrease of the jet penetration.

Small waves appear on the windward surface of the jet in the vicinity of the injector, and then they are amplified and accelerated downstream. The wave processes on the jet are clearly very important for the processes of breakup and atomization.

Figure 13.53 shows mass flow per unit area data taken at $x/d_j = 30$ for the 30 wt% loaded slurry jet from Thomas and Schetz (1985). The loading increases smoothly in the $+y/d_j$ direction and then jumps sharply at $y/d_j = 12$. The same trend is evident in the $+z/d_j$ direction. This increased loading is due to the solid particles with greater inertia separating from liquid, and it is consistent with the visual observations.

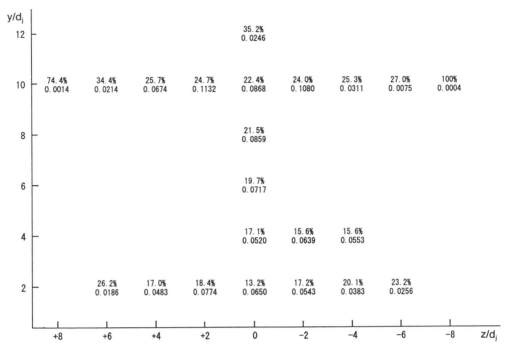

FIGURE 13.53 Spatial distribution of mass flux of droplets. (Reprinted with permission, Thomas, R. H. and Schetz, J. A., "Distributions Across the Plume of Transverse Liquid and Slurry Jets in Supersonic Airflow," *AIAA J.*, Vol. 23, No. 12, pp. 1892–1901, Copyright © 1985.)

Inamura *et al.* (1991a), (1991b), (1993a), and (1993b) have investigated the disintegration of water and slurry jets with 20, 40 wt% loading (2.33 μm aluminum) in subsonic airstreams. By their observations, the penetration of slurry jets is almost the same as that of water jets, and significant phase separation was never observed in subsonic airstream. From comparison with the observations by Schetz and coworkers, the phase separation generally seems to appear only under high shear conditions like in a supersonic airstream. It also depends, of course, on the diameter of solid particles in the slurry.

Figure 13.54 shows the disintegration phenomena of a water jet including a small amount of white color for better visualization and a slurry jet at the same air velocity and momentum flux ratio. The spray particles torn off from the side surface of the jet for the slurry are fewer than those for the water jet. Taking notice of the waves on the liquid surface in the vicinity of the injector, the waves of the slurry jet are smaller than those of the water jet. On the other hand, in the vicinity of the jet breakup point, the waves on the slurry jet are larger than those on the water jet. And, for both jets, a hollow can be seen on the stagnation line on the jet surface.

Empirical equations of jet penetration were deduced as follows from photographic techniques [Inamura *et al.* (1991b)] for water jets,

$$\frac{h}{d_j} = (1.18 + 240d)(\overline{q})^{0.36} \ln\left\{1 + (1.56 + 480d)\frac{x}{d_j}\right\} \qquad (13.166)$$

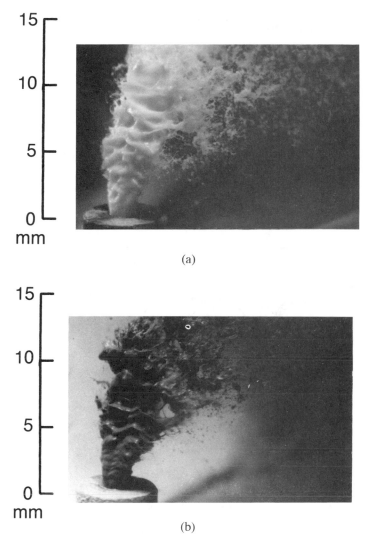

(a)

(b)

FIGURE 13.54 Disintegration phenomena of water jet including a small amount of white colors and slurry jet. (a) $U_\infty = 75$ m/s $U_j = 7.3$ m/s $\bar{q} = 9$, injectant: water $d_j = 2.0$ mm and (b) $U_\infty = 75$ m/s $U_j = 7.8$ m/s $\bar{q} = 9$, injectant: slurry $d_j = 2.0$ mm.

and for slurry jets,

$$\frac{h}{d_j} = (1.17 + 160d_j)(\bar{q})^{0.43} \ln\left\{1 + (0.75 + 950d)\frac{x}{d_j}\right\} \qquad (13.167)$$

Figure 13.55 depicts comparisons of jet penetration by measurements and by the empirical equations. The penetration of the slurry jet is a little bit larger than that of the water jet, however the differences are quite small. This trend agrees qualitatively with the observations.

Jet width was obtained by the following empirical equations from Inamura *et al.* (1991b) for water jets,

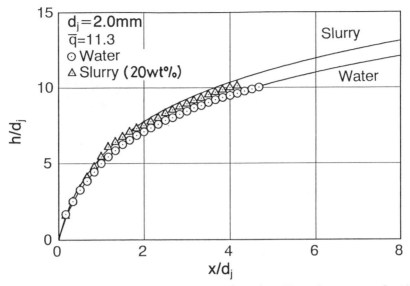

FIGURE 13.55   Penetrations of water jet and slurry jet. (From Inamura *et al.*, 1991a.)

$$\frac{w}{d_j} = 1.4(\overline{q})^{0.18} \left(\frac{x}{d_j}\right)^{0.49} \tag{13.168}$$

and for slurry jets (with 20 wt% loading, $d = 1.0$ mm),

$$\frac{w}{d_j} = 3.1(\overline{q})^{0.06} \left(\frac{x}{d_j}\right)^{0.32} \tag{13.169}$$

Figure 13.56 depicts the contours of the shape of water and slurry jet plumes from Inamura *et al.* (1993b). The values in percent in the figure stand for the fraction of time while the fixed probe placed in the plume does not sense contact with the injectant from the jets. The breakup lengths of the water jet in the airstream direction do not change with the momentum flux ratio, and the contour of the slurry jet is almost the same as that of the water jet with same momentum flux ratio. Some predictions using Eq. (13.166) for the top of the plume all also shown for comparison.

The velocity and frequency of the waves on the windward surface were measured by a newly developed laser instrument. Figure 13.57 presents the influence of the distance along the jet windward surface on the ratio of the wave velocity to air velocity from Inamura *et al.* (1993a). In the figure, s stands for the distance along the jet windward surface from the injector exit, and V_w stands for the wave velocity along the jet surface. There is no remarkable difference between the water jet and the slurry jet (20 wt% loading). This fact means that the effects of solid particles in the jet on the behavior of the jet are small in the range that solid particle loading is less than 20 wt%. The viscosity of the 20 wt% slurry is 22.5×10^{-3} Pa · s at low shear rate. However, at high shear rates of 10^6–10^7 such as in atomization process, the viscosity of the slurry decreases rapidly, and it approaches to that of water [see

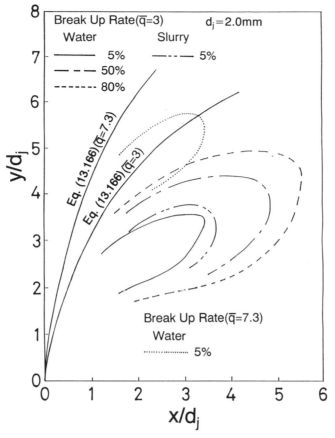

FIGURE 13.56 Contours of water jet and slurry jet. (From Inamura *et al.*, 1991b.)

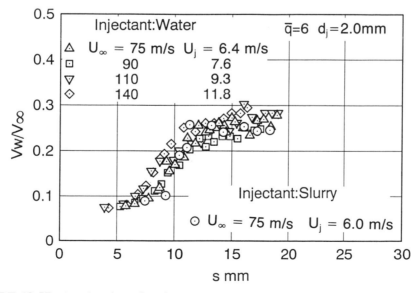

FIGURE 13.57 Acceleration of surface wave on water jet and slurry jet. (From Inamura *et al.*, 1993a.)

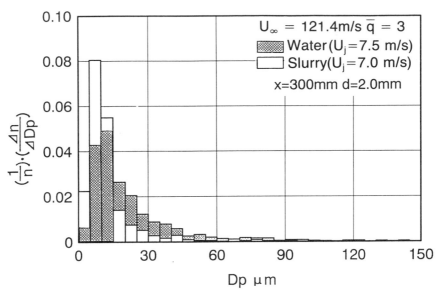

FIGURE 13.58 Spray particle size distributions. (From Inamura *et al.*, 1993b.)

Sakai *et al.* (1986)]. Consequently, at high shear rates a slurry jet behaves like a water jet.

Figure 13.58 shows the spray particle size distribution for the water and slurry (20 wt% loading) jets. In the figure, D_p is the particle diameter, and n is the number of particles. Sauter mean diameter for water is 75.7 μm, and for the slurry is 67.5 μm. There is not a large difference in the Sauter mean diameter. However, the particle size distribution of the slurry is quite different than that of water as shown in Fig. 13.58. The slurry spray includes much finer particles than the water spray. The slurry includes fine aluminum particles, and such fine particles play the role of nuclei for the atomization process. Also, the slurry shows a low viscosity similar to that of water at high shear rate like during the atomization process. This causes a much finer spray formation for slurry atomization. This phenomena can also be observed in the atomization of coal water mixtures [see Inamura *et al.* (1987)].

13.5 FLOW IN POROUS MEDIA
Robert A. Greenkorn

13.5.1 Introduction

A porous medium is a solid with holes in it. *Pores* which occupy a definite fraction of the bulk volume form a complex network of *voids*. In describing flow in porous media, interest is in interconnected pores, since these are the ones that affect flow. The *matrix* of a porous medium is the material in which the holes or pores are imbedded. The manner in which the pores are imbedded, how they are interconnected, and the description of their location, size, shape, and interconnection characterizes the porous medium.

The fraction of a porous medium that represents the connected pores is called the

TABLE 13.3 Typical Values of Porosity and Permeability [from Scheidegger (1974)]

Porous Solid	Porosity (Fraction)	Permeability (Darcy)
Sand	0.31–0.50	2–180
Sandstone	0.08–0.40	0.00000011–11
Limestone	0.015–0.20	0.000001–2
Brick	0.12–0.34	0.0048–0.022
Soil	0.43–0.54	0.29–14
Berl saddles	0.68–0.83	130,000–390,000
Wire crimps	0.68–0.76	3800–10,000
Silica powder	0.013–0.051	0.37–0.49

porosity. Porosity is an average value typical of the pore size distribution. It is measured by several techniques [see Bear (1972), Collins (1961), Corey (1977), Delhomme (1979), DeWiest (1965), Dullien (1992), Nunge (1970), Greenkorn (1972), (1981), (1983), Matherson and DeMarsily (1980), Muskat (1937), Poluborinova–Kochina (1962), and Scheidigger (1974)]. Table 13.3 shows some typical values of porosity for a variety of porous solids. An extremely large array of materials can be classified as porous media. Broadly speaking, porous media are classified as unconsolidated or consolidated and as ordered or random.

Porous media are characterized or described in terms of the properties of the matrix that affect flow. The flow properties which describe the matrix from the point of view of the contained fluid are pseudotransport properties: *permeability*, the *conductance* of the medium, *capillary pressure* (or relative permeability) caused by the interfacial forces due to surfaces, and *dispersion*, the mixing caused by the tortuous paths in the medium and other factors. These properties depend on the structure of the pores. They are bulk properties and have meaning only when applied to a medium having some minimum number of pores–a piece large enough to be volume-averaged.

Darcy related the volumetric flow rate Q of a fluid flowing linearly through a porous medium of cross sectional area, A, directly to the energy loss, $h_1 - h_2$, inversely to the length, L, of the medium, and proportional to a factor called the hydraulic conductivity, K. *Darcy's law* is expressed as

$$Q = \frac{KA\Delta h}{L} \tag{13.170}$$

where

$$\Delta h = \Delta z + \frac{\Delta p}{\rho} \tag{13.170a}$$

Darcy's law is empirical in that it is not derived from first principles; it is rather the result of experimental observation. Although Darcy's law is empirical, DeWiest (1965) has heuristically demonstrated that it is the empirical equivalent of the Navier–Stokes equations. Darcy's law is usually considered valid for creeping flow where the Reynolds number

$$\mathrm{Re}_p = \frac{D_p v_\infty \rho}{\mu(1 - \phi)} \tag{13.171}$$

is less than one, where D_p is equivalent particle diameter, v_∞ is the velocity of approach, ρ is fluid density, μ is fluid viscosity, and ϕ is porosity.

The hydraulic conductivity, K, defined by Darcy's law, Eq. (13.170), is dependent on the properties of the fluid as well as the pore structure of the medium. The hydraulic conductivity is temperature-dependent, since the properties of the fluid, density, and viscosity are temperature-dependent. Hydraulic conductivity can be written more specifically in terms of the *intrinsic permeability* and the properties of the fluids

$$K = \frac{k\rho g}{\mu} \tag{13.172}$$

where g is acceleration of gravity, k is the intrinsic permeability of a porous medium and is a function only of the pore structure. The intrinsic permeability is not temperature-dependent. Darcy's law is often written in differenial form, so that in one-dimension

$$\frac{Q}{A} = q = -\frac{k}{\mu}\frac{dp}{dx} \tag{13.173}$$

The minus sign results from the definition of Δp which is equal to $p_2 - p_1$, a negative quantity.

Permeability is usually measured by forcing a known volumetric flow through a porous medium of known length and area perpendicular flow. The pressure drop is measured and the permeability calculated from the integration of Eq. (13.173).

$$k = \frac{Q\mu\Delta x}{A\Delta p} \text{ darcies } \left(\frac{\mathrm{cm}^2 - cp}{\mathrm{atm} - \sec}\right) \tag{13.174}$$

Table 13.3 shows typical values of permeability.

It is usually assumed that Darcy's law is valid in three dimensions, that the permeability $\bar{\bar{k}}$ is a second-order tensor dependent on the directional properties of the pore structure, that $\bar{q}$ is a vector,

$$\bar{q} = -\frac{1}{\mu} (\bar{\bar{k}} \cdot \nabla p) \tag{13.175}$$

The main interest in describing porous media is to understand and to predict the passage of materials into and out of the media. The problem of the description of a porous medium is one of describing the geometrical or structural properties in some average fashion and relating these average structural properties to the flow properties. There are two levels of description. At the microscopic level, the description is statistical—in terms of *pore-size* distributions. At the macroscopic level, the description is in terms of the average or bulk properties and their variation at sizes or scales much larger than pores.

13.5.2 Single Fluid Flow

The parameters that relate a porous medium and single fluid flow in this case are porosity and permeability. These parameters are usually defined with reference to single fluid flow. The equation of continuity is

$$\phi \frac{\partial \rho}{\partial t} + (\nabla \cdot \rho \overline{q}) = 0 \tag{13.176}$$

where ϕ is the porosity of the medium, ρ is the density of the fluid in the pores. The equation of motion—Darcy's law—in three dimensions is given by Eq. (13.175).

For steady state linear flow, the equation describing the flow is the integral of Eq. (13.173)

$$q = \frac{kA}{\mu} \frac{p_1 - p_2}{L} \tag{13.177}$$

Example 13.1: What is the volumetric flow rate through a sandstone core that is 10-cm long and 2.5-cm in diameter and has a permeability of 2 darcies under a pressure drop of 1 atm? Using Eq. (13.177) and since $q = Q/A$

$$Q = \frac{kA}{\mu L} \Delta p = \frac{2 \left(\dfrac{cm^2 \; cp}{sec \; atm} \right) \pi (2.5)^2 \; (cm)^2}{1(cp)10(cm)} \; 1 \; cm = 1 \; (cm^3/sec) \tag{13.178}$$

13.5.3 Multifluid Immiscible Flow

The equations for multifluid immiscible flow are similar to those for single-fluid flow. However, the equations are written for each fluid and connected using the capillary pressure and saturation of the various fluids. The additional parameters, capillary pressure or equivalently relative permeability, are introduced.

Muskat (1937) assumed Darcy's law is valid for each flowing fluid. For fluids 1, 2.

$$q_1 = -\frac{kk_{r1}}{\mu} \frac{dp}{dx} \tag{13.179}$$

and

$$q_2 = -\frac{kk_{r2}}{\mu_2} \frac{dp}{dx} \tag{13.180}$$

The product kk_{r1} is the effective permeability of the medium to fluid 1 with fluid 2 present, k_1. The product kk_{r2} is the effective permeability of the medium to fluid 2 with fluid 1 present, k_2. Effective permeabilities depend on both pore structure and the fluids present in the porous medium. The values k_{r1} and k_{r2} are called the relative permeabilities. The relative permeabilities depend on saturation.

The measurement and calculation of capillary pressure and/or relative permea-

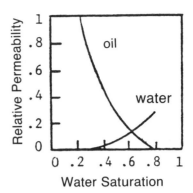

FIGURE 13.59 Relative permeability for oil and water flowing in sandstone as a function of water saturation.

bility as a function of saturation yield relative permeabilities for oil and water in sandstone as in Fig. 13.59, for example.

For flow of two immiscible fluids, the interfacial tension between the two causes a pressure discontinuity at the boundary separating the two fluids. This pressure is the capillary pressure and is related to saturation so that

$$p_c(S_1) = p_2 - p_1 \tag{13.181}$$

Calculations associated with flow for the case of immiscible flow can be found in the books cited in Sec. 13.5.1.

13.5.4 Multifluid Miscible Flow

The equations for multifluid miscible flow are the same as with single-fluid flow. However, if the two fluids begin flowing separately, a mixed region evolves between them. Much of the modeling for miscible flow is concerned with modeling the mixed region and the parameter associated with the mixing which is the *dispersion coefficient*. Dispersion results from the complex nature of the velocity in the flow paths in the media, although diffusion and turbulence also affect mixing.

To describe the mixing or dispersion of multifluid miscible flow, one must assume two fluids of equal viscosity and equal density, where one of the fluids is displacing the other fluid from a porous medium. If it is assumed that the flow is in one dimension, Darcy's law for the flow of each fluid and the sum of the flows is given by Eq. (13.173). The equation of continuity for the total flow is given by Eq. (13.176).

There is general agreement that the appropriate equation in one-dimension for the mixed region for a homogeneous medium is

$$\frac{\partial c}{\partial t} + v_x \frac{\partial c}{\partial x} = D \frac{\partial^2 c}{\partial x^2} \tag{13.182}$$

where D is the coefficient of dispersion in the direction of flow, the longitudinal dispersion, and c is concentration.

Solutions to Eq. (13.182) depend on the boundary conditions [see for example Greenkorn (1983)], but the most common solution in one-dimension for semi-infinite boundary conditions is

$$c = \frac{1}{2} c_0 \left(\operatorname{erfc} \frac{x - v_x t}{2\sqrt{Dt}} + e^{v_x x/D} \operatorname{erfc} \frac{x + v_x t}{2\sqrt{Dt}} \right) \qquad (13.183)$$

where c_0 is initial concentration.

Calculations for mixing are usually accomplished numerically. Longitudinal dispersion coefficients for sandstones are typically in the range of 10^{-3} to 10^{-5} cm^2/sec.

Currently there exists a dilemma concerning dispersion and the interpretation of laboratory measurements for use on a field scale [Matheron and DeMarsily (1980)]. Heterogeneous field dispersion may be a function of system size and solutions of Eq. (13.182) such as Eq. (13.183) do not correctly predict concentration.

13.6 INSTABILITIES OF SEPARATED/SWIRLING FLOWS AND FLOW-ACOUSTIC COUPLING
Donald O. Rockwell

13.6.1 Introduction

Although the designer is usually concerned with optimizing the steady performance characteristics of a flow system, a variety of possible instability mechanisms can give rise to intolerable noise and vibration in piping arrangements, reactor components, heat exchangers, turbomachinery exit chambers, and vehicular cutouts and protrusions [Naudascher and Rockwell (1980), (1994)]. In many cases, the magnitudes of local pressure fluctuations can be a substantial fraction of, or even the same order as, the mean dynamic pressure.

The onset of these instability mechanisms can be related to the mean characteristics of their corresponding flow fields, some of them having frequencies that scale with a characteristic length of the flow nozzle, duct, or chamber, while others scale with a characteristic thickness of the shear layer. It is important to recognize that certain of these instabilities do not necessarily involve transition from laminar to turbulent flow; for example, jets and mixing layers that are nominally-turbulent may produce strongly coherent self-sustained oscillations in the presence of acoustic resonators [Rockwell (1983) and Rockwell and Naudascher (1978)].

In this synopsis, we first examine the physical aspects of instabilities arising from mixing-layers, jets, thin trailing-edge wakes, bluff-body (cylinder and sphere) wakes, and swirling (confined-vortex) flows. Of primary interest will be definition of the possible flow regimes and the associated instability frequencies. Then, as a means of illustrating coupling between unstable flows and acoustic/free-surface resonators, these instability mechanisms will be linked to various resonator-controlled oscillations occurring in practice.

13.6.2 Stability of Thin Separated Shear Layers

A variety of mixing-layers, jets, and wakes, illustrated in Fig. 13.60, are highly receptive to disturbances from environmental noise, as well as extraneously-induced

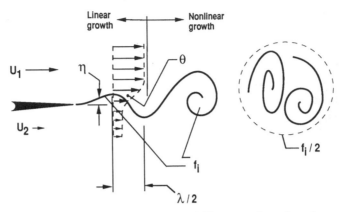

FIGURE 13.60   Growth of an instability wave in a shear layer.

perturbations from pumps, compressors, etc. In all of these shear-layer configurations, the growth of unstable disturbances leading to vortex formation is qualitatively similar. For example, for the free-shear layer configuration of Fig. 13.60, the displacement amplitude, η, of the separated shear layer may increase by several orders of magnitude as it develops at frequency, f_i, in the linear growth region downstream of separation. Once the disturbance amplitude reaches a limiting value, it experiences severe distortion in the nonlinear growth region, followed by roll-up into a discrete vortex. The neighboring vortices tend to successively coalesce or merge with one another, thereby lowering the predominant frequency in the shear layer; an exception is the planar wake. In addition to the fundamental frequency of the instability in the linear growth region, a number of additional frequencies can arise in the region of nonlinear disturbance growth; they are often accompanied by three-dimensional effects [Ho and Huerre (1984) and Rockwell (1983)].

To illustrate the process of disturbance amplification, and thereby frequency selection in the linear growth region, let us consider the case of the free-shear layer flow of Fig. 13.60. The momentum thickness, θ, of this layer is the appropriate length scale. The amplification factor, $-\alpha_i$, which is a negative quantity, indicates the rate at which the disturbance at a given frequency is amplified. It is related to the disturbance amplitude, η, by

$$\eta(x,\ t) \sim e^{-\alpha_i x} \cos{(\alpha_r x - \omega t)} \tag{13.184}$$

in which $\omega = 2\pi f_i$, where f_i is the initial instability frequency, and the wavenumber, α_r, is related to the wavelengh, λ, of the disturbance by $\alpha_r = 2\pi/\lambda$. In order to determine the amplification factor, $-\alpha_i$, and wavenumber, α_r, using linear stability theory, it is necessary to specify the mean velocity distribution; in fact, the following distribution serves as a good approximation

$$\frac{u}{U} = 1 + \Delta \tanh(y/2) \tag{13.185}$$

Here, $u(y)$ is the local velocity within the hear layer, $\Delta = (U_1 - U_2)/(U_1 + U_2)$ is the dimensionless magnitude of the velocity difference across the shear layer, and

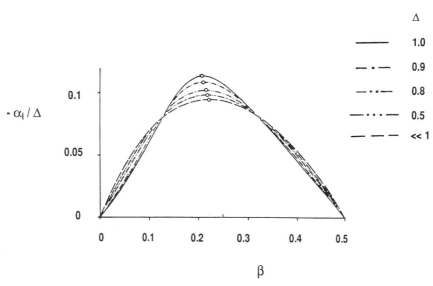

FIGURE 13.61 Amplification factor $-\alpha_i/\Delta$ of disturbance growing in a shear layer. (After Monkewitz and Huerre, 1982.)

$\overline{U} = (U_1 + U_2)/2$ is the average of the freestream velocities. When $\Delta = 0$, the flow is uniform and there is no shear layer, and when $\Delta = 1$, only the upper part of the stream is present. Remarkable is the fact that the frequency of the most amplified disturbance, and therefore the frequency of vortex formation, falls in the range $0.21 \lesssim \beta \le 0.225$, where $\beta \equiv 2\pi f_i\theta/\overline{U}$, for all values of Δ, as shown in Fig. 13.61.

Similar results for the amplification factor as a function of dimensionless frequency are available for the planar jet [Bajaj and Garg (1977)], axisymmetric jet [Michalke (1971)], and planar wake [Betchov and Criminale (1966) and Mattingly and Criminale (1972)].

The axisymmetric jet can also exhibit a *column instability* mode; it scales on jet diameter D: $0.2 \gtrsim fD/U \gtrsim 0.5$ [Ho and Huerre (1984)]. Even if the boundary layer is initially turbulent, and the foregoing thin shear layer instabilities are suppressed, this column instability can still occur.

Typically, the velocity/pressure fluctuations of the shear layer shown in Fig. 13.60 exhibit a number of frequency components in addition to the fundamental, β. This multiple frequency content is due to several sources [Rockwell (1983)]: the aforementioned successive vortex pairing that generates subharmonics of β even at high values of Re [Roshko (1976)], nonlinearities in the growth of the β component, generating its higher harmonics; and nonlinear interaction between two predominant frequency components (one of which may be β), giving rise to a number of sum and difference frequencies. All of this is associated with increasing three-dimensionality of the shear layer, energy exchange between frequency components, and broadening of the velocity/pressure spectra [Miksad (1972)].

At very low values of Reynolds number, all disturbances are damped as, for example, in a planar jet [Bajaj and Garg (1977)]. In such cases, the type of vortex formation shown in Fig. 13.60 will not occur.

The effect of compressibility on the stability characteristics is represented by the

Mach number, M. Michalke (1971) shows, for an axisymmetric jet, that as M increases, the flow becomes more stable. Also, for increasing M, the amplification factor, $-\alpha_i$, decreases, and the most unstable frequency, β, moves to lower values.

13.6.3 Stability of Flow Past Bluff Bodies

Flow past bluff bodies can give rise to two types of instabilities: an instability of the thin shear layer separating from the surface of the cylinder (see Sec. 13.6.2), and a global instability of the entire near-wake region giving rise to formation of the well-known Kármán vortices [Huerre and Monkewitz (1990)]. The coexistence of both of these types of instabilities at Re $= 10^4$ is shown in Fig. 13.62(a) [Rockwell *et al.* (1994)]; small-scale vortical structures arising from the shear-layer instability are embedded within the Kármán vortices. Depending upon the Reynolds number, Re $= \rho UD/\nu$, the bluff-body wake can be classified into several regimes, as described in Fig. 13.62(b). These regimes are summarized by Morkovin (1964) and Mc-

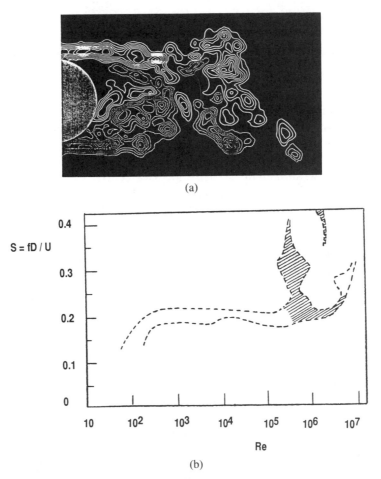

(a)

(b)

FIGURE 13.62 Unstable wake from a circular cylinder. (a) Instantaneous patterns of vorticity for flow at Re = 10,000. (From Lin, Towfighi, and Rockwell, 1995.) (b) Variation of Strouhal number $S = fD/U$. (After Morkovin, 1964 and McCroskey, 1977.)

Croskey (1977). The onset of well-defined unsteadiness sets in at Re $\sim$ 40 and pronounced shedding of the von Kármán type at Re $\sim$ 100. Alternate vortex shedding with varying degrees of background turbulence persists to very high Reynolds number.

The dimensionless shedding frequency, $S = fD/U$, is approximately $S = 0.21$ over the range $400 \gtrsim \text{Re} \gtrsim 2 \times 10^5$. At higher values of Re, the downstream movement of the separation points along the cylinder, and the consequent narrowing of the near-wake region, give a substantial increase in oscillation frequency, which is shown by the cross-hatched region in Fig. 13.62(b). The physical mechanisms leading to this wake narrowing and frequency increase are as follows. At a value of Re of the order of 2×10^5, the separated laminar shear layer quickly undergoes transition, experiences turbulent reattachment, followed by turbulent separation further downstream. In the range of roughly $2 \times 10^5 \gtrsim \text{Re} \gtrsim 3 \times 10^6$, there is an underlying coherence of the oscillations, though their spectral content is relatively broad band. As shown in Fig. 13.62(b), there is large scatter in the values of Strouhal number in this range, most likely due to effects such as free-stream turbulence and surface roughness. When Re reaches the order of 3×10^6, the boundary layer undergoes transition before separation, precluding formation of the laminar separation bubble. This is associated with an increased coherence or periodicity of the vortex shedding.

Bearman (1967) summarizes data for a variety of cylinders, wedges, plates, and ogive configurations, and defines a universal Strouhal number that accounts for the base pressure coefficient.

For the case of flow past spheres, the vortex shedding process is highly nonaxisymmetric, involving, at $\text{Re} = 10^3$ for example, the formation of vortex loops in a chain-like configuration [Achenbach (1978)]. Over the range $6 \times 10^3 \gtrsim \text{Re} \gtrsim 3 \times 10^5$, $0.125 \gtrsim S \gtrsim 0.20$. However, if we consider the range $200 \gtrsim \text{Re} \gtrsim 6 \times 10^3$ for several different experiments (and include circular flat plates as well as spheres), then the value of S rises rapidly with increasing Re, spanning the range $0.10 \gtrsim S \gtrsim 2.0$. At $\text{Re} \simeq 6 \times 10^3$, there is a change in the process of flow separation, accompanied by an order of magnitude drop in S.

13.6.4 Stability of Swirling Flows

Confined vortex or *swirling* flows occur in internal flow configurations such as vortex valves, swirl combustors, cyclone separators, and draft tubes of hydraulic machinery. They can give rise to a number of possible instability mechanisms; some are associated with cavitation in the core of such vortices, and many of them are triggered/enhanced by changes in cross-sectional area of the flow passage. Most of the observed instabilities of swirling flows include the phenomenon of vortex breakdown, reviewed by Leibovich (1984) and Escudier (1987), (1988).

Figure 13.63 illustrates vortex flow in a swirl chamber of diameter D; a volume flow, Q, is injected tangentially into the chamber, thereby producing a vortex with circulation number $\Omega = \Gamma/WD_e$, in which Γ = circulation of the vortex, W = mean axial velocity, D_e = inlet diameter of the flared pipe of total angle α, Q = volume flow injected tangentially, and t is the width of the tangential injection slot. Additional independent parameters describing the possible vortex breakdown regimes are Reynolds number, $\text{Re} = WD_e/\nu$, and the ratio of the radial to tangential velocities

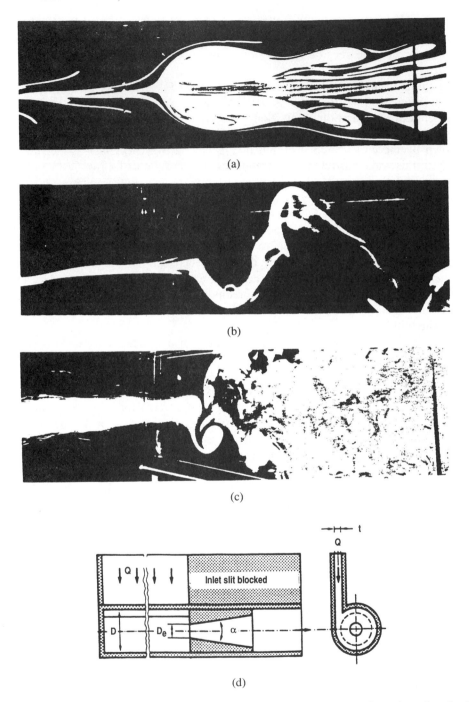

FIGURE 13.63 Typical instabilities for swirling (confined vortex) flows in a flared tube visualized by Escudier. (a) Re = 220, Ω = 7.25, L/D_e = 2.3, α = 20°; (b) Re = 510, Ω 4.28, L/D_e = 4.0, α = 10°; (c) Re = 3,800, Ω = 2.03, L/D_e = 8.3, α = 20°; (d) sketch of apparatus. (Reprinted with permission of Elsevier Science, Ltd., Escudier, M. P., "Vortex Breakdown: Observations and Explanations," *Progress in Aerospace Sciences*, Vol. 25, pp. 189–229, 1988.)

of the inflow, β. This parameter β, which represents the cotangent of the inflow angle, accounts for the noncylindrical influence of the flow inlet region. The Escudier–Zehnder [Escudier (1982)] criterion for the onset of vortex breakdown at a given location in a tube is, in terms of the pipe radius R, Reynolds number Re_B $\propto \Omega^{-3} R^{-1}$.

Figure 13.63 shows several basic breakdown mechanisms: 1) a quasi-axisymmetric bubble, 2) a nonaxisymmetric spiral, and 3) a spiral pattern associated with abrupt breakdown shortly after the swirling flow enters the flared section. There are several possible variations on these basic patterns [Escudier and Zehnder (1982) and Faler and Leibovich (1977)].

In some practical applications such as turbomachinery exit chambers, the mechanism of instability, associated with interaction of the two ends of the curved vortex as it enters the radial exit tube, is more complex than the well-posed experiments described above; Escudier and Merkli (1979) describe the basic characteristics of this type of oscillation.

In draft tubes of Francis turbines, pressure surges are due to swirling flow instabilities involving cavitation [Kubota and Aoki (1980)], the breakdown of the cavitating core at the elbow end of the draft tube substantially enhances the pressure fluctuation amplitude. In addition, Doerfler (1981) discusses types of flow instabilities occuring in draft tubes of Francis turbines; either a cavitated swirling flow or swirl about a cavitated zone attached to the hub of the rotor may occur.

13.6.5 Unstable Flow-Acoustic Coupling

Central to the effective coupling between one of the inherent instability mechanisms discussed above and a resonant acoustic mode of a duct, chamber, etc., is the coincidence, or near coincidence, of the instability and resonant mode frequencies. Although we have spoken of a single, most amplified frequency in conjunction with the shear layer instability of Fig. 13.61, the $-\alpha_i/\Delta$ vs. β diagram actually shows a band of possible frequencies that can undergo substantial amplification (i.e., large values of $-\alpha_i/\Delta$). This observation, along with the fact that resonators of the standing-wave type have n eigenfrequencies, make unstable-flow-acoustic resonator coupling a likely possibility in many flow systems. Details of this coupling for a range of situations are given by Naudascher and Rockwell (1994), Rockwell (1983), Rockwell and Naudascher (1978).

Bluff Body Wake-Resonant Duct Coupling. When a structure such as a flat plate, blade, or airfoil is mounted within a confined test section or channel, there exists the potential for coupling between the vortex shedding instability from the trailing-edge of the structure, and the acoustic (or free-surface) mode of the configuration. In fact, there are a number of possible eigenmodes [Parker and Pryce (1974)], so for systems with variable flow rates, there is a reasonable possibility that the flow-dependent shedding frequency will excite one of these modes. The flat plate with a cylindrically-shaped trailing-edge of Fig. 13.64 illustrates the basic features of this flow-acoustic coupling, usually called *lock-in*.

Let f_R designate the eigenfrequency of the plate-test section acoustic mode and f_i the vortex shedding frequency. For $f_i/f_R < 1$, the induced pressure fluctuation amplitude, $\tilde{p}_{rms}$, is due entirely to vortex shedding and has a reasonably constant, low

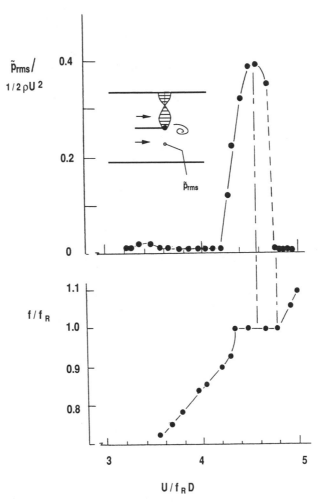

FIGURE 13.64 Bluff-body wake-acoustic mode coupling in a rectangular duct. (After Gaster, 1971.)

amplitude, however as $f_i \to f_R$, the dimensionless pressure amplitude amplitude, $\tilde{p}_{rms}/\frac{1}{2} pU^2$, increases rapidly, attaining a value of about 500 times that associated with the vortex shedding alone. Further increase of flow speed $U/f_R D$ to values outside the *lock-in range* again yields low values of pressure fluctuation amplitude.

Jet Cavity-Resonant Pipeline Coupling. In the case of pipeline systems where the piping components are sufficiently long to act as *organ-pipe resonators*, coupling between flow instabilities generated at tees, bends, expansions, or cavity configurations can produce strongly coherent oscillations, even if the time-averaged flow is turbulent. To illustrate this type of oscillation, consider the case of turbulent flow in a long pipe terminated by an axisymmetric cavity. Figure 13.65 reveals that the oscillation frequencies, corresponding to sharp peaks in the cavity pressure spectrum, are always *locked-in* to one of the organ-pipe modes. In this case, the $n = 3$, 4, 5 modes are excited, and there are frequency jumps between these modes to keep

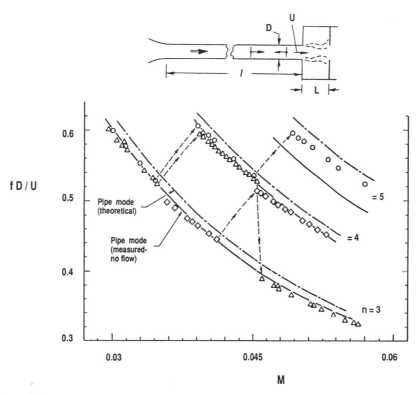

FIGURE 13.65 Jet-acoustic mode coupling in a cavity pipeline system. (After Rockwell and Schachenmann, 1982.)

the oscillation frequency within the upper and lower bounds, $3 \gtrsim S = (fD/U) \gtrsim 0.5$. As noted in Sec. 13.6.2, these limits approximately define the range of frequencies of the column type instability of the axisymmetric jet. At a given value of dimensionless cavity length, L/D, there is, for each lock-in condition, a distribution of cavity pressure amplitude similar to that of Fig. 13.65. Maximum amplitude occurs when the phase between the velocity fluctuations at the cavity exit and inlet $\phi_L - \phi_0 = 2n\pi$. In other words, there is a *phase-lock* between the separation and impingement locations of the separated flow zone, a criterion that holds for oscillations occurring in a variety of internal flow configuration [Rockwell (1983)].

In the case of flow past shallow cavities [Heller and Bliss (1975)], especially at higher values of Mach numbers, it is possible to have strongly coherent oscillations even if a normal mode of the cavity is not excited. As many as five frequencies can coexist.

There are a wide variety of resonators in engineering practice; in addition to standing-wave resonators (described in the foregoing) and Helmholtz resonators occurring in configurations with compressible gases, their free surface counterparts should also be accounted for in assessing flow-acoustic coupling phenomena driven by mixing-layers, jets, and wakes [Naudascher and Rockwell (1994)].

Swirling Flow-Resonant Exit Chamber Coupling. Interaction between an eigenfrequency of a turbomachine exit chamber and an instability of the vortex flow within

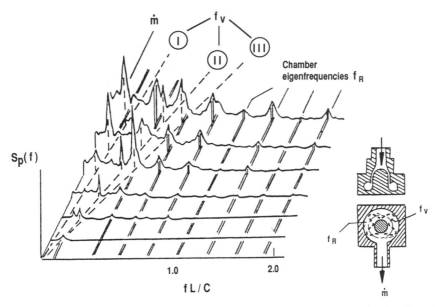

FIGURE 13.66 Swirling vortex flow-acoustic mode coupling in an exit chamber. (After Merkli, 1978.)

or at the exit of the chamber also can generate pronounced oscillations [Merkli (1978)]. Figure 13.66 shows a representation of an outlet chamber, having a ring (doughnut) shape, with its exit tube protruding outward in the radial direction. As discussed above, confined swirling (vortex) flows can yield periodic breakdown phenomena; in this case, the instability occurs at the junction between the exit chamber and the radial exit tube. The measured frequency spectra of Fig. 13.66 show that the predominant frequencies are a function of mass flow rate, $\dot{m}$, through the exit chamber of length L, with as many as nine spectral peaks occurring at the highest value of $\dot{m}$.

13.6.6 Fluid-Elastic Effects

In the interest of brevity, we have not addressed fluid-elastic effects of structures and structural components and their relation to the aforementioned flow instabilities. If the resonant frequency of the structure coincides with the flow-instability frequency and the acoustic-mode frequency, catastrophic vibration may set in [Naudascher and Rockwell (1980), (1994)].

13.7 STRATIFIED FLOW
Carl H. Gibson

13.7.1 Introduction

Fluids are often found with internal variations of density. When these density variations interact with the gravitational field, buoyancy forces arise that result in stratified flows. Stratification may produce or inhibit fluid motions. Unstable stratification produces flow and turbulence. Stable stratification inhibits and damps turbulence and rapidly converts the rare patches of turbulence into a particular class of internal wave motions termed fossil vorticity turbulence [see Gibson (1980), (1982), (1986), (1991a,b,c,d)]. Turbulence is defined as eddy-like motions where the inertial-vortex forces of the eddies are larger than any other forces that tend to damp them out, as discussed in Sec. 1.5. *Fossil turbulence* is a remnant fluctuation in any hydrophysical field produced by previous active turbulence that persists after the flow is no longer actively turbulent on the scale of the fluctuation. Examples of fossil turbulence are the *contrails* of jet aircraft, sky writing, oceanic temperature and salinity microstructure, astrophysical structures like stars, star clusters, galaxies, etc., cosmic rays, and remnant metallic eddies formed in explosive welding. *Fossil vorticity turbulence* means remnants of previous turbulence that persist in the vorticity fields as solitary waves, internal waves, or inertial-Coriolis waves.

Stratification profoundly affects the rates of diffusion and mixing of scalar fluid properties such as temperature and chemical species concentration, since these are generally dominated by turbulent diffusion. A familiar example is the atmospheric boundary layer where solar heating of the earth's surface produces density inversions that help to stir up accumulations of smog (and stir down plumes of smoke). Solar heating of the ocean surface layer produces the opposite effect, strongly stabilizing the upper layer and ultimately the entire water column, and strongly inhibiting vertical mixing.

In this section, we will examine some examples of stratified flows. However, it should be understood that this is an enormous area in fluid mechanics which can only be highlighted in such a limited space. Stratified flows may be laminar or turbulent, oscillatory or random, linear or nonlinear, and can cover a range of length and time scales from microscopic to astrophysical.

Several good reference books on stratified flows exist. Turner (1973), *Buoyancy Effects in Fluids*, is perhaps the most comprehensive, discussing linear internal waves, finite amplitude motions in stably stratified fluids, instability and the production of turbulence, turbulent shear flows in a stratified fluid, buoyant convection from isolated sources, convection from heated surfaces, double diffusive convection, mixing across density interfaces, and internal mixing processes, including twenty-four photographic plates illustrating stratified flow phenomena. Yih (1980) is more mathematical and includes discussion of flows in porous media, and the analogy between gravitation and acceleration. Tritton (1988) has well illustrated sections on convection in horizontal layers, thermal flows, free convection, and other stratified flows. Phillips (1977) gives a comprehensive treatment of surface waves, internal waves, and turbulence in the ocean. List and Jirka (1990) provides an excellent collection of theoretical and experimental papers from a major international symposium on the subject.

Both gravitational and rotational acceleration of density inhomogeneities are crucial to geophysical fluid mechanics. Pedlosky (1987) and Gill (1982) provide authoritative treatments of these aspects of stratified flows.

13.7.2 Equations of Motion

Stratified flows are subject to the usual conservation laws of mass, momentum, and energy. The conservation of mass leads to the general expression

$$\frac{1}{\rho}\frac{D\rho}{Dt} = -\nabla \cdot \bar{\mathbf{v}} \tag{13.186}$$

where ρ is the density, $\bar{\mathbf{v}}$ is the fluid velocity, and the operator $D/Dt = (\partial/\partial t + \bar{\mathbf{v}} \cdot \nabla)$ is the derivative following a fluid particle. Even though the density in a stratified flow is necessarily variable, as long as the density of individual fluid particles is constant then the *incompressibility condition* $\nabla \cdot \bar{\mathbf{v}} = 0$ will be satisfied.

If we assume that gravity is the only body force, the momentum conservation equations are

$$\rho\frac{D\bar{\mathbf{v}}}{Dt} = -\nabla p + \rho\bar{\mathbf{g}} + \mu\nabla^2\bar{\mathbf{v}} \tag{13.187}$$

where the molecular viscosity coefficient is assumed to be constant. Following Turner (1973), we can expand p and ρ about reference values p_0 and ρ_0 for which $\nabla p = \rho_0\bar{\mathbf{g}}$, and the resulting momentum equations are

$$\rho\frac{D\bar{\mathbf{v}}}{Dt} = -\nabla p' + \rho'\bar{\mathbf{g}} + \mu\nabla^2\bar{\mathbf{v}} \tag{13.188}$$

where $\rho' = \rho - \rho_0$ and $p' = p - p_0$. Therefore, for small departures of a stably stratified fluid from its equilibrium state, only density differences ρ' are of importance in affecting the flow, and gravitational accelerations are reduced by the factor ρ'/ρ.

Taking the curl of Eq. (13.188), assuming variable density, gives the conservation of angular momentum equation

$$\frac{D\bar{\omega}}{Dt} = (\bar{\omega} \cdot \nabla)\bar{\mathbf{v}} + \nu\nabla^2\bar{\omega} + \nabla p \times \nabla(1/\rho) \tag{13.189}$$

where the vorticity $\bar{\omega} = \nabla \times \bar{\mathbf{v}}$, and the kinematic viscosity $\nu = \mu/\rho$. According to this equation, the vorticity of a fluid particle can change by three mechanisms. The first term on the right hand side represents the rate of vortex line stretching. The second term is vorticity damping by viscous diffusion. The third term is unique to stratified flows, since it vanishes in the absence of density gradients. Since ∇p is nearly constant downward due to gravity, any tilting of strong density gradient surfaces from their equilibrium horizontal direction will result in the production of vorticity because the density and pressure gradients will cease to be collinear. Vorticity is produced and transmitted by this term as shear layers, turbulence, and internal wave motions that are clearly rotational, versus surface waves that are rotational only at the air-water interface.

Large Amplitude Fluctuations. Because the vorticity production rate increases with the magnitude of the density gradient, we can see from this term that sharp density gradients will tend to focus vorticity production to produce sharp shear layers. This is why stratified turbulence occurs in localized patches. Strong density gradients produce strong shears, but turbulence is strongly inhibited. Eventually, if tilting continues long enough for the local Froude number ($Fr = u/yN$) to exceed a critical

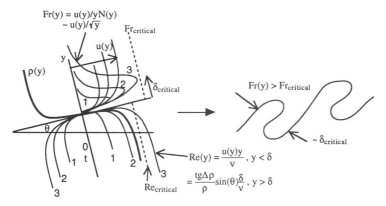

FIGURE 13.67 Formation of turbulence and a Kelvin–Helmholtz billow on a tilted density interface. The velocity, Fr, and Re profiles grow with time. If Re(y) exceeds Re$_{crit.}$ before Fr(y) exceeds Fr$_{crit.}$ then the boundary layers will be turbulent when the billow forms.

value Fr = Fr$_{crit}$, then turbulence develops to break down the density gradient and absorb and dissipate most of the accumulated kinetic energy on the shear layer. [See Eq. (13.190) and the accompanying discussion for a definition of N.] Figure 13.67 shows the formation of the first turbulent billow on a tilted density interface. The local Froude number, u/yN, is a function of y because both u and N are functions of the distance y from the density interface. For $y < \delta$, $u(y) \sim y$ and $N(y) \sim 1/\sqrt{y}$, so Fr(y) $\sim \sqrt{y}$, where δ is the thickness of the velocity boundary layer. For $y > \delta$, $u(y)$ = constant = $tg(\Delta\rho/\rho)\sin(\theta)$, where t is the time, g is gravity, $\Delta\rho$ is the density difference across the interface, and θ is the angle of tilt that occurs at time $t = 0$.

The first billows will be on scales proportional to the boundary layer thickness $\delta_{crit.}$ when Fr(y) exceeds Fr$_{crit.}$. If the boundary layers on the density interface before the first billow forms are turbulent, the dissipation rate, ϵ, within them will strongly increase due to vortex line stretching as the billow grows. Even if the boundary layers are laminar before the first billow forms, turbulence will form as the flow separates as it is distorted by the billow, and the dissipation will be amplified by vortex stretching within the billow. The turbulence eddies grow and pair in the usual cascade from small scales to large (see Sec. 1.5) until the buoyancy forces produced as heavy water is mixed up and light water is mixed down balance the inertial forces of the turbulence, converting the turbulence kinetic energy to that of a forced intenal wave that remembers the maximum size of the turbulence that produced it (a form of fossil turbulence), and splitting the original density layer into two.

Small Amplitude Fluctuations. If the vertical amplitude of stratified fluid motions is small compared to the horizontal wavelength, and if the density fluctuations ρ' are much less than ρ_0, then the equations of motion become

$$\frac{\partial \overline{\mathbf{v}}}{\partial t} = -\frac{1}{\rho_0} \nabla p' + \frac{\rho'}{\rho_0} \overline{\mathbf{g}} + \nu \nabla^2 \overline{\mathbf{v}} \tag{13.190}$$

which are known as the linearized Boussinesq equations if the viscous term is neglected. The density variation in the inertial term on the left has been neglected, which is the Boussinesq approximation, but we see the dominant effect of ρ'/ρ_0 on the gravitational term. The quantity $(\rho'/\rho_0)\overline{\mathbf{g}}$ is often called the *reduced gravity*, $\overline{\mathbf{g}}'$. Because of the assumption that vertical displacements are small, the nonlinear term in velocity has been dropped, and the equations are linear.

From Eq. (13.190), if we displace an element of inviscid fluid a distance η from its equilibrium position, then

$$\frac{\partial^2 \eta}{\partial t^2} = -\frac{g}{\rho_0} \frac{\partial \rho_0}{\partial z} \eta \tag{13.191}$$

will describe the displacement. Thus, the element will oscillate in simple harmonic motion with angular frequency

$$N = \left[\frac{g}{\rho_0} \frac{\partial \rho_0}{\partial z} \right]^{1/2} \tag{13.192}$$

where z is depth and N is the stratification frequency of the fluid, or the *Brunt–Väisälä frequency*. N is a very important parameter of stratified flows, as will be evident in the following discussion. For example, it appears in the *gradient Richardson number*, Ri

$$\mathrm{Ri} = \frac{N^2}{\left(\dfrac{\partial U}{\partial z} \right)^2} \tag{13.193}$$

where U is the horizontal velocity. Ri is a measure of the ratio of buoyancy forces to inertial forces in a shear flow, and it serves as the criterion for whether a shear flow can become turbulent. If Ri > 1/4 locally, then buoyancy forces prevent the development of turbulence.

The *flux Richardson number*, Rf, is defined as the ratio of the rate of removal of potential energy by buoyancy forces in a stratified flow to the rate of turbulent kinetic energy production and dissipation by shear.

$$\mathrm{Rf} = \frac{g\overline{\rho' w'}}{\rho_0 \epsilon} \tag{13.194}$$

where w' is the fluctuation of vertical velocity, and ϵ is the rate of viscous dissipation of turbulent kinetic energy per unit mass.

13.7.3 Convection in Horizontal Layers and Vertical Slots

If a thin layer of fluid is confined between parallel plates of different temperatures, the heat transferred to the fluid may set the fluid into convection motion. If the plates are horizontal and the hottest plate is below with temperature T_2, the induced density field is unstable. Convective motions develop if the *Rayleigh number*

$$\mathrm{Ra} = \frac{g\beta(T_2 - T_1)d^3}{\nu\alpha} \tag{13.195}$$

exceeds a critical value of about 1700, where β is the coefficient of thermal expansion, d is the separation of the plates, and α is the thermal diffusivity of the fluid. The convective motions consist of a set of organized rolling motions called *Benard cells*. For Ra values slightly above critical, the rolls are long and interlaced like fingerprints. For larger Ra values, the rolls shorten and form a honeycomb-like structure. The convective motions greatly increase the heat transfer rate, measured by the *Nusselt number*

$$\text{Nu} = \frac{qd}{k(T_2 - T_1)} = \frac{hd}{k} \tag{13.196}$$

where q is the rate of heat transfer per unit area, and k is the thermal conductivity of the fluid, h is the film heat transfer coefficient, and $(T_2 - T_1)$ is the temperature difference driving the flow.

If the plates are vertical, then convective motions occur for all values of Ra. Elongated vertical rolls develop that break up into smaller rolls as Ra increases above about 3×10^5. Above Ra $\approx 10^6$ the flow becomes turbulent.

The Rayleigh number is the product of the *Grashof number* and the Prandtl number. The Grashof number

$$\text{Gr} = \frac{g\beta q L^3}{\nu^2} \tag{13.197}$$

is the ratio of the inertial and viscous forces, under the condition that the buoyancy force and inertial force are of the same order of magnitude. Setting the inertial force per unit volume $f_I \approx U^2/L$ equal to the buoyancy force per unit volume $f_B \approx g\beta q$ gives a velocity scale $U \approx \sqrt{(g\beta q L)}$. The ratio of inertial to viscous forces is the Reynolds number Re $= UL/\nu$, which is thus

$$\text{Re} = \frac{UL}{\nu} = \left(\frac{g\beta q L}{\nu^2}\right)^{1/2} L = \left(\frac{g\beta q L^3}{\nu^2}\right)^{1/2} = \text{Gr}^{1/2} \tag{13.198}$$

when the Reynolds number is large.

When the buoyancy forces are comparable to the viscous rather than the inertial forces, the velocity scale is different since

$$|\nu\nabla^2\bar{\mathbf{v}}| \approx |g\beta q| \tag{13.199}$$

so the velocity scale for this flow is

$$U \approx g\beta q L^2/\nu \tag{13.200}$$

which gives

$$UL/\nu = \text{Gr} = \text{Re} \tag{13.201}$$

when the Reynolds number is small.

As we have seen, the criterion for the onset of Benard cell convection in horizontal layers is the magnitude of the Rayleigh number Ra $= \text{Gr} \cdot \text{Pr}$. Since this motion is at small Reynolds numbers, then Re $\approx$ Gr. The ratio of advective to conductive heat transfer is the *Peclet number*, Pe $\approx$ Re Pr, which in this instance is just the Rayleigh number

$$\frac{\text{advection}}{\text{conduction}} \approx \frac{|\bar{\mathbf{v}} \cdot \nabla T|}{\alpha \nabla^2 T} \approx \frac{UL}{\alpha} \approx \text{Gr Pr} = \text{Ra} \qquad (13.202)$$

for small Re.

Benard cell convection is thus driven by unstable stratification, but is stabilized by viscous forces and thermal diffusion which prevent the formation of turbulence as long as the Reynolds numbers are sufficiently low. Illustrations of these processes may be found in Titton (1988).

13.7.4 Viscous Convection and Diffusion in Fossil Turbulence Microstructure

A flow regime similar to Benard cell convection occurs in the restratification of remnant density microstructure produced by turbulence after the turbulence has been damped by buoyancy forces. Turbulence in stably stratified fluids tends to occur in isolated patches, as mentioned above, and produces a complex field of density microstructure through the combined action of advection and molecular diffusion during the overturning, or actively turbulent, phase of the turbulence event. Buoyancy forces rapidly convert the turbulence to internal wave motions, acting first on the largest eddies and progressing down to the viscous, or Kolmogorov, scale as the dissipation rate, ϵ, of the fluid motions decreases. Any density microstructure which remains in the fluid after the fluid is no longer overturning on the scale of the microstructure is known as fossil turbulence.

The density field consists of multiply connected isodensity surfaces, with a complex topology of extrema, saddle points and saddle lines which makes such microstructure of any scalar fluid property a unique signature of turbulence, either past or present. A local equilibrium is maintained at the smallest scales between the local rate of strain of the fluid, which streepens scalar gradients, and molecular diffusion, which smoothes them out. Universal similarity hypotheses based on the strain rate, molecular diffusivity of the scalar, and scalar dissipation rate have been proposed by Gibson (1968b) based on the kinematics of the scalar topology described by Gibson (1968a), and the similarity theory has been extended to include the fossil turbulence regime in Gibson (1980).

If the microstructure is not overturning on any scale, it is completely fossil turbulence microstructure. The geometry of the microstructure requires the existence of at least one density inversion per density extremum point, so each of these entities will drive a stratified viscous flow similar to the high Rayleigh number Benard cell in a horizontal layer. Such flows have not been well documented in the laboratory, and are the subject of active research [Gibson (1982)].

13.7.5 Stratified Turbulence

In nonstratified flows, turbulence occurs when the inertial forces of a perturbation become larger than the viscous forces which arise to damp out the perturbation. The characteristic inertial force of a perturbation of scale L with velocity U in a fluid of density ρ is

$$F_I \approx \rho U^2 L^2 \tag{13.203}$$

and the opposing viscous force in a fluid with viscosity μ is

$$F_V \approx \mu(U/L)L^2 \tag{13.204}$$

The ratio of the inertial to viscous forces is the Reynolds number, Re

$$F_I/F_V \approx \rho U^2 L^2 / \mu(U/L)L^2 = UL/\nu = \text{Re} \tag{13.205}$$

Where the Reynolds number is large, the nonlinear inertial forces can produce turbulence, otherwise the perturbation will be damped.

In a fully turbulent flow, the viscous forces become important only at small scales, where the Reynolds number is of order unity. According to the Obukhov–Kolmogorov law, velocity differences on large scales are given by

$$U(L) \approx (\epsilon L)^{1/3} \tag{13.206}$$

Substituting this expression into the definition of Re, setting Re = 1 and solving for the viscous length scale L_V gives

$$L_V = L_K = (\nu^3/\epsilon)^{1/4} \tag{13.207}$$

where L_K is the Kolmogorov length scale.

In a stably stratified flow, buoyancy forces F_B arise which also tend to damp out perturbations

$$F_B = g[(\partial\rho/\partial z)L]L^3 \tag{13.208}$$

where the term in brackets represents the mass difference between a volume of fluid of scale L displaced in the vertical direction z with a vertical density gradient $\partial\rho/\partial z$.

The length scale at which buoyancy forces equal inertial forces for a turbulent flow is found by setting $F_B \approx F_I$ in Eqs. (13.208) and (13.207), with the definition of N in Eq. (13.192), and solving for $L = L_R$

$$L_R = (\epsilon/N^3)^{1/2} \tag{13.209}$$

where L_R is called the Ozmidov length scale.

The qualitative criterion for the existence of active turbulence in a stratified flow is that

$$L_R \gg L \gg L_K \tag{13.210}$$

so that an eddy of scale L can overturn against the buoyancy forces which resist it at large scales and the viscous forces which resist it at small scales. Active turbulence is defined as an eddy-like state of fluid motion where the inertial-vortex forces of the eddies are larger than any other forces that tend to damp them out (see Sec. 1.5). Gibson (1980) has proposed a quantitative version of the preceding criterion in terms of the wavelength λ of turbulent Fourier components

$$1.2 L_{R \leq \lambda} \geq 15.2 \; L_K \tag{13.211}$$

which can be used to classify the hydrodynamic state of fluid motions in a stratified fluid as active turbulence, internal waves, or viscous laminar flow. Equating both sides of Eq. (13.211) gives the dissipation rate, ϵ_F, corresponding to the simultaneous equality of buoyant, inertial, and viscous forces on a turbulent eddy

$$\epsilon_F = 30 \; \nu N^2 \tag{13.212}$$

corresponding to the final decay of fossil turbulence in a stratified fluid. This criterion has been verified extensively by laboratory and field experiments [Gibson (1991a)]. If ϵ in a microstructure patch has $\epsilon < \epsilon_F$, then no turbulence exists. If the vertical scale, L_T, of the largest density overturns in a density microstructure patch are about half of one the wavelength, λ, then Eq. (13.211) provides a criterion for whether or not the microstructure is turbulent at all scales. Setting $2 L_T = 1.2 \; (\epsilon_0 / N^3)^{1/2}$ gives

$$\epsilon_0 = 2.8 \; L_T^2 N^3 \tag{13.213}$$

as a lower bound for the dissipation rate in a patch of microstructure that is turbulent at all scales. If the measured ϵ is larger than ϵ_0, then the patch is actively turbulent at all scales. If ϵ lies between ϵ_0 and ϵ_F, the microstructure is fossil turbulence for wavelengths $1.2 L_{R0 \leq \lambda} \geq 1.2 L_R$ and actively turbulent for $1.2 L_{R \geq \lambda} \geq 15.2 L_K$. Most of the microstructure patches detected in the ocean are in a mixed active-fossil hydrodynamic state by these criteria except near sources of active turbulence such as seamounts [Gibson *et al.* (1994)].

Self-gravitational stratified turbulent flow occurs in the formation of many astrophysical objects, such as stars, which form by gravitational condensation that is inhibited at small scales by turbulence. A minimum condensation length scale $\sim L_{ST}$ is possible in a gas of density ρ with turbulence dissipation rate ϵ, where $L_{ST} = \epsilon^{1/2}/(\rho G)^{3/4}$, G is $6.72 \times 10^{-11} \text{m}^3/\text{kg s}^2$, and L_{ST} is termed the turbulent *Schwarz Radius* of the condensation [Gibson (1988), (1995)].

13.7.6   Horizontal Turbulence

Horizontal perturbations in a stratified fluid are not constrained by buoyancy forces. Consequently, it is a commonly observed phenomenon that turbulent eddy motions occur in the ocean and atmosphere which have much larger horizontal scales than the vertical turbulence, and eddy diffusivities in the horizontal may be orders of magnitude larger than eddy diffusivities in the vertical. For example, the wake of a grounded tanker in a current is geometrically similar to the wake of a plate inclined at the same angle in a water tunnel [Van Dyke (1982), plates 172 and 173]. The inertial forces of the horizontal eddies are of order $F_I \approx \rho U^2 L h$ compared to the *viscous damping* forces on horizontal surfaces of $F_V \approx (\mu U/h) L^2$, where h is the thickness of the turbulent layer. Solving for L gives $L_{crit.} \approx h^2 U/\nu$, or hRe, where the Reynolds number is Uh/ν, and the *viscosity* depends on the vertical turbulence. Without vertical constraints of the turbulence such as the bottom and the air sea interface, or buoyancy forces, the *effective viscosity*, ν, would be of order Uh, giving $L_{crit.} \approx h$. However, the vertical constraints mentioned could reduce ν to nearly the molecular value. Therefore, the aspect ratio $L_{crit.}/h \approx$ Re of horizontal eddies can become very large without significant viscous restraints.

Horizontal eddies may be constrained at the largest scales by Coriolis forces, at an inertial-Coriolis scale similar to the Rossby radius of deformation $r_R \approx U/\Omega$ used by oceanographers. Horizontal turbulent eddies are converted into inertial-Coriolis eddy motions in a rotating system just as vertical turbulence is converted to internal waves, or fossil vorticity turbulence, when the turbulence is damped by buoyancy forces in a stably stratified fluid. As shown in Sec. 1.5, the inertial forces of growing horizontal turbulence eddies, $F_I \approx \rho U^2 Lh$, become equal to the Coriolis forces, $F_C \approx \rho U\Omega L^2 h$, at the Hopfinger scale $L_H \approx (\epsilon/\Omega^3)^{1/2}$, where $U \approx (\epsilon L)^{1/3}$ for the turbulent velocity at scale L, h is the thickness of the turbulent layer, and Ω is the vertical component of the angular velocity. Fossil two-dimensional turbulence remnants are produced when Coriolis forces convert the kinetic energy of two-dimensional turbulence to Coriolis-inertial eddies or waves, as discussed by Gibson (1991a,c).

13.8 MAGNETOFLUIDMECHANICS
S. T. Demetriades and C. D. Maxwell

13.8.1 Introduction

Any attempt to exceed the limits of specific energy (energy per unit mass) imposed by the chemical bond in the production, generation, containment, manipulation, conversion or utilization of energy or energetic fluids requires the application of the principles of magneto- and electrofluidmechanics. As examples, we may consider the generation of electricity from gases at temperatures above 2000 K, the acceleration of mass to velocities above 1600 m/sec and the manipulation of materials at temperatures above 3000 K. Although many other interesting applications of magneto- and electrofluidmechanics exist at lower temperatures (e.g., pumping of liquid metals), these other applications are technologically less demanding by comparison.

This is a very broad field of study, since it involves classical physics and thermodynamics (through the coupling of the Maxwell equations with the laws of motion and conservation), modern physics (e.g., through the equation of state and electron behavior) and chemistry (e.g., through the law of mass action). The degree of complexity of this field of study is commensurate with the number of disciplines involved, and, in fact, the pace of success in magnetofluidmechanics is intimately coupled to and exposes all the weaknesses of these disciplines.

13.8.2 Fundamental Theory

The Equations of Charged and Conducting Fluids. Magnetohydrodynamics and electrohydrodynamics are concerned with the mechanics of fluids which may support *electromagnetic body forces.* These forces arise from the presence in the fluid of a charge density, ρ_c, and a current density $\bar{\mathbf{J}}_0$. The total current density, $\bar{\mathbf{J}}_0$, consists of a convected portion, $\rho_c\bar{\mathbf{U}}_e$, and a conduction portion, $\bar{\mathbf{J}}$, which is carried by charge drifting at a relative velocity with respect to the macroscopic mean velocity, $\bar{\mathbf{U}}$. The electromagnetic force density which acts upon the fluid is given by

$$\bar{\mathbf{f}} = \rho_c\bar{\mathbf{E}} + \bar{\mathbf{J}}_0 \times \bar{\mathbf{B}} \tag{13.214}$$

where $\bar{\mathbf{E}}$ and $\bar{\mathbf{B}}$ are the electric and magnetic fields which are themselves determined by the charge and current present both in the fluid and its surroundings. The corresponding power flow, in watts/m^3, from the electromagnetic field to the fluid is given by

$$P = \bar{\mathbf{J}}_0 \cdot \bar{\mathbf{E}} \tag{13.215}$$

The *macroscopic* (as opposed to the *species*) equations for a fluid subject to electromagnetic forces are

$$D\rho/Dt + \rho\nabla \cdot \bar{\mathbf{U}} = 0 \tag{13.216a}$$

$$\rho D\bar{\mathbf{U}}/Dt + \nabla p - \nabla \cdot \bar{\pi} = \rho_c\bar{\mathbf{E}} + \bar{\mathbf{J}}_0 \times \bar{\mathbf{B}} \tag{13.216b}$$

$$\rho D(\epsilon + \tfrac{1}{2}U^2)/Dt + \nabla \cdot (p\bar{\mathbf{U}}) - \nabla \cdot (\bar{\pi} \cdot \bar{\mathbf{U}}) + \nabla \cdot \bar{\mathbf{q}} = \bar{\mathbf{J}}_0 \cdot \bar{\mathbf{E}} \tag{13.216c}$$

where ρ and e are the mass density and thermal energy per unit mass of the fluid. The pressure is p, $\bar{\pi}$ is the viscous stress tensor, and $\bar{\mathbf{q}}$, the heat flux vector. The electric field $\bar{\mathbf{E}}' \equiv \bar{\mathbf{E}} + \bar{\mathbf{U}} \times \bar{\mathbf{B}}$ may be thought of as the electric field in a frame of reference moving with the fluid velocity, $\bar{\mathbf{U}}$. The power flow $\bar{\mathbf{J}} \cdot \bar{\mathbf{E}}' \equiv \bar{\mathbf{J}} \cdot (\bar{\mathbf{E}} + \bar{\mathbf{U}} \times \bar{\mathbf{B}})$ is positive definite and dissipative (*Joule dissipation*) and represents the dissipation of electrical power into internal thermal energy as follows from the thermal energy equation which results from Eq. (13.216c) with U^2 eliminated by Eq. (13.216b).

The conservation laws, Eqs. (13.216a,b,c), must be supplemented with *thermodynamic state equations*

$$p = p(\rho, T), \qquad e = e(\rho, T) \tag{13.217}$$

and constitutive equations for the stress tensor, heat flux vector, and conduction current (see Chaps. 4 and 7). Fluids with permanent electric or magnetic dipoles possess thermodynamic and transport property functions which depend explicitly upon the fields $\bar{\mathbf{E}}$, $\bar{\mathbf{B}}$ [de Groot and Mazur (1962) and Penfield and Haus (1967)]. A typical form of the constitutive relationships consistent with the Navier–Stokes level of approximation is

$$(\bar{\pi})_{ij} = -\rho\nu[\tfrac{1}{2}(\partial U_i/\partial x_j + \partial U_j/\partial x_i) - \tfrac{1}{3}\partial U_k/\partial x_k\delta_{ij}] \tag{13.218a}$$

$$\bar{\mathbf{q}} = -\lambda\nabla T + \psi\bar{\mathbf{j}} \tag{13.218b}$$

$$\bar{\mathbf{J}} = \sigma(\bar{\mathbf{E}} + \bar{\mathbf{U}} \times \bar{\mathbf{B}}) + \phi\nabla T \tag{13.218c}$$

Equation (13.218) for the current density, J, which complements the Navier–Stokes relationship for $\overline{\pi}$ and $\overline{q}$ in a conducting fluid is called the *Ohm's Law*. The coefficients μ, λ, σ, ψ, and ϕ are the *scalar transport coefficients* of viscosity, heat conduction, and electrical conductivity. In general, the heat flow and conduction current are coupled. The cross coupling term $\phi \nabla T$ in the current equation may be neglected whenever the electric field $\overline{E}' = \overline{E} + \overline{U} \times \overline{B}$ is much greater in magnitude than $(k/e)\nabla T$, where k/e is the ratio of the Boltzmann constant to the electron charge.

Since each species (electrons, ions of different charge, atoms or molecules of different mass) in the fluid may have its own temperature, number density, etc., and its own constitutive or transport properties, Eqs. (13.216)–(13.218) are more properly written separately for each species and then appropriately summed. However, that treatment is beyond the scope of this article. It can be found in such sources as Demetriades *et al.* (1963), (1966), (1982), Argyropoulos *et al.* (1969), and Sutton and Sherman (1964).

In most engineering applications of MHD, there are important regions where the *electron temperature* is not the same as the gas temperature. The electron number density does not correspond to the gas temperature through some appropriate *Saha equation* (at the gas temperature) but is closer or equal to the number density obtained from a Saha equation at the electron temperature [Kerrebrock (1962)]. It is important to retain in the system of Eqs. (13.216)–(13.218) at least the separate electron energy equation which couples the local electron temperature to the electric and magnetic fields and local electron number density.

An adequate form of this electron energy equation is

$$\overline{J}_e \cdot \overline{E}' = \tfrac{3}{2} k n_e \nu \delta_{\text{eff}}(T_e - T_a) + \dot{R} + \sum_i s_i \epsilon_i \qquad (13.219)$$

where

$\dot{R}$ is the radiative losses from the heavy species at the electron temperature,

s_i is the rate of *chemical* reaction i of electrons and heavy particles per unit volume per unit time,

ϵ_i is the characteristic energy that is exchanged in each such reaction i in units of energy per reaction,

k is the Boltzmann constant,

n_e is the number density of electrons,

ν is the total electron collision frequency,

δ_{eff} is the effective energy loss factor per electron collision that involves elastic and inelastic energy exchange,

T_e is the electron temperature,

T_a is the heavy species (atoms, molecules or ions) temperature, and

m_e is the mass of the electron.

Usually, for diatomic or polyatomic gases and at temperatures of current interest in MHD, $\dot{R}$ and s_i are negligible compared to the inelastic collision term that contains $T_e - T_a$.

When a magnetic field is present, the transport properties become *tensorial*. This effect is known as the *Hall effect* and is described by the Hall vector, $\overline{\beta} = (e/m)\overline{B}/\nu$, where e/m is the charge to mass ratio of the conduction carriers and ν is their effective collision frequency for momentum scattering. When $|\overline{\beta}| \gtrsim 1$, terms proportional to $\nabla \overline{\mathbf{U}} \times \overline{\boldsymbol{\beta}}$, $\nabla T \times \overline{\boldsymbol{\beta}}$, and $\overline{\mathbf{E}} \times \overline{\boldsymbol{\beta}}$ must appear in Eqs. (13.218a, b, and c), respectively. In partially ionized gases, the Hall effect upon the stress tensor is often negligible, although its influence upon the heat flux and current may be substantial. The full *generalized Ohm's Law* in this case has been developed by Demetriades, Hamilton, Ziemer, and Lenn (1963), Demetriades and Argyropoulos (1966), and Schweitzer and Mitchner (1967). In the simplest case in which temperature gradients may be neglected, Eq. (13.218) becomes

$$\overline{\mathbf{J}}_{\perp} = [\sigma/(1 + \beta^2)] \, (\overline{\mathbf{E}}'_{\perp} - \overline{\mathbf{E}}'_{\perp} \times \overline{\boldsymbol{\beta}}) \tag{13.220a}$$

$$\overline{\mathbf{J}}_{\parallel} = \sigma \overline{\mathbf{E}}_{\parallel} \tag{13.220b}$$

where $\overline{\mathbf{J}}_{\perp}$, $\overline{\mathbf{E}}_{\perp}$ and $\overline{\mathbf{J}}_{\parallel}$, $\overline{\mathbf{E}}_{\parallel}$ are perpendicular and parallel to the magnetic field respectively. Using Eq. (13.220a), the expression $\overline{\mathbf{J}} \cdot \overline{\mathbf{E}}'$ may be expressed as $\overline{\mathbf{J}} \cdot \overline{\mathbf{E}}' = J^2/\sigma = [\sigma/(1 + \beta^2)]E'^2_{\perp} + \sigma E^2_{\parallel}$ showing that the *Joule dissipation* is a positive definite quantity.

The charge density ρ_c and current $\overline{\mathbf{J}}_0$ determine the field $\overline{\mathbf{E}}$, $\overline{\mathbf{B}}$ through the *Maxwell equations*

$$1/c^2 \partial \overline{\mathbf{E}}/\partial t - \nabla \times \overline{\mathbf{B}} = -\mu \overline{\mathbf{J}}_0 \tag{13.221a}$$

$$\partial \overline{\mathbf{B}}/\partial t + \nabla \times \overline{\mathbf{E}} = 0 \tag{13.221b}$$

$$\nabla \cdot \overline{\mathbf{E}} = \epsilon_0^{-1} \rho_c \tag{13.221c}$$

$$\nabla \cdot \overline{\mathbf{B}} = 0 \tag{13.221d}$$

Equations (13.216) and (13.221) with the state and constitutive relationships, Eqs. (13.217), (13.218), and (13.220) form a closed coupled system for the fluid variables ρ, $\overline{\mathbf{U}}$, e, ρ_c, and $\overline{\mathbf{J}}_0$ and the electromagnetic field variables $\overline{\mathbf{E}}$, and $\overline{\mathbf{B}}$.

Theorems of Bernoulli and Kelvin. From the identity $\overline{\mathbf{U}} \cdot \nabla \overline{\mathbf{U}} = \nabla \tfrac{1}{2} U^2 - \nabla \times \overline{\mathbf{U}} \times \overline{\mathbf{U}}$, the momentum equation may be expressed as

$$\rho(\partial \overline{\mathbf{U}}/\partial t - \overline{\boldsymbol{\Omega}} \times \overline{\mathbf{U}}) + \rho \nabla \tfrac{1}{2} U^2 + \nabla(p + w)$$
$$= -\nabla \cdot \overline{\boldsymbol{\pi}} + \mu_0^{-1} \overline{\mathbf{B}} \cdot \nabla \overline{\mathbf{B}} + \epsilon_0 (\overline{\mathbf{E}} \cdot \nabla \overline{\mathbf{E}} - \rho_c \overline{\mathbf{E}}) \tag{13.222}$$

where $\overline{\boldsymbol{\Omega}} = \nabla \times \overline{\mathbf{U}}$ is the vorticity, and $w = \tfrac{1}{2}(\mu_0^{-1} B^2 + \epsilon_0 E^2)$ is the electromagnetic energy density. Vorticity is discussed in Sec. 13.1. When the right hand side of Eq. (13.222) vanishes, *Bernoulli's theorem for electromagnetic flow* is obtained

$$\tfrac{1}{2} U^2 + (\tfrac{1}{2}\mu_0^{-1} B^2 + \tfrac{1}{2}\epsilon_0 E^2)/\rho + p/\rho = C(t) \tag{13.223}$$

Particular velocity fields which possess Bernoulli integrals, Eq. (13.223), include all velocity fields which may be represented as $\overline{\mathbf{U}} = \overline{\mathbf{k}}_1 \times \overline{\mathbf{E}} + \overline{\mathbf{k}}_1 \times \overline{\mathbf{B}}$ where $\overline{\mathbf{k}}_1$ and $\overline{\mathbf{k}}_2$ are arbitrary vectors. All electromagnetic fields which are free of tangential gradients similarly allow the existence of a Bernoulli integral. Most general flows of engineering interest do not satisfy these conditions.

In uniform density flow the curl of Eq. (13.216) yields the vorticity equation in electromagnetic flow

$$\rho(D\overline{\mathbf{\Omega}}/Dt - \overline{\mathbf{\Omega}} \cdot \nabla\overline{\mathbf{U}}) = -\nabla \times (\nabla \cdot \overline{\pi})$$

$$+ \overline{\mathbf{B}} \cdot \nabla\overline{\mathbf{J}}_0 - \overline{\mathbf{J}}_0 \cdot \nabla\overline{\mathbf{B}} + \rho_c\nabla \times \overline{\mathbf{E}} \quad (13.224)$$

When the right-hand side of Eq. (13.224) vanishes, *Kelvin's theorem* is obtained; the flux of vorticity through an elemental area $\overline{\mathbf{\Omega}} \cdot \delta\mathbf{A}$ is conserved. In conventional flow, viscous stresses act as the only mechanism for the creation of vorticity, however in electromagnetic flow, the curl of the Lorentz force, f, does not generally vanish even in the case of *ideal flow* in which $\sigma^{-1} \to 0$, and there is no Joule dissipation. Gradients of the current density in the direction of the magnetic field can represent an important source of vorticity in electromagnetic flow.

The Mangetohydrodynamic and Electrohydrodynamic Approximations. The divergence of Eq. (13.221) and use of Eq. (13.221) with $\overline{\mathbf{J}}_0 = \rho_c\overline{\mathbf{U}} + \overline{\mathbf{J}}$ from the Ohm's Law, Eq. (13.220), yields the equation governing the charge density

$$D\rho_c/Dt + \rho_c\nabla \cdot \overline{\mathbf{U}} = -(\epsilon_0^{-1}\sigma)\rho_c \quad (13.225)$$

From Eq. (13.225) the *decay time for charge*, τ_c, is given by $\tau_c = (\epsilon_0\sigma^{-1})$. If $t = L/U$ is a characteristic macroscopic time scale where L is the length scale, the condition $t/\tau_c = (L\tau_c/U) = 1$ divides conductors from non-conductors. If $t/\tau_c \ll 1$ the *electrohydrodynamic approximation* $\overline{\mathbf{J}} \simeq 0$ holds and

$$D\rho_c/Dt + \rho_c\nabla \cdot \overline{\mathbf{U}} = 0 \quad (13.225a)$$

Correspondingly, the Lorentz force becomes

$$\overline{\mathbf{f}} = \rho_c(\overline{\mathbf{E}} + \overline{\mathbf{U}} \times \overline{\mathbf{B}}) \quad (13.226)$$

For $t/\tau_c \gg 1$ there results $\rho_c \simeq 0$ and $\overline{\mathbf{J}}_0 = \overline{\mathbf{J}}$. Equation (13.225) then becomes

$$\nabla \cdot \overline{\mathbf{J}} = 0 \quad (13.227a)$$

which requires that Eq. (13.221) becomes

$$\nabla \times \overline{\mathbf{B}} = \mu_0\overline{\mathbf{J}} \quad (13.227b)$$

Correspondingly, the Lorentz force becomes

$$\overline{\mathbf{f}} = \overline{\mathbf{J}} \times \overline{\mathbf{B}} \quad (13.227c)$$

The approximation embodied in Eqs. (13.227), which stems from $\rho_c = 0$, is known as the *magnetohydrodynamic approximation*. Note particularly that the loss of $\partial\overline{\mathbf{E}}/\partial t$ from Eq. (13.221) in this approximation prohibits electromagnetic radiation.

In the presence of a uniform magnetic field, fluid vorticity is a source of electrical charging. From $\nabla \cdot \overline{\mathbf{J}}_0 = 0$ in steady flow, substitute $\overline{\mathbf{J}}_0 = \overline{\mathbf{J}}$ from Ohm's Law with uniform electrical conductivity and use Eq. (13.221) to obtain

$$\rho_c = -\epsilon_0 \, \overline{\mathbf{B}} \cdot \overline{\mathbf{\Omega}} \tag{13.228}$$

The Equation of Magnetic Induction. The elimination of $\overline{\mathbf{J}}$ and $\overline{\mathbf{E}}$ from Eq. (13.227b) and the Ohm's Law with Eq. (13.221b) results in the induction equation for $\overline{\mathbf{B}}$ for uniform density and electrical transport properties σ and $\overline{\beta}$

$$D\overline{\mathbf{B}}/Dt - \overline{\mathbf{B}} \cdot \nabla\overline{\mathbf{U}} = \eta\nabla^2\overline{\mathbf{B}} + \eta[(\overline{\beta} \cdot \nabla) \, \nabla \times \overline{\mathbf{B}} - (\nabla \times \overline{\mathbf{B}} \cdot \nabla)\overline{\beta}] \tag{13.229}$$

Equation (13.229) describes the convection and diffusion of the magnetic field with the *magnetic diffusivity*, $\eta = (\mu_0\sigma)^{-1}$. When $\eta \to 0$, the electrical forces are non-dissipative, and the magnetic field is convected as if the lines of force were bound to the fluid. The left-hand side of Eq. (13.229) is formally identical to the vorticity equation Eq. (13.224); $\overline{\mathbf{B}} \cdot \delta\overline{\mathbf{A}}$, which is the flux of $\overline{\mathbf{B}}$ through an elemental area $\delta\overline{\mathbf{A}}$, is conserved throughout the flow just as the flux of vorticity $\overline{\mathbf{\Omega}} \cdot \delta\overline{\mathbf{A}}$ is conserved when Kelvin's theorem holds. When $\overline{\beta} \equiv 0$, the diffusion is a scalar diffusion, but when $\overline{\beta} \neq 0$, the *induction equation* is a nonlinear tensor diffusion equation which has components which are quadratic in $\overline{\mathbf{B}}$.

The time scale, τ_B, for the field to diffuse over a length scale $L \sim |\nabla^{-1}|$ is given by

$$\tau_B = L^2/(\mu_0\sigma) \tag{13.230}$$

The convective time scale, L/U, may be compared to the diffusion time scale, τ_B, in a nondimensional parameter known as the *magnetic Reynolds number*, R_m

$$R_m = \tau_B/(L/U) = \mu_0\sigma UL \tag{13.231}$$

For $R_m \ll 1$, magnetic fields diffuse instantly through the region of interest. The regime of vanishing R_m has other important consequences. Consider a steady magnetic field $\overline{\mathbf{B}}_0$ which is established by currents external to the fluid. Let $\overline{\mathbf{J}}$ be the current present in the fluid and $\overline{\mathbf{B}}$ given by Eq. (13.227) which yields the field induced by the currents in the fluid. From Ohm's Law with $\overline{\mathbf{E}} \sim \overline{\mathbf{U}} \times (\overline{\mathbf{B}}_0 + \overline{\mathbf{B}})$, the current is given by $\overline{\mathbf{J}} \sim \sigma\overline{\mathbf{U}} \times (\overline{\mathbf{B}}_0 + \overline{\mathbf{B}})$. The field, $\overline{\mathbf{B}}$ induced by this current is from Eq. (13.227)

$$\nabla \times \overline{\mathbf{B}} \sim \mu_0\overline{\mathbf{J}} \sim \mu_0\sigma\overline{\mathbf{U}} \times (\overline{\mathbf{B}}_0 + \overline{\mathbf{B}}) \tag{13.232}$$

Since $|\nabla \times \overline{\mathbf{B}}| \sim B/L$, there results

$$\frac{|\overline{\mathbf{B}}/B_0|}{1 + |\overline{\mathbf{B}}/B_0|} \sim \mu_0\sigma UL \sim R_m \tag{13.233}$$

Thus when $R_m \gg 1$, the field satisfies $|\overline{\mathbf{B}}/B_0| \sim 1$, but for $R_m \ll 1$ this condition becomes $|\overline{\mathbf{B}}/B_0| \sim R_m \ll 1$ and the induced field $\overline{\mathbf{B}}$ may be neglected compared to the applied field $\overline{\mathbf{B}}_0$. From Eq. (13.221), $\partial\overline{\mathbf{B}}/\partial t \sim \nabla \times \overline{\mathbf{E}}$. For $E \sim UB$ and $\partial/\partial t$ of the order of the convective scale U/L, it will be true that

$$\frac{|\partial \overline{\mathbf{B}}/\partial t|}{|\nabla \times \overline{\mathbf{E}}|} \sim R_m \tag{13.234}$$

In the approximation of low magnetic Reynolds number, Ohm's Law is expressed only in terms of the applied field $\overline{\mathbf{B}}_0$

$$\overline{\mathbf{J}} = \sigma/(1 + \beta^2) \, [\overline{\mathbf{E}} + \overline{\mathbf{U}} \times \overline{\mathbf{B}}_0 - \overline{\boldsymbol{\beta}}_0 \times (\overline{\mathbf{E}} + \overline{\mathbf{U}} \times \overline{\mathbf{B}}_0)] \tag{13.235}$$

while the Maxwell equation given by Eq. (13.221b) becomes

$$\nabla \times \overline{\mathbf{E}} = 0 \tag{13.236}$$

and Eq. (13.221) in the MDH approximation becomes

$$\nabla \cdot \overline{\mathbf{J}} = 0 \tag{13.237}$$

Equations (13.235) to (13.237) serve to determine completely the fields $\overline{\mathbf{E}}$ and $\overline{\mathbf{J}}$ in the low magnetic Reynolds number and MHD approximations. The magnetic field $\overline{\mathbf{B}} \ll \overline{\mathbf{B}}_0$ induced by the current determined by Eqs. (13.235) to (13.237) may be determined from Eq. (13.227).

Characteristic Parameters. The characteristic parameters for the general fluid equations in the absence of body forces are three: 1) the Mach number, $M = U/c$ (where $c = \sqrt{\gamma p/\rho}$ is the speed of sound), 2) the Reynolds number, $R_e = UL\rho/\mu$, and 3) the Prandtl number, $P_r = C_p\mu/\lambda$ where C_p is the specific heat at constant pressure. Additional parameters are required in magnetohydrodynamic flow.

The first of these parameters is the magnetic Reynolds number, R_m, which describes the convection of magnetic energy compared to its diffusion. One other independent one-dimensional number is required for the description of electromagnetic effects in the MHD approximation. For $R_m \gg 1$, the ratio of magnetic energy density to the fluid energy density is an appropriate choice

$$I_U = \mu_0^{-1}B^2/\rho U^2 \tag{13.238}$$

A second choice is the ratio of magnetic pressure to fluid pressure

$$I_p = \mu_0^{-1}B^2/p \tag{13.239}$$

These parameters I_U, I_p are not independent but are related through the Mach number M

$$I_p = \gamma M^2 I_U \tag{13.240}$$

Under conditions where $R_m \ll 1$, magnetic diffusion effects dominate over convection, and a more appropriate choice of parameters are the *interaction parameters*, S_U and S_p. The parameter S_U is defined as the ratio of the Lorentz force (with $\overline{\mathbf{J}}$ expressed from Ohm's Law) to the change of fluid momentum in length L

$$S_U = \sigma B^2 L/\rho U \tag{13.241}$$

The parameter S_p is the ratio of the Lorentz force similarly expressed to the pressure gradient over length L

$$S_p = \sigma UB^2L/p \qquad (13.242)$$

The parameters S_U and S_p are not independent of I_U and I_p but are related through the magnetic Reynolds number

$$S_U = R_m I_U \qquad (13.243a)$$

$$S_p = R_m I_p \qquad (13.243b)$$

The *Hartmann number*, H_a, is defined as

$$H_a = (S_U R_e)^{1/2} \qquad (13.244)$$

and is a measure of the ratio of the Lorentz force to the viscous force.

For general magnetohydrodynamic flow, a typical independent and complete set of parameters are M, R_e, P_r, R_m, and S_U. When the electrical conductivity is tensorial due to the Hall effect, the *Hall parameter*, β, must be added to this set.

13.8.3   Simple Magnetohydrodynamic Processes

Local Currents and Fields. Consider the nature of the currents and fields $\bar{\mathbf{J}}$ and $\bar{\mathbf{E}}$ in a local region of flow. From the Ohm's Law, Eq. (13.220), it can be seen that the magnetic field effects involve $\bar{\mathbf{U}} \times \bar{\mathbf{B}}$ and $(\bar{\mathbf{U}} \times \bar{\mathbf{B}}) \times \bar{\boldsymbol{\beta}}$. These effects are confined to the plane perpendicular to $\bar{\mathbf{B}}$. Orient $\bar{\mathbf{B}}$ in the z direction, $\bar{\mathbf{B}} = \bar{\mathbf{B}}(0, 0, B_z)$; $\bar{\mathbf{U}}$ in the x (axial) direction, $\bar{\mathbf{U}} = \bar{\mathbf{U}}(U_x, 0, 0)$, and let $\bar{\mathbf{E}} = \bar{\mathbf{E}}(E_x, E_y, E_z)$, where y represents the transverse direction. The axial and transverse *load factors*, K_x, K_y, which serve to summarize the nature of the external connection to the fluid are defined as

$$K_x = \frac{-E_x}{\beta U_x B_z(1 - K_y)} \qquad (13.245a)$$

$$K_y = E_y/U_x B_z \qquad (13.245b)$$

The Ohm's Law, Eqs. (13.220), then become

$$J_x = (1 - K_x)(1 - K_y)\sigma/(1 + \beta^2)(\beta U_x B_z) \qquad (13.246a)$$

$$J_y = -(1 - K_y)(1 + K_x\beta^2)\sigma/(1 + \beta^2)(U_x B_z) \qquad (13.246b)$$

For $K_y = 1$, both the axial and transverse currents are *open circuited*, and from Eqs. (13.245), $E_x = 0$ and $E_y = U_x B_z$, its open circuit value. For $K_x = 1$ for any K_y, the axial current vanishes, and the transverse current is given by

$$J_y = -(1 - K_y)\sigma U_x B_z \qquad (13.247)$$

Thus, if J_x is prevented from flowing, the transverse current, J_y, achieves a level equal to that which it would have in the absence of the Hall effect; an axial or *Hall field*, E_x, will be present given, however, by $E_x = -\beta E_y$, where $E_y = (1 - K_y) U_x B_z$. Clearly for large Hall parameter β, the axial field will dominate the transverse field. On the other hand if $K_x = 0$, the axial field is shorted, and the transverse current from Eqs. (13.246) is reduced by the factor $1/(1 + \beta^2)$ from its value in the unshorted case given by Eq. (13.247).

Force, Power, and Dissipation. The local Lorentz force $\bar{\mathbf{f}}(f_x, f_y, 0)$ and power P are given from Eqs. (13.214), (13.215), and (13.246) as

$$f_x = -(1 - K_y)(1 + K_x\beta^2)/(1 + \beta^2)(\sigma U_x B_z^2) \tag{13.248a}$$

$$f_y = -(1 - K_y)(1 - K_x)/(1 + \beta^2)(\beta\sigma U_x B_z^2) \tag{13.248b}$$

$$P = -(1 - K_y)[K_y(1 + \beta^2 K_x)$$
$$- (1 - K_y)(1 - K_x)K_x\beta^2]/(1 + \beta^2)(\sigma U_x^2 B_z^2) \tag{13.249a}$$

and the dissipated power $P' = \bar{\mathbf{J}} \cdot \bar{\mathbf{E}}'$ is given by

$$P' = (1 - K_y)^2(1 + \beta^2 K_x^2)/(1 + \beta^2)(\sigma U_x^2 B_z^2) \tag{13.249b}$$

For $K_x < 1$ and $K_y < 1$, the force, f_x, is opposed to the flow velocity $\bar{\mathbf{U}}$, and power $P < 0$ is extracted from the flow. For $K_y > 1$, the force, f_x, is in the direction of the flow, and power $P > 0$ must be supplied to the fluid by the fields. For $K_x \neq 0$ a transverse force, f_y, will exist in addition to the axial force, f_x, for $\beta > 0$. For strong Hall effect $\beta \gg 1$, this force will dominate the axial force, f_y, creating significant transverse pressure gradients and, under some circumstances, secondary flows.

The total force, F_x, and power, P, which can be developed in a volume, V, are given by

$$F_x = -\int_V dV(1 - K_y)\sigma U_x B_z^2 \tag{13.250}$$

$$p = -\int_V dV(1 - K_y)K_y\sigma U_x^2 B_z^2 \tag{13.251}$$

In a uniform system, these expressions have the simple forms

$$F_x = -(1 - K_y)\sigma U_x B_z^2 V \tag{13.252}$$

$$p = -(1 - K_y)K_y\sigma U_x^2 B_z^2 V \tag{13.253}$$

It is important to observe that Eqs. (13.252) and (13.253) follow from Eqs. (13.250) and (13.251) only in the case that K_y, σ, U_x, and B_z are uniform over the volume V. In most realistic magnetohydrodynamic flows, these quantities are significantly non-uniform [Argyropoulos, Demetriades, and Kendig (1967), Demetriades and Argyropoulos (1969)] and the total force, F_x, or the total power, P, cannot be reliably estimated from Eqs. (13.252) and (13.253).

Magnetic Meter and Pump. Consider a channel aligned in the x direction with crossed electric $\overline{\mathbf{E}}(0, E_y, 0)$ and magnetic $\overline{\mathbf{B}}(0, 0, B_z)$ fields. The steady one-dimensional incompressible flow equations in the flow magnetic Reynolds number limit with f_x given by Eq. (13.248a) are

$$\partial U/\partial x = 0 \tag{13.254}$$

$$\partial(\tfrac{1}{2}\rho U^2 + p)/\partial x = -(1 - K_y)\sigma UB_z^2 \tag{13.255}$$

Since the condition $\nabla \times \overline{\mathbf{E}} = 0$ holds, we have $E_y = -\Phi/h$ where Φ is the voltage across the gap, h, thus, from Eq. (13.245) the voltage is given by $\Phi = -K_y UBh$. The pressure change, Δp, in length, L, of the channel from Eq. (13.255) is

$$\Delta p = (1 - K_y)\sigma UB_z^2 L \tag{13.256}$$

The total current per unit depth which flows in length, L, is given by

$$I = J_y L = -(1 - K_y)\sigma UB^2 L \tag{13.257}$$

The pressure change can thus be represented as

$$\Delta p = -IB \tag{13.258}$$

The power delivered to the fluid per unit depth is

$$P_L = J_y E_y L = -I\Phi \tag{13.259}$$

At open circuit $I = 0$, and from Eq. (13.259) the voltage at this condition is given by $\Phi = \Phi_{oc} = -UB_z h$. This voltage can, in principle, be used to meter the average velocity, U, which exists in the channel when a known magnetic field is applied over the gap of known dimension h. In practice, such meters must be corrected for leakage current losses and polarization voltages which exist on the electrodes for small devices. For a device of dimension $h = 10$ cm and imposed magnetic field $B_z = 0.1$ Tesla, a velocity of 10 m/s will register only 100 mV. Faraday attempted to measure the velocity of the Thames River using this principle. Although $h \simeq 100$ m for the river, the earth's magnetic field is of the order 10^{-4} Tesla, and thus $\Phi_{oc} \simeq 10$ mV for $U \simeq 1$ m/s which is in the noise range of the *spurious leakage* and *polarization effects*, not to say modern habitation noise.

To operate the channel as a *pump* ($\Delta p > 0$), a voltage, $\Phi < -UBh$, must be imposed across the gap. At the no-flow condition, $U = 0$, the pressure rise is maximum and is given by

$$\Delta p_{max} = (-\Phi)\sigma B(L/h) \tag{13.260}$$

For liquid metals, the pumping voltage is low because of the high electrical conductivity. For $L/h \sim 1$, $B \sim 10^{-1}$ Tesla, $\sigma \sim 10^7$ S/m, and $\Delta p_{max} \sim 1$ atm, the pumping voltage, Φ, is of the order $\Phi \sim 100$ mV. The corresponding current from Eq. (13.256) is $I \sim 10^6$ A/m, and the power is $P_L = 100$ kW/m. These voltage-current characteristics place a severe demand on the pump power supply.

13.8.4 Simple Magnetohydrodynamic Flows

Quasi-One-Dimensional Compressible Flow. In the quasi-one-dimensional description, the flow variables are averaged quantities over the cross section of a duct. The conservation laws, Eqs. (13.216), integrated over the cross section with the variables now representing averages over the cross section, are

$$\frac{\partial}{\partial t}(\rho A) + \frac{\partial}{\partial x}(\rho U A) = \sum_{\rho} \tag{13.261a}$$

$$\frac{\partial}{\partial t}(\rho U A) + \frac{\partial}{\partial x}(\rho U^2 A) = -A\left(\frac{\partial p}{\partial x} - f_x\right) - \tau_W C + \sum_U \tag{13.261b}$$

$$\frac{\partial}{\partial t}[\rho(e + \tfrac{1}{2}U^2)A] + \frac{\partial}{\partial x}(\rho U h_0 A) = PA + q_W C + \sum_e \tag{13.261c}$$

The dependent fluid variables are the pressure, p, enthalpy, $h = e + p/\rho$, and stagnation enthalpy, $h_0 = h + \tfrac{1}{2}U^2$. The local duct cross-sectional area is A, and the local perimeter is C. The functions Σ_ρ, Σ_U, Σ_e describe the sources of mass, momentum and energy due to mass addition (or depletion) within the flow train. In addition to the conservation laws the caloric and kinetic equations as given by Eqs. (13.217) complete the description. The average wall shear stress over the cross section is τ_W, and the average heat flux is q_W. These averages are composites of the electrode wall and insulating wall shear stresses and heat fluxes. These quantities may be computed from the boundary layer equations for each wall. The results of these boundary layer calculations may be summarized in the local, *dynamic friction factor*, $C_f(x, s, t)$, and Stanton number, $S_T(x, s, t)$, where s is a perimeter coordinate [Oliver, Crouse, Maxwell, and Demetriades (1980)].

The most fundamental solution of Eqs. (13.261) which illustrate electromagnetic effects are those for steady, constant area flow in the absence of viscous and heat conduction effects. Equations (13.261) are three simultaneous equations for the primitive variables ρ, U, and e. We may eliminate these variables in favor of one equation in the Mach number $M = U/\sqrt{\gamma p/\rho}$ using Eqs. (13.248) and (13.249) for f_x and P with $K_x = 1$

$$\frac{1}{M^2}\frac{dM^2}{dx} = \frac{s_p}{1 - M^2}\left[\frac{(1 - \gamma M^2)}{\gamma}(1 - K_y)^2\right.$$
$$\left. + (1 - K_y)\gamma M^2\left(1 + \frac{\gamma - 1}{2}M^2\right)\right] \tag{13.262}$$

In Eq. (13.262), $s_p \equiv \sigma U B^2/p$, which is the pressure interaction parameter per unit length. This equation defines the conditions which drive the flow to or from the *choke point*, $M = 1$. The term in $(1 - K_y)^2$ represents the effect of Joule dissipation, while that proportional to $(1 - K_y)$ represents the Lorentz force. When $(1 - K_y)$

<< 1 is satisifed, the Joule dissipation term is negligible compared to the force term. For all $K_y < 1$, the Lorentz force opposes the flow and drives the flow to the sonic point, $M = 1$. For all $K_y > 1$, the Lorentz force acts in the direction of the flow and drives the flow away from the sonic point. When $(1 - K_y) \sim 1$ holds, the effect of the Joule heating term is always to drive the flow to the sonic point; there are, however, values of K_y for which the bracketed term on the right hand side of Eq. (13.262) vanishes and continuous passage through $M = 1$ in an *electromagnetic throat* is, in principle, possible [Neuringer (1963)].

Exact integrals for steady flow in the absence of friction and heat transfer exist [Rosa (1968), pp. 51–54]. For the general case of flow which is unsteady and in which viscous and heat conduction effects are important, the solutions are best obtained by finite difference methods. Such solutions are presented in Sec. 3.8.5.

Viscous, Incompressible, Magnetohydrodynamic Flow. Equations (13.216) and (13.218) for steady, incompressible, low R_m flow in the x direction between the planes $z = \pm h$ in the presence of $\bar{\mathbf{B}} = \bar{\mathbf{B}}(0, 0, B_z)$, $\bar{\mathbf{U}} = \bar{\mathbf{U}}(U_x, 0, 0)$, $\bar{\mathbf{J}} = \bar{\mathbf{J}}(0, J_y, 0)$, and $\bar{\mathbf{E}} = \bar{\mathbf{E}}(0, E_y, 0)$ reduce to

$$\frac{\partial U_x}{\partial x} = 0 \tag{13.263a}$$

$$\frac{\partial p}{\partial x} = J_y B_z + \rho\nu \frac{\partial^2 U_x}{\partial z^2} \tag{13.263b}$$

$$J_y = -\sigma(E_y - U_x B_z) \tag{13.263c}$$

Since $U_x = U_x(z)$, define K_y as $K_y = E_y/U_{ave}B$ where U_{ave} is the average velocity between the planes $z = \pm h$. The Ohm's Law, Eq. (13.263c), then becomes

$$J_y = -\sigma(U_x - U_{ave}K_y)B \tag{13.264}$$

It can be seen that the current reverses at the point where $U_x = K_y U_{ave}$ with $J_y \lessgtr 0$ for $U_x \gtrless K_y U_{ave}$. Combining Eqs. (13.263b) and (13.264) and observing that $\partial^2 p/\partial x\, \partial z = 0$, since there is no motion in the transverse direction, there results

$$\frac{\partial^2 U_x}{\partial z^2} - \frac{H_a^2}{h^2} [U_x - (K_y + 1/S_p)U_{ave}] = 0 \tag{13.265}$$

where $H_a \equiv Bh\sqrt{\sigma/\rho\nu}$ is the Hartmann number and $S_p = \sigma B^2 U_{ave}/(-\partial p/\partial x)$ is the *pressure interaction parameter*.

The solution to Eq. (13.265) for $U_x(z)$ satisfying the boundary conditions $U_x(h) = U_x(-h)$ is

$$U_x(z) = (K_y + 1/S_p)U_{ave}\left[1 - \frac{\cosh\,(H_a z/h)}{\cosh\,(H_a)}\right] \tag{13.266}$$

By taking the average of Eq. (13.266), the interaction parameter, S_p, may be represented as

$$S_p = \frac{\sigma B^2 U_{ave}}{-\partial p/\partial x} = \frac{H_a - \tanh(H_a)}{(1 - K_y)H_a + K_y \tanh(H_a)} \qquad (13.267)$$

Equation (13.267) may be used to determine the average velocity in terms of the Hartmann number and the pressure gradient. For $H_a \gg 1$, $S_p \sim 1/(1 - K_y)$ with the Lorentz force balancing the pressure gradient. For $H_a \ll \tanh(H_a) = H_a - H_a^3/3$ and $S_p \sim H_a^2$ so that the pressure gradient is balanced by the viscous force.

The velocity profiles for insulating walls ($K_y = 1$) and a fixed interaction parameter S_p are shown in Fig. 13.68 as a function of Hartmann number. The effect of the magnetic field is to flatten the velocity profile in the core where $J_y < 0$ and $f_x < 0$ and to enhance the profile near the sidewalls where $J_y > 0$ and $f_x > 0$.

13.8.5 Large Magnetic Reynolds Number

Flow at large magnetic Reynolds number is characterized by the presence of magnetic fields induced by currents flowing in the fluid which are comparable to externally imposed magnetic fields. These internal fields, created by the internal currents tend to oppose the imposed field (*Lenz' Law*) thereby driving the total magnetic field

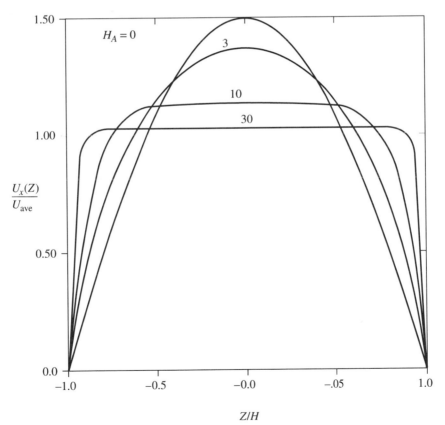

FIGURE 13.68 Viscous magnetohydrodynamic flow velocity profiles $U_x(z)$ as a function of Hartmann number H_a. Insulating walls ($K_y = 1$).

from the conducting fluid. At large magnetic Reynolds numbers, the currents and fields are confined to a thin zone at the boundaries of the conducting fluid.

Magnetic Field of a Conducting Fluid Slab. A fluid with a uniform velocity, $\overline{U} = \overline{U}(U_x, 0, 0)$, moves in a channel of dimension $a \ll L$ where L is the scale of variation of an imposed magnetic field $\overline{B}_0 = \overline{B}_0(0, 0, B_z)$. Scalar conductivity of magnitude σ is switched on between the planes $x = 0$ and a. The conductivity vanishes outside this slab.

The governing equation for $B_z(x)$ in steady flow is Eq. (13.229) (with $\overline{\beta} = 0$)

$$U \frac{\partial B_z}{\partial x} = \eta \frac{\partial^2 B_z}{\partial x^2} \tag{13.268}$$

while the current is determined from Eq. (13.227b)

$$J_y = -\mu_0^{-1} \frac{\partial B_z}{\partial x} \tag{13.269}$$

The boundary conditions are $B_z(a) = B_z^+$ and $B_z(0) = B_z^-$. The total current per unit depth flowing through the slab is given by

$$I = \int_0^a J_y \, dx = -\mu_0^{-1}(B_z^+ - B_z^-) \tag{13.270}$$

This current is also given in terms of the *load factor* $K_y \equiv E_y/UB_0$ from Eq. (13.247) as

$$I = -\int_0^a \sigma UB_0(B_z/B_0 - K_y) \, dx \tag{13.271}$$

In the limit in which we have $I = 0$, the field B_z must be equal to the applied field B_0. This is equivalent to demanding that the average of the edge fields be equal to the imposed field B_0; i.e., $(B_z^+ + B_z^-)/2 = B_0$.

The solutions to Eqs. (13.268) and (13.269) subject to these conditions are

$$B_z = B_0\{1 + (1 - K_y) [1 - e^{R_m(x/a - 1/2)}/\cosh (R_m/2)]\} \tag{13.272a}$$

$$J_y = (\mu_0^{-1}B_0/a) (1 - K_y)e^{R_m(x/a - 1/2)}/\cosh (R_m/2) \tag{13.272b}$$

From Eqs. (13.272a) and (13.272b), it can be seen that the field and current are progressively confined to a thin zone at the front of the conducting slab as the magnetic Reynolds number is increased (see Fig. 13.69). The total power extracted from the conducting slab per unit of cross-section flow area A may be expressed as

$$P/A = \int_0^a J_y E_y \, dx = -4K_y(1 - K_y) (B_0^2/2\mu_0)U \tanh (R_m/2) \tag{13.273}$$

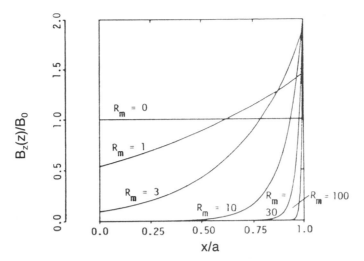

FIGURE 13.69 Variation of magnetic field as a function of nondimensional distance in flow direction with magnetic Reynolds number as a parameter.

As the magnetic Reynolds number is increased (e.g., by increasing the conductivity), $\tanh(R_m/2)$ approaches 1, and the maximum extracted power under load matching conditions ($K_y = 1/2$) for an ideal MHD generator is limited to the product of the magnetic energy of the applied field, B_0^2, and the flow velocity, U, as shown by Demetriades, Demetriades, and Demetriades (1985), who also introduce the concept of an optimum (or maximum allowable) conductivity.

Planar Flow in an Inhomogeneous Field. The behavior of the fields and currents in flow through an imposed localized magnetic field at large magnetic Reynolds number is inhernetly multi-dimensional. The imposed field $\overline{\mathbf{B}}_0 = \overline{\mathbf{B}}_0(0, 0, B_0)$ may be represented as

$$B_0(x) = B_0 f(x) \tag{13.274}$$

where B_0 is a constant amplitude and $f(x)$ is a shape function with a maximum at $f(0)$. The fluid is contained within the planes $y = \pm h$ and flows steadily at a uniform velocity $\overline{\mathbf{U}} = \overline{\mathbf{U}}(X_x, 0, 0)$. Conductors are arranged on the planes $y = \pm h$ over the region $|x| \leq a$. The region $|x| > a$ is an insulating surface.

The governing equation for $B_z(x, y)$ is the two-dimensional form of Eqs. (13.229)

$$U_x \frac{\partial B_z}{\partial x} = \eta \left(\frac{\partial^2 B_z}{\partial x^2} + \frac{\partial^2 B_z}{\partial y^2} \right) \tag{13.275}$$

and the current $\overline{\mathbf{J}} = \overline{\mathbf{J}}(J_x, J_y, 0)$ is given by

$$J_x = \mu_0^{-1} \frac{\partial B_z}{\partial y}, \qquad J_y = -\mu_0^{-1} \frac{\partial B_z}{\partial x} \tag{13.276}$$

The boundary conditions are that the current vanishes at $x = \pm\infty$, thus $\partial B_z/\partial x = 0$ at $x = \pm\infty$. On the conductors, the tangential field $E_x = \sigma^{-1}J_x$ must vanish, while the normal current J_y must vanish on the insulating surfaces.

Because of the nonuniform shape, $f(x)$, and the mixed boundary conditions on $y = \pm h$ the solutions of Eqs. (13.274) to (13.276) are obtained using finite-difference methods [Oliver, Swean, and Markham (1981)]. The appropriate non-dimensional parameters are the magnetic Reynolds number, $R_m = \mu_0\sigma Ua$, and the current load parameter, $\mu_0 I/B_0$, where the total current per unit depth flowing through the conductors to the external circuit is I.

In Fig. 13.70(a) the open circuit and in Fig. 13.70(b) the loaded condition for a magnetic Reynolds number $R_m = 1$ are exhibited. It can be seen that there is an onset of eddy current cells induced in the regions of imposed magnetic field gradients and a convection of the current pattern downstream of the conductors. The principal current flowing in the conductors, however, is still confined to the region between the electrodes.

In Fig. 13.71, the same situation is represented but for a magnetic Reynolds number $R_m = 7$. Considerable current flow induced by the magnetic field gradients exists under open circuit conditions. Note how the upstream eddy cell actually couples into the electrodes. Under load, the current exists completely downstream from the electrodes.

13.8.6   Magnetohydrodynamic Power Generators

Magnetohydrodynamic power generators are compact, high energy density converters of thermal energy to electricity. Such generators may function as *electromagnetic*

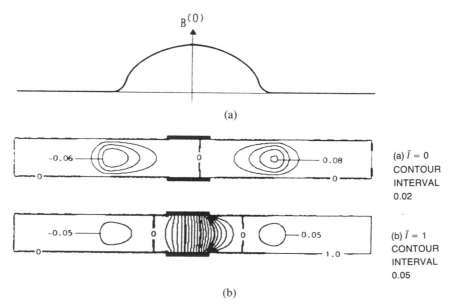

(a) $\bar{I} = 0$
CONTOUR
INTERVAL
0.02

(b) $\bar{I} = 1$
CONTOUR
INTERVAL
0.05

FIGURE 13.70   Induced magnetic field isolevels for $(B_z - B_0)/B_0$ for steady magnetohydrodynamic flow with profiles similar to those of Fig. 13.69 and $R_m = 1$. (a) The electrodes at open circuit ($\mu_0 I/B_0 = 0$), (b) the electrodes connected to a load ($\mu_0 I/B_0 = 1$).

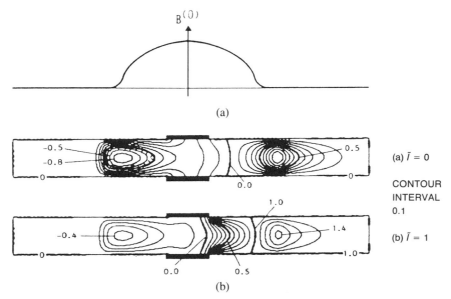

FIGURE 13.71 Induced magnetic field isolevels for $(B_z - B_0)/B_0$ for steady magnetohydrodynamic flow with the profiles of Fig. 13.69 and $R_m = 7$. (a) The electrodes are at open circuit ($\mu_0 I/B_0 = 0$), (b) the electrodes are connected to a load ($\mu_0 I/B_0$).

turbines in a complete power system cycle with either a fossil or nuclear heat source and a steam bottoming plant. In this function, the generators serve as high temperature turbines capable of handling a top cycle temperature in excess of 2500 K as well as slag-laden working fluid in the case of coal fired plants. Overall, MHD topped plant efficiencies should be in the range of 50 60% with the MHD generator extracting 20–25% of the input chemical energy [Heywood and Womack (1969)]. MHD generators may also function as portable devices with a rocket type combustor for short bursts of power [Maxwell and Demetriades (1986)], or the generators may be inherently unsteady, driven by high energy explosives [Bangerter, Hopkins, and Brogan (1975)].

Ideal Generators. The fundamental starting point of low-magnetic-Reynolds-number generator designs are the expressions for force and power in a basic crossed-field situation, which are Eqs. (13.248) and (13.247). Three principal generator types follow from these expressions—Faraday, Hall, and Diagonal. In the *Faraday generator* (Fig. 13.72), the electrode walls must be segmented and multiple loads attached to preserve the condition of vanishing Hall current ($J_x = 0$, $K_x = 1$). The load current in an axial length L is given by $I = J_y L$, and the voltage across the electrodes of separation, h, by $\Phi = -E_y h$. This configuration possesses only an axial Lorentz force and has a power density which is not degraded by the Hall effect. The large number of load circuits are a disadvantage in this configuration. In the *Hall generator* (Fig. 13.73), the Faraday circuits are shorted ($K_y = 0$), and the ends of the machine are connected to a single load. The load current for a machine of length, L, and transverse height, h, becomes $I = J_x h$, and the load voltage for this configuration is $\Phi = -E_x L$. This machine is only efficient at large Hall parameter

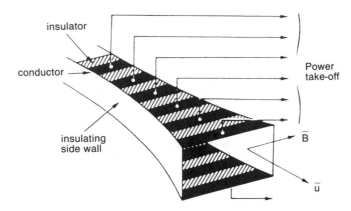

FIGURE 13.72   Magnetohydrodynamic Channel Configuration with the Faraday connection. Each electrode pair is individually loaded.

β; in addition, a large transverse Lorentz force, f_y, exists which exceeds the axial Lorentz force and develops secondary flows and transverse pressure gradients which degrade performance.

The *Diagonal generator* is constructed much like a Faraday generator (see Fig. 13.74). From Eqs. (13.245), the angle α of the equipotential lines which are orthogonal to field lines is given by

$$\tan \alpha = -\frac{E_y}{E_x} = \frac{K_y}{\beta(1 - K_y)K_x} \tag{13.277}$$

For a given Hall parameter, β, and operating load condition, K_y, the equipotentials which are orthogonal to the field lines given by Eq. (13.277) may be made of conductors and the power again extracted by a single load attached to the diagonal electrodes. This machine has the single load advantage of the Hall machine without its performance degradation. In this case, the load current into a diagonal conductor

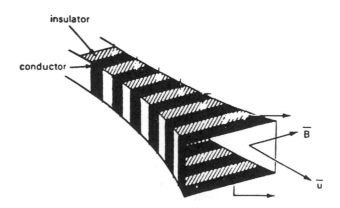

FIGURE 13.73   Magnetohydrodynamics Channel in the Hall Configuration. Faraday circuits are shorted, and power is extracted from the ends.

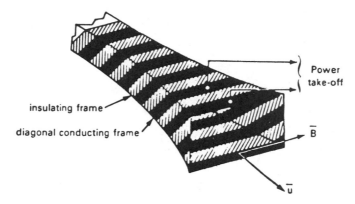

FIGURE 13.74 Magnetohydrodynamic Channel in Diagonal Conducting Wall Configuration. Diagonal conducting bars are aligned with design point equipotentials, and power is extracted from the ends.

is given by $I = h(J_x - \tan \alpha J_y)$, and the load voltage by $\Phi = -E_x L$. The local electrical efficiency, η, is defined as the ratio of the electrical power P to the mechanical power flow $\bar{f} + \overline{U}$. The force, power, efficiency of these configurations are given in Fig. 13.75. It will be noted that the diagonal configuration becomes inefficient at operating conditions away from its design point.

The maximum output power density for these machines is $(1/4)\sigma U^2 B^2$. For fossil fuel combustion products seeded with an easily ionized substance such as a potassium or cesium salt, the achievable conductivity is of the order of 10 S/m, the velocity of order 10^3 m/s, and the magnetic field of order 10 Tesla. The corresponding power produced by 1 m^3 of fluid is 10^3 MW.

The Magnetohydrodynamics of Real Generators. The magnetohydrodynamic flow in real generators is considerably more complex than the simple flows described in Secs. 13.8.3 and 13.8.4. MHD generators are principally configured in linear ducts in which a flow proceeds along the axis of a duct of general and variable cross sectional shape and size. The Mach numbers of interest are high subsonic ($M \gtrsim 1$) or moderately supersonic $M \gtrsim 1$. The MHD interaction parameter, S_U, will be of the order 6–10 for commercial scale (300 MWth–2000 MWth) systems. The typical duct L/D will be of the order $L/D \simeq 10$. The viscous Reynolds number will be of the order $R_e = 10^6$ for typical 1 m scale cross sections; the flow will therefore be turbulent in viscous regions. The turbulent MHD flow in such ducts will exhibit complex multi-dimensional fluid and electrical phenomena which are indicated in Fig. 13.76.

Boundary Layers: The fluid in the *anode wall layer* is subjected to a decelerating axial Lorentz force which is comparable to that in the core. It is also subjected to a transverse Lorentz force which results from the Hall current induced by conductivity nonuniformities between the boundary layer and the core. This force (and induced Hall current) exist even in a Faraday connection. The forces on the fluid in the anode boundary layer thus tend to lift the fluid off the wall and to induce a transverse pressure gradient. This creates an anode boundary layer which grows in thickness

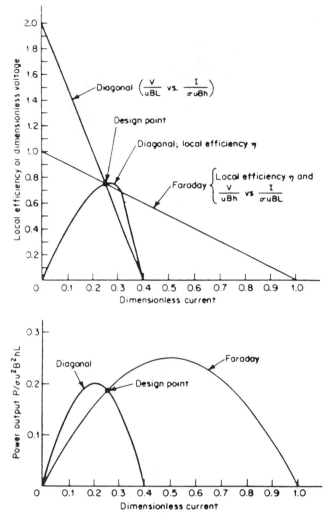

FIGURE 13.75 Comparison of the output characteristics of a diagonal and a Faraday generator. (From Rosa, 1968.)

faster than a conventional layer, has weaker turbulence production within it, a larger boundary layer shape factor, and a greater tendency to stall. Such boundary layers, computed with the multi-dimensional equations and an appropriate turbulence theory [Argyropoulos, Demetriades, and Lackner (1971)] are shown in Figs. 13.77 to 13.79 for a generator duct flow with an interaction parameter, S_U, of approximately 10.

Like the anode wall, the *cathode wall layer* is subjected to a decelerating Lorentz force which is comparable to that in the core. The nonuniformity induced transverse force, however, is in the direction to drive the cathode wall layer fluid into the wall. A corresponding transverse pressure gradient exists as shown in Fig. 13.77. In contrast to the anode wall layer, the transverse force on the cathode layer intensifies the turbulence production within the layer producing a thinner boundary layer with a lower shape factor and a diminished tendency to stall compared to a conventional

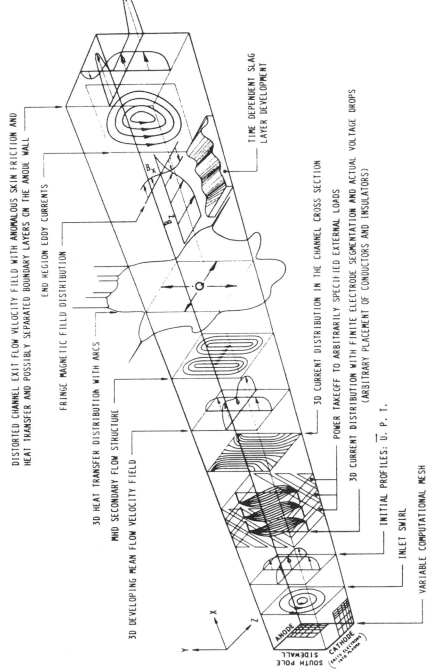

SELECTED CRITICAL PHENOMENA IN COAL FIRED MHD GENERATORS REVEALED BY STD/MHD CODES

DISTORTED CHANNEL EXIT FLOW VELOCITY FIELD WITH ANOMALOUS SKIN FRICTION AND
HEAT TRANSFER AND POSSIBLY SEPARATED BOUNDARY LAYERS ON THE ANODE WALL

END REGION EDDY CURRENTS

FRINGE MAGNETIC FIELD DISTRIBUTION

3D HEAT TRANSFER DISTRIBUTION WITH ARCS

MHD SECONDARY FLOW STRUCTURE

3D DEVELOPING MEAN FLOW VELOCITY FIELD

TIME DEPENDENT SLAG
LAYER DEVELOPMENT

3D CURRENT DISTRIBUTION IN THE CHANNEL CROSS SECTION

POWER TAKEOFF TO ARBITRARILY SPECIFIED EXTERNAL LOADS

3D CURRENT DISTRIBUTION WITH FINITE ELECTRODE SEGMENTATION AND ACTUAL VOLTAGE DROPS
(ARBITRARY PLACEMENT OF CONDUCTORS AND INSULATORS)

INITIAL PROFILES: $\vec{U}$, P, T,

INLET SWIRL

VARIABLE COMPUTATIONAL MESH

ANODE

SOUTH POLE
SIDEWALL

CATHODE
(EMITS ELECTRONS
INTO PLASMA)

B_x

B_z

X

Y

Z

FIGURE 13.76 Schematic representation of inherently three-dimensional fluid dynamic and electrical phenomena in linear MHD generator ducts. (Reprinted with permission from STD Research Corp.)

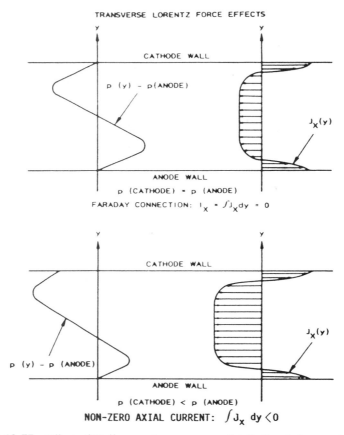

FIGURE 13.77   Effect of Hall current on pressure distribution between electrodes.

boundary layer. These general features of the cathode boundary layer for an $S_U \sim$ 10 scale machine may be seen in Fig. 13.79.

The decelerating Lorentz force which the fluid in the sidewall boundary layer experiences is significantly diminished compared to that in the core. In a uniform conductivity flow, the velocity profiles in the boundary regions are enhanced by the Lorentz force (Hartmann effect) as described in Sec. 13.8.3. When conductivity nonconformities are considered, the Hartmann effect becomes more pronounced; under circumstances of strong interaction ($S_U \gtrsim 1$), the boundary layers possess velocity overshoots. As a result, the sidewall boundary layers experience an accelerating pressure gradient compared to the core leading to the development of velocity overshoots and negative displacement thicknesses (Fig. 13.78). The turbulence production in the part of the boundary layer below the maximum velocity point is greatly intensified over that in the anode and cathode layers due to the enhanced shearing rate of the overshot velocity profile. The *skin friction and heat transfer are greatly enhanced on the sidewalls* (see Fig. 13.79). The sidewall layers are also free of Lorentz forces in the direction normal to their surfaces, since this is the magnetic field direction. The nature of the sidewall boundary layers in a 300 MWth machine can be seen in Fig. 13.78. Boundary layer profiles of this nature have been measured by Daily, Kruger, Self, and Eustis (1976).

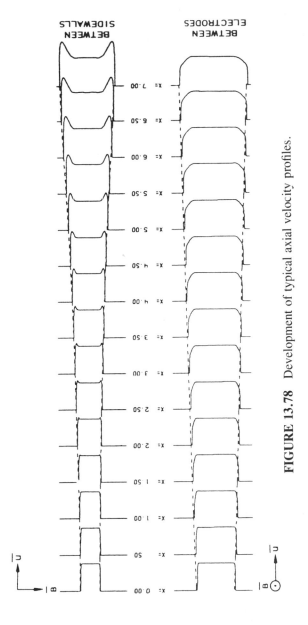

FIGURE 13.78 Development of typical axial velocity profiles.

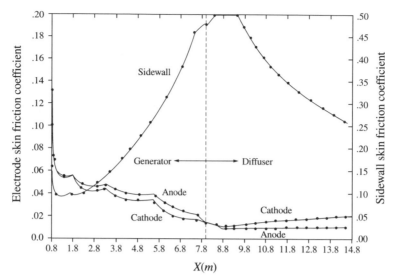

FIGURE 13.79 Skin-friction distributions on electrode and sidewalls for a typical high interaction MHD generator and diffuser.

In a conducting sidewall generator, the current is no longer constrained to enter the electrode across the magnetic field lines. Under some circumstances it is therefore possible for the *electrode wall* boundary layers to be relieved by their Lorentz forces as the current vector aligns itself with the magnetic field in the boundary layer region ($\overline{\mathbf{J}} \times \overline{\mathbf{B}} = 0$). In such a situation, the electrode wall layers are free to experience the analogous acceleration which the sidewalls experience in an insulating wall generator and to even develop velocity overshoots, correspondingly enhanced turbulence production, shearing rate, and enhanced heat transfer rates [Markham, Maxwell, Demetriades, and Oliver (1977)].

The actual structure of the flow in such cases will be complex, because zones of enhanced Lorentz force will develop just *outside* the boundary layer edge as the current density intensifies but still contains a significant component in the transverse direction. Further, secondary flows which transport high momentum fluid between the sidewalls, core, and electrode walls can be expected to modify the final balance of momentum in the boundary layer regions. *High-interaction conducting sidewall machines*, therefore, require the most careful analysis of multidimensional and secondary flow effects.

Secondary Flow: It was shown in Eq. (13.224) that gradients of the current density in the magnetic field direction are a source of vorticity. When such current gradients arise purely from velocity gradients at uniform conductivity, the vorticity production is negative, i.e., the vorticity is damped [Heywood (1968)]. On the other hand, when the current gradient is induced by an electrical conductivity gradient, the vorticity production is positive. Such nonuniformity induced gradients are of the essence in *MHD induced secondary flow*. Because of the low conductivity region near the electrodes, a Hall current, J_x, is induced over the electrodes [Demetriades, Argyropoulos, and Casteel (1970)]. This Hall current varies in the magnetic field direction due to the sidewall cooling, and this current density gradient becomes a source of axial vorticity. The feedback of this vorticity on the conductivity distribution can lead to amplification and an instability known as the *magnetothermal instability* [Demetrides, Oliver, Swean, and Maxwell (1981)].

A three-dimensional evolving flow which is *magnetoaerothermally unstable* is shown in Figs. 13.80 and 13.81. The case considered is for a subsonic segmented Faraday generator with a maximum magnetic field strength of 6 Tesla and an interaction $S_U = 10$. In Fig. 13.80, the structure of the flow after one meter of evolution down the channel is shown. It should be noted particularly that MHD forces have already dramatically altered the turbulently generated corner secondary flow cells transforming them to fill the full cross section of the channel. In Fig. 13.81, the inherently magnetohydrodynamic character of the flow is evident in all its characteristics including the secondary flow effects. In particular, the growing, intensifying current concentration in the central anode region is prominent. This magnetoaerothermal current concentration has brought about an extreme deceleration of the primary flow in this region and has correspondingly dramatically reduced the local skin friction in the central anode region and increased the local shape factor driving this region of the anode boundary layer towards separation [Maxwell, Swean, Vetter, Crouse, Oliver, Bangerter, and Demetriades (1981)].

The *anode–sidewall corner cells* have evolved such that there is one large slowly rotating cell which tends to convect hot fluid into the anode and relatively colder

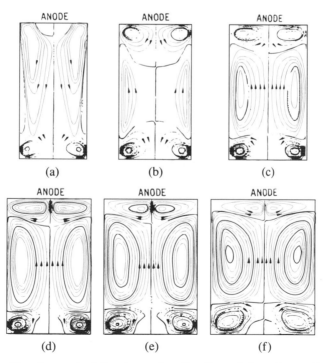

FIGURE 13.80 Development of the stream function formed from the vortical component of the secondary flows in the cross section at 1 m intervals in a 7 m long MHD generator duct. Station locations are relative to the first loaded electrode. (a) $x = 1$ m, (b) $x = 2$ m, (c) $x = 3$ m, (d) $x = 4$ m, (e) $x = 5$ m, and (f) $x = 6$ m.

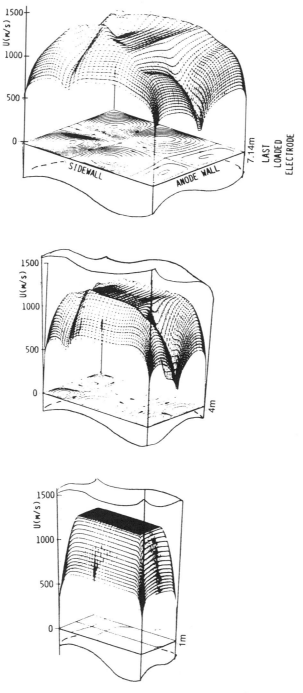

FIGURE 13.81   Velocity profiles and wall vorticity distributions for the case in Fig. 13.80. (Reprinted with permission from STD Research Corp.)

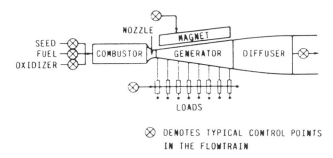

FIGURE 13.82 Transient MHD generator flow train.

fluid onto the cathode and a smaller more rapidly rotating cell which acts across the anode. In the cathode–sidewall corner, there is a large rapidly rotating cell which acts along the sidewall boundary layer region and a similar but smaller cell which acts over part of the cathode boundary layer region. Together, these cells are very efficient in bringing higher temperature fluid from the edge of the boundary layer into the corner region.

Joule heating participates with secondary flow convection in determining the temperature field. The locally enhanced temperature and conductivity produce regions of higher current densities J_y and J_x which, in turn, further heat the fluid through Joule heating. This is the result of intensified Joule dissipation in certain regions which is not convected away by the secondary flows or transferred to the walls by turbulent heat conduction.

Time Dependent Phenomena: Time dependent phenomena in a magnetohydrodynamic flow train as in Fig. 13.82, may be examined with the quasi-one-dimensional system of Eqs. (13.261) [Oliver, Crouse, Maxwell, and Demetriades (1980)]. A most basic time dependent event is the *start-up transient*. In Fig. 13.83, the time dependent start-up of a flow train after a steady oxidizer cold flow has been established is shown. This dynamic is exhibited in the space–time–amplitude plane (see Fig. 13.84) including the ignition in the combustor and the injection of the ionizable potassium carbonate which takes place approximately 8 milliseconds after ignition.

Of particular interest are the extreme excursions of the electric fields during this start-up transient. The axial field, E_x, experiences a severe overshoot as shown in Fig. 13.85, because the velocity in the channel is high before the seed comes on and the Lorentz force decelerates the flow to its design level.

13.8.7 Magnetohydrodynamic Accelerators

Steady Linear Accelerators. If power is supplied to the flow and $K_y > 1$ as shown in Sec. 13.8.2, the Lorentz force is in the direction of flow in an MHD channel, and the working fluid is accelerated [Demetriades (1964)]. Steady flow channel solutions for the exit state of the fluid in terms of the inlet state and the applied power may be obtained analogous to those for MHD power generators [Jahn (1968)]. The velocity enhancement U_e/U_0 (where U_e is the exit velocity of the channel, and U_0 is

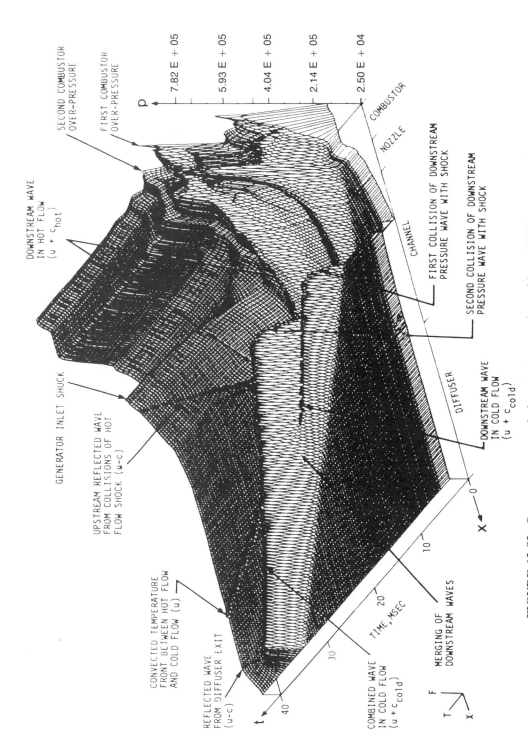

FIGURE 13.83 Pressure response during start-up transient of large MHD generator flow train.

868

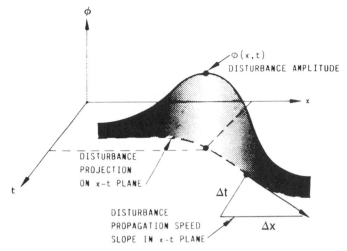

FIGURE 13.84 Amplitude-space-time representation of dynamic variables in transient flowtrain response. Any variable ϕ has an amplitude at any point in space x at time t which describes the strength of the disturbance. The projection of these amplitudes on the x–t plane defines a trajectory whose slope is the propagation speed of any disturbance in the variable ϕ.

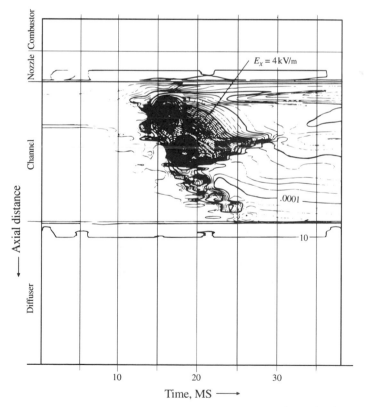

FIGURE 13.85 Hall field E_x response during the start-up transient of a large MHD flow train.

the inlet velocity) is a function of the interaction $S_U \sim \sigma B^2 L/\rho U$

$$U_e/U_0 = \left(1 + a \left\langle \frac{\sigma B^2}{\rho U} \right\rangle L\right)^\alpha \qquad (13.278)$$

where $\langle \ \rangle$ indicates an average over the channel and the pure numbers a and α have different order-unity values depending upon the particular assumptions of channel operation (isothermal, constant field, constant current, etc.).

The Magnetoplasmadynamic Accelerator. An accelerator of considerable interest is obtained by considering a steady flow accelerator of the type described in *Steady Linear Accelerators* in cylindrical coordinates with azimuthal symmetry (Fig. 13.86). For this configuration of anode and cathode in (r, ϕ, z) coordinates, we have $\overline{\mathbf{J}} = \overline{\mathbf{J}}(J_r, 0, 0)$ and $\overline{\mathbf{B}} = \overline{\mathbf{B}}(0, B_\phi, 0)$ where a current, I, is applied between the anode and the cathode, and the Hall current is suppressed. The magnetic field may be either applied or induced by the self-current ($R_m \gtrsim 1$) or both. With only variation in the r-direction, the current density is given in terms of the applied current by

$$J_r(r) = \frac{I}{2\pi r L} \qquad (13.279)$$

the field B_ϕ is determined from Eq. (13.227b)

$$\partial B_\phi/\partial z = -\mu_0 J_r \qquad (13.280)$$

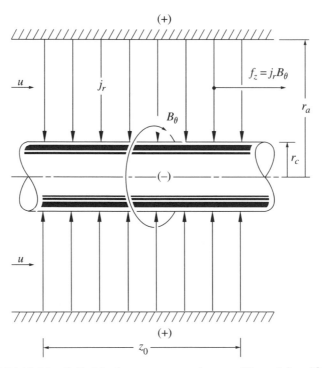

FIGURE 13.86 Cylindrical geometry accelerator. (From Jahn, 1968.)

and thus

$$B_\phi = -\frac{\mu_0 I}{2\pi r}(1 - z/L) \tag{13.281}$$

The Lorentz force from Eq. (13.227c) is given by $\bar{\mathbf{f}} = \bar{\mathbf{f}}(0, 0, f_z)$

$$f_z = J_r B_\phi \tag{13.282}$$

The total thrust force acting over the volume $0 \le z \le L$, $r_c \le r \le r_a$ is

$$F_z = \int f_z \, dV = \frac{\mu_0 I^2}{4\pi} \ln(r_a/r_c) \tag{13.283}$$

the thrust is thus quadratic in the applied current. For typical dimensions $r_c \sim 1$ cm, $r_a \sim 10$ cm, and $I \sim 1$ kAmp, a thrust of the order 1/4 N is generated. In general, the current pattern will possess a component in the z direction, particularly if the anode is placed downstream of the cathode. The Lorentz force component due to the presence of J_z is $f_r = -J_z B_\phi$ and represents a focusing (or inward pinching) of the jet. When the Hall effect is present, the azimuthal symmetry requires that the Hall field, E_ϕ, be shorted, and a Hall current will flow given by $J_\phi \sim \beta J_r$. Correspondingly components B_r and B_z of the magnetic field will be induced which will modify the focusing and thrust forces and also lead to a torque $r f_\phi = r(J_z B_r - J_r B_z)$ which will impart angular momentum to the fluid.

The MPD accelerator achieves a conducting fluid by the self-heating of the working fluid by the discharge itself. This machine operates most stably at low pressure, and in this regime, plasma kinetic effects such as ionization and recombination, charge exchange, and strongly non-Maxwellian behavior are most important.

REFERENCES

Abraham, G., "The Flow of Round Buoyant Jets Issuing Vertically into Ambient Fluid Flowing in a Horizontal Direction," Delft Hydraulic Laboratory, Publ. No. 81, 1971.

Abramovich, G. N., *The Theory of Turbulent Jets*, MIT Press, Cambridge, MA., 1960 (English edition).

Abramson, H. N. (Ed.), *The Dynamic Behavior of Liquids in Moving Containers*, NASA SP-106, 1966.

Achenbach, E., "Prepared Discussion," *Aerodynamic Drag Mechanisms*, Sovran, G., Morel, T., and Mason, W. T. (Eds.), Plenum Press, New York, pp. 153–157, 1978.

Adams, M. L. and Szeri, A. Z., "Incompressible Flow Between Finite Discs," *J. Appl. Mech.*, Vol. 49, pp. 1–9, 1982.

Adler, D. and Baron, A., "Prediction of a Three-Dimensional Jet in Cross Flow," *AIAA J.*, Vol. 17, No. 2, pp. 168–174, 1979.

Alvarez, J. and Jones, W. P., "Computation of a Jet Discharging into a Cross-Flow with a Second-Moment Turbulence Closure and a Low-Diffusive Convection-Discretization Scheme," *Engineering Turbulence Modelling and Experiments 2*, Rodi, W. and Martelli, F. (Eds.), Elsevier Publ., 1993.

Andreopoulos, J. and Rodi, W., "Experimental Investigation of Jets in a CrossFlow," *J. Fluid Mech.*, Vol. 138, pp. 93–127, 1984.

Andreopoulos, J., "On the Structure of Jets in a Crossflow," *J. Fluid Mech.*, Vol. 157, pp. 163–197, 1985.

Andreopoulos, J., "Measurements in a Pipe Flow Issuing Perpendicular into a Cross Stream," *J. Fluids Eng.*, Vol. 104, pp. 493–499, 1983.

Arai, T. and Schetz, J. A., "Penetration and Mixing of a Transverse Bubbling Jet into a Supersonic Flow," *Proc. 2nd Nat. (Japan) Symp. RAM/SCRAM JET*, Sendai, Japan, pp. 131–136, 1992.

Arai, T. and Schetz, J. A., "Injection of Bubbling Liquid Jets from Multiple Injectors into a Supersonic Air Stream," *J. Propulsion and Power*, Vol. 10, No. 3, pp. 382–386, 1994.

Argyropoulos, G. S. and Demetriades, S. T., "Influence of Relaxation Effects in Non-Equilibrium J × B Devices," *J. Appl. Phys.*, Vol. 40, No. 11, pp. 440x–4409, 1969.

Argyropoulos, G. S., Casteel, M. A., and Demetriades, S. T., "Two-Dimensional Distribution of Current Along Magnetohydrodynamic Channels," *Energy Conversion*, Vol. 10, pp. 189–192, 1970.

Argyropoulos, G. S., Demetriades, S. T., and Lackner, K., "Compressible Turbulent Magnetohydrodynamic Boundary Layers," *Phys. Fluids*, Vol. 11, No. 12, 1968.

Argyropoulos, G. S., Demetriades, S. T., and Kendig, A. P., "Current Distribution in Non-Equilibrium J × B Devices," *J. Appl. Phys.*, Vol. 38, No. 13, pp. 5233–5239, 1967.

Atkinson, K. M., Khan, Z. A., and Whitelaw, J. H., "Experimental Investigation of Opposed Jets Discharging Normally into a Cross-Stream," *J. Fluid Mech.*, Vol. 115, pp. 493–504, 1982.

Avrashkov, V., Baranovsky, S., and Levin, V., "Gasdynamic Feature of Supersonic Kerosene Combustion in a Model Combustion Chamber," AIAA Paper 90-5268, 1990.

Bajaj, A. K. and Garg, V., "Linear Stability of Jet Flows," *J. Appl. Mech.*, pp. 378–384, 1977.

Baker, A. J., Snyder, P. K., and Orzechowski, J. A., "Three Dimensional Nearfield Characterization of a VSTOL Jet in Turbulent Crossflow," AIAA Paper 87-0051, 1987.

Bangerter, C. D., Hopkins B. D., and Brogan, T. R., "Explosively Driven MHD Power Generation—A Progress Report," *Proc. 6th Int. Symp. MHD Power Generation*, Vol. IV, p. 155, Washington, D.C., 1975.

Baranovsky, S. I. and Schetz, J. A., "Effect of Injection Angle on Liquid Injection," *AIAA J.*, Vol. 18, No. 6, pp. 625–629, 1980.

Barata, J. M. M., Durão, D. F. G., Heitor, M. V., and McGuirk, J. J., "Impingement of Single and Twin Turbulent Jets Through a Crossflow," *AIAA J.*, Vol. 29, No. 4, pp. 595–602, 1991.

Batchelor, G. K., *An Introduction to Fluid Dynamics*, Cambridge University Press, Cambridge, 1967.

Bear, J., *Dynamics of Fluids in Porous Media*, Elsevier, New York, 1972.

Bearman, P. W., "On Vortex Street Wakes," *J. Fluid Mech.*, Vol. 24, Pt. 8, pp. 625–641, 1967.

Bennetts, D. A. and Hocking, L. M., "On Nonlinear Ekman and Stewartson Layers in a Rotating Fluid," *Proc. Roy. Soc. A*, Vol. 333, pp. 469–489, 1973.

Betchov, R. and Criminale, W. O., "Spatial Instability of the Inviscid Jet and Wake," *Physics of Fluids*, Vol. 9, pp. 356–362, 1966.

Bien, F. and Penner, S. S., "Velocity Profiles in Steady and Unsteady Rotating Flows for a Cylindrical Geometry," *Phys. Fluids*, Vol. 13, pp. 1665–1671, 1970.

Bödewadt, U. T., "Die Drehstremung uber festem Grunder," *Z. Angew. Math. Mech.*, Vol. 20, pp. 241–253, 1940.

Briggs, G. A., "Plume Rise Predictions," *Lectures on Air Pollution and Environmental Impact Analysis*, Chap. 3, American Meteorology Society, Boston, 1975.

Caldwell, D. R. and van Atta, C. W., "Characteristics of Ekman Boundary Layer Instabilities," *J. Fluid. Mech.*, Vol. 44, pp. 79–95, 1970.

Campbell, J. F. and Schetz, J. A., "Analysis of Injection of a Heated Turbulent Jet into a Cross Flow," NASA TR R-413, Dec. 1973b.

Campbell, J. F. and Schetz, J. A., "Flow Properties of Submerged Heated Effluents in a Waterway," *AIAA J.*, Vol. 11, No. 2, pp. 223–230, 1973a.

Catton, I., "Natural Convection in Enclosures," *Proc. 6th Int. Heat Transfer Conf.*, Vol. 6, pp. 13–31, 1978.

Cebeci, T. and Abbot, D. E., "Boundary Layers on a Rotating Disc," *AIAA J.*, Vol. 13, No. 6, pp. 829–832, 1975.

Cess, R. D., "Free-Convection Boundary-Layer Heat Transfer," *Handbook of Heat Transfer*, Sec. 6, McGraw-Hill, 1973.

Cham, T.-S. and Head, M. R., "Turbulent Boundary-Layer Flow on a Rotating Disc," *J. Fluid Mech.*, Vol. 37, pp. 129–147, 1967.

Chen, C. J. and Rodi, W., *Vertical Turbulent Buoyant Jets—A Review of Experimental Data*, Pergamon Press, New York, 1980.

Chen, C. L. H., "Aufrollung eines zylindrischen Strahles durch Querwind," doctoral dissertation, University of Göttingen, Göttingen, Germany, 1942.

Chew, J. W., "Development of Computer Program for the Prediction of Flow in a Rotating Cavity." *Int. J. Num. Methods in Fluids*, Vol. 4, No. 7, pp. 667–683, 1984.

Chien, J. C. and Schetz, J. A., "Numerical Solution of the Three-Dimensional Navier–Stokes Equations with Applications to Channel Flows and a Buoyant Jet in a Cross Flow," *J. Appl. Mech.*, Vol. 42, pp. 575–579, 1975.

Chorin, A. J., "Numerical Solution of the Navier–Stokes Equations," *Math. Comp.*, Vol. 22, pp. 745–762, 1968.

Claus, R. W. and Vanka, S. P., "Multigrid Calculations of a Jet in Crossflow," *J. Propulsion Power*, Vol. 8, pp. 185–193, 1992.

Clausing, "New Natural Convection Correlations for Isothermal Vertical Surfaces," ASME Paper 82-WA/HT-68, pp. 1–7, 1982.

Coelho, S. L. V. and Hunt, J. C. R., "The Dynamics of the Near Field of Strong Jets in Crossflows," *J. Fluid Mech.*, Vol. 200, pp. 95–120, 1989.

Collins, R. E., *Flow of Fluids through Porous Materials*, Van Nostrand Reinhold, New York, 1961.

Cooper, P. and Reshotko, E., "Turbulent Flow Between a Rotating Disc and a Parallel Wall," *AIAA J.*, Vol. 13, pp. 573–578, 1975.

Corey, A. T., *Mechanics of Heterogeneous Fluids in Porous Media*, Water Resources Publications, Fort Collins, CO, 1977.

Corrsin, S. and Uberoi, M., "Further Experiments on the Flow and Heat Transfer in a Heated Turbulent Air Jet," NACA TN 1985, 1949.

Crabb, D., Durão, D. F. G., and Whitelaw, J. J., "Round Jet Normal to a Cross Flow," *J. Fluids Eng.*, Vol. 103, No. 1, pp. 142–153, 1981.

Daily, J. M., Kruger, C. H., Self, S. A., and Eustis, R. A., "Boundary-Layer Profile Measurements in a Combustion Driven MHD Generator," *AIAA J.*, Vol. 14, pp. 997–1005, 1976.

Daly, B. J. and Harlow, F. H., "Transport Equations of Turbulence," *Physics Fluids*, Vol. B, pp. 2634–2649, 1970.

Danielsson, L. and Lundgren, S., "Measurements of Velocity Profile of Countercurrent Jets in a Rotating Cylinder Using Laser Doppler Anemometry," *Proc. 2nd Workshop on Gases in Strong Rotation*, Soubbaramayer (Ed.), Centre d'Etudes Nucleaires, France, pp. 175–190, 1977.

DeGroot, S. R. and Mazur, P., *Non-Equilibrium Thermodynamics*, North Holland, Amsterdam, 1962.

Delhomme, J. D., "Spatial Variability and Uncertainty in Groundwater Flow Parameters: A Geostatistical Approach," *Water Resources Res.*, Vol. 15, No. 2, p. 269, 1979.

Demetriades, S. T., "Momentum Transfer to Plasmas by Lorentz Forces," *Physico-Chemical Diagnostics of Plasmas*, Anderson, T. P., Springer, R. W., and Warder, R. C. (Eds.), Northwestern University Press, Evanston, IL, pp. 297–328, 1964.

Demetriades, S. T. and Argyropoulos, G. S., "Ohm's Law in Multicomponent Non-Isothermal Plasmas with Temperature and Pressure Gradients," *Phys. Fluids*, Vol. 9, No. 11, pp. 2136–2149, 1966.

Demetriades, S. T., Demetriades, J. T., and Demetriades, A. S., "Influence of Magnetic Reynolds Number on Power Generated by an Ideal MHD Device," *AIAA J.*, Vol. 23, No. 11, pp. 1813–1814, 1985.

Demetriades, S. T. and Maxwell, C. D., "Determination of Energy-Loss Factors for Slow Electrons in Hot Gases," STD Research Corp. Rept. STD-69-1, 1969.

Demetriades, S. T., Hamilton, G. L., Ziemer, R. W., and Lenn, P. D., "Three-Fluid Non-Equilibrium Plasma Accelerators, Part I," *Progress in Astronautics and Aeronautics*, Vol. 9, AIAA Series, pp. 461–611, Academic Press, New York, 1963.

Demetriades, S. T., Oliver, D. A., and Maxwell, C. D., "Brief Technical Discussion of the Consistent STD/MHD Code System: Presenting the Foundation, Formulations Assumptions and Constitutive Building Blocks of the STD/MHD Code System," STD Research Corp. Rept. No. STDR-82-7, 1982.

Demetriades, S. T., Oliver, D. A., Swean, T. F., Jr., and Maxwell, C. D., "On the Magnetoaerothermal Instability," AIAA-81-0248, 1981.

Demetriades, S. T., "Novel Method for Determination of Energy-Loss Factors for Slow Electrons in Hot Gases," STD Research Corp. Rept. No. STD-68-2, 1968.

Demuren, A. O. and Rodi, W., "Calculation of Turbulence-Driven Secondary Motion in Non-Circular Ducts," *J. Fluid Mech.*, Vol. 140, pp. 189–222, 1984.

Demuren, A. O. and Sarkar, S., "Perspective: Systematic Study of Reynolds Stress Closure Models in the Computations of Plane Channel Flows," *J. Fluids Eng.*, Vol. 115, pp. 5–12, 1993.

Demuren, A. O. and Wilson, R. V., "Estimating Uncertainty in Computations of Two-Dimensional, Separated Flows," *J. Fluids Eng.*, Vol. 116, pp. 216–220, 1994.

Demuren, A. O., "Modeling Turbulent Jets in Crossflow," *Encyclopedia of Fluid Mechanics*, Cheremisinoff, N. P. (Ed.), Vol. 2, Chap. 17, Gulf Publ., Houston, TX, 1985a.

Demuren, A. O., "Calculations of 3D Impinging Jets in Crossflow with Reynolds Stress Models," *Proc. Int. Symp. Heat Transfer in Turbomachinery*, Goldstein, R. J. (Ed.), Marathon, Greece, 1992, 1995.

Demuren, A. O., "False Diffusion in Three-Dimensional Flow Calculations," *Comp. and Fluids*, Vol. 13, pp. 41–65, 1985b.

Demuren, A. O., "Multigrid Acceleration and Turbulence Models for Computations of 3D Turbulent Jets in Crossflow," *Int. J. Heat and Mass Transfer*, Vol. 35, No. 11, pp. 2783–2794, 1992.

Demuren, A. O., "Numerical Calculations of Steady Three-Dimension Turbulent Jets in Cross Flow," *Comp. Meth. Appl. Mech. and Eng.*, Vol. 37, pp. 309–328, 1983.

DeWiest, R. J. M., *Geohydrology*, John Wiley & Sons, New York, 1965.

Doerfler, P., "Systemtheoretische Ansaetze fuer die Saugrohrschwingungen der Francisturbien," VDI (Society of German Engineers)—Berichte NR 424, pp. 211–222, 1981.

Drazin, P. G. and Reid, W. H., *Hydrodynamic Stability*, Cambridge University Press, London, 1981.

Dullien, F. A. L., *Porous Media Fluid Transport and Pore Structure*, 2nd ed., Academic Press, New York, 1992.

Ede, A. J., "Advances in Free Convection," *Advances in Heat Transfer*, Vol. 4, pp. 1–64, 1967.

Ekman, V. W., "On the Influence of the Earth's Rotation on Ocean Currents," *Arkiv for matematik, astronomi och fysik*, Vol. 2, pp. 1–52, 1905.

Erian, F. F. and Tong, Y. H., "Turbulent Flow due to a Rotating Disc," *Phys. Fluids*, Vol. 14, pp. 2588–2591, 1971.

Escudier, M. P. and Zehnder, N., "Vortex-Flow Regimes," *J. Fluid Mech.*, Vol. 115, pp. 105–121, 1982.

Escudier, M. P., "Vortex Breakdown: Observations and Explanations," *Progress in Aerospace Sciences*, Vol. 25, p. 189, 1988.

Escudier, M. P. and Merkli, B., "Observations of the Oscillatory Behavior of a Confined Vortex," *AIAA J.*, Vol. 17, pp. 253–260, 1979.

Escudier, M. P., "Confined Vortices in Flow Machinery," *Ann. Rev. of Fluid Mech.*, Vol. 19, p. 27, 1987.

Escudier, M. P., "Vortex Breakdown and the Criterion for Its Occurrence," *Topics in Atmospheric and Oceanographic Sciences: Intense Atmospheric Vortices*, Bengston, L. and Lighthill, M. J. (Eds.), pp. 247–258, 1982.

Faler, J. H. and Leibovich, S., "Disrupted States of Vortex Flow and Vortex Breakdown," *Physics of Fluids*, Vol. 20, No. 9, pp. 1385–1400, 1977.

Fan, L. N., "Turbulent Jets into Stratified or Flowing Ambient Fluids," Keck Laboratory of Hydraulics and Water Resources, California Institute of Technology, Report No. KII-R-15, 1967.

Fjortoft, R., "Applications of Integral Theorems in Deriving Criteria of Stability of Laminar Flow and for the Baroclinic Circular Vortex," *Geofys. Publ.*, Vol. 17, pp. 1–52, 1950.

Galperin, B. and Orszag, S. A., *Large Eddy Simulation of Complex Engineering and Geophysical Flows*, Cambridge University Press, Cambridge, 1993.

Gans, R. F., "On the Stability of Shear Flow in a Rotating Gas," *J. Fluid Mech.*, Vol. 68, pp. 403–412, 1975.

Garner, J. E., "A Review of Jet Efflux Studies with Application to V/STOL Aircraft," AEDC-TR-67-163, U.S. Air Force, Sept. 1967. (Available from DDC as AD 658 432.)

Gaster, M., "Some Observations on Vortex Shedding and Acoustic Resonances," Aeronautical Research Council Current Paper No. 1141, 1971.

Gebhart, B., "Buoyancy Induced Fluid Motions Characteristic of Applications in Technology," *J. Fluids Eng.*, Vol. 101, pp. 5–28, 1979.

Gebhart, B., "Natural Convection Flows and Stability," *Advances in Heat Transfer*, Vol. 9, pp. 273–348, 1973.

Gibson, C. H., "Fossil Two-Dimensional Turbulence in the Ocean," *Turbulent Shear Flows*, 7, Durst, F. and Reynolds, W. C. (Eds.), Springer-Verlag, Berlin, 1991c.

Gibson, C. H., Nabatov, V., and Ozmidov, R., "Measurements of Turbulence and Fossil Turbulence Near Ampere Seamount," *Dynamics of Atmospheres and Oceans*, pp. 175–204, 1994.

Gibson, C. H., "Alternative Interpretation for Microstructure Patches in the Thermocline," *J. Phys. Oceanogr.*, Vol. 12, pp. 374–383, 1982.

Gibson, C. H., "Fine Structure of Scalar Fields Mixed by Turbulence: I. Zero-Gradient Points and Minimal Gradient Surfaces," *Phys. Fluids*, Vol. 11, No. 11, pp. 2305–2315, 1968a.

Gibson, C. H., "Fine Structure of Scalar Fields Mixed by Turbulence: II. Spectral Theory," *Phys. Fluids*, Vol. II, No. 11, pp. 2316–2327, 1968b.

Gibson, C. H., "Fossil Temperature, Salinity, and Vorticity Turbulence in the Ocean," *Marine Turbulence*, Nihoul, J. (Ed.), Elsevier Publ., Amsterdam, 1980.

Gibson, C. H., "Internal Waves, Fossil Turbulence, and Composite Ocean Microstructure Spectra," *J. Fluid Mech.*, Vol. 168, pp. 89–117, 1986.

Gibson, C. H., "Oceanic and Interstellar Fossil Turbulence," *Radio Wave Scattering in the Interstellar Medium*, AIP Conference Proceedings 174, Lerner, R. G. (Ed.), Am. Inst. Phys., New York, pp. 74–79, 1988.

Gibson, C. H., "Kolmogorov Similarity Hypotheses for Scalar Fields: Sampling Intermittent Turbulent Mixing in the Ocean and Galaxy," *Turbulence and Stochastic Processes: Kolmogorov's Ideas 50 Years on Proc. Roy. Soc. (London)*, Vol. 433, No. 1890, pp. 149–164, 1991b.

Gibson, C. H., "Laboratory, Numerical, and Oceanic Fossil Turbulence in Rotating and Stratified Flows," *J. Geophys. Res.*, Vol. 96 (C7), pp. 12,549–12,566, 1991a.

Gibson, C. H., "Turbulence, Mixing, and Heat Flux in the Ocean Main Thermocline," *J. Geophys. Res.*, Vol. 96 (C11), pp. 20,403–20,420, 1991d.

Gibson, C. H., "Fossils of Primordial Turbulence and Non-Turbulence at the 'Schwarz Radii'—The Length Scales for Self-Gravitating Fluid Matter," submitted to *Physics of Fluids*, 1995.

Gill, A. E., *Atmosphere-Ocean Dynamics*, Academic Press, New York, 1982.

Greenkorn, R. A., "Steady Flow Through Porous Media," *A.I.Ch.E. J.*, Vol. 27, No. 4, p. 529, 1981.

Greenkorn, R. A., *Flow Phenomena in Porous Media*, Marcel Dekker, New York, 1983.

Greenkorn, R. A., "Matrix Properties of Porous Media," *Proc. 2nd Symp. IAHR-ISSS, Fundamentals of Transport Phenomena in Porous Media*, Goelph, 1992.

Greenspan, H. P., *The Theory of Rotating Fluids*, Breukelen Press, 1990.

Gusev, A. and Bark, F. H., "Stability of Rotation-Modified Plane Poiseuille Flow," *Phys. Fluids*, Vol. 23, pp. 2171–2177, 1980.

Hashimoto, K., "On the Stability of the Stewartson Layer," *J. Fluid Mech.*, Vol. 76, pp. 289–306, 1979.

Heller, H. H. and Bliss, D., "The Physical Mechanisms of Flow-Induced Pressure Fluctuations in Cavities and Concepts for Their Suppression," AIAA Paper No. 75-491, 1975.

Heywood, J. B. and Womack, G. J., *Open Cycle MHD Power Generation*, Pergamon Press, London, 1969.

Heywood, J. B., "The Effect of Swirl on MHD Generator Perfomance," *Adv. Energy Conversion*, 1968.

Hide, R. and Ibbetson, A., "On Slow Transverse Flow Past Obstacles in a Rapidly Rotating Fluid," *J. Fluid Mech.*, Vol. 32, pp. 251–272, 1968.

Hide, R. and Titman, C. W., "Detached Shear Layers in a Rapidly Rotating Fluid," *J. Fluid Mech.*, Vol. 29, pp. 39–60, 1967.

Higuera, F. J. and Martinez, M., "An Incompressible Jet in a Weak Crossflow," *J. Fluid Mech.*, Vol. 249, pp. 73–97, 1993.

Hirst, E. A., "Buoyant Jets with Three-Dimensional Trajectories," *J. Hydr. Div., Proc. ASCE*, HY 11, Vol. 98, pp. 1999–2014, 1972.

Ho, C.-M. and Huerre, P., "Perturbed Free Shear Layers," *Ann. Rev. Fluid Mech.*, Vol. 16, pp. 365–424, 1984.

Holodniok, M., Kubicek, M., and Hlavocek, V., "Computation of the Flow Between Two Rotating Coaxial Discs," *J. Fluid Mech.*, Vol. 81, pp. 689–699, 1977.

Holton, J. R., *An Introduction to Dynamic Meteorology*, Academic Press, New York, 1972.

Howard, J. H. G., Patankar, S. V., and Bordynuik, S. V., "Flow Prediction in Rotating Ducts Using Coriolis-Modified Turbulence Models," *J. Fluids Eng.*, Vol. 102, pp. 456–461, 1980.

Huerre, P. and Monkewitz, P. A., "Local and Global Instabilities in Spatially Developing Flows," *Ann. Rev. Fluid Mech.*, Vol. 22, p. 473, 1990.

Inamura, T., Nagai, H., and Tomita, M., "Atomization of Liquid Jets Including Solid Particles in a High-Speed Air Stream (2nd Rep.)," *Trans. JSME (Ser. B)*, Vol. 59, No. 557, pp. 296–303, 1993a (in Japanese).

Inamura, T., Nagai, H., Watanabe, T., and Yatsuyanagi, M., "Disintegration of Liquid and Slurry Jets Traversing Subsonic Airstreams," *Proc. 3rd World Conf. Experimental Heat Transfer, Fluid Mechanics, and Thermodynamics*, Honolulu, pp. 1522–1529, 1993b.

Inamura, T., Nagai, N., and Inagaki, H., "A Study on the Twin-Fluid Atomization of a Highly Concentrated Coal-Water Mixture," *JSME Int. J.*, Vol. 30, No. 269, pp. 1790–1796, 1987.

Inamura, T., Nagai, N., Hirai, T., and Asano, H., "Atomization of Liquid Jets Including Solid Particles in a High-Speed Air Stream (1st Rep.)," *Trans. JSME (Ser. B)*, Vol. 57, No. 541, pp. 3237–3243, 1991a (in Japanese).

Inamura, T., Nagai, N., Hirai, T., and Asano, H., "Disintegration Phenomena of Metalized Slurry Fuel Jets in High Speed Air Stream," *Proc. 5th Int. Conf. Liquid Atomization and Spray Systems*, NIST, Gaithersburg, MD, pp. 839–846, 1991b.

Isaac, K. M. and Schetz, J. A., "Analysis of Multiple Jets in a Cross Flow," *J. Fluids Eng.*, Vol. 104, pp. 489–492, 1982.

Jahn, R. G., *Physics of Electric Propulsion*, McGraw-Hill, New York, 1968.

Johnston, J. P., Halleen, R. M., and Lezius, D. K., "Effects of Spanwise Rotation on the Structure of Two-Dimensional Fully Developed Turbulent Channel Flow," *J. Fluid Mech.*, Vol. 56, pp. 533–557, 1972.

Joshi, P. B. and Schetz, J. A., "Effect of Injector Geometry on the Structure of a Liquid Jet Injected Normal to a Supersonic Airstream," *AIAA J.*, Vol. 13, No. 9, pp. 1137–1138, 1975.

Kamotani, Y. and Greber, I., "Experiments on a Turbulent Jet in a Cross Flow," *AIAA J.* Vol. 10, No. 11, pp. 1425–1429 (see also, NASA CR 72893, 1971), 1972.

Kamotani, Y. and Greber, I., "Experiments on Confined Turbulent Jets in Cross Flow," NASA Report, NASA CR-2392, 1974.

Keffer, J. F. and Baines, W. D., "The Round Turbulent Jet in a Cross-Wind," *J. Fluid Mech.*, Vol. 15, Pt. 4, pp. 481–496, 1963.

Kerrebrock, J. L., "Conduction in Gases with Elevated Electron Temperatures," *Engineering Aspects of Magnetohydrodynamics*, Columbia University Press, New York, 1962.

Kevorkian, J. and Cole, J. D., *Perturbation Methods in Applied Mathematics*, Springer-Verlag, Berlin, 1981.

Khan, Z. A., McGuirk, J. J., and Whitelaw, J. H., "A Row of Jets in a Cross Flow," AGARD CP 308, Paper 10, 1982.

Kim, S. W. and Benson, T. J., "Calculation of a Circular Jet in Crossflow with a Multiple-Time-Scale Turbulence Model," *Int. J. Heat Mass Transfer*, Vol. 35, No. 10, pp. 2357–2365, 1992.

Kubota, T. and Aoki, H., "Pressure Surge in the Draft Tube of a Francis Turbine," *Practical Experiences with Flow-Induced Vibrations*, Naudascher, E. and Rockwell, D. (Eds.), Springer-Verlag, Berlin, pp. 279–286, 1980.

Kush, E. A., Jr. and Schetz, J. A., "Liquid Jet Injection into a Supersonic Flow," *AIAA J.*, Vol. 11, No. 9, pp. 1223–1224, 1973.

Launder, B. E. and Spalding, D. B., "The Numerical Computation of Turbulent Flows," *Comp. Meths. Appl. Mech. Eng.*, Vol. 3, pp. 269–289, 1974.

Launder, B. E., Reece, G. J., and Rodi, W., "Progress in the Development of a Reynolds Stress Turbulence Closure," *J. Fluid Mech.*, Vol. 68, pp. 537–566, 1975.

Leibovich, S., "Vortex Stability and Breakdown: Survey and Extension," *AIAA J.*, Vol. 22, No. 9, p. 1192, 1984.

Lentini, M. and Keller, H. B., "The von Karman Swirling Flows," *SIAM J. Appl. Math.*, Vol. 38, pp. 52–64, 1980.

Lesieur, M., "Introduction á la Turbulence Bidimensionelle," *J. de Méc. Theor. et Appl.*, pp. 5–20, Numero Special 1983.

Less, D. and Schetz, J. A., "Quantitative Study of Time Dependent Character of Spray Plumes," *AIAA J.*, Vol. 24, No. 12, pp. 1974–1986, 1986.

Less, D. M. and Schetz, J. A., "Penetration and Breakup of Slurry Jets in a Supersonic Stream," *AIAA J.*, Vol. 21, No. 7, pp. 1045–1046, 1983.

Lewellen, W. S., "A Review of Confined Vortex Flows," Dept. of Aeronautics and Astronautics, MIT, Cambridge, MA, SPL-70-1, 1970.

Lilly, D. K., "On the Instability of the Ekman Boundary Layer," *J. Atmos. Sci.*, Vol. 23, pp. 481–494, 1966.

Lin, J.-C., Towfighi, J., and Rockwell, D., "Instantaneous Flow Past a Cylinder: Effect of Reynolds Number," *J. Fluids and Structures* (in press), 1995.

List, E. J., "Turbulent Jets and Plumes," *Ann. Rev. Fluid Mech.*, Vol. 14, pp. 189–212, 1982.

List, E. J., "Turbulent Jets and Plumes," *Mixing in Inland and Coastal Waters*, Chap. 9, Academic Press, 1979.

List, E. J. and Jirka, G. H. (Eds.), *Stratified Flows: Proceedings of the Third International Symposium on Stratified Flows*, Feb. 3–5, 1987, Pasadena, CA, American Society of Civil Engineers, New York, 1990.

Long, R. R., "Sources and Sinks at the Axis of a Rotating Liquid," *Quart. J. Mech. Appl. Math.*, Vol. 4, No. 2, pp. 385–393, 1956.

Lugt, H. J., *Vortex Flow in Nature and Technology*, Wiley–Interscience, New York, 1983.

Makihata, T. and Miyai, Y., "Prediction of the Trajectory of Triple Jets in a Uniform Cross Flow," *J. Fluids Eng.*, Vol. 105, pp. 91–97, 1983.

Markham, D. M., Maxwell, C. D., Demetriades, S. T., and Oliver, D. A., "A Numerical Solution to the Unsteady, Quasi-Three-Dimensional, Turbulent Heat Transfer Problem in an MHD Channel," Paper No. 77-HT-90 presented at AIChE-ASME Heat Transfer Conf., Salt Lake City, UT, 1977.

Matheron, G. and DeMarsily, G., "Is Transport in Porous Media Always Diffusive? A Counter Example," *Water Resources Res.*, Vol. 16, No. 5, p. 901, 1980.

Mattingly, G. E. and Criminale, W. O., "The Stability of an Incompressible Two-Dimensional Wake," *J. Fluid Mech.*, Vol. 51, Pt. 2, pp. 233–272, 1972.

Maxwell, C. D. and Demetriades, S. T., "Initial Tests of a Lightweight Self-Excited MHD Power Generator," *J. Propulsion and Power*, Vol. 2, No. 5, pp. 474–480, 1986.

Maxwell, C. D., Swean, T. F., Jr., Vetter, A. A., Crouse, R. D., Oliver, D. A., Bangerter, C. D., and Demetriades, S. T., "Three-Dimensional Effects in Large Scale MHD Generators," AIAA-81-1231m, CA, 1981.

McCroskey, W. J., "Some Current Research in Unsteady Fluid Dynamics—The 1976 Freeman Scholar Lecture," *J. Fluids Eng.*, pp. 8–44, 1977.

McGuirk, J. and Rodi, W., "Numerical Model for Heated Three-Dimensional Jets," *Heat Transfer and Turbulent Buoyant Convection, Studies and Applications for Natural Environment, Building, Engineering Systems*, Spalding, D. B. and Afgan, N. (Eds.), Hemisphere Publ., Washington, D.C., 1977.

Mellor, G. L. and Herring, H. J., "A Survey of Mean Turbulent Field Closure," *AIAA J.*, Vol. 11, pp. 590–599, 1973.

Merkli, P., "Acoustic Resonance Frequencies for a T-Tube," *J. Appl. Math. Physics (ZAMP)*, Vol. 29, pp. 486–498, 1978.

Michalke, A., "Instabilität eines Kompressiblen runden Freistrahls unter Berücksichtigung des Einflusses der Strahlgrenzschichtdiche," *Z. Flugwiss.*, Vol. 8, No. 9, pp. 319–328, 1971.

Miksad, R. W., "Experiments on the Nonlinear Stages of Free Shear Layer Transition," *J. Fluid Mech.*, Vol. 56, Pt. 4, pp. 694–719, 1972.

Monkewitz, P. A. and Huerre, P., "Influence of the Velocity Ratio on the Spatial Stability of Mixing Layers," *Physics of Fluids*, Vol. 25, No. 7, pp. 1137–1143, 1982.

Morkovin, M. W., "Flow Around a Circular Cylinder—A Kaleidoscope of Challenging Fluid Phenomena," *Symposium on Fully Separated Flows*, Hansen, A. G. (Ed.), ASME, New York, pp. 102–108, 1964.

Muskat, M., *The Flow of Homogeneous Fluids through Porous Media*, McGraw-Hill, New York, 1937.

Naudascher, E. and Rockwell, D., *Flow-Induced Vibrations: An Engineering Guide*, Balkema Press, Rotterdam, 1994.

Naudascher, E. and Rockwell, D., *Proc. IAHR/IUTAM Symp. Practical Experiences with Flow-Induced Vibrations*, Springer-Verlag, Berlin, 1980.

Naumann, R. J., "Susceptibility of Materials Processing Experiments to Low-Level Accelerations," *Spacecraft Dynamics as Related to Laboratory Experiments in Space*, NASA CP-2199, Nov. 1981.

Needham, D. J., Riley, N., and Smith, J. H. B., "A Jet in Crossflow," *J. Fluid Mech.*, Vol. 188, pp. 159–184, 1988.

Needham, D. J., Riley, N., Lyton, C. C., and Smith, J. H. B., "A Jet in Crossflow, Part 2," *J. Fluid Mech.*, Vol. 211, pp. 515–528, 1990.

Nejad, A. S. and Schetz, J. A., "Effects of Properties and Location in the Plume on Mean Droplet Diameter for Injection in a Supersonic Stream," *AIAA J.*, Vol. 21, No. 7, pp. 956–961, 1983.

Neuringer, J. L., "Optimum Power Generation Using a Plasma as the Working Fluid," *J. Fluid Mech.*, Vol. 7, pp. 287, 1960.

Northam, G. B., "Combustion in Supersonic Flow," *Proc. 21st JANNAF Combust. Meeting*, JPL, Pasadena, CA, p. 404, 1985.

Nunge, R. J. (Ed.), *Flow Through Porous Media*, Amer. Chem. Soc., Washington, D.C., 1970.

Oda, T., Hiroyasu, H., Arai, M., and Nishida, K., "Characteristics of Liquid Jet Atomization Across a High-Speed Airstream (1st Report, Experiment on Shape of Spray, Spatial Distribution of Injected Liquid and Sauter Mean Diameter)," *Trans. JSME (Ser. B)*, Vol. 58, No. 552, pp. 2592–2601, 1992 (in Japanese).

Oh, T. S. and Schetz, J. A., "Finite Element Simulation of Complex Jets in a Crossflow for V/STOL Applications," *J. Aircraft*, Vol. 27, No. 5, pp. 389–399, 1990.

Oliver, D. A., Crouse, R. D., Maxwell, C. D., and Demetriades, S. T., "Transient Processes in Large Magnetohydrodynamic Generator Flowtrains," *Proc., 7th Int. Conf. on Magnetohydrodynamic Electrical Power Generation*, MIT, 1980.

Oliver, D. A., Swean, T. F., Jr., and Markham, D. M., "Magnetohydrodynamics of Hypervelocity Pulsed Flows," *AIAA J.*, Vol. 19, No. 6, pp. 699–705, 1981.

Orth, R. C., Schetz, J. A., and Billig, F. S., "The Interaction and Penetration of Gaseous Jets in Supersonic Flow," NASA CR-1386, July 1969.

Ostrach, S., "Convection in Fluids at Reduced Gravity," *Spacecraft Dynamics as Related to Laboratory Experiments in Space*, NASA CP-2199, Nov. 1981.

Ostrach, S., "Fluid Mechanics in Crystal Growth," *J. Fluids Eng.*, Vol. 105, pp. 5–20, 1983.

Ostrach, S., "Low Gravity Fluid Flows," *Ann. Rev. Fluid Mech.*, Vol. 14, pp. 313–345, 1982a.

Ostrach, S., "Natural Convection Heat Transfer in Cavities and Cells," *Proc. 7th Int. Heat Transfer Conf.*, Vol. 1, pp. 365–379, 1982b.

Ostrach, S., "Natural Convection in Enclosures," *Advances in Heat Transfer*, Vol. 8, pp. 161–227, 1972.

Pao, H.-P. and Shih, H.-H., "Selective Withdrawal and Blocking Wave in Rotating Fluids," *J. Fluid. Mech.*, Vol. 57, pp. 459–480, 1973.

Parker, R. and Pryce, D. C., "Wake-Excited Resonances in an Annular Cascade: An Experimental Investigation," *J. Sound and Vibration*, Vol. 37, No. 2, pp. 247–261, 1974.

Patankar, S. V. and Spalding, D. B., "A Calculation Procedure for Heat, Mass and Momentum Transfer in Three-Dimensional Parabolic Flows," *Int. J. Heat Mass Transfer*, Vol. 15, pp. 2787–2805, 1972.

Patankar, S. V., Basu, D. K., and Alpay, S. A., "Prediction of the Three-Dimensional Velocity Field of a Deflected Turbulent Jet," *J. Fluids Eng.*, Vol. 99, pp. 758–762, 1977.

Pater, L. L., Crowther, E., and Rice, W., "Flow Regime Definition for Flow Between Corotating Discs," *J. Fluids Eng.*, Vol. 96, pp. 29–34, 1974.

Pearson, C. E., "Numerical Solutions for the Time Dependent Viscous Flow Between Two Rotating Coaxial Discs," *J. Fluid Mech.*, pp. 623–633, 1965.

Pedlosky, J., *Geophysical Fluid Dynamics* (2nd ed.), Springer-Verlag, New York, 1987.

Pedlosky, J., *Geophysical Fluid Dynamics*, Springer-Verlag, Berlin, 1979.

Penfield, P., Jr. and Haus, H. A., *Electrodynamics of Moving Media*, Research Monography No. 40, MIT Press, Cambridge, MA, 1964.

Perkins, S. C., Jr. and Mendenhall, M. R., "A Study of Real Jet Effects on the Surface Pressure Distribution Induced by a Jet in a Crossflow," NASA CR-166150, Mar. 1981.

Phillips, O. M., *The Dynamics of the Upper Ocean* (2nd ed.), Cambridge University Press, New York, 1977.

Poluborinova–Kochina, P. Ya., *Theory of Ground Water Movement*, Princeton University Press, Princeton, NJ, 1962.

Pratte, B. D. and Baines, W. D., "Profiles of the Round Turbulent Jet in a Cross Flow," *Proc. ASCE, J. Hydraulics Division*, pp. 56–63, Nov. 1967.

Rajaratnam, N., *Turbulent Jets*, Elsevier Scientific Publ., New York, 1976.

Rayleigh, Lord, "On the Dynamics of Revolving Fluids," *Proc. Roy. Soc. (London)*, Ser. A, Vol. 93, pp. 148–154, 1917.

Rockwell, D. and Naudascher, E., "Review—Self-Sustaining Oscillations of Flow Past Cavities," *J. of Fluids Eng.*, Vol. 100, pp. 152–165, 1978.

Rockwell, D. and Schachenmann, A., "Self-Generation of Organized Waves in an Impinging Turbulent Jet at Low Mach Numbers," *J. Fluid Mech.*, Vol. 118, pp. 79–107, 1982.

Rockwell, D., "Oscillations of Impinging Layers," *AIAA J.*, Vol. 21, No. 5, pp. 645–664, 1983.

Rogers, M. H. and Lance, G. N., "The Boundary Layer on a Disc of Finite Radius in a Rotating Fluid," *Quart. J. Mech. Appl. Math.*, Vol. 17, pp. 319–330, 1964.

Rosa, R. J., *Magnetohydrodynamic Energy Conversion*, McGraw-Hill, New York, 1968.

Roshko, A., "Structure of Turbulent Shear Flows: A New Look," *AIAA J.*, Vol. 14, pp. 1349–1357, 1976.

Rott, N. and Lewellen, W. S., "Boundary Layers and Their Interactions in Rotating Flows," *Prog. Aeronaut. Sci.*, Vol. 7, pp. 111–144, 1966.

Saffman, P. G. and Baker, G. R., "Vortex Interactions," *Ann. Rev. Fluid Mech.*, Vol. 11, Annual Reviews, Palo Alto, 1979.

Sakai, T., Sadakata, M., and Okamura, M., "Apparent Viscosity of CWS Through Over the Wide Shearing Rate Range," *Proc. 14th Conf. Liquid Atomization and Spray Systems in Japan*, pp. 109–114, 1986 (in Japanese).

Salzman, R. N. and Schwartz, S. H., "Experimental Study of a Solid-Gas Jet Issuing into a Transverse Stream," *J. Fluids Eng.*, Vol. 100, pp. 333–339, 1978.

Schatzmann, M., "An Integral Model of Plume Rise," *Atmospheric Environment*, Vol. 13, pp. 721–731, 1979.

Scheidegger, A. E., *The Physics of Flow through Porous Media*, 3rd ed., University of Toronto Press, Canada, 1974.

Schetz, J. A., *Injection and Mixing in Turbulent Flow*, AIAA, New York, 1980.

Schetz, J. A., "Interaction Shock Shape for Transverse Injection," *J. Spacecraft and Rockets*, Vol. 7, No. 2, pp. 143–149, 1970.

Schetz, J. A. and Padhye, A., "Penetration and Break-Up of Liquids in Subsonic Airstreams," *AIAA J.*, Vol. 15, No. 10, pp. 1390–1395, 1977.

Schetz, J. A., Hawkins, P. F., and Lehman, H., "The Structure of Highly Underexpanded Transverse Jets in a Supersonic Stream," *AIAA J.*, Vol. 5, No. 5, pp. 882–884, 1967.

Schetz, J. A., Jakubowski, A. K., and Aoyagi, K., "Jet Trajectories and Surface Pressures Induced on a Body of Revolution with Various Dual Jet Configurations," *J. Aircraft*, Vol. 20, No. 11, pp. 975–982, 1983.

Schetz, J. A., Jakubowski, A. K., and Aoyagi, K., "Surface Pressures on a Flat Plate with Dual Jet Configurations," *J. Aircraft*, Vol. 21, No. 7, pp. 484–490, 1984.

Schetz, J. A., Weinraub, R., and Mahaffey, R., "Supersonic Transverse Jets in a Supersonic Stream," *AIAA J.*, Vol. 6, No. 5, pp. 933–934, 1968.

Schlichting, H., *Boundary-Layer Theory*, McGraw-Hill, London, 1968.

Schweitzer, S. and Mitchner, M., "Electrical Conductivity of Partially Ionized Gases in a Magnetic Field," *Phys. Fluids*, Vol. 10, No. 4, p. 799, 1967.

Senodo, Y. and Haymai, H., "An Analysis of the Flow in a Casing Induced by a Rotating Disc Using a Four Layer Flow Model," *J. Fluids Eng.*, pp. 192–198, 1976.

Serrin, J., "Mathematical Principles of Classical Fluid Mechanics," *Handbuch der Physik*, Vol. VIII/1, Springer-Verlag, Berlin, pp. 125–350, 1959.

Shih, H.-H. and Pao, H.-P., "Selective Withdrawal in Rotating Fluids," *J. Fluid Mech.*, Vol. 49, pp. 509–527, 1971.

Smith, S. H., "The Nonlinear Split-Disc Problem," *Quart. J. Mech. Appl. Math.*, Vol. 29, No. 4, pp. 399–414, 1976.

Speziale, C. G., Sarkar, S., and Gatski, T. B., "Modeling the Pressure-Strain Correlation of Turbulence: An Invariant Dynamical Systems Approach," *J. Fluid Mech.*, Vol. 227, pp. 245–272, 1991.

Speziale, C. G., "Analytical Methods for the Development of Reynolds-Stress Closures in Turbulence," *Ann. Rev. Fluid Mech.*, Vol. 23, pp. 107–157, 1991.

Speziale, C. G., "Modeling of Turbulent Transport Equations," *ICASE/LaRC Short Course on Turbulent Flow Modeling and Prediction*, 1994.

Stark, J. A., Bradshaw, R. D., and Blatt, M. H., *Low-G Fluid Behavior Technology Summaries*, NASA CR-134746, Dec. 1974.

Stark, J. A., Blatt, M. H., Bennett, F. O., and Campbell, B. J., *Fluid Management Systems Technology Summaries*, NASA CR-134748, Dec. 1974.

Stewartson, K., "On Almost Rigid Rotations—Part 1," *J. Fluid Mech.*, Vol. 3, pp. 17–26, 1957.

Stewartson, K., "On Almost Rigid Rotations—Part 2," *J. Fluid Mech.*, Vol. 26, pp. 131–144, 1966.

Stewartson, K., "On Rotating Laminar Boundary Layers," *Boundary Layer Research*, Gortler, H. (Ed.), Springer-Verlag, Berlin, 1958.

Sutton, G. W. and Sherman, A., *Engineering Magnetohydrodynamics*, McGraw-Hill, New York, 1964.

Sykes, R. I., Lewellen, W. S., and Parker, S. F., "On the Vorticity Dynamics of a Turbulent Jet in a Crossflow," *J. Fluid Mech.*, Vol. 168, pp. 399–413, 1986.

Tatro, P. R. and Mollo–Christensen, E. L., "Experiments on Ekman Layer Instability," *J. Fluid Mech.*, Vol. 28, pp. 531–544, 1967.

Taylor, G. I., "Experiments on the Motion of Solid Bodies in Rotating Fluids," *Proc. Roy. Soc. A*, Vol. 104, pp. 213–218, 1923.

Thomas, R. H. and Schetz, J. A., "Distributions Across the Plume of Transverse Liquid and Slurry Jets in Supersonic Airflow," *AIAA J.*, Vol. 23, No. 12, pp. 1892–1901, 1985.

Tritton, D. J., *Physical Fluid Dynamics* (2nd ed.), Oxford [Oxfordshire]: Clarendon Press, 1988.

Turner, J. S., *Buoyancy Effects in Fluids*, Cambridge University Press, 1973.

Turner, J. S., "Buoyant Plumes and Thermals," *Ann. Rev. Fluid Mech.*, Ann. Reviews, Palo Alto, CA, 1969.

Van Dyke, M., *An Album of Fluid Motion*, Parabolic Press, Stanford, CA, 1982.

von Karman, Th., "Uber laminare and turbulente Reibung," *Z. angew. Math. Mech.*, Vol. 1, pp. 233–251, 1921.

Wright, S. J., "Mean Behavior of Buoyant Jets in a Crossflow," *ASCE J. Hydraulics Division*, Vol. 103, No. HY5, pp. 499–513, 1977.

Yatsuyanagi, N., "Combustion Characteristics of Metalized Hydrocarbon Fuels," *J. of J.S.A.S.S.*, Vol. 37, No. 427, pp. 393–399, 1989 (in Japanese).

Yih, C.-S., *Dynamics of Nonhomogeneous Fluids*, Collier–Macmillan Limited, London, 1965.

Yih, C.-S., *Stratified Flows* (2nd ed.), Academic Press, New York, 1980.

Zubkov, A. I. and Galgolev, A. I., "The Effect of Boundary Layer Thickness and Transverse Curvature of the Surface on the Geometry and Forces Acting in the Separation Zone Produced by Injection of a Jet into a Supersonic Flow over that Surface," *Fluid Mech-Soviet Res.*, Vol. 8, No. 1, pp. 69–79, 1979.

14 Multiphase Multicomponent Flows

CLAYTON T. CROWE
Washington State University
Pullman, WA

CONTENTS

Multicomponent–multiphase flows are encountered in a wide diversity of industries and applications ranging from the gas–particle flow in a rocket nozzle on a space craft to the motion of water through a subterranean aquifer. The study of these types of flows has also received considerable attention because of their importance in energy-related industries and applications.

Multicomponent–multiphase flows are identified according to the components and phases. A common example of a multicomponent single-phase fluid is air at standard temperatures and pressures. The air is a mixture of chemical species (components) with nitrogen being the major constituent. Typically, one treats the air as a single component fluid with appropriate properties such as gas constant, viscosity, thermal conductivity, etc. This approach becomes invalid at high temperatures where the

Handbook of Fluid Dynamics and Fluid Machinery, Edited by Joseph A. Schetz and Allen E. Fuhs
ISBN 0-471-12598-9 Copyright © 1996 John Wiley & Sons, Inc.

constituents dissociate selectively or at low temperatures where the water vapor will condense leaving water droplets in the flow. Another example of a multicomponent single-phase flow is the flow of two immiscible liquids such as oil and water, which is important to oil recovery processes. Generally, the mixture will be composed of droplets and one component immersed in a continuous conveying component such as oil droplets in water.

Multiphase flows are classified according to the phases in the mixture, namely, gas–liquid, gas–solid, and liquid–solid flows. These mixtures are usually referred to as two-phase. A mixture of gas, liquid, and solid flowing simultaneously constitutes a three-phase flow. Of course, a multiphase flow may also be a single or multicomponent flow. A common example of a single component two-phase flow is a stream-water flow found in nuclear power plants and other power systems. The combustion of droplets in a furnace, on the other hand, would be a multicomponent two-phase flow. Pneumatic transport of powder is a good example of a gas–solids flow. However, the flow of vapor in a porous media also constitutes a gas–solids flow. The flow of liquid with suspended solids is an example of a solids–liquid flow. These types of mixtures are generally referred to as slurries. Another example of a solids-liquid flow is the flow of mud. The movement of liquid in a solid, such as encountered in the drying of a porous material, is also a solids–liquid flow.

Multiphase–multicomponent flows represent a wide range of flow conditions and properties. This chapter will be developed according to the three combinations of two-phase flows. The chapter will first address gas–liquid flows, then gas–solid flows, and finally, liquid–solid flows. A separate section will treat liquid–liquid flows. The treatment of multicomponent gaseous flows can be found elsewhere in the handbook. A few examples of three-phase flows will complete the chapter.

14.1 GAS–LIQUID FLOWS

14.1.1 Flow Patterns

The most widely encountered examples of gas–liquid flows is the transport of gas-liquid mixtures in ducts. This section presents flow patterns and discusses pressure drop and void fraction calculations in ducts. The chapter concludes by discussing liquid sprays in a gas.

Gas–liquid flow in ducts is characterized by flow regimes ranging from bubble flow to annular-mist flow. The flow pattern also depends upon the orientation of the duct. The flow patterns identified for gas–liquid flow in a vertical duct are illustrated in Fig. 14.1 and are described as follows:

   a. *Bubble flow*: Flow of a liquid continuum with bubbles dispersed throughout the continuum.
   b. *Slug flow*: Bubbles coalesce and grow to dimensions comparable to the duct size. With this flow pattern, the bubble can rise in the center of the duct while the liquid film at the wall can fall downward. These types of flows are inherently intermittent and unsteady.
   c. *Churn flow*: The bubbles begin to form into irregular shapes with a more violent macroscopic mixing between phases. The liquid tends to move up and down in the tube in an oscillatory fashion.

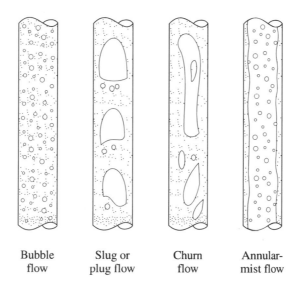

Bubble   Slug or   Churn   Annular-
flow   plug flow   flow   mist flow

FIGURE 14.1 Flow patterns for vertical gas–liquid flows.

    d. *Annular-mist flows*: A liquid flow is confined to a liquid layer moving along the wall and droplets (mist) in the core flow. A continuous exchange between the liquid layer and the mist flow occurs by entrainment of droplets from the liquid layer and deposition of core droplets on the wall.

    Early workers suggested the existence of a pure mist flow; that is, a flow with no liquid layer on the wall. However, it is unlikely except in heated systems that such a condition can exist due to the continual deposition of droplets from the core flow. In systems with heated walls, the liquid droplets approaching the wall may be vaporized before reaching the wall thereby maintaining a dry wall condition.

    Hewitt (1982) proposes another flow regime for vertical flows, namely, a *wispy annular flow* in which the droplets in the core region of the annular-mist flow agglomerate to form elongated segments of liquid (wisps).

    It is extremely important in the analysis of a gas–liquid flow in a duct to be able to predict the flow pattern in the duct. This is a very difficult task because of the complexity of these flows and the change in the controlling phenomena from one flow regime to the next. The search for a scheme to provide a reliable map of flow regimes continues. However, a scheme which appears to have wide acceptance is that proposed by Hewitt and Roberts [pp. 2–18 in Hewitt (1982)], shown in Fig. 14.2, in which the flow is characterized by the superficial momentum fluxes of each phase, $\rho_L U_L^2$ and $\rho_G U_G^2$. The *superficial velocities*, U_L and U_G, are the velocities of the liquid and gaseous phases, respectively, if that phase were to flow through the duct alone. One notices that the annular mist flow is achieved for high gas flow rates and low liquid flow rates, while bubble flows correspond to high liquid flow rates and low gas flow rates.

    It is also important to distinguish between the upward cocurrent and countercurrent flows. For low upward gas velocities in annular mist flows, the liquid layer may actually be falling. However, with increasing velocity within the core, the shear stress on the liquid film may reverse the flow direction of the liquid film. The point

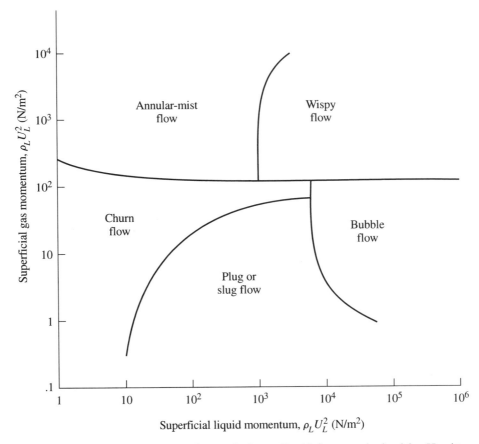

FIGURE 14.2 Flow pattern map for vertical gas–liquid flow as obtained by Hewitt and Roberts. (From Hewitt, 1982.)

where the shear stress merely supports the liquid layer on the wall leads to a *flooding* condition characterized by a disruptive liquid layer.

Downward vertical gas–liquid flows are much different. In this flow orientation, there could be an annular liquid flow on the wall with no gas flow in the core. In fact, annular flow predominates in this flow condition. Very little information is available for flow regime maps for vertical downward gas–liquid flows.

The flow of a gas–liquid mixture in a horizontal duct is strongly influenced by gravity, leading to stratification of the flow. The various flow regimes for horizontal gas–liquid flow in a duct are illustrated in Fig. 14.3 and are described below.

a. *Dispersed bubble flow*: Flow in which bubbles are dispersed in the liquid continuum. The bubbles tend to migrate and concentrate towards the upper half of the tube.

b. *Plug flow*: Large bubbles of dimensions comparable to tube size occupy the top part of the pipe leading to an unsteady, intermittent flow. These large gas bubbles may contain droplets. Also, small bubbles may appear in the liquid. The large bubbles appear as slugs which travel through the pipe. This regime is sometimes referred to as *slug flow*.

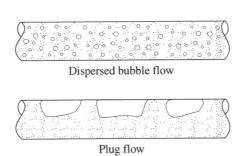

Dispersed bubble flow

Plug flow

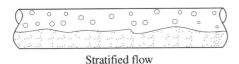

Stratified flow

Annular-mist flow

FIGURE 14.3 Flow patterns of horizontal gas–liquid flows.

c. *Stratified flow*: A flow in which there is a complete gravitational separation with the liquid on the bottom and the gas on the top. The occurrence of waves on the surface of the liquid due to the gas motion gives rise to a wavy, stratified flow leading to droplet formation and entrainment into the gas flow.

d. *Annular mist flows*: This regime is characterized by a liquid layer on the wall and a droplet flow in a gaseous core. It is similar to the annular mist flow in a vertical duct with the exception of asymmetry of the liquid layer due to gravitational effects.

The most widely used generalized flow map for horizontal gas–liquid flows is the Baker chart, deveoped by Baker (1954). Baker chose to map the flow regimes using the coordinates $\dot{m}_G/\lambda_B$ and $(\dot{m}_L/\dot{m}_G)\psi_B\lambda_B$, where $\dot{m}_G$ is the mass flow rate per unit duct area (kg/m^2 s) of the gas and $\dot{m}_L$ is the corresponding rate of the liquid. The nondimensional parameters λ_B and ψ_B were introduced to account for variations in the density, surface tension, and dynamic viscosity of the flowing media. These parameters are functions of the fluid properties normalized with respect to the properties of water and air at standard conditions. Both λ_B and ψ_B reduce to unity for air–water flows at standard conditions.

More recent studies have shown that the Baker chart is not entirely adequate. Mandhane *et al.* (1974) have proposed that the superficial phase velocities be used as the mapping coordinates. The superficial phase velocity is the velocity which would exist if that phase alone occupied the entire cross-sectional area of the duct. The superficial velocities of each phase are obtained by

$$U_G = \dot{m}_G/\rho_G \text{ (m/s)} \qquad (14.1a)$$

$$U_L = \dot{m}_L/\rho_L \text{ (m/s)} \qquad (14.1b)$$

where ρ_G and ρ_L are the densities (kg/m^3) of the gas and liquid phases, respectively. The flow pattern map, shown in Fig. 14.4, represents the correlation of nearly 6000 data points. The basic map is based on the observed flow regimes for air–water systems. It was found, however, that the map is reasonably accurate for other systems as well. A computer program is provided by Mandhane *et al.* (1974) for locating the flow regime for general gas–liquid systems.

Example 14.1   An air–water flow exists in a horizontal 5 cm-diameter pipe. The mass flow rate of the water is 1 kg/s and the mass flow rate of the air is 0.1 kg/s. The density of the water is 1000 kg/m^3 and the air, 1.2 kg/m^3. Using Mandhane's flow pattern map, find the flow regime for this system.

The cross-sectional area of the pipe is

$$A = (\pi/4)(0.05)^2 = 0.00196 \text{ m}^2$$

The superifical velocity of the water is

$$U_W = \dot{m}_W/\rho_W A = 1/(1000 \times 0.00196) = 0.51 \text{ m/s}$$

and the superficial velocity of the air is

$$U_A = \dot{m}_A/\rho_A A = 0.1/(1.2 \times 0.00196) = 42.5 \text{ m/s}$$

This point lies just inside the annular mist flow regime on the flow pattern map (Fig. 14.4).

Studies on the effect of pipe diameter and fluid properties on flow regimes are reported by Weisman *et al.* (1979).

Taitel and Dukler (1976) have attempted to develop an analytic approach to predict the transition between flow regimes and have had a measure of success in correlating Mandane's data. Weisman and co-workers [Choe *et al.* (1978)] have highlighted some shortcomings in the analytic model relating to the effects of liquid viscosity and surface tension. The complexity of the gas–liquid flows in ducts will preclude for some time the development of reliable flow mapping schemes.

The above discussion pertains to vertical and horizontal ducts. Additional flow regime maps must be generated for inclined ducts or ducts of nonuniform cross-sectional area.

14.1.2   Pressure Drop Predictions in Gas–Liquid Flows

Pressure drop predictions are important to the design of any flow system, because these predictions dictate the power requirements or the flow rate achievable in the system for the power available. This information is essential to the design of heat exchangers, condensers, or other elements in a flow loop.

Gas–liquid flows are so complicated and cover such a wide range of flow patterns that the majority of studies on pressure drops have been empirical. The empirical correlations developed are far from universal and may lead to significant errors when

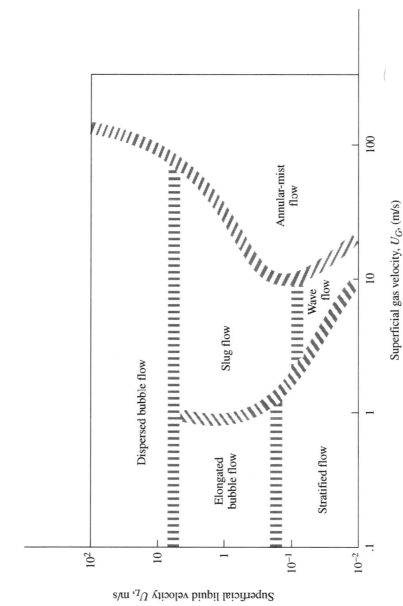

FIGURE 14.4 Flow pattern map for horizontal gas–liquid flow. (Reprinted with permission from Elsevier Science, Ltd., Mandhane, J. M., *et al.* "A Flow Pattern for Gas Liquid Flow in Horizontal Pipes," *Intl. J. Multiphase Flow*, Vol. 1, pp. 537–553, 1974.)

applied to flow regimes different from those for which they are developed. In this section, the classic approach to predicting pressure drop will be discussed, with reference being made to more recent directions and developments.

The *void fraction* of a gas–liquid mixture is the ratio of the time averaged volume of the vapor to the volume of the mixture, or

$$\alpha_G = V_G/V_M \qquad (14.2)$$

It is also the fraction of the cross-sectional area of the duct occupied by the gas. The *volume fraction* of the liquid is related to the void fraction by

$$\alpha_L = 1 - \alpha_G \qquad (14.3)$$

The *quality* is related to mass fluxes of both phases by

$$x = \dot{m}_G/(\dot{m}_G + \dot{m}_L) \qquad (14.4)$$

where $\dot{m}_G$ (kg/m^2 s) is the mass flux of the gas, and $\dot{m}_L$ (kg/m^2 s) is the mass flux of the liquid.

The *phase velocities*, which are the actual velocities of each phase, are given by

$$u_G = U_G/\alpha_G \text{ (m/s)} \qquad (14.5a)$$

$$u_L = U_L/\alpha_L \text{ (m/s)} \qquad (14.5b)$$

where U_G and U_L are the superifical velocities defined by Eq. (14.1).

The density of the two-phase mixture is related to the phase densities and void fractions by

$$\rho_M = \alpha_G\rho_G + \alpha_L\rho_L \text{ (kg/m}^3) \qquad (14.6)$$

By the assumption of a homogeneous flow, the mixture is treated as a single phase with the velocities and temperatures of both phases being equal. The continuity equation for steady, quasi one-dimensional, homogeneous flow through a duct is given by

$$d(\rho_M u_H A)/dx = 0 \qquad (14.7)$$

where u_H is a homogeneous flow velocity, and x is the distance along the duct axis.

Application of the momentum equation to a steady, homogeneous, quasi one-dimensional flow yields

$$d(\rho_M u_H^2 A)/dx = -A\, dp/dx - \rho_M g\, A \sin\theta - \tau_0 P \qquad (14.8)$$

where θ is the angle of inclination of the duct from the horizontal, p is the pressure, P is the duct perimeter, and τ_0 is the local shear stress at the wall. Subtracting the continuity equation from the momentum equation and dividing by the duct area gives

$$\rho_M u_H du_H/dx = -dp/dx - \rho_M g \sin\theta - P\tau_0/A \qquad (14.9)$$

This equation can be rewritten for the sum of a series of pressure gradients by

$$dp/dx = -\rho_M u_H du_H/dx - \rho_M g \sin \theta - P_{T0}/A \qquad (14.10a)$$

$$= dp/dx|_a + dp/dx|_b + dp/dx|_f \qquad (14.10b)$$

In Eq. (14.10), the first term is the pressure gradient required to change the flow velocity, the second term is the pressure gradient required to balance the body forces due to gravity and, finally, the last term is the pressure gradient needed to overcome the frictional forces on the wall.

Empirical correlations are needed to evaluate the frictional pressure gradients. It is common practice to introduce frictional multipliers which are the ratios of the actual frictional pressure drop to the pressure drops that would exist if each phase flowed separately, namely,

$$\phi_G^2 = \frac{dp/dx|_f}{dp/dx|_G} \qquad (14.11a)$$

$$\phi_L^2 = \frac{dp/dx|_f}{dp/dx|_L} \qquad (14.11b)$$

where ϕ_G and ϕ_L are the nondimensional frictional multipliers.

In addition, it is convenient to define a frictional multiplier based on the pressure gradient that would exist if the total mass flow consisted of the liquid phase alone, namely,

$$\phi_{L0}^2 = \frac{dp/dx|_f}{dp/dx|_{L0}} \qquad (14.12)$$

The frictional pressure gradient, $(dp/dx)_{L0}$, can be related to the Darcy–Weisbach friction factor in the usual way,

$$-(dp/dx)_{L0} = \dot{m}^2 f_{L0}/2\rho_L D \qquad (14.13)$$

where f_{L0} is the Darcy–Weisbach friction factor obtained through the standard procedures for single phase flows [see Roberson and Crowe (1993), for example]. For laminar flow, f_{L0} is 64/Re. For turbulent flow in a smooth pipe (Re > 2000), the friction factor can be represented empirically by the Blasius equation

$$f_{L0} = .316(\dot{m}D/\mu_L)^{-1/4} \qquad (14.14)$$

Thus, if the value of the frictional multiplier is available, the pressure gradient is obtained from

$$-\frac{dp}{dx}\Big|_f = \phi_{L0}^2 \dot{m}^2 f_{L0}/2\rho_L D \qquad (14.15)$$

The frictional multipliers can be predicted for homogeneous flows. By this approach, the two-phase mixture is treated as a single phase flow with the properties

TABLE 14.1 Selected Values for φ_{L0}^2 for the Homogeneous Steam-Water System

Pressure (MPa)	Steam Quality (%)			
	10	50	80	100
0.101	121.2	435	623	738
0.689	21.8	80.2	115.7	137.4
3.44	1.44	17.45	25.1	29.8
10.3	1.95	5.08	7.08	8.32
20.7	1.0	1.0	1.0	1.0

of the mixture. In order to determine the friction factor, it is necessary to obtain the Reynolds number which, in turn, requires a value for the mixture viscosity. Several equations have been proposed for the two-phase viscosity but the one most commonly used is that due to McAdams *et al.* (1942).

$$1/\mu_M = x/\mu_G + (1 - x)/\mu_L \tag{14.16}$$

The two-phase Reynolds number can then be written as

$$\text{Re}_M = (\dot{m}Dx/\mu_G)\,(1 + (1 - x)\mu_G/x\mu_L) \tag{14.17}$$

Using the Blasius equation for pressure drop in a smooth pipe yields a two-phase multiplier of

$$\phi_{L0}^2 = \left(1 + x\,\frac{\rho_L - \rho_G}{\rho_G}\right)\left(1 + x\,\frac{\mu_L - \mu_G}{\mu_G}\right)^{-1/4} \tag{14.18}$$

This multiplier can be evaluated for any quality, temperature, and pressure condition for which viscosity data are available.

Some values calculated for steam–water flows are shown in Table 14.1. One notes that ϕ_{L0}^2 decreases with increasing pressure at a given quality reaching unity at the critical point.

The frictional pressure drop is then calculated by Eq. (14.15). This method works quite well for dispersed phase flows (bubbly flows) but tends to underpredict the pressure drop for separated flows.

Example 14.2   The mass flow rate of a steam/water mixture in a 10 cm diameter pipe is 2 kg/s. The pressure is 0.689 MPa (100 psia) and the quality is 50%. The pressure gradient is desired. From Chap. 2 in which a set of property tables for steam and water are given one finds

$$\mu_L = 1.64 \times 10^{-4} \text{ Ns/m}^2$$

$$\rho_L = 904 \text{ kg/m}^3$$

From Table 14.1, the value of the frictional multiplier for this condition is

$$\phi_{L0}^2 = 80.2$$

The Reynolds number of the liquid is found to be

$$\mathrm{Re} = \frac{\dot{m}D}{\mu_L} = 1.55 \times 10^5$$

The corresponding fraction factor from Eq. (14.14) is

$$f_{LO} = 0.016$$

Finally, the pressure gradient is obtained from Eq. (14.15) and is

$$\left.\frac{dp}{dx}\right|_f = -461 \text{ N/m}^3$$

The most widely used scheme to correlate pressure drop in separated flows is that due to Lockhart and Martinelli (1949). In a separated flow, the pressure drop through the liquid and gas phases must be the same so

$$(dp/dx)_f = \phi_G^2(dp/dx)_G = \phi_L^2(dp/dx)_L \tag{14.19}$$

Lockhart and Martinelli (L–M) chose to define a new parameter, X, which is the ratio of the liquid and gas frictional multipliers, namely,

$$X^2 = \phi_G^2\phi_L^2 = (dp/dx)_L/(dp/dx)_G \tag{14.20}$$

The correlations for ϕ_L and ϕ_G were presented in graphical form as a function of X and separate curves were identified for the laminar and turbulent conditions of the two phases. Chisholm (1967) proposed the following algebraic representation for the L–M correlation.

$$\phi_L^2 = 1 + C/X + C/X^2 \tag{14.21a}$$

$$\phi_G^2 = 1 + CX + CX^2 \tag{14.21b}$$

where the constant, C, depends on the turbulent or laminar state of the separated phases. The values for C proposed by Chisholm are given in Table 14.2. The laminar or turbulent classification is based on the Reynolds number with the superficial velocity of each phase.

TABLE 14.2   Value of C for Separated Laminar-Turbulent Flows

Liquid	Gas	C
turbulent	turbulent	20
laminar	turbulent	12
turbulent	laminar	10
laminar	laminar	5

The L–M correlations are not universal and lead to large errors in pressure drop predictions in some flow regimes. For example, the L–M correlation overpredicts the pressure drop for bubbly flows. Work has continued to find a better scheme to predict gas–liquid pressure drops. Baroczy (1965) proposed an empirical correlation that incorporates mass flow effects. More recently, Freidel [Hewitt (1982)] developed a new correlation and fit it with 25,000 data points. His correlation includes both gravitational and surface tension effects and is given by

$$\phi_{LO}^2 = E + 3.24FH/Fr^{0.045}We^{0.035} \tag{14.22}$$

where

$$E = (1 - x)^2 + x^2\rho_L f_{G0}/\rho_G f_{L0}$$

$$F = x^{0.78}(1 - x)^{0.24}$$

$$H = (\rho_L/\rho_G)^{0.91}(\mu_G/\mu_L)^{0.39}(1 - \mu_G/\mu_L)^{0.7}$$

$$Fr = (\dot{m}/\rho_M)^2/gD$$

$$We = \dot{m}^2D/\rho_M\sigma$$

and where f_{G0} and f_{L0} are the friction factors corresponding to the total mass flow being either the gas or the liquid. Freidel claims a standard deviation of 30% for single component flows and up to 50% for two-component flows.

The scatter in the data underscores the need to develop more phenomenological models for the pressure drop in gas–liquid flows. Work is currently proceeding in this direction. The development of a physical model for an annular type of mist flow is contingent on quantifying mechanisms for droplet formation, entrainment rate, and average droplet velocity in the gaseous core. This observation results from a simple analysis of annular mist flow in a duct as shown in Fig. 14.5. Droplets are formed at the liquid surface and become entrained in the gaseous core. At the same time droplets are deposited on the liquid layer from the gaseous core. A frictional force also acts on the gaseous flow at the liquid layer. Expressing the momentum equation for the control volume shown in the figure, one has for fully developed flow

$$A(p_1 - p_2) - \tau_f \pi D\Delta L = \dot{m}_d u_d - \dot{m}_e u_l \tag{14.23}$$

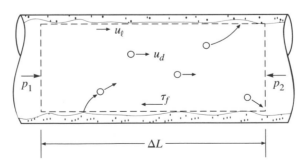

FIGURE 14.5 Control volume for annular mist flow in duct.

where $\dot{M}_d$ is the mass rate of droplet deposition on the liquid layer, u_d is the droplet velocity in the core, $\dot{M}_e$ is mass rate of droplet entrainment, and u_l is the velocity of the surface of the liquid layer. Neglecting u_l compared to u_d, the equation for pressure drop becomes

$$p_1 - p_2 = \tau_f \pi D \Delta L / A + \dot{m}_d u_d / A \qquad (14.24)$$

For a fully developed flow, the entrainment rate will equal the deposition rate. Expressing the entrainment rate as the flux based on the surface area of the liquid layer, the equation for pressure gradient can be expressed as

$$-\Delta p / \Delta L = 4\tau_f / D + 4\dot{m}_e u_d / D \qquad (14.25)$$

where $\dot{m}_e$ is the droplet entrainment flux (kg/m^{2}s). The first term on the right-hand side of Eq. (14.25) is the component of the pressure drop due to friction while the second term is the component due to droplet–liquid layer interaction.

The entrainment rate depends on the flow rates and viscosities of the gas and liquid as well as on the surface tension of the liquid. The droplet velocity will depend on the droplet size and effective residence time of the droplet in the gaseous core. Fundamental quantitative information of this nature will ultimately lead to more reliable physical models for the pressure drop in annular mist flows.

The flow of gas–liquid mixtures in variable area ducts has been the subject of several studies in multiphase flows. Of specific interest is the flow of gas–liquid mixtures in nozzles and ventures. Several experiments have been reported on the measurement of pressure drop of steam–water and air–water mixtures across venturi meters. Also, many empirical correlations have been developed for the pressure drop. One correlation which has been used extensively is that of Chisholm (1967) which is expressed as

$$\Delta p_M / \Delta p_G = 1 + C(\Delta p_L / \Delta p_G)^{1/2} + (\Delta p_L / \Delta p_G) \qquad (14.26)$$

where Δp_M is the pressure drop with the two-phase mixture, and Δp_L and Δp_G are the pressure drops if the liquid or gaseous phases were to flow through the venturi alone. The constant C, as given by Chisholm, is a function of the density ratio, namely

$$C = (\rho_L / \rho_H)^{1/2} + (\rho_H / \rho_L)^{1/2} \qquad (14.27)$$

where ρ_H is the density of the mixture if the flow were homogeneous. This empirical correlation was developed assuming no mass exchange between the core and liquid layer on the wall. Recent experiments have shown that this correlation is not universally valid. Further work in this area has been reported by Martindale and Smith (1981) and El Haggar and Crowe (1983).

14.1.3 Void Fraction

A fundamental flow parameter in gas–liquid flow is the void fraction or the volume of the gaseous phase per unit volume of mixture. This void fraction can be regarded as the cross-sectional area occupied by the gaseous phase in the flow of a gas–liquid mixture in a duct.

A model used to predict void fraction, which works well for bubbly flows is the drift flux model [Wallis (1969)]. In this model, the flux of each phase is referenced to the superficial velocity of the mixture which is defined as

$$U_M = (Q_L + Q_G)/A = U_L + U_G \tag{14.28}$$

where Q_L and Q_G are discharge (m³/s) of the liquid and gas phases. The flux of the gas phase, with respect to the mixture velocity, is

$$j_{GM} = \alpha_G(u_G - U_M) \tag{14.29}$$

and the flux of the liquid phase with respect to the mixture velocity is

$$j_{LM} = \alpha_L(u_L - U_M) = (1 - \alpha_G)(u_L - U_M) \tag{14.30}$$

The sum of these two velocities is zero, because there is no net discharge with respect to a plane moving at the superficial mixture velocity.

$$\begin{aligned} j_{GM} + j_{LM} &= \alpha_G u_G + \alpha_L u_L - (\alpha_L + \alpha_G)U_M \\ &= U_M - U_M = 0 \end{aligned} \tag{14.31}$$

so

$$j_{GM} = -j_{LM} \tag{14.32}$$

Substituting the relationship between phase velocity and superficial velocity into the equation for j_{GM}, Eq. (14.29), yields

$$j_{GM} = (1 - \alpha_G)U_G - \alpha_G U_L \tag{14.33}$$

which shows that j_{GM} varies linearly with α_G, the void fraction.

The drift flux must also be a function of the properties of the system such as void fraction, flow pattern, and properties of the fluid. Wallis (1969) has found that the variation of j_{GM} with void fraction for vertical bubbly flow is represented well by

$$j_{GM} = u_\infty \alpha_g(1 - \alpha_g)^n \tag{14.34}$$

where the exponent n depends on the bubble Reynolds number and the Galileo number,* and u_∞ is a terminal velocity of the bubble in an infinite fluid. For small bubbles, the exponent is two and the rise velocity is given by

$$u_\infty = 2r_b^2(\rho_L - \rho_G)g/9\mu_L \tag{14.35}$$

where r_b is the bubble radius. The solution of Eqs. (14.33) and (14.34) will yield the void fraction for a vertical bubbly flow. The graphical solutions for cocurrent,

*The Galileo number, sometimes also referred to as the Archimedes number is defined as $\rho_L g d_b^3(\rho_L - \rho_G)/\mu_L^2$, where d_b is the bubble diameter.

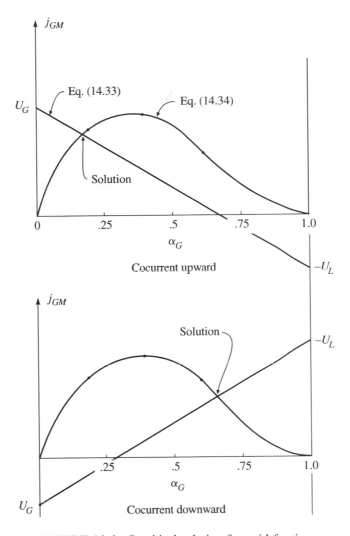

FIGURE 14.6 Graphical solution for void fraction.

vertical upward and vertical downward flows are shown in Fig. 14.6. The same technique can be applied for counter-current flows. There will obviously be no solution for liquid upward, gas downward flows.

Other correlations have been developed for predicting void fraction. Lockhardt and Martinelli (1949) proposed an empirical variation for void fraction based on the Martinelli parameter, X. Other correlations have been developed for other flow regimes such as stratified, slug, and annular flows.

An application of bubble flows appears in Sec. 23.13.

14.1.4 Droplet Sprays

Droplet sprays represent an important area of gas–liquid flows because of their application to processes such as combustion of liquid fuels, spray drying of food stuffs and chemicals and spray cooling. A nozzle issuing a spray into a gas flow is shown

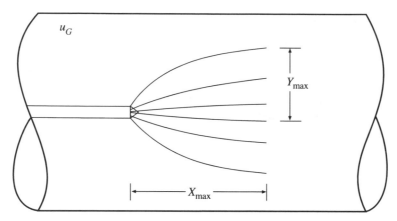

FIGURE 14.7 Spray into gas flow in a duct.

in Fig. 14.7. Spray patterns are identified as hollow cone or solid cone sprays, depending on the droplet flux distribution within the spray.

A variety of atomizer designs are available to generate different spray patterns and droplet size distributions. These atomizers are generally classified as wheel atomizers, two-fluid atomizers, and pressure atomizers. Sonic and vibratory atomizers are less common.

The extent of penetration of a spray into the local gas-flow field is an important design consideration. A simple analysis, assuming that the gas-flow field is unaffected by the spray, yields the extent of spray penetration and is useful for preliminary design of spray systems. Consider the droplet in the gas flow shown in Fig. 14.8. If Stokes law is valid for the aerodynamic drag, the equation of motion for the droplet is

$$m_D \frac{du_D}{dt} = 3\pi\mu_g d_D(u_G - u_D) \tag{14.36}$$

where m_D is the droplet mass, d_D is the droplet diameter, μ_G is the gas viscosity, and u_G and u_D are the gas and droplet velocities. It is assumed that the drag is unaffected by droplet evaporation and that the mass flux from the droplet surface is uniform. Taking a spherical droplet, the equation becomes

$$\frac{du_D}{dt} = \frac{18\mu_G}{\rho_D d_D^2}(u_G - u_D) \tag{14.36a}$$

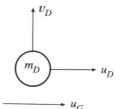

FIGURE 14.8 Droplet in gas flow field.

If the droplets are evaporating (or burning) a reasonable empirical equation for the rate of droplet evaporation is the *diameter-squared law* which states that the square of the droplet diameter decreases linearly with time

$$d_D^2 = d_0^2 - \lambda t \tag{14.37}$$

where d_0 is the initial diameter and λ is the evaporation rate constant. Substituting this relation into Eq. (14.36) gives

$$\frac{du_p}{dt} = \frac{1}{\tau_A} \frac{(u_G - u_D)}{(1 - t/\tau_B)} \tag{14.38}$$

where $\tau_A = \rho_D d_0^2/18\mu_G$ is the aerodynamic response time of the droplet, and $\tau_B = d_0^2/\lambda$ is the droplet evaporation, or burning, time. Integrating this equation twice and setting t equal to τ_B gives the penetration distance of the spray. The maximum penetration is

$$\frac{S_{max}}{\tau_A u_0} = \frac{\tau_B/\tau_A}{1 + \tau_B/\tau_A} [(u_G/u_0)(\tau_B/\tau_A) + 1] \tag{14.39}$$

where u_0 is the initial droplet velocity. This equation is useful to estimate spray penetration.

Example 14.3 A fuel spray with 100 μm droplets is injected at an angle of 30° with a velocity of 60 m/s into an airflow moving at 10 m/s. The viscosity of the air is 1.8×10^{-5} Ns/m^2 and the evaporation constant is 10^{-6} m^2/s. The density of the liquid fuel is 800 kg/m^3. The aerodynamic response times and evaporation times are

$$\tau_A = 25 \text{ ms}$$

$$\tau_B = 10 \text{ ms}$$

The initial droplet velocity in the lateral direction is 60 (sin 30°), or 30 m/s. The component of the gas velocity in the lateral direction is zero, so

$$Y_{max} = 0.21 \text{ m}$$

Correspondingly, the maximum travel in the gas flow direction before complete evaporation is

$$X_{max} = 0.40 \text{ m}$$

The original analyses of spray systems were based on the assumption, as used above, that the spray did not affect the gas flow field, and the droplet dynamics were unaffected by the presence of neighboring droplets. More accurate analyses, however, must include the multiphase transfer processes taking place in the spray. For example, with a cold liquid sprayed into a hot gas fume, transfer of heat from the gas to the liquid occurs with the associated cooling of the gas and reduced evaporation rate of the droplets. The mass released by droplet evaporation represents an additional gaseous mass to the conveying gas flow. The deceleration of the droplets gives rise to a momentum increment in the gas. All of these exchange mechanisms will affect the gas flow and droplet motion.

The availability of computers and the development of computer models [Crowe (1991)] have enabled the prediction of spray patterns and properties including the two-way coupling effects: that is, not only the effect of the gas on the droplet parameters, but also the effect of the droplet on the gas flow field. Developments are continuing in both experimental studies and numerical modeling.

14.2 GAS–SOLID FLOWS

The motion of particles in gas streams has been of interest to engineers for many years in the design of pneumatic conveying systems. Since the early 1900's, there has been a continuing effort to improve the design of such systems. After the development of reliable pumping equipment, pneumatic conveying has found application in transporting powdered cement, grains, metal powders, ores, coal, and so on. The major advantage in pneumatic conveying over other systems such as conveyor belt, is the continuous operation, the relative flexibility of the pipe line location to avoid obstacles or to save space, and the capability to tap the line at any location to remove all or a portion of the powder. The book by Klinzing (1981) offers an excellent background in the principles of pneumatic transport.

14.2.1 Flow Patterns

Like gas–liquid flows, the patterns of gas–solid flows in a pipe will depend on many factors, such as solids/gas ratio, Reynolds number, and other properties of the particulate material. The flow patterns shown in Fig 14.9 represent a few patterns that have been observed [Wen (1979)] for gas–particle flow in a horizontal duct. Homogeneous flow represents a situation where the gas velocity is sufficiently high and the solids loading is sufficiently low that the turbulence within the duct maintains good mixing and property uniformity across the duct area. Reducing the gas velocity leads to a concentration of particles on the lower section of the wall, and the homogeneous flow begins to degenerate. As the particles build up a layer on the bottom wall, ripples are formed by the moving gas stream, and the flow is known as a *dune flow*. As the powder continues to fill the pipe, there will be alternate regions where the particles have settled out and others will remain in suspension, giving rise to an intermittent flow pattern called a *slug flow*. Finally, with increased loading, the powder plugs the pipe completely and gas flow then represents a flow through a packed bed. Besides the flow patterns shown above, there are also other patterns. There is no chart, however, equivalent to the Baker chart for gas–particle flow in a horizontal duct.

Homogeneous flow

Dune flow

Slug flow

Packed bed

FIGURE 14.9 Flow patterns for horizontal gas–solid flow.

14.2.2 Pressure Drop in Constant-Area Ducts

A typical variation in pressure drop due to gas–particle flow in a horizontal duct as a function of superficial gas velocity is shown in Fig. 14.10. The various curves represent increased solids loading. At low solids loading and higher velocity, the flow pattern is homogeneous, and the pressure decreases with decreasing gas velocity. With decreased gas velocity, particles start to deposit on the pipe wall. Deposition continues until the reduction in the cross-sectional area for the gas flow produces a gas velocity high enough to reentrain the particles from the surface. This velocity is called the *saltation velocity*. The pressure drop is higher because of the increased gas velocity. A further reduction in superficial velocity leads to further deposition, reduced gas flow area, increased gas phase velocity, and attendant increased pressure drop. The same trends exist for higher loadings with correspondingly higher pressure drops.

Most of the analytic models that have been developed for pressure drop due to gas–particle flow in horizontal pipes are based on homogeneous flow. The mechanics of the other flow regimes are very complex and only recently have been addressed through numerical simulation [Tsuji *et al.* (1992)]. The pressure drop in a fully developed flow is attributed to two sources: the frictional forces on the gas due to the walls and the drag acting on the particles. As the particles impact the wall and rebound or are deposited on the wall and reentrained by the gas flow, the gas flow must reaccelerate these particles toward the mean flow velocity in the duct. The force required for particle acceleration is the pressure drop due to the particles in the duct.

Because of the technological significance of this problem, innumerable studies

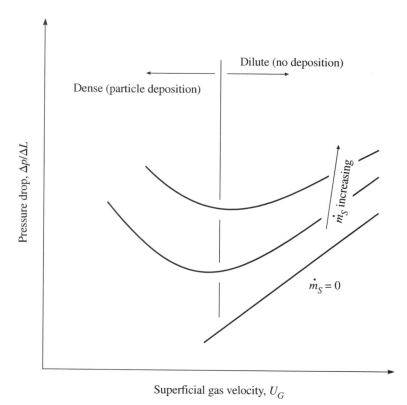

FIGURE 14.10   Dependence of pressure drop on superficial gas velocity for gas–particle flow in a horizontal duct.

have been done to measure the pressure drop in pneumatic conveyance systems and develop a correlation to the data. Barth (1958) suggested that the appropriate non-dimensional parameters for correlation of the pressure drop are the Froude numbers based on the gas velocity and the terminal velocity of the particles. He obtained a good correlation for his data with the Froude number based on the flow velocity. It was found, however, that the correlation was not valid for other systems. Several attempts were made to extend Barth's analysis to make the correlation more general. One of the difficulties of trying to correlate the numerous data available is the uncertainty of the data itself. There is some concern that many of the data reported did not adequately account for acceleration effects, that is, the reestablishment of uniform flow downstream of a flow disturbance. It is recommended that one seeking data for design purposes should seek experimental results in the literature that best correspond to the specific design conditions of interest.

There have been some attempts at developing more phenomenological models for the pressure drop in pneumatic transport systems. These models are based on the premise that the pressure drop in the duct is due to the aerodynamic force needed to maintain the particle's velocity. Ths pressure gradient is given by Brothroyd (1971)

$$\frac{\Delta p_S}{\Delta L} = -n \frac{\pi d_P^2 C_D}{8} \rho_G (u_G - u_P)^2 \tag{14.40}$$

where n is the particle number density, (m^{-3}) d_P is the particle diameter (m), C_D is the drag coefficient, and ρ_G is the gas density (kg/m^3). The ratio of the pressure drop due to the solids to the pressure drop due to gas alone can be written as

$$\frac{\Delta p_S}{\Delta p_G} = \frac{3}{2} \frac{C_D \rho'_P D (u_G - u_P)^2}{\rho_P d_P u_G^2 C_f} \tag{14.41}$$

where ρ'_P is the bulk density of the particles (kg/m^3), D is the pipe diameter (m), ρ_P is the material density of the particles (kg/m^3), and C_f is the local shear stress coefficient at the wall due to the gas flow alone. This expression can be used only if the particle velocity (particle–gas velocity difference) is known.

Some correlations for particle velocity have been developed. Schuchart (1968) proposed that the ratio of particle velocity to gas velocity be expressed as

$$\frac{u_P}{u_G} = \left[1 + C \left(\frac{\rho_P}{\rho_G} - 1 \right)^{2/3} \left(\frac{d_P}{D} \right) \left(1 + \frac{200}{\text{Fr} - \text{Fr}_0} \right) \right]^{-1} \tag{14.42}$$

where Fr is the Froude number of flow in the pipe and Fr_0 is the Froude number when the particles begin to settle out. The constant C varies between 0.014 and 0.05. Other correlations for slip velocity have been developed. However, based on the apparent inconsistencies in the data, it is recommended that only correlations be used which closely match the conditions of interest to the designer. Extrapolation with data correlations to other flow conditions is not recommended.

There is an obvious need for more descriptive models for particle motion in a duct. One important difference in model development is the distinction between coarse and fine particle suspensions. Gravity plays a significant role in particle motion for coarse particle suspensions. Owen (1969) has developed a model in which it was assumed that the particle rebounds in the direction normal to the wall and reaccelerates toward the gas velocity. The time of flight before another rebound with the wall is related to gravity and duct dimensions. Using this approach, Owen developed the following expression for the solids/gas pressure drop ratio.

$$\frac{\Delta P_S}{\Delta P_G} = \frac{\dot{m}_S g}{\rho_G R u_G u_P u_t} \tag{14.43}$$

where $\dot{M}_S$ is the mass flow rate of the particles (kg/s), and R is the duct radius. The predictions provided by this model appear to agree quite closely with experimental observations. However, some aspects of the model require further study. For example, the inverse relationship between pressure drop and terminal velocity is questionable. Still, this approach may serve to be fruitful.

In fine suspension flow, the terminal velocity of the particles is small compared to the flow velocity, and the particles tend to remain in suspension. In this case, the presence of the particles plays a more important role on the effective viscosity of the fluid. The pressure drop of the two-phase mixture is a function of the effective viscosity.

Considerably more experimental work and model development is needed to permit reliable predictions of pressure drop in horizontal pneumatic conveyance ducts.

The application of more sophisticated instrumentation and experimental techniques to measure particle gas flow properties will lead to improved understanding and, ultimately, more descriptive analytic and numerical models. A recent review of numerical models has been published by Sommerfeld and Zivkovic (1992).

14.2.3 Gas–Solid Flows in Variable-Area Ducts

Several studies have been reported in the literature on the flow of gas–particle mixtures in venturi meters and rocket nozzles. The venturi meter studies have been concerned with metering the flows of gas–solid suspensions, while the study of gas–particle flows in rocket nozzles has been related to nozzle performance loss.

An important definition in the development of the equations for gas–particle flow in variable area ducts is the definition of *dilute* and *dense* flows. The definition used here describes a phenomenon different than that applied to pneumatic transport in which dilute and dense refer to the flow regimes above and below the saltation velocities. In this instance, *dilute* refers to a gas–particle flow in which the particle motion is established by aerodynamic forces and not by particle–particle collisions. A dilute or dense flow is characterized by the ratio of the aerodynamic response time (τ_A) to the time between particle–particle collisions (τ_C). For a definition of τ_A, see the discussion in connection with Eq. (14.38). If τ_A/τ_C is much less than unity, then the particles have sufficient time to respond to the local aerodynamic flow field before the next collisoin. This gas–particle flow is controlled by particle aerodynamics and is referred to as dilute. On the other hand, if τ_A/τ_C is much greater than unity, then the particle motion is controlled by collision, and the flow is *dense*.

The governing equations for steady, dilute, nonreacting gas–particle flow in the variable area duct can be expressed in terms of the governing equations for each phase. The continuity equation for the gas phase is

$$\frac{d}{dx}(\alpha_G \rho_G u_G A) = 0 \qquad (14.44a)$$

and, that for the particulate phase is

$$\frac{d}{dx}(\alpha_P \rho_P u_P A) = 0 \qquad (14.44b)$$

where A is the cross-section area of the duct (m^2). In many gas–particle flows, the volume fraction of the gas is very close to unity, so α_G is set equal to unity and the product, $\alpha_P \rho_P$, is replaced by the bulk density of the particles, ρ_P' (kg/m^3).

The momentum equation for the gas can be written as

$$\frac{d}{dx}(\alpha_G \rho_G u_G^2 A) = -A\frac{dp}{dx} - f_D - \tau_G P \qquad (14.45a)$$

where f_D is the aerodynamic force per unit length of duct (N/m) on the gas due to the particles and τ_G is the shear stress on the gas at the wall. The momentum equation for the particulate phase is

$$\frac{d}{dx}(\rho_P' u_P^2 A) = f_D - \tau_P P \qquad (14.45b)$$

where τ_P is the effective shear stress due to the particle–wall interaction.

The term f_D is the coupling force between the gas and the particulate phases in the momentum equation and can be expressed as

$$f_D = nAF_D \qquad (14.46)$$

where n is the particle number density (m^{-3}), and F_D is the drag on an individual particle. The aerodynamic drag includes the contributions due to pressure gradient and unsteady effects as well as the steady state drag. For gas-particles flows where the density ratio of the gas to the particles (ρ_G/ρ_P) is small (about 10^{-3}) the drag on a particle can be expressed as

$$F_D = 3\pi\mu_G d_P\lambda(u_G - u_P) \qquad (14.47)$$

where λ is ratio of the drag coefficient to Stokes drag. Similar equations can be written for the energy of the gas and particulate phases where the coupling between the phases is represented by heat transfer.

Two important scaling parameters for gas–particle flow evolve from nondimensionalization of the gas-phase momentum equation. Substituting Eq. (14.47) into Eq. (14.45a) and using Eq. (14.44a) gives

$$\dot{m}_G \frac{du_G}{dx} = -A \frac{dp}{dx} - \frac{\lambda\rho_P'A}{\tau_A}(u_G - u_P) - \tau_G P \qquad (14.48)$$

where $\dot{m}_G$ is the gas mass flow rate (kg/s) and τ_A is the aerodynamic response time of the particle. Recognizing that $\rho_P' Au_P$ is the mass flow rate of the particular phase, $\dot{m}_S$ (kg/s), the gas-phase momentum equation becomes

$$\frac{du_G}{dx} = -\frac{A}{\dot{m}_G}\frac{dp}{dx} - \lambda\frac{\dot{m}_S}{\dot{m}_G}\frac{1}{\tau_A}\left(\frac{u_G}{u_P} - 1\right) - \frac{\tau_G P}{\dot{m}_G} \qquad (14.49)$$

Nondimensionalizing the velocity with respect to a characteristic velocity of the gas, U_0, and the distance with respect to a characteristic length of the flow system, L, results in

$$\frac{d\bar{U}_G}{d\bar{x}} = -\frac{d\bar{p}}{d\bar{x}} - \left(\frac{\dot{m}_S}{\dot{m}_G}\right)\frac{L}{U_0\tau_A}(\bar{u}_G/\bar{u}_P - 1) - \tau_G\frac{PL}{A} \qquad (14.50)$$

where the bars over the variables indicate nondimensional variables.

The two important nondimensional parameters which appear are $\dot{m}_S/\dot{m}_G$ and $\tau_A U_0/L$. The first parameter is the *loading*, Z. As the loading approaches zero, the effect of the particles on gas flow properties is minimized, and the flow can be treated as a single phase gas flow.

The parameter $\tau_A U_0/L$ is the *Stokes number* (St), and it can be considered as the ratio of particle response time, τ_A, to flow residence time, L/U_0. If the Stokes number is large, the particles do not have time to respond to changes in the gas flow velocity. This is known as the *frozen flow* condition. On the other hand, if the Stokes number is small, the particles have ample time to respond to changes in the gas flow field and to maintain velocities close to those of the gas. In the limit of zero Stokes number, the particle and gas velocities are equal (*equilibrium flow*).

Adding the particle and gas–phase momentum equations, Eqs. (14.45a) and (14.45b), and setting the phase velocities equal gives

$$\alpha_G \rho_G (1 + Z) u_G \frac{du_G}{dx} = -A \frac{dp}{dx} - P(\tau_G + \tau_P) \tag{14.51}$$

This equation represents the flow of a fluid with density $\rho_G(1 + Z)$ and shear stress $(\tau_P + \tau_G)$. Thus, the equilibrium flow condition allows one to treat the two-phase mixture as a single phase fluid with modified properties.

Experimental studies of the pressure drop in a Venturi meter with gas–particle flow have shown that the pressure drop varies nearly linearly with loading. Typical data from Lee and Crowe (1982) are shown in Fig. 14.11 where Δp_M is the pressure drop with gas–particle flow, and Δp_G is the pressure drop for gas flowing alone. The equilibrium flow curve corresponds to the pressure drop which would occur if the gas were inviscid and incompressible and the Stokes number were zero. Increasing the Stokes number reduces the pressure drop, because the gas is less effective in accelerating the particles. As a rule of thumb, equilibrium flow conditions can be assumed for Stokes numbers less than 1/10.

Example 14.4   Fifty μm coal particles are conveyed by air at standard conditions through a Venturi meter. The throat diameter of the meter is 10 cm, and the flow velocity at the throat is 30 m/s. The density of the coal is 1300 kg/m^3. Can one treat the flow as an equilibrium flow?

The approach to equilibrium flow is quantified by the Stokes number. The aerodynamic response time of the coal particles is

$$\tau_A = \frac{(1300)\,(50 \times 10^{-6})^2}{(18)(1.8 \times 10^{-5})} = 10 \text{ ms}$$

The system residence time is

$$\tau_S = L/U_0 = 10 \times 10^{-2}/30 = 3.3 \text{ ms}$$

The Stokes number is $10/3.3 = 3$. The flow will not be in velocity equilibrium, and the pressure drop will be smaller than that for an equilibrium flow.

Several numerical solutions for the gas–particle flow equations have appeared in the literature. The analyses of gas particle flows in a rocket nozzle were among the

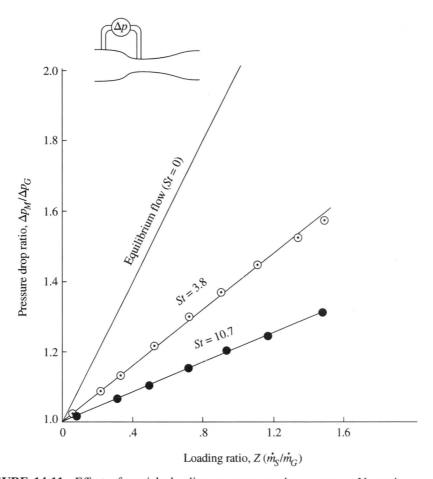

FIGURE 14.11   Effect of particle loading on pressure drop across a Venturi meter with gas–particle flow. (Reprinted with permission of ASME Lee J. and Crowe, C. T., 1982, "Scaling Laws for Metering the Flow of Gas-Particle Suspensions Through Venturis," *J. of Fluids Engr.*, Vol. 104, No. 1, pp. 88–91, 1982.)

first numerical studies that included the effect of the particles on the gas-flow field as well as the effect of the gas on the particles (two-way coupling). The numerical model originally developed by Kliegel (1960) was based on a quasi one-dimensional flow that showed the interesting feature of the sonic line occurring downstream of the geometric throat. The need for more accurate numerical models in the expansion section of the nozzle led to the application of the *method of characteristics* (see Sec. 8.7) to supersonic gas–particle flows. These models have met with considerable success in predicting performance of rocket nozzles.

14.2.4   Erosion

One of the major concerns with pneumatic transport and the operation of turbomachinery in a dusty environment is erosion. Because of particle inertia and gas-phase turbulence, particles do not follow the gas path lines, and they may impact on walls or other surfaces. This impact leads to metal removal from the surface or erosion.

Several studies of the erosive nature of particle-laden flow have appeared in the literature [Sheldon and Finnie (1966) and Tabakoff (1982)]. One important aspect of erosion is the effect of particle impact velocity and the orientation of the velocity vector to the surface on the erosion rate. A typical variation of erosion rate with the angle of approach of the particle to the surface is shown in Fig. 14.12. It is noted that the erosion rate appears to go through a maximum at an approach angle of approximately 30°. It is also found that the erosion rate increases with increased approach velocity and increased material temperature.

The accurate prediction of erosion is a difficult task, because the velocity and angle of approach of the particles at impact requires a reliable description of the gas-flow field. Also, information is needed on the velocity and angle at which the particle is reflected from a surface after impact, so the particle trajectory can be calculated for the next impact. In addition, there is the possibility of particle fracture after impact which further complicates the analysis. Several numerical studies have been undertaken in this area, many of which can be found in Tabakoff (1982).

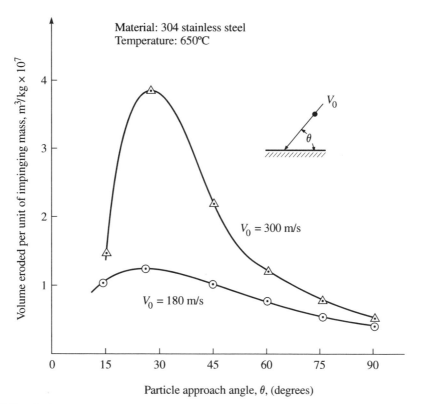

FIGURE 14.12 Dependence of erosion rate on particle approach angle and material temperature. (Reprinted with permission from ASME, Tabakoff, W., "Performance Deterioration on Turbomachinery with Presence of Solid Particles," *Particle Laden Flow in Turbomachinery*, pp. 3–22, 1982.)

14.3 LIQUID–LIQUID FLOWS

The flow of two immiscible liquids is encountered in many industries, especially the chemical and petroleum industries. The flow of oil and water in pipe lines is important to the transport of crude oil to treatment facilities. A major interest has been the introduction of water in specified amounts to pipelines conveying heavy, viscous crude oil for the purpose of reducing the head loss and pumping power requirements. In this case, the water forms an annular region at the wall with a core flow of the more viscous fluid.

The flow patterns associated with liquid–liquid flows are as diverse as those with gas–liquid flows. In general, the immiscible fluids tend to stratify according to their density ratio. However, various flow patterns appear to be a function of superficial velocities as well as the fluid properties. In general, the flow patterns in oil/water mixtures with the same densities can be classified as follows and as in Fig. 14.13.

a. Water drops in oils: Water drops are suspended in the oil where the oil forms a continuous media.

b. Oil in water annulus: Water forms an annular region around the core flow of oil.

c. Oil slugs in water: The oil forms large oblong slugs with an annular water region between the slug and the wall.

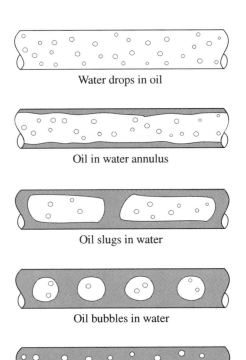

Water drops in oil

Oil in water annulus

Oil slugs in water

Oil bubbles in water

Oil drops in water

FIGURE 14.13 Flow patterns in equal density flowing oil–water mixtures.

 d. Oil bubbles in water: The slugs become more symmetric and appear as large bubbles with diameters comparable to the tube diameter.

 e. Oil droplets in water: The oil droplets are suspended in the water as the water flows as the continuous medium.

Studies of the flow patterns of oil/water mixtures flowing in horizontal one-in. pipe lines have been reported by Charles *et al.* (1961). The experiments were done with oils that had the same density as water but with viscosities ranging from 6.3–65 cp at 25°C. The various flow patterns observed were plotted on a map using superficial velocities as coordinates as shown in Fig. 14.14. At large superficial oil velocities and low water velocities, the water droplets-in-oil pattern evolves, while oil droplets-in-water results from the large superficial water velocities and low superficial oil velocities. The flow pattern map was somewhat different for high viscosity oil, in which case the patterns were less stable and there seemed to be a condition whereby water droplets in oil would change directly to oil slugs in water.

Measurements [Charles *et al.* (1961)] of the pressure drop along the pipe with oil–water mixtures showed different trends as a function of the oil/water flow rate

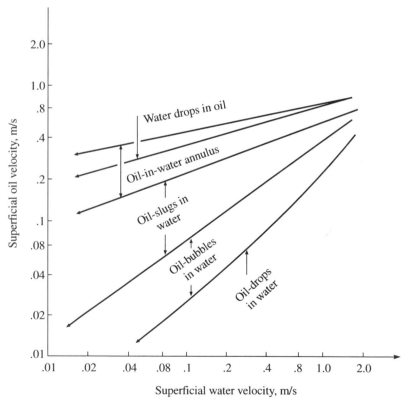

FIGURE 14.14   Flow pattern map for equal density oil–water mixture with $6 < \mu_{oil}/\mu_{water} < 17$. (Reprinted with permission from Charles, M. E., Govier, G. W., and Hodgson, G. W., "Horizontal Pipeline Flow Density Oil/Water Mixture," *Can. J. Chem. Engr.*, Vol. 39, No. 1, pp. 27–36, 1961.)

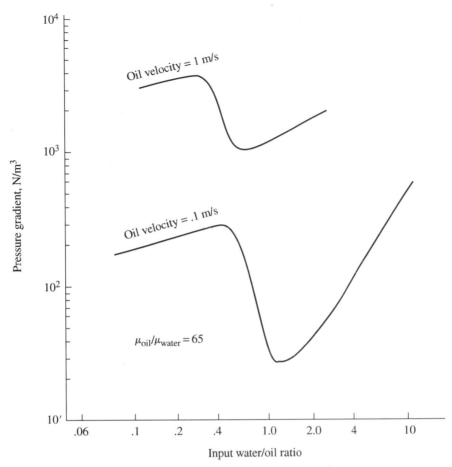

FIGURE 14.15   Dependence of pressure gradient on water/oil ratio. (Reprinted with permission from Charles, M. E., Govier, G. W., and Hodgson, G. W., "Horizontal Pipeline Flow Density Oil/Water Mixture," *Can. J. Chem. Engr.*, Vol. 39, No. 1, pp. 27–36, 1961.)

ratio depending on the viscosity of the oil. The pressure gradient for the most viscous oil/water mixture as a function of the water/oil flow rate ratio is shown in Fig. 14.15 for constant oil velocities. One notices a sharp drop in the pressure gradient beginning at a water/oil ratio of approximately 0.5 and reaching values as much as a factor of 10 less than those for pure oil flow alone. The technological significance of this is obvious from the point of view of power pumping requirements. The pressure drop curves for the less viscous oil/water mixtures show the same trends, but the changes are less pronounced.

14.4   SOLID–LIQUID FLOWS

Solid–liquid flows are encountered in the transport of slurries and in the motion of liquid in aquifers. This section will be concerned with the flow of slurries. A discussion of flows in porous media can be found in Sec. 13.5. The book by Shook and Roco (1991) provides a comprehensive discussion of the principles and current practice of slurry flows. Slurry pipelines, constructed to transport a variety of ores and coals, have proven to be economically viable.

14.4.1   Homogeneous or Heterogeneous Flows

Slurries in horizontal pipelines are classified as either *homogeneous* or *heterogeneous*. A heterogeneous slurry is composed of large particles which tend to settle out on the lower side of the pipe line. In such systems, the pressure loss variation with velocity shows a minimum value where a bed starts to accumulate on the pipe bottom. The velocity at which this occurs is called the *deposition velocity*. In a homogeneous slurry, the fluid mixing is sufficient to keep the particles in suspension. In this case, there is no minimum pressure drop but only a change of pressure drop gradient as the flow changes from laminar to turbulent.

Of course, no horizontal pipe flow would be completely homogeneous. A measure of homogeneity is the variation of particulate concentration across the pipe section. The rule of thumb is that the deviation in particle concentration be less than 20% to be classified as a homogeneous flow. Studies have shown that the particle concentration variation depends on both particle size and flow velocity. Wasp *et al.* (1970), using coal–water slurries with 50% by weight coal moving at 2 m/s in a 30-cm pipe, found that homogeneous flow could be maintained with particles as large as 600 μm in diameter. Larger particles could be homogeneously suspended at larger velocities. There is no well-established technique, given the liquid and particle parameters, flow velocities, and loading, with which one can assess the homogeneity of the flow.

14.4.2   Pressure Drop in Homogeneous Slurry Flows

The most common approach to analyzing homogeneous slurry flows is the application of models for non-Newtonian fluids. Chapter 3 focuses on non-Newtonian flows. Particularly, the slurry is treated as a *Bingham plastic* fluid or as a *power law* fluid. The variation of shear stress with velocity for each type of fluid is shown in Fig. 14.16. A Bingham fluid is represented by a threshold shear stress below where there is no motion and above where the shear stress varies linearly with the strain rate. The power law fluid is one in which the shear stress varies with the power of the strain rate.

$$\tau = k \left(\frac{du}{dy} \right) \left| \frac{du}{dy} \right|^{n-1}$$
(14.52)

where n and k are empirical parameters* which, for a Newtonian fluid, would reduce to unity and the dynamic viscosity of the fluid.

To predict the pressure drop of homogeneous slurry flow in pipes, it is imperative to know if the flow is laminar or turbulent. The usual single phase procedure does not apply here because of the non-Newtonian character of the fluid. In the case of a

---

*For $n > 1$, the fluid is *pseudoplastic* and for $n < 1$, *dilatant*.

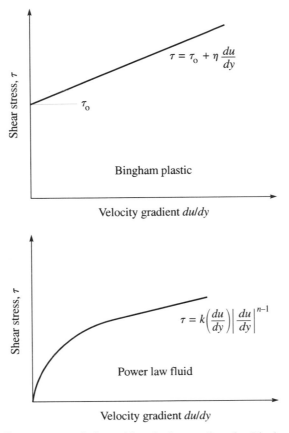

FIGURE 14.16 Shear stress variation with velocity gradient for Bingham plastic and pseudoplastic power law fluid.

Bingham plastic, the most significant nondimensional parameter is the *Hedstrom number* defined as

$$\mathrm{He} = \tau_o \rho_M D^2 / \eta^2 \qquad (14.53)$$

where τ_o is the threshold yield stress, and η is the slope of shear stress versus the strain rate in the linear region. Hanks and Pratt (1967) correlated a large amount of data and found the result shown in Fig. 14.17 where the critical Reynolds number is a function of Hedstrom number. In this case, the Reynolds number is defined as

$$\mathrm{Re} = \frac{u_M D \rho_M}{\eta} \qquad (14.54)$$

This technique appears to give reliable predictions of transition velocity.

Example 14.5 A homogeneous slurry flow in a 5-cm pipe, and the density of the slurry is 1500 kg/m^3. The slurry is modeled as a Bingham plastic with a threshold

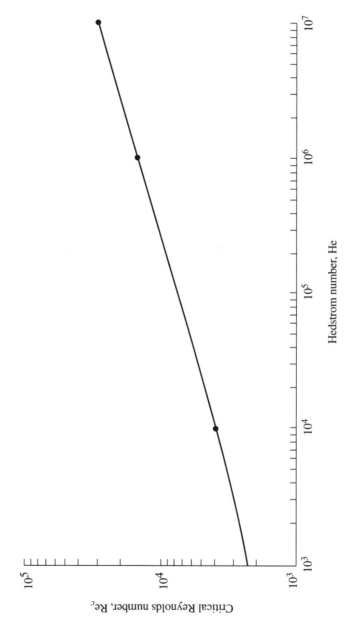

FIGURE 14.17 Critical Reynolds number versus Hedstrom number for a slurry modeled as a Bingham plastic. (Reprinted with permission from Hanks, R. W. and Pratt, D. R., "On the Flow of Bingham Plastic Slurries in Pipes and Between Parallel Plates," *Soc. of Petroleum Engr.*, pp. 342–346, 1967.)

yield stress of 10 Pa and a slope of 0.01 Ns/m^2. The transition velocity between laminar and turbulent flow is to be calculated.

The Hedstrom number is evaluated first and found to be

$$H_e = (10)(1500)(0.05)^2/0.01^2 = 3.75 \times 10^5$$

Referring to Fig. 14.17, the critical Reynolds member is

$$\text{Re}_c = 10^4$$

The transition velocity is

$$u_M = \eta \, \text{Re}_c/\rho_M D = 0.01 \times 10^4/(1500 \times 0.05)$$

$$= 1.33 \text{ m/s}$$

Laminar flow of a Bingham plastic in a circular pipe yields the following implicit relationship for shear stress at the wall.

$$\tau_W = \frac{8\eta u_M}{D} \left[1 - \frac{4}{3} \frac{\tau_o}{\tau_W} + \frac{1}{3} \left(\frac{\tau_o}{\tau_W} \right)^4 \right]^{-1} \tag{14.55}$$

Neglecting the fourth-order term, the shear stress can be expressed as

$$\tau_W = \eta \left(\frac{8u_M}{D} \right) + \frac{4}{3} \tau_o \tag{14.56}$$

For a power law fluid, the shear stress at the wall assumes the form

$$\tau_W = k \left(\frac{8u_M}{D} \right)^n \tag{14.57}$$

where k and n are empirical parameters.

For turbulent flows, the best-known correlation for friction factor is that of Dodge and Metzner (1959) which assumes the slurry can be treated as a power law fluid. A chart for Darcy–Weisbach friction factor versus Reynolds number was developed where Reynolds number was defined as

$$\text{Re} = \rho_M u_M^{2-n} D^n / k 8^{n-1} \tag{14.58}$$

The Darcy–Weisbach friction factor versus Reynolds number is shown in Fig. 14.18. The simplicity of this approach makes it attractive. Caution is recommended, however, in using a universal approach to treat non-Newtonian fluids; many fluids may not be modeled as power-law fluids.

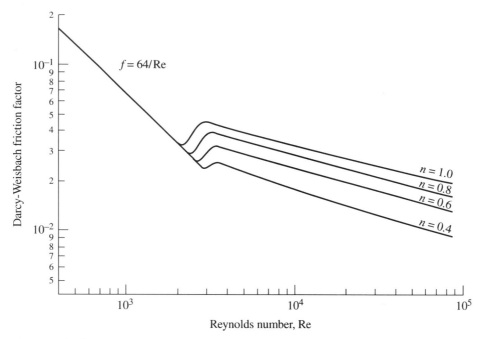

FIGURE 14.18   Darcy–Weisbach friction factor versus Reynolds number for slurry modeled as power law fluid. (Reprinted with permission from Dodge, D. W. and Metzner, A. B., 1959. ''Turbulent Flow of Non-Newtonian Systems,'' *AIChE J.*, Vol. 5, No. 2, pp. 182–204, 1959.)

14.4.3   Pressure Drop in Heterogeneous Slurry Flows

The heterogeneous slurry in a horizontal pipe is characterized by a large concentration of particles near the bottom of the pipe. The velocity at which the particles begin to deposit on the duct wall is known as the deposition velocity. The design of systems to transport heterogeneous slurries should ensure operational velocities higher than the deposition velocity to reduce erosion effects and the likelihood of plugging. Thus, the deposition velocity is an important design parameter.

Several correlations have been proposed to relate deposition velocity to the particle and fluid properties. The best-known is that of Durand [Govier and Aziz (1972)] which is expressed as

$$V_D = \kappa[2gD(\rho_P/\rho_L - 1)]^{1/2} \tag{14.59}$$

where D is the pipe diameter, ρ_P is the solids material density, ρ_L is the fluid density, and κ is a constant depending on the particle diameter and solids concentration. The value of κ ranges from 0.8 to 1.5. Wasp *et al.* (1979) proposed an extension of Durand's correlation to include particle size effects and applied it to sand–water slurries. The correlation can be well-represented by the expression

$$V_D = 3.0 \, (\alpha_P)^{0.2} \, [2gD(\rho_P/\rho_L - 1)]^{1/2} \, (d_P/D)^{1/6} \tag{14.60}$$

where α_P is the average volume fraction of the particles and d_P is the particle diameter. This correlation tends to underpredict the deposition velocities for coal–

water and iron–water slurries. A more detailed analysis of deposition velocity can be found in the work of Thomas (1964).

Example 14.6   A heterogeneous slurry is to be transported in a 10-cm pipe. The fluid is water. The particles are 500 μm in diameter and have a material density of 2000 kg/m^3. The volume fraction of the particles is 30%. The deposition velocity is needed.

Using the correlation represented by Eq. 14.60, one finds

$$V_D = 3(0.3)^{-2}[(2)(9.81)(0.1)(2 - 1)]^{1/2}(0.0005/0.1)^{1/6}$$

$$= 1.37 \text{ m/s}$$

The velocity must be larger than 1.37 m/s to prevent deposition of the particles.

The head loss of heterogeneous slurries flowing in horizontal pipes was studied extensively by Durand. The correlations that he developed remain the accepted practical empirical relationships for pipeline design. Durand correlated the friction factor by

$$\frac{f - f_L}{f_L} = C\alpha_P \left[(gD/u_M)^2 \left(\frac{\rho_P}{\rho_L} - 1 \right) C_D^{-1/2} \right]^{3/2} \tag{14.61}$$

where f_L is the friction factor corresponding to the flow of the liquid alone and C_D is the drag coefficient of a particle at terminal velocity in the pure liquid. The coefficient, C, varies from 80–150. Some adjustment of Eq. (14.61) has to be done to account for a particle size distribution.

Obviously, the analysis of slurry flows depends on identifying a model (Bingham plastic or power law fluid) to characterize the rheological properties. Of course, one is unable to assume that the same model and parameters are going to apply in all flow conditions. Thus, one is cautioned against extrapolation of the models to predict flow properties for flow conditions much removed from those for which the models were adapted. Considerably more work is needed to develop more fundamental models from which reliable slurry flow parameters can be predicted.

14.5 THREE-PHASE FLOWS

Three-phase flows are not frequently encountered in multiphase systems. However, one system which involves a three-phase flow is the air-lift pump in which air is injected near the bottom of a pipe filled with a liquid–solid mixture, and the bubbles rise through the mixture giving rise to a pumping action. This concept has applications in well-drilling and pumping coal in shafts [Weber and Dedegil (1976)].

Another example of three-phase flow is injection of air into slurries for the purpose of drag reduction. Experimental work by Heywood and Richardson (1978) has shown that very significant reduction in pressure gradient can be achieved by this method. These studies will become of increasing technological importance with the use of coal/water slurries as fuels for power plants.

REFERENCES

Baker, O., "Simultaneous Flow of Oil and Gas," *Oil Gas J.*, Vol. 53, pp. 185, 1954.

Baroczy, C. J., "A Systematic Correlation for Two-Phase Pressure Drop," *Chem. Engr. Prog. Sym. Series*," Vol. 62, pp. 232–249, 1965.

Barth, W., "Stromunsvorgange beim Transport von Festteilchen und Flussigkeitsteilchen in Gasen," *Chemie-Ing. Tech.*, Vol. 30, pp. 171–180, 1958.

Brothroyd, R. G., *Flowing Gas-Solids Suspensions*, Chapman and Hall, Ltd., 1971.

Charles, M. E., Govier, G. W., and Hodgson, G. W., "Horizontal Pipe-Line Flow of Equal Density Oil/Water Mixture," *Can. Chem. Engr.*, Vol. 39, No. 1, pp. 27–36, 1961.

Chisholm, D., "A Theoretical Basis for the Lockhart-Martinelli Correlation for Two-Phase Flow," *Int. J. Heat Mass Transfer*, Vol. 10, pp. 1767–1778, 1967.

Chisholm, D., "Pressure Gradients During the Flow of Incompressible Two-Phase Mixtures Through Pipes, Venturis, and Orifice Plates," *Brit. Chem. Engr.*, Vol. 12, No. 9, p. 455, 1967.

Choe, W. G., Weinberg, L., and Weisman, J., "Observation and Correlation of Flow Pattern Transitions in Horizontal Cocurrent Gas-Liquid Flows," *Two Phase Flow and Reactor Safety*, Veziroglu, T. N. and Kakac, S. (Eds.), Hemisphere Publishing, pp. 1357–1393, 1978.

Crowe, C. T., "The State-of-the-Art in the Development of Models for Dispersed Phase Flows," *Proc. Intl. Conf. Multiphase Flows*, Tsukuba, Japan, Vol. 3, pp. 49–60, 1991.

Dodge, D. W. and Metzner, A. B., "Turbulent Flow of Non-Newtonian Systems," *AIChE J.*, Vol. 5, No. 2, pp. 189–204, 1959.

El-Haggar, S. and Crowe, C. T., "Numerical Model for Annular-Dispersed Air-Water Flow in a Venturi," Int. Conf. on Physical Modeling of Multiphase Flow, BHRA Fluid Engr., pp. 41–52, 1983.

Govier, G. W. and Aziz, K., *The Flow of Complex Mixtures in Pipes*, Van Nostrand Reinhold Co., New York, pp. 648, 1972.

Hanks, R. W. and Pratt, D. R., "On the Flow of Bingham Plastic Slurries in Pipes and Between Parallel Plates," *Soc. Petroleum Engrs. J.*, pp. 342–346, 1967.

Hewitt, G. F., "Liquid-Gas Systems," Chapter 2 of *Handbook of Multiphase Systems*, Hetroni, G. (Ed.), Hemisphere Publishing, 1982.

Heywood, N. I. and Richardson, J. F., "Head Loss Reduction by Gas Injection for High Shear-Thinning Suspensions in Horizontal Pipe Flow," *5th Int. Conf. on the Hydraulic Transport of Solids in Pipes*, BHRA Fluid Engr., Cranfield, U.K., pp. 1–22, 1978.

Kliegel, J. R., 1960, "One-Dimensional Flow of a Gas-Particle System," Paper No. 60-S, 28th IAS Meeting, New York.

Klinzing, G. A., *Gas-Solid Transport*, McGraw-Hill, New York, 1981.

Lee, J. and Crowe, C. T., "Scaling Laws for Metering the Flow of Gas-Particle Suspensions Through Venturis," *J. Fluids Eng.*, Vol. 104, No. 1, pp. 88–91, 1982.

Lockhart, R. W. and Martinelli, R. C., "Proposed Correlation of Data for Isothermal Two-Phase, Two-Component Flow in Pipes," *Chem. Engr. Prog.*, pp. 39–48, 1949.

Mandhane, J. M. *et al.*, "A Flow Pattern Map for Gas-Liquid Flow in Horizontal Pipes," *Int. J. Multiphase Flow*, Vol. 1, pp. 537–533, 1974.

Martindale, W. R. and Smith R. V., "Two-Phase, Two-Component Interface Drag Coefficients in Separated Phase Flows," *Int. J. Multiphase Flow*, Vol. 7, pp. 211–219, 1981.

McAdams, W. H., Woods, W. K., and Heroman, L. C., "Vaporization Inside Horizontal Tubes, 2: Benzene-Oil Mixtures," *Trans. ASME*, Vol. 64, pp. 193–200, 1942.

Owen, P. R., "Pneumatic Transport," *J. Fluid Mech.*, Vol. 39, No. 2, pp. 407–432, 1969.

Roberson, J. A. and Crowe, C. T., *Engineering Fluid Mechanics*, Houghton–Mifflin, 1993.

Schuchart, P., "Widerstandsgezetze beim pneumatischen Transport in Rohrkummern," *Chemie-Ingr. Tech.*, Vol. 40, pp. 1060–1067, 1968.

Sheldon, G. L. and Finnie, I., "The Mechanism of Material Removal in the Erosive Cutting of Brittle Materials," *J. Engr. Industry, Trans. ASME, Series B*, Vol. 4, pp. 393–399, 1966.

Shook, C. A. and Roco, M. C., 1991, *Principles and Practice of Slurry Flow*, Butterworth.

Sommerfeld, M. and Zivkovic, G., "Recent Advances in the Numerical Simulation of Pneumatic Conveying through Pipe Systems," *Comp. Meth. Appl. Sci.*, pp. 201–212, 1992.

Tabakoff, W., "Performance Deterioration on Turbomachinery with Presence of Solid Particles," *Particle Laden Flow in Turbomachinery*, ASME/AIAA Joint Meeting, St. Louis, MO, pp. 3–22, 1982.

Taitel, Y. and Dukler, A. E., "A Model for Predicting Flow Regimes Transitions in Horizontal and Near-Horizontal Gas-Liquid Flows," *AIChE J.*, Vol. 22, pp. 47–55, 1976.

Thomas, D. G., "Transport Characteristics of Suspensions: Part IX, Representation of Periodic Phenomena on a Flow Diagram for Dilute Suspension Transport," *AIChE J.*, Vol. 10, No. 3, pp. 303–308, 1964.

Tsuji, Y., Tanaka, T., and Ishida, T., "Lagrangian Numerical Simulation of Plug Flow of Cohesionless Particles in a Horizontal Pipe," *Powder Tech.*, Vol. 71, pp. 239–250, 1992.

Wallis, G. B., *One-Dimensional Two-Phase Flow*, McGraw-Hill, New York, 1969.

Wasp, E. J., Kenny, J. P., and Gandhi, R. L., *Solid-Liquid Flow Slurry Pipeline Transportation*, Gulf Publishing Co., Houston, TX, 1979.

Wasp, E. J., *et al.*, "Deposition Velocities, Transition Velocities and Spatial Distribution of Solids in Slurry Pipe Lines," *1st Int. Conf. on Hydraulic Transport in Slurry Pipelines*," BHRA Fluid Engr., Cranfield, U.K., Paper H4, 1970.

Weber, M. and Dedegil, V., "Transport of Solids According to the Air-Lift Principle," *4th Int. Conf. on Hydraulic Transport of Solids in Papers*, BHRA Fluid Engr., Cranfield, U.K., pp. 1–24, 1976.

Weisman, J., Duncan, D., Gibson, J., and Crawford, T., "Effects of Fluid Properties and Pipe Diameter on Two-Phase Flow Patterns in Horizontal Lines," *Int. J. Multiphase Flow*, Vol. 5, pp. 437–462, 1979.

Wen, C.-Y., *Coal Conversion Technology*, Addison–Wesley, New York, 1979.

Index